Proteomics of Spermatogenesis

Proteomics of Spermatogenesis

Dr. G.S. Gupta, Ph.D.
Former Professor and Chairman
Department of Biophysics
Panjab University
Chandigarh-160014
India

Springer

A C.I.P. Catalogue record for this book is available from the Library of Congress.

ISBN: 0-387-25398-X

Printed in the United States of America.

9 8 7 6 5 4 3 2 1

springeronline.com

PREFACE

Although morphological events in mammalian spermatogenesis have been known for many years, it is only through recent development of experimental techniques in cellular and molecular biology that made it possible to understand molecular biology of male gametogenesis in sufficient detail. However, despite considerable research over past several decades, there has been no systematic attempt to organize protein sequences/structures involved in spermatogenesis. The **PROTEOMICS OF SPERMATOGENESIS,** first of its kind, is the first ever effort to describe proteomics of an organ system such as male reproduction and deals with germ cell specific proteins from the point of view of their structures and functions as well as their clinical applications. However, the subject focuses mainly on the description of protein isoforms, which have been either considered specific to- or dominantly expressed in germ cells and finally localized in spermatozoa. The book has been written keeping in mind that the subject may be beneficial not only to students of reproductive biology in understanding spermatogenesis, but may be useful in understanding the causes of genetic infertility in human males and in other mammalian species, although few examples on proteins during spermatogenesis of non-mammalian species also have been cited. The salient feature of proteomics of spermatogenesis is the compilation of up to date information, based on the available data in literature, which has been interpreted and described in the words of original researchers. More importantly, each chapter in relation to a group of proteins has been properly introduced, although the classification of these proteins is arbitrary and based on their cellular localization or their functions. The knowledge of germ cell specific protein isoforms and understanding of sperm specific proteins and polypeptides acquired during maturation in epididymis offers potential application for targeted intervention in testis without generalized effects on stages of spermatogenesis, and in the development of a contraceptive vaccine in males and females. Although each chapter is unique, but under a broader base the book can be classified into sections such as (i) Spermatogenesis (Chapters 1-5), (ii) Cytoskeleton proteins (Chapters 6-10, and Chapter 25), (iii) Proteins involved in the regulation of gene expression, and in the transcriptionl and translational activity (Chapters 11-17), (iv) Informational macromolecules and their relevance in cell communication during spermatogenesis and sperm-oocyte interactions (Chapters18-22), (v) Proteins participating in cell adhesion and fertilization (Chapters 23-27, and Chapter 34), (vi) Is-proteins which regulate sperm motility (Chapters 28-29) and participate in quality control of sperm functions (Chapters 30 and 31). In addition, the discovery of association of germ cell specific isoforms with non-germ cell/ somatic cell tumors has opened new challenges for their application in the diagnosis and prognosis of oncogenesis and immunotherapy of variety of malignancies (Chapter 32). Therefore, the main objectives of proteomics of spermatogenesis were to acquaint the reproductive biologists and andrologists with the current status of basic and applied research on specialized proteins of mammalian germ cells, their role in spermatogenesis, and to help in identifying research strategies that might yield information useful in the design of male anti-fertility agent, and antigenic peptides as future perspectives for development of contraceptive-cum-cancer vaccines in males and females. Since each topic has been properly introduced, proteomics of spermatogenesis may be referred to as a text book for students undergoing advanced training in reproductive biology and as a guide for Research and Development by pharmaceutical industries.

The author acknowledges the financial grant from Department of Science and Technology under USERS scheme and Emeritus Fellowship from University Grants Commission during the tenure of this project. The author also appreciates the cooperation of various research investigators for providing scientific literature and various copyright owners such as publishing agencies and scientific societies, and authors for granting copyright permissions to reproduce figures and other illustrations from original journals and books.

The author wishes to express his personal gratitude to Professor K.N.Pathak, the Vice-Chancellor and Dr. S.N. Sanyal, the Chairman, Department of Biophysics, Panjab University, Chandigarh for providing space and office facilities of the department during the entire tenure of writing of the manuscript. I am also thankful to my colleagues and friends, and office staff for their support and co-operation during this task. My special appreciation goes to Kishori (wife), Rajesh (son), and other family members who had been supporting this venture with great patience and their personal inconveniences.

G.S.Gupta

CONTENTS

Chapter 1

SPERMATOGENESIS

1.1. TESTIS COMPARTMENTS

The main function of testis is to produce the male gametes and steroid hormones. Spermatogenesis and steroidogenesis take place in two different compartments: seminiferous tubules and interstitium respectively that are morphologically and functionally distinguished from each other. Although anatomically divided, both compartments are functionally connected to each other, and their integrity is essential for normal germ cell production. The functions of the testis and thereby also the functions of its compartments are primarily regulated by the hypothalamus and the pituitary gland, whereas at the testicular level various local regulatory molecules modulate the endocrine hormone actions in somatic and germ cells directly. In mammalian species, the testicular tubular compartment consists of a variety of cell types. The Sertoli cells comprise the main structural component of the seminiferous epithelium. They are responsible for the physical support of the germ cells, in addition to providing nutrients and growth factors. The germ cells are sequentially organized into several layers signifying the respective mitotic or meiotic processes and spermatid development. The presence of distinct germ cell associations allowed stages of the spermatogenic cycle of the seminiferous epithelium to be described on the basis of morphological changes in spermatid morphology. Although the staging is arbitrary, it is of great help in describing structural and physiological changes in the seminiferous epithelium. Each seminiferous tubule is surrounded by mesenchymal cells, which comprise the peritubular myoid cells whose contractile elements generate peristaltic waves along the tubules, but do not present a tight diffusion barrier. The interstitium, the other compartment is populated by androgen producing Leydig cells, which are heterogeneous in respect to their physiological and structural features. Vascular smooth muscle cells, macrophages and endothelial cells are also located in the interstitial space of the testis. The physiological function of macrophages has not been well studied. However, their presence is crucial for (re)population of Leydig cells during development and after experimental depletion. Immune cells, known to secrete a number of growth factors and cytokines, are part of the intra-testicular communication pathways. In addition, neuronal connections also influence cellular interaction in the testis.

1.1.1. Intra-Testicular Communication

At the organ and cell levels, a number of signaling factors operate for rapid communication and responsiveness. Presently, cellular communication is categorized into endocrine, paracrine and autocrine signaling. A signaling molecule can functionally cover more than one category.

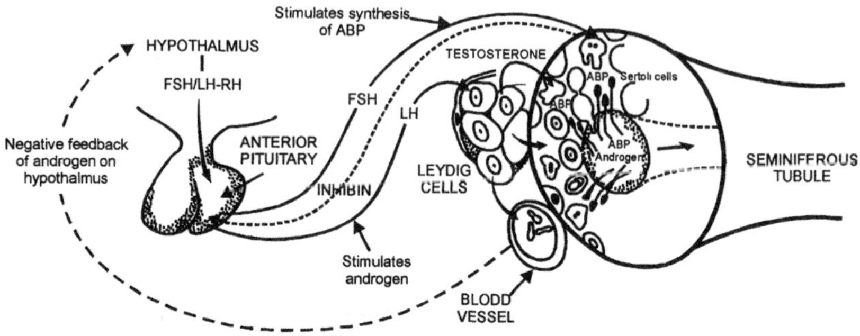

Fig.1.1: Hypothalmus-pituitary-testis axis in mammalian male reproduction.

To ensure coordinate organ function, response and activation, cells in one organ synthesize and release signaling molecules that act on distant organs. This signaling mechanism, termed endocrine signaling is mediated by hormones. Hormones are transported via the bloodstream from the site of production, and reach their cellular target through diffusion or mediated by receptor. Hence, endocrine communication, albeit indispensable and highly effective, is relatively slow **(Fig 1.1)**.

In addition to the regulation of testis function by hypothalamus and the pituitary gland, another level of interaction exists between the neighbouring cellular elements within each of two testicular compartments. While paracrine factors secreted from cells act through diffusion on neighbouring cells, the secreted molecules that act back on the cells from which they originate are referred to autocrine factors. Intracrine signaling has occasionally been used to describe factors that are produced and active within the same cell. In juxtacrine signaling, exoplasmic components of plasma membrane bind and act on adjacent cells through direct cell contact. However, same molecule can work for endocrine, paracrine and autocrine functions. Within testis, paracrine communication comprises not only signaling between neighbouring cells but also between the testicular compartments and among cells far from being in close proximity to each other. In testis, the paracrine mechanisms occur between immune cells, fibroblasts and Leydig cells in the interstitium, between interstitial cells and peritubular cells, between peritubular cells and Sertoli cells, between Sertoli cells and germ cells and among germ cells themselves. Sertoli cells are closely linked by tight and gap junctions from puberty onwards. This structure is known as the 'blood-testis barrier', which represents a tight diffusion barrier dividing the testis into two functional compartments (basal and adluminal) within each seminiferous tubules. Sertoli cells, the only cell type extending into two compartments have the important role of coordinating the secretion of signaling factors into tubular compartments **(Fig. 1.1)**. Sertoli cells also are endowed with a variety of structural features, which enable them to establish and maintain contact with the adjacent germ cells. It is evident that communication between the compartments is essential for functioning of the testis although the precise mechanisms of these interactions are less evident. Presently, more than 100 local factors have been identified and considered to be important for the testis function, but little is known concerning the relevance of these factors in human male infertility. Hence it would seem necessary to distinguish between local factors that are essential for spermatogenesis and those that show redundancies.

1.1.2. Seminiferous Tubules

Seminiferous tubules are enclosed by one or more layers of adventitial cells derived from primitive connective tissue elements of the interstitium. In rodents, a single layer of polygonal cells form a continuous epitheloid sheet surrounding the tubule. Because of their atypical shape and epitheloid organization, they are referred to as myoid cells or peritubular cells. In larger species, ram, bull, boar, man and monkey the adventitial cells form multiple layers. The properties of these cells differ from species to species. In adult mammals, the seminiferous tubules are lined by a complex stratified epithelium composed of two major categories of cells, supporting cells and spermatogenic cells. Supporting cells of single kind, called Sertoli cells were earlier believed to form a syncytium. But later they were shown to be individual cells uniformly spaced on the basal lamina with germ cells occupying expanded intercellular spaces between them. Sertoli cells cease to divide at the time of puberty but persist for whole life of an individual. The three dimensional configuration of Sertoli cell is extraordinarily complex.

The spermatogenic cells include severed morphologically defined cell types: spermatogonia, primary spermatocytes, secondary spermatocytes, spermatids and spermatozoa. Ontogenetically these spermatogenic cell types are not distinct but are successive stages of a process which after proliferation and differentiation lead to formation of mature differential spermatozoon. The proliferative activity in epithelium is confined to spermatogonia and spermatocytes near the base. The earlier cells, speramatogonium rests on the basal lamina propria or boundary tissue. Thus seminiferous epithelium in adults consists of a fixed population of non-proliferating supporting cells and a highly proliferating and differentiating population of germ cells with their stem cells at the base of the epithelium. As they develop, the germ cells are displaced upward along the sides of supporting cells. The topographical relations between germ cells and Sertoli cells change as germ cells move upwards from the base to the lumen, and have important implications in cell adhesion and communications. In seminiferous tubules, typical gap junctions or desmosomes are not found between Sertoli cells and germ cells in the upper two thirds of epithelium, due to free movement of germ cells to move upward as seen in other somatic tissues. However, specialized junctions (occluding junctions), described between adjacent Sertoli cells near the base of the epithelium, form the morphological basis of blood testis permeability barriers. These junctions divide the epithelium into basal compartment containing stem cells of spermatogenesis and adluminal compartment consisting of more advanced stages of spermatogenesis. In addition, these junctions regulate the permeability selective molecules necessary for spermatogenesis, without interruption of the permeability barrier. A number of studies have suggested that the basement membrane (BM) around seminiferous tubules has an important role in supporting testis differentiation, influencing in particular the differentiation of peritubular cells and the proliferation and differentiation of Sertoli cells, and their interaction with germ cells. In addition to compartmentalization, BM of seminiferous tubules acts as substrate for cells in contact and also provides important signals for differentiation, maintenance, and remodeling of tissues.

1.2. SPERMATOGENESIS

Spermatogenesis is a process by which spermatozoa are formed from spermatogonial stem cells during adult's reproductive phase. The process of sperm formation is initiated in the mouse embryo at around day 11.5 postcoitum (pc), when primordial germ cells (PGC) colonise the genital ridge. Under the influence of Y chromosome bearing Sertoli cells, the PGCs proliferate, some of which undergo apoptosis, while the remainder convert to gonocytes. The gonocytes proliferate for a few days and then arrest in G_0/G_1 phase of cell cycle. After remaining quiescent until after birth, gonocytes are reactivated, and differentiate into spermatogonia to initiate the process of spermatogenesis. In rat and mouse the gonocytes resume proliferation (first wave of proliferation) within a few days after birth to form adult type spermatogonia. In mice the first wave of spermatogenesis occurs on day 5 after birth and at 6 months of age after birth in men. While some spermatogonia become self-renewing spermatogonial stem cells, most of them differentiate into spermatocytes, and meiosis begins at approximately day 10 pp in mice and at puberty in man. In the mouse, haploid spermatids are generated by day 20, and spermatozoa first appear in seminiferous tubules by approximately day 35. The onset of puberty and associated increase in gonadotrophin and androgen levels result in the progression of spermatocytes and the appearance of haploid spermatids. Thus, the entire process of spermatogenesis occurs in three sequential phases of cell proliferation and differentiation called: i) mitotic phase, ii) meiotic phase, and iii) post-meiotic phase, which involves stepwise progression of morphologically undifferentiated spermatids to highly differentiated spermatozoa. In mouse spermatogenesis, the mitotic phase lasts for 10 days, meiotic phase for 11 days, while post-meiotic phase lasts for 14 days. The final division produces preleptotene spermatocytes, which begin meiotic phase and undergo last cell cycle S-phase of spermatogenesis..

1.2.1. Mitotic Phase

In mitotic phase, also called spermatocytogenesis, primitive spermatogonia proliferate by mitosis to give rise to several successive generations of spermatogonia, each generation being more differentiating than the preceding one. Traditionally spermatogonia have been divided into two types of spermatogonia (A and B type). These can be distinguished with little difficulty. The type A spermatogonia do not have heterochromatin, whereas type B spermatogonia possess abundant heterochromatin in their nuclei. In human testis, the type A spermatogonium has a spherical or ellipsoid nucleus and one or two nucleoli attached to inner part of nuclear envelope. The type B spermatogonium has spherical nucleus containing a single nucleolus and the chromatin of varying size (heterochromatin), many of which are distributed along the nuclear envelope. In rats and mice intermediate type spermatogonia can also be observed. The A type spermatogonia undergo a series of divisions that result into other type A spermatogonia. During spermatogonial division, A single (A_s), a paired (A_p) and A aligned (A_{al}) spermatogonia can be seen according to their arrangement on the basal side of seminiferous tubules. Single spermatogonium (A_s) is the stem cell for spermatogenesis. On division, A_s produces two new stem cells, where as A_p spermatogonia are connected through intercellular cytoplasmic bridges, the functional significance of which is not clear. The paired spermatogonia (A_p) further divide to form chains of 4, 8 and 16 A_{al} spermatogonia **(Fig.1.2)**. The A_{al} spermatogonia undergo five successive divisions giving rise to A2, A3, A4, intermediate and finally B spermatogonia. B spermatogonia further divide to give primary spermatocytes, which are produced by last mitotic division during spermatogenesis (de Rooij and Grootegoed, 1998). Type A spermatogonia express a very high level of telomerase (Ravindranath et al., 1997). The expression of telomerase decreases with further stages of spermatogenesis and disappears in late spermatids.

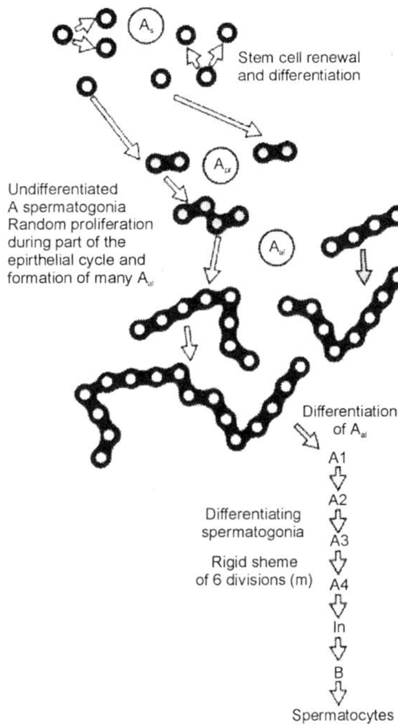

Fig.1.2. Scheme of spermatogonial multiplication and stem cell renewal, which probably applies to all mammals except humans. Stem cells (A_s) proliferate, renewing the stem cell pool and also producing undifferentiated A type paired spermatogonia (A_{pr}), joined together by intercellular cytoplasmic bridges. Further division of A_{pr} produce chains of aligned spermatogonia (A_{al}), which differentiate through six mitotic divisions into A1, A2, A3, A4, intermediate (In), and B spermatogonia to become primary spermatocytes. Reproduced with permission from de Rooij DG, Grootgoed JA.Curr Opin Cell Biol 10; 694-701: 1998 © Elsevier

1.2.2. Meiotic Phase

Most of the somatic cells contain chromosomes in pairs and hence called diploid, while gametes (sperm and ovum) possess only one of each pair. Such cells are called haploid. Haploidy of mammalian gametes is essential, since after fertilization, the zygote establishes the diploid character of chromosome number. The special type of nuclear division, which forms haploid gametes, is termed 'Meiosis'. The meiotic phase terminates at the primary spermatocytes, which at first resembles the cytological characteristics of spermatogonia from which they arise. Primary spermatocytes enter into prophase I of maturation or meiotic division. Their chromatin reorganizes into thread like chromosomes, characteristic of leptotene stage of meiosis. During meiotic phase (leptotene, zygotene, pachytene, diplotene and diakinesis) chromosomes condense. Two important events in meiosis are: linear pairing of chromosomes and interchange of genetic segments between homolgous chromatids during zygotene stage through formation of synaptonemal complex. This is followed by two meiotic divisions that occur in rapid succession without DNA replication to produce spermatids, which are re-modeled into spermatozoa. The process of meiosis and formation of synaptonemal complex has been discussed in more details in Chapter 6. During meiosis a wide variety of genes are up-regulated in spermatocytes. Some of these genes are transcribed only in spermatogenic cells, whereas others produce transcripts specific or unique to spermatocytes. The expression and regulation of several of these genes during meiosis has been recorded during last decade (McCarrey, 1998; Eddy and O'Brien, 1998) (see Chapter 14). The RNA synthesis is low at preleptotene, leptotene, zygotene and early pachytene spermatocytes. However, RNA synthesis increases rapidly in pachytene spermatocytes of mouse, rat, hamster, and human testes. Nuclear RNA synthesis is highest at zygotene stage in both mouse and human spermatocytes suggesting that RNA synthesis occurs during meiosis (Eddy and O'Brien, 1998).

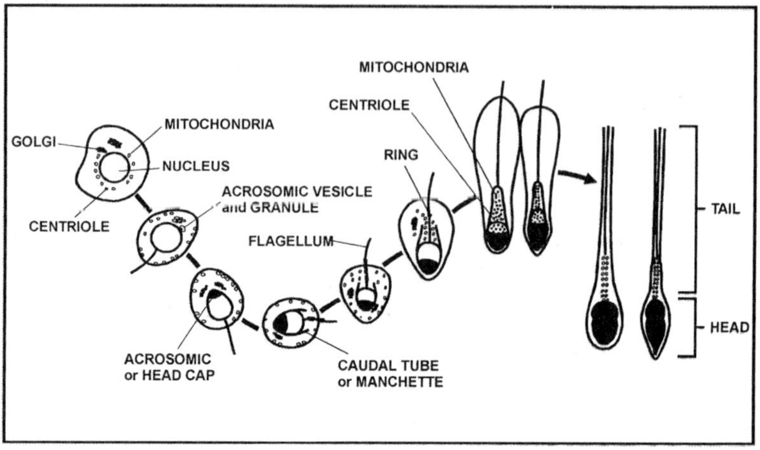

Fig.1.3. Various steps of human spermiogenesis. Reproduced with permission from Hamilton, D.W. in: Cellular and Molecular Events in Spermiogenesis 1990 © WHO.

1.2.3. Post-meiotic Phase (Spermiogenesis)

Spermiogenesis is the final phase of differentiation of male germ cells prior to their release from seminiferous epithelium. During this phase the round spermatids, which are relatively less undifferentiated haploid germ cells, undergo complex morphological, biochemical and physiological changes that result in the formation of asymmetrical flagellated spermatozoa. During last decade it has been possible to study spermiogenesis in greater details. Formation of acrosome and formation of sperm tail are two major transforming events occurring during spermiogenesis. Spermiogenesis is relatively a long process in mammals occupying time duration of about 14 days in mice and 22 days in humans. Though spermiogenesis is a continuous process, it has been divided into a number of morphological events. In most of the species these events can be grouped into Golgi phase, cap phase, acrosome phase and maturation phase of spermiogenesis (Clermont et al., 1990)**(Fig. 1.3)**.

Golgi phase: The earliest spermatids, somewhat smaller than secondary spermatocytes from which they are formed are round and have a nucleus of 5-6 μm in diameter. The cytoplasm contains abundant organelles as seen in electron microscope. A small Golgi apparatus is seen in juxta-nuclear cytoplasm. The Golgi-lysosomal complex in spermatids is more complex than in other cells. The first sign of differentiation of spermatozoon is the formation of several granules within Golgi. These pro-acrosomal granules are rich in carbohydrates and enclosed within a membrane vesicle. These granules, as well as vesicles, coalesce to form acrosomal vesicle. At certain time the membrane of vesicles attaches to the nuclear envelope. As the acrosomal vesicle begins to develop, the centrioles found close to Golgi, translocate to the pole opposite to the developing acrosome, and establish the longitudinal polarity of the cell. The distal centriole becomes oriented parpendicular to the cell surface and elaborates the formation of flagellum. The proximal centriole also attaches to the nuclear envelope and eventually forms the connecting piece in the neck region of the sperm. At the same time cytoplasmic tubules arise and form cylindrical structure called manchette. The stimulus and the mechanism of translocation of centriole and its attachment to the nucleus are unknown.

Manchettes are sleeve like structures and composed of large number of parallel arrays of microtubules extending caudally from spermatid nucleus and attached to cell membrane located at the posterior margin of acrosomal cap. The elongation of spermatid is attributed to flow of cytoplasm back along the manchette. The manchette is one of the several transient organelles that develops during spermiogenesis and disappears without leaving any remanant in spermatozoa. Another morphological event during Golgi phase of spermiogenesis involves clustering of mitochondria. At the beginning, mitochondria are around the nucleolus. During Golgi phase, these disperse close to periphery of the cell and close to plasma membrane. The cause and the mechanisms of morphological changes in mitochondrial redistribution are unknown (Hamilton, 1990).

Cap phase: In cap phase of spermiation, just after the acrosomal vesicle attaches to the nuclear envelope, the attachment site begins to extend distally towards the caudal end of the cell resembling a partially inflated balloon, until it covers half of the nuclear surface (Hamilton, 1990). The cap phase is followed by acrosome phase during which acrosome completes its differentiation; mitochondria migrate from cell periphery and come closer to the developing flagellum. Associated with the movement of mitochondria towards developing axoneme is the development of annulus / and elaboration of outer dense fibres associated with the axoneme and fibrous sheath. The development of annulus delimits the boundary of mitochondrial sheath. Followed by mitochondrial sheath formation, the cell nucleus is condensed and spermatids begin to elongate, a process which is characteristic of mature sperm. The final stage of spermatid terminal differentiation involves the removal of their bulk of cytoplasm in a process known as spermatid individualization.

At cap phase flagellum consists only of axial filament-complex called axoneme. Axoneme is made of two central fibrous and nine peripheral doublets, which are in continuation with the wall of distal centriole. During further differentiation of the tail, nine longitudinally oriented segmented columns arise around the central doublet. These are connected to each other proximally and to the base of the condensed nucleus to form chromatid body, which appears and disappears quickly. As manchette disappears, mitochondria gather around the segment of the flagellum between annulus and the nucleus and arranged helically around it to complete differentiation of middle piece (Fig. 1.3). While these developmental events are in progress, a succession of circumferentially oriented ribs are deposited around the tail, distal to the annulus to form fibrous sheath of principal piece (Bloom and Fawsett, 1971).

Maturation: With the completion of formation of tail, spermatozoa are separated from the excess cytoplasm, which remains attached to the epitehlium in the form of residual body, rich in fine granules, lipid droplets and degenerating excess organelles. For all other purposes spermatid at this stage appears like a fully formed sperm. Chromatin remodeling continues during sperm maturation (Manfredi Romanini et al., 1986). It is possible to distinguish and analyze the two processes involved in sperm maturation, i.e. chromatin condensation (compact structure) and chromatin stabilization (a tough structure, which protects the genomic DNA). The first process, involving a reduction in the nuclear size and a decrease in the amount of sperm DNA reaches its maximum in caput epididymidis sperm, in which the sheared chromatin was mainly organized into 120 Å thick knobby fibers. On the contrary, chromatin stabilization, the onset of which occurs in the testis (at the late spermatid stage) via the formation of –S-S- cross links, is completed in the vas deferens, where chromatin has a superstructure consisting of thicker fibers, with diameters of 210 and 350 Å. Associated with these morphological charges is a remarkable transformation in basic proteins, in which histones are replaced by protamines during spermiogenesis. Protamines are arginine rich small mol wt proteins, which are synthesized and associate with condensing chromatid. The shaping of sperm nucleus occurs due to aggregation of DNA protein complexes. Pericentromeric regions of 9 chromosomes of the rat show few tandem associations in spermatids at steps 1-8 of spermiogenesis. At step 9, pericentromere regions are associated to form an elongated cluster in the spermatid nucleus and continue to be so in the sperm nucleus. Telomere signals showed a tendency for pair wise association, which was more pronounced in elongated spermatid and epididymal sperm nuclei. Spermatozoa showed telomeres at the periphery and the pericentromeres located in the nuclear interior. Chromosome 2 and 12 predominantly occupied compact and variably shaped territories during spermatid maturation. It appeared that the associations of pericentromeres during step 9 render a well-defined nuclear topology, which facilitates the ordered compaction of the genome at subsequent stages (Meyer-Ficca et al., 1998). Moreover, apoptotic proteins, including the activation of caspases play an essential role during spermatid individualization in *D. melanogaster* (Arama et al., 2003). However, testicular spermatozoa are not fertile and further maturation of sperm occurs in epididymis in order to produce a specialized cell, which can fertilize the egg (see Chapter 34).

Centromere Protein B and Spermiogenesis: The centromere protein B (CENP-B) is a centromeric DNA binding protein. It recognizes a 17-bp sequence motif called the CENP-B box, which is found in the centromeric region of most chromosomes. It binds DNA through its amino terminus and dimerizes through its carboxy terminus. Localization of centromere proteins allowed us to follow their movements during the different phases of spermiogenesis. Since the number of centromeres remained close to the number of chromosomes until the cap phase of spermatid differentiation, it has been hypothesized that the labeling of young spermatids corresponds to centromeric proteins associated with their specific DNA counterparts, while the centromere proteins, possibly detached from their DNA loci, were released from nuclei of old spermatids in the same way as are histones and transition proteins (Courtens et al., 1992).

1.3. ENDOCRINE HORMONES AND SPERMATOGENESIS

Leutinizing hormone (LH)/testosterone and follicle stimulating hormone (FSH) are the prime regulators, which control spermatogenesis. However, androgens are indispensable for initiation and maintenance of spermatogenesis. Although testosterone feeds back on both gonadotrophic hormones, an additional feedback loop exists between the testis and the brain for FSH. Inhibin, activin and follistatin are involved in this regulatory system. While inhibin functions to suppress FSH secretion, the homodimer activin stimulates FSH release, whereas follistatin antagonizes activin action. Testicular receptors for LH, FSH, and androgens are confined to somatic cells. a) Testosterone, but not FSH, is required for Sertoli cell proliferation during fetal and early neonatal life; b) FSH and testosterone both regulate the late stages of Sertoli cell proliferation; c) FSH has a general trophic effect on Sertoli cell activity in the pubertal and adult mouse; and d) androgens are required for specific transcript expression during prepubertal development (Johnston et al., 2004) **(Fig.1.1).**

1.3.1. Action of LH

The LH stimulates testosterone synthesis in the Leydig cells, whereas FSH governs spermatogenesis through paracrine action of the Sertoli cells. In adult mammals, including man, early deprivation of gonadotrophin and testosterone results into a stage specific degeneration of germ cells in the testis. The earliest morphological signs of germ cell degeneration involved preleptotene and pachytene spermatocytes, step 7 and step 19 spermatids. The germ cell degeneration was accompanied by a significant increase in apoptosis. It suggested that germ cells die in the testis in response to a lack of hormonal stimulation (Hikim et al., 1995). The stimulatory effects of LH or human chorionic gonadotrophin (hCG) on male gametogenesis are mediated by testosterone, through the expression of the key genes of 3 β HSDI in adult Leydig cells, which are dependent on LH signaling (Zhang et al., 2004). In the majority of the species both LH/testosterone and FSH are needed to ensure normal production of germ cells in testis. In certain animal species, e.g. Djungarin hamsters, FSH is the only hormone responsible for spermatogenesis. Although LH and testosterone are necessary for the development of androgen dependent organs, and sexual behaviour, both gonadotrophins are necessary for spermatogenesis in primates. Immunization either against LH or FSH affects spermatogenesis in most rodents and in primates, and the inactivation of LH/FSH receptors does interfere with testicular function and spermatogenesis, although to varying degrees. It is also clear from many studies that gonadotrophic hormones have primacy over local factors since lack of these endorcrine hormones cannot be compensated for by local factors. In addition, compelling evidence suggests that the activity of endocrine factor / receptors is controlled at the testicular level by local factors (Weinbauer and Wessels, 1999). The testicular production of androgens is dependent on hypothalamus and the pituitary, and the interrelationship between the hypothalmo-hypophyseal system and the gonad is coordinated through feedback systems. Effect of hypophysectomy and subsequent testosterone administration on germ cells indicates that cell-specific mRNAs appear concomitantly with germ cell reappearance in a time-dependent manner in the testes of testosterone treated hypophysectomized adult rats (Trasler et al., 1992). However, the intrauterine sex differentiation of each sex is independent of LH action, which plays a crucial role in attaining sexual maturity post-natally. This was demonstrated in LH receptor (LHR) knockout (LuRKO) mice, which delivered phenotypically normal males and females. However, postnatal growth and maturation of testes and accessory sex organs were blocked in LuRKO delivered males, and their spermatogenesis was arrested at the round spermatid stage, in association with low number of Leydig cells and sex hormone, but elevated gonadotrophin levels (Zhang et al., 2001). Expression of gonadotropin receptors and aromatase transcripts in pigs were detected as early as 25 dpc for the LH receptor and aromatase, and 28 dpc for the FSH receptor in testis (Parma et al., 1999).

1.3.2. FSH Action

In response to FSH, the Sertoli cells express a large number of differentiated gene products, such as transferrin, kit ligand, androgen binding protein and several others (see Chapter 2). Transgenic mice, crossed into the hypogonadal (hpg) mouse genotype showing absence of LH but secreting FSH alone showed increase in testis weights up to 5-fold relative to non-transgenic hpg controls (Allen et al, 2001).

1.3.3. FSH Receptor

The FSH receptor (FSH-R) belongs to a superfamily of receptors that act through interactions with G-proteins, as do receptors for LH/hCG (chorionic gonadotrophin) and thyroid stimulating hormone (TSH). These receptors in the superfamily contain seven transmembrane domains and a large glycosylated extracellular domain required for interaction with the complex

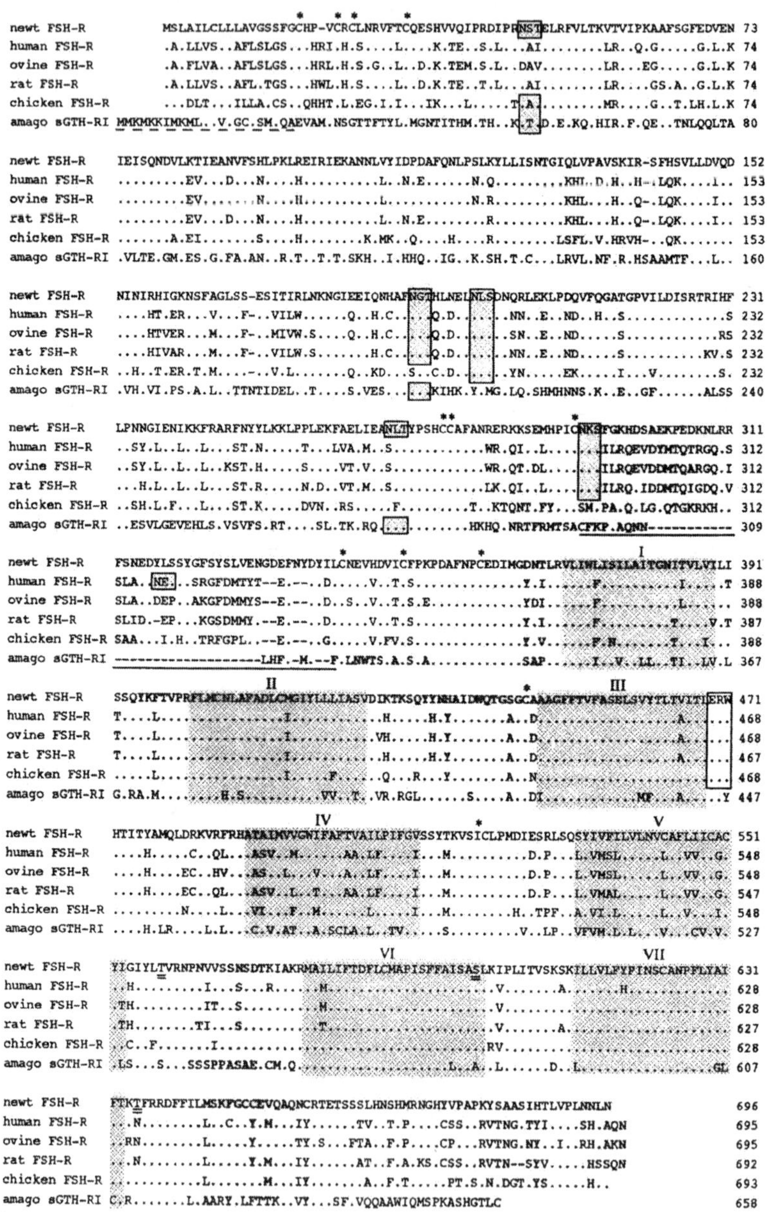

Fig.1.4. Aligned amino acid sequences of FSH receptors (FSH-R) from newt, human, ovine, rat, and chicken and GTH-RI of amago salmon. Cysteine residues are marked by (*).Transmembrane domains (I - VII) are light shaded with roman numerals. N-glycosylation sites are shown by stippled boxs. Acidic-Arg-aromatic motif (ERW) is shown by an open box. The signal peptide is underlined with dashed line. Solid line indicates heterologous region. Potential phosphorylation sites in newt FSH-R are double underlined. Reprinted with permission from Y. Nakayama et al, Biochem Biophys Res Commun 275; 121-28 : 2000 © Elsevier

heterodimeric structure formed by α and β subunits of these hormones. Binding of the hormones to the receptor activates a G-protein, which then stimulates the membrane bound adenylyl cyclase, resulting in the elevation of the intracellular cAMP concentration. The biochemical effects of FSH in mammals are mediated through its interaction with its receptor that is specifically expressed in Sertoli Cells. FSH action on SC is mediated via a specific Sertoli cell G protein coupled-receptor. In a transgenic model the action of a mutant human FSH receptor ($FSHR^{+}$) containing a single amino acid substitution (Asp567Gly) revealed that transgenic $FSHR^{+}$ could stimulate a constitutive FSH-like Sertoli cell response in gonadotropin-deficient testes through pathways that involved LH-independent testicular steroidogenesis (Haywood et al., 2002). The FSH-R cDNAs have been cloned from several species of mammals such as human (Minegishi et al., 1991; Yarney et al., 1993), sub-human primates (Gromoll et al., 1993), equine (Robert et al., 1994), porcine (Remy et al., 1995), ovine, bovine and other species including amphibians (see Nakayama et al., 2000) **(Fig.1.4)**.

The amino acid residues 9-30 in the extracellular domain of rat FSH-R comprise a specific FSH binding domain. The corresponding sequence in newt FSH-R is highly homologous with those in mammals. However, in rats there is an additional site (residues 300-315) for interaction with FSH, but this sequence is highly heterologous between mammals and newt (see c/r Nakayama et al., 2000). The FSH receptor is modulated by cAMP-related transcription factors, CREB and CREM. Protease activity in Sertoli cells might be involved in the control of protein half lives (Monsees et al., 1998) and in determining stage dependent effects on spermatogenesis. The germ cell stage–specific expression of the *FSH-R* gene in the prepubertal and adult rat seminiferous epithelium suggests that the onset of stage-specific *FSHR* gene expression is concomitant with maturation of the Sertoli cell population and completion of the first generation of spermatocytes indicating that spermatogonia and spermatocytes may be involved in the regulation of *FSHR* gene expression (Rannikko et al., 1996). However, the trophic actions of gonadotrophic hormones within the testis are subject to modulation by local growth factors (review Weinbauer and Wessels, 1999).

Activation of Protein Kinase-A Pathway: Activation of protein kinase A (PKA) by FSH and cAMP in Sertoli cells plays an important role in the regulation of testicular functions. The regulatory subunit, RIIβ of PKA is transcriptionally induced in Sertoli cells by cAMP. The basic helix-loop-helix (bHLH) family of transcription factors has been shown to influence FSH-mediated gene expression in Sertoli cells. Knutsen et al, (1997) localized a footprint, which overlaps one of the major transcription initiation sites in the basal promoter (-293 to -123). The protein binding this sequence belongs to the NF-1 family of transcription factors and to a bHLH response element. An upstream region (-723 to - 395, cAMP-responsive region) was found to enhance cAMP responsiveness in Sertoli cells but not in peritubular cells (Knutsen et al., 1997). It appears that the functions of the bHLH proteins are modulated by Id (inhibitor of differentiation) proteins, which lack the DNA binding basic domain. The Id protein forms transcriptionally inactive dimer with bHLH proteins and thus regulates the cell proliferation and differentiation. Since there inhiitors of differentiation Sertoli cells co-express all four isoforms of Id (Id1, Id2, Id3, and Id4), it was suggested that these inhibitors of differentiation are differentially regulated and may have distinct functions (Sablitzky et al., 1998; Chaudhary et al., 2001). In addition to PKA activation, Crepieux et al, (2001) demonstrated that FSH activates the ERK in MAP-kinase pathway following dual coupling of the FSH-R both to G_s and to G_i heterotrimeric proteins, in a PKA- and also Src-dependent manner.

c-Fos Gene and Sertoli Cell Activation by FSH: The expression c-fos is known to be associated with the transduction of cell surface stimuli into changes in nuclear function. The FSH induces a transient expression of c-fos in cultured Sertoli cells. This induction is probably mediated by cAMP and likely involves an increased transcription of the c-fos gene (Hall et al, 1988). Although the action of FSH on Sertoli cells is considered to be mediated by cAMP, dibutyryl cAMP (dbcAMP), an analogue of cAMP, induces much less c-fos mRNA expression than FSH, suggesting that additional cAMP independent mechanisms may mediate the effect of FSH, on c-fos. Specific intracellular inhibitors of PKC and enhancer of $(Ca^{2+})_i$ demonstrated that Sertoli cell c-fos mRNA expression is under multifactorial regulation such as cAMP, calcium and PKC (Jia et al, 1996). Sertoli cell differentiation of basement membrane is mediated by the c-fos protooncogene. To examine the role of c-fos in Sertoli cell attachment, spreading, and differentiation on extracellular matrix, Papadopoulos and Dym (1994) suggested that c-fos might mediate the events involved in Sertoli cell attachment and spreading upon contact of the cells with extracellular matrix.

The regulation of c-fos by FSH was investigated with various CAT constructs containing segments of the c-fos promoter, such as SRE, CRE, and AP1/phorbol ester/TPA response element (TRE). Transfection experiments indicated that the FSH can stimulate all three response elements, as well as a whole *c-fos* promoter construct. More dramatic effect was seen on the SRE-CAT than on a cAMP analogue. Nuclear extracts of FSH-stimulated Sertoli cells caused labeled AP1 oligonucleotide to form a DNA/protein complex indicating activation of the *c-fos* gene and binding of the c-fos/jun complex. The induction of Sertoli cell differentiation appears to involve the serum response element (SRE) of the c-*fos* promoter to activate c-*fos* and intermediate bHLH factor(s) that regulate down-stream Sertoli cell-differentiated genes. The SRE of the c-*fos* promoter is influenced through serum response factor (SRF). E-box nucleotide sequence is present within the SRE. The bHLH proteins may directly influence the SRE of the c-*fos* promoter. The activation of the c-*fos* promoter in Sertoli cells was found to be inhibited with the over-expression of the inhibitory HLH protein, ld. The Id specifically inhibited the activation of SRE in Sertoli cells and no other elements tested. This provides a direct role of bHLH proteins in regulating SRE activity in Sertoli cells. The SRF and bHLH proteins appear to bind the SRE and activate the c-*fos* promoter in Sertoli cells. Observations provide evidence that a bHLH protein can interact with the SRE of the c-*fos* promoter to influence hormone-induced promoter activation. Cross-talk between these nuclear transcription factors appears to be instrumental in the control of Sertoli cell-differentiated functions (Chaudhary et al, 1996; Chaudhary and Skinner, 1999). Therefore, FSH appears to act through multiple transcriptional activation pathways. The first involves cAMP and the CRE at both early-event genes (e.g., *c-fos*) and downstream genes (e.g., transferrin). It is likely that other pathways involve alternate signal transduction events (e.g., calcium mobilization) and promoter response elements (e.g., SRE).

FSH Stimulation of LH Receptor in Leydig Cells: Although earlier reports suggested a stimulatory effect of FSH on Leydig cell function, controversy existed due to unavailability of FSH preparations free of contaminating LH. Availability of recombinant human FSH preparations made it possible to reinvestigate the role of FSH on leydig cells. Immature male rats were hypophysectomized and implanted with osmotic minipumps releasing recombinant FSH (rFSH) or pituitary FSH (hpFSH). After treatment, increase in testicular weight was found associated with the increase in LH receptor number and with the increase in LH receptor mRNA levels. It seems that rFSH, like purified pituitary FSH, is capable of increasing the LH receptor content and steroidogenic responsiveness of Leydig cells through paracrine mechanisms together with a stimulatory effect on spermatogenesis (Vihko et al, 1991).

1.3.5. Local Factors in Trophic Hormone Action

The LH-induced testosterone synthesis is modulated by local factors in testis. Mice expressing a mutant form of colony stimulating factor (CSF)-1 lack tissue macrophages and produce low levels of testosterone and low fertility, which is restored after administration of CSF-1 to these animals. A macrophage derived factor was identified that stimulated testosterone production (Lukyanenko et al., 1998). Various Sertoli cell factors also are known to influence testosterone synthesis and secretion by Leydig cells following hormonal activation (Saez, 1994). Insulin like growth factor (IGF)-1, activin and inhibin increase LH responsiveness whereas transforming growth factors (TGF-α, TGFβ) inhibit LH responsiveness and number of LH receptors. The stimulatory effects of IGF-I are mediated through prolonged half life of mRNA (Zhang et al., 1998). Progesterone also has been reported to regulate LH receptor expression negatively (El-Hafnawy and Huthaniemi, 1998). Co-culture of human Leydig cells and Sertoli cells prolong testosterone synthesis and enhance the responsiveness in terms of testosterone release to hCG. Anti-Muellerian hormone (AMH) inhibits LH receptors expression in vitro and prevents Leydig cell differentiation (Josso et al., 1998) and suppresses LH induced testosterone production and Sertoli cell aromatase activity (Rouiller-Fabre et al., 1998). Platelet derived growth factor A (PDGF-A) deficient mice showed that PDGF-A gene is essential for development of Leydig cell lineage and that PDGF-A may play a role in cascade of genes involved in testis differentiation (Gnessi et al., 2000) (see Chapter 3).

1.4. TESTOSTERONE AND SPERMATOGENESIS

Testosterone is the classical hormone that initiates androgen dependent functions in the entire organism. While the androgen receptor is ubiquitously expressed in the entire organism, testosterone plays specific functions in the testis. Androgen receptor is a ligand dependent transcription factor whose properties have been well characterized (Quigley, 1998). The intra-testicular importance of testosterone is well established in maintaining and resorting quantitative spermatogenesis in all species. In addition, substantial restoration of spermatogenesis can be achieved in regressed testis after administration of testosterone in GnRH immunized rats or gonadotrophin deficient mice (McLachlan et al., 1994; Singh et al., 1995).

The testosterone, produced by Leydig cells in interstitial compartment, acts on different cell types present in both the interstitial and the tubular compartments. During postnatal development testosterone promotes the functional maturation of myoid cells and Sertoli cells. The stimulatory effects of testosterone on gametogenesis are indirect since testosterone only binds to somatic testicular cells. It is likely that as yet unknown local mediators of testosterone action could exist (Saez, 1994). Testosterone acts within the testis following conversion into dihydrotestosterone (DHT) by 5α-reductase activity. At comparable dose, DHT is as effective as testosterone in stimulating spermatogenesis. 5α- reductase of testosterone by 5α-reduction is particularly important for progression through mid-spermiogenesis, since this phase of germ cell development is more sensitive to withdrawl of androgens (O'Donnell et al., 1999). Evidence also suggests that local aromatization of testosterone into oestrogen might be needed for intact spermatogenesis. Apart from the trophic action of testosterone on the germinal epithelium, testosterone also plays an important role in the morphological and presumably functional differentiation of the somatic testicular cell types prior to the onset of spermatogenic activity (Quigley, 1998). Testosterone is also capable of modifying Leydig cell function. It is well known that the differentiation of Leydig cells from their fibroblast precursors is under the control of LH/androgens. This view is

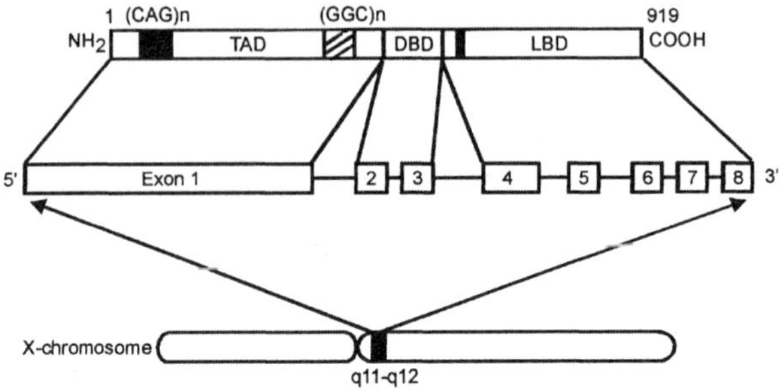

Fig.1.5. Structural organization of androgen receptor (AR) gene and the androgen receptor protein. The AR gene, located on X-chromosome, consists of 8 exones, which encode AR protein with 919 amino acid residues. The three major domains of AR (TAD = Transactivation domain, DBD = DNA-binding domain, and LBD = ligand-binding domain) are marked. The two homopolymeric regions encoding glutamine (CAG)n and glycine (GGC)n repeats are shown in the figure. Bipartite signal is marked by (*).

supported by the demonstration that exogenous administration of testosterone increases androgen receptor expression on Leydig cells, and that the functional maturation of Leydig cells is controlled by LH/hCG. Hence, testosterone is an example of a molecule involved in endocrine, paracrine and autocrine signaling in the male. Observations suggest also an FSH dependent local interaction between Sertoli cells and peritubular cells during androgen action in the primate testis (Weinbauer and Wessels, 1999).

1.4.1. Androgen Receptors

One potential mechanism for the tubular specificity of response to androgens is differential expression of the androgen receptor (AR). The testis contains at least fivefold more testosterone than needed for the maintenance of spermatogenesis and for saturation of total receptor binding sites. The site(s) of androgen action within the testis, i.e. the cellular distribution of the androgen receptor, is not very clear. Early studies favored the hypothesis that both germ cells and somatic cells express an AR moiety different from androgen binding protein. In contrast, other investigators were unable to measure androgen receptors in germ cells. In contrast to its established effects, the mechanisms of testosterone action are poorly understood. Nonetheless, when androgen receptor was studied by immuno-chemical techniques, conflicting results appeared. Within the testis, androgen receptors are found in Leydig cells, peritubular cells and Sertoli cells in rat and human. Germ cells are devoid of androgen receptors although the presence of receptors on elongating spermatids of steps X-XII has been suggested. However, androgen effects on germ cells are conveyed by the somatic cells of the testis. Both FSH and testosterone influence the expression of the androgen receptor, which is stage specific phenomenon (Suarez–Quian et al., 1999). The molecular cloning of rat androgen receptor allowed deduction of the amino acid sequence of the receptor and the synthesis of unique peptide segments of this protein (Bremner et al., 1994; Ornerier et al., 1995; Vornberger et al., 1994). Mutations in androgen receptor may result in complete feminization to idiopathic male infertility. Studies confirmed that missense amino acid substitutions in the ligand binding domain of androgen receptor leads to infertility through defective protein-protein interactions between receptor domains and coactivator proteins (Yong et al., 2003) **(Fig.1.5)**. Expression of androgen receptor in Sertoli cells is highly dependent on the androgenic milieu in adult animals but less so in immature animals.

The androgen receptor binds its ligand in the cytoplasm, then translocates to the nucleus and thus stimulates the transcription of specific genes in rat and man. Localization of androgen receptor remains unchanged regardless of the status of spermatogenesis indicating that changes in the expression of androgen receptor are not the major cause for defects of spermatogenesis. Although many efforts have been made, the expression of any specific gene product or protein could not be correlated with the early effects of testosterone withdrawal. However, during the first few days of androgen deficiency, an overall stimulation of gene expression and protein production is dependent on the presence of all germ cell types and appears to be mediated through androgen binding protein (ABP). The ABP reaches its peak in the spermatogenic cycle, in which the first effects of androgen deficiency are observed at stages, VII VIII. (review-Schlatt et al., 1997). This observation speculates on possible non-androgen receptor mediated testosterone action. Using Cre/loxP technology, De Gendt et al., (2004) generated mice with a ubiquitous knockout of androgen receptor as well as mice with a selective androgen receptor knockout in Sertoli cells (SC) only. Expression of the homeobox gene *Pem*, which is androgen-regulated in SC, was severely decreased in association with decrease in testis weight. This observation established the absolute requirement of androgen receptor in SC for androgen maintenance and spermatogenesis (De Gendt et al., 2004).

After various considerations, implications for androgen regulation of spermatogenesis suggest: 1) The presence of androgen receptor (AR) in adult rat testis Leydig cells is consistent with the hypothesis that androgens modify Leydig cells activity in an autocrine fashion. Since, all Leydig cells did not exhibit AR immunostaining at steady state a differential functional activity of these cells within the population can be suggested. 2) The intense AR immunostaining of smooth muscle cells present in the interstitium indicates that these cells are targets for androgens. 3) Presence of AR in both Sertoli and peritubular myoid cells suggests their indirect involvement in the androgenic control of spermatogenesis. 4) The stage specific AR immunoreactivity in Sertoli cells may, however, be more indicative of a specific androgen response during these stages, whereas, peritubular myoid cells and Leydig cells may participate in the total maintenance of spermatogenesis. Moreover AR is detectable by day 5 and becomes stage specific between days 21-35 at the stage of elongated spermatids. The specific presence of AR in elongated spermatids indicated that androgens could act directly on germ cells to regulate spermatogenesis (Vornberger et al., 1994; Bremner et al., 1994). Observations led to evolve into current view that androgens control spermatogenesis indirectly by manifesting their action solely in somatic cells of the testis, principally in Sertoli, Leydig, and pertibular cells. In primates, androgens exert important effects during prepuberal development by stimulating the proliferation and differentiation of peritubular cells and Sertoli cells. In adult rats, androgen withdrawal leads to germ cell depletion with the first detrimental effects observed in spermatids after only 3 days that showed signs of degeneration and subsequent release into the lumen.

1.4.2. Androgen Binding Protein

Androgen binding protein (ABP) and sex hormone binding globulin (SHBG) are secretory glycoproteins that bind dihydrotestosterone (DHT), testosterone, and estradiol with high affinity. They share the same amino acid sequence but differ slightly in oligosaccharide content. The ABP is produced by the Sertoli cells in the seminiferous tubules and is secreted into the luminal fluid. After transport to the epididymis with maturing sperm, it is internalized by the epithelium in the initial segment and proximal caput epididymis possibly by a receptor mediated process. Nonetheless, reports also suggested that ABP is produced by epididymis where it is present in high concentration. It shows that secretion of ABP occurs along the entire epididymis, whereas endocytosis is region specific and suggests an important physiological role of this region in the induction of sperm maturation (Hermo et al., 1998).The ABP is produced by Sertoli cells in many mammals, including humans, rats, and mice. The livers of most adult animals, but not of rodents, secrete SHBG into the blood, where it circulates as the major sex steroid binding protein. Although the best-characterized ABP was isolated from the rat epididymis, the mouse also synthesizes ABP mRNA in the testis and transports ABP to the epididymis.

ABP is thought to regulate spermatogenesis and sperm maturation by maintaining high androgen levels in the testis and epididymis. The identification of ABP/SHBG membrane receptors in several potential target tissues and increase in cellular cAMP levels in response to SHBG and sex steroid administration suggests that these proteins have a much broader function possibly as hormones (Joseph et al., 1997). The SHBG is endocytosed by germ cells in monkeys and rodents that suggests that sex hormone binding protein is required for spermatogenesis. In monkeys, it has been suggested that steroid binding proteins may be required for spermatogenesis by acting at the germ cell lineage site either by themselves or by serving as steroid transmembrane carriers. The ABP can influence the testicular and intra-tubular distribution of androgens. The ABP levels are highest during the final stages of spermatid maturation (stages VIII-XII), and its production is controlled by testosterone and FSH; its content in testis is lowered by gonadotrophin deficiency. Dihydrotestosterone binding activities in the testis and epididymis of the homozygous transgenic mice were elevated as compared to wild animals. Increased ABP levels in the transgenic mice are associated with structural and functional abnormalities in the testis. The variable spermatogenic disorder was associated with variable male fertility. The homozygous transgenic male and female mice also had a serious motor dysfunction affecting their hind limbs (Joseph et al., 1997; Reventos et al., 1993). Testicular and epididymal ABP levels can be maintained in hypophysectomized rats with pregnenolone treatment alone, and that sperm fertilizing ability can be maintained when intraluminal androgen levels are low and ABP levels are near normal. This suggests that ABP may be important in the maintenance of normal sperm fertilizing ability. A postnatal study revealed that ABP secretion is regulated by different factors.

The rat and human ABP/SHBG genes have been cloned and their organizations characterized (Joseph, 1996b; Joseph et al., 1997; Sullivan et al., 1993). The rat ABP is regulated by two promoter genes: P_A and P_I. While promoter P_I regulates the synthesis of an mRNA that encodes the secreted Sertoli cell form of ABP/SHBG, P_A regulates an mRNA that encodes an alternate form of ABP/SHBG with a different N-terminal sequence, which targets ABP/SHBG to the nucleus (Joseph et al., 1996a). To study the expression of the rat ABP/SHBG gene, transgenic mice were developed with a genomic DNA fragment, containing 1.5 kb upstream of the P_I transcription start site and all the exons, capable of directing tissue specific gene expression in the testis and ovary.

1.4.3. Estradiol

The presence of the enzyme aromatase in sperm initiated the idea that conversion of androgens into oestrogens in the duct system is used as a physiological signal for the quantity of sperm released from the testis and might act as a signal for the regulation of organ function. Now estradiol has become a 'male' hormone over the past few years since the discovery that male patient lacking estrogen receptor or aromatase activity show incomplete bone maturation. This steroid has been considered as a new local regulator of spermatogenesis: mice lacking estrogen receptor exhibited severe impairment of spermatogenesis and were infertile (Eddy et al., 1996) (see Chapter 4).

1.5. PROLACTIN

Prolactin (PRL, 23000 M_r) is an anterior pituitary hormone present in many species and it expresses various functions. In mammals, PRL stimulates gonadal tissues and mammary glands and exerts immunomoduatory effects on the mammalian body. In vertebrates, PRL acts as an osmoregulatory hormone. The effects of PRL on mammalian testes have been studied with various experimental models such as PRL deficient hereditary dwarf mice, hypophysectomized rats and/or dopamine agonist treated animals. The PRL reduces testicular shrinkage in short term hypophysectomized mice, while it increases testicular weight in long term treatment in hypophysectomized mice. Although it is known that PRL binds to Leydig cells and increases the binding activity of hCG to Leydig cells, the exact role of PRL in the regulation of testicular function is controversial.

Prolactin receptor (PRL-R) mRNA has been detected in rabbit and rat testis. In situ hybridization shows that mRNA is expressed not only in Leydig cells but also in the first layer of the seminiferous epithelium (Ouhtit et al., 1993). However, in contrast to report of Ouhtet et al, (1993), Hondo et al, (1995) detected PRL-R on the surface of Leydig cells, Sertoli cells, all phases of spermatogonia, elongated spermatids and spermatocytes, and spermatozoa, where as mRNA is expressed in all stages of spermatogenic cycle. Results indicated that PRL-R mRNA expression is almost consistent with PRL binding sites except for elongated spermatids and spermatozoa, and suggested that PRL may have direct effects on spermatogenic cells.

Prolactin modulates also testicular function by increasing testicular LH receptors and therefore responsiveness to LH. Seasonal changes in testicular function are not only the result of changing serum concentrations of LH, FSH, and PRL, but also concomitant changes in testicular responsiveness to these hormones. For example testicular regression induced by short days in the golden hamster is associated with a progressive loss of testicular LH, FSH, and PRL receptors, whereas recrudescence induced by long days is associated with a progressive increase of LH, FSH, and PRL receptors. Howell-Skalla et al., (2000) investigated the expression of testicular LH, FSH, and PRL receptor mRNA during the non-breeding and breeding seasons and cloned and sequenced a fragment of the extra-cellular portion of the (LH-R: 852 bp) and (PRLR: 680 bp) in the black bear. Comparisons of the partial cDNA and predicted amino acid sequences of these receptors from many species suggested that the LH-R and FSH-R are highly conserved (LH-R: 87.1 - 93.7%; FSH-R: 86.0 - 92.7%) whereas the PRL-R is less well conserved (81- 87%) and that the expression of testicular gonadotropin and PRL receptor mRNA is seasonally regulated (Howell-Skalla, 2000).

1.6. SEROTONIN RECEPTOR IN SERTOLI CELLS

The Sertoli cell-germ cell co-cultures showed the presence of a cDNA of 355-nt that is up-regulated in Sertoli cells by pachytene spermatocytes. The mRNA sequence revealed a homology of 91% (nt) and 86% (aa) to a serotonin receptor present in Sertoli cells. Its induction did not require cell contact, as germ cell conditioned medium also induced the mRNA. The search of germ cell factor(s) inducing the serotonin receptor mRNA revealed it a factor of more than 10-kDa that survived freezing and thawing, and was heat sensitive. The induction of a serotonin receptor mRNA in Sertoli cells by pachytene spermatocytes demonstrates that meiotic germ cells induce mRNA encoding this important receptor in Sertoli cells by a distinct pathway (Syed et al., 1999).

Conclusion: Although either of the LH/testosterone and FSH is capable of stimulating all phases of the spermatogenic cycle including the formation of sperm, the combination of testosterone and FSH is necessary in most instances to achieve quantitatively normal germ cell numbers. Evidence for differential effects of testosterone and FSH on male gametogenesis, however, is lacking and a synergistic effect is observed when they are combined.

1.7. REFERENCES

Allan CM, Haywood M, Swaraj S, Spaliviero J, Koch A, et al. A novel transgenic model to characterize the specific effects of follicle-stimulating hormone on gonadal physiology in the absence of luteinizing hormone actions. Endocrinology 2001; 142; 2213-20.

Arama E, Agapite J, Steller H. Caspase activity and a specific cytochrome C are required for sperm differentiation in *Drosophila.* Dev Cell 2003; 4: 687-97.

Bloom W, Fawsett DW. A text book of histology, WB Saunders Co, Philadelphia, 1978

Bremner WJ, Millar MR. Sharpe RM, Saunders PTK. Immunohistochemical localization of androgen receptors in the rat-testis: evidence for stage-dependent expression and regulation by androgens. Endocrinology 1994; 135: 1227-34.

Chaudhary J, Johnson J, Kim G, Skinner MK. Hormonal regulation and differential actions of the helix-loop-helix transcriptional inhibitors of differentiation (Id1, Id2, Id3 and Id4) in Sertoli cells. Endocrinology 2001; 142: 1727-36.

Chaudhary J, Skinner MK. Basic Helix-Loop-Helix proteins can act at the E-Box within the serum response element of the c-fos promoter to influence hormone-induced promoter activation in sertoli cells. Mol Endocrinol 1999; 13: 774-86.

Chaudhary J, Whaley PD, Cupp A, Skinner MK. Transcriptional regulation of Sertoli cell differentiation by follicle-stimulating hormone at the level of the c-fos and transferrin promoters. Biol Reprod 1996; 54: 692-9.

Clermont Y, Hermo L, Rambourg A, Thorne-Tjomsland G. Structural and Cytochemical characteristics of the Golgi apparatus during the formation of acrosome system in spermatids, In: Cellular and molecular events in spermiogenesis (Hamilton DW, Waites GMH eds), Cambridge University Press, Cambridge, pp 17-39,1990.

Crepieux P, Marion S, Martinat N, Fafeur V, Vern YL, Kerboeuf D, Guillou F, Reiter E. The ERK-development signaling is stage-specifically modulated by FSH, during primary Sertoli cell maturation. Oncogene 2001; 20: 4696-709.

Courtens JL, Biggiogera M, Rothfield NF, Burnier M, Faken S. Migration of centromere proteins in rabbit spermatids. Mol Reprod Dev 1992; 32: 369-77

De Gendt K, Swinnen JV, Saunders PT, Schoonjans L, Dewerchin M, Devos A, Tan K et al. A Sertoli cell-selective knockout of the androgen receptor causes spermatogenic arrest in meiosis. Proc Natl Acad Sci USA. 2004; 101: 1327-32.

De Rooij DG, Grootgoed JA. Spermatogonial stem cells. Curr Opin Cell Biol 1998; 10: 694-701.

Eddy EM, O'Brien DA. Gene expression during mammalaion meiosis. Curr Topics Dev Biol 1998;37:141-99.

Eddy EM, Washburn TF, Bunch DO, Goulding EH, Gladen BC, Lubahn DB, Korach KS. Targeted disruption of the estrogen receptor gene in male mice causes alteration of spermatogenesis and fertility. Endocrinology 1996; 137: 4796-4805.

El-Hafnawy T, Huthaniemi I. Progesterone can participate in down-regulation of the luteinizing hormone receptor gene expression and function in cultured murine Leydig cells. Mol Cell Endocrinol 1998; 137: 127-38.

Gnessi L, Basicni S, Mariani S, Arizzi M, Spera G, Wang C, Bondjers C, Karlsson L, Betsholtz C. Leydig cell loss and spermatogenic arrest in platelet-derived growth factor (PDGF)-A deficient mice. J Cell Biol 2000 1019-26

Gromoll J, Dankbar B, Sharma RS, Nieschlag E. Molecular cloning of the testicular follicle stimulating hormone receptor of the non human primate Macaca fascicularis and identification of multiple transcripts in the testis. Biochem Biophys Res Commun 1993; 196: 1066-72.

Hall SH, Joseph DR, French FS, Conti M. Follicle stimulating hormone induces ransient expression of the

protooncogene c-fos in primary Sertoli cell cultures. Mol Endocrinol 1988; 2: 55-61.
Hamilton DW. Spermiogenesis: An itroduction to the symposium. In: Cellular and molecular events in spermiogenesis (Hamilton DW, Waites GMH eds), Cambridge University Press, Cambridge, pp 3-14,1990.
Haywood M, Tymchenko N, Spaliviero J, Koch A, Jimenez M, Gromoll J et al. An activated human follicle-stimulating hormone (FSH) receptor stimulates FSH-like activity in gonadotropin-deficient transgenic mice. Mol Endocrinol 2002; 16: 2582-91.
Hermo L, Barin K, Oko R. Androgen binding protein secretion and endocytosis by principal cells in the adult rat epididymis and during postnatal development. J Androl 1998; 19: 527-41.
Hikim AP, Wang C, Leung A, Swerdloff RS. Involvement of apoptosis in the induction of germ cell degeneration in adult rats after gonadotropin-releasing hormone antagonist treatment. Endocrinology 1995; 136: 2770-5.
Hondo E, Kurohmaru M, Sakai S, Ogawa K, Hayashi Y. Prolactin receptor expression in rat spermatogenic cells. Biol Reprod 1995; 52: 1284-90.
Howell-Skalla L, Bunick D, Bleck G, Nelson RA. Bahr JM. Cloning and sequence analysis of the extracellular region of the polar bear (Ursus maritimus) luteinizing hormone receptor (LHr), follicle stimulating hormone receptor (FSHr), and prolactin receptor (PRLr) genes and their expression in the testis of the black bear (Ursus americanus). Mol Reprod Dev 2000; 55: 136-45.
Jia MC, Ravindranath N, Papadopoulos V, Dym M. Regulation of c-fos mRNA expression in Sertoli cells by cyclic AMP, calcium, and protein kinase C mediated pathways. Mol Cell Biochem 1996;156: 43-9.
Johnston H, Baker PJ, Abel M, Charlton HM, Jackson G, Fleming L, Kumar TR, O'Shaughnessy PJ. Regulation of Sertoli cell number and activity by follicle-stimulating hormone and androgen during postnatal development in the mouse. Endocrinology 2004; 145: 318-29.
Joseph DR, Becchis M, Fenstermacher DA, Petrusz P. The alternate N-terminal sequence of a rat androgen-binding protein/sex hormone-binding globulin contains a nuclear targeting signal. Endocrinology 1996a; 137: 1138-43.
Joseph DR, O'Brien DA, Sullivan PM, Becchis M, Tsuruta JK, Petrusz P. Overexpression of androgen binding protein/sex hormone globulin in male transgenic mice. Tissue distribution and phenotypic disorders. Biol Reprod 1997; 56: 21-32.
Joseph DR, Power SGA, Petrusz P. Expression and distribution of androgen-binding protein/sex hormone-binding globulin in the female rodent reproductive system. Biol Reprod 1996b; 55: 14-20.
Josso N, di Clemente N, Gouedard L. Anti-mullierian hormone and its receptors. Mol Cell Endocrinol 2001; 179: 25-32.
Kasahara M, Gutknecht J, Brew K, Spurr N, Goodfellow PN. Cloning and mapping of a testis-specific gene with sequence similarity to a sperm-coating glycoprotein gene. Genomics 1989; 5: 527-34.
Knutsen HK, Tasken K, Eskild W, Richards JS, Kurten RC, Torjesen PA, Jahnsen T, Hansson V, Guerin S, Tasken KA. Characterization of the 5'-flanking region of the gene for the cAMP-inducible protein kinase A subunit, RIIβ, in Sertoli cells. Mol Cell Endocrinol 1997; 129: 101-14.
Lukyanenko YO, Carpenter AM, Brigham DE, Stocco DM. Regulation of Leydig cells through a steriodogenic acute regulatory protein-impendent pathway by a lipophilic factor from macrophages. J Endocrinol 1998; 156: 267-75.
Manfredi Romanini MG, Biggiogera M, Formenti D, Franschini A, Garagna S, Pellicciari C, Redi CA. Sperm-chromatin maturation in the mouse. A cytochemical approach. Histochemistry 1986; 84: 484-91
Meyer-Ficca M, Muller-Navia J, Scherthan H. Clustering of pericentromeres initiates in step 9 of spermiogensis of the rat (Rattus norvegicus) and contributes to a well-defined genome architecture in the sperm nucleus. J Cell Sci 1998: 111(pt10): 1363-70
McLachlan RI, Wrford NG, Tsonis C, de Kretser DM, Robertson DM. Testosterone effects on spermatogenesis in the gonadotropin-releasing hormonr-immunized rat. Biol Reprod 1994; 50: 271-80.
McCarrey Jr. Spermatogenesis as a model system for analysis of regulatory mechanisms associated with tissue-specific gene expression. Cell Dev Biol 1998; 9: 459-66
Minegishi T, Nakamura K, Takakura Y, Ibuki Y, Igarashi M. Cloning and sequencing of human FSH receptor cDNA. Biochem Biophys Res Commun 1991; 175: 1125-30.
Monsees TK, Goring M, Schill WB, Miska W. Possible roles of proteases in the regulation of spermatogenesis. Andrologia 1998; 30: 185-91.
Nakayama Y, Yamamoto T, Oba Y, Nagahama Y, Abe SI. Molecular cloning, functional characterization and gene expression of a follicle – stimulating hormone receptor in the testis of newt cynops prrrhogaster. Biochem Biophys Res Commun 2000; 275: 121-28.
O'Donnell L, Pratis K, Stanton PG, Robertson DM, McLachlan RI. Testosterone-dependent restoration of spermatogenesis in adult rats is impaired by a 5α-reductase inhibitor. J Androl 1999; 20: 109-17.
Ornerier W, Prins G, Musto NA, Suarez-Quian CA. Androgen receptor distribution in rat testis: new implications for androgen regulation of spermatogenesis. Endocrinology 1995; 123: 2307-16.

Ouhtit A, Morel G, Kelly PA. Visualization of gene expression of short and long form of prolactin receptor messenger RNAs encoding a unique precursor. Biol Reprod 1993; 49: 528-36.
Quigley CA. The androgen receptor. Physiology and pathophysiology. In: Testosterone: action, deficiency substitution (Nieschlag E, Behre HM, eds) Springer Verlag, Berlin. pp 33-107; 1998.
Parma P, Pailhoux E, Cotinot C. Reverse transcription-polymerase chain reaction analysis of genes involved in gonadal differentiation in pigs. Biol Reprod 1999; 61: 741-8.
Papadopoulos V, Dym M. Sertoli cell differentiation on basement membrane is mediated by the c-fos, protooncogene. Proc Natl Acad Sci USA 1994; 91: 7027-31.
Rannikko A, Penttila L, Zhang FP, Toppari J, Parvinen M, Huhtaniemi I. Stage-specific expression of the FSH receptor (FSHR) gene in the prepubertal and adult rat seminiferous epithelium. J Endocrinol 1996; 151: 29-35.
Ravindranath N, Dalal R, Solomon B, Djakiew D, Dym M. Loss of telomerase activity during male germ cell differentiation. Endocrinology 1997; 138: 4026-30.
Remy JJ, Lahbib-Mansais, Yerle Y, Bozon M, Couture L, Pasjot E, Grebert D, Salesse R. The porcine follitropin receptor: cDNA cloning, functional expression and chromosomal localization of the gene. Gene 1995; 163: 257-61.
Reventos J, Sullivan PM, Joseph DR, Gordon JW. Tissue-specific expression of the rat androgen-binding protein/sex hormone-binding globulin gene in transgenic mice. Mol Cell Endocrinol 1993; 96: 69-73.
Robert P, Amsellem S, Christophe S, Benifla J, Bellet D, Koman A, Bidart J. Cloning and sequencing of equine testicular follitropin receptor. Biochem Biophys Res Commun 1994; 201: 201-07.
Rouiller-Fabre V, Carmona S, Merhi RA, Cate R, Habert R, Vigier B. Effect of anti-Mullerian hormone on Sertoli and Leydig cell functions in fetal and immature rats. Endocrinology 1998; 139: 1213-20.
Sablitzky F, Moore A, Bromley M, Deed RW, Newton JS, Norton JD. Stage-and subcellular-specific expression of Id proteins in male germ and Sertoli cells implicates distinctive regulatory roles for Id proteins during meiosis, spermatogenesis, and Sertoli cells function. Cell Growth Differ 1998; 9: 1015-24.
Saez JM. Leydig cells: Endocrine, paracrine, and autocrine regulation. Endocrine Rev 1994; 15: 574-94.
Schlatt S, Meinhardt A, Nieschlag E. Paracrine regulation of cellular interactions in the testis: factors in search of a function. Eur J Endocrinol 1997; 137: 107-17.
Singh J, O'Neill C, Handelsman DJ. Induction of spermiogenesis by androgens in gonadotrophin-deficient mice. Endocrinology 1995; 136: 5311-21.
Suarez-Quian CA, Martinez-Garcia F, Nistal M. Regadera J. Androgen receptor distribution in adult human testis. J Clin Endocrinol Metab 1999; 84: 350-58.
Sullivan PM, Wang YM, Joseph DR. Identification of an alternate promoter in the rat androgen-binding protein/sex hormone-binding globulin gene that regulates synthesis of a mRNA encoding a protein with altered function. Mol Endocrinol 1993; 7: 702-15.
Syed V, Gomez E, Hecht NB. Messenger ribonucleic acids encoding a serotonin receptors and a novel gene are induced in Sertoli cells by a secreted factor(s) from male rat meiotic germ cells. Endocrinology 1999; 140: 5754-60.
Trasler JM, Alcivar AA, Awoniyi CA, Santulli R, Zirkin BR, Hecht NB. Temporal gene expression is restored concomitantly with germ cells in the experimentally regressed rat testis. Endocrinology 1992; 131: 297-4.
Vihko KK, LaPolt PS, Nishimori K, Hsueh AJ. Stimulatory effects of recombinant follicle-stimulating hormone on leydig cell function and spermatogenesis in immature hypophysectomized rats. Endocrinology 1991; 129; 1926-32.
Vornberger W, Prins G, Musto NA, Suarez-Quian CA. Androgen receptor distribution in rat testis: new implications for androgen regulation of spermatogenesis. Endocrinology 1994; 134: 2307-16.
Weinbauer GF, Wessels J. Paracrine control of spermatogenesis. Andrologia 1999; 31: 249-62.
Yong EL, Loy CJ, Sim KS. Androgen receptor gene and male infertility. Human Reprod Update 2003; 9:1-7
Yarney TA, Sairam MR, Khan H, Ravindranath N, Payne S, Seidah NG. Molecular cloning and expression of the ovine testicular follicle stimulating hormone receptor. Mol Cell Endocrinol 1993; 93: 219-26.
Zhang FP, El-Hafnawy T, Huthaniemi I. Regulation of luteinizing hormone receptor gene expression by insulin like growth factor-I in an immortalized murine Leydig tumor cell line (BLT-1). Biol Reprod 1998; 59: 1116-23.
Zhang FP, Pakarainen T, Zhu F, Poutanen M, Huhtaniemi I. Molecular characterization of postnatal development of testicular steroidogenesis in luteinizing hormone receptor knockout mice. Endocrinology 2004; 145: 1453-63.
Zhang FP, Poutanen M, Wilbertz J, Huhtaniemi I. Normal prenatal but arrested postnatal sexual development of luteinizing hormone receptor knockout (LuRKO) mice. Mol Endocrinol 2001; 15: 172-83.

Chapter 2

PARACRINE ROLE OF SERTOLI CELL

2.1. SERTOLI CELL

Enrico Sertoli in 1865 first proposed that Sertoli cells in testis function as "nurse cells". In other words, these cells provide nutrients and regulatory factors for the sustenance of germ cells. Since the time of E. Sertoli (1865), number of morphological studies showed intimate relationships between the plasma membrane of Sertoli cells and different categories of germ cells. As stated in Chapter 1, the seminiferous tubules are separated into compartments by tight-junctional complexes between adjacent Sertoli cells. The timely opening and closing of inter-Sertoli cell tight junctions in the testis are essential cellular events in the completion of spermatogenesis. Blood macromolecules are effectively excluded from the inner or adluminal compartment, and the milieu in this compartment is composed of secretions from Sertoli cells and germ cells. Germ cell development is dependent on the interaction of groups of germ cells with Sertoli cells. The first constructed Sertoli cell from a rat displayed contact with nearly different germ cells of 4 different cell types (type B spermatogonia, pachytene spermatocytes, step 5 spermatids and step 17 spermatids). While assessing the effect of various inhibitors of protein phosphatases and kinases on the inter-Sertoli tight junction permeability barrier, it appeared that the interplay of protein kinases and phosphatases is very crucial in modulating the assembly and maintenance of inter-Sertoli tight junctions in the testis (Li et al., 2001).

Sertoli cells divide during the prepuberal period until the blood testis barrier is established. Sertoli cells coordinate germ cell proliferation and development, and determine adult testis size and sperm production. The evidence for production of sperm cells is compelling since under normal and pathological conditions, the number of Sertoli cells is correlated precisely with the number of sperm produced. Sertoli cell secretion is influenced by the type of germ cells, which are present in the seminiferous epithelium. In the rat, FSH establishes the competence of Sertoli cells to bind round spermatids. In these studies, while testosterone alone stimulated spermatid binding once the Sertoli cells had been primed with FSH, the FSH deprived Sertoli cells lost the ability to bind spermatids. But testosterone could restore the spermatid binding only in presence of FSH. Studies on heterologous germ cell transplantation also suggested that mouse testis could support the development of rat germ cells including the formation of spermatozoa in association with mouse Sertoli cells.

2.1.1. Co-culture Experiments

Reports from several laboratories strongly suggest that Sertoli cells exert a paracrine control of the two main testicular functions: androgen secretion and spermatogenesis. Supporting evidence of this role of Sertoli cells was obtained by co-culture of Sertoli cells with other testicular cells. i) Co-culture of pig or rat Sertoli cells with pig Leydig cells produces an increase in the hCG receptor number and an increase in steroidogenic activity of Leydig cells. Pretreatment with FSH further increased the values of these two parameters. The effect of Sertoli cells on Leydig cells depends upon the ratio of the two cells and on the substrate in which the cells are cultured. Moreover, Leydig cells produce an increase in the FSH receptor number and an increase in the FSH stimulation of plasminogen activator by Sertoli cells. ii) Co-culture of rat or pig Sertoli cells with rat germ cells induces an increase in the RNA and DNA biosynthetic activities of germ cells. Most of the stimulatory effects seemed to be mediated by diffusible factors, secreted by Sertoli cells, but full expression of the stimulatory action was observed when germ cells were in contact with other cells. In this co-culture system, a fraction of rat germ cells containing mainly mature forms of spermatocytes inhibited rat Sertoli cell RNA and DNA synthesis, but had no effect on pig Sertoli cells. iii) On the contrary, a fraction of rat germ cells richer in spermatogonia and preleptotene spermatocytes stimulated rat Sertoli cells DNA synthesis but was without effect on pig Sertoli cell (reviewed in Saez et al., 1986). iv) Sertoli cells conditioned medium derived from 12 days old rats possess factors, which show natural killer cell activity of mammalian lymphocytes and age related stimulation of quiescent rat pre-spermatogonia. A dose dependent inhibition of T and B lymphocyte proliferation has also been shown by pre-pubertal Sertoli cell secreted protein (pSCSP). These results suggested that by the time the blood testis barrier had been formed, Sertoli cell in rat testis had already synthesized immuno-regulatory proteins (Martinova et al., 1993). v) Sertoli cells provide regulatory factors for the sustenance of germ cells. The promoter sequences from the *Pem* homeobox gene that directs Sertoli cell specific expression have been identified. A 0.6-kb 5'-flanking sequence directs transgene expression specifically in the testis and the epididymis but not in any other tissue, has been tested. There are at least two regulatory regions in the *Pem* proximal promoter: one that directs androgen receptor-dependent expression specifically in Sertoli cells and another that confers stage-specific expression in neonates and adults by acting as a negative regulator (Rao et al., 2003).

2.1.2. Dependence of Sertoli Cell Products on Stage of Spermatogenesis

Large percentage of protein synthesis by Sertoli cells is related to the secretion of glycoproteins, which can be classified in categories as: i).The transport or bioprotective proteins that are secreted in relative high abundance. These include metal ion transport proteins such as transferrin and cerulloplasmin. ii). The second category of secreted proteins includes proteases and protease inhibitors, which play important role in tissue remodeling that occurs during spermiation and movement of pre-leptotene spermatocytes into the adluminal compartment. iii). Another category of Sertoli cells secretion includes the glycoproteins that form the basement membrane between the Sertoli cells and the peri-tubular cells. iv) Sertoli cell regulatory glycoproteins are produced in very low abundance but sufficient to carry out their biochemical roles. These categories of glycoproteins function as growth factors or paracrine factors and include product such as Mullerian inhibiting substance, inhibin and others. Secretory products of Sertoli cells depend on stages of spermatogenesis. According to Griswold (1995), cycle of the seminiferous epithelium in the rat consists of a series of 14 columns that describes a different group of cellular associations found in the testis. In rodents, the stages of cellular associations are arranged in a consecutive and linear fashion along the seminiferous tubule. An alternative way to depicting this cycle is the expression in the form of circular representation where the duration of the stages is proportional to the arc of the circle. Germ cell development

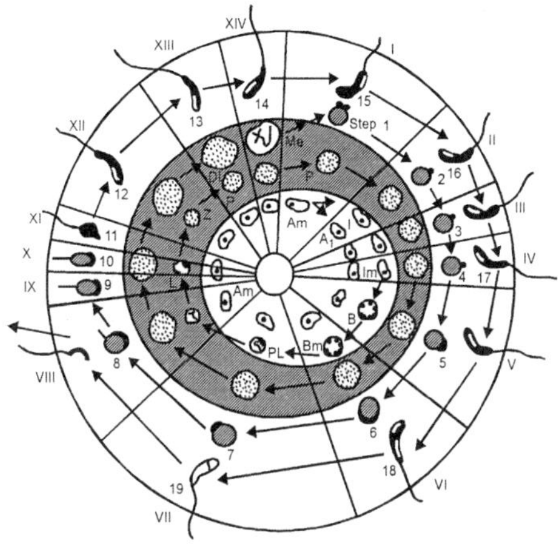

Fig.2.1. Diagramatic representation of 14 stages of cycle of seminiferous epithelium in rat. Cycle is pictured as a circle with each stage defining an arc proportional to the duration of that stage. Germ cell development is pictured as a spiral from the center circle of A_1 spermatogonia. While oval in the center of diagram encloses developmental events that occur in basal compartment, shaded region encloses pre-meiotic events within adluminal compartment. Out side shaded area are post-meiotic stages that also occur in adluminal compartment. Am - mitosis of A spermatogonia; B-B spermatogonia; I-zygotene; Me - meiotic divisions I and II. Spermatid steps are numbered 1-19. Reprinted with permissiom from M.D. Griswold. Biol Reprod.52; 221-16:1995 © American Society for Study of Reproduction.

can be depicted as a spiral beginning from the A1 spermatogonia to the spermatids. Several molecular events taking place in Sertoli cells and germ cells have shown that the secretions by Sertoli cells and thus the immediate environment in the seminiferous epithelium varies as a function of the stage of the cycle. Sertoli cells associated with germ cells at one stage of the cycle express gene different from those in other parts of the cycle. This activity is exemplified by cathepsin, which is secreted from Sertoli cells at stages VI and VII, where as cystatin, a protease inhibitor specific for cathepsin L, is secreted primarily at stages XII-I. The secretion of these two proteins clearly varies in different parts of the cycle. Similar analysis has been done for number of proteins, which suggest a dual mode of Sertoli cell function. One mode includes the stages of meiotic division, in which the secretion of a group of Sertoli cell products is maximal and the other mode includes the spermiation stages (Griswold, 1995) **(Fig.2.1)**.

2.2. BASEMENT MEMBRANE COMPONENTS OF SEMINIFEROUS TUBULES

It is now clear that a large degree of molecular heterogeneity exists in basement membranes (BMs) of different tissues and at different stages of development. Basement membrane function is altered in acquired and genetic diseases, such as Good pasture syndrome, Alport syndrome, and diffuse leiomyomatosis. Therefore, it is important to analyze the molecular changes in BMs that are associated with key cellular changes in the developing testis (Loveland et al., 1998; Kahsai et al., 1997). Sertoli and myoid cells are known to produce several components of basement membrane. A cooperative interaction between Sertoli and peritubular cells is necessary for the synthesis and deposition of the seminiferous tubule BM, which is influenced by locally produced factors such as platelet derived growth factor. Abnormal thickening of the seminiferous tubule BM is frequently associated with reduced spermatogenic function in humans, and it has been suggested that deregulated production of BM components directly influences the phenotype of Sertoli and peritubular myoid cells.

2.2.1. Collagen Type IV and Type II

Type IV collagen, the major constituent of BMs has been linked to the pathogenesis of many disorders. Type IV collagen comprises a family of triple-helical isoforms consisting of six genetically distinct chains designated α1(IV) to α6(IV). The existence of six α(IV) chains allows for as many as 56 different kinds (isoforms) of triple helical molecules, which differ in type and stoichiometry of various chains. Evidence for heterotrimers having composition of $[\alpha 1\ (IV)]_2$ α2 (IV) and $[\alpha 3(IV)]_2$ α4 (IV) has been obtained. The six α(IV) chains differ considerably with respect to tissue distribution. The α1 (IV) and α2(IV) chains have a ubiquitous distribution whereas the α3(IV), α4(IV) and α5(IV) chains have a restricted distribution in both human and rodent tissues. Seminiferous tubule basement membrane (STBM) in testis is composed of all six α-chains of type IV collagen. The content of α3(IV) chain (40%) and the α4(IV) chain (18%) in STBM is substantially higher than in any other basement membrane collagen. During structural reorganization of BMs of normal testicular development, Collagen VI is initially restricted to the interstitial matrix between seminiferous cords of the fetal and early postnatal rat testis, and subsequently it is observed to be associated with the BM around Sertoli cells and peritubular cells by Day 14, when these cell types are beginning to differentiate. In the mouse, collagen α3(IV) chains may have a role in spermatogonial proliferation. In addition, three major hexamer populations were found that represent the classical network of the α1(IV) and α2(IV) chains and two novel networks, one composed of the α1(IV)-α6(IV) chains and the other composed of the α3(IV)- α6(IV) chains. The results established a structural linkage between the α3(IV) and α5(IV) chains, suggesting a molecular basis for the conundrumin in which mutations in the gene encoding the α5(IV) chain cause defective assembly of the α5(IV) chain in the glomerular basement membrane of patients with Alport syndrome (Kahsai et al., 1997).

The gene for type II collagen, *Col2α1*, is known as a target of Sox9 in mouse chondrocytes and *Col9α3* as a Sox9 target in testis. The nonchondrocytic collagen transcript isoforms are also expressed in the early male mouse gonad. Male-specific, gonadal expression of nonchondrocytic *Col2α1* was first seen at 11.5 dpc and undetectable by 13.5 dpc. This was accompanied by increasing expression of nonchondrocytic *Col9α1, Col9α2,* and *Col9α3,* first detected at 11.5 dpc. Germ cells depleted testes showed that *Col9α3* and *Col2α1* are expressed in Sertoli cells within the developing testis cords. Nonchondrocytic type II collagen contains a cysteine-rich domain that binds members of the transforming growth factor β superfamily of signaling molecules. This interaction may play a role in the development of testis (McClive et al., 2003).

2.2.2. Laminin

Sertoli cells alone produced laminin and collagen IV but not fibronectin, while myoid cells produced all three proteins. In Sertoli-myoid co-cultures, a sequential deposition of the components into extracellular fibers has been noted during 5 days of culture. Analysis revealed that mRNA levels for the laminin B1 chain and collagen IV increased from Day 3 to 5 in Sertoli cell monocultures. By contrast, the levels of laminin B1, collagen IV, heparan sulfate, proteoglycan and fibronectin decreased in the co-cultures. Transcripts for the laminin A chain were not detected in the myoid cell; instead these cells produced the mRNA for the laminin homologue merosin. This provides the basis for the mechanisms that regulate the expression of the basement membrane genes by Sertoli and peritubular myoid cells in the testis (Richardson et al., 1995). Rat Sertoli cells respond to extra-cellular matrix macromolecules by increase of Ca^{2+}. Laminin dependent Ca^{2+} increase was down-regulated by FSH (Taranta et al., 2000).

2.2.3. Fibulins and Other Components

Fibulins are a new class of matrix glycoproteins characterized by the presence of 9-11 epidermal growth factor like repeats that include consensus sequences for Ca^{2+} binding. Two members have been identified: fibulin1 and fibulin 2. Fibulin 1 is a 90kDa monomer protein of fibulins which has four alternate splice variants. Fibulin 2 has a larger N-terminal domain and it is a dimmer of two disulfide bonded 195 kDa monomers. The tissue expression pattern of fibulin1 and fibulin2 are broad in interstitial matrices and BMs of embryonic and adult tissues. Analyses of interactions with other extracellular matrix components have shown that both fibulin–1 and fibulin–2 bind strongly to nidogen and fibronectin and more weakly to some other components, by Ca^{2+-} dependent mechanisms in most cases. This suggests a potential role for fibulins in matrix assembly. Fibulins may be involved in the structural reorganization of BMs during testis development. Localization of fibulins has been compared with localization of other components (Laminin, nidogen, perlecan, collagen VI, and fibronectin) alongwith propyl-thiouracil (PTU) treated hypothyroid rats, in which testis development is significantly delayed (Loveland et al., 1998). The delay in maturation of the seminiferous tubule BM in the testes of PTU-treated rats demonstrates a correlation between changes in the composition of the tubule BM and cellular development of the testis (Loveland et al., 1998).

2.3. SERTOLI CELL PRODUCTS IN GERM CELLS DEVELOPMENT

Sertoli cells synthesize and secrete a large variety of factors including steroids, proteins, growth factors, opioids, proteases and prostaglandins. With this scenario the Sertoli cells control the spermatogenic process in all aspects and guide the germ cells along their long and complex development from a stem cell spermatogonium into elongated spermatid. Sertoli cells and germ cells are in fact intimately associated on a functional and morphological basis, and for that Sertoli cells possess specialized processes for interaction with germ cells. Strong evidence comes from transplantation studies of germ cells, which suggested that rat spermatozoa are produced in mouse testes along with mouse germ cell production. Morphologically rat and mouse sperm were nourished by mouse Sertoli cells maintaining 8 days spermatogenic cycle in mice and 12 days in rats (Franca et al., 1998). Cell-cell interaction between germ and Sertoli cells controls the duration of cell cycles and cellular organization. Rat germ cells that were transplanted and supported by mouse Sertoli cells always differentiated with the cell cycle timing characteristic of the rat and development spermatogenic pattern of rat, demonstrating that germ cell differentiation during spermatogenesis is regulated by germ cells alone (Franca et al., 1998). The obvious conclusion from various experiments is that the number of germ cells appears to be directly related to the number of Sertoli cells and probably to their synthetic capability. Notwithstanding, the fundamental notion that androgens and FSH are required for the onset and the maintenance of spermatogenesis is relatively unchallenged. It appears that Sertoli cells probably involve the expression of a number of genes unique to prenatal differentiation. The Tyro3, Axl and Mer receptors are normally expressed in Sertoli cells during postnatal development whereas their ligands, Gas6 and protein S are produced by Leydig cells before sexual maturity, and by both Leydig and Sertoli cells thereafter. The concerted activation of Tyro 3, Axl and Mer in Sertoli cells is critical to the role that these cells play as nurturers of developing germ cells.

The expression of germ cell specific genes in co-cultures of Sertoli cells with either pachytene spermatocytes (PS) or round spermatids (RS) showed that the mRNAs encoding phosphoprotein p19 and the testis specific histone TH2B were specifically expressed in pachytene spermatocytes whereas those encoding the transition proteins TP1 and TP2 were specific to round spermatids (Weiss et al., 1997). Differential display gene expression induced by co-culturing rat Sertoli cells and germ cells identified cDNAs that are either down regulated or up-regulated in co-cultures of germ cells and Sertoli cells. In conclusion, it showed that a number of specific genes are important in cell-cell communication during spermatogenesis (Syed and Hecht., 1997). Syed and Hecht (2001) concluded that the Sertoli cells from aged regressed testes are unable to respond to selective signals from germ cells of young rats; germ cells from regressed testes showed a similar selective loss. Among these most notable was the expression by Sertoli cells of Serotonin gene expressed, which failed to express in germ cell regressed testes. Such disruptions in communication between Sertoli cells and germ cells likely contribute to germ cell loss during aging (Syed and Hecht, 2001). Besides several growth factors, ABP, FSH regulatory factors (inhibin family) (discussed in chapter 3), secreted by Sertoli cells, and some of the recently discovered factors need attention for future study. N-cadherin has also been suggested as a candidate for mediating the endocrine controlled interaction between Sertoli cells and round spermatids (Perryman et al., 1996).

2.3.1. Cytokines

Cytokines are small secretory cellular proteins that attach to cell surface receptors. Cytokines trigger cell proliferation and differentiation. A role for TNFα in Sertoli-germ cell communication and leukaemia inhibitory factor has been implied in the autocrine control of Sertoli cell proliferation (Hara et al., 1998). Cytokines show discrepant effects on tumour cells varying from anti- to pro-apoptotic activities. Testicular germ cell tumors (TGCT) express a functional form of the pro-inflammatory cytokine interferon-γ (IFNγ). Analysis of TGCT cell lines demonstrated that they were resistant to the endogenous IFNγ. In order to explain tumour resistance to endogenous IFNγ, Schweyer et al., (2003) suggested that despite secreting biologically active IFNγ, the concentration of the endogenous IFNγ is too low to stimulate the IFNγR/STAT signaling pathway in TGCT in an autocrine and/or paracrine manner.

Associated molecule with SH3 domain of STAM (AMSH) is critical in the cytokine-mediated intracellular signal transduction downstream of the Jak2/Jak3-STAM complex. A related molecule of AMSH in the mouse brain is AMSH-FP (AMSH-Family Protein), which may act as an adaptor of gene transcription. AMSH-FP has two splicing forms: one is ubiquitous, whereas the other one is restricted to testis. The abundant type is named as AMSH-FPα and the testis form as AMSH-FPβ. The AMSH-FPβ is a variant lacking N-terminal 166 amino acid residues of AMSH-FPα. Analysis of the 5'-untranslated regions in AMSH-FPβ and AMSH-FPβ mRNAs and exon-intron structure of AMSH-FP gene suggested that testis-specific transcripts are generated due to alternative promoter usage and/or alternative splicing. Importantly, AMSH-FPβ mRNA was not detected in juvenile and infertile mouse testis but was restrictively expressed in the haploid stage of testicular germ cells in the normal mature testis. The AMSH-FPβ appears to have a functional role in the spermiogenesis (Kitajima et al., 2003).

IL-1 and IL-6: Among several other growth factors in the testis, interleukin-1 (IL-1) is particularly significant. IL-1α is present in rat and human testis, where it may promote DNA synthesis and differentiation of spermatogonia and preleptotene spermatocytes in the rat testis, whereas mice lacking IL-1 type-1 receptor were fully fertile. Sertoli cells produce IL-1α, whose secretion is enhanced by known activators of macrophages/monocytes, such as LPS and latex beads. Constitutive expression of IL-1αmRNA in rat Sertoli cells is dependent on interaction of germ cells (Jonsson et al., 1999). Other IL-6 family members, which are important in cardiac differentiation and pleiotropic functions in the hematopoietic and immune systems are also produced by Sertoli cells. In contrast to IL-1, IL-6 may be an inhibitor of pre-meiotic DNA synthesis within the seminiferous epithelium. IL-1 and IL-6 productions are stimulated by residual bodies, which are cytoplasmic fragments of spermatids that are shed at the time of sperm release in testis. It is suggested that IL-1 produced by Sertoli cells in response to LPS or residual bodies induces IL-6 through the lipoxygenase pathway (Syed et al., 1995). In addition to growth factors, Sertoli cell produces energy substrates such as lactate. Lactate represents a preferential energy substrate for germ cells. The inability of germ cells to use glucose to energize their metabolism, their preference for lactate as an energy source have generated a concept related to Sertoli cells-germ cell metabolic cooperation, with lactate playing a pivotal role. Assuming that IL-1α is produced by both Sertoli and germ cells, which utilize lactate for their energy metabolism, Nehar et al, (1998) suggested that IL-1α may represent a signal in the metabolic cooperation existing between Sertoli cells and germ cells, resulting into a redistribution of LDH isoforms.

Tumor Necrosis Factor α (TNF-α): Tumor necrosis factor α (TNF-α) released by spermatids is detected by Sertoli cells and may serve as a paracrine factor, regulating as yet a defined process in spermatogenesis (De et al., 1993; Benahmed, 1997)). The presence of TNF-α appears to provide TM4 Sertoli cells with the active cofactor for NO production. Although the mechanism of action is unknown, the TNF-α-induced signaling for induction of NO synthesis is not dependent on protein kinase C activation (Chung et al., 2000). The bioactive plasminogen activator inhibitor-1 (PAI-1) protein and PAI-1 mRNA are stimulated by TNF-α. It was suggested that some of the biological effects of TNF-α on PAI-1 might be secondary to de novo synthesis of EGFR and may occur in the context of physiological interactions between the testis and the immune system (Le Magueresse-Battstoni et al., 1997).

Interferons: Interferon (IFN) proteins are known for their antiproliferative and immunoregulatory activities and their crucial involvement in cellular antiviral action. There are two types of IFNs : type I IFNs (IFNsα, IFNsβ, IFNΩ, and IFNδ and type II IFN (IFN-γ). Peritubular and Sertoli cells in testes secrete substantial amount of type I IFNs. In contrast, the meiotic pachytene spermatocytes and the postmeiotic haploid early spermatids produce low levels of type I IFNs constitutively. In the search of the testicular antiviral defense system, Dejucq et al (1998) studied IFN production by Leydig cells, testicular macrophages, and spermatogonia after exposure to Sendai virus and found that spermatogonia do not constitutively express IFNs and give a very poor response to the virus. In contrast, testicular macrophages constitutively produce type I IFNs, which was markedly stimulated by Sendai virus. Leydig cells produced twice as much type I IFNs as testicular macrophages after viral exposure, producing both IFNα and -γ. The Sendai virus stimulated testosterone production by Leydig cells. Thus, the potential antiviral system is represented by Leydig cells and, also by macrophages, which play a key role in protecting both androgen production and spermatogenesis (Dejucq et al., 1998). Targeted gene mutations studies suggested that IFNs can alter the development of testicular germ cells. In transgenic mice over-expressing either the IFNα or IFNβ gene, the process of spermatogenesis is disrupted with concomitant destruction of germ cells. Such effects of IFNs and those of Dejucq et al (1998) are suggestive of specific receptor molecules as targets interacting with cytokine ligands within the seminiferous tubule (Sertoli and/or germ cells) or interstitium (e.g. Leydig cells). Interferon-γ (IFN-γ) like IL-6 is a multifunctional cytokine that regulates cell function through tyrosine phosphorylation of latent proteins. It was shown that IFN-γ transmits its signal through a cell surface receptor (IFNγ-R), which consists of a primary ligand binding α-chain (IFNγRα) and a signaling β-chain (IFNγRβ) (see Kanzaki and Morris, 1998). Kanzaki and Morris (1998) studied the expression of (IFNγRα) and (IFNγRβ)subunits in purified rat testicular cells and their differential regulation by cytokines in Sertoli cell cultures, as well as the cell signaling pathway of testicular IFNγR, and its functional role using primary Sertoli cells. It was demonstrated that the mRNAs and their proteins for both chains of IFNγR subunits are present in cultured primary Leydig and Sertoli cells of testis of immature rats. The expression of both IFNγR component mRNAs in Sertoli cells was increased by its specific ligand (IFNγ), as well as IL-1α and TNF-α. IFNγ-activation of the Janus tyrosine kinases (JAK) indicated that IFNγR, expressed in the Sertoli cell, is functional. Moreover, IFNγ modulates the expression of interferon regulatory factor (IRF)-1 and IL-1β converting enzyme genes in Sertoli cells. Data suggested a role(s) for IFN-γ in the regulation of distinct gene expression and cell-specific sensitivity to apoptosis in the testis (Kanzaki and Morris, 1998).

The binding of IL-6 and IFN-α to distinct membrane receptors activates several members of the Jak family and with subsequent tyrosine phosphorylation of the receptor and its associated binding component, gp 130 mediated in part by activation of kinases and STAT3 phosphorylation. Jenab and Morris (1997) showed that IL-6 and IFN-α induce nuclear factor STAT-3 and STAT-1 DNA-binding activity to the cis-inducible element of c-fos in a genistein –dependent pathway. Quantitative analyses showed that differential induction of c-fos, junB and c-myc mRNA by these cytokines occurs at transcriptional levels. Genistein treatment blocks the induction of *c-fos* and *junB* gene expression demonstrating that tyrosine phosphorylation of STAT proteins is involved in the cytokine regulation of the Sertoli immediate early genes. Thus transcriptional regulation of Sertoli cell immediate early genes by IL-6 and IFN-γ is mediated through phosphorylation of STAT-3 and STAT-1 proteins (Jenab and Morris, 1997).

Leukemia Inhibitory Factor: Leukemia inhibitory factor (LIF), a member of IL-6 ciliary neurotrophic and gp130 pleiotropic family of growth factors, inhibits differentiation of embryonic stem cells, induces tubulogenesis in the embryonic kidney, and regulates sperm differentiation. Leukemia inhibitory factor has been shown to enhance the survival of rat testicular gonocytes and Sertoli cells (De Miguel et al., 1998). The gp130 and LIF mRNAs expressed in the somatic (the Sertoli and Leydig cells) and specific germ cells (spermatogonia, pachytene, round, and elongated spermatids) of rodent testes are suggestive of cell specific LIF mediated functions (Jenab and Morris, 1998) (see Chapter 3).

2.3.2. Growth Factors

Growth factors are mainly polypeptide molecules that upon binding to surface receptors in target cells elicit the activation of signal transduction pathways involved in cells division and

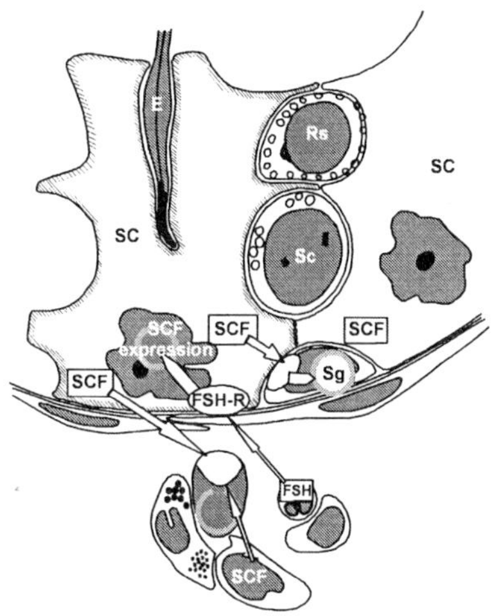

Fig.2.2. Contact of Serfoli cell with all other germ cells. Stem cell factor (SCF) produced by Sertoli cells and Leydig cells interacts at FSH-R in Sertoli cell. The SCF binds stem cell/spermatogonial c-kit receptor with protein kinase activity, and initiates spermatogonial proliferation. SC, Serfoli cell cytoplasm; Sg, spermatogonia; Sp, spermatocyte; Rs, round spermatid; E, elongating spermatid.

cell specific differentiation. The action of growth factors can be stimulatory or inhibitory and this has also been shown for many factors using cultured testicular cells. Several of these factors (FGF, IGF) have been found in Sertoli cells and germ cells. On the other hand, EGF/TGF have been specifically implicated in the survival and proliferation of rat type A-spermatogonia. Nerve growth factor (NGF) is a protein by which germ cells communicate with Sertoli cells. The fibroblast growth factor (FGF) family represents signal molecules of another group. Sertoli cells and germ cells produce FGF-like proteins. FGF receptors are present in Sertoli cells. A similar category of FGFs function has been postulated for the insulin like growth factor (IGF) family, which includes three structurally related peptides: insulin, IGF-I (also called somatomedin C) and IGF-II. Another molecule probably acting mainly as an environmental factor is transforming growth factor (TGF)-α. It is synthesized and secreted by Sertoli cells, Leydig cells and peritubular cells while Sertoli cells express the TGF/epidermal growth factor receptor. The large family of TGF-β growth factors, which also includes the inhibins and activins, is the most versatile group with the largest multifunctional spectrum of effects among all known growth factors. The proteins or mRNAs for several growth factors and growth factor receptors have been identified in the testis. The EGF receptors are present on Sertoli, Leydig and peritubular cells (see Chapter 3 for details) (Schlatt et al., 1997; Weinbauer and Wessels, 1999).

2.3.3. Stem Cell Factor

Unlike cytokines, the need for stem cell factor (SCF) for spermatogenesis is clearly established (Loveland and Schlatt, 1997) **(Fig.2.2)**. Sertoli cell is the only cell in seminiferous tubules that produce SCF. Mutant mice lacking either c-kit or SCF are sterile and their gonads are deficient in germ cells. Testicular expression of SCF is under control of FSH. Investigations have yielded compelling evidence that the c-kit/SCF system is pivotal for spermatogonial differentiation in rodents and humans. The stem cell factor is an important local regulator and mediator of spermatogonial development in vivo, which is synthesized and secreted by somatic cells, acts on germinal cells and whose expression is under endocrine control. In human testis, c-kit is present in spermatogonia and SCF in Sertoli cells (Strohmeyer et al., 1995; Sandlow et al., 1997). A correlation between spermatogonia and seminal SCF in men has been suggested (Fujisawa et al., 1998) (see Chapter 17).

2.3.4. Sertoli Cell's Cystatins

The cystatins are a superfamily of cysteine proteinase inhibitors that consist of three related families. The in vivo function of cystatins is not understood. Sertoli cells synthesize and secrete cystatin C, a potent inhibitor of cysteine proteases. Cultured Sertoli cells secrete three forms of cystatin C: a predominant pair of proteins of 13-14-kDa and a less abundant 20-kDa protein. Rat testis does not show cystatin C in spermatogonia or spermatocytes. Sertoli cells contain a 700-nt cystatin C transcript. Population of spermatocytes contained a 550 nt transcript whereas RNA from purified spermatids contains a small 550-nt transcript. The 700-nt transcript is present in testes of rats of 5-79 days of age, the 500-nt transcript is detected initially in testis from 24 days old rats, and the 550-nt transcript is detected initially at 35 days of age. The RNA from stage synchronized testes showed that steady state level of both the 550 and 700-nt transcripts were lowest in stages VI-VII of the cycle. The cystatin C in the testis may inhibit the proteolytic activity of the cysteine protease, cathepsin L in all stages except stages VI-VII (Tsuruta et al., 1993). Two new members of mouse cystatin multigene family called cystatin SC (cystatin-related gene expressed in Sertoli cells) and cystatin TE-1 (cystatin-related gene highly expressed in testis and epididymis) have been characterized recently (Li et al., 2002). The full length cDNA sequences of cystatin SC contains an open reading frame that enodes a putative signal peptide of 20 amino acids and a mature protein of 110 amino acids, whereas that of cystatin TE-1 encodes a protein of 128 amino acids. Both of these cystatins sequences contain four highly conserved cysteine residues in precise alignment with other cystatins. The cystatin SC mRNA was detected only in testis, whereas cystatin TE-1 was highly expressed in testis and epididymis. The functions of these proteins in vivo are unknown (Li et al., 2002).

2.3.5. Mannose–6-phosphate Receptor

Sertoli cells secrete glycoproteins containing mannose-6-phosphate. Analysis of the conditioned medium for lysosomal enzyme precursors, which typically bear the M-6-P recognition marker, indicated that Sertoli cells selectively secreted β-N-acetylhexosaminidase and α-mannosidase, but not β-glucuronidase or β-galactosidase. Thus it provides evidence that mannose-6-phosphate receptors (MPRs) on the surface of spermatogenic cells endocytose secreted glycoproteins that are likely to be present in the seminiferous epithelium (O'Brien et al., 1993; Gamzu et al, 1998). The relative expression of mannose receptors mRNAs in spermatogenic and Sertoli cells, poly (A) RNAs were examined by using cDNA probes specific for the cation independent (CI-) and cation dependent (CD) MPRs. A single CI-MPR transcript, of 10 kb in size, ubiquitously present in all tissues or cell types is more abundant in Sertoli cells than in spermatogenic cells of adult testes. Multiple CD-MPR are synthesized by pachytene spermatocytes and round spermatids including a 2.4-kb CD-MPR mRNA that is indistinguishable from small CD-MPR transcripts of somatic tissues and Sertoli cells. Smaller

CD-MPR mRNAs of 1.4 and 1.6 kb are prominent in pachytene spermatocytes and round spermatids, respectively, but undetectable in somatic tissues. It appeared that alternate polyadenylation signals are used to produce multiple CD-MPR transcripts in spermatogenic cells (O'Brien et al., 1994). The significance of MPR in spermatogenesis needs to be investigated in details.

2.3.6. Other Sertoli Cell Products

It seems Sertoli cells synthesize and secrete many serum proteins behind blood testis barrier that are also produced by hepatocytes (Braghiroli et al., 1998). Sertoli cells provide proteoglycans needed for structural support of germ cells. Cell surface proteoglycans are mostly heparan sulfate (HSPG) produced by immature rat Sertoli cells. Expression of three HSPG, syndecan–1, syndecan–4 and glypican–1 in Sertoli cells of 20 day old rat suggested a regulation of their expression by the phosphatidyl inositol pathway. This over expression could be related with paracrine factors secreted by germ cells (Brucato et al., 2000). Among these proteoglycans glypican-1, syndecans-1 and -4 mRNAs were expressed and differentially regulated. Glypican-1 and syndecan-1 mRNA expression was up-regulated under PKC activation in contrast to syndecan-4 mRNA, which was not affected (c/r Brucato and Villers, 2002). Proteoglycans synthesis is regulated by insulin growth factor-I (see chapter 3). In addition, a physiological effector of the PKC such as ATP gave similar effects.

It has been proposed that stage-specific gene expression in Sertoli cells results from sequential activation and repression of transcription. However, the exact molecular mechanisms are unknown. By use of gene trapping strategy, one can identify genes, which in Sertoli cells are either up-or down-regulated by signals emitted by germ cells. Using this technique gene encoding Fra1, a component of AP1 transcription complex was identified. Accumulation of Fra1 mRNA was induced in Sertoli cells by addition of either round spermatids or nerve growth factor (NGF). The effect of NGF was mediated by the TrkA receptor through ERK1-ERK2 kinase pathway. Results suggested that AP1 activation might be an important relay in the Sertoli cell-germ cell cross-talk (Vidal et al., 2001).

A new gene product SERT shows no homology to any of the known gene sequences in the GenBank database. But some homology was found with short sequences in EST database. It is a Sertoli cell product which shows no homology in Genbank database. The SERT is an RNA binding protein motif involved in *Drosophila* sex determination (Chaudhary et al., 2004). A 3-kb genomic fragment immediately upstream of the rat cathepsin L translation start site has been identified in Sertoli cells of transgenic mice, in a manner comparable to that of the endogenous gene (predominantly in stages VI-VIII tubules).

Transferrin : In vivo evidence for the role of several proteins in spermatogenesis is lacking in most of the cases. One exception is the putative role of transferrin (Tf), an iron transport protein. Transferrin seems to play complex physiological roles related to cell function, differentiation and proliferation. The Sertoli cell makes transferrin as part of a proposed iron shuttle system that effectively transports iron around the tight junction complexes to the developing germ cells. However, existence of an iron shuttle system does not prove that iron transport is necessary for germ cell development. The protein is essentially synthesized in hepatocytes, but also in Sertoli cells, in the epithelial cells of the choroid plexus in rodents and in oligodendrocytes in all species analyzed (Zakin et al., 2002).

Nuclear Factor kβ (NF-kβ): The Rel/nuclear factor (NF) κβ family of transcription factors is important intracellular conveyors of extra-cellular signals in a number of systems. NF-κβ p50 and p65 proteins have been found to be constitutively present and active in the nucleus of Sertoli cells, of rat testis. In vivo, NF κβ proteins are present in the nucleus of Sertoli cells during all 14 (I-XI) cyclical stages of spermatogenesis with elevated level of expression in stage X IV. In contrast, NF κβ p50 and p65 subunits are transiently expressed in the nuclei of germ cells with peak levels found in pachytene spermatocytes during stages VII-XI and at lower levels in stage I-VII spermatids. The TNF-α, which is produced by round spermatids in the testis, increased nuclear NF-κβ binding activity to Sertoli cells. Stimulation of Sertoli cells with activators of PKA signaling pathway such as forskolin or FSH also increased NF-κβ DNA binding activity. Addition of TNF-α to Sertoli cells further stimulated κβ enhancer mediated transcription. Thus, NF-κβ proteins are stage specifically localized to Sertoli cell and spermatocytes nuclei and play a role in the regulation of stage specific gene expression during spermatogenesis (Delfino and Walker, 1998; Lilienbaum et al., 2000).

2.4. SERTOLI CELL JUNCTION ADHESIONS

2.4.1. Neural Cell Adhesion Molecule (NCAM) at Sertoli Cell-Gonocyte Junction

During neonatal development gonocytes resume mitotic activity and or migrate to basal lamina. These crucial events occur without added factors or hormones and are hence likely to depend on interaction with adjacent Sertoli cells. It seems likely that changing intercellular adhesion is a function of progessive development of Sertoli cells. Coupling of the kit receptor protein on gonocytes to stem cell factor from Sertoli cells is vital for successful migration by gonocytes. During the neonatal period, neural cell adhesion molecule (NCAM) is the main adhesive molecule expressed and functioning at birth, with a progressive decline as development proceeds. This factor is recognized in supporting of Sertoli cell differentiation (Newton and Millette, 1992). Gonocytes or Type A spermatogonia, persist in long-term co-culture maintained for 15 days and that NCAM continues to be expressed at high levels in these cultures. Other avenues including secretion of growth factors and other peptides also exist in the makeup of the extracellular matrix, which Sertoli cells and gonocytes contact (Orth et al., 2000). Gonocyte-Sertoli cell adhesion is very susceptible to phospholipase C in co-cultures isolated from newborns. The way in which gonocytes adhered to Sertoli cells appeared to change during the immediate postnatal period, as observed by change in phospholipase sensitivity. This perhaps indicates production of a phospholipase C-resistant NCAM isoform following several days after birth. The NCAM is found ubiquitously at cell-cell interfaces within the seminiferous cord from birth through day 10 and thereafter is restricted to interstitial cells. Moreover, only polysialic acid-negative 140-kDa NCAM is expressed in the testis and supported the view that 140-kDa NCAM is prominent in neonatal testes but is down- regulated by as yet unidentified mechanisms thereafter (Li et al., 1998). Although down-regulation of NCAM-based intercellular adhesion during postnatal maturation is likely to be important for appropriate differentiation of testicular cells, such down-regulation of NCAM fails to occur in hormone-and serum free Sertoli cell-germ cell cocultures. Besides NCAM, P-cadherin is also present in neonatal testicular cords, and seems to disappear from the seminiferous epithelium after the first postnatal week. Thyroid hormone T3 is involved in control of Sertoli cell proliferation and differentiation. The effects of T3 on adhesive factors showed down-regulation of NCAM expression in vitro, in contrast to the continued expression of NCAM in cultures without added T3. This suggested that T3 acts as a regulator of NCAM expression in neonatal testicular cells and as a modifier of gonocyte/Sertoli cell adhesion in vitro (Laslett et al., 2000).

2.4.2. α6β1 Integrin

On the basis of various evidences, it has been suggested that Sertoli cells express β1 and α6 integrin chains, which are associated to form α6β1 dimers. The receptor α6β1 is localized within the seminiferous epithelium suprabasally, at a less defined level possibly at the point of contact between Sertoli cells and basal germ cells and/or inter-Sertoli cell junctions. The pattern of expression strongly suggests that α6β1 integrin localizes in Sertoli cell at the level of ectoplasmic specialization (ES) containing cell-to-cell contacts, and that the receptor belongs to the Sertoli cell lateral domains engaged in the ES junctional structures at the level both of inter-Sertoli junctions and at the contacts between Sertoli cells and elongating spermatids. The receptor belongs to the inter-Sertoli junctional machinery and its expression requires FSH stimulation (Salanova et al, 1998).

2. 4.3. Connexin 43 and Zonaoccludin-1

Connexin43 (Cx43), the gap junction protein encoded by the *Gja1* gene, is expressed in several cell types of the testis. Cx43 at gap junctions couple Sertoli cells with each other, Leydig cells with each other, and spermatogonia/spermatocytes with Sertoli cells. Cx43 based cell-cell communication is regulated by growth factors and oncogenes. Ultrastructure analysis of Sertoli cell line (42GPA9) revealed that 42GPA9 cells form gap junctions as small electron-dense bridges that connect the plasma membranes of adjacent cells. These gap junctions contained protein connexin 43 (Lablack et al., 1998). The underlying protein interacting with Cx43 is zonaoccludin–1 protein (Zo-1). The Zo-1 is a 220 kDa peripheral membrane protein having multiple protein interaction domains including 3 PDZ domains and a SH3 (Src homology-3) domain. The interaction of Cx43 with Zo1 occurs through second PDZ domain of Zo-1. The Cx43 and Zo-1 are co-localized at gap junctions (Giepmans and Moolenaar, 1998). Histological examination in mice lacking Cx43 indicated that the development of grafted testes kept pace with that of non-grafted testes in terms of the onset of meiosis, but this development required the presence of the host gonads. Cx43 null mutant neonates had a germ cell deficiency that arose during fetal life, and this deficiency persists postnatally, giving rise to a "Sertoli cell only" phenotype. Thus intercellular communication via Cx43 channels is required for postnatal expansion of the male germ line (Roscoe et al., 2001).

2.4.4. Tpx-1

The Tpx-1 is a testicular cell adhesion molecule responsible for the specific interaction between spermatogenic and Sertoli cells. Tpx-1 is produced and secreted from spermatogenic cells at various differentiation stages and mediates the interaction of these cells with Sertoli cells (Maeda et al., 1998; 1999). As Sertoli cells interact with spermatogenic cells and eliminate degenerating cells by phagocytosis, it is possible that Tpx-1 might play some role in this phenomenon. The Tpx-1 has been reported to exist in the acrosome (Foster and Gerton, 1996) or the outer dense fiber of sperm (O'Bryan et al., 1998). It mediates the binding of spermatogenic cells to Sertoli cells of the rat in primary culture. Structure function analysis revealed that the 101-amino terminal residues were sufficient for cell adhesion activity, whereas the carboxyl-terminal cysteine-rich region was dispensable.

The TPX-1 (or AA1, the guinea pig homologue of TPX 1) is abundantly present in the mouse sperm acrosome. The synthesis of the mRNA and protein of AA1 becomes detectable at the pachytene spermatocyte stage, reaching a peak in round spermatids, and thereafter stably present in spermatogenic cells. However, reports also suggest that TPX-1 may not be solely responsible for the adhesion of spermatogenic cells to Sertoli cells, since an anti-TPX-1 polyclonal antibody inhibited it by only 50%. Other testicular proteins e.g. 1,4 galactosyltransferase and N-cadherin are likely to be also involved in the specific interaction between the two cell types. In addition to 1,4 galactosyl transferase, multiple fucosyl transferases and their lectin/selectin ligands are involved in mediating germ cell-Sertoli cell adhesion to form a cohesive epithelium and thus help in germ cell adluminal translocation within the seminiferous epithelium (Raychoudhury and Millette, 1997).

Characterization: The Tpx-1 consists about 243 amino acids with a cluster of hydrophobic amino acid residues near the amino terminus and a cysteine rich region in the carboxyl-terminal half, with no sequences identified that are considered to be transmembrane regions or N-linked glycosylation site. Within the Cys-rich region, five Cys residues appear every nine amino acids near the C-terminus. These characteristics of Tpx-1 classify it to be a member of the CRISP family of proteins and epididymal protein called acidic epididymal glycoprotein (AEG) that is synthesized in epididymal epithelium. Since more than 55% of the amino acid sequence of Tpx-1 is identical to that of AEG, these two proteins could possess similar biological functions (Maeda et al., 1998; 1999)) **(Fig.2.3)**.

The specific gene *Tpx-1* is located between *Pgk-2* and *Mep-1* on mouse chromosome 17. The predicted coding sequence of Tpx-1 from BALB/c mice showed 64.2% nucleotide and 55.1% amino acid sequence similarity with that of a rat sperm coating glycoprotein gene. The mouse Tpx-1, and human counter part TPX1 showed 77.8% nucleotide and 70% amino acid sequences similarity. The *Tpx-1* from mouse is more similar to *TPX1* from man than it is to a rat sperm coating glycoprotein gene. Thus *Tpx-1 (TPX1)* and a sperm coating glycoprotein gene are closely related, but distinct gene products belonging to the same gene family. The *TPX-1* was mapped to chromosome 6 p 21 qter (Kasahara et al., 1989). The mouse *Tpx-1* gene was found to contain ten exons, with an unusually large intron (approximately 17.0-kbp) between exons 8 and 9.

2.4.5. Clusterin

Sulfated glycoprotein-2 (SGP-2, Clusterin) is a heterodimeric glycoprotein produced by a wide array of tissues and found in most biological fluids. Clusterin was originally isolated from ram rete testis fluid on the basis of its ability to elicit aggregation of Sertoli cells from testes of immature rats. Clusterin constitutes ~ 18% of the total protein of ram testis fluid, The main source of clusterin in the testis appears to be the Sertoli cells but testicular epithelial cells can also secrete clusterin in vitro, and also produced by epididymal cells (see Chapter 34). The concentration of clusterin in testis is several told higher than in serum. Clusterin is localized mainly in the adluminal region of the seminiferous epithelium and the epithelial cells of the rete testis and caput epididymis (Ackland et al., 1992; O'Bryan et al., 1994). Clusterin also aggregates erythrocytes from several species and TM 4 Sertoli cells derived from mouse testis. Clusterin is implicated in many biological processes including cell adhesion, apoptosis and transformation. Clusterin expression was demonstrated during sperm maturation and is over-expressed in different malignancies. Disulfide linked dimeric clusterin from ram rete testis fluid has a molecular mass of ~ 80-kDa under non-reducing conditions and comprises two monomers of 40kDa that have identical pI of 3.7 and are rich in carbohydrates. Testicular clusterin has two molecular forms that have almost identical amino acid compositions and identical N-terminal amino acid sequences. Each form consists of two subunits that are non-

```
A
   1 CTC CTG ATC TTC AAA CAT CAG AAG AAA GGA CAA GAT AAG GCA GAT ATT TCA ACT   54
  55 GTC AAA TCA ACA CTT CCA GCC ATG GCT TGG TTC CAG GTG ATG CTG TTT GTC TTT  108
   1                             M   A   W   F   Q   V   M   L   F   V   F    11
 109 GCT GTG CTG CTA CCA TTG CCA CCC ACA GAA GGA AAG GAT CCA GAC TTC GCT ACT  162
  12  A   V   L   L   P   L   P   P   T   E   G   K   D   P   D   F   A   T    29
 163 TTG ACA ACC AAC CAA ATA CAA GTT CAA AGA GAG ATC ATA GCT AAA CAC AAT GAA  216
  30  L   T   T   N   Q   I   Q   V   Q   R   E   I   I   A   K   H   N   E    47
 217 CTG AGG AGA CAA GTT AGC CCC CCT GGC AGC AAC ATA CTA AAA ATG GAA TGG AAC  270
  48  L   R   R   Q   V   S   P   P   G   S   N   I   L   K   M   E   W   N    65
 272 GTA CAA GCA GCA GCA AAT GCT CAA AAG TGG GCT AAT AAC TGT ATT TTA GAA CAC  324
  66  V   Q   A   A   A   N   A   Q   K   W   A   N   N   C   I   L   E   H    83
 325 AGT AGT ACA GAA GAC CGG AAA ATC AAT ATA AAA TGT GGC GAG AAT CTC TAT ATG  378
  84  S   S   T   E   D   R   K   I   N   I   K   C   G   E   N   L   Y   M   101
 379 TCG ACT GAC CCT ACA TCC TGG AGA ACC GTA ATT CAA AGC TGG TAT GAA GAA AAT  432
 102  S   T   D   P   T   S   W   R   T   V   I   Q   S   W   Y   E   E   N   119
 433 GAA AAC TTC GTT TTC GGC GTA GGA GCT AAA CCC AAT TCC GCT GTC GGA CAC TAC  486
 120  E   N   F   V   F   G   V   G   A   K   P   N   S   A   V   G   H   Y   137
 487 ACT CAG CTT GTT TGG TAT TCA TCT TTC AAA GTT GGA TGT GGA GTT GCT TAC TGT  543
 138  T   Q   L   V   W   Y   S   S   F   K   V   G   C   G   V   A   Y   C   155
 544 CCC AAT CAA GAT ACC CTG AAA TAC TTC TAT GTT TGC CAT TAC TGT CCT ATG GGT  594
 156  P   N   Q   D   T   L   K   Y   F   Y   V   C   H   Y   C   P   M   G   173
 595 AAC AAC GTG ATG AAA AAG AGT ACC CCA TAT CAT CAA GGG ACA CCT TGT GCT AGT  648
 174  N   N   V   M   K   K   S   T   P   Y   H   Q   G   T   P   C   A   S   191
 649 TGT CCC AAT AAC TGT GAT AAT GGA TTG TGC ACC AAT AGC TGT GAT TTT GAA GAT  702
 192  C   P   N   N  (C)  D   N   G   L   C   T   N   S  (C)  D   F   E   D   209
 703 TTA CTT AGT AAC TGT GAA TCC TTG AAG AGT TCA GCA GGC TGT AAA CAT GAG TTG  756
 210  L   L   S   N  (C)  E   S   L   K   S   S   A   G  (C)  K   H   E   L   227
 755 CTC AAG GCA AAG TGT GAG GCT ACT TGC CTA TGT GAA GAC AAA ATT CAT TAA CAT  810
 228  L   K   A   K  (C)  E   A   T   C   L   C   E   D   K   I   H   *       243
 811 GCC CAG CGT GCA GCA TGA CAG ACT ACA TGA GAA GGG GTA CAG ACT TAG TTG AGA  864
 865 CAT GAC AGG GAA AAC CTA TAG GAG AGT AGT GAA ACA GTG CAT CCC AAA TGA CAA  918
 919 GGC TTC TTT CCT TCC TGG ATT TAT ATA GAA ATG TCT TTC ATA CAG CCA TTA AGA  972
 973 AAG GTG TCA TTT AGG ATA ACA ACT CTG GAT TTT GAC CAA CTT TGC TGC TTC AAA 1026
1027 TGT AGT GAA GCG AAT CAA GTG GAG AAT TTT GAA AGT TGT ACC ATA ACT GGT CAT 1080
1081 TCA CCT CTA GAA CTT TGA AAA GGA GAG AAC TGT TTG TGT CCT AAA CCA ACC TGC 1134
1135 AAT GGA AGA ATG GGC TGT AGT TAC ATC ACC ATC AAC CTA CTT CAT AGT GCC TAC 1188
1189 CAG GAT GAA TCT TGA CAT CTA GAT TTG TCT TAT GTC TTC TTA CTT TAA CAC AAA 1242
1243 TGA TCA TCT TTT CCA ATA AAG AAT TCA AGC TAC CAC AAA AAA AAA AAA AAA AAA 1296
1297 AA
```

```
B
                          **   *  *     **              **** *     *  * *** **   *** **
# 97 (rat)           1 MAWFQVMLFVFAVLLP/LP/P/TEGKDPDFAT/LTTNQIQVQREIIAKHNELRR  50
Tpx-1 (mouse)        1 ------------L--RS///-L---------TSL--/----------VN-------  50
AA-1 (guinea pig)    1 --LLP-VV-LITM---CVL///-N----A-TALI--/-S---N---N---Q--K  50
TPX-1 (human)        1 --LLP-/--LVT----S--///A-----A-TAL---/-L-------VN------K  49

                         * *  *  *****         *** *** * * **     ***      ******** *
# 97 (rat)          51 QVSPPGSNILKMEWNVQAAANAQKWANNCILEHSSTEDRKINIKCGENLYMSTD 104
Tpx-1 (mouse)       51 S-N-T--D------SI--TT-------K-------KD------R---------- 104
AA-1 (guinea pig)   51 S-T--A--M-----SRE--V-------R-T-V--NPD---TST---------S- 104
TPX-1 (human)       50 A----A--M-----SREVTT---R---K-T-Q--DP----TSTR--------S- 103

                       *  *   ****   *   *  *** *  ** ************      *** *****
# 97 (rat)         105 PTSWRTVIQSWYEENENFVFGVGAK/PNSAVGHYTQLVWYSSFKVGCGVAYCPN 157
Tpx-1 (mouse)      105 --L-S-------N---D--Y-----/------------------I---I----- 157
AA-1 (guinea pig)  105 -S--SDA----FD-SQD-T----P-SH-AV------------YL----I----- 158
TPX-1 (human)      104 ----SSA-----D-ILD--Y---P-S--AV-----------TYQ----I----- 157

                      ** *** *** *** ***     * *** ** *** **  *  *******     ***
# 97 (rat)         158 QDTLKYFYVCHYCPMGNNVMKKSTPYHQGTPCASCPNNCDNGLCTNSCDFEDLL 211
Tpx-1 (mouse)      158 --N-----------------------Q------------E--------------- 211
AA-1 (guinea pig)  159 --S---Y---Q---A----YT-N---K--I------GH-E--------EY---- 212
TPX-1 (human)      158 --S---Y---Q---A---MNR-N---Q------G--DD--K-------QYQ--- 211

                      *** ***  *** * **  ** *** ** **
# 97 (rat)         212 SNCESLKSSAGCKHELLKAKCEATCLCEDKIH                       243
Tpx-1 (mouse)      212 -------T----------T--Q----------                       243
AA-1 (guinea pig)  213 -------NT---E-Q--VE--K---R-----Y                       244
TPX-1 (human)      212 ---D---NT---E-----E--K------N--Y                       243
```

Fig.2.3. (A). Nucleotide and predicted amino acid sequence of rat Tpx-1 (cDNA clone #97). The N-terminal hydrophobic sequence and a presumed polyadenylation signal are underlined. The five Cys residues near C-terminus are circled. (B). Amino acids similarity among rat Tpx-1 (clone #97), mouse Tpx-1, guinea pig auto antigen AA-1 and human TPX-1. Gaps are shown by slashes. Identical amino acid residues are indicated by horizontal bars and those conserved amino acids are asterisked (*).Reprinted with permission from Maeda et al, Biochem Biophys Res Commun 248; 140-46: 1998 © Elsevier.

identical to each other in their amino acid composition. Deglycosylation of testicular clusterin also leads to a reduced molecular mass with no biological activity, suggesting that the serum form is nonglycosylated clusterin and that glycosylation is necessary for biological activity. Rat clusterin isolated from culture medium of Sertoli cell enriched cultures also comprises of two subunits of 43- and 38-kDa. The subunits, though not identical to each other, show marked similarities, with the respect to ram subunits. (Law and Griswold, 1994; O'Bryan et al., 1994). The disulfide linked dimeric protein secreted by Sertoli cells and found in seminiferous tubule fluid is composed of monomers of M_r 47000 and 34000 whereas the epididymal protein exhibits monomers of M_r 40000 and 29000. (see Chapter 34). Clusterin is associated with normal germ cell development and the loss of clusterin expression might play a role in the malignant transformation of germ cells (Behrens et al., 2001).

2.4.6. Osteopontin

A bone cell adhesion molecule, osteopontin (OPN) is present in the rat testis, epididymis and rat sperm. The OPN is a multifunctional phosphoprotein, which is regulated by circulating androgens in epididymis. A 30kDa band from epididymis and a 60kDa band from testis, OPN stains in spermatogonia and early spermatocytes, suggesting its role in adhesion of these cells wth adjacent Sertoli cells (Luedtke et al., 2002). Sequence analysis of one of four positive cDNA clones, designated as pREPs, revealed identity with rat OPN cDNA. The partial cDNA clone pREP5 encompasses 64% of the 1,457 residues. The OPN is detected in rat testis as well as in epididymal fluid and sperm extracts. The OPN appears to be a Sertoli cell product. Though the function of OPN in testis is uncertain but the characterization of this protein suggests that OPN could play a role in testicular cell adhesion during spermatogenesis (Siiteri et al., 1995).

2.4.7. Other Junction Proteins

Oligodendrocyte Specific Protein: Among unknown FSH targets, oligodendrocyte-specific protein (OSP) mRNA was identified by Hellani et al. (2000). The OSP was initially shown being produced in oligodendrocytes, which are supporting cells for specialized neurons. The OSP has been identified as a new member of claudin family (claudin 11). The key role of OSP in central nervous system myelin and Sertoli cell tight junction strands was demonstrated in *osp/claudin* 11 null mice. The testicular expression of OSP is dependent upon male gonad development and systemic and local signaling molecules. The OSP is expressed early in fetal development in Sertoli cells, immediately after the peak of SRY (sex-determining region, Y gene: Chapter 10) expression, but just before that of the anti-Mullerian hormone. The OSP expression, regulated by FSH and TNFα starts from day 3 postnatally and reaches a plateau between days 6 and 16. It appears that OSP expression during fetal life is under SRY control during male sex differentiation (Hellani et al., 2000).

Testin: Sertoli cells express testin, which is tightly associated with the disruption of Sertoli-germ cell junctions (Grima et al., 1997). Treatment of rats with CdCl2, which disrupted the inter-Sertoli tight junctions, failed to induce any change in testicular testin expression. In constrast, lonidamine, an antispermatogenic drug that rearranges the Sertoli cell adhesion junctions, induced a drastic increase in testicular testin expression (Grima and Cheng, 2000).

2.5. INTRACELLULAR BRIDGES/ ECTOPLASMIC SPECIALIZATIONS

In areas of Sertoli cell crypts attached to the heads of elongate spermatids occurs a class of actin-related adherent junctions (AJs) termed ectoplasmic specializations (ESs). Each of these structures consists of the plasma membrane in regions adherent to the spermatid head, a cistern of endoplasmic reticulum, and an intervening layer of hexagonally packed actin filaments. The three elements of the plaque occur as a structural unit that remains attached to the spermatid when the latter is mechanically separated from the epithelium. Actin filaments of 6-7 nm in diameter are concentrated in specific regions of spermatogenic cells and Sertoli cells. In spematogenic cells actin filaments occur in intercellular bridges and in the sub-acrosomal space. In Sertoli cells they are abundant in ectoplasmic specializations and in regions adjacent to tubulobulbar processes of spermatogenic cells. Fluorescence generated by labeled actin in ectoplasmic specializations shows linear tracts that follow the outer contour of spermatid heads, demonstrating that Sertoli cell ectoplasmic specializations remain attached to spermatids, mechanically dissociated from the seminiferous epithelium. Vinculin, also present in ectoplasmic specializations, is co-distributed with actin bundles within each ectoplasmic specialization. In vivo, the endoplasmic reticulum component of the plaque is related, on its cytoplasmic face, to microtubules (Pfeiffer et al., 1991). During spermatogenesis, actin-based ectoplasmic specializations, between Sertoli and germ cells undergo extensive restructuring in the seminiferous epithelium to facilitate germ cell movement across the epithelium. One of these events is the movement or translocation of spermatids to the base and then back to the apex of the epithelium. The physiological significance of this translocation event is not known. This change in position occurs after spermatids have acquired an elongate shape and while they are attached to apical invaginations (crypts) of Sertoli cells.

The espins are actin-binding and -bundling proteins localized to parallel actin bundles. The 837- amino-acid "espin" of Sertoli cell-spermatid junctions (ES) and the 253-amino acid "small espin" of brush border microvilli are splice isoforms that share a C-terminal 116-amino-acid actin-bundling module but contain different N-termini. Properties of espin suggested that it is a major actin-bundling protein of the Sertoli cell-spermatid junction complexes. Scinderin is detected in a region corresponding with the sub-acrosomal space in the round spermatids and with the acrosome in the elongated spermatids. In Sertoli cells, scinderin may contribute to the regulation of tight junctional permeability and to the release of the elongated spermatids by controlling the state of peri-junctional actin (Pelletier et al., 1999).

The AJ assembly is associated with a transient induction of β1-integrin, vinculin, p-FAK-Tyr(397), and phosphatidylinositol 3-kinase (PI3K) but not the nonphosphorylated form of focal adhesion kinase (FAK), paxillin, and p130 Cas. However, phsphorylated form of focal adhesion kinase (p-FAK) is involved in the events of Sertoli-germ cell AJ assembly and disassembly. In addition, AJ dynamics in the testis, particularly at the site of ES, are regulated by proteins that are found in the focal adhesion complex (FAC) in other epithelia, such as β1-integrin, vinculin, and nonphosphorylated form of focal adhesion kinase (FAK) utilizing the integrin/pFAK/PI3K/p130 Cas signaling pathway (Siu et al., 2003). The action, though, mediated by the effects of TGF-β3 involved integral membrane proteins and adaptors of TJ and AJs, it did not involve protease inhibitors (Wong et al., 2004).

Though the mechanism(s) that regulates AJ dynamics in the testis is virtually unknown, Rho GTPases have been implicated in the regulation of these events in other epithelia. The in vitro assembly of the Sertoli-germ cell AJs but not of TJs is associated with a transient but significant induction of RhoB. RhoB displays a cell-specific association during the epithelial cycle and localizes in the seminiferous epithelium in stage specific manner, being lowest prior to spermiation. In conclusion, it was confirmed that Sertoli-germ cell AJ dynamics are regulated, at least in part, via the integrin/ROCK/LIMK/cofilin signaling pathway (Lui et al., 2003a).

2.6. SERTOLI CELL TIGHT JUNCTIONS DYNAMICS

Several molecules have been shown to regulate Sertoli cell tight junctions (TJ) dynamics; they include, for example, transforming growth factor β3 (TGFβ3), occludin, protein kinase A, protein kinase C, and signaling pathways such as the TGFβ3/p38 mitogen-activated protein kinase pathway. Yet the mechanisms that regulate these events are essentially unknown (Lui et al., 2003b, c). cAMP and cGMP are known modulators of Sertoli cell TJs. The assembly of TJ barrier is associated with a significant plummeting in the levels of iNOS (NO synthase) and eNOS, suggesting that their presence might perturb TJ assembly through production of NO. Subsequently it was confirmed that iNOS (no synthase) and eNOS are linked to TJ-integral membrane proteins, such as occludin, and cytoskeletal proteins, such as actin, vimentin, and α-tubulin. In addition inhibition of protein kinase-G facilitated the Sertoli cell TJ barrier assembly. Cytokines, such as TGF-β and TNF-α, known to perturb the Sertoli cell TJ barrier, also stimulated Sertoli cell iNOS and eNOS expression in vitro. These results suggested NOS as an important physiological regulator of TJ dynamics in the testis, exerting its effects via the NO/soluble guanylate cyclase/cGMP/protein kinase-G signaling pathway (Lee et al., 2003).

The significance of TGFβ3 in the regulation of Sertoli cell TJ dynamics was evident possibly via inhibitory effects of TGFβ on the expression of occludin, claudin-11, and zona occludin-1 (ZO-1). The TGFβ3-induced effect on Sertoli cell TJ dynamics appears to be mediated via the p38 MAP-kinase pathway (Lui et al., 2003a). The blood-testis barrier (BTB) functions, studied by the $CdCl_2$-induced damage in rats demonstrated a significant reduction in testicular occludin along with a loss of immunoreactive occludin in the seminiferous epithelium at the site of the BTB. This was followed with a surge in testicular TGF-β3, as well as p-p38 MAP kinase (the phosphorylated/activated form of p38), but not p38 MAP kinase, nor ERK/p-ERK. It appeared that BTB dynamics in vivo are regulated by the TGF-β3/p38 MAP kinase pathway, which in turn determines the level of occludin at the site of Sertoli cells TJs (Lui et al., 2003c).

2.7. SPERMIATION

Mechanisms leading to the control of spermiation process, and the proteins that interact to adhere mature spermatids to Sertoli cells, are poorly understood. The tubulin, actin, paxillin and vinculin are present at the luminal edge of the seminiferous tubule (Wine and Chapin, 1999). These proteins have been co-localized with β- integrins in other cells. Clear evidence was also found for the presence of N-cadherin and its associated intracellular proteins: β-catenin, pp120, desmogein, pp60 (src), and Csk. In addition, N-cadherin and desmogein were found around spermatids retained by epithelium. This suggests a mechanism for spermatid adhesion and the control of spermiation in rat testes (Wine and Chapin, 1999). Observations related to the spatial and temporal distribution of integrin-related signaling molecules at ESs, and the presence of β1 integrin/integrin related kinase (ILK) interaction in non-cultured epithelium suggested their role in Sertoli-spermatid junctions. Pan-cadherin and β-catenin and focal adhesion kinase (FAK), a known integrin-associated molecule, did not associate with these adhesion molecules present there in or with ES. Rather phosphotyrosine residues were found to be located in ES particularly during sperm release and turnover of basal junction complexes. These observations indicated that ESs share cell signaling features both of cell-cell junctions and of cell-extracellular matrix junctions (Mulholland et al., 2001). Chapin et al (2001) identified specific proteins linked together as part of multiprotein complex, as well as several additional proteins (cortactin, ERK1/2, and 14-3-3 zeta) that may be functioning in both structural and signal transduction roles. The data suggested that protein phosphorylation is central to the control of spermiation (Chapin et al., 1996; Chapin et al., 2001). Since the effect of NGF is mediated by the TrkA receptor and the ERK1-ERK2 kinase pathway, it was suggested that AP1 activation may be an important relay in the Sertoli cell-germ cell cross-talk (Vidal et al., 2001). Nevertheless, a phospholipid kinase is known to regulate actin organization and intercellular bridge formation during germline cytokinesis (Brill et al., 2000).

Microtubules are prominent elements of the Sertoli cell cytoskeleton. They surround apical crypts containing elongate spermatids, generally arranged parallel to the axis of spermatid translocation, and have their positive ends positioned at the base of the epithelium. The nature of motor proteins involved in transport of the spermatid/junction complexes along microtubules in not well known. Specialized junction plaques between Sertoli cells and spermatids in the rat testis support microtubule translocation in vitro. It has been proposed that motor proteins on the endoplasmic reticulum in ES interact with adjacent microtubules to translocate the junction plaques, and hence the attached spermatids, within the epithelium. The microtubule-based motility of spermatid translocation has been consistent with several observations (Beach and Vogl, 1999). Besides in spermatid manchettes and in Sertoli cell cytoplasm, dynein has been also found in ectoplasmic specializations associated with spermatids during early stages of differentiation. Kinesin has been detected in the spermatids manchette and the Sertoli cell trans-Golgi network. In addition, the heterotrimetic motor protein kinesin-II has been localized in the mid-piece and flagellum of echinoderm sperm. The subunits of this protein show considerable sequence homology with the mouse kinesin-like proteins KIF3A and KIF3B. KIF3B is present in high concentration in testis. In order to determine the presence of a microtubule-dependent motor proteins in Sertoli cell region involved with spermatid translocation, it was shown that an isoform of cytoplasmic dynein may be responsible for the apical translocation of elongated spermatids that occurs before sperm release whereas kinesin-II was associated with spermatid tails (Miller et al., 1999) (see chapters 9 and 25).

Gelsolin is concentrated in ES adhesion complexes, besides phosphatidylinositol 4,5-bisphosphate and phosphoinositide specific phopholipase C. Treatment of spermatid - Sertoli cells junction complex with phospholipase (PLC) resulted in release of gelsolin and loss of actin from the complex. It seems that the hydrolysis of phosphatidylinositol 4, 5 bisphosphate may result in a local Ca^{2+} surge via the action of inositol triphosphate on junctional endoplasmic reticulum. This Ca^{2+} surge would facilitate the actin severing function of gelsolin within the adhesion complex (Guttman et al., 2002).

Interaction between protease and protease inhibitors seems to play an important role in remodeling and restructuring of the seminiferous tubule during spermatogenesis. Controlled proteolytic activity is also involved in the migration of germ cells from the basal compartment to the lumen of the seminiferous epithelium, and in the release of spermatids during spermiation. The reported occurrence of Sertoli cell membrane – associated proteases indicates the possible involvement of regulatory peptide systems within these paracrine systems, the kallikrein-kinin system, in cells of the seminiferous tubule (Monsees et al., 1998).

Suppression of testicular testosterone in adult rats by low-dose testosterone and estradiol (TE) treatment causes the premature detachment of step 8 round spermatids from the Sertoli cell. At higher doses of testostesterone, newly elongating spermatids associated with ES proteins within 2 days. Thus it suggested that Sertoli cell ES structure was qualitatively normal in TE-treated rats, and thus the absence of this structure is unlikely to be the cause of round spermatid detachment. The study suggested that defects in adhesion molecules between round spermatids and Sertoli cells are likely to be involved in the testosterone dependent detachment of round spermatids from the seminiferous epithelium.

2.8. GERM CELL–SERTOLI CELL INTERACTIONS

Sertoli cells and germ cells are in fact intimately associated on a functional and morphological basis, and for that Sertoli cells possess specialized processes for interaction with germ cells. On the other hand, evidence has also accumulated to support the view that the germ cells also control the activity of the Sertoli cell, since elimination of a specific germ cell type by specific testicular toxins, provoked stage dependent changes in the secretion of inhibin from Sertoli cells (Sharpe, 1994). Germ cells composition and their number have considerable influence on Sertoli cell function. The induction of a serotonin receptor mRNA in Sertoli cells by pachytene spermatocytes demonstrated that meiotic germ cells induce mRNA encoding an important receptor in Sertoli cells (Chapter 1).

It is hypothesized that germ cells, in particular pachytene spermatocytes and early spermatids, may influence Sertoli cell function during sexual development in the rat. The ability of pachytene spermatocyte proteins to induce specific Sertoli cell secretion indicated that the pachytene spermatocytes are able to influence their surrounding milieu, and provided further support to the concept of a paracrine interaction between germ cells and Sertoli cells during spermatogenesis (Djakiew and Dym, 1988; Tanaka et al, 1999). A 24.5kDa protein and haspin (a nuclear protein kinase) derived from rat spermatids are associated with Sertoli cell secretory function. Cleaved internal fragment of 24.5 kDa protein participates in the negative regulation of Sertoli cell function during spermatogenesis. Besides 24.5-kDa protein and haspin, another 29-kDa protein present in cytosolic fraction of round spermatids also influences Sertoli cell functions (Onoda and Djakiew, 1993). Testicular germ cells secretion can stimulate Sertoli cell α2–macroglobulin (α_2-MG) expression. Enriched populations of germ cells from adult rats has been found to influence 20 day old rat Sertoli cell secretory activity by stimulating ABP and inhibiting oestradiol 17β production in the presence of FSH as well as of dibutyryl cyclic AMP (dbcAMP). The androgen dependent changes in secretion of proteins, which occur specifically at stage VI-VIII of the normal spermatogenic cycle, are largely dependent on a normal germ cell complement, with each germ cell type regulating the secretion of proteins essential for its own development (Mckinnell and Sharpe, 1992). Information on co-cultures of rat pachytene spermatocytes and round spermatids suggested that germ cells interact with Sertoli cells, possibly by a protein that acts as a short loop paracrine factor, which regulates the expression of proenkephalin gene in Sertoli cells and that the stage specific regulation of proenkephalin levels in Sertoli cells may occur in vivo (Fujisawa et al., 1992). Spermatogenic immunoglobulin superfamily (SgIGSF), a mouse protein belonging to the Ig superfamily is expressed in the spermatogenic cells of seminiferous tubules. The SgIGSF on the surface of spermatogenic cells binds to some membrane molecules on Sertoli cells in a heterophilic manner and thereby may play diverse roles in the spermatogenesis (Wakayama et al., 2003). An upstream domain that mediated an inhibitory effect by male germ cells was found within this 3-kb promoter region, which is associated with the regulatory elements for transcription (Charron et al., 2003). A 120-bp region that spans the transcription start site of the rat cathepsin L gene was sufficient to activate transcription in Sertoli cells. Within this region, Sp3 specifically bound to a GC-box, where as Sp1-binding activity was absent from nuclear extracts of Sertoli cells. The GC-box was found to be essential for promoter activity. It supported that the GC-

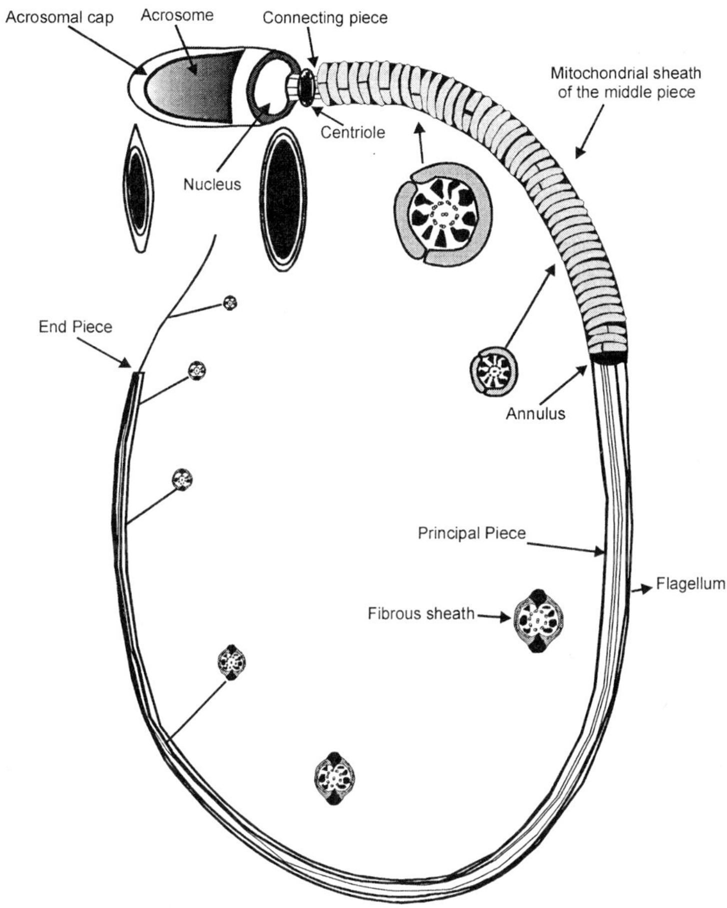

Fig. 2.4. Mature human spermatozoon.

box is a critical regulatory element for the cathepsin L promoter in mature Sertoli cells (Charron et al., 2003). The mature human sperm after passing through male ejaculatory tract is shown in **Fig. 2.4.**

Basigin is a transmembrane glycoprotein, ablation of which leads to azoospermia in mice. In mouse testis basigin is localized not only on the plasma membrane of spermatocytes and spermatids, but also on the plasma membrane of the Sertoli cell processes, which contact the spermatocytes and spermatids. Experimental cryptorchid testes, which contained only spermatogonia and Sertoli cells did not show basigin activity. The concomitant expression of basigin with appearance of spermatocytes suggested that basigin is involved in the interaction between Sertoli cells and germ cells at specific stages of spermatogenesis (Maekawa et al., 1998). Allurin represents the first vertebrate sperm chemo-attractant, which belongs to CRISP family to be found in the female reproductive tract (Olson et al., 2001) (see Chapter 34)

2.9. P-MOD-S FROM PERITUBULAR CELLS

P-Mod-S, a paracrine and a mesenchymal factor modulates the functions of Sertoli cells. Skinner and Fritz (1985; 1986) isolated a factor from conditioned medium of pertitubular cells that stimulated Sertoli cell production of transferrin, ABP, and several other proteins. This factor did not have any mitogenic activity and was found to be trypsin sensitive but was heat and acid stable and bound to heparin. The protein was called P-Mod-S for the factor from peritubular (P) cells that modulates (Mod) Sertoli (S) cell functions. Further characterization showed that there are two proteins of 54 and 56-kDa, which have been termed P-Mod- S A and -B (Skinner et al., 1988). Norton and Skinner showed that P-Mod-S increases transferrin and ABP mRNA levels in cultured Sertoli cells and elevated intracellular cGMP but not cAMP. P-Mod-S is more active than FSH alone and was almost as effective as a combination of FSH, testosterone, retinol, and insulin in stimulating Sertoli cell function (Ackland et al., 1992). The P-Mod-S can activate the c-fos promoter through the SRE and cis-inducible element, CRE and TPA response element (TRE). P-Mod -S does not act through stimulation of cAMP and PKA pathways. P-Mod-S can activate c-fos promoter through SRE and SIE and can cause activating protein 1 (AP1) oligonucleotide to form a DNA-protein complex, indicating activation of the *c-fos* gene and binding of the c-fos-jun complex. When transfected into Sertoli cells, the 581-bp transferrin minimal promoter (Tf) responded to P-Mod-S slightly, but was activated by FSH. However, the 2.6-kb Tf promoter construct responded to P-Mod-S. This shows that response elements of c-fos promoter involve intermediate transcription factors in induction of Sertoli cells differentiation by P Mod S (Whaley et al, 1995).

2.10. REFERENCES

Ackland JF, Schwartz NB, Mayo KE, Dodson RE, Nonsteroidal signals originating in gonads. Physiol Rev 1992; 72: 731-87.

Beach SF, Vogl AW. Spermatid translocation in the rat seminiferous epithelium: coupling membrane trafficking machinery to a junction plaque. Biol Reprod 1999; 60: 1036-46.

Behrens P, Jeske W, Wernert N, Wellmann A. Downregulation of clusterin expression in testicular germ cell tumours. Pathobiology 2001; 69: 19-23.

Benahmed M. Role of tumor necrosis factor in the male gonad. Contracept Fertil Sex 1997; 25: 569-71.

Braghiroli L, Silvestrini B, Sorrentino C, Grima J, Mruk D, Cheng CY. Regulation of α2-macroglobulin expression in rat Sertoli cells and hepatocytes by germ cells in vitro. Biol Reprod 1998; 59: 111-23.

Brill JA, Hime GR, Scharer-Schuksz M, Fuller MT. A phospholipid kinase regulates actin organization and intrecellular bridge formation during germline cytokinesis. Development 2000; 127: 3855-64.

Brucato S, Harduin-Lepers A, Godard F, Bocquet J, Villers C. Expression of glypican-1, syndecan-1 and syndecan-4 mRNAs protein kinase C-regulated in rat immature Sertoli cells by semi-quantitative RT-PCR analysis. Biochim Biophys Acta 2000; 1474: 31-40.

Brucato S, Villers C. Protein kinase C regulation of glypican-1, syndecan-1 and syndecan-4 mRNAs expression during rat Sertoli cell development. Biochimie 2002; 84: 681-6.

Chapin C, Mousseau B, Ramsay K, Duddy S, Li Y, Kiley SC and Jaken S. identification of a major protein kinase C-binding protein and substrate in rat embryo fibroblasts. Decreased expression in transformed cell. J Biol Chem 1996; 271: 6417-22

Chapin RE, Wine RN, Harris MW, Borchers CH, Haseman JK. Structure and control of a cell-cell adhesion complex associated with spermiation in rat seminiferous epithelium. J Androl 2001; 22: 1030-52.

Charron M, DeCerbo JN, Wright WW. A GC-box within the proximal promoter region of the rat cathepsin L gene activates transcription in Sertoli cells of sexually mature rats. Biol Reprod 2003; 68: 1649-56.

Chaudhary J, Sadler-Riggleman I, Skinner MK. Identification of a novel Sertoli cell gene product SERT that influences follicle stimulating hormone actions. Gene 2004; 324: 79-88.

Chen B, Li A, Wang D, Wang M, Zheng L, Bartles JR. Epsin contains an additional actin-binding site in its N-terminus and is a major actin binding protein of Sertoli cell spermatid ectoplasmic specialization

junctional plaque. Mol Biol Cell 1999; 10: 4327-39.
Chung CK, Chung BH, Shin T, Chae HJ, et al. Protein kinase C independent activation of inducible nitric oxide synthase by tumor necrosis factor -α in TM4 Sertoli cells. Immunopharmacol 2000; 22: 49-59.
De Miguel PD, Boer-Brouwer MD, Paniagua R, et al. Leukemia inhibitory factor and ciliary neurotropic factor promote the survival of Sertoli cells and gonocytes in a coculture system. Endocrinology 1996; 137: 1885-96.
De SK. Chen HL, Pace JL, Hunt JS, Terranova PF, Enders GC. Expression of tumor necrosis factor-α in mouse spermatogenic cells. Endocrinology 1993; 133: 389-96.
Dejucq N, Lienard MO, Guillaume E, Dorval I, Jegou B. Expression of interferons-α and-γ in testicular interstitial tissue and spermatogonia of the rat. Endocrinology 1998; 139: 3081-87.
Delfino F, Walker WH. Stage-specific nuclear expression of NF-κB in mammalian testis. Mol Endocrinol 1998; 12: 1696-07.
Djakiew D, Dym M. Pachytene spermatocyte proteins influence Sertoli cells function. Biol Reprod 1988; 39: 1193-205.
Foster JA, Gerton GL. Autoantigen 1 of the guinea pig sperm acrosome is homologue of Tpx-1 and human Tpx-1 and is a member of the cysteine secretary protein (CRISP) if any. Mol Reprod Dev 1996; 44: 221-29.
Franca LR, Ogawa T, Avarbock MR, Brinster RL, Russell LD. Germ cell genotype controls cell cycle during spermatogenesis in the rat. Biol Reprod 1998; 59: 1371-77.
Fujisawa M, Bardin CW, Morris PL. A germ cell factor (S) modulates preproenkephalin gene expression in rat Sertoli cells. Mol Cell Endocrinol 1992; 84: 79-88.
Fujisawa M, Kanzaki M, Okuda Y, Okada H, Arakawa S, Kamidono S. Stem cell factor in human plasma as a marker for spermatogenesis. Urology 1998; 51: 460-63.
Gamzu R, Yogev L, Kleiman S, Botchan A, et al. Expression of mannose-ligand receptors on human spermatozoa: effect of lecithin and association with sperm binding to the zona pellucida. Fertil Steril 1998; 70: 766-70.
Giepmans BN, Moolenaar WH. The gap junction protein connexin43 interacts with the second PDZ domain of the zona occludens–1 protein. Curr Biol 1998; 8: 931-4.
Gizang-Ginsberg E, Wolgemuth DJ. Expression of the proopiomelanocortin gene is developomentally regulated and affected by germ cells in the male mouse reproductive system. Proc Natl Acad Sci U S A 1987; 84: 1600-4
Grima J, Cheng CY. Testin induction: the role of cyclic 3', 5'-adenosine monophosphate/protein kinase A signaling in the regulation of basal and lonidamine-induced testing expression by rat sertoli cell. Biol Reprod 2000; 63: 1648-60.
Grima J, Zhu LJ, Cheng CY. Testin is tighly associated with testicular cell membrane upon its secretion by Sertoli cells whose steady state mRNA level in the testis correlates with the turnover and integrity of inter-testicular cell junctions. J Biol Chem 1997; 272: 6499-509.
Griswold MD. Interactions between germ cells and Sertoli cells in the testis. Biol Reprod 1995; 52: 221-16.
Guttman JA, Janmey P, Vogl AW. Gelsolin-evidence for a role in turnover of junction-related actin filaments in Sertoli cells. J Cell Sci 2002; 115: 499-505.
Hara T, Tamura K, de Miguel MP, Mukouyama Y, Kim HJ, Kogo H, Donovan PJ, Miyajima A. Distinct roles of oncostatin M and leukemia inhibitory factor in the development of primordial germ cells and Sertoli cells in mice. Dev Biol 1998; 201: 144-58.
Hellani A, JI J, Maudutt C, Deschildre C, Tabone E, Benachmed M. Developmental and hormonal regulation of the expression of oligodendrocytes-specific protein/claudin 11 in mouse testis. Endocrinology 2000; 141: 3012-19.
Jenab S, Morris PL. Testicular leukemia inhibitory factor (LIF) and LIF receptor mediate phosphorylation of signal transducers and activators of transcription (STAT-3) and STAT-1 and induce c-fos transcription and activator protein-I activation in rat Sertoli but not germ cells. Endocrinology 1998; 139: 1883-90.
Jenab S, Morris PL. Transcriptional regulation of Sertoli cell immediate early gene by interleukin-6 and interferon-γ is mediated through phosphorylation of STAT-3 and STAT-1 proteins. Endocrinology 1997; 138: 2740-97.
Jonsson CK, Zetterstrom RH, Hoist M, Parvinen M, Soder O. Constitutive expression of interleukin 1alpha messenger ribonucleic acid in rat Sertoli cells is dependent upon interaction with germ cells. Endocrinology 1999; 140; 3755-61.
Kahsai TZ, Enders GC, Gunwar S, Brummark C, Wieslander J, Kalluri R, Zhou R, Noelken ME, and Hudson BG. Seminiferous tubule basement membrane composition and organization of type IV collagen chains, and the linkage of a3(IV) and a5(IV) chains. J Biol Chem 1997; 272: 17023-32.
Kasahara M, Gutknecht J, Brew K, Spurr N, Goodfellow PN. Cloning and mapping of a testis-specific gene with sequence similarity to a sperm-coating glycoprotein gene. Genomics 1989; 5: 527-34.

Kanzaki M, Morris PL. Identification and regulation of testicular interferon-γ (IFNγ) receptor subunits: IFNγ enhances interferon regulatory factor-1 and interleukin-1β converting enzyme expression. Endocrinology 1998; 139: 2636-44.

Kitajima K, Matsumoto K, Tahara M, Takahashi H, Nakamura T, Nakamura T. A newly identified AMSH-family protein is specifically expressed in haploid stages of testicular germ cells. Biochem Biophys Res Commun 2003; 309: 135-42.

Lablack A, Bourdon V, Defamie N, Batias C, Mesnil M, Fenichel P, Pointis G. Segretain D. Ultrastructural and biochemical evidence for gap junction and connexin 43 expression in a clonal Sertoli cell line: a potential model in the study of junctional complex formation. Cell Tissue Res 1998; 294: 279-87.

Laslett AL, Li LH, Jester WF Jr, Orth JM. Thyroid hormone down-regulates neural cell adhesion molecule expression and affects attachment of gonocytes in Sertoli cell-gonocyte cocultures. Endocrinology 2000; 141: 1633-41.

Law GL. Griswold MD. Activity and form of sulfated glycorprotein 2 (clusterin) from cultered Sertoli cells, testis, and epididymis of the rat. Biol Reprod 1994; 50: 669-79.

Le Magueresse-Battistoni B, Pernod G, Kolodie L, Morera AM, Benahmed M. Tumor necrosis factor-α regulates plasminogen activator inhibitor-1 in rat testicular peritubular cells. Endocrinology 1997; 138: 1097-105.

Lee NP, Cheng CY. Regulation of Sertoli cell tight junction dynamics in the rat testis via the nitric oxide synthase/soluble guanylate cyclase/3',5'-cyclic guanosine monophosphate/protein kinase G signaling pathway: an in vitro study. Endocrinology 2003; 144: 3114-29.

Li JC, Mruk D, Cheng CY. The inter-Sertoli tight junction permeability barrier is regulated by the interplay of protein phosphatases kinases : an in vitro study. J Androl 2001; 22: 847-56.

Li LH, Jester WF, Jr, Orth JM. Expression of 140kDa neural cell adhesion molecule in developing testes in vivo and in long term cell genocyte cocultures. J. Androl 1998; 19: 365-73.

Li Y, Friel PJ, Robinson MO, McLean DJ, Griswold MD. Identification and characterization of testis- and epididymis-specific genes: Cystatin SC and Cystatin TE-1. Biol Reprod 2002; 67: 1872-80.

Lilienbaum A, Sage J, Memet S, Rassoulzadegan M, Cuzin F, Israel A. NF-κB developmentally regulated during spermatogenesis in mice. Dev Dyn 2000; 219: 333-40.

Loveland K, Schlatt S, Sasaki T, Chu ML, Timpl R, Dziadek M. Developmental changes in the Basement membrane of the normal and hypotheyroid postnatal rat testis: segmental localization of fibulin-2 and fibronectin. Biol Reprod 1998; 58: 1123-1130.

Luedtke CC, McKee MC, Cyr DG, Gregory M, Kaartinen MT, Mui J, Hermo L. Osteopontin expression and regulation in the testis, efferent ducts, and epididymis of rats during postnatal development through to adulthood. Biol Reprod 2002, 66: 1437-48.

Lui WY, Lee WM, Cheng CY. Sertoli-germ cell adherens junction dynamics in the testis are regulated by RhoB GTPase via the ROCK/LIMK signaling pathway. Biol Reprod 2003a; 68: 2189-206.

Lui WY, Lee WM, Cheng CY. Transforming growth factor beta3 regulates the dynamics of Sertoli cell tight junctions via the p38 mitogen-activated protein kinase pathway. Biol Reprod 2003b; 68: 1597-612.

Lui WY, Wong CH, Mruk DD, Cheng CY.TGF-beta3 regulates the blood-testis barrier dynamics via the p38 mitogen activated protein (MAP) kinase pathway: an in vivo study. Endocrinology 2003c; 144: 1139-42.

Maeda T, Nishida J, Nakanishi Y. Expression pattern, subcellular localization and structure-function relationship of rat Tpx-1, a spermatogenic cell adhesion molecule responsible for association with Sertoli cell. Dev Growth Differ 1999; 41: 715-22.

Maeda T, Sakashita M, Ohba Y, Nakanishi Y. Molecular cloning of the Rat Tpx-1 responsible for the interaction between spermatogenic and Sertoli cells. Biochem Biophys Res Commun 1998; 248: 140-46.

Maekawa M, Suzuki-Toyota F, Toyama Y, Kadomatsu K, Hagihara M, Kuno N, Muramatsu T, Dohmae K, Yuasa S. Stage specific localization of basigin, a member of the immunolobulin superfamily, during mouse spermatogenesis. Arch Histol cytol 1998; 61: 405-15.

Martinova YS, Kancheva LS, Nikolova DB, Georgieve VD. Differential effects of prepubertal rat Sertoli cell secreted proteins on somatic testicular and nontesticular cells. Mol Cell Endocrinol 1993; 98: 75-9.

McClive PJ, Sinclair AH. Type II and type IX collagen transcript Isoforms are expressed during mouse testis development. Biol Reprod 2003; 68:1742-7.

McKinnell C, Sharpe RM. The role of specific germ cell types in modulation of the seceration of androgen-regulated proteins (APRs) by stage VI-VIII seminiferous tubules from the adult rat. Mol Cell Endocrinol 1992; 83: 219-31.

Monsees TK, Goring M, Schill WB, Miska W. Possible roles of proteases in the regulation of spermatogenesis. Andrologia 1998; 30: 185-91.

Miller MG, Mulholland DJ, Vogl AW. Rat testis motor proteins associated with spermatid translocation

(dynein) and spermatid flagella (kinesin-II). Biol Reprod 1999; 60: 1047-56.
Mulholland DJ, Dedhar S, Vogl AW. Rat seminiferous epthelium contains a unique junction (Ectoplasmic specialization) with signaling properties both of cell/cell and cell/matrix junctions. Biol Reprod 2001; 64: 396-407.
Nehar D, Mauduit C, Boussouar F, Benahmed M. Interleukin 1α stimulates lactate dehydrogenease A expression and lactate production in cultured procine Sertoli cells. Biol Reprod 1998; 59: 1425-32.
Newton SC, Millette CF. Sertoli cell plasma membrane polypeptides involved in spermatogenic cell-Sertoli cell adhesion. J Androl 1992; 13: 160-71.
O'Brien DA, Gabel CA, Eddy EM. Mouse Sertoli cells secrete mannose 6-phosphate containing glycoproteins that are Endocytosed by spermatogenic cells. Biol Reprod 1993; 49: 1055-65.
O'Brien Da, Welch JE, Fulcher KD, Eddy EM. Expression of mannose 6-phosphate receptors messenger ribonucleic acids in mouse spermatogenic and Sertoli cells. Biol Reprod 1994; 50: 429-35.
O'Bryan MK, Mallidis C, Murphy BF, Baker HW. Immunohistological localization of clusterin in the male genital tract in humans and marmosets. Biol Reprod 1994; 50: 502-9.
O'Donnell L, Stanton PG, Bartles JR, Robertson DM. Sertoli cell ectoplasmic specializations in the seminiferous epithelium of the testosterone-suppressed adult rat. Biol Reprod 2000; 63: 99-108.
Olson JH, Xiang X, Ziegert T, Kittelson A, Rawls A, Bieber AL, Chandler DE. Allurin, a 210kDa sperm chemoattractant from Xenopus egg jelly, is related to mammalian sperm-binding proteins. Proc Natl Acad Sci (U S A) 2001; 98: 11205-10.
Onoda M, Djakiew D. A 24,500 Da protein derived from rat germ cells is associated with Sertoli cell secretory function. Bichem Biophys Res Commun 1993; 197; 688-95.
Orth JM, Jester WF, Li LH, Laslett AL. Gonocyte-Sertoli cell interactions during development of the neonatal rodent testis. Curr Top Dev Biol 2000; 50: 103-24.
Pelletier R, Trifaro JM, Carbajal ME, Okawara Y, Vitale ML. Calcium-dependent actin filament-severing protein scinderin levels and localization in bovine testis, epididymis, and spermatozoa. Biol Reprod 1999; 60: 1128-36.
Perryman KJ, Stanton PG, Loveland KL, McLachlan RI, Robertson DM. Hormonal binding decency of neural cadherin in the binding of round spermatids to Sertoli cells in vitro. Endocrinology 1996; 137: 3877-83.
Pfeiffer DC, Vogl AW. Evidence that Vinculin is co-distributed with actin bundles in ectoplasmic : junctional") specializations of mammalian Sertoli cells. Anat Rec 1991; 231: 89-100.
Quill TA, Garbers DL. Sperad is a novel sperm-specific plasma membrane protein homologous to a family of cell adhesion proteins. J Biol Chem 1996; 271: 33509-14.
Rao MK, Wayne CM, Meistrich ML, Wilkinson MF.Pem homeobox gene promoter sequences that direct transcription in a Sertoli cell-specific, stage-specific, and androgen-dependent manner in the testis in vivo. Mol Endocrinol. 2003; 17: 223-33.
Raychoudhury SS, Millette. Multiple fucosyltransferases and their carbohydrate ligands are involved in spermatogenic cell-Sertoli cell adhesion in vitro in rats. Biol Reprod 1997; 56: 1268-73.
Richardson LL, Kleinman HK, and Dym M. Basement membrane gene expression by Sertoli and peritubular myoid cells in vitro in the rat. Biol Reprod 1995; 52: 320-30.
Roscoe WA, Barr KJ, Mhawi AA, Pomerantz DK, Kidder GM. Failure of spermatogenesis in mice lacking connnexin43. Biol Reprod 2001; 65: 829-38.
Saez JM, Tabone E, Perrard – Sapori MH. Rivarola MA. Paracrine role in Sertoli cells. Med Biol 1986; 63: 225-36.
Salanova M, Ricci G, Boitani C, Stefanini M, Grossi SD, Palombi F. Junctional contacts between Sertoli cells in normal and aspermatogenic rat seminiferous epithelium contain á1â5 integrins, and their formation is controlled by follicle stimulating hormone. Biol Reprod 1998; 58: 371-78.
Sandlow JI, Feng HL, Sandra A. Localization and expression of the c-kit receptor protein in human and rodent testis and sperm. Urology 1997; 49: 494-500.
Schlatt S, Meinhardt A, Nieschlag E. Paracrine regulation of cellular interactions in the testis: factors in search of a function. Eur J Endocrinol 1997; 137: 107-17.
Schweyer S, Soruri A, Peters J, Wagner A, Radzun HJ, Fayyazi A. Malignant germ cell tumours of the testis express interferon-gamma, but are resistant to endogenous interferon-gamma. Br J Cancer. 2003 ; 89: 915-21.
Silteri JE, Ensrud KM. Moore A, Hamilton DW. Identification of osteopontin (OPN) mRNA and protein in the rat testis and epididymis and on sperm. Mol Reprod Dev 1995; 40: 16-28.
Siu MK, Mruk DD, Lee WM, Cheng CY. Adhering junction dynamics in the testis are regulated by an interplay of beta 1-integrin and focal adhesion complex-associated proteins. Endocrinology 2003; 144: 2141-63.
Skinner MK, Fetterolf PM, Anthony CT. Purification of a paracrine factor, P-Mod-S, produced by testicular

peritubular cells that modulates Sertoli cell function. J Biol Chem 1988; 263: 2884-90.

Skinner MK, Fritz IB, Testicular peritubular cells secrete a protein under androgen control that modulates Sertoli cell functions. Proc Natl Acad Sci USA 1985; 82: 114-18.

Skinner MK, Fritz IB. Identification of a non-mitogenic paracrine factor involved in mesenchymal-epithelial cell interactions between testicular peritubular cells and Sertoli cells. Mol Cell Endocrinol 1986; 44: 85-97.

Strohmeyer T, Reese D, Press M, Ackermann R, Hartmann M, Slamon D. Expression of the c-kit proto-oncogene and its ligand stem cell factor (SCF) in normal and malignant human testicular tissue. J Urol 1995; 153: 511-15.

Syed V, Hecht NB. Selective loss of Sertoli cell and germ cell function leads to a disruption in Sertoli-cell germ cell communication during aging in the Brown Norway rat. Biol Reprod 2001; 64: 107-12.

Syed V, Hecht NB. Up-regulation and down-regulation of genes expressed in cocultures of rat Sertoli cells and germ cells. Mol Reprod Dev 1997; 47: 380-9.

Syed V, Stephan JP, Gerard N, Legrand A, Parvinen M, Bardin CW, Jegou B. Residual bodies activate Sertoli cell interleukin-1 α (IL-1α) release, which triggers IL-6 production by an autocrine mechanism, through the lipoxygenase pathway. Endocrinology 1995; 136: 3070-8.

Sylvester SR, Morales C, Oko R, Grishwold MD. Localization of sulfated glycoprotein-2 (clusterin) on spermatozoa and in the reproductive tract of the male rat. Biol Reprod 1991; 45: 195-207.

Tanaka H, Yasuhide Y, Masami N, et al. Identification and characterization of a Haploid germ cell-specific nuclear protein kinase Haspin in spermatid nuclei and its effects on somatic cells. J Biol Chem 1999; 274: 17049-57.

Taranta A, Teti A, Stefanini M, D'Agosino A. Immediate cell signal induced by laminin in rat sertoli cells. Matrix Biol 2000; 19: 11-8.

Tsuruta JK, O'Brien DA, Griswold MD. Sertoli cell and germ cells cystatin C: stage-development expression of two distinct messenger ribonucleic acid transcripts in rat testes. Biol Reprod 1993; 49: 1045-54.

Vidal F, Lopez P, Lopez-Fernandez LA, Ranc F, Scimeca JC, Cuzin F. Rassoulzadadegan M. Gene trap analysis of germ cell singaling to Sertoli cells: NGF-TrkA mediated induction of Fra1 and Fos by post-meiotic germ cells. J Cell Sci 2001; 114 (pt2): 435-43.

Wakayama T, Koami H, Ariga H, Kobayashi D, Sai Y, Tsuji A, Yamamoto M, Iseki S. Expression and functional characterization of the adhesion molecule spermatogenic immunoglobulin superfamily in the mouse testis. Biol Reprod. 2003; 68: 1755-63.

Weinbauer GF, Wessels J. Paracrine control of spermatogenesis. Andrologia 1999; 31: 249-62.

Weiss M, Vigier M, Hue D, Perrard-Sapori MH, Marret C, Avallet O, Durand P. Pre-and postmeiotic expression of male germ – specific genes throughout 2-week cocultures of rat germinal and Sertoli cells. Biol Reprod 1997; 57: 68-76.

Whaley PD, Chaudhary J, Cupp A, Skinner MK. Role of specific response elements of the c-fos promoter and involvement of intermediate transcription factor(s) in the induction of Sertoli cell. Differentiation (transferrin promoter activation) by the testicular paracrine factor PmodS. Endocrinology 1995; 136: 3046-53.

Wine RN, Chapin RE. Adhesion and signaling proteins spatiotemporally associated with spermiation in the rat. J Androl 1999; 20: 198-213.

Wong CH, Mruk DD, Lui WY, Cheng CY. Regulation of blood-testis barrier dynamics: an in vivo study. J Cell Sci. 2004;117(Pt 5): 783-98.

Zakin MM, Baron B, Guillou F. Regulation of the tissue-specific expression of transferrin gene. Dev Neurosci. 2002; 24: 222-6.

Chapter 3

NON-STEROIDAL SIGNAL MOLECULES IN SPERMATOGENESIS

Non-steroidal growth factors are mainly polypeptides that upon binding to surface receptors in target cells elicit the activation of signal transduction pathways involved in cell division and cell specific differentiation. The action of growth factors can be stimulatory or inhibitory as shown using cultured testicular cells. Since growth factors elicit diverse effects in variety of organ systems during development, it has been difficult to establish and to assign a specific function to a specific growth factor during the process of spermatogenesis. Situation is further complicated by the fact that several of these factors (FGF, IGF) have been found in Sertoli cells and germ cells. On the other hand, EGF and TGF have been specifically implicated in the survival and proliferation of rat type A-spermatogonia.

3.1. INHIBIN FAMILY

Inhibin and activin are dimeric glycoproteins sharing a common-β subunit. Inhibin contains an inhibin specific α-subunit and one of the two closely related-β subunits (β_A or β_B), whereas activin is a homo- or heterodimer of β-subunits. Immunoreactive and bioactive inhibin, and three inhibin/activin subunit mRNAs (α, β_A, and β_B) were identified in Sertoli cells. Small amounts of immunoactive inhibin, inhibin α-subunit proteins and activin, and the subunit mRNAs were also detected in the Leydig cells of rat testis and in Leydig derived cell lines.

3.1.1. Inhibin

Inhibin isolated from follicular fluid is a heterodimeric glycoprotein that exists in multiple biologically active forms. Porcine inhibin is a 32-kDa protein, which is composed of two disulfide bound subunits of 18kDa (α) and 14 kDa (β). Two related, yet distinct forms of the smaller β subunit, termed β_A and β_B have been described, and both these subunits form complex with the α-subunit to form biologically active inhibin (termed inhibin A and inhibin B). The amino acid sequences of inhibins for porcine, bovine, human, and rat sources have been described by molecular cloning of the corresponding cDNAs. Structural analysis has revealed that each of the three inhibin subunits is encoded by a separate gene and that each mature subunit resides at the carboxy terminal end of a much larger precursor protein. In addition, the subunits of inhibin share a substantial sequence identity with an emerging family of proteins with growth regulating properties, the prototype of which is transforming growth factor-β (TGF-β).

Expression and Regulation: In males, the Sertoli cell appears to be the major producer of inhibin. However, reports also suggest that the α-subunit and possibly inhibin itself are produced by Leydig cells also, whereas β_A is present in the nuclei of spermatocytes (Ackland et al., 1992). Inhibin, originally considered only as gonadal hormone preferentially suppresses FSH secretion in the pituitary in a classic negative feedback loop with FSH. FSH has been shown to be a major regulator of inhibin. In males inhibin and its mRNA levels are positively regulated by FSH in vivo and in vitro, but regulation of its α and β chains is independent. In addition to FSH, adenosine and EGF stimulate the production of inhibin by Sertoli cells, whereas endorphin is inhibitory. The IGF-I has no influence on the level of inhibin in Sertoli cells. The role of inhibin in adult males is less clear except in seasonal breeders where it appears to regulate FSH during testicular growth. Inhibin is also present in different regions of human epididymis. The epididymal inhibin was found to be bioactive, since it suppressed specifically the FSH levels of rat pituitary in vitro.

Analysis by *in situ* hybridization in rat testis showed that inhibin β_A mRNA expresses in Sertoli cells in a highly stage specific manner. The mRNA levels started to accumulate in Sertoli cells at stage VIII of the cycle and were highly expressed during stages IX-XI. Follistatin mRNA expression was identical to that of inhibin β_A, while inhibin β_B mRNA expressed maximally in Sertoli cells at stages XIII-III. Coexpression of inhibin β_A and follistatin mRNAs in stage IX-XI Sertoli cells suggested a close interplay between these molecules. The pattern of inhibin β_B mRNA expression was similar to that of inhibin–α mRNA. Localization of activin A receptor mRNA in spermatogenic cells suggested that activin might influence meiotic division and early stage of spermiogenesis (Kaipia et al., 1992).

The evidence for the presence of immunoreactive inhibin and its regulation by specific germ cell type in rat testis has been presented by various authors. Pineau et al., (1990) provided evidence for the involvement of germ cells, particular of early spermatids, in the local testicular regulation of inhibin gene expression and its production in the rat. This may be of crucial importance for the ontogeny of this parameter of Sertoli cell function, and may have important implications with regard to postulated roles of inhibin. Exogenously administered inhibin impairs spermatogenesis and alters steroid production. In this regard inhibin appears to be highest in mature animals and appears to depress following the initiation of spermatogenesis and formation of the blood testes barrier. This pattern has also been seen with both α and β inhibin mRNA in rat testes. However, Bhasin et al., found that inhibin mRNA content within seminiferous tubules varied with the stage of the cycle (review Ackland et al., 1992).

3.1.2. Activins

Activins are involved in the regulation of several reproductive processes in higher vertebrates. The active form of activin is composed of two β subunits covalently attached by a single disulphide link. So far, five β subunits have been identified, but only β_A and β_B subunits have been shown to dimerize, giving three possible types of activin species, which are named according to the subunit species of which they are composed (activin A, B and AB). The activin-inhibin subunits are structurally related to the transforming growth factor-β (TGF-β) family of peptides, which also include bone morphogenetic proteins (BMP), Mullerian inhibiting substance (MIS) and growth and differentiation factor (GDF). Originally, activins were considered as regulatory proteins mainly in the gonadal-pituitary axis since they were capable of positive regulation of pituitary FSH secretion. Several studies have confirmed this activity of activin on FSH production as well as other pituitary produced proteins, such as follistatin

(Phillips and de Kretser, 1998). Subsequently it was accepted that activin could modulate many ovarian cell functions, at least in vitro.

Inhibin and Activin as Paracrine/Autocrine Factors: Activin, which is composed of disulfide bonded homodimeric-β subunits $\beta_A\beta_A$ or $\beta_B\beta_B$ (activin A and B, respectively) or heterodimeric β-subunits (activin AB), stimulates FSH secretion. Activin like inhibin has been suggested, from their expression profiles, to act as paracrine and autocrine regulators of spermatogenesis. The Sertoli cells are the major sites of activin and inhibin production although the Leydig cells may produce low levels of the α-subunit. Activin appears as an inhibitor of LH/hCG action on androgen biosynthesis at all stages of Leydig cell differentiation whereas inhibin facilitates LH action only on immature Leydig cells. Activin also stimulates the proliferation of spermatogonial cells in vivo and in vitro, where as inhibin decreases spermatogonial proliferation when injected locally to hamster testis (reviewed in Tanimoto et al., 1999). Now inhibin and activins are as important for males as they are for female reproductive processes. Activin A stimulates proliferation of testicular cells in immature rats. However, activin actions on testicular cells are age specific. In the presence of FSH, activin A stimulates the proliferation of Sertoli cells from 9 days old rats but not from 18 days old rats (Boitani et al, 1995). Transgenic mice, over-expressing the A subunit in the testis showed that the control of the expression of the subunits is critical for male gonadal functions (Tanimoto et al., 1999). The localization of activin receptor mRNA in rat seminiferous epithelium by Kaipia et al. (1993) provided evidence, which suggests that inhibin and activin are paracrine/autocrine regulators in the testis. The genes encoding the inhibin/activin subunits are widely expressed in many tissues suggesting that inhibin and activin may also act as paracrine / autocrine factors to regulate cellular functions in tissues other than testis and ovary.

Distribution of Activins in Testis: In rat testes, the mRNAs expression encoding the β-subunits localized in the Sertoli cells is dependent on the seminiferous epithelial cycle. The β_A-subunit mRNA starts to appear at stage VIII, reaches a maximum at stages IX-XI, whereas maximum β_B-subunit mRNA concentrations are observed at stage XIII (Kaipia et al., 1992). However, despite the presence of its mRNA, the subunit protein was not detected in the rat testes, although the activin dimmer was detected in Leydig and Sertoli cells (Wada et al., 1996). The protein distribution and the developmental expression of β_B subunits in the testes appear to be different from those of the β_A subunit, at least in rats. During development β_A subunit synthesis was immunochemically detected in fetal Leydig cells but the signal decreased markedly after birth, whereas the β_B subunit was detected in the Sertoli cells and its abundance increased after birth (Majdic et al., 1997) indicating different role for the subunits during the testicular maturation.

Blauer et al., (1999) provided evidence for intra-nuclear activin/inhibin β_A subunit of 24 kDa and showed its distribution in the rat seminiferous epithelium. β_B subunit precursor contains a functional nuclear localization signal within the lysine rich sequence corresponding to amino acids 231-244. In all stages of the rat seminiferous epithelial cycle, an intense activity of nuclear β_A was demonstrated in intermediate or type B spermatogonia or primary spermatocytes in their initial stages of the first meiotic prophase, as well as in pachytene spermatocytes and elongating spermatids. Results suggested that activin/ inhibin β_A may elicit its biological functions through two parallel signal transduction pathways, one involving the dimeric molecules and cell surface receptors and the other an alternately processed β_A sequence having molecular weight 1.5 times that of present in dimeric molecule and acting directly within the nucleus. β_A may play a significant role in the regulation of nuclear functions during meiosis and spermiogenesis (Blauer et al., 1999).

Expression of activin receptors in testes: The expression of ActR IIB gene in Sertoli cells around meiotically dividing spermatocytes suggested a possible role of activin during meiotic maturation in addition to stimulatory effect on spermatogonial proliferation. Like most members of the TGF-β family, activins mediate their action by binding to a complex of transmembrane serine and threonine kinase receptors. So far, four activin receptors have been identified by their ability to bind activin. These proteins can be classified into two main categories namely the type I receptor group, which comprises activin receptor like kinase ALK2 and ALK4, and the type II receptor group, comprising the activin type II and activin type IIB receptor (ActRII and ActRIIB). In addition, five isoforms of ActRIIB, generated by alternative splicing of its mRNA, have been identified in mice and bovine species. Thus, activins have the potential to act through several forms of receptors. The ALK2 gene is expressed in bovine testes, and there is a change in mRNA profile during testicular maturation. The cDNAs encoding two type II activin receptors (ActR II and ActR IIB) have been cloned. The mRNAs coding these receptors are detected in the specific populations of germ cells. Activin–A receptor II mRNA was localized mainly in spermatogenic cells with maximal expression in late primary spermatocytes at stages XIII-XIV and in early round spermatids at stages I-IV, with a low level of expression by Sertoli cells. The ActRIIB mRNA was localized to spermatogonia. In rats ActR II appears to be expressed mainly in spermatogenic cells whereas the ActR II mRNA is present at a very low level or absent. The ActR II B mRNA has also been detected in bovine and mouse testes (Ethier et al., 1997; Wu et al., 1994). The examination of isoform mRNA expression of both activin receptor type II (ActR- II) and type II B (ActR- II B) genes showed high levels of the 6.0-kb mRNA in overies, whereas the 3.0-kb transcript was major mRNA found in the testis. The isoforms II B2 and II B4 are the major mRNA species expressed in both female and male gonads and extra-gonadal reproductive tissues (Wu et al., 1994). Mice with inhibin α subunit gene deletion develop testicular tumors. These mice were associated with a 200 folds increase in testicular expression of activin β_A subunit mRNA form of 6.5 kb and a minor form of 4.5 kb. Testicular expression of the type II activin receptors was decreased 3 fold in these mice. Therefore, increased expression of activin is accompanied by a tissue specific reduction in the expression of its own receptors mRNA in vivo (Ethier and Findlay, 2001).

Cell Signaling and transcriptional activation by activins: Activins have the potential to act through their receptors. It is thought that activin binds first to an activin type II receptor and then an activin type I receptor is recruited into the complex. The formation of this ligand receptor complex induces the activation of the type I receptor by the kinase activity of the nearby type II receptor (Mathews, 1994). Once activated, the kinase of activin type I receptor stimulates the down stream pathways. The direct substrates for the kinase type I receptor to form a novel family of proteins are called Smads.

Smad proteins: The Smad proteins, which are transcription factors found in vertebrates, insects and nematodes, are the only TGF-β receptor substrates, that possess the ability to propagate signals. These proteins can be grouped in three subclasses: the receptor activated Smads, the common Smads, and the inhibitory Smads. They are downstream mediators for the family of TGFβ growth factors and associated with both TGFβ and activin signaling. However, very little is known about the expression and regulation of the Smad signaling proteins in the testis. Smad2 and Smad3 proteins are expressed in the postnatal testes of rats from 5 days to 60 days of age. Their mRNA levels run parallel to protein synthesis. These proteins are mainly localized in the cytoplasm of meiotic germ cells, Sertoli cells, and Leydig cells, whereas Smad3 protein is localized to the nucleus of preleptotene to zygotene primary spermatocytes in young rats. The presence of these downstream mediators in these cell types supports a role

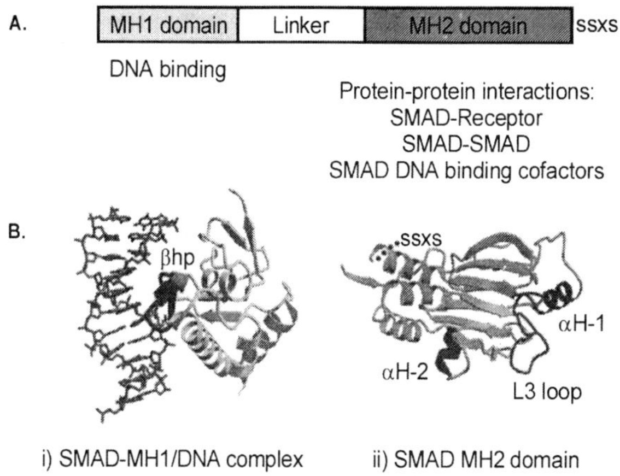

Fig.3.1. (A) Smad structural domains and their functions. (B) Three dimemsional structure of (i) Smad3 MH1 domain bound to AGAC in SMAD-MH1/DNA complex and (ii) Smad2 MH2 domain. Reprinted with permission from J. Massague and D. Wootton, The EMBO Journal 19; 1745-54: 2000 © http://www.nature.com/

for TGFβ and activin during spermatogenesis. The difference between the expression of Smad2 and Smad3 suggests that they may have different functions within the testis (Xu et al., 2003). Activin can signal specifically through Smad2 and Smad3. Once Smad2 or Smad3 are activated by phosphorylation, they associate with Smad4 and the dimer migrates to the nucleus. The Smad complex can then bind to specific promoters and associate with transcriptional co-factors and co-activators to induce a transcriptional response (Massague and Chen, 2000; Ethier and Findlay, 2001) **(Fig.3.1)**.

3.1.3. Follistatin

Follistatin (FS) is a monomeric glycosylated polypeptide originally isolated from follicular fluid as an inhibitor of FSH secretion. It is also produced in a wide range of non-reproductive tissues and is a high affinity binding protein of activin and to a lesser extent of inhibin. Shimasaki et al., (1988) and Robertson et al., (1987) independently reported the isolation of three single chain proteins with FSH release inhibitory activity from bovine follicular fluid, all having same amino terminal amino acid sequence as porcine FS. The cDNA cloning of the porcine ovarian FS precursor revealed two populations of cDNAs, which differed at the 3'-region of the open reading frames; one population encodes a precursor of 317 amino acids while the other encodes another precursor having the same 317 amino acids but with an additional 27 amino acids at the carboxy terminal; the two populations of FS mRNAs are generated by alternative splicing. The FS gene is approximately 6 kb long and consists of six exons separated by five introns. The first exon encodes the putative signal sequence, followed by four exons, which encode the four domains of FS, three of which are highly homologous to the other. The last exon encodes the extra 27 amino acid carboxy terminal domain of the residual precursor (Shimasaki et al., 1988; Houde et al., 1994).

Follistatin is present in the cytoplasm and nucleus of late pachytene spermatocytes. Although the reaction in the cytoplasm disappeared after meiosis, it continued to be intense in the nucleus from pachytene spermatocytes to round spermatids (Ogawa et al., 1997). A role for follistatin in male germ cell differentiation was studied at the time when spermatogonial stem cells and committed spermatogonia first appear in the developing testis. Germ cells have the potential to regulate their own maturation through production of endogenous activin A and follistatin, whereas Sertoli cells, which produced the activin/inhibin β_A subunit, the inhibin α subunit, and follistatin, have the potential to regulate germ cell maturation as well as their own development (Meehan et al., 2000). It was suggested that the main function of follistatin is to act as a regulator of bioactivity rather than as a carrier protein. It could be possible that the relative abundance of follistatin and activin in a cell may determine cell proliferation (Mather et al, 1993). In the rat testis, follistatin mRNA is expressed in Sertoli cells and the level of expression varies in a stage specific manner. The protein has been localized to Sertoli and Leydig cells (Majdic et al., 1997). In addition to Leydig and Sertoli cells of the testis, the prostate gland and seminal vesicle may form the predominant source of follistatin in seminal plasma (Anderson et al., 1998).

3.2. GROWTH FACTORS

3.2.1. Insulin Like Growth Factors

The insulin like growth factor (IGF) system is well defined, with profound effects on the growth and differentiation of normal and malignant cells. The established components of the IGF system include IGFs (IGF-I and IGF-II), type I and type II IGF receptors, IGF binding proteins (IGFBPs), and IGFBP proteases. The IGF-I (also called somatomedin C) and IGF-II, which are structurally similar to insulin, are two highly homologous hormone peptides of approximately 70-kDa molecular mass. The IGFs interact with specific cell surface receptors, designated type-I and type-II IGF receptors, and can also interact with the insulin receptor. The type-I IGF receptor, like the insulin receptor, is a receptor with tyrosine kinase activity. The type II IGF receptor is structurally distinct, binds primarily IGF-II but also serves as a receptor for mannose-6-phosphate containing ligands. There are at present, six well characterized mammalian IGFBPs, designated IGFBP-1 through–6. Of particular interest is the discovery of several groups of cysteine rich proteins with discrete but striking, structural and functional similarities to the IGFBPs (Hwa et al., 1999).

IGF and IGF Receptors in Testis: Insulin and its receptors have been extensively studied in the gonads and is the subject of several reviews. Based on the sequence analysis of the complete IGF-1 mRNA, it seems that the protein is synthesized as a precursor containing a signal peptide of 22 or 25 amino acid residues. In addition, the sequence extended in the 3'-direction showed the presence of multiple polyadenylation sites in IGF-1 message (Le Bouc et al., 1986). Insulin like growth factor-I has been found in all cell types in rat testis. In addition to its finding in Sertoli cells, IGF-I is present over peritubular cells. The IGF-I was apparent in spermatogenic cells, Sertoli cells, and Leydig cells only during the first 2 wk after birth, and declined thereafter in adults (Hansson et al., 1989).

Anterior pituitary hormones regulate IGF-1 release and synthesis in the testis. Testicular IGF was decreased after hypophysectomy, and LH and FSH could restore levels of IGF-I mRNA but not IGF-II. In addition to gonadotrophins, stroids have a stimulatory effect on IGF-I in the testis. Growth hormone (GH) is important in stimulating IGF in liver and its secretion from the gonads. A slight increase is found in IGF-I RNA after GH exposure, and expression was greatly enhanced by addition of DES with GH (review Ackland et al., 1992). In contrast, IGF-II mRNA expression was detected in all testicular cell types irrespective of the hormonal treatment (Dombrowicz et al., 1992).

The IGF-1 receptors have been identified on all cell types in testis. Receptors for IGF-1 have been found on Leydig cells and pachytene spermatocytes. IGF receptors are regulated both developmentally and hormonally, and hence IGF-I may be more important during developmental stages. The FSH, cAMP analogues, and a calcium ionophore reduce the production of IGF binding protein-1 by rat Sertoli cells *in vitro*. In situ hybridization studies revealed that IGF-I mRNA was not detected in human testis, whereas IGF-I mRNA was abundant in testicular blood vessels and peritubular connective tissue. Both IGF-I and IGF-II receptors mRNAs were most abundant in the germinal epithelium. Insulin like growth factors binding protein-1 mRNA was not detected while IGFBP 2-6 mRNAs were expressed abundantly.

Functions of IGF in Testis and Sperm: Insulin and the IGFs have a variety of effects on gonadal cells, including increasing proliferation, affecting steroidogenesis, and influencing other aspects of differentiation. As IGF-I and IGF-II are locally produced in the testis these factors have a selective paracrine or autocrine role in the regulation of spermatogonial proliferation during spermatogenesis. The IGF-I and IGF-II and insulin promote differentiation of spermatogonia to primary spermatocytes in organ culture of newt testes (Nakayama et al., 1999; Tajima et al., 1995). The growth-promoting role of the IGFs has been demonstrated conclusively from the dwarfing phenotypes observed after targeted mutagenesis of the mouse *Igf1, Igf2,* and *Igf1r* genes. The testes are reduced in size and sustain spermatogenesis only at 18% of the normal level. Despite the mutational impact on the epididymis, capacitated sperm are able to fertilize wild type eggs in vitro. The infertility of male mutants is due to failure of androgen synthesis resulting in drastically reduced levels of serum testosterone (Baker et al., 1996). The IGF-I deficiency may play a pathogenetic role in hypogonadism of cirrhosis. Low doses of IGF-I reverse testicular atrophy and appear to improve hypogonadism in advanced experimental cirrhosis (Castilla-Cortazar et al., 2000).

Evaluation of the transgene expression during testicular development revealed that expression begins between 10 and 15 days of development, coinciding with the appearance of the zygote and pachytene primary spermatocytes during early spermatogenesis. The over-expression of IGF-I induces increased IGFBP localization in the testis. Inhibition of IGF activity by IGFBP would explain the lack of dramatic physiological effects in IGF-I transgenic mice, despite the presence of elevated testicular IGF-I. Thus induced IGFBP in testis due to over-expression of IGF-I confirms the existence of a well-regulated testicular IGF system in testicular function (Dyck et al., 1999; Suh et al., 1997).

Insulin and IGF-1 are found to inhibit germ cell synthesis of sulfogalactosylglycolipid (SGG), which is a differentiation marker in spermatogenesis (Lingwood et al., 1994). Soluble factors have been shown to potentially mediate regulatory interactions between peritubular cells and Sertoli cells. IGF-I ubiquitously produced in the testis stimulates proteglycans (PG) synthesis in several cell types. The exposure of peritubular cells from immature rat testis to IGF-I induces time dependent increase of newly synthesized proteoglycans. The stimulation of PG synthesis by IGF-I appeared to be due, at least partially, to an increase of glycosylation process, and appeared to be mediated by the classical tyrosine kinase-signaling pathway. Although the mechanism of hyaluronic acid synthesis was completely different from those of other glucosaminoglycan (GAG), IGF-I dramatically enhances its production by PT cells (Thiebot et al., 1997).

The presence of IGF-IR on sperm and the presence of IGF-I in semen, and the ability of IGF-I to stimulate sperm motility provide evidence that the IGF system is involved in the fertilization in the bovine species (Henricks et al., 1998). IGF-IR is present primarily in the equatorial region, along with a weak binding in the acrosomal region of human spermatozoa. Since the IGF-IR has tyrosine kinase activity and its ligand IGF-I is present in the seminal plasma, the IGF-I system may be involved in the signal transudation leading to sperm capacitation and acrosome exocytosis (Naz and Padman, 1999).

3.2.2. Fibroblast Growth Factors

Fibroblast growth factors (FGFs) are a family of heparin binding polypeptides involved in an array of biological processes during development and adult life. The FGF family consists of at least 12 members. Heparan sulfate proteoglycans (HSPGs) are low-affinity binding sites for FGF ligands and are present in the extracellular matrix or as cell surface-bound molecules. Fibroblast growth factor exists in two highly homologous forms, acidic (aFGF) and basic (bFGF) that can bind to the same receptor. The aFGF has a limited tissue distribution and is not found in the gonads. Basic FGF has been isolated from testis and has been shown to be bioactive. Basic fibroblast growth factor of 30-kDa restricted to the tubular compartment is a 1146-amino acid peptide composed of a single chain, which can exist in a truncated form that has the first 15 amino acids from the amino terminus deleted. Basic FGF is a potent mitogen but also has differentiating capabilities. It has been implicated in many functions (et al., 1991).

In the male gonads, FGFs are implicated in spermatogonial and Sertoli cell mitosis and in Leydig cell function. Sertoli cell extracts were shown to contain bFGF. Observations suggest that the autocrine/paracrine actions of FGFs take place in both the compartments of testis. The FGF-1, FGF-2 and FGFR-I have been localized in adult testis. Immature Leydig cells bind FGF-1 and FGF-2 in vitro through a specific FGFR. The FGFR-I mRNA has been detected in the immature rat testis, and levels were found to decrease during development. It can stimulate c-fos mRNA in Sertoli cells in vitro, and this induction appears to be independent of cAMP and Ca^{2+}/phospholipids pathways. In immature pig Sertoli cells in culture, bFGF stimulates cell proliferation and DNA synthesis. In addition, bFGF could increase FSH receptors on Sertoli cells and could act synergistically with EGF to increase PA. It has been shown that bFGF affects steroidogenesis in various testicular cells. Basic FGF regulates aromatase activity in porcine Leydig cells. It was found to inhibit LH- and forskolin induced testosterone production in rat testicular cell cultures but did not decrease progesterone production. It also inhibited conversion of exogenous progesterone and testosterone, indicating that bFGF may suppress 17β-hydroxylase activity. However, in immature porcine Leydig cells, bFGF augmented hCG-stimulated testosterone production.

Four high-affinity FGF receptor (FGFR) genes have been characterized; these encode transmembrane protein tyrosine kinase receptors. The FGF-2 is involved in the autocrine and paracrine regulation of proliferation and differentiation of spermatogonia and spermatocytes via the receptors FGFR-1, FGFR-3 and FGFR-4, which were localized to germ cells of human testis (Steger et al., 1998; Resnick et al., 1998). In the fetal testis, reactivity for FGFR-1 was seen in gonocytes, Sertoli cells, Leydig cells, and mesenchyme, and FGFR-3 was localized in gonocytes. The FGF-2 stimulates steroidogenesis by immature Leydig cells in the absence of LH, mediated through HSPGs. The HSPGs are also essential for the autocrine growth factor regulation of adult Leydig cell steroidogenesis. In the immature testis, FGFR-1 was localized to spermatogonia, and all four FGFRs were localized in pachytene spermatocytes, immature Leydig cells, and peritubular cells. In the adult testis Sertoli cells were immunoreactive for FGFR-4, and germ cells were immunoreactive for all four FGFRs (Cancilla and Risbridger, 1998).

3.2.3. Epidermal Growth Factor

Epidermal growth factor (EGF) is a single polypeptide of 53 amino acids that was first isolated and purified from the submandibular glands of male mice and contains three disulphide bonds.

The EGF is shown to be equivalent to urogastrone, which was isolated from human urine. The mature EGF (EGF_m) is a proteolytic processed polypeptide of EGF precursor (EGF_p), an integral membrane protein of 140,000 M_r that exhibits eight EGF like repeats in addition to the EGF_m. Although the normal physiological action of EGF_m remains obscure, several biological effects have been attributed to it, including growth and differentiation effects on the cells in vitro. The EGF has been implicated in the regulation of spermatogenesis. For example, ablation of the submandibular gland (sialoadenectomy) in adult male mice led to a marked decrease in circulating levels of immunoreactive EGF, without diminishing testosterone and FSH, but correlated with decrease of epididymal sperm. These effects were reversed by administration of EGF_m.

The EGF is involved in differentiation of the male reproductive system through modulation of androgen receptor activity. It was suggested that proper EGF expression is important for completion of spermatogenesis (Wong et al., 2000). EGF administration was required to reverse the infertility. It appears that the effects are mediated via the testicular somatic cells. Epidermal growth factor like activity has been detected in human seminal plasma. Specific EGF binding activity was detectable in testicular membrane preparations but not in membranes prepared from human prostate, seminal vesicle, epididymis, or spermatozoa (Gupta, 1996).

Epidermal growth factor can stimulate Sertoli cell proliferation and has an additive effect with other growth factors. In Leydig cells, EGF_m also affects androgen production. Consistent with studies was the demonstration that both Leydig cells and Sertoli cells in rodent, nonhuman primates, and human testes exhibit EGF receptor. The EGF has been observed in the mouse and human testis and produced specifically in the mature germ cells residing in the adluminal compartment of the blood testis barrier. It has been suggested that the testis must be included in the list of tissues capable of synthesizing EGF_P. It induces differentiation of type A spermatogonia in cryptorchid testis at 100-200 ng/ml, whereas EGF at lower concentrations (1 to 100 ng/ml) inhibited mitotic activity in spermatogonia. The action on EGF receptor (EGFR) expression and inhibition of EGFR results in inhibition of Wolffian duct differentiation. A zinc finger protein, ZPR1 binds to the cytoplasmic tyrosine kinase domain of the epidermal growth factor receptor (EGFR). This interaction is mediated between zinc fingers of ZPR1 and sub-domains X and XI of the EGFR tyrosine kinase. Treatment of mammalian cells with EGF caused decreased binding of ZPR1 to the EGFR and the accumulation of ZPR1 in the nucleus. The ZPR1 represents a prototype class of molecule that binds to the EGFR and is released from the receptor after activation (Galcheva-Gargova et al., 1996). EGF-receptor is localized in the head of bull spermatozoa and can induce the acrosome reaction in its typical dose dependent manner. Epidermal growth factor also stimulates human sperm capacitation by activating the tyrosine kinase of the EGF receptor, which regulates phosphorylation at multisites.(Lax et al, 1994).

3.2.4. Transforming Growth Factor-α

Transforming growth factor α is biologically and structurally related to EGF. From the expression patterns of EGF, TGF-α and their common receptor, TGF was suggested to act during early pubertal stages to support the active somatic cell growth in testis. However, its role in spermatogenesis is not evident. Transforming growth factor α is a 50 amino acid polypeptide and is highly homologous to EGF, sharing ~21 residues with mouse EGF, including three intra-chain disulfide bonds. Both TGF-α and EGF bind to the same receptor and have similar action in many cell systems. But they can also have distinct effects. In males, TGF-α mRNA (but not EGF) is found in peritubular cells and Sertoli cells, and cultures of Sertoli cells secreted immunoreactive TGF-α. While TGF-α is essentially observed in somatic cells, specifically in perinatal Leydig cells and in mature Sertoli cells, EGF is localized both in germ cells and in somatic cells with a preferential tubular expression. EGF receptor was localized to peritubular cells, but none was found on Sertoli cells consistent with the finding that TGF-α stimulated proliferation of peritubular cells but not Sertoli cells (Caussanel et al., 1996). A study by Levine et al., (2000) was an investigation of the hypothesis that TGF-α is involved in regulating embryonic testis growth. Characterization of cell proliferation in the rat testis throughout embryonic and postnatal development indicated that each cell type has a distinct pattern of proliferation. Both TGF-α and the EGFR were found to express in the embryonic and postnatal rat and mouse testes. Perturbation of TGF-α function by anti-TGF-α Abs on testis organ cultures dramatically inhibited the growth of both embryonic and neonatal testis. Therefore, TGF-α appears to influence embryonic testis growth but not morphogenesis (i.e. cord formation). Treatment of embryonic testis organ cultures with exogenous TGF-α also perturbed development, leading to an increased proliferation of unorganized cells. TGF-α knockout mice had no alterations in testis phenotype, while EGFR knockout mice had a transient decrease in the relative amount of interstitial cells. This suggests that there may be alternate or compensatory factors that allow testis growth to occur in the apparent absence of TGF-α actions in the mutant mice. It seems that TGF-α is an important factor in the regulation of embryonic testis growth, although other factors may also be involved in the process.

3.2.5. Transforming Growth Factor-β

The transforming growth factor-β (TGF-β) family of secretory polypeptides includes various forms of TGF-β, the bone morphogenetic proteins (BMPs), the Nodals, the activins, the anti-Mullerian hormone, and many other structurally related factors in vertebrates, insects and nematodes. More than 30 proteins have been identified in this family. These factors regulate cell migration, adhesion, multiplication, differentiation and apoptosis throughout the life span of various cells. Transforming growth factor-β is a 25000 Da peptide composed of two identical subunits. Each subunit contains nine cysteine residues, seven of which are conserved throughout the TGF-β family. The protein is highly conserved among species; the human, bovine, and porcine sequences are identical and differ from that of the rat by only one amino acid. Most cell types contain TGF-β and its receptor, and it may act in an autocrine fashion. Transforming growth factor is secreted in an inactive form and can be activated with acid or plasmin. Unlike classical hormones, whose actions are few and concrete, the members of the TGF-β family have many different effects depending on the type and state of the cell. For example, in the same healing wound TGF-β may stimulate or inhibit cell proliferation depending on whether the target is a fibroblast or a keratinocyte. Hence TGF-β signaling is regulated positively or negatively. Positive regulation amplifies signals to a level sufficient for biological action. Negative regulation occurs at the extracellular, membrane, cytoplasmic and nuclear levels. TGF-β signaling is often regulated through negative feedback mechanism, which limits the magnitude of signals and terminates signaling. Regulation of TGF-β signaling might be tightly linked to tumor progression, since TGF-β is a potent growth inhibitor in most cell types (Miyazono, 2000). The TGF-β family members are multifunctional hormones, and hence the TGF-β signal transduction pathways may be numerous and complex in order to account for diversity of responses. However, mediation of diverse TGF-β responses, like activins, is explained by the presence of a family of receptor substrates, the Smad proteins that move into the nucleus where they act as transcription factors. The TGF-β (ligand) assembles the receptor complex that activates Smads, and the Smads assemble multi-subunit complexes that regulate transcription.The developmental expression of the multiple forms of TGF-β (TGF-β1, β2, β3) genes in whole testis and isolated somatic cell type showed that TGF-β1 and TGFβ2 mRNA expression was predominant in the immature testis and decreased at the onset of puberty. The TGF-β has dramatic effects on peritubular cell differentiation and the production of specific proteins. TGF-β3 mRNA expression peaked at an early pubertal stage, coincident with the initiation of spermatogenesis. Peritubular and Sertoli cells expressed each isoform of TGF-β during development. Peritubular cells mRNA expression of TGF-β1, β2, and β3 decreased transiently at the onset of puberty. Both cultured peritubular and Sertoli cells can produce the proteins for TGF-β1, β2 and β3.The FSH causes a dramatic decrease in Sertoli cell TGF-β2 expression, and has no effect on TGF-β1 or TGF-β3 expression. TGF-β1 has no effect on Sertoli cell proliferation (Mullaney and Skinner, 1993). Germinal cells express a unique transcript of TGF-β1 and respond to TGF-β1. This suggests that TGF may play an important role in morphogenesis and differentiation of the spermatogenic cells. Local production of TGF-β may also act to limit the action of growth stimulators in the tubule. Three TGF-β genes have been cloned, designated TGF-β_1, TGF-β_2 and TGF-β_3 (Derynck et al., 1988).

Wang and Zhao (1999) examined the expression of Smad2, the signal transducer of the TGF-β in mouse testes. Both Smad2 mRNA and protein are present in meiotic germ cells from preleptotene to pachytene spermatocytes but not in postmeiotic germ cells. Smad2 expression was also observed in interstitial cells and Sertoli cells. Presence of Smad2 in testis provides molecular evidence for TGF-β signal transduction during spermatogenesis. The DNA-Smad interaction are shown in **Fig.3.1** (Massague and Wootten, 2000).

In addition to the role of gonadotrophins in testis, it is now clear that TGF-β is another important factor for spermatogenesis. Both TGF-β mRNA and TGF-β like immunoreactive material are present in the testis of several mammals. In rats, Sertoli cells and peritubular cells express TGF-β mRNA and secrete TGF like peptides in vitro. Leydig and Sertoli cells from immature pig testis express TGF-β mRNA and protein both in vivo and in vitro. In adult mouse testis TGF-β transcript of small size was also found in germ cells. Isolated testicular cells, including germ cells and Sertoli cells, contain mRNAs for the three types of TGF-β receptors (Avallet et al., 1997). Sertoli cells express a single TGF-β transcript of 2.5-kb, but no TGF-β protein could be detected in Sertoli cell conditioned medium. Neither TGF-β mRNA nor protein could be detected in either pachytene spermatocytes or early spermatids. Co-culture of Sertoli cells with either spermatocytes or spermatids resulted in a 2 folds increase of TGF-β protein. Thus Sertoli cell-germ cells interactions regulate TGF-β expression (Avallet et al., 1997).

TGF-β Receptors: The mechanism of the TGF-β receptor activation involves two different transmembrane protein serine/threonine kinases, receptor types-I and -II (TβRI and TβRII) brought together by the ligand that acts as a receptor assembly factor (Massague and Wotton, 2000). Both (TβRI and TβRII) type of receptors are present in the rat testis. In *situ*, Leydig cells showed a strong TβRII immunoreactivity whereas the type-1 receptor (TβRI) activity was weak. The TβRII was detectable in Sertoli cells and in germ cells. The expression of both receptors depended on the seminiferous cycle stage. TβRI first appeared in pachytene spermatocytes and was absent in elongated spermatids from stage XIV onwards, whereas TβRII was observed as early as the spermatogonia stage; it increased in pachytene spermatocytes at the onset of TβRI and disappeared in elongating spermatids from stage XI onwards. This showed that TGFβ could affect somatic cells functions and exert a direct effect on germ cells (Olaso et al., 1998).

3.2.6. Anti-Mullerian Hormone

During male sexual development in mammals, anti-Mullerian hormone (AMH) induces the regression of the Mullerian ducts that form the primordia of the female reproductive tract. Anti-Mullerian hormone, also known as Mullerian-inhibiting substance (MIS), a member of the transforming growth factor β family produced by immature Sertoli cells and, to a lesser degree, by granulosa cells from birth to the end of reproductive life, does not affect gonadal determination but has a negative effect upon gonadal development in both sexes. The AMH represents a useful marker of androgen and FSH action within the testis, as well as of the onset of meiosis. It blocks meiosis in fetal ovaries and inhibits the transcription of aromatase and LH receptor. AMH also affects the development and function of the adult testis by blocking the differentiation of mesenchymal into Leydig cells and by independently decreasing the expression of steroidogenic enzymes (Lyet et al., 1995; Josso et al., 1998). Like other members of TGF-β family, AMH signals through two related but distinct receptors, both serine/threonine kinases with a single transmembrane domain, type I and type II receptors. The type II receptor has been cloned and is expressed solely in AMH target organs. AMH appears to involve the type I receptor BMPR-IB and downstream effector Smad. However, its role in AMH biological actions remains to be seen (Josso et al., 2001). Clinical studies demonstrated a negative correlation between serum AMH and testosterone in puberty but not in the neonatal period. In normal mice, intratesticular testosterone repressed AMH synthesis, but not by serum testosterone, explaining why AMH is down-regulated in early puberty when serum testosterone is still low. In neonatal mice, AMH is not inhibited by intratesticular testosterone due to deficient expression of androgen receptors by Sertoli cell. Androgen insensitive patients exhibit elevated AMH in coincidence with gonadotrophin activation. In immature normal and in androgen insensitive Tfm mice, FSH administration gives elevation of AMH levels, indicating a role of androgens. The role of meiosis on AMH was investigated in Tfm and pubertal XXSxrb mice, in which germ cells degenerate before meiosis. It was found that meiotic entry acts in synergy with androgens to inhibit AMH (Al-Attar et al., 1997).

Sequence comparisons of MIS type II receptor gene demonstrated that MIS gene (*Amhr)* is highly conserved (Mishina et al., 1997). GATA-4, a zinc finger transcription factor and nuclear receptor steroidogenic factor (SF-1) are early markers of Sertoli cells that have been shown to regulate MIS transcription. GATA and SF-1 binding sites are adjacent to one another in the MIS promoter. Co-expression of both factors lead to a strong synergistic activation of the MIS promoter. GATA-4 enhances *MIS* transcription by binding at its DBD in promoter region and its direct interaction with SF1. GATA-4 can also synergize SF1 promoter by direct protein – protein interaction with SF-1 mediated through the zinc finger region of GATA-4 (Tremblay et al., 1999) **(Fig.3.2)**. An orthologue of AMH from the tammar wallaby (Macropus eugenii) is highly conserved with the eutherian orthologues within C-terminal mature domain. The N-terminus of marsupial AMH is divergent and larger than that of eutherian species. It is located on chromosome 3/4, showing autosomal localization in other species. Like eutherian AMH genes, the wallaby 5' regulatory region contains binding sites for SF1, SOX9, and GATA factors, in addition to a putative SRY-binding site. In the developing testis, AMH is localized in the cytoplasm of the Sertoli cells but is lost by adulthood (Pask et al., 2004).

3.2.7. Platelet Derived Growth Factor

Platelet derived growth factor (PDGF), initially isolated from platelets stimulates fibroblast growth. The PDGF plays paracrine and autocrine roles in the testis, and may mediate

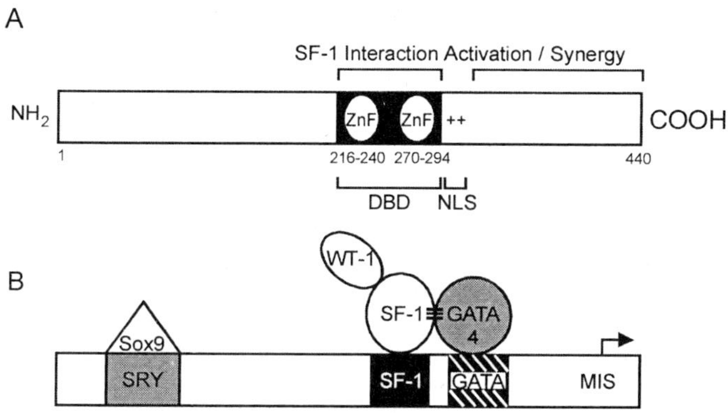

Fig.3.2: Role of GATA-4 in Mullerian inhibiting substance (*MIS*) gene expression. (A). Schematic diagram of GATA-4 protein showing DNA binding domain (DBD), nuclear localization signal (NLS), and zinc finger (ZnF) domain. (B). Control of *MIS* transcriptional regulation by different transcriptional factors (GATA-4, Sox9, WT1 and SF1) (see text).

communication between the interstitial compartment and the seminiferous epithelium. The PDGF is composed of two homopolymer chains of 31 kDa. The direct action of PDGF is on spermatogenesis and its local production in testis is not known. However, PDGF-A-deficient male mice are associated with progressive loss of testicular size, Leydig cells loss, and spermatogenic arrest. The testicular defects seen in PDGF-A-/- mice, combined with the normal developmental expression of PDGF-A and PDGF-Rα indicated that the PDGF-A gene is essential for the development of the Leydig cell lineage. The Leydig cell loss and the spermatogenic impairment in the mutant mice are reminiscent of cases of testicular failure in man (Gnessi et al., 2000). Platelet derived growth factor ligand and receptor (PDGF-R) α, β subunits mRNA express in the Sertoli and Leydig cells of the rat testis. The Leydig cells contained predominantly mRNA encoding PDGF-β subunits, while the Sertoli cells contained abundant mRNAs encoding both ligand subunits. None of the PDGF subunit mRNA was detected in the germ cells (Loveland et al., 1995; Luconi et al., 1995).

3.2.8. Nerve Growth Factor

Nerve growth factor (NGF) is a protein by which germ cells communicate with Sertoli cells. The NGF is known to have important functions in the development and maintenance of sensory and sympathetic neurons. In the adult rat testis, the NGF and its gene are localized to spermatocytes and early spermatids of all stages, while NGF receptor mRNA was detected in Sertoli cells of testis of immature and mature cells and is down-regulated by testosterone. The NGF immunoreactivity and mRNA have been detected in the testis of the adult mouse, rat and human. Hypophysectomy increased both NGF-R mRNA in testis and the number of NGF-R hybridizing cells in seminiferous tubules. This was suppressed by treatment with hCG or testosterone, but not with FSH. The NGF-R mRNA also increases after destruction of Leydig cells or blocking of the androgen receptor. This suggests that NGF produced by male germ cells regulates testicular function in an androgen-modulated fashion by mediating an interaction between germ and Sertoli cells. NGF stimulates in vitro DNA synthesis of seminiferous tubule segments with preleptotene spermatocytes at the onset of meiosis suggesting that NGF is a meiosis growth factor that acts through Sertoli cells (Parvinen et al., 1992).

The neurotropin receptors consist of a low-affinity p75 nerve growth factor receptor (p75NGF-R) and a family of high-affinity tyrosine receptor kinases (trk). Both the p75NGF-R gene product and the trk receptor gene product were detected in immature rat testis, with maximal expression found in stages VII and VIII of the cycle, the sites of onset of meiosis in rats. The low-affinity NGF receptor protein was in the plasma membrane of the apical Sertoli cell processes as well as in the basal plasma membrane of these cells at stages VII to XI. Tyrosine kinase receptor (trk) mRNA encoding an essential component of the high- affinity NGF receptor is also present at all stages. Expression of the testicular p75NGF-R and the trk receptor progressively decline in older animals so that they are barely detectable in 90-day-old adult rats. The p75NGFR was detected in membrane fractions of Sertoli cells, and not in membrane fractions of round spermatids and primary spermatocytes, while microsomal fractions of peritubular myoid cells were immunoreactive for a 65-kDa band (Djakiew et al., 1994). The expression of p75NGF-R during early gonadal development has been demonstrated in mesenchymal cells of the embryonic mouse and rat testes. The expression in the prepuberal and adult mouse and rat testis, of an abundant and shorter transcript of 3.2kb that cross-hybridizes to receptor mRNA (3.7-kb) also has been shown. This new mRNA species, which appears at the beginning of spermatogenesis, is expressed in pachytene spermatocytes and round spermatids (Russo et al., 1995).

3.2.9. Neurotrophic Factors

Normally neurotrophic factors (NTs) are defined as molecules that maintain neuronal cells. However, they possess a range of other functions, which are performed outside nervous system. For example, glial cell line derived neurotophic factor, leukemia inhibitory factor, and IL-6 all belonging to neurotrophic factor family, play non-neural functions in stem cell differentiation and kidney morphogenesis during embryogenesis, maintenance of immune cells, sperm cell differentiation and others. The expression of NTs and their receptors in the human fetal testis during the second trimester indicated the possible roles of neutrophins in the proliferation and survival of germ cells and peritubular cells (Robinson et al., 2003).

Neurotropin-3: The neurotropins are involved in embryonic morphological events (cord formation; NT3) and in growth of the perinatal testis (NT3 and NGF on postnatal Day 0). Nerve growth factor and NT3 alone or in combination inhibited expression of mRNA for TGF-α while NT3 increased mRNA expression of EGFR. The combination treatment of neurotropins inhibited expression of TGF-β1 and increased expression of TGF-β1. This suggests that NT3, NGF, trk A and trk C are localized to cells critical to seminiferous cord formation and appear to be important regulators of morphological sex determination. In addition to these morphological effects, both NT3 and NGF stimulate postnatal Day 0 testis growth and may elicit their action through altering the expression of locally produced growth factors such as TGF-β and TGF-β (Cupp et al., 2000).

Glial Cell Line-derived Neurotrophic Factor: Glial cell line-derived neurotrophic factor (GDNF) is essential for ureteric branching in kidney morphogenesis and regulating the fate of stem cells during spermatogenesis. The expression of GDNF family ligands and related receptors are present in germ cells of adult mice. Though the GDNF mRNA was detected in ovary, it was not detectable in testis. Neuroturin (NTN), another ligand in this family, gave rise to strong mRNA hybridization signals in testis at stages IX-XII and I-II of the spermatogenic cycle. In testis, the transducing receptor, RET as well as GDNF receptor α-1 (GFR) α-1 and GFR α-2 are distributed in complementary and overlapping patterns, the former at stages XI-XII-I and the latter at stages VII and VIII, while (GFR) α-3 could not be detected (Widenfalk et al., 2000). In another study, GFR α-1 and GFR α-2 were found to be expressed in germ cells while GDNF and NTN were expressed by Sertoli cells. Thus, GDNF-family ligands may act as paracrine factors in spermatogenesis and the circuit may be active in germ cell tumors (Viglietto et al., 2000). Transgenic loss of function and over-expression models showed that the dosage of GDNF produced by Sertoli cells regulates cell fate decisions of undifferentiated spermatogonia that include the stem cells for spermatogenesis. Gene-targeted mice with one *GDNF*-null allele showed depletion of stem cell reserves, whereas mice over-expressing GDNF showed accumulation of undifferentiated spermatogonia. They were unable to respond properly to differentiation signals and undergone apoptosis upon retinoic acid treatment. Non-metastatic testicular tumors were regularly formed in older GDNF-over expressing mice. Thus, GDNF contributes to paracrine regulation of spermatogonial self-renewal and differentiation (Meng et al., 2000).

3.2.10. Vascular Endothelial Growth Factor

Vascular endothelial growth factor (VEGF), also known as vascular permeability factor (VPF), and its receptors fms-like tyrosine kinase (Flt-1) and fetal liver kinase (KDR in humans and FIk-1 in mouse) are present in the human testis. The VEGF mRNA was detected in the human testicular tissue and in fragments of seminiferous tubules. VEGF antibody revealed two protein bands corresponding to 24 and 49kDa in testis. Studies suggested that VEGF acts as a paracrine, mitogenic and angiogenic factor, responsible for modulating the capillarization of the human testicular tissue and maintaining the functions of testicular microvasculature. The VEGF may also influence the permeability of capillaries passing through the group of Leydig cells and those localized within the lamina propria of human seminiferous tubules. The differences in the expression pattern of the VEGF receptors in the human testicular tissue probably reflect different VEGF effects in different compartments of human testis (Ergun et al., 1997). While VEGF and VEGF receptors mRNA are present in the human epididymal tissue, VEGF protein was localized in peritubular and basal cells of the epididymal duct (Ergun et al., 1997). However, Korpelainen et al., (1998) produced evidence that over-expression of VEGF in the testis and epididymis of transgenic mice causes infertility. Findings suggested that the VEGF transgene has non-endothelial target cells in the testis and that VEGF may regulate male fertility.

3.2.11. Hepatocyte Growth Factor/Activator-Inhibitor

The hepatocyte growth factor (HGF) is known to regulate cell growth, movement and cell differentiation, and is involved in embryonal morphogenesis. The hepatocyte growth factor receptor (HGFR, c-MET), a tyrosine kinase receptor protein transduces multiple effects in mammalian tissues, and is widely expressed in different cell types including the male reproductive tract. Both c-MET mRNA and protein are expressed in prebuberal rat testis. The interstitial tissue is also c-MET positive. The protein encoded by the *c-MET* proto-oncogene is present in myoid cells. HGF induces morphological changes in myoid cells and in c-MET-expressing Sertoli cells where it is developmentally regulated and functionally active in postpuberal Sertoli cells. In rat testis c-MET is expressed during postnatal life until the sexual maturity. The c-MET expression persists in myoid cells during postnatal testis development and its expression in Sertoli cells correlates over time with germ cell de-differentiation and lumen formation The co-expression of factor and receptor in the myoid cells suggests a role for HGF as autocirne regulator of myoid cells and, possibly, as regulator of mammalian testicular function (Catizone et al., 2001).

Mouse hepatoma-derived growth factor (HDGF)-related protein (HRP)-1 and –2 genes are mouse homologues of the HDGF gene isolated from a mouse testis cDNA library. The HRP-1, HRP-2 and HDGF, which comprise an HDGF gene family, have a highly conserved sequence in their N-terminal region, called the HATH region (homologous to the amino terminus of HDGF). All of these genes are highly expressed in the testis. Although HRP-2 and HDGF genes are ubiquitously expressed in many organs, the HRP-1 gene is exclusively expressed in the testis. The HRP-1 gene is not expressed in the ovary but its product was detected in the nuclei of germ cells, not in somatic cells. The HRP-1 gene is expressed through pachytene spermatocyte to round spermatid (Kuroda et al., 1999).

Hepatocyte growth factor activator inhibitor type 1 (HAI-1) and type 2 (HAI-2) are newly identified integral membrane Kunitz-type proteinase inhibitors. These inhibitors have important regulatory roles in pericellular activation of hepatocyte growth factor/scatter factor (HGF/SF). HGF/SF is also involved in testicular development and spermatogenesis. In human testis, HAI-2 was strongly expressed whereas HAI-1 mRNA was hardly detectable. The mRNA size of HAI-2 was shorter in the testis (1.2 kb) than those in the other tissues such as placenta (1.5 kb). Two major transcription start sites in the HAI-2 gene are located at -30 bp and -360 bp upstream from the translation initiation codon. The site at –360 appeared to be mainly used in the placenta and the other non-testicular organs, where as site at –30 is used in testis, resulting in the -300 bp shorter mRNA. HAI-2 is expressed exclusively in primary spermatocytes (Yamauchi et al., 2003).

3.3. PLASMINOGEN ACTIVATORS

It has been well established that limited proteolysis plays a role in spermatozoal maturation and capacitation. Plasmin, a tryptic protease, is thought to play an important role in extracellular proteolysis. It is generated from plasminogen by limited proteolysis catalyzed by plasminogen activators, i.e. tissue plasminogen activator (tPA) and urokinase plasminogen activator (uPA). Binding of uPA to a high affinity, specific plasma membrane receptors (uPAR) that is present on a variety of different cell types localizes the generation of plasmin to the close environment of the cell surface. Ejaculated mouse spermatozoa have cell surface associated uPA, which is synthesized by epithelial cells of the caudal part of epididymis and the vas deferens. Binding of uPA to spermatozoa suggested the presence of uPA receptors on sperm surface. Surface bound uPA of mouse sperm appears to be acquired from genital tract secretions at ejaculation. In situ hybridization demonstrated the presence of uPA-R mRNA in germ cells. Binding of uPA, but not of a mutant enzyme lacking the receptor binding domain, indicated the presence of uPA-R on spermatids and sperm (Zhou and Vassali, 1997).

Both uPA and tPA-like activities are found in seminiferous tubules and are secreted by Sertoli cells, and are under the control of FSH (Vihko et al., 1986). Sertoli cells secrete PA, which is under the regulation of FSH, c-AMP and neighbouring spermatogenic cells. Pre-leptotene spermatocytes are important regulators of uPA secretion. Cyclic secretion of PA starts at the age of 28 days, and from 40 days onwards. The high secretion rates are localized in stages VII and VIII of the cycle of the seminiferous epithelium. The PA secretion in vitro by Sertoli cells is strongly enhanced by those Sertoli cells that had phagocytosed *E. Coli.* Residual bodies (RB) play a role in the regulation of proteolytic system in testis and in coordination of spermatogenesis. Addition of RBs to Sertoli cells in culture and their phagocytosis by Sertoli cells leads to rapid stimulation of Sertoli cell IL-1α, which regulates spermatogenesis. It seems that RBs stimulate PA, which is regulated by IL-1α (Sigillo et al., 1998).

3.3.1. Ly-6/Urokinase-Type PlasminogenActivator Receptor

A member of the Ly-6/urokinase-type plasminogen activator receptor (uPA-R) superfamily of receptors, SAMP14 has been identified in testis. The SAMP14 is retained on the inner acrosomal membrane of the human spermatozoa following the acrosome reaction and may play a role in fertilization. The SAMP14 sequence predicted a glycosylphosphatidylinositol (GPI)-anchored protein with a signal peptide, a transmembrane domain near the carboxyl terminus, and a putative transamidase cleavage site in the proprotein. The SAMP14 has a single functional domain similar to the Ly-6 urokinase plasminogen activator receptor superfamily proteins, and the gene mapped to 19q13.33, near the PLAUR locus for uPAR at 19q13.2. The SAMP14 was localized to outer and inner acrosomal membranes as well as the acrosomal matrix of ejaculated human sperm. Some SAMP14 is loosely associated with the plasma membrane. SAMP14 represents a GPI-anchored putative receptor in the Ly-6/uPAR family that is exposed on the inner acrosomal membrane after the acrosome reaction (Shetty et al., 2003).

3.3.2 Plasminogen Activator Inhibitor Proteins

Sertoli cells secrete PAs on both sides of the blood testis barrier, i.e. in the basal and apical compartments of the seminiferous tubules, whereas peritubular cells secrete plasminogen activator inhibitor-1 (PAI-1). Both types of PA are inhibited specifically by PAI-1. The PAI-1 mRNA and bioactive PAI-1 protein are stimulated by TNF-α. The biological effects of TNF-α on PAI-1 might be secondary to de novo synthesis of EGFR and may occur in the context of physiological interactions between the testis and the immune system (Le Magueresse-Battstoni et al., 1997). The signal intensity of PAI–1 increases in the course of differentiation until a maximum is reached at stages VII-VIII and remains associated with germ cells after spermiation and during maturation in the epididymis. In mature caudal sperm, PAI-1 is localized on the plasma membrane surrounding the head. In addition the activity was detectable in the acrosome and in the tail (Manske et al., 1994). It appeared that there exists a direct linkage between molecular events triggered by phagocytosis and the regulation of tPA and PAI-1 in Sertoli cells (Bourdon et al, 1999).

Since Sertoli cells secret PA, primarily of uPA type, and under FSH stimulation, of the tPA type, whereas peritubular cells secrete PAI-1, it was hypothesized that cell-cell interactions between these two cell types may regulate net protease activity within the seminiferous tubules. While it is possible that PAI-1 produced by peritubular cells counteracts the basal secretion of PA, the nature of the PA inhibitor acting in the apical compartment was a PAI-1 protein that was secreted by Sertoli cells. Sertoli cells contain a transcript of 3-3.2-kb, which hybridizes specifically to a PAI-1 cDNA. The secreted protein complexed with t-PA was biologically active. This protein co-migrated with purified PAI-1 as a doublet of 46 and 49 kDa and is under a complex hormonal (FSH) and paracrine and/or autocrine control exerted at least by bFGF and TGF-β1 (Le Magueresse-Battstoni, et al., 1998).

3.4. PEPTIDE HORMONES

3.4.1 Rennin–Angiotensin System

The rennin angiotensin system (RAS) has long been considered to be a circulating hormonal system responsible for the regulation of blood pressure and electrolyte balance. Presently there has been a shift in emphasis from endocrine to paracrine and autocrine functions of the RAS in tissues. Key components of the RAS, notably angiotensinogen and rennin, need to be present for local RAS. These components have been demonstrated in a number of tissues with physiological relevance. In the reproductive tissues, RAS may be important in the regulation of reproductive functions (Vinson et al., 1997).

3.4.2. Angiotensin Converting Enzyme

Angiotensin I-converting enzyme (ACE, EC 3.4.15.1) is primarily a dipeptidyl carboxy peptidase, responsible for removing dipeptides from the C-terminus of circulating angiotensin I (A1, DRVYIHPFHL), bradykinin (RPPGFSPFR), and the hemo-regulatory peptide, N-acetyl SDKP (AcSDKP) and participates in blood pressure regulation. Human ACE exists in two isoforms: somatic ACE (sACE, M_r 150-180 x 10^3 Da) and testicular or germinal ACE (gACE, M_r 100-110 x 10^3 Da). The ACE is expressed in somatic tissue (vascular endothelial cells and at the brush border of renal proximal convoluted tubule, jejunal villus, and epididymal duct epithelia), whereas the germinal form is expressed uniquely in germinal cells with a precise stage specific pattern, starting in round spermatids and finishing in spermatozoa. In situ hybridization demonstrated that there was no direct correlation between the levels of ACE mRNA and the enzyme it encodes for, i.e. ACE in a given epithelium (c/r Sibony et al., 1994). The larger sACE is anchored to the plasma membrane by a hydrophobic sequence towards the C-terminus and has two catalytically active domains. The two domains have a high level of sequence homology. The structure of the mammalian ACE gene has clearly arisen from a gene duplication event during the course of evolution. Amino acid sequence of the C-domain is identical with gACE, apart from a 68 amino acid N-terminal sequence. The predicted pro-protein has a signal sequence for secretion, and when a full length cDNA for human gACE is expressed in transfected Chinese hamster ovary cells, gACE is expressed as both a membrane bound and a soluble secreted enzyme. In rats, the link of gACE in maturing spermatids and spermatozoal protein differs from plasma membrane sACE in that the germinal enzyme is processed to a soluble protein with an intracellular localization (Isaac et al., 1997).

Ontogeny: In rat and mouse, testis specific ACE mRNA and its gene product are present only after completion of meiosis. The ACE mRNA and its corresponding protein are first synthesized during the cap phase (steps 4-7), followed by maximum expression during the acrosome phase (Steps 8-12). The ACE mRNA is not detectable in spermatids beyond step 14, whereas its gene product is expressed until the end of spermatid maturation (Sibony et al., 1994). However, Langford er al, (1993) noted that although testis ACE protein is first detected in step 10 spermatids, testis ACE mRNA is first detected in a developmentally younger cell population, late pachytene spermatocytes. Thus, testis ACE mRNA is translationally arrested for a period of several days until late in spermiogenesis. The localization of gACE to the acrosome and the Golgi apparatus close to the acrosome, at a stage when this organelle is very active in secretion led to consider the possibility that gACE is involved in the intracellular processing of spermatid peptide pro-hormones. The peptide hormone intermediates with Lys/Arg-Arg at the C-terminus are high affinity substrates for human gACE. The gACE from human sperm cleaves Arg-Arg from the C-terminus of the CCK5-GRR (GWMDFGRR), a peptide corresponding to the C-terminus of CCK-gastrin pro-hormone intermediate. It was suggested that gACE might be an alternative for carboxypeptidase E for trimming of basic dipeptides from the C-terminus of intermediates generated from pro-hormones by subtilisin like convertase in human male germ cells (Isaac et al., 1997).

Characterization: Testis ACE (gACE) is produced in large quantities during mammalian spermatogenesis and the testis is considered a tissue with the highest levels of gACE in an adult male animal. Germinal and somatic forms of ACE are encoded by a single gene. Testis ACE cDNA contains 2,435-bp and encodes a protein of 732 amino acids. The N-terminal 66 amino acids are unique to testis isozyme while remaining 666 are identical to carboxyl half of mouse somatic ACE (Ehlers et al., 1989) **(Fig.3.3)**. The overall conservation of amino acid sequence between the testis isozymes of the mouse, rabbit, and human is 78 to 84%. The conservation of amino acids for the N-terminal domain uniquely expressed within the testis is 63 to 67% between these species. RNA transcription of the testis ACE isozyme begins 16 or 17 bases upstream from the translation start site. To create the unique isozyme of gACE, the testis begins mRNA transcription in the middle of the exonic-intronic structure of somatic ACE, within a sequence treated as an intron by somatic tissues. Testis ACE is not the result of alternative RNA splicing but seems due to the start of transcription at a unique site within the ACE gene (Howard et al., 1990). Both somatic and testicular ACE isoenzymes are anchored by a single hydrophobic transmembrane polypeptide near the C-terminus. The majority of the enzyme's mass including its catalytic site(s) is exposed to the extracellular milieu. Therefore, ACE fulfils the criteria of an ectoenzyme (Kohn et al., 1998). A protein of 68-70-kDa showing strong homology with number of ACE is present in equine sperm. An 84% identity is found between testis specific and somatic forms of human and mouse ACE. The stallion protein is specific for the peri-acrosomal membrane of ejaculated, epididymal, and testicular spermatozoa (Dobrinski et al., 1997).

Gene organization: The transcription of testis ACE is initiated in the middle of this gene, within introns 12 of somatic ACE. This organization into two transcription units reflects the presence of a sperm specific promoter within the ACE gene. Using transgenic mice, Langford et al., (1991) have localized this tissue-specific promoter of a 700 bp genomic region, which induces high levels of reporter gene transcription only within developing spermatozoa. In order to define the minimal testes ACE promoter and to characterize DNA protein interactions mediating germ cell specific expression, studies showed that cell and stage specific expression of testis ACE requires only a small portion of the immediate upstream sequence extending from 91 to – 9 relative to transcriptional start site. A critical motif within this core promoter is a CRE sequence that interacts with a testis specific transactivating factor. Since this putative CRE has been conserved within the testis ACE promoter of different species and found at the same site in other genes that are expressed specifically in the testis, it is likely that CRE may provide a common mechanism for the recognition of sperm specific promoters. Howard et al., (1993) have suggested two important DNA regulatory elements starting at positions – 55 and –32. DNA constructs were made in which these motifs were either eliminated or substituted. Disruption of either motif reduced in vitro transcription to about 30% while mutation of both elements abolished transcription. The DNA element at –55, TGAGGTCA, is homologous to a consensus CRE. The motif at – 32, TCTTAT, is located at a position analogous to a TATA box. Substitution of the –32 motif with a consensus TATA box sequence (TATAAA) stimulated transcriptional activity about 3-fold. As measured by gel mobility shift, oligonucleotides

1	MGQGWATPGL	PSFLFLLLCC	GHHLLVLSQV	ATDHVTANQG	ITNQATTRSQ	TTTHQATIDQ
61	TTQIPNLETD	EAKADRFVEE	YDRTAQVLLN	EYAEANWQYN	TNITIEGSKI	LLEKSTEVSN
121	HTLKYGTRAK	TFDVSNFQNS	SIKRIIKKLQ	NLDRAVLPPK	ELEEYNQILL	DMETTYSLSN
181	ICYTNGTCMP	LEPDLTNMMA	TSRKYEELLW	AWKSWRDKVG	RAILPFFPKY	VEFSNKIAKL
241	NGYTDAGDSW	RSLYESDNLE	QDLEKLYQEL	QPLYLNLHAY	VRRSLHRHYG	SEYINLDGPI
301	PAHLLGNMWA	QTWSNIYDLV	APFPSAPNID	ATEAMIKQGW	TPRRIFKEAD	NFFTSLGLLP
361	VPPEFWNKSM	LEKPTDGREV	VCHPSAWDFY	NGKDFRIKQC	TSVNMEDLVI	AHHEMGHIQY
421	FMQYKDLPVT	FREGANPGFH	EAIGDIMALS	VSTPKHLYSL	NLLSTEGSGY	EYDINFLMKM
481	ALDKIAFIPF	SYLIDQWRWR	VFDGSITKEN	YNQEWWSLRL	KYQGLCPPVP	RSQGDFDPGS
541	KFHVPANVPY	VRYFVSFIIQ	FQFHEALCRA	AGHTGPLHKC	DIYQSKEAGK	LLADAMKLGY
601	SKPWPEAMKL	ITGQPNMSAS	AMMNYFKPLT	EWLVTENRRH	GETLGWPEYN	WAPNTARAEG
661	STAESNRVNF	LGLYLEPQQA	RVGQWVLLFL	GVALLVATVG	LAHRLYNIRN	HHSLRRPHRG
721	PQFGSEVELR	HS				

Fig.3.3. Amino acid sequence of angiotensin-converting enzyme (ACE) {peptidyl-dipeptidase A (EC 3.4.15.1) }from mouse testis. Source: http//www. ncbi.nlm.nih.gov (accession number A35655)

encompassing the – 32 motif and the consensus TATA box formed different DNA protein complexes. However, the –32 nt motif was recognized by nuclear liver or testis proteins (Zhou et al., 1995).

Sperm ACE: ACE is mainly located at the plasma membrane of the acrosomal region, equatorial segment, postacrosomal region and mid-piece. In contrast, only weak ACE like immunoreactivity was found at the flagellum. In case of cells with missing plasma membranes, ACE seems to be located at the surface of the outer acrosomal membrane. Kohn et al., (1998) did not indicate direct effects of ACE on the capacitation process or acrosome reaction. ACE is important for achieving in vivo fertilization and that sperm from mice lacking both ACE isozymes showed defects in transport within the oviducts and in binding to zonae-pellucidae. Evidence for the importance of ACE in fertilization has been obtained from mice carrying an insertional mutation resulting in the inactivation of both sACE and gACE. Male mice homozygous for the deficient mutation are infertile. The effect of the absence of gACE on male fertility is not due to the failure to generate ACE, since male mice that do not express the angiotensinogen gene are fertile. Males generated by gene targeting that lack sACE but retained gACE were normally fertile. Moreover, angiotensin I was not essential substrate for testis ACE (Hagaman et al., 1998; Krege et al., 1995; c/r Isaac et al., 1997).

Effect of Angiotensins (Angiotensin Receptors on Sperm): Males lacking somatic ACE but retaining testis ACE are normally fetile, establishing that somatic ACE in males is not essential for their fertility. Furthermore, male and female mice lacking angiotensinogen have normal fertility, indicating that angiotensin I (AT1) is not a necessary substrate for testis ACE. Moreover, angiotensin II induced the acrosome reaction whereas angiotensin I had no effect. The effect of angiotensin II seems to be calcium dependent and mediated by protein kinases (Kohn et al., 1998). On the other hand Gur et al (1998) showed involvement of AT1-R in acrosome reaction induced by AT2. In capacitated sperm, AT2 and AT1-R are localized in the head and tail, whereas in non-capacitated cells, the receptors are localized in the tail only. Angiotensin II markedly stimulates acrosomal exocytosis of capacitated bovine sperm in vitro, while AT2 did not show any effect on non-capacitated sperm. Ability of AT2 to stimulate the acrosome reaction depends on the presence of calcium ions. AT2 via activation of AT1-R may play a regulatory role in the induction of the acrosome reaction (Gur et al., 1998). Motility data on stimulated and unstimulated human sperm showed that AT2 may increase both the percentage of motile sperm and their linear velocity, while the AT1-R (AT1-angiotensin type

I receptor) antagonist DuP753 inhibits the action of AT2. The AT1-Rs are present in primary spermatogonia and in spermatid tails, but immunoreactivity is not seen in sperm contained in caput or cauda epididymis, showing that AT1 receptor function is regulated during transit through the reproductive tract. Since local tissue rennin angiotensin systems are present in both male and female tracts, it was suggested that AT2 has a role in the maintenance of sperm function and fertility (Vinson et al., 1997). Angiotensin II induces contraction, growth, and increase in intracellular calcium in rat peritubular myoid cells via angiotensin II type I receptors, and suggests that Ang II is involved in the paracrine regulation of the seminiferous tubule function (Rossi et al., 2003).

Angiotensin System in Epididymis: The rennin-angiotensin system (RAS) has been implicated in epididymis on the basis of the presence of immunoreactive angiotensin I and II, the rennin and ACE like activities in epididymal cells. Angiotensin II (Zhao et al., 1996) and its receptor subtypes, AT1-R and AT2-R have been localized to the epithelial cells of the rat epididymis, indicating that a tissue RAS may play a role in electrolyte and fluid secretion (Leung et al., 1999).

3.4.3. Kallikrein-Kinin System

Four components of the kallikrein-kinin system kininogens, tissue kallikreins, kinins, and kininases have been found to be present in human male genital secretions. Kinins are continuously released from seminal plasma kininogens through limited proteolytic action of kininogenases on tissue kallikrein from prostate and sperm acrosin. Kinins are the terminal effectors of the kallikrein-kinin system that increase sperm motility and sperm metabolism at nanomolar concentration. These effects are possibly mediated by a specific sperm membrane integrated, subtype B2 receptors, which exhibit high affinity for bradykinin and kallidin (lys-bradykinin). B2 receptor expresses in Sertoli cell and peritubular cells of mature and immature rat testes. Specific staining of B2 receptors occurred on spermatocytes and spermatids. The B_1 sub-type receptor is almost absent (Monsees et al., 2002). The two major kinases that are present in seminal plasma are kinase II and neutral metallo-endopeptidase. Kininase II identical with ACE, is also involved in the rennin-angiotensin system as it converts angiotensin I into angiotensin II, and thus it is the connecting enzyme of both systems. Apart from the observed effects of kinins on sperm motility, the kallikrein-kinin system is thought to be involved in the regulation of spermatocytes, and the ^{3}H- thymidine incorporation of testicular tissue enhances glucose intake, and can increase testicular blood flow. Clinical trials suggest that systemic administration of kallikrein is particularly useful for treatment of infertile men suffering from asthenozoospermia and/or oligozoospermia. The B2R mRNA and protein were localized in mature testis exclusively on pachytene spermatocytes and round and elongated spermatids besides in blood vessels. B2R mRNA and protein were additionally located in peritubular cells in pre-pubertal rat testis. Tissue kallikrein (tK) was localized to acrosomal cap of round and elongated spermatids in addition to endothelial cells of blood vessels (Monsees et al., 2003).

Specific immunostaining demonstrated that the kininogenase tissue kallikrein is present in round and elongated spermatids and absent from Leydig cells, Sertoli cells, peritubular cells, spermatogonia and spermatocytes. Bradykinin in the supernatant of Sertoli cell cultures was effectively degraded. Specific protease confirmed the occurrence of several metalloproteases on Sertoli cell membranes, including neutral metalloendopeptidases kininase type II (ACE) and kininase type I (metallocarboxypeptidase) as well as bradykinin B2 receptor mRNA in testis and Sertoli cell extracts.

3.4.4. Proopiomelanocortin Peptides

Three classes of opioid receptors μ, δ and κ mediate physiological and pharmacological functions of the endogenous opioid peptides and exogenous opioid compounds in the central nervous system as well as in peripheral tissues, and the action of each is mediated by PKC pathway. κ-opioid receptor gene expression in Sertoli cells is under intereukin-6 regulation. The prototype peptides first identified in other tissues and later found localized in reproductive tissues are the proopiomelanocortin (POMC)-derived peptides, including ACTH, MSH, and β-endorphin. The ACTH and endorphin are found in the male reproductive tract of several species. In rats, ACTH and β-endorphin like material were described in Leydig cells, epididymis, seminal vesicle, vas deferens, and prostate. β-endorphin was localized in human and rat Leydig cells. Characterization of the POMC derived peptides produced in gonadal tissues has indicated that processing of the precursor to MSH and β-endorphin occurs in the testes, as occurs in the intermediate pituitary. A POMC like mRNA of ~200-nt shorter than the authentic 1,050-nt pituitary POMC mRNA was found in the testis. It now appears that the germ cells also produce lesser amounts of POMC mRNA as shown by in situ hybridization and by analysis of germ cell deficient mutant mice. The small sized POMC mRNA present in pachytene spermatocytes has also been found in the ovaries of rats, mice, and monkeys. In testis, POMC mRNA levels can change in response to maturational or hormonal signals and developmentally regulated in mice. The expression of the POMC gene in testis is regulated by gonadotrophins but not by glucocorticoids, which regulate the gene in the anterior pituitary. Opiate receptors have been found in Sertoli cells but not in Leydig cells, suggesting that β-endorphin produced by Leydig cells, acts in a paracrine fashion on Sertoli cells. The POMC mRNAs are most abundant in a subpopulation of Leydig cells associated with tubules in specific stages of the cycle of the seminiferous epithelium. The POMC transcripts of approximately 800-nt are present in enriched populations of meiotic prophase spermatocytes and in caput epididymis but are absent in cauda epididymis and vas deferens (Gizang-Ginsberg and Wolgemuth, 1987).

3.4.5. Proenkephalin and Prodynorphin peptides

Two precursor proteins encoding opioid peptides are proenkephalin and prodynorphin (proenkephalin β). These two genes like POMC are expressed in testes. The gene for the opioid precursor proenkephalin is expressed in both spermatogenic and somatic cells. Proenkephalin and POMC derived peptides produced by germ cells may function at different stages of spermatogenesis. In the mouse and rat, spermatogenic cells produce a 1700-nt proenkephalin RNA, while somatic cells selectively express a 1450-nt form. The mRNA size of the spermatids pre-proenkephalin is about 1900 bases, which is larger than the other pre-proenkephalin genes. The spermatids contained small amounts of enkephalin protein, but little or no met–enkephalin. The proenkephalin gene is developmentally regulated during spermatogenesis, with the highest expression occurring in late pachytene spermatocytes and postmeiotic round spermatids. The production of proenkephalin by rat Sertoli cells and Leydig cells is stimulated by FSH or cAMP. Prodynorphin mRNA likewise is also found in testis. A 200 nt mRNA shorter than hypothalamic counter-part is present in Leydig cells and is stimulated by cAMP (Ackland et al., 1992).

cDNA: The rat and mouse proenkephalin genes are selectively expressed from an alternate, germ cell specific promoter in meiotic and early haploid cells. Germ cells interact with Sertoli cells, possibly by a protein that acts as a short loop paracrine factor, which regulates the expression of proenkephalin gene in Sertoli cells. Sequence analysis of proenkephalin cDNA from the testis has revealed that the germ cell derived transcript contains a novel 5'-terminal region in which the somatic exon 1 sequence is replaced with a portion of the first intron (intron A), which separates the somatic exons 1 and 2. Although the testis contains relatively high levels of proenkephalin mRNA, only small quantities of proenkephalin or proteolytic products are found in testis extracts. Rat testis cDNA library showed two types of proenkephalin cDNA, corresponding to somatic and germ cells specific transcripts. The germ cell specific transcripts translate in vtiro with lower efficiency than do somatic transcripts. Translational inefficiency is mediated primarily by the presence of four upstream AUG codons followed by short open reading frame within the leader sequence. Secondary structure in the form of a stem-and-loop upstream of the initiator codon for proenkephalin was found to have a inhibitory effect on the rate of translation (Rao and Howells, 1993). Two distinct forms of proenkephalin mRNA are present in the murine testis, a family of 1.7-kb germ cell specific transcript and a 1.45 kb form that is also found in somatic tissues. Both forms of proenkephalin mRNA transcripts are present in cells undergoing meiosis and spermiogenesis, whereas the 1.45-kb mRNA was detected primarily in type B spermatogonia. In contrast, in situ hybridization analysis did not detect significant amount of the 1.45 kb transcript in any of the spermatogenic cells types. The two mRNAs expressed differently during development. The 1.45 kb transcript was the only form present in the prepubertal testis and localized mainly in interstitial cells. In contrast, the 1.7-kb transcripts were the major mRNAs observed in the adult testis and were localized to spermatogenic cells. A transition from the prepubertal to the adult pattern occurred, when proenkephalin expressing pachytene spermatocytes began to populate the seminiferous tubules. Proenkephalin gene expression in mutant (at/at) mice, which lacked germ cells, was identical to that observed in the early prepubertal testis. This suggested that the 1.45 kb proenkephalin mRNA is developmentally down-regulated in mouse interstitial cells and that this process requires ongoing spermatogenesis (Mehta et al., 1994).

Promoter: Using cDNA sequencing, the rat and mouse germ cells proenkephalin RNAs were generated by alternative transcriptional initiation. Transcription of the rat and mouse spermatogenic cell RNAs is initiated downstream from the proenkephalin somatic promoter in the first somatic intron (intron As). In both species, the germ cell cap site region consists of multiple start sites distributed over a length of approximately 30bp. Within rat and mouse intron A_s, the region upstream of the germ cell cap site is GC rich and lacks TATA sequences. A consensus binding site for the transcription factor Sp1 was identified in introns A_s downstream of the proenkephalin germ cell cap site region. These features are characteristic of several promoters that lack TATA sequences. Homologies were also identified between the proenkephalin and rat cytochrome c spermatogenic cell promoters, including the absence of a TATA box, a multiple start site region, and several common sequences. This promoter motif thus may be shared with other genes expressed in male germ cells. In proenkephalin transgenic mice, a short promoter confers high testis expression and reduced fertility. Studies support a role for specific promoter sequences in testis expression (Kilpatrick et al., 1987; O'Hara et al., 1994).

The rat minimal proenkephalin germ line promoter comprises to a 116-bp region encompassing the transcriptional start site. Further, a proximal 51-bp sequence located in the 5' flanking region is absolutely required for promoter activity. This 51-bp sequence corresponds to a binding element (GCP1) that forms cell specific complexes with rat germ cell nuclear factors distinct from CREB proteins. Further, GCP1 contains direct repeat sequences required for factor binding and transgenic expression in spermatogenic cells. These repeat elements are highly similar to sequences with the active regions of other male germ line promoter expressed during meiosis. The GCP1 may, therefore contain transcriptional elements that participate during meiosis in the differentiation of spermatocytes and early haploid spermatids (Liu et al., 1997).

3.4.6. Action of Other Peptide Hormones

Corticotropin Releasing Factor: Corticotropin releasing factor (CRF) is a 41 amino acid peptide originally isolated from sheep hypothalamus. Hypothalamic CRF increases secretion and synthesis of ACTH and stimulates release of β-endorphin from the pituitary. It may have a similar function of regulating POMC gene products in the gonads. Corticotropin releasing factor of testicular origin is identical to hypothalamic CRF and has been localized on the Leydig cells of several species and to the germ cells of rats. Local synthesis of CRF has been confirmed in rat testis, which is developmentally regulated. Corticotropin releasing factor had no effect on basal testosterone level produced by Leydig cells but decreased hCG stimulated production. The CRF also acted to decrease c-AMP, and these effects appeared to be mediated via PKC-mediated mechanism, since CRF effects can be mimicked by phorbol esters (Ackland et al., 1992).

Growth Hormone Releasing Hormone: A growth hormone-releasing hormone (GHRH) mRNA has been identified in placenta and testicular germ cells besides in hypothalamus. GHRH, secreted from rat adult Leydig cells, stimulates basal and LH-induced cAMP formation and steroidogenesis. In some studies cAMP production in Sertoli cells was stimulated by GHRH. Fabbri et al., (1995) described a potential paracrine action of GHRH in the Sertoli cells, with stimulation of extracellular cAMP formation which was more prominent in pubertal than in adult cultures. The GHRH like immunoreactivity was present in rat interstitial cells and in the acrosomal region of early and intermediate spermatids at stages III-VI of the seminiferous epithelium cycle. It seems that GHRH acts synergistically with FSH to promote cAMP production in Sertoli cells in culture. Testicular GHRH of Leydig and germ cell origin may be an important paracrine regulator of Sertoli cell function. Alternatively, GHRH present in germ cells may exert stage specific intracrine functions (Fabbri et al., 1995). The GHRH mRNA produced by spermatogenic cells is approximately 1700 nt in length, whereas GHRH transcripts in hypothalamus and placenta are 750 nt. The 5'end of germ cell GHRH mRNA and its comparison with the genomic sequence shows that GHRH transcription in testis initiates approximately 700 bp 5' to transcription initiation in placenta and 10.7-kb pairs 5' to that in hypothalamus. The 5'-flanking region of the testicular GHRH gene contains a TATA-like motif and sequences homologous to spermatogenic specific cis-acting elements. Analysis of rat liver DNA suggested that just one GHRH gene is present in rat and that both alternative transcription initiation and splicing of the GHRH gene exist in rat testicular germ cells (Srivastava et al., 1995). Expression of the GHS-R gene in rat testis takes place in a developmental, stage-specific, and hormonally regulated manner (Barreiro et al., 2003).

Thyroid Hormone: Thyroid hormone (TH) plays a pivotal role in the regulation of prepuberal testis development with specific influence on the differentiation of Sertoli cells, the only cell type that expresses thyroid hormone receptors in testes. The H directly affects the development of prepuberal testes and the regulation of FSH-R and ABP gene expression in Sertoli cells, and LH-R mRNA levels in Leydig cells. Resulting action may lead to further modulation of the effect of gonadotropins on testes function (Rao et al., 2003). Though TRH mRNA is expressed in testis, it could not be detected as early as day 8 post-natal. TRH mRNA signals were detected on day 15 and progressively increased till day 90. It seems that the transcription of TRH-R gene in the rat testis was development-dependent (Liu et al., 2003).

Gondotrophin Releasing Hormone: Gondotrophin releasing hormone (GnRH), a decapeptide originally discovered from hypothalamus, acts on anterior pituitary and secretes LH and FSH, which regulate gonadal functions. However, in surprising results, GnRH agonists were found to have antigonadal effects in both sexes leading to reduction of estrogen, progesterone and LH receptors. This led to suggest that GnRH had direct gonadal site for action. Hence receptors for GnRH were observed in testis. Several reports suggest that GnRH like activity is present in Sertoli cells and in the nuclei of germ cells. However, gonadal GnRH is not the same product as produced by hypothalamus. The GnRH receptors with high or low affinity for GnRH have been observed in several species. In rat GnRH-R have been noticed in Leydig cells and appear to be coupled to PKC, PLC and Ca^{2+} turnover (Ackland et al., 1992 and other cited reviews).

Neurohypophyseal Hormones: The presence of oxytocin and vasopressin in the gonads has been reviewed and the reader is referred to Wathes (1989) and Ackland et al., (1992) for a comprehensive treatment of two hormones in reproductive tissues.

Oxytocin: Immunoreactive oxytocin (OT) is found in testis of many species, including rats. The presence of neurophysin and the finding of oxytocin mRNA suggests a local site of synthesis but this has not been confirmed. Oxytocin in the efferent ducts and the epididiymis is apparently not synthesized locally but may be taken up by endocytosis produced from other sources. The oxytocin receptor (OTR) gene and protein is expressed in the human epididymis and stimulates in vitro contractility. The receptor is not only present in the smooth muscle cells of the human epididymis but also in the epithelial compartment. The OT induces the release of another potent stimulator of epididymal contractility, endothelin-1 (ET-1). Thus OT seems to control epididymal motility and sperm progression through the male genital tract (Filippi et al., 2002).

Vasopressin: Wathes (1989) reported the prsence of arginine vasopressin (AVP) in the human and the bovine male reproductive tissues. Immunoreactivity has been reported in both Sertoli cells and Leydig cells of rats. Similar to oxytocin, AVP exerts paracrine effects on gonadal function. Vasopressin produces a dose dependent decrease in gonadotropin stimulated androgen biosynthesis in cultured testicular cells. Members of the vasopressin family were also shown to increase progesterone synthesis by the Leydig cell. Both AVP and AVT appear to act by deceasing levels of 17-α hydroxylase and 17, 20 desmolase. AVP also increases basal level of testosterone. But in presence of hCG, AVP decreases its production. The effects of AVP on Leydig cells are mediated by the V_1-receptor in rats and humans. After hypophysectomy these receptors in the testis are decreased but can be restored with the administration of exogenous LH or GH. (Ackland et al., 1992).

Calcitonin Related Peptide: Calcitonin gene-related peptide (CGRP) is a 37-amino acid neuropeptide with widespread expression and a wide variety of biological effects. It influences many aspects of mammalian development and affects the function of male and female reproductive tissues. The major cell response to CGRP is an increase in the levels of intracellular cAMP. The CGRP-receptor component (RCP) proteins are highly homologous among species and present in many tissues. Particularly striking is the high levels of expression in murine testis, where CGRP-RCP are present in the head region of spermatozoa. The high concentration of CGRP-RCP in acrosome of murine spermatozoa suggests that this protein plays an important function in reproduction. After acrosome reaction *in vitro*, spermatozoa no longer exhibited CGRP-RCP immunoreactivity. This loss indicated that the release or breakdown of this protein occurs upon acrosomal discharge (Balkan et al., 1999). CGRP is present in many regions of the male reproductive tract, where it appears to function as a modulator of secretion. Although the role of CGRP in sperm is unclear, other neuropeptides are known to affect the function of these cells. There is also some evidence that calcitonin affects sperm function. Semen contains a high concentration of calcitonin, and human spermatozoa possess calcitonin receptors. Although human calcitonin has no effect on the motility of human sperm, salmon calcitonin does inhibit motility (Balkan et al., 1999).

Atrial Natriuretic Peptides: Atrial natriuretic peptides (ANPs), originally isolated from heart muscle, have been shown to have potent diuretic and hypotensive effects. Immunoreactive ANP receptors are present on Leydig cells but not in the tubular compartment of rat testis, and discussed in Chapter 18.

3.5. REFERENCES

Ackland JF, Schwartz NB, Mayo KE, Dodson RE. Non-steroidal signals originating in the gonads. Physiol Rev 1992; 72: 731-87.

AL-Attar L, Noel K, Dutertre M, Belville C, Forest MG et al. Hormonal and cellular regulation of Sertoli cells anti-Mullerian hormone production in the postnatal mouse. J Clin Invest 1997; 100: 1335-43.

Anderson RA Jr, Feathergill KA, Rawlins RG, Mack SR, Zaneveld LJ. Atrial natriuretic peptide: a chemoattractant of human spermatozoa by a guanylate cyclase-dependent pathway. Mol Reprod Dev 1995; 40: 371-8.

Anderson RA, Lrvine DS, Balfour C, et al. Inhibin B in seminal plasma: testicular origin and relationship to spermatogenesis. Hum Reprod 1998; 13: 920-26.

Avallet O, Gomez E, Vigier M, Bernard J, Saez JM. Sertoli cell-germ cell interactions and TGF-β1 expression and secretion in vitro. Biochem Biophys Res Commun 1997; 238: 905-9.

Baker J, Hardy MP, Zhou J, Bondy C, Lupu F, Bellve AR, Efstratiadis A. Effects of an Igf1 null mutation on mouse reproduction. Mol Endocrinol 1996; 10: 903-18.

Balkan W, Oates EL, Howard GA, Roos BA. Testes exhibit elevated expression of calcitonin gene-related peptide receptor component protein. Endocrinology 1999; 140: 1459-69.

Barreiro ML, Suominen JS, Gaytan F, Pinilla L, Chopin LK, Casanueva FF et al., Developmental, stage-specific, and hormonally regulated expression of growth hormone secretagogue receptor messenger RNA in rat testis. Biol Reprod 2003; 68: 1631-40.

Blauer M, Husgafvel S, Syvala H, Tuohimaa P, and Ylikomi T. Identification of a nuclear localization signal in activin/inhibin β_A subunit; intranuclear β_A in rat spermatogenic cells. Biol Reprod 1999; 60: 588-93.

Boitani C, Stefanini M, Fragale A, Morena AR. Activin stimulates Sertoli cell proliferation in a defined period of rat testis development. Endocrinology 1995; 136: 5438-44.

Bourdon V, Defamie N, Fenichel P, Pointis G. Regulation of tissue-type plasminogen activator and its inhibitor(PAI-I) by lipopolysaccharide-induced phagocytosis in a Setoli cell line. Exp Cell Res 1999; 15: 367-72.

Cancilla B, Risbridger GP. Differential localization of fibroblast growth factor Receptor – 1, -2, -3, and –4 in fetal, immature, and adult rat testes. Biol Reprod 1998; 58: 1138-45.

Castilla-Cortazar I, Garcia M, Quiroga J, Diez N, et al. Insulin-like growth factor-I reverts testicular atrophy in rats with advanced cirrhosis. Hepatology 2000; 31: 592-600.

Catizone A, Ricci G, Galdieri M. Expression and functional role of hepatocyte growth factor receptor (C-MET) during postnatal rat testis development. Endocrinology 2001; 142: 1828-34.

Caussanel V, Tabone E, Mauduit C, Dacheux F, Benahmed M. Cellular distribution of EGF, TGFα and their receptor during postnatal development and spermatogenesis of the boar testis. Mol Cell Endocrinol 1996; 123: 61-9.

Cupp AS, Kim GH, Skinner MK. Expression and action of neurotropin-3 and nerve growth factor in embryonic and early postnatal rat testis development. Biol Reprod 2000; 63: 1617-28.

Derynck R, Lindquist PB, Lee A, Wen D, Tamm J, Graycar JL, Rhee L, Mason AJ, Miller DA, Coffey RJ, et al. A new type of transforming growth factor β, TGF-β3. EMBO Journal. 1988; 7: 3737-43.

Djakiew D, Pflug B, Dionne C, Onoda M. Postnatal expression of nerve growth factor receptors in the rat testis. Biol Reprod 1994; 51: 214-21.

Dobrinski I, Ignotz GG, Fagnan MS, Yudin SI, Ball BA. Isolation and characterization of a protein with homology to angiotensin converting enzyme from the periacromsomal plasma membrane of equine spermatozoa. Mol Reprod Dev 1997; 48: 251-60.

Dombrowicz D, Hooghe-Peters EL, Gothot A, Sente B, Vanhaelst L, et al. Cellular localization of IGFI and IGF-II mRNAs in immature hypophysectomized rat testis and epididymis after in vivo hormonal treatment. Arch Intern Physiol Biochem Biophys 992; 100: 303-8.

Dyck MK, Ouellet M, Gagn M, et al. Testes-specific transgene expression in insulin-like growth factor-I transgenic mice. Mol Reprod Dev 1999; 54: 32-42.

Ehlers MR, Fox EA, Strydom DJ,, Riordan,JF. Molecular cloning of human testicular angiotensin-converting enzyme: the testis isozyme is identical to the C-terminal half of endothelial angiotensin-converting enzyme. Proc Natl Acad Sci USA 1989; 86: 7741-45.

Ergun S, Kilic N, Fiedler W, Mukhopadhyay AK. Vascular endothelial growth factor and its receptors in normal human testicular tissue. Mol Cell Endocrinol 1997; 131: 9-20.

Ethier JF, Findlay JK. Roles of activin and its signal transduction mechanisms in reproductive tissues. Reproduction 2001; 121: 667-675.

Ethier JF, Lussier JG, Silversides DW. Bovine activin receptor type IIB messenger ribonucleic acid displays alternative splicing involving a sequence homologous to Src-homology 3 domain binding sites. Endocrinology 1997; 138: 2425-34.

Fabbri A, Ciocca DR, Ciampani T, Wang J, Dufau ML. Growth hormone-releasing hormone in testicular interstitial and germ cells: potential paracrine modulation of follicle-stimulating hormone action on Sertoli cell function. Endocrinology 1995; 136: 2303-8.

Filippi S, Vannelli GB, Granchi S, Luconi M, Crescioli C, Mancina R, Natali A, Brocchi S, Vignozzi L, Bencini E, Noci I, Ledda F, Forti G, Maggi M. Identification, localization and functional activity of oxytocin receptors in epididymis. Mol Cell Endocrinol. 2002; 193: :89-100.

Galcheva-Gargova Z, Konstantinov KN, Wu IH, Klier FG, Barrett T, Davis RJ. Binding of zinc finger protein ZPR1to the epidermal growth factor receptor. Science 1996; 272: 1797-02.

Gizang-Ginsberg E, Wolgemuth DJ. Expression of the proopimelanocortin gene is developmentally regulated and affected by germ cells in the male mouse reproductive system. Proc Natl Acad Sci U S A 1987; 84: 1600-4.

Gnessi L, Basciani S, Mariani S, Arizzi M, Spera G, Wang C, Bondjers C, Karlosson L, Betzholtz C. Leydig cell loss and spermatogenic arrest in platelet-derived growth factor (PDGF)-A-deficient mice. J Cell Biol 2000; 149: 1019-26.

Gupta C. The role of epidermal growth factor receptor (EGFR) in male reproductive tract differentiation: stimulation of EGFR expression and inhibition of Wolffian duct differentiation with anti-EGFR antibody. Endocrinology 1996; 137: 905-10.

Gur Y, Breitbart H, Lax Y, Rubinstein S, Zamir N. Angiotensin II induces acrosomal exocytosis in bovine spermatozoa. Am J Physiol 1998; 275: 87-93.

Hagaman JR, Moyer JS, Bachman ES Sibony M, et al. Angiotensin-converting enzyme and male fertility. Pro Natl Acad Sci USA, 1998; 95: 2552-57.

Hanssan HA, Billig H. Isgaard J. Insulin-like growth factor I in the developing and mature rat testis: immunohistochemical aspects. Biol Reprod 1989; 40: 1321-28.

Henricks DM, Kouba AJ, Lackey BR, Boone WR, Gray SL. Identification of insulin-like growth factor 1 in bovine seminal plasma and its receptors on spermatozoa: Influence on sperm motility. Biol Reprod 1998; 59: 330-37.

Houde A, Lambert A, Saumande J, Silversides DW and Lussier JG. Cloning and tissue expression of bovine follistatin cDNA. Mol Reprod Dev 1994; 39: 127-35.

Howard T, Balogh R, Overbeek P, Bernstein KE. Sperm-specific expression of angiotensin-converting enzyme (ACE) is mediated by a 91-base-pair promoter containing a CRE-like elment. Mol Cell Biol 1993; 13: 18-27.

Howard TE, Shai SY, Langrord KG, Martin BM, Bernstein KE, et al. Transcription of testicular angiotensin-converting enzyme (ACE) is initiated within the 12th intron of the somatic ACE gene. Mol Cell Biol 1990; 10: 4294-4302.

Hwa V, Oh Y, Rosenfeld RG. The insulin like growth factor-binding protein (IGFBP) superfamily. Endcor Rev 1999; 20: 761-87.

Isaac RE, Williams TA, Sajid M, Corvol P, Coates D. Cleavage of arginyl-arginine and lysyl-arginine from the C-terminus of pro-horomone peptides by human germinal angiotensin I-converting enzyme (ACE) and the C-domain of human somatic ACE. Biochem J 1997; 328: 587-91.

Josso N, di Clemente N, Gouedard L. Anti-mullierian hormone and its receptors. Mol Cell Endocrinol 2001; 179: 25-32.
Josso N, Racine C, Di Clemente N Rey R, Xavier F. The role of anti-mullerian hormone in gonadal development. Mol Cell Endocrinol 1998; 145: 3-7.
Kaipia A, Parvinen M, Toppari J. Localization of activin receptor (Act R-IIB2) mRNA in rat seminiferous epithlium. Endocrinology 1993; 132: 477-79.
Kaipia A, Penttila TL, Shimasaki S, Ling N, Parvinen M, Toppari J. Expression of inhibin βA and βB, follistatin and activin-A receptor messenger ribonucleic acids in the rat seminiferous epithelium. Endocrinology 1992; 131: 2703-10.
Kilpatrick DL, Borland K, Jin DF. Differential expression of opioid peptide genes by testicular germ cells and somatic cells. Proc Natl Acad Sci U S A 1987; 84: 5695-9.
Kohn FM, Muller C, Drescher D, Neukamm C, el Mulla KF, Henkel R, Hagele W, Hinsch E, Habenicht UF, Schill WB. Effect of angiotensin converting enzyme (ACE) and angiotensins on human sperm functions. Andrologia 1998; 30: 207-15.
Korpelainen KI, Karkkainene MJ, Tenhunen A, Lakso M, et al. Overexpression of VEGF in testis and epididymis causes infertility in transgenic mice: evidence for nonendothelial targets for VEGF. J Cell Biol 1998; 143: 1705-12.
Krege JH, John SWM, Langenback LL et al. Male female differences in fertility and blood pressure in ACE deficient mice Nature 1995; 375: 146-48.
Kuroda T, Tanaka H, Nakamura H, Nishimune T, Kishimoto T. Hepatoma-derived growth factor-related protein (HRP)-1 gene in spermatogenesis in mice. Biochem Biophys Res Commun 1999; 262: 433-37.
Langford KG, Zhou Y, Russell LD, Wilcox JN, Bernstein KE. Regulated expression of testis angiotensin-converting enzyme during spermatogenesis in mice. Biol Reprod 1993; 48: 1210-8.
Lax Y, Rubinstein S, Breitbart H. Epidermal growth factor induces acrosomal exocytosis in bovine sperm. FEBS Lett 1994; 339: 234-8.
Le Bouc Y, Dreyer D, Jaeger F, Binoux M, Sondermeyer P. Complete characterization of the human IGF-I nucleotide sequence isolated from a newly constructed adult liver cDNA library. FEBS Lett 1986; 196: 108-12.
Le Magueresse-Battistoni B, Pernod G, Sigillo F, Kolodie L. Plasminogen activator inhibitor-1 is expressed in cultured rat Sertoli cells. Biol Reprod 1998; 59: 591-98.
Leung PS, Wong TP, Sernia C. Angiotensinogen expression by rat epididymis evidence of an intrinsic, angiotensin-generating system. Mol Cell Endocrinal 1999a; 155: 115-22.
Levine E, Cupp AS, Miyashiro L, Skinner MK. Role of transforming growth factor-α and the epidermal growth factor receptor in embryonic rat testis development. Biol Reprod 2000; 62: 477-90.
Li Y, Lingwood CA, Sakac D, Saltiel A. Developmentally regulated testicular galactolipid sulfotransferase inhibitor is a phosphoinositol glycerolipid and insulin-mimetic. Mol Reprod Dev 1994; 37: 462-6.
Liu F, Tokeson J, Persengiev SP, Ebert K, Kilpatrick DL. Novel repeat elements direct rat proenkephalin transcription during spermatogenesis. J Biol Chem 1997; 272: 5056-62.
Liu XP, Li Z, Zhang YQ, Xu RJ. [Expression of thyrotropin-releasing hormone receptor mRNA in the rat testis development] Shi Yan Sheng Wu Xue Bao 2003; 34: 109-14. [Article in Chinese]
Loveland KL, Zlatic K, Stein-Oakley A, Risbridger G, deKretser DM. Platelet-derived growth factor ligand and receptor subunit mRNA in the Sertoli and Leydig cells of the rat testis. Mol Cell End 1995; 27: 108: 155-9.
Luconi M, Bonaccorsi L, Krausz C, Gervasi G, Forti G, Baldi E. Stimulation of protein tyrosine phosphorylation by platelet-activating factor and progesterone in human spermatozoa. Mol Cell Endocrinol 1995; 27: 35-42.
Lyet L, Louis F, Forest MG, Jasso N, Behringer RR, Vigier B. Ontogeny of reproductive abnormalities induced by deregulation of anti-mullerian hormone expression in transgenic mice. Biol Reprod 1995;52:444-54.
Majdic G, McNeilly AS, Sharpe RM, et al. Testicular export of inhibin and activin subunits and follistatin in the rat and human and neonate and during postnatal development in the rat. Endocrinology 1997; 138: 2136-47.
Manske M, Welte B, Hitschke K, Wack A, Bade EG. Plasminogen activator inhibitor type 1 in rat spermatozoa: localization in the tail, in the acrosome and on the surface of the head. Zygote 1994; 2: 243-52.
Massague J, Wotton D. Transcriptional control by the TGF-β/Smad signaling system. The EMBO Journal 2000; 19: 1745-54.
Mather JP, Roberts PE, Krummen LA. Follistatin modulates activin activity in a cell- and tissue specific menner. Endocrinology 1993;132:2732-4.
Mathews LS. Activin receptors and cellular signaling by the receptor serine kinase family. Endocrine Rev 1994; 15: 310-25.
Mayerhofer A, Russell LD, Grothe C, Rudolf M, Gratzl M. Presence and localization of a 30-kDa basic

fibroblast growth factor – like protein in rodent testes. Endocrinology 1991; 129: 921-24.
Meehan T, Schlatt S, O'Bryan MK, de Kretser DM, Loveland KL. Regulation of germ cell and Sertoli cells development by activin, follistatin and FSH. Dev Biol 2000; 220: 225-37.
Mehta ND, Don J, Zinn SA, Millette C, Wolgemuth DJ. Kilpatrick DL. Proenkephalin gene expression in testicular interstitial cells is down – regulated coincident with the appearance of pachytene spermatocytes. Endocrinology 1994; 135: 1543-50.
Meng X, Lindahl M, Hyvonen ME, Parvinen M, et al. Regulation of cell fate decision of undifferentiated spermatogonia by GDNF. Science 2000; 287: 1489-93.
Mishina Y, Tizard R, Deng JM, Pathak BG et al. Sequence, genomic organization, and chromosomal location of the mouse Mullerian-inhibiting substance type II receptor gene. Biochem Biophys Res Commun 1997; 237: 741-46.
Miyazono K. Positive and negative regulation of TGF-β signaling. J Cell Sci 2000; 113: 1101-9.
Monsees TK, Blocher S, Heidorn F, Winkler A, Siems WE, et al. Expression and location of the bradykinin B_2 receptor in rat testis. Biol Reprod 2002; 67: 1832-39.
Monsees TK, Blocher S, Loddo C, Steger K, Schill WB. Tissue kallikrein and bradykinin B2 receptors in the reproductive tract of the male rat. Andrologia 2003; 35: 24-31.
Mullaney BP, Skinner MK. Transforming growth factor-β (β1, β2 and β3) gene expression and action during pubertal development of the seminferous tubule: potential role at the onset of spermatogenesis. Mol Endocrinol 1993; 7: 67-76.
Nakayama Y, Yamamoto T, Abe SI. IGF-I, IGF-II and insulin promote differentiation of spermatogonia to primary spermatocytes in organ culture of newt testes. Int J Dev Biol 1999; 43: 343-7.
Naz RK, Padman P. Identification of insulin-like growth factor (IGF)-1 receptor in human sperm cell. Andrology 1999; 43: 153-9.
O'Hara SF, Donovan DM, Lindberg I, Brannock MT, et al. Proenkephalin transgenic mice: a short promoter confers high testis expression and reduced fertility. Mol Reprod Dev 1994; 38: 275-284.
Ogawa K, Hashimoto O, Kurohmaru M, Mizutani T, Sugino H, Hayashi Y. Follistatin-like immunoreactivity in the cytoplasm and nucleus of spermatogenic cells in the rat. Eur J Endocrinol 1997; 137: 448-9.
Olaso R, Pairault C, Habert R. Expression of type I and II receptors for transforming growth factor β in the adult rat testis. Histochem Cell Biol 1998; 110: 613-8.
Parvinen M, Pelto-Huikko M, Soder O, Schultz R, Kalpia A. Expression of β-nerve growth factor and its receptors in rat seminiferous epithelium: specific function at the onset of meiosis. J Cell Biol 1992; 117: 629-41.
Pask AJ, Whitworth DJ, Mao CA, Wei KJ, Sankovic N, Graves JA, Shaw G, Renfree MB, Behringer RR. Marsupial anti-Mullerian hormone gene structure, regulatory elements, and expression. Biol Reprod 2004; 70: 160-7.
Philips DJ, de Kretser DM. Follistatin: a multifunctional regulatory protein. Front Neuroendocrinol 1998; 19: 287-322.
Pineau C, Sharpe RM, Saunders PT, Gerard N, Jegou B. Regulation of Sertoli cell inhibin production and of inhibin α-subunit mRNA levels by specific germ cell types. Mol Cell Endocrinol 1990; 72: 13-22.
Rao JN, Liang JY, Chakraborti P, Feng P. Effect of thyroid hormone on the development and gene expression of hormone receptors in rat testes in vivo. J Endocrinol Invest. 2003; 26: 435-43.
Rao SM, Howells RD. cis-acting elements in the 5'-untranslated region of rat testis proenkephalin mRNA regulate translation of the precursor protein. J Biol Chem 1993; 268: 22164-9.
Resnick JL, Ortiz M, Keller JR, Bonovan PJ. Role of fibroblast growth factors and their receptors in mouse primordial germ cell growth. Biol Reprod 1998; 59: 1224-29.
Robertson DM, Klein R, de Vos FL et al. The isolation of polypeptides with FSH suppressing activity from bovine follicular fluid, which are structurally different to inhibin. Biochem Biophys Res Commun 1987; 149: 744-49.
Robinson LL, Townsend J, Anderson RA. The human fetal testis is a site of expression of neurotrophins and their receptors: regulation of the germ cell and peritubular cell population. J Clin Endocrinol Metab 2003; 88: 3943-51.
Rossi F, Ferraresi A, Romagni P, Silvestroni L, Santiemma V. Angiotensin II stimulates contraction and growth of testicular peritubular myoid cells in vitro. Endocrinology 2002; 143: 3096-104.
Russo MA, Odorisio T, Fradeani A, Rienzi L, De Felici M, Cattaneo A, Siracusa G. Low-affinity nerve growth factor receptor is expressed during testicular morphogenesis and in germ cells at specific stages of spermatogenesis. Mol Reprod Dev 1994; 37: 157-66.
Shetty J, Wolkowicz MJ, Digilio LC, Klotz KL, Jayes FL, et al. SAMP14, a novel, acrosomal membrane-associated, glycosylphosphatidylinositol anchored member of the Ly-6/urokinase-type plasminogen activator receptor superfamily with a role in sperm-egg interaction. J Biol Chem. 2003; 278: 30506-15.
Shimasaki S, Koga M, Esch F, Mercado M, Cooksey K, Koba A, and Ling N. Porcine follistatin gene structure supports two forms of mature follistatin produced by alternative splicing. Biochem Biophys Res Commun

1988; 152: 717-23.

Sibony M, Segretain D, Gasc JM. Angiotensin-converting enzyme in murine testis: step-specific expression of the germinal isoform during spermiogenesis. Biol Reprod 1994; 50: 1015-26.

Sigillo F, Pernod G, Kolodie L, Benahmed M, LeMagueresse Battistoni B. Residual bodies stimulate rat Sertoli cell plasminogen activator activity. Biochem Biophys Res Commun 1998; 250: 59-62.

Srivastava CH, Monts BS, Rothrock JK, Peredo MJ, Pescovitz OH. Presence of a spermatogenic-specific promoter in the rat growth hormone-releasing hormone gene. Endocrinology 1995; 136: 1502-8.

Steger K, Tetens F, Seitz J, Grothe C, Bergmann M. Localization of fibroblast growth factor 2 (FGF-2) protein and the receptors FGFR 1-4 in normal human seminiferous epithelium. Histochem Cell Biol 1998; 110: 57-62.

Suh DS and Rechler MM. Hepatocyte nuclear factor 1 and the glucocorticoid receptor synergistically activate transcription of the rat insulin-like growth factor binding protein-1 gene. Mol Endocrinol 1997; 11: 1822-31.

Tajima Y, Watanabe D, Koshimizu U, Matsuzawa T, Nishimune Y. Insulin-like growth factor-1 and transforming growth factor-α stimulate differentiation of type A spermatogonia in organ culture of adult mouse cryptorchid testis. Int J Androl 1995; 18: 8-12.

Tanimoto Y, Tanimoto K, Sugiyama F, Horiguchi H. Male sterility in transgenic mice expressing activin βA subunit gene in testis. Biochem Biophys Res Commun 1999; 259:699-705.

Thiebot B, Bichoualne L, Langris M, et al. IGF-I stimulates synthesis of undersulfated proteoglycans and of hyaluronic acid by peritubular cells from immature rat testis. Biochim Biophys Acta 1997;1358:127-41.

Tremblay JJ, Viger RS. Transcription factor GATA-4 enhances Mullerian inhibiting substance gene transcription through a direct interaction with the nuclear receptor SF-1. Mol Endocrinol 1999; 13: 1388-101.

Trudeau VL, Matzuk MM, Hache RJ, Renaud LP. Overexpression of activin βA subunit mRNA is associated with decreased activin type II receptor mRNA levels in the testes of α-inhibin deficient mice. Biochem Biophys Res Commun 1994; 203: 105-12.

Viglietto G, Dolci S, Bruni P, Baldasrarre G, et al. Glial cell line-derived neutrotrophic factor and Neurturin can act as paracrine growth factors stimulating DNA synthesis of Ret-expressing spermatogonia. Int J Oncol 2000; 16: 689-94.

Vihko KK, Tappari J, Saksela O, Suominen JJ, Parvinen M. Testicular plasminogen activators during postnatal development in the rat. Acta Endocrinol 1986; 112: 431-5.

Vinson GP, Saridogan E, Puddefoot JR, Djahanbakhch O. Tissue renin-angiotensin systems and reproduction. Human Reprod 1997; 12: 651-62.

Wada M, Shintani Y, Kosaka M, Sano T, Hizawa K, Saito S. Immunohistochemical localization of activin A and follastatin in human tissues. Endocrine J 1996; 43: 375-85.

Wang RA, Zhao GQ. Transforming growth factor β signal transducer Smad2 is expressed in mouse meiotic germ cells, Sertoli cells, and Leydig cells during spermatogenesis. Biol Reprod 1999; 61: 999-1004.

Wathes DC, Oxytocin and vasopressin in the gonads. In: Oxford Reviews of reproductive biology (ed SR Milligan). Oxford Univ Press 1989; Vol. II, p. 226-83.

Widenfalk J, Parvinen M, Lindqvist E, Olson L. Neurturin, RET, GFRRα-1 and GFRα-2, but not GFR-α3, mRNA are expressed in mice gonads. Cell Tissue Res 2000; 299: 409-15.

Wong RW, Kwan RW, Mak PH, Sham MH, Chan SH. Expression of epidermal growth factor induced hypospermatogenesis in transgenic mice. J Biol Chem 2000; 275: 18297-301.

Wu, TC, Jih MH, Wang L, Wan YJ. Expression of activin receptor II and IIB mRNA isoforms in mouse reproductive organs and oocytes. Mol Reprod Dev 1994; 38: 9-15.

Xu J, Beyer AR, Walker WH, McGee EA. Developmental and stage-specific expression of Smad2 and Smad3 in rat testis. J Androl. 2003; 24: 192-200.

Yamauchi M, Itoh H, Naganuma S, Koono M, Hasui Y, Osada Y, Kataoka H et al., Expression of hepatocyte growth factor activator inhibitor type 2 (HAI-2) in human testis: identification of a distinct transcription start site for the HAI-2 gene in testis. J Biol Chem 2003; 383: 1953-7.

Zhao W, Leung PY, Cheng-Chew SB, Chan HC, Wong PYD. Localization and distribution of angiotensin II in the rat epididymis. J Endocrinol 1996; 149: 217-22.

Zhou H, Vassalli JD. The receptor for urokinase-type plasminogen activator is expressed during mouse spermatogenesis. FEBS Letters 1997; 413: 11-15.

Zhou Y, Delafontaine P, Martin BM, Bernstein KE. Identification of two positive transcriptional elements within the 91-base promoter for mouse testis angiotensin converting enzyme (Testis ACE) Dev Genet 1995; 16: 201-9.

Chapter 4

SPECIFICITY OF RETINOL, ESTROGEN AND STEROID LINKED PROTEINS

4.1. ACTION OF RETINOIDS

Retinoids and estrogens are small lipophilic compounds, which play important roles in verteberate development, reproduction and homeostasis. Both ligands are activators of their own receptors, which belong to nuclear superfamily and act as transcription factors. The requirement of vitamin A in spermatogenesis became clear from vitamin A deficient (VAD) mice, which showed the presence of only Sertoli cells and undifferentiated A spermatogonia in testis, with loss of subsequent stages of spermatgenesis. Vitamin A deficient mice showed germ cell arrest at the end of S phase of cell cycle. H1 histone kinase, the marker of M phase, is decreased in vitamin A deficient testis but restored or increased following retinol treatment (Wang and Kim, 1993). Retinoic acid receptors have been identified in testis germ cells by many investigators. In part of this section, our main concern would be to describe the characteristic features of the retinol binding proteins (the transport proteins of retinol), specific enzymes involved in retinol metabolism, the retinol receptors and their nuclear receptor homologues found in testis as orphan receptors. The molecules that received considerable attention during recent years, comprise derivatives of vitamin A (retinol) called retinoids. Retinol is converted to active metabolites including all-trans-retinoic acid (RA) through multiple metabolic pathways. Retinoids are important molecules whose action is mediated through receptors during development and differentiation.

4.1.1. Retinol Binding Proteins

As hydrophobic nutrients, retinoids and fatty acids exist in aqueous environments bound to specific carrier proteins. These proteins belong to a superfamily of cellular lipophilic transport proteins that are believed to be involved in the uptake, transport, and metabolism of their ligands. This superfamily includes retinoid-binding protein family, the fatty acid binding protein family, adipocyte lipid binding protein and myelin P2. These proteins are small (14-15kDa) in molecular size and have high degree of sequence conservation across species. They have strikingly similar tertiary structures and many physical characteristics in common.

Cellular retinol–binding protein (CRBP) has been known in testis and epididymis since long. The CRBP was found in the seminiferous tubules, with a striking localization within Sertoli cells. A distinct cyclic variation of specific staining for CRBP within Sertoli cells is observed during the spermatogenic cycle. Within the epididymis CRBP is selectively localized to the proximal portion of the caput epididymis. In vitamin A-deficient rats the CRBP is markedly reduced in both testes and epididymis (Kato et al., 1985). The distribution studies of CRBP and cellular retinoic acid-binding protein (CRABP) in rat testis indicated that the level of CRABP in Sertoli cells preparations containing germ cells was approximately five times greater than that observed in Sertoli cells free from germ cells (Blaner et al., 1987). The presence of lecithin retinol acetyltransferase activity within adluminal germ cells suggests that germ cells receive vitamin A as retinol and synthesize the retinyl esters that have been shown to be present in mature sperm (Schmitt and Ong, 1993). Retinyl esters are converted to retinoic acids.

The CRABP has been isolated from male germ cells. The cDNA clone of 561-bp of CRABP containing a full open reading frame codes a protein of 132 amino acids. The protein has 58% amino acid sequence identity to bovine myelin P2, 58% identity to urine adipocyte lipid-binding protein, and 40% identity to rat CRABP. Binding of carboxyl groups suggested fatty acid or an acidic retinoid as ligands, and appears to be restricted to late germ cells. Immunostaining was first detectable in mid-pachytene spermatocytes and increased in intensity as these cells progressed to elongated spermatids, suggesting that this testis lipid binding protein has a specific role in sperm development (Schmitt et al., 1994).

4.1.2. Retinol to Retinoic Acids

Retinol is converted to several forms of retinoic acids (4 oxo-RA). Hydroxylation into 4-hydroxy retinoic acids is the main route. The formation of 4 oxo-RA is catalyzed by cytochrome P450 (CYP). The CYP enzymes are active in hydroxylation of endogenous and exogenous (xenobiotic) compounds. Retinoic acid is a potent inducer of in vivo proliferation of growth-arrested A spermatogonia in VAD mouse testis (Gaemer et al., 1996). A specific RA-hydroxylase has been cloned from regenerating dorsal zebrafish fin, representing a new CYP subfamily designated as CYP26. Subsequent cloning of CYP26 from mouse, *Xenopus* and humans has shown that CYP26 is highly conserved among these species. All CYP26 homologues possess specific properties of the CYP superfamily, such as a transmembrane anchoring domain, a proline-rich domain, a steroid and oxygen binding site, and a conserved heme-binding domain including the highly conserved cysteine residue at position 442. Amino acid sequence comparison of CYP26 revealed 93% identity between human and mouse, 66% between human and *Xenopus*, and the human and zebrafish sequences were 65% identical. The CYP26 is rapidly induced by retinoic acid in a variety of cell lines (van der Burg et al., 1999).

4.1.3. *Stra 8-A*: A Retinoic Acid Inducible Gene

The full-length cDNA corresponding to *Stra8,* a gene inducible by retinoic acid in embryonal carcinoma cells, encodes a 45-kDa protein. Both *Stra8* mRNA and protein are induced in cell treated with all *trans* and *9-cis* retinoic acids. The two stereoisomers of 4 oxo-RA differentially regulate the phosphorylation status of the Stra8 protein, which exists in differentially phosphorylated forms. Stra 8 protein is a cytoplasmic protein. During mouse embryogenesis, *Stra8* expression is restricted to male gonadal development where as in adult male mice Stra8 is restricted to the pre-meiotic germ cells in testis. Thus, the mouse Stra8 protein may play a role in the pre-meiotic phase of spermatogenesis (Miyamoto et al., 2002; Oulad-Abdelghani et al., 1996).

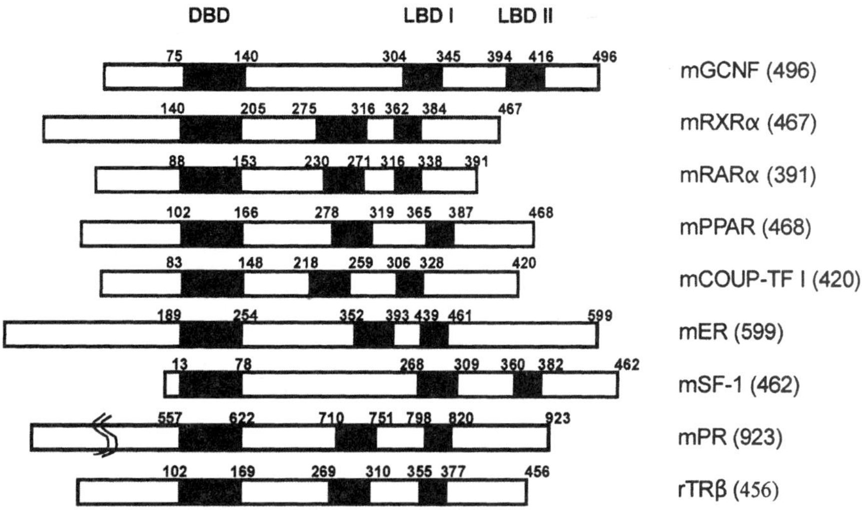

Fig.4.1. Comparison of several members of nuclear receptor superfamily. Mouse (m) germ cell nuclear factor (mGCNF), retinoid-X-receptor (mRXRα), retinoic acid receptor-α (RARα), peroxisome proliferators-activated receptor (PPAR), progesterone receptor (PR), estrogen receptor (mER), steroidogenic factor (SF-1), and rat thyroid receptor β (rTRβ). Number of amino acids in each receptor are mentioned on top of each bar. DNA binding domain (DBD) and Ligand binding domain (LBD) I and LBD II are shown.

4.2. NUCLEAR RECEPTORS

The nuclear receptor superfamily comprises a group of ligand dependent transcription factors that include the receptors that bind to and activated by retinoids, steroids, and thyroid hormones. Ligand activated receptors regulate the expression of target genes by binding *cis*-acting sequences. The common structural feature of nuclear receptor superfamily is a tripartite domain structure consisting: i) a hypervariable N-terminus that contributes mainly to the transactivation function, ii) a highly conserved DNA binding domain, which is responsible for DNA recognition and dimerization and iii) the conserved C-terminus, which contains the sub-domains II and III and is involved in nuclear localization, ligand binding, receptor dimerization, silencing and transactivation. The most conserved DNA binding domain contains 65-68 amino acid residues. Eight of the nine non-variant cysteins form two type-II zinc modules. The sequence identity in the DNA binding domain of any member to the rest of the family ranges between 40-99%. The DNA binding domain contains two zinc finger motifs and mediates the interaction of the receptor with specific DNA sequences (hormone response elements HRE). These response elements consist of either a direct or inverted repeat or a palindrome. The ligand-binding domain contains a ligand inducible transactivation function and a heptad repeat motif implicated in the dimerization function of nuclear receptors. Based on the sequence and structural similarities with known members of the steroids hormone superfamily, a variety of orphan receptors have been cloned for which the ligands are yet to be identified. The nuclear receptors modulate transcription by interacting with those DNA regulatory elements, which control RNA polymerase complex (see Chapter 1 for androgen receptor and Chapters 14 and 16 for transcription factors). Although progesterone receptors have been identified on sperm surface, they are different from nuclear receptors and have not been related with transcriptional activity.

Many of the orphan receptors with apparent no known ligands have homology with retinoic acid receptors (**Fig. 4.1**).

4.3. RETINOID RECEPTORS AND THEIR HOMOLOGUES

4.3.1. Retinoid Receptors

The retinoid receptors belong to the steroid/vitamin D (VDR), thyroid (TRs) hormone nuclear receptor superfamily. Two distinct classes of retinoid receptors, RARs (retinoic acid receptors) and RXRs (retinoid X receptors) have been described. They are characterized by their ligand- and DNA-binding abilities, and by their possible dimerization partners. Retinoic acid receptors (RARα) are present in spermatogenic cells while RXRβ receptors are located in Sertoli cells. A receptor related to retinoids called GCNF or RTR is present in round spermatids and developing oocytes. Each class is composed of three genes (RARα, -β, -γ, and RXRα, -β, -γ), the transcription of which results in several isoforms because of alternative promoter usage and mRNA splicing. Both, all trans-retinoic acid (ATRA) and 9-cis-retinoic acid (9-cisRA) can bind to RARs, but only 9-cis-RA is capable of binding to RXRs. In addition to the existence of different ligands and receptors, the complexity of retinoid signaling is increased by the possible formation of different receptors homo- and heterodimers. These dimers can bind to different response elements in the promoters of certain genes and act as transcription factors. The actions of the RXRs are very diverse as they are involved in several signaling pathways as heterodimeric partners involved in DNA binding. The RXRs can form heterodimers with RARs and a number of other members of the steroid hormone receptor family and, upon binding of 9-cis-RA, RXRs can also form homodimers that bind to response elements other than the heterodimers (Brocard et al., 1996). An important role for RARs and RXRs during development was clear from their specific expression patterns, coinciding with most of the target tissues affected in vitamin A-deficient (VAD) mouse embryos.

The expression patterns of RARs and RXRs have been studied in a number of tissues. In the mouse testis, all three RARs are present. Both RARα and RARγ null mutant mice were found to be sterile. In contrast, mice lacking all RARβ isoforms develop normally and are fertile. This does not necessarily mean that RARβ is not involved in spermatogenesis, since there is evidence of functional redundancy among the RARs. On the contrary, RXRβ null mutant mice are sterile because of abnormal spermiogenesis (Gaemers et al., 1998). Receptor knockouts of members of the RAR/RXR provided extensive support that RARs and RXRs are indeed the mediators of known functions of retinoids during development (Kastner et al., 1995, 1997). It is postulated that retinoic acid may regulate the differentiation of spermatogenic cells either directly or through its actions on Sertoli cells. Furthermore the differential effects of testosterone on the steady-state levels of mRNA of testicular retinoic acid receptors-α (RARα) and -γ suggested that a RAR-mediated mechanism may be involved in the testosterone regulation of some Sertoli cell functions (Huang et al., 1994).

Both Sertoli cells and germ cells are capable of expressing different retinoic acid receptors. The RAR-α mRNAs are present in both Sertoli and germ cells of the testis. A 2.7-kb mRNA was expressed solely in Sertoli cells, whereas a 3.4-kb mRNA was distributed in both Sertoli and germ cells. In addition, two minor germ cell-specific mRNAs are detected primarily in the pachytene spermatocytes. By contrast, only one transcript for RARβ was found exclusively in Sertoli cells, suggesting that each mRNA may have specific functions in mediating the effects of retinoids during spermatogenesis. The expression of RAR-α mRNAs was regulated during the spermatogenic cycle, which supported that α-mRNA transcription is necessary in advanced germ cells than preleptotene spermatocytes. The highest level of retinoic acid receptor (RAR-α) transcripts occur in round spermatids at stage VIII of the spermatogenic cycle (Akmal et al, 1997). The precise regulation of receptor by retinol suggests that its synthesis is required before it can be used to modulate the transcription of retinoid-inducible genes. While Sertoli cells and spermatocytes attained high level of 3.4-kb RARγ, trace amount was also present in spermatids. Testosterone administration resulted in an increase of both 3.4 and 2.7-kb RARγ (Huang et al., 1994). Analysis in the developing testes revealed that the mRNA level was high from 10-15 days of age, both in Sertoli cells and in germ cells, and then declined in 20 day-old rats. These results suggested a role for RAR-α during meiosis, at the transition from round to elongating spermatids and in Sertoli cells of developing testis.

All-trans-retinoic acid (ATRA) and FSH modulate RAR-α receptor sub-cellular localization, leading to changes in its transcriptional activity and protein expression in mouse Sertoli cell. It was shown that ATRA induced the nuclear localization of RAR-α within 30 min and that exposure for longer term increased the receptor transcriptional activity and RAR-α protein expression. Conversely, FSH suppressed the ATRA-induced nuclear localization, transcriptional transactivation, and protein expression of RAR-α. Treatment with protein kinase A-selective antagonists suggested the involvement of protein kinase A in mediating the inhibitory effects of FSH. This is an example of a unique signaling convergence between the RAR-α and the FSH mediated signaling pathways, which may have significant implications in the testis since both are critical regulators of testis physiology (Braun et al., 2000).

4.3.2. Germ Cell Nuclear Factor (GCNF)

The study of orphan receptors has expanded our view of the super family of nuclear receptors. It has facilitated the discovery of new ligands. For example 9-cis-retinoic acid has been identified as a ligand for the retinoid-X-receptor (RXR). Chen et al., (1994), cloned a novel member of the nuclear receptor super family that was designated as germ cell nuclear factor (GCNF). The GCNF is a putative transcription factor, based on homology to the nuclear receptor super family of ligand-activated transcription factors. The GCNF binds specifically to DR0 as a homodimer and does not bind with high affinity to DR1-DR6 sequences (Cooney et al., 1998). The deduced protein sequence of GCNF contains 495 amino acids and consists of the characteristic DNA binding and ligand binding domains of the nuclear receptor super family. The primary sequence of this new orphan is distinct from those of previously cloned members and subgroups. It showed maximum 61% identity in amino acids with murine RXRα in DNA binding domain I. This orphan receptor binds to the sequence TCAAGGTCA that includes the steroidogenic factor 1 half site and direct repeat with base pair spacing elements. The receptor was predominantly expressed in the testis. In situ hybridization experiments showed it to be located in the spermatogenic cells. High expression of this gene is restricted to developing germ cells, the occytes and spermatogenic cells. It was speculated that this orphan receptor is a molecule involved in regulating some aspect of meiosis, and in the regulation of gene expression in germ cells during gametogenesis. In situ hybridization showed that GCNF mRNA is lacking in the testis of hypogonadal mutant mice, which lacked developed spermatids, as against in wild type testis. The GCNF mRNA was most abundant in stage VII round spermatids in wild type testis, and its expression first occurred in testis of 20-day old mice, when round spermatids first emerged. Therefore, GCNF expression, in male, occurs post-meiotically and may participate in the morphological changes of the maturing spermatids. In contrast, female expression of GCNF was shown in growing oocytes that had not completed the first meiotic division. Although

```
  1 MERDERPPSG GGGGGGSAGF LEPPAALPPP PRNGFCQDEL AELDPGTNGE TDSLTLGQGH
 61 IPVSVPDDRA EQRTCLICGD RATGLHYGII SCEGCKGFFK RSICNKRVYR CSRDKNCVMS
121 RKQRNRCQYC RLLKCLQMGM NRKAIREDGM PGGRNKSIGP VQISEEEIER IMSGQEFEEE
181 ANHWSNHGDS DHSSPGNRAS ESNQPSPGST LSSSRSVELN GFMAFRDQYM GMSVPPHYQY
241 IPHLFSYSGH SPLLPPQARS LDPQSYSLIH QLMSAEDLEP LGTPMLIEDG YAVTQAELFA
301 LLCRLADELL FRQIAWIKKL PFFCELSIKD YTCLLSSTWQ ELILLSSLTV YSKQIFGELA
361 DVTAKYSPSD EELHRFSDEG MEVIERLIYL YHKFHQLKVS NEEYACMKAI NFLNQDIRGL
421 TSASQLEQLN KRYWYICQDF TEYKYTHQPN RFPDLMMCLP EIRYIAGKMV NVPLEQLPLL
481 FKVVLHSCKT STVKE
```

Fig.4.2: Amino acid sequence of orphan nuclear receptor RTR homologue from mouse testis. Source: http//www.ncbi.nlm.nih.gov (accession number Q64249).

mRNA messages contained the DNA binding domain in both, only the larger message was recognized from the extreme 3'-untranslated region. In addition the coding region and portion of the 3'-untranslated region of the GCNF cDNA are conserved in the rat (Katz et al., 1997). The GCNF was highly expressed with two transcripts (7.4 and 2.1-2.3-kb) in mouse testis and with only one transcript (7.4-kb) slightly expressed in brain, liver, and kidney. Two *Gcnf* transcripts, encoded by a single copy gene, are expressed in the testis of several mammalian species. The 3'-UTRs of two transcripts are quite distinct. The 7.4-kb transcript, which appears earlier in spermatogenesis, has a very long 3'-UTR of 445-nt. In contrast, the 2.1-kb transcript, which is expressed predominantly during the haploid phase of spermatogenesis, has a 3'-UTR of only 202-nt in length. Both transcripts share the same coding region and are associated with polysomes. The GCNF is localized in the nuclei of pachytene spermatocytes and round spermatids, and was first detectable in early pachytene spermatocytes (stage II) until spermatids began to elongate in stage IX. The GCNF is distributed throughout the nucleus, more prominently in heterochromatic regions at some stages and in condensed chromosomes undergoing the meiotic divisions. This expression profile suggested that GCNF plays a role in transcriptional regulation during meiosis and the early haploid phase of spermatogenesis (Yang et al., 2003). It also expresses in epididymis, albeit at a very low level. The expression of GCNF transcripts in epididymis were down regulated by androgen in mice. The two epididymal mRNAs (7.4 and 3.1-kb) encode for the same protein as in testis. Its expression was localized in the principal cell in the corpus region (Hu et al., 2003).

4.3.3. Retinoid Testis Receptor (RTR)

Hirose et al., (1995a) identified and cloned another member of the nuclear receptor super family from murine testis, referred to as retinoid receptor related testis associated receptor or RTR. The amino acid sequence of RTR is most closely related to that of the mouse RXR with an overall identity of 32-34%. The highest similarity (61%) is observed in the DNA binding domain **(Fig.4.2)**. The RTR is predominantly expressed in the testis and RTR-mRNA is not expressed in early germ cells or Sertoli cells but is most abundant in round spermatids. This putative transcription factor was suggested to play a role in the regulation of gene expression particularly during the post-meiotic phase of spermatogenesis.

4.3.4. Tr 2–11 Homologue

Lee et al (1996) suggested that a Tr 2-11, a putative DNA binding protein of the zinc finger type from mouse testis is highly homologous with the human and rat orphan receptor Tr2-11 isolated from prostate. This transcript was detected in early to mid-gestation embryos and was seen to level off during later stages of development. In adult animals, a high expression was detected only in the testis starting at postnatal day 18, a stage when active meiosis begins to occur. In mice fed a vitamin A depleted diet, where the testes were depleted of advanced germ cells, expression of this protein could not be detected, suggesting a biological relation of this receptor in male germ cell differentiation. Using a retinoic acid response element (RARE), the expression of this protein dramatically repressed both the basal and the retinoic acid (RA) regulated promoter activities of this reporter. Thus, this orphan receptor appears to play a role in modulating both the basic transcription machinery and the RA signaling pathway during embryogenesis and male germ cell differentiation (Lee et al., 1994).

4.3.5. hTAK1

A new member of orphan receptor family is TAK1, which is highly expressed in several tissues including testis. A 9.4-kb TAK1 transcript is expressed ubiquitously in human tissues but a 2.8-kb mRNA was largely restricted to the testis. Most of the sequences of TAK1 are identical to the sequence of the DNA binding domain of known members of the nuclear receptor family and included retinoic acid receptor-α (RAR-α), retinoid Xβ (RXR-β) receptor, thyroid hormone receptor and the human homologue of Nurr1. However, one fragment constituted a unique sequence not previously known. The predicated amino acid sequence suggested that this DNA fragment encodes the DNA binding domain of a novel member of the steroid hormone receptor super family, which was named TAK1.

Analysis of the *TAK1* gene revealed a long ORF that starts with a putative initiation codon at nucleotide 241 and stop codon at nucleotide 2029. TAK1 encodes a protein of 596 amino acid residues with a predicted mol wt of 66-kDa. In vitro translation of a full-length coding region of TAK1 cDNA revealed that the major protein migrated at approximately 65-kDa band on SDS-PAGE. Like other members of this family, the putative DNA binding domain (amino acid residues 116-182) of TAK1 contains nine cysteine residues, eight of which participate in the formation of two zinc fingers. The carboxyl terminus of the protein may contain a ligand-binding domain. The TAK1 exhibits a 64% identify with the human Tr2-II receptor from a human testis cDNA library. The greatest similarity between TAK1 and Tr2-11 occurs in the putative DNA binding domain in the region I, where 82% of 66 amino acids are identical in the ligand- binding domain. The amino acid residues in regions II and III between TAK1 and Tr2-II are respectively, 81% and 91% identical. Outside these three regions much homology between TAK1 and Tr2-11 does not exist. The N-terminal region and the region between the DNA and ligand binding domains exhibit only a 36% and 42% identity, respectively. Although the similarity between TAK1 and Tr2-11 is substantial, it is much lower than the similarity among members of the retinoic acid receptor family. The next best homology was observed between the DNA binding domains of TAK1 and COUPTFs (73% identity) and the RXR receptor (70% identity) (Hirose et al., 1994) **(Fig.4.3)**. The amino acid sequence of mouse TAK1 (mTAK1) is highly homologous to that of human TAK1, with an overall identity of 98%. The TAK1 transcript is predominantly expressed in pachytene spermatocytes, at a low level in round spermatids, but not in germ cells at earlier phases of spermatogenesis or in Sertoli cells. The TAK1 gene is located on human chromosome 3 (Hirose et al., 1995b).

4.3.6. Nuclear Receptor Co-activator

Transcriptional activities of nuclear receptors are modulated by co-activators and co-repressors. The amplified in breast cancer-3 protein (AIB3, also known as ASC-2, RAP250, PRIP, TRBP,

```
  1 MTSPSPRIQI ISTDSAVASP QRIQIVTDQQ TGQKIQIVTA VDASGSSKQQ FILTSPDGAG
 61 TGKVILASPE TSSAKQLIFT TSDNLVPGRI QIVTDSASVE RLLGKADVQR PQVVEYCVVC
121 GDKASGRHYG AVSCEGCKGF FKRSVRKNLT YSCRSSQDCI INKHHRNRCQ FCRLKKCLEM
181 GMKMESVQSE RKPFDVQREK PSNCAASTEK IYIRKDLRSP LIATPTFVAD KDGARQTGLL
241 DPGMLVNIQQ PLIREDGTVL LAADSKAETS QGALGTLANV VTSLANLSES LNNGDASEMQ
301 PEDQSASEIT RAFDTLAKAL NTTDSASPPS LADGIDASGG GSIHVISRDQ STPIIEVEGP
361 LLSDTHVTFK LTMPSPMPEY LNVHYICESA SRLLFLSMHW ARSIPAFQAL GQDCNTSLVR
421 ACWNELFTLG LAQCAQVMSL STILAAIVNH LQNSIQEDKL SGDRIKQVME HIWKLQEFCN
481 SMAKLDIDGY EYAYLKAIVL FSPDHPGLTG TSQIEKFQEK AQMELQDYVQ KTYSEDTYRL
541 ARILVRLPAL RLMSSNITEE LFFTGLIGNV SIDSIIPYIL KMETAEYNGQ ITGASL
```

Fig.4.3: Amino acid sequence of orphan nuclear receptor TAK homologue from mouse testis. Source: http//www.ncbi.nlm.nih.gov (accession number JC 4299).

and NCR) is a newly identified nuclear receptor co-activator that is amplified and over-expressed in breast cancers. The concentrations of AIB3 mRNA differ substantially in different tissues in a descending order from the following: testis, brain, thymus, white fat, pituitary, ovary, adrenal gland, lung, uterus, kidney, heart, skeletal muscle, liver, and virgin mammary gland. The AIB3 mRNA level in the testis is 165-fold higher than that in the virgin mammary gland (Zhang et al., 2003).

4.4. ACTION OF ESTROGENS

It is widely accepted that "estrogen" is a "female" hormone. However, it has been reported that estrogen is present in some male vertebrates and that it's receptors are expressed in the male reproductive organs. In mammals, estrogen appears to regulate the re-absorption of luminal fluid in the epididymis and its effect on sexual behavior has been a noticeable. The fact that the rete testis and the efferent ducts are target organs for oestrogen action led to the hypothesis that efferent duct ligation under oestrogen deficiency is the primary cause for testicular fluid accumulation and subsequent damage of the seminiferous epithelium. Estrogen inhibits expression of aquaprorin–1, a water channel protein, confirming the role of estrogen in fluid re-absorption. Hence it appears that estrogen does not act on the gametogenic epithelium directly. The newly discovered role of estrogen in the regulation of fluid re-absorption in the efferent ductules of testis has been reviewed (Hess, 2000). In vitro effects on spermatogonial stem cells and in vivo studies on Japanese eel of estradiol-17β (E_2) and spermatogenic cells in diethylstilbestrol (DES) treated newborn rats confirmed that estrogen is an indispensable "male hormone" in the early spermatogenetic cycle (Miura et al., 1999; Toyama et al., 2001). Further studies revealed the stimulation of rat gonocyte proliferation by estorgens in vitro (Li et al., 1997) and that aromatase inhibitors impaired the development of spermatids in monkeys (Shetty et al., 1997, 1998). Observations suggested that the mode of testosterone action on spermatogenesis / sperm transport is not only mediated by DHT but also by estradiol as a local factor involved in male mouse gametogenesis. Whether this also holds true for other species remains to be seen (Weinbauer and Wessels, 1999). Growth factors have been shown to synergize in estrogen signaling pathway, although the mechanism largely remains unknown. Kato et al., (2000) reviewed that the MAP-kinase activated by growth factors phosphorylates the Ser118 residue of the human ERα A/B domain and this phosphorylation potentiates the N-terminal trans-activation function (AF-1) of human ERα, indicating the possible molecular mechanism of a novel cross-talk between ER and growth factor signaling pathways.

4.4.1. Estrogen Receptors

Estrogen receptors (ERs) have been identified in Sertoli and Leydig cells but not in myoid and endothelial cells of rat and mouse testes. Less clear is the expression of ER in the human testis (Saati et al., 1993; c/r Ciocaa et al., 1995). Most of the ER-regulated genes are present in classic estrogen target tissues. The estrogen receptor(s) protein functions as a trans-acting regulatory factor (transcription factor) interacting with specific cis-acting regulatory sequences termed estrogen-responsive elements (EREs), which are located in the promoter (5'-flanking) regions of the target genes. Most of the genes regulated by estrogens respond within one or more hours after estrogen administration increasing their mRNA levels. Only a few, such as c-Myc and pS2 found in breast cancer cells, respond within 15 min after steroid administration. There are several other proteins, mainly enzymes that are modulated by estrogens, but at present we do not know with certainty whether they are activated via ER or by post-translational modifications. Ciocca et al (1995) implicated estrogen, estrogen receptor and hsp27 in a short feedback control mechanism of androgen production.

Estrogen receptor exists in two forms. ERα is found in all reproductive tissues and most abundant. However, ERβ is the product of a different gene, which does not bear much similarity with ERα. The ERα and ERβ both bind with oestradiol and interact with ERE sequences in DNA in target tissues. Both receptors show different transcriptional activity, ERβ being less active than ERα. Nonetheless, ERα is most abundant than ERβ. The ERβ transcripts have been identified in breast, ovary, uterus, thymus, lung, intestine and kidney as well as in male reproductive tissues including testis (reviewed in Hewitt and Korach, 2003). The ERβ has high homology and resembles many of the physiological characteristics of ERα. Both forms are present in epididymis. Estrogen receptor-α is expressed in Leydig cells and ERβ in Sertoli cells and in mitotic and meiotic germ cells (Saunders, 1998). Lambard et al., (2004) observed an ERα isoform variant that lacks exon 4 in human germ cells and P450 arom as a single band of 49-kDa in germ cells, as was reported for human ejaculated spermatozoa. The ERβ was seen at 50 and 60-kDa, which could correspond to the long and short forms of ERβ formed from the use of alternative start sites. Two proteins were also observed in human immature germ cells: one of the expected size (66-kDa) and a second one of 46-kDa. In mature spermatozoa, only the 46-kDa protein was observed (Lambard et al., 2004). Presence of receptor β in germ cells suggests an important role during gametogenesis. However, elimination of the receptor by gene targeting failed to alter fertility of male mice but provoked signs of prostate and bladder hyperplasia.

The gonocytes of rat fetal testis at 16 days post-coitum and testes of 4-day old animals, contained the ER-β mRNA in their cytoplasm and the ER-β protein in their nucleus. In testes of 11- and 15-day-old rats, ERβ mRNA and protein were detected in Sertoli cells and type A spermatogonia. No signal was found in other types of germ cells. In the adult testes, expression of ERβ mRNA as well as ERβ protein was found in pachytene spermatocytes from epithelial stages VII-XIV and in round spermatids from stages I-VIII. Low ERβ expression was observed in all type A spermatogonia, including undifferentiated A spermatogonia, whereas no expression was found in type B spermatogonia and early spermatocytes. At all ages, Sertoli cells showed a weak hybridization signal as well as weak immunoreactivity for ERβ. ERβ mRNA or protein was not detected in the interstitial tissue of adult rats indicating that Leydig cells as well as Sertoli cells and peritubular cells either do not express or express weak activity of ERβ (van Pelt et al., 1999).

Estrogen receptor α is more abundant in the efferent ductules of the testis than in any other tissue of the male or female. The structure of the efferent ductules is important, as these tubules are responsible for the re-absorption of almost 90% of the luminal rete testis fluid. It was reasonable to hypothesize that oestrogen receptors play a role in the regulation of fluid re-absorption in males. The oestrogen receptor α knockout mice indicated complete infertility in both male and female αERKO and αβERKO mice, whereas βERKO males exhibited normal fertility, and βERKO females exhibited overall decreased fertility, with three subsets of phenotypes in a continuous mating study (Hewitt and Korach, 2003). In this model, the efferent ductules are altered markedly from a re-absorptive epithelium to squamous epithelium devoid of lysosomes and endocytotic organelles. Although the separate roles for oestrogens and androgens in the regulation of fluid re-absorption are controversial and remain to be resolved, it is now evident that structure and function of efferent ductules are essential for fertility. Mechanisms involved in the regulation of water and ion transports in the male reproductive tract by estrogens need to be investigated (Hess, 2000).

4.4.2. P450 Aromatase

Early work on estrogen synthesis indicated that Leydig and Sertoli cells were the only important cells in the production of this steroid in the adult testis. Now it is clear that germ cells and spermatozoa, besides Leydig cells also contain aromatase and produce oestrogen. Testicular expression of P450 aromatase (P450arom), the enzyme that converts androgens into estrogens, has been shown in both somatic and germ cell types in several species. Human spermatozoa are the potential sites of estrogen biosynthesis (Aquila et al., 2002). P450arom is highly conserved among mammals and all vertebrates, and demonstrates 50-90% peptide sequence identity between fish and mammalian forms of the enzyme despite differences in gene size (Conley and Hinshelwood, 2001) **(Fig.4.4)**. The P450arom activity is present in germ cells (pachytene spermatocytes, and round and elongated spermatids), and that germ cells are a source of testicular estrogens (Levallet et al., 1998a,b). The sequence of PCR products from spermatozoa shares 98% identity with published human p450arom sequence. On Western blot with a mAb directed against aromatase, Lambard et al., (2003) detected two bands (53 and 49-kDa) in microsome preparations from purified spermatozoa, whereas in total extracts of purified spermatozoa (with and without cytoplasmic droplets), authors found the aromatase corresponding to a 49-kDa band with a stronger intensity when cytoplasmic droplets were present. The presence of aromatase in sperm initiated the idea that conversion of androgens into oestrogens in the duct system is used as a physiological signal for the quantity of sperm released from the testis and might act as a signal for the regulation of organ function. Aromatase deficiency was associated either with macro-orchidism or with smaller testes and impaired fertility. Aromatase activity is much more abundant in younger than in mature germ cells and the activity is two to four fold greater in spermatozoa.

The P450arom complex consists of two components: aromatase cytochrome P450 and, coupled to it, a ubiquitous flavoprotein, NADPH-cytochrome P450 reductase. P450arom is a member of P450 super family of enzymes, in a subgroup referred to as the steroid hydroxylases. The reaction mechanism of P450arom has been reviewed (Conley and Hinshelwood 2001). Typically cytochromes P450 incorporate molecular oxygen into substrates by reactions that are dependent on efficient electron transfer from donor molecules. However, the catalytic process leading to the aromatization of androgens is unusually complex. The reaction is thought to involve two consecutive hydroxylations at the C19 methyl group (19-hydroxylase activity) of the steroid substrate, forming 19-hydroxy and eventually 19-oxo-intermediates. A third oxidative event, which is still debated culminates in the cleavage of the angular C19 methyl group (by desmolase activity) and estrogen formation. Androstenedione and testosterone are the most common, physiologically important substrates for P450arom.

```
  1 MVLEMLNPIH YNITSIVPEA MPAATMPVLL LTGLFLLVWN YEGTSSIPGP GYCMGIGPLI
 61 SHGRFLWMGI GSACNYYNRV YGEFMRVWIS GEETLIISKS SSMFHIMKHN HYSSRFGSKL
121 GLQCIGMHEK GIIFNNNPEL WKTTRPFFMK ALSGPGLVRM VTVCAESLKT HLDRLEEVTN
181 ESGYVDVLTL LRRVMLDTSN TLFLRIPLDE SAIVVKIQGY FDAWQALLIK PDIFFKISWL
241 YKKYEKSVKD LKDAIEVLIA EKRRRISTEE KLEECMDFAT ELILAEKRGD LTRENVNQCI
301 LEMLIAAPDT MSVSLFFMLF LIAKHPNVEE AIIKEIQTVI GERDIKIDDI QKLKVMENFI
361 YESMRYQPVV DLVMRKALED DVIDGYPVKK GTNIILNIGR MHRLEFFPKP NEFTLENFAK
421 NVPYRYFQPF GFGPRGCAGK YIAMVMMKAI LVTLLRRFHV KTLQGQCVES IQKIHDLSLH
481 PDETKNMLEM IFTPRNSDRC LEH
```

Fig.4.4. Amino acid sequence of human aromatase.Source: http//www.ncbi.nlm.nih.gov (accession number 112503).

4.4.3. P450 Arom Gene (*CYP19)*

The aromatase gene is unique and its expression is regulated in a tissue and a cell-specific manner via the alternative use of various promoters located in the first exon I. Aromatase enzyme is the product of *CYP19* gene, which encode aromatase in most of the mammals. Although there may be multiple first exons that are involved in tissue-specific expression, these exons are untranslated in most species. Therefore, P450arom expressed in different tissues represents only one protein with one amino acid sequence. High conservation is particularly evident in the gonad- and brain-specific first exons and promoters for which the cDNAs and genes from many mammalian species have been cloned (Hinshelwood et al., 1997; 2000). The human *CYP19* gene extends over 70-kb, whereas the medaka *CYP19* gene only 2.6-kb. Partial sequence analysis indicated that the pig gene and other mammalian genes use the same splice sites as the human gene, even in the 5'-untranslated region of the transcript. Expression of P450arom in adult rat germ cells gave cDNA products of 334-bp in length that corresponded to the size expected from the final nested amplification. Rat genomic DNA yielded a product for P450arom of approximately 2000-bp, suggesting that the rat gene contains an intron in the amplified region, as in human P450arom gene. The mouse *Cyp19* gene includes three different promoters that specifically direct expression in ovary, testis, and brain (Golovine et al., 2003).

Testosterone (T) enhances *CYP19* gene expression in Leydig cells and germ cells, and augments the estradiol outputs. The non-aromatizable androgen 5α-DHT induces the same effect as T on P450 aromatase (P450arom) gene expression but was inefficient on the estradiol output. In rat germ cells, the aromatase gene expression is not under androgen control but subjected to cytokine, TNFα and growth factor, TGFβ regulation (Carreau et al., 2001). In germ cells, TNFα likely enhances expression of aromatase through promoter PI.4 in primary spermatocytes, possibly via an AP1 site upstream the GAS element, while in round spermatids TNFα requires glucocorticoids as a co-stimulator to increase CYP19 gene expression. Using an in vitro model of mature rat Leydig cells, pachytene spermatocytes and round spermatids, it has been suggested that several factors direct the expression of the aromatase gene and it is obvious that not only promoter PII but also promoter PI.4 are concerned (Bourguiba et al., 2003).

Two Isoforms of P450arom in Rat Germ Cells: Two unusual isoforms (Ex10-S and INT) of P450arom are expressed in rat germ cells. These sequences demonstrated that an alternative splicing event occurred first at the exon-intron boundary of the GT consensus sequence of the last coding exon, and second in the internal 5' donor inside exon 9 used as a minor cryptic splicing site. These isoforms lacked the last coding exon, which contained the heme-binding domain; in addition, for the Ex10-S transcript, the catalytic domain was also absent because of a frame shift in the open reading frame. The truncated P45-arom polypeptides without the heme-binding domain were unable to convert androgens into estrogens. Adult rat germ cells are able to express P450arom mRNA, which is then translated into a biologically active enzyme, involved in estrogen production. Alternative splicing events of P450arom mRNA exist in pachytene spermatocytes and round spermatids, which probably cannot encode functional aromatase molecules (Levallet et al., 1998).

4.4.4. P450arom Deficiency

Human aromatase deficiency is a rare disorder and is usually caused by single base-pair changes resulting in amino acid substitution or premature stop codons resulting in virilization in the third trimester of pregnancy. The mice deficient in aromatase developed disruptions to spermatogenesis despite normal levels of androgens. Spermatogenesis primarily was arrested at early spermatogenic stages, as characterized by an increase in apoptosis and the appearance of multinucleated cells. There was a significant reduction in round and elongated spermatids with out any change in Sertoli cells and early germ cells. It appeared that local expression of aromatase is essential for spermatogenesis and provides evidence for a direct action of estrogen on male germ cell development (Robertson et al., 1999). The aromatase knockout male mouse (ArKO) has shortened femur length and bone mineralization. Male ArKO mice testes demonstrated arrest of spermatogenesis at the level of round spermatids and Leydig cell hyperplasia. ArKO mice also exhibited evidence of insulin resistance and visceral adiposity (Bilinska et al., 2000; Simpson, 2000; Conley and Hinshelwood, 2001).

4.5. OTHER STEROIDS LINKED ENZYMES IN GERM CELLS

Normal reproductive development in mammals is critically dependent on the regulated biosynthesis of sex steroids. Many diseases and conditions affecting fertility and general health are accompanied by aberrations in androgen or estrogen metabolism. Therefore, the balance between androgen and estrogen production is not only essential for normal sexual development and reproduction, but also for the normal growth and physiological well being of both sexes.

4.5.1. Testis-Specific Lanosterol 14α-Demethylase (CYP51)

In contrast to the majority of tissues in which cholesterol is the main product, male and female gonads over-produce two late intermediates of the cholesterol biosynthetic pathway: follicular fluid meiosis activating sterol (FF-MAS; 4,4-dimethyl-5α-cholesta-8, 14, 24-triene-3β-ol) and testis meiosis activating sterol (T-MAS; 4,4-dimethyl-5α-cholesta-8, 24-diene-3β-ol) (Byskov et al. 1999; Yoshida et al., 1996). Both tissue-isolated FF-MAS, the product of lanosterol 14α-demethylase (CYP51) and chemically synthesized FF-MAS (Grondahl et al. 1998), and T-MAS, the product of sterol 14α-reductase, stimulate reinitiation of meiosis in mouse oocytes in vitro and are believed to have important, but not yet fully understood roles in fertilization (Byskov et al. 1999; Grondahl et al. 1998; Yoshida et al., 1996; Debeljak et al., 2003). The CYP51 removes 14α methyl group from lanosterol forming 4,4 dimethyl 5α-cholesta-8, 14, 24-triene 3β-ol (FF-MAS). At least six enzymatic steps after CYP51 are needed for the production of cholesterol from its precursors **(Fig.4.5)**. Human sperm lysates can synthesize cholesterol in vitro but the activity is too low to contribute free cholesterol content of the sperm (Cross, 1998).

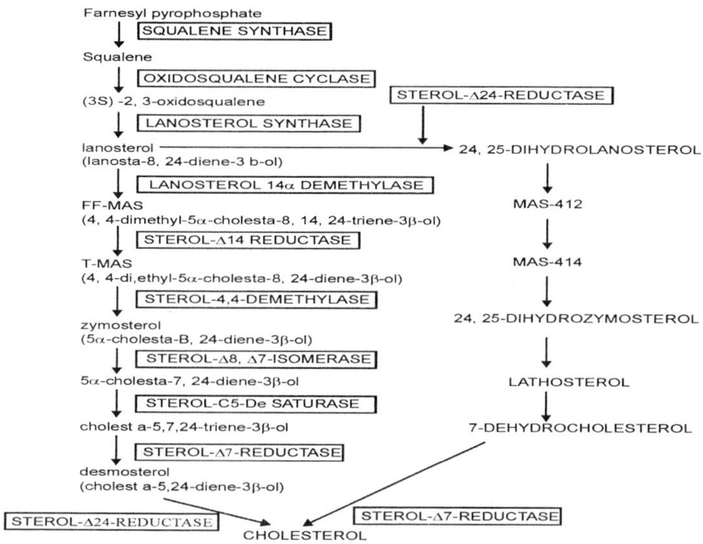

Fig.4.5. Formation of meiosis-activating sterol (FF-MAS), T-MAS, MAS-412, and MAS-414 from lanosterol and relating enzymes during cholesterol biosynthesis in post-meiotic male germ cells.

Lanosterol 14α-demethylase is a cytochrome P450 (CYP51) enzyme of the cholesterol biosynthetic pathway belonging to the *CYP51* gene family, which is evolutionarily conserved member of the CYP super family. Mammalian *CYP51* genes are unique in sharing several common characteristics: highly conserved exon/intron borders and proximal promoter structures, ubiquitous expression at the highest level in the testis, and appearance of testis-specific transcripts that arise from differential polyadenylation site usage. The CYP51 enzyme of eukaryotes catalyzes the 14α-demethylation of 14-methylsterol, which is an essential step in sterol synthesis. The CYP51 oxidatively demethylates substrates in the presence of NADPH, molecular oxygen and the microsomal NADPH-cytochrome p450 reductase (Shyadehi et al. 1996). Lanosterol is synthesized from squalene through sequential epoxidation and cyclization reaction that are catalyzed by squalene epoxidase and oxidosqualene cyclase **(Fig.4.5)**.

Ontogeny and Localization in Testis: Testis-specific *CYP51* mRNA arises from the use of an upstream polyadenylation site and is restricted to germ cells, being most abundant in elongating spermatids in stages VII-XIV whereas somatic *CYP51* transcripts are present in all cells. In contrast, the mRNA levels of squalene synthase (an enzyme preceding CYP51 in the pathway) are maximal in round spermatids, and no germ cell-specific transcript is observed. The rat male germ cell specific *CYP51* transcript is translated in vitro to two proteins of 55 and 53.5-kDa. The CYP51 activity is higher in protein extracts of testes and germ cells of sexually mature rats than in prepubertal animals, in which postmeiotic germ cells are not yet present. This shows increased capacity for the production of T-MAS sterols by male germ cells that have already completed meiosis, suggesting that they serve a role different from meiosis activation (Stromstedt et al., 1998). A comparative study of CYP51, NADPH P450-reductase and squalene synthase was made during spermatogenesis. The CYP51 was detected in step 3-19 spermatids where P450 reductase was also present. Squalene synthase was present at stage 2-15 spermatids indicating that squalene synthase and CYP51 proteins are not equally expressed in same stages of spermatogenesis. The study provided evidence that step 19 spermatids and residual bodies of the rat testis have the capacity to produce T-MAS sterols in situ (Majdic et al., 2000).

The CYP51 has been unequivocally localized to acrosomal membranes of male germ cells in mouse, bull, and ram, in which it synthesizes FF-MAS in the presence of the acrosomal form of $NADPH_2$-P450 reductase. The CYP51 is localized on inner and outer acrosomal membranes of male germ cells, the round and elongated spermatids. Significance of CYP51 localization on the acrosome, which is a Golgi-derived organelle, is not known. But it was proposed that CYP51-formed FF-MAS could function as a signaling sterol during fertilization (Cotman et al., 2001). In the mouse, CYP51 (53-kDa) resides in endoplasmic reticulum (ER) and Golgi during all phases of acrosome development, indicating an intracellular transport from ERs through the Golgi to the acrosome. Demethylation of lanosterol to FF-MAS by the acrosomal lanosterol 14α-demethylase enzyme complex demonstrated the ability of ejaculate sperm to synthesize meiosis-activating sterols. While sterol regulatory element binding protein (SREBP)-dependent transcriptional regulation of CYP51 contributes to synthesis of cholesterol, the germ-cell-specific cAMP/CREMτ -dependent upregulation might contribute to increased production of MAS (Debeljak et al., 2003).

Characteization and Regulation of CYP51: The amino acid sequences of the rat and human CYP51 enzymes exhibit 93% identity and 35-42% homology to those of lower eukaryotic CYP51s. Human *CYP51* mRNA is expressed in all tissues, with particularly high levels in testis. In addition to the functional *CYP51* gene, intronless *CYP51* pseudogenes are found in human and rat. Processed pseudogenes arise from reverse transcription of mRNA in germ cells followed by their random incorporation into the genome. Only genes highly expressed in germ cells have such intronless gene copies, and discovery of *CYP51*-processed pseudogenes suggested that the high level of CYP51 expression in testis includes expression in germ cells.

In contrast to the liver, regulation of pre- and post MAS gene differs in the testis. Genes encoding pre-MAS enzymes [HMG-CoA synthase (SYN), HMG-CoA reductase (RED), farenesyl diphosphate synthase (FPP), squalene synthase (SS), and CYP51] are upregulated during sexual development of the testis, although not all genes are turned on at the same time. Furthermore, two post-MAS genes, C-4 sterol methyl oxidase and sterol Δ7-reductase, are expressed at low levels and are not upregulated either in rat or human. As a cholesterogenic gene, *CYP51* is regulated by a sterol/sterol regulatory element binding proteins (SREBPs) dependent pathway in somatic tissues. The SREBPs are basic helix-loop-helix (bHLH)-Zip proteins, which are synthesized as membrane-bound precursors, are cleaved to form a soluble, transcriptionally active mature SREBP that regulates the promoters for genes involved in lipid synthesis. Homeostasis is controlled by sterol feedback inhibition of this maturation process. Although, the expression of SREBP genes occurs in male germ cells, the role of these proteins in the control of sterol regulatory elements during spermatogenesis has been unclear. In contrary to SREBP-dependent gene regulation in somatic cells, expression of some of these genes in male germ cells does not depend on sterols (Wang et al., 2002). A novel isoform of SREBP2, called SREBP2gc is highly enriched in rat and mouse spermatogenic cells. The SREBP2gc is expressed in a stage-dependent manner as a soluble, constitutively active transcription factor that is not under feedback control of sterols. These findings likely explain the apparent sterol-insensitive expression of lipid synthesis genes during spermatogenesis (Wang et al., 2002).

In addition, male germ cells contain alternative routes to control expression of cholesterogenic genes (Rozman et al., 1999). Promoters of the human and rat *CYP51* genes also contain SRE and CRE elements, which indicates the presence of two main regulatory routes:

```
1     GCGGCGGCTGTTTGCTAGTGTACAGCAGCCGGCCTGTAAACTAAGTGTTCTGCAGGTGTTGGTTGACTGAGTGGAGTTTAAGGTTCCTGG
90    AACAGAGGTTAACTTTATGCTGTACGGTCAGCATAAAGTCTTTCTACAGTTGTCTTGGCAAAGTTCTACTTGTGCAGGGCTGGACTGTAC

1                                                                       M  I  W  F  S  T  A  V  L  I  G
180   TACCTTGCAGAGCTGATTGAAGAGTACACGGTGGGCACCAGCAGAATCATCAAATACATGATCTGGTTCTGCACAGCAGTGCTGATTGGC

12    L  Y  V  F  E  R  F  P  T  S  M  I  G  V  G  L  F  T  N  L  V  Y  F  C  L  L  Q  T  F  P
270   CTCTACGTCTTTGAGCGCTTCCCCACCAGCATGATTGGCGTGGGCCTTTTCACCAACCTGGTCTACTTTGGCTTCTCCAGACCTTCGGC

42    F  I  M  L  T  S  P  N  F  I  L  S  C  G  L  V  V  V  N  H  Y  L  A  F  Q  F  F  A  E  E
360   TTCATCATGCTGACATCACCTAACTTCATCCTGTCATGCGGGCTAGTGGTGGTGAACCATTACCTGGCATTTCAGTTTTTTGCGGAAGAA

72    Y  Y  P  F  S  E  V  L  A  Y  F  T  F  C  L  C  A  G  S  S  Y  R  H  M  R  N  M  K  G  L
450   TATTATCCTTTCTCTGAGGTGCTGGCCTACTTCACATTCTGCCTGTGTGCTGGGAGCTCCTACAGACATATGCGGAACATGAAAGGACTG

102   R  H  Q  A  V  L  A  I  G  Q  E  L  N  R  R  A  L  G  D  P  S  P  G  W  M  G  Q  V  R  R
540   AGGCATCAAGCTGTGCTAGCCATTGGCCAAGAGCTGAACCGGAGAGCCCTAGGGGACCCCAGTCCTGGGTGGATGGGTCAGGTCCGGCGT

132   R  S  S  L  L  G  S  Q  L  E  A  T  L  Y  S  D  Q  E  L  S  Y  I  Q  Q  G  E  E  A  M  Q
630   CGGAGCTCTCTACTTGGTTCTCAACTGGAAGCAACACTCTACAGTGACCAGGAGCTGTCCTACATCCAGCAAGGAGAGGAAGCTATGCAA

162   K  A  L  G  I  L  N  N  Q  E  G  W  K  K  E  S  Q  Q  E  N  G  D  E  V  L  S  K  V  V  P
720   AAGGCCTTGGGCATACTCAACAACCAGGAAGGCTGGAAGAAGGAAAGCCAGCAGGAGAATGGAGATGAAGTGCTAAGTAAGGTGGTGCCA

192   G  V  G  K  V  F  R  L  E  V  L  L  D  Q  P  M  D  R  L  Y  E  E  L  V  D  R  M  E  A  M
810   GGTGTGGGCAAGGTGTTCCGACTGGAGGTGCTGCTAGACCAGCCCATGGACAGACTCTATGAAGAACTGGTGGACCGCATGGAGGCCATG

222   G  E  W  N  P  N  V  K  E  I  K  V  L  K  K  I  G  K  D  T  V  I  T  H  E  L  A  A  A  A
900   GGAGAGTGGAACCCAAATGTCAAGGAAATCAAGGTCCTGAAGAAGATTGGAAAAGACACAGTCATCACGCATGAGCTGGCTGCAGCAGCA

252   R  G  N  L  V  G  P  R  D  F  V  S  V  R  C  T  K  R  R  G  S  T  C  V  L  A  G  M  A  T
990   CGAGGCAACCTGGTGGGGCCGCGAGACTTCGTAAGCGTACGCTGTACCAAGCGTAGAGGTTCCACCTGTGTGCTGGCAGGCATGGCCACA

282   H  F  G  E  M  P  E  Q  R  G  V  I  R  A  E  H  G  P  T  C  M  V  L  H  P  L  A  G  S  P
1080  CACTTTGGGGAGATGCCTGAGCAAAGAGGTGTCATCAGAGCTGAACACGGTCCCACCTGCATGGTGCTTCATCCACTGGCTGGAAGTCCC

312   S  K  T  K  L  T  W  L  L  S  I  D  L  K  V  W  L  P  K  T  I  I  N  Q  V  L  S  Q  T  Q
1170  TCAAAGACCAAACTCACGTGGCTGCTCAGTATTGACCTCAAGGTGTGGCTGCCAAAGACCATCATCAACCAAGTCTTATCACAGACCCAG

342   I  E  F  A  S  H  L  R  K  R  L  E  S  S  P  A  S  E  A  Q  C  *
1260  ATAGAGTTCGCCAGCCACTTGCGCAAGCGCTTGGAGTCCAGCCCTGCCTCTGAGGCCCAGTGTTAAGGACTGCCCACCACATCTACCTGC
1350  AGGCCATTGGAAGTTCTCACAGGAAGTCTGCAAGTCTGTTCACCTTCAGCCAAGGACAACGAGAGGGGTAGTAGTCATTAGACACTAGGA
1440 .CTGACTGGTAAAAACGTACTACAGGAAAAACAGAACTGAGGCTTACAATAAAGGTTCTGCAGTGTCTAAAAAAAAAAAAAAAAAAAAAA
1530  AAAAAAAAAAAAAAAAAAAAAAAAAAAAAA
```

Fig.4.6. cDNA and deduced amino acid sequences of rat StAR from rat testis. The polyadenylation signal is underlined. Reproduced with permission from Lee et al, Biochem Biophys Res Commun 230; 528-32; 1997 © Elsevier.

the sterol dependent regulation and the cAMP-dependent regulation of CYP51 genes. While feedback regulation by sterols is characteristic for all genes involved in cholesterol biosynthesis and homeostasis, the cAMP-dependent regulation is unique, indicating that CYP51 may play tissue-specific roles distinct from cholesterol biosynthesis (Rozman, 2000). In testis, cAMP/CREMτ-dependent regulation of *CYP51* predominates, leading to increased levels of shortened *CYP51* mRNA transcripts. CREM -/- mice lack the abundant germ cell-specific *CYP51* mRNAs in testis while expression of somatic *CYP51* transcripts is unaffected. The mRNA levels of squalene synthase in testis of CREM-/- mice remained unchanged as compared with wild-type animals, showing that regulation by CREMτ is not characteristic for all cholesterogenic genes expressed during spermatogenesis. The –334/+314-bp CYP51 region can mediate both sterol/SREBP dependent as well as cAMP/CREMτ-dependent transcriptional activation. The cAMP-dependent transcriptional activator CREMτ from germ cell nuclear extracts binds to a conserved CYP51 CRE2 element while no SREBP-1 binding is observed in germ cells. The two regulatory pathways, mediating expression of *CYP51* describe this gene as a cholesterogenic gene and also as a haploid expressed gene. Besides CREM, other unknown transcription factors control expression of individual cholesterogenic genes during spermatogenesis. Since T-MAS in testis increases 8 fold during development where as MAS is hardly detected in liver, it seems that the lack of a coordinate transcriptional control over the cholesterol biosynthetic pathway

```
1                                                      Met Trp Phe Met Tyr Val
1                         CGCGCAGGCCCGTCCCGGCGGCCCGGGGGGCCC ATG TGG TTC ATG TAC GTG

7    Leu Ser Trp Leu Ser Leu Phe Ile Gln Val Ala Phe Ile Thr Leu Ala Val Ala Ala
52   CTG AGC TGG CTG TCG CTG TTC ATC CAG GTG GCA TTC ATC ACT CTG GCC GTC GCG GCT

26   Gly Leu Tyr Tyr Leu Ala Glu Leu Ile Glu Glu Tyr Thr Val Ala Thr Ser Arg Ile
109  GGA CTG TAC TAC CTT GCA GAG CTG ATA GAA GAG TAC ACG GTG GCC ACC AGC CGA ATC

45   Ile Lys Tyr Met Ile Trp Phe Ser Thr Ala Val Leu Ile Gly Leu Tyr Val Phe Glu
166  ATC AAA TAC ATG ATC TGG TTC TCC ACA GCA GTG CTG ATC GGC CTC TAC GTC TTT GAG

64   Arg Phe Pro Thr Ser Met Ile Gly Val Gly Leu Phe Thr Asn Leu Val Tyr Phe Gly
223  CGT TTC CCC ACC AGC ATG ATT GGC GTG GGC CTT TTC ACC AAC CTC GTC TAC TTT GGC

83   Leu Leu Gln Thr Phe Pro Phe Ile Met Leu Thr Ser Pro Asn Phe Ile Leu Ser Cys
280  CTT CTC CAG ACC TTC CCC TTC ATC ATG CTG ACG TCA CCT AAC TTC ATC CTG TCA TGC

102  Gly Leu Val Val Val Asn His Tyr Leu Ala Phe Gln Phe Phe Ala Glu Glu Tyr Tyr
337  GGG CTA GTG GTG GTG AAT CAT TAC CTG GCA TTT CAG TTT TTT GCG/GAA GAG TAT TAT

121  Pro Phe Ser Glu Val Leu Ala Tyr Phe Thr Phe Cys Leu Trp Ile Ile Pro Phe Ala
394  CCT TTC TCT GAG GTC CTG GCC TAC TTC ACA TTC TGC CTG TGG ATA ATC CCG TTT GCT

140  Phe Phe Val Ser Leu Ser Ala Gly Glu Asn Val Leu Pro Ser Thr Met Gln Pro Gly
451  TTC TTC GTG TCA CTT TCG GCT GGG GAG AAT GTC CTG CCC TCC ACC ATG CAG CCA GGC

159  Asp Asp Val Val Ser Asn Tyr Phe Thr Lys Gly Lys Arg Gly Lys Arg Leu Gly Ile
508  GAT GAC GTG GTC TCC AAT TAC TTC ACC AAA GGC AAG CGA GGC AAG CGC TTA GGC ATC

178  Leu Val Val Phe Ser Phe Ile Lys Glu Ala Ile Leu Pro Ser Arg Gln Lys Ile Tyr
565  CTG GTT GTC TTC TCC TTC ATC AAA GAG GCC ATC CTA CCC AGT CGG CAG AAG ATA TAC

662  TGA CCCTTTGGGGAGGGCATGTGGGGGGCAAGACCAAGAGAGCCAGGCCCCTGGGACTCTGGGCTAATCGTGCC
696  TGGGAGCCTGGAAGGTGTCTTCTGCCAGCCCAGGTCTCCCTCCCCTCTGTTCTGAAGCCTCGTTCCCCACCTCCT
771  CAGGAGGCTTGAATGCTGCCAGGGGCACCAGAGAGGGTAGCAGAGCCTGCTATGCTAGGAGGCTATGTCCCCAGC
846  CCTCCCTTTCCCCAGAGTTGGGTCTGGTTGGATGGGGCAGGCAGGGAAGCTGGTATGAAGCTGGGGAACTGACAG
921  CAAGTTCTCAGGCAGGTGACCGTAGGGAGGAGCTGAGGGTTGGCACTTCCAGGAAAAGTGCATTTTGCTGTTGAA
996  ATGTGGGTCCTGTCCAATGCTGATTCCCTTTCGAATAAAGATTCTAGGTGGCACTGTTTGCCTTAATACCCCAGG
1071 AGTTCATCTTCTTCTCTGCCTGCTAACCTTCTGCCTAGACTAGCTAGCCTGGCTCTAGGGACTCTTCTGCTCTTG
1146 CCCTTCTGGAGGGATTCTCATGGCCCTAAGTGGAAGGGAACAGGGATGAGAATGAGTGGGCAAAGGCAGTTTTT
1221 CTGAGGCTTCTCCTCCTCTCCGTGTCTCCTTCCCCAGACTGTGATAGTCCAGGGTGGGGAGGGGTAGAAAGTCT
1296 GTAGCTGGGTTAGAGATGTAGACCATCTACAATACCCTATTTATGTCTTCCTTTGAGAAAAAGGAGAGATTTCTT
1371 GGTAAATTCTATATCAGGCTCGGTGAAAAAAAAAAAAAAA
```

Fig.4.7. Nucleotide and deduced amino acid sequences of mouse *Tex261*. The region of homology with rat StAR cDNA is at nt106 to 433. The polyadenylation signal is underlined. Reproduced with permission from Lopez-Fernandez et al, Biochem Biophys Res Commun 242; 565-69; 1998 © Elsevier.

contributes importantly to over-production of the signaling sterol T-MAS in testis (Tracer et al., 2002).

4.5.2. Steroidogenic Acute Regulatory (StAR) Protein and Homologues

In steroidogenic cells, during acute regulation of steroid hormone biosynthesis in response to the trophic hormone, the immediate effect is to transport cytosolic cholesterol to the matrix side of mitochondrial inner membrane where cholesterol is transformed into pregnenolone by a cholesterol side chain cleavage enzyme. A 30-kDa phosphoprotein is candidate protein responsible for acute control of steroid hormone biosynthesis. The cDNA for the 30-kDa-proteins, now called steroidogenic acute regulatory (StAR) protein, has been cloned first in mouse and later in human, bovine, ovine and rat (c/r Lee et al., 1997). The multiple alignment of these deduced amino acid sequences revealed that they have 85%-88% identity and more than 90% similarity. The deduced amino acid sequence from rat StAR clone has an additional 86 amino acid stretch at amino terminus when it was compared with those sequences in other species. The other part of the amino acid sequence has 94% identity to mouse StAR protein sequence. Three transcripts (1.3-b, 1.6-kb, and 3.5-kb), which hybridized to the clone, were detected in testis, ovary and adrenal gland. The cDNA expressed 30-kDa and 47-kDa proteins in COS1 cells (Lee et. al., 1997) **(Fig. 4.6)**. It has been proposed that IL-1α, which is constitutively produced by the rat testis, is an important paracrine regulator of postnatal Leydig cell maturation. Regulation of StAR protein expression is one of the possible mechanisms by which IL-1α contributes to the differentiation of immature Leydig cells into adult cells (Svechnikov et al., 2003).

Tex261: The *Tex261* is another *StAR* like gene cloned from a subtractive cDNA library from 10-day postnatal mouse testis. Tex261 and StAR are differentially localized presumably with different functions. The *Tex261* transcribes three mRNAs of 3.5, 1.6, and 1.4-kb. The 3.5 and 1.4-kb transcripts are expressed in different gonadal and somatic tissues where as the 1.6-kb transcript is only detected in testis and differentially regulated during development. This 1.6-kb mRNA is highly expressed in adult testis, beginning at 15 days of postnatal life, which coincides with the presence of pachytene cells in prepuberal mouse. A region of 86 amino acids from the predicted Tex261 has been reported as a part of the steroidogenic acute regulatory protein *StAR* gene by its sequence identity to the rat *StAR* cloned cDNA **(Fig.4.7)**. Lopez-Fernandez et al., (1998) demonstrated that, in the mouse, *StAR* and *Tex261* are two different genes with different functions, yet a high identity (43%) at amino acid level is detected in a region of 153 amino acids corresponding to a transmembrane protein.

4.5.3. T-StAR/ETOILE

RNA binding motif (RBM) is an RNA-binding protein encoded on the Y-chromosome in mammals and is expressed only in the nuclei of male germ cells (see chapters 10 and 14). Of a number of potential partners for RBM identified, the most frequent isolate encoded is a RNA-binding protein, termed T-StAR that is closely related to SAM68, a Src-associated protein of unknown function. The mouse homologue was also cloned and designated ETOILE. It mapped to chromosome 15, while T-StAR mapped to the syntenic region on human chromosome 8. The T-StAR/etoile is expressed primarily in the testis. In rat germ cells, the expression of both T-StAR/etoile and SAM68 is regulated during meiosis. Transfection of T-StAR/etoile caused an accumulation of protein in the compartment of the nucleus, adjacent to the nucleolus but distinct from the peri-nucleolar compartment. The RBM and other hnRNP G family members are candidate downstream targets for regulation by T-StAR/etoile and SAM68 (Venables et al., 1999).

4.6. REFERENCES

Akmal KM, Dufour JM, Kim KH. Retinoic acid receptor α gene expression in the rat testis: Potential role during the prophase of meiosis and in the transition from round to elongating spermatids. Biol Reprod 1997; 56: 549-56.

Aquila S, Sisci D, Gentile M, Middea E, Siciliano L, Ando S. Human ejaculated spermatozoa contain active P450 aromatase. J Clin Endocrinol Metab 2002; 87:3385-90.

Bilinska B, Schmalz-Rraczek B, Sadowska J, Carreau S. Localization of cytochrome P450 aromatase and estrogen receptors α and β in testicular cells—an Immunohistochemical study of the bank vole. Acta Histochem 2000; 102: 167-81.

Blaner WS, Galdieri M, Goodman DS. Distribution and levels of cellular retinol and cellular retinoic acid-binding protein in various types of rat testis cells. Biol Reprod 1987; 36: 130-7.

Bourguiba S, Genissel C, Lambard S, Bouraima H, Carreau S. Regulation of aromatase gene expression in

Leydig cells and germ cells. J Steroid Biochem Mol Biol 2003; 86: 335-43.
Braun KW, Tribley WA, Griswold MD, Kim KH. Follicle stimulating hormone inhibits all-trans-retinoic acid-induced retinoic acid receptor α nuclear localization and transcription activation in mouse Sertoli cell lines. J Biol Chem 2000; 275: 45-51.
Brocard J, Kastner P, Chambon P. Two novel RXRα isoforms from mouse testis. Biochem Biophy Res Comm 1996; 229: 211-18.
Byskov AG, Andersen CY, Nordholm L, Thogersen H, Guoliang X, Wassman O, Guddal JVAE, Roed T. Chemical structure of sterols that activate oocyte meiosis. Nature 1995; 374: 559-562.
Carreau S, Bourguiba S, Lambard S, Galeraud-Denis I, Genissel C, Levallet J.Reproductive system: aromatase and estrogens. Mol Cell Endocrinol 2002;193:137-43.
Chen F, Conney AJ, Wang Y, Law SW. and Malley BW. Cloning of a novel orphan receptor (GCNF) expressed during germ cell development. Mol Endocrinol 1994; 8: 1434-44.
Ciocca DR, Vargas Roig LM. Estrogen receptors in human non-target tissues: biological and clinical implications. Endocrine Rev 1995; 16: 35-62.
Conley A, Hinshelwood M. Mammalian aromatases. Reproduction 2001; 121: 685-95.
Cooney AJ, Hummelke GC, Herman T, Chen F, Jackson KJ. Germ cell nuclear factor is a response element-specific repressor of transcription. Biochem Biophys Res Commun 1998; 245: 94-100.
Cotman M, Jezek D, Fon Tacer K, Frangez R, Rozman D. A functional cytochrome P450 lanosterol 14 alpha-demethylase CYP51 enzyme in the acrosome: transport through the Golgi and synthesis of meiosis-activating sterols. Endocrinology. 2004;145:1419-26.
Cotman M, Rozma D, Banek L, Jezek D. Localization of Lanosterol 14α-demethylase in round and elongated spermatids of the mouse testis: an immunoelectron microscopic and stereological study. Pflugers Arch 2001; 442 (6 Suppl 1): 167-8.
Cross NL. Role of cholesterol in sperm capacitation. Biol Reprod 1998; 59: 7-11.
Debeljak N, Fink M, Rozman D. Many facets of mammalian lanosterol 14alpha-demethylase from the evolutionarily conserved cytochrome P450 family CYP51. Arch Biochem Biophys. 2003;409:159-71.
Debeljak N, Horvat S, Vouk K, Lee M, Rozman D. Characterization of the mouse lanosterol 14α-demethylase (CYP51), a new member of the evolutionarily most conserved cytochrome P450 family. Arch Biochem Biophys 2000; 379: 37-45.
Gaemers IC, Van Pelt AMM, van der Saag SPT, De Rooiij DG. All-trns-4-oxo-retinoic acid: a potent inducer of in vivo proliferation of growth arrested A spermatogonia in the vitamin A deficient mouse testis. Endocrinology 1996; 137: 479-85.
Gaemers IC, van Pelt AMM, van der Saag, PT, Hoogerbrugge JW, Themmen APN, de Rooij DG. Differential expression pattern of retinoid X receptors in adult murine testicular cells implies varying roles for these receptors in spermatogenesis. Biol Reprod 1998; 58: 1351-56.
Golovine K, Schwerin M, Vanselow J. Three different promoters control expression of the aromatase cytochrome p450 gene (cyp19) in mouse gonads and brain. Biol Reprod. 2003; 68 : 978-84.
Grondahl C, Ottesen JL, Lessl M, Faarup P, Murray A, Gronvald FC, Hegelehartung C, Ahnfeltronne I. Meiosis–activating sterol promotes resumption of meiosis in mouse oocytes cultured in vitro in contrast to related oxysterols. Biol Reprod 1998; 58: 1297-1302.
Hess RA. Oestrogen in fluid transport in efferent ducts of the male reproductive tract. Rev Reprod 2000;5: 84-92.
Hewitt SC, Korach KS. Oestrogen receptor knockout mice: roles for oestrogen receptors α and β in reproductive tissues. Reproduction 2003; 125: 143-49.
Hinshelwood MM, Michael MD, Simpson ER. The 5'-flanking region of the ovarian promoter of the bovine CYP19 gene contains a deletion in a cyclic adenosine 3'-5' monophosphate-like responsive sequence Endocrinology 1997; 138: 3704-10.
Hinshelwood MM, Smith ME, Murry BA, Mendelson CR. A 278 bp region just upstream of the human CYP19 aromatase gene mediates ovary-specific expression in transgenic mice Endocrinology 2000;14:2050-53.
Hirose T, Fujimoto W, Yamaai T, Kim KH, Matsuura H, Jettten AM. Tak1: molecular cloning and characterization of a new member of the nuclear receptor superfamily. Mol Endocrinol 1994;8:1667-94.
Hirose T, O'Brien DA, Jetten AM. RTR: a new member of the nuclear receptor superfamily that is highly expressed in murine testis. Gene 1995a; 152: 247-51.
Hirose T, O'Brien DA. Jetten AM. Cloning of the gene encoding the murine orphan receptor Tak1 and cell-type-specific expression in testis. Gene 1995b; 152: 239-242.
Hu Y, Zhou Z, Xu C, Shang Q, Zhang YD, Zhang YL. Androgen down-regulated and region specific expression of germ cell nuclear factor in mouse epididymis. Endocrinology 2003; 144: 1612-9.
Huang HFS, Li MT, Pogach LM, Qian L. Messenger ribonucleic acid of rat testicular retinoic acid receptors

: developmental pattern, cellular distribution and testosterone effect. Biol Reprod 1994; 51: 541-50.
Inkster S, Yue W, Brodie A. Human testicular aromatase : Immunocytochemical and biochemical studies. J Clin Endocrinol Metab 1995; 80: 1941-7.
Jiang Y, Kuo CL, Pernecky SJ, Piper WN. The detection of cytochrome P450 2E1 and its catalytic activity in rat testis. Biochem Biophys Res Comm 1998; 246: 578-83.
Kastner P, Mark M, Chambon P. Nonsteroid nuclear receptors: what are genetic studies telling us about role in real life? Cell 1995; 83: 859-69.
Kastner P, Mark M, Ghyselinck N, Krezel W, Dupe V Grondona JM, Chambon P. Genetic evidence that the retinoid signal is transduced by heterodimeric RXR/RAR functional units during mouse development. Development 1997; 124: 313-26.
Kato M, Sung WK, Kato K, Goodman DS. Immunostochemical studies on the localization of cellular retinol-binding protein in rat testis and epdidiymis. Biol Reprod 1985; 32: 173-89.
Kato S, Masuhiro Y, Watanabe M, Kobayashi Y, Takeyama KL, Endo H, Yanagisawa J. Molecular mechanism of a cross-talk between oestrogen and growth factor signaling pathways. Genes Cells 2000; 5: 593-601.
Katz D, Niederberger C, Slaughter GR, Conney AJ. Characterization of germ cell-specific expression of the orphan nuclear receptor, germ cell nuclear factors. Endocrinology 1997; 138: 4364-72.
Kojima M, Morozumi T, Onishi A, Mitsuhashi T. Structure of the pig sterol 14α-demethylase (CYP51) gene and its expression in the testis and other tissues. J Biochem 2000; 127: 805-11.
Lambard S, Galeraud-Denis I, Bouraima H, Bourguiba S, Chocat A, Carreau S. Expression of aromatase in human ejaculated spermatozoa: a putative marker of motility. Mol Hum Reprod 2003; 9: 117-24.
Lambard S, Galeraud-Denis I, Saunders PT, Carreau S. Human immature germ cells and ejaculated spermatozoa contain aromatase and oestrogen receptors. J Mol Endocrinol. 2004; 32: 279-89.
Lee CH, Chang L, Wei LN. Molecular cloning and characterization of a mouse nuclear orphan receptor expressed in embryos and testes. Mol Reprod Dev 1996; 44: 305-14.
Lee H-K, Ahn RS, Kwon H-B, Soh J. Nucleotide sequence of rat steroidogenic acute regulatory protein complementary DNA. Biochem Biophys Res Comm. 1997; 230: 528-32.
Levallet J, Bilinska B, Mittre H, Genissel C, Fresnel J, Carreau S. Expression and immunolocalization of functional Cytochrome P450 aromatase in mature rat testicular cells. Biol Reprod 1998; 58: 919-26.
Levallet J, Mittre H, Delarue B, Carreau S. Alternative splicing events in the coding region of the cytochrome P450 aromatase gene in male germ cells. J Mol Endocrinol 1998; 20: 305-12.
Li H, Papadopoulos V, Vidic B, Dym M, Culty M. Regulation of rat testis gonocytes proliferation by platelet derived growth factor and estradiol: identification of signaling mechanisms involved. Endocrinology 1997; 138: 1289-1298.
Lopez-Fernandez LA, Parraga M, Mazo JD. Tex261, a novel gene presumably related but distinct from steroidogenic acute regulatory (StAR) gene, is regulated during the development of germ cells. Biochem Biophys Res Comm 1998; 242: 565-69.
Majdic G, Parvinen M, Bellamine A, Harwood Jr AB, et al. Lanosterol 14α-demethylase (CYP51) NADPH-cytochrome P450 reductase and squalene synthase in spermatogenesis: Late spermatids of the rat express proteins needed to synthesize follicular fluid meiosis activating sterol. J Endocrinal 2000; 166: 463-74.
Miura T, Miura C, Ohta T, Nader MR, Todo T, Yamauchi K. Estradiol -17β stimulates the renewal of spermatogonial stem cells in males. Biochem Biophys Res Comm 1999; 264: 230-34.
Miyamoto T, Sengoku K, Takuma N, Hasuike S, Hayashi H, Yamauchi T, Yamashita T, Ishikawa M. Isolation and expression analysis of the testis-specific gene, STRA8, stimulated by retinoic acid gene 8. J Assist Reprod Genet. 2002; 19: 531-5.
Oulad-Abdelghani M, Bouillet P, Decimo D, Gansmuller A, et al. Characterization of a premeiotic germ cell-specific cytoplasmic protein encoded by Stra8, a novel retinoic acid-responsive gene. J Cell Biol 1996; 135: 469-77.
Robertson KM, O'Donnell L, Jones ME, Meachem SJ, Boon WC, et al. Impairment of spermatogenesis in mice lacking a functional aromatase (cyp19) gene. Proc Natl Acad U S A 1999; 96: 7986-91.
Rozman D, Fink M, Fimia GM, Sassone-Corsi P, Waterman MR. Cyclic adenosine 3, 5'-monophosphàte (cAMP) responsive element modulator (CREM)-dep Endocrinologyent regulation of cholesterogenic lanosterol 14α-demehtylase (CYP51) in spermatids. Mol Endocrinol 1999; 13: 1951-62.
Rozman D. Lanosterol 14α-demethylase (CYP51)- a cholesterol biosynthetic enzyme involved in production of meiosis activating sterols in occytes and testis a mini review. Pflugers Arch 2000; 439: 56-7.
Saunders PTK. Oestrogen receptors β (ERβ). Rev Reprod 1998; 3: 164-71.
Schmitt MC, Jamison RS, Orgebin-Crist MC, Ong DE. A novel, testis specific member of the cellular lipophilic transport protein superfamily, deduced from a complimentary deoxyribonucleic acid clone. Biol Reprod 1994; 51: 239-45.
Schmitt MC, Ong DE. Expression of cellular retinol-binding protein and lecithin-retinol acyltransferase in

developing rat testis. Biol Reprod 1993; 49: 972-9.
Shetty G, Krishnamurthy H, Krishnamurthy HN, Bhatnagar S, Moudgal RN. Effect of estrogen deprivation on the reproductive physiology of male and female primates. J Steroid Biochem Mol Biol 1997; 61: 157-66.
Shetty G, Krishnaumurthy H, Krishnamurthy HN, Bhatnagar AS, Moudgal NR. Effect of long-term treatment with aromatase inhibitor on testicular function of adult male bonnet monkeys, *M. radiata*. Steriods 1998; 63: 414-20.
Shyadehi AZ, Lamb DC, Kelly SL, Kelly DE et al. The mechanism of acyl carbon bond cleavage reaction catalyzed by recombinant sterol 14α demethylase of candida allacans. J Biol Chem 1996; 27: 12445-50.
Simpson ER. Genetic mutations resulting in loss of aromatase activity in humans and mice. J Soc Gynecol Investig 2000; 7: 18-21.
Stromstedt M, Waterman MR, Haugen TB, Tasken K, Parvinen M, Rozman D. Elevated expression of lanosterol 14α-demethylase (CYP51) and the synthesis of oocyte meiosis-activating sterols in postmeiotic germ cells of male rats. Endocrinology 1998; 139: 2314-21.
Svechnikov K, Stocco DM, Soder O. Interleukin-1alpha stimulates steroidogenic acute regulatory protein expression via p38 MAP kinase in immature rat Leydig cells. J Mol Endocrinol. 2003; 30: 59-67.
Tacer KF, Haugen TB, Baltsen M, Debeljek N, Rozman D. Tissue-specific transcriptional regulation of the cholesterol biosynthetic pathway leads to accumulation of testis meiosis-activating sterol (T-MAS). J Lipid Res 2002; 43: 82-9.
Toyama Y, Ohkawa M, Oku R, Maekawa M, Yaasa S. Neonatally administered diethylstilbestrol retards the development of the blood-testis barrier in the rat. J Androl 2001; 22: 413-23.
van der Burg B, Sonneveld E, Lemmen JG, van der Sagg PT. Morphogenetic action of retinoids and estrogens. Int J Dev Biol 1999; 43: 735-43.
van Pelt AM, de Rooij DG, van der Burg B, Vander Sao T, Gustaffsson JA, Kuiper GG. Ontogeny of estrogen receptors-β expression in rat testis. Endocrinology 1999; 140: 478-83.
Venables JP, Vernet C, Chew SL Elliott DJ, Gansmuller A, et al. T-STAR/ETOILE: a novel relative of SAM68 that interacts with an RNA-binding protein implicated in spermatogenesis. Hum Mol Genet 1999; 8: 959-69.
Wang H, Liu F, Millette CF, Kilpatrick DL. Expression of a novel, sterol-insensitive form of sterol regulatory element binding protein 2 (SREBP2) in male germ cells suggests important cell- and stage-specific functions for SREBP targets during spermatogenesis. Mol Cell Biol 2002; 22: 8478-90.
Wang Z, Kim KH. Vitamin A-different testis germ cells are arrested at the S phase of the cell cycle: a molecular study of the origin of synchronous spermatogenesis in regenerated seminiferous tubules. Biol Reprod 1993; 48: 1157-65.
Weinbauer GF, Wessels J. Paracrine control of spermatogenesis. Andrologia 1999; 31: 249-62.
Yang G, Zhang YL, Buchold GM, Jetten AM, O'Brien DA. Analysis of germ cell nuclear factor transcripts and protein expression during spermatogenesis. Biol Reprod. 2003; 68: 1620-30.
Yoshida Y, Yamashita C, Noshiro M, Fukuda M, Aoyama Y. Sterol 14-demehtylase P450 activity expressed in rat gonads: contribution to the formation of mammalian meiosis-activating sterol. Biochem Biophys Res Comm. 1996; 223: 534-8.
Zhang H, Liao L, Kuang SQ, Xu J. Spatial distribution of the messenger ribonucleic acid and protein of the nuclear receptor coactivator, amplified in breast cancer-3, in mice. Endocrinology. 2003; 144: 1435-43.

Chapter 5

HOMEOSTASIS OF GERM CELLS AND APOPTOSIS

5.1. APOPTOSIS

During development of most metazoan animals, many more cells are produced than are eventually needed. If cells are no longer needed, they commit suicide by activating an intracellular death program. The pattern of events in death by suicide is so orderly that the process is often called programmed cell death or PCD or apoptosis. Apoptosis or programmed cell death is a morphologically distinct form of cell death, which is designed to remove unwanted and potentially dangerous cells. Apoptosis is the orchestrated collapse of cells staging membrane fragmentation, cell shrinkage, mitochondrial breakdown, release of cytochrome C, protein fragmentation, chromatin condensation and DNA degradation followed by rapid engulfment of corpses by neighboring phagocytic cells. Thus apoptosis plays a key role in removing surplus cells and sculpting the developing embryo. Regulation of apoptosis is associated with a variety of diseases including cancer, AIDS, neuro-degenerative diseases and ischaemic stroke. Apoptosis represents an active, gene directed mechanism and offers a possible method to control diseases by therapeutic purposes. Several molecules that are responsible for the regulation and execution of apoptosis have been identified during last decade. Now it is clear that an unusual class of cysteine proteases, termed caspases is responsible for execution of cell death. The basic pathways of apoptosis have been conserved throughout animal evolution, although nature of molecules/molecular complexes has changed with evolution of species. The basic molecules involved in mammalian apoptosis are listed in **Table-5.1** (reviews - Song and Steller, 1999; Lawen, 2003)). Thus the cellular machinery of programmed cell death turns out to be as intrinsic to the cell as, mitosis. Programmed cell death is needed to destroy cells that represent a threat to the integrity of the organism. For example cells infected with viruses and cells with damaged DNA are needed to be destroyed by apoptosis. The early steps in apoptosis are reversible - at least in C. elegans. In some cases, final destruction of the cell is guaranteed only with its engulfment by a phagocyte.

5.1.1. Death Signals

The apoptosis occurs when the balance between the positive signals, which are needed for continued survival, and the receipt of negative signals which are responsible for cell loss, is lost. Some examples of positive signals include growth factors for neurons and interleukin-2 (IL-2), an essential factor for the mitosis of lymphocytes. The continued survival of most cells requires that they receive continuous stimulation from other cells and, for many, continued

Table 5.1. Basic Components of Apoptotic Pathways in Mammalian Cells.

Component	Intrinsic Pathway	Extrinsic Path way
1. Apoptosis Promoter	Bax/ BH3 only proteins	Fas/FasL, TNFR1/TNFα
2. Apoptosis Inhibitor	Bcl-2, Bcl-x_L	FLIP
3. Adaptor	Apaf-1	FADD, TRADD
4. Initiator Caspase	Caspase-9	Caspase-8
5. Caspase Inhibitor	IAP	IAP
6. IAP Inhibitor	Smac/Diablo	Smac/Diablo
7. Effector Caspase	Caspase-3, Caspase-7	Caspase-3, Caspase-7

adhesion to the surface on which they are growing. Negative signals include: i) increased levels of oxidants within the cell, ii) damage to DNA by these oxidants or other agents like ultraviolet light , x-rays, chemotherapeutic drugs, iii) accumulation of proteins that fail to fold properly into their proper tertiary structure, and iv) molecules that bind to specific receptors on the cell surface and signal the cell to begin the apoptosis. These death activators include: tumor necrosis factor-α (TNF-α) that binds to the TNF receptor; lymphotoxin that also binds to the TNF receptor; Fas ligand (FasL), a molecule that binds to a cell-surface receptor Fas (also called CD95).

5.1.2. Caspases

Caspases are a family of proteases that have cysteine at the active site (cysteine proteases) and function during apoptosis or cytokine processing, or both. They get their name because they cleave proteins - mostly each other - at aspartic acid (Asp) residues. They are synthesized as pro-enzymes (or zymogens) and remain inactive in most healthy cells. Upon activation by different death signals, the single chain pro-caspase is cleaved at specific aspartic acid residue by other caspases in proteolytic cascade to remove an inhibitory N-terminal pro-domain and to generate two distinct subunits. These subunits assemble into a heterotetramer to form the active protease. Once activated, caspases are thought to cleave many important cellular proteins and there by bring about the characteristic apoptotic morphology (Song and Steller, 1999). Caspase 9 is one of a family of over a dozen caspases. Caspase 9 cleaves and, in so doing, activates other caspases. The sequential activation of one caspase by another creates an expanding cascade of proteolytic activity (rather like that in blood clotting and complement activation), which leads to digestion of structural proteins in the cytoplasm, degradation of chromosomal DNA, and phagocytosis of the cell.

5.2. THE MECHANISMS OF APOPTOSIS

The initiation of apoptosis is regulated by many signals, which can originate either from within the doomed cell or from its extracellular environment. It shows that multiple mechanisms are used to control the conversion of the caspase zymogene to the active protease. Pro-caspase

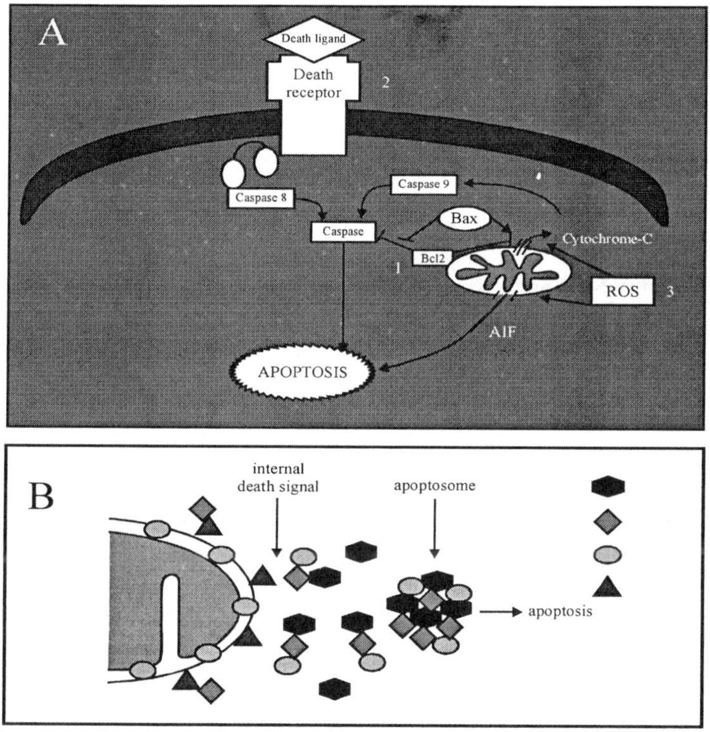

Fig.5.1. A. A model showing three mechanisms of apoptosis: (1) The Intrinsic or Mitochondrial Pathway; (2) The Extrinsic or Death Receptor Pathway, and (3) Apoptosis-Inducing Factor (AIF) Release Pathway. B. The Intrinsic or Mitochondrial Pathway, showing release of cytochrome C and formation of apoptosome.

activation of target cell can be triggered from out side by activation of cell receptor present on the cell surface. For example, killer lymphocyte produces Fas ligand (FasL), which triggers target cell by binding with Fas receptor (FasR) present on target. Intrinsic pathway is generated by signals arising within the cell. Presently, there are 3 different mechanisms by which a cell commits suicide by apoptosis **(Fig.5.1A)**.

5.2.1. The Intrinsic or Mitochondrial Pathway

Intrinsic pathway is generated by signals arising within the cell. When cells are damaged or stressed from within the cell, mitochondria are induced to release cytochrome C into the cytosol. i). In a healthy cell, the outer membranes of its mitochondria express the protein Bcl-2, which is bound to the protein Apaf-1 (apoptotic protease activating factor-1). The adapter protein, Apaf-1 is represented by CED-4 in *C. elegans*. ii). Internal damage to the cell (e.g., from reactive oxygen species) causes Bcl-2 to release Apaf-1, and a related protein, Bax to penetrate mitochondrial membranes, causing cytochrome C to leak out. The released cytochrome C and Apaf-1 bind to molecules of caspase 9. The resulting complex of cytochrome C, Apaf-1, caspase 9, (and ATP) is called the apoptosome. These proteins form aggregates in the cytosol **(Fig.5.1 B)**.

The CED-4 or Apaf-1 regulates the activation of a cytoplasmic caspase cascade by complex formation, possibly in a manner similar to the model described for CD-95/Fas system. In this process, the release of cytochrome C is a crucial step. However, the precise circumstances during which cytochrome C is released in situ, and its role in the regulation of apoptosis in vivo, remain to be determined. Nevertheless, it appears that Apaf-1 can bring pro-caspase molecules into close proximity, so that they can activate one another to form apoptosome **(Fig.5.1B)**. Apaf-1 activity has been identified in testis and its role in germ cell apoptosis is being appreciated.

5.2.2. The Extrinsic or Death Receptor Pathway

Extrinsic pathway is triggered by death activators by binding to receptors at the cell surface. The death activators include: TNF-α, lymphotoxin, and Fas ligand (FasL). Fas and the TNF receptor are integral membrane proteins with their receptor domains exposed at the surface of the cell. Binding of the complementary death activator (FasL and TNF respectively) to membrane receptor transmits a signal to the cytoplasm that leads to activation of caspase 8 **(Fig.5.1A)**. Caspase 8 (like caspase 9) initiates a cascade of caspase activation leading to phagocytosis of the cell. The best understood pathway for activating caspases through induced proximity is the CD-95/Fas/Apo-1 system. The CD-95/Fas belong to the tumor necrosis factor receptor (TNFR) family and function in the removal of activated cytotoxic T cells at the end of the immune response. Binding of extra-cellular ligands, such as the Fas ligand (FasL), to these receptors induces trimerization of the receptor. This trimerization, in turn, recruits the adaptor molecule FADD and pro-caspase-8 into a multimeric complex termed DISC (death inducing signal complex) in which caspase-8 is activated. Activated caspase goes on to cleave downstream caspase zymogens, such as caspase-3. In addition to activating caspase-3, caspase-8 also cleaves a pro-apoptotic member of the Bcl-2 family that amplifies the death signal received from the cell surface. Several of these steps are mediated by NO actions (Chung et al., 2001).

5.2.3. Apoptosis-Inducing Factor (AIF) Release Pathway

A third mechanism may be triggered by dangerous reactive oxygen species. Neurons, and perhaps other cells, have another way to self-destruct that - unlike the two paths described above - does not use caspases. In this mechanism, apoptosis-inducing factor (AIF) is released. The AIF is a protein that is normally located in the intermembrane space of mitochondria. When the cell receives a signal telling it that it is time to die, AIF is released from the mitochondria (like the release of cytochrome C in the first pathway), migrates into the nucleus, binds to DNA, which triggers the destruction of the DNA and cell death **(Fig.5.1A)**.

Oxidative Stress and Apoptosis Signal-Regulating Kinase 1 (ASK1)

Apoptosis signal regulating kinase (ASK1) mediated apoptosis MAP-kinase cascade is evolutionarily well conserved in all eukaryotic cells and consists of three kinases that establish a sequential activation pathway (see Chapter 19). The ASK1 is a member of the MAPKKK family and activates both the SEK1 INK and MKK3/MKK6-p38 signaling cascades. The ASK1 is activated in cells after treatment with inflammatory cytokines and to various types of stresses. Over-expression of wild ASK1 is a key element in cytokine and stress induced apoptosis. Moreover apoptosis is reduced in ASK1 knockout mice (review-Matsuzawa and Iehijo, 2001). Oxidative stress induced activation of ASK1 results in apoptosis. Thioredoxin (Trx) has been identified as a negative regulator of ASK1/JNK-p38 pathway (Saitoh et al., 1998). The Trx is also present in testicular germ cells, although its associating protein ASK1 is yet to be known (see Chapter 29). The Trx is redox regulatory protein, which has two redox sensitive cysteine residues at the active center. Only a reduced form of Trx is associated with the N-terminal regulatory domain of ASK1 and silences the activity of ASK1; oxidization of Trx results in the dissociation of ASK1 to active kinase. The ASK1-Trx complex is thus thought to be a redox sensor, which functions as a molecular switch of external and internal redox status for the kinase-signaling module. The 14-3-3 protein is another class of proteins that directly bind to ASK1 and that over-expression of 14-3-3 proteins blocked ASK1 induced apoptosis. Recently, protein serine/threonine phosphatase-5 (PP5) was identified as a negative regulator for activated ASK1 in response to various types of stresses. Though, these components have been identified in testis, their role and correlation with apoptosis of germ cells in testis needs exploration (review-Matsuzawa and Iehijo, 2001).

NO as Regulator of Apoptosis

Nitric oxide (NO) can promote apoptosis (pro-apoptosis) in some cells, whereas it inhibits apoptosis (anti-apoptosis) in other cells. This complexity is due to the rate of NO production and its interaction with biological molecules such as iron, thiols, proteins, and reactive oxygen species. Long-lasting steady state level of NO acts as a proapoptotic modulator by activating caspase family proteases through the release of mitochondrial cytochrome C into the cytosol, upregulation of p53 expression, activation of JNK/SAPK, and altering the expression of apoptosis associated proteins including Bcl-2 family of proteins. However, physiological concentration of NO prevents cells from apoptosis induced by Fas, TNFα, and lipopolysaccharide. The anti-apoptotic mechanism can be understood via expression of protective genes such as heat shock proteins, Bcl-2 as well as direct inhibition of apoptotic caspase family proteases by S-nitrosylation of the cysteine thiol (Chung et al., 2001).

NO as a Pro-apoptotic Inducer: High concentrations of NO or peroxynitrite induce apoptotic cell death, in several cell types. Most of the proapoptotic effects of NO on these cells seem to be independent (but not all) of cGMP. The NO mediated apoptosis can be influenced by the redox state of the transition metal complexes within cells and expression of survival genes as: (a) NO can directly induce cytochrome C release through mitochondrial membrane potential loss by binding to iron of cytochrome c oxidase (complex IV) in mitochondrial electron transfer chain. Under this condition, superoxide generated from mitochondria interacts with NO to form peroxynitrite, which induces cytochrome C release. The effect of NO on cytochrome C release depends on intracellular redox potential and levels of target molecules, such as iron, glutathione, and superoxide. (b) NO mediates p53 accumulation, which induces cell cycle arrest by p21 upregulation or apoptosis by increase in the ratio of Bax/Bcl-x_L, cytochrome C release, and caspase activation. (c) NO activates the c-Jun N-terminal kinase (JNK)/stress activated protein kinase (SAPK) group (MAPKs) which is involved in NO-mediated apoptotic cell death. NO produced either by NO donors or by NO synthase induces apoptotic cell death with the activation of JNK/SAPK and p38 MAPK and caspase-3 activation. JNK/SAPK and p38 MAPK may be a critical mediator for NO-induced caspase-3 activation. (d) NO donors increase cellular ceramide level. Ceramide formation can induce several apoptotic signal pathways including the release of mitochondrial cytochrome C into cytosol, the activation of caspases-9 and –3, JNK/SAPK activation the inhibition of protein kinase B/Akt and the suppression of Bcl-2 expression (Chung et al., 2001).

NO as Anti-apoptotic Modulator: The NO can protect some cells from apoptosis induced by many different types of stimuli such as TNFα and oxidative stress. The mechanism underlying NO-mediated anti-apoptotic effects may depend on cell type specific with multiple signal pathways (review Chung et al., 2001). Antiapoptoic effects of NO can be cGMP-dependent and cGMP-independent. In some cells such as hepatocytes, antiapoptotic effects of NO are associated with cGMP production, which suppresses mitochondrial cytochrome C release, ceramide generation and caspase activation. Moreover, both NO and cGMP protect splenic B lymphocytes from apoptosis by increasing the expression of Bcl-2 and the activation of Akt/PKB, which induces phoshorylation of Bad and procaspase-9. Inhibition of Caspase Activity is caused by S-Nitrosylation.

Nitric oxide potentially induces the expression of several cytoprotective genes inducing HSP70 and HSP32 (Heme oxygenase), which are protectors of apoptosis. The molecular mechanism underlying the antiapoptotic effect of NO-induced HSP70 may be associated with two possibilities: i) HSP70 inhibits oligomerization of Apaf-1 by associating with the caspase-recruitment domain (CARD) of Apaf-1 resulting in the suppression of the apoptosome formation. ii) HSP70 involves the chaperone-mediated import of precursor proteins into the mitochondria, thus inhibiting cytochrome C release. NO can also regulate the Bcl-2 family proteins. NO donor elevates Bcl-2 expression both at mRNA and protein levels and prevents apoptotic cell death (Chung et al., 2001).

5.3. INHIBITORS OF APOPTOSIS PROTEINS (IAPS)

Several inhibitors of apoptosis proteins (IAPs) have been discovered during their ability to inhibit apoptosis. The characteristic structural motif of all IAPs is the baculovirus repeat (BIR) of 70 amino acids. Besides the BIR domains, some IAPs also contain additional structural motifs such as a RING and a CARD domain. Homologs of IAPs have been identified in mammals and even in yeast. The IAPs appear to inhibit cell death through direct interactions with caspases. The well-known human anti-apoptotic factors directly bind to and inhibit caspases and pro-caspases, which are the major executioners of apoptosis. In turn, Smac/DIABLO, a recently identified pro-apoptotic molecule is released from the mitochondria and promotes apoptosis by binding to and inhibiting IAPs. The CAD (caspase-activated deoxyribonuclease) causes nuclear DNA fragmentation by its apoptosis signal specific DNase activity (Matsuzawa and Ichijo, 2001; Song and Steller, 1999).

5.4. APOPTOSIS DURING SPERMATOGENESIS

Fate of male germ cells in testis is determined by a complex network of external and internal signals, which include stem cell factor (SCF), leukemia inhibitory factor (LIF) and Desert Hedgehog (Dhh), as well as endocrine signals such as pituitary gonadotropins and testosterone. Like various other cell types, male germ cells also respond to external signals, and to their internal milieu, by activating intracellular signaling pathways that ultimately control their fate and maintain their homeostasis. The FasL/Fas-R, p53 and proteins of the Bcl-2 family provide few of the signaling apoptotic pathways that appear to be essential for male germ cells homeostasis. The competitive interactions of the pro- and anti-survival Bcl2 family proteins regulate the activation of the proteases (caspases), which dismantle the cell

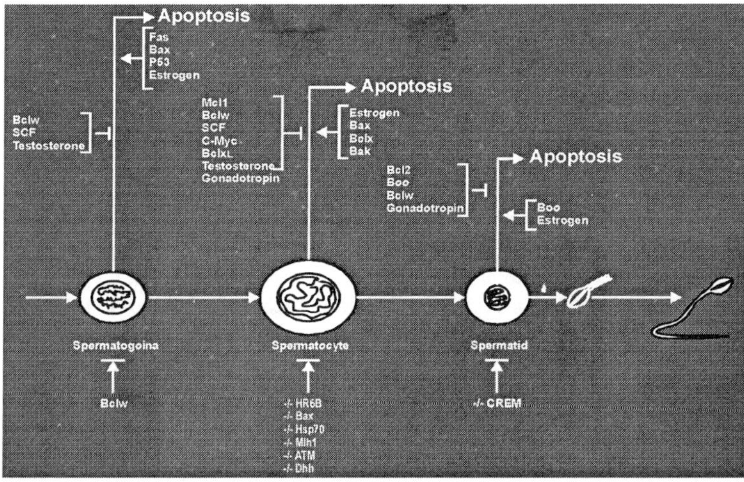

Fig.5.2. Apoptosis of different germ cell types during spermatogenesis. Signals that regulate germ cell apoptosis are given by arrows indicating promotion of apoptosis and by —I indicating inhibition of apoptosis. Genetic modifications of apoptosis regulators and the effects of mutations in genes encoding apoptosis regulators are indicated below the appropriate cell type. (BMP, bone morphogenetic protein; SCF, stem cell factor, IL, interleukin; LIF, luekaemia inhibitory factor, FGF, fibroblast growth factor) (Print and Loveland, 2000)

by the mechanisms that are not fully understood. The subject has been extensively reviewed by Print and Loveland (2000) **(Fig.5.2)**. In order to achieve homeostasis of each germ cell type in the adult, germ cell renewal, proliferation, export and apoptosis must be finely balanced. The exact incidence of adult male germ cell apoptosis remains unclear, since not all-degenerating germ cells display the classical morphology of apoptosis. Spermatogonia and round spermatids almost certainly die by apoptosis, since they display many of its classical morphological and biochemical features. Apoptotic germ cells are either sloughed into the tubule lumen, or phagocytosed by Sertoli cells (Print and Loveland, 2000) (see Chapter 2).

5.4.1. FAS System in Testis

The Fas (CD95) is a transmembrane receptor protein that belongs to the tumor necrosis factor-related type-II transmembrane receptor protein (TNFR). It contains a "death domain" and is capable of initiating apoptosis when stimulated by receptor cross-linking or binding to its ligand, FasL (CD95L). The FasL-Fas interaction triggers the death of cells expressing Fas, a process best studied in lymphoid cells. The biological importance of the FasL-Fas interaction is understood by abnormalities seen in several strains of mutant mice, in which the Fas system is not functional. In testis, the Fas system has been implicated in maintaining immune privilege. By transplanting Sertoli cells lacking functional FasL across immunological barriers, Bellgrau *et al.* showed that FasL was critical to prevent immune rejection (Lee et al., 1997). In testis, FAS system is the key regulator of spermatogenesis. Both Fas ligand (FasL) and Fas receptor (Fas) are expressed post-natally in rat testis with peak expression associated with the high levels of germ cell apoptosis found during the first wave of spermatogenesis. The Fas has been localized to germ cells where as FasL to Sertoli cells. FasL expressed by Sertoli cells initiates the apoptotic death of germ cells expressing Fas. The expression of Fas occurs on germ cells including spermatogonia, spermatocytes, and spermatids. The Fas expression on germ cells was increased following cryptorchidism. But the expression of Fas on germ cells does not correlate with spontaneous apoptosis or apoptosis induced by cryptorchidism (Ogi et al., 1998). During photoperiod-induced testicular regression, the Fas system is activated (Young and Nelson, 2001). The *Fas* and *FasL* genes are dramatically up-regulated after exposure to mono-(2-ethylhexl) phthalate (MEHP) and 2, 5-hexanedione, two widely studied Sertoli cell toxicants known to induce germ cell apoptosis. The testicular expression of RIP and FAP-1, components of the Fas activating complex, increased after exposure to MEHP, which induces massive germ cell death through Sertoli cells. Finally, the expression of additional apoptosis-inducing genes, including tumor necrosis factor receptor (TNFR), FADD, TRAIL, and DRS, were detected in mammalian testis. These results provide strong support to the concepts: (1) Sertoli-germ cell interactions are important in the control of germ cell apoptosis; and (2) the Fas system and similar paracrine systems are important modulators of testicular germ cells homeostasis (Boekelheide et al., 1998 ; Lee et al., 1997). The Fas appears to mediate germ cell apoptosis via caspase path way, since caspase inhibitor, benzyloxycarbonyl-Val-Ala-Asp (OMe) fluoromethylketone inhibited germ cell death (Pentikainen et al., 1999). It is thus possible that apoptosis of spermatogenic cells is induced by the binding of Fas L to Fas through interaction between Sertoli cells and spermatogenic cells. If this is the case, Sertoli cells are presumably responsible for both apoptosis induction and heterophagic elimination of spermatogenic cells. In this action of phagocytosis, class B scavenger receptor type1 (SR-BI) functions at least partly as a PS receptor, enabling Sertoli cells to recognize and phagocytose apoptosis spermatogenic cells at all stages of differentiation (Shiratsuchi et al., 1997; 1999).

5.4.2. Apaf-1 and Cytochrome C in Germ Cell Apoptosis

A complex composed of Apaf-1, dATP, and cytochrome C activates cytoplasmic caspases leading to apoptotic cell death. In Apafl (-/-) knockout model, most Apaf-1 mutants died pre-natally and frequently exhibited neural lesions. The neural lesions that developed in the knockout were due to an excess of neural progenitor cells that manifested as early as embryonic day 9.5 in development. In contrast to previous reports on the Apaf-1 knockout mice, 5% of the mutants successfully survived to adulthood. In these survivors, the brain develops normally, but in males, there is degeneration of spermatogonia resulting in the virtual absence of sperm. Thus, it was suggested that cytochrome C-mediated apoptosis is not absolutely required for normal neural development, but is essential for spermatogenesis. These findings suggested that alternative apoptotic pathways are present in conjunction with and parallel to Apaf-1, that modify effects on programmed cell death (Honarpour et al., 2000).

5.4.3. p53 Induced Pathway

DNA damage can trigger apoptosis. This response requires p53, which can activate transcription of genes that encode proteins participating in the release of cytochrome C from mitochondria. The p53, cell cycle regulator appears to be required for radiation induced apoptosis of differentiating spermatogonia. There are also p53 independent mechanisms. For example, p53 may not be required for radiation induced apoptosis of stem spermatogonia, which are more radio-resistant, while p53 and p21 and other mechanisms contribute to spermatocyte apoptosis in ATM deficient mice (reviewed in Print and Loveland, 2000) **(Fig.32.1).** The p53 is also a tumor suppressive factor, which has important functions in the control of somatic cell growth and differentiation (see Chapter 32). During normal spermatogenesis, absence of p53 from spermatogonia, but its presence at unusual high concentration in tetraploid pachytene spermatocytes at both mRNA and protein level are suggestive of its role during the meiotic prophase. In p53-deficient mice, however, the testis contains about 50% more A_1 spermatogonia than in wild-type mice. This indicates that either p53 is involved in cell cycle regulation of undifferentiated spermatogonia or p53 normally induces apoptosis in some of these cells. However, strong immunohistochemical staining for p53 seen in spermatogonia following 4Gy of X-irradiation indicated its role in extensive apoptosis of differentiating spermatogonia. Apparently, in differentiating spermatogonia, increased p53 expression correlates with increased apoptosis (de Rooij and Grootgoed, 1998). The p53 is required for radiation-induced apoptosis in many cells. But in radiation induced damage of germ cell loss, it has been found that differentiating spermatogonia died of apoptosis that is dependent on p53 where as stem spermatogonia, which are more radio-resistant did not (Hasegawa et al., 1998). In a recent study, however, apoptosis and up-regulation of Fas following radiation have been linked to p53-dependent events. Surprisingly, Fas is necessary, in part, for radiation-induced p53-dependent apoptosis, where as FasL is not (Embree-ku et al, 2002).

A growing body of evidence demonstrates that germ cell death during normal spermatogenesis and that induced by increased scrotal temperature occurs via apoptosis. However, it is unknown that the germ cell loss associated with cryptorchidism occurs by apoptosis. But presence of p53 in tetraploid pachytene spermatocytes, a germ cell stage most vulnerable to heat stress, suggests that p53 is involved in apoptosis. Testicular p53 is associated with the nuclear envelope and translocated into the nucleus in response to heat stress. In experimental unilateral cryptorchidism, the testicular germ cell loss, and DNA fragmentation all began in the cryptorchid testes on Day 6-7 in wild-type mice. In contrast, these changes were delayed by 3 days in p53-/- mice. This suggested that abdominal heat stress induces germ cell loss through two apoptosis pathways: a p53-dependent pathway responsible for the initial phase of germ cell apoptosis, and a p53-independent pathway that accounts for subsequent apoptosis (Yin et al., 1998) (see Chapter 32).

The p63, a member of p53 gene family, is required for p53-dependent apoptosis and can induce apoptosis in the absence of p53 through the activation of p53-target genes. Nakamuta and Kobayashi (2003) examined the expression pattern of p63 in the mouse testis from embryonic day (Ed) 13.5 to Ed 18.5 to clarify their possible role in germ cell apoptosis. The p63 was found in the nucleus of germ cells at Ed 13.5, and continuously observed until Ed 18.5. The RT-PCR using specific primers for the p63 isotype showed that both transcripts were expressed in the fetal gonads. Thus, p63 appears to play a role in the embryonic testes (Nakamuta and Kobayashi, 2003).

5.4.4. Bcl-2 Protein Family

Male germ cells respond to the various external signals and to their internal milieu, by activating internal signaling mechanisms that ultimately determine their fate. Proteins, which arbitrate germ cell apoptosis include B-cell lymphoma/leukemia-2 (Bcl-2). In *C.elegans,* the *Ced-9* encodes a protein homologous to mammalian Bcl-2-like molecule that regulates apoptosis in mammals. A connection has been established between members of the CED-9/Bcl-2 family and caspases. Bcl2 proteins characterized by presence of Bcl2 homology (BH)1 and BH2 can either induce or inhibit apoptosis. Hence Bcl-2 family proteins are divided into three categories: the anti-apoptotic members such as Bcl-2, Bcl-X_L, Bcl-w, Mcl-1, and Bfl1/A1; the pro-apoptotic members such as Bax, Bak, and Bok; and the pro-apoptotic BH3-only proteins such as Bad, Bid, Bik, Bim, Blk, Hrk and Noxa (Srivastava et al., 1999). The mechanisms by which Bcl-2 family proteins regulate mitochondrial dependent apoptosis are still being debated. The voltage dependent anion channel (VDAC) has been suggested as one of the targets (Shimizu et al., 1999) for Bcl-2 family and thus each Bcl-2 family protein functions as a gatekeeper of the PT pore by either promoting or preventing the release of cytochrome. Because Bcl-2 is present in the outer mitochondrial membrane, it has been suggested that it blocks apoptosis by inhibiting the release of apoptosis inducing factors, such as cytochrome C from mitochondria **(Fig 5.1A)**. One important property of Bcl-2-like proteins is their ability to form homo- and heterodimers with other family members to modulate their activity. Therefore, the battle between these antagonistic Bcl-2 family members could provide a mechanism to fine tune the levels of protection against cell death.

Presence of Bcl-2 itself in spermatogonia is not well established (Furuchi et al 1996; Sugiyama et al., 2001 and c/r). Nevertheless, Bcl-2 over expression has dramatic effects on spermatogenesis, associated with a massive accumulation of spermatogonia. Hence, members of the Bcl-2 family may control this aspect of spermatogenesis, although the Bcl-2 protein itself does not seem to be involved. Probably, the *Bcl-x* gene encodes the relevant factor. In the testis, at least two splice variants of Bcl-x are expressed: Bcl-x_S and Bcl-x_L. Whereas Bcl-x_L suppresses apoptosis, Bcl-x_S has on opposing effect. Both forms of Bcl-x, Bcl-x_S and/or Bcl-x_L were found strongly expressed in spermatogonia in the human testis. These observations suggested that pro-survival, Bcl-x_L is important in germ cell density regulation. Since Bcl-x_L knockout mice die during embryogenesis, however, conditional gene knockout studies are required to dissect Bcl-x_L's function after birth.

Another member of the Bcl-2 family such as the pro-apoptotic Bax protein plays a fundamental role. Since male Bax knockout mice possess germ cells, Bax is not essential for germ cell survival in the embryo. The Bax protein appears localized to those germ cells, which are under-going apoptosis during the first spermatogenic wave. The Bax, however, appears to be essential for germ cell survival during the first wave of spermatogenesis. The Bax may trigger the physiological apoptosis of spermatogenesis and spermatocytes. Genetically modified mice also revealed a critical influence of Bax on germ cell homeostasis. Testes of adult *bax* knockout mice contained excessive numbers of spermatogonia and pre-leptotene spermatocytes, consistent with failed apoptosis during the first wave of spermatogenesis (Print and Loveland, 2000). The Bax is a potent inducer of apoptosis in adult male mice as well. The *Bax* gene knockout adult male mice are sterile. The histological features of the testes of *Bax* knockout mice resemble those of the *Bcl-2* and *Bcl-x_L* transgenic testes, showing an accumulation of spermatogonia. This indicated that Bax is an important factor in the control of spermatogonial apoptosis. Probably the balance between the expression and action of Bcl-x_L and Bax control the cell-density-related apoptosis of spermatogonia (review- de Rooij and Grootgoed, 1998).

Male germ cell apoptosis, of which the default pathway is similar to that of the female, is likely to be influenced by male gonadal environments (Kasai et al., 2003). Similar to rodents, mild testicular hyperthermia results in azoospermia and oligozoospermia in monkeys through increased germ cell apoptosis with minimal effect on the hormonal milieu. Short-term exposure of rat testis to mild heat results, within few hours, cell-specific activation of germ cell apoptosis. Among initial events was the redistribution of Bax from a cytoplasmic to paranuclear sites in heat-susceptible germ cells. The relocation of Bax was followed by sequestration of mitochondria and endoplasmic reticulum (ER) into paranuclear areas, cytosolic translocation of cytochrome c in association with activation of the initiator caspase 9 and the executioner caspases 3, 6, and 7, and cleavage of PARP. Furthermore, though the Bax was co-localized with ER in the susceptible germ cells, the Fas system was dispensable for heat-induced germ cell apoptosis in these studies. The mitochondria- and possibly also ER-dependent pathways could be the key apoptotic pathways for heat induced germ cell death in the testis (Lue et al.,

2003; Hikim et al., 2003; Sinha Hikim et al., 2003).

The *Bcl-w* is expressed in spermatids and Sertoli cells and does not seem to function in spermatogonia. The pro-survival protein Bcl-w plays a different role. The Bcl-w appears to be dispensable in testis during embryogenesis and during the first two weeks after birth. The incidence of germ cell apoptosis in *Bcl-w* knockout mice, however, becomes dramatically elevated between two and four weeks of age. The testes of six weeks old *Bcl-w* knockout mice contained numerous apoptotic cells, many of which remained linked as multinucleate symplasts. Testicular degeneration was observed in *Bcl-w deficient* mice, however (Print et al., 1998; Ross et al., 2001).

Among other members of the Bcl-2 family that are expressed in testis but their role in spermatogenesis has not been investigated, include Bfl-1 (A1) whose cellular distribution is unknown. The Bad, Bok and Bim are also expressed in adult testis. The DIVA or Boo, a new member of the Bcl-2 family, is apparently expressed only in reproductive tissues; its role, however, is controversial. Inohara et al (1998 c/r Song et al 1999) observed this protein in spermatids, where they suggest it to perform a pro-apoptotic role, while Song et al, (1999) suggested that it is in fact anti-apoptotic, and observed transcripts in epididymis but not testis. The expression of Boo was highly restricted to the epididymis implicating it in the control of sperm maturation. Boo inhibits apoptosis, homodimerizes or heterodimerizes with some death promoting and suppressing Bcl-2 family members. More importantly, Boo interacts with Apaf-1 and forms a multimeric protein complex with Apaf-1 and caspase-9 (Song et al., 1999). The Mcl-1, another member of Bcl-2, has been detected in Leydig cells and residual bodies but not germ cells.

5.5. FACTORS CONTROLLING APOPTOSIS IN SPERMATOGENESIS

5.5.1. Paracrine Control of Apoptosis

Several paracrine and endocrine signals are prime regulators of germ cell fate, and may be particularly important during embryonic development and PP spermatogenesis. For example, PGC express leukemia inhibitory factor (LIF) receptors and the survival of PGC in culture is promoted by LIF. The role played by LIF in vivo is unclear however, since mice lacking the LIF gene are fertile. The expression patterns, and the finding that Gas 6 promotes PGC survival in vitro, suggest that Gas 6 may promote PGC survival in the embryo. Other factors, which promote PGC survival in vitro include: interleukin-4, basic fibroblast growth factor (bFGF), the stem cell factor (SCF or Steel factor) (Chapter 17), and bone morphogenetic protein (BMP)-4. In contrast, TGFβ promote gonocyte apoptosis in vitro. In the adult, members of the BMP family promote germ cell survival in vivo. BMP-8A and BMP-8B, secreted by round spermatids, appear to provide survival signals to spermatocytes. Dhh is a paracrine signal, which appears to promote germ cell survival indirectly. The Dhh, encoded by hedgehog gene *Dhh* and secreted by Sertoli cells appears to activate transcription of the *patched gene* in Leydig cells. Mice lacking Dhh exhibited accelerated spermatocyte apoptosis, possibly due to abnormal Leydig cell indicating its function in germ cell homeostasis (Print and Loveland, 2000) **(Fig.5.2)**

5.5.2. Endocrine Control of Apoptosis

Endocrine hormones such as testosterone and the FSH and LH have been known to influence germ cell apoptosis. Withdrawl of gonadotropin and testosterone causes apoptosis of spermatogonia. But this occurs through indirect effects, since hormone receptors are present on somatic but not germ cells. Accordingly, in both breeding and non-breeding state, the fate of male germ cells is dependent on extracellular signals and on the quality of internal environment, which influence endocrine state of the organisms (Young and Nelson, 2001). Oestrogen treatment also induced apoptosis of all germ cells including elongating spermatids. The mechanism by which androgen withdrawal induces germ cell death remains unclear. Possibly androgen withdrawal alters expression of the Bcl-2 family proteins in germ cells, since Bcl-x_L and Bcl-2 expression in the testis was altered following long term anti-androgen treatment for prostate cancer. The relationship between apoptosis and apoptosis related genes was studied for a short term using EDS to withdraw androgen support. Though the apoptotic index enhanced, 8 days after EDS administration as demonstrated by fragmented DNA, there was no significant change in the levels of clusterin, Bcl-xl, Bak, and Bad, where as the expression of Bcl-2 and Bax was up-regulated at 8 days after EDS administration. The induction of Bax at this time suggests that it may play a role in germ cell apoptosis following androgen withdrawl. The concomitant elevation in Bcl-2 expression and also a decline in the expression of Fas-L and Fas-R in the pachytene spermatocytes and spermatids may represent a survival mechanism for the remaining germ cells (Woolveridge et al., 1999). It appears that a network of signals, including the endocrine, paracrine and direct cell contact signals described combine to maintain homeostasis of each germ cell type at each developmental stage.

5.5.3. Selective Apoptosis of Damaged Germ Cells

Apoptosis of male germ cells occurs when spermatogenesis cannot proceed beyond certain points due to molecular insufficiencies. For example, spermatocytes of mice lacking the heat shock protein 70.2 or the DNA mismatch repair protein Mih1 could not complete meiosis and were subsequently deleted by apoptosis. The high incidence of spermatocyte and round spermatid apoptosis has been described in infertile men. In addition, men with ataxia telangiectasia and mice lacking the ATM gene are infertile as a result of extensive spermatocyte apoptosis following failed meiosis. Apoptosis may also selectively remove damaged spermatogonia and damaged haploid spermatids. Vasectomy induces a large amount of germ cell apoptosis in association with activation of MAPKs. In contrast to the delayed phase up to 24 weeks after vasectomy, Shiraishi et al., (2002) observed hyperdynamic cellular turnover, spermatocyte loss through apoptosis and enhanced germ cell proliferation transiently at the early phase after vasectomy (Shiraishi et al., 2002).

Spermatids of mice lacking the CREM protein showed spermatogenic arrest in the first step of spermiogenesis and appeared to be subsequently removed by apoptosis. A variety of extracellular stimulii seem to mediate the selective apoptosis of damaged germ cells. Disruption of spermatogenesis and germ cell apoptosis in cryptorchid testis probably results from abnormal expression of temperature related genes. Three temperature-related expressed sequence (TRS) tags (ESTs), TRS1, TRS3, and TRS4, expressed specifically in scrotal testis have been related to apoptosis. The TRS1 was mainly expressed in the spermatocytes and the round spermatids, while TRS1, TRS2, and TRS3 were found to be temperature related ESTs during spermatogenesis (Cai-xia et al., 1999). Liu et al., (2003) cloned a gene, called TSARG3 gene, which is related to human testis spermatogenesis apoptosis. A mouse homolog of this gene was also identified. Galectin-1, a conserved beta-galactoside-binding protein, induces apoptosis of activated T cells and suppresses the development of autoimmune responses and chronic inflammation. In rat testes, expression of galectin-1 was developmentally regulated, and was present throughout the spermatogenic cycle including sperm, which were strongly labelled. During spermiation (stages VI-VIII), a strong labeling was observed at the luminal pole of seminiferous epithelium, and localized on apical stalks of Sertoli cells, on heads of mature spermatids, and in the cytoplasm of residual bodies (Dettin et al., 2003).

5.5.4. Sertoli Cell-Germ Cell Contact

Signals that promote proliferation can also promote apoptosis. Thus germ cells apoptosis also depends upon signals delivered from Sertoli cells by direct membrane contact. Of these, stem cell factor (SCF) has received the most attention. The mechanism though not fully understood but SCF is also required during the first wave of spermatogenesis (see chapter 17). The membrane bound SCF is expressed on the basal regions of Sertoli cells, while c-kit is expressed on the corresponding surface of spermatogonia. When SCF/c-kit interaction in adults was blocked in vivo, the incidence of apoptosis in spermatogonia and spermatocytes was increased (Packer et al., 1995). As stated earlier, Sertoli cells also depend on the Fas system to regulate germ cell fate, since Fas-L (ligand) is expressed on Sertoli cells, while Fas is expressed on the surface of adjacent spermatogonia. In rodent testes, germ cell apoptosis was accompanied by increased expression of Fas on the germ cell surface, and by increased expression of Fas ligand on Sertoli cells. Male mice that completely lack Fas are fertile, however, arguing that while Fas may contribute to germ cell homeostasis, it is not essential.

5.5.5. Phagocytosis of Apoptotic Cells

Apoptotic cells are rapidly removed by phagocytosis by cells in direct contact with apoptotic bodies. The phosphatidylserine (PS), which is normally hidden within the plasma membrane, is exposed on the surface during apoptosis. The PS is bound by receptors on phagocytic cells like macrophages and dendritic cells, which then engulf the cell fragments. The phagocytic cells secrete cytokines that inhibit inflammation (e.g., IL-10 and TGF-α). Among other molecules, which can recognize apoptotic cells are: lectins, $\alpha v\beta 3$ intergrin (vitronectin receptor)/ CD36 complexes, and scavenger receptor like molecule, all of which most likely capture target cells by binding to sugars, thrombospondin, and ABC transporter ABC1. Sertoli cells phagocytose degenerating spermatogenic cells and are thus likely to be involved in the elimination of apoptotic spermatogenic cells in the testis. However, little is known about the regulation of phagocytic function of Sertoli cells. From primary co-cultures of spermatogenic and Sertoli cells of the rat, it appeared that upon induction of spermatogenic cell apoptosis, PS and probably other acidic phospholipids, which are normally localized in the inner leaflet of the plasma membrane of Sertoli cells, translocate to the outer leaflet and serve as a signal for phagocytosis by Sertoli cells. The PS thus most likely serves as a common phagocytosis signal for several different phagocytes such as macrophages, vascular smooth muscle cells, and Sertoli cells, but the modes of recognition of target cells by these phagocytes are presumably somewhat different (Shiratsuchi et al., 1997; 1999).

5.6. REFERENCES

Boekelheide K, Lee J, Shipp EB, Richburg JH, Li G. Expression of Fas system-related genes in the testis during development and after toxicant exposure. Toxicol Lett 1998; 28: 503-8.

Cai-xia G, Tie-shan T, Xiao-ming M, et al. Cloning of novel temperature related expressed sequnce tag in rat testis during spermatogenesis. Biochem Biophys Res Comm 1999; 258: 401-6.

Chung HT, Pae HO, Choi BM, Billiar TR, Kim Y -M. Nitric Oxide as a Bioregulator of apoptosis. Biochem Biophys Res Comm 2001; 282: 1075-79.

deRooij DG, Grootegoed JA. Spermatogonial stem cells. Curr opin cell Biol 1998; 10: 694-701.

Dettin L, Rubinstein N, Aoki A, Rabinovich GA, Maldonado CA. Regulated expression and ultrastructural localization of galectin-1, a proapoptotic beta-galactoside-binding lectin, during spermatogenesis in rat testis. Biol Reprod. 2003; 68: 51-9.

Embree-ku M, Venturini D, Boekelheide K. Fas is involved in the p53-dependent apoptotic response to ionizing radiation in mouse testis. Biol Reprod 2002; 66: 1456-61.
Furuchi T, Masuko K, Nishimune Y, Obinata M, Matsui Y. Inhibition of testicular germ cell apoptosis and differentiation in mice misexppressing Bcl-2 in spermatogonia. Dev 1996; 122: 1703-9.
Hasegawa MH, Zhang Y, Niibe H, Terry NHA, Meistrich ML. Resistance of differentiating spermatogonia to radiation-induced apoptosis and loss in p53-deficient Mice. Radiat Res 1998; 149: 263-70.
Hikim AP, Lue Y, Yamamoto CM, Vera Y, Rodriguez S, Yen PH, Soeng K, Wang C, Swerdloff RS. Key apoptotic pathways for heat-induced programmed germ cell death in the testis. Endocrinology. 2003; 144: 3167-75.
Honarpour N, Du C, Richardson JA, Hammer RE, Wang X, Herz J. Adult Apaf-1-deficient mice exhibit male infertility. Dev Biol 2000; 218: 248-58.
Kasai S, Chuma S, Motoyama N, Nakatsuji N. Haploinsufficiency of Bcl-x leads to male-specific defects in fetal germ cells: differential regulation of germ cell apoptosis between the sexes. Dev Biol 2003; 264: 202-16.
Lawen A. Apoptosis-An introduction. Bioessays 2003; 25: 888-96
Lee J, Richburg JH, Younkin SC, Boekelheide MB. The Fas system is a key regulator of germ cell apoptosis in the testis. Endocrinology 1997; 138: 2081-88.
Liu G, Lu G, Fu J, Liu S, Xing X, Li L. Molecular cloning of TSARG3 gene related to apoptosis in human spermatogenic cells [Article in Chinese]. Zhonghua Yi Xue Yi Chuan Xue Za Zhi. 2003;20:
Lue YH, Lasley BL, Laughlin LS, Swerdloff RS, Hikim AP, Leung A, Overstreet JW, Wang C. Mild testicular hyperthermia induces profound transitional spermatogenic suppression through increased germ cell apoptosis in adult cynomolgus monkeys (Macaca fascicularis). J Androl. 2002; 23: 799-805.
Matsuzawa A, Iehijo H. Molecular Mechanisms of the decisions between life and death regulation of Apoptosis by apoptosis signal regulating kinase I. J. Biochem 2001; 130: 1-8.
Nakamuta N, Kobayashi S. Expression of p63 in the testis of mouse embryos. J Vet Med Sci 2003;65:853-6.
Ogi S, Tanji N, Yokoyama M, Takeuchi M, Terada N. Involvement of Fas in the apoptosis of mouse germ cells induced by experimental crytorchidism. Urol Res 1998; 26: 17-21.
Packer AI, Besmer P, Bacharova RF. Kit ligand mediates survival of type A spermatogonia and dividing spermatocytes in postnatal mouse testes. Mol Reprod Dev 1995; 42: 303-10.
Pentikainen V, Erkkila K, Dunkel L. Fas regulates germ cell apoptosis in the human testis in vitro. Am J Physiol 1999; 276: 310-16.
Print CG, Loveland KL, Gibson L, Meehan T, Stylianou A et al. Apoptosis regulator bcl-w is essential for spermatogenesis but appears otherwise redundant. Proc Natl Acad Sci USA 1998; 95: 12424-31.
Print CG, Loveland KL. Germ cell suicide: New insights into Apoptosis during spermatogenesis. Bioessays 2000; 22: 423-29.
Ross AJ, Amy SP, Mahar PL, Lindsten T, Knudson CM, Thompson CB, Korsemeyyer SJ, MacGregor GR. Bcl_w mediates survival of postmitotic Sertoli cells by regulating BAX activity. Dev Biol 2001; 239: 295-308.
Saitoh M, Nishitoh H, Fujii M, Takeda K, Tobiume K et al., Mammalian thioredoxin is a direct inhibitor of apoptosis signal regulating kinase (ASK) The EMBO Journal 1998; 17: 2596-06.
Shimizu S, Narita, Tsujimoto Y. Bcl-2 family proteins regulate the release of apoptogenic cytochrome c by the mitochondrial channel VDAC. Nature 1999; 399: 483-87.
Shiraishi K, Yoshida K, Fujimiya T, Naito K. Activation of mitogen activated protein kinases and apoptosis of germ cells after vasectomy in the rat. J Urol. 2002; 168:1273-8.
Shiratsuchi A, Kawasaki Y, Ikemoto M, Arai H, Nakanishi Y. Role of class B scavenger receptor type I in phagocytosis of apoptoic rat spermatogenic cells by Sertoli cells. J Biol Chem 1999; 274: 5901-8.
Shiratsuchi A, Umeda M, Ohba Y, Nakanish Y. Recognition of phosphatidylserine on the surface of apoptosis spermatogenic cells and subsequent phagocytosis by Sertoli cells of the rat. J Biol Chem 1997;272:2354-58.
Sinha Hikim AP, Lue Y, Diaz-Romero M, Yen PH, Wang C, Swerdloff RS. Deciphering the pathways of germ cell apoptosis in the testis. J Steroid Biochem Mol Biol 2003;85:175-82.
Song Q, Kuang Y, Dixit VM, Vincenz C. Boo, a novel negative regulator of cell death, interacts with Apaf-1. The EMBO Journal 1999; 18: 167-78.
Song Z, Steller H. Death by design: Mechanism and control of apoptosis. TIBS 1999; 24: 968-4.
Srivatsava RK, Mi QS, Hardwick JM, Longo DL. Deletion of the loop region of bcl-2 completely blocks paclitaxel-induced apoptosis. Proc Natl Acad Sci USA 1999; 96: 3775-80.
Sugiyama N, Obinata M, and Matsui Y. Bcl-2 inhibits apoptosis of spermatogonia and growth of spermatogonial stem cells in a cell-intrinsic manner. Mol Reprod and Dev 2001; 58: 30-8.
Woolveridge I, de Boer Brouwer M, Taylor MF, Teeds KJ, Wu FC, Morris ID. Apoptosis in the rat spermatogenic epithelium following androgen withdrawl: changes in apoptosis-related genes.Biol Reprod1999;60:461-70.
Yin Y, DeWolf WC, Morgentaler A. Experimental crytorchidism induces testicular germ cell apoptosis by p53-dependent and independent pathways in mice. Biol Reprod 1998; 58: 492-96.
Young KA, Nelson RJ. Mediation of seasonal testicular regression by apoptosis. Reproduction 2001; 122: 677-85.

Chapter 6

NUCLEAR SKELETON PROTEINS: NON-HISTONES

6.1. CHROMOSOMAL ORGANIZATION

The function of chromosomes is to carry the genetic information from one cell generation to another. The replication of chromosome depends on the precise mechanism of replication. The major chemical components of chromosomes include DNA (35%), RNA (5%) and proteins (60%), which comprise histone proteins and non-histone proteins. In situ hybridization to telomeric cDNA showed that chromosomes are arranged tandemly and in a defined order in the sperm nucleus. Among five types of histones, which have been differently designated, H1 histone is most easily removed and so is least tightly bound. Histones are rich in basic amino acids and hence histones are positively charged and combine with negatively charged DNA to form DNA-Histone complex. Histone proteins play a structural role rather than a regulatory role. However, chromatin reconstitution and other experiments established that histones do play a regulatory role as well. This regulatory role of histones is more of general than of nature specific and is exercised by repressing the activity of genes. The basic chromatin structure involving DNA and histones was proposed by R.D. Kornberg and J.O. Thomas in 1974. In this model DNA interacts with a tetramer ($H3_2$-$H4_2$) and two molecules of an oligomer (H2A-H2B), so that a tetramer involving two molecules each of the histones H3 and H4, is associated with two molecules each of the histones H2A and H2B and with 200 base pairs of DNA making a repeating unit. One molecule of H1 is also associated with each repeating unit. P. Oudet et al., (1975) proposed the term nucleosome for repeating units, which were observed as beads on strings in electron microscope. The non-histone proteins display more but still limited diversity and include RNA polymerase. Heterogeneity of these proteins indicates that these proteins are not as conserved in evolution as histones. The non-histone proteins differ between different tissues of the same organism suggesting that they regulate the activity of specific genes. Chromatin reconstitution experiments confirmed and established in a number of cases that specific non-histone proteins switch on specific genes. Certain non-histones induce RNA synthesis in vitro.

6.2. CELL DIVISION

Spermatogenesis can be divided into three principal phases. In the first phase called spermatocytogenesis, the most primitive spermatogonia proliferate by mitotic division to replace

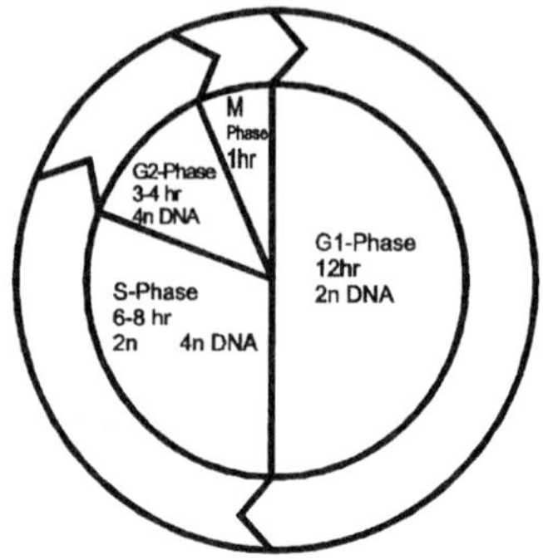

Fig 6.1. Cell cycle of a diploid cell.

themselves by successive generations of spermatogonia, each more differentiated than the preceding one. During last stage of mitosis, type B spermatogonia yield preleptone spermatocytes. The second phase of spermatogenesis is marked by meiosis in which spermatocytes undergo maturation divisions in which chromosome number is reduced to half and produce a cluster of cells called spermatids. During last phase of spermatogenesis, called spermiogenesis, the spermatids undergo a series of morphological transformation leading to formation of spermatozoa. Therefore brief description of arrangement of chromosomes during mitosis and meiosis will be necessary before taking up molecular basis of spermatogenesis.

6.2.1. Mitosis

In continuously dividing cells, an individual cell passes through four phases, shown in figure (**Fig.6.1**). The figure depicts that prior to mitotic phase cell prepares for cell division, marked by interphase. Interphase can be divided by three phases G_1, S and G_2 where G_1 is a resting phase, followed by a synthetic S phase. During S phase, DNA synthesis takes place and G_2 phase is again a resting phase following DNA synthesis. The three phases viz. G_1, S and G_2 constitute interphase. The mitotic division takes place during M phase. During interphase there is little microcopically resolvable structure. The mitotic phase consists of the following stages (Gupta, 1999; Bloom and Fawsett,1970).

Prophase: Although the DNA synthesis is already completed during interphase, at prophase of mitosis, the cell is still preparing for division. During initial stage of prophase chromosomes appear as thin, filamentous uncoiled structures. As the time passes, chromosomes continue to condense, and become thicker and distinct. Each chromosome is double consisting of two sister chromatids. Chromatids are functional units of chromosomes. An important characteristic of mitotic prophase is longitudinal splitting of each chromosome into two sister chromatids. Double structure of each chromosome thus may be conspicusous at late prophase. Sister chromatids at this stage are attached only at centromere. At this position two chromatids will remain attached to spindle tubule. Soon after, nucleolous diminishes in size and disappears before the cell enters mitotic metaphase and the nuclear membrane disappears at late prophase. The disappearance of nuclear envelope brings the end of prophase.

Metaphase: The mitotic spindle tubules start appearing and chromosomes begin to gather in the same plane at late prophase or early metaphase. The spindle tubules get attached at primary constriction of chromosomes called centromeres. The chromosomes begin active movement leading to arrangement of chromosomes in the center or at equatorial plate. Spindle is an organelle responsible for orderly arrangement and separation of chromosome strands late during cell division. The spindle apparatus, which helps in arrangement of chromosomes at the plate, is formed with the help of centrosome particularly in case of animal cells. Centrosome comprises of two centrioles, which separate and move to opposite poles of the spindle at the time of spindle formation. As the two centrioles separate, astral rays appear radiating out into the surrounding cytoplasm from each centriole. These rays when join form spindle fibres. However, in most of the plant cells, centrosomes are missing, but spindle is still formed.

Anaphase: Following spindle formation and arrangement of chromosomes on equatorial plate at metaphase, chromosomes separate and split at centromere also. The two sister chromatids appear as a separate structures, which are called chromosomes. At the close of anaphase the two groups of chromatids, now called daughter chromosomes are separated and move towards opposite poles where they gather near spindle poles. How chromosomes move at anaphase, is still a matter of debate. It seems that there is repulsion between centrosomes and that there is contraction of spindle fibres, which help the movement. Kinesin and cytoplasmic dynein ATPases seem to be involved in this process (Chapter 9).

Telophase: After anaphase, chromosomes reach the poles, the discontinuous segments of nuclear membrane is reconstructed around chromosomes grouped at the poles giving rise to one nucleus at each pole. The nucleoli also reappear at this stage. With these events karyokinesis is complete. Karyokinesis is followed by cytokinesis in which cytoplasm is divided leading to the formation of two cells. The two daughter nuclei are genetically identical and attain interphase condition.

6.2.2. Meiosis

All somatic cells divide by mitosis. But in reproductive cells, cell division occurs through a process, called meiosis. In sexual method of reproduction, fusion of sex cells-spermatozoa and ovum is essential. Meiotic division consists of two successive divisions of a cell, so that one complete meiotic division results into four daughter cells. The first meiotic division results in separation of two members of each pair of homologous chromosomes followed by reduction in number of chromosomes to half without any division of chromosomes, while the second meiotic division involves separation of chromatids of the chromosomes. Consequently, number of chromosomes, which is reduced during first meiotic division, remains constant (haploid) during second division. Therefore, meiosis is described in two parts: (i) first meiotic division, and (ii) second meiotic division. Preceding meiosis, there is also an interphase just like the one described in mitosis, consisting of G1 phase, S phase and G2 phase. However, in meiosis G2 phase is of short duration or altogether absent, so that meiotic division takes over just after DNA synthesis is complete. Gametes formed after meiosis consist half the number of chromosomes as compared to somatic cells and are described as haploid. The male and female haploid cells after fertilization, restore the normal chromosome number in fertilized egg (zygote).

First Meiotic Division

First meiotic division is more important than the second meiotic division, since it is the reduction division, while second division is like a mitotic division and is called equational division. Like any cell division, first meiotic division also consists of prophase, metaphase, anaphase and telophase.

(a). First Prophase: First prophase is of a very long duration and is also complex, differing from mitotic prophase in several respects. For convenience, first prophase has been subdivided into five stages, namely, leptotene, zygotene, pachytene, diplotene and diakinesis.

(i). Leptotene (leptonema).This is the earliest stage of meiosis following interphase. At this stage chromosomes appear as long thin, thread-like structures, which are loosely interwoven. On these thread like chromosomes, bead-like structures called chromomeres are found all along the length of chromosomes. In some animals, particularly in insects, chromosomes have their ends drawn together at a point on nuclear membrane, near centriole. The polarization of chromosome ends is not seen in plants, and appears to be due to absence of centrosome.

(ii). Zygotene (zygonema). Zygotene is characterized by pairing of homologous chromosomes (synapsis). This pairing occurs between homologous segments even if they are present in non-homologous chromosomes (e.g. in case of translocations). The pairing is brought about in a zipper-like fashion and may start at centromere, at chromosome ends or at any other position. Another important feature of this pairing is that pairing is allowed only between two chromosomes in one region. For instance, in an autotetraploid, where there are four homologous chromosomes, only two chromosomes in a particular region will pair. However, the pairing between giant chromosomes in somatic cells of salivary glands of *Drosophila* is exception and may not result into meiotic pairing. Moreover, the factors, which bring about pairing of chromosomes in meiotic prophase remain unresolved.

(iii). Pachytene (pachynema). After chromosomes pairing at zygotene, the chromosomes begin to contract and the cell enters the stage of pachytene, where chromosomes are shortened and coiled, and chromosomes appear as thickened thread-like structures, haploid in number. Each thread, however, if carefully examined has two homologous chromosomes closely appressed against each other. Such appressed pairs of homologous chromosomes are called bivalents. Each chromosome in a bivalent form has two chromatids, and each bivalent consists of four chromatids, called a tetrad. Crossing over or exchange of segments of chromatids is brought about at this stage. The nucleolus still persists. Attached to nucleolus can be seen the nuclear organizing bivalent.

(iv). Diplotene (diplonema). At diplotene thickening and shortening of chromosomes take place further and paired chromosomes start separating from one another along their length. This separation starts at centromere, and travels towards the ends. This type of separation from centromere towards the ends is known as teminalization. After separation, dual nature of a bivalent becomes distinct and hence the name diplotene. Homologous half chromosomes make contact across one another and are held together only at certain points along the length. Such points of contact between homologous chromosomes are known as chiasmata (singular, chiasma) and represent the places of crossing over or exchange of chromosomal segments leading to a genetic phenomenon of "crossing over". As terminalization occurs, these chiasmata move towards the ends of chromosomes. During crossing over at one point, only two chromatids, one from each of the two homologous chromosomes, participate. Chiasmata are however, not the cause but are the consequence of crossing over. The mechanisms of crossing over and chiasma formation have been reviewed (Gupta, 1999).

(v) Diakinesis. At diakinesis, chromosomes continue to undergo further contraction, which is the only distinction between diplotene and diakinesis. Nucleolus disappears at this stage. Due to further terminalization and contraction, bivalents appear as rounded bodies darkly stained and evenly scattered throughout the cell. As a result chiasmata are mainly terminal and chromosome counting becomes easy at this stage.

(b) First Metaphase: At metaphase chromosomes are most condensed and have a smooth outline. The spindle apparatus starts appearing and bivalents become attached to spindle through centromeres as in mitosis. Bivalents then appear in the form of metaphase plate, due to the movement known as congression.

(c) First Anaphase: The movement of chromosomes of a bivalent from equatorial plate to poles constitutes first anaphase. In first anaphase of meiosis, sister chromatids do not separate but go to the same pole and thus differs from mitotic anaphase. In other words, when sister chromatids go to same pole, it is called a reductional or disjunctional division; whereas when they separate and go to two poles it is an equational division. After anaphase I, each pole has a haploid number of chromosomes and thus the chromosome number is reduced. The meiotic division is also called a reduction division, due to reduction in chromosome number.

(d) First Telophase and Interphase : At first telophase, nuclear membranes are formed around the groups of chromosomes at the two poles. In mammals, nuclei are reconstituted for a brief time but in other forms the chromosomes may proceed directly into second division without full reconstitution of nucleus. After formation of nuclei, chromosomes pass into a small interphase before the second meiotic division will start. First meiotic division, which is completed at first telophase, may be followed by cytokinesis giving rise to a dyad. Such a division is called successive division. However, cytokinesis may be postponed till the end of second division, when four cells are formed due to simultaneous division.

Second Meiotic Division

First meiotic division is followed by a second meiotic division with or without intervening interphase. The second meiotic division is identical to that seen in mitotic division and is sometimes referred to as meiotic mitosis. At second prophase, chromosomes are already double, each having two sister chromatids with a single functional centromere. These chromosomes soon arrange at metaphase plate during second metaphase. The centromere, then splits and two chromatids, which may now be called chromosomes pass to two poles during second anaphase. This is soon followed by second telophase and cytokinesis.

6.3. THE SYNAPTONEMAL COMPLEX

Synaptonemal complex (SC) is a feature of meiotic prophase. Synaptonemal complex is a tripartite structure usually found between the two-paired homologous chromosomes of each bivalent in all animal and plant nuclei undergoing meiosis. This is believed to be a physical structure, which is associated with synapsis of homologous chromosomes. The synaptonemal complex is a configuration in which the 10nm fibres of the chromosomes are arranged into a superstructure that can be viewed under the electron microscope. It is composed of three parallel, electron dense elements that are separated by less dense areas. The two lateral elements seem to be composed of fibers that are slightly wider than 10 nm and are called synaptomeres. They vary in structure at different stages of meiotic prophase 1 within a species. The central element is ladder like configuration in the center of SC. In some species it is comparatively more pronounced

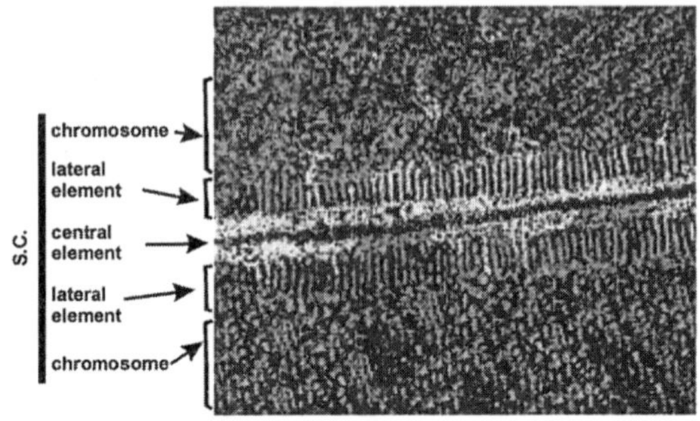

Fig.6.2. The electron micrograph of a bivalent with a synaptonemal complex showing central and lateral elements.

(Fig. 6.2). The transverse elements are electron dense filaments that interconnect the central element with the lateral elements. The lateral regions establish contacts with DNA of the sister chromatids of each homologue, and the central region connects the two lateral regions through an array of transverse filaments. The lateral elements may be spaced at a distance ranging from 20-30nm to as much as 100-125nm. The lateral elements are rich in DNA, RNA and proteins, but the central element contains mainly RNA, protein and some DNA. Sometimes SC like structures are not associated with synapsed chromosomes. The lateral elements of these anomalous complexes may differ in size and density from those in the true synaptonemal complexes of that species. However, the true autosomal synaptonemal complex is always a unit tripartite complex but anomalous complexes are often multiple stacks of alternating lateral elements and central elements.

The synaptonemal complex is more directly related to the process of recombination. The synaptonemal complex appears mainly to be a protein framework that is responsible for proper alignment of the homologous chromosomes. Inhibition of DNA synthesis, which occurs at very low level, at meiotic prophase, can arrest the functions of SC. However, during recombination by crossing over, it is necessary that DNA of the paired chromatids should reach the central component of SC within a distance of $\simeq 1.0\ \mu m$ for the recombination to occur. At diplonema the SC is shed from the bivalents with the exceptions of the regions in which the repelling homologues are held together each by a chiasma. Thus, a chiasma contains a piece of synaptonemal complex that will finally disappear and would be replaced by a chromatin bridge.

Unlike eutherian males, pairing of the sex chromosomes in marsupial males during the first meiotic prophase is not mediated by a synaptonemal complex. Instead, a specific structure, the dense plate, develops during pachytene between the sex chromosomes. It was proposed that the unique modifications of the composition and structure of the axial elements of the sex chromosomes in meiotic prophase might result in the prescription of synaptonemal complex formation between male marsupial sex chromosomes, where the dense plate is an extension of the axial elements of sex chromosomes. This replaces synapsis to maintain X and Y association during first meiotic prophase (Page et al., 2003).

6.4. SYNAPTONEMAL COMPLEX PROTEINS

The transverse filaments mediate synapsis and contain a 125-kDa protein, termed SYN1 in hamster (synapsis; Dobson et al., 1994) or SCP1 in rat/mouse (synaptonemal complex protein 1. The SYN1/SCP1 is recruited to the SC at zygotene and it fully assembled at pachytene, when complete synapsis is detected. By late pachytene, the gradual displacement of SYN1/SCP1 from the chromosomes initiates, marking the beginning of SC dissolution (Dobson et al., 1994). The first identified component of the axial elements (cores) of the lateral regions of the SC was termed SCP3 in rat/mouse (synaptonemal complex protein 3), and COR1 in hamster. A second component of the meiotic chromosome cores, termed SCP2 in rat and human has been identified (Offenberg et al., 1998; Schalk et al., 1998). The displacement and degradation of SYN1/SCP1 and COR1/SCP3 are precisely programmed. Both proteins with ubiquitin conjugating enzyme Ubc9 indicated that ubiquitination could be the signal for initiation of their degradation (Tarsounas et al., 1999).

Monoclonal antibodies against synaptonemal complex proteins recognized (i) a 190-kDa polypeptide, (ii) a 30 and a 33-kDa polypeptide, (iii) two polypeptides with molecular weights of about 120-kDa and (iv) polypeptides with molecular weights of 66-55-kDa within spermatocytes. The 66 to 55-kDa polypeptides were not confined to SC; rather, these polypeptides appeared to be chromosomal components. The 190, 30, and 33-kDa polypeptides were part of the lateral elements of paired as well as unpaired segments of SCs. The 120-kDa polypeptides were localized on the inner edge of the lateral elements, specifically in paired segments of SC. The 190, 120, 30 and 33-kDa polypeptides turned out to be specific for nuclei of zygotene up to and including diplotene spermatocytes. Only in some early spermatids, the 190-, 30-, and 33-kDa polypeptides could be detected presumably in remnants of synaptonemal complexes and seemed to arise by rearrangements of pre-existing components in the nucleus but their major components were synthesized in meiotic prophase (Heyting et al., 1989). A 65-kDa protein is present in rat SC called SC65 (Chen et al., 1992). The longest open reading frame, initiating at an ATG codon, encodes a protein of 431 amino acids, with a relative molecular mass of 50kDa. The SC65 was located on the SC between the pairing faces of the parallel-aligned cores of homologues chromosomes in spermatocytes. However, the product was not specific since SC65 gene is transcribed in testis, brain, and heart at similar levels, and in the liver at a much lower level and could be a similar product of 65-kDa shown by Heyting et al, (1989). Another 220-kDa protein is present during meiosis and spermiogenesis. Its localization in the nuclear matrix suggests the participation of pre-existing nucleoskeletal proteins in the molecular organization of the SC and its dynamics during meiotic prophase (Gil-Alberdi and del Mazo, 1992).

6.4.1. Synaptonemal Complex Protein-1

In yeast, several defective mutants of meiosis have been shown to be defective in SC assembly or disassembly. The transverse filaments mediate synapsis and contain a 125-kDa protein, termed SYN1 in hamster or SCP1 in rat/mouse (synaptonemal complex protein 1). The SCP1 protein contains a long central coiled-coil motif and the molecules are probably organized as dimmers, each forming a coiled-coil fiber. The N-terminal end of the SCP1 protein is located within the central element of the synaptonemal complex, whereas the C-terminal end is close to or within the lateral element of the synaptonemal complex. This supports the notion that SCP1 is an extended filamentous protein and that the two molecules of SCP1 dimmer are likely to have the same polarity. SCP1 dimmers, anchored in opposite lateral elements, could establish contact with each other in the central element via their N-termini. Liu et al., (1996) found that the N-terminal end of the SCP1 protein indeed interacted with itself, but not with other protein domains tested. It was suggested that a transversal filament consists of one or more pairs of SCP1 dimmers, each pair being organized in a head to head arrangement with the C-termini anchored in the lateral elements and the two N-termini being joined in the central element (Liu et al., 1996). The SYN1/SCP1 is recruited to the SC at zygotene and it fully assembled at pachytene, when complete synapsis is detected. By late pachytene, the gradual displacement of SYN1/SCP1 from the chromosomes initiates, marking the beginning of SC dissolution (Dobson et al., 1994). A cDNA encoding SCP1 has been isolated from the rat and human and is believed to be a major component of the transverse filaments of SCs. The human homologue of SCP1 is mapped to human chromosome 1p13 (Kondoh et al., 1997). The *Scp1* gene probe hybridized with two sequences on different chromosome; Scp-rs2 was mapped to Chr3, whereas Scp1-rs3 was mapped to Chr7 (Taketo et al., 1997).

The SCP1 consists of three domains: i) a short, proline rich N-terminal part, ii) a stretch of 700 amino acid residues capable of forming an amphipathic α–helix, and iii) a C-terminal domain of 240 amino acid residues, that is capable of binding to DNA. Polyclonal antibodies against three non-overlapping fragments of SCP1 were used to study the orientation of SCP1 in sections of rat testis. On the basis of the distribution of immunogold label and the predicted secondary structure and dimensions of SCP1 molecules, it was suggested that: the C-terminus of SCP1 lies in the inner half of the LE, the molecules protrude from the LE through the central region into the CE, and end up with their N-terminus between the center of the CE and the opposite LE, so that the N-termini of SCP1 molecules from opposite LEs overlap. The model has several implications for the assembly of SCs and the possible functions of SCP1 (Schmekel et al., 1996). The SCP1, evolved by specialization of a nuclear matrix protein (Meuwissen et al., 1992), consists of 946 amino acid residues and has a molecular weight of 111-kDa. It shares several features with nuclear lamins and other nuclear matrix proteins. The major part of SCP1 consisting of long stretches capable of forming amphipathic helices shows amino acid similarity to coiled coil region of myosin heavy chain. A leucine zipper is included in this region. The carboxy terminus has two small basic domains and several S/T-P-X-X motifs, which are characteristic of DNA binding proteins. One of these motifs is a potential target site for $p34^{cdc2}$ protein kinase. The amino terminus is acidic and relatively proline rich, but does not contain the S/T-P-X-X motif (**Fig. 6.3**). The transcription of the gene encoding SCP1 is restricted to zygotene diplotene spermatocytes.

Constructs with *Scp1* 5' upstream sequences directed the expression of reporter genes to pachytene spermatocytes in transgenic mice. A short fragment encompassing the transcription start (nt -54 to +102) was sufficient for stage specific expression and for temporal regulation during development. Upstream enhancer element(s) quantitatively regulating expression were localized in the region between –54 and –260. The gene is normally expressed both in the male and female gonads, but none of the promoter sequences active in the testis allowed the expression of reporter genes during meiosis in the ovary (Sage et al., 1999).

SCP1 as Cancer-testis Antigen (CTA): The *Scp-1* is also expressed selectively in a variety of neoplastic tissues and tumor cell lines. The *Scp-1* differs from other members of the class of CTA by its localization on chromosome-1 and its frequent expression in malignant gliomas. The aberrant expression of SCP-1 in tumors might contribute to their genomic instability and suggests that the functional role of other CTA might also relate to meiosis (Tureci et al., 1998) (see Chapter 32).

Ott: Ott (ovary-testis transcribed) homologul of the rat SC protein SCP1 is expressed in embryonic ovary, adult brain and testis. At least seven *Ott* genes are transcribed specifically during meiosis and are predicted to encode "pioneer" proteins with an unusual structure, containing

```
GGAATTCGGCACGAGTCGGGCTTCTAGGAGGTTCTGAGGCAGAAGCTCTCAGTTCCATTCGATGTCTTCATAAAGAAGCGCTCGCGGGCACGGAGCCTGCCATCGGCCCATGGAGAAGC
AGAAGCGGAGCCTGCCATCGGGCCATGGAGAAGCAGAAGCCTTCACGTTGTTCGTCCCACCGACACTGAGCAGCAGTCAAGTGTCGGCTGTGAAACCTCAGACGGCGGGAGGAGACTCC
AACTACTTCAAGACTGTCAACAAATGCACAGAAGGTGATTTTGGTGTTCCATTGACA

   1 ATG TCC AGT CTA TCA AAA AAT CGG GAA AAC ATT GAT ACA GAT CCC GCT TTT CAA AAA CTT AGC ATT TTG CCC ATG CTT GAA CAG GTT GCA
   1 M   S   S   L   S   K   N   R   E   N   I   D   T   D   P   A   F   Q   K   L   S   I   L   P   M   L   E   Q   V   A

  91 AAT TCT GGC AGT TGC CAC TAT CAG GAA GGA GTA AAT GAC TCT GAT TTT GAG AAT TCA GAG CCA ATG AGC AGA CTG TAC TCA AAG CTG TAT
  31 N   S   G   S   C   H   Y   Q   E   G   V   N   D   S   D   F   E   N   S   E   P   M   S   R   L   Y   S   K   L   Y

 181 AAA GAG GCT GAA AAG ATA AAA AAG TGG AAA GTG AGC ATA GAG TCT GAA CTG AAG CAG AAA GAA AAT AAG TTG CAA GAA AAC AGA AAG ATA
  61 K   E   A   E   K   I   K   K   W   K   V   S   I   E   S   E   L   K   Q   K   E   N   K   L   Q   E   N   R   K   I

 271 ATT GAA GCC CAG CGA AAA GCC ATT CAG GAA CTT CAG TTT GAA AAT GAA AAA GTA AGC TTG AAA TTA GAA GAA GAA ATT CAA GAA AAT AAA
  91 I   E   A   Q   R   K   A   I   Q   E   L   Q   F   E   N   E   K   V   S   L   K   L   E   E   E   I   Q   E   N   K

 361 GAT TTA ATC AAG GAG AAT AAT GCT ACA AGA CAT TGG TGT AAT TTA CTC AAG GAA ACC TGT GCT AGA TCT GCA GAA AAG ACA AGT AAA TAT
 121 D   L   I   K   E   N   N   A   T   R   H   W   C   N   L   L   K   E   T   C   A   R   S   A   E   K   T   S   K   Y

 451 GAA TAT GAG CGA GAA GAA ACC AGA CAA GTT TAT GTG GAT CTA AAT AAT AAC ATT GAG AAA ATG ATA CTA GCT TTT GAG GAA CTT CGT GTG
 151 E   Y   E   R   E   E   T   R   Q   V   Y   V   D   L   N   N   N   I   E   K   M   I   L   A   F   E   E   L   R   V

 541 CAA GCT GAG AAT GCC AGG CTG GAA ATG CAC TTT AAG TTA AAG GAA GAT CAT GAA AAA ATC CAA CAT CTT GAA GAA GAA TAT CAG AAG GAA
 181 Q   A   E   N   A   R   L   E   M   H   F   K   L   K   E   D   H   E   K   I   Q   H   L   E   E   E   Y   Q   K   E

 631 GTA AAC AAC AAG GAA AAC CAG GTA TCA CTA CTA TTG ATC CAA AGT ACT GAG AAA GAA AAT AAA ATG AAA GAT TTA ACA TTT CTG CTA GAG
 211 V   N   N   K   E   N   Q   V   S   L   L   L   I   Q   S   T   E   K   E   N   K   M   K   D   L   T   F   L   L   E

 721 GAA TCC AGA GAT AAA GCT AAT CAA TTA GAG GAA AAA ACA AAA TTA CAA GAT GAA AAC TTA AAA GAA TTA AAT GAA AAG AAG GAT CAT TTA
 241 E   S   R   D   K   A   N   Q   L   E   E   K   T   K   L   Q   D   E   N   L   K   E   L   N   E   K   K   D   H   L

 811 ACA TCA GAA CTT GAA GAT ATT AAA ATG TCC ATG CAA AGA AGT ATG AGC ACT CAG AAG ACT TTA GAG GAA GAT TTA CAG ATA GCA ACA AAA
 271 T   S   E   L   E   D   I   K   M   S   M   Q   R   S   M   S   T   Q   K   T   L   E   E   D   L   Q   I   A   T   K

 901 ACG ATT TAT CAG CTC ACT GAA GAA AAA GAA GCT CAA ATG GAA GAA CTC AAC AAA GCT AAA ACT ACT CAC TCA CTT GTG GTG ACT GAA CTT
 301 T   I   Y   Q   L   T   E   E   K   E   A   Q   M   E   E   L   N   K   A   K   T   T   H   S   L   V   V   T   E   L

 991 AAA GCC ACT ACA TGT ACC TTG GAG GAA TTA CTG AGA ACA GAA CAG CAA AGA TTG GAA AAT AAT GAG GAT CAA CTG AAA CTG ATT ACT ATG
 331 K   A   T   T   C   T   L   E   E   L   L   R   T   E   Q   Q   R   L   E   N   N   E   D   Q   L   K   L   I   T   M

1081 GAG CTC CAG AAG AAA TCA AGT GAA CTA GAA GAG ATG ACT AAA TTT AAA AAT AAC AAA GAA GTG GAA CTT GAA GAA TTA AAA ACC ATA TTG
 361 E   L   Q   K   K   S   S   E   L   E   E   M   T   K   F   K   N   N   K   E   V   E   L   E   E   L   K   T   I   L

1171 GCA GAA GAC CAA AAA CTT TTA GAT GAA AAG AAA CAA GTT GAG AAG CTT GCT GAA GAA TTA CAA GGG AAA GAA CAA GAA CTA ACT TTC CTT
 391 A   E   D   Q   K   L   L   D   E   K   K   Q   V   E   K   L   A   E   E   L   Q   G   K   E   Q   E   L   T   F   L

1261 TTG CAA ACC AGA GAG AAA GAA ATC CAT GAT TTG GAA GTA CAA GTA ACT GTC ACT AAA ACA AGT GAA GAA CAT TAT TTA AAA CAG GTT GAA
 421 L   Q   T   R   E   K   E   I   H   D   L   E   V   Q   V   T   V   T   K   T   S   E   E   H   Y   L   K   Q   V   E

1351 GAA ATG AAA ACT GAG CTT GAA AAA GAG AAA CTT AAG AAT ATT GAA TTA ACT GCA AAC TCT GAC ATG CTT TTG CTT GAG AAC AAA AAA TTG
 451 E   M   K   T   E   L   E   K   E   K   L   K   N   I   E   L   T   A   N   S   D   M   L   L   L   E   N   K   K   L

1441 GTA CAA GAA GCA AGT GAT ATG GTC CTA GAA CTC AAG AAA CAT CAA GAA GAT ATC ATT AAT TGC AAA AAG CAA GAA GAA AGG ATG TTG AAA
 481 V   Q   E   A   S   D   M   V   L   E   L   K   K   H   Q   E   D   I   I   N   C   K   K   Q   E   E   R   M   L   K

1531 CAA ATA GAA ACT TTG GAA GAA AAA GAA ATG AAT TTA AGG GAT GAA CTG GAA TCA GTA AGA AAA GAG TTC ATA CAG CAA GGA GAT GAA GTT
 511 Q   I   E   T   L   E   E   K   E   M   N   L   R   D   E   L   E   S   V   R   K   E   F   I   Q   Q   G   D   E   V

1621 AAA TGT AAA TTG GAC AAG AGT GAA GAA AAT GCT CGA AGC ATT GAA TAT GAA GTT TTA AAG AAA GAA AAG CAG ATG AAG ATA TTA GAA AAT
 541 K   C   K   L   D   K   S   E   E   N   A   R   S   I   E   Y   E   V   L   K   K   E   K   Q   M   K   I   L   E   N

1711 AAG TGT AAT AAT TTA AAG AAA CAA ATC GAA AAT AAA AGC AAG AAT ATT GAA GAG CTT CAC CAG GAG AAT AAA GCC TTG AAA AAA AAA AGT
 571 K   C   N   N   L   K   K   Q   I   E   N   K   S   K   N   I   E   E   L   H   Q   E   N   K   A   L   K   K   K   S

1801 TCA GCA GAA AAC AAA CAA CTG AAT GCA TAT GAG ATA AAG GTC AAT AAA TTA GAG TTG GAA TTA GCA AGT ACC AAG CAA AAA TTT GAA GAA
 601 S   A   E   N   K   Q   L   N   A   Y   E   I   K   V   N   K   L   E   L   E   L   A   S   T   K   Q   K   F   E   E

1891 ATG ATT AAC AAC TAC CAG AAA GAA ATT GAG ATA AAA AAG ATT TCA GAA GAA AAG CTT TTG GGA GAG GTT GAG AAA GCC AAA GCA ACA CTT
 631 M   I   N   N   Y   Q   K   E   I   E   I   K   K   I   S   E   E   K   L   L   G   E   V   E   K   A   K   A   T   V

1981 GAT GAA GCC GTA AAG TTA CAG AAA GAA ATT GAT TTA CGA TGC CAA CAT AAA ATA GCT GAG ATG GTA GCA CTT ATG GAA AAA CAT AAG CAC
 661 D   E   A   V   K   L   Q   K   E   I   D   L   R   C   Q   H   K   I   A   E   M   V   A   L   M   E   K   H   K   H

2071 CAA TAT GAT AAG ATT GTT GAA GAA AGA GAC TCA GAA TTA GGA CTT TAT AAA AAC AGA GAA CAG GAA CAG TCT TCA GCA AAG GTT GCT TTG
 691 Q   Y   D   K   I   V   E   E   R   D   S   E   L   G   L   Y   K   N   R   E   Q   E   Q   S   S   A   K   V   A   L

2161 GAG ACT GAA TTA TCT AAT ATC AGA AAC GAA CTT GTA TCC CTT AAG AAG CAA CTT GAA GTA GAA AAA GAA GAA AAA GAG AAA TTA AAA ATG
 721 E   T   E   L   S   N   I   R   N   E   L   V   S   L   K   K   Q   L   E   V   E   K   E   E   K   E   K   L   K   M

2251 GAA CAA GAA AAC ACA GCT ATT CTC ACA GAT AAA AAA GAC AAG AAA ATA CAG GCA TCT TTG CTG GAA TCA CCT GAA GCC ACT AGT TGG AAA
 751 E   Q   E   N   T   A   I   L   T   D   K   K   D   K   K   I   Q   A   S   L   L   E   S   P   E   A   T   S   W   K

2341 TTT GAT TCT AAA ACA ACT CCC TCA CAA AAT ATA TCT CGG CTT TCC TCA TCA ATG GAT AGT GGC AAA TCC AAA GAT AAC AGA GAT AGT CTG
 781 F   D   S   K   T   T   P   S   Q   N   I   S   R   L   S   S   S   M   D   S   G   K   S   K   D   N   R   D   S   L

2431 CGG GCA TCT GCC AAA AGC ATT TTA TCT ACA ACA GTT ACA AAG GAA TAT ACA GTG AAG ACA CCA ACT AAA AAG AGC ATA TAT CAA AGA GAA
 811 R   A   S   A   K   S   I   L   S   T   T   V   T   K   E   Y   T   V   K   T   P   T   K   K   S   I   Y   Q   R   E

2521 AAC AAG TAT TTA CCT ACT GGA GGA AGT AAT AAA AAG AGA AAA ACT GTC TTT GAA TTT GAT GTT AAT TCA GAT AGT TCA GAA ACT ACT GAT
 841 N   K   Y   L   P   T   G   G   S   N   K   K   R   K   T   V   F   E   F   D   V   N   S   D   S   S   E   T   T   D

2611 CTT TTG AGC TTG GTT TCA GAG GAA GAT ATA TCA AAC AGG ATT TAT AAT AAT AAT ACA CCA GAT TCT CAT CTA TTA GTC AAA ACT CCC AAA
 871 L   L   S   L   V   S   E   E   D   I   S   N   R   I   Y   N   N   N   T   P   D   S   H   L   L   V   K   T   P   K

2701 CAG ACT CCT TTA TCT TTA TCA ACT CCT GCA TCT TTT ACG AAG TTT GGA AGT CTG AAA AAA ATG AGA GAA GAC CGT TGG GCA ACG ATT GCT
 901 Q   T   P   L   S   L   S   T   P   A   S   F   T   K   F   G   S   L   K   K   M   R   E   D   R   W   A   T   I   A

2791 AAA ATT GAT AGG AAA AGA AGA CTA AAG GAA GCA GAA AAG TTA TTT ACT TAA
 931 K   I   D   R   K   R   R   L   K   E   A   E   K   L   F   T   *

TTTCAGAAAATCAATGTTGGTTATAGAGACTAATAGTTAACTTTATTATTTCCCAGAGAGCCAAACTTTACTGAGGAATTCAGACTTTAAAATTAATTACATAAATAATTTGTTTGTTAT
GTTTGTGGTCTAAATATAAAGTAACTATATTTGAATTTTTTTATTGTATTCAGATACTTAGATTTTGTTGCTTTTACTTGCAATGGATATTGAAATTCAAAGCTTTTTTTTGTTTTTGTT
TTTTTTTTTTTTTGTCTTCTCCTCCCTACCCCCAAGAGCTGAGGACCAAACCCAGTGCTTTGTGCTTGCTAGGCAAGCGCTCTACCACTGAGCTTAATCCCTAACCCTGAAGTTCAAATT
TTTTATTTTGAGCTTTTGACCATTAGAAATGCATTATCGAGACTTATAATTTATTCTTTATTATTAAAATATTTTGGATGCAAAAAAAAAAAAAAAAA
```

Fig.6.3: The nucleotide and amino acid sequence of rat synaptonemal complex protein 1 (SCP-1). Reprinted with permission from Meuwissen et al., The EMBO Journal 11; 5091-5100: 1992© http://www.nature.com/

```
   1 MPIRPDLQQL EKCIDDALRK NDFKPLKTLL QIDICEDVKI KCSKQFFHKV DNLICRELNK
  61 EDIHNVSAIL VSVGRCGKNI SVLGQAGLLT MIKQGLIQKM VAWFEKSKDI IQSQGNSKDE
 121 AVLNMIEDLV DLLLVIHDVS DEGKKQVVES FVPRICSLVI DSRVNICIQQ EIIKKMNAML
 181 DKMPQDARKI LSNQEMLILM SSMGERILDA GDYDLQVGIV EALCRMTTEK QRQELAHQWF
 241 SMDFIAKAFK RIKDSEFETD CRIFLNLVNG MLGDKRRVFT FPCLSAFLDK YELQIPSDEK
 301 LEEFWIDFNL GSQTLSFYIA GDNDDHQWEA VTVPEEKVQI YSIEVRESKK LLTIILKNTV
 361 KISKREGKEL LLYFDASLEI TNVTQKIFGA TKHRESIRKQ GISVAKTSLH ILFDASGSQI
 421 LVPESQISPV GEELVSLKEK SKSPKEFAKP SKYIKNSDKG NRNNSQLEKT TPSKRKMSEA
 481 SMIVSGADRY TMRSPVLFSN TSIPPRRRRI KPPLQMTSSA EKPSVSQTSE NRVDNAASLK
 541 SRSSEGRHRR DNIDKHIKTA KCVENTENKN VEFPNQNFSE LQDVIPDSQA AEKRDHTILP
 601 GVLDNICGNK IHSKWACWTP VTNIELCNNQ RASTSSGDTL NQDIVINKKL TKQKSSSSIS
 661 DHNSEGTGKV KYKKEQTDHI KIDKAEVEVC KKHNQQQNHP KYSGQKNTEN AKQSDWPVES
 721 ETTFKSVLLN KTIEESLIYR KKYILSKDVN TATCDKNPSA SKNVQSHRKA EKELTSELNS
 781 WDSKQKKMRE KSKGKEFTNV AESLISQINK RYKTKDDIKS TRKLKESLIN SGFSNKPVVQ
 841 LSKEKVQKKS YRKLKTTFVN VTSECPVNDV YNFNLNGADD PIIKLGIQEF QATAKEACAD
 901 RSIRLVGPRN HDELKSSVKT KDKKIITNHQ KKNLFSDTET EYRCDDSKTD ISWLREPKSK
 961 PQLIDYSRNK NVKNHKSGKS RSSLEKGQPS SKMTPSKNIT KKMDKTIPEG RIRLPRKATK
1021 TKKNYKDLSN SESECEQEFS HSFKENIPVK EENIHSRMKT VKLPKKQQKV FCAETEKELS
1081 KQWKNSSLLK DAIRDNCLDL SPRSLSGSPS SIEVTRCIEK ITEKDFTQDY DCITKSISPY
1141 PKTSSLESLN SNSGVGGTIK SPKNNEKNFL CASESCSPIP RPLFLPRHTP TKSNTIVNRK
1201 KISSLVLTQE TQNSNSYSDV SSYSSEERFM EIESPHINEN YIQSKREESH LASSLSKSSE
1261 GREKTWFDMP CDATHVSGPT QHLSRKRIYI EDNLSNSNEV EMEEKGERRA NLLPKKLCKI
1321 EDADHHIHKM SESVSSLSTN DFSIPWETWQ NEFAGIEMTY ETYERLNSEF KRRNNIRHKM
1381 LSYFTTQSWK TAQQHLRTMN HQSQDSRIKK LDKFQFIIIE ELENFEKDSQ SLKDLEKEFV
1441 DFWEKIFQKF SAYQKSEQQR LHLLKTSLAK SVFCNTDSEE TVFTSEMCLM KEDMKVLQDR
1501 LLKDMLEEEL LNVRRELMSV FMSHERNANV
```

Fig.6.4. Amino acid sequence of human synaptonemal complex protein 2 (SCP-2). Source: http/ www.ncbi.nlm.nih.gov (accesson number NP_055073).

tandam arrays of a degenerate eight amino acid repeat (Kerr et al., 1996). One gene is stringently testis specific and another is expressed exclusively in testis and embryonic ovary. The latter clone is not expressed in the testes of adult sex reversed mice, which lack germ cells, and therefore represents a meiosis specific gene. Steady state levels of a 2.3-kb polyadenylated *Ott* mRNA are high through out meiotic prophase in the testis when the X chromosome is generally transcriptionally inactive. A second transcript of 1-kb was also detectable at 4 weeks age and onwards. The two mRNAs have different 3'-ends and contain different protein coding information.

6.4.2. Synaptonemal Complex Protein-2

A second component of the meiotic chromosome cores, termed SCP2 in rat and human, has been identified (Offenberg et al., 1998; Schalk et al., 1998). The major protein components of SC have electrophorectic mobilities that correspond to (Mr) 30000 – 33000 and 190000-Da. The nucleotide of the cDNA corresponding to 196-kDa-protein was called *SCP2*. The SCP2 is a basic protein (pI, 8.0) with a molecular mass of 173-kDa. At the C-terminus, a stretch of approximately 50 amino acid residues is capable of forming coiled-coil structures. The SCP2 contains two clusters of serine/threonine-P motifs, which are common in DNA binding proteins. These clusters flank the central, most basic part of the protein (pI, 9.5). Three of the serine/ threonine-P motifs are potential target sites for p34 (Cdc2) protein kinase. In addition, SCP2 has eight potential cAMP/cGMP dependent protein kinase target sites. The gene encoding SCP2 is transcribed specifically in the testis, in meiotic prophase cells. At the amino acid sequence and secondary structural levels SCP2 shows some similarity to the Rad1 protein, which is involved in meiotic recombination and the assembly of axial elements of SCs in yeast. It seems that SCP2 is a DNA binding protein involved in the structural organization of meiotic prophase chromosomes (Offenberg et al., 1998) **(Fig.6.4)**.

6.4.3. Synaptonemal Complex Protein-3

The first identified component of the axial elements (cores) of the lateral regions of the SC was termed SCP3 in rat/mouse (synaptonemal complex protein 3); and COR1 in hamster. Two-hybrid analysis showed that the interaction between COR1(SCP3) molecules is mediated by the coiled-coil domains present in the carboxyl half of the molecule. It is likely that COR1/SCP3 is involved with the attachment of the DNA to SCs, and based on its persistence along the separated chromosomal axes after dissolution of synapsis. It has been postulated that it is involved in sister chromatid cohesiveness at meiosis. At metaphase II, COR1 or SCP3 localizes to the sister kinetochores, suggesting a role in their cohesiveness until the final meiotic segregation. The SCP3 is one of the major protein components of the lateral elements (LEs) of SCs of the rat, with Mr of 30-kDa. It has significant amino acid similarity to the pM1 protein, which is one of the predicted products of an X-linked lymphocyte regulated gene family of the mouse: there are 63% amino acid sequence similarity and 35% amino acid identity between the SCP3 and pM1 proteins. However, the two proteins still differ (Lammers et al., 1994). Di Carlo et al., (2000) observed that male primordial germ cells also express SCP3 for a short time before undergoing G1 arrest. This supports the hypothesis that primordial germ cells are programmed to enter meiosis irrespective of the sex and that foetal testis produces a factor that inhibits such programme. The chromosome behaviour accompanying the transition to metaphase I in rat spermatocytes was studied using antibodies against COR1/SCP3 and against the synaptic protein SYN1/SCP1. An interesting aspect related to the dynamics of SCP1 and SCP3 on the meiotic chromosomes is the precise timing of their departure from the chromosomes: SYN1/SCP1 displacement coincides with the repulsion of homologue at diplotene, while COR1/SCP3 disappears at metaphase II synchronously with sister chromatid separation. This suggests that the displacement and degradation of SYN1/SCP1 and COR1/SCP3 are precisely programmed. Both proteins with ubiquitin conjugating enzyme Ubc9 in a two-hybrid system indicated that ubiquitination could be the signal for initiation of their degradation (Moens and Coworkers-reviewed in Tarsounas et al., 1999). Okadaic acid, an inhibitor of the protein, induces a rapid SC dissolution and bivalent separation, followed by chromosome condensation and chiasmata formation, similar to the succession of events in untreated germ cells in cultures. This cell-culture system can be used in the study of this transition at the molecular level from pachytene to metaphase and provides a model for in vivo transition. The effect of okadaic acid is most likely mediated by the activation of tyrosine kinases. The okadaic acid induced progression to the metaphase 1 arrest is not affected by the inhibition of protein synthesis. However, pachytene spermatocytes, in the presence of inhibitors, showed loss of synapsis, which is abnormal in that it is not accompanied by chiasmata formation. The two meiotic specific proteins (SCP1/SCP3) were phosphorylated in vitro by extracts isolated from pachytene cells. The hypothesis proposes that phosphorylation of SYN1/SCP1 and COR1/SCP3 targets their removal from the chromosomes and that activity of the kinases involved correlates with the presence of these two proteins on the chromosomes (Tarsounas et al., 1999).

A null mutation in the SCP3 gene of homozygous mutant males lead to sterility due to massive apoptotic cell death during meiotic prophase. The SCP3 deficient male mice failed to form axial/lateral elements and SCs, and the chromosomes in the mutant spermatocytes failed to synapse. While the absence of SCP3 affected the nuclear distribution of DNA repair and recombination proteins Rad51 and RPA, as well as SCP1, a residual chromatin organization remained in the mutant meiotic cells (Yuan et al., 2000). The human testis-specific SYCP3 is associated with human non-obstructive azoospermia. Miyamoto et al., (2003) suggested that SYCP3 has an essential meiotic function in human spermatogenesis that is compromised by the mutant protein via dominant negative interference.

6.4.4. Other SC Proteins

Stag: The Stag3, a member of the stromalin protein family, is expressed specifically in testis and associates to the synaptonemal complex. The mammalian STAG3 is expressed in germinal cells, although its function in meiosis is not well known. STAG3 has a role in sister chromatid arm cohesion during mammalian meiosis. Studies in prophase I cells suggest that STAG3 is a component of the axial/lateral element of the SC. In metaphase I, STAG3 is located at the inter-chromatid domain and is absent from the chiasma region. In late anaphase I and at the later stages of meiosis, STAG3 is not detected. The STAG3 interacts with the structural maintenance chromosome proteins SMC1 and SMC3, which have been reported to be subunits of the mitotic cohesion complex. Human STAG3 gene is mapped at Tq22 of chromosome 7. The STAG3 is involved in chromosome pairing and maintenance at SC structure, and The STAG3 duplications predispose to germline chromosomal rearrangement within SC (Pezzi et al., 2000; Prieto et al., 2001). Mouse *Stag3* gene has been assigned to cM 78 on chromosome 5. The rat *Stag3* cDNA is 4181 nt long, contains a highly polymorphic hexanucleotide repeat in the coding region, and encodes a 1256 amino acid protein with 93 and 77% sequence identity to mouse and human *Stag3* respectively. *Stag3* is not responsible for the aspermic phenotype in as/as males nor suggests that the SC structures are primarily affected in these rats (Bayes et al., 2001).

FK506 Binding Protein (Fkbp6): The FK506 binding protein (Fkbp6) localizes to meiotic chromosome cores and regions of homologous chromosome synapsis. Targeted inactivation of Fkbp6 in mice results in aspermic males and the absence of normal pachytene spermatocytes. Loss of Fkbp6 results in abnormal pairing and misalignments between homologous chromosomes, nonhomologous partner switches, and autosynapsis of X chromosome cores in meiotic spermatocytes. Fertility and meiosis are normal in Fkbp6 mutant females. Thus, Fkbp6 is a component of SC essential for sex-specific fertility and for the fidelity of homologous chromosome pairing in meiosis (Crackower et al., 2003).

CPEB: CPEB is a sequence specific RNA binding protein that regulates translation during vertebrate oocyte maturation. Male CPEB null mice also contained germ cells arrested at pachytene. The germ cells from the knockout mice harbored fragmented chromatin, suggesting a possible defect in homologous chromosome adhesion or synapsis. Two CPE containing SC protein mRNAs interacting with CPEB contained shortened poly (A) tails and mostly failed to sediment with polysomes in the null mice. Synaptonemal complexes were not detected in these mice. The CPEB, therefore, controls germ cell differentiation by regulating the formation of the synaptonemal complex (Tay and Richter, 2001).

6.5. STRUCTURAL MAINTENANCE CHROMOSOME PROTEINS

Establishing and maintaining proper sister chromatid cohesion throughout the cell cycle are essential for maintaining genome integrity. Cohesins are mutli-subunit protein complexes involved in sister chromatid cohesion. The association / dissociation of these complexes are the key steps in chromosome segregation. The relatively easy procedure developed for the purification of proteins specifically associated with chromosomes made possible the identification of proteins associated with them (Hirano et al., 1997). Structural maintenance chromosome (SMC) proteins are components of cohesion complexes that function in chromosome cohesion. The SMC proteins exist in the form of two high molecular weight complexes (8S and 13S), termed condensins. In somatic cells, the heterodimeric SMC proteins are involved in chromosome condensation and gene dosage compensation (SMC2 and -4), and sister chromatid cohesion and DNA recombination (SMC1 and -3). The SMC1α and SMC3 are localized to wild-type mouse meiotic chromosomes, but with distinct differences in their patterns and hence these proteins are involved in meiosis. The SMC protein encoded by the fission yeast *rad18* gene is involved in several DNA repair processes and has an essential function in DNA replication and mitotic control. It has a hetero-dimeric partner SMC protein, like Spr18, with which it forms the core of a multi-protein complex. In testis, intense SMC chromatin association occurs in meiotic prophase cells, weaker staining in round spermatids and absence of the SMC proteins in elongated spermatids. In spermatocytes nuclei spreads, the SMC1 and SMC3 proteins localize in a beaded structure along the axial elements of synaptonemal complexes of pachytene and diplotene chromosomes. Both SMC proteins are present in rat spermatocytes and SC where SMC proteins interact with SCP2 and SCP3. This study suggests a model for the arrangement of SMC proteins in mammalian meiotic chromatin (Eijpe et al., 2001).

The hSMC6 and hSMC5 proteins interact to form a tight complex analogous to the yeast Rad18/Spr18 hetero-dimer. In proliferating cells these proteins are bound to both chromatin and the nucleocytoskeleton. A phosphorylated form of hSMC6 localizes to interchromatin granule clusters. Total level of hSMC6 and its phosphorylated form remain constant through the cell cycle. The hSMC5 and hSMC6 proteins are expressed at extremely high levels in the testis and associate with the sex chromosomes in the late stages of meiotic prophase, suggesting a possible role for these proteins in meiosis (Taylor et al., 2001). Lee et al., (2003) suggested that mammalian Rec8 protein, in association with SMC3 and SMC1β but not SMC1α, is involved in meiosis-specific chromosome behavior, and that homologous chromosome separation is triggered by selective loss of Rec8 from chromosome arms in meiosis I, while sister chromatid cohesion is maintained until metaphase II/anaphase II transition by centromeric Rec8 during mammalian meiosis.

The SMC protein (SMC1β) with a unique basic-DNA binding C-terminal motif is highly homologous to SMC1 (now called SMC1α) and is not present in the yeast genome. The SMC1 is specifically expressed in testes and coimmunoprecipitates with SMC3 from testis nuclear extracts, but not from a variety of somatic cells, showing tissue specific nature of SMC protein isoforms. Various stages of meiosis revealed localization of SMC1β along the axial elements of SCs in prophase 1. Most SMC1β dissociates from the chromosome arms in late-pachytene diplotene cells. However, SMC1β but not SMC1α, remains chromatin associated at the centromeres up to metaphase II. Thus, SMC1β and not SMC1α is likely involved in maintaining cohesion between sister centromeres until anaphase II (Revenkova et al., 2001). Anti-SMC3 coincided with axial elements of the synaptonemal complex, while SMC1α was found mainly in regions where homologues were synapsed. This pattern was especially visible in pachytene sex vesicles where SMC1α localized only weakly to the asynapsed regions. At diplotene, SMC3, but not SMC1α, remained bound along axial elements of desynapsed chromosomes. The study suggested that two cohesin proteins might not always be associated in a dimer and may function as separate complexes in mammalian meiosis, with SMC1α playing a more specific role in synapsis. In addition, results indicated that cohesin cores can independently form double-strand break and homologous pairing (James et al., 2002).

Darwiche et al (1999) cloned mouse orthologues of proteins involved in sister chromatid cohesion in other organisms. The mouse orthologues, of *SMC1S.c.* and *SMC3S.c.*, mSMCB and mSMCD respectively are transcribed at similar levels in adult mouse tissues except in testis, which has an excess of mSMCB transcripts. The bulk of the mSMCB, mSMCD and PW29 (an orthologue of Med1p S.c.) proteins undergo redistribution from the chromosome vicinity to the cytoplasm during prometaphase and back to the chromatin in telophase. This pattern of localization suggests a complex role for this group of SMC proteins in chromosome dynamics. The PW29 protein and PCNA, which have both been implicated in sister chromatid cohesion, indicated that these proteins may not function in the same cohesion pathway. Over-expression of a PW29-GFP fusion protein in mouse fibroblasts leads to inhibition of proliferation, implicating this protein and its complex with SMC proteins in the control of mitotic cycle progression (Darwiche et al., 1999).

6.6. T-COMPLEX POLYPEPTIDES

Complete t haplotype chromosomes contain a number of genes that cause sterility in compound t^x/t^y males and affect transmission ratios of sperm in heterozygote, tx/+ males. However, Silver et al., (1983) described a group of polypeptides, five of which are certainly encoded by genes within the t complex and are expressed in mouse testis. The mouse t complex polypeptides, TCP-3 and TCP-7, are synthesized in a testes-specific manner with highest levels of expression during haploid stages of spermatogenesis. A third, TCP-1, is also expressed at highest levels in haploid cells, and continues until the last residual body stage of spermatogenesis. Some of these t complex polypeptides (TCPs), such as TCP-1 may be correlated with transmission ratio distortion factor; others, TCP-3 and TCP-7, may be correlated with a factor (tcs-2) involved in the sterility of compound t^x/t^y males (Silver et al., 1987).

6.6.1. TCP-1

The TCP-1 is one of the best studied of t complex polypeptides and is apparently a non-glycosylated external membrane protein. It exists in two electrophoretic forms: TCP-1A and TCP-1B. The TCP-1A is encoded by all complete t haplotype chromosomes, where as all inbred strains encode TCP-1B. The TCP-1A has the same molecular weight as TCP-1B but has a more acidic isoelectric point. Some of these t-complex polypeptides are abundantly present in testes. Willison (1986) cloned the gene for germ cell variant form, TCP-1A. The cDNA clone, pB1.4, which hybridizes to a 19S mRNA is abundant in haploid cells during mouse spermatogenesis and derives from the 3'end of the mRNA encoding TCP-1B. The *Tcp-1* gene shows multiple changes between the predicted amino acid sequences of TCP-1B and TCP-1A. An additional Taq-1 site is created by a T to C transition in the predicted ORF of the Tcp-1^a gene **(Fig. 6.5)**. The resultant RFLP has allowed typing of the *Tcp-1* gene cluster in 54 complete and partial t haplo-type chromosomes. DNA sequence comparison of the *Tcp-1* gene suggests that the t haplotype chromosome arose within the genus Mus more than million years ago (Willison et al., 1986).

6.6.2. Tcte 2

Tcte2 (t complex testes expressed 2) is a male meiosis-specific gene that maps to band 3.3 of mouse chromosome 17. Two distinct male fertility defects, hybrid sterility and transmission ratio distortion, have been mapped to this region. Hybrid sterility arises in crosses between different mouse species and the F1 generation males have defects in the first meiotic division and are sterile. Transmission ratio distortion is shown in males heterozygous for the t haplotype form of chromosome 17. The *Tcte 2* gene expresses a coding mRNA and a number of putative non-ORF transcripts in meiosis. A deletion of the 5' part of the locus abolishes *Tcte2* expression

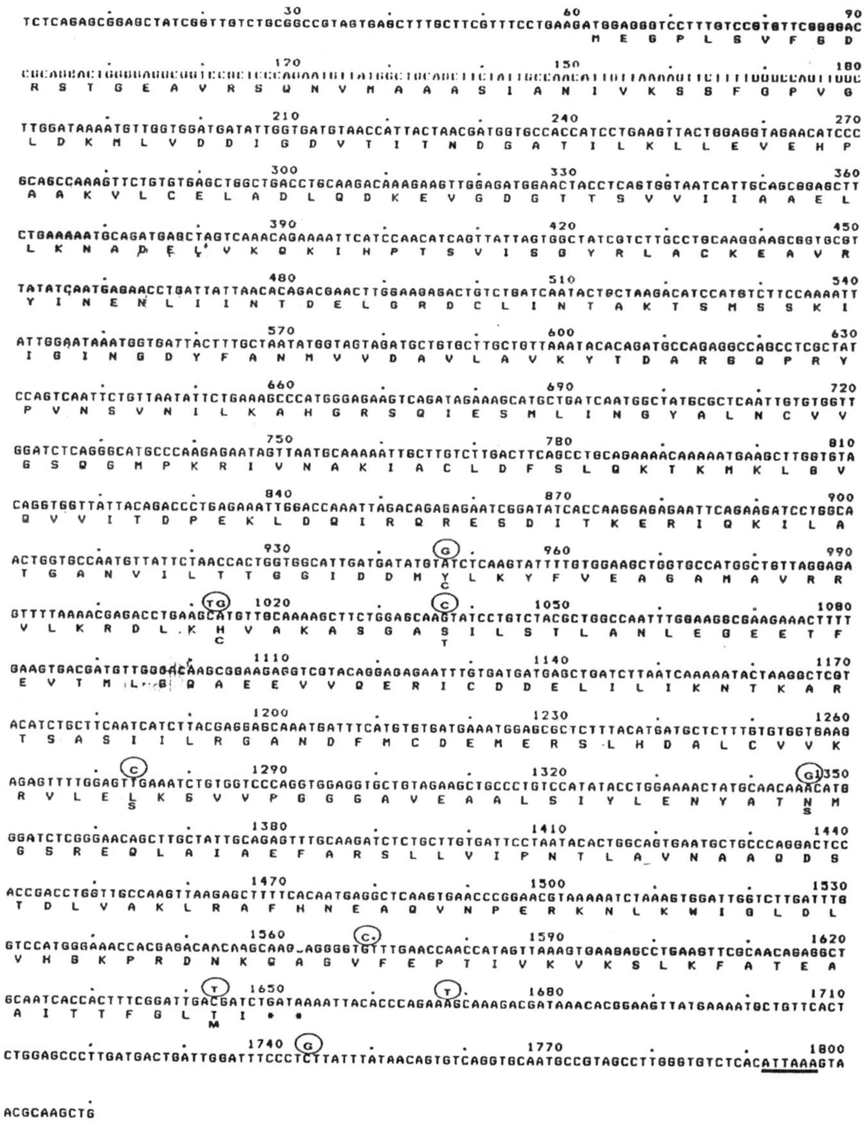

Fig.6.5. Nucleotide and deduced amino acid sequences of *Tcp-1[b]* and partial nucleotide sequence of *Tcp-1[a]* cDNA (874 bases). Reprinted with permission from K.R. Willison et al. Cell 44; 727-38: 1986 © Elsevier.

on the t haplotype form of chromosome 17. Additionally, the series of putative non-ORF RNAs at the *Tcte2* locus are differentially spliced in species that show hybrid sterility when crossed to laboratory mice. Polymorphisms in t haplotype and in different mouse species allow alleles of Tcte2 to be proposed as candidates for loci, which contribute to both meiotic drive and hybrid sterility phenotypes. The *Tcte2* is the candidate gene to support this link at a molecular level (Bradotti and Barlow, 1997).

6.7. CENTROMERE PROTEIN-B

The centromere protein B (CENP-B) is a centromeric DNA binding protein. It recognizes a 17-bp sequence motif called the CENP-B box, which is found in the centromeric region of most chromosomes. It binds DNA through its amino terminus and dimerizes through its carboxy terminus (Fig.6.6). Localization of centromere proteins allowed us to follow their movements during the different phases of spermiogenesis. Since the number of centromeres remained close to the number of chromosomes until the cap phase of spermatid differentiation, it has been hypothesized that the labeling of young spermatids corresponds to centromeric proteins associated with their specific DNA counterparts, while the centromere proteins, possibly detached from their DNA loci, were released from nuclei of old spermatids in the same way as are histones and transition proteins (Courtens et al., 1992). Though CENP-B protein has been proposed to perform a vital role in organizing chromatin structures at centromeres, other evidences do not agree with this view. For example, CENP-B found at inactive centromeres on stable dicentric chromosomes, and also mitotically stable chromosomes lacking alpha satellite DNA have been reported. Moreover, male and female *CenpB* null mice had normal body weights at birth and at weaning, but these subsequently lagged behind those of the heterozygous and wild type animals. The weight and sperm content of these testes of *CenpB* null mice were also reduced. Otherwise, null mutant mice lacking CENP-B were viable and fertile indicating that mice without CENP-B undergo normal somatic and germline development. These results suggested that *CenpB* is not essential for mitosis or meiosis, although the observed weight reduction raised the possibility that *CenpB* deficiency may subtly affect some aspects of centromere assembly and function (Hudson et al., 1998; Perez Castro et al., 1998). The crystal structure of the complex of the DNA-binding region (129 residues) of CENP-B and CENP-B box DNA at 2.5 Å indicated involvement of two helix-turn-helix domains, which are bound to adjacent major grooves of DNA. The DNA is kinked at the two-helix contact sites; the DNA region between the kinks is straight. The overall bending angle of 59° was suggested to be important for centromere-specific chromatin structure (Tanaka et al., 2001) (**Fig.6.6**).

6.8. NUCLEAR LAMINS

The nuclear lamina is a karyoskeletal structure located at the periphery of cell nuclei in the form of nuclear envelope. The major constituents are the lamins, which belong to the evolutionary conserved multigene family of intermediate filament proteins. Lamins show a conspicuous cell type-specific expression pattern.(Benaventa and Krohne, 1985) Somatic cells of vertebrates express A-type (lamins A and C) as well as B-type (lamins B1 and B2) lamins. Spermatogenic lamin isoforms B3 and C2 are shorter splicing variants of their somatic counterparts. Although a lamina structure has been demonstrated to be a ubiquitous component of somatic nuclei, its existence in certain meiotic stages during spermatogenesis has been a matter of debate. A lamin protein of 52kDa selectively expressed during spermatogenesis, appears to belong to nuclear family selectively expressed during meiotic stages of spermatogenesis in rat. Vester et al., (1993) suggested that the nuclear lamina structure is present in rat pachytene spermatocytes and the expression pattern is similar to lamin B1.

However, anti-lamin B has been found to bind to the nuclear periphery of all cell types examined, including Sertoli cells, primitive type A spermatogonia, preleptotene, leptotene, zygotene and pachytene spermatocytes, and round spermatids in testis. In sperm nuclei, the antigenic determinants were localized to a narrow domain of the nucleus, where as, after removing the perinuclear theca, anti-lamin B localizes to the entire nuclear periphery in a

A

B

C

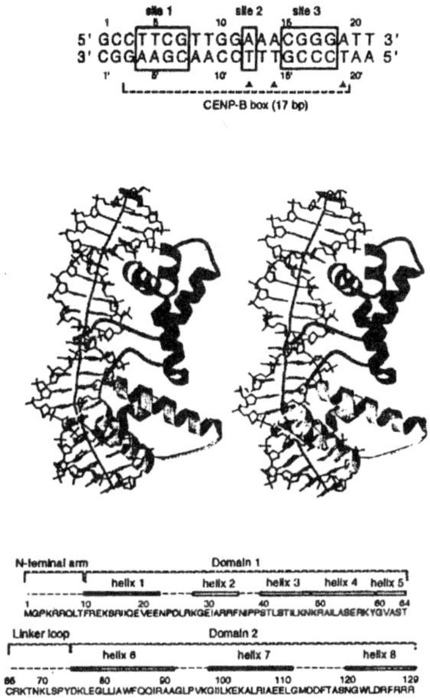

Fig.6.6. The crystal structure of the CENP-$B_{1\text{-}129}$ complexed with 21 mer DNA. (A) The 21 mer DNA sequence with CENP-Box. The three boxes marked 1, 2, and 3 indicate the essential bases for CENP-B binding to CENP-B box DNA. Closed triangles indicate the thymine residues, each of which was replaced by 5'-iodouracil for phasing. (B) Overall structure of CENP-$B_{1\text{-}129}$ complex. (C). Amino acid sequence and secondary structure of CENP-$B_{1\text{-}129}$. Reprinted with permission from Y. Tanaka et al, The EMBO Journal 20; 6612-18: 2001© http://www.nature.com/.

punctate pattern, to determinants previously covered by the theca constituents. Anti-lamin B reacts with approximately 68-kDa polypeptide in all germ cells and, to a lesser extent, with four additional polypeptides present only in meiotic and post-meiotic nuclear matrices. In contrast to anti-lamin B, anti-lamin A and C did not bind to the germ cells at any stage of spermatogenesis (Moss et al., 1993). Furukawa and Hotta (1993) reported the cDNA of a 53-kDa lamin from mouse spermatocytes, termed lamin B3, and restricted to spermatogenic cells. The *Lamin B3* gene is generated from *lamin B2* gene by differential splicing and alternative polyadenylation. When lamin B3 was introduced into somatic cells in culture, their nuclear morphology was transformed from spherical to hook shaped. This suggested that the germ cells specific lamin B3 **(Fig. 6.7)** is involved in the reorganization of nuclear and chromosomal structures during meiotic division.

A-type lamin of 52-kDa from mouse pachytene spermatocytes, termed lamin C2 because of its similarities with lamin C, has a sequence identical to that of lamin C except that the N-terminal segment, containing the head and the α helical coil1-A domain is replaced with a short non-α-helical stretch of amino acids. In mice, lamin C2 is specifically expressed in germ cells and arises from differential splicing of somatic type A lamin gene. The specific expression and unique structure suggests a role for lamin C2 in determining the organization of nuclear and chromosomal structures during spermatogenesis (Furukawa et al., 1994). Lamin C2 is selectively expressed in spermatocytes and regulated at the transcriptional level. This is in contrast to the situation during amphibian oogenesis, where lamin expression is largely post-transcriptionally

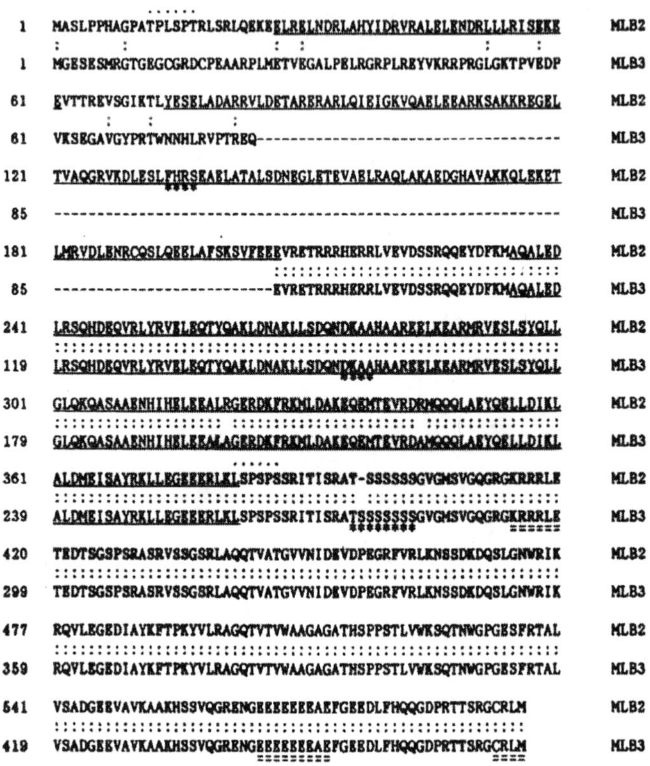

Fig,6.7. Amino acid sequence comparison of murine lamins B2 and B3. Alpha helical coils are underlined. Identical amino acids are marked with double dots. Asterisks indicate three areas of lamin B2 specific sequences. Two sequence motifs that are phosphorylated by Cdc2 kinase are indicated by dotted lines. Reprinted with permission from K. Furukawa and Y. Hotta, The EMBO Journal 12; 97-106: 1993© http://www.nature.com/.

regulated. Moreover, the expression pattern of lamin C2 temporally coincides with that of structural protein components of the synaptonemal complex. This indicated that pairing and recombination of homologous chromosomes during meiotic prophase is accompanied by significant changes in the organization of the karyoskeleton of SC, which are accomplished by the expression of stage-specific proteins (Alsheimer and Benavente, 1996). Wang et al, (1992) identified a cDNA fragment from *Drosophila* testis that exhibited a complete sequence similarity with the cDNA of the laminin B2 chain. Transcripts of laminin B2 were detected in the RNA of male germ cells. Antigenic sites against laminin B2 were found in the lamp brush loops in primary spermatocytes, in nuclei of spermatids and in heads of spermatozoa.

6.9. NUCLEAR ASSOCIATED PROTEIN (NASP)

Among sperm specific proteins that function in fertilization, an autoantigenic post-acrosomal sperm protein has been found to originate in the testis as a nuclear associated protein (NASP) and has several structural features of nuclear proteins. In testis NASP was first detected in the nuclear area of primary spermatocytes. During the subsequent meiotic divisions, NASP was partitioned into the cytoplasm and then re-associated with the reforming nucleus. During spermiogenesis, NASP became restricted to the post-acrosomal region of the spermatozoon, although some labeling appeared in residual bodies and subsequently in the lumen of the tubule. The detection of NASP mRNA transcripts in primary spermatocytes supported by in situ hybridization indicated that NASP expression was under transcriptional control (Welch and O'Rand, 1990). The NASP contains a c-terminal nuclear translocation signal and has structural similarities to the lamins and other nuclear proteins; its 2.5-kb mRNA is apparently tissue, but not species specific. The polyadenylated mRNA has 39 bases of 5' untranslated sequence, an open reading frame of 2043 bases encoding 680 amino acids, and a 104 base 3'untranslated region (2186). The encoded polypeptide has a calculated molecular weight of 73533Da and a pI = 4.06, containing 25% acidic residues. Secondary structure predictions for NASP showed that 69% of the molecule had a high probability of forming α-helices and that several α-helical regions had a characteristic repeating heptad pattern that in the intermediate filaments and nuclear lamins is involved in coiled-coil interactions with other molecules. In addition to the nuclear translocation signal common to many nuclear proteins, NASP also showed homology with the *Xenopus* histone-binding protein, N1/N2 (Welch et al., 1990). Batova and O'Rand (1996) determined the number, location and activity of the histone–binding domains on the primary sequence of human NASP (hNASP). Recombinant polypeptides expressing the full-length hNASP were tested for the polypeptides spanning the N-terminal region (amino acids 32-192) and two additional regions (amino acids 193-352 and aa 353-572). The lack of binding to the expressed C-terminal (amino acid 573-787), which also contains polyacidic amino acids, suggested that the binding of hNASP to the somatic core histones is a sequence-specific and also involves electrostatic interaction. Thus, there are at least three functional histone–binding domains in hNASP, two of them encompassing the predicted histone binding sites, homologous to the N1/N2 protein, and a third novel domain. Therefore, hNASP may be defined as a nuclear histone-binding protein found in human testis. A somatic form of NASP (sNASP) is also present in mitotic cells (Richardson et al, 2000). The NASP from myeloma cells is complexed only with H1, linker histones. The sNASP is a shorter version of testicular NASP (tNASP) with two deletions in the coding regions and differs from tNASP in its 5' untranslated regions. Expression of sNASP mRNA is regulated during cells cycle and consistent with a role as a histone transport protein. Expression of NASP mRNA parallels with the expression of histone mRNA.

6.10. SPERM CYLICIN

In the mammalian sperm head the nucleus, in its posterior part, is tightly associated with a large and dense non-filamentous cytoskeletal structure, the calyx, whose major proteins are basic, representing a novel category of cytoskeletal element. One of these calyx proteins, termed cylicin II (calicin), has been characterized from bull and man. The polypeptide of 588 amino acids (Mr of 66,889; pI = 8.1), very similar in the two species, is encoded by a approximately 2.2-kb mRNA that has been detected only in testis but not in any other tissue or cell cultures. Cylicin is homologous to the kelch protein of the ring canal structure of *Drosophila* ovaries. In particular, cylicin contains three consecutive repeating units of 48 amino acids each, which are homologous to the so-called "β-strand folds" occurring in proteins of the kelch family, including the scruin of Limulus sperm (Chapter 24) and a series of the eukaryotic, bacterial, and viral proteins. The amino terminal domain of cylicin contains a region of about 100 amino acids homologous to an extended motif shared by the kelch protein as well as various zinc finger and poxvirus proteins (Hess et al., 1993; 1995). Cylicin seems to function as a morphogenic cytoskeletal element in spermatogenic differentiation besides its relation to the demonstrated absence or altered arrangement in human tertozoospermia such as "round-headed" or other post-acrosomal sheath defect- sperm malformations (van Bulow et al., 1995).

6.11. NUCLEAR PORE ASSOCIATED PROTEINS

A cDNA (BS-63) was found to consist of 1933-bp with an open reading frame of 1824 bp. The deduced polypeptide consisted of 608 amino acid residues containing XFXFG or FG motifs that are characteristic of nuclear pore complex (NPC) proteins and act as potential binding sites for Ran. The N-terminal end has high homology with Ran BP2/Nu 358. The BS–63 is transcribed in two transcripts i.e. 6.0 and the 8.5kb. The 8.5kb transcript was present in low amounts in somatic tissues whereas the 6.0kb transcript is expressed only in testis. In situ hybridization analysis of human testis showed that BS-63 mRNA is expressed only in germ cells at all stages of spermatogenesis but not in Sertoli cells (Wang et al 1999).

An mRNA with a substantial similarity to the rat p62 mRNA that encodes a nucleoporin, was cloned from the rat testis. Nucleoporin related (NPR) cDNA revealed an mRNA of 1.3kb, different from the 2.7–kb transcript attributed to the p62 gene and found primarily in the haploid germ cells of the adult testis but not in Sertoli cells. The central region of the NPR cDNA sequence was identical to the 3'portion of the p62 cDNA containing heptad repeat sequences. However, the 5'region and the extreme 3'region of the NPR cDNA sequence were different from the p62 cDNA. The extreme 3' untranslated region (UTR) contained a 212-bp inverted repeat of a sequence located in the middle of the NPR cDNA that is identical to the p62 sequence. The inverted repeats of the NPR sequence could hybridize, leading to the formation of circular transcripts (Wang et al., 1994).

A 3kb cDNA, encoding a rat nuclear pore- associated protein *(Npap60)* was cloned and characterized. The predicted NPAP60 protein contains 381 amino acids with a composition of 25.6% charged residues and is highly hydrophilic. The Npap60 gene appears to be conserved in mouse, rat, and human. The NPAP60 colocalizes with nuclear pore complexes in RAT1A cells. The expression of *Npap60* is about 10.20 times higher in rat testis than in somatic tissues. The sub-cellular localization of NPAP60 protein changes dramatically during male germ cell differentiation, from nuclear pore complex-like staining in spermatocytes to whole nucleus staining in mature spermatozoa. These changes are temporally and spatially related to nuclear reorganization during male germ cell differentiation (Fan et al., 1997).

6.12. HIGH MOBILITY GROUP (HMG) PROTEINS

The testis contains a substantially higher level of HMG2 than any other rat tissue. High levels of HMG2 in the testis were due to pachytene spermatocytes and early spermatids. Spermatogonia, early primary spermatocytes, late spermatids, Sertoli and Leydig cells showed very low levels of HMG2 similar to those in non-proliferating somatic tissues. HMG proteins are synthesized in spermatogonia and primary spermatocytes, but not in spermatids. Rat testis HMG2 exhibited two bands, A "slow" form co-migrated with somatic cell HMG2, while the other "fast" band migrated ahead of the somatic form and appeared to be testis specific accounting high level of HMG2. The very high level of HMG2 in testis is not associated with proliferative activity (Bucci et al., 1984).

A cDNA encoding a DNA-binding protein has been isolated by screening a mouse testicular expression cDNA library with a concatemer of a 12-bp putative protein-binding element present in the promoter of the testis-specific gene PGK-2. Sequence analysis of the isolated cDNA indicated the presence of an open reading frame that encodes a protein with two-conserved DNA–binding motifs of high–mobility group. Expression of the gene is restricted to the post-pubertal testis and located predominantly in the nuclei of elongated spermatids at step 9 and 10. The HMG box protein gene may be involved in the regulation of gene expression of the haploid male genome. The gene has been termed testis – specific HMG (tsHMG) (Boissonneault and Lau 1993).

The testis specific high mobility group (tsHMG) protein, similar to the human mitochondrial transcription factor-1 and to the linker associated protein δ of *Tetrahymena thermophila* micronuclei, is thought to play a structural role. TsHMG is restricted to quiescent elongating and condensing spermatids and displays preferential binding to supercoiled plasmid DNA and protects DNA from relaxing activity. The tsHMG protein can also introduce negative supercoils into a relaxed plasmid substrate in a topoisomerase 1-dependent manner. The tsHMG protein was substrate of a Ca^{2+} phospholipid dependent protein kinase C present in testis of adult mice and that the phosphorylation by PKC is required for both DNA binding and topoisomerase I dependent supercoiling activities of tsHMG. This supports that the spermatid tsHMG protein is a topological factor (transition protein) that can modulate the activity of topoisomerase 1 and can contribute to the important transition in chromatin structure, which leads to the decrease in transcriptional activity, which is observed at the early stages of spermatid elongation (Alami-Ouahabi et al., 1996).

The Sox-5 is one of a family of genes, which shows homology to the HMG box region of the testis-determining gene SRY. The Sox-5 protein is localized on the nucleus of post-meiotic round spermatids in the mouse testis. The Sox-5 binds specifically on the sequence AACAAT with moderately high affinity. Moreover, interaction of Sox-5 with its target DNA induces a significant bend in the DNA, characteristic of HMG box proteins. Circular dichroism spectroscopy of the Sox-5 HMG box and its specific complex with DNA shows an alteration in the DNA spectrum, perhaps as a consequence of DNA binding. Analysis of the structure of the Sox-5 HMG box by 2D NMR suggests that both the location of helical secondary structure as well as the tertiary structure is similar to that of HMG1 box 2 (Connor et al, 1994) (Chapter 10).

6.13. OTHER NUCLEAR ANTIGENS

Germ Cell Nuclear Antigens: A mouse germ cell nuclear antigen (GCNA1) is present in pre-spermatogonia (gonocytes) in males and in oogonia and oocytes of females within the gonadal ridge from embryonic day 11.5 onward, but rarely in primordial germ cells prior to their arrival at the gonadal ridge. The GCNA1 is abundantly present in spermatogonia, spermatocytes but decreases subsequently in spermatids. The antigen is approximately 80-110kDa on immunoblots of isolated pachytene spermatocytes and round spermatids. The GCNA1 serves as a common marker of the germ cell lineage in male and female mice after primordial germ cells arrive in the gonadal ridge until they reach the diplotene/dictate stage of the first meiotic division (Enders and May, 1994). A series of antigens, specific to germ cells have been described through use of mAb(s) (Tsuchida et al. 1995; Tanaka et al. 1997; Ling et al., 1998; Periera et al 2000). TRA 54 mAb recognized antigen in the range of Mr of 200-, 190-, and 85-kDa whereas mAb TRA55 showed positive bands with molecular weights of 43-, 46-, 49- and 55-kDa. A monoclonal antibody (mAb TRA 104) raised against mouse testicular germ cells was able to recognize the nuclei of testicular germ cells at all the stages of differentiation from embryonic gonocytes to spermatids and did not react with any somatic cells. The antigen(s), referred to as a germ cell specific nuclear antigen(s) (GENA) is first detected in the genital ridge at around 12 days of gestation and increases gradually until adulthood. The GENA may be a useful marker to study the mechanism of germ cell differentiation in testis (Tanaka et al, 1997).

Another germ cell nuclear antigen with approximately 44-kDa was identified in the nuclei of early meiotic cells from pre-leptotene to early pachytene spermatocytes exhibiting a stage specific appearance in the cycle of the seminiferous epithelium. The activity was associated with the meiotic chromosomes but absent from late pachytene spermatocytes or more advanced stages of spermatogenesis and also absent from spermatogonia, Sertoli and Leydig cells. The pattern of expression of the antigen during early meiotic stages of spermatogenesis could strengthen its possible role in meiotic division (Atanassova et al., 2000).

The cDNA cloning of a novel 60kDa skeletal protein from mouse spermatocytes, termed MNS1 (meiosis-specific nuclear structural protein), whose computer-predicted protein configuration indicates long α-helical coiled-coil domains flanked by non-helical terminal domains.The MNS1 protein is specifically expressed at the pachytene stage during spermatogenesis, so that its function may involve the determination and maintenance of the appropriate nuclear morphology during meiotic prophase (Furukawa et al., 1994b).

Nicolin 1: A mammalian gene, termed *nicolin 1* gene (NICN1) is present in human, dog and mouse and absent from the available genome sequences of non-mammalian organisms. The *NICN1* gene consists of six exons and spans about 6-kb of genomic DNA. It encodes a 213 amino acid protein that does not belong to any known protein family. Nicolin 1 is a nuclear protein, which is expressed from a 2.5-kb *NICN1* mRNA in a tissue-specific manner. The highest NICN1 expression levels were found in brain, testis, liver, and kidney with weak expression in spleen, leukocytes, small intestine and colon (Backofen et al., 2002).

Nucleoplasmin : Nucleoplasmin, one of the most abundant proteins in *Xenopus laevis* oocytes, has been involved in the chromatin remodeling that takes place immediately after fertilization. This molecule is responsible for the removal of the sperm-specific proteins and deposition of somatic histones onto the male pronuclear chromatin. The pentameric form of nucleoplasmin binds to a histone octamer equivalent consisting equal amounts of the four core histones, H2A, H2B, H3, and H4, without any noticeable preference for any of them. Interactions other than those electrostatic in nature (likely hydrophobic) probably also play a critical role in the formation of the complex between the negatively charged nucleoplasmin and positively charged histones (Arnan et al., 2003).

SP17: Sp17 is mammalian testis and sperm protein that has been characterized from human, rabbit and mouse testis and spermatozoa. In the rabbit, Sp17 is a member of the RSA family (17-kDa) of rabbit sperm-specific autoantigen that binds rabbit zona pellucida as well as sulfated carbohydrates. In other species (human, pig, mouse), RSA-like molecules have been identified. The mouse mRNA for Sp17 encodes a 149-amino acid protein with a predicted molecular weight of 17296Da. The mouse Sp17 (MSp 17) cDNA sequence is 82% identical to the rabbit Sp17 cDNA sequence while the MSp17 protein sequence is 74% identical to the rabbit protein sequence. Mouse native Sp17 has an apparent molecular mass of 24kDa on SDS-PAGE. In the absence of proteolytic inhibitors, part of the C-terminal of native MSp17 is cleaved, giving rise to an 18-kDa band. The SP17 is present in spermatocytes in the testis. In spermatozoa, Sp17 antibody is not able to bind acrosome of intact spermatozoa, but it is present on the equatorial surface of live-acrosome reacted spermatozoa (Kong et al., 1995) (Chapter 24). Recent data

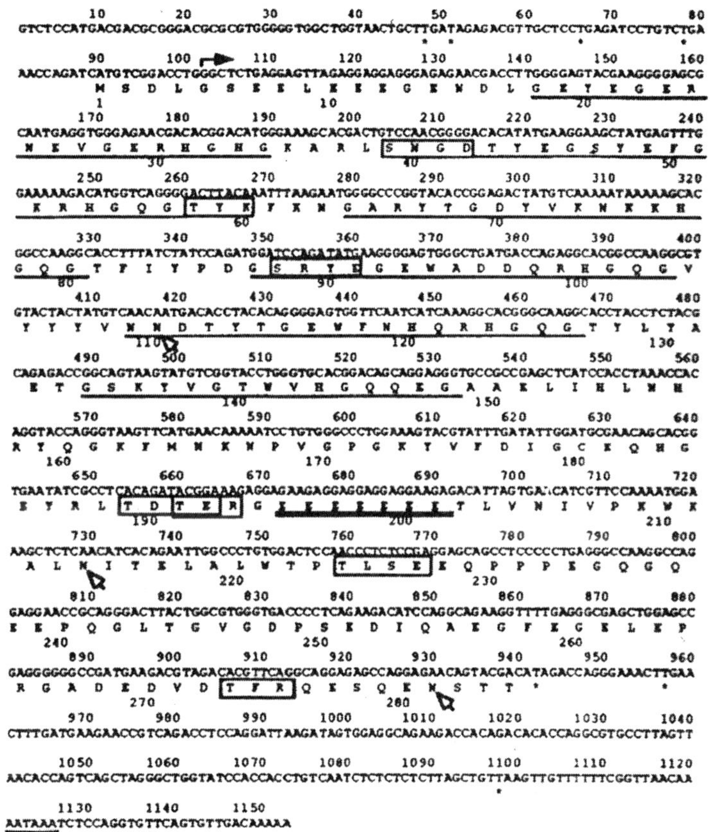

Fig.6.8. The nucleotide and deduced amino acid sequence of meichroacidin cDNA. The deduced amino acid sequence of longest ORF is given under DNA sequence. Stop codons are shown by asterisks. The putative polyadenylation signal is underlined in bold face and phosphorylation signals are boxed. The glutamic acid rich region is double- underlined. An open arrow indicates a potential glycosylation site (N-X-S/T). Closed arrow shows the 5' end of the A-2 cDNA fragment. Reprinted with permission from J. Tsuchida et al. Dev Biol 197; 67-76: 1998© Elsevier.

indicated that Sp17 might be expressed more widely, both in tumours and in normal somatic tissues albeit at a much lower expression than in the testis. On the basis of the extremely high sequence conservation throughout the N-terminal half of Sp17, and the presence within this region of A-kinase anchoring protein (AKAP)-binding motif, it has been postulated that the proposed role of Sp17 in zona pellucida binding is unlikely to be its principal function (Frayne and Hall, 2002).

Meichroacidin: The deduced amino acid sequence of a nuclear protein consists of 284 aa residues, including a nominal repeat structure in the N-terminal region and a transcript of 1.3-kb exclusively expressed in the testis and ovary with a molecular weight of approximately 40-kDa and pI of 4.9 **(Fig 6.8).** The protein was predominantly present in the cytoplasm of pachytene spermatocytes to round spermatids. However, during the disappearance of the nuclear envelope at both the first and second meiotic divisions, the protein was localized around the metaphase chromosomes and spindles, hence named meichroacidin, which stands for male meiotic metaphase chromosome-associated acidic protein (Tsuchida et al., 1998).

Tsg 118 protein: A murine gene encoding a nucleolar protein called Tsg118, with a predicted molecular mass of 59.4-kDa and a high content of basic amino acids has been shown to be associated with condensed chromosome in mitotic cells. Homologous human gene was localized to chromosome 16p12.3. The Tsg118 protein is predominantly expressed in proliferating somatic cells and in male germ cells. Tsg118 shows localization to the dense fibrillar component of the nucleolus. The nucleolar localization of the Tsg118 protein appears to be temporally restricted to the interphase stages of the somatic cell cycle and to the meiotic phase of spermatogenesis. As the nucleolar signal disappears in mitotic cells, the Tsg118 protein instead becomes associated with the surface of the condensed chromosomes (Larsson et al., 1999).

Tstis-Specific Protein-57 (Tsp57) : A nuclear receptor-associated protein-80 (RAP80) is highly expressed in spermatocytes and appears to have a role in regulating gene expression. Protein interacting with RAP80 was named as testis-specific protein (Tsp), referred to as Tsp57. Tsp57 encodes a basic protein with a mass of 56.8-kDa. The amino acid sequence of Tsp57 is highly conserved (87%) between mouse and human (Kim et al., 2004). The mouse and human Tsp57 genes map to chromosomes 9A1 and 11q21, respectively. The expression of Tsp57 mRNA was highly restricted to the testis and temporally regulated during testicular development, and greatly induced between Day 21 and Day 25 of postnatal testicular development. The hybridization signal for Tsp57 mRNA was strongest in seminiferous tubules at stages VI-VIII of spermatogenesis, consistent with the idea that Tsp57 is most highly expressed in round spermatids. However, Tsp57 expression was absent from testes in vitamin A-deficient mice, which do not have round spermatids. Thus Tsp57 protein plays a role in the postmeiotic phase of germ cell differentiation. The Tsp57 contains two putative nuclear localization signals: NLS1 and NLS2. The cellular localization showed that the Tsp57 functions as a nuclear protein (Kim et al., 2004).

6.14. REFERENCES

Alami-Ouahabi N, Veilleux S. Meistrich ML, Boissonneault G, The testis specific high-mobility-group protein, a phosphorylation-dependent DNA-packaging factor of elongating and condensing spermatids: Mol Cell Biol 1996; 16: 3720-9

Alsheimer M, Benavente R. Change of karyoskeleton during mammalian spermatogenesis: expression pattern of nuclear lamin C2 and its regulation. Exp Cell Res 1996; 228: 181-8

Arnan C, Saperas N, Prieto C, Chiva M, Ausio J. Interaction of nucleoplasmin with core histones. J Biol Chem 2003; 278: 31319-24.

Atanassova NN, Russinova AI, Kancheva LS, Valkova CA. Stage-specific nuclear antigen is expressed in rat male germ cells during early meiotic prophase. Mol Reprod Dev 2000; 56: 45-50

Backofen B, Jacob R, Serth K, Gossler A, Naim HY, Leeb T. Cloning and characterization of the mammalian-specific nicolin 1 gene (NICN1) encoding a nuclear 24 kDa protein. Eur J Biochem 2002;269:5240-5.

Batova I, O'Rand MG. Histone-binding domains in a human nuclear autoantigenic sperm protein. Biol Reprod 1996; 54: 1238-44.

Bayes M, Prieto I, Noguchi J, Barbero JL, Perez Jurado LA. Evaluation of the Stag3 gene and the synaptonemal complex in a rat model (as/as) for male infertility. Mol Reprod Dev 2001; 60: 414-7

Benavente R, Krohne G. Change of karyoskeleton during spermatogenesis of Xenopus: expression of lamin LIV, a nuclear lamina protein specific for the male germ line. Proc Natl Acad Sci (U S A) 1985; 82: 6176-80

Bloom W, Fawsett DW. A text book of histology, WB Saunders Co, Philadelphia, 1978

Boissonneault G, Lau YF. A testis specific gene encoding a nuclear high-mobility group box protein located in elongating spermatids. Mol Cell Biol 1993; 13: 4323-30.

Braidotti G, Barlow DP. Identification of a male meiosis-specific gene, Tcte2, which is differentially spliced in species that form sterile hybrids with laboratory mice and deleted in t chromosomes showing meiotic drive. Dev Biol 1997; 186: 85-99

Bucci LR, Brock WA, Goldknopf IL, Meistrich ML. Characterization of high mobility group protein levels during spermatogegesis in the rat: J Biol Chem 1984; 259; 8840-6

Chen Q, Pearlman RE, Moens PB. Isolation and characterization of a cDNA encoding a synaptonemal complex protein. Biochem Cell Biol 1992; 70: 1030-8.

Connor F, Cary PD, Read CM, Preston NS, Driscoll PC, Denny P, Crane-Robinson C, Ashworth A. DNA binding and bending properties of the post-meiotically expressed Sry-related protein Sox-5. Nucleic Acids Res 1994; 22: 3339-46.

Courtens JL, Biggiogera M, Rothfield NF, Burnier M, Faken S. Migration of centromere proteins in rabbit spermatids. Mol Reprod Dev 1992; 32: 369-77

Crackower MA, Kolas NK, Noguchi J, Sarao R, Kikuchi K, et al. Essential role of Fkbp6 in male fertility and homologous chromosome pairing in meiosis. Science 2003; 300(5623): 1291-5.

Darwiche N, Freeman LA, Strunnikov A. Characterization of the components of the putative mammalian sister chromatid cohesion complex. Gene 1999; 233: 39-47.

Di Carlo AD, Travia G, De Felici M. The meiotic specific synaptonemal complex protein SCP3 is expressed by female and male primordial germ cells of the mouse embryo. Int J Dev Biol 2000; 44: 241-4.

Dobson MJ, Pearlman RE, Karaiskakis A. Spyropoulos B, Moens PB. Synaptonemal complex proteins. Occurrence, epitope mapping, and chromosome disjunction. J Cell Sci 1994; 107: 2749-60.

Eijpe M, Heyting C, Gross B, Jessberger R. Association of mammalian SMC1 and SMC3 proteins with meiotic chromosomes and synaptonemal complexes. J Cell Sci 2000; 113(pt 4): 673-82.

Enders GC, May JJ 2nd. Developmentally regulated expression of a mouse germ cell nuclear antigen examined from embryonic day 11 to adult in male and female mice. Dev Biol 1994; 163: 331-40

Fan F, Liu CP, Korobova O, Heyting C, Offenberg HH, Trump G, Arnheim N. cDNA cloning and characterization of Npapa60: a novel rat nuclear pore-associated protein with an unusual subcellular localization during male germ cell differentiation. Genomics 1997; 40: 444-53.

Furukawa K, Hotta Y. cDNA cloning of a germ cell specific lamin B3 from mouse spermatocytes and analysis of its function by ectopic expression in somatic cells. The EMBO Journal 1993; 12: 97-106

Furukawa K, Inagaki H, Naruge T, Tabata S, Tomida T, Yamaguchi A, Yoshikuni M, Nagahama Y, Hotta Y. cDNA cloning and functional characterization of a meiosis-specific protein (MNS1) with apparent nuclear association. Chromosome Res 1994a; 2: 99-113.

Furukawa J, Inagaki H, Hotta Y. Identification and cloning of an mRNA coding for a germ cell-specific A-type lamin in mice. Exp Cell Res 1994b; 212, 426-430.

Gil-Alberdi L, del Mazo J. Microtubule associated proteins during mouse spermatogenesis: localization of a protein immunochemically, related to brain MAP1b protein in the synaptonemal complex.Cytogenet Cell Genet 1992;59:1-5.

Gupta PK. Genetics. Rastogi Publications, Meerut, 2000.

Heyting C, Dietrich AJ, Dettmers RJ, Offenberg RH, Redeker EJ, Vink AC. Synaptonemal complex proteins. Genome 1989; 31: 81-7.

Hirano T, Kobayashi R, Hirano M. Condensins, chromosome condensation protein complexes containing XCAP-C, XCAP-E and a *Xenopus* homolog of *Drosophila* Barren protein. Cell 1997; 89: 511-21.

Hudson DF, Fowler KJ, Earle E, Saffery R, Kalitsis P, Trowell H, Hill J, Wreford NG, deKretser DM, Candcilla MR, Howman E, Hii L, Cutts SM, Irvina DV, Choo KH. Centromere protein B null mice are mitotically and meiotically normal but have lower body and testis weights. J Cell Biol 1998; 141: 309-19.

James RD, Schmiesing JA, Peters AH, Yokomori K, Disteche CM. Differential association of SMC1alpha and SMC3 proteins with meiotic chromosomes in wild-type and SPO11-deficient male mice. Chromosome Res. 2002;10: 549-60.

Kerr SM, Taggart MH, Lee M, Cooke HJ. Ott, a mouse X-linked multigene family expressed specifically during meiosis. Hum Mol Genet 1996; 5: 1139-48.

Kim YS, Nakanishi G, Oudes AJ, Kim KH, Wang H, Kilpatrick DL, Jetten AM. Tsp57: a novel gene induced during a specific stage of spermatogenesis. Biol Reprod. 2004; 70: 106-13.

Kondoh N, Nishina Y, Tsuchida J, Koga M, Tanaka H et al. Assignment of synaptonemal complex protein 1 (SCP1) to human chromosome 1p13 by fluorescence in situ hybridization and its expression in the testis. Cytogent Cell Genet 1997; 78: 103-4.

Lammers JH, Offenberg HH, van Aalderen M, Vink AC, Dietrich AJ, Heyting C. The gene encoding a major component of the lateral elements of synaptonemal complex of the rat is related to X-linked lymphocyte-regulated genes. Mol Cell Biol 1994; 14: 1137-46.

Larsson M, Brundell, E, Jorgensen PM, Stahl S, Hoog C. Characterization of a novel nucleolar protein that transiently associates with the condensed chromosomes in mitotic cells. Eur J Cell Biol 1999; 78:382-90

Lee J, Iwai T, Yokota T, Yamashita M.Temporally and spatially selective loss of Rec8 protein from meiotic chromosomes during mammalian meiosis. J Cell Sci 2003;116(Pt 13): 2781-90.

Liu JG, Yuan I, Brundell E, Bjorkroth B, Daneholt B, Hoog C. Localization of the N-terminus of SCP1 to the central element of the synaptonemal complex and evidence for direct interactions between the N-termini of SCP1 molecules organized head-to-head. Exp Cell Res 1996; 10: 11-9.

Meuwissen RLJ, Offenberg HH, Dietrich AJJ, Riesewijk A, van Lersel M, Heyting C. A coiled-coil related protein specific for synapsed regions of meiotic prophase chromosome. The EMBO Journal 1992; 11: 5091-100.

Miyamoto T, Hasuike S, Yogev L, Maduro MR, Ishikawa M, Westphal H, Lamb DJ. Azoospermia in patients heterozygous for a mutation in SYCP3. Lancet. 2003; 362(9397): 1714-9.

Moss SB, Bumhan BL, Bellve AR. The differential expression of lamin epitopes during mouse spermatogenesis. Mol Reprod Dev 1993; 34: 164-74

Offenberg HE, Schalk JA, Meuwissen RL, van Aalderen M, Kester HA, Dietrich AJ, Heyting C. SCP2: a major protein component of the axial elements of synaptonemal complexes of the rat. Nucleic Acids Res 1998; 26: 2572-9.

Page J, Berrios S, Rufas JS, Parra MT, Suja JA, Heyting C, Fernandez-Donoso R. The pairing of X and Y chromosomes during meiotic prophase in the marsupial species Thylamys elegans is maintained by a dense plate developed from their axial elements. J Cell Sci. 2003; 116(Pt 3): 551-60.

Perez- Castro AV, Shamanski FL, Meneses JJ, Lovato TL, Vogel KG, Moyzis RK, Pedersen R. Centromeric protein B null mice are viable with no apparent abnormalities. Dev Biol 1998; 201: 135-43.

Pezzi N, Prieto I, Kremer L, Perez Jurado LA, Valero C, Del Mazo J, Martinez-A C, Barbero JL. STAG3 a novel gene encoding a protein involved in meiotic chromosome pairing and location of STAG3 related genes flanking the Williams-Beuren syndrome deletion. FASEB J 2000; 14: 581-92.

Prieto I, Suja JA, Pezzi N, Kremer L, Martinez-AC, Rufas JS, Barbero JL. Mammalian STAG3 is a cohesion specific to sister chromatid arms in meiosis I. Nature Cell Biol 2001; 3: 761-6.

Revenkova E, Ejpe M, Heyting C, Gross B, Jessberger R. Novel meiosis-isoform of mammalian SMC1. Mol Cell Biol 2001; 21: 6984-98.

Richardson RT, Batova IN, Widgren EE, Zheng LX, Whitfield M, Marzluff WF, O'Rand MG. Characterization of the histone H1 binding protein, NASP, as a cell cycle regulated somatic protein. J Biol chem. 2000; 275: 30378-86.

Sage J, Martin L, Meuwissen R, Heyting C, Cuzin F, Rassoulzadegan M. Temporal and spatial control of the *Sycp1* gene transcription in the mouse meiosis regulatory elements active in the male are not sufficient for expression in the female gonad. Mech Dev 1999; 80: 29-39.

Schalk JA, Dietrich AJ, Vink AC. Offenberg HH, Van Aalderen M, Heyting C. Localization of SCP2 and SCP3 protein molecules within synaptonemal complexes of the rat. Chromosoma 1998; 107: 540-8.

Schmekel K, Meuwissen RL, Dietrich AJ, Vink AC, van Marle J, van Veen H, Heyting C. Organization of SCP1 protein molecules within synaptonemal complexes of the rat. Hum Mol Genet 1996; 10: 20-30.

Silver LM, Kleene KC, Distel RJ, Hecht NB. Synthesis of mouse t complex proteins during haploid stages of spermatogenesis. Dev Biol 1987: 119: 605-8.

Silver LM, Uman J, Danska J, Garrels JI. A diversified set of testicular cell proteins specified by genes within the mouse t-complex. Cell 1983; 35: 35-45.

Taketo MM, Araki Y, Matsunaga A, Yokoi A, Tsuchida J, Nishina Y, Nozaki M, Tanaka H, Koga M, Uchida K, Matsumiya K, Okuyama A, Rochelle JM, Nishimune Y, Matsui M, Seldin MF. Mapping of eight testis-specific genes to mouse chromosomes. Genomics 1997, 46: 138-42

Tanaka H, Pereira LA, Nozaki M, Tsuchida J, Sawada K, Mori H, Nishimune Y. A germ cell-specific nuclear antigen recognized by a monoclonal antibody raised against mouse testicular germ cells. Int J Androl 1997; 20: 361-6.

Tanaka Y, Nureki O, Kurumizaka H, Fukai S, Kawaguchi S, Ikuta M, Iwahara J, Okazaki T, Yokoyama S. Crystal structure of the CENP-B protein-DNA complex: the DNA binding domains of CENP-B induce kinks in CENP-Box DNA. The EMBO Journal 2001; 20: 6612-18.

Tarsounas M, Morita T, Pearlman RE, Moens B. RAD51 and DMC1 form mixed complexes associated with mouse meiotic chromosome cores and synaptonemal complexes. J. Cell Biol 1999; 147: 207-19.

Tarsounas M, Pearlman RE, Moens PB. Meiotic activation of rat pachytene spermatocytes with okadaic acid: the behavior of synaptonemal complex components SYN1/SCP1and COR1/SCP3. J Cell Sci 1996; 112(pt4): 423-34.

Tay J, Richter JD. Germ cell differentiation and synaptonemal complex formation are disrupted in CPEB knockout mice. Dev Cell 2001; 1: 201-13.

Taylor EM, Moghraby JS, Lees JH, Smit B, Moens PB, Lehmann AR. Characterization of a novel human SMC heterodimer homologous to the Schizosaccharomyces pombe Rad18/Spr18 complex. Mol Biol Cell 2001; 12: 1583-94.

Tsuchida J, Nishina Y, Wakabayashi N, Nozaki M, Sakai Y, Nishimune Y. Molecular cloning and characterization of meichroaciadin male meiotic metaphase chromosome-associated acidic protein. Dev Biol 1998; 197: 67-76

Tureci O, Sahin U, Zwick C, Koshowski M, Seitz G, Pfreundschuh M. Identification of a meiosis specific protein as a member of clas of cancer/testes antigens. Proc Natl Acad Sci USA 1998; 95: 5211-6.

Vester B, Smith A, Krohne G, Benavente R. Presence of a nuclear lamina in pachytene spermatocytes of the rat. J Cell Sci 1993: 104(pt2): 557-63

von Bulow M, Heid H, Hess H, Franke WW. Molecular nature of calicin, a major basic protein of the mammalian sperm head cytoskeleton. Exp cell Res 1995: 219: 407-13

Wang LF, Zhu HD, Miao SY, Cao DF, Wu YW, Zong SD, Koide SS. Molecular cloning and characterization of a novel testi-specific nucleoporin-related gene. Arch Androl 1999; 42: 71-84.

Wang ZQ, Akmal KM, Kim KH. An unusual nucleoporin-related messenger ribonucleic acid is present in the germ cells of rat testis. Biol Reprod 1994; 51: 1022-30.

Welch JE, O'Rand MG. Characterization of a sperm-specific nuclear autoantigenic protein II. Expression and localization in the testis. Biol Reprod 1990; 51: 1022-30.

Welch JE, Zimmerman LJ, Joseph DR, O'Rand MG. Characterization of sperm specific nuclear autoantigenic protein. I. Complete sequence and homology with the xenopus protein N1/N2. Biol Reprod 1990; 43, 559-68.

Willison KR, Dudley K, Potter J. Molecular cloning and sequence analysis of a haploid expressed gene encoding t Complex polypeptide 1. Cell 1986; 44: 727-38.

Yuan L, Liu JS, Zhao J, Brundell E, Daneholt B, Hoog C. The murine SCP3 gene is required for synaptonemal complex assembly, chromosome synapsis, and male fertility. Mol Cell 2000; 5: 73-83.

Chapter 7

NUCLEAR SKELETON PROTEINS:
CHROMOSOMAL BASIC PROTEINS

7.1. MAMMALIAN TESTIS HISTONES

During spermatogenesis, there are dramatic changes in basic nuclear proteins as the germinal cells progress through spermatogenesis. DNA is associated with five types of histone proteins in eukaryotes. The structural unit of chromatin (the nucleosome) consists of a core containing two copies of each histone H2A, H2B, H3, and H4 with approximately 146 bp of DNA coiled around the core (Richmond, et al., 1984). Very little work has been conducted on the average spacing between nucleosomes during the spermatocytes and early spermatid stage of spermatogenesis. It is accepted that two classes of histone H1 genes are present in most eukaryotes: 1) Replication-dependent histones are expressed during the S-phase of the cell cycle and further characterized by genes that do not have introns and have corresponding mRNAs that lack polyadenylation. 2) Histone genes of the second class can have intrones and can have corresponding mRNAs that are polyadenylated and expressed at a low but constant level throughout the cell cycle. But a third class of histones is the testis specific histone subtypes expressed in the germinal cells during spermatogenesis (Meistrich, 1989; Wolfe and Grimes, 1999). The levels and synthesis of histone variants have been measured in spermatogonia and in various stages of primary spermatocytes. The results showed a difference in regulation of the synthesis and accumulation of testis – specific histones H1t, TH2A, TH2B, and TH3. The TH3 is present and actively synthesized in A and B spermatogonia. While no detectable amounts of H1t were found, only low levels of TH2A and TH2B could be found in spermatogonia, TH2A and TH2B are already present and actively synthesized in early primary spermatocytes (around the preleptotene stage), and H1t does not accumulate until the pachytene stage.

In mammals, core and H1 histones function not only to package DNA into the 10 nm fiber containing nucleosomes and into the 30 nm coiled fiber but apparently they also function to regulate transcription of some genes, possibly by interfering with binding of essential transcription factors. A high degree of conservation among H1t genes from different species supports an important role for this specific histone that cannot be filled by other H1 variants. Using *Xenopus* egg extracts, while the somatic H1s significantly inhibited DNA replication in *Xenopus* sperm nuclei, little or no inhibition was seen with H1t. The differences in H1-chromatin interactions might explain some of the diversity in H1 function. Using a series of domain-switch mutants of H1(0) and H1t, De et al., (2002) identified the H1 carboxyl-terminal domains as the domains responsible for the differential affinity for chromatin and for the differential effects of H1 variants upon DNA replication.

Nevertheless, approximately 15% of the DNA remains associated with histone in a sequence specific manner in human sperm. H1 is absent while histone H2 takes the minor form of H2AX and H2AZ variants. The H3 and H4 are highly acetylated. Since remodeling of chromatin structure contributes to various aspects of gene activity, such as transcription, replication, repair and recombination, it can be suggested that histones in sperm could influence which genes are transcribed after fertilization. Additional factors involved in chromatin remodeling include histone modifying enzymes, multi-subunit ATP dependent chromatin remodeling factors and histone chaperones. The TAF-1 has histone–binding and nucleosome assembly and remodeling activities, and is a novel histone chaperone. Recombinant human and *Xenopus* TAF-1ß decondenses Xenopus sperm chromatin. It was suggested that TAF-1 associates with the chromatin through its interaction with histones H3/H4 (Matsumoto et al., 1999). Gardiner et al (1998) suggested that sperm chromatin structures are generally similar in different men but the length of histone-associated regions can vary. The association of sperm DNA with histones or protamines sometimes changes within as little as 400 bp of DNA suggesting that there is a fine control over the retention of histones (Gardiner et al., 1998).

7.2. HISTONE H2A/TH2A

7.2.1. Somatic Variants

The H2A.1 and H2A.2 (somatic H2A variants) and H2A.X (a germ cell enriched variant) seems to be the most abundant of the H2A histones, within spermatogonia. The mouse histone H2A.X gene promoter is regulated by the transcription factor E2F and CCAAT binding protein (Yagi et al., 1995). The promoter region of the mouse H2A.X gene showed that maximal promoter activity is present in the construct containing up to –282 bp H2A.X upstream region. Within this region, two sequences regulating the promoter activation were detected. One was an E2F site and another was a CCAAT box. These sequences are also required for the DNA protein binding activities. Thus the promoter of the H2A.X gene was controlled by both the transcription factor E2F and H1TF2 through the E2F and CCAAT element (Yagi et al., 1995).

7.2.2. Testis Specific TH2A Variant

The testis-specific histone TH2A is synthesized and accumulates in spermatocytes. The synthesis of TH2A begins in early spermatocytes and accumulates to about 25% of the H2A complement in pachytene spermatocytes. TH2A synthesis is not detected in early spermatids but the histone is present in these cells. As will be found later that one H2A variant gene is paired with the somatic H2B gene in the rat. In addition, the testis-specific TH2A variant gene is paired with the testis-specific TH2B variant gene. The genes encoding rat somatic and testis–specific H2A and TH2A histones share 300 bp of 5'upstream region with somatic H2B genes and testis specific TH2B genes respectively (Huh et al, 1991). These genes are transcribed divergently from promoter regions that overlap. The deduced amino acid sequences showed that H2A and TH2A histones have eight amino acid difference in the first half of the molecules and three consecutive changes in the C-terminal region. The TH2A gene is expressed only in testis, independent of DNA replication. Nevertheless, the promoter of the TH2A gene contains the octamer-like element "ATTTGCCAT", an S-phase specific transcription regulatory element responsible for the S-phase specific transcription of the human H2B gene. Although synthesis of TH2A and TH2B histones is independent of DNA replication and insensitive to inhibitors of DNA synthesis in testis, the regulatory region shared by the two genes contained a bi-directional S phase – specific transcription regulatory element. In addition, TH2A gene, like TH2B gene, contains the consensus sequence elements in the 3'non-coding region, which is involved in the S phase – specific stabilization of histone mRNA.

7.3. HISTONE H2B/TH2B

7.3.1. Somatic H2B Variants

There are at least two variants of H2B in rat testis, a somatic variant of H2B and a testis specific variant TH2B. The somatic variant of H2B appears to predominate in spermatogonia. The synthesis of testis-specific variant TH2B begins in spermatocytes and accumulates in pachytene primary spermatocytes (Meistrich, 1989). Synthesis of H2B histones has not been detected in spermatids but H2B and TH2B are present in early spermatids (step 1-8 in the rat). However, in mouse testis, the H2B transcription occurs in round spermatids. The sequence of H2B cDNA clone predicts a protein that is extremely similar to other mouse H2B proteins, except at the carboxyl-terminus where the testes H2B contains an additional 12 amino acids, seven of which are hydrophobic. In contrast to the replication-dependent histone mRNAs, the 3'-untranslated region of this cDNA contains the poly (A) addition sequence (AAUAAA) upstream of a poly (A) tract. Furthermore, the conserved hairpin structure immediately upstream of replication-dependent histone mRNA termini is not present. H2B mRNA transcript is spermatid specific and it is translated during spermiogenesis (Moss et al, 1989). Three sperm H2B sub-types have been identified. Messenger RNAs encoding the sperm specific H2B and H1 proteins have been identified in testes of three different sea urchin species. These transcripts are found only in testis and not in any other adult tissue or embryonic stage tested (Lai and Childs, 1986).

A spermatid-specific H2B (ssH2B) variant has been identified in mouse round spermatids that has an additional 12 amino acids at its carboxyl terminus as compared to somatic H2Bs. The mRNA encoding this protein was also present in rat round spermatids (Unni et al, 1995b). The ssH2B was detected in tubule sections at stages II-VI. It reached a maximum level at stages VII-VIII and then decreased to minimum level at stages XIII-I. Since ssH2B was not detected in pachytene spermatocytes, the stage-specific levels of ssH2B corresponded to the respective steps of spermatids present in the tubules. While ssH2B constituted a relatively small amount (approximately 2%) of total H2B protein in round spermatids, its presence in mouse and rat suggests that its function may be conserved in mammalian spermatogenic cells (Unni et al, 1995b). Although the S-phase specific regulatory elements (SPRE) for several histone genes have been characterized, none of the genes seems to have any common sequence elements for the S-phase specific regulation. This suggests that the S-phase specific transcription of different histone genes may be regulated by different mechanisms or by different trans-acting factors. A nuclear protein factor, which interacts with the human H2B SPRE containing an octamer motif (ATTTGCAT) has been purified from HeLa cell nuclear extracts and seems to stimulate transcription of the H2B gene in vitro.

7.3.2. Testis TH2B

Rat TH2B: The rat testis specific TH2B histone gene is restricted to pachytene spermatocytes. The gene has conserved the promoter elements involved in the S-phase specific expression of somatic H2B histone genes as well as the S-phase specific stabilization of histone mRNA.

Therefore, the gene has the potential to be expressed during the S phase in all tissues as well as during early stages of spermatogenesis in which spermatogonia undergo rapid cell divisions. One of the basic questions is how a gene, which does not have any cell type specific regulatory element can be expressed in a cell type-specific manner during spermatogenesis. In TH2B gene, this appears to be achieved at least at two levels: i) assembly of germ cell specific chromatin structure as evidenced by the hypomethylation of the gene from the beginning to the end of spermatogenesis, and ii) repression of the gene transcription in non-expressing spermatogenic cells (Hwang et al., 1990).

In contrast to somatic histones, the TH2B gene contains S-phase specific regulatory elements (SPRE). Since the gene is not expressed except in pachytene spermatocytes, the SPRE are most probably repressed in somatic cells. In pachytene spermatocytes, the SPRE could be either repressed or involved in the cell – type specific expression of TH2B gene. The DNA sequence elements involved in the S_0 phase-dependent transcription of the cloned TH2B gene in somatic cells were characterised (Hwang et al., 1990). The cloned TH2B gene has two DNA sequence elements, which stimulate transcription of the cloned gene in an S-phase dependent manner when introduced into somatic cells. The factors interacting with the two elements, CCAAT at –127 bp and octamer ATTTGCAT at –93 bp, interact with each other to bring about a maximum stimulation of S-phase dependent-transcription. The level of CCAAT and octomer-binding proteins is unchanged during the cell cycle, and the S-phase dependent transcription of TH2B and endogenous mouse H2B genes does not require synthesis of new proteins during the S phase. Thus, cell cycle specific postranslational modification of regulatory proteins may be responsible for the S-phase-dependent transcription of H2B histone genes (Hwang et al, 1990).

Another study revealed that DNA methylation plays a key role in the transcriptional repression of the germ cell-specific TH2B gene (Chapter 14). The CpG methylation sites within the promoter of the TH2B gene are hypomethylated in germinal cells but they are methylated in somatic tissues. This is consistent with the hypothesis that DNA methylation inhibits gene activity by preventing the binding of transcription factors to their recognition sequences. The binding of ubiquitous transcription factors to the promoter region of TH2B gene may be blocked in nuclei of liver and DNA methylation can directly interfere with the binding of transcription factors recognizing a hexamer (ACGTCA) motif. In vitro DNA methylation and transfection experiments demonstrated that expression of TH2B gene is inhibited by DNA methylation in vivo (Choi and Chae, 1991). Although the gene is hypomethylated at all stages of spermatogenesis, it is not expressed in germinal cells before the pachytene spermatocytes stage. The low level of TH2B mRNA in these early spermatogenic cells may be a result of transcriptional repression of the gene by a pre-meiotic cell-specific protein, which binds to a site between the TATA element and the transcription initiation site of the TH2B gene (Lim and Chae. 1992). The regulation expression of the rat TH2B gene appears to be quite different from the regulation of expression of the sea urchin sperm H2B-1 gene where a protein factor from sea urchin embryos binds over a CCAAT box preventing interaction with a positive–active CCAAT-binding factor resulting in a block to transcription. In sperm cells the CCAAT-binding factor binds resulting in transcription of the sperm H2B-1 gene.

Human TH2B: In contrast to spermatozoa from other species, human spermatozoa still contain a significant amount of histone, including testis – specific histone 2B (hTH2B). It has been shown that an antibody targeting tyrosine hydroxylase, which cross reacts with rat TH2B, specifically reacts with human TH2B on Western blots due to homology at N-terminus. In human testis, hTH2B immunostaining, first apparent in spermatogonia shows marked variation, especially at the pachytene spermatocyte state and then reached an intense signal in round spermatids. During condensation of the spermatid nucleus, the immunodetectability of TH1B disappears gradually and is completely absent at end of spermatogenesis. The immunostaining of hTH2B was observed only when spermatozoa obtained from semen were decondensed to make nuclear proteins accessible to the antibody (van Roijen et al., 1998). It has been suggested that the N-terminus of rat TH2B is less tightly bound to DNA or to other proteins at this time in preparation for the removal of TH2B and other histones.

The amino acid sequence of TH2B deduced from cDNA sequences showed extensive sequence divergence at the N-terminal end among H2B and TH2B histones; the rest was highly conserved. TH2B contains one cysteine whereas no cysteine is present in somatic histones. The TH2B in testis is detected 14 days after birth, reaching a maximum at Day 20. The level of H2B mRNA showed a reciprocal pattern. This contrasting pattern can be due to the gradually changing proportion of cell types with testicular maturation (Kim et al., 1987). The human testis/sperm-specific histone hTH2B is 85% homologous to somatic H2B and has over 93% homology with the testis H2B of rodents. Two genetic alleles of hTH2B were found in the human population. The hTH2B gene is transcribed exclusively in testis, and the corresponding protein is also present in mature sperm. The hTH2B was concentrated in spots located at the basal nuclear area of a subpopulation (20% of cells) of mature sperm. This fact may be of particular importance, because the hTH2B "positive" and "negative" sperm cells may undergo significantly different decondensation processes following fertilization (Zalensky et al., 2002).

7.4. HISTONE H3 /TH3 VARIANTS

The germ cell-specific histone TH3 is the only testis-specific histone to appear in spermatogonia (Meistrich, 1989). A somatic variant, most likely H3.2, is most abundant at this stage. Histone TH3 synthesis is not detected in spermatocytes but its presence is found in spermatocytes and early spermatids. TH3 from rat testis has been characterized by amino acid analysis, tryptic peptide mapping, labelling with cystine, and by electrophoretic mobility as a variant of somatic H3. Its mobility on gels is retarded more than that of the somatic variants H3.2 and H3.3, but less than that of H3.1; it migrates between the H2As and H1s. The TH3 is found in significant amount in spermatogonia and in similar or slightly higher amount in spermatocytes and round spermatids. Synthesis of TH3 takes place in the spermatogonia but not in spermatocytes, in contrast to the other testis-specific histones (Trostle–Weige et al, 1984).

In an attempt to define patterns of replacement of histone H3.3 gene expression during male germ cell differentiation, Bramlage (1997) isolated cDNAs corresponding to the murine H3.3A and H3.3B genes. RNase protection assays showed that H3.3A mRNA is present in pre- and post-meiotic cells, whereas expression of the H3.3B gene is essentially restricted to cells of the meiotic prophase. The gene structure of human H3.3B showed characteristic features of H3.3 gene. It contains an intron of about 0.5-kb in the 5'untranslated region and two smaller introns within the coding gene portion. Secondly, no histone gene-specific dyad symmetry element was found in the 3'untranslated region, but three putative polyadenylation signals were detected downstream of the gene. The human H3.3B gene was mapped to the telomeric region of chromosome 17 (17q25). This localization of the H3.3B gene and its solitary arrangement contrast with the majority of the replication-dependent histone genes, which form a large cluster on chromosome 6 and a second cluster on chromosome 1 (Albig et al., 1995).

```
   Coding Sequence

(S)TCTGGACGAGGGAAAGGCGGCAAAGGTCTGGGTAAAGGTGGCGCCAAGCGACACCGCAAGGTTCTCCGA
           *  *     *  *     *  *  *           *     *  *     *  *  *  *
(G)TCTGGACGTGGTAAAGGTGGTAAAGGGCTTGGGAAAGGTGGCGCTAAGCGTCATCGCAAAGTCCTGCGC
   SerGlyArgGlyLysGlyGlyLysGlyLeuGlyLysGlyGlyAlaLysArgHisArgLysValLeuArg
                                                            20

   GATAACATCCAGGGCATTACCAAGCCCGCTATCCGGCGTTTGGCTCGGCGCGGCGGTGTCAAGCGTATC
     *              *  *        *     *  **    *        *  *  *     *
   GACAACATCCAGGGCATCACAAAGCCCGCCATCCGCCGCCTGGCCCGGCGCGGAGGAGTGAAGCGCATC
   AspAsnIleGlnGlyIleThrLysProAlaIleArgArgLeuAlaArgArgGlyGlyValLysArgIle
                                                    40

   TCGGGTCTTATCTACGAGGAGACTCGCGGCGTCCTGAAGGTTTTCCTTGAAAATGTGATTCGCGATGCT
     *  *  *              *     *  *        *     *  *  *     *     *  *
   TCCGGCCTCATCTACGAGGAGACCCGCGGTGTGCTGAAGGTGTTCCTGGAGAACGTGATCCGCGACGCC
   SerGlyLeuIleTyrGluGluThrArgGlyValLeuLysValPheLeuGluAsnValIleArgAspAla
                                           60

   GTCACCTACACCGAGCACGCCAAGCGCAAGACGGTCACGGCTATGGACGTGGTGTACGCGCTCAAGCGC
              *                          *  *           *        *  *
   GTCACCTACACGGAGCACGCCAAGCGCAAGACGGTCACTGCCATGGACGTGGTCTACGCGCTTAAACGC
   ValThrTyrThrGluHisAlaLysArgLysThrValThrAlaMetAspValValTyrAlaLeuLysArg
                                  80

   CAGGGCCGCACGCTCTATGGCTTCGGCGGC
              *     *        *
   CAGGGCCGCACACTCTACGGCTTCGGTGGC
   GlnGlyArgThrLeuTyrGlyPheGlyGly
                        100
```

Fig.7.1. Comparison of the coding and non-coding regions of the rat somatic (S) and germinal (G) histone H4 genes. Asterisks mark nucleotides which differ within the coding regions. Reproduced with permission from S. A. Wolfe et al. Biochim Biophys Acta 1007; 140-50: 1989 © Elsevier.

7.5. HISTONE H4t

The germinal histone H4t and H1t genes are expressed during spermatogenesis in rat pachytene spermatocytes, and the somatic histone H4 gene expressed only in nongerminal cells of rat and somatic tissues. Only one histone H4t polypeptide has been identified in rats. The H4t is a highly conserved polypeptide having 102 amino acids in length (Grimes et al., 1987; Wolfe, et al. 1989) (**Fig.7.1**). Multiple unique rat histone H4t genes have been sequenced and there may be about 6 or 7 different H4t genes in mammals. Apparently these different genes are differentially expressed so that only a sub-set is expressed in testis germinal cells like the other testis histone genes. High synthesis of H4t in pachytene spermatocytes occurs at a time when there is no DNA synthesis, and therefore at least one of the H4t genes is transcribed in male germinal cells that are not replicating DNA. Although the H4t gene is closely linked to the testis-specific histone H1t gene that is transcribed only in testis, the H4t gene is transcribed in both testis and non-testis tissues. The deduced amino acid sequence of the rat H4 histone is identical to that of the sequence of human histone H4 but the nucleotide sequence of the coding region differs significantly from the coding region of the human histone H4. The S1-nuclease protection analyses revealed that mRNAs of both species are present in a fraction of testis cells highly enriched in pachytene spermatocytes, while only the H4 mRNA species was present in a rat myeloma cell line (Grimes et al, 1987).

Although the predicted amino-acid sequences of testis specific and somatic H4 were identical, 49 out of 102 codons differed. The leader sequence of the germinal histone H4t mRNA was 17 bases compared to 40 bases for the somatic histone H4 mRNA, and the 3' terminal sequence of the germinal histone H4t mRNA was 52 bases compared to 75 bases for the somatic histone H4 mRNA. The germinal histone H4t gene also lacked a consensus purine-rich motif, which was present in the 5' noncoding region of the somatic histone H4 gene (Wolfe et al 1989). The histone H4 gene promoter is functional in rat testis. Examination of the proximal promoter of the histone H4 gene indicated that specific protein-DNA interactions occur when pachytene spermatocyte nuclear proteins were mixed with promoter fragments essential for regulating transcription of the histone H4 gene (Wolfe and Grimes, 1991). Mouse and human H4t genes, associated with H1t gene, were sequenced by Drabent et al (1995). The deduced amino acid sequences were identical to the other mouse or human H4t histone but the gene differed significantly in their nucleotide sequences. Both human and mouse genes were located on the same DNA strand compared with H1t gene. In contrast to this identical transcriptional orientation of H1t and its neighbouring H4t gene in mouse and man, the H4t gene with the opposite orientation has been described in the vicinity of rat H1t gene. It was concluded that both murine and human H4t gene like the H1t gene are expressed in testicular cells whereas H4t gene, in contrast to H1t gene is expressed in non-testicular human and mouse culture cells.

Acetylation of H4t: Mouse spermatogonia and preleptotene spermatocytes contain acetylated core histones H2A, H2B and H4, whereas no acetylated histones were observed throughout meiosis in leptotene or pachytene spermatocytes or in most round spermatids. Acetylated forms of H2A and H2B, H3 and H4 reappeared in step 9 to 11 elongating spermatids, and disappeared later in condensing spermatids. Observations suggested that deacetylases are responsible for maintaining a deacetylated state of histones in these cells. The regulation of histone deacetylase activity / concentration appears to play a major role in controlling histone hyperacetylation and probably histone replacement during spermiogenesis (Hazzouri et al., 2000; Wade et al., 1997). During early spermatids when the histones are beginning to be replaced by the transition proteins, the core histone H4 becomes hyperacetylated (Meistrich, et al., 1992). Though most of the testicular cells were acetylated, histone H4t from fractions enriched in elongated spermatids was hyperacetylated (Grimes and Henderson, 1984). It appeared that histone hyperacetylation is a physiological process that precedes and facilitates the removal of the core histones from the nucleosome during this time period in mammals as found in trout testis. As such, it is an obligatory step in the removal of the histones. Alternatively, the process of acetylation/deacetylation of histones may be one of the mechanisms responsible for initiation of transcription at the sites of acetylation of DNA-histone complex, and thus controlling stage specific expression of several proteins during meiosis.

7.6. HISTONE H1 AND TESTIS VARIANT H1t

7.6.1. Histone H1

Each cell contains a complement of H1 family histones, which can vary during development and differentiation. Some H1 subtypes are present in only a specific cell lineage such as the H1t subtype found in only germinal cells of the mammalian testis and the H5 subtype found in nucleated erythrocytes. The H1 family of histones is the largest and most variable of the histones. The H1 linker histones are components of chromatin associated with the 10nM nucleosome containing DNA fiber, which changes into a higher ordered structure, the 30nm solenoid. This transition results in sequestering of genes into heterochromatin. The primary function of H1 appears to be structural in chromatin but it has also been noted to be involved in the repression of transcription of some genes, possibly by the displacement of essential transcription factors. Members of the H1 family exhibit the common characteristics of having a short, randomly coiled basic amino-terminal arm and a long, randomly coiled, basic carboxy-terminal arm, which flanks a central, trypsin-resistant, globular domain. The basic N- and the C-terminal arms are thought to interact with linker DNA while the globular domain is essential for H1 interaction with the nucleosome. Cross species comparisons have shown that the properties of most of the subtypes are conserved. Vertebrates have seven known different subtypes of the histones H1 gene including histone H1°. The period of H1.1 gene transcription is restricted to the proliferative phase in which cells undergo successive divisions, whereas the H1.1 protein is a component of the proliferative, meiotic, and spermatogenic phases (Franke et al., 1998). The relative amounts of histone subtypes H1.2 is high in dividing cells, while subtypes H1.3 and H1° are higher in non-dividing cells. Subtypes H1.3, H1.4, H1.5 and H1°, which constitute the largest amount of histone H1 in somatic tissues, are much less abundant in germinal cells. During spermatogenesis somatic histone subtypes initially make up about 50% of the H1 histone and drop off to 1% by the late pachytene stage. While H1.1 mRNA and protein was already present in spermatogonia of the prepuberal testis, the H1.1 mRNA remained confined to the most peripheral layer of gem cells of 20-day-old and adult mice. In contrast to the H1.1 mRNA, the H1.1 protein persisted in the germ cell chromatin with decreasing concentration throughout meiosis and in postmeiotic cells. The somatic histones are first replaced by H1.1 and H1.2 in the spermatogonial stage where they are most abundant accounting of about 80% of H1. These two histones are then replaced by the testis-specific histone H1t in the primary spermatocytes stage where it constitutes about 60% of the H1 complement in late pachytene spermatocytes.

Yan et al., (2003) identified a spermatid-specific linker histone H1-like protein (termed HILS1) in the mouse and human. Both mouse and human HILS1 genes are located in intron 8 of the alpha-sarcoglycan genes. HILS1 is highly expressed in nuclei of elongating and elongated spermatids (steps 9-15). The HILS1 displays several biochemical properties that are similar to those of linker histones. Because HILS1 is expressed in late spermatids that do not contain core histones, HILS1 may participate in spermatid nuclear condensation through a mechanism distinct from that of linker histones. HILS1 may also regulate gene transcription, DNA repair, and/or other chromosome processes during mammalian spermiogenesis (Yan et al., 2003)

Functions of histone H1: Histone H1 participates in transcriptional regulation by stabilization of nucleosomes and conversion of DNA into a higher ordered structure. Recently, histone H1 has been shown to compete with transcriptional activators as measured by in vitro competition for binding to specific sites on DNA. Histone H1 added to naked DNA was shown to inhibit transcription of a test gene flanked by binding sites for Sp1, GAL4-VP16, or the GAGA factor. Addition of the appropriate factor was shown to counteract the H1 mediated repression. The results from these experiments provide evidence that histone H1 can act as a general repressor that can be selectively removed or depleted by specific transcriptional factors. The repressor function of histone H1 may not be a generalized mechanism whereby the whole chromosome is covered nonspecifically by H1, but occurs in specific localized regions (Croston et al., 1991; c/r Wolfe et al., 1999). Although assumed to mediate chromatin condensation, H1t binds DNA with lower affinity than other H1 variants, rendering it relatively more sensitive to DNase I. Studies leading to these conclusions [De Lucia et al, 1994; Khadake et al, 1994] provide compelling evidence supporting a suggestion that H1t maintains chromatin in a relatively decondensed state, facilitating events such as recombination (De Lucia et al, 1994; see vanWert et al., 1996).

```
                                                           TCAGGTGG   AATCCAAGTT
-180 CCCGTGGGTC  ACCTGTGTCA  TAACCTGAGC  GATTCTGCAG  AGAGCACACA  TGCAAGTTTC
-120 TACTTTTTGG  GGGAAAAGAA  ACACAAATGC  CCCTTCCCCA  GGGGCGGGGG  AGGCGCCTAG
 -60 GGATGCACCA  ATCACAGCGC  GCCCTGCTCT  ATATAAGGCC  CCCCCCCGGA  CCCCCTCTCT
  +1 ACTCCAGCGC  TGTTCGCTCT  CCTGTACTTT  TTTTTTTAAT  TGGGTGTAGT  TCTTGACCAG
  61 CTCTTGACT   ATG TCG GAA ACG GCT CCT GCA GCC TCA AGT ACT CTT GTT CCA GCT
                 Met Ser Glu Thr Ala Pro Ala Ala Ser Ser Thr Leu Val Pro Ala
                      1                                  10
 115 CCT GTA GAG AAA CCT GCA ACT AAA AGG CGA GGG AAG AAG CCT GGC ATG GCC ACT
     Pro Val Glu Lys Pro Ala Thr Lys Arg Arg Gly Lys Lys Pro Gly Met Ala Thr
                         20                                      30
 169 GCT CGC AAA CCT CGG GGT TTC TCG GTT TCC AAG TTG ATT CCT GAG GCC CTT TCC
     Ala Arg Lys Pro Arg Gly Phe Ser Val Ser Lys Leu Ile Pro Glu Ala Leu Ser
                                 40                                      50
 223 ATG TCT CAG GAA CGG GCA GGA ATG TCC CTT GCT GCC CTG AAG AAA GCC CTG GCT
     Met Ser Gln Glu Arg Ala Gly Met Ser Leu Ala Ala Leu Lys Lys Ala Leu Ala
                                         60
 277 GCG GCT GGC TAT GAC GTG GAG AAG AAC AAC AGT CGT ATC AAG CTG GCC CTC AAG
     Ala Ala Gly Tyr Asp Val Glu Lys Asn Asn Ser Arg Ile Lys Leu Ala Leu Lys
          70                                      80
 331 AGA CTT GTG AAT AAG GGA GTC CTG GTG CAG ACC AAG GGC ACC GGA GCC TCA GGC
     Arg Leu Val Asn Lys Gly Val Leu Val Gln Thr Lys Gly Thr Gly Ala Ser Gly
                 90                                  100
 385 TCC TTC AAG CTT AGC AAG AAG GCA GCT TCA GGC AAC GAC AAG GGC AAG GGC AAG
     Ser Phe Lys Leu Ser Lys Lys Ala Ala Ser Gly Asn Asp Lys Gly Lys Gly Lys
                         110                                     120
 439 AAA TCT GCT TCT GCC AAG GCT AAG AAA CTG GGC TTG TCC AGG GCC TCG AGA TCC
     Lys Ser Ala Ser Ala Lys Ala Lys Lys Leu Gly Leu Ser Arg Ala Ser Arg Ser
                                 130                                     140
 493 CCC AAG AGT AGT AAG ACC AAG GTT GTC AAG AAG CCA AAA GCT ACG CCC ACA AAG
     Pro Lys Ser Ser Lys Thr Lys Val Val Lys Lys Pro Lys Ala Thr Pro Thr Lys
                                         150
 547 GGT TCT GGG AGC AGA AGG AAG ACC AAA GGG GCC AAG GGC TTG CAA CAG CGC AAA
     Gly Ser Gly Ser Arg Arg Lys Thr Lys Gly Ala Lys Gly Leu Gln Gln Arg Lys
          160                                     170
 601 AGC CCC GCC AAA GCG AGG GCA ACC AAC TCC AAC TCT GGG AAG TCA AAG ATG GTC
     Ser Pro Ala Lys Ala Arg Ala Thr Asn Ser Asn Ser Gly Lys Ser Lys Met Val
                 180                                     190
 655 ATG CAG AAG ACC GAC CTA AGG AAG GCA GCA GGA AGA AAG TGA GTTT   CAAAGCCAGT
     Met Gln Lys Thr Asp Leu Arg Lys Ala Ala Gly Arg Lys  *
                         200
 711 TTTCAAAAAC  CCAAAGGCTC  TTTTAAGAGC  CACTTACATA  CTTCTTAAAA  TGGCCAAACA
 771 CTGAGCAAAA  GTTAGAGGTG  GGCAGTCACT  TAGGTCGA
```

Fig.7.2. Nucleic acid sequence of the rat *H1t* gene and the predicted amino acid sequence of ORF of H1t. Underlined regions show homology with other H1 variants. Reproduced with permission from K.D.Cole et al: J Biol Chem 261; 7178-83: 1986 © The American Society of Biochemistry and Molecular Biology.

7.6.2. Testis Specific Histone-1 (H1t)

Germinal cell histone, H1t, is found in most mammalian species and in most vertebrates. Translation of rat histone H1t occurs exclusively in pachytene primary spermatocytes and it is the most abundant H1 variant in these meiotic cells. Synthesis of H1t is dependent upon the steady-state level of H1t mRNA. The H1t mRNA does not accumulate in the testis before the spermatocytes stage. Moreover, there is no detectable H1t mRNA in early spermatids or in nongerminal testis cells such as Leydig cells. Examination of enriched populations of the various cell types from testis showed transcriptional initiation in both the pachytene spermatocytes enriched fraction. It is presumed that the testis H1t may be essential for the unique chromatin structure and for the unique transcription pattern in primary spermatocytes, although these possibilities have not been adequately tested. However, Drabent et al (1998) showed that mouse H1t gene begins to express at premeiotic stages of spermatogenesis and that H1t mRNA is detectable even in spermatogonia, albeit at a low level. In contrasts to H1t mRNA, the H1t protein has not been detected in spermatogonia (Drabent et al., 1998). However, the H1t is detected in the chromatin of mouse meiotic prophase chromosomes, only after synapsis and synaptonemal complex (SC) assembly is completed and before core separation is initiated. The H1t protein is evenly distributed over euchromatin, heterochromatin and SC (Moens, 1995).

Mice homozygous for the mutated H1t gene locus developed normally and showed no anatomic abnormalities until the adult stage. The H1t deficient mice were fertile and reproduced as wild-type mice. The process of spermatogenesis and the testicular morphology remained unchanged in the absence of H1t. RNase protection analysis demonstrated that H1.1, H1.2 and H1.4 histone gene expression is enhanced during spermatogenesis in H1t-deficient mice (Drabent et al, 2000; Fantz et al., 2001; Rabini et al, 2000).

Aminoacid Sequence of H1t: The rat H1t gene encodes 207-amino-acid protein (excluding the initiating methionine) that perfectly matches with amino-acid sequence determined directly. The gene lacks introns and has good matches to all the consensus sequences known to lie upstream from a variety of H1 genes from diverse organisms. It has also the standard downstream palindromic sequence that specifies the 3' end of most histone messages (Cole et al, 1986a) (**Fig. 7.2**). The amino acid sequence determined for H1t isolated from boar testes showed that the protein is composed of 211 amino acids with the composition having 3 aspartic acids, 6 asparagines, 13 threonines, 21 serines, 7 glutamic acids, 7 glutamines, 15 prolines, 14 glycines, 37 alanines, 11 valines, 1 methionine, 3 isoleucines, 14 leucines, 1 tyrosine, 1 phenylalanine, 42 lysines, 15 arginines and a calculated molecular weight of 22,059 disregarding post-translational acetylation of the amino-terminal alanine. The carboxyl-terminal half of H1t is distinguished from the standard somatic family by being somewhat shorter and by the presence of 10 arginine residues. In contrast to many H1 proteins, the carboxy-terminal region of H1t does not show an obvious pattern of peptide repeats (Cole et al. 1984). There is a high degree of conservation in amino acid sequence of histone H1t from different mammalian species. This is in contrast to the high degree of variability between the seven subtypes of H1t, which have a lower degree of conservation than the core histones (Koppel et al, 1994). The H1t, like other H1 variants, has a variable N-terminal domain, a conserved central globular domain, and a basic and highly variable C-terminal domain. Globular domains of H1 histones including H5, a variant found in nucleated erythrocytes, are evolutionarily conserved. This domain binds to DNA and two probable DNA binding sites have been identified [Ramakrishnan et al, 1993]. The H1t globular domain may have similar DNA binding sites.

The sequence of the first 108 residues of rat H1t has been compared against boar H1t and also with a consensus sequence, characteristic of standard somatic H1 variant (Cole et al, 1986b). The two versions of H1t share a common pattern of divergence from the standard somatic consensus. For example, they both share many substitutions within the H1 globular region (residues 40-110), a portion of the molecule that is virtually invariant among the common somatic variants. Within the entire region, there are 28-shared locations, where both forms of H1t differ from the somatic consensus. However, identical substitutions occur at only 15 of these sites, and each protein also differs from the consensus at five (boar) or ten (rat) additional locations that are not shared. These results established that H1t from diverse sources shows a characteristic pattern of divergence from the sequence of standard somatic H1 proteins (Cole et al, 1986b).

Promoter of H1t / CCTAGG element: The human H1t gene is located on chromosome 6 (Koppel et al., 1994). Understanding the regulation of H1t gene is challenging since its promoter shares all its control elements of somatic or early spermatogenic cell H1 genes. Cis acting elements sufficient for regulating testis specific transcription, have been found within 2.4-kb upstream and 3.78-kb down stream of H1t gene. Understanding transcriptional regulation of H1t gene has been recently reviewed (Grimes et al., 2003). A unique 40-base pair promoter element designated H1t/TE is essential for spermatocyte-specific expression. The H1t/TE

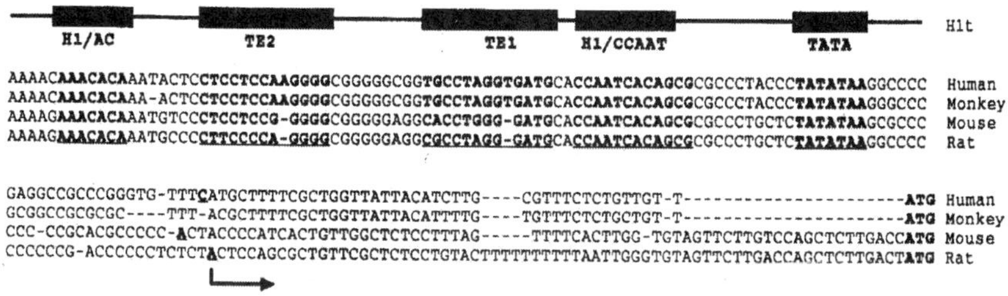

Fig.7.3. Conserved promoter regions of mammalian H1t genes. The H1t/TE region is conserved in all the promoters. The transcription start site of rat gene is indicated by an arrow and those of rat, mouse and human by bold character. Reproduced with permission from S. A. Wolfe et al. Biochemistry 34; 12461-69: 1995 © The American Chemical Society.

element contains three subelements designated TE2, GC-box, and TE1 based on in vitro footprinting and electrophoretic mobility shift assays. The TE1 sub-element is bound to a protein complex. Mutation of TE1 leads to a drop in H1t promoter activity in germinal GC-2spd cells as well as in nongerminal cell lines. The rat TE1 but not TE2 contains a CpG dinucleotide, which is methylated in liver but not in primary spermatocytes. Methylation of the cytosine at this site almost eliminates nuclear protein binding. Thus, there are significant functional differences in the TE2 and TE1 sub-elements of the H1t promoter with TE1 serving as a transcriptional activator-binding site and TE2 serving as a repressor- binding site in some cell lines (Wilkerson et al., 2003) **(Fig. 7.3)**.

A unique 18-bp sequence element is present between the H1/GC box and H1/CCAAT box seen in the rat and the human H1t promoters (Wolfe and Grimes, 1993). Testis nuclear proteins to the unique base sequence element within the H1t promoter showed relatively tight binding with an excess of homologous DNA but not with a mutated element. Testis proteins from prepubertal animals did not bind to the 18-bp promoter element but proteins from enriched populations of primary spermatocytes did bind. Therefore, the temporal correlation between onset of transcription of the H1t gene and the time when the specific H1t promoter-binding proteins detected in primary spermatocytes suggested that the DNA-binding proteins could be germinal cell-specific transcription factors that participate in the formation of an active H1t transcription initiation complex.

Transgenic mice carrying a 6.8-kb fragment of rat genomic DNA encompassing the H1t gene expressed rat H1t at high levels in the testis and in no other organ examined. H1t fragments truncated to within 141 bp of the gene in the 5' direction or within 837 bp in the 3' direction retained testis specificity. Expression of rat H1t protein was also evident in the testes of the transgenic mice. The stage of spermatogenesis of transgene expression was restricted to germinal cells in three different constructs, and H1t (-948/+71)-lacZ fusion was expressed specifically in the spermatocytes and round spermatids of a transgenic line, confirming that sequences sufficient for correct, tissue and developmental expression lie within this 1019-bp segment of the gene (Bartell et al., 1996). *H1t* gene for the testis-specific linker histone did not demonstrate an intragenic activating region (IAR). While H1t has a similar α sequence and did bind YY1, it lacks the Ω homologies of H1d (Horvath et al., 2003).

Examination of the H1t promoter revealed a sequence element located between the H1/GC box and the H1/CCAAT box homologous to an element that directs the testis specific expression of ß-tubulin in *Drosophila*. The central portion of the element contains the palindromic sequence CCTAGG. Additionally, this palindrome is found in the human, monkey, and rabbit H1t promoters in the same location. The H1t/CCTAGG palindrome is actually the center of a large 18-19 bp element. The nucleotide sequence within this element is highly conserved with only two base differences being noted upon comparison of five species of mammals. Thus far, this element has been found only in histone H1t promoters. The H1t/CCTAGG containing sequence element currently appears to be an excellent candidate for interaction with a transcriptional regulatory protein. Since this element is highly conserved in mammalian H1t promoters, it potentially represents a binding site for a transcriptional factor (Clare et al,1997b; Wolfe et al, 1995) **(Fig. 7.3)**.

TE Elements: Histone H1t genes from both Sprague-Dawley rats and humans are very similar, but they differ substantially from nongerminal H1 gene (Cole et al., 1986; Drabent et al., 1991; Wolfe and Grimes, 1993). The appearance of testis-specific nuclear proteins that bind to a unique promoter sequence element designated H1t/TE located between the H1/AC box and the H1/CCAAT box correlates with the onset of transcription of the H1t and H4t genes during the meiotic cell cycle. In order to determine whether sequences flanking the rat H1t gene are sufficient to confer tissue-specific expression in vivo, a 6859 bp EcoRI restriction fragment of genomic DNA containing the rat histone H1t gene was microinjected into mouse embryos. The descendent of resulting transgenic mice expressed the rat gene in proper tissue and at proper meiotic cell cycle stage. The appearance of appropriate meiotic cell cycle-specific transcription indicated the importance of the conserved promoter sequence elements between the two species (vanWert et al., 1995 ; Bartellet al., 1996).The Results from transgenic mice in other experiment showed that the truncated form of 6859 bp rat genomic Eco R1 shortened H1t promoters extending 210 bp upstream from H1t gene ATG start codon, was sufficient to generate high levels of rat H1t mRNA in testis of transgenic mice. But northern blots failed to detect rat H1t mRNA in other tissues (Bartell et al, 1996).

Comparison of DNA sequences of human, monkey, mouse and rat revealed the presence of almost identical sequences in the proximal promoter region extending from the H1/AC box through the TATAA box. In addition to conserved elements common to replication-dependent H1 promoters, the H1t promoter contains a unique TE element, and sequences within this element may contribute to enhanced expression of the gene in primary spermatocytes. Two imperfect inverted repeat sequences designated TE1 and TE2, that are located within the large TE element, overlap a central GC-rich region and bind specifically to nuclear proteins derived from primary spermatocytes where H1t gene is transcribed. Binding activity was also present in spermatids. Protein interactions experiments indicated that a complex of proteins with a molecular mass of approximately 180 kDa binds TE1. The GC-rich region in H1t and in some replication dependent histone H1 promoters contains an Sp1 consensus sequence. Although the H1t/TE element that contains the GC-rich region binds nuclear proteins, it does not appear to bind Sp1 obtained from cell populations enriched in primary spermatocytes (Wolfe et al. 1995; vanWert et al, 1996).

Although, subtle differences are found in the H1t promoter compared to the other H1 promoters, lines of evidence support the hypothesis that the sequence element TE located within the H1t promoter is essential for enhanced testis-specific transcription of this gene. Transgenic mice bearing a rat H1t transgene, which contains a replacement of the TE element with stuffer DNA, failed to express rat H1t mRNA. In addition, an upstream sequence appeared to function as a silencer element that leads to transcriptional repression of the H1t gene in nongerminal cells. Thus, multiple promoter elements appear to contribute to the regulation of H1t gene (Grimes et al, 1997).

When the native rat H1t gene along with flanking sequences, including 2453 bp upstream and 3784 bp downstream from the coding region, was microinjected into mouse embryos, the offspring of the resulting transgenic mice transcribed the transgene in a tissue specific manner and only in primary spermatocytes. In one study the TE promoter element was deleted and replaced with a heterologous stuffer DNA fragment. When the mutant rat DNA fragment was used to create transgenic mice, off springs of the mice bearing the promoter mutation did not transcribe the rat H1t gene in any tissue. On the other hand, transcription of the rat H4t transgene, which is located approximately 1.5kb downstream from the H1t gene, occurred in these animals. These studies support the hypothesis that the TE element is essential for enhanced testis – specific transcription of the H1t gene in primary spermatocytes (van Wert et al, 1998).

Transcriptional Repression: Recently there has been a focus upon transcriptional repressors. RNA polymerase II requires a set of proteins (general transcription factors and others) to assemble around the start point of transcription guided in part by the TATA box (Chapter 14). Efficient transcription may require activator proteins, which may require additional proteins termed mediator components that associate with RNA polymerase II. Eukaryotic repressors work at least in three ways: 1) they bind to DNA by interferring with the binding of activator. 2) They bind with DNA alongwith activator and quench the activity of activator. 3) In third mechanism, repressor acts on transcription machinery. Therefore, H1t may have multiple key functions during spermatogenesis. At present, it is not yet clear that histone H1t is essential to the process of spermatogenesis. However, subtle changes in the H1 promoter lead to spermatocyte-specific regulation. Studies summarized by Grimes et al (1997) have indicated that a sequence upstream from H1t/ AC box may act as a silence element and may be involved in repression of transcription of H1t gene in non-germinal cells. Clare et al (1997a) indicated that H1t gene is strongly repressed by a G/C-rich region downstream of the TATA box.

Studies with proliferating cell lines revealed several upstream elements that may contribute to control of testis specific histone H1t gene transcription (Drabent and Doenecke, 1997). H1t transcription as detected with reporter constructs driven by the H1t promoter is low compared to transcription from somatic H1t constructs in cells undergoing mitosis. Low transcriptional activity seen with the H1t promoter indicated the presence of silencer elements and of significant differences within the proximal promoter that render H1t less efficient in interacting with transcriptional factors utilized by the somatic variants. Prior work suggested that regulation of H1t gene transcription during the mitotic cells cycle might be influenced by sequences upstream from the H1t proximal promoter. The question that needs to be addressed is whether there are active mechanisms that silence transcription of H1t gene or they maintain its transcription at a low level in somatic cells. The transcription of the H1t gene appears to be reduced by sequences between –1999 and 1506. But no regulatory effect could be shown for the H1 gene specific for TG box, 452bp upstream of transcription start-site in the expression of the mouse H1t gene (Drabent and Doenecke, 1997).

To assess the contribution of the upstream DNA sequence to H1t transcription silencing in non-expressing cells, a set of histone H1t-promoted reporter vectors was constructed. Transient transfection of mouse C1271 cells with these reporter vectors allowed Wolfe et al (1999) to identify a transcriptional silencer located between 948bp and 780bp upstream from the H1t transcriptional initiation site. Histone H1t promoted luciferase activity increased when the region between 948 bp and 875 bp upstream from the transcriptional initiation site was eliminated. Addition of a 73bp rat H1t promoter fragment (-948 to -875, containing the 5' portion of the silencer region) to a site immediately upstream from the histone H1t proximal promoter led to significantly reduced luciferase expression upon transient transfection. Nuclear proteins were found to bind DNA within the H15 silencer region when assayed by in vitro DNase 1 foot printing. Thus inactive transcriptional silencer mechanism involving a specific and autonomous H1t promoter element (nt –948/-875) may be operative to minimize expression of the H1t gene in non-testicular cells. Further examination of nuclear binding protein to this DNA regulatory region by EMSA using extracts revealed that non-expressing cells might contain proteins that repress H1t gene transcription (Wolfe and Grimes, 1999).

H1t promoter shares all of the recognized control elements with somatic H1 genes. To investigate the mechanism of this apparent repression, inhibitory sequences were located immediately downstream of the TATA box in H1t gene. A series of deletions and short oligonuclotide mutations, scanned across the region between the TATA box and cap site, identified two tracts of C (GC box 2) as the inhibitory sequences. While both Sp1 and Sp3 bind to this region weakly in vitro and may not be responsible for the inhibitory effect of GC box 2, additional binding proteins (CTB-4 and CTB-5) were identified as better candidates for mediating the repressive effect. Repression of the H1t promoter was relieved by mutation of GC box 2, whereas additional mutations in GC box 1 upstream of the CAAT box led to a large decrease in activity. This indicated that these two G/C-rich elements, GC Box 1/GC Box 2, have opposite effects on promoter activity of H1 (Grimes et al, 1992; Wolfe et al, 1995; Clare et al, 1997). GC-box is a consensus site for binding of Sp transcription-factors family members among which Sp1 and Sp3 are present in testis of 9-day-old and adult rats and in pachytene primary spermatocytes and early spermatids. A 95- to 105-kDa form of Sp1 is most abundant in the tissues and cell lines examined, but a 60-kDa form of Sp1 is the most abundant species in spermatocytes and early spermatids. Sp1 and Sp3 from adult testis, primary spermatocytes, and early spermatids can bind to the H1t/TE element. Thus Sp transcription factors might be involved in transcription of the H1t gene and the GC-box and the TE1 subelement are required for activation of the H1t promoter (Wilkerson et al., 2002).

This supports that multiple mechanisms are involved in transcriptional repression of the H1t gene in other cell types. 1) Transcriptional repression of H1t gene is mediated in part by specific proximal and distal promoter elements, and 2) in some cell types by methylation of CpG dinucleotides within the promoter. A distal promoter element located between 948 and 780 bp upstream from the transcription initiation site and a GC-rich region between the TATA box and transcription initiation site are known to contribute to repression. In addtion, transcriptional repression of the histone H1t gene is regulated by an element within the proximal promoter. This element was designated H1t promoter repressor element (RE) located between -130 and -106 bp. Transcriptional repression mediated by the RE element in NIH 3T3 cells differs significantly from the mechanism mediated by the GC-rich region. Furthermore, binding proteins that form the RE complex are not present in rat testis where the gene is actively transcribed (Wolfe and Grimes, 2003).

7.7. HISTONES IN CHROMOSOME ASSEMBLY

Phosphorylation of Histones: During mitosis, the relaxed interphase chromatin undergoes dramatic changes resulting in the formation of highly condensed mitotic chromosomes. Though the process of chromosome assembly is still poorly understood, the SMC (structural maintenance of chromosomes) family proteins seem to play an essential role in chromosome condensation. The SMC proteins exist in the form of two high molecular weight complexes (8S and 13S), termed condensins. The 13S condensin, phosphorylated by Cdc2 in vitro, seems to be responsible for the triggering of chromosome condensation. Chromosome assembly is also accompanied by phosphorylation of histones H1 and H3. Linker histone H1 is highly phoshyorylated at the beginning of mitosis in cells in culture and is rapidly dephosphorylated after anaphase, suggesting a role for these modifications in chromosome condensation.

However, other experiments showed that the absence of linker histones does not affect either chromosome condensation or nuclear assembly, thus arguing against a role for histone H1 hyperphosphorylation in these processes. On the contrary, histone H3-Ser10 phosphorylation is tightly correlated with chromosome condensation during both mitosis and meiosis. In addition, this modification of histone H3 seems to be required for the initiation but not the maintenance of the chromosome condensed state. It was also shown that H3-Ser10 phosphorylation is linked to chromosome condensation and segregation in vivo and is required for proper chromosome dynamics (Wei et al., 1999). The mRNAs encoding the testis specific histone TH2B show temporal expression with the phosphoprotein p19 and the transition proteins TP1 and TP2 during pubertal development and throughout the spermatogenic cycle of rat (Marret et al., 1998).

An important contribution to study chromosome assembly was made by the use of mitotic extracts isolated from *Xenopus* eggs: incubation of demembranated *Xenpous* sperm nuclei or somatic nuclei in such extracts resulted in the assembly of mitotic chromosomes. These in vitro assembled-chromosomes exhibited identical physical properties to those of native chromosomes (Houchmandzadeh and Dimitrov, 1999). The use of mitotic extracts demonstrated that topoisomerase II is required for assembly and condensation, but not for structural maintenance of mitotic chromosomes. Both SMC proteins and topoisomerase II interact with DNA:13S condensin and inducing an ATP-dependent positive supercoiling in closed circular DNA.

Histone Tails: In another study (reviewed in de la Barre et al., 2000), the effect of histones was investigated on the interactions of SMC protein and topoisomerase II with nucleosomal DNA keeping in mind that the nucleosome is composed of two superhelical turns of DNA wrapped around an octamer of core histones consisting of two each H2A, H2B, H3 and H4. The structure of the histone octamer determined by X-ray crystallography represents a tripartite assembly with a centrally located (H_3-H_4) tetramer, flanked by H2A-H2B dimers. All four histones within the octamer consist of an elongated folded domain (the 'histone-fold') and extremely located N-terminal tails with a large number of positively charged residues. The N-terminal tails are highly flexible and they contain the sites for different histone post-translational modifications (Luger and Richmond, 1998). The tails are found to be active players in numerous vital functions in eukaryotic cells. For example, acetylation of the tails seems to be involved in transcriptional regulation. It was also shown that the N-terminal tails interact with repressor proteins and that they are essential for the assembly of specific protein complexes required for the establishment of chromatin silencing. Moreover, acetylation at specific tail lysines is involved in chromatin assembly in vivo. The histone tails also seem to be implicated in nucleosome positioning in some yeast gene promoter. Evidence has been presented that shows that flexible histone tails are essential players in the chromosome condensation process. Based on these results it has also been shown that chromosome assembly is correlated with phosphorylation of histone H3 at Ser10 by a specific kinase, which is involved in chromosome assembly (reviewed in de la Barre et al., 2000). De la Barre (2000) studied the role of core histone tails in the assembly of mitotic chromosome. Incubation of sperm nuclei in the *Xenopus* egg extracts led to the formation of mitotic chromosomes, a process, which correlated with phosphorylation of the N-terminal tail of histone H3 at Ser10. When the extracts were supplemented with H1 depleted oligosomes, they were not able to assemble chromosomes. Selective elimination of oligosome histone tails by trypsin digestion resulted in dramatic decrease in their ability to inhibit chromosome condensation. The chromosome assembly was also inhibited by each of the histone tails with differing efficiency. In addition, it was found that nucleosomes were recruiting through the flexible histone tails some chromosome assembly factors, different from topoisomerase II and 13S condensin. Thus it was demonstrated that histone tails play an essential role in chromosome assembly. Evidence indicates that the nucleosomes, through physical association, were able to deplete the extracts from the kinase phosphorylating histone H3 at Ser10, suggesting the role of a kinase in chromosome condensation (de le Barre et al., 2000).

7.8. UBIQUITINATION OF HISTONES

Although much is not known regarding the levels of ubiquitinated histones during mammalian spermatogenesis, studies in other vertebrate groups have established that they vary during development. In the rooster, the levels of uH2A increase markedly from an undetectable level in pachytene spermatocytes to the highest level in late spermatids, just before the histones are replaced by protamines. In contrast, the level of uH2A remains unchanged from early to late stage germ cells in trout testes, whereas there is a decline in uH2B and an increase in u2H2B (deubiquitinated form). The major role for ubiquitination in cells is the targeting of proteins for degradation by polyubiquitination. In contrast, some proteins, particularly the histones H2A and H2B are often mono-ubiquitinated, which appear to be involved in processes other than degradation. The role of histone mono-ubiquitination is currently not clear, but it has been shown that uH2A and uH2B are associated with transcriptionally active chromatin. Mono-and polyubiquitinated H2A and H2B were more readily dissociated from chromatin by salt, indicating that they destabilize the nucleosome. Because of potential role of histone ubiquitination in altering chromatin structure, Chen et al (1998) characterized the levels of ubiquitination of specific histones in meiotic and postmeiotic germ cells in rat testes and observed that the levels of the major ubiquitinated histone forms, mono and poly–ubiquitinated H2A, were highest in the pachytene spermatocyte stage, declined or undectable thereafter through the round spermatid stage, and reached their lowest levels with increase in elongating spermatids. Three additional ubiquitinated histone species, besides H2A in the fraction enriched in elongating spermatids corresponded to uH3, uTH3, and uH2B. From these studies, it was speculated that ubiquitination of histones could loosen the nucleosome structure in preparation for histone removal, be a consequence of nucelosome relaxation or disruption caused by other means, or target H3 for degradation (Chen et al, 1998) (Chapter 31). In the HR6B knockout mice lacking ubiquitin enzyme, an overt defect in the overall pattern of histone ubiquitination was not detected. The most intense uH2A signal in pachytene spermatocytes was detected in the sex body that contains the heterochromatic X and Y-chromosomes. The postmeiotic uH2A immunoexpression in elongating spermatids indicated that nucleosome destabilization induced by histone ubiquitination might play a facilitating role during histone to protamine replacement (Barrends et al, 1999).

7.9. TRANSITION PROTEINS

Transition proteins (TPs) are basic proteins, which participate in remodeling of chromatin by replacement of histones during spermiogenesis. The transition proteins TP1 and TP2 are the most abundant basic transition proteins, which alongwith their genes have been studied in more details. The genes of TP1 and TP2 are transcribed and their mRNAs accumulate in early spermatids (steps 1-8) but the mRNAs are not translated until about step 9 when the TP proteins begin to accumulate (Hecht, 1989; Kistler, 1989). Although other TPs exist, TP1 and TP2 are the predominant ones found in the nuclei of rodent spermatids. The TP1, a 6.2-kDa protein, consists of ~20% arginine and 20% lysine and lacks cysteine. The TP2, a 13-kDa protein, consists of ~10% each arginine and lysine and 5% cysteine. TP1 is expressed abundantly in most mammals and is highly conserved showing cDNA nucleotide and amino acid sequence homologies of 90% across species. The TPs in the rat constitute >90% of basic chromosomal protein of spermatid nuclei. A cDNA clone for the smallest TP1 showed that its message first appears postmeiotically in late round spermatids. The message remains translationally inactive for some 3-4 days before translation occurs in early elongating spermatids. While translationally repressed, TP1 message is nonpolysomal and has a discrete size of about 590 bases, including a 140-residue poly (A) tail. In contrast, polysome-associated message is of heterogeneous size due to variability of poly (A) lengths (Heidaran and Kistler, 1987). Sequential expression of nucleoproteins during spermiogenesis has been studied (Kistler et al, 1996; Oko et al, 1996 Steger et al, 1998; Siffroi et al, 1999). Transition protein 2 was the first to appear as a faint band at stages IX-XI, followed by high levels at stages XII-XIV of the cycle. TP1 showed a low expression at stage XII of the cycle and peaked at stages XIII-1, whereas protamine 1 first appeared at stage 1 of the cycle and remained high throughout the rest of the spermiogenesis. It appeared that TP2 is the first nucleoprotein that replaces histones from the spermatid nucleus, and its appearance is associated with the onset of nuclear elongation. The TP1 shows up along with the compaction of the chromatin. Thus, the two transition proteins seem to have distinct roles during transformation of the nuclei and compaction of spermatid DNA (Kistler et al, 1996; Oko et al, 1996).

Transition protein-2 (TP2), which is expressed during stages 12-15 of mammalian spermiogenesis, undergoes phosphorylation immediately after its synthesis. The TP2 is phosphorylated in vitro at threonine 101 and serine 109 by spermatid nuclei with the help of sperm-specific isoform of protein kinase A (Cs-PKA). The Cs-PKA was present in the haploid spermatids and absent from premeiotic germ cells. The sequence of rat Cs-PKA at the N terminus differs from mouse and human by one amino acid. The cytoplasmic factors and ATP are absolute requirements for translocation of TP2 into the nucleus. Phosphorylation up-regulates the NLS dependent import of TP2 into the nucleus (Ullas and Rao, 2003).

In vitro, TP1 decreases the melting temperature of DNA and relaxes the DNA in nucleosomal core particles, which led to the proposal that TP1 reduces the interaction of DNA with the nucleosome core. In contrast, TP2 increases the melting temperature of DNA and compacts the DNA in nucleosomal cores, suggesting that it is a DNA-condensing protein. These apparently distinct functions of TP1 and TP2, as well as the absence of TP1 in human with spermatid arrest, suggest that TP1 is essential for histone displacement and other aspects of spermatid development. These predictions were tested in mice lacking TP1 protein generated by targeted deletion of the TP1 gene in embryonic stem (ES) cells. Surprisingly, testis weights and sperm production were normal in the mutant mice, and only subtle abnormalities were observed in sperm morphology. Electron microscopy revealed large rod-like structures in the chromatin and mutant step 13 spermatids, in contrast to the fine chromatin fibrils observed in wild type. Steps 12-13 spermatid nuclei from the testis of *Tp1*-null mice contained, in place of TP1, elevated levels of TP2 and some Prm2 precursor. Most of the precursor was processed to mature Prm2, but high levels of incompletely processed forms remained in epididymal spermatozoa. Sperm motility was reduced severely, and ~60% of *Tp-1* null males were infertile. Thus, TP1 is not essential histone displacement or chromatin condensation. The absence of TP1 may partially be compensated for by TP2 and Prm2 precursor, but this deregulation of nucleoprotein replacement results in an abnormal pattern of chromatin condensation and in reduced fertility (Yu et al., 2000). These observations opened new suggestions aimed at the identification of conserved upstream sequence elements required for testis-specific transcriptional activation. The report by Caron et al suggests a repair role for TP-1 in DNA single-strand breaks (sSB) both in vivo and in vitro. The sSB may be the consequence of topoisomerase II activity present in spermatids. Essentially, the basic TP-1 has the ability to interact with the phosphate groups of the disrupted single-strand DNA and enables a still unknown ligase to repair the gap. This TP-1/DNA interaction is facilitated by a loss of the supercoiled nature of DNA. Chromatin condensation can proceed once the repair is completed. This is supported by the observation that chromatin condensation is abnormal in TP1 null mice (Kierszenbaum, 2001).

During human and mouse spermiogenesis, the transition proteins (TPs) are replaced by two protamines, protamine 1 (Prm1) and protamine 2 (Prm2). Whereas Prm1 is synthesized as a mature protein, Prm2 is synthesized as the precursor. In mouse, mature Prm2 of 63 residues is derived from the precursor, pPrm2, of 106 residues by sequential proteolyses that produces at least six identifiable intermediates. The genomic loci for the protamines are found in the same gene cluster as TP2 in all mammals examined, suggesting some functional relationship among three proteins. The TP1, however, is on a separate chromosome and is not clearly related to the other three proteins. Steger et al (1998) studied the sequential expression of H1t and the TP1 and TP2 during human spermatogenesis. While TP-1 mRNA was present in spermatids from step 2 to early step 4, the TP1 protein occurred, with temporal delay, in the nuclei of step 3 and step 4 spermatids. The TP2 protein was observed in the nuclei of spermatids from step 1 to step 5. Since the TP1 protein appeared with temporal delay, it could be assumed that the corresponding TP-1 mRNA is translationally delayed (Steger et al, 1998). Siffroi et al (1999) showed that, at the end of spermatid elongation phase, the disappearance of TP1 and Prm 1 transcripts may be related to the transcriptional activity while the deposition of transition proteins and protamines occurs successively within spermatid nuclei. Thus, TPs may be implicated in arrest of transcriptional activity at mid-spermiogenesis besides their role in histone to protamine replacement such as replacement of Prm1 and Prm2, Prm3 and Prm4 (c/r Siffroi et al, 1999). Studies on the gene expression of H1t, CREM, and TP1 in testicular biopsies from men with normal spermatogenesis and with round spermatid maturation arrest indicated that TP1 is an important parameter in the histone to protamine exchange and in the initiation of spermatid elongation. The CREM is involved in the regulation of TP1 gene expression and consequently plays a vital role in the correct differentiation step from round spermatids to mature spermatozoa (Steger et al, 1999).

Reinhart et al (1991) studied the nucleotide sequence and organization of the gene for TP2. The *Tp2* gene, isolated from a bull cosmid library, was found to contain a single intron of 910 bp. The coding sequence consists of 390 bp and has a similarity of about 70% to that of the TP2 c-DNA of mouse and rat. On amino-acid sequences basis, the bull TP-2 is 14 and 15 amino acids longer than that of mouse and rat, respectively, with 42-45% similarity between bull and rodents. However, the evolutionary divergence does not occur at the cost of basic amino acids, which are of functional importance in DNA-protein interaction in the condensing spermatid nucleus. The TP2 gene is closely linked to the protamine genes in the bull genome (Reinhart, et al 1991). Nayernia and colleagues pursued the identification of the regulatory region of TP-2 directing the expression of the CAT reporter gene in spermatids. They find that a 525-bp stretch of the 5'-flanking sequence of rat TP-2 gene regulates CAT gene expression in round spermatids and that the translation of the mRNA is delayed by 6 days. This temporal expression pattern is similar to that reported for TP-2 during spermatogenesis.

TP-DNA Interaction: The apparent association constant for binding of TP2 to these nucleic acids was found to be 1.63. Thermal denaturation studies of calf thymus DNA and its complexes with TP2 showed that at 1mM NaCI, TP2 shifted the Tm from 53° C to 62-67° C while at 60 mM NaCl, the Tm shifted from 72 to 78°C - suggesting that TP2 is a DNA stabilizing protein. Circular dichroism studies of TP1-DNA and TP2-DNA complexes revealed that TP2 has a better DNA condensing property than TP1 (Bhaskaran and Rao, 1990). Rat TP1, having higher affinity for single-stranded DNA, induces local melting of DNA. Rat TP2 with two possible zinc finger motifs prefers and stabilizes a GC-rich sequence, a phenomenon, which is zinc-dependent. The transition in nuclear proteins requires changes of activities of topology-modifying enzymes such as topoisomerase II. The testis-specific high-mobility-group protein from mouse late spermatids has been reported to be a DNA-packaging factor that can modulate the activity of topoisomerase I (c/r Akama et al, 1999). Akama et al (1999 c/r Akama et al, 2000) developed methods for isolating intact boar TP1-4, and reported that boar TP1 and TP3, having higher affinity for single-stranded DNA, and boar TP4, having higher affinity for double-stranded DNA, are DNA-melting proteins, mediated through the stacking of Tyr-32, Trp-18 and Trp-126 with nucleic acid bases, respectively.

7.10. PROTAMINES

7.10.1. Characterization and Functions

The nuclear compaction of the sperm nucleus during spermiogenesis involves the sequential replacement of histones by transition proteins and then by protamines. These chromation changes contribute to the cessation of transcription in the maturing spermatids. Since transcription of the two mouse protamine genes is temporally coordinated it is likely that similar sequences in the 5'-flanking regions of the protamine genes participate in their regulation (Yiu et al, 1997). Protamines (Prms) are small, arginine-rich proteins expressed only during the postmeiotic stages of spermatogenesis. In most mammals there are two distinct protamine (Prm) proteins and two corresponding Prm genes (Hecht, 1989). In rats both Prm genes are expressed in appropriate amount but protamine 2 does not accumulate in rat spermatids (Hecht, 1989). Protamine-2 in mouse and other mammals is translated as a precursor molecule requiring post-translational processing. The absence of mature rat Prm2 protein in spermatozoa may partly be due to a reduced level of Prm2 precursor mRNA and perhaps also due to the inability of the rat testis to process Prm2 precursor protein completely. These proteins have been isolated from large number of vertebrates including fish, domestic fowl and several mammals. Prms associate ultimately with DNA through transition of several nuclear basic proteins including somatic histones, testis-specific histones and spermicidal basic proteins. Thereafter, the cysteine residues in the nucleoprotamine complex are completely cross-linked to form a tight network structure during transit of spermatozoa in epididymis.

By the criterion of protein gel electrophoresis, the spermatozoa of most mammals have only one protamine. Exceptions are the mouse and hamster, those have two Prms, while the humans have four Prms. Prm2 is the main component of proteins composed of Prm2, Prm3 and Prm4, which differ only by an amino terminal extension of one to four residues. Prm1, Prm2, Prm3, and Prm4, have 50, 57, 54 and 58 amino acids respectively. The amount of Prm 2 in spermatozoa varies from about 70% of the protamine in the mouse to an undetectable level by gel electrophoresis in the rat and most other mammals. Although the reason for this variability is not understood, apparently there is much flexibility in the number and/or type of proteins needed to compact the DNA in sperm. The genes for the mouse and human protamines are closely linked and present as single copy gene on chromosome 7 and 16 respectively.

```
mP1                                                                                          1
                                                                                         Ala Arg
 GACCCCTGCTCACAGGTTGGCTGGCTCGACCCAGGTGGTGTCCCCTGCTCTGAGCCAGCTCCCGGCCAAGCCAGCACC ATG GCC AGA
-78             -60                                  -30                            1
                                10                                          20
Tyr Arg Cys Cys Arg Ser Lys Ser Arg Ser Arg Cys Arg Arg Arg Arg Arg Arg Cys Arg Arg Arg Arg
TAC CGA TGC TGC CGC AGC AAA AGC AGG AGC AGA TGC CGC CGT CGC AGA CGA AGA TGT CGC AGA CGG AGG
                             30                                         60
                30                                  40
Arg Arg Cys Cys Arg Arg Arg Arg Arg Arg Cys Cys Arg Arg Arg Arg Ser Tyr Thr Ile Arg Cys Lys
AGG CGA TGC TGC CGG CGG AGG AGG CGA AGA TGC TGC CGT CGC CGC CGC TCA TAC ACC ATA AGG TGT AAA
             90                                         120
    50
Lys Tyr ***
AAA TAC TAG ATGCACAGAATAGCAAGTCCATCAAAACTCCTGCGTGAGAATTTTACCAGACTTCAAGAGCATCTCGCCACATCTTGAA
  150                                180                                210

AAATGCCACCGCCGATGAAAAACAGGAGCCTGCTAAGGAACAATGCCACCTGCCAATAAATGTTGAAAACTCAAAAAAAAAAAAAAAAAAA  3'
  240                                270                                300                   326

mP2         1                                 10
         Met Val Arg Tyr Arg Met Arg Ser Pro Ser Glu Gly Pro His Gln Gly Pro Gly Gln Asp
 CGACCC ATG GTT CGC TAC CGA ATG AGG AGC CCC AGT GAG GGT CCG CAC CAG GGG CCT GGA CAA GAC

20                                30                                             40
His Glu Arg Glu Glu Gln Gly Gln Gly Gln Gly Leu Ser Pro Glu Arg Val Glu Asp Tyr Gly Arg
CAT GAA CGC GAG GAG CAG GGG CAG GGG CAA GGG CTG AGC CCA GAG CGC GTA GAG GAC TAT GGG AGG

       ↓                       50                                      60
Thr His Arg Gly His His His His Arg His Arg Arg Cys Ser Arg Lys Arg Leu His Arg Ile His
ACA CAC AGG GGC CAC CAC CAC CAC AGA CAC AGG CGC TGC TCT CGT AAG AGG CTA CAT AGG ATC CAC

                        70                                      80
Lys Arg Arg Arg Ser Cys Arg Arg Arg Arg Arg His Ser Cys Arg His Arg Arg Arg His Arg Arg
AAG AGG CGT CGG TCA TGC AGA AGG CGG AGG AGA CAC TCC TGC CGC CAC AGG AGG CGG CAT CGC AGA

                90                                      100
Gly Cys Arg Arg Ser Arg Arg Arg Arg Arg Cys Arg Cys Arg Lys Cys Arg Arg His His His ***
GGC TGC AGA AGA TCC CGA AGG AGG AGG AGA TGC AGG TGC AGG AAA TGT AGG AGG CAC CAT CAC TAA

GCCTCCCCAGGCCTGTCCATTCTGCCTGGAGCCAAGGAAGTCACTTGCCCAAGGAATAGTCACCTGCCCAAGCAACATCATGTGAGGC

CACACCACCATTCCATGTCGATGTCTGAGCCCTGAGCTGCCAAGGAGCCACGAGATCTGAGTACTGAGCAAAGCCACCTGCCAAATAAA

GCTTGACACGAGAAAAAAAAAAAAAAAAA   3'
```

Fig.7.4. Nucleotide and corresponding amino acid sequences of the cDNA of mouse Prml and Prm2. The deduced amino acid sequence is presented above the nucleotide triplets. Amino acids are numbered beginning with alanine for for Prml and valine for Prm2 because methionine does not appear at the amino terminus of protamines. The termination codon, TAG for Prml and TAA for Prm2, are denoted by asterisks. The canonical hexanucleotide, AATAAA, is underlined. Reproduced with permission from N.B.Hechet: In: Cellular and Molecular Events in Spermiogenesis, 1989 © WHO

Intermediate basic proteins HPI-1 and HPI-2 are common precurssors of Prm2 protamine family. These proteins are synthesized in large amounts in human spermatids during elongation phase and disappear almost totally in mature spermatids when deposition of protamines is completed in condensed nuclei (Prigent et al, 1998).

Ontogeny: During spermiogenesis, round spermatids morphologically differentiate into mature spermatozoa. The protamine gene in the bull is postmeiotically expressed and the mRNA is synthesized as a 680 nt long molecule (Lee et al., 1987). Protamine transcripts of 0.6-kb and 90.9-kb in lengths for Prml and Prm2, respectively, were detected only in testicular RNA. When cDNA clones from boar (BPrm-1 and BPrm-2) were used for hybridization experiments with the testicular RNA of those mammalian species, which lack protamine 2 in their spermatozoa, the presence of transcripts for both protamines was detected. Protamines were not observed in association with spermatogonia, spermatocytes, Sertoli cells, or interstitial cells. Thus, the human Prml and Prm2 transcripts are expressed postmeiotically in round and elongating spermatids. (Wykes et al, 1995). The genes for these basic nuclear proteins exist as a single linear array of *Prml, Prm2,* and also *TP2* on human chromosome 16p 13.2.

Characterization: The Prml has a molecular weight of 7400 and is characterized by the presence of arginine, cysteine, lysine, and tyrosine. By contrast, Prm2 is unusual in containing an abundance of arginine, histidine, lysine, and cysteine, but no tyrosine. The primary structure of Prm2 with a molecular weight of 8841-Da contains 62 amino acids and includes six clusters of arginine and histidine, distributed throughout the polypeptide, each ranging from five to eight amino acids in length. Sequence comparisons of mouse and human protamines revealed greater homology between human Prm2 and mouse Prm2 than within the Prm1 family from the two mammalian species. The Prm2 family proteins are synthesized as a larger precursors having between 66 and 101 residues from a single copy gene present on chromosome 16 in humans and processed to matured form of the polypeptide within the condensing spermatids nucleus (Alimi et al., 1993). In mouse, Prm2 is synthesized as a precursor of 106 aminoacids, which finally matures into 63 amino acid protein found in sperm (**Fig.7.4**). It has been demonstrated that the bull, boar and rat lack Prm2 in the sperm nucleus but still contain the Prm2 gene in their respective genomes. The lack of Prm2 in the boar and bull has been attributed to mutations in their genes. However, evidence for the presence of Prm2 precursors in the mature rat sperm has been presented (Bunick et al., 1990; Dominjoud et al., 1990).

Complete amino acid sequences for members of the Prm1 family have been determined for bulls, boars, rams, and humans and based on cDNA sequencing, sequence for mouse Prm1 has been predicted. Primary sequences for members of the Prm2 family have been determined for human by protein sequencing and for mouse by cDNA sequencing. The Prm1s from these mammalian species have proven to be comparable in length (50 amino acids), each consisting of a central core of three arginine clusters and the less basic amino and carboxyl-terminal regions.The protamines of mouse, bull, and boar are organized into three domains, a central highly basic core consisting of clusters of arginines and two less basic regions at the amino and carboxy termini. The conserved domains of arginine are considered to interact tightly with DNA to yield highly condensed nucleoprotamine, with the cysteine groups stabilizing the complex through the formation of covalent disulfide bonds. The amino acid sequence of boar protamine is given as:

Ala-Arg-Tyr-Arg-Cys_2-Arg-Ser-His-Ser-Arg-Ser-Arg-Cys-Arg-Pro-Arg_4-Cys-Arg_4-Cys_2-Pro-Arg_5-Ala-Val-Cys_2-Arg_2-Tyr-Thr-Val-Lle-Arg-Cys-Arg_2-Cys.

Both Prms shared 80% homology between boar and bull protamines. The 404-bp cDNA encoding the cysteine-rich tyrosine-containing mouse Prm1 encodes a polypeptide of 50 amino acids of which 28 are arginine, 9 are cysteine, and 3 are tyrosine. The insert contains the complete 3'non-coding region of 151 bases and most of the 5'noncoding region. The predicted amino acid sequence of mouse Prm1 is about 80% homologous to boar Prm and 67% homologous to bull Prm and contains the central, highly basic domain of four arginine clusters found in the trout protamines (Kleene et al, 1985). From the kinetics of DNA condensation and decondensation by Prm1 and synthetic peptides corresponding to specific segments of the bull *Prm1* DNA binding domain, it appeared that the number of clustered arginine residues present in the DNA binding domain is the most important factor affecting the condensation and stability of the DNA-protamine complex prior to the formation of inter-protamine disulfide cross-links. The use of Arg, rather than Lys residues, and the inclusion of Tyr or Phe residues in the hinge regions between anchoring domains provide additional stability to the complex.

7.10.2.Protamine Genes

Protamine Genes: The first initiation codon (AUG) is located 78 bases from the 5'end of Prm-1 and appears in the context of a sequence (GCACCAUGG) that closely matches the consensus sequence (CCA/GCCAUGG) for initiation of protein synthesis in eukaryotes. Mouse Prm1, a polypeptide of about 55 amino acids contains 4 tyrosines, 9 cysteines and no histidine, whereas mouse Prm2, a polypeptide of 65 amino acids includes 12 histidines, 3 cysteines, and no tyrosines. The predicted amino acid composition of the protamine encoded by mouse *Prm1* agrees almost perfectly well with the composition of protamine 1 in the number of arginine (28 vs 32), cysteine (9 vs. 9), tyrosine (3 vs 4), serine (4 vs 4) and alanine, threonine, and isoleucine (1 vs 1). The 3'-untranslated sequence of mouse Prm1-mRNA is 151 bases excluding the poly (A) tail. The total length of the 5'-untranslated sequence of the mRNA for protamine was deduced to contain 113 bases from the combined length of the coding and 3' untranslated region of *mPrm-1* excluding the poly (A) and the size of protamine mRNA after deadenylation with RNAse H, 420-307 bases. Hence, *mPrm1-1* lacks about 30 bases at the 5'-end (Bellve et al., 1988; Kleene et al, 1985).

While comparing the rat and mouse *Prm2* gene sequences, Bunick et al, (1990) showed that rat gene sequence encompasses 435 nt of the coding region, which includes an intron of 120 nt, 461nt 5' to the coding sequence and 181 bases 3' to it. In the mouse the Prm2 gene is abundantly transcribed and translated. There are two deletions of 38 and 20 nt in length and three insertions of 8, 13 and 6 nt in length in the rat r*Prm2* gene. The m*Prm2* protein is initially synthesized as a precursor and then proteolytically processed to yield the mature protein. In contrast, the rat *Prm2* transcripts are present at 2-5% that found in the mouse. The mature protein was not detected in rat spermatozoa. Although there is 92% nucleotide similarity between rat and mouse genes and 91% similarity of the predicted amino acids sequences, the *rPrm2* promoter was only 30% as efficient as the *mPrm2* promoter. Analyses of total sperm basic nuclear proteins using a monoclonal antibody specific for Prm2 suggested that the rat Prm2 mRNA is translated in vivo but is not properly processed.

Each of the human *Prm1* and *Prm2* genes contains a single intron consisting of 91 and 163bp, respectively. The 5'-noncoding region of Prm1 contained 664 bp, while the 5'-noncoding region of Prm2 showed 902bp. Both genes possess typical TATAA and CAAT boxes at conventional distances from the transcription start points, which could be assigned to nucleotides -91 and -110 for *Prm1* and *Prm2* genes, respectively. Comparison of the 5'-noncoding region of *Prm1* and *Prm2* genes reveals 12 different motifs in common, 8 of which are clustered in both genes and could reflect regulatory elements for testis-and spermatid-specific gene expression. Both human genes are found to be clustered at a distance of 4.8 kb. Comparison of the genomic organization of human and mouse protamine genes revealed greater similarities between the two in the 5'- noncoding region (Domenjoud et al, 1990).

Choudhary et al (1995), addressed the mechanism governing transcriptional potentiation, of a region of approximately 40kb of the human genome encompassing Prm(s) genes. The region containing these genes was introduced into the germ line of mice. In situ hybridization and Southern analysis showed that this segment of the human genome integrated into independent chromosomal sites while maintaining its fidelity. The expression of the endogenous mouse protamine Prm1 and Prm2 genes as well as the mouse, TP2 gene were expressed along with their human transgene counterparts. In sperm the human Prm1 → Prm2 → TNP2 genomic domain was contained as 28.5-kb contiguous segment bounded by an array of nuclear matrix associated topoisomerase II consensus sites.

Cis-acting Elements: Protamine gene expression may be controlled at many levels. To identify cis-acting sequences that regulate the stage and cell-specific expression of the two coordinately regulated protamine genes in the mouse, the nucleotide sequences of the 5'- flanking regions and coding regions were compared. Unlike most histone genes and the multigene family of trout protamine genes, which are intronless, each mouse protamine gene has a single, short intervening sequence. Although the coding regions do not share significant nucleotide homology, the 5'-flanking regions contain several short homologous sequences that may be involved in gene regulation. An additional shared sequence is present in the 3'-untranslated region surrounding the poly (A) additional signal in both genes.

Johnson et al (1991) identified short nucleotide sequences, which are present in the mouse *Prm1* and *Prm2* and other testes specific genes, which may be necessary for their testis specific expression.The genes for these two protamines are coordinately regulated and are transcribed only in the round spermatids. The poly (A) tails of both mRNAs are shortened, concurrent with their appearance on polysomes. Specific fragments of the 5'-flanking regions of both the *mPrm1* and *mPrm2* genes were shown to be sufficient to confer testis-specific expression on marker genes in transgenic mice. The nucleotide sequences of these 5'-flanking regions contain several shared sequences, which were called the X, A, F, C, D, E and B boxes.

In order to detect regulatory conserved DNA elements within the *Prm1*-encoding gene (P10 promoter), Queralt and Oliva (1993) have sequenced this region from the rat, guinea pig, gorilla, orangutan, anubis-baboon and red monkey *Prm1* genes and compared it to the homologous human, bull, boar, and mouse nt sequences. Authors demonstrated the presence of a consensus sequence, HSMCYTCAYAAT (Prot1C: protamine 1 consensus), from nucleotide position -64 to position -53 in all *Prm1* genes whose promotor sequences were known. Sequences similar to Prot1C are found in the promoter region of other testis-specific genes, such as the transition protein 1-encoding gene promoter, which is thought to have derived from the Prm1 genes. The relevance of this conserved element in the expression of *Prm1* genes is strongly supported by the demonstration of a mouse testis trans-acting factor (Tet-1) which binds and matches, in the mouse, the first 11-bp of the corresponding consensus Prot1C sequence reported. Another highly conserved element (TGTGAGG) has also been identified at 203-nt upstream from Prot1C. The sequence forms a perfect palindrome with the central 7-nt of Prot1C and is absent in the homologous region of other genes. Further upstream, at positions -113 to -132, a third highly conserved region is present (MATGCCCATATWTGGRCAYG), which is similar to the c-fos SRE (serum-response element) and contains the central core common to all SREs. This element has not been found in the homologous region of other sperm-specific genes (Queralt and Oliva, 1993).

Zambrowicz et al (1993) from transgenic mice concluded that a 113-bp region from –15 to –37 could direct spermatid-specific transcription, which has defined sequences that are essential for proper function. A number of ubiquitous and testis-specific proteins bind within this region including nuclear protein Tet-1, which binds to an 11-mer sequence at –64. Tet-1 has been proposed to activate the transcription of mPrm1. In vitro DNA-protein interaction, using rat testis nuclear extracts suggested that Tet-1 is a tissue specific trans-acting factor. The Tet-1 recognizing sequence was delineated to 11-mer TGACTTCATAA at position – 64. Tet-1 appears to be distinct from known cAMP responsive element binding factors (Tamura et al, 1992). Likewise, a 859-bp region of the mPrm2 gene immediately upstream of the transcription start site confers specific expression in round spermatids. Analyses of the mPrm2 promoter by in vitro transcription assays identified a potential positive regulatory region from nucleotide -10 to -29. Mobility shift assays of mPrm2 promoter region –140 to –23 detected the binding of ubiquitous and testis-specific proteins (c/r Yiu et al., 1997).

mPrm2 Gene-Protein Interactions: Studies analyzing protein-DNA interactions within the proximal *mPrm2* promoter have identified five protein-binding sites (Ha et al, 1997). Site-1 (-64 / -48) contains the single core motif AGGTCA recognized by orphan nuclear receptors that are believed to be important for gonadal and brain development (Enmark et al., 1995). Site-1 also contains a half-site of CRE that binds CREB or CREM and a sequence similar to CRE element. Site 2 has a putative Y-box at –83 /-72 nt that binds a family of sequence-specific DNA-binding proteins; one of them capable of activating germ cell-specific transcription (c/R Yiu et al, 1997). A testicular Y-box protein has also been shown to bind to a putative Y-box at –489/ -478 of the mPrm2 promoter.

The functional significance of protein-DNA instructions within 180-bp upstream of mPrm2 was studied by mutational analysis. Deletion and mutational analyses revealed two positive regulatory sequences for *mPrm2* transcription at positions –59/-47 and –83/-72 of the mPrm2 promoter. The proximal element at –59 / -47 binds to a testis-specific protein that was named protamine-activating factor-1 (PAF-1). The PAF-1 reaches high levels in round spermatids at the time of mPrm2 transcription. Deletion of the –59/-47 sequence resulted in about a 3-fold reduction of mPrm2 transcription in vitro. Although the PAF-1 binding site (PAF –responsive element, PAF-RE) contains the sequence GTCA present in the CRE and is very similar to the estrogen responsive element, mobility shift assays revealed that neither the cAMP-responsive element modulator nor the estrogen receptor is the protein(s) binding to PAF-RE. Competition mobility shift assays showed that the second positive regulatory element at –83/ -72 binds a Y-box-binding protein. It suggested that the testis-specific PAF-1 and a Y-box binding protein are needed to activate *mPrm2* transcription in postmeiotic male germ cells (Yiu et al, 1997). The coordinate transcription of the mouse protamine genes suggested that they might contain common regulatory elements responsible for their expression.

Translational regulation of Prm gene: The synthesis of protamines is regulated during germ cell development by mRNA storage in the cytoplasm of differentiating spermatids. Two highly conserved sequences, the Y and H elements present in the 3'-untranslated regions (UTRs) of all known mammalian protamine mRNAs, form RNA-protein complexes and specifically bind a protein of 18-kDa. Kwon and Hecht (1993) showed that translation of fusion mRNAs was markedly repressed in reticulocyte lysates supplemented with a mouse testis extract enriched for the 18-kDa protein when the mRNAs contained the 3'UTR of Prm2 or Y and H elements of Prm2. The 18-kDa-protein is developmentally regulated in male germ cells and requires phosphorylation for RNA binding. It was proposed that a phosphorylated 18-kDa protein plays a primary role in repressing translation of mPrm2 mRNA by interaction with the highly conserved Y and H elements (Chapter 14).

7.11. NATURE OF BASIC PROTEINS IN OTHER SPECIES

During the final stages of spermatogenesis, the compaction of DNA in many organisms is accomplished by the replacement of histones with transition proteins TP1 and TP2 and finally with a class of arginine-rich proteins called protamines. In other organisms, however, condensation of sperm DNA can occur with comparable efficiency in the presence of somatic-type histones or, alternatively, an intermediate class of proteins called protamine-like (PL) proteins. Lewis and Ausio (2002) proposed a model for a chromatin structure in the sperm of closely related species that is mediated by somatic-type histones, which are frequently found associated with these proteins. This structure supports the concept that the protamine-like (PL) proteins may represent various evolutionary steps between a sperm-specific histone H1 precursor and true protamines. There is an unequivocal support to the notion that vertebrate protamines evolved from histones (Lewis et al., 2004). The *S. solidissima* sperm has an SNBP

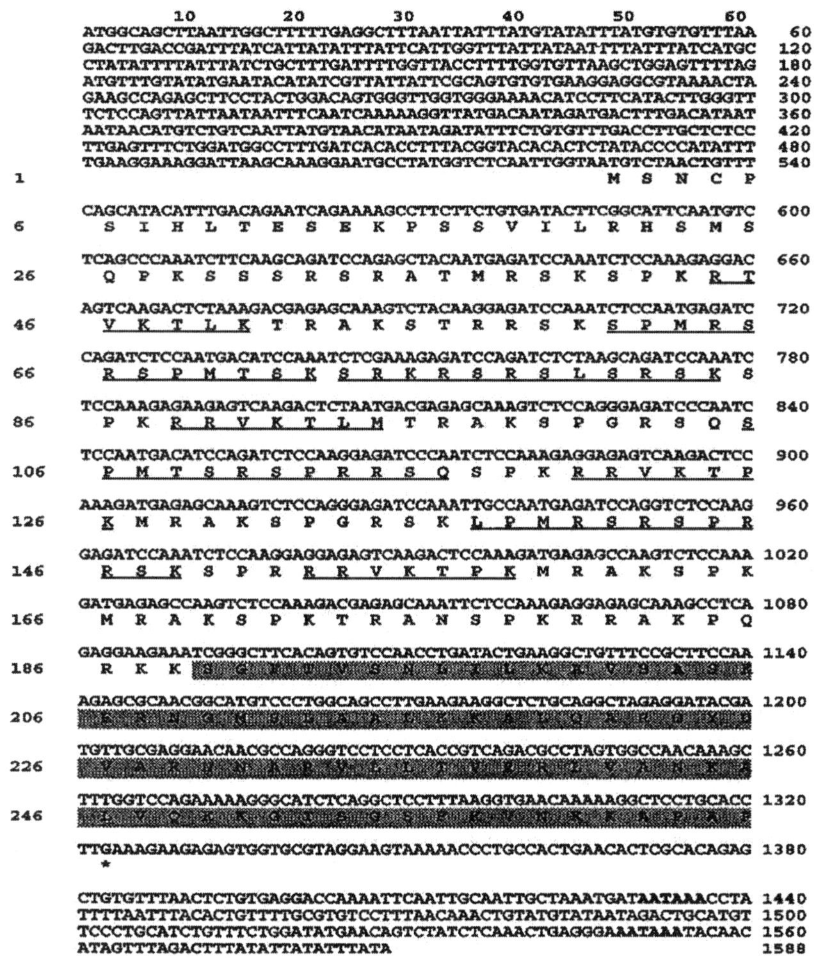

Fig.7.5. Nucleotide and predicted amino acid sequence of a genomic clone 2B of a high molecular weight chromatin protein from winter flounder sperm. The translation stop site is shown by asterisk. Matches to HM_rBNP hepatapeptide and dodecapeptide repeats are underlined. The histone H1-like globular core region begins at residue 189 and ends at 266. Reproduced with permission from C.E. Watson and P.L. Davies: J Biol Chem 273; 6157-62: 1998 © The American Society of Biochemistry and Molecular Biology.

composition in which about 80% of a PL-1 protein coexists with 20% sperm specific histones. The PL-1 is H1 like protein and contains cysteine. This protein is responsible for aggregation behaviour. The cysteine residue is located in the globular part of the protein, and is consistently present in PL-1 proteins from sperm of other bivalve mollusks. It seems that cysteine is mainly present in invertebrate H1, with the exception of H5 isoform of Japanese quail (Zhang et al., 1999).

7.11.1. Basic Proteins of Winter Flounder

The winter flounder is a fish that does not replace its testis histones with protamines. On the contrary, it retains somatic histones throughout spermatogenesis and does not synthesize significant quantities of sperm – specific histone variants. Instead, this fish produces a unique group of high molecular mass basic nuclear proteins (HM$_r$BNPs) in mid-spermatids that are retained in the mature sperm, and make up ~ 25% of the total acid-soluble nuclear proteins. These HM$_r$BNPs have Arg-, Ser-, Lys-, and Pro- repeat lengths rich in apparent molecular mass ranging from 80-200-kDa. They do not contain cysteine, and they have only very limited quantities of hydrophobic amino acids. When the HM$_r$BNPs first appears in mid-spermatid nuclei they are extensively phosphorylated, approximately one in five residues is phosphoserine. Dephosphorylation of the HM$_r$BNPs in late spermatids coincides with a major re-arrangement of the chromatin, in which the nucleosome repeat length increases from 195 to 220bp. A similar change in nucleosome repeat length is also seen in developing sea urchin sperm, where the repeat increases from 220 to 250bp, and in chicken erythrocytes, where it increases from 190 to 212bp. Based on their size and amino acid composition, the HMrBNPs do not fit into either the histone or protamine category. The HM$_r$BNPs appear to have evolved from an extreme H1 variant that has an N-terminal tail of HM$_r$BNP-like sequence linked to an H1 globular region. Based on sequences of the most abundant HM$_r$BNP cDNAs, and the lack of hybridization between HM$_r$BNP mRNAs and a DNA probe for the H1 globular region, the latter domain appears to have been lost during expansion and amplification of the HM$_r$BNP-like repeats. Transcripts of the HM$_r$BNP and Hq variant gene are present in testis RNAs only during the mid-spermatid stage of spermatogenesis, at the same time that HM$_r$BNPs in their highly phosphorylated form first appear in the nucleus. Judging by the lack of a lag between HM$_r$BNP mRNA synthesis and translation, the mRNAs for these highly basic proteins are not stored for any length of time. Instead, the deposition of HM$_r$NNPs onto DNA, which coincides with the major reorganization and silencing of the chromatin, may be controlled by dephosphorylation (Watson et al, 1999).

Although the HMrBNPs have no obvious homology to histones, protamines, or other sperm-specific chromatin proteins, isolation of a clone (2B) from a winter flounder establishes a link between the HM$_r$BNPs and histone H1 (Watson and Davies, 1998; Eatson et al, 1999 c/r Watson et al., 1999). The 2B sequence contains an open reading frame, which encodes a 265-residue protein. At its N terminus the translation product contains numerous simple repeats that match the oligopeptides contained within the HM$_r$BNPs **(Fig 7. 5)**. Unexpectedly, the C terminus of the putative protein shows 66% identity and 76% conservation to the histone H1 globular domain. This connection suggests that the HMrBNPs may have originated from the extended N-terminal tail region of a testis- specific, H1-like linker histone (Watson and Davies, 1998).

7.12. REFERENCES

Akama TO, Nakagawa H, Sugihara K, Narisawa S, Ohyama C, Nishimura S, O'Brien DA, Moremen KW, Millan JL, Fukuda MN. Science, 2002;295(5552):53-54.

Albig W, Bramlage B, Gruber K, Klobeck HG, Kunz J, Doenecke D. The human replacement histone H3.3B gene. Genomics 1995; 30: 264-72.

Alimi E, Martinage A, Arkhis A, Belaiche D, Sautiere P, Chevaillier P. Amino acid sequence of the human intermediate basic protein 2. (HPI2) from sperm nuclei. Structural relationship with protamine P2. Eur J Biochem 1993; 214: 445-50.

Baarends WM, Hoogerbrugge JW, Roest HP, Qoms M, Vreeburg J, Hoeijmakers JH, Grootegoed JA. Histone ubiquitination and chromatin remodeling in mouse spermatogenesis. Dev Biol 1999; 207: 322-33.

Bartell JG, Davis T, Kremer EJ , Dewey MJ, Kistler WS. Expression of the rat testis-specific histone H1t gene in transgenic mice. One kilobase of 5'-flanking sequence mediates correct expression of a lacZ fusion gene. J Biol Chem 1996; 271: 4046-54.
Baskaran R, Rao MR. Interaction of spermatid-specific protein TP2 with nucleic acids, in vitro. A comparative study with TP1. J. Biol Chem 1990; 265: 21039-47.
Bellve AR, Purification and characterization of mouse protamine P1 and P2. Amino acid sequence of P2. Biochemistry 1988; 27: 2890-97.
Bramlage B, Kosciessa U, Doenecke D. Differential expression of the murine histone genes H3. 3A and H3. 3B. Differentiation 1997; 62: 13-20.
Bunick D, Balhorn R, Stanker LH, Hecht NB. Expression of the rat protamine 2 gene is suppressed at the level of transcription and translation. Exp Cell Res 1990; 188: 147-52.
Chen HY, Sun JM, Zhang Y, Davie JR, Meistrich ML. Ubiquitination of histone H3 in elongating spermatids of rat testis. J Biol Chem 1998; 273: 13165-69.
Choi YC, Chae CB. DNA hypomethylation and germ cell specific expression of testis specific H2B histone gene. J Biol Chem. 1991; 266: 504-11.
Clare SE, Fantz DA, Kistler WS, Kistler MK. The testis-specific histone H1t gene is strongly repressed by a G/C-rich region just down stream of the TATA box. J Biol Chem 1997a; 272: 33028-36.
Clare SE, Hatfield WR, Fantz DA, Kistler WS, Kistler MK. Characterization of the promoter region of the rat testis-specific histone H1t gene. Biol Reprod 1997b; 56: 73-82.
Cole KD, Kandala JC, Kistler W. Isolation of the gene for the testis-specific H1 histone variant H1t. J Biol Chem 1986a; 261: 7178-83.
Cole KD, York RG, Kistler WS. Sequence of the amino terminal half of rat testis-specific histone variant H1t. Biochim Biophys Acta 1986b; 869: 223-229.
Cole KD, York RG, Kistler WS. The Amino Acid Sequence of Boar H1t, a Testis-specific H1 Histone Variant. J Biol Chem. 1984; 259: 13695-702.
Croston GE, Kerrigan LA, Lira LM, Marshak DR, Kadonaga JT. Sequence specific antirepression of histone H1 mediated inhibition of basal RNA polymerase II transcription. Science 1991; 251: 643-49.
de la Barre AE, Gerson V, Gout S, Creaven M, Allis CD, Dimitrov S. Core histone N-termini play an essential role in mitotic chromosome condensation. The EMBO Journal 2000; 19: 379-91.
De Lucia F, Faraone – Menella MR, D' Erme M, Quesada P, Caiapa P, Farina B. Histone induced condensation of rat testis chromatin : testis specific Hit versus somatic H1 variants. Biochem Biophys Res Commun 1994; 198: 32-39.
De S, Brown DT, Lu ZH, Leno GH, Wellman SE, Sittman DB. Histone H1 variants differentially inhibit DNA replication through an affinity for chromatin mediated by their carboxyl-terminal domains. Gene 2002; 292: 173-81.
Doenecke D, Albig W, Bode C, Drabent B, Franke K, Gavenis K, Witt O. Histones genetic diversity and tissue-specific gene expression. Histochem Cell Biol 1997; 107: 1-10.
Domenjoud L, Nussbaum G, Adham IM, Greeske G, Engel W. Genomic sequences of human protamines whose genes, PRM1 and PRM2, are clustered. Genomics 1990; 8: 127-33.
Drabent B, Bode C, Miosge N. Expression of the mouse histone gene H1t begins at premeiotic stages of spermatogenesis. Cell Tissue Res 1998; 291: 127-132.
Drabent B, Bode C. Doenecke D. Structure and expression of the mouse testicular H1 histone gene (H1t). Biochim Biophys Acta 1993: 1216: 311-313.
Drabent B, Doenecke D. Testis-specific expression of the mouse histone gene H1t is regulated by several promoter elements. Biol Reprod 1997; 57: 1375-83.
Drabent B, Kardalinou E, Bode C, Doenecke D. Association of histone H4 genes with the mammalian testis-specific H1t histone gene. DNA Cell Biol 1995; 14: 591-597.
Drabent B, Kardalinou E, Doenecke D. Structure and expression of the human gene encoding testicular H1 histone (H1t). Gene 1991; 103: 263-269.
Drabent B, Saftig P, Bode C, Doenecke D. Spermatogenesis proceeds normally in mice without linker histone H1t. Histochem Cell Biol 2000; 113: 433-442.
Enmark E, and Gustaffsson J. Mol. Endocrinol. (1996) 10: 1293-07. Ha H, van Wijnen AJ, Hecht NB. J. Cell Biochem 1997; 64: 94-105.
Fantz DA, Hatfield WR, Horvath WR, Kistler MH, Kistler WS. Mice with a targeted disruption of the H1t gene are fertile and undergo normal changes in structural chromosomal proteins during spermiogenesis. Biol Repord 2001; 64: 425-31.
Franke K, Drabent B, Doenecke D. Testicular expression of the mouse histone H1.1 gene. Histochem Cell Biol 1998; 109: 383-90.
Gardiner M, Ballesteros M, Gordon M, Tam PPL. Histone and protamine-DNA association: conservation

of different patterns within the β-globin domain in human sperm. Mol Cell Biol 1998; 18: 3350-56.
Grimes SR Jr, Henderson N. Hyperacetylation of histone H4 in rat testis spermatids. Exp. Cell Res 1984; 152: 91-7.
Grimes SR Jr, van Wert J, Wolfe SA. Regulation of transcription of the testis-specific histone H1t gene by multiple promoter elements Mol Biol Rep 1997; 24: 175-184.
Grimes SR, Weisz-Carrington P, Daum H 3rd, Smith J, Green L, Wright K, Stein G, Stein J. A rat histone H4 gene closely associated with the testis-specific H1t gene. Exp. Cell Res 1987; 173: 534-45.
Grimes SR, Wilkerson DC, Noss KR, Wolfe SA. Transcriptional control of the testis-specific histone H1t gene. Gene. 2003; 304: 13-21.
Ha H, Van Wijnen AJ, Hecht NB. Tissue-specific protein-DNA interactions of the mouse protamine 2 gene promoter. J Cell Biochem 1997; 64: 94-105.
Hazzouri M, Pivot-Pajot C, Faure AK, Usson Y, Pelletier R, Sele B, Khochbin S, Rousseaux S. Regulated hyperacetylation of core histones during mouse spermatogenesis: Involvement of histone deacelylases. Eur J Cell Biol 2000; 79: 950-60.
Hecht N. Gene expression during spermiogenesis In: Cellular and molecular events in spermiogenesis (Hamilton DW and waites GMH eds), Cambridge University Press, 1990, pp 265-83.
Hecht N. In: Adolph KW (ed). Molecular Biology of Chromosome function. Springer – Verlag, New York 1989; pp. 396-420.
Heidaran MA, Kistler WS. Transcriptional and translational control of the message for transition protein 1, a major chromosomal protein of mammalian spermatids. J Biol Chem 1987; 262: 13309-15.
Horvath GC, Dasgupta A, Kistler MK, Kistler WS. The rat histone H1d gene has intragenic activating sequences that are absent from the testis-specific variant H1t. Biochim Biophys Acta. 2003; 1625: 165-72.
Houchmandzadeh B, Dimitrov S. Elasticity measurements show the existence of a thin rigid core inside mitotic chromosome. J Cell Biol 1999; 145: 215-23.
Huh NE, Hwang I, Lim K, You KH and Chae CB. Presence of a bi-directional S phase-specific transcription regulatory element in the promoter shared by testis specific TH2A and TH2B histone genes. Nucleic Acids Res 1991; 19: 93-98.
Hwang I, Lim K, Bom, Chae C. Characterization of the S-phase-specific transcription regulatory elements in a DNA replication-independent Testis-specific H2B (TH2B) histone gene. Mol Cell Biol 1990; 10: 585-592.
Johnson PA, Bunick D, Hecht NB. Protein binding regions in the mouse and rat protamine-2 Genes. Biol Reprod 1991; 44: 127-34.
Khadake JR, Markae ER, Rao MRS. Testis specific histone (H1t) is not phosphorylated with chromatin. Ind J Biochem Biophys 1994; 31: 335-38.
Kierszenbaum AL. Transition nuclear proteins during spermiogenesis: unrepaired DNA breaks not allowed. Mol Reprod Dev 2001; 58: 357-58.
Kim YJ, Hwang I, Tres LL, Kierszenbaum AL, Chae CB. Molecular cloning and differential expression of somatic and testis-specific H2B histone genes during rat spermatogenesis. Dev Biol 1987; 124: 23-24.
Kistler WS, Henriksen K, Mali P, Pavinen M. Sequential expression of nucleo proteins during rat spermiogenesis. Exp Cell Res 1996; 225: 374-81.
Kistler WS. In : Hinilica LS (ed). Histones and other basic nuclear proteins. CRC Press, Boca Raton FL. 1989; 331-45.
Kleene KC, Distel RJ, Hecht NB, Nucleotide sequence of a cDNA clone encoding mouse protamine 1. Biochemistry 1985; 24: 719-22.
Koppel DA, Wolfe SA, Fogelfeld LA, Merchant PS, Prouty L, Grimes SR. Primate testicular histone H1t genes are highly conserved and the human H1t gene is located on chromosome 6. J. Cell Biochem 1993; 55: 1-12.
Kwon YK, Hecht NB. Binding of a phosphoprotein to the 3' untranslated region of the mouse protamine 2 mRNA temporally represses its translation. Mol Cell Biol 1993; 13: 6547-57.
Lai JC, Childs G. Isolation and characterization of the gene encoding the testis specific histone protein H2B-2 from the *Urchin lytechinus pictus*. Nucleic Acids Res 1986; 14: 6845-56.
Lee CH, Bartels I, Engel W. Haploid expression of a protamine gene during bovine spermatogenesis. Biol Chem Hoppe Seyler 1987; 368: 807-11.
Lewis JD, Ausio J. Protamine-like proteins: evidence for a novel chromatin structure. Biochem Cell Biol 2002; 80: 353-61.
Lewis JD, Saperas N, Song Y, Zamora MJ, Chiva M, Ausio J. Histone H1 and the origin of protamines. Proc Natl Acad Sci USA. 2004; 101: 4148-52.
Lim K, Chae CB. Presence of a repressor protein for testis-specific H2B (TH2B) histone gene in early stages of spermatogenesis. J Biol Chem 1992; 26: 15271-3.

Luger K, Richmond TJ. The histone tails of the nucleosome. Curr Opin Genet Dev 1998; 8: 140-46.

Marret C, Avallet O, Perrard-Sapori MH, Durand P. Localization and quantitative temporal expression of mRNAs encoding the testis specific histone TH2B, the phosphoprotein p19, the transition proteins 1 and 2 during pubertal development and throughout the spermatogenic cycle of the rat. Mol Reprod Dev 1998; 51: 22-35.

Meishtrich ML. In: Hnilica LS (ed). Histones and other basic nuclear proteins. CRC Press, Boca Raton, 1989; 165-82.

Moens PB. Histones H1 and H4 of surface – spread meiotic chromosomes. Chromosoma 1995; 104: 169-74.

Moss SB, Challoner PB, Groudine M. Expression of a novel histone 2B during mouse spermiogenesis. Dev Biol 1989; 133: 83-92.

Moss SB, Orth JM. Localization of a spermatid specific histone 2B protein in mouse spermiogenic cells. Biol Reprod 1993; 48: 1047-56.

Oko RJ, Jando V, Wagner CL, Kistler WS, Hermo LS. Chromatin reorganization in rat spermatids during the disappearance of testis-specific histone, H1t, and the appearance of transition proteins TP1 and TP2. Biol Reprod 1996; 54: 1141-57.

Perez- Castro AV, Shamanski FL, Meneses JJ, Lovato TL, Vogel KG, Moyzis RK, Pedersen R. Centromeric protein B null mice are viable with no apparent abnormalities. Dev Biol 1998; 201: 135-43

Prigent Y, Troalen F, Dadoune JP. Immunoelectron microsopic visualization of intermediate basic proteins HPI1 and HPI2 in human spermatids and spermatozoa. Reprod Nutr Dev 1998; 38: 417-27.

Queralt R, Oliva R. Identification of conserved potential regulatory sequences of the protamine-encoding P1 genes from ten different mammals. Gene 1993; 133: 197-204.

Rabini S, Franke K, Saftig P, Bode C, Doenecke D, Drabent B. Spermatogenesis in mice is not affected by histones H1. 1 deficiency. Exp Cell Res 2000; 255: 114-24.

Ramakrishnan V, Finch JT Graziano V, Lee PL, Sweet RM. Crystal structure of globular domain of histone H5 and its implications for nucleosome binding. Nature 1993; 362: 219-223.

Reinhart N, Kremling H, Luerssen H, Adham IM, Engel W. Characterization of a gene encoding a basic protein of the spermatid nucleus, TNP2, and its close linkage to the protamine gens in the bull. Biol Chem Hoppe Seyler 1991; 372: 431-6.

Richmond TJ, Finch JT, Rushton B, Rhodes D, Klug A. Structure of nucleosome core particle at 7A resolution. Nature 1984; 311: 532-37.

Siffroi JP, Alfonsi MF, Dadoune JP. Colocalization of HP1 and TP1. Int J Androl 1999; 22: 83-90.

Steger K, Klonisch T, Gavenis K, Behr R, Schaller V, Drabent B, Doenecke D, et al. Round spermatids show normal testis-specific H1t but reduced cAMP-responsive element modulator and transition protein1 expression in men with round spermatid maturation arrest. J Androl 1999; 20: 747-54.

Steger K, Klonisch T, Gavenis K, Drabent B, Doenecke D, Bergmann M. Expression of mRNA and of protein in nucleoproteins during human spermiogenesis. Mol Hum Reprod 1998; 4: 939-45.

Stewart TA, Hecht NB, Hollingshead PG, Johnson PA, Leong. JAC, Pitts SL. Mol Cell Biol 1998; 8, 1748-55.

Tamura T, Makino Y, Mikoshiba K, Muramatsu M. Demonstration of a testis specific transacting factor Tet-1 in vitro that binds to the promoter of mouse protamine 1 gene. J Biol Chem 1992; 267: 4327-32.

Trostle-Weige PK, Meistrich ML, Brock WA Nishioka K. Isolation and characterization of TH3, A germ cell-specific variant of Histone 3 in rat testis. J. Biol Chem 1984; 259, 8769-76

Ullas KS, Rao MR. Phosphorylation of rat spermatidal protein TP2 by sperm-specific protein kinase A and modulation of its transport into the haploid nucleus. J Biol Chem 2003; 278: 52673-80.

Unni E, Mayerhofer A, Zhang Y, Bhatnagar YM, Russell LD, Meistrich ML. Increased accessibility of the N-terminus of testis-specific histone TH2B to antibodies in elongating spermatids. Mol Reprod Dev 1995a; 42: 210-9

Unni E, Zhang Y, Kangasniemi M, Saperstein W, Moss SB, Meistrich ML. Stage-specific distribution of the spermatid-specific histone 2B in the rat testis. Biol Reprod 1995b; 53: 820-6

van Roijen HJ, Ooms MP, Spaargaren MC, Baarends WM, Weber RF, Grootegoed JA, Vreeburg JT. Immunoexpression of testis-specific histone 2B in human spermatozoa and testis tissue. Hum Reprod 1998; 13: 1559-66.

van Wert JM, Panek HR, Wolfe SA, Grimes SR. The TE promoter element of the histone H1t gene is essential for transcription in transgenic mouse primary spermatocytes. Biol Reprod 1998; 59: 704-10.

van Wert JM, Wolfe SA, Grimes SR, Binding of nuclear proteins to a conserved histone H1t promoter element suggests an important role in testis-specific transcription. J Cell Biochem 1996; 60: 348-62.

van Wert JM, Wolfe SA, Grimes SR. Testis-specific expression of the rat histone H1t gene in transgenic mice. Biochemistry 1995; 34: 8733-43.

Wade PA, Pruss D, Wolfe AP. Histone acetylation-chromatin in action. Trends Biochem Sci. 1997; 22: 128-132.

Watson CE, Davies PL. The high molecular weight chromatin proteins of winter flounder sperm are related to an extreme histone H1 variant. J Biol Chem 1998; 273: 6157:62.

Watson CE, Gauthier SY, Davies PL. Structure and expression of the highly repetitive histone H1-related sperm chromatin proteins from winter flounder. Eur. J Biochem 1999; 262: 258-67.

Wei Y, Lanlan Y, Brown J, Gorovsky MA, Allis CD. Phosphorylation of histone H3 is required for proper chromosome condensation and segregation Cell 1999; 97: 99-109.

Wilkerson DC, Wolfe SA, Grimes SR. H1t/GC-box and H1t/TE1 element are essential for promoter activity of the testis-specific histone H1t gene. Biol Reprod. 2002; 67: 1157-64.

Wilkerson DC, Wolfe SA, Grimes SR. TE2 and TE1 sub-elements of the testis-specific histone H1t promoter are functionally different. J Cell Biochem 2003 15;88:1177-87.

Wolfe S A, Anderson JV, Grimes SR, Stein GS, Stein JS. Comparison of the structural organization and expression of germinal and somatic rat histone H4 genes. Biochim Biophys Acta 1989; 1007: 140-150.

Wolfe SA, Grimes SR. Binding of nuclear proteins to an upstream element involved in transcriptional regulation of the testis-specific histone H1t gene. J Cell Biochem 1999; 75: 555-65.

Wolfe SA, Grimes SR. Histone H1t: a tissue-specific model used to study transcriptional control and nuclear function during cellular differentiation. J. Cell Biochem 1993; 53: 156-60.

Wolfe SA, Grimes SR. Protein-DNA interactions within the rat histone H4t promoter. J Biol Chem 1991; 266: 6637-43.

Wolfe SA, Grimes SR. Transcriptional repression of the testis-specific histone H1t gene mediated by an element upstream of the H1/AC box. Gene. 2003; 308:129-38.

Wolfe SA, Mottram PJ, vanWert JM, Grimes SR. Localization of upstream element involved in transcriptional regulation of the rat testis-specific histone H1t gene in somatic cells. Biol Reprod 1999; 61: 1005-11.

Wolfe SA, van Wert JM, Grimes SR. Expression of the testis-specific histone H1t gene: evidence for involvement of multiple cis-acting promoter elements. Biochemistry 1995; 34: 12461-69.

Wykes SM, Nelson JE, Visscher DW, Djakiew D, Krawetz SA. Coordinate expression of the PRM1, PRM2, and TNP2 multigene locus in human testis. DNA Cell Biol 1995; 14: 155-61.

Yagi H, Kato T, Nagata T, Habu T, Nozaki M, Matsushiro A, Nishimune Y, Morita T. Regulation of the mouse histone H2A.X promoter by the transcription factor E2F and CCAAT binding protein. J Biol Chem 1995; 270: 18759-65.

Yan W, Ma L, Burns KH, Matzuk MM. HILS1 is a spermatid-specific linker histone H1-like protein implicated in chromatin remodeling during mammalian spermiogenesis. Proc Natl Acad Sci USA. 2003; 100: 10546-51.

Yiu GK, Hecht NB. Novel Testis-specific protein-DNA interactions activate transcription of the mouse protamine 2 gene during spermatogenesis. J Biol Chem 1997; 272: 26926-32.

Yu YE, Zhang Y, Unni Y, Shirley CR, Deng JM, Russel LD, Weil MM, Behringer RR, Meistrich ML. Abnormal spermatogenesis and reduced fertility in transition nuclear protein 1 deficient mice. Proc Natl Acad Sci USA 2000; 97: 4683-88.

Zalensky AO, Siino JS, Gineitis AA, Zalenskaya IA, Tomilin NV, Yau P, Bradbury EM. Human testis/sperm-specific histone H2B (hTSH2B). Molecular cloning and characterization. J Biol Chem 2002; 277: 43474-80..

Zambrowicz BP, Harendza CJ, Zimmermann JW, Brinster RL, Palmiter RD. Analysis of the mouse protamine 1 promoter in transgenic mice. Proc Natl Acad Sci USA 1993; 90: 5071-5.

Zhang F, Lewis JD, Ausio J. Cysteine-containing histone H1-like (PL-1) proteins of sperm. Mol Reprod Dev 1999; 54: 402-9.

Chapter 8

MICROTUBULES

8.1. INTRODUCTION

Microtubules (MT) in eukaryotes are made on a common frame-work. They are about 25 nm in diameter and appear as hollow cylinders with a central lumen of about 15 nm in width. The wall of the microtubule is formed from a circle of subunits, each 4 nm in diameter and 8 nm long. Microtubules have specific functions in the mitotic and meiotic spindle, in the centriole, in the manchette and flagellar axoneme of spermatozoa, in secretion and transport in many cell types and may act as tracks for vesicles and organelle transport. Microtubules interact with force-generating proteins to generate a variety of intracellular movements, including intracellular particle transport, ciliary-flagellar beating, and chromosome-spindle movements during mitosis-meiosis. The core of the eukaryotic flagellum is the axoneme, a complex motile organelle composed of approximately 200 different polypeptides. The most prominent components of the axoneme are the central pair and nine outer doublet microtubules. Each doublet microtubule contains an A and a B tubule; these are composed, respectively, of 13 and 10-11 protofilaments. Microtubules are assembled from heterodimers of two closely related proteins called α- and β-tubulins along with microtubules-associated proteins such as tektins. Each α and β subunit, in α- and β-tubulins contains a dynein binding domain located near the C-termini of their respective tubulin subunits. Blocking either α and β tubulin binding domains of dynein attenuates motility in demembranated sea urchin sperm up to 50% whereas blocking of both α-and β-tubulin binding domains on dynein produces greater decrease in motility. Among the most interesting proteins associated with microtubules are the energy-transducing ATPases, known as dyneins, which act as microtubule motors. These motor proteins, which are able to produce sliding between adjacent microtubules as well as transport of cytoplasmic particles along a single microtubule, occur widely in many types of cells. Flagellar dynein binds to both subunits of the MT polymer, α-and β-tubulins, although the two subunits appear to contribute equivalent, but functionally separate roles to flagellar motility (Goldsmith et al., 1995).

The direction of movement on microtubules is determined by their structural polarity, which derives from the head to tail organization of their constituent α, β-tubulin heterodimers. In the cytoplasm of many cells, microtubules exist in a state of dynamic instability at both ends, with phases of growth by polymerization of tubulin alternating stochastically with phases of shrinkage by relatively rapid depolymerization. The net growth of the microtubule occurs more rapidly at one end, known as the plus end, than at the other end, known as the minus end, and it can thus be used to obtain a marker of microtubules in a variety of cells, the polarity that is visible by light microscopy. Study of the polarity of microtubules in a variety of cells has shown that they are most often oriented uniformly with their minus ends lying toward the centrosome and their plus ends toward the cell periphery. In spite of their generally constant structure, microtubules are capable of playing diverse roles in several forms of cell

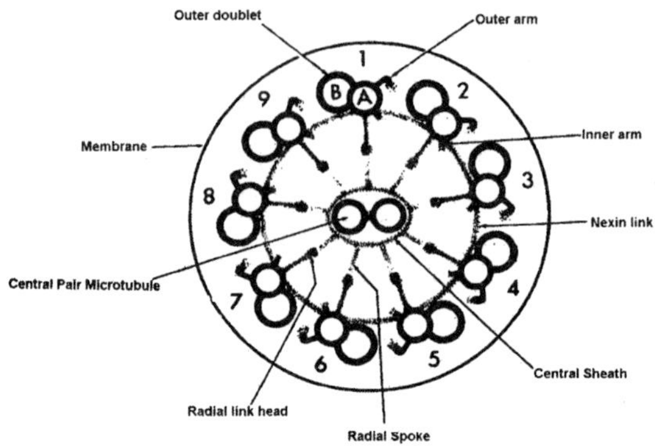

Fig.8.1. Schematic diagram of a transverse section of a flagellum or cilium, illustrating the 9 + 2 arrangement of microtubules. The microtubules link together at regular intervals along the length of the axoneme, through various projections from microtubules.

motility, as well as fulfilling a more static skeletal function in maintaining cell shape and asymmetry.

8.2. FLAGELLAR STRUCTURE AND DOUBLET SLIDING

Cilia and flagella are unique motile systems that can be isolated with great ease from the rest of the cell and reactivated. After mild treatment with detergent, flagella can then be purified, leaving an axoneme, a cylinder of 9 doublet microtubules (which consist of 13 complete protofilament A-tubules and 11 protofilament B-tubules) surrounding 2 singlet microtubules, all interconnected by numerous microtubule-associated proteins and protein complexes. Dyneins are attached to outer doublets in two rows of projections or "arms" along each A-tubule. These arms extend toward the B-tubule of an adjacent doublet, where they form rigor-like cross-bridges in the absence of ATP, and transient, force-generating attachments when Mg-ATP is present. The outer row consists of a single variety of dynein with a 24-nm periodicity, whereas the inner row contains several dyneins arrayed in a 960nm repeat pattern. Brief exposure of axonemes to protease before addition of Mg-ATP prevents beating, but releases the potential for doublet microtubules to slide, one along the next, until all nine doublets lie end to end. This simple observation reveals the basic sliding mechanism that underlies flagellar motility. Axonemes can be structurally divided into two systems: the central pair complex and the outer doublet complex. The two single-micotubules of the central pair are inter-connected by short bridges at a periodicity of 16nm. Other structures project off of these microtubules with periods of 16 or 32 nm to encircle the microtubules in a cylindrical cage. In some organisms the entire central pair complex has been observed to rotate relative to the surrounding cylinder of doublet microtubules. The central pair complex may regulate dynein activity but may not be essential for doublet sliding or bend propagation (Mitchell 1994) **(Fig. 8.1)**. Doublets are interconnected at 96 nm by bridges called nexin links located in the same region as inner row /arm dyneins. The nexin link is a major part of dynein regulatory complex (Woolley, 1997). Nexin links have been distinguished from dyneins by their resistance to conditions that extract dyneins. It seems that nexin links maintain interdoublet connections without generating any sliding resistance, and that stiftners of ATP depleted axoneme is entirely due to rigor dynein cross-bridges (Bozkurt and Woolley, 1993).

Radial spokes project from each outer doublet toward the central pair complex, where they appear to form contacts with central pair projections. Spokes are found in groups of two or three (depending upon the organism) at a repeat interval of 96 nm, with the S2 spoke in each group located just proximal. Numerous spoke assembly mutations have been isolated; 17 proteins have been identified as spoke components, and S3 have already been cloned and sequenced. Defects in radial spoke assembly inhibit flagellar motility, but bypass suppressor mutations have been characterized that can restore motility without restoring the spoke structures (Mitchell, 1994). All flagellar dyneins tested are minus-end directed motors and induce sliding of an adjacent doublet toward the tip of the axoneme.

8.3. TUBULINS

In most eukaryotes the tubulin genes comprise small multigene families with approximately equal numbers of genes for α- and β-tubulin, the structural proteins of microtubules. Tubulin dimers, formed from globular α and β-subunits, and the tektins, three equimolar α-helical proteins that form filaments, mutually associate to form the junctional regions of doublet and triplet microtubules. The α- and β-tubulins are tightly bound together by noncovalent bonds. Each tubulin monomer has a binding site for GTP. It appears that microtubules, regardless of source, are the only structural constituent of the spindle apparatus essential for cleavage furrow induction (Alsop and Zhang, 2003).

The properties of the microtubule network are regulated at various levels including tissue-dependent isotype switching, post-translational modification of α– and β-tubulin, and by a variety of microtubule-associated molecules. The azh/azh (for abnormal sperm head shape) mouse mutant is an ideal model for analyzing tubulin isotypes and microtubule-associated proteins of the manchette and axoneme in light of a potential role of the manchette in the shaping of the sperm head and formation of the tail. The cationic dye, stains-all, is known to stain brain β-tubulin blue and α-tubulin red. This stain can also be applied to detect β-tubulin in axonemal tubulins from various sources such as cilia of protozoa and sperm flagella. Furthermore, it selectively stains isoforms of axonemal β-tubulin blue following isoelectric focusing, whereas those of α-tubulin are stained red. Stains-all staining is a useful tool for electrophoretic analysis of axonemal tubulins (Nakamura et al., 1990). While searching possible differences in tubulin isotype variants in stage specific manchettes and axonemes of wild type and azh/azh mutant mice, Mochida et al., (1999) found that (1) manchettes of azh/azh abnormal mutants are longer than in wild type mice; (2) manchette and sperm tail axonemes display a remarkable variety of post-translationally modified tubulins (acetylated, glutamylated, tyrosinated, a-3/7 tubulins). (3) an acidic 62-kDa protein was identified as the main component of the peri-nuclear ring of the manchette in wild type and azh/azh mice; (4) bending of looping of the mid-piece of the tail of azh/azh sperm, accompanied by a dislocation of the connecting piece from head attachment sites, were visualized in about 35% of spermatids/sperm; and (5) a lasso-like tail configuration was predominant in epididymal sperm of azh/azh mutants. It appeared that spermatid and sperm tail abnormalities in the azh/azh mutant reflected structural and/or assembly deficiencies of peri-axonemal proteins responsible for maintaining a stiffened tail during sperm maturation (Mochida et al., 1999).

Results in *Drosophila* also indicated that the post-translational modification occurs (PTM) as a very late event, after complete assembly of axonemal microtubules, and that the axonemal tubulin becomes modified when axonemal microtubules become coupled with the membrane, suggesting that the modification may in some way be induced by the microtubule membrane interaction (Bressac et al., 1995). The PTM occurring in the C-terminal tail of axonemal tubulin from sea urchin, *Paracentrotus lividus,* spermatozoa suggested that the majority of the isolated C-terminal peptides were unmodified and that polyglycylation and polyglutamylation could occur simultaneously on α-tubulin (Mary et al., 1997).

8.4. MICROTUBULES IN SPERMATOGENIC CELLS

In Sertoli cells microtubles are composed of tubulin subunits, which preferentially polymerize to one end of microtubules giving them fast growing [+] and slow growing [-] ends. The majority (greater than 93%) of microtubules in Sertoli cells are oriented with their [+] ends towards the cell nucleus and their [-] ends toward the lumen of the seminiferous tubule. Thus, the bulk transport of seminiferous tubule constituents must be towards the [-] ends of the microtubules. In contrast to Sertoli cells, axons have microtubules oriented [-] ends towards the nucleus. Microtubules dependent transport in Sertoli cells probably involves molecules that can bind to and move along microtubules. The translocation of maturing spermatids to the seminiferous tubule lumen from a basal location during stages V-VI has been hypothesized to involve cytoplasmic specialization and microtubules.

Another prominent microtubule structure in the testis is the manchette, which consists of parallel, cross-linked arrays of microtubules extending from the nuclear ring of the maturing elongate spermatids towards the lumen of the seminiferous epithelium. Manchettes first form within the spermatid cytoplasm as they begin to elognate and begin to break up as spermatids are translocated through the seminiferous epithelium during stages IV-VI. During disassembly, fragments of manchettes have been observed in the cytoplasm of horse and cat spermatids. Ultrastructural observations from mutant mice and chemically disrupted manchette models indicated that manchettes might be involved in spermatid nuclear reshaping although chromatin condensation may also be important in this process. The observed association of cytoplasmic vesicles with manchette microtubules has led to an additional hypothesis that manchettes "may serve as a track or conveyor along which the cytoplasm is translocated from the anterior to the caudal region of the cell".

8.4.1. Testicular-α Tubulin

The α- and β-tubulins exist in a number of different isotypes with distinct expression patterns during development. The appearance of tubulin mRNA transcripts in meiotic and post-meiotic testicular cells has been studied by isolation of a mouse testis α-tubulin cDNA clone, the 3' end to which is homologous to atleast two different α-tubulins. These transcripts are desirable only during the haploid phase of spermatogenesis. The un-translated 3'-end of this DNA is homologous to two RNA transcripts present in post-meiotic cells of the testis but absent from meiotic cells and from several tissues including brain. The temporal expression of this α - tubulin cDNA provides evidence for haploid expression of a mammalian structural gene (Distel et al., 1984). The absence of transcripts homologous to pRDaTT.3 in mouse brain suggests that the testicular and brain α-tubulins are from different genes. Multiple α-tubulin mRNAs are found in sea urchin testes and a testis specific β-tubulin is present in *Drosophila*. However, in contrast to the *Drosophila* tubulin, the transcript from mouse α-tubulin gene is detected only in haploid cells. The appearance of transcripts of this α-tubulin in round spermatids coincides with the formation of the manchette and flagellar axoneme. Thus, the haploid-

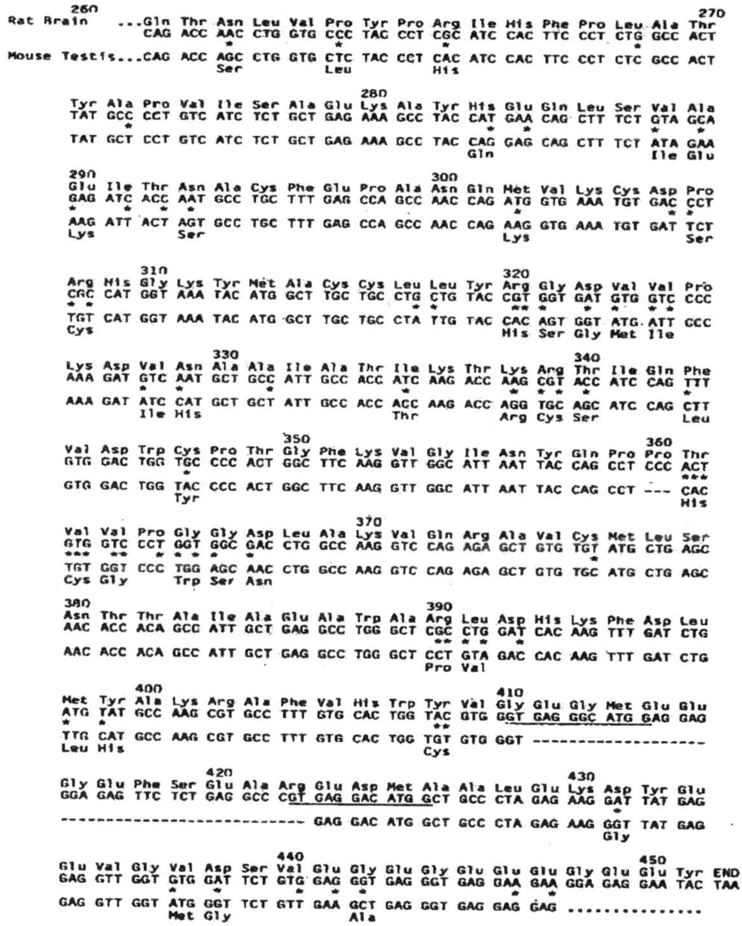

Fig.8.2. Nucleotide and predicted amino acid sequences of the cDNAs of rat brain and mouse testis α-tubulins. Reproduced from N.B.Hechet: In: Cellular and Molecular Events in Spermiogenesis, 1990 © WHO.

specific tubulin mRNA may be involved in these structures, both of which are unique to spermatogenesis (Distel et al., 1984; Hecht, 1990) **(Fig. 8.2)**.

Five mouse isotypes of α tubulin described are distinguished by the presence of unique amino acid substitution within the coding region. In most, but not all, their specific amino acids are clustered at carboxy-terminus. One of the tubulin epitope is testis specific and is coded by two closely related genes (Ma3 and Ma7), which have homologous 3'-untranslated regions but differ at multiple third codon positions and in their 5'-untranslated regions. Same family of α tubulin of testis specificity is present in human testis. Another α-tubulin gene Ma6 is expressed ubiquitously whereas a third gene Ma4 is unique in that it does not encode a carboxy terminal tyrosine residue and yields two transcripts of mRNA: one 1.8-kb mRNA that is abundantly present in muscle and the other transcript of 2.1-kb, which is abundant in testis (Villasante et al., 1986) **(Fig.8.3)**.

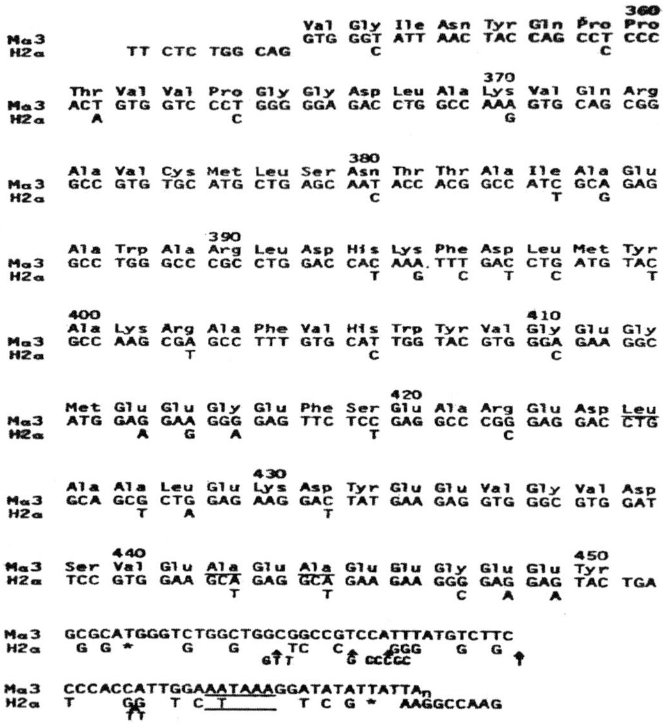

Fig.8.3. A human α-tubulin gene, H2α encoding a isotype similar to Ma3/Ma7. Figure shows the amino acid differences between the two. Deletions are marked by asterisks. Reproduced with permission from A. Villasante et al, Mol Cell Biol 6; 2409-19: 1986 © American Society for Microbiology

8.4.2. Cell Specific Expression of β-Tubulin Isoforms

The β1, β2 and β3 tubulins are distributed very specifically in *Drosophila* testes.The β3-tubulin is present exclusively in cytoplasmic microtubules of somatic cells, while the β1 isotype is localized in the somatic cells and in early germ cells of both the microtubules of the cytoskeleton as well as in the mitotic spindle. In constrast, β2-tubulin is present in all microtubular arrays of germ cells from meiotic prophase onward, though not detectable in somatic cells. Thus, a switch of β tubulin isotype from β1 to β2 occurs during male germ cell differentiation. The co-polymerization properties of the different tubulin isotypes revealed that neither β1 nor β3 is detectable in the axoneme in the wild-type situation. Analysis of transgenic flies revealed that both β1 and β3 tubulin isotypes have the potential to co-incorporate with β2-tubulin into microtubules of the sperm axoneme. Male flies homozygous for the fusion genes (β2-β1 or β2-β3) remain fertile, despite the mixture of β tubulin isotypes in the axoneme (Kaltschmidt et al., 1991).

The β-Tub85D gene encodes the testis-specific β2-tubulin isoform in *D melanogaster*.The functional capacity of different β-tubulin isoforms was tested in vivo by expressing β3-tubulin either in place of or in addition to β2-tubulin in the male germ line of *D. melanogaster*. The testes-specific β2-isoform is conserved relative to major metazoan β-tubulins, while the developmentally regulated, β-3 isoform is considerably divergent in sequence. β3-tubulin is normally expressed in discrete subsets of cells at specific times during development but not expressed in the male germ line. β2-Tubulin is normally expressed only in the post-mitotic germ cells of the testis, and is required for all microtubule based functions is these cells. Functions of β2-tubulin include assembly of meiotic spindles axonemes, and at least two classes of cytoplasmic microtubules, including those associated with the differentiating mitochondrial derivatives. Study shows that β3-tubulin can support only a subset of the multiple functions normally performed by β2, and that the microtubules associated with the mitochondrial derivatives mediate their elongation. Data suggested that the *Drosophila* β tubulin isoforms β2 and β3 are not equivalent in intrinsic functional capacity, and that the assembly of the doublet tubules of the axoneme imposes different constraints on β tubulin function than does assembly of singlet microtubules (Kemphues et al., 1982; Hoyle and Raff, 1990; Hutchens et al., 1997; Raff et al., 1997).

Mutant flies: Mutations responsible for different classes of functional phenotypes are distributed throughout the β2-tubulin molecule. The amino acid sequence of the β2-tubulin internal variable regions is required for generation of correct axoneme morphology and not for microtubule functions. The β2-tubulin carboxy-terminal isotype-defining domain is required for sub-structural organization of the axoneme. The β2 variant lacking the carboxy-terminus and the B2t6 variant (an allele within an internal cluster of variable amino acids in vertebrate β-tubulins) complement each other for mild-to-moderate meiotic defects but do not complement for proper axonemal morphology. It was proposed that the integrity of this structure in the Drosophila testis β2-tubulin isoform is required for proper axoneme assembly but not necessarily for general microtubule functions (Fackenthal et al., 1995). A recessive male sterile mutation (B2t8) that encodes a stable variant of the testis-specific β-2-tubulin of *Drosophila* causes the assembly of aberrant microtubules both in vivo and in vitro. The B2t8 mutation appears to cause defects in the formation of inter-protofilament bonds. In homozygous mutant males, the most commonly observed aberrant structures are sheets of protofilaments curved to form an S in cross-section rather than a normal, closed microtubule (Fuller et al., 1987).

Regulation of β1-tubulin gene in Drosophila: The β1 isotype is localized in microtubles of somatic cells and early germ cells. A switch of tubulin isotypes from β1 to β2 occurs during male germ cell differentiation. β1-tubulin gene is regulated by a short upstream sequence while intron elements guides, expression in somatic cells. Within the male gonads, expression of the β1-tubulin gene is under cell-type-specific control mediated by independent cis-acting elements. Therefore in the germ line, control of β1-tubulin expression is strictly governed by promoter-proximal elements, while for the somatic parts of the testis enhancer elements confer less stringent expression control (Buttgereit and Renkawitz-Pohl, 1993). In mammals, like in *Drosophila*, the two βIV isotypes (βIVa and βIVb) contain a specific acidic seven amino acid sequence in its carboxyl terminus (Jensen-Smith et al., 2003). Using different mAbs, immunocytochemical staining discriminated between various parts of a spermatozoon epitopes of class III β-tubulin present in the flagellum of boar sperm. A tubulin epitope from the C-terminal domain of β-tubulin was detected in the triangular segment of the post-acrosomal part of the sperm head. Ionophore induced acrosome reaction indicated its participation in early fertilization. Three mAbs directed against epitopes on the C-terminal end of neuron-specific class III β-tubulin stained the flagella. This suggested that various tubulins are involved in the functional organization of the mammalian sperm flagellum and head (Peknicova et al., 2001).

8.4.3. Glutamylated and Glycosylated Tubulins (Post-Translational Modifications)

Microtubule diversity can be maintained not only by differential genetic expression of tubulin polypeptides but also by tubulin monomer acetylation and detyrosination, and post-translational modifications of α-tubulin. The distribution of post-translationally modified microtubules in the cells of the seminiferous epithelium showed a distinct pattern of staining of anti-Glu antibody. In mammalian sperm flagellum glutamylated and glycylated tubulin isoforms are detected according to longitudinal gradients and preferentially in axonemal doublets 1-5-6 and 3-8, respectively. This suggested a role for these tubulin isoforms in the regulation of flagellar beating (Hermo et al., 1991). Glutamylated tubulin in the flagellum of mouse and man spermatozoa were detected by using two mAbs: TAP 8\952 and AXO 49 glycosylated tubulin was not found in centrioles and cytoplasmic microtubules (manchette) of germ cells. In human mature sperm axonemal microtubules are rich in monoglycosylated tubulin from base to tip of the flagellum. In mouse sperm flagellum a similar gradient of mono-glycosylated tubulin was present in addition to an opposite gradient to poly-glycosylated tublulin. In both species monoglycosylated tubulin predominated in doublet 3-8 whereas glutamylated tubulin occurred in doublet 1-5-6. The distribution of these tubulin isoforms and the known inhibition of motility by their specific antibodies are consistent with a complementary role of tubulin glycylation and glutamylation in the regulation of flagellar beating in mammalian spermatozoa (Kann et al., 1998; Fouquet et al., 1994).

Using two mAbs GLU-1 and A1-6, raised against γ-L-glutamyl-glutamic acid dipeptide (Glu-Glu) and Ca^{2+} dependent ATPase from *Paramecium*, respectively, it was suggested that GLU-1 and A1.6 are able to recognize the glutamylated form of β-tubulin. The C-terminal Glu-Glu sequence is present in the C terminus of the detyrosinated form of α-tubulin and the glutamlyl side chain of β-tubulin, and that stable axonemal microtubules appeared to be rich in post-translationally modified tubulin subunits (Kuriyama et al., 1995). In mouse and man, the glutamylated tubulin could be involved in a functional heterogeneity of microtubules in peripheral doublets of the sperm flagellum. It was suggested that the glutamylated tubulin interact with other axonemal and/or periaxonemal proteins, which could be involved in flagellar beating and its regulation. Microtubules of some Sertoli cell branches were not acetylated but glutamylated. Sperm extracts showed a high level of α- and β-tubulin heterogeneity, comparable to that found in brain. The major testis-specific tubulin isotypes (murine α 3/7 and murine β 3) are also glutamylatable. These results showed a sub-cellular sorting of post-translationally modified tubulin isoforms in spermatids and the glutamylation being associated with the most stable microtubule structures (Fouquet et al., 1994).

In contrast to mammalian sperm flagellum in rodent lung ciliated cells, all tubulin isoforms are uniformly distributed in all axonemal microtubules with a unique deficiency of glutamylated tubulin in the transitional region. A similar distribution of tubulin isoforms is observed in cilia of Paramecium, except for a decreasing gradient of glutamylated tubulin in the proximal part of axonemal microtubules. In the sea urchin sperm flagellum, predominant labeling of tyrosinated and detyrosinated tubulin in 1-5-6 and 3-8 doublets, respectively, were observed together with decreasing proximo-distal gradients of glutamylated and polyglycylated tubulin labeling and an increasing gradient of monoglycylated tubulin labeling. In flagella of *Chlamydomonas*, the glutamylated and glycylated tubulin isoforms are detected at low levels.

The sea urchin sperm comprising the nine outer doublets, differ strongly in tubulin variants. A tubule contains over 95% unmodified, tyrosinated α β-tubulin. In β tubules, α-tubulin is approximately 65% detyrosinated and both α- and β-tubulins are 40-45% polyglycylated. This raises the possibility that post-trasnlational modifications of tubules reflect or specify structurally and functionally distinct microtubules (Multigner et al., 1996).

From the panel of mAbs against sea urchin sperm axonemal proteins, the antitubulin mAb B3, which inhibited the flagellar motility of sperm, recognizes predominantly α-tubulin of sperm axoneme as well as brain α- and β-tubulins. The B3 epitope was restricted to the last 13 amino acid residues of the C-terminal domain of α-tubulin and recognized polyglutamylated motifs in the C-terminal domain of α-tubulin. With the exception of the anti-polyglutamylated mAb GT335 and mAb B3 other mAbs had no effect on motility. The B3 and GT335 acted by decreasing the beating amplitude without affecting the fallgellar beat frequency. The antigens corresponding to these mAbs revealed their presence along the whole axoneme of sea urchin spermatozoa. This indicated that the polyglutamylated lateral chain of α-tubulin plays a dynamic role in a dynein-based motility process (Gagnon et al., 1996). To sum up, these results show different composition and organization of tubulin isoforms of cilia and flagella, suggesting various models of functional organization and beating regulation of the axoneme (Pechart et al., 1999).

γ-Tubulin: γ-tubulin is an essential component of link between the centrosome and the microtubule suggesting a direct role in microtubule nucleation in *Xenopus* eggs (Stearns and Kirschner, 1994). Human centrosome γ-tubulin requires at least three proteins h102p, h104p and h76p for microtubules nucleation (Fava et al., 1999). Recently several novel proteins complexes found in centrosome have been implicated in the role of γ-tubulin with the nucleation of microtubules (see Centrosome).

8.4.4. δ-Tubulin

A screen for flagellar mutants in the green alga *Chlamydomonas reinhardtii* has led to the identification of a fourth member of the tubulin gene superfamily, δ-tubulin. The deficiency in a functional δ-tubulin gene copy causes aberrant numbers of flagella, depending on the age of the corresponding basal bodies; mutants also show abnormal ultrastructure of the basal bodies and a misplacement of the cleavage furrow at mitosis. The mouse δ-tubulin homolog gene is highly expressed in testis. In the elongating spermatid, δ-tubulin associated with the manchette is responsible for reshaping of the sperm head. The protein is specifically localized at the peri-nuclear ring of the manchette, at the centriolar vaults and along the principal piece of the sperm flagellum. In somatic cell lines, unlike most other tubulins, mammalian δ-tubulin is both cytoplasmic and nuclear and does not co-localize with microtubules. The protein was enriched at the spindle poles during mitosis. The role of mammalian δ-tubulin appears to be distinct from other tubulins (Smrzka et al., 2000).

8.5. MICROTUBULE ASSOCIATED PROTEINS

8.5.1. Tektins

A tektin heteropolymer forms a unique protofilament of flagellar microtubules. Contrary to what is generally assumed, at least one protofilament in the wall of the A tubule is not composed of tubulin. This nontubulin protofilament is primarily composed of tektins, which show some structural similarity to intermediate filament proteins. A 480Å axial periodicity within these ribbons, can be related to the structure of tektin and may determine the large-scale structure of the axoneme in terms of the binding of dynein, nexin and radial spokes to the doublet microtubules (Nojima et al., 1995). Tektins comprise a family of filament forming proteins that are known to co-assemble with tubulins to form ciliary and flagellar microtubules.

They do not show obvious similarity with α and β tubulins. Tektins contain higher cysteine and tryptophan content than desmin and vimentin and have twice α-helical content as tubulin. Besides microtubules, tektins are associated with several mammalian cell lines (Steffen and Linck, 1992). The protofilament ribbon (pf-ribbon) from ciliary/flagellar axonemes can be obtained with treatment of detergents. The Pf ribbons consisting of 3-4 longitudinal protofilaments, are composed of α/β tubulin, tektins A (55-kDA), B (51-kDa), and C (47-kDa), and two polypeptides with molecular masses of 77- and 83-kDa. Although axonemal tektins possess biochemical and immunological properties similar to intermediate filament protein (IFs), sequence analysis of cDNAs encoding the tektins from sea urchin and mammals has revealed that they represent an independent class of filamentous proteins, distinct from the IFs. Analysis of isolated pf-ribbons has shown that the tektins assemble as a heteropolymeric filament within the wall of the axonemal A-tubule. Tektin filaments contribute to the stability and three-dimensional form of doublet microtubules. In addition to their association with doublet microtubules, the tektins have been localized to the basal body, the interphase centrosomes/centrioles and the mitotic/meiotic spindle poles of species ranging from clam to human (Hinchcliffe and Linck, 1998).

The sequences of tektin A (~53-kDa), B (~51-kDa) and C (~47-kDa), predicted from cDNAs from sea urchin (*Strongylocentrotus purpuratus*) sperm flagellar microtubules, and partial sequences from mouse and human were compared (Norrander et al., 1996) The predicted amino acid sequence identities/similarities are: for tektins A and C, 42/54%, for tektins A and B, 34/51%, for tektins B and C, 29/42%, for tektins C and a partial cDNA clone from mouse testis, 55/65%, and for tektins B and a partial cDNA clone from the human brain, 45/47%. The three tektins and human clone possess a sequence repeat RPNVELCRD. The structural pattern of all three tektins is similar to intermediate protein filament proteins. Tektins are predicted to form extended rods composed of two α-helical segments (approximately 180 residues long) capable of forming coiled-coils, which are interrupted by short non-helical linkers. Along each tektin rod, cysteine residues occur with a periodicity of approximately 8 nm, coincident with the axial repeat of tubulin dimers in microtubules. The segment length of tektins AB heterodimers is likely to be 16 nm. Both segments of tektins C may be 24 nm long but one may be 16 nm. On the basis of the available evidence, it was proposed that co-assembly of tektin AB heterodimers with tektins C dimers produces filaments with over all repeats of 8, 16, 24, 32, 40, 48, 96 nm (Norrander et al., 1996).

During development, mammalian *TEKT1* RNA expression is detected in spermatocytes and in round spermatids of mouse testis. Anti-Tekt1 antibodies showed no distinct labeling of any subcellular structure in spermatocytes, whereas in round spermatids anti-Tekt1 antibodies co-localized with anti-ANA antibodies to the centrosome. At a later stage, elongating spermatids displayed a larger area of anti-Tekt1 staining at their caudal ends; and as spermiogenesis proceeds, the anti-Tekt1 staining disappears. These results provide the intra-species evidence to suggest that Tekt1 is transiently associated with the centrosome, and that Tekt1 is one of several tektins to participate in the nucleation of the flagellar axoneme of mature spermatozoa perhaps being required to assemble the basal body (Larsson et al., 2000). Tektin-t from mouse haploid germ cells encodes a protein of 430 deduced amino acids possessing RSNVELCRD, the conserved sequence of tektin family proteins, having a molecular weight of 86-kDa. Tektin-t is localized in the flagella of elongating spermatids from developmental step 15 to maturity (Iguchi et al., 1999) **(Fig.8.4A and B)**.

Another product of *tektin* gene family from the human fetal brain, named as human Tektin1 gene consists of 1375-bp and has a putative open reading frame encoding 418 amino acids. The predicted protein is 48.3-kDa in size, and its amino acid sequence is 82% identical to that of the mouse, rat and dog. One conserved peptide RPNVELCRD is observed at 323-331

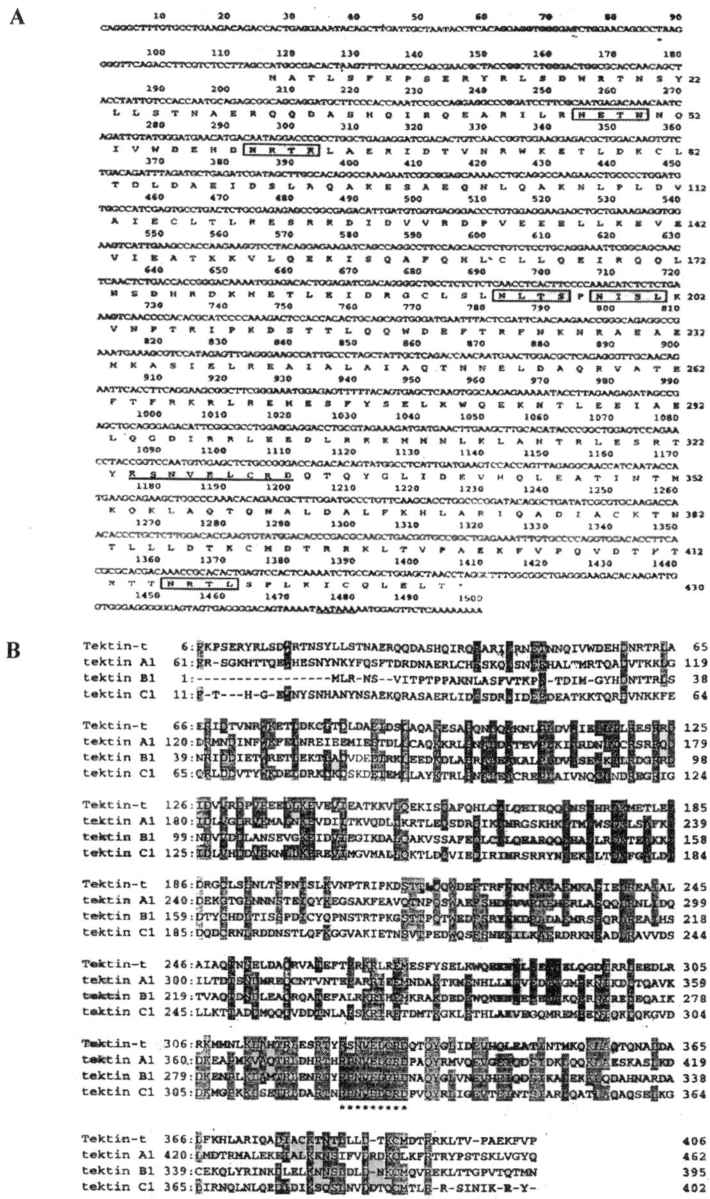

Fig.8.4 (A). The nucleotide and amino acid sequence of germ cell specific tektin-t from mouse testis. Asterisks at stop codons (nt 87-89 and 1407-1409) are indicated. The polyadenylation signal is underlined. Sequence at amino acid 324-332 corresponds to RPNVELCRD repeat present in many tektins. Potential glycosylation sites are boxed. (B). Maximum alignment of deduced amino acid sequence of tektin-t with other tektins from sea urchin sperm flagella. Asterisks indicate the conserved RPNVELCRD repeat. Reproduced with permission from N. Iguchi et al, FEBS Lett 456; 315-21: 1999 © Elsevier.

of the amino acid sequence, which is a prominent feature of tektins and is likely to represent a functionally important protein domain. *Tektin1* gene was mapped to the human chromosome 17-bp and is predominantly expressed in human testis. Its mRNA was localized to spermatocytes and round spermatids in the seminiferous tubules of the mouse testis, indicating that the homologous protein may play a role in spermatogenesis (Xu et al., 2001). A recently discovered member of the *Tektin* gene family from mice is *Tektin3,* which encodes a 1.7-kb transcript in the testes of adult mice with *Tektin3* mRNA expression in late pachytene spermatocytes and early round spermatids. The putative mouse TEKTIN3 protein shares 83.5% overall sequence identity with the human ortholog and includes a fully conserved carboxy terminal nonapeptide signature sequence present in all TEKTIN family members. The *Tektin3* gene maps to murine chromosome 11 in a region that is syntenic to the human 17p12 chromosomal region containing the human *TEKTIN3* gene. Studies demonstrated that TEKTIN3 is a novel evolutionarily conserved male germ cell-enriched protein that might perform important roles in male reproductive development (Roy et al., 2004)

8.5.2. Microtubule Associated Protein-2 like Proteins

Microtubules, composed of α-and β-tubulin subunits, are polymerized and stabilized by a variety of microtubule-associated proteins (MAPs). Several MAPs have been identified in the testis. MAPs from mammalian brain stimulated outer doublet tubulin assembly, decorating the microtubules with fine filamentous projections. Sertoli cell microtubules have also been implicated in directing spermatid translocation in the seminiferous epithelium. In relation to structural elements, tubulin is localized to the region of the axoneme, the centriolar adjunct and the equatorial part of the acrosome, while microtubule associated protein-1 and –2 (MAP1 and MAP2) were localized to the post-nuclear region and in the fibrous sheath of the principal piece of the flagella. A putative MAP2 protein of 255-kDa has been identified in bull sperm. A MAP2-like protein is present in both somatic and germ cells, with a particularly distinct localization within the cytoplasm of primary and secondary spermatocytes during meiotic metaphase. The MAP2 mRNA is processed into at least two variants encoding proteins designated MAP2a, MAP2b, and MAP2c. Of the 5.7-kb of coding sequence in the 9-kb mRNA that encodes MAP2a and MAP2b, a deletion of approximately 4-kb produces mRNA encoding MAP2c, which consists of only the N- and C-terminal regions of MAP2b. Using polyclonal and monoclonal antibodies reactive with MAP2, a 74-kDa protein corresponding to MAP2c was detected in the rat testis. The predominant mRNAs in testis of 6-kb and 2.5-3.5-kb corresponded to MAP2c. A single 6-kb mRNA with the potential to encode MAP2c was detected in enriched preparations of immature Sertoli cells and adult Leydig cells. Round spermatids contained at least two MAP2 mRNA between approximately 2.5 and 3.5-kb size that displayed a stage-specific pattern of expression. In general the MAP2 protein contains domains that mediate binding to microtubules, to calmodulin, and the regulatory subunit of the cAMP-dependent kinase (PKA). Involvement of MAP2 in microtubule bundling is suggested by the formation of stable bundles of microtubules (Loveland et al., 1996).

The phosphorylation state of bovine sperm MAP2 is uniquely sensitive to regulation by intracellular pH (pHi), calcium, isobutyl-3-methylxanthine (MIX), H-8, and fluoride. The phosphoprotein co-sediments along with the particulate sperm heads during sub-cellular fractionation, and is not detectable in other sperm fractions. Sperm MAP2 is phosphorylated on serine residues, resulting in change in electrophoretic mobility. The co-localization of sperm MAP2 with the head fraction and the unique sensitivity of its phosphorylation level to modulators, which are known to regulate capacitation and the acrosome reaction, suggest that sperm MAP2 phosphorylation may be an intermediate step in the regulation of one or both of these sperm processes (Carr and Acott, 1990).

Tau: Tau is a heat stable MAP previously believed to be limited in its expression in mammals to the nervous system. It has been identified in rat and bovine testis, a unique non-neuronal location. Total rat testis RNA showed a 1.7-kb that hybridizes with a 51-nucleotide oligomer complementary to a conserved portion of the tau transcript. In the bull testis, tau is localized to the spermatid manchette. Tau most likely plays a structural role in the manchette; however, tau immunoreactivity also is observed in late stage 1 spermatids prior to manchette formation, suggesting that tau may serve a function in manchette assembly (Ashman et al., 1992).

E-MAP77: In echinoderms the major microtubule-associated protein is also a 77-kDa, WD repeat protein, called EMAP. The EMAP-related proteins have been identified in sea urchins, starfish, sanddollars, and humans. Unlike MAP-2, MAP-4, or tau proteins, EMAP binding to microtubules is not lost by cleavage of tubulin with subtilisin. In addition to binding to the microtubule polymer, EMAP binds to tubulin dimmers in a 1:1 molar ratio. These results indicated that EMAP, unlike brain microtubule-associated proteins promotes microtubule dynamics (Hamill et al., 1998).

The 77/83-kDa polypeptides, termed Sp77 and Sp83 from sperm flagella of the sea urchin *Stronglyocentrotus purpuratus*, are structural proteins associated with stable doublet microtubules, and may be components of basal bodies and centrioles of sea urchins and mammalian cells (Hinchcliffe and Linck, 1998). Studies have shown that like tektins, present in the basal bodies and centrosomes/centrioles of cells ranging from clam to human, the Sp83 decorates the spindle poles in sea urchin zygotes, and the interphase centrosome and spindle poles in CHO cells. The 80-kDa-polypeptide is associated with the centrioles themselves. Thus Sp77 and Sp83 can be considered structurally defined proteins within the axoneme, exclusive to the stable region of the flagellar A tubules. The Sp77/83 are completely solubilized from the taktin filament by treatment with 0.5% sarkosyl and 2M-urea and are distinct from the tektin proteins.

8.5.3. E-MAP-115/ MTEST 60

A microtubule-associated protein E-MAP-115, originally characterized from HeLa cells, is predominantly expressed in cells of epithelial origin. It is involved in the regulation of cell polarization. The presence of E-MAP-115 in the mouse and rat seminiferous epithelium indicates its distinct association with the spermatid manchette. At steps 15-16 when manchette had been disassembled, the E-MAP-115 disappeared. The E-MAP-115 is a single polypeptide whose gene appears in two mRNA species (3.4 and 2.4-kb). The MTEST 60, a spermatid-specific transcript, showed a 100% homology over region of 68-193-bp of E-MAP-115 sequence. The reported specific localization of E-MAP-115 to the spermatid manchette strongly supports its role as a regulator of cell polarization (Penttila et al., 2003).

8.5.4. TBP-1-Like Subfamily with ATPase and Protease Domains

Rivkin et al (1997) cloned a TBP-1 (tat-binding protein-1; designated rat testis TBP-1 (rtTBP-1), a new member of the family of putative ATPases associated with the 26S proteasome complex. The 1.63-kb rtTBP-1 cDNA encoded a 49-kDa protein with 99% amino acid identity to human TBP-1 protein. The RtTBP-1 protein present in the rat spermatid manchette and outer dense fibers of the developing sperm contains a heptad repeat of six leucine-type zipper

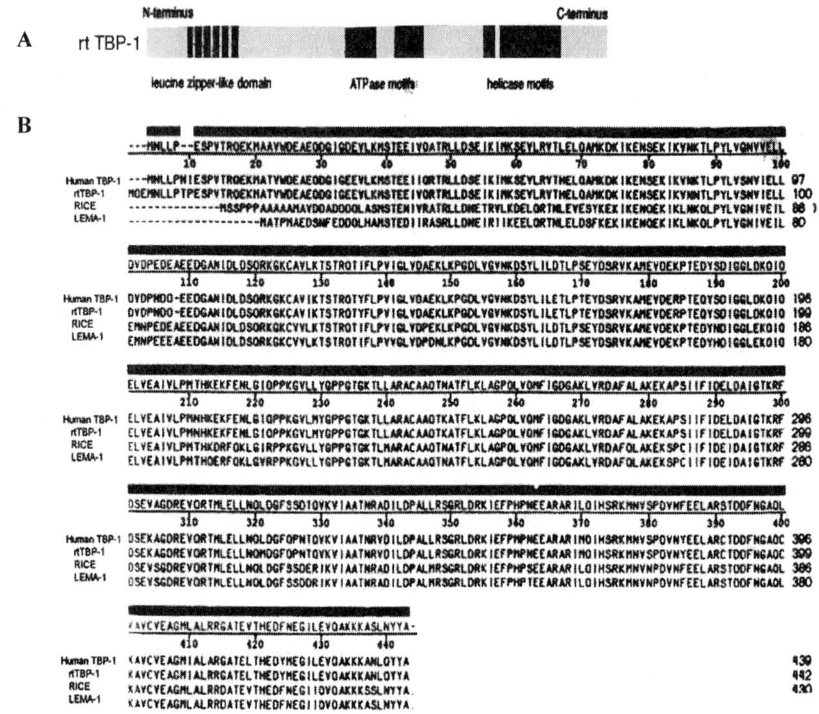

Fig.8.5. (A) Schematic presentation of rat tat binding protein-1 (rt TBP-1) with various motifs. (B) Amino acid sequence of rt TBP-1 and its comparison with human TBP-1 and other proteins. Reproduced with permission from E. Rivkin et al, Mol Reprod Dev 48;77-89:1997© John Wiley & Sons. Inc http://www3.interscience.wiley.com/cgi-bin/jabout/37692/ProductInformation.html.

fingers at the amino terminal end and highly conserved ATPase and DNA/RNA helicase motifs towards the carboxyl terminal region. Rat testis TBP-1 consists of a triplet with a molecular mass of 52-48-kDa and acidic pI (5.0-5.9) **(Fig.8.5)**. A identical immunoreactive triplet was detected in pachytene spermaocytes, round spermatids and epididymal sperm. RtTBP-1 immunoreactive sites co-localize with α-tubulin-decorated manchettes of elongating spermatids. The TBP-1 is one of the 20 subunits forming the 19S regulatory complex of the 26S proteasome (see Chapter 31) (Mochida et al., 2000).

8.5.5. CLIP50

Another protein component of peri-nuclear ring termed CLIP-50 because of its high similarity at the C-terminal region of the microtubule-binding protein CLIP-170/restin was characterized (Tarsounas et al., 2001). The CLIP-50 lacks the characteristic microtubule-binding motif, but retains a portion of the predicted coiled-coiled domain and the metal-binding motif. The CLIP-50 transcript and protein are abundant in testis and also expressed in heart, lung, kidney, brain and skin. In the spermatids, CLIP-50 protein localizes specifically to the centriolar region where the sperm tail originates, and to the peri-nuclear ring from which the manchette emerges. The CLIP-50 staining is retained in the ring throughout its migration over the surface of the nucleus, which accompanies the nuclear shaping into its characteristic sperm configuration. The localization pattern indicates a specific function for this CLIP derivative during mouse spermiogenesis.

8.5.6. CAS

Human CAS cDNA encodes a 971-aa open reading frame that is homologous to the yeast gene CSE1. The CSE1 is involved in chromosome segregation and is necessary for B-type cyclin degradation in mitosis. The CAS levels are high in proliferating and low in non-proliferating cells. The CAS is an approximately 100-kDa protein present in the cytoplasm of proliferating cells at levels between $2X10^5$ and $1X\ 10^6$ molecules per cell. The intracellular distribution of CAS resembles that of tubulin. In interphase cells, anti–CAS antibody shows microtubule-like patterns and in mitotic cells it lebels the mitotic spindle. The CAS is diffusely distributed in the cytoplasm with only traces present in tubulin paracrystals or bundles. Thus, CAS appears to be associated with but not to be an integral part of microtubules. Elevated level of CAS is observed in proliferating cells such as testicular spermatogonia, in the basal layer cells of the colon, and in axons and purkinje cells in the cerebellum, which contain many microtubules. The cellular location of CAS is consistent with an important role in cell division as well as in ciliary's movement and vesicular transport (Scherf et al., 1996).

8.6. RNA BINDING PROTEINS IN MICROTUBULES

In addition to Spnr, TB-RBP and other RNA binding proteins present in sperm (Rohwedder et al., 1996), human spermatozoa are positive for HLA class I mRNA for β-actin, retinoblastoma (RB), CD4, or kappa light chain genes in the sperm. Physiological role of HLA transcripts in sperm is unknown (Chiang et al., 1994).

8.6.1. Spermatid Perinuclear RNA Binding Protein (Spnr)

Murine RNA-binding protein is localized to cytoplasmic microtubules. Six independent clones that encode Spnr, a spermatid peri-nuclear RNA binding protein, have been observed in sperm. Spnr is a 71-kD protein that contains RNA binding domains. The Spnr mRNA is expressed at high levels in the testis, ovary, and brain, and is present in multiple forms in those tissues. In testis it is expressed exclusively in post-meiotic germ cells and localized to the manchette of developing spermatids (Schumacher et al, 1995; Schumacher et al., 1998). Despite the highly abnormal manchettes and microtubule aggregates formed in *azh*, *hop-sterile*, *tw2*, and *tw8* mutants, Spnr remains associated with the manchettes. Localization of Spnr to the abnormal manchettes suggested that Spnr is tightly bound to the manchette. Spnr could bind manchette microtubules directly, or it could bind indirectly via an interaction with a microtubule-associated protein (MAP). In addition to its function as an RNA-binding protein, Spnr is also a bona fide MAP. Spnr (GT) mutants show a high rate of mortality, reduced weight, and an abnormal clutching reflex. In addition to minor anatomical abnormalities in the brain, males exhibit defects in spermatogenesis. The sperm from mutant males display defects in the flagellum and consequently show decreased motility and transport within the oviducts. Thus Spnr (GT/GT) mutant male mouse is a unique model for some human male infertility cases (Pires-daSilva et al., 2001).

8.6.2. Testis-Brain RNA Binding Protein

Numerous functions have been proposed for the testis-brain RNA-binding protein and its human homolog, Translin, ranging from mRNA transport and translational regulation to DNA rearrangement and repair. The TB-RBP interacts with the TER-ATPase in vitro and in vivo. TB-RBP co-localizes with actin in the cytoplasm of male germ cells and serves as a link in attaching specific mRNAs to cytoskeletal structures. It suggests an involvement for the ubiquitously expressed TER ATPase in intracellular and intercellular mRNA transport (Hecht, 2000; Wu et al., 1999). The intracellular association between TB-RBP and specific target mRNAs suggests an involvement of TB-RBP in microtubule-dependent mRNA transport in the cytoplasm of cells (Morales et al., 1998; Wu and Hecht 2000). The TB-RBP gene encodes three mRNAs of 3.0, 1.7, and 1.0-kb, which differ only in their UTRs. The 1-kb mRNA predominates in testis, while somatic cells preferentially express the 3.0-kb TB-RBP mRNA. The 1-kb mRNA is translated several-fold more efficiently than the 3-kb TB-RBP in rabbit reticulocyte lysates. Polyadenylation factor, CstF 64 is abundant in the testis and preferentially binds to a distal site in the TB-RBP pre-mRNA that produces the 3-kb TB-RBP. Moreover, up-regulation or over-expression of CstF 64 increases the poly (A) site selection for the 1-kb TB-RBP mRNA. It has been proposed that the level of CstF 64 modulates the level of TB-RBP synthesis in male germ cells by an alternative processing of the TB-RBP pre-mRNA (Chennathukuzhi et al., 2001; Wu et al., 1999).

8.6.3. Fragile X Mental Retardation-1 Protein

Fragile-X-mental retardation-1 protein (FMRP) is the archetype of a class of cytoplasmic mRNA-binding protein that includes the fragile X-related 1 and 2 proteins (FXR1P and FXR2P). In mouse the highest level of FXR1P is found in the adult testis. In young animals, changes in mRNA spliced variants and their corresponding protein isoforms occur during spermatogenesis. The FXR1P is associated with microtubule elements. Since the cytoskeletal framework is implicated in cellular plasticity as well as in mRNA transport new possibilities for the function(s) of the FXR proteins exist (Huot et al., 2001).

8.7. OTHER MICROTUBULE ASSOCIATED PROTEINS

RS20: RS20 is a 20-kDa microtubule-associated protein found in several human tissues. The intercellular distribution of RS20 in spermatozoa from both human and rat shows that RS is mainly a flagellum protein, but in rats, it is also abundant in the sperm head. In the sperm tail, RS20 was co-localized with β-tubulin, the major component of the axoneme, suggesting that RS20 is also associated with the flagellum axoneme (Whyard et al., 2000).

MAST205: MAST205 colocalizes with the spermatid manchette. Sequencing of full-length cDNA clones encoding MAST205 revealed it to be a novel serine/threonine kinase with a catalytic domain related to those of the A and C families. The testis specific MAST205 mRNA increases during prepuberal testis development, peaking at the spermatid stage. The MAST205 protein complex may function to link the signal transduction pathway with the organization of manchette microtubules (Walden and Cowan, 1993) (see protein kinases).

Sperm contain two isozymes of creatinine kinase (Crk). In sea urchin sperm, one enzyme is located in sperm head and another in the tail. Tail isozyme exists along the entire flagellum.

This unusual Crk isozyme of Mr 145 kDa, is a component of the flagellar axoneme. Thus, although the two sperm isozymes have similar kinetic properties, they differ in affinity for microtubules, a characteristic that may determine the regional differentiation needed for establishing a phosphocreatine shuttle (Tombes et al., 1988).

VCY2: The function of VCY2 is still unknown. An interacting partner of VCY2 is VCY2IP-1 that encodes an ORF of 1059 amino acids. The amino acid sequence of VCY2IP-1 shows 59.3% and 41.9% homology to two human microtubule-associated proteins (MAPs), MAP1B and MAP1A, respectively. The VCY2IP-1 has an extensive homology to the N-terminus and C-terminus regions of MAP1B and MAP1A, placing it within a large family of MAPs. The VCY2IP-1 is localized to chromosome 19p13.11. The VCY2IP-1 gene spans 15-kb and consists of seven exons. A 3.2-kb VCY2IP-1 transcript is predominantly expressed in human testis. The VCY2IP-1 transcripts are present in germ cells whereas VCY2 and VCY2IP-1 transcripts are present in human ejaculated spermatozoa (Wong et al., 2004).

Bub1: The cDNA sequences from several organisms with highly significant predicted protein sequence homologies to *S. Cerevisiae* Bub1p, a protein required for function of the spindle assembly checkpoint in budding yeast have been characterized. Multiple-tissue Northern analysis indicated conservation of expression pattern in mouse and human with markedly high mRNA levels in testis. Checkpoint protein gene products from human BUB1 and MAD2 reveals an expression pattern with common tissue distribution, consistent with their roles in a common pathway. In addition an mRNA from fibroblast transformation system encodes rat BUB1 (Pangilinan et al., 1997).

Mitotic Spindle Associated Protein 126: The cDNA of hMAP126 mitotic spindle associated protein has been obtained from human testis. This gene encodes a protein of 1120 amino acids. The hMAP1 is abundantly expressed in the testis. Furthermore, hMAP 126 was identified to be post-translationally modified and phosphorylated by $p34^{cdc2}$ kinase in vitro. The hMAP-126 may be involved in the functional and dynamic regulation of mitotic spindles (Chang et al, 2001).

Glycolytic Enzymes: Mammalian sperm is active in glycolysis and uses energy substrates required during sperm motility. Hence several unique isoforms of proteins of glycolytic pathway and respiratory energy metabolism are present in sperm flagella. Phosphoglycerate hydrolase is most abundant in mature spermatozoa and in residual cytoplasmic bodies detached from elongating spermatids. Enolase is localized mostly to the tail of mature spermatozoa and only α isoform of derived cytosolic enolase (4.9%) co-sediments with microtubules stabilized in the presence of taxol (Gitlits et al., 2000). Other isozymes, which are highly specific to sperm and present in sperm flagella include LDH-C_4, PGK2, PDH2 and sperm specific hexokinase (see Chapter31).

8.8. CENTROSOME

Microtubules are generally nucleated from a specific intracellular site, known as microtubule-organizing center (MTOC). The MTs are nucleated at their minus end. In animal cells, most microtubules are anchored at one end in a structure called the centrosome. During mitosis, the centrosome duplicates to form the spindle poles which is the single well-defined MTOC located near the nucleus. Electron microscopy revealed that the centrioles are composed of nine triplet microtubules arranged to form a short cylinder, and are surrounded by a dense amorphous material (the pericentriolar material, or centrosome matrix). Cells lacking centrioles contain a focus of microtubule growth, with antigenic determinants common to those found in the centrosome. The centrosome nucleates the growth of microtubules and thereby determines the number, and the distribution of hundreds of microtubules in an animal cell. The centrosome and its associated microtubules direct the events of mitosis, and they play a central role in the organization of the interior of the cell during interphase. The centrosome undergoes characteristic structural changes during the cell cycle. Understanding the molecular events that underlie the observed function of the centrosome is matter of future investigations.

Centriole and Spindle Pole Body: Many eukaryotic cells do not contain the classic centrosome structure found in animal cells, although they contain functionally analogous organells. In *S. Cerevisiae*, for instance, microtubules are organized by a structure called spindle pole body embedded in the nuclear envelope and contain no centrioles. The centriole bears a structural resemblance to the basal body. Functions of centrioles remain unknown. Many of the spindle poles formed do not contain centrioles, but appear to function normally, suggesting that centrioles do not play an essential role. The animal cell centrosome and the spindle pole body of yeast are structurally distinct, but functionally similar.

Centrosome Assembly and Duplication: Earlier it was thought that the centrosome is a relatively static structure that simply nucleates microtubules. Now, it is becoming clear that the centrosome is a dynamic organelle that is constantly changing during the cell cycle. The centrosome must duplicate once and only once in order to assemble a bipolar spindle and to segregate the chromosomes accurately. The mechanism by which centrosome duplication occurs is unknown. Formation of a new centrosome normally occurs in association with a pre-existing centrosome. But in some cases formation of new centrosome can occur independently of a pre-existing centrosome. Evidences suggest that proteins present in the cytoplasm can assemble to form a new MTOC (Kellogg et al., 1994). Mutation in Cdc31 of budding yeast affects both mitotic and meiotic division and the defect lies at cytoplasmic side of the half bridge of the spindle body responsible for spindle pole body duplication.

Centosome in Microtubule Nucleation: The role of centrosome is MT nucleation is largely unknown. In all cells, the centrosome catalyzes the formation of small microtubule seeds that are capable of microtubule nucleation at tubulin concentrations where individual microtubules are unstable and can exist only transiently (Mitchison and Kirschner, 1984). The centrosome appears to have distinct microtubule-nucleating sites that are regulated during the cell cycle. Entry into mitosis in animal cells is associated with a several fold increase in the number of microtubules nucleated by the centrosome.

In vertebrate cells, the increase in the number of microtubules begins at about the time of prophase, and is accompanied by a dramatic increase in both the amount of pericentriolar material and the level of phosphorylation of centrosomal proteins. Since cyclin-dependent protein kinase (Cdk) complex can induce mitosis it is likely that they are responsible for causing the increased ability of the centrosome to nucleate microtubules. It is, however, not clear whether these Cdks directly phosphorylate centrosomal proteins or instead initiate a cascade of other events that lead to the observed changes in the centrosome. However, cyclin A has been localized to the centrosome during prophase, and cyclin B and protein kinase $P34^{cdc2}$ are localized at or near centrosome during G2. These observations suggested that cyclin-dependent kinase complexes might directly phosphorylate centrosomal proteins. However, GTP-bound -Ran induces microtubule and pseudo-spindle assembly in mitotic egg extracts in the absence of chromosome and centrosomes, and that chromosomes induce the assembly of spindle microtubules in these extracts through generation of Ran-GTP. The Ran-GTP has independent effects on both the nucleation activity of centrosomes and the stability of centrosomal microtubules, and the inhibition of Ran-GTP production prevents assembly of a bipolar spindle in M-phase extracts (Carazo-Salas et al, 2001). A complete understanding of microtubule nucleation requires the identification of the proteins that are directly involved. Several lines of evidence support the view that γ-tubulin is involved in microtubule nucleation. However, like other cellular process, it is necessary to reconstitute a system in vitro with purified components to justify the function. During recent years different centrosomal proteins have been identified in different species. It will be important to identify the homologues of centrosomal proteins in diverse species, and allow their functions to be characterized in organism, experimentally accessible.

8.9. CENTROSOME PROTEINS

While the development of protocols for centrosome isolation has not been successful but has helped in characterization and in identification of a number of centrosomal proteins. Although a number of centrosomal proteins have been identified a total of about 10 proteins, from a wide range of organisms were reviewed, and the list is growing (Kellogg et al., 1994). Sequence analysis indicates that the Cdc31 protein belongs to a family of Ca^{2+} binding proteins that includes centrin and calmodulin. The Cdc31 is 50% identical to centrin, a highly conserved centrosomal protein. Other genes involved in pole body duplication include *KAR1*, *mps1* and *NDC1* gene. Pericentrin, a centrosomal protein from human is probably the best characterized of the centrisomal proteins. The DMAP190 is a 190kDa protein from *Drosophila* microtubule. Another protein that interacts with microtubules is LK6. The LK6 antibodies recognize a 180-kDa protein in crude extracts and an approximately 200-kDa protein in the fraction of proteins that bind to a microtubule affinity column and localized to centrosome. The sequence of the LK6 has shown it to be a kinase that phosphorylates itself in vitro.

8.9.1. γ-Tubulin in MT Nucleation

γ-Tubulin: γ-Tubulin is an essential component that links the centrosome and the microtubule and suggests a direct role in microtubule nucleation in Xenopus eggs (Stearns and Kirschner, 1994). The role of γ-tubulin in the nucleation of microtubules was first suggested by inferences drawn from genetic and immunocytochemical observations (Joshi et al., 1992). γ-tubulin in humans centrosome requires at least threee proteins h102p, h104p and h76p for microtubules nucleation (Fava et al., 1999). Recently several novel proteins complexes found in centrosome have been implicated in the role of γ-tubulin with the nucleation of microtubules. γ-tubulin is a member of the tubulin family whose sequence is highly conserved from yeast to man. This motor microtubule protein is localized to the microtubule organizing centers. A mutation in the gene encoding γ-tubulin produces a microtubuleless mitotic arrest in *Aspergillus nidulans*. A polyconal antibody that binds to a highly conserved segment of γ-tubulin revealed that the this antibody binds to native centrosomes at all phases of the cell cycle. In the presence of the γ-tubulin antibody, microtubules fail to regrow into cytoplasmic arrays after depolymerization induced by nocodazole or cold. Moreover, cells injected with Ab immediately before or during mitosis failed to assemble a functional spindle. This shows that γ-tubulin is required for microtubule nucleation throughout the mammalian cell cycle.

Two genes corresponding to γ-tubulin have been identified in *Drosophila*: γ tubulin 23C (γ-*Tub23C*) and γ-*Tub37CD* (Wilson et al., 1997). Analysis showed that γ-Tub37CD and γ-Tub23C are differentially expressed during gametogenesis and development. During oogenesis, γ-Tub23C was detected by centrosomes and in the cytoplasm of mitotic germ cells, but not in germ cells following completion of mitosis. In contrast, γ-Tub37CD was not detected in proliferating germ cells, but appeared to accumulate in germ cells during egg chamber development and none of the γ-tubulin isoform was detected at the anterior or posterior poles of developing oocytes. During spermatogenesis, only γ-Tub23C was detected at centrosomes, where it showed cell cycle- and differentiation dependent organization. Following reconstitution of a canonical centrosome at fertilization, only γ-Tub37CD was detected at centrosomes in syncytial embryos, but both γ-Tub37CD and γ-Tub23C were detected at centrosomes in cellularized embryos. Co-localization of these two isoforms suggests that γ-Tub23C and γ-Tub36CD both contain structural features of γ-tubulins essential for localization to centrosomes (Wilson et al., 1997). To assess the role of γ-tubulin in spindle assembly in vivo, Sampaio et al (2001) followed meiosis progression in γ-Tub23(PI) mutant spermatocytes. Study suggested that microtubule polymerization or stabilization from the centrosome might be possible in a γ-tubulin independent manner in *Drosophila* spermatocytes. However, γ-tubulin seemed to be essential for spindle assembly in these cells (Sampaio et al., 2001).

Xgrip210 and γ-Tubulin Ring Complex (γ-TuRC): The γ-tubulin ring complex (γ-TuRC) isolated from the cytoplasm of vertebrate and invertebrate cells is a microtubule nucleator in vitro. The γ-TuRC is a lock-washer shaped structure and topped with a cap. Microtubules are nucleated from the uncapped side of the γ-TuRC. Hence, the cap structure of the γ-TuRC is distal to the base of the microtubules, giving the end of the microtubule the shape of pointed cap. The Xgrip210, a subunit of *Xenopus* γ-TuRC is a conserved centrosomal protein that is essential for the formation of γ-TuRC. The Xgrip210 is localized to the ends of microtubules nucleated by the γ-TuRC, which is localized more distally toward the tip of the γ-TuRC cap structure, than that of γ-tubulin. Immunodepletion of Xgrip 210 blocks not only the assembly of the γ-TuRC, but also the recruitment of γ-tubulin and its interacting protein, Xgrip109, to the centrosome. Thus Xgrip 210 is a component of the γ-TuRC cap structure that is required for the assembly of the γ-TuRC (Martin et al., 1998; Zhang et al., 2000).

8.9.2. Centrin and Other Proteins in Centrosome

The *Vfl2* gene in *Chlamydomonas* encodes centrin (also known as caltractin), a protein originally identified in basal body in the alga, *Tetraselmis* but later found in all organisms. Centrin is a member of a family of EF-hand-Ca^{2+} binding proteins that includes calmodulin and Cdc31. In *Drosophila* several proteins form centrosomal complexes. Two proteins have been named DMAP60 and DMAP85, respectively. The cDNA that encodes DMAP60 has been cloned. It contains six consensus sites for phosphorylation by cyclin-dependent kinases. The DMAP60 and γ-tubulin interact with each other. The cDNA encoding human centrin contains the open reading frame of 516 bp and predicts a product of 172 amino acids with a molecular mass of 19,528Da and a pl 4.61. The human centrin sequence has four putative calcium binding domains as defined by the EF-hand consenus. Human centrin is a protein of approximately 20,000 M(r) as predicted from the cDNA clone. In HeLa cells centrin is localized at the centrosome of intephase cells and redistributes to the region of the spindle poles during mitosis. Thus centrin is a ubiquitous component of centrosomes and mitotic spindle poles of diverse organisms and suggests that centrin plays a role in centrosome separation at the time of mitosis (Errabolu et al., 1994).

h76p: The γ-tubulin and other associated proteins h103p (hGCP2), h104p (hGCP3) and h76p (hGCP4) besides recently discovered protein complexes seem to be essential in MT nucleation. These proteins are also present in soluble complexes. Several pieces of evidence indicate that h76p, besides h103p and h104p is involved in microtubule nucleation. γ-tubulin complexes containing h76p bind to microtubules. The h76p is recruited to the spindle poles and to Xenopus sperm basal bodies and is necessary for aster nucleation by sperm basal bodies. The h76p also shares partial sequence identity with human centrosomal proteins h103p and h104p, suggesting a common protein core. Hence, human γ-tubulin appears associated with at least three evolutionary related centrosomal proteins, whose functions at the molecular level are unknown (Fava et al., 1999).

Spc98: Since crucial functions such as microtubules nucleation can be maintained by common mechanism, hence components Spc98p and Spc97p present in trimeric complex formed by γ-tubulin of spindle pole body in yeast must be present in divergent species. The deduced protein sequence of human Spc98p showed 22% identity and 45% similarity with yeast homologue, and is generated by alternate splicing at 3'end. Human Spc98p like yeast is concentrated at centrosome although also present in cytosol complex. Both human γ-tubulin and human Spc98p are in same complex. Anti-Spc98p antibody inhibited microtubule nucleation suggesting that Spc98p is required for microtubule nucleation (Tassin et al., 1998).

Abnormal Spindle (Asp): Abnormal spindle (Asp) is a 220-kD microtubule-associated protein from *Drosophila* that has been suggested to be involved in microtubule nucleation from the centrosome. The Asp is enriched at the poles of meiotic and mitotic spindles and localizes to the minus ends of central spindle microtubules. Localization of Asp to these structures is independent of a functional centrosome and Asp is not an integral centrosomal protein. In *asp* mutants, microtubule nucleation occurs from the centrosome, and γ-tubulin localizes correctly however, spindle pole focussing and organization are severely affected. Wakefield et al (2001) determined the role of Asp in the absence of centrosomes. Phenotypic analysis in *asp* and asterless double mutants shows that Asp is required for the aggregation of microtubules into focused spindle poles, reinforcing the conclusion that its function at the spindle poles is independent of any putative role in microtubule nucleation. Study also suggested that Asp has a role in the formation of the central spindle. The inability of Asp mutants to correctly organize the central spindle leads to disruption of the contractile ring machinery and failure in cytokinesis (Wakefield et al., 2001).

8.10. REFERENCES

Alsop GB, Zhang D. Microtubules are the only structural constituent of the spindle apparatus required for induction of cell cleavage. J Cell Biol 2003;162:383-90.

Ashman JB, Hall ES, Eveleth J, Boekeheide K. Tau the neuronal heat-stable microtubule-associated protein, is also present in the cross-linked microtubule network of the testicular spermatid manchette. Biol Reprod 1992; 46: 120-9.

Bozkurt HH, Woolley DM. Morphology of nexin links in relation to interdoublet sliding in the sperm flagellum. Cell Motil Cytoskeleton 1993; 24: 109-18.

Bressac C, Bre MH, Darmanaden-Delorme J, Laurent M, Levilliers N, Fleury A. A massive new posttranslational modification occurs on axonemal tubulin at the final step of spermatogeneis in *Drosophila*. Eur J Cell Biol 1995; 67 : 346-55.

Buttgereit D, Renkawitz-Pohl R. Expression of β1 tubulin (β Tub56D)in *Drosophila* testis stem cells is regulated by a short upstream sequence while intron elements guide expression in somatic cells. Mole Gen Genet 1993; 241: 263-70.

Carazo-Salas RE, Gruss OJ, Mattaj IW, Karsenti E. Ran-GTP coordinates regulation of microtubules nucleation and dynamics during mitotic spindle assembly. Nat Cell Biol 2001; 3: 228-34.

Carr DW, Acott TS. The phosphorylation of a putative sperm microtubule-associated protein 2 (MAP2) is uniquely sensitive to regulation. Biol Reprod 1990; 43: 795-805.

Chang LC, Yew PR. Regulation of nuclear transport and degradation of the Xenopus cyclin dependent kinase inhibitor, p27Xic1. J Biol Chem 2001; 276: 1610-17.

Chennathukuzhi VM, Lefrancois S, Morales CR, Syed V, Hecht NB. Elevated levels of the polyadenylation factor CstF 64 enhance formation of the 1 kb testis brain RNA-binding protein (TB-RBP) mRNA in male germ cells. Mole Reprod Dev 2001; 58: 460-9.

Chiang MH, Steuerwald N, Lambert H, Main EK, Steinleitner A. Detection of human leukocyte antigen class I messenger ribonucleic acid transcripts in human spermatozoa via reverse transcription polymerase chain reaction. Fertil Steril 1994; 61: 276-80.

Distel RJ, Kleene KC, Hecht NB. Haploid expression of a mouse testis α-tubulin gene. Science 1984; 224: 67-70.

Errabolu R, Sanders MA, Salisbury JL. Cloning of a cDNA encoding human centrin, an FF-hand protein of centrosomes and mitotic spindle poles. J Cell Sci 1994; 107(ptl): 9-16.

Fackenthal JD, Hutchens JA, Turner FR, Raff EC. Structural analysis of mutations in the *Drosophila* β2 tubulin isoform reveals regions in the β-tubulin molecule required for general and for tissue-specific microtubule functions. Genetics 1995; 139: 267-86.

Fava F, Raynaud-Messina B, Leung-Tack J, Mazzolini L, Li M, guillemot JC, Cachot D, Tollon Y, Ferrara P, Wright M. D, Tollon Y, Ferrara P, Wright M. Human 76p: a new member of the γ-tubulin associated protein family. J Cell Biol 1999; 147: 857-68.

Fouquet JP, Edde B, Kann ML. Wolff A, Desbruyeres E, Denoulet P. Differential distribution of glutamylated tubulin during spermatogenesis in mammalian testis. Cell Motil Cytoskeleton 1994; 27: 49-58.

Fuller MT. Caulton JH, B. Hutchens JA, Kaufman TC, Raff EC. Genetic analysis of microtubule structure: a β-tubulin mutation causes the formation of aberrant microtubules in vivo and in vitro. J Cell Biol 1987; 104: 385-94.

Gagnon C, White D, Cosson J, Huitorel P, Edde B, Desbruyeres E, Lafanechere L, Multigner L, Job D, Cibert C. The polyglutamylated lateral chain of α-tubulin plays a key role in flagellar motility. J Cell Sci 1996; 109 (Pt 6): 1545-53.

Gitlits VM, Toh BH, Loveland KL, Sentry JW. The glycolytic enzyme enolase is present in sperm tail and displays nucleotide-dependent association with microtubules. Eur J Cell Biol 2000; 79: 104-11.

Goldsmith M, Yarbrough L, van der Koov D. Mechanics of motility: distinct dynein binding domains on α- and β–tubulin. Biochem Cell Biol 1995; 73: 665-71.

Hamill DR, Howell B, Cassimeris L, Suprenant KA. Purification of a WD repeat protein, EMAP, that promotes microtubule dynamics through an inhibition of rescue. J Biol Chem 1998; 273: 9285-91.

Hecht N. Gene expression during spermiogenesis In: Cellular and molecular events in spermiogenesis (Hamilton DW and Waites GMH eds), Cambridge University Press, 1990, pp 265-83.

Hecht NB. Intracellular and intercellular transport of many germ cell mRNAs is mediated by the DNA- and RNA-binding protein, testis brain-RNA binding protein (TB-RBP). Mol Reprod Dev 2000; 56(2 Suppl): 252-3.

Hermo L, Oko R, Hecht NB. Differential post-translational modifications of microtubules in cells of the seminiferous epithelium of the rat: a light and electron microscope immunocytochemical study. Anat Rec 1991; 229: 31-50.

Hinchcliffe EH, Linck W. Two proteins isolated from sea urchin sperm flagella: structural components common to the stable microtubule of axonemes and centrioles. J Cell Sci. 1998; 111: 585-595.

Hoyle HD, Raff EC. Two Drosophila β tubulin isoforms are not functionally equivalent. J Cell Biol 1990; 111: 1009-26.

Huot ME, Mazoui R, Leclerc P, Khandjian EW. Developmental expression of the fragile X-related 1 proteins in mouse testis: association with mcirotubule elements. Hum Mol Genet 2001; 10: 2803-11.

Hutchens JA, Hoyle HD, Turner FR, Raff EC. Structurally similar Drosophila α-tubulins are functionally distinct in vivo. Mol Biol Cell 1997; 8: 481-500.

Iguchi N, Tanaka H, Fujii T, Tarmura K, Nojima H, Nishimune Y. Molecular cloning of haploid germ cell specific tektin cDNA and analysis of the protein in mouse testis. FEBS Lett 1999; 456: 315-21.

Jensen-Smith HC, Luduena RF, Hallworth R. Requirement for the beta-I and beta-IV tubulin isotypes in mammalian cilia.Cell Motil Cytoskeleton. 2003; 55: 213-20.

Joshi HC, Palacios MJ, McNamara L, Cleveland DW. γ-tubulin is a centrosomal protein required for cell cycle-dependent microtubule nucleation. Nature 1992; 356: 80-82.

Kaltschmidt B, Glatzer KH, Michiels F, Leiss D, Renkawitz-Pohl R. During Drosophila spermatogenesis β1, β2 and β3 tubulin isotypes are cell-type specifically expressed but have the potential to co-assemble into the axoneme of transgenic flies. Eur J. Cell Biol 1991; 54: 110-20.

Kann ML, Prigent Y, Levilliers N, Bre MH, Fouquet JP. Expression of glycylated tubulin during the differentiation of spermatozoa in mammals. Cell Motil Cyoskeleton 1998; 41 : 341-52.

Kellogg DR, Moritz M, Alberts BM. The centrosome and cellular organization. Annu. Rev. Biochem. 1994; 63: 639-74.

Kemphues KJ, Kaufman TC, Raff RA, Raff EC. The testis-specific β-tubulin subunit in *Drosophila Melanogaster* has multiple functions in spermatogenesis. Cell 1982; 31 (3 Pt 2): 555-70.

Kuriyama R, Levin A, Nelson D, Madl J, Frankfurter A, Kimble M. Monoclonal anti-dipeptide antibodies cross-react with detyrosinated and glutamylated forms of tubulin. Cell Motil Cytoskeleton 1995; 30: 171-82.

Larsson M, Norrander J, Graslund S, Brundell E, Linck R, Stahl S, Hoog C. The spatial and temporal expression of Tektl, a mouse tektin C homologue, during spermatogenesis suggest that it is involved in the development of the sperm tail basal body and axoneme. Eur J Cell Biol 2000; 79: 218-25.

Loveland KL, Hayes TM, Meinhardt AS. Zlatic, KS, Parvinen M, de Kretser DM, McFarlane JR. Biol Reprod 1996; 54: 896-04.

Martin OC, Gunawardane RN, Iwamatsu A, Zheng Y. Xgrip 109: a γ-tubulin-associated protein with an essential role in γ-tubulin ring complex (γ-TuRC) assembly and centrosome function. J Cell Biol 1998; 141: 675-87.

Mary J, Redeker VJ, Le Caer JP, Rossier J, Schmitter JM. Posttranslational modifications of axonemal tubulin. J Protein Chem 1997; 16: 403-7.

Mitchell DR. Cell and Molecular Biology of flagellar dyneins. Intern Rev Cyt 1994; 155 : 141-80.

Mitchison T, Kirschner M. Dynamic instability of microtubule growth. Nature1984; 312: 237-242.

Mochida K, Tres LL, Kierszenbaum AL. Structural features of the 26S proteasome complex isolated from rat testis and sperm tail. Mol Reprod Dev 2000; 57: 176-84.

Mochida K, Tress LL, Kierszenbaum AL. Structural and biochemical features of fractionated spermatid manchettes and sperm axonemes of the azh/azh mutant mouse. Mol Reprod Dev 1999; 52: 434-44.

Morales CR, Wu XQ, Hecht NB. The DNA/RNA-binding protein, TB-RBP moves from the nucleus to the cytoplasm and through intercellular bridges in male germ cells. Dev Biol 1998; 201: 113-23.

Multigner L, Pignot-Paintrand I, Saoudi Y, Job D, Plessmann U, Rudiger M, Weber K. The A and B tubules of the outer doublets of sea urchin sperm axonemes are composed of different tubulin variants. Biochemistry 1996; 35: 10862-71.

Nakamura K, Masuyama E Wada S, Okuno M. Application of 'stains all' staining to the analysis of axonemal tubulins: identification of β-tubulin and β- isotubulins. J. Biochem Biophys Methods 1990; 21: 237-45.

Nielsen MG, Turner FR, Hutchens JA, Raff EC. Axoneme-specific β-tubulin specialization: a conserved C-terminal motif specifies the central pair. Curr Biol 2001; 11: 529-33.

Nojima D, Linck RW, Egelman EH. At least one of the protofilments in flagellar microtubules is not composed of tubulin. Curr Biol 1995; 5: 158-67.

Norrander JM, Perrone CA, Amos LA, Linck RW. Structural comparison of tektins and evidence for their determination of complex spacings in flagellar microtubules. J Mol Biol 1996; 257: 385-97.

Pangilinan F, Li Q, Weaver T, Lewis BC, Dang CV, Spencer F. Mammalian BUB1 protein kinases: map positions and in vivo expression. Genomics 1997; 46: 379-88.

Pechart I, Kann ML, Levilliers N, Bre MH, Fouquet JP. Composition and organization of tubulin isoforms reveals a variety of axonemal models. Biol Cell 1999; 91: 685-97.

Peknicova J, Kubatova A, Sulimenko V, Draberova E, Viklicky V, Hozak P, Draber P. Differential subcellular distribution of tubulin eptiopes in boar spermatozoa: recognition of class III b-tubulin epitope in sperm tail. Biol Reprod 2001; 65: 672-9.

Pires-daSilva A, Nayemia K, Engle W, Torres M, Stoykova A, Chowdhury K, Gruss P. Mice deficient for spermatid perinuclear RNA-binding protein show neurologic spermatogenic and sperm morphological abnormalities. Dev Biol 2001 15; 233: 319-28.

Raff EC, Fackenethal JD, Hutchens JA, Hoyle HD, Turner FR. Microtubules architecture specified by a β-tubulin isoform. Science 1997 3; 275(5296): 70-3.

Rivkin E, Cullinan EB, Tres LL, Kierszenbaum AL. A protein associated with the manchette during rat spermiogenesis is encoded by a gene of the TBP-1 like subfamily with highly conserved ATPase and protease domains. Mol Reprod Dev 1997; 48: 77-89.

Rohwedder A, Liedigk O, Schaller J, Glander HJ, Werchau H. Detection of mRNA transcripts of beta integrins in ejaculated spermatozoa by nested reverse transcription-polymerase chain reaction. Mol

Hum Reprod 1996; 2: 499-505.
Roy A, Yan W, Burns KH, Matzuk MM. Tektin3 encodes an evolutionarily conserved putative testicular microtubules-related protein expressed preferentially in male germ cells. Mol Reprod Dev. 2004; 67:295-302.
Sampaio P, Rebollo E, Varmark H, Sunkel CE, Gonzalez C. Organized microtubule arrays in γ-tubulin depleted Drosophila spermatocytes. Curr Biol 2001; 11: 1788-93.
Scherf U, Pastan J, Willingham MC, Brinkmann U. The human CAS protein, which is homologous to the CSE1 yeast chromosome segregation gene product is associated with microtubules and mitotic spindle. Proc Natl Acad Sci (U S A) 1996; 93: 2670-4.
Schumacher JM, Artzt K, Braun RE. Spermatid perinuclear ribonucleic acid binding protein binds microtubules in vitro and associates with abnormal manchettes in vivo in mice. Biol Reprod 1998; 59: 69-76.
Schumacher JM, Lee K, Lee K, Edelhoff S, Braun RE.Spnr, a murine RNA-binding protein that is localized to cytoplasmic microtubules. J Cell Biol 1995 129:1023-32.
Smrzka OW, Delgehyr N, Bornens M. Tissue-specific expresson and subcellular localization of mammalian δ-tubulin. Cur Biol 2000; 10: 413-6.
Stearns T, Kirschner M. In vitro reconstitution of centrosome assembly and function: the central role of gamma-tubulin. Cell 1994; 76: 589-91.
Steffen W, Linck RW. Evidence for a non-tubulin spindle matrix and for spindle components immunologically related to tektin filaments. J Cell Sci 1992; 101 (Pt 4): 809-22.
Tarsounas M, Pearlman RE, Moens PB. CLIP-50 immunolocalization during mouse spermiogenesis suggests a role in shaping the sperm nucleus. Dev Biol 2001; 236: 400-10.
Tassin AM, Celati C, Moudjou M, Bornens M. Characterization of the human homologue of the yeast spc98p and its association with γ-tubulin. J Cell Biol 1998; 141: 689-701.
Tombes RM, Farr A, Shapiro BM. Sea urchin sperm creatine kinase : the flagellar isozyme is a microtubule-associated protein. Exp Cell Res 1988; 178: 307-17.
Villasante A, Wang D, Dobner P, Dolph P, Lewis A, Cowan NJ. Six mouse -tubulin mRNAs encode five distinct isotypes: Testis-specific expression of two sister genes. Mole Cell Biol 1986; 6: 2409-19.
Wakefield JG, Bonaccorsi S, Gatti M. The *Drosophila* protein asp is involved in microtubule organization during spindle formation and cytokinesis. J Cell Biol 2001; 153: 637-48.
Walden PD, Cowan NJ. A novel 205-kilodalton testis specific serine/threonine protein kinase associated with microtubules of the spermatid manchette. Mol Cell Biol 1993; 7625-35.
Whyard TC, Cheung W, Sheynkin Y, Waltzer WC, Hod Y. Identification of RS as a flagellar and heat sperm protein. Mol Reprod Dev 2000; 55: 189-96.
Wilson PG, Zheng Y, Oakley CE, Oakley BR, Borisy GG, Fuller MT. Differential expression of two γ-tubulin isoforms during gametogenesis and development in Drosophila. Dev Biol 1997; 184: 207-21.
Wong EY, Tse JY, Yao KM, Lui VC, Tam PC, Yeung WS. Identification and characterization of human VCY2-interacting protein: VCY2IP-1, a microtubule-associated protein-like protein. Biol Reprod. 2004;70 : 775-84.
Woolley DM. Studies on the eel sperm flagellum.I. The structure of the inner dynein arm complex. J Cell Sci 1997; 110(Pt 1): 84-94.
Wu XQ, Hecht NB. Mouse testis brain-ribonucleic acid-binding protein/translin colocalizes with microtubules and is immunoprecipitated with messenger ribonucleic acids encoding myelin basic protein, α calmodulin kinase II, and protamines 1 and 2. Biol Reprod 2000; 62: 720-5.
Wu XQ, Lefrancois S, Morales CR, Hecht NB. Protein-protein interactions between the testis brain RNA-binding protein and the transitional endoplasmic retriculum ATPase, a cytoskeletal γ actin and Trax in male germ cells and the brain. Biochemistry 1999; 38; 11261-70.
Xu M, Zhou Z, Cheng C, Zhao W, Tan R, Huang Y, Wang W, Xu J, Zeng L, Xie Y, Mao Y. Cloning and characterization of a novel human TEKTIN1 gene. Int Biochem Cell Biol 2001: 1172-82.
Zhang L, Keating TJ, Wilde A, Borisy GG, Zheng Y. The role of Xgrip 210 in γ-tubulin ring complex assembly and centrosome recruitment. J Cell Biol 2000; 151: 1525-36.

Chapter 9

MICROTUBULE BASED MOTOR PROTEINS: Kinesins and Dyneins

Motor proteins use energy of ATP hydrolysis to move along microtubules or actin filaments. The first motor protein identified was skeletal muscle myosin, which generates force for muscle contraction and moves on tracks of actin filaments. The motor proteins that move on microtubules belong to either kinesin superfamily or dynein family. Dyneins, kinesins and myosins represent the three major classes of molecular motors that translocate along cytoskeletal elements. Recent Studies show that dyneins exhibit much greater molecular complexity than the other cytoskeletal motors: For example, the kinesin is a 360-kDa tetramer whereas the outer dynein arm has a molecular mass of 1.9-MDa, having 13 different polypeptide components. Unlike in myosin and kinesin, the microtubule-binding domain in dynein is located far from the ATP-hydrolysis site. The in vitro microtubule-translocation parameters of dynein also are quite different from those of kinesin. Studies indicate that the dynein motor unit is constructed around a series of AAA (ATPase Associated with diverse/cellular Activities) domains, which suggest that this enzyme class is fundamentally distinct from kinesin and myosin at both the structural and mechanistic levels.

9.1. KINESIN MOTOR

Kinesin was first isolated from squid axoplasm and has since been isolated from various sources. Kinesin is a unidirectional motor moving towards the [+] end of microtubules. Kinesin may be involved in fast axonal transport in axons, mitotic spindle movement in sea urchins and the formation of tubulovesicular networks. Cytoplasmic dynein first isolated from sea urchins, moves unidirectionally towards the [-] end of microtubules and has been implicated in organelle transport and mitotic spindle rearrangements. Hirose et al (1995) re-examined the polarity of microtubules and sheets decorated with kinesin motor domain. A conflicting conclusion was that the main mass of the kinesin head is associated with the tubulin closer to the plus end of a microtubule.

The kinesin superfamily of motor proteins comprises of more than 100 proteins that participate in a wide variety of movements within cells. Many kinesin-related proteins (KRPs) participate in spindle morphogenesis and chromosome movement in cell division. Kinesins, like myosin II, are tetrameric proteins made of two heavy chains (KHCs) and two light chains (KLCs) of 124 and 64-kDa, respectively, with a total length of around 80 nm, and involved in transport of cargo across microtubules. All kinesins have a conserved motor domain containing the sites for ATP hydrolysis and microtubule binding required for the generation of force against microtubules. In addition to variation in motor domain placement, the homology of superfamily members to one another decreases dramatically outside the motor domain. The

A

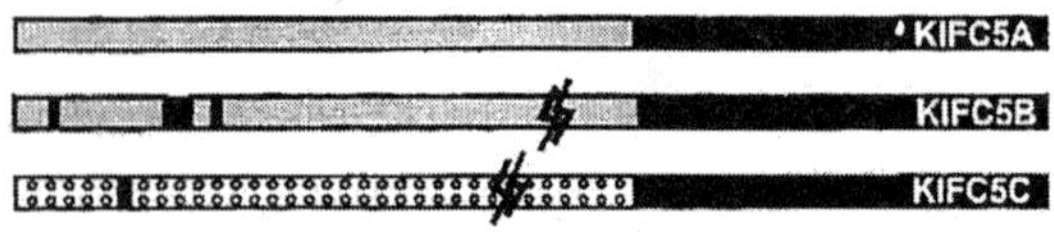

B

```
KIFC5A   1    MDVQAQRPPLLEVKRNVELKAALVKSSSRVPLSASRLKRGPDQMEDALEPAKKRTRVMGA
KIFC5B   1    MDVQAQRK-GREGKRNVELKAALVKSSSRLPLSASSLKRGPDQMEDALEPAKKRTRVMGA
KIFC5C   1    MDVQAQRPPLLEVKRNVELKTSLVKFSSRVPRQQAGSRVGPDQMEDALEPVKKRTRGMGA

KIFC5A  61    VTKVDTSRPRGPLLSTVSQTQGHTAAQKGPKKTGPRGCSAVGTVLRSQKPAPAAPAQKPG
KIFC5B  60    VTKVDTSRPRGPLLSTVSQTQGHNAV------------------LRSQKPAPAAPA-KPG
KIFC5C  61    VTKVDISRPRGPLFSTVSQTQGHTAAQKGPKKQDLLGCSTVGTVLRNQT-TPTVPAQKPG

KIFC5A 121    TSTAPVVVGKRAGKRPAWDLKGQLCDLNEELKRYREKTQMLELENRGLREQLREVQEQAT
KIFC5B 101    TSTAPLVVGKRAGKRPAWDLKGQLCDLNEELKRYREKTETLELENRGLREQLREVQEQAT
KIFC5C 120    TSTAPVVVRKRTGKRLAWELKGQLFELSEELKRYREKNQTLERENRGLRKHLREVQEQAT

KIFC5A 181    TLGTERNTLEGELASVRSRAEQDQQRLETLSARVLELEECLGTRERLLQELQGERL
KIFC5B 161    TLGTERNTLEGELASVRSRAEQDQQRLETLSARVLELEECLGTRERLLQELQGERL
KIFC5C 180    TLGTEQSTLEGELANVRTRAEQDQQRLEMLSACVLELEEYLGTTERLVQELQGERL
```

Fig.9.1. Comparison of KIFC5-like Isoforms. (A) A schematic presentation of KIFC5 like c-DNAs. The head domain (black) is located at C-terminus (The KIFC5B and KIFC5C clones are partial as shown by zigzag lines and contain tail domains only). Small black boxes indicate areas of sequence differences. The circled region in KIF5C shows multiple nucleotide differences in this region. (B) The aligned amino acid sequences of three KIFC5 molecules (Navolanic and Sperry, 2000).

divergent "tail" domains of these proteins are thought to specify the cargo to which they attach and move. The presence of kinesin-like motor protein in the testis was demonstrated in the spermatid manchette and the Sertoli cell trans-Golgi network (Yoshida et al., 1994). In addition, the heterotrimeric motor protein kinesin-II of 85-kDa and 115-kDa subunits has been localized in the mid-piece and flagellum of echinoderm sperm. The kinesin-II localization pattern is punctate and discontinuous. In flagella it is quite distinct from the continuous lebeling present in sperm labeled with anti-flagellar dynein. In the mid-piece the kinesin-II staining is similar to the pattern present in sperm labeled with an anti-centrosomal antibody (Henson et al., 1997). The subunits of this protein show considerable sequence homology with the mouse kinesin-like proteins KIF3A and KIF3B (Kinesin family like). The KIF3B is present in high concentration in testis. Although alternative splicing of mRNA is common mechanism, which generates diversity of proteins, this mechanism does not seem to operate in kinesin superfamily. Only kinesin-associated proteins (KAP), such as KAP3, associated with kinesin family like 3A (KIF3A), KIF3B, and the kinesin light chains (KLCs) have been shown to be the products of alternatively spliced genes.

9.1.1. Kinesin Related Proteins

There are at least ten families of kinesin related proteins (KRPs). Most of them have motor domain at the N-terminus of heavy chain and walk toward the [+] end of the microtubule. But one family has the motor domain at the C-terminus and walks in opposite direction, toward [-] end of the MT. Many of the kinesin family members have specific roles in mitotic and meiotic spindle formation and chromosome separation during cell division. Six members of the kinesin superfamily (KRP1-KRP6), those are expressed in the seminiferous epithelium of the rat were reported by Sperry and Zhao (1996), and one member of this group from the mouse has been cloned by Navolanik and Sperry, (2000) (**Fig. 9.1**). Members of this family share a highly homologous head domain responsible for force generation attached to a divergent tail domain thought to couple the motor domain to its target cargo. Among six (KRP1-KRP6) kinesin related proteins, three are expressed in seminiferous epithelial regions rich in meiotic cells and KRP2 showed to be promising for a motor in meiosis and homologous to mitotic apparatus.

Spermatogenesis Related Factor-1 (SRF-1): The *SRF-1* is a new gene product that may function as a molecular motor in meiosis in rat testis. The gene with testis specific expression coinciding with spermatogenesis was cloned and appeared spermatogenesis related factor-1 (SRF-1) specific gene. Expression was detected from 5 weeks of age and increased up to 15 weeks; the 750-bp cloned gene was coded for an open reading frame of 202 amino acids and expressed mainly in spermatocytes. The deduced amino acid sequence of SRF-1 showed partial homology with kinesin-related proteins (Yamano et al., 2001).

A Spermatogenesis-associated Factor (SPAF): The SPAF is aberrantly expressed at the malignant conversion stage in a clonal epidermal model of chemical carcinogenesis. Sequence analysis revealed two ATPase modules, classifying this protein as a new member of the AAA-protein family. The expression of AAA protein occurs in mouse spermatogonia and early spermatocytes in the seminiferous tubules. In view of mitochondrial localization, it seems that SPAF may be involved in morphological and functional transformations of mitochondria during spermatogenesis (Liu et al., 2000).

9.1.2. Kinesin Family C-Terminal 5A Gene (KIFC5A)

The full-length coding sequence for mouse KIFC5A (kinesin family carboxy-terminal 5A) cDNA, encoding a motor protein found in the testes is homologous to a group of carboxyl-terminal motors, including hamster CHO2, human HSET, and mouse KIFC1 and KIFC4. The KIFC 5A and KIFC1 cDNAs are nearly identical except for the presence of two additional sequence blocks in the 5'end of KIFC5A and a number of single base-pair differences in their motor domains. PCR amplification and sequencing of the 5'end of KIFC5A identified 3 distinct RNA species in testes and other tissues. In order to examine the possible role of these motors in germ cells of the testes, experiments suggested that KIFC5-like motor proteins are associated with multiple microtubule complexes in male germ cells, including the meiotic spindle, the manchette, and the flagella (Navolanic and Sperry, 2000).

Kinesin light chain-3: The function of light chains may be in cargo binding and regulation of kinesin activity. In mouse, gene for *KLC1* is expressed in neuronal tissue and plays a role in transport whereas *KLC2* is more widely expressed in other tissues. In mouse, an additional gene *KLC3*, codes for another isoform of KLC. The *KLC3* gene is located in close proximity to the *ERCC2* gene. The KLC3 can be classified as a genuine light chain: it interacts in vitro with the kinesin heavy chain (KHC); the interaction is mediated by a conserved heptad repeat sequence, and it associates in vitro with microtubules. In mouse and rat testis, KLC3 protein expression is restricted to round and elongating spermatids, and KLC3 is present in sperm tails. In contrast, KLC1 and KLC2 are detected before meiosis. However the expression profiles of the three known KHCs and KLC3 differ significantly: *Kif5a* and *kif5b* are not expressed after meiosis, and *kif5c* expressed slowly at spermatid, is not detectable in sperm tails. The *KLC3* gene characterization shows that it carries out a unique and specialized role in spermatids (Junco et al., 2001).

Function of Heavy Chain: Kinesin and myosin have been proposed to transport intracellular organelles and vesicles to the cell periphery in several cell systems. In search of a function local burst of Ca^{2+} regulated exocytosis was triggered by wounding the cell membrane and the resulting individual exocytosis events in real time were visualized depending on the sensitivities to reagents targeting different motor proteins. The function blocking anti-kinesin antibody SUK4 as well as the stalk-tail fragment of kinesin heavy chain specifically inhibited a slow phase, while myosin ATPase inhibitor inhibited both the slow and fast phases. The blockage of Ca^{2+} calmodulin-dependent protein kinase II with auto-inhibitory peptide also inhibited the slow and fast phases, consistent with disruption of a myosin-actin dependent step of vesicle recruitment. These direct observations provide evidence that in intact living cells, kinesin and myosin motors may mediate two sequential transport steps that recruit vesicles to the release sites of Ca^{2+} regulated exocytosis, although the identity of the responsible myosin isoform is not yet certain. These results provide in vivo evidence for the cargo-binding function of the kinesin heavy chain tail domain (Bi et al., 1997). In addition to motor function of kinesin, Macho et al., (2002) suggested that kinesin controls the activity of a transcriptional co-activator by a tight regulation of its intracellular localization. Functional relation between CREM and ACT, present in round spermatids showed that ACT selectively interacts with KIF17b, a kinesin mainly expressed in male germ cells. The ACT-KIF17b interaction is restricted to specific stages of spermatogenesis and directly determines the intracellular localization of ACT.

9.1.3. Kinesin Motor in *Drosophila*

Drosophila kinesin heavy chain alone, without the light chains and other eukaryotic factors, is able to induce microtubule movement in vitro, and that a fragment likely to contain only the kinesin head is also capable of inducing microtubule motility. It was shown that the amino terminal of 450 amino acids of kinesin contains all the basic elements needed to convert chemical energy into mechanical force (Yang et al., 1990). In *Drosophila*, kinesin like protein 3A (KLP3A) localizes to the equator of the central spindle during late anaphase and telophase of male meiosis. Mutations in the KLP3A gene disrupt the interdigitation of microtubules in spermatocytes central spindles. Findings suggested that the KLP3A presumptive motor protein is a critical component in the establishment or stabilization of the central spindle. These results imply that the central spindle is the source of signals that initiate the cleavage furrow in higher cells (Williams et al., 1995). *Drosophila melanogaster* females homozygous for mutations in the gene encoding the KLP3A are sterile. Williams et al (1997) investigated the basis of this sterility. The major defect caused by depletion of the KLP3A protein was either in specification of the female pronucleus, or in migration of the male and female pro-nuclei toward each other.

9.1.4. Structure-Function Relation

Kinesin has two identical 7-nm long motor domains, which it uses to move between consecutive tubulin binding sites spaced 8-nm apart along a microtubular protofilament. The molecular mechanism of this movement may be common to all families of motor proteins. A region adjacent to kinesin's motor catalytic domain (the neck) contains a coiled-coil that is sufficient for motor dimerization and has been proposed to play an essential role in processive movement. However, our understanding on-how kinesin generates force and motion remains incomplete. The domain organization of kinesin, its docking on microtubule protofilament and the comparison of kinesin motor (a plus end directed motor) with ncd (a minues end directed motor invoved in chromosome segregation) are shown in **Fig. 9.2.** In contrast to continuous coiled-coil of ncd neck, the kinesin neck consists of two short β-strands (termed the neck linker) followed by coiled-coil helilx (neck-coiled-coil), all of which are connected by short

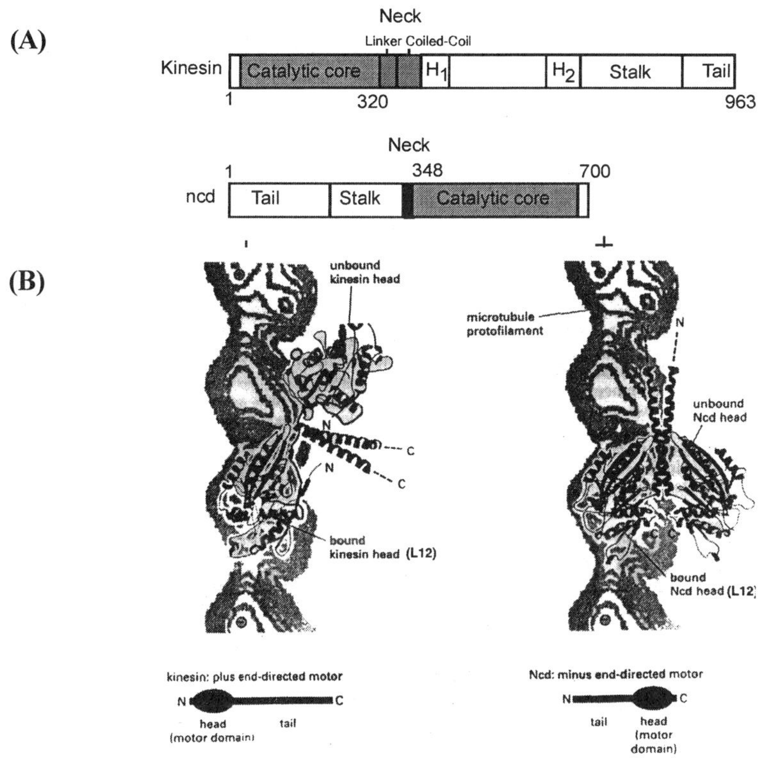

Fig. 9.2. (A) Domain organization of kinesin and ncd homodimers. The polypeptide chain of the monomer is organized into four domains (left to right in kinesin, right to left, C to N terminus, in ncd) I. A globular catalytic core, which contains the ATP- and microtubule-binding sites and is conserved throughout the kinesin superfamily. II is a 'neck' region, which is adjacent to the catalytic core and is defined by the class-specific sequence conservation. The neck of ncd contains 13 class conserved amino acids (R335-G347) that precede the first β-strand of the catalytic core. Kinesin's conserved neck (35 residues) emerges from the C terminus of the catalytic core and consists of two distinct regions. The first ten residues, termed the 'neck linker', are highly conserved among all plus end directed motors and interact with the catalytic core. The subsequent region form a coiled-coil('neck coiled-coil'), does not interact with the core, and is conserved among conventional kinesin subfamily. III. An α-helical 'stalk' domain that enhances dimer formation through an extended coiled-coil. H1 and H2 indicate flexible hinge region. IV. A small C-terminal 'tail" or attachment domain. (B) Model showing the ncd and kinesin dimer structures docked onto a tubulin protofilament. The bound ncd and kinesin heads are positioned similarly, with loop L12 docked onto the tubulin (background). Because of the distinct architectures of the kinesin and ncd necks, the unbound kinesin head points towards the plus end, whereas the unbound ncd head is tilted towards the minus end of the protofilament. Reproduced from E.P. Sablin et al Nature 395, 813-16;1998© nature.com/nature.

loops **(Fig 9.2.A)** (Sablin et al, 1998). Models have suggested that the neck enables head-to-head communication by creating a stiff connection between the two motor domains, but also may unwind during the mechano-chemical cycle to allow movement to new tubulin binding sites. To test these ideas the neck coiled-coil was mutated in a 560-amino acid (aa) dimeric kinesin construct and then the processivity of molecules moving along a microtubule was determined. Results showed that replacing the kinesin neck coiled-coil does not greatly reduce the processivity of the motor. This result argues against models in which extensive unwinding of the coiled-coil is essential for movement. Furthermore, deleting the neck coiled-coil decreases the processivity but does not abolish it. It was also demonstrated that processivity is increased by three-fold when the neck helix is elongated by seven residues. These structural features of the neck coiled-coil, although not essential for processivity, but seemed to tune the efficiency of single molecule motility (Romberg et al., 1998).

Kawaguchi and Ishiwata (2000) examined the effect of temperature on the gliding velocity, force, and processivity of single kinesin molecule interacting with a microtubule between 15 and 35°C. The average run length, (a measure of processivity of kinesin), and the gliding velocity increased with temperature with Arrhenius activation energy of 50 KJ/mol, showing temperature dependence of the microtubule-based ATPase. It appeared that the force generation is attributable to the temperature-insensitive nucleotide-binding state(s) and/or conformational change(s) of kinesin-microtubule complex, whereas the gliding velocity is dependent on ATP hydrolysis.

9.2. DYNEIN MOTORS

Dyneins are large, multi-subunit energy transducing ATPases that interact with microtubules to generate force and the sliding movement between tubules that underlies the beating of cilia and flagella of eukaryotes. They are a family of minus-end-directed microtubule motors, but not related to kinesin superfamily. Dyneins are present in eukaryotic cilia and flagella and are present in the cytoplasm, where they are involved in the transport of particles and organelles along microtubules and in the transport of condensed chromosomes during mitosis. In addition, these motor proteins are found in association with many forms of cytoplasmic motility, including axonal transport centripetal localization of Golgi vesicles and are present in simple non-flagellated organisms as *D. discoideum* and *S. cerevisisae*. As molecular motors, dyneins are unique for their large size, variety, and subunit complexity. Dynein motor proteins comprise a single homologous family with three main branches, cytoplasmic dynein, axonemal dynein, and a third branch represented by DYH1B that lies between the other two. The cytoplasmic dyneins are heavy chain homodimers, with two large motor domains as heads. The axonemal heavy chain dyneins include heterodimers and hetrotrimers, with two or three motor domain heads, respectively. Axonemal outer arm dyneins are more abundant than inner row dyneins and have higher ATPase activity. Genetic and biochemical dissection of flagellar dyneins has revealed that several similar but unique types are present in any given cilium or flagellum, and that some dyneins contain as many as 3 catalytic heavy chains, each greater than 500-kDa, and up to 12 smaller proteins, while others appear to have a single heavy chain and 2 or 3 smaller subunits. In just one organelle, the *Chlamydomonas* flagellum, at least 11 dynein heavy-chain isoforms are present, distributed among 5 or more unique enzyme complexes.

In all branches of the family, the dynein heavy chain has four copies of the P-loop motif for a nucleotide-binding site spaced approximately 300 residues apart in its mid-region, with amino acid sequence GPAGTGKT in the P-loop of the hydrolytic ATP-binding site. Heavy chains are encoded by multiple genes, and their non-hydrolytic P-loop motifs are much more divergent with little trace of their origin by gene duplication. The DYH1B subfamily is more closely related to the cytoplasmic dyneins in sequence, but appears related to axonemal dyneins in function and has not been found in organisms, such as yeast and *Dictyostelium*, that are totally without cilia or flagella (Gibbons, 1995). Axonemal dyneins have two or three globular heads joined by flexible tails to a common base, with each head/tail unit consisting of a single heavy chain polypeptide of relative molecular mass greater than 400-kDa. Defects in human axonemal dynein complexes have been shown to be associated with Kartagener's syndrome.

Cytoplasmic heavy chains and heavy chains from the outer arms of axonemes are the best characterized of the dynein genes. These share extensive sequence similarity and are conserved throughout species. In recent years several genes encoding intermediate and light chains have been identified; these have a remarkable diversity of products, which also seem to be highly conserved between species, although they fall into several complex groups (Milisav 1998). Knowledge of flagellar motors and their potential significance for flagellar function is important for understanding of sperm motility relating to fertility.

9.3. AXONEMAL DYNEINS

9.3.1. Doublet Tubules of Sperm Flagella

The propagated bending waves of cilia and flagella are the result of localized sliding between doublet tubules in the axoneme. The doublet tubules of cilia and sperm flagella contain three forms of dynein cross-bridges, termed "arms", with one form of outer arm and two forms of inner arm being disposed along the length of the tubule in a regular pattern **(Fig. 9.1)**. Each form has a unique polypeptide composition, with a total Mr of 1,200,000-1,900,000. The structure and function of the outer arms are better known than those of the inner arms. Outer arm dyneins each contain two or three distinct heavy chain polypeptide subunits of Mr>350,000, termed α and β (and γ if required) as well as a number of subunits of lesser weight. Solubilized outer arm dynein consists of two or three globular heads joined by flexible stems to a common base; in favorable instances it is possible to see also a projection about 10-nm long extending from each of the globular heads at an angle relative to the flexible stem. Each of the dynein ATPase subunits isolated from inner arms appears to contain a single subunit of actin, the function of which remains obscure as actin normally associates with myosin. In situ, the ATP cyclic cross-bridging of the dynein arms appears to involve the interaction of the projections on the globular heads with tubulin subunits in the B-tubule of the adjacent doublet. The angle of attachment of globular heads to B tubules depends on the presence and absence of ATP.

Axonemal dyneins show extreme polymorphism, with all the ATPase subunits of the outer arms and the two species of inner arms differing from each other in both structure and polypeptide composition. While the enzymatic regions of the dynein subunit ATPases are highly conserved, their non-enzymatic regions appear to be less highly conserved than other proteins associated with motility. The low level of immunological cross-reactivity among the isoforms of dynein in a given species, considered in conjunction with the nearly uniform diversity of dynein arm structure in different species, strongly suggests that the different isoforms of the dynein heavy chain are encoded by a highly conserved multigene family, rather than being the result of post-translational modification or of alternative RNA splicing.

9.3.2. Outer Row Dyneins

Axonemal outer arm dyneins are more abundant and better studied than inner row dyneins and have higher ATPase activity. The subunit composition of outer row dynein varies slightly among organisms, but all contain 10-15 proteins spanning over a range of molecular weights. Outer row dyneins of *Chlamydomonas* contain three 500-kDa catalytic heavy chains (α, β, and γ), two intermediate chains of 78-kDa and 70-kDa (IC78 and IC70), and eight light chains of 22-kDa to 11-kDa. Both intermediate chains and two of the three heavy chains have been cloned and sequenced. *Paramecium* outer row dynein, like that of *Chlamydomonas,* has also three heavy chains, two intermediate chains (88-kDa and 76-kDa) and eight light chains. *Tetrahymena* outer row dynein differs from *Chlamydomonas* and *Paramecium* in having three IC chains rather than only two. In sea urchin sperm, outer row dynein consists of only two heavy chains (α and β), three intermediate chains of 112-kDa, 79-kDa, and 70-kDa (IC1, IC2, and IC3), and at least six light chains of 23-kDa to 6-kDa. Sea urchin β-chain was the first dynein heavy chain to be cloned and sequenced (Mitchell, 1994). Information on mammalian sperm dynein chains from different species is scanty. Our main aim would to concentrate the knowledge available on mammalian species. For comparison, however, limited data available on species other than mammals would also be reviewed.

Light chains: Little is known about the structure and function of the dynein light chains. In *Chlamydomonas*, high salt solution dissociates a smaller complex containing the γ heavy chains and two light chains (22-kDa and 18-kDa) from a complex containing the remainder of the enzyme. Further treatment with detergent or low ionic strength dialysis can be used to purify sub-complexes of the α-chain and 16-kDa and 18-kDa light chains. In similar dissociation studies, sea urchin outer row dynein can be split into three fractions consisting of the α-chain alone; the β-chain and IC1; and a complex IC2, IC3, and several light chains. There is no evidence for tight association of heavy chains and light chains in sea urchin dyneins.

The mAb (D405-14) inhibiting the motility of demembranated-reactivated sperm models at low concentrations recognized a single polypeptide of 33-kDa (p33) in sea urchin axonemal proteins. A full-length cDNA clone encoding the 33-kDa protein predicted a polypeptide of 260 amino acids having a mass of 29730-Da and a pI = 9.3. Sequence comparison indicated that p33 is 66% identical (74% similar) to the p28 light chain of axonemal inner dynein arm of *Chlamydomonas reinhardtii* and suggested the possibility of a p28 light chain homolog in sea urchin sperm axoneme (Gingras et al., 1996).

Human Inner-Dynein Light Chain Gene: Splice-site mutations in the *Chlamydomonas* inner dynein arm gene (p28) are associated with impaired flagellar motility. The human homolog of the *Chlamydomonas* dynein p28 gene is an attractive candidate gene for patients with ICS. The human homolog (hp28) of *Chlamydomonas* p28 is coded by a transcript of 921-bp with an open reading frame of 257 amino acids. The hp28 gene was mapped to human chromosome 1p35.1. The hp28 cDNA probe hybridizes to sequences in species from yeast to human. Two hp28 gene transcripts of 0.9 and 2.5-kb are present in many tissues. The 0.9-kb transcript is expressed at a 20-fold higher level than the 2.5- kb transcript in the testis (Kastury et al., 1997).

Intermediate chains: Sequence analysis showed that outer row dynein intermediate chains from many organisms retain at least limited homology. Antibodies against the smaller *Chlamydomonas* (IC70) react with sea urchin IC3 and trout sperm IC2. A mAb specific to *Chlamydomonas* IC78 also reacts with trout sperm IC2, indicating that the two *Chlamydomonas* sequences are themselves related. A low similarity between IC78 and IC70 can be observed in sequence alignment. The IC70 also shows homology to a 74-kDa intermediate chain of mammalian cytoplasmic dynein. Based on comparison with the rat brain cytoplasmic dynein intermediate chain, the *Chlamydomonas* IC70 protein can be divided into three regions: an N-terminal region with no obvious sequence similarity, a central region that retains approximately 28% similarity to the C-terminal portion of the cytoplasmic dynein subunit, and a short C-terminal region that has a high probability of forming an α-helical coiled coil. In 74-kDa brain dynein, a coiled coil structure is predicted as an N-terminal rather than C-terminus domain. The functional significance of these predicted α-helical structures seems to participate in subunit interactions, perhaps between the two intermediate chains or between an intermediate chain and a heavy chain.

To assess the microtubule translocation ability of individual ATPase subunits of outer arm dynein, it was demonstrated that a single sub-domain in dynein, the b/1C1 ATPase, is sufficient for microtubule sliding activity (Sale and Fox, 1988). To investigate the role of axonemal components in the mechanics and regulation of flagellar movement, Gagnon et al generated a series of monoclonal antibodies (mAb) against sea urchin *(Lytechinus pictus)* sperm axonemal proteins. One of these mAb, D1 recognizes an antigen of 142-kDa axonemal proteins and a purified outer arm dynein, suggesting that it acts by binding to the heaviest intermediate chain (IC1) of the dynein arm. Antigen revealed its presence on the axonemes of flagella from sea urchin spermatozoa and O. marina but not on the microtubule network of the dinoflagellate (Gagnon et al., 1996). Newt *Cynopspyrrohogaster* spermatocyte dynein IC cDNA was found to encode a polypeptide consisting of 694 amino acids residues with 66.8% and 45.8% amino acid similarity to sea urchin dynein IC3 and *Chlamydomonas* respectively. The predicted protein contains five WD repeats and a novel repeated motif in the C-terminal region. Newt dynein IC mRNA was expressed in the spermatocytes and round spermatid stages, suggesting that dynein IC plays a role in formation of flagella as well as in meiotic events (Yamamoto et al., 2001).

Heavy chains: Comparison of cleavage maps with maps based on the predicted amino acid sequence has provided an initial framework for proposing relationships between chemically reactive sites and conserved sequence motifs common to the catalytic sites of other nucleotide metabolizing enzymes, and between quaternary structures observed by electron microscopy and secondary structure predicted from the sequence. Regions predicted to form α-helical coiled-coils and an N-terminal domain sufficient to support dynein assembly of *Chlamydomonas* β heavy chain.

Two flagellar dynein heavy chain sequences, published in early 1990s (Gibbons et al., 1991; Ogawa 1991), belong to outer row dynein β-chains from sea urchin embryonic cilia (see Mitchell, 1994). Most of the work appearing from Gibbons laboratory was based on use of vanadate and UV radiation. The cDNA encoding the β heavy chain of a dynein motor shows the predicted amino-acid sequence with four ATP-binding consensus sequences in the central domain. The dynein β heavy chain is thought to associate transiently with a microtubule during ATP hydrolysis, but the ATP-dependent microtubule binding sequence common to the kinesin super-family is not found in the dynein β heavy chain. These unique features distinguish the dynein β heavy chain from other motor protein super-families (Ogawa, 1991). β-heavy chain of sea urchin has 4466 residues and contains the consensus motifs for five nucleotide binding site. The probable hydrolytic ATP-binding site can be identified by its location close to or at the V1 site of vanadate-mediated photo-cleavage. The predicted secondary structure of the β heavy chain consists of an α/β type pattern along its whole length. The two longest regions of potential α helix, with unbroken heptad hydrophobic repeats of 120 and 50 amino acids length, may be of functional importance. But dynein does not seem to contain an extended coiled-coil tail domain (Gibbons et al., 1991).

These two sequences showed dyneins to be unrelated to other motor enzymes such as myosin and kinesin, except for the presence of P-loop motifs. Two cytoplasmic heavy chain sequences, from rat brain cDNAs and from the slime gave some insights into regions that may be common to all dyneins and regions that may be specific to dyneins designed for flagellar motility rather than cytoplasmic transport. Further insights into specialization within outer row dyneins were made available from the sequences of the three *Chlamydomonas* outer row dynein heavy chains. Amino acid sequences of α, β and γ chains showed a striking feature of four nucleotide binding consensus sequences within the central third of the molecule among all dynein heavy chains. While the first of these four sites (P1) is thought to be the catalytic site for ATP hydrolysis, based on its location at the approximate site of V1 photocleavage, functions for the remaining three sites were not completed (Mitchell, 1994). Like kinesin and myosin, dynein heavy chains appear as globular heads attached to extended tails. Unlike those other motor enzymes, dynein tails do not dimerize through formation of α-helical coiled coils. Many regions of each heavy chain are predicted to exist in α-helical structures, but few possess the heptad hydrophobic repeat typical of coiled-coils and none of these are very extensive.

The 19S outer arm dynein of trout sperm flagella is made of two heavy chains, five intermediate chains and at least six light chains. β-heavy subunit antibody recognized the dynein intermediate chains and tubulins. A Mg^{2+} dependent ATPase activity from trout sperm axonemes cross-reacted with a heterologous anti-sea urchin dynein antiserum. The RNA transcripts which were detected had a size (7.5-12 kb) consistent with those expected for the dynein heavy chains. All the dynein heavy chain (DHC) cDNAs encode portions of a highly unusual DNA coding sequence comprised of 21-bp direct repeats. The predicted open reading frame of this repeat is ILe/Leu-His-Val-ILe-Gln-Tyr-Ser and is characteristic of an extensive α-helical coiled-coil domain. The presence of an in-frame translation termination codon indicated that this domain is located at the carboxyl-terminus of the DHC (Garber et al., 1989). The specific anti-trout β dynein heavy chain (DHC) reacted with the DHC from the sea-urchin sperm 21S dynein and with heavy chain (>400-kDa) of demembranated ram sperm but did not react with heavy chains in trout brain or liver extracts, sheep brain extract or purified brain and testicular cytoplasmic dyneins. Thus the specific eptitope of one of the sperm outer arm dynein heavy chain is conserved throughout evolution and this epitope is not present on cytoplasmic dynein (Gatti and Dacheux, 1996). Marchese-Ragona et al (1987) used STEM to elucidate the structures and masses of 12S and 19S dynein from bull sperm flagella. The 12S particle was a single globular particle, whereas the 19S dynein particle consisted of globular heads joined to a common base. The average mass of the 19S particle was 1.6±0.04. X 10^6Da. Thus, with the exception of the larger mass, the bull sperm, 19S dynein molecule resembles the two-headed 21S dynein from sea urchin sperm flagella and the 18S dynein from *Chlamydomonas* with the possibility of a third head giving rise to the 12S particle. The structure, mass and polypeptide composition of bull sperm flagella dynein were compared with outer arm dyneins from *Chlamydomonas, Tetrahymena,* and *Sea urchin* sperm flagella (Eyer et al., 1990; Inaba, 1995).

A dynein related human transcript (designated DNEL1) can encode a protein with a size of 91087-Da. The DNEL1 mRNA is expressed specifically in testis. Analysis of whole genome has placed DNEL1 in 17q distal to the ERBA2L locus. The DNEL1 shares a high sequence identity and amino acid similarity with the C-terminal region of the outer arm axonemal dynein β-heavy chains derived from sea urchin and other species, but not to any gene encoding dynein intermediate or light chains. The similarity in C-terminal region of DNEL-1 described suggests an origin from a common progenitor gene and seems to play a possible role in sperm development or motility (Milisav et al., 1996).

9.3.3. Outer Arm Dynein Structure

Outer arm dyneins are better studied than the inner arm dyneins. They have the property of adhesion hence could be exploited for EM preparation. Upon adhesion to the surface they usually spread into a fan shaped bouquet in which globular domains are joined at bases through extended structure of heavy chain. The different globular heads with unique shapes and a short thin, knobbed antinna projects from each head in unique orientation. The heavy chain domains cannot be deduced from amino acid sequence. However, the isolated structure based model does not reconcile with views is situ outer arm structure as seen in EM. Many discrepancies seem to exist possibly due to preparation artifacts or change in conformation in isolated preparation (Mitchell, 1994).

Sequence of IC1 of the outer arm dynein of sea urchin sperm axoneme shows a unique primary structure: the N-terminal part is homologous to the sequence of thioredoxin: the middle part consists of three repetitive sequences, homologous to the sequence of nucleoside diphosphate kinase, and the C-terminal part contains a high proportion of negatively charged glutamic acid residues. Thus IC1 dynein intermediate chain is distinct from IC2 and IC3 and may be a multifunctional protein. The thioredoxin-related part of IC1 is more closely related to those of two redox-active *Chlamydomonas* light chains than thioredoxin. It is proposed that the three intermediate chains are the basic core units of sperm outer arm dynein because of their ubiquitous existence. The recombinant thioredoxin-related part of IC1 and outer arm dyneins from sea urchin and distantly related species were specifically bound to- and eluted from a phenylarsine oxide affinity column with 2-mercaptoethanol, indicating that they contain vicinal dithiols competent to undergo reversible oxidation/reduction (Ogawa et al., 1996).

9.3.4. Tctex-1: A Cytoplasmic LC Dynein in Inner Dynein Arm

The Tctex1 (Tcd1) has been identified as a light chain (LC) of cytoplasmic dynein, an inner arm dynein of *Chlamydomonas*. The murine t complex (haplotype) is an alternate form of the proximal 30-40 Mb of chromosome 17. The t form of this chromosome contains a series of inversions that serve to suppress recombination with wild type and, therefore, the t complex is normally inherited intact. The most striking property of this famous genetic unit is that heterozygous (+/t) males pass the t-bearing copy of the chromosome to > 95% of their progeny. This phenomenon is referred to as transmission ratio (TR) distortion or meiotic drive. Transmission ratio distortion (TRD) in mouse t-haplotypes remains the most significant example of meiotic drive in vertebrates. While the underlying mechanism that fuels it is still mysterious, TRD is clearly a complex multi-gene phenomenon and a combined effect of 4-5 gene products. All of these genes encoding distorters and responder proteins (expressed during late spermiogenesis) are located within t complex. Although much is known about genetics of t complex, molecular basis involved in the effects of spermiogenesis and neural tube formation remains obscure (King et al., 1996).

Unexpectedly, the deduced amino-acid sequence of LC3 (12.3-kDa) of outer arm dynein of sea urchin sperm flagella is highly homologous to that of mouse and human Tctex1, a protein encoded by a member of the multi-gene family in the t complex region involved in spermiogenesis. The cytoplasmic dynein light chain Tctex1 is a candidate for one of the distorter products involved in the non-Mendelian transmission of mouse t haplotypes. It has been unclear, however, how the t-specific mutations in this protein, which is a cytoplasmic dynein in many tissues, could result in a male germ cell-specific phenotype. It was demonstrated that Tctex1 is not only a cytoplasmic dynein component, but is also present both in mouse sperm and *Chlamydomonas* flagella. Genetic and biochemical dissection of the *Chlamydomonas* flagellum revealed that Tctex1 is a component of inner dynein arm II. Combined with the identification of another putative t complex distorter, Tctex2, within the outer dynein arm, results support the hypothesis that transmission ratio distortion (Meiotic drive) of mouse t haplotypes involves dysfunction of both flagellar inner and outer dynein arms but does not require the cytoplasmic isozyme (Harrison et al., 1998; Kagami et al., 1998) **(Fig.9.3)**.

```
   1 GGCACGAGCTAGTATCTTATCACGTTGCTCTGCTCCTCCAATAGCCCTGGAAAGGACCTC   60
   1                                                         M  E    2
  61 GAAGCTCTAGACTTGTTTCGCGCAGAACCTTGCCTTTCATTGGGGGCCTGCAGCCATGGA  120
   3  G  V  D  P  A  V  E  E  A  A  F  V  A  D  D  V  S  N  I  I    22
 121 GGGGGTAGATCCCGCTGTGGAGGAGGCGGCCTTCGTGGCAGACGACGTTAGCAACATCAT  180
  23  K  E  S  I  D  A  V  L  Q  N  Q  Q  Y  S  E  A  K  V  S  Q    42
 181 CAAGGAGTCAATTGATGCAGTGCTGCAAAACCAGCAATACAGCGAGGCGAAGGTTAGCCA  240
  43  W  T  S  S  C  L  E  H  C  I  K  R  L  T  A  L  N  K  P  F    62
 241 ATGGACGTCGAGCTGTCTAGAACATTGCATAAAGCGCCTCACAGCGTTGAATAAGCCGTT  300
  63  K  Y  V  V  T  C  I  I  M  Q  K  N  G  A  G  L  H  T  A  A    82
 301 CAAGTACGTGGTGACCTGCATCATCATGCAGAAGAACGGCGCGGGCCTGCACACGGCGGC  360
  83  S  C  W  W  D  S  T  T  D  G  S  R  T  V  R  W  E  N  K  S   102
 361 ATCATGCTGGTGGGACAGCACGACCGACGGGAGCCGCACCGTGCGGTGGGAGAACAAGAG  420
 103  M  Y  C  I  C  T  V  F  G  L  A  I  *                         114
 421 CATGTACTGCATCTGCACGGTGTTTGGTCTGGCGATCTGAGGCGCGCTGTCGAACGGCAG  480
 481 GGAAGGATTGACGTCTCGGAAGAGGAGAAGTGGTCCGAGGAGAAGAGCTGTGGCGCAGGG  540
 541 CGCCCTGGTAGCGGTTTGTGAGGCCTTCTGCACGATGGACTGTGCTAAACGCTGAGCCAA  600
 601 GCAGAGCGCAGTGGCGAACTCCGCGCTACGGTATTCACGGCAAGGAAGCGCGGGGAATGG  660
 661 AGGTACGCTGTGCCGTTGCAAGGAACACGCGGCGTCGGCAAACAGCACGGCTCCAGCCGG  720
 721 GGTGCTGCTGGGTCGACACACACGCAGATGGGGGCAGCATGCTTGATTGGAGGCAGATGC  780
 781 GTGAATCTGAATGCGCGCTGTACGGGTGTCATGTGAGAGCATGTGTATGACGGCAGGCTT  840
 841 CGGGATTACCTTTGCGTTGACAATCACGTCTGCCAGGATAGTATTGTGCAGTAAGGCTGG  900
 901 CGGCAGGGCCATGGGTGTGTGGGGGCGTGGGGATGCAGCAATTGTTTTTCCCCGAGTGCG  960
 961 CTAGGTGGAATGAGTGCAGATGGGTGGTGGGGTAGGTGCGACCGCATTCGTGGGCGGAAA 1020
1021 TGGTGCAACAGGAGGTGCAGGAGGTGTACAGAAGAGAGCGTGATGAGCGCTTCCAGTGTG 1080
1081 CCAGGTTTTCGGCGTGCAAAACGGGTTGGCAGAATGGCAGTTGGGATCAGAGAGTCTGCG 1140
1141 GAACGTCTTCTTAAAGAGCTTCAGCGGAAAAAAAAAAAAAAAAAAAAAAAA          1190
```

Fig.9.3. The cDNA and putative amino acid sequence of *Chlmydomonas* Tctex1. The 5' untranslated region contains three in-frame stop codons. Reprinted with permission from A. Harrison et al. J Cell Biol 140; 1137-47: 1998 © The Rockefeller University Press.

9.3.5. C and A Heavy Chain Dynein (C/A Dynein)

The C/A dynein is present in sea urchin sperm flagellar axoneme and composed of C and A heavy chains and some low molecular mass polypeptide chains. The ATPase activity of C/A dynein is about one half that of outer arm 21 S dynein at 25^0. The C/A dynein was able to translocate the microtubules towards its plus end. Cross bridges were formed between adjacent microtubules in the bundle with an axial periodicity of about 24 nm. The ATPase activity of C/A dynein was enhanced several fold by the microtubules at concentration as low as 1 mg/ml. On the other hand, 21 S dynein bound to microtubules showed 24-nm axial periodicity only in the absence of ATP, without activating its ATPase activity by the microtubules. The C/A dynein is considered to be a component of inner arms in the axonemes of sea urchin sperm and binds to- and bundles the microtubules in the absence of ATP. In contrast to outer arm 21S dynein C/A dynein is not released from the microtubules in the presence of ATP. Thus the mode of interaction of C/A dynein with microtubules in the presence of ATP was distinct from that of flagellar or ciliary outer arm dyneins. Thus C/A dynein, as a component of inner arms plays a distinct role in the flagellar movement of sea urchin sperm (Yokota and Mabuchi, 1994; 1997).

Homozygosity for the t haplotype allele of the testis-specifically expressed axonemal dynein heavy chain (axDHC) gene, *Dnahc8*, has been linked to male sterility resulting from aberrant sperm motility. However, the near absence of *Dnahc8* expression has been associated with male sterility resulting from an early breakdown in sperm flagellar development. Samant et al., (2002) studied the organization and expression of full-length *Dnahc8(+) and Dnahc8(t)* transcripts. At least two alternatively spliced, testis-specific *Dnahc8* mRNAs transcribed from both the + and t alleles have been obtained. A highly expressed isoform encodes a protein with significant homology with the γ heavy chain of the *Chlamydomonas* axonemal outer arm dynein, while a more poorly expressed isoform codes for a protein, which diverges from that of other axDHCs at both its N and C termini. While both mRNA species accumulate exclusively in mid- to late spermatocytes, each isoform showed spatial independence. The *Dnahc8* may have acquired functional plasticity in the testis through the tightly controlled expression of both typical and unusual isoforms (Samant et al., 2002).

9.4. RADIAL SPOKE PROTEINS

The radial spokes of sea urchin sperm axoneme appear to be arranged in triplets with intra-triplet distance of 40 nm, distance between spoke 1 and 2 of 32 nm and between spoke 2 and 3 of 24 nm. Spoke 1 is thicker than others; the complex of the triplets around the 9 microtubules doublets forms a helix of 96 nm pitch. Nexin links connect the A tubule of a doublet with the B of the adjacent one, and also form a helix of 96 nm pitch (Baccetti et al., 1985). Radial spokes have been analyzed genetically through mutations. Several radial spoke subunits have been cloned and sequenced and several bypass suppressor mutations have been isolated that allow motility in the absence of spokes. One suppressor (*sup-pf-1*) is an outer arm β-dynein mutation, whereas mutations at five other suppressor loci were all found missing a subset of same six axonemal proteins. These proteins, which range in size from 108-kDa to 29-kDa were called as the dynein regulatory complex or DRC. There are evidences to support that both radial spokes and the DRC play important role in dynein regulation but the mechanisms through which that control is exerted remain largely unknown. The tentative position at spoke S2 has been ascertained. Although flagella missing radial spokes do no beat, their doublet microtubules can still slide if they are treated with trypsin in the presence of ATP: Comparison of sliding rates of axonemes in absence of radial spokes consistently suggests that dyneins are only partially inhibited by the absence of radial spokes (Mitchell, 1994).

Perturbing effects of anti-body D-316 resulted in isolation of a 90-kDa protein from sea urchin sperm. The 90-kDa protein exists as a complex in its native form. A full length cDNA encoding the 90-kDa protein predicts a protein of 552 amino acids (pI=4.0) with a mass of 62.72-kDa (p63). Sequences in databases indicated that p63 is related to radial spoke proteins 4 and 6 (RSP4 and RSP6) of *Chlamydomonas* reinhardtii, which share 37% and 35% similarity respectively with p63. But sea urchin protein has features distinct from RSP4 and RSP6 such as presence of three major acidic stretches, which contain 25, 17 and 12 aspartate and glutamate residues of 34-, 22- and 14- amino acid long stretches respectively that are predicted to form α-helical coiled-coil secondary structures. A major role for p63 in the maintenance of a planer form of sperm flagellar beating or radial spoke heads has been predicted (Gingras et al., 1998).

9.5. FORCE GENERATION BY DYNEIN ARMS AND BEAT RHYTHMICITY

Eukaryotic flagella beat rhythmically. Dynein powers flagellar motion, and oscillation may be inherent to this protein. When the dynein arms on the doublet microtubule make a contact, a singlet microtubule and activated by photolysis of caged ATP, they generate a peak force of approximately 6pN and move the singlet microtubule over the doublet microtubule in a processive manner. The force and displacement oscillate with a peak-to-peak force and

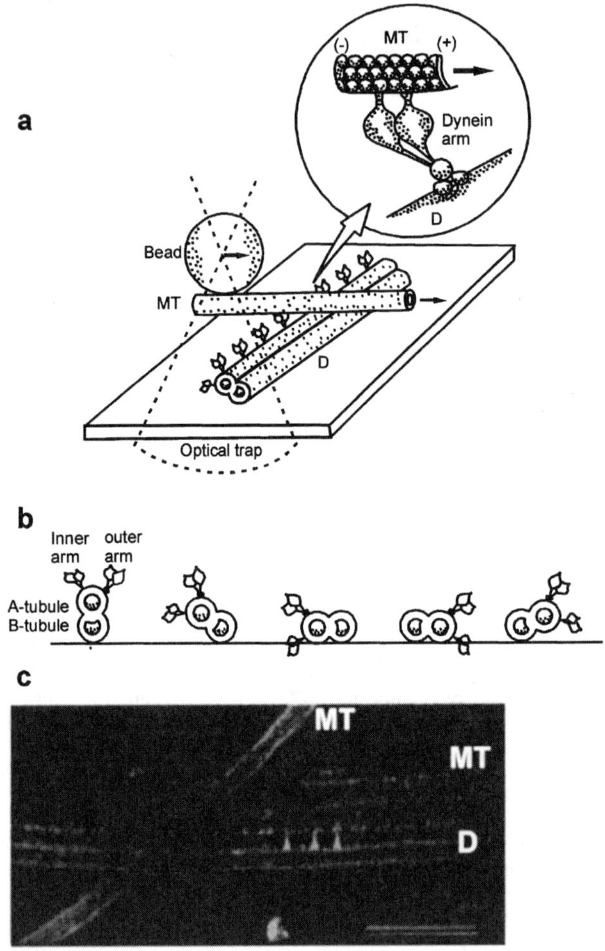

Fig.9.4. Interaction of singlet microtubule with dynein arms present on doublet microtubules. **a**. Schematic diagram of exposed dynein arms on a doublet microtubule (D), interacting with a singlet microtubule (MT). The singlet MT is manipulated by means of an optically trapped streptavidine-coated bead. A two-headed arm pulls the singlet MT in the direction of its plus (+) end, causing the bead to move away from the centre of the trap force (black arrows). **b**. Possible orientations of doublet MT (D) on the glass surface (bottom) with outer and inner dynein arms. Dynein arms pointing upwards may be capable to interact with a singlet MT. **c**. Electron micrograph showing two singlet MTs interacting with dynein arms (white arrowheads) on a doublet MT (D). Scale bar, 100 nm. Reproduced with permission from C. Shigyoji et al. Nature 393; 711-14: 1998 © Macmillan Magazines, Ltd. http://www.nature.com/Nature.

amplitude of approximately 2pN and approximately 30nm, respectively. The geometry of the interaction indicates that very few (possibly one) dynein arms are needed to generate the oscillation. The maximum frequency of the oscillation at 0.75 mM ATP is approximately 70-Hz. This frequency decreases as the ATP concentration decreases. A similar force was generated by inner dynein arms on doublet-microtubules that were depleted of outer dynein arms. Thus it showed that the oscillation of the dynein arm may be a inherent basic mechanism underlying flagellar beating (Shingyoji et al., 1998) (**Fig. 9.4**). However, the force generated by a detergent-extracted reactivated bull sperm flagellum during and isometric stall was 2.5 0.7 x 10^{-5} dyne (2.5 0.7 x 10^{-10}N). Schmitz et al (2000) suggested that more than 90% of the sustained force component is generated by the part of the flagellum between the probe and the flagellar base. Based on this premise, the isometric stall force per dynein head estimated to be 5.0 x 10^{-7} dyne (5 pN). This equals approximately 1.0 x 10^{-6} dyne (10 pN) per intact dynein arm. These values are close to the isometric stall force of isolated dynein.

9.6. CYTOPLASMIC DYNEINS

Cytoplasmic dynein is a complex of two catalytic heavy chains and at least seven co-purifying polypeptides of unknown function. The most prominent of these is a 74-kDa electrophoretic species, which can be resolved as two to three bands. Paschal et al., (1992) selected the rat brain cDNAs encoding the 74-kDa species. Deduced sequence of a full-length cDNA predicts a 72,753 Da polypeptide determined by microsequencing. Analysis suggested the existence of at least three isoforms of 74-kDa cytoplasmic dynein subunit and revealed the carboxyl-terminal half of the polypeptide to be 26.4% identical and 47.7% similar to the product of the *Chlamydomonas* outer dynein arm6 (ODA6) gene, a 70-kDa intermediate chain of flagellar outer arm dynein. The 74-kDa peptide is widespread in tissue distribution, as expected for a cytoplasmic dynein subunit. Nonetheless, the antibody recognized a 67-kDa species in ram sperm flagella and pig tracheal cilia, supporting the existence of a distinct but related cytoplasmic and axonemal polypeptides in mammals. The 74kDa polypeptide, perhaps mediates the interaction of cytoplasmic dynein with membraneous organelles and kinetochores (Paschal et al., 1992).

Under electron microscopy, solubilized dynein appears as a set of globular heads joined by flexible stems to a common base, with the number of heads being equal to the number of ATPase subunits present. Dynein ATPase activity is inhibited by low concentrations of vanadate, with concomitant loss of function and ATPase activity. There have been several reports of high molecular weight ATPase in sea urchin eggs that have properties resembling axonemal dynein. The closest homology between axonemal and cytoplasmic dyneins seems likely to occur at the energy-transducing level at which the ATP-driven cycling domains, putatively located in the microtubule being moved along.

9.6.1. Tctex-2 : An Analogue of Outer Dynein Arm LC2

The Tctex2 (Tcd3) is one of the LCs of axonemal outer dyneins from *Chlamydomonas* and sea urchin and murine sperm flagella. As stated, *Tctex-2* is also a distorter gene of mouse t haplotypes. King and associates by identifying *Chlamydomonas* insertional mutants with deletions in the gene encoding LC2 dynein demonstrated that the LC2 gene is the same as the outer dynein arm 12 (*ODA12*) gene. Complete deletion of the *LC2/ODA12* gene causes loss of all outer arms and rescues the motility phenotype. Therefore, LC2 is required for outer arm assembly. The fact that LC2 is an essential subunit of flagellar outer dynein arms led to propose a mechanism whereby transmission ratio distortion is explained by the differential binding of mutant (t haplotype encoded) and wild-type dyneins to the axonemal microtubules of t-bearing or wild-type sperm, with resulting differences in their motility (Pazour et al., 1999).

The *Tctex2* maps to the t-complex and encodes a membrane-associated protein found exclusively on the sperm tail. The t-haplotype form of Tctex2 is aberrant in both the levels of

```
Chlamy    ------------------------------------------------------------
Mouse     MERRGRMAKTPTGQTHQSPVSKRERKPSMFEKESYAQILRERLRESFHDVQYVEPPFDDS
Trout     ------------------------------------------------------------
SU        ------------------------------------------------------------

Chlamy    --------------------------------------------------------MDDM
Mouse     MERRGRMAKTPTGQTHQSPVSKRERKPSMFEKESYAQILRERLRESFHDVQYVEPPFDDS
Trout     ---------------------MSERRGSIMKG-IRKDSHTDRLRGSTMDIEMHDD-DDSH
SU        -------------------MNRRATLQVENLNMPQRSSLFDLTARRPHEMGQSYLALGCA

Chlamy    PFQSGGIKEVVYENTFITDPEGYGPNTKFERHKVQAVLKQVLKERIEKQKYDPVKGAQIS
Mouse     IADVGKEWKSALAKLKFANSYRMEPLKKFQAHLVETKIQQILKDSLKDVKYD-DKAPHLS
Trout     PVKRGSSVWKSGFKGKLANTYRLGPKQKFLPHLIQKKGEELMNQAFGELAYDHDNCRDVA
SU        PSDHEIEEYHVKHTVEFENTYRMEPKNRFVAERVEPIMKEVMEKHLEKVSYEPTQCGHLA
                        .  .     *  .*    ..  .....  .    *.  .   ..

Chlamy    KQLADDLREKVKALGYDRYKLVIQVTVGQKQGQAMRIISRCLWDQTNDNFASEYYENESM
Mouse     LELADGILAAVKEFAYHRYKFIIQVLFIQKTGQAINIASRWIWDVAWDNWVEAKHETESY
Trout     DKVAADVLAFCKEQVFDRYRYVVRVVVGDKKGQTVKIASRTLWDDEKDNFLTLNFENRHL
SU        KTIVSEIKTQVKELEFDRYKIICLVDIGEKRDQDIHMGSRCLWDHQRDTYASATYENHSL
            .   .   *   . **. .   *   .*  *  .  ** .**   * .    *

Chlamy    YCVCQVYGLYYE
Mouse     VVLALVFALYCE
Trout     FAVGMVFAIYFE
SU        FAAAIVFAVYFE
             *. .* *
```

Fig.9.5 Amino acid comparison of mouse Tctex2 dynein light chain with 19kDa LC dynein from *Chlmydomonas* (Chlasy), trout LC2, and sea urchin LC1 (SU). Identical amino acids are marked by asterisks. Phosphorylation sites by PKA are boxed. Reprinted with permission from K. Inaba et al, Biochem Biophys Res Commun 256; 177-83; 1999 © Elsevier.

its expression and its primary amino acid sequence. The multiple amino acid changes make it unlikely that it can function normally and, since it is found on sperm tails, suggest that it may actively interfere with the development of normal gamete function in males and sterility (Huw et al., 1995). Molecular analysis of a 19,000-Mr protein from the *Chlamydomonas* flagellum reveals that it is homologous to the t complex-encoded protein Tctex-2. The 19,000-Mr protein extracted from the axoneme co-migrates with the outer dynein arm in sucrose density gradients. This protein also is specifically missing in axonemes prepared from a mutant that does not assemble the outer arm. These results raise the possibility that Tctex-2 is a sperm flagellar dynein component (Patel-King et al., 1997).

A dynein-mediated TRD model has been proposed to the mechanism for how the +/- sperm become defective during spermatogenesis in +/t males, in motility by the action of Tcds. It has been concluded that the mutations of Tcds resulting into the defective sperm motility depends on abortive loss of cAMP-dependent phosphorylation of sperm axonemal components. The major target of cAMP–dependent protein phosphorylation is the structural component of dynein. Although OAD generally contains six to eight LCs, only a few, or in most cases only one LC can be phosphorylated by PKC. The dynein LCs known to be phosphorylated in a cAMP-dependent manner include a 29-kDa LC in *paramecium* cilia, a 20-kDa LC in *Spisula gill* cilia and a 18-20kDa LC in *Ciona* sperm. The 22-kDa OAD LC in salmonid fish is phosphorylated in a cAMP-dependent manner in parallel to the activation of sperm motility. In a similar finding, Inaba et al (1999) identified a Tctex2-related LC of OAD from sea-urchin sperm as the only LC to be phosphorylated. Molecular analyses of the phosphorylated LCs from sperm flagella of the salmonid fish and sea urchin revealed that the light chains are homologs of the mouse Tctex2, a putative t complex distorter. Thus mouse Tctex2 might also be a light chain of flagellar outer arm dynein and that the abortive phosphorylation of Tctex2/OAD light chain might be related to the loss of progressive movement of sperm (Inaba et al., 1999) **(Fig. 9.5)**.

9.6.2. Cytoplasmic Dynein in Sertoli Cells and Germ Cells

The Sertoli cells cytoskeleton is complex and morphologically related to a number of events regulating spermatogenic cells. One of these events is the movement or translocation of spermatids to the base and then back to the apex of the epithelium. This change in position occurs after spermatids have acquired an elongate shape, and while they are attached to apical invaginations (crypts) of Sertoli cells. Microtubules are prominent elements of the Sertoli cell cytoskeleton. They surround apical crypts containing elongate spermatids, generally arranged parallel to the axis of spermatid translocation, and have their positive ends positioned at the base of the epithelium. To determine whether motor proteins are present in Sertoli cell, it was shown that besides presence of dynein, kinesin II Ab reacted with sperm tail regions involved with spermatid translocation. Some of the dynein was found associated with Sertoli cell junction plaques attached to spermatids concluding that cytoplasmic dynein may be responsible for apical translocation of elongate spermatids that occurs before sperm release (Miller et al., 1999). Criswell and Asai (1998) presented evidence for four cytoplasmic dynein heavy chain isoforms in rat testis. Cytoplasimic dynein was observed in Sertoli cells during all stages of spermatogenesis, with a peak in apical cytoplasm during stages IX-XIV. Cytoplasmic dynein was also localized within Sertoli cells to steps 9-14 (stage IX-XIV) germ cell-associated ectoplasmic specialization. Yoshida and Coworkers (1994) localized cytoplasmic dynein, to spermatid manchettes and also noted a moderately intense staining in Sertoli cells.

In germ cells, cytoplasmic dynein was present in manchettes of steps 15-17 (stages I-IV) spermatids followed by observation of small, hollow circular structures in the cytoplasm of step 17 and step 18 spermatids. The stage-dependent apical Sertoli cell cytoplasmic dynein immunofluorescence was consistent with the hypothesis that cytoplasmic dynein is involved in Sertoli cell protein transport and secretion. The localization of cytoplasmic dynein and kinesin to manchettes is consistent with current hypotheses concerning manchette function. Cytoplasmic dynein is the only known kinetochore protein capable of driving chromosome movement toward spindle poles. In grasshopper dynein is a transient component of spermatocytes and binding is sensitive to presence of microtubules. As a result the bulk of dynein leaves the kinetochore very early in mitosis soon after kinetochores begin to attach microtubules. Thus the possible function of this dynein fraction is limited to the initial attachment and movement of chromosomes and/or to a role in the mitotic checkpoint (Hall et al., 1992; King et al., 2000b).

9.7. STRUCTURE-FUNCTION RELATIONS IN DYNEINS

Unfortunately much of the work on dyneins relating to motility with outer arms in situ involves *Chlamydomonas* mutants, while work on purified fractions was concentrated on sea urchin sperm dyneins. Hence making co-relation is difficult. The questions to be answered include whether both (all three) outer arm heavy chains participate equally in force generation, and whether any of the smaller chains serve specific regulatory roles in modulating heavy chain activity. Sequence data showed that three outer arm heavy chains retain only about 45% sequence identity suggesting functional diversity within outer arm, where each heavy chain seems to contribute differently to flagellar motility (Mitchell, 1994). For example mutations in α and β subunits yield different results functionally. Sea urchin α and β heavy chains differ in their microtubule binding properties as shown by an assay based on bridging of microtubules. The α subunit of sea urchin sperm outer arm dynein mediates structural and rigor binding to microtubules (Moss et al., 1992). The pH dependent changes in normal and lacking outer

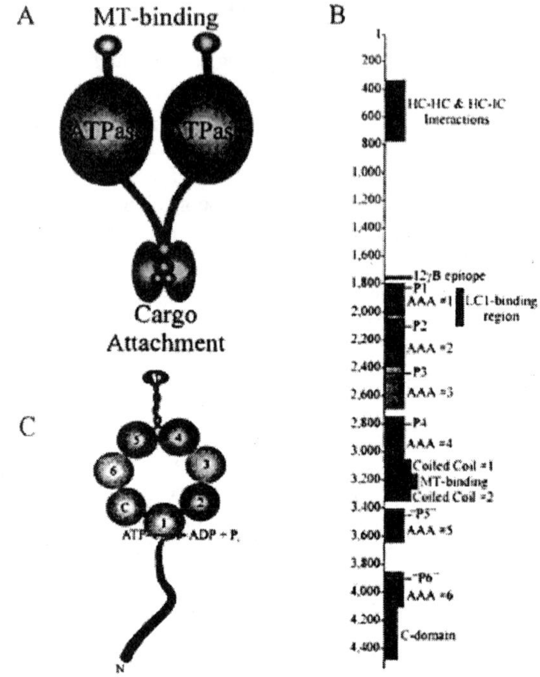

Fig.9.6. Organization of the dynein motor domain. (A) Generic model of a dynein particle containing two heavy chains (HCs). C-terminal portion of each HC forms a globular head containing ATPase sites and has a small stalk-like structure that terminates in a microtubule-binding globular unit. The base of dynein particle consists of N-terminal end of HCs and other accessory proteins, which help in cargo binding besides other putative functions. (B) Map of gamma HC from *Chlmydomonas* outer arm dynein having six AAA#1-AAA#-6 domains (both evident and cryptic P loops), microtubule binding region and coiled-coil segments. It illustrates the regions involved in HC-HC and HC-IC interactions besides a unrelated C-terminal sub-domain. (C) Model of each HC illustrating heptameric structure of the head. Reprinted with permission from S.M King. J Cell Sci 113; 2521-26: 2000 © The Company of Biologists Ltd.

dynein arm (LODA) sperm population suggested a role for outer dynein arms in the initiation of the propagation of flagellar bends at alkaline pH depending on the Ca^{2+} in the medium. Thus a functional difference between outer and inner dynein arms of spermatozoa depends on pH sensitivity. It was concluded that the dynein arms generate force in both directions and this feature of the dynein arms arises from at least the inner dynein arms. In addition to hydrolyzing ATP to generate the driving force necessary for microtubule sliding within the axoneme, 19S dynein may also interact with ATP to polymerize tubulin into microtubules (Eyer et al., 1987).

9. 7.1. AAA Domains and Organization of the Dynein Motor Unit

Present studies indicate that the dynein motor unit is constructed around a series of AAA (ATPase Associated with diverse/cellular Activities) domains, which suggests that this enzyme class is fundamentally distinct from kinesin and myosin at both the structural and mechanistic levels. The AAA domain is found in a very wide variety of proteins ranging from bacterial protease regulators and metal chelatases to transcriptional regulators and the microtubule-severing protein katanin. The EM and other structural analysis of 2-3 proteins having AAA domain have suggested the present dynein domain organization **(Fig. 9.6)**. This domain consists of an ATP-binding motif and P-loop combined with a core α/β structure on which protein-specific modules are added (King, 2000b). In combination with the recent sequence comparisons revealing dynein as a member of this group, the new structural data allow us to draw conclusions about the organization of this motor and the mechanisms of dynein function. A spermatogenesis-associated factor (SPAF) is aberrantly expressed at the malignant conversion stage in a clonal epidermal model of chemical carcinogenesis. Sequence analysis revealed two ATPase modules, classifying this protein as a new member of the AAA-protein family.

Dyneins consist of a large number of globular motor domains depending on number of heavy chains present in each enzyme. Globular motor domains are attached to a cargo-binding unit, through stems. These enzymes are built around 1-3 heavy chains of 520kDa each, which contributes an N-terminus, a stem and a C-terminus globular head that contains a ATP hydrolysis sites. The microtubule–binding region is located at small globular domain emanate at the tip of stalk that protrudes from the main head. A generic model of dynein motor containing two heavy chains is shown in Fig. 9.5 (King, 2000a).

9.8. REFERENCES

Baccetti B, Porter KR, Ulrich M. High voltage electron microscopy of sperm axoneme. J Sub Microsc Cytol 1985, 17: 171-6.

Bi GQ, Morris RL, Liao G, Alderton JM, Scholey JM, Steinhardt RA. Kinesin-and myosin driven steps of vesicle recruitment for Ca^{2+} regulated exocytosis. J. Cell Biol 1997; 138; 999-1008.

Criswell PS,Asai DJ.Evidence for four cytoplasmic dynein heavy chain Isoforms in rat testis.Mol Biol Cell1998;9:237-47

Eyer J, White D, Gagnon C. Bull sperm 19S dynein polymerizes brain tubulin into microtubules. Biochem Biophys Res Common 1987; 148: 218-24.

Eyer J, White D, Gagnon C. Presence of a new microtubule cold-stabilizing factor in bull-sperm dynein preparations. Biochem J 1990; 270: 821-4.

Gagnon C, White D, Cosson J, Huitorel P, Edde B, Desbruyeres E, Lafanechere L, Multigner L, Job D, Cibert C. The polyglutamylated lateral chain of α-tubulin plays a key role in flagellar motility. J Cell Sci 1996; 109 (Pt 6): 1545-53.

Garber AT, Retief JD, Dixon GH, Isolation of dynein heavy chain cDNAs from trout testis which predict an extensive carboxyl-terminal a-helical coiled-coil domain. EMBO J 1989; 8: 1727-34.

Gatti JL, Dacheux JL, Immunological cross-reaction between sperm dynein heavy chains from different species. Reprod Nutr Dev 1996; 36:213-20.

Gibbons IR, Gibbons BH, Mocz G, Asai DJ. Multiple nucleotide-binding sites in the sequence of dynein β heavy chain. Nature 1991; 352 6336): 640-3.

Gibbons IR. Dyenin family of motor proteins: present status and future questions. Cell Motil Cytoskeleton 1995; 32: 136-44.

Gingras D, White D, Garin J, Cosson J, Huitorel P, Zingg H, Gagnon C. Molecular cloning and characterization of a radial spoke head protein of sea urchin sperm axonemes: involvement of the protein in the regulation of sperm motility. Mol Biol Cell 1998; 9: 513-22.

Gingras D, White D, Garin J, Multigner L, Job D, Cosson J, Huitorel P, Zingg H, Dumas F, Gagnon C. Purification, cloning, and sequence analysis of a Mr = 30000 protein from sea urchin axonemes that is important for sperm motility. Relationship of the protein to a dynein light chain. J Biol Chem 1996; 271: 12807-13.

Hall ES, Eveleth J, Jiang C, Redenbach DM, Boekelheide K. Distribution of the microtubule-dependent motors cytoplasmic dynein and kinesin in rat testis. Biol Reprod 1992; 46: 817-28.

Harrison A, Olds-Clarke P, King SM. Identification of the t complex-encoded cytoplasmic dynein light chain tctex1 in inner arm 1 supports the involvement of flagellar dyneins in meiotic drive. J. Cell Biol 1998; 140: 1137-47.

Henson JH, Cole DG, Roesener CD, Capuano S, Mendola RJ, Scholey JM. The heterotrimeric motor protein kinesin II localizes to the midpiece and flagellum of sea urchin and sand dollar sperm. Cell Motil Cytoskeleton1997;38:29-37.

Hirose K, Fan J, Amos LA. Re-examination of the polarity of microtubules and sheets decorated with kinesin motor domain. J Mol Biol 1995; 251: 329-33.

Huw LY, Goldsborough AS, Willison K, Artzt K. Tctex2: a sperm tail surface protein mapping to the t-complex. Dev Biol 1995; 170:183-94.

Inaba K, Kagami O, Ogawa K. Tctex2-related outer arm dynein light chain is phosphorylated at activation of sperm motility. Biochem Biophys Res Commun 1999; 256: 177-83.

Inaba K. ATP-dependent conformational changes of dynein: evidence for changes in the interaction of dynein heavy chain with the intermediate chain 1: J Biochem (Tokyo) 1995; 117: 903-7.

Junco A, Bhullar B, Tamasky HA, van der Hoorn FA. Kinesin light chain KLC3 expression in testis is restricted to spermatids.

Boil Reprod 2001; 64: 1320-30.

Kagami O, Gotoh M, Makino Y, Mohri H, Kamiya R, Ogawa K. A dynein light chain of sea urchin sperm flagella is a homologue of mouse tctex 1, which is encoded by a gene of t-complex sterility locus. Gene 1998; 211: 381-86.

Kastury K, Taylor WE, Shen R, Arver S, Gutierrez M, Fisher CE, Coucke PJ, Van Hauwe P, Van Camp G, Bhasin S. Complementary deoxyribonucleic acid cloning and characterization of a putative human axonemal dynein light chain gene. J Clin Endocrinol Metab 1997; 82: 3407-53.

Kawaguchi K, Ishiwata S. Temperature dependence of force, velocity, and processivity of single kinesin molecules. Bichem Biophys Res Commun 2000; 272:895-99.

King SM. AAA domains and organization of the dynein motor unit. J Cell Sci 2000a; 113 : 2521-26.

King JM, Hays TS, Nicklas RB. Dynein is a transient kinetochore component whose binding is regulated by microtubule attachment, not tension. J Cell Biol 2000b;15 : 739-48.

King SM, Dillman JF III, Benashski SE, Lye RJ, Patel-King RS, Pfister KK. The mouse t-complex-encoded protin Tctex-1 is a light chain of brain cytoplasmic dynein. J Biol Chem 1996; 271:32281-87.

Liu Y, Black J, Kulessz-Martin MF. SPAF, a new AAA-protein specific to early spermatogenesis and malignant conversion. Oncogene 2000; 19: 1579-88.

Macho B, Brancorsini S, Fimia GM, Setou M, Hirokawa N, Sassone-Corsi P. CREM-dependent transcription in male germ cells controlled by a kinesin. Science 2002; 298(5602): 2388-90.

Marchese-Ragona SP, Gagnon C, White D, Isles MB, Johson KA. Structure and mass analysis of 12S and 19S dynein obtained from bull sperm flagella. 143: Cell Motil Cytoskeleton 1987; 8: 368-74.

Milisav I, Jones MH, Affara NA. Characterization of a novel human dynein-related gene that is specifically expressed in testis. Mamm Genome 1996; 7: 667-72.

Milisav I. Dynein and dynein-related genes. Cell Motil Cytoskeleton 1998; 39:261-72.

Miller MG, Mulholland DJ, Vogl AW. Rat testis motor proteins associated with spermatid translocation (dynein) and spermatid flagella (kinesin-II). Biol Reprod 1999; 60: 1047-56.

Mitchell DR. Cell and Molecular Biology of flagellar dyneins. Intern Rev Cyt 1994; 155 : 141-80.

Moss AG, Sale WS, Fox LA, Witman GB. The α-subunit of sea urchin sperm outer arm dynein mediates structural and rigor binding to microtubules. J Cell Biol 1992; 118: 1189-200.

Navolanic PM, Sperry AO. Identification of isoforms of a mitotic motor in mammalian spermatogenesis. Biol Reprod 2000; 62:1360-69.

Ogawa K, Takai H, Ogiwara A, Yokota E, Shimizu T, Inaba K, Mohri H. Is oute arm dynein intermediate chain 1 multifunctinoal? Mol Biol Cell 1996; 1895-907.

Ogawa K. Four ATP-binding sites in the midregion of the heavy chain of dynein. Nature 1991; 352(6336): 643-5.

Paschal BM, Mikami A, Pfister KK, Vallee RB. Homology of the 74-kD cytoplasmic dynein subunit with a flagellar dynein polypeptide suggests an intracellular targeting function. J Cell Biol 1992; 118: 1133-43.

Patel-king RS, Benashski SE, Harrison A, King SM. A *Chlamdomonas* homologue of the putative murine t complex distorter Tctex-2 is an outer arm dynein light chain. J Cell Biol 1997; 137: 1081-90.

Pazour GJ, Koutoulis A, Benashski SE, Dickeri BI, Sheng, H, Patel-King RS, King SM, Wiman G. LC2, *Chlamdomonas* homologue of t complex encoded protein Tctex 2, is essential for outer dynein arm assembly. Mol Biol Cell 1999; 10: 3507-20.

Romberg L, Pierce DW, Vale RD. Role of the kinesin neck region in processive microtubules-based motility. J Cell Biol 1998; 140: 1407-16.

Sale WS, Fox LA. Isolated β-heavy chain subunit of dynein translocates microtubules in vitro. J Cell Biol 1988; 107:1793-97.

Samant SA, Ogunkua O, Hui L, Fossella J, Pilder SH. The T complex distorter 2 candidate gene, Dnahc8, encodes at least two testis-specific axonemal dynein heavy chains that differ extensively at their amino and carboxyl termini. Dev Biol. 2002; 250: 24-43.

Shingyoji C, Higuchi H. Yoshimura M, Katayama E, Yanagida T. Dynein arm oscillating force generations. 348: Nature 1998; 393(6686): 711-4.

Sperry AO, Zhao LP. Kinesin related proteins in the mammalian testes: candidate motors for meiosis and morphogenesis. Mole Biol Cell 1996; 7: 289-305.

Williams BC, Dernburg AF, Puro J, Nokkala S, Goldberg ML. The Drosophila kinesin-like protein KLP3A is required for proper behavior of male and female pronuclei at fertilization. Development 1997; 124: 2365-76.

Williams BC, Riedy ME, Williams EV, Gatti M, Goldberg ML. The Drosophila kinesin-like protein KLP3A is a midbody component required for central spindle assembly and initation of cytokinesis. J Cell Biol1995;129:709-23.

Yamamoto T, Matsuda Y, Hikino T, Nakayama Y, Abe S. Cloning of dynein intermediate chain cDNA of the newt *cynops pyrrhogaster.* DNA Res 2001; 8: 81-4.

Yamano Y, Ohyama K, Sano T, Ohta M, Shimada A, Hirakawa Y, Sugimoto M, Morishima I. A novel spermatogenesis related factor-1 gene expressed in maturing rat testis: Biochem Biophys Res Commun 2001; 289: 888-93.

Yang JT, Saxton WM, Stewart RJ, Raff EC, Goldstein LS. Evidence that the head of kinesin is sufficient for force generation and motility in vitro. Science 1990; 249 (4964): 42-7.

Yokota E, Mabuchi I. Interaction of flagellar inner arm dynein isolated from sea urchin sperm with microtubules in the presence of ATP. Eur J Cell Biol 1997; 72 : 214-21.

Yokota E, Mabuchi I. Isolation and characterization of a novel dynein that contains C and A heavy chains from sea urchin sperm flagellar axonemes: J Cell Sci 1994a; 107 (Pt 2): 345-51.

Yoshida T, Ioshii SO, Imanaka-Yoshida K, Izutsu K. Association of cytoplasmic dynein with manchette microtubule and spermatid nuclear envelope during spermiogenesis in rats. J Cell Sci 1994; 107: 625-33.

Chapter 10

SEX CHROMOSOMAL PROTEINS AND AUTOSOMAL HOMOLOGUES

Abnormal spermatogenesis is caused due to multiple reasons. Survey suggests that among infertile couples, male infertility is responsible for ~40% out of which 10.7% are azoospermic or oligospermic. The problem in these men appeared to be related to the failure of spermatogenesis, the cause of which is unknown. In many cases the failure of spermatogenesis is due to genetic defects. Although the process of spermatogenesis has been intensively studied, our knowledge on the genetic basis of spermatogenesis is meager. Chromosomal abnormalities constitute an important factor, which can cause spermatogenic breakdown at various points resulting in chromosomally derived sterility (review Ma et al., 2000; Shah et al., 2003; Harley et al 2003). In recent years deficiency of certain proteins associated with genes on sex chromosomes has been ascribed as the cause of oligo- or azoospermia. Y chromosome plays significant role in such infertility. Now it is clear that the Y chromosome is necessary not only for sexual development but also for spermatogenesis. Y-chromosome controls the sex in *Homo sapiens* and other mammals like *Melandrium.* Turner's syndromes (XO) are sterile female individuals having certain abnormalities. Similarly, Klinefelter's syndromes (XXY) are males, despite presence of two X-chromosomes. Klinefelter syndrome, with an incidence of 1:600 male newborns, is the most frequent form of male hypogonadism. However, despite its relatively high frequency, the syndrome is often overlooked (Kamischke et al., 2003). In contrast to this, a *Drosophila* fly with 2A + XXY, which is exactly like Klinefelter's syndrome of human beings in chromosome constitution, is a normal female individual and XO *Drosophila* fly is male. This suggests that in human beings sex determining mechanism is like that in plants and not like that in *Drosophila.* Genetic studies of male infertility have demonstrated that the long arm of the Y chromosome (Yq) harbors at least 15 gene families, some of which are necessary for spermatogenesis. With the advent of present molecular techniques, a variety of Y-deletions of different locations are being identified in infertile patients with different phenotypes (**Fig.10.1**). Chromosomal aberrations may be of two types (i) numerical and (ii) structural. The former type-1 aberrations arise either because of missing or due to addition of accessory chromosome due to meiotic non-disjunction in paternal or maternal germ cells. Accessory chromosome may be an autosome or sex chromosome or both leading to somatic or reproductive cell defects. However, structural chromosome aberrations may involve loss or duplication of genetic information from one chromosome to another. Such aberrations can be at microscopic level or may be located at molecular level due to point mutations. Such aberrations may be in hundreds due to phenotypic abnormalities. The sex-linked sequences generally are less nuclease sensitive than are autosomal sequence. Interestingly, a hot spot of recombination (within the *Eb* gene) showed a high level of nuclease sensitivity, while a cold spot of recombination (centromeric satellite region) exhibited lower sensitivity, more similar to that of

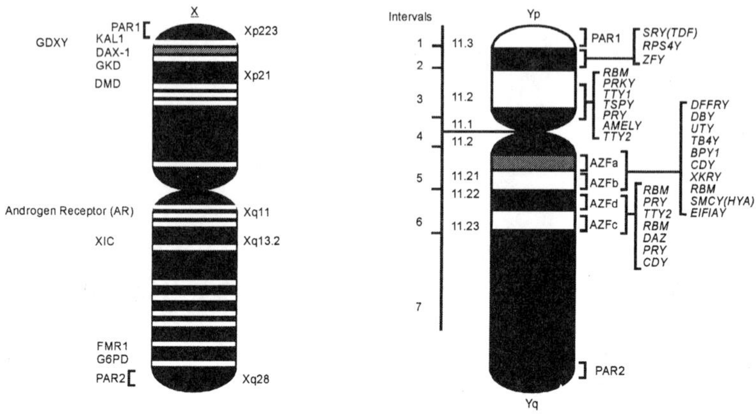

Fig.10.1. Diagrammatic representation of sex chromosomes showing genes on the X and Y chromosomes that have been found to be associated with either spermatogenesis or the differentiation of the male phenotype. The Y chromosome is divided into AZFa, AZFb, AZFc, and AZFd regions and mapped to intervals 5 and 6. Pseudoautosomal regions on X and Y (PAR1 and PAR2) chromosomes are shown. Genes located on AZF regions are also indicated (Ma et al 2000).

sex-linked sequences. Thus, in pachytene spermatocytes the XY chromosome pair is more condensed and inaccessible to enzymatic digest, whereas the autosomal chromatin is in a more open configuration and more accessible for enzyme digestion. In addition, while autosomal gene nuclease sensitivity was equivalent to that at the pachytene stage, X-linked sequences were more nuclease sensitive. These differences in chromatin nuclease sensitivity correlate with differences in meiotic recombination activity and may be mechanistically related (Wiltshire et al., 1998).

10.1. X-CHROMOSOME ABERRATIONS AND SPERMATOGENESIS

Infertility is caused by minor deletions or translocations of gene on the X chromosome. Pronounced effects are seen after deletion of major portion of X chromosome.

10.1.1. Xp22 Contiguous Gene Syndrome

The Xp22 contiguous gene syndrome is a rare disorder. De-novo deletion of Xp22 pter from the X chromosome affects several genes, including those responsible for glycerol kinase deficiency and Kallmann's syndrome (KAL1). KAL-1 is deleted in ~50% of the patients with the Xp22 contiguous gene syndrome, causing a primary deficiency in GnRH resulting into failure of sperm production (Diemer and Desjardins, 1999).

10.1.2. Translocation of X-Chromosome Genes to Autosomes and Y-Chromosome

Normal spermatogenesis is dependent on X chromosome inactivation, a process directed by an X-linked gene during the spermatocytes stage of germ cell development. The pairing and recombination of X and Y-chromosomes, followed by X inactivation, is limited to the pseudo-autosomal regions and occurs when both chromosomes are transcriptionally inactive.

Translocation of genes on the X chromosome to either autosomes or the Y chromosome is not a common cause of male infertility. But balanced translocations of genes on the X chromosome to either autosome or Y chromosome interfere with male sexual differentiation or spermatogenesis or both. Such translocations typically occur at break points in the pseudo-autosomal boundary region of the X chromosome (Xp-22-pter) and in the Yq11 region of the Y chromosome. Sexual differentiation is affected by translocations in Xp21.2-pter *(GDXY)* and this region appears to be crucial for the differentiation of the male phenotype. The *DAX-1,* a gene located at Xp21, is a decisive X gene for sexual differentiation. Expression of *DAX-1* in mice indicates that it plays a critical role in early differentiation of adrenal and Sertoli cells in the genital ridge but prolonged expression of DAX-1 interferes with testis development (Swain et al., 1998 c/r Diemer and Desjardins, 1999; Shah et al., 2003).

10.2. X-CHROMOSOME LINKED PROTEINS

Certain X-chromosomal genes remain transcriptionally silent in spermatids. In contrast, other X-chromosomal genes such as *Ubelx* and *MHR6A* are post-meiotically transcribed genes (see Chapter 31). The inactivation of *Pgk-1* and *Pdha-1* in spermatocytes and spermatids is compensated by testis specific expression of the autosomal genes *Pgk-2* and *Pdha-2,* which encode corresponding isozymes. The *Pgk-2* and *Pdha-2* are intron-less retroposons that originated from reverse transcription of mRNA molecules. The X-chromosomal gene glucose-6-phosphate dehydrogenase *(G6pd)* is expressed in most cell types of mammalian species. In the mouse, the chromosomal *G6pd-2* encoding a G6PD isoenzyme also does not contain introns and appears to represent a retroposed gene (see Chapter 28).

10.2.1. Pro-mAKAP82

The fibrous sheath pro-AKAP82 is an X-linked gene that is mapped to human chromosome Xp11.2, and is localized to sperm flagellum. The cDNA sequence of human sperm pro-hAKAP82 is highly homologous to the mouse pro-mAKAP82 sequence, and the functional domains of the pro-hAKAP82 protein, the protein kinase A binding, and the pro-hAKAP82/hAKAP82 cleavage sites are identical to those of the mouse protein. Compared with pro-mAKAP82, pro-hAKAP82 was considerably less processed in human sperm. Although sperm pro-mAKAP localizes only to the proximal portion of the principal piece of the flagellum, pro-hAKAP82 localized to the entire length of the principal piece. The pro-hAKAP82 gene, mapped to human chromosome Xp11.2, indicated that defects in this gene are maternally inherited. It was suggested that hAKAP82 plays several roles in sperm motility, including the regulation of signal transduction pathways (Turner et al., 1998). The primary structure of bovine bAKAP82 is highly conserved in amino acid sequence corresponding to the region of mAKAP82 responsible for binding to regulatory subunit of protein kinase A (see Chapter 29).

10.2.2. SPAN-X

Testis-specific protein designated SPAN-X is a sperm specific protein associated with the nucleus on the X chromosome. SPAN-X sequences showed no significant similarity with known peptide sequences. Its peptide sequences contained three overlapping consensus nuclear localization signals, a high percentage (33%-37%) of charged amino acid residues, and a relatively acidic isoelectric point (pI, 4.88-6.05). The SPAN-X mRNA is expressed in haploid, round and elongating spermatids. The SPAN-X gene was mapped to chromosome

Xq271. Anti-recombinant SPAN-X antibodies reacted with broad bands migrating between 15-20-kDa. Human spermatozoa demonstrated SPAN-X localization to nuclear craters and cytoplasmic droplets. Expression of SPAN-X, exclusively in haploid spermatids leads to interesting questions regarding the transcription of sex-linked genes during spermiogenesis (Westbrook et al., 2000). Wang et al., (2003) demonstrated that the X-linked gene encoding SPAN-Xb is expressed in multiple myeloma and other hematologic malignancies such as chronic lymphocytic leukemia (CLL), chronic myeloid leukemia (CML), and acute myeloid leukemia (AML). Thus, SPAN-Xb is a cancer/testis antigen that shows a restricted normal tissue expression (Wang et al., 2003).

10.2.3. Cleavage Stimulation Factor Like Protein

To determine whether the 64-kDa protein of the cleavage stimulation factor (CstF-64) is altered in male germ cells, Wallace et al., (1999) examined its expression in mouse testis. In addition to the 64-kDa protein a 70-kDa related protein was abundant in testis and undetectable in other tissues except in brain at low level. The CstF-64 was limited to meiotic spermatocytes and post-meiotic spermatids. In contrast, the 64-kDa form was absent from spermatocytes, suggesting that the testis specific CstF-64 might control expression of meiosis-specific genes. The CstF-64 may be absent in spermatocytes because the X chromosome is inactivated during male meiosis. Wallace et al, (1999) suggested that testis specific CstF-64 may be expressed from an autosomal homologue of the X chromosomal gene.

10.3. Y-CHROMOSOME ABERRATIONS

The human Y chromosome contains over 60 million nucleotides, but least number of genes compared to any other chromosome and acts as a genetic determinant of the male characteristic features. Thus far, 156 transcription units, 78 protein-coding genes and 27 distinct proteins of the Y chromosome have been identified. The male specific region, MSY, comprising 95% of the Y chromosome represents a mosaic of heterochromatic and three classes of euchromatic (X-transposed, X-degenerate and ampliconic) sequences. The MSY euchromatic sequences show frequent gene conversion. Of the eight massive palindromes identified on the human Y chromosome, six harbor vital testis specific genes (Ali and Hasnain, 2003; Skaletsky et al., 2003) (**Fig.10.1**). The human male infertility has been attributed to mutations in the genes on Y chromosome and autosomes and failures of several physical and physiological features including paracrine controls. In addition, deletion of any one or all the three azoospermia (AZFa, AZFb or AZFc) factor(s) and several unidentified regulatory elements located elsewhere in the genome result in infertility. The Yq11 region is an important region that contains several genes that regulate transcription and are required for spermatogenesis such as AZF (Vogt, 1998). Characterization of palindromic complexes on the long arm of Y chromosome encompassing AZFb and AZFc regions and identification of some of endogenous retroviruses close to AZFa region have facilitated our understanding on the organization of azoospermia factors. Considerable overlap has been shown to exist between AZFb and AZFc regions encompassing a number of genes and transcripts. However, with the exception of AZF, information on the exact number of genes or the types of mutations prevalent in the infertile male is not available. Similarly, roles of sizable body of repetitive DNA present in close association with transcribing sequences on the Y chromosome are yet not clear (Ali and Hasnain, 2003; Skaletsky et al., 2003).

10.3.1. Y-Chromosome and Sex Reversal

Most direct evidence for the presence of the spermatogenetic factor on the Y chromosome came from the study of Sxr^a gene, a dominant mutation causing sex reversal of female mice. The sex reversed XX males in mice resulted due to transfer of a sex reversing factor (Sxr) from a Y chromosome carrying a duplication for a region of its minute short arm translocated to the distal end of its long arm (this Y chromosome was designated Y^{sxr}). This duplicated segment was transferred to X-chromosomes due to crossing over during male meiosis and then was found to be on the distal arm of X-chromosome, which presumably pairs with a segment of Y chromosome. In situ hybridization to mitotic chromosomes of XY^{sxr} males showed that the rearranged chromatin is located distal to the telomeric signal and that the rearrangement arose from a recombination event involving the distal Y telomere sequences, i.e., within the telomere, a structure historically assumed to be incapable of participating in chromosome rearrangements (Ashley et al., 1995).

Cases of XX sex reversal were also reported in goats and humans. It has been found that these XX sex reversed males may result from either of the following two reasons: (i) due to transfer of a segment (called TDF = testis determining factor) of Y chromosome to X due to abnormal crossing over, in the same manner as the Sxr is transferred from Y to X in mice or (ii) due to mutation in a locus (which represses testis development), so that XX individuals now develop testes. Most of XX males are found to possess Y derived DNA sequences thus supporting the first model.

10.3.2. The SRY: a Sex-Determining Region on YGene in Mammals

In 1980s gene *ZFY* (ZF= zinc finger) was found on human Y chromosome, which was once believed to be TDF (testis determining factor) and was found in sex reversed XX males, but absent in sex reversed XY females. We now know that *ZFY* does not produce the functional transcripts, playing a role in sex determination. On the other hand, another gene *SRY* in humans (or Sry in mouse) has been located in sex determining region of the Y chromosome in 1990s. This gene has many properties expected for TDF. In some sex reversed XY human females, a de novo mutation found in the *SRY* gene, confirmed that *SRY* is required for male sex determination.

Deletions and Mutations of the SRY Gene: The human sex-determining region *SRY* located on the Y chromosome seems to be conserved and specific to the Y chromosome of all mammals examined. In pig genital ridges SRY transcripts were first detectable from 23 days post-coitum (dpc), then declined sharply after 35 dpc. The SRY is required for the regression of Mullerian duct and the subsequently formation of testes, the two sequential step in male sexual differentiation. The *SRY* gene is located on the short arm of the Y chromosome at Yp11.3, adjacent to the pseudo-autosomal boundary on interval 1. Over 26 inactivating point mutations in *SRY* have been discovered so far, and all are associated with gonadal dysgenesis or sex reversal syndromes or both. Point mutations in *SRY* produce structural and functional effects similar to those seen following deletion or translocation. The *SRY* gene spans over 35-kb pairs with an HMG box region of 80 amino acids, crucial for gene function. Mutations in HMG box of 240-bp conserved-motif of *SRY* have most deleterious effects. The 5'-regions of *SRY* gene from several mammalian species including human and sub-human primates have been identified. Highly conserved potential regulatory elements in 10 species are: AP1, Barbie, GATA, Gfi1, cMyb, NF1, Oct1, Sp1, and SRY (Margarit et al., 1998). The *SRY* is absent in some patients due to deletion or translocation to X-chromosome during paternal meiosis. Translocation of *SRY*

```
  1 MQSYASAMLS VFNSDDYSPA VQENIPALRR SSSFLCTESC NSKYQCETGE NSKGNVQDRV
 61 KRPMNAFIVW SRDQRRKMAL ENPRMRNSEI SKQLGYQWKM LTEAEKWPFF QEAQKLQAMH
121 REKYPNYKYR PRRKAKMLPK NCSLLPADPA SVLCSEVQLD NRLYRDDCTK ATHSRMEHQL
181 GHLPPINAAS SPQQRDRYSH WTKL
```

Fig.10.2. Amino acid sequence of sex determining region on Y chromosome (SRY) protein. Source: http// www.ncbi.nlm.nih.gov (Accession no. 005066).

to the X chromosome during meiosis results in either XY female or XX male genotypes in the off spring. The *SRY* encodes for a DNA binding protein SRY and is transcribed as a single exon (Cameron and Sinclair, 1997). This protein functions as a transcription factor regulating the formation of the testis in fetal development. The binding of SRY protein produces a bending to target DNA, and is crucial for male development. However, more recent studies revealed that testes might fail to differentiate in patients with a 46, XY complement of chromosomes and normal *SRY* gene expression, indicating that genes from other chromosomes are required for spermatogenesis and the differentiation of the male phenotype. In mice *Sry* is expressed just before the overt differentiation of the bipotential gonad and for a brief period in somatic cells, to differentiate Sertoli cells (c/r Harley et al., 2003). In addition, the family of SRY/sex proteins seems to be involved in other function such as in the down regulation of mouse Msh4 (MutS homologue 4) gene expression through regulation of promoter transcriptional activity (Santucci – Darmanin et al., 2001).

In the absence of *Sry* expression, neither XX nor XY develop testes, where as alterations in *Sry* expression are often associated with abnormal sexual differentiation. Nef et al., (2003) showed that the insulin receptor tyrosine kinase family, comprising Ir, Igf1r and Irr, is required for the appearance of male gonads and thus for male sexual differentiation. XY mice that are mutant for all three receptors develop ovaries and show a completely female phenotype. Reduced expression of both Sry and the early testis-specific marker Sox9 indicates that the insulin signaling pathway is required for male sex determination (Nef et al., 2003). It is now assumed that SRY acts upstream of SOX9 in the sex determination cascade, but the regulatory link, which should exist between these two genes remains unknown (Pannetier et al., 2003).

Characterization of SRY: The ORF of human SRY is contained within a single exon and encodes a protein of 204 amino acids. The protein chain can be divided into three regions, a central domain and two nuclear localization signals. The central domain contains 79 amino acids, which forms HMG box and functions as a DNA-binding and DNA-bending domain. The C-terminal domain of 7 amino acid have the potential to interact in vitro with PDZ proteins, whereas the N-terminus can enhance DNA binding activity after phosphorylation within this domain. The HMG box within SRY is conserved motif, which can recognize DNA. This motif has the ability to bind specifically to A/TAACAAT/A sequences with high affinity, and non-specifically to cruciform DNA **(Fig10.2)**. The affinities and specificities of full-length hSRY and the hSRY-HMG domain for 20-bp DNAs containing cisplatin, target sites in the CD3 epsilon gene enhancer (AACAAAG) suggested that SRY offers a role in the study of cytotoxicity and organotropic specificity of cisplatin and hence therapeutic target for testicular tumors (Trimmer et al., 1998).

The NMR structure of HMG box in SRY-DNA complex revealed that HMG box consists of three alpha helixes forming an L-shape structure. Helix-1 and helix-2 run anti-parallel and form short arm. The longer helix-3 runs anti-parallel to the N-terminal sequence strand and forms long arm. The structure is stabilized by conserved aromatic amino acids, which form hydrophobic core of the protein **(Fig.10.3A)** Binding of SRY causes DNA to unwind and

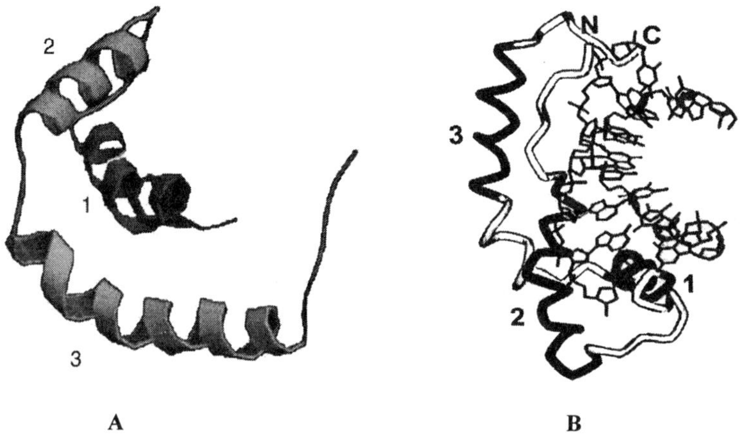

Fig. 10.3. (A) The L shaped structure and three alpha-helices (1, 2 and 3) of SRY derived by NMR. Adapted from http:ncbi.nlm.nih.gov. (B) The HMG box of SRY bound to the minor groove of DNA causes DNA to bend and unwind. Adapted with permission from M.H. Werner et el, Cell 81;706-14:1995b © Elsevier.

bending of DNA by about 80 degrees. The bending of DNA presumably results into a particular DNA architecture in chromatin responsible for displacing of a nucleosome at fra-2 promoter (Ng et al., 1997; Werner et al., 1995a,b) (**Fig.10.3 B**). The SRY and other HMG-box proteins can also bind cruciform DNA with normal affinity as compared to its 50-fold less binding affinity with Ile68Thr SRY protein variant (Rev: Harley et al., 2003).

SRY Interacting Proteins: SRY protein contains a bipartite nuclear localization signal (NLS) at the N-terminal end and a simian virus 40-type at C-terminal end of the HMG box (Argentaro et al., 2003)(**Fig. 10.4.**). The N-terminal NLS forms part of CaM-binding domain, which helps in the nuclear import of other proteins through CaM. The C-terminal of SRY binds importin beta. Another protein, SIP1 interacts with SRY via a PDZ- type domain. The biological significance of these interactions is unknown.

Other Sex Determining Genes: In addition to Y linked testis-determining gene *SRY*, a number of other sex determining genes have been identified, which are associated with a failure of sex determination. These include *WT1, DAX1, SOX9* and *ATRX*. Most cases of human sex reversal, XY females and XX males, need not necessarily be caused by mutations in these genes (McElreavey and Fellous, 1997). Among other sex determining genes, the key stage seems to be 28-dpc, in which *SOX9*, an autosomal gene expression switches between the male and female, and anti-Muellerian hormone (AMH) expression begins to attest Sertoli cell differentiation and the corresponding seminiferous cord formation in the male (see autosomal genes).

10.4. Y CHROMOSOME AND SPERMATOGENESIS

Translocations between the Y chromosome and autosomes in men have shown adverse effects and severe disorders in spermatogenesis. Analysis of testicular biopsies suggests that translocations of the Y chromosome explain multiple disorders in spermatogenesis ranging from failure of spermatogonial differentiation to formation of post-meiotic cells. Many effects are due to interference with sex chromosome pairing and incomplete defective inactivation of X chromosome (Diemer and Desjardins, 1999; Ma et al. 2000).

Drosophila: The Y chromosome has long been considered to be important in *Drosophila* spermatogenesis since lack of the Y chromosome would result in XO male sterility. The spermatogenesis in XO males of *D. melanogaster* is caused due to meiotic arrest (Ma et al., 2000). It was later demonstrated that deletions of the regions containing kl-5, kl-3 and ks-1 would prevent the formation of the outer dynein arm of the axoneme, leading to the collapse of spermatogenesis while deletion of ks-2 would result in a complex phenotype with nuclear crystal formation and abnormal meiosis. The presence of lampbrush loop like structure formed from the Y chromosome in the primary spermatocyte suggested that a lampbrush loop is a male fertility gene, and it contains only one complementation group (Hackstein and Hochstenbach; 1995). Subsequent study revealed that repetitive DNA sequences of fertility gene of *D. Hydei* are conserved on the Y chromosome in humans. More over seven of these DNA sequences, known as the pY6H were located to interval 6 of human Y chromosome. Three members of pY6H family, pY6HP35, pY6HP52 and pY6BS65/E were found missing in oligospermic and azospermic men.

In primary spermatocytes, three of the loci in lampbrush loops contain RNA transcripts and associated proteins. The identities and functions of these Y chromosome loop-binding proteins are largely unknown. The RB97D protein, which is essential for spermatogenesis and bound to a specific lamp brush loop, contains two copies of a well characterized RNA binding domain, the RNA recognition motif, followed by a proline-glutamine rich domain and found in only spermatocytes nuclei. The protein and its loop binding function are evolutionarily conserved and can be essential for fertility (Heatwole and Haynes, 1996). Another gene, *Dhmst 101* is a member of small gene family specifically expressed in D. *hydei* and encodes a protein of 344 aa with a Mr of 37793 Da. The main portion of the protein contains repetitive highly charged amino acids. The protein is localized in elongated spermatids (Neesen et al., 1994). The *boule (bol)* gene has been identified in *Drosophila. Boule* cDNA codes a protein of 228 amino acids that contained a single RNP type RNA binding domain. The putative RNA binding motif of Boule protein is most similar (42%) to that of human *DAZ* and mouse protein Dazl1. The *DAZ* maps to Yq in humans, but *boule* and mouse Dazl1 are autosomal (see section 10.4.2)

Mouse: Observations from different laboratories suggested the existence of a gene or genes involved in spermatogenesis in the Sxr^a region, since the germ cells entered meiosis normally (Burgoyne, 1979). Another mutant called Sxr^b discovered in 1984 also showed complete block of spermatogenesis at pre-pubertal stage at the onset of meiosis in adult testis. It has been found that Sxr^b is derived from Sxr^a gene, which is a deleted portion of DNA that contains genes for spermatogenesis (review Ma et al., 2000). However, the genetic background has a significant influence on the sperm morphology and fertilization efficiency in mice after Y chromosome deletion. Spermatozoa from CBA mice with B10.BR-Ydel strain showed deformation in acrosome region suggesting that Y chromosome has a regulatory role in spermiogenesis (Styrna et al., 2003). The deletion of Y chromosome also affects the epididymal maturation of sperm and their motility (c/r Styrna et al., 2003).

Human: Translocation of a part of the Y chromosome to autosomes typically occurs in chromosomes 1, 3 and 11 and impairs spermatogenesis. Breakpoints in the Y chromosome frequently arise in the Yq11 or Yq12 region. Genes affected by these translocations have not been identified but probably include those in the azoospermia factor (*AZF*) region of the Y chromosome. This region has been divided into three sub-regions (AZFa, AZFb, AZFc; Table 10.1) where genetic defects have been associated with different grades of impairment and arrest of spermatogenesis due to chromosomes pairing. A new locus, *AZFd* locus was discovered below *AZFb* and *AZFc* (Kent-First, 1999). Infertile men subsequently showed the identification of four gene families for *AZF*. They are RNA binding motif (RBM) (previously named Y-linked RNA recognition motif (*YRRM*)), deleted in azoospermia *(DAZ)*, chromodomain Y (*CDY*) and *Drosophila* related fat facets related Y *(DFFRY)* (**Fig. 10.1**). Many effects are due to interference with sex chromosome pairing and incomplete or defective inactivation of X chromosome.

10.4.1. AZFa Region

Deletion analysis of Yq11 region has revealed more than 15 genes, which have been identified as AZF candidates, and several studies are in progress. Three different regions of Yq11.21-23 within interval 5 and 6 (AZFa, AZFb, AZFc; **Fig.10.1.**) have been mapped. The *DFFRY* and *DFFRX* are homologues of the *Drosophila* developmental gene fat facets *(faf)*, which have been mapped to the proximal Yq11.2 and Xp11.4 respectively in humans. The *faf* gene is a member of the gene family encoding deubiquitinating enzymes, which remove ubiquitin from protein ubiquitin conjugates (Chapter 31). The mouse *Dffry* homologue has been characterized and mapped to the *Sxrb* region on the short arm of the Y chromosome. Deletion of *Sxrb* has been shown to be associated with early spermatogenic failure with an almost total loss of germ cells in meiosis. The expression of mouse *Dffry* is restricted to the testis and can be first detected between 7.5 and 10.5 days after birth, when type A, type B spermatogonia, preleptotene and leptotene spermatocytes are present (Brown et al., 1998).

The human *DFFRY* gene is expressed ubiquitously in adult and embryonic tissues, including the testis (Ma et al., 2000). The *DFFRY* gene is mapped to AZFa and deleted in azospermic and oligospermic men. De novo deletion of genes in the Yq11.21 (AZFa) causes the complete loss of germ cells production. Deletions within AZFa- region occur at much lower frequency than for AZFb and AZFc. AZFa deletions are generally associated with Sertoli cell only syndrome. Candidate genes for infertility within this region include *DFFRY*, *DBY* (Dead box Y gene) and *UTY* (ubiquitous transcribed repeat gene on Y chromosome). The *DFFRY* encodes an ubiquitin-specific hydrolase, which is involved in oocyte development in *Drosophila* and also plays a role in ubiquitin-dependent degradation of proteins; it specifically targets D-jun (see chapter 17) and other members involved in the regulation of D-Jun activity. The DBY encodes a protein that contains the DEAD box motif generally found in RNA helicases (see Chapter 13). The role of *UTY* is currently unknown (c/r Shah et al 2003).

10.4.2. ZFb Region

Within the Yq11.22-23 region (AZFb; **Fig.10.1.**), four genes (*RBM1, SMCY, TSPY and E1F1AY*) with suspected implications for spermatogenesis have been identified. *RBM1* (RNA binding motif 1) and multiple repeats of the RBM gene family are distributed on the Y chromosome. Deletion of regions containing either *RBM1*, *SMCY* or the other genes results in early arrest of spermatogenesis at the level of pre- and post-meiotic spermatocytes in testis. Ejaculates from patients with deletions in the AZFb region are most commonly accompanied with azoospermia, or severe oligozoospermia. Some of these genes encode RNA binding proteins, which are structurally similar to other human RNA binding proteins, while others function as translation initiators and proteases. Certain genes, such as RBM1, are only transcribed in the testis at specific stages of germ cell development (Delbridge et al., 1997; c/r Diemer and Desjardins, 1999). HSFY (heat shock transcription factor, Y chromosome) has been mapped in the AZFb region of the Y chromosome, whose deletion results in severe male infertility. It belongs to the heat shock factor family that is being implicated in spermatogenesis. Alternative splicing generates three different transcripts and proteins, each containing an HSF domain typical of HSF proteins. Two identical and functional full-length copies of HSFY map in palindrome P4 of AZFb, whereas four similar sequences mapping in two clusters in palindrome P1 of AZFc and P3 seem to represent pseudogenes. Sequences similar to few HSFY exons are also located in the short arm of chromosomes Y, X and 22. The three HSFY transcripts are differentially expressed; transcript 1 being present in many tissues including testis and ejaculated sperm, and transcripts 2 and 3 being testis-specific (Tessari et al., 2004).

10.4.3. AZFc Region

The Yq11.22-23 region also contains *AZFc* that regulates spermatogenesis related *DAZ* and *SPGY* genes (spermatogenesis gene on Y also referred to as *DAZ2*) and located distal of *DAZ*. Two new candidate genes *TTY1* and *TTY2* have also been known but their translation in germ cell development has not been confirmed. The *DAZ* and *SPGY (DAZ2)* share high structural homologies, by having an identical 72 base pair exons, but appear to function as independent genes of the same family. Deletions of *DAZ* and *DAZ2* interfere with the maturation of spermatids (**Table 10.1**). Genes in the Yq11.21-23 region (**Fig.10.1**) share similar functions and regulate the transcription of other genes involved in the germ cell development. In recent years RNA binding motif (RBM) proteins similar to DAZ in humans have been located on autosomes (see autosomal DAZ like proteins). Vogt and Fernandes (2003) discussed also the polymorphic DAZ gene family and disclosed putative origins of its molecular heterogeneity in fertile and infertile men identified by the analyses of Single Nucleotide Variants (SNVs) in this AZFc gene locus.

10.4.4. RBM Gene Family

RNA recognition motif of RBM encodes approximately 90 amino acids, a hallmark of a super-family of RNA binding proteins, the heterogenous nuclear ribonucleoprotein (hnRNP) family. Y chromosome *Rbm* genes are found in all mammals and are related to the X chromosome gene, which encodes hnRNPG. Its closest member hnRNPG RNA binding protein is widely expressed in the human, and mapped to the short arm of chromosome 6p12. The RBM is a nuclear protein expressed in fetal, pre-pubertal, and adult male germ cells. *Rbm* sequences found on both arm of the Y chromosome make genotype phenotype correlations difficult for this gene family. In the absence of detectable *Rbm* expression, one can see stages of germ cell development up to early meiosis, but not past this point into the haploid phase of spermatogenesis (Elliott et al., 1997). In man, deletion of the functional copies of *Rbm* is associated with meiotic arrest rather than sperm anomalies (Mahadevaiah et al., 1998). The expression of the *Rbm* and RBM proteins is found exclusively in germ cells in the testis. The deletion of RBM member(s) in the *AZFb* region appears to be associated with spermatogenic arrest at meiosis (Elliot et al., 1997; 1998). The RBM protein consists of four tandem repeats at 37 residue peptide, named the 'SRGY box' because of its high content of Ser-Arg-Gly-Tyr amino acids. The *Rbm* is a multi-copy gene family with an estimated 30-40 members (some of

Table 10.1. Sex Chromosomal of Gene Products in sex differentiation and testis development

Gene	Protein Type	Comments
Sry	Sex related Y protein (SRY).	Transcription factor, determines sex in mammals, located on Y chromosome; related to sex differentiation and spermatogensis in males, mutation in SRY gene leads to sex reversal in females.
Sox 9	SRY related HMG Box (SOX) protein	Transcription factor located on Y chromosome; regulates sex differentiation and bone morphogenesis; involved in sex reversal in females.
DAX1	DAX1	Nuclear hormone receptor, located on genital ridge of both male and females; Deficiency causes congenital adrenal hypoplasia and dosage dependent sex reversal.
WT1	WT1 Zinc finger protein.	Transcription factor; XY male develope into females.
GATA 4	GATA 4	Embryonic development, expresses in both sexes, relates to testis differentiation through AMH.
WTN4	WTN4: Cysteine rich Transcription factor.	Dose dependent sex differentiation; Regulator ofDAX1
DMRT1	Male specific protein with DNA binding domain (DBD)	Expresses in genital ridge in both sexes; Mutation in DBD shows incomplete testis development
RBM	RBM containing protein	Regulate spermatogenisis; deficiency causes oligospermia in men.
DAZ	DAZ and SPGY RNA binding proteins	Regulate spermatogenisis and matulation spermatids;deficiency causes oligospermia in men.
SPANX	SPANX	Expresses in spermatogenisis and in leukemic patients (Cll, CML, AML)
HR6A	HR6A	X chromosome protein, defficiency causes male infertility.
HR6B	HR6B	Y chromosomal and autosomal protein, located on elongating spermatids involved in ubiquitin conjugation.

which are pseudogenes), spread over both arms of the Y chromosome but mainly in intervals 5 and 6 (Delbridge et al., 1999; Mazeyrat et al., 1999; Elliott et al., 2000b).

Interestingly, the *Rbm* genes in mice and marsupials contain only one SRGY box, which is also the case in the *hnRNPG* gene. This led to the proposal that the Y-linked RBM family may have resulted from a transposition of an *hnRNPG* like ancestral gene to the Y chromosome.

The mouse *Rbm* gene was initially thought to be a single one. However, it is now known that this has multiple copies, some of which have been mapped to the Sxr^b region. At least 25 and probably > 50 copies of *Rbm* are present on the mouse Y chromosome short arm located between *Sry* and the centromere. As in human, RBM plays a role in spermatogenesis in mouse testis, but there are some differences in pattern of expression in two species. Mice carrying deletion Yd1 that maps on proximal Y short arm are female due to position effect resulting in non-expression of *Sry*.

The functions of RBM and hnRNPG are unknown. The RBM has a different expression profile from its closest homologue hnRNPG. Despite its ubiquitous expression in all transcriptionally active germ cell types, the proportion of RBM, distributed between punctuate-nuclear structure and the remainder of the nucleoplasm, is dynamically modulated over the course of germ cell development. Moreover, pre-mRNA splicing components are targeted to the same punctate nuclear regions as RBM during the early stages of germ cell development, but this spatial association breaks down in late meiosis. After meiosis, pre-mRNA splicing components are differently targeted to a specific region of the nucleus, suggesting dynamic and possibly multiple functions for RBM in germ cell development (Chai et al., 1997; Elliot et al., 1998).

Some RBM homologue has been identified on the X chromosome and chromosome 11. The RBM member of *hnRNP* gene is identified as a single copy gene on chromosome 11. The gene *hnRNP G-T* contains an uninterrupted open reading frame and no introns, consistent with derivation from a retroposon. However, unlike many retroposon-derived genes, *hnRNP G-T* is not a pseudogene. It is highly expressed in germ cells and in particular in the nuclei of meiotic spermatocytes. Human like hnRNP G-T protein is present in the testis of several mammals and is highly conserved (Elliot et al., 2000b). Within germ cell nuclei, protein-protein interactions showed that RBM protein directly interacts with members of the SR family of splicing factors and in addition, strongly interacts with itself. The SR proteins appear the only splicing factors bound by RBM, which are required for splicing of panel of pre-mRNAs and that RBM protein is a conserved mammalian splicing regulator, which operates as a germ cell-specific cofactor for more ubiquitously expressed pre-mRNA splicing activators (Elliott et al., 2000a).

10.5. THE DAZ GENE FAMILY AS AZF CANDIDATE

It is now clear that *DAZ* is a multiple gene family with at least seven copies clustered within a 1-Mbp region in the distal interval 6. The *DAZ* family shares certain characteristics with the *RBM* family, namely it also encodes RNA binding protein and expressed in germ cells only. Another member of this family and related to *DAZ*, named *SPGY* was identified and the complete sequence of *SPGY* has been released. The *DAZ* family contains 7 to 24 tandem repeats (termed *DAZ* repeat) of 72 bp with homology to the DYS1 repeat. Y-linked DAZ is seen only in old world monkeys but not in other mammals. The Y-linked *DAZ* was believed to be the best candidate for *AZF* by some investigators because of its deletions found in a number of azoospermic or oligospermic men. However, individuals lacking *DAZ* show various phenotypes ranging from azoospermia, oligospermia and some are even fertile. These observations imply that the gene is not required for the completion of normal spermatogenesis. Further studies reveal that the sequence of the *boule* gene is more like that of the autosomal *DAZL1* and *Dazl1* than that of the Y-linked *DAZ* (Ma et al., 2000).

Proteins that interact with DAZ proteins have been identified. The PUM2, a human homologue of Pumilio, a protein required to maintain germ line stem cells in *Drosophila* and *C.*

elegans, forms a stable complex with DAZ through the same functional domain required for RNA binding, protein-protein interactions and rescue of Pumilio mutations in flies. The PUM2 is expressed predominantly in human embryonic stem cells and germ cells and co-localizes with DAZ and DAZL in germ cells. It seems that PUM2 is a component of conserved cellular machinery that may be required for germ cell development (Moore et al., 2003).

10.6. THE CDY GENE FAMILY

The human chromodomain Y(*CDY*) is a gene family with at least three members identified and named as *CDY1* major, CDY1 minor and CDY2. The CDY family encodes a protein containing a chromatin-binding domain and a catalytic domain. The predicted coding regions of *CDY1* and *CDY2* were 98% identical in amino acid sequence of the predicted proteins. Most of the putative protein encoded by the *CDY1* minor transcript is identical to that encoded by the major transcript except that its carboxy terminus is divergent. The *CDY1* genes are mapped to intervals 6F and the *CDY2* to 51 of the Y chromosome. Like the *DAZ* family, the human *CDY* has an autosomal homologue, referred to as *CDY*-like *(CDYL)*, mapped to the distal short arm of chromosome 6. The Y-linked *CDY* genes are restricted to primates, while the autosomal *CDYL* homologues are widely present in other mammals. In mice, the *CDYL* homologue, known as *Cdyl* has been mapped to chromosome 13. The predicted mouse CDYL and human CDYL proteins have 93% overall amino acid identity, while the predicted human CDYL and CDY proteins have only 63% identity. Location of *CDY* on the region important for spermatogenesis and its testis-specific expression has made it a potential candidate for *AZF* (Ma et al., 2000).

10.7. OTHER SPERMATOGENESIS-RELATED GENES ON Y-CHROMOSOME

H-Y Antigen: The H-Y antigen is male specific antigen, which was first described during skin graft experiments. The mouse H-Y antigen gene *(Hya)* locus (termed *Smcy*) has been cloned from the region encoding *Spy* and *Hya* (Agulnik et al., 1994). The human homologue (*SMCY)* has also been mapped to the same deletion interval as human H-Y antigen locus, *HYA* (Agulnik et al., 1994). Now several Hys have been identified, including MHC class I-restricted and MHC class II-restricted epitopes.

TSPY: The TSPY, "the testis-specific protein, Y-encoded", is the product of a tandem gene cluster on human proximal Yp. Testis specific protein Y is a human Y-chromosome derived gene with numerous functional and non-functional copies. The TSPY has homology with other human and non-human proteins, including SET and NAP, factors that are suggested to play a role in DNA replication. A set of distinct *TSPY* transcripts has been identified with diverse exon compositions. The TSPY occurs mainly in a modified, putatively phosphorylated form. By immunostaining it was detected in distinct subsets of spermatogonia, and also strongly immunostained in early testicular carcinoma in situ, while seminomatous tumour cells showed low activity. The data point to a phosphorylation-dependent TSPY-function in early spermatogenesis, immediately prior to the spermatogonia-to-spermatocytes transition, and in early testicular tumorigenesis (Schnieders et al., 1996). Specific expression patterns in testis and testicular tumors, as in prostate cancer samples and cell lines led to the postulation of a potential role in cell proliferation, supported by the presence of a suppressor of variegation, enhancer of zeste and Trithorax/nucleosome assembling protein (nucleosome assembly protein) domain in the mature protein. TSPY expresses in two transcripts of variable length, termed TSPY-S and -L, which differ in their 3'-translated region due to alternative splicing, and in the level of transcripts; TSPY-S being 3-4-fold more abundant than TSPY-L. Surprisingly, no evidence of a G-G-18 haplotype was found for the TSPY-L transcript, while this haplotype makes up almost 50% of all TSPY-S transcripts. This excludes the corresponding TSPY-1 locus from alternative splicing. The only significant differences between the TSPY-1 locus and eight other loci were located in the promotor region as revealed by sequence comparisons (Krick et al., 2003).

PRY Genes: PRY (PTP-BL related on the Y chromosome) has been proposed as a candidate spermatogenesis gene. Comparison of the cDNA sequence with the genomic sequence revealed five exons. The Y chromosome revealed the presence of two full-length copies in AZFb *(PRY1* and *PRY2)* and two shorter versions of the *PRY* gene containing exons 3, 4 and 5 in AZFc (*PRY3* and *PRY4*). A clone containing exons 3, 4, 5 is located in area 5L (between AZFa and AZFb), a clone containing a sequence homologous to exon 5 is located in the area 5M (in AZFb) and a clone containing a fragment homologous to exon 3 is located in 6F. A repeat structure of exons 1 and 2 is present on the short arm of the Y chromosome as well as on the long arm. The *PRY1* and *PRY2* gene copies that are located in AZFb, are known to be expressed in the testis (Stoufs et al., 2001).

VCX/Y Proteins: Expression and localization of VCX/Y proteins and their possible involvement in regulation of ribosome assembly during spermatogenesis have been identified recently (Zou et al., 2003). The VCY2 is a testis-specific protein with unknown function. It is located in the AZFc region on chromosome Yq and is frequently deleted in infertile men with severe oligozoospermia or azoospermia. Ubiquitin-protein ligase E3A (UBE3A) contains a HECT domain that binds VCY2. Two UBE3A transcripts of 1.4 and 2kb are abundantly expressed in human testis. Both VCY2 and UBE3A mRNAs are expressed in ejaculated human spermatozoa, indicating that both genes localize in the germ cell compartment. This suggests that UBE3A ubiquitination may be required for VCY2 function (Wong et al., 2002). An interacting partner of VCY2 is VCY2IP-1 that encodes an ORF of 1059 amino acids. The amino acid sequence of VCY2IP-1 shows 59.3% and 41.9% homology to two human microtubule-associated proteins (MAPs), MAP1B and MAP1A, respectively. The VCY2IP-1 has an extensive homology to the N-terminus and C-terminus regions of MAP1B and MAP1A, placing it within a large family of MAPs. VCY2IP-1 is localized to chromosome 19p13.11. The VCY2IP-1 gene spans 15-kb and consists of seven exons. A 3.2-kb VCY2IP-1 transcript is predominantly expressed in human testis. The VCY2IP-1 transcripts are present in germ cells, while VCY2 and VCY2IP-1 transcripts are present in human ejaculated spermatozoa (Wong et al., 2004).

10.8. AUTOSOMAL GENES PRODUCTS

Cystic fibrosis (CF) is most common fatal autosomal recessive disorder, found among Caucasians. It arises because of mutations in cystic fibrosis transmembrane conductance regulator gene (*CFTR*) (Hargreave, 2000). The *CFTR* codes for a membrane protein that functions as an ion channel and also influences the formation of male reproductive tract. The CFTR protein is expressed in human epididymis, which generates proper environment for sperm maturation and sperm transport (see Chapter 32).

```
  1 MNLLDPFMKM TDEQEKGLSG APSPTMSEDS AGSPCPSGSG SDTENTRPQE NTFPKGEPDL
 61 KKESEEDKFP VCIREAVSQV LKGYDWTLVP MPVRVNGSSK NKPHVKRPMN AFMVWAQAAR
121 RKLADQYPHL HNAELSKTLG KLWRLLNESE KRPFVEEAER LRVQHKKDHP DYKYQPRRRK
181 SVKNGQAEAE EATEQTHISP NAIFKALQAD SPHSSSGMSE VHSPGEHSGQ SQGPPTPPTT
241 PKTDVQPGKA DLKREGRPLP EGGRQPPIDF RDVDIGELSS DVISNIETFD VNEFDQYLPP
301 NGHPGVPATH GQVTYTGSYG ISSTAATPAS AGHVWMSKQQ APPPPPQQPP QAPPAPQAPP
361 QPQAAPPQQP AAPPQQPQAH TLTTLSSEPG QSQRTHIKTE QLSPSHYSEQ QQHSPQQIAY
421 SPFNLPHYSP SYPPITRSQY DYTDHQNSSS YYSHAAGQGT GLYSTFTYMN PAQRPMYTPI
481 ADTSGVPSIP QTHSPQHWEQ PVYTQLTRP
```

Fig.10.4. Amino acid sequence of SOX9 [SRY-related high-mobility group (HMG) box 9]. Source: http// www.ncbi.nlm.nih.gov (accession number P48436).

10.8.1. SOX9 and Other SOX Proteins

The Sox (Sry-type HMG box) group of transcription factors, which is defined by a high-mobility group (HMG) DNA-binding domain, is categorized into six subfamilies. The effects of autosomal translocations on spermatogenesis involves on the formation of synaptonemal complexes. More important is the formation of asynaptical and hetero-synaptical complexes formed between non-homologous chromosomes and still more important with sex chromosome bivalent. Such interactions may lead to interfere in the inactivation of X chromosome, and hence meiosis cycle.

Disorders in SOX-9 (SRY type HMG box gene 9), originating at any level, cause a semi-lethal, recessive disorder with a range of skeletal malformation syndrome known as campomelic dysplasia (CD) and the inability to develop testis. The protein encoded by this gene recognizes the sequence CCTTGAG along with other members of the HMG-box class DNA-binding proteins. It acts during chondrocyte differentiation and, with steroidogenic factor 1, it regulates transcription of the anti-Muellerian hormone (AMH) gene. The *SOX-9* is the first sex-determining gene to be discovered on autosomes. Deletion, translocation and inactivating mutations involving SOX-9 cause CD, including sex reversal in individuals with a normal male karyotype, 46, XY (Pringle and Page, 1997). These features are associated with translocations at breakpoints located at q24.3-25.1 region of chromosome 17. All these chromosomal changes appear to involve a defect in the promoter element for *SOX-9*.

The human *SOX9* gene encodes a protein of 509 amino acids. The HMG box of SOX9 shares 70% homology with HMG of SRY protein. Hence SOX9 is SRY-related transcription factor and

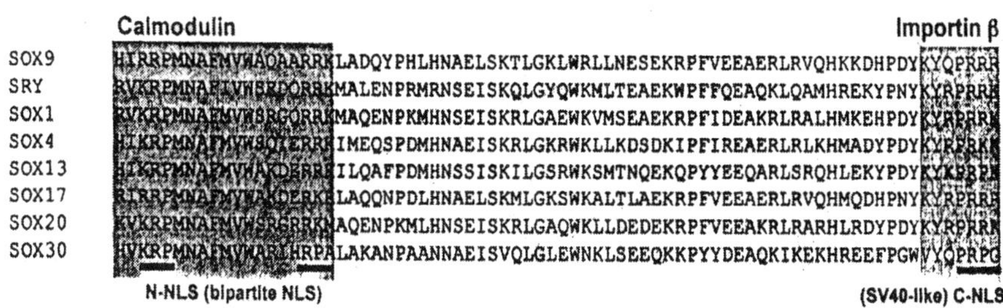

Fig.10.5. Calmodulin and importin beta binding domains of SRY HMG doman. An alignment of the HMG box of SOX protein (representative from each gr oup). The shaded region represents the CaM binding domain mapped in SRY. The N-andC-terminal NLS sites are underlined. Note lack of conservation of CaM and importin beta domains between SOX and non-SOX proteins. Adapted from A Argentaro et al. J Biol Chem 278;33839-47;2003 © ASBMB.

Table 10.2. Autosomal Genes and Their Products in Gametogenesis

Gene	Protein Type	Comments
Dazl	RNA binding protein	Located in cytoplasm of premeiotic cells, deficiency causes male and female infertility
HR6B	Ubiquitin-conjugating enzyme	Y-chromosomal homologue, present in spermatids, deficiency causes male infertility.
BMP 8B	Signaling Protein (TGF-β-like)	Present in round spermatids, causes germ cell development.
Dhh	Signaling	Present in Sertoli cells, causes germ cells development
Hsp 70-2	Heat-shock protein	Present on synaptonemal complex, chaperone protein, responsible for male fertility.
Pms2	PMS mismatch repair Protein	Related to DNA recombination repair, responsible for male infertility.
Mlh1	MLH1 mismatch repair Protein	Present on synaptonemal complex, related to DNA recombination repair, responsible for male infertility, and growth retardation.
Msh4	MSH4 mismatch repair Protein	Causes male infertility, related to DNA recombination repair, responsible for male infertility.
Fertilin β	Fertilin β	Sperm acrosomal protein involved in fertilization of ovum. Deficiency causes failure of fertilization.
AKAP 82	Protein kinaseA anchoring protein 82	Present on fibrous sheath of sperm, responsible for capacitation and fertilization.
ATM/ ATR	Check point kinases	Present on synaptonemal complex, deficiency causes male and female infertility.

binds with high affinity at SRY binding site AACAAT and TCF1 site, AACAAAG **(Fig. 10.4).** In addition, SOX9 contains other protein domains including two translational activation domains. SOX9 is more conserved protein than SRY protein. Mutations in *SOX9* gene have been found in individuals who are chromosomally males but phenotypically females. Studies in transgenic mice showed that female-to-male sex reversal occurs when female mice carry *Sox9*. Studies also suggest that the expression of Sox9 is up-regulated in the Sertoli cells shortly after *Sry* expression in mice, and that the *Sox9* is activated by Sry (Clarkson and Harley, 2002; Harley et al 2003). Apart from sharing a high level of homology with orthologs, the C-terminal domain of SOX9 also shares homology with other SOX proteins, such as SOX8 and SOX10. SOX9 contains several putative sites for phosphorylation including seven sites for casein kinase II, five for PKC and two for PKA. The regulatory region of *SOX9* promoter seems to be very large and the expression of *SOX9* is likely to be complex. Analysis of 6.8-kb fragment of mouse *Sox9* promoter in testis and overy from 13.5 dpc embryos showed that a minimal interval located between –193 and –73 from Sox9 transcription start site was sufficient to direct maximal promoter activity in male and female gonadal cells (c/r Harley et al., 2003) **(Fig 10.5)**.

Sox8: *Sox8* is a member of the Sox family of transcription factor genes and is closely related to *Sox9*. Like *Sox9, Sox8* is expressed in the developing mouse testis around the time of sex determination, suggesting that it might play a role in regulating the expression of testis-specific genes. The expression of anti-Mullerian hormone (AMH) in Sertoli cells requires the interaction of several transcription factors, including SF1, SOX9, GATA4, WT1, and DAX1. Expression of Sox8 begins just prior to that of *Amh* at 12 dpc in mouse testes and continues beyond 16 dpc in Sertoli cells. In vitro, SOX8 binds specifically to SOX binding sites within the *Amh* minimal promoter and, like SOX9, acts synergistically with SF1 through protein-protein interaction to enhance *Amh* expression at lower levels compared with SOX9. This provides evidence that SOX8 may partially compensate for the reduced SOX9 activity in campomelic dysplasia and substitute for *Sox9* where *Sox9* is either not expressed or expressed too late to be involved in sex determination or regulation of *Amh* expression (Schepers et al., 2003).

Sox5 and Sox6: Sox5 and Sox6 belong to the group D subfamily, which is characterized by conserved N-terminal domains including a leucine-zipper, a coiled-coil domain and a Q-box. These genes are expressed as long and short transcripts that exhibit differential expression patterns. In mouse, the long form of Sox5, L-Sox5, is co-expressed and interacts with Sox6. Together, these two proteins appear to play a key role in chondrogenesis and myogenesis. The SoxLZ/Sox6 contains a leucine zipper motif in addition to an HMG box, which is its DNA binding domain. A SoxLZ/Sox6 binding protein, termed Solt, is present in mouse testis. Like SoxLZ/Sox6 mRNA, Solt mRNA was exclusively expressed in the testis of mouse. Solt contains an unusual leucine zipper, which bound to the leucine zipper region of *SoxLZ/Sox6* in vitro. In transient transfection assays with SoxLZ/Sox6 expression of a reporter gene that carries a cis-binding region for Sox proteins was enhanced co-expressing Solt and Ca^{2+}/calmodulin-dependent protein kinase IV (Yamashita et al., 2000).

In humans, unlike the short form, which shows testis-specific expression, L-SOX5 is expressed in multiple tissues. L-SOX5 shows strong expression in chondrocytes and striated muscles, indicating a likely role in human cartilage and muscle development, like SOX6. The human L-SOX5 cDNA encodes a 763-amino-acid protein that is 416 residues longer than the short form and contains all of the characteristic motifs of group D Sox proteins. The predicted human L-SOX5 protein shares 97% amino acid identity with its mouse counterpart and 59% identity with human SOX6. The L-SOX5 gene contains 18 exons and shows similar genomic structure to SOX6. Two transcription start sites in L-SOX5 with multiple alternatively spliced mRNA variants that are distinct from the short form, have been observed (Ikeda et al., 2002).

10.8.2. Autosomal DAZ like (DAZL) Proteins

Autosomal gene *DAZL* is a family of genes (*DAZL, DAZ, BOULE*), all of which contain a consensus RNA binding domain and are expressed in germ cells. Although the *DAZ* homologues are present on the Y chromosomes of only great apes and old world monkeys, yet all mammals contain a single copy *DAZ*- like gene, *DAZL1*, on their autosomes. Autosomal *DAZ* homologous genes have been characterized and mapped to chromosome 17 in mice and named *Dazl1* (formerly *DAZla)* and *DAZL1 (DAZh)* mapped to chromosome 3p24 in humans. It has been proposed that the human Y-linked *DAZ* family is derived from a *DAZL1-like* autosomal ancestral gene. Like the Y-linked *DAZ, DAZL1* and *Dazl1* are cytoplasmic proteins expressed exclusively in testis and ovary. The *Dazl1* knockout mice express reduced number of germ cells and complete absence of gamete production in both males and females. These observations suggested that *Dazl1* is required for gametogenesis. In absence of Dazl protein, germ cells can complete mitosis, and embark on functional differentiation but progression through meiotic prophase requires this RNA binding protein in both male and females (Saunders et al., 2003). The *DAZL1* like DAZ encodes proteins with an RNA recognition motif RRM/RBM and varying numbers of copies of *DAZ,* and repeat DAZL1 has a single copy. The *DAZ* and *DAZL1* have 85% sequence similarity over most of their lengths, including the RBM domain and the *DAZ* repeats, but they have different C terminal sequences due to a frame shift downstream of the DAZ repeat region. A role for DAZ and DAZL1 in spermatogenesis is supported by their exclusive expression in germ cells, their homology to a *Drosophila* male infertility gene, *boule,* and the sterility of *Dazl1* knock out mice. The tubules contain only a few premeiotic spermatogonia that rarely progress into meiosis suggesting that *Dazl1* is required for both the development and the maintenance of the germ cells.

In the mouse 3.5-kb and 4.5-kb mRNA transcripts of Dazl1 are detected in male and female embryonic gonads at 12.5 dpc (days post coitum). During this period, the only germ cells present in the gonad are primordial germ cells. The *Dazh* transcripts were not detected in embryonic gonads of mice that lack germ cells because of mutation in W gene, suggesting that expression is limited to germ cells (Seligman and Page, 1998). Though DazL protein is strictly cytoplasmic and that human DAZ protein is restricted to post-meiotic cells, by contrast human DAZ and human and mouse DAZL1 proteins are present in both the nuclei and cytoplasm of fetal gonocytes and in spermatogonial nuclei. The proteins reallocate to the cytoplasm during male meiosis. In situ hybridization shows that Dazla (DAZL1) expression in mouse testis is restricted to pre-meiotic stage IV-VI of spermatogenesis (Neiderberger et al., 1997). Unlike DAZ, human DAZL protein persists in spermatids and even spermatozoa. The DAZL transcripts are reduced in men in association with spermatogenic cells (Reijo et al., 2000, Lin et al., 2001). While in other reports, the DAZL1 has been found abundantly in the cytoplasm of pachytene spermatocytes and to a lesser degree in the cytoplasm of type B spermatogonia and preleptotene and zygotene spermatocytes (Ruggiu et al., 1997), whereas DAZ is localized either to late spermatids and sperm tails (Habermann et al., 1998) or to spermatogonia to spermatocytes (Dorfman et al., 1999). The DAZ proteins function in the RNA metabolism of late spermatids, presumably in the storage or transport of testis specific mRNA, the translation of which is repressed until the formation of mature spermatozoa. Deletion of DAZ genes does not interfere with human sperm maturation but results in a gradual reduction of mature spermatozoa (Habermann et al., 1998). Tsui et al (2000) found that DAZL1 and DAZ bound to various homopolymers in vitro. The sedimentation profile in sucrose gradient indicated that DAZL1 is associated with polyribosomes. Further capture of DAZL1 on oligo (dT) beads demonstrated that association is mediated through binding of DAZL1 to poly (A) RNA and suggests that DAZL1 is involved in germ cell specific regulation of mRNA translation (Tsui et al., 2000).

10.8.3. Murine DazL1 Binding mRNAs

Jiao et al (2002) identified mRNAs from testes that were specifically bound by Dazl1. One mRNA encoded the Tpx-1 protein, a germ cell adhesion protein essential for the progression of spermatogenesis (see Chapters 1 and 34). A 26 nt region necessary and sufficient to bind Dazl1 was found within additional mRNAs isolated. These included (i) mRNA encoding Pam, a protein associated with myc, (ii) GRSF1, an mRNA binding protein involved in transcriptional regulation, and (iii) TRF2, a TATA box binding protein like protein involved in transcriptional

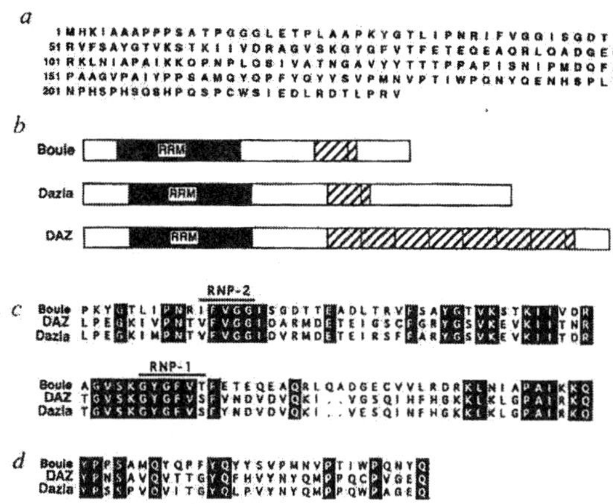

Fig.10.6. Comparison of Boule and DAZ proteins. (a) Sequence of Boule protein. (b) Domain organization of boule and DAZ proteins (mouse DAZ like = Dazla and human DAZ). RNA recognition motif (RRM), RNA binding domains and DAZ motifs are indicated by black and shaded boxes respectively. (c) and (d) Alignment of RRM (c) and DAZ (d) domains of boule, DAZ and Dazla. The RNP domains are indicated by lines in (c). (d) Alignment of 24 amino acid DAZ motif and 5 amino acid extension found in Boule, DAZ and Dazla. Reproduced with permission from C.G. Eberhardt et al. Nature 381: 783-5: 1996 © Macmillan Magazines Ltd. http://www.nature.com/Nature.

regulation. Each mRNA containing the Dazl1 binding site was specifically bound by DazL1 protein. A similar sequence is also present in the Cdc25A, a threonine/tyrosine phosphatase involved in cell cycle progression. The Dazl1 and Cdc25A homologues are functionally linked in *Drosophila* and are necessary for spermatogenesis. These results demonstrated that murine Dazl regulates a subset of mRNAs necessary for germ cell development and cell cycle progression (Jiao et al., 2002).

10.8.4. Boule and DAZ

The *boule* (bol), the infertility gene was identified in *Drosophila. Boule* cDNAs encoded a protein of 228 amino acids that contained a single RNP type RNA-binding domain. The putative RNA binding domain of the Boule protein is most similar (42% identify) to that of the human DAZ protein and a closely related to mouse protein, Dazl1. Boule also has 33% sequence similarity to a second region of DAZ and Dazl1. This region contains a motif, termed a DAZ repeat that is present once in the mouse sequence and seven times in the human protein. The positions of the RNP domain and the first DAZ repeat are conserved among the fly, mouse and human proteins. Male flies with the boule mutation have morphologically normal primary spermatocytes, which fail to enter into meiotic division, suggesting that boule is required for the G2/M, transition. Northern analysis of RNA from adult male and female flies demonstrated that *boule* expression is limited to male. In *bol* homozygotes, this transcript, which is normally 3.0kb in length, is truncated to just 1.1kb and the RNA is severely reduced in contents (**Fig.10.6**). The defect in spermatogenesis seen with either allele in *trans* to a deletion is indistinguishable from that seen with *boule* homozygotes, suggesting that the phenotype of the boule gene could be brought about by a strong or complete loss of gene function (Eberhardt et al., 1996). Loss of boule function results in azoospermia; meiotic divisions are blocked, although limited spermatid differentiation occurs. Histological examination of boule testis with cell cycle markers indicates that the primary defect is at the meiotic G2/M transition (Eberhardt et al., 1996).

10.8.5. MORC Gene

Inoue et al., (1999) described the autosomal recessive mouse *morc* TgN (Tyr) 1Az (microchidia) mutation, which causes arrest of spermatogenesis prior to the pachytene stage of meiosis prophase I, and characterization of the *morc* locus and positional cloning of a gene disrupted by the *morc* TgN (Tyr) 1 Az mutation. This gene termed *Morc*, encodes a 108-kDa protein expressed specifically in male germ cells. The transgene integrated within the first intron of *morc* and was accompanied by an intragenic deletion of approximately 13-kb of genomic sequences, removing exons 2-4 and abrogating expression of the wild type transcript. MORC protein sequences revealed putative nuclear localization signals, two predicted coiled-coil structural motifs and limited homology to GHL (GyraseB, Hsp90, Mut1) ATPase. The human MORC homologue showed that it too is testis-specific, but closely related human genes, and transcribed in multiple somatic tissues.

9.8.6. Other Gene Products in Infertility

A male-enhanced antigen-1 (MEA1) appears to be serologically detectable male antigen and associated with chromosome 17 (Ohinata et al., 2002). The *Zfy* in mice (*ZFY* in humans), once considered to be the best candidate for the testis-determining gene (*Tdf*) (Page et al., 1987), is now suggested to be involved in germ cell proliferation and development. The Y353/B, originally isolated from a mouse Y chromosome-enriched DNA library and mapped to the long arm of the Y chromosome (Bishop et al., 1985), has recently been proposed as a candidate multiple-copy spermiogenesis gene (Conway et al., 1994; Burgoyne et al., 1992). Lahn and Page (1997) isolated twelve new genes or gene families mapped to the non-recombination region of the Y chromosome, or NRY. Seven of these genes are Y chromosome specific and only expressed in the testis (Lahn and Page, 1997). The *DAX1* codes a nuclear hormone receptor protein, which is related to the congenital adrenal hypoplasia and sex reversal in humans. *Dax1* proteins are present in the early genital ridge of both males and females. Since *Dax1* is expressed at the same time as *Sry*, it is found that Sry and Dax1 are antagonistic to each other (Veitia et al., 2001). The WTN4 gene encodes cysteine rich glycoproteins that regulate transcriptional control during male sexual development. Its over-expression in humans is considered to result in XY sex-reversal. During sex determination, WNT4 up-regulates the expression of *DAX1* (Jordan et al., 2001 c/r Shah et al., 2003)

10.9. XY BODY

During the first meiotic prophase of mammalian spermatogenesis, the sex chromosomes X and Y show characteristic allocyclic behaviour with respect to autosomes. This is particularly evident during pachytene stage when X and Y sex chromosomes condense to form the so-called sex vesicle, sex body or XY body, which is morphological manifestation of the process of meiotic sex chromosome inactivation (MSCI) and is visually distinct from transcriptionally active autosomes. This structure is characterized by the condensed state of chromatin, transcriptional inactivity, and the limited extension of chromosome pairing, which is usually restricted to a short segment of sex chromosome axial elements. The molecular basis and functional significance of sex vesicle formation particularly heterochromatinization during mammalian spermatogenesis remains obscure. A large number of sex body located proteins are being identified but their functions are unclear. Meiosis-specific sex vesicle protein called XY40 (40 kDa; pl 5.7. 5.8) can be extracted from rat pachytene spermatocytes and recovered in particles of 9.5 S with a native molecular mass of approximately 152 kDa. It was proposed that protein XY40 may be involved in the allocyclic behavior of sex chromosomes during male meiotic prophase (Smith and Benavente, 1992a, b).

X inactive specific transcript *(Xist)* plays a crucial role during X-chromosome inactivation. Xist is exclusively transcribed from inactive X chromosome. In eutherian mammals, the Xist RNA is concentrated in the XY body. This suggests that Xist RNA is involved in the inactivation of the male X chromosome, and that it has spreading capability, not only in cis but also in a quasi-cis mode, to juxtaposed non-X chromosomes (Ayoub et al., 1997). "Asynaptin" (ASY), detected in male meiosis exclusively in association with the X and Y chromatin of sex body, is also expressed in pachytene oocytes of XY females where it coats the chromatin of the asynapsed X in the absence of MSCI. Furthermore, in pachytene oocytes of females carrying a reciprocal autosomal translocation, ASY associates with asynapsed autosomal chromatin. Thus the location of ASY to the sex body during male meiosis is likely to be a response to the asynapsis of the non-homologous regions of the heteromorphic X-Y bivalent, rather than being related to MSCI. In contrast to ASY, another sex body protein XY77 proved to be male sex body specific (Turner et al., 2000).

The HP1 class of chromobox genes is thought to encode proteins involved in the packaging of chromosomal DNA into repressive heterochromatin domains. Two proteins that have been implicated in effecting chromosome condensation through meiotic prophase are the mammalian HP1-like protein M31 (HP1b or MOD1) and the unusual core histone marcoH2A1.2. Analyses of M31 and macroH2A1.2 localization in mouse testis indicated that both proteins are components of meiotic centromeric heterochromatin and of the sex body. The presence of M31 and H2A1.2 in sex body raised the possibility that these proteins co-operate in meiotic sex chromosome inactivation. In order to investigate the roles of M31 and macroH2A1.2 in meiosis, Turner et al., (2001) examined the localization patterns of these proteins in surface spread meiocytes from male and female mice. Study of the distribution of a murine M31 during spermatogenesis revealed spreading from the tip of the XY body in mid stage pachytene spermatocytes to include the whole of the XY body in late pachytene spermatocytes. Moreover, the formation of the XY body during spermatogenic progression in neonatal mice coincides with the expression of nuclear isoform of M31, M31 (p21). In addition to their previous patterns, both proteins localize to a focus within the portion of the pseudoautosomal region (PAR) that contains the steroid sulphatase (Sts) gene (Turnor et al., 2001). It supports the view that a common mechanistic basis exists for heterochromatin induced repression, homeotic gene silencing, sex chromosome inactivation and complete desynapsis of the terminally associated X and Y-chromosomes prior to anaphase during mammalian spermatogenesis (Motzkus et al., 1999; Turner et al., 2001).

10.10. REFERENCES

Agulnik AI, Mitchell MJ, Lerner JL, Woods DR, Bishop CE. A mouse Y chromosome gene encoded by a region essential for spermatogenesis and expression of male-specific minor histocompatibility antigens. Human Mol Genet 1994; 3: 873-878.

Ali S, Hasnain SE. Genomics of the human Y-chromosome. 1. Association with male infertility. Gene 2003; 321: 25-37.

Argentaro A, Sim H, Kelly S, Preiss S, Claton A, Jans DA, Harley VR. A sox9 defect of calmodulin-dependent nuclear import in campomelic dysplasia / autosomal sex reversal. J Biol Chem 2003; 278:33839-47.

Ashley, T, Lieman J, Ward DC. Multicolor FISH with a telomere repeat and Sry sequence shows that Sxr (Sex reversal) in the mouse is a new type of chromosome rearrangement. Cytogenet Cell Genet 1995; 71: 217-22.

Ayoub N, Richler C, Wahrman J. Xist RNA is associated with the transcriptionally inactive XY body in mammalian male meiosis. Chromosoma 1997; 106: 1-10.

Bishop CE, Boursot P, Baron B, Bonhomme F, Hatat D. Most classical Mus musculus domestcus laboratory mouse strains carry a Mus musulus Y chromosome. Nature 1985; 315: 70-72.

Brown GM, Furlong RA, Sargent CA, Erickson RP, Longrepied G, Mitchell M et al. Characterization of the coding sequence and fine mapping of the human DFFRY gene and comparative expression analysis and mapping to the Ssrb interval of the mouse Y chromosome of the Dffry gene. Human Mole Genet 1998; 7: 97-107.

Burgoyne PS, Mahadevaiah SK. Sutchliffe MJ, Palmer SJ. Fertility in mice requires X-Y pairing and a Y-chromosomal 'spermiogenesis' gene mapping to the long arm. Cell 1992 71: 391-98.

Burgoyne PS. Evidence for an association between univalent Y chromosomes and spermatocyte loss in XYY mice and men. Cytogenet Cell Genet 1979; 23: 84-89.

Cameron FJ, Sinclair AH. Mutations in SRY and SOX9: testis determining genes. Human Mutation 1997; 9: 388-95.

Chai NN, Salido EC, Yen PH. Multiple functional copies of the RBM gene family, a spermatogenesis candidate on the human Y chromosome. Genomics 1997; 45: 355-61.

Clarkson MJ, Harley VR. Sex with SOX on: SRY and SOX9 in testis development. Trends Endocrinol Metabol 2002; 13: 106-11.

Conway SJ, Mahadevaiah SK, Darling SM, Capel B, Rattigan AM, Burgoyne PS. Y353/B : a candidate multiple copy spermiogenesis on mouse Y chromosome. Mamm Genome 1994; 5: 203-10.

Delbridge ML, Lingenfelter PA, Disteche CM, Graves JA. The candidate spermatogenesis gene RBMY has a homologue on the human X chromosome. Nature Genetics 1999; 22: 223-224.

Diemer T, Desjardins C. Developmental and genetic disorders in spermatogenesis. Human Reprod Update 1999; 5: 120-140.

Dorfman DM, Genest DR, Pera RAR. Human *DAZL1* encodes a candidate fertility factor in woman that localizes to the prenatal and postnatal germ cells. Hum Reprod 1999; 14: 2531-36.

Eberhardt CG, Maines JZ, Wasserman SA. Meiotic cycle requirement for a fly homologue of human Deleted in Azoospermia. Nature 1996; 381: 783-5.

Elliott D, Bourggois CF, Klink F, Stevenin J, Cooke H. A mammalian germ cell-specific RNA-binding protein interacts with ubiquitously expressed proteins involved in splice site selection. Proc Natl Acad Sci USA 2000b; 97: 5717-22.

Elliott DJ, Millar MR, Oghene K, et al Expression of RBM in the nuclei of human germ cells is dependent on a critical region of the Y chromosome long arm. Proc Natl Acad Sci 1997; 94: 3848-53.

Elliott DJ, Oghene K, Makarov G, Makarova O, Hargreave TB, Chandley AC, Eperon IC, Cooke HJ. Dynamic changes in the subnuclear organization of pre-mRNA splicing proteins and RBM during human germ cell development. J Cell Sci 1998; 111 (pt9): 1255-65.

Elliott DJ, Venables JP, Newton CS, Lawson D, Boyle S, Eperon IC, Cooke HJ. An evolutionarily conserved germ cell-specific hnRNP is encoded by a retrotransposed gene. Hum Mol Genet 2000b; 9: 2117-24.

Habermann B, Mi HF, Edelmann A, Bohring C, Backert IT, Kiesewetter F, Aumuller G, Vogt PH. DAZ (Deleted in Azoospermia) genes encode proteins located in human late spermatids in sperm tails. Hum Reprod 1998; 13: 393-9.

Hackstein J Hochstenbach R. The elusive fertility genes of Drosophila: the ultimate heaven for selfish genetic elements. Trends in Genetics 1995; 11: 195-200.

Hargreave T. Genetic basis of male infertility. Br Med Bull 2000;3: 650-71.

Harley V R, Clarkson M J, Argentaro A: The molecular action and regulation of the testis determining factors, SRY (sex-deteremining region on the Y chromosome) and SOX9 (SRY-related high-mobility group (HMG) box 9). Endocr Rev 2003; 24: 466-87.

Heatwole VM, Haynes SR. Association of RB97D , an RRM protein required for male fertility, with a Y chromosome lampbrush loop in Drosophila spermatocytes. Chromosoma 1996; 105: 285-92.

Ikeda T, Zhang J, Chano T, Mabuchi A, Fukuda A, Kawaguchi H, Nakamura K, Ikegawa S. Identification and characterization of the human long form of Sox5 (L-SOX5) gene. Gene. 2002; 298: 59-68.

Inoue N, Hess KD, Moreadith RW, Richardson LL, Handel MA, Watson ML, Zinn AR. New gene family defined by MORC, a nuclear protein required for mouse spermatogenesis. Hum Mol Genet 1999; 8 : 1201-7

Jiao X, Trifillis P, Kiledjian M. Identification of target messenger RNA substrates for the murine azoospermia-like RNA-binding protein. Biol Reprod 2002; 66: 475-85.

Kamischke A, Baumgardt A, Horst J, Nieschlag E. Clinical and diagnostic features of patients with suspected

Klinefelter syndrome. J Androl 2003; 24: 41-8.
Kent-First M, Muallem A, Shultz J, Pryor J, Roberts K. Solten W et al. Defining regions of the Y-chromosome responsible for male infertility and identification of a fourth AZF region (AZFb) by Y-chromosome microdeletion detection. Mole Reprod Develop 1999; 53: 27-41.
Krick R, Jakubiczka S, Arnemann J. Expression, alternative splicing and haplotype analysis of transcribed testis specific protein (TSPY) genes. Gene 2003; 302: 11-9.
Lahn BT, Page DC. Functional coherence of the human Y chromosome. Science 1997; 278: 675-680.
Lin YM, Chen CW, Sun HS, Tsai SJ, Hsu Cc, Teng YN, Lin JS, Kuo PL. Expression patterns and transcript concentrations of the autosomal DAZL gene in testes of azoospermic men. Mol Hum Reprod 2001; 7: 1015-22.
Ma K, Mallidis C, Bhasin S. The role of Y chromosome deletions in male infertility. Eur J Endocrnol 2000; 142: 418-30.
Mahadevaiah SK, Odorisio T, Elliott DJ, Rattigan A, Szot M, Laval SH, Washburn LL, McCarrey et al. Mouse homologues of the human AZF candidate gene RBM are expressed in spermatogonia and spermatids, and map to a Y chromosome deletion interval associated with a high incidence of sperm abnormalities. Hum Mol Genet 1998; 7: 715-27.
Margarit E, Guillen A, Rebordosa C, V-dal-Taboada J, Sanchez M, Ballesta F, Oliva R. Identification of conserved potentially regulatory sequences of the SRY gene from different species of mammals. Biochem Biophys Res Commun 1998; 245: 370-77.
Mazeyrat S, Saut N, Mattei MG, Mitchell MJ. RBMY evolved on the Y chromosome from a ubiquitously transcribed X-Y identical gene. Nature 1999; 22: 224-26.
McElreavey K, Fellous M. Sex-determining genes. Trends Endocrinol Metab 1997; 9: 342-45.
Motzkus D, Singh PB, Hoyer-Fender S. M31, a murine homolog of Drosophila HP1, is concentrated in the XY body during spermatogenesis. Cytogenet Cell Genet 1999; 86: 83-8.
Moore FL, Jaruzelska J, Fox MS, Urano J, Firpo MT, Turek PJ, Dorfman DM, Pera RA. Human Pumilio-2 is expressed in embryonic stem cells and germ cells and interacts with DAZ (Deleted in AZoospermia) and DAZ-like proteins. Proc Natl Acad Sci U S A 2003; 100: 538-43.
Niederberger C, Agulnik AI, Cho Y, Lamb D, Bishop Ce. In situ hybridization shows that Dazla expression in mouse testis is restricted to premeiotic stage IV-VI of spermatogenesis. Mamm Genome 1997:8:277-8.
Neesen J, Bunemann H, Heinlein UA. *The drosophila hydei* gene Dhmst101(1) encodes a testis-specific, repetitive, axoneme associated protein with differential abundance in Y chromosomal deletion mutant flies. Dev Biol 1994; 162: 414-25.
Nef S, Verma-Kurvari S, Merenmies J, Vassalli JD, Efstratiadis A, Accili D, Parada LF. Testis determination requires insulin receptor family function in mice. Nature. 2003; 426(6964): 291-5.
Ng KW, Ridgway P, Cohen DR, Tremethick DJ. The binding of a Fos/Jun heterodimer can completely disrupt the structure of a nucleosome. The EMBO Journal 1997; 16: 2072-85.
Ohinata Y, Sutou S, Kondo M, Takahashi T, Mitsui Y. Male enhanced antigen-1 gene flanked by two overlapping genes is expressed in late spermatogenesis. Biol Reprod 2002; 67: 1824-31.
Page DC, Mosher R, Simpson EM, Fisher EM, Mardon G, Pollack J et al. The sex-determining region of the human Y chromosome encodes a finger protein. Cell 1987; 5: 1091-1104.
Pannetier M, Servel N, Cocquet J, Besnard N, Cotinot C, Pailhoux E. Expression studies of the PIS-regulated genes suggest different mechanisms of sex determination within mammals. Cytogenet Genome Res. 2003;101:199-205.
Pringle MJ, Page DC. Somatic and germ cell sex determination in the developing gonad. In:Infertility in the male (Lipschultz LL and Howards SS, eds),1997, pp 3-22, Mosbey-Year Book, St Louis.
Rejio RA, Dorfman DM, Slee R, Renshaw AA, Loughlin KR, Cooke H, Page DC. DAZ family proteins exist throughout male germ cell development and transit from nucleus to cytoplasm at meiosis in humans and mice. Biol Reprod 2000; 63: 1490-6.
Ruggiu M, Speed R, Taggart M, McKay SJ, et al. The mouse Dazla gene encodes a cytoplasmic protein essential for gametogenesis. Nature 1996; 389: 73-77.
Santucci-Darmanin S, Vidal F, Scimeca JC, Turc-Carel C, Paquis-Flucklinger V. Family of SRY/Sox proteins is involved in the regulation of the mouse Msh4 (MutS homolog 4) gene expression. Mol Reprod Dev 2001; 60; 172-80.
Saunders PTK, Turner JMA, Ruggiu M, Taggart M, Burgoyne PS, Elliot D, Cooke HJ. Absence of m*Dzl* produces a final block on germ cell development at meiosis. Reproduction 2003; 126: 589-97.
Schmidt EE, Hanson ES, Capecchi MR. Sequence independent assembly of spermatid mRNAs into messenger ribonucleoprotein particles. Mol Cell Biol 1999; 19: 3904-15.
Schnieders F, Dork T, Arnemann J, Vogel T, Werner M, Schmidtke J. Testis-specific protein, Y-encoded (TSPY) expression in testicular tissues. Hum Mol Genet 1996; 5: 1801-7.
Seligman J, Page DC. The Dazh gene is expressed in male and female embryonic gonads before germ cell sex differentation. Biochem Biophys Res Commun 1998; 245: 878-82.

Shah K, Sivapalan G, Gibbons N, Tempest H, Griffin DK. The genetic basis of infertility. Reproduction 2003; 126: 13-25
Skaletsky H, Kuroda-Kawaguchi T, Minx PJ, Cordum HS, Hillier L, Brown LG, Et al. The male specific region of the human Y chromosome is a mosaic of discrete sequence classes. Nature 2003; 423(6942): 825-37. Comment in: Nature. 2003; 423(6942): 810-1, 813.
Smith A, Benavente R. Identification of a short nuclear lamin protein selectively expressed during meiotic stages of rat spermatogenesis. Differentiation 1992a; 52: 55-60.
Smith A, Benavente R. Meiosis-specific protein selectivily associated with sex chromosomes of rat pachytene spermatocytes. Proc Natl Acad Sci (U S A) 1992b; 89: 6938-42.
Schepers G, Wilson M, Wilhelm D, Koopman P. SOX8 is expressed during testis differentiation in mice and synergizes with SF1 to activate the Amh promoter in vitro. J Biol Chem 2003; 278: 28101-8.
Stouffs K, Lissens W, Van Landuyt L, Tournaye H, Van Steirteghem A, Liebaers I. Characterization of the genomic organization, localization and expression of four PRY genes (PRY1, PRY2, PRY3 and PRY4). Mol Hum Reprod 2001; 7: 603-10.
Styrna J, Kilarski W, Krzanowska H. Influence of the CBA genetic background on sperm morphology and fertilization efficiency in mice with a partial Y chromosome deletion. Reproduction 2003; 126:579-88.
Tessari A, Salata E, Ferlin A, Bartoloni L, Slongo ML, Foresta C. Characterization of HSFY, a novel AZFb gene on the Y chromosome with a possible role in human spermatogenesis. Mol Hum Reprod. 2004; 10: 253-8.
Trimmer EE, Zamble DB, Lippard SJ, Essigmann JM. Human testis-determining factor SRY binds to the major DNA adduct of cisplatin and a putative target sequence with comparable affinities. Biochemistry 1998; 37: 352-62.
Tsui S, Dai T, Warren ST, Salido EC, Yen PH. Association of the mouse infertility factor DAZL1 with actively translating poyribosomes. Biol Reprod 2000; 62: 1655-60.
Turner JM, Burgoyne PS, Singh PB. M31 and macroH2A1.2 colocalise at the pseudoautosomal region during mouse meiosis. J Cell Sci 2001; 114 (pt18): 3367-75.
Turner JM, Mahadevaiah SK, Benavente R, Offenberg HH, Heyting C, Burgoyne PS. Analysis of male meiotic "sex body" proteins during XY female meiosis provides new insights into their functions. Chromosome 2000; 109: 426-32.
Turner RM, Johnson LR, Haig-Ladewig L, Gerton GL, Moss SB. An X-linked gene encodes a major human sperm fibrous sheath protein, hAKAP82. J Biol Chem 1998; 273: 32135 – 41.
Vogt PH. Edelmann A. Kirsch S, Henegarin O. Hirschmann P, Kiesewetter F et al, Human Y chromosome azoospermia (AZF) mapped to different subregions in Yq11. Genetics 1996; 5: 933-41.
Vogt PH, Fernandes S. Polymorphic DAZ gene family in polymorphic structure of AZFc locus: Artwork or functional for human spermatogenesis? APMIS. 2003;111: 115-26.
Wallace AM, Dass B, Ravnik SE, Tonk V, Jenkins NA, Gilbert DJ, Copeland NG, MacDonald CC. Two distinct forms of the 64,000 Mr protein of the cleavage stimulation factor are expressed in mouse male germ cells. Proc Nat Acad Sci (U S A) 1999; 96: 6763-8
Wang Z, Zhang Y, Liu H, Salati E, Chiriva-Internati M, Lim SH. Gene expression and immunologic consequence of SPAN-Xb in myeloma and other hematologic malignancies. Blood. 2003; 101: 955-60.
Werner MH, Bianchi ME, Gronenborn AM, Clore GM. NMR spectroscopic analysis of the DNA conformation induced by the human testis determining factor SRY. Biochemistry 1995; 34: 11998- 12004.
Werner MH, Huth JR, Gronenborn AM, Clore GM. Molecular basis of human 46X,Y sex reversal revealed from the three dimensional solution structure of thr human SRY-DNA complex. Cell 1995; 81: 705-14.
Westbrook VA, Dickman AB, Klotz KI, Khole VV, von Kap-Herr C, Golden WL, Eddy RL, Shows TB, Stoler MH, Lee CY, Flickinoer CI Herr JC. Spermatid-specific expression of the novel X Linked gene product SPAN-X localized to the rucleus of human spermatozoa. Biol Reprod 2000; 63: 469-81
Wiltshire T, Park C, Handel MA. Chromatin configuration during Meiosis I prophase of spermatogenesis. Mol Reprod Develop 1998; 49: 70-80.
Wong EY, Tse JY, Yao KM, Lui VC, Tam PC, Yeung WS. Identification and characterization of human VCY2-interacting protein: VCY2IP-1, a microtubule-associated protein-like protein. Biol Reprod 2004; 70 : 775-84.
Wong EY, Tse JY, Yao KM, Tam PC, Yeung WS. VCY2 protein interacts with the HECT domain of ubiquitin-protein ligase E3A. Biochem Biophys Res Commun 2002; 296: 1104-11.
Yamashita A, Ito M, Takamatsu N, Shiba T. Characterization of Solt, a novel SoxLZ/Sox6 binding protein expressed in adult mouse testis. FEBS Lett 2000; 481: 147-51.
Zou SW, Zhang JC, Zhang XD, Miao SY, Zong SD, Sheng Q, Wang LF. Expression and localization of VCX/Y proteins and their possible involvement in regulation of ribosome assembly during spermatogenesis. Cell Res 2003; 13: 171-75.

Chapter 11

CELL CYCLE COMPONENTS

11.1. CELL CYCLE

In continuously dividing cells, an individual cell passes through four phases, shown in **Fig. 11.1.** G1 is the resting phase. During S phase DNA synthesis occurs. The S phase is followed by G2 phase, which is again the resting phase after DNA synthesis. The three phases viz. G1, S and G2 constitute interphase, while the main mitosis takes place during M phase. The duration of different phases of cell cycle depends not only in different organisms but also varies in different tissues of same organism. The crucial features in cell cycle include the existence of two transition control points, at G2/M boundary and during G1 phase. The M phase is characterized by activation of a kinase, which is described by various names depending on the assay such as MPF or H1 kinase or by neutral name M phase kinase. The identification of multiple forms of protein kinases in cell cycle and cell cyclins in spermatogenesis indicates that mitotic and meiotic phases of spermatogenesis are regulated at multiple points and many fold more complicated than the cell cycle thought in yeast or Xenopus oocytes.

11.2. CELL CYCLE GENES IN YEAST

Earlier investigations on cell division conducted on budding yeast (Saccharomyces cerevisiae) and the fission yeast (Schizosaccharomyces pombe) were responsible for major breakthrough in the discovery of present day research in cell cycle. Budding yeast is an oval shaped cell that divides by forming a bud. The fission yeast is a rod shaped cell, which grows by elongation at its ends. Despite of differences in their shapes, they share a number of features, which are useful for genetic studies. A large number of mutants, which helped in the discovery of cell division cycle (cdc) were isolated. These mutants arrest the cell cycle at specific checkpoints. The products of mutant genes were analyzed and compared with the proteins isolated and studied through the biochemistry of the cell division. Cell cycle components controlling cell cycle are highly conserved from yeast to mammalian cells.

11.2.1. Cdc2, Cdc28 and Cdc13 Genes

The study on yeast mutants suggested that the Cdc2 gene in fission yeast and Cdc28 gene in budding yeast were responsible both for passage through start (i.e. transition from G1 to S phase), and for transition from G2 to M phase. Therefore, these two genes are considered to be equivalent to each other. The protein product of these genes and of their homologs in other organisms, named p34cdc2 (Mr 34kDa, p34cdc2), is found in all eukaryotic cells. Another

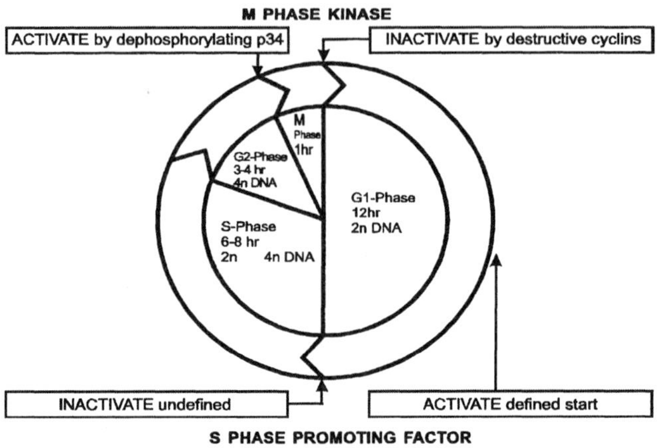

Fig 11.1. Cell cycle of a diploid cell showing effects of M-Phase Kinase and of S-Phase Promoting factor.

fission yeast gene Cdc13, which produces a cyclin or its homolog, is also required for the induction of mitosis. The products of Cdc2 and Cdc13 (p34cdc2 and cyclin) have been demonstrated to participate in cell division **(Fig.11.2)**.

11.2.2. Cdc25 and Weel Genes

The coordinated activity of Cdc2 and Cdc13 is regulated by two additional genes: Cdc25 (stimulates entry into mitosis), and Wee1 (inhibits entry into mitosis), which act antagonistically to control and regulate the entry of cell into mitosis. Another gene nim1 exercises a negative control on Wee1. While nim1 and Wee1 produce kinases that cause phosphorylation, the Cdc25 produces a phosphatase, which causes dephosphorylation of phosphate group at a tyrosine residue of p34cdc2. Thus, these genes act through regulation of protein phosphorylation and dephosphorylation. An increase in the ratio of activities of two genes, namely that of Cdc25 to Wee 1, increases the cell size required for entry into mitosis, while a decrease in this ratio leads to a decrease in the critical cell size. The over-expression of Cdc25 in a mutant for Wee1, causes premature entry into mitosis resulting into lethality. Several nim mutants such as nim A (nim= never-in-mitosis; bim= blocked-in- mitosis) arrest cell cycle in G2 phase in Aspergillus. A Drosophila gene named string and and temperature sensitive nim T in Aspergillus are homologs of Cdc25 of fission yeast **(Fig.11.2)**.

11.2.3. Cyclin and Other Genes in Budding Yeast

Two types of cyclins are mainly involved in mitotic cell division in budding yeast : G1 cyclins and M (mitotic) cyclins (called cyclin B). G1 cyclins play an essential role in start. Originally three cyclin genes (Cln1, Cln2, Cln3) were identified in budding yeast. The levels of cyclin 1 (CLN1) and CLN2 undergo periodic changes with the phase of cell cycle, peaking in late G1, whereas CLN3 is low throughout the cell cycle. In addition, two more genes SWI4 and SWI6 (and also Cdc 28) (SWI = for switching) are required for transcription of HO gene coding an endonuclease. The products of SWI4 and SWI6, required for transcription of Cln1 and Cln2 are components of a transcription factor (SBF) that binds the promoter element responsible for

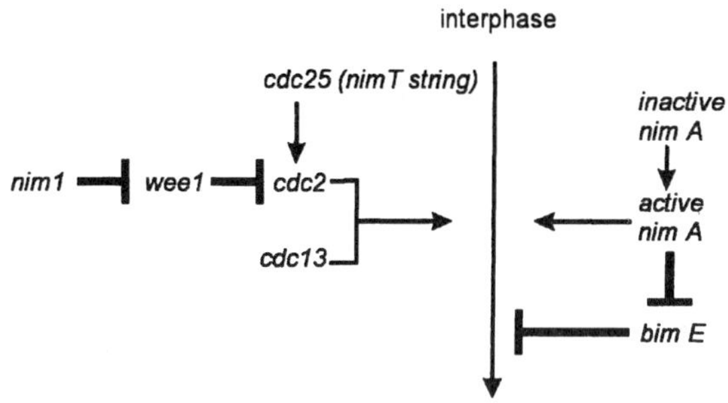

Fig. 11.2. Induction of mitosis in the fission yeast. Stimulation (→)and inhibition(⊣) of genes by gene products are shown. cdc2 and cdc13 gene products are shown acting in concert of each other to induce mitosis.

activation of HO, Cln1 and Cln2 genes. If these genes are absent, cancer like situation may arise, and if positive feedback loop is dampened, the cells stop undergoing rapid cell divisions. In mammalian cells several classes of genes have been identified, that regulate cell cycle (see section 11.6). In addition to cyclin genes, a gene for duplication of centrosome and a number of genes required for spindle function have been identified in budding yeast.

11.3. DEPENDENCE OF MITOSIS ON DNA SYNTHESIS

In several organisms, mitosis can be arrested by the inhibitors or by inactivation of replication enzymes due to mutations. Temperature sensitive mutants defective in DNA synthesis machinery have been isolated, which do not normally undergo mitosis at the restrictive temperature. These mutations, induced by irradiation in fission yeast, involve genes codng for DNA polymerases, viz., Cdc17 (DNA polymerase I), Cdc9 (DNA polymerase II) and cdc2 (DNA polymerase III); the corresponding genes in budding yeast are POL1, POL2 and POL3. However, the dependence of mitosis on DNA synthesis in these mutants is relieved by a complete deficiency of RAD9. It suggests that RAD 9 gene inhibits mitosis in these mutants. It is speculated that RAD 9-gene product may interact with MPF or other components, which are essential for the initiation of mitosis.

The cell cycle controls the initiation of DNA replication once per cycle by an efficient mechanism. Only G1 cells are competent to initiate DNA replication. Molecular approaches revealed that DNA replication begins at origins of replication in chromosomes. In budding yeast, several gene products and their regulators are controlled by a common upstream regulatory element, which represents a recognition site. The origin recognition complex (ORC), a multi-protein complex has been identified in budding yeast. These complexes bind the replication origins through out cell cycle. One of these regulatory proteins is Cdc6. During most part of the cell cycle Cdc6 remains at very low level, but increases transiently at early G1. The Cdc6 binds at ORC and assembles related proteins of a pre-replicative complex (pre-RC). At this point the S-phase cyclin-cdk complex triggers the S-phase DNA replication by phosphorylating proteins at DNA replication origin. The mechanism ensures that DNA is duplicated only once per cycle.

11.4. BIOCHEMISTRY OF CELL DIVISION: AN OVERVIEW

11.4.1. MPF and Kinase Activity

A significant progress was made in understanding the biochemisrtry of cell division from *Xenopus* eggs. The transition from G1 into M phase requires the higher eukaryotic cells to initiate complex processes including nuclear envelope breakdown, chromatin condensation, and reorganization of the cytoskeleton, in a highly coordinated manner. The onset of M phase is only normally permitted after completion of a round of DNA replication. The major biochemical event that controls cell division was the discovery of maturation promoting factor or MPF (a protein) from immature frog eggs. The MPF causes the eggs to undergo meiosis and divide, and thus prepares the egg for fertilization. Subsequently, it was shown that MPF could also induce mitosis. The MPF induces chromosome condensation, nuclear envelope breakdown and assembly of spindle. MPF also activates enzymes that conjugate ubiquitin to cyclin leading to cyclin degradation, which render MPF inactive. Cyclin degradation and MPF activation lead to chromosome segregation, chromosome decondensation, nuclear deformation and cytokinesis. Now it has become apparent that a central and rate limiting function in the transition from G2 into M is performed by a protein kinase that is highly conserved on its primary amino acid sequence and completely conserved in its M-phase-promoting activity throughout the eukaryotes. It is now known that MPF activity depends on two protein species present in equimolar amounts in purified MPF preparations. The first component is the protein kinase catalytic subunit p34cdc2, independently identified genetically in yeast as having a function required for the onset of M phase. The second component of MPF is a B cyclin protein, the biochemical role of which remains unclear, though potential functions include specification of localization of the MPF protein kinase heterodimer. It was also shown that MPF activity is not constant but fluctuates during the cell cycle, i.e. it rises when the cell enters cell division and drops sharply after the division is complete. Presence of a similar cyclin protein was observed in sea urchin eggs and clam eggs. The cyclin activity also fluctuated like MPF. Later, cyclin was shown to be a part of MPF. Mutation analysis in fission yeast showed that cdc2 gene codes for p34cdc2, the kinase responsible for phosphorylation of cyclin. The predicted ATP binding region of the protein is phophorylated during interphase, resulting in inactivation of the pre-MPF complex until M phase onset. The protein MPF was isolated and purified and was shown to consist of a 45 kDa cyclin protein and a 34 kDa cdc2 kinase (p34cdc2) the catalytic subunit of MPF. These two proteins (particularly p34cdc2) are highly conserved and have been found in a wide range of organisms ranging from yeast to humans. Though the level of p34cdc2 is constant during the cell cycle, the level of cyclin rises in the cell cycle at two points (G1 cyclins accumulate upto 'start' point and M cyclins accumulate upto metaphase). After reaching their peak values, cyclins undergo degradation, failing which the cell division cannot proceed further. It was later shown that cyclin synthesis itself is sufficient to activate the p34cdc2 component of MPF and makes frog eggs undergo mitosis. Further, the cyclins accumulate in each interphase and their destruction was necessary (but not sufficient) to shut off p34cdc2 component of MPF thus permitting completion of cell division through inactivation of MPF.

In *Xenopus* egg extracts, three mitotic MPF (Cdc2-cyclin) complexes have been identified; they are activated and inactivated sequentially, beginning with Cdc2-cyclin A1, followed by Cdc2-cyclin B1, and finally Cdc2-cyclin B2. The last of the three, Cdc2-cyclinB2 is inactivated just after nuclear envelope breakdown. Chromatin condensation and nuclear envelope break down persist throughout the remainder of M phase in the absence of active Cdc2. This persistence could be the result of slow reversal of the effects of Cdc2. Alternatively, these aspects of mitosis could be actively maintained by some regulatory protein other than Cdc2 (Guadagno and Ferrel, 1998).

With the help of certain mutants, which replicate their DNA twice in fission yeast, it was also concluded that immediately after M phase, Cdc2 molecule (or p34cdc2 / Cdk1) is in a form, described as S-form, which can be converted into active SPF (due to linkage with G1 cyclins) and induce DNA replication. At some point after start, but before the completion of DNA replication, the Cdc2 molecule is converted into a form described as M form, which can be converted into active MPF and not into active SPF. The formation of active MPF guarantees completion of S phase. During the passage through mitosis, M form of Cdc2 again gets converted into S form. This is possibly brought about by a separate Cdk-activating kinase (CAK). In mitotic cells, this is accomplished by a complex of cyclin H and Cdk7. Cyclin H/Cdk7 complexes, present during meiosis, form active complexes in testicular cells and are strong candidates for the activating kinase for cyclin A1-associated kinase (Kim et al., 2001). Thus S and M forms of Cdc2 are distinguished from each other due to post-translational modification of this protein.

11.4.2. Phosphorylation / Dephosphorylation of Cdc2

During the cell cycle, besides the reversible phosphorylation of cyclin protein, Cdc2 also undergoes phosphorylation and dephosphorylation, which regulate its activity. After passing through 'start', the cells synthesize mitotic cyclins (cyclin B), which accumulate and bind to Cdc2. This complex is initially inactive, due to associated phosphorylation of tyrosine-15 (Tyr-15) of Cdc2 (Tyr-15 phosphorylation inhibits kinase activity). This inactive complex undergoes further phosphorylation at threonine-160 (Thr-160), which is necessary for MPF activity, but is not sufficient to overcome the inhibitory effect of tyrosine phosphorylation. Nonetheless, the phosphorylation of nuclear lamins to promote nuclear disassembly, and the phosphorylation of vimentin and caldesmon, substrates potentially involved in cytoskeletal rearrangement, are reasonably well-substantiated roles for Cdc2 at M phase onset. While protein phosphorylation is clearly of fundamental importance for initiation of M phase, protein phosphatases also have important roles to play, not only in the activation of the pre-MPF complex, but also in removing MPF-driven phosphorylation to restore the interphase state on completion of M phase. One interesting possibility, suggested initially by analogy with yeast systems, is that Cdc2 is required for G1 commitment in animal cells, as well as for M phase onset. Subsequent removal of tyrosine phosphate from Cdc2 during G2 activates MPF and leads to induction of mitosis (M phase). In fission yeast, tyrosine kinase is the product of Wee1 gene, whereas phosphatase is the product of Cdc25 gene. Active MPF activates tyr-phosphatase and inhibits tyr-kinase. This leads to entry into mitosis. The corresponding genes (Wee1, and Cdc25) and their protein products have been identified in mammalian male germ cells . During G2, the cyclin B-Cdc2 complex accumulates in the oocyte. But it is kept inactive through phosphorylation of Thr-14 and/or Tyr-15 by the Wee1 kinase(s). Inactivation of Wee1 kinase and/or activation of phosphatase Cdc25 triggers dephosphorylation of Thr-14 and Tyr-15 and cause activation of MPF. A possible link between activation of MAPKs and MPF has been proposed on the basis of the finding that MAPK activated p90rsk binds to and phosphorylates the Wee1 homolog, Myt1. This phosphorylation inhibits Myt1 activity and interferes with inhibition of MPF activity in vitro (see MAPK signaling pathways).

11.5. MPF ACTIVITY IN SPERMATOGENESIS

While the role of MPF in controlling meiosis in oocytes has been well documented, very little is known about its function in male germ cells. The localization of cyclin B1 and Cdk1 within the testis showed that both cyclin B1 and Cdk1 were present at highest levels in pachytene spermatocytes, with lower levels in the post-meiotic cells. Zhu et al., (1997) suggested the requirement of HSP 70-2 during Cdk1 activity. The Hsp 70-2 gene is expressed only in spermatogenic cells at a significant level. This provides an evidence for a link between an HSP70 molecular chaperone and Cdk1 activity, essential for meiosis during spermatogenesis (Zhu et al., 1997). While Cdk1 and cyclin B1-associated kinase activities were present in meiotically dividing pachytene spermatocyte, with no activity in post-meiotic spermatids, PCNA was detected in the mitotically proliferating spermatogonia, but not in spermatocytes, which had just entered meiosis. PCNA staining reappeared in later stages of meiotic prophase, in particular zygotene and pachytene spermatocytes. As these cells are undergoing meiotic recombination, the presence of PCNA in these pre-meiotic cells could suggest a second function such as in DNA excision repair (Chapman and Wolgemuth, 1994a, 1994b).

11.5.1. Cdk activity in Spermatogenic Cells

Besides the immature frog eggs, spermatocytes also produce MPF required for mitotic cell division, suggesting that MPF activity is needed for mitosis as well as for meiosis, and is associated with metaphase state. A line of research established that a 34-kDa protein kinase for entry into mitosis in yeast is functionally conserved in human cells. In mouse, cyclin dependent kinase (Cdk) family genes are expressed not only in dividing but also in terminally differentiated mouse germ cells, suggesting their possible function during both cell division and differentiation. Whereas in yeast cell cycle control is achieved by a single Cdk p34cdc2 or p34cdc28, in higher eukaryotes, multiple genes appear to function. These various Cdks have high homology in amino acid sequence, and the PSTAIRE motif, where cyclins bind is highly conserved. The genes whose products are known to interact with cyclins or to be functionally homologous to the yeast Cdcs are designated as Cdks (Cdk1 to Cdk6; Pines, 1993). Other structurally and functionally related kinases whose cyclin partners are not known have been named by virtue of their amino acid homologies in a conserved domain (Pines1993; Rhee and Wolgemuth, 1995). The roles of Cdk family in murine germ cell development have been examined by studying the expression of five Cdk family genes (Cdc2, Cdk2, Cdk4, Pctaire-1, and Pctaire-3) in mouse reproductive organs. Analysis revealed distinctive expression pattern of these genes with striking cellular lineage, and developmental stage specificity. In premeiotic spermatocytes in the testis, Cdc2 and Cdk2 transcripts were most abundant in late pachytene to diplotene spermatocytes soon to undergo meiosis. Expression of Cdk family genes has been also observed in non-proliferating cell types. All five Cdk family genes examined are expressed in Sertoli cells of the adult testis; Pctaire 1 and Pctaire 3 showed highest expression in post-meiotic spermatids with high levels of Pctaire1 in post-meiotic germ cells. This showed that Cdk family kinases may exhibit various functions in germinal and somatic cells during spermatogenesis, not only in the cell cycle but also in other regulatory processes, including differentiation (Rhee and Wolgemeuth, 1995).

11.5.2. Multiple Forms of Cdks in Spermatogenesis

During last decade, several Cdks genes have been cloned which are related to amino acid homologies but their functions in cell cycle are unknown due to want of cyclins as partners;

```
1    AATTCGGCACGAGCGCCGTTGTCTGCGCTGCGCTGCGCTGCGCTGGACCAGTTTCGCGAAGTTGCCGGGGCCCTACGCCGCCCAGATGTGCGACCTCATT
1                                                                                         M  C  D  L  I
101  GAACCGCAGCCGGCCGAGAAGATCGGCAAGATGAAGAAGTTGAGGAGAACTTTGTCCGAGAGTTTCAGCCGCATCGCTCTGAAGAAAGAGGACACCACCT
6     E  P  Q  P  A  E  K  I  G  K  M  K  K  L  R  R  T  L  S  E  S  F  S  R  I  A  L  K  K  E  D  T  T

201  TTGATGAGATATGTGTCACAAAGATGTCTACCCGGAACTGCCAGGGGACAGATTCAGTGATCAAGCACCTGGACACAATTCCTGAAGACAAGAAAGTCAG
39   F  D  E  I  C  V  T  K  M  S  T  R  N  C  Q  G  T  D  S  V  I  K  H  L  D  T  I  P  E  D  K  K  V  R

301  GGTTCAGAGGACGCAGAGCACTTTTGACCCATTTGAGAAACCAGCCAACCAAGTCAAAAGGGTCCATTCTGAGAACAATGCATGCATTAACTTTAAATCC
73    V  Q  R  T  Q  S  T  F  D  P  F  E  K  P  A  N  Q  V  K  R  V  H  S  E  N  N  A  C  I  N  F  K  S

401  TCCTCTGCTGGCAAAGAGTCACCTAAAGTTCGGCGGCACTCCAGCCCCAGCTCGCCAACGAGTCCCAAATTTGGAAAAGCTGACTCATACGAAAAACTGG
106  S  S  A  G  K  E  S  P  K  V  R  R  H  S  S  P  S  S  P  T  S  P  K  F  G  K  A  D  S  Y  E  K  L

501  AAAAACTGGGGGAAGGATCTTATGCAACAGTGTACAAAGGGAAAAGCAAAGTGAATGGGAAGCTGGTGGCTCTAAAGGTGATCCGGCTGCAGGAAGAAGA
139  E  K  L  G  E  G  S  Y  A  T  V  Y  K  G  K  S  K  V  N  G  K  L  V  A  L  K  V  I  R  L  Q  E  E  E
                      I                                           II
601  GGGCACACCTTTCACAGCCATCAGGGAAGCTTCCCTGTTGAAAGGACTAAAGCACGCCAACATCGTGTTGCTTCACGACATCATCCACACTAAGGAAACC
173   G  T  P  F  T  A  I  R  E  A  S  L  L  K  G  L  K  H  A  N  I  V  L  L  H  D  I  I  H  T  K  E  T
                III                                                IV
701  CTGACCCTTGTCTTTGAATACGTGCACACTGATTTATGTCAGTACATGGAACAGCACCCTGGAGGACTCCATCCAGATAATGTGAAGTTGTTTTTATTTC
206  L  T  L  V  F  E  Y  V  H  T  D  L  C  Q  Y  M  E  Q  H  P  G  G  L  H  P  D  N  V  K  L  F  L  F
                          V
801  AGCTGCTGCGAGGACTGTCTTACATCCACCAGCGTTATATTTTGCACAGAGACCTGAAACCGCAGAACCTTCTCATCAGCGATACGGGGGAGTTGAAGCT
239  Q  L  L  R  G  L  S  Y  I  H  Q  R  Y  I  L  H  R  D  L  K  P  Q  N  L  L  I  S  D  T  G  E  L  K  L
                                                     VI
901  GGCAGATTTCGGTCTGGCAAGAGCAAAATCCGTCCCTAGCCACACATACTCCAATGAAGTGGTTACCTTGTGGTACAGACCTCCAGATGTTCTTCTGGGC
273   A  D  F  G  L  A  R  A  K  S  V  P  S  H  T  Y  S  N  E  V  V  T  L  W  Y  R  P  P  D  V  L  L  G
            VII                                                          VIII
1001 TCTACAGAATATTCCACCTGCCTTGACATGTGGGGAGTTGGCTGTATCTTCGTTGAGATGATCCAAGGAGTTGCTGCGTTTCCAGGAATGAAAGACATTC
306   S  T  E  Y  S  T  C  L  D  M  W  G  V  G  C  I  F  V  E  M  I  Q  G  V  A  A  F  P  G  M  K  D  I
                                IX
1101 AGGATCAACTTGAACGGATATTTCTGGTTCTTGGAACACCGAATGAGGACACGTGGCCTGGAGTTCATTCTTTACCACATTTTAAGCCAGAACGCTTTAC
339  Q  D  Q  L  E  R  I  F  L  V  L  G  T  P  N  E  D  T  W  P  G  V  H  S  L  P  H  F  K  P  E  R  F  T
                     X
1201 CGTGTACAGCTCTAAAAGCCTTAGACAAGCATGGAATAAGCTCAGCTATGTAAATCATGCTGAAGACTTGGCCTCCAAGCTTCTCCAGTGTTCCCCAAAG
373   V  Y  S  S  K  S  L  R  Q  A  W  N  K  L  S  Y  V  N  H  A  E  D  L  A  S  K  L  L  Q  C  S  P  K
                                                                                     XI
1301 AACAGGCTATCAGCACAGGCCGCCTTGAGCCATGAGTATTTCAGCGATCTGCCTCCACGGCTATGGGAGCTGACTGATATGTCTTCTATTTTTACCGTCC
406   N  R  L  S  A  Q  A  A  L  S  H  E  Y  F  S  D  L  P  P  R  L  W  E  L  T  D  M  S  S  I  F  T  V

1401 CAAATGTGAGATTGCAACCAGAAGCTGGAGAGAGCATGAGGGCCTTTGGAAAAAACAATAGTTATGGAAAAGCCTATCGAACAGCAAACACTGACAAGC
439  P  N  V  R  L  Q  P  E  A  G  E  S  M  R  A  F  G  K  N  N  S  Y  G  K  S  L  S  N  S  K  H  •

1501 CCGGCACTCTCAGGAGAGCACAGGATTAAGTTGTCATCATTCTGGGAAGAAGAAGAAGAGATAAAAATAAATCAATCAATAAAAACCACTAATGAAGTGC
1601 CTCAGTGCGAGGGCCCTGAATCGGTTCTCTTCTGCCCGGTGCAGTGGATTTCACTGACATGAAAATGGAAGCTGGCAAGAACCTGTTTCCTCTGCAATTT
1701 ATTTAAAACCTTGCACGCATTTGGATACCTTGTGACTTCCAAGAACTACGTTACGATTAAGCTTTGCTTACTGCTACATGGCATGTATTCTTTTCCGTCT
1801 TTTGTGTCTGGTTCTGTTTGCTTCCCCCTGCCGCACAGTGTCCATCGCACGGTGTCTCTGTGAAGGGTTTTGCTGCTTTCAGCAGCCTCGTCTTAAGTAC
1901 ATTAAGACAACACATTTGGTGTTCACACTTCCTCAGCCATGTCTGCACTTGAAAGCCACACAGCGGCAGGAGACAATGTGTGTTTTCCCTGAGAGCAGAG
2001 TGCATTCTGCAACCTCCAGGGAAGAACATTCTTTTTGCTAAAACAAACTCCATGCTTTCTGACAGACTAGACAGCGTGGTCCCAGACAGTGGGGCCCTAC
2101 CTGTTGCCTGCCTTGGAGGACATTTAGAAATGTAAGGGTTTTTTTGTTTTGTTTTGTTTTTCGTTTTTTATTGGAAATGAGCACACATTTACACAGGAGA
2201 CCAAAATCCACTTGCTATTTCAATAGCCTGACTGTTTCTGAGCAGGTAGAGAAATAAACATGGAAAACAAAGACAGGGTGACTGAAGGTTAATAAGCATT
2301 ACTATCATTTTTTTTAGCCTGTAGAAAGCAAGGCAACTTTTTCATGTGTCTTGTCCACATCTTACCTTTTCCCATCAGCTCTGTATTATTAAACTGTTAC
2401 ACTCTATTTTTAAACTGGAGAAAATAGTAGCAAATATCATTCTTAAAACCTGGGTACATAGAAAAAAAAAAAATCTGTCTGTCAGGGTAGGTCCTGAATGC
2501 AGCCTTGGCTGATTAAAGCTTAGAAATCACATTTTATAATTATCCAGACTTTAAAATGTGCTTATTTACGACAAAGGACCTTTGAATTTAATTCGATGTT
2601 CAGAAACATTCCAGGCCGTTCGGAAGGCATCACTGGGTACCAGACTTGAAACCAGTCTTCAGTAACAAGTGTTCAAAGGCAGATCTTACAAGCAATGCAG
2701 GAATTTTGCTTTAAGAATAAAAATGGCCAAGAAAGGCCCTTGTTTTAAGGACCTTTGGGACCTTGCAAAGCGTAGGCAAGTCTGCTCTAGCATACATTGA
2801 ACCTTGATCTGGGGCAGCCGCCCAGTGCTGGCTACTTTCATGGACAGCCTCACTGCAGAATCCCAGGGGCAGTTAGGAAGCTATGTGAAGCATGCCCTTG
2901 CTCCTCCTGACATAAATCCCCATGGAAACCCAACACATCTGTGTACGGTGGTCTGTGTTTTACTCCTCCTGAGAAAAAGGAGACGCAGGCTACAAGCCTG
3001 AAATGTGACACAATTTTAGGACCAAGACTAGAAATCACAAGGCTGATTTGACAAGGGCTTTCTGATTCTCCAATGAGGGGCAAGCTTCCTTATACTCAAA
3101 CCCAGTGTATCCCTGGAAGTAGACCCAGGTTTCCAAGTCTTCAAAGGCACTGTGTCTGAGTAAGAGGTGTAACCCAGCCCCCAGCCCCTCCGAGCGGGTC
3201 ACTCACTGCAGCCTCACAAACACTGAGAAAGGAACTGCTTACAGGATGGTTTGGGGAAAAAAGCATAAGATACCCAAGGTAACAATGTTTATTCTTCACT
3301 AGCTGGGAATCGGTCTCCATCCCTCTTTTGTACATTCCTTCCATGGAAGTCTTTACAGGTATCTTGATTTAAAGTATAATTAAATTGGGAGTTCTTTAA
3401 GTGGGGAGAAAACTTTATGTGTAATTAAAAACAGTCTCCCTGTGATTTGCTTAGAATCCTCCATACATCTGGGGACGGGGGAAGATTATGTTTGTTTTCA
3501 CTAAGTAGAAGTCAGGAGCTCACAGGAATGCTGGGAGGGTTTCAGTAAGAACATAAAGGCAACCCCATGTCCTTGCCCTCGTCTAGACAGCTTCTCCTGG
3601 GAAGCAAGACATAACACTGCTCAGTGTGCGGTCTGCTTCACGACAGGGACAGGCACACCTCTCAAGTTTAAAGGGTTTCAGTTTCCTGTCCTTGTCACCA
3701 GTGAAGGAAATGCAAAGTGTCTGCCCCCACCCCTCCACCCAACACAGTCCCCCTTCTCTGGCTTTTGCTCTCCTGGCCTCTCCAAGGGTGTCATGAAAAT
3801 CAGTATAGACTTTGTTTCATTAACATGTCATATGTTGAGAGAGCACTGGAATGTTTGGTGAGTTTATTTTTCTATGAACGCGTAGCTTGAGTTTGGACAA
3901 CTAGTAAGGAATGCACCAACACTTTGGCATGACAATTGTATGAAAGCAACCTCACTGCACAAAGTATTTGGAAAAGAAGTATGGGACTTAGTTTCTATCT
4001 TCTTAAGGGGAGCCGTTTAATTAAATACAACATCCATGGGACAGGAAATGTGTTTGCTATAAAATTAGAGATATAAGGCAGGGGCAGCAATCTGCCCACC
4101 ATGCAAGGATGAAAATCTGTTTCCAGCTCTCAAGTGGCTAAAAGTTTCCAAGAGAGTCCGTGGGATTATAGGAAAACCAGGGCCTCTTGGTACCCTGTCC
4201 TCAGAGTGTGTGACAGCCACTGTAATCAGTGCTCTATGTATGCGTCCATAGCAGACACTCATCTGTGAATATCGCTGGGTTTGTAATCTTTTGTTATATA
4301 AGAGGATGTTTAGGCTGTACATACTGGGGTAGATCATTGCCTGCCCCTATACATAGAAATATGCTGCATAATTGTGCATAACTTCCTTCTCCGTGGCTGG
4401 CTTGTAGACAGAGGAAACTGTATGTTAGCCTTCAGTTTCCTCATATACCAAACTGAAAACGTTTAAATCTCAAATGAAGTAAGCAAGGTTCTGTTCTCAC
4501 TGCCCACTTTCTCAGCAGCATATATTAGA
```

Fig. 11.3. Nucleotide and deduced amino acid sequence of murine Pflaire-1.The Kinase domains are underlined and numbered. The polyadenylation signal is shown in bold letters. Reproduced with permission from V. Besset et al. Mol Reprod Dev 50;80-29:1998 © John Wiley & Sons Inc. http://www3.interscience.wiley.com/cgi-bin/jabout/37692/ProductInformation.html.

some of these kinases with PSTAIRE and Ckd5 are not directly related to mitosis. The PCTAIRE genes from human, mouse and lower organisms have been cloned. Gene PCTAIRE1 is expressed abundantly in highly differentiated tissues. Mouse Pctaire1 encoded two major proteins of Mr~62,000 and ~68000, found predominantly in testis and brain. Within these two tissues, Pctaire 1 was most abundant in the cytoplasm of terminally differentiated cells, notably, the pyramidal neurons in brain and elongated spermatids in testis. A kinase activity toward myelin basic protein was associated with Pctaire 1 in the adult testis and brain and that its activity was potentially regulated through association with regulatory partner(s) (Besset et al., 1999).

The murine Pftaire-1 has 50% and 49% amino acid identity with Cdk5 and Pctaire-3, respectively, and contains the eleven sub-domains characteristic of the protein kinases. It has two transcripts of approximately 5.5 and 4.9kb in size, which are expressed at low level in all murine tissues, except in the brain, testis and embryo, where high expression was detected. The mRNAs analysis shows that Pftaire-1 is expressed in late pachytene spermatocytes in the testis and in post-mitotic neuronal cells both in the brain and embryo suggesting a role of Pftaire 1 in the process of meiosis as well as neuron differentiation and/or neuron function (Besset et al., 1998) **(Fig. 11.3)**. The putative Cdk1 (Cdc2) related serine/threonine protein kinase isolated from a Hela cell cDNA library encodes a protein of 469 amino acids, sharing 95% identities with the mouse PFTAIRE1 throughout the entire protein sequence. This gene, designated human hPFTAIRE1, is located at human chromosome 7q21.13. Another 6kb transcript is detected with varied levels of expression of hPFTAIRE1 in 16 human tissues. The hPFTAIRE1 was highly expressed in brain, pancreas, kidney, heart, testis, and ovary and exhibits cytoplasmic distribution (Yang and Chen, 2001).

11.5.3. Human Wee1

Regulatory phosphorylation of the Cdc2p kinase by Wee type kinases prevents eukaryotic cells from entering mitosis or meiosis at an inappropriate time. This suggests that a novel, sperm-specific pathway negatively regulates WEE-1.3 to allow the G2/M transition of male meiosis I, and that dominant wee-1.3 mutants prevent this negative regulation. In S.pombe, Wee1 shows a genetic control of integeration between growth and cell division. Nakanishi et al. (2000) identified human Wee1B (a new member of Wee1 family of Cdk-inhibitory kinases) protein comprising of 561 amino acids. Human Wee1B mRNA is particularly abundant in testis. GFP-Wee1B showed a predominantly nuclear localization in Hela cells. Recombinant human Wee1B effectively phosphorylated cyclin B associated Cdk1 on tyrosine-15, resulting in the inactivation of the kinase activity of Cdk1. The identification of Wee1B suggests that inhibition of Cdk1 is mediated at multiple levels in mammals. All metazoa also have a membrane-associated Wee1p-like kinase named Myt1, which is less well characterized. The canonical Wee1p kinase is a soluble protein that functions in the eukaryotic nucleus. The C elegans Myt1 ortholog is encoded by the wee-1.3 gene, and six dominant missense mutants prevent primary spermatocytes from entering M phase. These mutants revealed that WEE-1.3 protein is required for embryonic development, germline proliferation and initiation of meiosis during spermatogenesis (Lamitina and L'Hernault 2002).

11.5.4. Other Protein Kinases in Cell Cycle

A heterodimeric surface receptor of 65-kDa (p65) and 95-kDa (p95) is expressed on the surface of proliferating cells such as activated T lymphocytes and other mitotic cells. The p65/p95 receptor associated with kinase activity is involved in the regulation of mitotic cell division. All cells undergoing mitotic or meiotic division in the rat testis express the p65/p95 receptor.

The p65/p95 receptor is localized on cell surface as well as in the nucleus on chromosomes associated with mitotic and meiotic divisions. Cells that do not divide, did not express this receptor. It seems that further studies on this receptor and its ligand should provide a better understanding of the control of cell division (Roux et al., 1998).

The 95-kDa nuclear protein (NP95), which contains a leucine zipper motif, a zinc finger motif, a putative cyclin A/E-cyclin dependent protein kinase 2 phosphorylation site, and retinoblastoma protein binding motifs, is associated with S-phase progression of mouse cells. It is suppressed during G1 and G2/M phases in normal thymocytes but expressed at a constantly high level irrespective of cell stage in mouse T cell lymphoma cells. The NP95 is known to be expressed strongly only in proliferative tissues and cells, and is localized in S-phase nuclei. Double immunostaining of NP95 and PCNA showed that NP95 was co-localized with PCNA. Construction of 3-D images indicated that NP95 was localized with PCNA in replication sites in a somewhat distinct temporal manner. During meiosis, NP95 was present not only in proliferating spermatogonia but also in meiotic spermatocytes and differentiating spermatids, which were not proliferating (Uemura et al., 2000). Association of PCNA and NP95 suggests that NP-95 is involved in mitosis and meiosis in testis.

11.6. CELL CYCLINS IN SPERMATOGENIC CELLS

Cyclins are a family of proteins implicated in the induction and control of mitosis and were first described as proteins periodically synthesized during early embryonic development of murine invertebrates. They are characterized by their accumulation in the interphase and dramatic destruction at the end of each cell cycle via the ubiquitin pathway. Cyclic changes in levels of cyclins result in cyclic assembly and activation of cyclin-Cdk complexes. Cyclins described above can be of three types: 1) M-cyclins (cyclin B) found during mitotic phase; 2) S-cyclins (cyclin A) found during S-phase and 3) G1/S cyclins (called Cln in yeast and cyclin E in vertebrates) found at the end of G1 phase of the cell cycle and commit the cell for DNA replication. In combination with mitotic cyclins, p34cdc2 can form active protein kinase, known as growth associated H1 kinase (Histone H1 as substrate) or MPF, which induces the events of mitosis. In addition, G1 cyclins (cyclin D), a fourth class of cyclins, are associated to promote the passage through start or restriction point in late G1. Thus, Cdk can combine with G1 cyclins, and catalyze passage through start (an event that commits the cell to initiate DNA replication). The Cdk-G1 cyclin complex is described as start promoting factor (SPF). Thus the alternation of replication (S phase) and segregation (M phase) is actually brought about by alternating activation of Cdk component of the SPF and MPF activities.

However, presently ten classes of cyclins (designated cyclin A through I and cyclin K) have been reported in animal cells; several of them have multiple members. The cyclins have been found to contain a well conserved amino acid sequence known as the cyclin box. All proteins designated as cyclins contain this structural motif, which has necessary information for binding of Cdks (Zhu and Naz, 1998). Three forms of cyclin D (D1-D3) have been described for the mouse testis, but none of them appear to be associated specifically with dividing germ cells (Ravnik et al., 1995). Moreover, cyclin D1 was found in the Sertoli cells, which are non-dividing in the adult, and cyclin D3 (also non-dividing) was found in round spermatids. Cyclin D2 has been seen in both spermatogonia and in Sertoli cells (Nakayama et al., 1996); testis size is largely reduced in cyclin D2-deficient mice (Gromoll et al., 1997).

11.6.1. Cyclin B: A Component of MPF

During cell division, cyclin levels were seen to peak at each M phase, with two types of cyclins A and B, being distinguished by their different gel mobilities and the slightly earlier appearance and disappearance of the A type. Molecular cloning of cyclin genes from a variety of eukaryotes confirmed the distinction between the A and B type cyclins at the amino acid level. The description of cyclin behaviour in the early embryo set the stage for the discovery that the 45-kDa component of purified MPF was a B cyclin. No enzymic activity has been attributed to the cyclin B moiety, while p34cdc2 (Cdc2) has protein kinase activity, which is essential for initiation of M phase. It was concluded from several studies that cyclin B has an accessory function for MPF and has no catalytic activity of its own.

In constrast to other cyclins, which are nuclear, cyclin B accumulates in the cytoplasm of interphase cells and stays there until the beginning of mitosis. Only a limited number of studies have addressed the role of cyclins in spermatogenesis, and a general scheme has yet to emerge. In the rat testis, cyclin B and its kinase are associated with spermatogonial/early spermatocyte proliferation whereas in the mouse testis, mRNA expression of two cyclins (B1 and B2) correlates with second meiotic division and early round spermatid development. For mouse cyclin A1, expression is testis specific and is confined to meiotic divisions (Sweeny et al., 1996).

Characterization: Sequence analysis of the isolated cyclin B1 clones obtained from a rat testis library revealed sequence identity to a rat cyclin B1 that has been isolated from a rat embryo fibroblast cell line. The 3'-untranslated region in cDNA clones (37 bp) was significantly shorter than the sequences from the rat (~840 bp) and the mouse (~720 bp). When the amino acid sequences of cyclin B1 of different species were compared, characteristic elements such as the cyclin destruction boxes, the typical cyclin B1 motif, and the cytoplasm retention signal displayed very high homology. Trembley et al., (2000) cloned 1.8 kb of DNA sequence upstream of the rat cyclin B1 gene translation start site and confirmed the mRNA transcription start point (tsp) at approximately 100bp upstream of the translation start site. A second potential tsp was also located at approximately 32bp downstream from the first. Like many other cyclin promoters, there was no apparent TATA box upstream of the transcription initiation sites. However, computer analysis of the promoter region identified a group of consensus transcription factor binding sites, some of which are also reported in other cyclin promoters. These include those for p53, p21, Ap1, Ap-2, Ets-1, CAATT, E-Box and Y1. The E-Box and/or CCAAT binding sites appeared to be important for transcription. In addition, there may be negative regulatory elements present between 1800 and 1100bp upstream of the translation start sites (Trembley et al., 2000).

Localization in testis : A single 1.7 kb cycB2 expression was most abundant in the germ cells, specifically in pachytene spermatocytes. This is in constrast to the highest levels of expression of cycB1 being present in early spermatids. There are lineage and developmental specific differences in the pattern of the B cyclins in mammalian germ cells, in constrast to the co-expression of B cyclins in the early conceptus (Chapman and Wolgemuth, 1993) **(Fig. 11.4).** Therefore, B1 and B2 cyclins exhibit different patterns of expression during murine spermatogenesis. Cyclin B2 transcripts are most abundant in pachytene spermatocytes whereas the highest levels of expression of cyclin B1 are present in early spermatids. However, cyclin B1 protein appears more abundant in pachytene spermatocytes than in the post-meiotic round spermatids. Moreover, cyclin B2 null mice developed normally and were fertile, whereas cyclin B1-null mice died in utero (Brandeis et al., 1998). This indicates that cyclin B2 is dispensable for

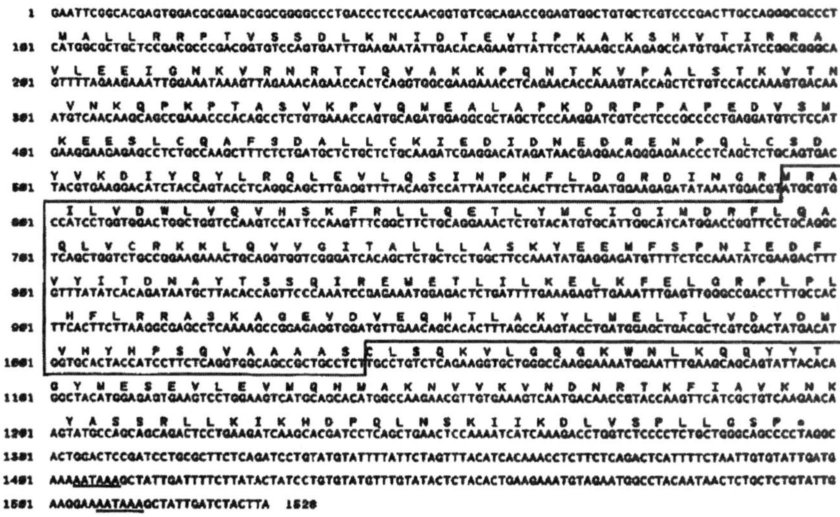

Fig 11.4. Nucleotide sequence and deduced amino acid sequence of mouse cycB2 cDNA are shown. A termination codon is shown at the end with an asterisk(*).The boxed region (called cyclin box) of amino acids is most highly conserved among cyclin. The polyadenylation signal in 3' untranslated region underlined. Reproduced with permission from D.L. Chapman and D.J. Wolgemuth, Development 118;229-240:1993 © The Company of Biologists, Ltd.

murine spermatogenesis whereas cyclin B1 protein has a role in this process.

Gromoll et al., (1997) found B1 mRNA in cytoplasm of primary and secondary spermatocytes and round spermatids but cyclin B1 protein was present only in round spermatids, leading these authors to conclude that "a role for this cyclin in the second meiotic division is unlikely". In situ hybridization of Cdk1 mRNA in the mouse testis has shown that this gene is expressed primarily in meiotic cells of the testis (Rhee and Wolgemuth, 1995). However, immunocytochemical detection of Cdk1 has revealed a lower staining of meiotically dividing spermatocytes than of early pachytene spermatocytes (Ravnik and Wolgemuth, 1999). This later result challenges somewhat the widely recognized role of Cdk1 in the G2/M transition of both mitotically and meiotically dividing cells. By constrast, Cobb et al., (1999) observed similar amount of Cdk1 (and cyclin B1) in young and late spermatocytes (c/r Godet et al., 2000). In rat, cyclin B1 and Cdk1 were high in young pachytene spermatocytes just before the first meiotic division. The relative levels of the two proteins remained high in secondary spermatocytes, then decreased in round spermatids at the exit of meiosis. H1 kinase activity was higher in pachytene spermatocytes than in round spermatid fractions from both adult and young rats. This suggested that the meiotic G2/M transition is associated to high levels of cyclin B1 and Cdk1 indicating a high MPF activity irrespective of the age of the animals (Godet et al., 2000).

Functions :The similarity of the spermatid-specific expression patterns in the mouse and rat, and the observation that cyclin B1 protein is present in these spermatids, lent support to the view that B-type cyclins could be involved in spermatid differentiation. Although rat and mouse cyclin B1 are highly homologous (93%), the timing of their appearance during the spermatogenic cycle is different in the two species. Rat cyclin B1 mRNA is already expressed in late pachytene spermatocytes and during the second meiotic division, but since the protein was not found in these cells, a role for this cyclin in the second meiotic division is unlikely. In

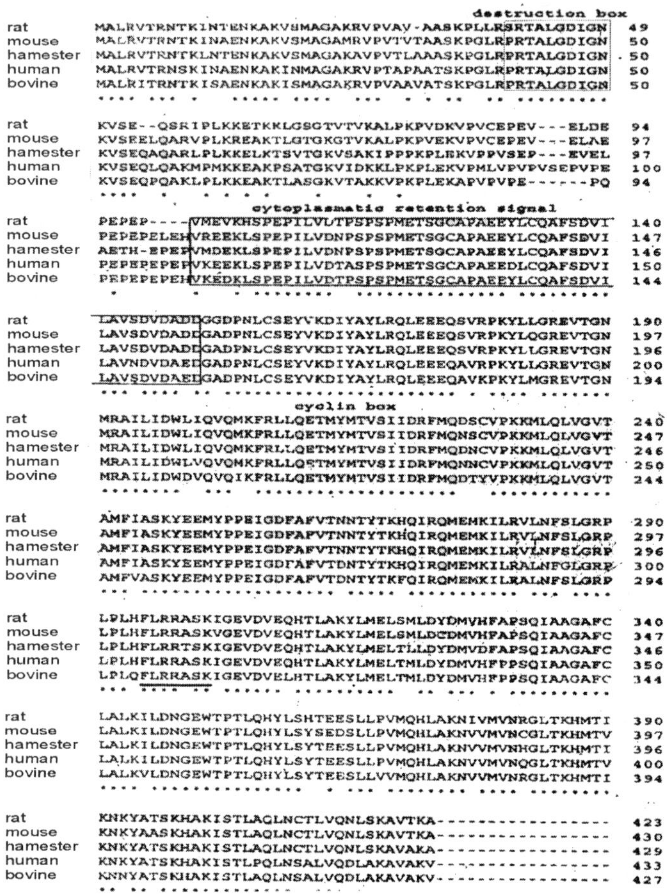

Fig.11.5. Comparison of amino acid sequences of rat, mouse, hemester, human, and bovine cyclin B1. The destruction box and cyclin box are shown. The conserved moti for B-type cyclins is underlined. Reprduced with permission from J.Gormoll et al, Biol Reprod 57;1312-19:1997 © Society for the Study of Reproduction.

vitro studies have shown that human cyclin B1 is cytoplasmic throughout G2-phase but that it translocates into the nucleus at the beginning of mitosis. The translocation from the cytoplasm into the nucleus takes place as a complex consisting of cyclin B1 and Cdc2 during mitosis. This complex is associated with the spindle caps and the main spindle and probably, directly interacts with microtubules. However, in the study of Gromoll et al. (1997), cyclin B1 protein could be detected only in the cytoplasm and not in the nucleus of germ cells. Since cyclin B1 has to combine with Cdc2 to form the MPF complex, it may be that cyclin B1 remains cytoplasmic due to the lack of Cdc2 expression. The presence of Cdc2 in mouse Sertoli cells and finding of the nuclear localization of cyclin B1 in the rat Sertoli cell, a location in which cyclin B1 should normally be detected together with Cdc2 during mitotic divisions in somatic cells, might indicate other functions of the MPF, e.g. cell differentiation. Cyclin B1 protein was seen only in round spermatids, in constrast to cyclin B1 mRNA expression in late meiotic spermatocytes and early round spermatids. Similar to other gene products whose RNA is transcribed and stored and

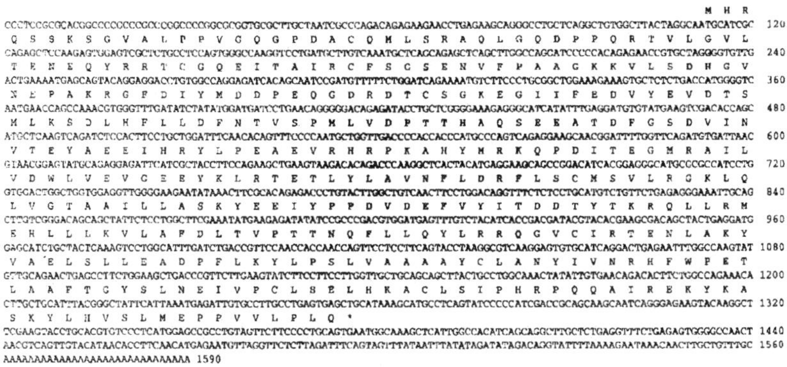

Fig. 11.6. Nucleotide sequence of the cDNA clones encoding murine cyclin. Al from adult testis library. The amino acid sequence of encoded polypeptide is shown above the nucleic acid sequence. Reproduced with permission from C. Sweeney et al. Development 122; 53-64: 1998 © The Company of Biologists, Ltd.

translated several days later since spermatids have no proliferative activity, the presence of regulatory and stimulatory components of the cell cycle implies that cyclins serve other functions than control of cell division in these cells. However, the functional significance of cyclin expression in spermatids remains unknown (Gromoll et al., 1997) **(Fig. 11.5).**

Factors Controlling B Cyclin Activity : At M phase of the cell cycle, the H1 histone kinase activity associated with the Cdc2 kinase/cyclin B complex is an excellent indicator of mitosis (M phase). Various factors control cell cycle in testes. Depletion of gonadotropins did not change the expression pattern of cyclin B1 in the testis after treatment with antagonist of GnRH. This indicated that cyclin B1 is not influenced by gonadotropins and that it might be regulated by paracrine factors deriving from a Sertoli cell-germ cell or a germ cell-germ cell interaction. This provided evidence that the signals triggering the onset of cyclin B1 expression does not originate from the pituitary within the germinal epithelium.

The germ cells of vitamin A deficient testis begin to enter the M phase around 4 h after retinol injection, with the highest percentage of cells entering at 12 h. These results are consistent with the suggestion regarding origin of synchronous spermatogenesis in regenerated seminiferous tubules at the end of the S phase of the cell cycle (Wang and Kim, 1993). The abdominal temperature though has no significant influence on the transcription of Cdc2 and Cyclin B1 in the spermatogonia and pachytene/diplotene primary spermatocytes, but it blocks the translation of both of them. Due to the deficiency of p34cdc2 or Cdc2 and Cyclin B1, the spermatogonia and pachytene/diplotene primary spermatocytes failed to form MPF and hence could not undergo karyokinesis. The development of primary spermatocytes was arrested at the G2/M phase transition. Moreover, testosterone could regulate the cyclin B1 expression in spermatogenic cells since testosterone could recover spermatogenesis in the unilateral scrotal testis, which was influenced by the contralateral cryptorchid testis. However, testosterone could not salvage the spermatogenic block in the cryptorchid testis (Kong et al., 2000).

11.6.2. Cyclin A1 in Male Germ Cells

The mammalian A-type cyclin family consists two members, cyclin A1 (encoded by Ccna1) and cyclin A2 (encoded by Ccna2). Cyclin A2 promotes both G1/S and G2/M transitions, and targeted deletion of Ccna2 in mouse is embryonic lethal. In contrast to cyclin A1, cyclin A2,

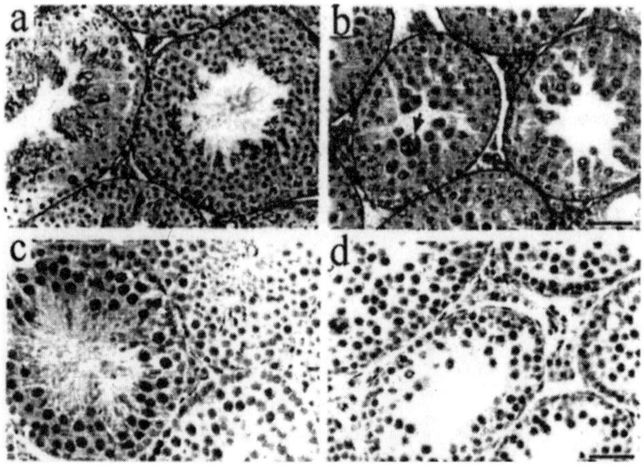

Fig.11.7. Spermatogenesis abnormalities in Cenal-/-mice. b and d indicate metaphase of first mciotic division in Cenal-/- mice and b show immunohisto chemical localizationof cyclin A1 indicating immediately proceeding the first meiotic division in wild type (c) and in mutant mice (d) Arrow shows the appearance of multinucleated giant cells. Reproduced with permission from D. Liu et al, Nature Genet 20;377-80:1998 © www/ http.nature.com/

which has greatest sequence identity with human cyclin A and Xenopus cyclin A2, is expressed in embryos and all tissues analyzed. Sweeney et al., (1996) showed the existence of two A- type cyclins in the mouse germ cells **(Fig.11.6).** In the adult mouse, the expression of cyclin A1, which has greatest sequence identity with Xenopus cyclin A1, is restricted to germ cell lineage and expressed in humans at highest levels in the testis and brain among normal tissues, in addition to certain myeloid leukemia cells and cyclin A1 induced leukemia in a transgenic mouse model. The Ccna1-/- males are sterile due to a block of spermatogenesis before the first meiotic division, whereas females were normal. Arrest of meiosis in Ccna1-/- males was associated with increased germ cell apoptosis, desynapsis abnormalities and reduction of Cdc2 kinase activation at the end of meiotic prophase (Liu et al., 1998) **(Fig. 11.7).** Cyclin A1 expression was very low in breast cancer, non-small cell lung cancer and in cervical carcinoma. However, substantial expression of cyclin A1 was found in testicular and ovarian cancer and in endometrial cancer.

Cyclin A1 gene, a human cyclin A like gene has 48% identity with human cyclin A and is more related to the cloned murine cyclin A1. The expression of cyclin A1 mRNA is differentiation and cell cycle regulated in the ML-1 cells. Its antibody recognizes the Mr 65 kDa cyclin A1, along with the Mr 33 kDa Cdk and other proteins at Mr 39, 42, 45, 95 and 110 kDa. In an in vitro assay Cdk cyclin A1 revealed activities against histone H1. Cyclin A1 binds to Cdk2 but not to Cdc2, Cdk4 and Cdk5. Human cyclin A1 gene was mapped to chromsome 13q12.3-q13 approximately 1000 kb from the sequence tagged site marker WI-3374 (Yang et al., 1997; 1999). Cyclin A1 could bind to cell cycle regulators such as the transcrption factor E2F-1, and the p21 family of proteins. The in vitro interaction of cyclin A1 with E2F-1 is greatly enhanced when cyclin A1 is complexed with Cdk2. Associations of cyclin A1 with Rb and E2F-1 in vivo with several cell lines suggested that the Rb family of proteins and E2F-1 may be important targets for phosphorylation by the cyclin A1-associated kinase and thus cyclin A1 may function in the mitotic cell cycle in certain cells and possibly in testis (Yang et al., 1999).

The levels of cyclin A1 mRNA rise dramatically in late pachytene spermatocytes and become undetectable soon after completion of the meiotic divisions, suggesting its expression to be cell cycle regulated. In lysates of germ cells from adult testes, cyclin A1 is present in p13suc1 precipitates and possesses H1 kinase activity. Three kinase partners of cyclin A1 have been identified: p34cdc2, a polypeptide of 39 kDa that is related to p33cdk2 and, in lesser quantities, p33cdk2. Cyclin A1 is also present in metaphase I and metaphase II oocytes, where a proportion of the cyclin A1 co-localizes with the spindle. This indicates that mammalian germ cells contain cyclin A1-dependent kinases that either act as a substitute for, or in addition to the cyclin A2-dependent kinases characterized in somatic tissues (Sweeney et al., 1996) **(Fig.11.6).**

11.6.3. Cyclin A2

The cyclin A2 expresses mainly in stem germ cells, the spermatogonia, and at highest levels in preleptotene spermatocytes, in which premeiotic DNA synthesis occurs. The concurrent localization of cyclin A2 mRNA and protein suggested that cyclin A2 is regulated at the level of transcription in these cells. The observed cellular specificity of cyclin A2 expression is consistent with its function during mitosis in the stem cell stage of this lineage, while the restricted localization at meiotic stage suggests its function in G1/S or S but not in the meiotic divisions per se (Ravnik and Wolgemuth, 1996). The cellular distribution of the cyclin A1 and cyclin A2, and their Cdk partners, Cdk1 and Cdk2 in the spermatogenic lineage showed that cyclin A1 is present only in male germ cells just prior to or during the first, but not the second, meiotic division. By constrast, cyclin A2 was expressed in spermatogonia and was most abundant in preleptotene spermatocyte cells, which were ready to enter the meiotic pathway. The Cdk1 was most apparent in early pachytene spermatocytes, followed by declie in diplotene and meiotically dividing spermatocytes, the cells in which cyclin A1 expression was strongest, whereas Cdk2 highly expressed in all spermatocytes. Notably, in cells undergoing the meiotic reduction divisions, Cdk2 appeared to localize specifically to the chromatin. This was not the case for spermatogonia undergoing mitotic divisions. In contrast to binding of cyclin A1 to both Cdk1 and Cdk2 in earlier studies, Ravnik and Wolgemuth (1999) showed that cyclin A2 binds only Cdk2, and suggested that the A-type cyclins and their associated kinases have different functions in the initiation and passage of male germ cells through meiosis (Ravnik and Wolgemuth, 1999). Role of cyclin A1 in G2/M transition during meiosis of germ cells was confirmed in mice in which targeted disruption of A-type cyclin gene, results in block of spermatogenesis prior to the entry into metaphase I. The meiotic arrest is accompanied by a defect in Cdc2 kinase activation at the G2/M transition, raising the possibility that a cyclin A1-dependent process dictates the activation of MPF. Subsequently it was shown that like Cdc2, the expression of B-type cyclins is retained in cyclin A1-deficient spermatocytes, while their associated kinases remained at inactive states. Treatment of arrested germ cells with the protein phosphatase type-1 and type2A inhibitor okadaic acid restored the MPF activity and induced entry into M phase and the formation of normally condensed chromosome bivalents, concomitant with hyperphosphorylation of Cdc25 proteins. Conversely, inhibition of tyrosine phosphatases, including Cdc25(s) by vanadate, suppresses okadaic acid induced metaphase induction. Further more, the protein kinase complex of cyclin A1 and Cdk2 phosphorylated Cdc25A and Cdc25C in vitro. These studies suggest that in normal meiotic male germ cells, cyclin A1 participates in the regulation of other protein kinases or phosphatases critical for the G2/M transition and may be directly involved in the initial amplification of MPF through activating phosphorylation of Cdc25 phosphatases (Liu et al., 2000).

11.6. 4. D-Type Cyclins

The D-type cyclins have been shown to be differentially expressed in a number of isolated cell types and cell lines, suggesting distinct roles in cell cycle regulation in particular cell lineages. The D-type cyclins (D1, D2 and D3) are critical governors of the cell cycle during G1 phase of the mammalian cell cycle. These D type cyclins are expressed in overlapping, apparently redundant fashion in the proliferating tissues. To investigate why mammalian cells need three distinct D-type cyclins, Sicinski et al., (1996) generated mice bearing a disrupted cyclin D gene by gene targeting in embryonic stem cells. Cyclin D2 deficient females were sterile owing to the inability of ovarian granulose cells to proliferate normally in response to FSH, whereas mutant males displayed hypoplastic testes. Despite a low size of testis, microscopic picture was normal and there was no sign of impairment of fertility. In ovarian granulose cells, cyclin D2 specifically induced by FSH via a c-AMP-dependent pathway indicated that expression of the various D-type cyclins is under control of distinct intracellular signaling pathways. The hypoplasia seen in cyclin D2 ovaries and testes prompted to examine human cancers deriving from corresponding tissues, and found that some human ovarian and testicular tumors contain high levels of cyclin D2 mRNA (Sicinski et al., 1996).

Major transcripts of cyclin D1, cyclin D2, and cyclin D3, respectively of 4.2kb, 6.8 kb, and 2.3-kb have been detected in the adult mouse testis. Additional transcripts of 1.8 and 2.7-kb were detected for cyclin D3 in the testis. Other tissues, did not have additional transcripts, which were limited to germ cells cyclin. Cyclin D1 was expressed in a pattern consistent with expression of non-dividing Sertoli cells, while cyclin D2 levels appeared slightly enriched in germ cell deficient testes as compared to intact testis (Musa et al.,1998; Ravnik et al., 1995). However, cyclin D3 expression was highest in the non-dividing, haploid, round spermatids (Ravnik et al., 1995; Zhang et al., 1999).

Temporal and spatial expression studies of cyclin D3 during rat testis development showed no detectable cyclin D3 in Sertoli cells, interstitial cells, or fibroblasts within seminiferous tubules or blood vessels whereas known cyclin D3 partner, Cdk4 was located mainly in nuclei of spermatogonia and in early stage of primary spermatocytes. But cyclin D3 was detectable in meiotically active male PCNA-positive germ cells. Moreover, in contrast to in vitro reports, cyclin D3 was not located in the nucleus, but rather in the cytoplasm of male germ cells in vivo. The presence of cyclin D3 in spermatocytes and its location in the cytoplasm seems to speculate that cyclin D3 may have functions in male germ cells other than mitosis (Kang et al., 1997; Zhang et al., 1999). Feng et al., (2000) suggested that SCF induced cyclin D3 expression and phosphorylation of retinoblastoma protein through pathway that is sensitive to both wortmannian and rapamycin.

In contrast to adult testis in the early stages of postnatal testicular development, cyclin D3 protein was detected in spermatogonia and Leydig cells. Cdk-4 and p27 were expressed in a similar subset of testicular cells, suggesting that they may regulate cyclin D3 function during testicular development in a cell type specific manner. Cyclin D3 with associated kinase activity, detected in immature, but not adult testes suggests unique role for cyclin D in the control of cell division and differentiation in the germ line (Zhang et al., 1999).

Cyclin D2 Eexpression in GCTs: Human male germ cell cyclin D2 expression in germ cell tumors (GCTs) arises in the spermatocytic lineage, and subsets display embryonal like differentiation. Virtually all human male germ cell tumors exhibit multiple copies of the short arm of chromosome 12. Among the candidate amplified genes mapped to 12p, expression of the cyclin D2 gene was deregulated in a panel of GCT cell lines, with the relative level of steady state mRNA and protein inversely correlated. The GCT cell lines with a more differentiated

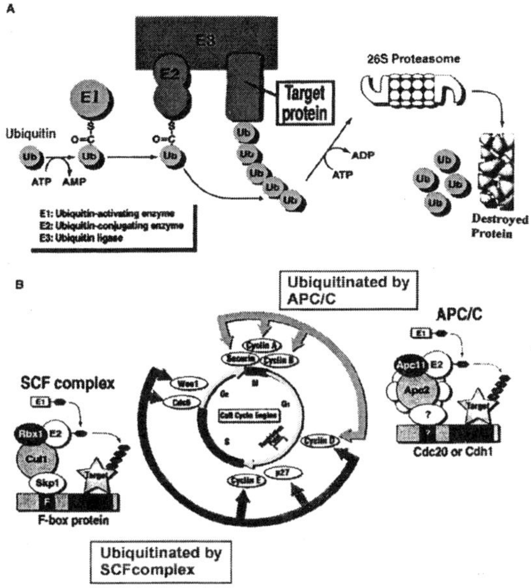

Fig. 11.8. Cell cycle regulation by ubiquitin proteasome pathway. (A) Ubiquitin polymerization on the target protein and its subsequent degradation by 26 Sproteasome. (B) Roles of two distinct ubiquitin ligases in regulation of cell cycle. Reproduced with permission of K. Nakayama et al, Biochem Biophys Res Commun 282;853-60;2001© Elsevier.

phenotype displayed lower cyclin D2 expression with a concurrent increase in expression of the cell cycle inhibitor p21. In the GCT cell lines in which cyclin D2 was highly expressed, cyclin D2 was in complex with its expected catalytic partners (Cdk4 and Cdk6). Whereas no detectable cyclin D2 expression was evident in normal human germ cells, cyclin D2 was expressed in the abnormal germ cells of all carcinoma in situ. In GCT specimens that displayed no evidence of differentiation (seminoma) or primitive differentiation (embryonal carcinoma), cyclin D2 expression was detected. However, in tumor specimens with certain patterns of differentiation (teratoma and yolk sac tumor), expression was down or up-regulated depending on the pattern. Thus, aberrant cyclin D2 expression is an early event in germ cell tumorigenesis (Houldsworth et al, 1997).

Cyclin D and its Partner Cdk4: The mechanism by which cyclin D/Cdk4 complex regulates cell cycle progression is not clear. Although several factors have been implicated in the regulation of Cdk4 activity, little is known regarding the contribution of cyclin dependent kinase inhibitors (CKIs) in Cdk4 activation in the mid-G1 phase. Cyclin D/Cdk4 appears to initiate phosphorylation of retinoblastoma protein (Rb) leading to inactivation of the S-phase inhibitory action of Rb. However, cyclin D/Cdk4 has been postulated to act in a non-catalytic manner to regulate the cyclin E/Cdk2-inhibitory activity of p27 kip1 by sequestration. Cdk4 (-/-) mice survived embryogenesis and showed growth retardation and reproductive dysfunction associated with hypoplastic seminiferous tubules in the testis and perturbed corpus luteum formation in the ovary. These phenotypes appear to be opposite to those of p27-deficient mice such as gigantism and gonadal hyperplasia. A majority of Cdk4 (-/-) mice developed diabetes mellitus associated with degeneration of pancreatic islets, and quiescent fibroblast exhibited substantial delay in

S-phase entry after serum stimulation. The cell cycle delay was due to increased binding of p27 to cyclin E/Cdk2 and diminished activation of Cdk2 accompanied by impaired Rb phosphorylation. Importantly, fibroblasts from Cdk4 (-/-) p27 (-/-) embryos displayed partially restored kinetics of the G/S transition, indicating the significance of the sequestration of p27 by Cdk4. These results suggest that at least part of Cdk4's participation in the rate limiting mechanism for the G/S transition consists of controlling p27 activity (Tsutsui et al., 1999 ; Nakayama et al., 2001) **(Fig 11.8).** Takahashi et al., (2000) found that most of Cdk4 remained bound to p15 when cells were in a quiescent state. Following CSF-1 stimulation, Cdk4 bound to cyclin D1 and then to p21, concomitant with the dissociation of p15 from the complexes. The activation of Cdk4 correlated well with p21 binding to the complexes, and the majority of active Cdk4 complexes contained p21. Taken together, it was suggested that p21 and cyclin D1 act cooperatively as activators of Cdk4 through the release of CKIs of the INK4 family.

11.6.5. Cyclin G Associated Kinase (GAK)-Cdk5

Human cDNA encoding GAK, an association partner of cyclin G and CDK5 encodes GAK of 1311 amino acids and shows all the unique motifs that characterize rat GAK, such as the presence of a Ser/Thr kinase domain, a tensin/auxilin homologous domain and a Tyr phosphorylation target site. The expression profiles of GAK and cyclin G during the Hela cell cycle showed that GAK expression oscillates slightly, peaking at G1phase, although the histone H1 kinase activity remains constant throughout the cell cycle. Also the kinase activity in cyclin G-antibody complex fluctuated during the cell cycle with a peak at G1 phase, although the expression level of cyclin G remained constant. The GAK is expressed ubiquitously, with the highest level of expression being observed in the testis. The chromosomal localization of GAK has been assigned to 4p16 (Kimura et al., 1997).

11.6.6. Cyclin H / Cdk7 Complex

Cell division requires that Cdks be activated by phosphorylation. In mitotic cells, this is accomplished by the Cdk activating kinase (CAK), which is a complex of cyclin H and Cdk7. Kim et al., (2001) have shown that cyclin A1 is meiosis specific and forms an active kinase with Cdk2. Because cyclin A1 is required for meiosis and its associated kinase must be phosphorylated, it was proposed that cyclin H/Cdk7 functions to activate cyclin A1/Cdk2 in meiotic cells. Cyclin H and Cdk7 have been found during meiosis and the mRNAs encoding cyclin H and Cdk7 are abundant in the nucleus of spermatocytes in stages IV to XII of the spermatogenic cycle, overlapping the same stages that express cyclin A1-associated kinases. Finally, H1-kinase assay in immunoprecipitates of cyclin H and Cdk7 from testicular extracts showed that these proteins interact to form an active kinase. This indicated that cyclin H/Cdk7 complexes, present during meiosis, form active complexes in testicular cells and are strong candidates for the activating kinase for cyclin A1-associated kinase (Kim et al., 2001).

11.6.7. Sperm Cyclin I

The presence of cell cyclins has also been demonstrated in spermatozoa. Besides the presence of A and B cyclins, the human sperm showed the presence of an isoform of cyclin I, designated cyclin ITI. The cyclin ITI cDNA is 1443 bp long and has an ORF of 178 amino acids with the first ATG Met start codon at nt 1. It has a well conserved sequence of cyclin box. The 5' flanking sequence of 728-pb has a high similarity (99.6%) with the 5' introns of the human -polymerase gene. The nucleotide sequence of cyclin ITI cDNA has high similarity (99.3%) with the cyclin I gene sequence (Fig.11.10). The cyclin ITI was found to have three amino acids mutations at nt 25, 172, 223, respectively, including one in the well conserved sequence of cyclin box. The cyclin ITI is expressed in human testis at an increased level compared to other tissues, indicating its presence in human sperm. This new I isoform may have a role in spermatogenesis or human sperm especially in capacitation and or acrosome reaction (Zhu and Naz, 1998).

11.6.8. Cyclin K

Human cyclin K is a RNA polymerase II associated cyclin having both carboxy terminal domain kinase and Cdk-activating kinase activity. The cyclin K gene encodes a 357-amino-acid protein most closely related to human cyclins C and H, which have been proposed to play a role in regulating basal transcription through their association with and activation of cyclin dependent kinases that phosphorylate the carboxyl-terminal domain (CTD) of the large subunit of RNA polymerase II (RNAPII). Murine and Drosophila melanogaster homologues of cyclin K have also been identified. Cyclin K mRNA is ubiquitously expressed and most abundant in the developing germ cells of the adult testis and ovaries. Cyclin K is associated with potent CTD kinase and Cdk activating kinase (CAK) activity in vitro and coimmunoprecipitates with the large subunit of RNAP II. Thus, cyclin K represents a new member of the transcription cyclin family, which may play a dual role in regulating Cdk and RNAP II activity (Edwards et al., 1998).

11.7. CDC25 PHOSPHATASES IN MALE GERM CELLS

The Cdc25 family of protein phosphateses positively regulate the cell division cycle by activating cyclin-dependent protein kinases. The Cdc25 belong to protein threonine/tyrosine phosphatases that activate cyclin dependent kinases. In humans and rodents, three Cdc25 family members denoted Cdc25A, B, and C have been identified. The murine forms of cdc25 exhibit different pattern of expression during development and in adult mouse tissues. Cdc25A and Cdc25C were differentially localized in rat germ cells during spermatogenesis. Cdc25A protein was distributed in the cytoplasm of germ cells, whereas Cdc25C was localized to the nucleus of late spermatocytes and round spermatids. Among two Cdc25A transcripts, one transcript of 3.8 kb was expressed in the early germ cells including spermatogonia, and the other of 3.3-kb was highly expressed in spermatocytes and round spermatids. The mRNA levels for Cdc25A were highest in diplotene spermatocytes and round spermatids but undetectable in Sertoli cells. Both transcripts for Cdc25A increased with retinol treatment in the vitamin A-deficient testis. Results suggest that cdc25A may function during mitosis of spermatogonia. A cDNA encoding the murine homologue of the human Cdc25C gene was isolated. Murine Cdc25C transcripts of 2.1kb were detected in midgestation embryos and in several adult tissues, including testis and ovary; the highest levels were detected in the testis, with abundant level of a 1.9-kb transcript that was not detected in the other tissues. In the testis, Cdc25C expression was localized in germ cells, specifically in late pachytene diplotene spermatocytes and round spermatids, whereas Cdc25B expression was detected in the somatic cells. The study not only indicates the expression of the murine Cdc25 genes to be lineage and developmental stage-specific in the mouse testis but also suggests that the Cdc25 genes may have different functions in the germinal and somatic compartments some of which are involved in cellular proliferation (Wu and Walgemuth, 1995 b; Mizoguchi and Kim, 1997). Among adult tissues in which Cdc25C is detected, its transcripts are most abundant in testis, followed by thymus, ovary, spleen, and intestine. Mice lacking Cdc25C were fertile, indicating that Cdc25C does not contribute an essential function during spermatogenesis or oogenesis in the mouse. Cdc25C(-/-) mice are viable and do not display any obvious abnormalities. T- and B-cell development was also normal in Cdc25C(-/-) mice, and Cdc25C(-/-) splenic cells exhibited normal proliferative responses in vitro. Findings indicated that Cdc25A and/or Cdc25B may compensate for loss of Cdc25C in the mouse (Chen et al., 2001).

Twine, the homolog of gene Cdc25 has been found in Drosophila. String but not twine transcripts express in the proliferating cells of newly cellularized embryos. Both genes are abundantly expressed in nurse cells during oogenesis. In the testis, twine transcripts are seen in the growing stage of premeiotic cysts. Analysis of a twine mutant suggests a requirement for the gene during oogenesis, during syncytial embryonic development, and for male meiosis (Alphey et al., 1992). Sigrist et al., (1995) analyzed the requirement for Drosophila cdc2 kinase during spermatogenesis in temperature sensitive mutant lines (Dmcdc2ts). While meiotic spindles and metaphase plates were never formed in Dmcdc2ts mutants at high temperature, chromosomes still condensed in late spermatocytes and spermatid differentiation (sperm head and tail formation). The same phenotype was also observed in twine and twine/ Dmcdc2ts double mutant testes, consistent with the idea that the cdc2 kinase activity required for meiotic divisions is activated by the Twine/cdc25 phosphatase. Confirming this notion, it was found that entopic expression of the string/cdc25 phosphatase, which is known to activate the cdc2 kinase before mitosis, results a partial rescue of meiotic divisions in twine mutant testis (Sigrist et al., 1995). However, in comparison to fission yeast, in mammalian male germ cells, the functions of cdc25 isoforms are not well defined since cdc25c(-/-) male mice are viable and fertile (Chen et al., 2001) (Chapter 20).

11.8. CYCLIN DEPENDENT KINASE INHIBITORS

The activity of Cdks, which play a key role in cell cycle progression in eukaryotic cells is regulated through a number of mechanisms. One of the regulatory mechanism has emerged with the discovery of the Cdk inhibitors (Cyclin dependent Kinase Inhibitors, CKIs). Two families of CKIs, the INK4 family and the Cip/Kip family, have been reported in mammalian cells. These two classes are structurally distinct, and their members interact with cyclins and Cdks in completely different ways. These CKIs are able to modulate Cdk activity during G1/S phase transition. Members of INK4 family specifically bind and inhibit cyclin D-dependent kinases, whereas Cip/Kip family proteins bind and regulate A-, D-, and E-dependent kinases. Members of the INK4 family comprising four polypeptides of Mr 15-19 kDa (p16 or INK4a; p15 or INK4b; p18 or INK4c and p19 or INK 4d) share characteristic four fold ankyrin repeats and bind specifically to Cdk4 and its close homolog Cdk6. In contrast to INK family Cip/Kip proteins (p21, p27, and p57) share a homologous amino terminal domain that contains contiguous cyclin and Cdk-binding regions, and bind to a wide range of cyclin/ Cdk complexes. Another characteristics of the Cip/Kip family of CKIs is that the p21 molecule can be a component of catalytically active enzymes, raising the possibility that p21 may have roles to play in cell cycle regulation of other than as a CKI.

11.8.1. INK4 Family

By disrupting cyclin D-dependent holoenzymes, INK4 proteins prevent phosphorylation of the retinoblastoma (Rb) protein and block entry into the DNA synthetic phase of the cell division cycle. The p16 (INK4a) is a potent tumor suppressor in human, whereas involvement, if any, of other INK4 proteins in tumor surveillance is less well documented. p18 (INK4c) and p19 (INK4d) are expressed during mouse embryogenesis in stereotypic tissue specific pattern and are also detected, together with p15 (INK4b), in tissues of young mice. The p16 is expressed neither before birth nor readily at appreciable levels in young animals, but its increased expression later in life suggests that it plays some checkpoint function in response to cell stress, genotoxic damage, or aging per se.

Mice lacking p19 gene developed into adulthood, had a normal life span, and did not spontaneously develop tumors. Tumors did not arise at increased frequency in animals neonatally exposed to ionizing radiation or the carcinogen, dimethylbenzanthracene. Mouse embryo fibroblasts, bone marrow-derived macrophages, and lymphoid T and B cells from these animals proliferated normally and displayed typical lineage specific differentiation markers. Males exhibited marked testicular atrophy associated with increased apoptosis of germ cells. The absence of tumors in p19 (INK4d) deficient animals demonstrated that, unlike p16, p19 is not a tumor suppressor but is instead involved in spermatogenesis (Zindy et al., 2000). The p19 negatively regulates the proto-oncogenic cyclin D/Cdk4 complexes whose ability to phosphorylate the tumour suppressor promotes G1/S transition. Bartkova et al (2000a) examined the presence of p19 (INK4d) in the human testis, at various stages of germ cell tumour pathogenesis. It was found that the p19 protein is abundant in spermatocytes of normal human adult testes, whereas p19 was not detectable in testicular cancer, including the preinvasive carcinoma in situ stage. Together with the lack of p19 in human foetal germ cells, these results supported the concept of foetal origin of the testicular germ cell tumors, and helped to understand better the emerging role of the Rb pathway in spermatogenesis and tumorigenesis in the human testis (Bartkova et al., 2000).

Male mice lacking both the p18 (INK4c) and p19 (INK4d) genes are infertile, whereas female fecundity is unaffected. Whether or not p19 is present or absent, animals lacking p18 develop hyperplasia of interstitial testicular Leydig cells, which produce reduced levels of testosterone. The anterior pituitary of fertile mice lacking p18 or infertile mice doubly deficient for p18 and p19 produces normal levels of LH. By contrast, p19 null or double null mice produce elevated levels of FSH. Because p19-null mice are fertile, increased FSH production by the anterior pituitary is also unlikely to contribute to the sterility observed in p18/p19 double null males. Thus p18 and p19 are essential for male fertility. These two Cdk inhibitors collaborate in regulating spermatogenesis, helping to ensure mitotic exit and the normal meiotic maturation of spermatocytes (Zindy et al., 2001).

11.8.2. Cip/Kip Family of Cdk Inhibitors

In contrast to INK4 family Cip/Kip proteins (p21, p27, and p57) share a homologous amino terminal domain that contains contiguous cyclin and Cdk-binding regions, and bind to a wide range of cyclin Cdk complexes. The p24 has homology to dual specificity protein phosphatase but its substrate is unknown. p27 blocks the cells cycle in response to transforming growth factor . But p21 has attracted the most attention. p21 was first identified as a component, along with the proliferating cell nuclear antigen (PCNA), of almost all Cdk complexes in normal but not in transformed cells. Another characteristics of the Cip/Kip family of CKIs is that the p21 molecule can be a component of catalytically active enzymes, raising the possibility that p21 may have roles to play in cell cycle regulation of other than as a CKI. It has been proposed that members of the Cip/Kip family of CKIs act as assembly factors of cyclin D1/Cdk4 (Takahashi et al., 1999).

Waga et al., (1994) showed that p21 is able to inhibit DNA replication directly, quite independently of the cyclin dependent kinases. This finding has profound implications for understanding the regulation of both DNA synthesis and DNA repair. The p21 is involved in inhibiting the cell cycle in response to DNA damage. It is induced in -irradiated cells that have wild type p53, binds to the putative S-phase promoting kinase, cyclin E-CDK2, and so prevents cells from initiating DNA replication in vivo and in vitro. The p21 seems to inhibit elongation rather than involved in initiation events, and that its inhibition can be relieved by adding excess of PCNA. Therefore, p21 can inhibit the DNA replication machinery directly, and Waga et al, (1994) presented data to show that it does so by binding to PCNA (Pines, 1994).

The p21Cip1/WAF1 expression was studied before and after irradiation in the adult mouse testis. In normal testis a weak staining for p21 is found in pachytene spermatocytes in epithelial stage V up to step 5 spermatids. X-irradiation at 4 Gy resulted in a transient increase of p21 despite the fact that the apoptotic spermatocytes did not increase, whereas number of apoptotic spermatogonia increased to 10 fold and spermatogonial staining for p21 was absent. This suggested that these cells do not undergo a p21 induced G1 arrest prior to DNA repair or apoptosis. This implies that p21 is a factor, which could be important during the meiotic prophase in spermatocytes and repair mechanisms in these cells, but not in spermatogonial cell cycle delay or apoptosis induction (Beumer et al., 1997).

During mouse testicular development, p27kip1 is induced when the fetal germ cells, gonocytes become quiescent on day 16 postcoitum suggesting that p27kip1 is an important factor for the G1/G0 arrest in gonocytes. But proliferating spermatogonia among different types of germ cells were not stained for p27kip1. Although during development, Sertoli cells have proliferating activity and only are occasionally lightly stained for p27kip1 in the adult testis, the terminally differentiated Sertoli cells are heavily stained for p27kip1. Since p27kip1 is only expressed in Sertoli cells, the role of p27kip1 in spermatogonia and preleptotene spermatocytes must be indirect. Hence supporting role of Sertoli cells in spermatogenic process depends on the expression of p27kip1in these cells. Since the expression of p27kip1 is transiently increased after x-irradiation in whole testicular lystates, hence, p27kip1seems to be involved in the cellular response after DNA damage (Beumer et al., 1999). The p27kip1 knockout mice are associated with aberrations in the spermatogenic process and multiple organ hyperplasia and have 2-fold larger testes than their wild-type litter mates. A higher pituitary tumor incidence was reported in p27kip1 knockout mice indicating that loss of p27kip1 may contribute to oncogenesis and tumor progression (Beumer et al., 1999). These observations indicated that p27kip1 has a role in the regulation of spermatogonial proliferation, or apoptosis, and the onset of the meiotic prophase in pre-leptotene spermatocytes.

Role of p27 and p57 has been studied in p27-/- and p57-/- knock out mice. Phenotypes of p27-/- mice are substantially different from those of p57-/- mice, suggesting that spatial and temporal expression patterns of p27kip1 and p57kip2 are distinct. Whereas certain organs and tissues (brain, lens, ganglion, lung, heart, liver, skin and kidney) expressed both proteins, other expressed only p27kip1 or p27 (thymus, spleen, retina, testis and ovary) or only p57kip2 or p57(gut, palate, pancreas, cartilage and skeletal muscle). In addition, some organs expressed both p27 and p57 but showed mutually exclusive patterns of distribution among tissues restricted to embryogenesis. Expression of p27kip1 in many tissues was maintained in adult animals. The expression of p27kip1 and p57kip2 was inversely correlated with cell proliferation, suggesting that p27kip1 and p57kip2 are expressed exclusively in postmeiotic cells. These complex spatial and temporal patterns of expression are consistent with the phenotypes of mice deficient in p27kip2 or p57kip2 and they suggest that these proteins might play important roles in tissue development (Nagahama et al., 2001).

Evidences suggest that HSPs are likely involved in cell cycle control under normal conditions without stress. 73hsc (heat shock cognate protein), which belongs to the HSP70 family of molecular chaperones, interacts with p27kip1 inhibitor of Cdk during G1/S transition. This indicated that 73hsc might be involved in the cell cycle progression through the regulation of cell cycle regulators such as p27Kip1 (Nakamura et al., 1999). The Xenopus Levies p27Xic1 or Xic1 is a member of Cip/Kip CKI family and inhibits Cck2-cyclin E in vitro as well as DNA replication in Xenopus egg extracts. Ubiquitination of Xic1 occurs exclusively in the nucleus and that nuclear localization of Xic1 is necessary for its degradation. However, binding of Xic1 to Cdk2-cyclin E is essential for Xic1 ubiquitination and degradation. The amino acids 180-83 of Xic1 are critical determinants for recognition and degradation of Xic1 by the ubiquitin machinery (Chuang and Yew, 2001).

p27 as tumor suppressor: Absence of the tumor suppressor p27 (kip1) results in increased body size, hyperplasia of several organs including the testis, and cancer in mice. Similarly, lack of inhibin causes testicular and ovarian tumors of the granulosa/Sertoli cell lineage beginning at 4 weeks of age and adrenal tumors in gonadectomized mice. In inhibin knockout mice Inh(-/-), Cdk4 and cyclin D2 mRNA levels are elevated, and p27 levels are decreased in testicular tumors from Inh(-/-) mice. These findings suggested that increased Cdk4/cyclin D2 (positive) activity and decreased p27 (negative) activity is causal for gonadal tumor formation. To test this hypothesis, double mutant mice lacking both p27 and inhibin were generated to determine whether the tumor suppressors p27 and inhibin have additive suppressor activity in the gonads. Like Inh(-/-) mice, p27(-/-) Inh(-/-) mice (double mutant) demonstrated elevated serum activin levels, ovarian and testicular tumors, and a resultant lethal cachexia like syndrome. However, double mutant mice died earlier than single mutant mice. Moreover, tumor foci in double mutant mice could be observed as early as 2 weeks of age in males and as early as 4 weeks in females. These findings demonstrated that absence of both Inhibin and p27 in mice causes earlier development of ovarian and testicular tumors and earlier death compared with absence of inhibin alone (Cipriano et al., 2001). Bartkova et al., (2000b) examined cyclin A, B1 and E along with Cdk inhibitors, p18, p21 and p27, in human testicular tumors, including carcinomas in situ (CIS) and germ cell tumours (GCTs). A correlation seemed to exist between proliferation and abundance of the cyclin proteins, and abundance of p18, and the lack of p21 and p27 in normal spermatogenesis. Increased abundance of cyclin E, and down regulation or loss of p18 were the features that correlated with progression from CIS to invasive germ cell tumours (Bartkova et al., 2000).

11.9. ACTIVATION OF MAPK PATHWAY DURING MEIOSIS

MAP-kinases are inactivated by dephosphorylation either at threonine or at tyrosine. The MAP kinase (ERK) activation occurs via a complex cascade of kinases that include, immediately upstream, MAP kinase-kinase (MAPKK called MEK in mammalian Ras signaling pathway), which in turn is phosphorylated by one of the MAP kinase-kinase-kinase (MAPKKK called Raf in Ras signaling pathway). The Raf kinase is activated by activated Ras. Once activated, the MAP kinase relays the signal down stream by phosphorylating various proteins within cell including protein kinases and regulatory proteins. On entry into the nucleus, it can phosphorylate one or more component of gene regulatory complex, and may activate the transcription of immediate early genes. Among the genes activated by Ras pathway are the genes encoding G1 cyclins required for cell division. Activation of the ERK subfamily of MAP kinases is a well-characterized response to both receptor tyrosine kinases (RTKs) and integrin activation. Although there are many routes to ERK activation, GRB2 pathway is the best understood and provides a paradigm for understanding receptor mediated signal transduction. The duration, as well as the magnitude of the ERK signal are critical factors in determining the proliferative response of cells to changes in their extracellular environment (Roovers and Assoian, 2000).

MAPK pathway seems to facilitate or regulate not only mitosis and meiosis but also plays important role in sperm maturation through capacitation and or acrosome reaction. Several lines of approaches raise the possibility that p42 MAPK (ERK2) pathway participates in mitosis. A series of events can be artificially reproduced in mouse spermatocytes by treatment with the serine threonine/phosphatase inhibitor, okadaic acid (OA). The OA overcomes the checkpoints that normally delay the progression of the meiotic cycle of mid- and late-pachytene spermatocytes and induces nuclear envelope breakdown and chromosome condensation resembling a situation that was observed during G2/M transition after prophase-1. The meiotic progression of mouse spermatocytes induced by OA is accompanied by an increase in H1 kinase activity, which is considered a sign of MPF activation. The increase in H1 kinase activity during OA induced G2/M transition in mouse spermatocytes is due to a concurrent activation of a cyclin/Cdk activity, which suggests that activation of MPF actually occurs, under these experimental conditions (Sette et al., 1999). In addition to MPF, ERK1 is also required for G2/M progression during meiosis in mouse spermatocyte. Immunoprecipitation with ERK antibodies showed that p44/ERK1 but not p42/ERK2 is stimulated in OA treated spermatocytes. It was suggested that ERK1 is specifically activated during G2/M transition in mouse spermatocytes, that contributes to the mechanisms of MPF activation, and essential for chromosome condensation associated with progression to meiotic metaphase (Sette et al.,1999).

11.10. REFERENCES

Alphey L, Jimenez J, White- Cooper H, Dawson I, Nurse P, Glover DM. Twine, a cdc25 homolog that functions in the male and female germ line of Drosophila. Cell 1992; 69: 977-88.

Bartkova J, Thulberg M, Rajpert De Meyts E, Skakkebaek NE, Bartek J. Lack of p19INK4d in human testicular germ cell tumours contrasts with high expression during normal spermatogenesis. Oncogene 2000; 19: 4146-50.

Bartkova J, Thullberg M, Rajpert De Meyts E, Skakkeback NE, Bartek J. Cell cycle regulators in testicular cancer: loss of p18INK4C marks progression from carcinoma in situ to invasive germ cell tumours. Int J Cancer 2000, 85: 370-5

Besset V, Rhee K, Wolgemuth DJ. The cellular distribution and kinase activity of the Cdk family member Pctaire1 in the adult mouse brain and testis suggest function in differentiation. Cell Growth Diff. 1999; 10: 173-81.

Besset V, Rhee K, Wolgemuth DJ. The identification and characterization of expression of Pftaire-1, a novel Cdk family member, suggest its function in the mouse testis and nervous system. Mol Reprod Dev 1998; 50: 18-29.

Beumer TL, Kiyokawa H, Roepers-Gajadien HL. Van Den Bos LAC, Lock TMTW, Gademan IS, Rutgers DH, Koff A, and De Rooij DG. Regulatory role of p27kip1 in the mouse and human testis. Endocrinology 1999; 140: 1834-40.

Beumer TL, Roepers-Gajadien HL, Gademan LS, Rutgers DH, de Rooij DG. 21(Cip1/WAF1) expression in the mouse testis before and after X irradiation. Mol Reprod Dev 1997; 47: 240-7.

Brandeis M, Rosewell I, Carrington M, Crompton T, Jacobs MA, Kirk J, Gannon J, Hunt T. CyclinB2-null mice develop normally and are fertile whereas cyclin B1-null mice die in utero. Proc Natl Acad Sci (USA) 1998; 95: 4344-49.

Chapman DL, Wolgemuth DJ. Isolation of the murine cyclin B2 cDNA and characterization of the lineage and temporal specificity of expression of the B1 and B2 cyclins during oogeneis, spermatogenesis and early embryogenesis. Development 1993; 118: 229-40.

Chapman DL, Wolgemuth DJ; Expression of proliferating cell nuclear antigen in the mouse germ line and surrounding somatic cells suggests both proliferation dependent and independent modes of function. Int J Dev Biol 1994b, 38: 491-7.

Chapman DL, Wolgemuth DJ. Regulation of M-phase promoting factor activity during development of mouse male germ cells. Dev Biol 1994a; 165: 500-6.

Chen MS, Hurov J. White LS, Woodford-Thomas T, Piwnica-Worms H. Absence of apparent phenotype in mice lacking Cdc25C protein phosphatase. Mol Cell Biol 2001; 21: 3853-61.

Chuang LC, Yew PR. Regulation of nuclear transport and degradation of the Xenopus cyclin dependent kinase

inhibitor, p27Xic1. J Biol Chem 2001; 276: 1610-17.
Cipriano SC, Chen L, Burns KH, Hoff A, Matzuk MM. Inhibin and p27 interact to regulate gonadal tumorigenesis. Mol Endocrinol 2001; 15: 985-96.
Cobb J, Cargile B, Handel MA. Acquisition of competence to condense metaphase I chromosomes during spermatogenesis. Dev Biol 1999; 205: 49-64.
Edwards MC, Wong C, Elledge SJ. Human Cyclin K, a novel RNA polymerase II associated cyclin possessing both carboxy-terminal domain kinase and Cdk-activating kinase activity. Mol Cell Biol 1998; 18: 4291-3000.
Feng LX, Ravindranath N, Dym M. Stem cell factor/c-kit upregulates cyclin D3 and promotes cell cycle progression via phosphoinositide 3-kinase/ p70 S6 kinase pathway in spermatogonia. J Biol Chem 2000;275:25572-6.
Godet M, Tomas A, Rudkin BB, Durand P. Developmental changes in cyclin B1 and cyclin dependent kinase 1 (CDK1) levels in the different populations of spermatogenic cells of the post-natal rat testis. Eur J Cell Biol 2000; 79: 816-23.
Gromoll J, Wessels J, Rosiepen G, Brinkworth MH, Weinbauer GF. Expression of mitotic cyclin B1 is not confined to proliferating cells in the rat testis. Biol Reprod 1997; 57: 1312-19.
Houldsworth J, Reuter V, Bosl GJ, Chaganti RS. Aberrant expression of cyclin D2 is an early event in human male germ cell tumorigenesis. Cell Growth Differ 1997; 8: 293-9.
Kang MJ, Kim MK, Terhune A, Park JK, Kim YH, Koh GY. Cytoplasmic localization of cyclin D3 in seminiferous tubules during testicular development. Expt Cell Res 1997; 234: 27-36.
Kim JM, McGaughy JT, Bogle RK. Ravnik SE. Meiotic expression of the cyclin H/Cdk7 complex in male germ cells of the mouse. Biol Reprod 2001; 64: 1400-8.
Kimura SH, Tsuruga H, Yabuta N, Endo Y, Nojima H. Structure, expression, and chromosomal localization of human GAK. Genomics 1997; 44: 179-87.
Kong WH, Zheng G, LU JN, Tso JK. Temperature dependent expression of cdc2 and cyclin B1 in spermatogenic cells during spermatogenesis. Cell Res 2000; 10: 289-302.
Lamitina ST, L'Hernault SW. Dominant mutations in the Caenorhabditis elegans Mytl ortholog wee-1.3 reveal a novel domain that controls M-phase entry during spermatogenesis. Development 2002; 129: 5009-18.
Lewin B. Driving the cell cycle: M Phase kinase, its partners, and substrates.Cell 1990; 61: 743 -52.
Liu D, Liao C, Wolgemuth DJ. A role for cyclin A1 in the activation of MPF and G2-M transition during meiosis of male germ cells in mice. Dev Biol 2000; 224: 388-400.
Liu D, Matzuk MM, Sung WK, Guo Q, Wang P, Wolgemuth DJ. Cyclin A1 is required for meiosis in the male mouse. Nature Genet 1998; 20: 377-80.
Luconi M, Barni T, Vannelli GB, Krausz C, Marra F, Benedetti PA, Evangelista V, Francavilla S, Properzi G, Forti G, Baldi E. Extracellular signal regulated kinases modulate capacitation of human spermatozoa. Biol Reprod 1998; 58: 1476-89.
Mizoguchi S, Kim KH. Expression of cdc25 phosphatases in the germ cells of the rat testis. Biol Reprod 1997; 56: 1474-81.
Musa FR, Tokuda M, Kuwata Y, Ogawa T, Tomizawa K, Konishi R, Takenaka I, HataseO. Expression of Cyclin dependent kinase 5 and associated cyclins in Leydig and Sertoli cells of the testis. J Androl 1998; 19: 657-66.
Nagahama H, Hatakeyama S, Nakayama K, Nagata M, Tomita K, Nakayama K. Spatial and temporal expression of patterns of cyclin dependent kinase (CDK) inhibitors p27kip1 during mouse development. Anat Embryol (Ber) 2001; 203: 77-87.
Nakamura S, Tatuno I, Noguchi Y, Kitagawa M, Kohn LD, Suito Y, Hirai A. 73-kDa heat shock cognate protein interacts directly with P27Kip1, a cyclin. Biochem Biophys Res Comm 1999; 257: 340-43.
Nakanishi M, Ando H, Watanabe N, Kitamura K, Ito K, Okayama H, Miyamoto T, Agui T, Sasaki M. Identification and characterization of human Wee1B, a new member of the Wee1 family of Cdk-inhibitory kinases. Genes Cells 2000; 5: 839-47
Nakayama K, Ishida N, Shirane M, Inomata A, Inoue T, Shishido N, Horii I, Loh Dy, Nakayama K-I. Mice lacking p27kip1 display increased body size, multiple organhyperplasia, retinal dysplasia and pituitary tumors. Cell 1996; 85: 707-20.
Nakayama K-I, Hatakeyama, Nakayama K. Regulation of cell cycle G1-S transition in proteolyses of cyclin E and p27kip1 Biochem Biophys Res Commun 2001; 282: 853-60.
Norbury C, Nurse P. Animal cell cycles and their control. Ann Rev Biochem 1992; 61: 441-70.
Pines J, Cyclins and cyclin dependent kinases: Take your partners. Trends Biochem Sci 1993; 18: 195-97.
Pines J. p21 inhibits cyclin shock. Nature 1994; 369: 520-21.
Sicinski P, Donaher JL, Geng Y, Parker SB, Gardner H, Park MY, Robker RL, Richards JS, McGinnis LK, Biggers JD, Eppig JJ, Weinberg RA. Cyclin D2 is an FSH responsive gene involved in gonadal cell proliferation and oncogenesis. Nature 1996; 384: 470-74.

Ravnik SE, Rhee K, Wolgemuth DJ. Distinct patterns of expression of the D-type cyclins during testicular development in the mouse. Dev Genet 1995; 16: 171-8.

Ravnik SE, Wolgemuth DJ. Regulation of meiosis during mammalian spermatogenesis: The A-type cyclins and their associated cyclin dependent kinases are differentially expressed in the germ cell lineage. Dev Biol 1999; 207: 408-18.

Ravnik SE, Wolgemuth DJ. The developmentally restricted pattern of expression in the male germ line of a Murine cyclin A, Cyclin A2, suggests roles in both mitotic and meiotic cell cycles. Dev Biol 1996; 173: 69-78.

Rhee K, Wolgemuth DJ. Cdk family genes are expressed not only in dividing but also in terminally differentiated mouse germ cells, suggesting their possible function during both cell division and differentiation. Dev Dyn 1995; 204: 406-20.

Roovers K, K. Assoian. Integrating the MAP kinase signal into the G1 phase cell cycle machinery. Bioessays 2000; 22: 818-26.

Roux E, Robaire B, Saragovi HU. A cell cycle regulating receptor is localized on cell surface and in nuclei of mitotically and meiotically dividing cells. DNA Cell Biol 1998; 17: 237-47.

Sette L, Barchi M, Bianchini A, Conti M, Rossi P, Geremia R. Activation of the mitogen-activated protein kinase ERK1 during meiotic progression of mouse pachytene spermatocytes. J Biol Chem 1999; 274: 33571-79.

Sigrist S, Ried G, Lehner CF.; Dmcdc2 kinase is required for both meiotic divisions during Drosophila spermatogenesis and is activated by the Twine/cdc25 phosphatase. Mech Dev 1995; 53: 246-60.

Sweeney C, Murphy M, Kubelka M, Ravnik SE, Hawkins CF, Wolgemuth DJ, A distinct cyclin A is expressed in germ cells in the mouse. Development 1996; 122; 53-64.

Takahashi H, Menjo M, Kaneko Y, Ikeda K, Matsushime H, Nakanishi M. Cdk4 Activation is Dependent on the Subunit Rearrangement in the Complexes. Biochem Biophys Res Comm 2000; 267: 388-93.

Trembley JH, Chen Z, Rodrigues CM, Kren BT, Steer CJ. Genomic organization and promoter characterization of the rat cyclin B1 gene. Gene 2000; 255: 93-104.

Tsutsui T, Hesabi B, Moons DS, Pandolfi PP, Hansel KS, Koff A, Klyokawa H. Targeted disruption of CDK4 delays cell entry with enhanced p27(kip1) activity. Mol Biol Cell 1999; 19: 7011-9.

Uemura T, Kubo E, Kanari Y, Ikenmura T, Tatsumi K, Muto M. Temporal and spatial localization of novel nuclear protein NP95 in mitotic and meiotic cells. Cell Struct Funct 2000; 25: 149-59.

Wang Z, Kim KH. Vitamin A-deficient testis germ cells are arrested at the end of S phase of the cell cycle: a molecular study of the origin of synchronous spermatogenesis in regenerated seminiferous tubules. Biol Reprod 1993; 48: 1157-65.

Wu S, Wolgemuth DJ; The distinct and developmentally regulated patterns of expression of members of the mouse Cdc25 gene family suggest differential functions during gametogenesis. Dev Biol 1995; 170: 195-206.

Xue L, Lucocq M. Low extracellular pH induces activation of ERK 2, JNK, and p38 in A431 and swiss 3T3 cells. Biochem Biophys Res Common 1997; 241,236-242.

Yang R, Morosetti R, Koeffler HP; Characterization of a second human cyclin A that is highly expressed in testis and in several leukemic cell lines. Cancer Res 1997; 57: 913-20.

Yang R, Muller C, Huynh V, Fung YK, Yee AS, Koeffler HP. Functions of cyclin A1 in the cell cycle and its interactions with transcription factor e2F-1 and Rb family of proteins. Mol Cell Biol 1999; 19: 240-7.

Yang T, Chen JY; Identification and cellular localization of human PFTAIRE1. Identification and cellular localization of humand PFTAIRE1. Gene 2001; 267: 165-72.

Zhang Q, Wang X, Wolgemuth DJ. Developmentally regulated expression of cyclin D3 and its potential in vivo interacting protein during Murine gametogenesis. Endocrinology 1999; 140: 2790-2800.

Zhu D, Dix DJ, Eddy EM. HSP70-2 is required for CDC2 kinase activity in meiosis I of mouse spermatocytes. Development 1997; 134: 3007-14.

Zhu X, Naz RK. Expression of a novel isoform of cyclin I in human testis. Biochem and Biophy Res Comm 1998; 248: 56-60.

Zindy F, den Besten W, Chen B, Rehg JE, Alters E, Barbacid M, Pollard JW, Sherr CJ, Cohen PE, Roussel MF. Control of spermatogenesis in mice by the cyclin D-dependent kinase inhibitors p18 (Ink4c) and p19 (Ink4D). Mol Cell Biol 2001; 21: 3244-55.

Zindy F, van Deursen J, Grosveld G, Sherr CJ, Roussel MF INK4d-deficient mice are fertile despite testicular atrophy. Mol Cell Biol 2000; 20: 372-78.

Chapter 12

ISOPROTEINS IN DNA SYNTHESIS

DNA replication takes place at Y shaped structure, called a replication fork. During DNA replication in vivo, each of the DNA strand acts as a template for formation of a new strand, through semi-conservative mechanism by the action of DNA polymerase. In contrast to DNA replication in prokaryotes, eukaryotes require two different DNA polymerases (DNA polymerase-α and DNA polymerase-δ). During 1990s, a study based on SV40 genome in cell free system suggested that eukaryotic DNA replication depends on T antigen (tumor antigen) derived from SV40 and on other mammalian proteins including DNA polymerases. The DNA polymerase, with self correcting property catalyses nucleotide polymerization in 5'--> 3'direction copying the DNA template strand with great fidelity. The replication fork is asymmetric in nature and hence two strands of DNA double helix are antiparallel. The synthesis in 5'--> 3' direction is continuous on one of the arms of replicating fork (leading strand). The synthesis on the other strand (lagging strand) is discontinuous and hence lagging strand produces short segments (Okazaki fragments) by discontinuous backstitching mechanism. The DNA fragments synthesized in pieces are primed by short RNA primer molecules, which require a special kind of DNA polymerase-primase. The DNA primase joins 3' end of newly synthesized DNA to 5' end of the previous one. Besides DNA polymerase and DNA primase, DNA synthesis requires DNA helicases and single strand DNA binding (ss B) protein (also called replication factor-1) in opening of double helix. The RNA primers are replaced by DNAs in lagging strand by another enzyme. The two ends of DNA pieces are sealed by DNA ligase, which joins 3' and 5' end. During course of DNA unwinding, the torsion developed in two strands, is relieved by topoisomerases, I and II by making a nick in one of the DNA strands. Many of these proteins associate with each other at replication fork, which helps in the movement of individual components in coordinated fashion. Thus various protein components that regulate DNA synthesis in eukaryotes include i) T antigen, ii) replication factor A (RF-A) (also called eukaryotic ssB), iii) topoisomerase I iv) topoisomerase II, v) DNA polymerase (Poly-α or α Pol) which is associated with eukaryotic primase to form poly–primase complex, vi) DNA polymerase-δ (Pol-δ), vii) proliferating cell nuclear antigen (PCNA) also known as cyclin, viii) replication factor C (RF-C). Of these components, T antigen is derived from SV40, whereas other components are derived from host mammalian system. The T antigen has a DNA binding domain, a ATPase domain and helicase activity. Since 1990, several DNA repair poteins and proteins involved in DNA recombination have been added in the list (see Chapter 13).

12.1. EUKARYOTIC DNA POLYMERASES

Five different classes of DNA polymerases have been identified, in different type of eukaryotic cells. These include: (i) DNA polymerase (Pol)-α, a relatively high molecular weight enzyme, also called *cytoplasmic polymerase,* or appropriately called *large polymerase*, since it is also found in the nucleus and since another small nuclear polymerase is also available in cytoplasm; (ii) DNA polymerase-ß or *nuclear polymerase* (also called *small polymerase*), found only in vertebrates, and not found either in budding yeast nor in other lower eukaryotes; (iii) DNA polymerase-γ is a *mitochondrial polymerase*, but encoded in the nucleus; (iv) DNA polymerase-δ is a novel enzyme present in mammalian cells, but is PCNA dependent for DNA synthetic processivity; (v) DNA polymerase-ε (earlier described as DNA polymerase-δ II), is PCNA independent. In some other cases, a distinction between DNA polymerase-δ and DNA polymerase-ε has not been possible. Another calf thymus enzyme designated as DNA polymerase-δ I represents DNA polymerase α with 3'-5' exonuclease activity. The large DNA polymerase α is the major DNA polymerase in eukaryotic cells and was believed for a long time to be the only enzyme involved in DNA replication. DNA polymerases can also be classified on the basis of amino acid sequence similarity into at least four different families: (i) *Escherichia coli* Pol I-type DNA polymerases as mitochondrial Pol γ and cellular Pol δ, (ii) eukaryotic-type DNA polymerases, as Pol-α, Pol-δ, Pol-ε and Pol-ξ; (iii) homologous to the α subunit of *E Coli* Pol III with not known homologues in eukaryotic cells; and (iv) Pol X, homologous to nuclear Pol-β. The Pol X family of DNA polymerases is a peculiar group that includes vertebrate Pol-β, involved in DNA repair.

Three of the nuclear DNA polymerase-α, -δ and -ε belong to a group of α like DNA polymerases that share many structural and catalytic properties. The DNA polymerase-β and -δ are sufficient for replication of the circular DNA of the mammalian virus SV-40 in vitro and also in vivo. In this model, DNA polymerase-α primase is responsible for the synthesis of short RNA-DNA precursor chains at the replication origin and prime synthesis of both the leading and lagging strands. These RNA-DNA precursors are extended by DNA polymerase-δ. The switch from DNA primer synthesis to DNA synthesis by DNA polymerase-δ depends on replication factor C and PCNA, which replace the DNA polymerase α primase complex with DNA polymerase-δ. Consequently, DNA polymerase-δ is responsible for extending RNA-DNA precursors into Okazaki fragments during the discontinuous synthesis of the lagging strand and the processive elongation of the leading strand. The components for in vitro replication of SV 40 DNA have been identified. The structure and catalytic properties of human DNA polymerase ε have been thoroughly studied. Although this enzyme does not seem to be involved in the replication of SV40 DNA, a study showing cross linking of DNA polymerase-α, -δ and -ε to nascent cellular DNA in replicating chromosomes suggests that DNA polymerase ε plays a role in the replication of cellular DNA in a process closely associated with replication. The DNA polymerase ε has also been shown in the repair of UV damage in human fibroblasts (Kamel et al., 1997).

12.1.1. DNA Polymerases in Spermatogenesis

Spermatogenesis offers a model for assessing the involvement of proteins in DNA replication, in meiotic recombination, and in DNA repair processes that occur during spermatogenesis (Monesi et al, 1962). Nuclear DNA polymerase from mouse sperm is stimulated by ATP and Mg^{2+} dependent but inhibited by low and high concentrations of KCI (Philippe and Chevaillier, 1976). Mouse testes, DNA polymerase- α is a high molecular weight enzyme, which sediments at 6-8 S and has maximal activity at a pH near neutrality. Its activity is severally depressed by KCI concentrations above 50 mM and by sulfhydryl group inhibitors. The velocity sedimentation at unit gravity allows to compare the relative specific activity of DNA polymerase-α and DNA polymerase-β during spermatogenesis (Hecht et al., 1979). As expected, DNA polymerase-α and -δ are highly expressed in the testes of young animals, which contain a high proportion of proliferating cells. The steady state levels of mRNAs encoding DNA polymerase-ε and the Rad51 remained constant during testis development, whereas the mRNA levels of DNA polymerase-α and -δ declined from birth until sexual maturity. As expected, DNA polymerase-α and PCNA had relatively strong activity in mitotically proliferating spermatogonia and even stronger staining in preleptotene cells undergoing meiotic DNA replication. In contrast, DNA polymerase-ε was detectable in mitotically proliferating spermatogonia but not in the early stages of meiotic divisions, and was present in haploid spermatids up to the stage at which the flagellum starts developing. Thus, DNA polymerase-ε functions in mitotic replication, in the completion of recombination in late pachytene cells and in the repair of DNA damage in round spermatids. In contrast, DNA polymerase-α and -δ appeared to be involved in meiotic DNA synthesis, which occurs early in meiotic prophase, in addition to functioning in DNA replication in proliferating spermatogonia (Kamel et al., 1997).

12.1.2. DNA Polymerase-β

DNA polymerase-β (or β-Poly) is one of the constitutively expressed 'housekeeping' enzyme in mammalian cells. Structural conservation of this polymerase suggests that this enzyme is essential for survival. The enzyme is involved in cellular maintenance e.g., DNA synthesis during DNA repair and recombination and in some type of DNA repair, particularly in DNA synthesis in small gaps formed during excision repair (Singhal et al., 1995). DNA polymerase-β is a low molecular weight enzyme of sedimentation coefficient 3-4 S, which has a pH optimum around 8.8 and is stimulated by KCI. β-Pol is a monomeric protein of 39-kDa, folded in two independent domains, an N-terminal domain of 8kDa and a polymerization domain of 31-kDa. The latter domain has the general right hand fold of polymerases, defined by the sub-domains named fingers, palm and thumb. In addition to its polymerization activity, Pol-β can incise the sugar-phosphate bond 3' to either an intact AP site (AP lyase activity) or to a 5' pre-incised site by an AP endo-nuclease, releasing a 2-deoxyribose 5-phosphate (dRPase activity). These Pol-β activities are located in the repair 8-kDa domain and contributes to base excision repair (BER), the major pathway that removes a damaged base residue in DNA (reviewed in Garcia-Diaz et al., 2000). Evidences are suggestive that β-Pol plays a role in mammalian meiotic events associated with synapsis and recombination (Plugg et al., 1997). The DNA polymerase-β is located preferentially in the nucleus.

Less complex species of eukaryotes are shown to possess only one type of DNA polymerase. The DNA polymerase-α and -β can be distinguished from one another by the differential effects of *N*-ethylmaleimide, KCI, ara-CTP and temperature, as well as on the basis of sedimentation. The sensitivity of DNA polymerase-β to elevated temperatures as compared to DNA polymerase α provides a new means of distinguishing between these two enzymes (Philippe and Chevaillier, 1976; Wilson et al., 1988). The cDNAs for rat and human DNA polymerases-β have been cloned. With the aid of these cDNAs, human cells are shown to contain a single copy gene encoding DNA polymerase-β. In most adult rat tissues the abundance of DNA polymerase-β mRNA is low. In contrast, young brain and testes show β-polymerase mRNA levels 5- and 15 fold higher, respectively. Four different (4.0, 2.5, 2.2, 1.4-kb) transcripts hybridizing to β-polymerase probe were found in all tissues examined. The 40-kb transcript was dominant for young and adult brain, where as the 1.4-kb transcript was dominant for testes (Nowak et al., 1989). Another human isoform, β-Pol 2 is closely related to β-Pol. The predicted β-Pol2 protein contains a BRCA1 C-terminus (BRCT) motif in its N-terminus. The -COOH terminal region of β-Pol2 shows 33% amino acid sequence identity with human β-Pol.

The β-Pol2 is expressed exclusively in testis. The truncated form of β-Pol2 lacking BRCT motif retains polymerase activity. Thus, β-Pol2 is related to DNA β-Pol, with a BRCT motif, which is dispensable for Pol activity (Nagasawa et al., 2000).

β-pol promoter gene: The promoter for the human β-pol gene has been found to be TATA less, but it does have multiple GC boxes and one ATF/CRE binding site located within 50 residues 5' of the major mRNA start site. The ATF/CRE binding site has been found to be essential for activity of the cloned promoter. The bovine testes DNA binding protein with specificity for the β-pol promoter ATF/CRE binding site is phosphorylated in vivo and contains several phosphorylation sites. Phosphorylation systems may change binding of this ATF/CRE binding protein to the β-pol promoter and in turn modulate the promoter. Co-transfection with a protein kinase A expression plasmid to elevate phosphorylation was found to strongly reduce promoter activity (Englander et al., 1991).

A protein fraction with binding affinity for an ATF/CREB like binding element GTGACGTCAC, at –49 to –40 in the core β-polymerase promoter was purified from a nuclear extract of bovine testes. This protein activates transcription and has a M_r of about 100-120kDa. SDS-PAGE revealed that it contains several polypeptides in the range of M_r 30-52kDa, yet two of these peptides (Mr 49-kDa and 52-kDa) are predominant. Specific binding to the palindrome is salt sensitive and is consistent with the formation of nine ion pairs and has a Ka of 5.8×10^{11} M^{-1} at 200-mM KCI. Kinetic studies with synthetic oligonucleotides as binding ligands indicated that the purified protein could bind tighter to or discriminate between the β-polymerase ATF/CREB element and similar elements derived from somatostatin and chronic gonadotropin genes (Widen and Wilson, 1991).

YY1 transcription factor in germ cells: The YY1 is the abbreviation of yin and yang 1 and reflects the dual function of this ubiquitously expressed factor to act under various conditions, either as a transcriptional repressor or as transcriptional activator. Although YY1 from testicular nuclear extract was found to bind to a promoter element of the p53 gene and to an initiator motif of the DNA polymerase-β gene, it was proposed that YY1 could be an important transcription factor for the regulation of male germ cell-specific genes (Schulten et al., 1999). The putative YY1 binding sites were also found in the promoters of other testis-specific gene, such as pro-acrosin gene (Schulten et al., 1999).

β-poly in testicular tumors: Presence of very high level of DNA polymerase β mRNA in pachytene spermatocytes points to the role of this enzyme in mammalian meiosis. Analysis of telomeric length and DNA polymerase-β expression in testicular germ cell tumors suggested that seminoma and nonseminoma, two main histological types of testicular germ cell tumors are derived independently from transformed foetal primordial cells. If these tumour cells originate from meiotically arrested pachytene spermatocytes, it should be expected that DNA polymerase-β mRNA is over-expressed in them (Nowak et al., 2002).

12.1.3. Pol λ

A new gene encoding a DNA polymerase-λ has been identified at mouse chromosome 19. Murine Pol-λ consisting of 573 amino acid residues has a 32% identity to Pol β involved in nuclear DNA repair in eukaryotic cells. Pol-λ contains all the critical residues involved in DNA binding, nucleotide binding and selection, and catalysis of DNA polymerization, that are conserved in Pol-β and other DNA polymerases belonging to family X. The Pol -λ also conserves

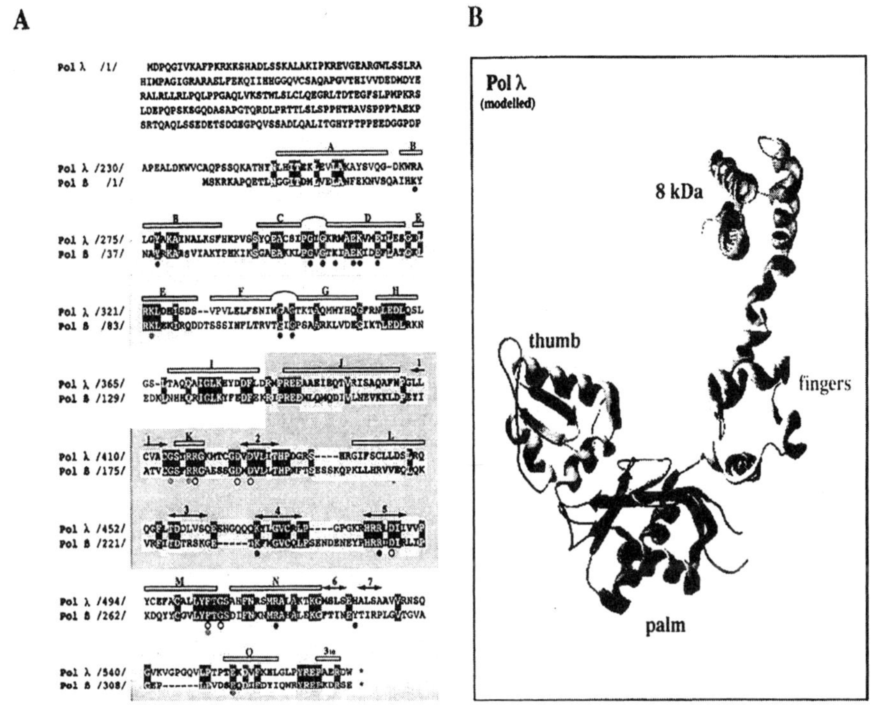

Fig.12.1 (A) Pair-wise homology between mouse testis polymerase λ and rat polymerase-β. The rat poly β has been divided into four sub-domains: (i) 8-kDa domain (aa residues 1-91), (ii) fingers (aa residues 92-148), (iii) palm (aa residues 149-260), and (iv) thumb (aa residues 261-335) (Pelletier et al 1994; Sawaya et al, 1994). α-helix (lettered A, B, ---) and β-strands (numbered 1, 2, 3, ---) are shown at the top of the alignment. Two hair-pin-helix motifs are located between C and D (8-kDa sub-domain), and between F and G (finger sub-domain). **(B)** 3D-Structure model of mouse Pol λ, showing four distinguishable sub-domains on the pattern of Pol β, within a region of 271-573 amino acid residues. The 8-kDa domain possesses dRPase activity and also provides an additional DNA binding capacity. While fingers and palm sub-domains provide contact point for DNA strand, and the surface for substrate binding respectively, thumb sub-domain helps in template selectivity and reaction cycling. Reproduced with permission from M.Garcia-Diaz et al. J Mol Biol 30; 851-67: 2000 © Elsevier; H.Pelletier et al. Science 264;1891-1903:1994© AAAS; MR Sawaya et al. Science 264;1930-35:1994© AAAS.

the critical residues of Pol-β required for its intrinsic deoxyribose phosphate lyase (dRPase) activity. The first 230 amino acid residues of Pol-λ, that have no counterpart in Pol-γ, contain a BRCT domain, present in a variety of cell cycle check point control proteins, responsive to DNA damage and proteins involved in DNA repair. The Pol-λ mRNA is highly expressed and developmentally regulated in testis. A major transcript of 2.4-kb mRNA accumulates at high level in testis on day 16 postnatal (pn), but not detected on day 7 pn suggesting that λ-Pol is expressed in mid-pachytene spermatocytes and early round spermatids of mouse testis. The high level of Pol-λ specifically expressed and mainly associated to pachytene spermatocytes, suggests a potential role of Pol-λ in DNA repair synthesis associated with meiosis (Gargcia-Diaz et al., 2000). A binary alignment of Pol-λ and Pol-β based on crystal structure of Pol-β suggests a nuclear localization signal at sequence PKRKKSH, located at residues 11 to 17 of Pol-λ. Leaving first 239 aa residues of Pol-λ, that have no counterpart in Pol-β, the remaining

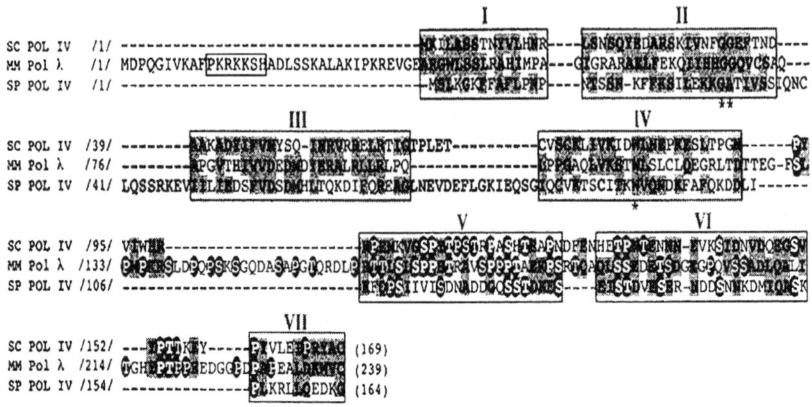

Fig.12.2. Amino acid similarity in N-terminus region of mouse (MM 239 aa) Pol λ with POL IV from *S. cerevisiae* (SC; 169 aa) and *S. pombe* (SP; 164aa). The main segments (I to VII) of amino acids are boxed. The region from segment I to IV forms a BRCT domain. Reproduced with permission from M. Garcia-Diaz et al. J Mol Biol 30; 851-67: 2000 © Elsevier.

residues could be aligned with Pol-β with four sub-domains. The alignment also shows that Pol-λ shares 25 out of 27 aa that are invariant among DNA polymerase X family members **(Fig. 12.1 and Fig. 12.2)**.

Human and mouse Pol-λ has homology with Pol-β, while another polymerase, Pol-μ is closer to terminal deoxynucleotidyl transferase. However, Pol-λ and Pol-μ share similar structural organization including a BRCT domain, helix-loop-helix DNA binding motif and Pol X domain. The mRNA expression of Pol-λ is highest in testis and in fetal liver, while expression of Pol-μ is maximum in lymphoid tissues (Auofouchi et al., 2000).

12.1.4. DNA Polymerase ξ

The *REV3* gene of *S. cerevisiae* encodes the catalytic subunit of DNA polymerase-ξ which is involved in translesion synthesis. The mouse homologue of this gene, Rev3L, encodes a putative protein of 3122 amino acids. The sequence conservation to its yeast counterpart is restricted to several regions. In the carboxy terminal part of the protein all six domains are present that are characteristic for DNA polymerase-α. In the amino terminal part of the protein two regions can be identified with considerable similarity to the NT boxes of mouse polymerase -δ. In addition, a region of 60 residues unique for the REV3 homologues is found in the middle part of the protein. Transcription of the *Rev3L* gene was highest in brain, ovary and testis. Mouse *REV3L* shows strong conservation with the human *REV3L* gene, which was localized to the long arm of chromosome 6, region 21-22. The mouse equivalent gene maps to chromosome 10, distal to the *c-myb* gene, close to the *Macs* gene (Van Sloun et al., 1999).

12.2. DNA LIGASES

The pleiotropic effects of mutations in the DNA ligase gene of prokaryotes include conditional lethality, sensitivity to DNA damage and hyper recombination. In contrast to prokaryotes, multi-cellular eukaryotes contain more than one species of DNA ligase. DNA ligase I is required for DNA replication involved in DNA repair. Pol-β and DNA ligase-1 seem to be components of a multi-protein complex that performs base excision repair (Prasad et al., 1996). The activity of DNA ligase II activity, the major DNA joining enzyme in the liver is increased following treatment with DNA damaging agents, suggesting a role for this enzyme in DNA repair. DNA ligase III is a component of a calf thymus recombination complex and has been found associated with a human DNA strand break repair protein, Xrcc1, suggesting roles for this enzyme in genetic recombination and DNA repair. The 70-kDa ligase II and 100-kDa DNA ligase III from bovine liver and bovine testis respectively revealed that these polypeptides share extensive regions of identity, which indicated that they are probably encoded by the same gene. A comparison of the peptide sequences from DNA ligases II and III with the predicted amino acid sequences of other DNA ligases revealed that these enzymes are more highly related to the DNA ligases encoded by cytoplasmic poxviruses than to mammalian DNA ligase I and other reflective DNA ligases. DNA ligase may be required to perform the meiosis specific DNA transactions. A characteristic feature of meiosis is the high frequency of homologous recombination that leads to the production of genetically reassorted haploid gametes. These recombination events appear to occur within the synaptonemal complex and presumably are completed by DNA synthesis and DNA ligation prior to the first meiotic division (Chen et al., 1995).

The human DNA ligase III cDNA encodes a polypeptide of 862 amino acids, whose sequence is more closely related to the DNA ligases encoded by poxviruses than to replicate DNA ligases, such as human DNA ligase I. In vitro transcription of the cDNA produced a catalytically active DNA ligase similar in size and substrate specificity to the purified bovine enzyme. The DNA ligase III gene was localized to human chromcsome 17. DNA ligase III is ubiquitously expressed at low levels, except in the testes, in which the steady state levels of DNA ligase III mRNA are at least 10 fold higher than those detected in other tissues and cells during spermatogenesis. DNA ligase I mRNA expression correlating with the contribution of proliferating spermatogonia cells in the testes suggested its role in DNA replication. In contrast, elevated levels of DNA ligase III mRNA were observed in primary spermatocytes undergoing recombination prior to the first meiotic division. Therefore, it was suggested that DNA ligase III seals DNA strand breaks that arise during the process of meiotic recombination in germ cells and as a consequence of DNA damage in somatic cells (Chen et al., 1995) **(Fig 12.3).**

Two forms of DNA ligase III are produced by paired (duplex DNA paired with a single stranded DNA) structures. Another ligated product of DNA ligase andenylate adduct of 100kDa was also formed. This showed that DNA pairing co-purifies with DNA ligase from extract of rat testes (Acharya et al., 1996).

12.3. DNA HELICASES

The unwinding of parent DNA strands is a prerequisite to basic genetic processes including DNA replication, DNA repair, recombination, and transcription. In each of these processes, unwinding of duplex DNA is catalyzed by a DNA helicase, which functions to destabilize hydrogen bonds between complementary base pairs in duplex DNA. The energy necessary for this reaction is provided by hydrolysis of nucleosides and deoxyncleoside 5'-triphosphates. DNA helicase generally possesses an intrinsic ATPase activity. DNA helicases are ubiquitously present from prokaryotes to eukaryotes. Helicases have several characteristic motifs and some of them are functionally important. Two sequence signatures, so called Walker A and B motifs, have been identified in all helicases so far examined. There are at least 41 helicase genes in the

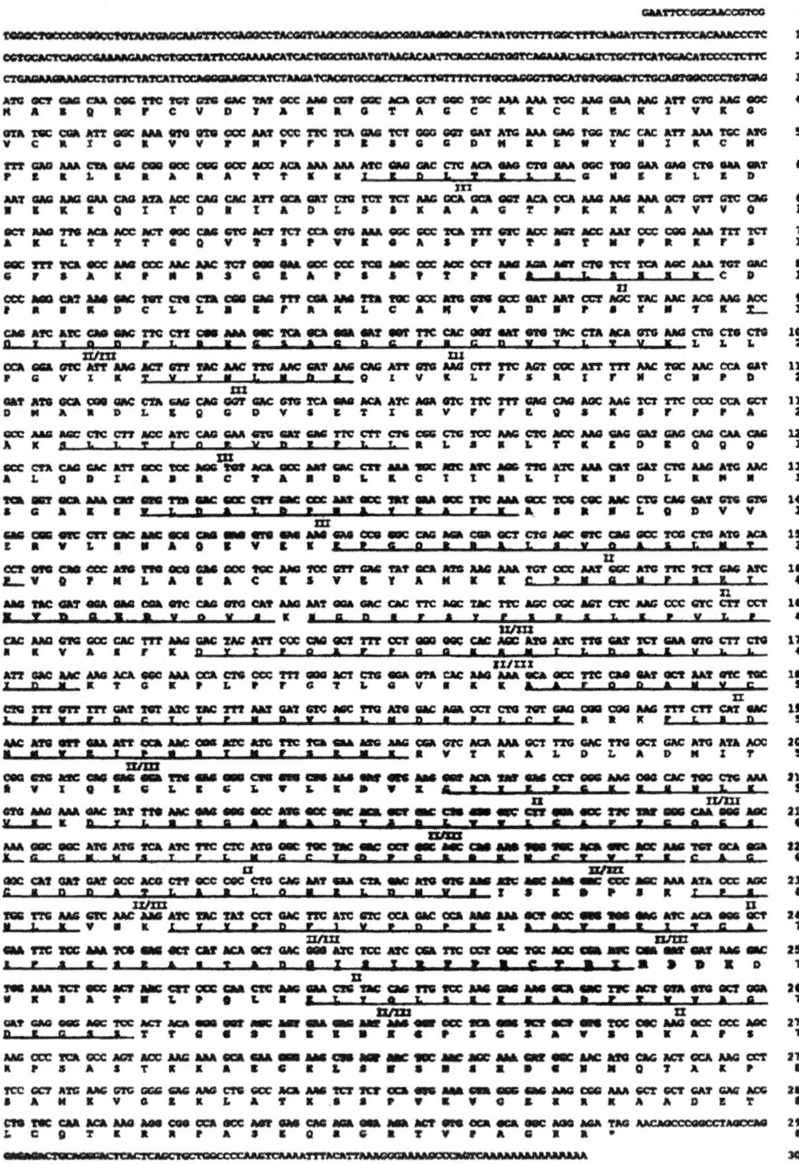

Fig.12.3. cDNA and predicted amino acid sequences of human DNA ligase III. Regional homologies to bovine DNA lagases II and III, indicated by II and III respectively, are underlined. Reproduced with permission from J. Chen et al, Mol Cell Biol 15:5412-221: 1995 © The Society of Microbiologists.

genome. Inspite of so much diversity, most of our biochemical knowledge about DNA helicases comes from analyses of bacterial and phage enzymes because little is known about the biochemical properties of most of the eukaryotic DNA helicases. However, studies demonstrated that six putative helicases are mutated in the human diseases such as Xeroderma pigmentosum,

A

```
  1 MASVSALTEE LESVASELHA IDIQIQELTE RRQELLQRKS VLTGKIKQYL EDSSAEASSD
 61 LDTSPAAWNK EDFPWFGKVK DVLQNVFKLQ KFRPLQLETI NVTMARKDIF LVMPTGGGKS
121 LCYQLPALCS DGFTLVICPL ISLMEDQLMV LKQLGISATM LNASSSKEHV KWVHAEMVNK
181 NSQLKLIYVT PEKIAKSKMF MSRLEKAYEA GRLTGVAVDE VHCCSQWGHD FRPDYKALGI
241 LKRQFPNASL MGLTATATNH VLKDVQKILC VGKCLTFTAS FNRPNLFYEV RQKPSSAEDF
301 TEDIVKLING RYKGQSGIIY CFSQKDSEQI TISLQKLGIH AGTYHANMEP EDKTKVHTQW
361 SANELQVVVA TVAFGMGIDK PDVRFVIHHS MSKSMENYYQ ESGRAGRDDS RADCILYYGF
421 GDIFRISSMV VMENVGQQKL YEMVSYCQNV SKCRRVLIAQ HFDEVWNADA CNKMCDNCCK
481 DVSFEKKNVT QHCRDLIKIL KQAEGLNEKL TPLKLIDTWM GKGAPKLRVA GVVAPALPRE
541 DLERIVAHAL LQQYLKEDYS FTAYATISYL KVGPRACLLS NEAHAVTMQV KKSAQSSVRG
601 APSEARQVEQ VDSKGEEQSS GNSQKSKSRL QPSGSKNAGA KKRKLDDA
```

B

```
  1 MASVSALTEE LESVASELHA IDIQIQELTE RRQELLQRKS VLTGKIKQYL EDSSAEASSD
 61 LDTSPAAWNK EDFPWFGKVK DVLQNVFKLQ KFRPLQLETI NVTMARKDIF LVMPTGGGKS
121 LCYQLPALCS DGFTLVICPL ISLMEDQLMV LKQLGISATM LNASSSKEHV KWVHAEMVNK
181 NSQLKLIYVT PEKIAKSKMF MSRLEKAYEA GRLTGVAVDE VHCCSQWGHD FRPDYKALGI
241 LKRQFPNASL MGLTATATNH VLKDVQKILC VGKCLTFTAS FNRPNLFYEV RQKPSSAEDF
301 TEDIVKLING RYKGQSGIIY CFSQKDSEQI TISLQKLGIH AGTYHANMEP EDKTKVHTQW
361 SANELQVVVA TVAFGMGIDK PDVRFVIHHS MSKSMENYYQ ESGRAGRDDS RADCILYYGF
421 GDIFRISSMV VMENVGQQKL YEMVSYCQNV SKCRRVLIAQ HFDEVWNADA CNKMCDNCCK
481 DVSFEKKNVT QHCRDLIKIL KQAEGLNEKL TPLKLIDTWM GKGAPKLRVA GVVAPALPRE
541 DLERIVAHAL LQQYLKEDYS FTAYATISYL KVGPRACLLS NEAHAVTMQV KKSAQSSVRG
601 APSEARQVEQ VDSKGEEQSS GSHAADTAAN V
```

Fig.12.4. Amino acid sequences of human testis DNA replicases. (A) Q1α,(B) Q1β derived from cDNA. Source: http//www. ncbi.nlm.nih.gov. (A) (Accession number BAA75085) and (B) (Accession number BAA75086).

Bloom's syndrome, Cockayne's syndrome, trichothiodystrophy, Werner's syndrome, and α thalassemia. At least five human RecQ related genes exist: RecQ1, BLM, WRN, RecQ4 and RecQ5. Mutations in BLM, WRN and RecQ4 are associated with Bloom, Werner and Rothmund-Thomson syndromes, respectively, involving a predispostion to malignancies and a cellular phenotype that includes increased chromosome instability. Studies indicate an intimate association of Rec Q homologues in DNA replication. Genetic and biochemical analysis have suggested that these Rec Q homologues interact with proteins involved in DNA replication and function in a pathway from DNA replication check point to homologous recombination (Enomoto, 2001). Thus, identification of a novel helicase is medically important.

DNA Helicase Q1 (Rec QL) : Wang et al., (1998) cloned cDNA encoding mouse homologue for the human DNA helicase Q1 Rec Q like /RecQL (human helicase Q1), which has homology with E. Coli RecQ protein and encodes two isoforms. The two isoforms are identical in the carboxyl terminal sequence spanning over less than 30 amino acids. One of the two isoforms Q1α contains a sequence KKRK in the carboxyl terminus, which is also contained in human helicase Q1 and was confirmed to function as a nuclear localization signal. The other Q1β form does not contain such a sequence. Expression of mouse helicase Q1 mRNA is extremely and relatively high in the testis and the thymus, respectively. Helicase Q1 α is expressed in all tissues whereas the Q1β form expressed only in the testis. A survey of expression of Q1 β mRNA in the testis after birth revealed that Q1 α mRNA is expressed in all testes of mice aged from 7 days to 8 weeks, and the expression of Q1β mRNA begins 14 days after birth, corresponding to the appearance of cells in the pachytene stage (Wang et al., 1998) (**Fig 12.4**).

RecQ5: Rec-5 is small, containing only a core part of the RecQ helicase, but three isomer transcripts code for small RecQ5 a (corresponding to the original RecQ5 with 410 amino acid), new large RecQ5 b (991 amino acids) and small RecQ5 g (435 amino acids) proteins that contain the core helicase motifs. Genomic structure of three isoforms generated by differential splicing from the RecQ5 gene showed it to contain at least 19 exons. Northern blot analysis using a RecQ5 specific probe indicates that RecQ5 mRNA is expressed strongly in the testis. Three N-terminally tagged RecQ5 isomers expressed in 293EBNA cells showed that RecQ5b migrates to the nucleus and exists exclusively in the nucleoplams, while the small RecQ5 a and RecQ5g proteins stay in the cytoplasm. The nucleoplasmic RecQ5 like yeast Sgs1 DNA helicase binds to topoisomerases 3a and 3b but not to topoi 1. These results predict that RecQ5 may have an important role in DNA metabolism and may also be related to a distinct genetic disease (Shimamoto et al., 2000).

TIP49a: A 49-kDa TATA binding protein (TBP)- interacting protein (TIP49a; termed TIP49 from rat liver nuclear extracts and a general transcription factor is needed for all classes of RNA polymerases. A computer search revealed a high similarity between TIP49a and a prokaryotic DNA helicase, RuvB, which is involved in branch migration of the Holliday junction. It is noteworthy that TIP49a contains both the Walker A and B motifs of the bacterial RuvB homologous regions, suggesting that TIP49a is a putative DNA helicase. The nuclear protein TIP49a from rat testis possessed ATPase activity that was stimulated by single stranded DNA but not by double stranded DNA or by RNA polymers tested. The TIP49a specifically interacts with ATP, unwinds DNA duplex in the presence of ATP or dATP. The TIP49a is abundantly expressed in testes and moderately in spleen, thymus, and lung. In mouse seminiferous tubules, the protein was restricted to germ cell lineage from late pachytene spermatocytes to round spermatids. The TIP49a is a DNA helicase and may play a role in recombination and transcription (Makino et al., 1999).

SNF2: The SNF2 gene family consists of a large group of proteins involved in transcriptional regulation, maintenance of chromosome integrity, and various aspects of DNA repair. An SNF2 family human cDNA, with identity to the *E coli* RNA polymerase binding protein HepA and named the human hepA related protein (HHARP/SMARCAL1) has been cloned. In addition the mouse and *C. elegans* orthologues have been identified. Phylogenetic analysis indicated that the HARP proteins shared a high level of sequence similarity to the seven-motif helicase core region (SNF2 domain) with identifiable orthologs in other eukaryotic species, except for yeast. Purified His tagged HARP/SMARCAL1 protein exhibits single stranded DNA dependent ATPase activity. Both the human and the mouse genes consist of 17 exons and 16 introns. The human gene maps to chromosome 2q34-q36 and mouse gene is localized to syntenic regions of chromosome 1. HARP/SMARCAL1 transcripts are ubiquitouly expressed in tissues, with testis presenting the highest level of mRNA expression in humans (Coleman et al., 2000).

12.4. DNA TOPOISOMERASES

A DNA topoisomerase can be viewed as a reversible nuclease that adds itself to a DNA backbone phosphate there by breaking a phosphodiester bond in DNA strand. This reaction is reversible and phosphodiester bond reforms as protein leaves. DNA topoisomerase I produces a single strand nick whereas DNA topoisomerase II forms a covalent linkage to both the strands of DNA helix making a double strand break (dsb) in the helix. Both topoisomerases are activated by sites on chromosomes where two helices cross-over each other. The DNA topoisomerases are localized in the nucleus, where they remove torsional stress in DNA. Their function is important for replication, transcription, chromosome condensation, and chromosome segregation during mitosis and meiosis.

DNA Topoisomerase I: The transcript of the topoisomerase (topoi) I gene was highest in somatic cells of the testis and in the mitotically proliferating spermatogonia and meiotic prophase spermatocytes, with the level of transcript decreasing dramatically in postmeiotic spermatids. In contrast, the levels of topoi II α transcript were negligible in germ cell free testes and highest in late meiotic prophase cells and round spermatids. Enzyme activity for both topoi I and topoi II was detected in both pachytene spermatocytes and in rounds spermatids; topoi II exhibited a higher level of activity in meiotic spermatocytes than in round spermatids. Camptothecin, inhibitor of topoi I, caused some abnormalities of paired meiotic homologs. In contrast, inhibitors of topoi II dramatically inhibited the formation of metaphase chromosomes in cells induced to progress from prophase to metaphase. However, the disassembly of the synaptonemal complex was not inhibited, indicating that this process could be uncoupled from condensation of chromatin to form chromosomes.

DNA Topoisomerase II: Type II DNA topoisomerases (topoi II) are a class of ubiquitous enzymes indispensable for a number of nuclear processes. These enzymes alter topological properties of DNA by creating transient dsb and transport an intact segment of DNA through the gap and finally religate the cleaved strand. In vertebrates, topoi II is thought to play fundamental role in transcription, genome stability, DNA condensation, sister chromatid separation, segregation in mitosis, and disjunction of homologue during meiosis. Topoi II is concentrated along a dense axial core of metaphase chromosomes and is therefore implicated to play a scaffolding role. In vertebrate cells it has been identified as a marker for cell proliferation. Two isoforms of topoi II have been identified in cultured murine and human cells. These two isoforms topoi II α, a 170 kDa protein and topoi II β, a 180 kDa protein are encoded by separate genes and differ in several biochemical properties (Galande and Muniyapa et al., 1996).

Cobb et al., (1997) presented an evidence for a functional requirement of topoi II activity in the transition from meiotic prophase to meiotic metaphase I in mammalian spermatocytes. The presence of topoi II α and histone H3 at the centromeric heterochromatin was studied with the process of chromosome condensation and individualization of chiasmate bivalents in murine spermatocytes. Evidence showed that activity of topoi II is required for condensation of chromatin at the end of meiotic prophase. Histone H3 becomes phosphorylated at the end of prophase, beginning with its phosphorylation at the centromeric heterocheromatin in the diplotene stage (Cobb et al., 1999).

Topoi II has been purified and characterized from rat testis and its catalytic activity was compared with that of the purified enzyme from liver. The purified enzymes appeared to be monomers under denaturing conditions; however they differed in their relative molecular mass. Topoi II from testis and liver have apparent molecular masses of 150 and 160 kDa, respectively, optimally active between pH 6.5 and 8.5. Both enzymes are able to promote decatanation and relax supercoiled DNA substrates in an ATP and Mg^{2+} dependent manner. Intriguingly, the relaxation activity of liver topoi II was inhibited by potassium glutamate at 1 M, whereas testis enzyme required about half its concentration. These findings argue that topoi II from rat testis is structurally distinct from that of its somatic form (Galande and Muniyappa, 1996).

Topoi II is inhibited by merbarone (a cancer chemotherapeutic drug) during male mouse meiotic divisions in vivo, with increase in polyploid and hypoploid metaphase II spermatocytes at time intervals corresponding to the treatment of the first meiotic divisions and diplotene diakinesis with a clear increase in the frequency of arrested divisions. Thus, merbarone severely damages normal meiotic processes through inhibition of topoi II (Kallio and Lahdetie, 1997). The Topoi II inhibitor, etopoiside exhibits significant mutagenicity in primary spermatocytes. The effect of etopoiside on crossing-over in male mice, resulted in decreased crossing-over in the p-Tyr(c) interval of mouse chromosome 7. The temporal and spatial protein patterns suggested the formation of multiple lesions in the DNA after recombination nodules (MLH1 foci) had disappeared from spermatocytes. Since etopoiside blocks religation of the cut made by topoi II, repair of DNA damage may result in rejoining of the original DNA strands, undoing the reciprocal exchange that had already occurred, resulting in reduced crossing-over despite a normal frequency of MLH1.Three periods in spermatogenesis respond to etopoiside in different ways. Exposure of (a) late differentiating spermatogonia (and, possibly, preleptotene spermatocytes) results in cell death; (b) early to mid pachytene induces specific locus deletions and crossover reduction; and, (c) late pachytene through diakinesis leads to genetically unbalanced conceptuses as a results of clastogenic damage (Russel et al., 2000).

DNA topoisomerase IIIβ: The disruption of the mouse *topoi IIIβ* gene encoding DNA topoisomerase IIIβ leads to a progressive reduction in fecundity. The litter size in crosses of *topoi IIIβ* (-/-) mice decreases with passage of time and through successive generations. This decrease seems to reflect embryonic death rather than impaired fertilization. These observations are suggestive of a gradual accumulation of chromosomal defects in germ cells lacking DNA topoisomerase IIIβ. Results were interpreted in terms of a specific role of a type IA DNA topoisomerase in the resolution of meiotic double-Holliday junctions without crossing over (Kwan et al., 2003).

Spo11 : A Variant of Topoisomerase: In *S cerevisiae, S pombe, C elegans* and *Drosophila melanogaster*, Spo11 is required for meiotic recombination, suggesting a conserved mechanism for the initiation step of this process. The Spo11 initiates meiotic recombination by garneting DNA double strand breaks (dsbs) and is required for meiotic synapsis in *S. cerevisisae*. The protein SPO11 from *archaebacterium sulfulobus shibatae,* is the catalytic subunit of topo VI DNA topoisomerase. Cloning showed that the putative mouse and human Spo11 proteins are 82% identical and share approximately 25% identity with other family members. Mouse Spo11 was localized to chromosome H2H4, and human SPO11 was localized to chromosomes 20q13.2.q13.3. Surprisingly, disruption of mouse *Spo11* leads to infertility. The *Spo11* localizes to discrete foci during leptotene and to homologously synapsed chromosomes. Other mouse mutants that arrest during meiotic prophase (ATM-/-, Dmc1-/-, mei1-/-, and Morc (-/-)) showed altered Spo11 protein localization and expression. It seems that there is an additional role for SPO11, after it generates dsb in synapsis. The cDNAs resulting from alternative splicing suggests that there are potential variants of the mouse SPO11 protein. Expression of the mouse *Spo*11 gene was detected only in the testis, in agreement with its predicted function in the initiation of meiotic recombination (Romanienko and Camerini-Otero, 1999; 2000; Metzler - Guillemain and deMassy, 2000). But RT-PCR results showed ubiquitous expression of at least a portion of Spo11 in mouse. Human SPO11 was also present in several somatic tissues.

12.5. REVERSE TRANSCRIPTASE

The reverse transcriptase (RT) activity is present in murine epididymal spermatozoa. Sperm cells incubated with human poliovirus RNA can take up exogenous RNA molecules and internalize them in nuclei. The released hypersensitive DNA from murine spermatozoa has been cloned. A high proportion of sequences were found to be of retroposon origin from a variety of families, particularly of the LINE/L1 group. The findings that these sequences, organized in nucleosomes are hypersensitive to nuclease cleavage and that many of them contain uninterrupted LINE-derived ORFs suggest that they may derive from LINE elements potentially coding for RT in male germ cells. In search of endogenous RT activity in mature murine spermatozoa, PCR analysis of two-cell embryos showed that poliovirus RNA-challenged spermatozoa transferred retrotranscribed cDNA molecules into eggs during in vitro fertilization. The RT molecules can be visualized on sperm nuclear scaffolds by immunogold electron microscopy. These observations, together with DNA-dependent DNA and RNA polymerases in mammalian spermatozoa support the view that potentially active domains for RT exist in the sperm genome. The potential ability of LINE-L1 to reshuffle the genome has been recently demonstrated by the finding that L1 elements integrate into transcribed genes by retrotransposing their 3'sequences into new genomic locations (Moran et al., 1999). The results reported raise the possibility that RT is involved in the reshuffling of genetic material in a subfraction of sperm chromatin (Giordano et al., 2000) (see Chapter13).

12.6. TELOMERE PROTEINS

The physical termini of chromosomes in eukaryotes are composed of a specific DNA-protein complex called telomere. The structure of telomeres has been studied in a wide variety of organisms to demonstrate that they are highly conserved elements throughout the eukaryotes, both in structure and function. Telomeric DNA has been shown to consist of simple randomly repeated nucleotide bases characterized by clusters of G residues in one strand and C residues in the other. Another feature of telomere is a 3'overhang sequence of 12-16 nucleotides in length of the G-rich strand. The same repeated sequence is found at the ends of all chromosomes in a species and the same telomere sequence may occur in widely divergent species, as in humans, and some acellular slime molds (trypanosomes) and fungi like *Neurospora.* At every telomere, as much as 10-kb of this repeat sequence may occur. The telomeric DNA is also complexed with non-histone proteins and the complex structure being associated with nuclear lamina, as shown in *Oxytricha*, a ciliated protozoan. The telomeric DNA is synthesized under the influence of telomerase, which is a ribonucleoprotein, whose RNA component works as a template for the synthesis of telomeric DNA repeats. Telomeres have evolved to fulfill several essential functions: 1) they protect chromosomes from end degradation and fusion; 2) they facilitate complete replication of chromosomal DNA molecules; and 3) they participate in chromosome positioning and proper segregation during mitosis and meiosis. Telomere DNA in vertebrates is represented by a tandemly repeated double-stranded hexanucleotide $[(TTAGGG)(AATCCC)]_N$ and a single-stranded overhang of the G-strand at the very end.

12.6.1. Telomere Binding Proteins

Both double-stranded and single-stranded telomere DNAs are associated with various proteins; double-stranded telomere-binding proteins (dsTrBP) and single stranded telomere-binding proteins (ssTrBP), first described in ciliated S. *cerevisiae.* In vertebrates, the first double-stranded ds TrBP characterized at the molecular level was human telomere repeat factor (hTrRF). The 60-kDa hTrRF1 isolated from HeLa cells has been found in nuclear extracts from monkey, rodent, and chicken cells, whereas TrRF1 mRNA has been detected in all human tissues analyzed. Therefore, all vertebrates have similar dsTrBP, and the corresponding gene is expressed in every somatic tissue. A second human dsTrBP-hTrRF2- has been also isolated. Sequence analysis indicated that TrRF1 and TrRF2 are distinct proteins sharing a common DNA-binding domain that has strong sequence homology to the DNA-binding domains of *Myc* protooncogene. Important telomere function has been associated with TrRFs. The hTrRF1 indirectly regulates telomerase activity while hTrRF2 prevents chromosome end-to-end fusion (Zhang et al.,2001).

Among ssTrBP are XTEF in *Xenopus laevis* egg and ovaries, qTrBP42 single-stranded and quardruplex telomeric DNA-binding protein from rat hepatocytes, and avian cells protein recognizing hairpin DNA structure. Peptide sequencing and mobility shift experiments using (UUAGGG) as substrate showed that these proteins are highly homologous to heterogeneous nuclear ribonucleoproteins. Protein binding of double stranded telomere DNA has been identified in human sperm (Kozik et al., 2000). Kozik et al (2000) described another protein from bovine sperm nuclii designated bovine sperm TrBP (bs-ssTrBP) that specifically interacts with single-stranded $(TTAGGG)_N$ DNA. The native size of the protein was in the range of 20-40kDa. Cross linked complexes between bs-ssTrBP and telomere DNA indicated that several polypeptides are involved in complex formation. Bovine s-ssTrBP had high specificity toward nucleotide sequence, since single nucleotide substitutions in the $(TTAGGG)_4$ substrate suppressed binding. The minimal number of (TTAGGG) repeats required for binding of bs-ssTrBP was 3. The protein recognized linear DNA but not folded DNA structures. It seems that the bs-ssTrBP participates in telomere-telomere interactions. Germ cell-specific telomere-binding proteins may participate in down-regulation of telomerase activity during spermatogenesis, and in establishing telomere-telomere associations localized at nuclear membrane. It has been proposed that the latter is important for telomere functions during meiosis and after fertilization bs-ssTrBP could contribute to one of these functions.

During spermatogenesis, chromatin structure is reorganized, DNA becomes super-condensed, and genetic activity is completely shut down. Specific and random nuclear architecture has been demonstrated as in mammalian sperm cells in particular. This architecture is characterized by telomere localization at the extreme nuclear periphery where telomeres interact in dimers and tetramers. Telomere localization and interaction observed in mature mammalian sperm are different from those in somatic nuclei. Distinct sperm-specific features of telomeres are established during spermatogenesis. In spermatogonia, telomeres are randomly scattered throughout the nucleoplasm; in spermatocytes, all telomeres re-localise towards nuclear membrane and form associations. Telomeres associations during meiosis have been described in a variety of organisms, where they are involved in initiation of synapsis or crossing over. The meiotic telomere complex is similar to its somatic counterpart and contains significant amounts of TrRF1, TrRF2, and hRap1, while tankyrase, a poly –(ADP-ribose) polymerase at somatic telomeres and nuclear pores, form a small signal at ends of human meiotic chromosome cores (see section 12.6.2.). Analysis of rodent spermatocytes reveals TrRF1 at mouse, TrRF2 at rat, and mammalian Rap1 at meiotic telomeres of both rodents. It has also been demonstrated that telomere repositioning during meiotic prophase occurs in sectors of the nuclear envelope that are distinct from nuclear pore dense areas. The latter form during preleptotene/leptotene stage and are present during entire prophase (Scherthan et al., 2000).

12.6.2. Telomerase and Telomere Length

Telomeric DNA is not completely synthesized by conventional DNA polymerase, and in the absence of the enzyme telomerase (a ribonucleoprotein), chromosomes lose 50-200-bp per mitotic division. Cellular aging is characterized by a decrease in telomere length, and this has been implicated as a mitotic clock that signals cells to stop division when telomeres reach a critically short length. Telomerase extends telomeres by the addition of six bp direct repeats

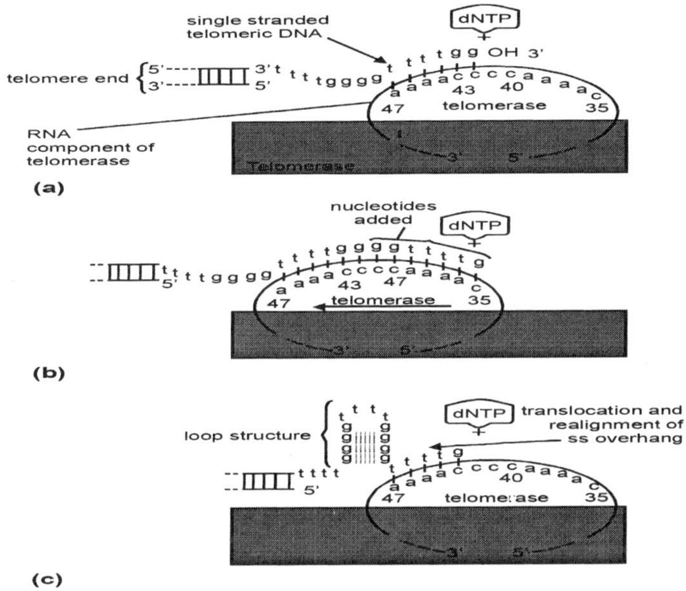

Fig.12.5. Action of *Euplotes* telomerase. (a) Binding of overhang of G- rich-single strand DNA primer at telomere end with telomerase (A ribonucleprotein) RNA template according to Watson-Crick base pairing rule, (b) polymerization of telomeric DNA on RNA template of telomerase till its end, and (c) backward translocation of elongated-telomeric end and realignment with the template according to Watson-Crick rule at 3' end. The newly synthesized DNA primer forms a stem-loop (hair pin like) structure due to telomeric repeat aided by G:G pairing as shown by shaded area. The cycle repeats with 3'overhang of telomeric DNA bound to telomerase.

onto chromosome. All tissues do not necessarily show evidence of active telomerase even after screening with PCR mediated telomeric amplification protocol. In most normal human somatic cells or in epithelial tissues, which contain reproductive somatic cells, telomerase is absent. In contrast, telomerase activity has been observed only in germ cells, with some activity in bone marrow and peripheral blood leukocytes (Fujisawa et al., 1998). Conversely, maintenance of telomere length appears necessary for cellular immortality. The telomere hypothesis suggests that telomerase activity is high in embryonic cells and that it decreases in somatic tissues during development and differentiation. Telomerase is expected to remain active in germline cells to ensure the transmission of full-length chromosomes to progeny. Consistent with this hypothesis, telomerase activity has been detected in extracts of human ovaries and testis. The presence of telomerase activity in testis was anticipated from the long telomeres in sperm and the longer telomere length in sperm of older men. Similarly, mouse testis telomere lengths are 1-3pairs longer than those in somatic tissues. Reports on sub cellular localization of telomecase in dividing germ cells are not very clear. The observation that mouse testicular telomerase activity cannot be detected until 42 days of age suggests that telomerase activity is low until advanced stages of spermatogenesis. Telomerase activity may be important during meiosis (Eisenhauer et al., 1997). The progressive reduction of telomere lengths to critically short sizes has been correlated with the cessation of cell division and the onset of senescence **(Fig.12.5).**

The activity of telomerase in fetal and adult testis of human and mouse has been demonstrated. Telomerase activity is present at birth, increases progressively until the 4th week after birth, and declines to a relatively low level by the 10th week in the rat testis. In isolated germ cell tubes, a very strong activity was consistently observed in type A spermatogonia compared to a lower level of activity in populations of pachytene spermatocytes and round spermatids. Telomerase activity was totally absent from epididymal spermatozoa (Ravindranath et al., 1997). In association, telomere length was increased during development of male germ cells from spermatogonia to spermatozoa and was inversely correlated with the expression of telomerase activity (Achi et al., 2000).

Telomerase in Pathalogical conditions: Fujisawa et al (1998) showed that lower activities of DNA polymerases were not the sole cause of hypospermatogenesis in the testis of oligozoospermic men with varicocele. Since telomeric DNA is not completely synthesized by conventional DNA polymerases, it is telomerase that is involved in the elongation of telomere, and thus enables germ cells to divide continuously. Telomerase activity is present in the pachytene spermatocytes and round spermatic of rat, but not in spermatozoa from either the caput or the cauda epididymis, and that the telomerase activity in spermatids is higher than in pachytene spermatocytes. Since it is necessary for these germ cells in the human testis to divide for much of the individual's life, the presence of telomerase in human testis is essential. The activity of telomerase in the human testis with idiopathic azoospermia does not change between the testis with maturation arrest and those with hypospermatogenesis in obstructive azoospermia or oligozoospermia. Testis with Sertoli cell-only , having no germ cells showed no telomerase activity. This showed that the telomerase activity was associated with germ cells.

Telomerase activity and expression of human telomerase RNA (hTR) and human telomerase reverse transcriptase (hTERT) in human testis was studied in biopsy samples of patients showing Sertoli cell only syndrome, patients with maturation arrest and patients showing obstructive azoospermia. All 12 biopsy specimens from patients with obstructive azoospermia were positive for telomerase activity, hTR, and hTERT. 8/9 patients with maturation arrest were positive for telomerase, 9 for hTR and 5 cases for hTERT, where as none of the patients showing only Sertoli cell syndrome were positive with telomerase or hTERT but all of them showed hTR activity. Telomerase activity and evidence of hTERT in testicular tissue are highly sensitive and highly specific markers of gametogenesis, which could gain an importance as part of the fertility workup before microinjection procedures (Schrader et al., 2000). Telomerase uses a special mechanism for DNA synthesis at telomeric ends **(Fig 12.5)**.

Telomere maintenance plays an important role in cell proliferation and tumor survival. Nowak et al (2000) compared telomeric length and telomerase activity between two major tumors. Fifteen out of 16 seminoma samples revealed telomeric restriction fragment (TrRF) length below 13 kb; the remaining seminoma showed TRF length $\geq$ 23kb, which is similar to that in human sperm. Nine out of 11 seminoma specimens and six out of seven nonseminomas showed moderate to high telomerase activity. The major difference in telomeric length between seminomas and nonseminomas, is apparently unrelated to the presence of telomerase activity, which suggests that a germline-like homeostasis of telomeric length is preserved in human nonseminomas (Nowak, 1997; Nowak et al., 2000).

12.6.3. Tankyrase

Smith et al (1998) described the discovery of a protein present in human cells that controls telomerase activity or blocks the access of telomerase to chromosome ends. The new protein was called Tankyrase. Tankyrase has 24 so called ankyrin repeats, which in other proteins are involved in protein-protein interactions. The other section of tankyrase looks like catalytically active region of an unusual enzyme called PARP [(poly-adenosine diphosphate-ribose) polymerase]. The PARP plays a role in DNA repair (Pennisi, 1998). It is yet to be seen if tankyrase is present in testicular germ cells, where it controls the telomerase activity.

12.7. REFERENCES

Acharya S, Choudhury NR, Rao MR. characterization of a DNA pairing activity copurifies with DNA ligase activity from partially purified extract of rat testis. Biochim Biophys Acta 1996; 1309: 131-46.

Achi MV, Ravindranath N, Dym M. Telomere length in male germ cells is inversely correlated with telomerase activity. Biol Reprod 2000; 63: 591-8.

Aoufouchi S, Flatter E, Dahan A, Faili A, Bertocci B, Storck S, Belbos F et al. Two novel human and mouse DNA polymerases of the pol X family. Nucleic Acids Res 2000; 28: 3684-93.

Chen J, Tomkinson AE, Ramos W, Mackey ZB, Danehower S et al. Mammalian DNA Ligase III: molecular cloning, chromosomal localization, and expression in spermatocytes undergoing meiotic recombination. Mol Cell Biol 1995; 15: 5412-22.

Cobb J, Miyaike M, Kikuchi A, Handel MA. Meiotic events at the centromeric heterochromatin: histone H3 phosphorylation, topoisomerase II α localization and chromosome condensation. Chromosoma 1999; 108: 412-25.

Cobb J, Reddy RK, Park C, Handle MA. Analysis of expression and function of topoisomerase I and II during meiosis in male mice. Mol Reprod Dev 1997; 46: 489-98.

Coleman MA, Eisen JA, Mohrenweiser HW. Cloning and characterization of BARP/SMARCAL1: a prokaryotic HepA-related SNF2 helicae protein from human and mouse. Genomics 2000; 65: 274-82.

Eisenhauer KM, Gerstein RM, Chiu CP, Conti M, Hsueh AJW. Telomerase activity in female and male rat germ cells undergoing meiosis and in early embryos. Biol Reprod 1997; 56: 1120-25.

Englander EW, Widen SG, Wilson SH. Mammalian β-polymerase promoter: phosphorylation of ATF/CRE-binding protein and regulation of DNA binding. Nucleic Acids Res 1991; 19: 3369-75.

Enomoto T. Functions of RecQ family helicases : possible involvement of bloom's and werner's syndrome gene products in guarding genome integrity during DNA replication. J Biochem 2001; 129: 501-7.

Fujisawa M, Tanaka H, Tatsumi N, Okada H, Arakawa S, Kamidono S. Telomerase activity in the testis of infertile patients with delected causes. Human Reprod 1998; 13: 1476-79.

Galande S, Muniyappa K. Purification and functional characterization of type II DNA topoisomerase from rat testis and comparison with topoisomerase II from liver. Biochim Biophys Acta 1996; 1308: 58-66.

Garcia-Diaz M, Dominguez O, Lopez-Fernandez LA, de lera LT, Saniger ML et al. DNA polymerase Lambda (Pol λ), a novel eukaryotic DNA polymerase with a potential role in meiosis. J Mol Biol 2000; 301: 851-67.

Giordano R, Magnano AR, Zaccagnini G, Pittoggi C, Moscufo N, Lorenzini R, Spadafora C. Reverse transriptase activity in mature spermatozoa of mouse. J Cell Biol 2000; 148: 1107-13.

Hecht NB, Farrell D, Williams JL. DNA polymerases in mouse spermatogenic cells separated by sedimentation velocity. Biochim Biophys Act 1979; 561: 358-68.

Kallio M, Lahdetie J. Effects of the DNA topoisomerase II inhibitor merbarone in male mouse meiotic divisions in vivo: cell cycle arrest and induction of aneuploidy. Environm Mole Mut 1997; 29: 16-27.

Kamel D, Mackey ZB, Sjoblom T, Walter CA, McCarrey JR, Uitto L, Palosaari H, Lahdetie J, Tomkinson AE, Syvaoja JE. Role of deoxyribonucleic acid polymerase ε in spermatogenesis in mice. Biol Reprod 1997; 57: 1367-74.

Kozik A, Bradbury EM, Zalensky AO. Identification and characterization of a bovine sperm protein that binds specifically to single-stranded telomeric deoxyribonucleic acid. Biol Reprod 2000; 62: 340-69.

Kwan KY, Moens PB, Wang JC. Infertility and aneuploidy in mice lacking a type IA DNA topoisomerase III β. Proc Natl Acad Sci U S A 2003; 100: 2526-31.

Mackey ZB, Ramos W, Levin DS, Walter CA, McCarrey JR, Tomkinson AE. An alternative splicing event which occurs in mouse pachytene spermatocytes generates a form of DNA ligase III with distinct biochemical properties that may function in meiotic recombination. Mol Cell Biol 1997; 17: 989-98.

Makino Y, Kanemaki M, Kurokawa Y, Koji T, Tamura Ta. Rat RuvB-like protein, Tip49a, is a germ cell-enriched novel DNA helicase. J Biol Chem 1999; 274: 15329-35.

Metzler-Guillemain C, de Massy B. Identification and characterization of an SPO11 homolog in the mouse. Chromosoma 2000; 109: 133-8.

Monesi V. Autoradiographic study of DNA synthesis and the cell cycle in spermatogonia and spermatocytes

of mouse tetis using tritiateed thymidine. J Cell Biol 1962; 14: 1-18.
Moran JV, DeBerardinis RJ, Kazazian HHJr. Exon shuffling by L1 retrotransposition. Science 1999; 283: 1530-34.
Nagasawa K, Kitamura K, Yasui A, Nimura Y, Ikeda K, Hirai M, Matsukage A. Nakanishi M. Identification and characterization of human DNA polymerase β2, a DNA polymerase β related enzyme. J Biol Chem 2000; 275: 31233-8.
Nowak R, Grzybowska EA, Wilczynska A, Pykalo R, Siedlecki JA. Low expression of DNA polymerase β in human testicular germ cell tumours. Acta Oncologica 2002; 41: 188-91.
Nowak R, Siedlecki JA, Kaczmarek L, Zmudzka BZ, Wilson SH et al. Levels and size complexity of DNA polymerase β mRNA in rat regenerating liver and other organs. Biochim Biophys Acta 1989; 1008: 203-7.
Nowak R, Sikora K, Pietas A, Skoneczna I, Chrapusta SJ. Germ cell-like telomeric length homeostasis in nonseminomatous testicular germ cell tumors. Oncogene 2000; 19: 4075-8.
Nowak R. Telomeric length variation in human testicular germ cell tumours : Very long and homogeneous telomeres in spermatocytic seminoma. Oncology Reports 1997; 4: 1099-1101.
Pelletier H, Sawaya MR, Kumar A, Wilson SH, Kraut J. Structures of Tenary Complexes of Rat DNA Polymerase β| a DNA Template-Primer, and ddCTP. Science 1994;264:1891-1903.
Pennisi E. A possible new partner for telomerase. Science 1998; 282: 1395-96.
Philippe M , Chevaillier P. Further characterization of a DNA polymerase activity in mouse sperm nuclei. Biochim Biophys Acta 1976; 447: 188-202.
Plug AW,Clairmont CA, Sapi E, Ashley T, Sweasy JB. Evidence for a role for DNA polymerase β in mammalian meiosis. Proc Natl Acad Sci USA 1997; 94: 1327-31.
Prasad R, Singhal RK, Srivastava DK, Molina JT, Tomkinson AE, Wilson SH. Specific interaction of DNA polymerase β| and DNA ligase 1 in a multiprotein base excision repair complex from bovine testis. J Biol Chem 1996; 271: 16000-7.
Ravindranath N, Dalal R, Solomon B, Djakiew D, Dym M. Loss of telomerase activity during male germ cell differentiation. Endocrinology 1997; 138: 4026-30.
Romanienko PJ, Camerini-Otero RB. The mouse Spo11 gene is required for meiotic chromosome synapsis. Mol Cell 2000; 6: 975-87.
Romanienko PJ, Camerini-Otero RD. Cloning, characterization and localization of mouse and human SPO11. Genomics 1999; 61: 156-69.
Russell LB, Hunsicker PR, Hack AM, Ashley T. Effect of the topoisomerae-II inhibitor etopoiside on meiotic recombination in male mice. Mutat Res 2000; 464: 201-12.
Sawaya MR, Pelletier H, Kumar A Wilson SH, Kraut J. Science 1994;264:1930-35.
Schulten H J , Engel W , Nayernia K , Burfeid P : Yeast one-hybrid assay identifies YY1 as a binding factor for a proacrosin promoter element. Biochem Biophys Res Commun 1999; 257: 871-73.
Scherthan H, Jerratsch M, Dhar S, Wang YA, Goff SP, Pandita TK. Meiotic telomere distribution and Sertoli cell nuclear architecture are altered in Atm- and Atm-p53-deficient mice. Mol Cell Biol 2000;20:7773-83.
Schrader M, Muller M, Heicappell R, Krause H, Schulze W, Miller K. Telomerase activity and expression of telomerase subunits in the testicular tissue of infertile patients. Fertil Steril 2000; 73: 706-11.
Shimamoto A, Nishikawa K, Kitao S, Furuichi Y. Human RecQ5β a large isomer of RecQ5 DNA helicase, localize in the nucleoplasm and interacts with topoisomerases 3α and 3β. Nucleic acids Res 2000; 28: 1647-55.
Singhal RK, Prasad R, Wilson SH. DNA polymerase β conducts the gap filling step in uracil-initiated base excision repair in a bovine testis nuclear extract. J Biol Chem 1995; 270: 949-57.
Smith S, Giriat I, Schmitt A, de Lange T. Tankyrase, a poly (ADP-ribose) polymerase at human telomeres. Science 1998;282:1484-87.
Van Sloun PP, Romeijn RJ, Eeken JC. Molecular cloning, expression and chromosomal localization of the mouse Rev 31 gene, encoding the catalytic subunit of polymerase zeta. Mutat Res 1999; 433: 109-16.
Wang WS, Seki M, Yamaoka T, Seki T, Tada S, Katada T, Fujimoto H, Enomoto T. Cloning of two isoforms of mouse DNA helicase Q1/RecQL cDNA; α form is expressed ubiquitously and β form specifically in the testis. Biochim Biophys Acta 1998; 198-202.
Widen SG, Wilson SH. Mammalian β-polymerase promoter: large-scale purification and properties of ATF/CREB palindrome binding protein from bovine testes. Biochemistry 1991; 30: 6296-305.
Wilson S, Abbotts J, Widen S. Progress toward molecular biology of DNA polymerase β. Bichim Biophys Acta 1988; 949: 149-57.
Zhang D, Penttila TL, Morris PL, Roeder RG. Cell and stage-specific high-level expression of TBP-related factor 2 (TRF2) during mouse spermatogenesis. Mech Dev 2001; 106: 203-5.

Chapter 13

DNA REPAIR AND RECOMBINATION

Repair of DNA damage is a fundamental process for all living cells. Persistence of DNA damage can result in mutagenesis and carcinogenesis. Defects in mismatch repair and nucleotide excision repair are associated with hereditary non-polyposis colorectal cancer and Xeroderma pigmentosum, respectively. In somatic cells, nucleotide excision (NER) is a major pathway to counteract the mutagenic effects of DNA damage. Two NER sub-pathways have been identified: global genome repair (GGR) and transcription coupled repair (TCR). In contrast to somatic cells, little is known regarding the expression of these pathways in germ cells. Mixed spermatogenic germ cell nuclear extracts exhibit high NER activity. The order of tissues/ cells exhibiting the greatest to lowest activity was: mixed germ cells>Sertoli cells> thymocytes> small intestine> liver> brain (Intano et al., 2001). However, Repair in spermatogenic cells is inefficient in the overall genome in transitionally active genes indicating non-functional GGR and TCR. Incision activities of extracts from diplotene cells and round spermatids are low, indicating a stage dependent expression of incision activity. Sequestering of NER proteins by misrepaired regions in DNA involved in synapsis and recombination may underline the lack of NER activity in premeiotic cells (Jansen et al., 2001).

13.1. MISMATCH REPAIR

In copying the total genome before cell division, the DNA replication machinery makes errors, approximately one error per 10^7 incorporated nucleotides. These errors are corrected by DNA mismatch repair. DNA mismatch repair plays a vital role in the faithful replication of DNA, and its inactivation leads to a mutator phenotype that has been associated with the common cancer susceptibility syndrome: Hereditary Non-Polyposis Colorectal Cancer (HNPCC). The mismatch-repair (MMR) system plays a central role in maintaining genetic stability and requires evolutionarily conserved protein factors. Mismatch repair, homologous recombination repair, and post-replication repair in the context of mammalian gametogenesis have been reviewed (Barrends et al., 2001). Attention was also given to DNA damage induced cell cycle checkpoint control. Among various repair mechanisms, DNA mismatch repair plays a prominent role in the correction of replicative mismatches that escape DNA polymerase proof reading, and other mismatches, which arise owing to spontaneous deamination of 5-methylcytosine and those which are caused due to mispairs during genetic recombination. Most of the proteins, named in mismatch repair in eukaryotes are based on the homology with proteins involved in the repair mechanism in prokaryotes. Microsatellite instability is characteristic of certain types of cancer, and is present in rodents lacking specific DNA mismatch repair proteins.

```
  1 MEQTEGVSTE CAKAIKPIDG KSVHQICSGQ VILSLSTAVK ELIENSVDAG ATTIDLRLKD
 61 YGVDLIEVSD NGCGVEEENF EGLALKHHTS KIQEFADLTQ VETFGFRGEA LSSLCALSDV
121 TISTCHGSAS VGTRLVFDHN GKITQKTPYP RPKGTTVSVQ HLFYTLPVRY KEFQRNIKKE
181 YSKMVQVLQA YCIISAGVRV SCTNQLGQGK RHAVVCTSGT SGMKENIGSV FGQKQLQSLI
241 PFVQLPPSDA VCEEYGLSTS GRHKTFSTFR ASFHSARTAP GGVQQTGSFS SSIRGPVTQQ
301 RSLSLSMRFY HMYNRHQYPF VVLNVSVDSE CVDINVTPDK RQILLQEEKL LLAVLKTSLI
361 GMFDSDANKL NVNQQPLLDV EGNLVKLHTA ELEKPVPGKQ DNSPSLKSTA DEKRVASISR
421 LREAFSLHPT KEIKSRGPET AELTRSFPSE KRGVLSSYPS DVISYRGLRG SQDKLVSPTD
481 SPGDCMDREK IEKDSGLSST SAGSEEEFST PEVASSFSSD YNVSSLEDRP SQETINCGDL
541 DCRPPGTGQS LKPEDHGYQC KALPLARLSP TNAKRFKTEE RPSNVNISQR LPGPQSTSAA
601 EVDVAIKMNK RIVLLEFSLS SLAKRMKQLQ HLKAQNKHEL SYRKFRAKIC PGENQAAEDE
661 LRKEISKSMF AEMEILGQFN LGFIVTKLKE DLFLVDQHAA DEKYNFEMLQ QHTVLQAQRL
721 ITPQTLNLTA VNEAVLIENL EIFRKNGFDF VIDEDAPVTE RAKLISLPTS KNWTFGPQDI
781 DELIFMLSDS PGVMCRPSRV RQMFASRACR KSVMIGTALN ASEMKKLITH MGEMDHPWNC
841 PHGRPTMRHV ANLDVISQN
```

Fig.13.1. Amino acid sequence of mouse PMS2. Reproduced with permission from S.M.Baker et al. Cell 82; 309-19: 1995 © Elsevier (Source: http//www. ncbi.nlm.nih.gov) (Accession number P54279).

Microsatellite instability and DNA mismatch repair protein defects are present in some azoospermic men, predominantly in Sertoli cell-only patients.

13.1.1. Mismatch Repair Genes

Three genes, *mutS, mutL*, and *mutH* are central to the correction of replication errors in E. coli. In mammals, there are many MutS and *MutL* homologues such as MSH1-6, MLH1 and MLH3, and PMS1 and PMS2 in humans and their counterparts in mice **(Fig.13.1)**. Since the discovery of a link between the malfunction of post-replicative MMR and human cancers, a number of reports have appeared on the function of MutS and MutL homologues in the correction of replication errors. In addition, several MutS-like and MutL-like proteins also participate in meiotic recombination. Studies indicate that the MutS protein recognizes the mismatch and that the MutH protein acts in a complex with MutS and MutL as an endonuclease to induce an initiating single strand break on the newly synthesized strand (Arnheim and Shibata, 1997). The *mutS* and *mutL* homologues with role in DNA mismatch repair have been identified in yeast (as PMS) and mammals (Baker et al., 1995). The pathway in eukaryotes consists of the DNA-binding components, which are the homologues of the bacterial MutS protein (MSH 2, 6), and the MutL homologues, which bind to the MutS homologues and are essential for the repair process. Three of the six homologues of MutS that function in these processes, Msh2, Msh3 and Msh6, are involved in the mismatch repair of mutations, frameshifts and replication errors, and two others, Msh4 and Msh5, have specific roles in meiosis.

Most of the information about various components of the mismatch repair system in spermatogenesis has been gained from mouse gene knockout models, containing a homozygous null mutation of a gene encoding one of these components. Knockout mice for one of the genes *MSH2*, *MSH3* and *MSH6* showed pronounced mismatch repair deficiency for somatic cells, but fertility appeared to be normal (Arnheim and Shibata, 1997; de Wind et al., 1999). However, several other MLH-MSH-PMS homologues seemed to be involved in gametogenesis, with or without an additional role in mismatch repair in somatic cells. The phenotype of *Mlh1* knockout mice also showed male and female infertility, owing to pachytene arrest (Baker et al., 1996; Edelmann et al., 1996). Meiotic recombination is severally impaired in *Mlh1* knockout mice. Mice that are deficient in either the *Pms2* or *Msh2* DNA mismatch repair genes have microsatellite instability and a predisposition to tumors. The *PMS2* deficient males display sterility associated with abnormal chromosome pairing in meiosis. Mice deficient in another mismatch repair gene, *Mlh1*, possess not only micro satellite instability but are also infertile (both males and females). The *Mlh1* deficient spermatocytes exhibit high levels of prematurely separated chromosomes and arrest in first meiotic division. The *Mlh1* appears to localize to sites of crossing over on meiotic chromosomes. Thus Mlh1 is involved in DNA mismatch repair and meiotic crossing over. The MLH1 protein has been localized to "early" and late recombination nodules on the synaptonemal complex (Baker et al., 1996; Wang et al., 1999).

The *PMS2* knockout mice have a phenotype associated with overall genomic instability. However, *PMS2* knockout mice develop male-restricted infertility. In the spermatocytes of these mice, pairing of homologous chromosomes is impaired, but there is no complete meiotic arrest and some malformed spermatids are still produced. PMS2 deficient animals appear prone to sarcomas and lyphomas. PMS2 deficient males are infertile, producing only abnormal spermatozoa. Analysis of axial element and synaptonemal complex formation during meiotic prophase-1 indicated abnormalities in chromosome synapsis. These observations suggested links among mismatch repair, genetic recombination, and chromosome synapsis in meiosis (Baker et al., 1995). In MSH4 and MSH5 knockout mice, meiosis proceedes upto zygotene stage, and pairing of homologous chromosomes is greatly diminished. Human MSH4 has been shown to interact with MLH1, and the two proteins partially co-localize on the synaptonemal complex during spermatogenesis in mice (Knetiz et al., 2000; Santucci-Darmanin et al., 2001; Barrends et al., 2001).

The full-length cDNA clone for rat MLH1 has an open reading frame of 2274 nt for a protein of 757 amino acids. Northern blot analysis of MLH1, MSH2, MSH3, and MSH6 in the testes of rats of different ages showed differential expression of these genes as a function of developmental maturation of the testis. The expression analysis suggested that MSH3 might have a more predominant role in the meiotic recombination (Geeta Vani et al., 1999). The rat MSH2 cDNA has an open reading frame of 2802 nucleotides coding for a protein of 933 amino acids (100-kDa) (**Fig.13.2**). Rat MSH-2 protein shows significant homology to human and mouse MSH2. The MSH2 in the testes of rats of different ages showed maximum expression at 20 days, at a time when the germ cells undergo premeiotic DNA replication, followed by down-regulation in the MSH 2 expression beyond 25 days, by the time the germ cells have entered meiotic prophase (Geeta Vani and Rao, 1997).

In addition to MLH1 and PMS2, there are two other MutL homologues in mice and humans: PMS1 and MLH3. The PMS1 is most likely involved in the repair of mitochondrial DNA mismatches. Targeted inactivation of either one of these genes results in a gametogenesis specific phenotype in both cases, displaying male and female meiotic arrest (de Vries et al., 1999; Edelmann et al., 1999; Kneitz et al., 2000). Of the four MutL homologues, Mlh1, Mlh3, Pms1and Pms2, three are involved in mismatch repair and at least two, Pms2 and Mlh1, are essential for meiotic progression in both yeast and mice. *Mlh3(-/-)* mice are viable but sterile. The Mlh3 is required for Mlh1 binding to meiotic chromosomes and localizes to meiotic chromosomes from the mid-pachynema stage of prophase I (Lipkin et al., 2002). A heterotetrameric complex of two heterodimer of different MutL proteins can recognize a single base pair mismatch, or a loop of single stranded DNA, depending on the composition of the complex.

The MutL homologue MLH3 has been recently identified in mammals. The MutL homologue 3 (Mlh3) is a member of a family of proteins conserved during evolution and having dual roles in DNA mismatch repair and meiosis. Evidences support a role for this

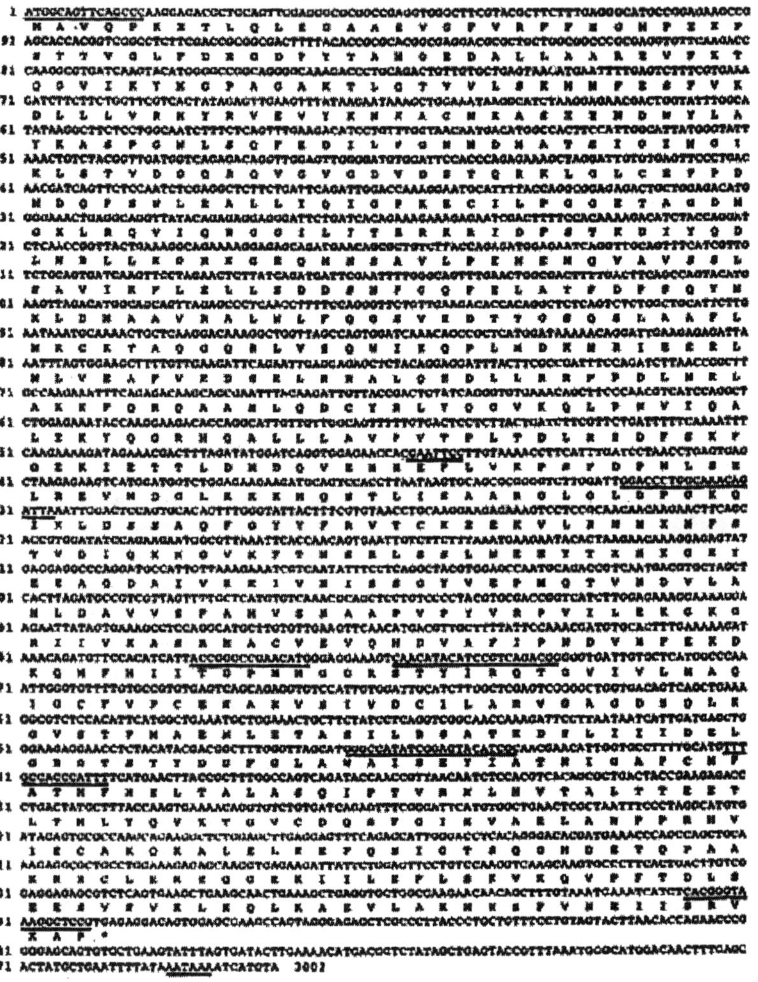

Fig.13.2. cDNA and deduced amino acid sequence of rat MSH2. The cDNA has an ORF from nt 1-2802, coding a protein of 933aa. The 3'-UTR extends from nt 2803 to 3002. The putative polyadenylation signal AATAAA is underlined. The unique EcoR1 present in rMSH2 is also underlined. Reprinted with permission from R. Geeta Vani and M.R.S. Rao. Gene 185; 19-26: 1997 © Elsevier.

protein in post-replicative MMR. The *MLH3* gene is expressed in mouse meiotic cells and in human testis, and the MLH3 protein is found in mouse spermatocytes. The meiosis-specific MSH4 protein, known to participate to meiotic recombination, is co-immunoprecipitated with MLH3 from mouse meiotic cell extracts. Report suggests that MLH3 is associated with MSH4 in mammalian meiotic cells, and possibly plays a role in mammalian meiotic recombination (Santucci-Darmanin et al., 2002).

A RT-PCR approach was used to identify germ cell transcripts for PMS2, and MSH2 and MSH3. Both the *PMS2* and *MSH2* genes are highly expressed in mitotically proliferating spermatogonia, and early in meiotic prophase in the leptotene and zygotene spermatocytes.

Thereafter, their expression declines in early and mid-pachytene spermatocytes, and in postmeiotic spermatids. In constrast, expression of *MSH3* was at its highest in pachytene spermatocytes. These patterns of expressions of mismatch repair enzymes are consistent with the proposed roles of the products in mismatch repair during both DNA replication and recombination (Richardson et al., 2000).

The MSH4 is a meiosis specific protein that is necessary for meiotic recombination in *S. cerevisiae*. The human MSH4 protein in testis is involved first in synapsis and second during recombination together with MLH1. The mouse *Msh4* gene is located on chromosome 3. The pattern of expression during spermatogenesis is consistent with a role for MSH4 both during zygonema and pachynema. Promoter activity of *mMsh4* 5'-flanking region was found with several SRY/Sox binding sites (chapter 10). Co-transfection experiments showed that SRY could down-regulate *mMsh4* promoter transcriptional activity. It is proposed that the regulation of *mMsh4* expression could be one of the reasons for the persistence of SRY and/or SRY related proteins in adult testis (Santucci-Darmanin et al., 2001). The human *MSH4* homologous gene shows 28.7% identity with the *S. cerevisiae* MSH4 protein and only detectable in testis and in ovary with a lower level of expression. The MSH4 is mapped to human chromosome 1 at band p31 (Paquis-Flucklinger et al., 1997).

A human orthologue of MSH5-cDNA, as well as the human gene of MSH5 contains a 2505-bp ORF that encodes an 834 amino acid polypeptide with a predicted molecular mass of 92.9 kDa. The amino acid sequence includes sequence motifs that are conserved in all known MutS homologues existing in bacteria to humans. The cDNA appears to be a member of the *MutS* family and shares 30% amino acid sequence identity with that of *S. Cerevisiae* MSH5, that plays a critical role in facilitating crossover during meiosis. The 2.9-kb human MSH5 mRNA species is present in all human tissues with highest expression in testis. The expression pattern of MSH5 resembled that of other human MutS homologues *MSH2, MSH3*, and *MSH6* genes. The *MSH5* gene spans approximately 25 kb and contains 26 exons that range in length from 36 bp to 254 bp for exon 8-25. *MSH5* has been mapped to human chromosome band 6p21.3 (Her and Dogget, 1998).

13.1.2. Proteins with Nuclease Activity

Apurinic / Apyrimidinic Endonuclease: The multifunctional mammalian apurinic/apyrimidinic endonuclease (APE) is responsible for the repair of AP sites in DNA. In addition, this enzyme has been shown to function as a redox factor facilitating the DNA binding capability of Jun-Jun homodimers and Fos-Jun heterodimers. The distribution of the APE transcript has shown that this gene is expressed in all tissues at relatively similar levels. Examination of the adult testis indicated that the expression of the transcript varies with the stage of spermatogenesis with the highest levels being present over round spermatids. Evidences show that the APE gene is not homogeneously expressed, but rather is found in subpopulations of cells in the brain and testes during development (Wilson et al., 1996).

Exonuclease 1: The human *EX01* gene product, exonuclease1, shares 26% amino acid sequence identity with the *S. cerevisiae* EX01. The human and *S. cerevisiae* proteins showed a similar ability to complement the mutator phenotype of *S. cerevisiae rad27* mutants indicating that the two proteins are functionally similar. There appear to be two forms of *hEX01* that differ by the COOH terminal 1 and 44 amino acids, respectively, and these appear to result from alternative RNA splicing. The hEX01 gene consists of 14 exons and is transcribed to yield a 3 kb mRNA. The human gene is located at 1q42.2-qter and expressed in high levels in testis. Elevated expression was also observed in thymus and colon and to a lesser extent in small intestine, placenta, spleen, and ovary (Lee et al., 1999). A human exonuclease *(hExo1)* cDNA, related to the yeast exonuclase 1, comprises 2541 bp, which code for a Mr 94,000 protein that appears to be highly expressed in testis and at very low levels in other tissues. The *hExo1* gene has 14 exons and is located on chromosome 1q43. The hEXO1 interacts strongly with the human mismatch repair protein MSH2, suggesting its involvement in the mismatch repair process and/or DNA recombination (Schmutte et al., 1998; Tishkoff et al., 1998). The mouse gene, *mEX01* exonuclease 1, encodes a approximately 92 kDa protein that shares homology to proteins of the RAD2 nuclease family, most notably human 5' to 3' exonuclease Hex1/hExo1, yeast (EX01) proteins and *Drosophila melanogaster* Tosca. The *mExo1* gene maps to distal chromosome 1 and is expressed prominently in testis, an area of active homologous recombination, and spleen, a prominent lymphoid tissue. An increased level of *mExo1* mRNA is observed during a stage of testis development where cells that are actively involved in meiotic recombination arise first and represent a significant proportion of the germ cell population (Lee et al., 1999).

ERCC4: The *ERCC4* is an essential human gene in the nucleotide excision repair (NER) pathway. The ERCC4 and ERCC1 are also uniquely involved in removing DNA inter strand cross-linking damage. The ERCC1-ERCC4 heterodimer, like the homologous Rad10-Rad1 complex, has been found to possess an endonucleolytic activity that incises on the 5'-side of damage. The *ERCC4* gene, assigned to chromosome 16p 13.1-p13.2, corrects the defect in CHO mutants of NER complementation group 4 and is implicated in complementation group F of the human disorder Xeroderma pigmentosum (XP-F). The ERCC4 gene structure and functional cDNA sequence encoding a 916 amino acid protein (104 kDa), has substantial homology with the eukaryotic DNA repair and recombination proteins MEI-9 (melanogaster), Rad16 (pombe), and Rad1 (cerevisiae). The ERCC4 cDNA efficiently corrected mutants in rodent NER complementation groups 4 and 11. In cells of XP-F patients, the ERCC4 protein level was reduced to less than 5% consistent with XPF being the *ERCC4* gene. The considerable identity (40%) between *ERCC4* and *MEI-9* suggests a possible involvement of ERCC4 in meiosis. In baboon tissues, ERCC4 is expressed weakly and was not significantly higher in testis than in nonmeiotic tissues (Brookman et al., 1996).

RAD23A/RAD23B: The *S. cerevisiae RAD23* gene is involved in nucleotide excision repair (NER). Two human homologues of HHR23A and HHR23B (HGMW-approved RAD23A and RAD23B), were isolated. The HHR23B protein is complexed with the protein defective in the cancer prone repair syndrome *Xeroderma pigmentosum*, complementation group C, and is specifically involved in the global genome NER sub-pathway. Overall structure for all RAD23 homologues shows an ubiquitin like N-terminus followed by a strongly conserved 50 amino acid domain that is repeated at the C-terminus. The MHR23A was assigned to mouse chromosome 8C3 and MHR23B to 4B3. Because of the close chromosomal proximity of human XPC and HHR23B, the mouse XPC chromosomal location was determined at 6D. Both MHR23 genes are expressed in all tissues with elevated RNA expression of both MHR23 genes in testis. Although the RAD23 equivalents are well conserved during evolution, the mammalian genes did not express the UV-inducible phenotype of their yeast counterpart (van der Spek et al., 1996).

XPA/XAB1: The *Xeroderma pigmentosum* group A protein (XPA) plays a central role in nucleotide excision repair (NER). A cytoplasmic GTP binding protein that binds to XPA and designated XPA binding protein 1 (XAB1) was identified. The deduced amino acid sequence of XAB1 consisted of 374 residues with a molecular weight of 41 kDa and a pI 4.65. The XAB1 has four sequence motifs G1-G4 of the GTP binding proteins family in the N-terminal half, and also contains acidic region in the C-terminal portion. The XAB1 mRNA is expressed ubiquitously, Consistent with the GTP binding motif, recombinant XAB'1 protein showed intrinsic GTPase activity and binds to the N-terminal region of XPA. The deletion of five amino acid residues, 30-34 of XPA required for nuclear localization of XPA abolished the interaction with XAB1. This suggested that XAB1 is a cytoplasmic GTPase involved in nuclear localization of XPA (Nitta et al., 2000).

XPF: The human XPF protein, an endonuclease subunit essential for DNA excision repair, may also function in homologous recombination. The predicted mouse XPF protein, encoded by a 3.4 kb cDNA, contains 917 amino acids and is 86% identical to human XPF. Appreciable similarity also exists between mouse XPF and homologous proteins in budding yeast (Rad 1), fission yeast (Rad16), and fruit fly (Mei-9), all of which have dual functions in excision repair and recombination. Sequence analysis of the 38.3 kb Xpf gene localized to a region in proximal mouse chromosome 16, revealed more than 72% identity to human XPF in 16 regions. Of these conserved elements, 11 were exons and 5 were noncoding sequences within introns. The *Xpf* transcript and protein levels are specifically elevated in adult mouse testis. Increased levels of *Xpf* and *Ercc1* mRNAs are correlated with meiotic and early post-meiotic spermatogenic cells and support a distinct role for the XPF/ERCC1 junction specific endonuclease during meiosis (Shannon et al., 1999).

13.2. PROTEIN SPECIFICITY IN RECOMBINATION REPAIR

DNA recombination plays an important role in the acquirement of gene diversity and in the survival of cells. Genetic recombination in meiosis produces genetic diversity. The term recombination has often been used as a synonym to crossing-over. However, it should be recognized that recombination covers a much wider spectrum of events including, (i) crossing -over, both meiotic and mitotic (reciprocal exchange of segments), (ii) gene conversion (non-reciprocal event), (iii) exchange between sister chromatids (even though this does not lead to a change in genotpye), (iv) repair of DNA damage, etc.

Recombination nodules are important structures, associated with paired pachytene chromosome considered to be involved in meiotic recombination. A correspondence between meiotic exchange events and the numbers and locations of recombination nodules has been observed in a variety of materials suggesting the possible role of these structures in recombination. In *Drosophila* females, these nodules can be spherical (larger) or ellipsoidal (smaller). Nodules play a positive role during recombination in SC. In *Drosophila* mutants, with defective recombination, a study suggests that nodules help in choice of number and location of recombination sites and in recombination itself. Meiotic recombinations in the proximal region of the mouse MHC are clustered within certain segments of chromosome, known as hotspots. One of such hotspots, is mapped between the Pb and Ob genes, located very close to the 3' end of the Lmp2 gene, which encodes a subunit of a proteolytic proteasome (Wagner et al., 2001). Ohinata et al., (2003) cloned human and mouse *Peas* cDNAs (*hPEAS/ mPeas*) and analyzed their stage-specific expressions. Peas protein contains six repeated kelch motifs, structurally similar to RAG2, a V(D)J recombination activator. Peas is evolutionarily conserved among mammals, birds, insects, and nematodes and is specifically transcribed in testis at pachytene spermatocytes in which it is localized to the cytoplasm and meiotic chromatin. It seems to be involved in meiotic recombination (Ohinata et al., 2003). Recombinational repair removes DNA damage and suppresses oncogenesis and cell death. DNA recombination has been extensively studied in prokaryotes and lower eukaryotes. E. coli RecA has been shown to be major protein involved in DNA recombination. This protein recognizes homologous regions of double stranded DNA and promotes DNA strand exchange. In *Saccharamyces cerevisisae* two classes of genes are required for meiotic recombination. One class includes Rad52 (*RAD50-57*) epistasis group of genes involved in both mitotic DNA repair and meiotic recombination through interaction with double strand breaks (dsbs) in DNA. Among them, Rad51 protein of yeast is structurally and functionally similar to RecA protein of *E.coli*. The genes belonging to the other class are *HOP1*, *SPO11*, *RED1*, *REC102*, *ZIP1* and *Dmc1*, which are essential only for meiotic recombination. Four genes encoding proteins homologous to RecA that have been isolated are: *RAD51, RAD55, RAD57*, and *DMC1*. The Rad51 protein plays a crucial role in both mitotic and meiotic recombination and in the repair of DNA double strand breaks. In mammals, seven *recA*-like genes, RAD51, *DMC1/LIM15, REC2/R51H3/RAD51B, RAD51C, XRCC2, XRCC3,* and *TRAD/R51H3/ RAD51D*, have been reported independently by three groups (reviewed in Kawabata and Saeki et al., 1999). With the mouse gene, one group showed a human orthologue, *HsR51H3*, of mouse *TRAD* and another revealed an alternative transcript, *HsRAD51D*, encoding a truncated protein with a partial N-terminal region (~14%) of the orthologue and lacking two nucleotide binding motifs conserved in the *recA/RAD51* family.

13.2.1. RAD52

RAD52 is required for recombination repair of double strand breaks in DNA. The RAD52 homologues from man and mouse were cloned by the PCR. DNA sequence analysis revealed an open reading frame of 418 amino acids for the human RAD52 homologs and of 420 amino acid residues for the mouse counterpart. The identity between the two proteins is 69% and the overall similarity 80%. The homology of the mammalian proteins with their counterparts from yeast is primarily concentrated in the N-terminal region. A relatively high level of gene expression was observed in testis and thymus, and at low levels in other tissues suggesting that the mammalian RAD52 protein, like its homologue from yeast, plays a role in recombination. The mouse RAD52 gene is located near the tip of chromosome 6 in region G3. The human equivalent maps to region p13.3 of chromosome 12 (Muris et al., 1994).

Shen et al, (1996) identified a 120kDa protein that associates with the human RAD51 and RAD52 proteins. This protein shares significant amino acid homology with the yeast protein SMT3, which functionally associates with the yeast mitosis fidelity protein MIF2. It also shares moderate homology with ubiquitin and several other proteins, including the N-terminus of the RAD23 protein and an ubiquitin cross-reacting protein. The gene is tentatively designated UBL1 for ubiquitin like 1. The UBL1 mRNA is expressed in many human tissues, but at high level in testis. The *UBL1* gene is mapped to chromosome 2q32.2-q33, and a related sequence may be located on chromosome 1q23-q25 (Shen et al., 1996).

13.2.2. RAD 51

The expression of mouse *Rad51* mRNA transcript was observed from late G1 phase through M phase. During the period of late G1-S-G2, the RAD51 proteins are observed exclusively in nuclei. The Rad51 and Lim15 proteins are present in chromatin loops when chromosomes form SC. The proteins may promote pairing of homologous DNA sequences that would lead to formation of SC. The Rad51 protein in the SC cores may be involved in chiasma formation in late stages. The Lim 15 instead, may be involved in recombination in the telomeric region or in cohesion of sister chromatids for segregation (Barlow et al., 1997a,b; Ikeya et al., 1996). The *Rad51* gene is expressed not only in meiotic tissues but also in somatic tissues dependent on a cell cycle (late G1-S-G2M), suggesting involvement in double strand break repair of DNA in replication. Gene targeting studies of mouse *Rad51* revealed that the absence of *Rad51* results in early embryonic lethality. However, Yamamoto et al., (1996) detected RAD51 in proliferating cells including spermatogonia in testis. It also expressed in spermatocytes during early and mid-prophase of meiosis and resting oocytes before maturation. RAD51 recombinase has been also detected in human and mouse spermatocytes (Barlow et al., 1997b). Thus, it appeared that mouse Rad51 expression is closely related to the state of cell proliferation and is presumably involved in DNA repair coupled with DNA replication, as well as in meiotic DNA recombination in spermatocytes (Yamamoto et al., 1996). On the assumption that Rad51 protein plays a role in early meiotic chromosomal events, Moens et al., (1997) examined the location and time of appearance of Rad51 protein in meiotic prophase chromosomes. The Rad51 foci in mouse spermatocytes appeared after the emergence of, and attached to, short chromosomal core segments. These foci increased in number to about 250 per nucleus at the time when core formation is extensive. In the male mouse foci decreased in number to approximately 100 while chromosome synapsis was in progress. When synapsis was completed, the numbers of autosomal foci declined to near 0 while X chromosome retained about 15 foci throughout this time. This stage coincides with the appearance of testis specific histone H1t at mid to late pachytene. Electron microscopy revealed association of 100 nm nodules associated with core, which was rich in Rad51. The absence of Rad51 foci in the chromatin loops suggests that in wild type mice Rad51/DNA filaments are restricted to DNA at the cores/synaptonemal complexes. However, the expected association of Rad51 could not be verified immunocytochemically (Moens et al., 1997).

In vitro, the human Rad51 protein (hRad51) promotes homologous pairing and strand exchange reactions suggestive of a key a role in genetic recombination. In human spermatocytes, hRad51 was found to form discrete nuclear foci from early zygotene to late pachytene. The foci always co-localized with lateral element proteins, components of the SC. During zygotene, the largest foci were present in regions undergoing synapsis, suggesting that Rad51 is a component of early recombination nodules. Pachytene nuclei showed a greatly reduced level of Rad51 labeling with the exceptions of any asynapsed autosomes and XY segments, which were intensely labeled. The distribution of Rad51 in mouse spermatocytes was similar to that found in human spermatocytes except that in this case Rad51 was detectable at leptotene. These results suggested that the Rad51 has a role in the interhomologue interactions that occur during meiotic recombination. These interactions are spatially and temporally associated with synapsis during meiotic prophase I (Ashley et al., 1995; Baumann et al., 1996; Barlow et al., 1997a; Haff et al., 1995).

Using sequence homologies from eukaryotic members, Cartwright et al., (1998) identified fragments of two additional mammalian genes with homology to RAD51. Cloning of full-length cDNAs for both human and mouse genes showed that the sequences are highly conserved, and that the predicted proteins have characteristic features of this gene family. One of the novel genes (R51H2) occurs in two forms in human cDNA, differing extensively at the 3' end, probably due to an unusual form of alternative splicing. The new genes (R51H2 and R51H3) were mapped to human chromosomes 14q23-24 and 17q1.2, respectively. The combination of gene structure conservation and the transcript expression patterns suggest that these new members of the recA/RAD51 family may also function in homologous recombination repair pathways (Cartwright et al., 1998).

13.2.3. RAD51 Paralogs

Cells defective in any of the RAD51 paralogs (RAD51B, RAD51C, RAD51D, XRCC2, and XRCC3) are sensitive of DNA cross-linking agents and to ionizing radiation. Because the paralogs are required for the assembly of DNA damage-induced RAD51 foci, and show genomic instability, their defect is thought being caused by an inability to promote efficient recombinational repair. These five paralogs exist in two distinct complexes in human cells: one contains RAD51B, RAD51C, RAD51D, and XRCC2 (defined as BCDX2), whereas the other consists of RAD51C with XRCC3. Both protein complexes have been investigated (Masson et al., 2001). The BCDX2 binds single stranded DNA and single-stranded gaps in duplex DNA, in accord with the proposal that the paralogs play an early role in recombinational repair. Moreover, BCDX2 complex binds specifically to nicks in duplex DNA. It was suggested that the extreme sensitivity of paralog-defective cell lines to cross-linking agents is owing to defects in the processing of incised cross-links and the consequential failure to initiate recombinational repair at these sites (Masson et al., 2001).

RAD51C: The full length of approximately 1.3-kb cDNA clone of RAD51C encodes a protein of 276 aa, having a 18-26% aa identity with other human Rad51 family members. The RAD51C includes mapped sequences tagged site location near the end of chromosome 17q. The RAD51C transcript is expressed in various human tissues, with highest level of expression in testis and it binds to two other members of the Rad51 protein family (Xrcc3 and Rad51B) but not to itself. Findings suggest that Rad51C may function similarly to the yeast Rad55 or Rad57 proteins, rather than as a Rad51 functional homologue (Dosanjh et al., 1998).

In humans, its interaction with proteins like BRCA1 and BRCA2 has provided an insight into the mechanism of how these molecules function as tumor suppressors. Several, members of the Rad51-like family have been identified, including RAD51I2. This gene has been found to be amplified in breast tumors suggesting its role in tumor progression. The cloning of the murine homologue of the human RAD51I2/RAD51C gene revealed that murine RAD51I2 protein is 86% identical and 93% similar to human homologue. In spite of such high sequence conservation, the murine protein lacks the first nine amino acids present in the human protein, which by the sequence of the 5' end of the murine cDNA-Rad51I2 is expressed in several adult tissues as well as in embryos at various developmental stages Rad51I2 gene maps to chromosome 11 and is located in the systemic region of human chromosome 17q22-23, where the human RAD51I2 is present (Leasure et al., 2001). A full-length 1.5-kb mouse cDNA cone of Rad51d that encodes a predicted 329-amino-acid protein was isolated. The Rad51d mRNA was present in every mouse tissue examined. Four different transcript sizes of Rad51d mRNA were detected, one of which was specific to testis. Human cDNA clones that predicted 71% amino acid identity to the mouse protein and their sequences provided evidence for alternative splicing. These mRNAs predict proteins that are truncated relative to the mouse and lack the ATP-binding motif characteristic of RecA-related proteins. The Rad51d was mapped to mouse chromosome 11, 48.5cm from centromere. The human orthologue RAD51D was mapped to chromosome 17q11, which is a region syntenic to mouse Chromosome 11. Due to its expression pattern and sequence similarity to other RAD51 family members, it is likely that Rad51d is required for DNA repair and meiotic recombination (Pitman et al., 1998).

Unique multiple alternative transcripts are present in human *TRAD/R51H3/RAD51D* gene. One of the transcripts encoded a 328-amino-acid protein with 83.0% overall amino acid identity and 98.2% similarity with the mouse *TRAD* gene and had two nucleotide binding consensus sequences, motif A and motif B, conserved among members of this family. Other transcripts encode truncated proteins with a partial N-terminal region of the orthologue or short proteins lacking internal sequences, which contain nucleotide-binding motifs. Northern blot revealed that multiple transcripts of the human *TRAD* gene were expressed in various tissues and their distribution was not ubiquitous (Kawabata and Salki, 1999).

DMC1: *DMC1* (disrupted meiotic cDNA) is a gene, which plays a role specifically in meiotic recombination in yeast and is another homologue of RecA in eukaryote. Yeast *dmc1* mutants fail in reciprocal recombination and accumulate dsb recombination intermediates. They are incapable of forming normal synaptonemal complex, thus causing an arrest of late meiotic prophase at a meiosis specific cells cycle checkpoint that monitors various defects of chromosome metabolism. The Dmc1 protein is structurally and evolutionally related to the gene products of yeast *RAD51* and *E.coli RecA*. In yeast, Dmc1 and Rad51 proteins have been shown to co-localize within spread meiotic nuclei, suggesting their co-operative function in meiotic recombination. Moreover, Dmc1 and Rad51 form complex to promote recombination events. However, Dmc1 and Rad51 exhibit different expression patterns during meiosis even though they are structurally similar to each other. In higher eukaryotes *RAD51* homologues have been isolated from the chicken, human and mouse. They encode proteins sharing 80% similarity to the yeast Rad51 protein and ~ 50% similarity to E.coli *RecA* protein. (reviewed in Habu et al., 1996) **(Fig.13.3)**. Habu et al., (1996) cloned the mouse and human homologues of the yeast *DMC1* gene. The predicted human DMC1 proteins showed 54.1% sequence identity with yeast Dmc1 protein. The domain II region, highly conserved in the *E.coli* RecA-like protein family, was also found in the mammalian DMC1 proteins, including the two ATP binding motifs and DNA binding sites within the region. Since the alternatively spliced *Dmc1-d* transcript was detected in both male and female germ cells of man and mouse, the encoded protein DMC1-D may have some important role in mammalian genetic recombination in meiosis.

Thus, it is not easy to evaluate the importance of Rad51 in meiosis because homozygotes die before gametes are produced. In contrast to the *Rad51* the *Dmc1* gene is expressed only during meiosis. Both RAD51 and DMC1 proteins have structures similar to the central core region of the RecA protein (domain II), which includes two nucleotide-binding motifs. Purified human DMC1 has DNA dependent ATPase activity, single stranded DNA binding activity, and strand exchange activity Therefore, the DMC1 protein is expected to catalyze the DNA mediated recombination reaction in meiosis, as observed with RecA and RAD51 proteins. Homologues associate with each other in the zygotene stage and synapse along their lengths to form synaptonemal complexes in the pachytene stage of meiosis (Yoshida et al., 1998).

The precise immunolocalization of RAD51 and DMC1 proteins on the meiotic chromosomes of plants and animals has been complicated by their high degree of identity at the amino acid level. Antibodies that have been immunodepleted of cross-reactive epitopes RAD51 and DMC1 have shown identical distribution patterns in mouse spermatocytes in successive prophase I stages, and colocalization of the two proteins on the meiotic chromosome cores at early prophase indicated that mouse RAD51 and DMC1 establish protein-protein interactions with each other and with the chromosome core component COR1 (SCP3) in a two-hybrids system. These results suggest that the formation of a multiprotein recombination complex associated with the meiotic chromosome cores is essential for the development and fulfillment of the meiotic recombination process (Tarsounas et al., 1999) **(Fig.13.3)**. During mouse spermatogenesis DMC1 protein was detected in leptotene to zygotene spermatocytes, when homologous pairing likely initiates. Gene disruption results in an arrest of meiosis of germ cells at the early zygotene stage, followed by apoptosis. Meiotic chromosome analysis of *Dmc1* deficient mouse spermatocytes revealed random spread of univalent axial elements without correct pairing between homologues. Thus, the mouse *Dmc1* gene is required for

```
  1 MKEDQVVAEE PGFQDEEESL FQDIDLLQKH GINVADIKKL KSVGICTIKG IQMTTRRALC
 61 NVKGLSEAKV DKIKEAANKL IEPGFLTAFE YSEKRKMVFH ITTGSQEFDK LLGGGIESMA
121 ITEAFGEFRT GKTQLSHTLC VTAQLPGAGG YPGGKIIFID TENTFRPDRL RDIADRFNVD
181 HDAVLDNVLY ARAYTSEHQM ELLDYVAAKF HEEAGIFKLL IIDSIMALFR VDFSGRGELA
241 ERQQKLAQML SRLQKISEEY NVAVFVTNQM TADPGATMTF QADPKKPIGG HILAHASTTR
301 ISLRKGRGEL RIAKIYDSPE MPENEATFAI TAGGIGDAKE
```

Fig.13.3. Amino acid sequence of meiotic recombination protein DMC1/LIM15 homolog. Source: http// www. ncbi.nlm.nih.gov (Accession number Q14565).

homologous synapsis of chromosomes in meiosis (Yoshida et al., 1998).

MmLim15: Using the sequence information on RecA like genes isolated from meiosis specific libraries of yeast and plant cell, a new gene was identified from a mouse testis cDNA library. The gene product termed, MmLim15 had a molecular mass of 37.88 kDa consisting of 340 amino acids. The predicted amino acid sequence showed 72-78% similarly to those of Lim15 from Lilium and other meiosis specific RecA-like proteins and RAD51 and its homologues. MmLim15 is testis specific and participates in meiosis (Sato et al., 1995a). Meiosis specific RecA like gene homologous cDNA clone, isolated from a cDNA library of human testis, had the coding capacity of a protein consisting of 340 amino acid residues, with the average size of putative eukaryotic RecA-like proteins. Expression of gene named *HsLIM15* gave products of two different sizes in testis: while the longer was predominantly seen in the testis, the shorter was commonly present to all the tissues including the testis. Analysis of the sequences indicated that the longer product corresponded to the above cDNA clone where as the shorter one was its deletion product missing an internal 165 bp portion. The result suggests that the mRNA species coding for the putative meiosis specific RecA like protein in human is predominantly expressed in testis possibly as an alternative splicing product of a ubiquitously expressed gene (Sato et al., 1995b). A testis specific 1730-bp fragment of the novel mouse gene specifically expresses a transcript of about 10kb in size. The presence of AATAAA polyadenylation site at its end indicated that the cloned fragment represented the end part of the corresponding gene. The plus chain of the cloned fragment contained motifs of the mouse RAD50 gene; the 387 bp fragment of the minus chain showed 96% homology with the Homo sapiens Hd741-f cDNA fragment; and the whole minus chain was 50% homologous to the prokaryotic FTSZ/A genes. The presumptive 44-kDa protein encoded by the plus chain (sense) open reading frame contained serine, proline and threonine rich regions and was 25 to 30% homologous to a number of nuclear DNA binding proteins of eukaryotes. Although analysis of the similarly between the 44-kDa proteins and the E. coli RecA protein did not show any significant homology between them, it revealed their identify of five amino acid residues involved in the formation of the epitope common to the paratope of the RecA protein (Loseva et al., 2000).

Pir 51: A human protein, named Pir51, strongly interacts with human Rad51 recombinase. The Pir51 mRNA is expressed in a number of human organs, most notably in testis, thymus, colon and small intestine. Gene locus was mapped to chromosome 12q13.1-13.2. Pir51 protein binds both single-and double-stranded DNA, and is capable of aggregating DNA. The protein also binds RNA. Pir may represent a member of the multiprotein complexes postulated to carry out homologous recombination and DNA repair in mammalian cells (Kovalenko et al., 1997).

13.2.4. Excision Repair Cross Complimenting Genes

XRCC1: The use of repair deficient rodent cell mutants has proved to be the most successful approach to the molecular cloning of human DNA repair genes. Elevated rates of sister chromatid exchange, reduced efficiency of homologous recombination, slow rates of DNA single strand break repair, and hypersensitivity to X-rays and radiomimetic drugs are hallmark features of the CHO cell line, EM9 and thus have provided a tool for isolating the human genes *ERCC1, ERCC2, ERCC3, ERCC5,* and *ERCC6* on the basis of functional complementation of the mutations by transfections of human genomic DNA, followed by vector cloning. These five *ERCC* (excision repair cross complementing) genes correct mutations that produce a common phenotype of hypersensitivity to UV radiation, which is analogous to the defect in cells derived from humans with the disease Xeroderma pigmentosum. A human gene that efficiently corrects that defect in EM9 cells was identified and named *XRCC1* (X-ray repair cross complementing). The *XRCC1* gene is assigned to human chromosome 19 and was shown to lie on the long arm in the region 19q13.2-13.3 (Thompson et al., 1990). The XRCC1 is implicated in DNA repair on the basis of its ability to restore the parental phenotype when transfected into the repair deficient EM9 cell. Although it has been suggested that XRCC1 is involved in base excision repair, its precise biochemical function is not known. However, histidine tagged XRCC1 protein copurifies with DNA ligase III. The recombinant protein designated pcD2EHX, fully corrected the EM9 phenotype of high sister chromatid exchange, indicating that the histidine tag was not detrimental to XRCC1 activity. Extract of EM9 cells transfected with pcD2EHX resulted in copurification of XRCC1 and DNA ligase III activity. The co-purification of DNA ligase III activity with histidine tagged XRCC1 suggests that the two proteins are present in the cell as a complex (Caldecott et al., 1994). EM9 cells, deficient in XRCC1 have reduced levels of DNA ligase II suggesting association of XRCC1 for DNA ligase III and further substantiating that XRCC1 is involved in DNA strand break repair. Analyses of *Xrcc-1* steady state RNA levels in murine and baboon demonstrated that *Xrcc-1* is a low abundance message with the highest expression detected in testis. Interestingly, DNA ligase III activity is highest in testis. The *Xrcc-1* message was significantly higher than found from W/W^v mice, which are deficient in germ cells and that found in somatic tissues (Walter et al., 1996). The Xrcc-1 expression is most abundant in pachytene spermatocytes and round spermatids with low expression in Sertoli cells, types A and B spermatogonia, preleptotene spermatocytes, and leptotene plus zygotene spermatocytes. The relatively abundant XRCC-1 expression in pachytene spermatocytes and round spermatids suggest that XRCC-1 is involved in DNA strand break repair associated with meiotic recombination in addition to its implicated role in strand break repair associated with base excision repair (Walter et al., 1996).

A nearly full-length clone obtained from the pcD2 human cDNA expression library gave ~80% correction of EM9, as analysed by the level of sister chromatid exchange. Based on the analysis of the nucleotide sequence of the cDNA insert compared with that of the 5' end of the gene from a cosmid clone, the cDNA clone appeared to be deficient in ~100bp of transcribed sequence, including 26 nucleotides of coding sequence. The cDNA probe detected a single transcript of ~2.2kb in HeLa polyadenylated RNA. From the open reading frame and the positions of likely start sites for transcription and translation, the size of the putative XRCC1 protein is 633 amino acids (69.5 kDa) and the size of the XRCC1 gene as 33 kb. The deduced amino acid sequence did not show significant homology with any protein in the protein sequence databases.

The 5' region of baboon XRCC1 was cloned and characterized. The 400 bp-5'-flanking region showed the greatest promoter activity, while -194 to –8 bp of the 5'flanking region displayed core promoter activity in transient transfection assays. A comparison between baboon and human 5' flanking sequences in the core promoter region revealed a potential CAAT-box, an imperfect CREB-binding site and two putative Sp1-binding sites. Transient transfection assays indicated that the distal Sp1-binding site has a functional role in transcription and that Sp1 or an Sp1 like protein is bound to the DNA fragment in vitro (Zhou and Walter 1998).

XRCC 2: The cDNA for *XRCC2* gene from both human and mouse shows that it is a highly conserved member of the recA/RAD51 recombination repair gene family. The cDNA is able to complement significantly the phenotype of a unique cell line, irs1, which shows extreme sensitivity to DNA cross-linking agents and genetic instability. This phenotype is consistent with a role for the *XRCC2* gene in recombination repair in somatic cells, suggesting that in addition to RAD51, other members of this gene family have an important function in high fidelity repair processes in mammals. Despite this function, the *XRCC2* gene transcript is expressed at a very low level in somatic tissues, but is elevated in mouse testis, suggesting an additional role in meiosis (Cartwright et al., 1998b).

13.2.5. BLM Helicase

Bloom's syndrome (BS) is rare autosomal recessive human genetic disorder characterized by a number of abnormalities including short stature, immunodeficiency, male sterility and cancer predisposition. Cells from BS have genomic instability and an increased frequency of sister chromatid exchange. The gene mutated in BS, *BLM*, encodes a 1417-amino-acid residue protein with a central domain that is homologus to a subfamily of helicases, including E *coli* RecQ, which have a characteristic DEXH box in one of the helicase motifs. N- and C- terminal domains have little homology as in Werners Syndrome (WS) related protein, WRN which has RecQ helicase activity. Both BLM and WRN proteins have been shown to possess 3'-5' helicase activity. Despite detailed investigations into BLM and BS cells, the precise physiological function(s) for BLM in humans is unclear (Moens et al., 2000). Human males homozygous for BLM mutations are infertile and heterozygous individuals display increased frequencies of structural chromosome abnormalities in their spermatozoa. Also, mutations in the *S. cerevisiae* homologue of BLM, *Sgs1*, cause a delay in meiotic nuclear division and a reduction in spore viability. The association of BLM protein with the recombinases and with a specific crossover site supports a role for this protein in meiotic recombination. The colocalization of RAD51/DMC1 with BLM and significant excess of BLM signals in the synapsed pseudoautosomal region of the X-Y chromosomes (a recombinational hot spot), provide indications that BLM protein may function in the meiotic recombination process (Moens et al., 2000).

13.3. POST-REPLICATION REPAIR GENE PRODUCTS

Post-replication DNA repair pathway is very complicated. This pathway does not exert DNA repair activity, but rather allows mitotic somatic cells in S phase to proceed with DNA replication over a DNA template containing damaged bases, in an error free or error prone manner. This mechanism is important for rescuing the cells from premature termination of DNA replication, which would otherwise lead to unscheduled cell death (Baarends et al., 2001). At least five multienzyme DNA repair pathways are known; one of these is termed post-replication repair (PRR). The PRR is not really a repair system, as the name suggests but it facilitates tolerance of base damage during replication, overcoming termination of replication at the sites of damage (Friedberg and Gerlach, 1999). The genes involved in PRR are placed in the very heterogenous RAD6 epitasis group (Friedberg and Gerlach, 1999). Yeast *rad6* mutants display the most severe phenotype. Other prominent members are *RAD6, RAD9, RAD18, REV1, REV3, REV7* and *RAD30* (McDonald et al., 1997). Two mammalian homologues of the yeast *RAD6* gene (*HR6A, HR6B*) and their role in the mouse spermatogenesis by targeted gene disruption have been discussed in a number of studies. Other possible homologues of PRR genes are: Yeast *RAD18,* which encodes a 55.5 kDa protein with a C_3HC_4-RING-Zn^{2+}-finger and C_2HC-Zn^{2+}-finger.

13.3.1. RAD6 and its Mammalian Homologues: HR6A and HR6B

In S. *cerevisiae*, genes involved in post-replication DNA repair are members of the RAD6 group. In mammalian cells, two conserved homologues of this gene have been identified: the X-chromosomal HR6A and autosomal HR6B gene. The *mHR6B* gene is expressed in many cell types throughout the body, but the only pronounced phenotypic expression of the loss of function in gene mutation is male infertility (Roest et al., 1996). The *mHR6B* gene is an autosomal mouse homologue of the *S..cerevisiae* Rad6 gene. Another mouse homologue, the *mHR6A* gene, is located on the X chromosome. *RAD6* homologues of several higher eukaryotes including two closely related human homologues *hHR6A* and *hHR6B* have been cloned. The encoded human proteins are structurally and functionally highly conserved: they share 70% sequence identity with *S. cerevisiae RAD6* and are able to ubiquitinate histones *in vitro*. Both human gene products can substitute for the mutagenesis and UV resistance function of the yeast protein but not for its role in sporulation (Roest et al., 1996). Protein targets for the RAD6-dependent ubiquitination include histones H2A and H2B, and possibly other chromosomal proteins (Baarends et al, 2001). The conservation of Rad6 from yeast to mammals is striking, showing upto 70 % amino acid sequence identity; and from mouse to human, the conservation of amino acid identity is 100% for both *HR6A* and *HR6B.* The HR6A and HR6B proteins are not completely identical to each other, but show 96% amino acid sequence identity. In HR6A and HR6B knockout mice no major phenotypes are observed for somatic tissues. With regard to a role for RAD6 homologue in mammalian gametogenesis, the *HR6A* knockout mice showed maternal effect infertility due to block in embryonic development with normal male fertility. In contrast, *HR6B* knockout mice demonstrated impaired spermatogenesis and male restricted infertility. Increased apoptosis in HR6B knockout primary spermatocytes was detected during the first wave of spermatogenesis, indicating that HR6B performs a primary role during the meiotic prophase, with major changes in the synaptonemal complexes, which were found to be longer. Complex proteins were depleted from near telomeric regions in the HR6B knockout pachytene nuclei. A specific requirement for the ubiquitin-conjugating activity of HR6B in relation to dynamic aspects of the SC and meiotic recombination in spermatocytes could be a possible role (Baarends et al., 2003). There are indications that the ubiquitin pathway plays a very active role during spermatogenesis (see Chapter 31). The *mHR6A* and *mHR6B* mRNAs are found in brain, heart, and testis in relatively high amounts. A similar expression pattern has been seen for two other genes involved in the ubiquitin pathway, *Ube1x* and *Ube1y*. Both genes located on X and Y-chromosomes respectively are meiotically suppressed, but reactivated in spermatids. In the series of reactions of the ubiquitin pathway, HR6A and HR6B are two of the many enzymes with ubiquitin conjugating activity. It is unknown whether RAD6 is involved in this posttranslational modification (Grootgoed et al., 1998).

In the *mHR6B* knockout mice, there is a reduced number of elongating spermatids, which also show an irregular pattern of nuclear condensation, resulting in the presence of many immature germ cells in the epididymis. Findings provide a parallel between yeast sporulation and mammalian spermatogenesis and strongly implicate hHR6 dependent ubiquitination in chromatin remodeling since heterozygous male mice and even knockout female mice are completely normal. A preferential localization of hHR6 in euchromatin areas, suggests that the protein is associated with transcriptional active regions. Thus both hHR6 genes have overlapping, constitutive functions related to chromatin conformation and that they have a specific role in spermatogenesis, involving Ub mediated histone degradation (Grootgoed et al., 1998; Koken et al., 1996).

13.3.2. Mammalian Homologues of Rad 18

The yeast *RAD18* gene encodes a protein with a RING zinc finger domain and a classical DNA binding zinc finger. Both domains are highly conserved among RAD18 of lower eukaryotes Nuv A *(Neurospora crassa)* and Uvs-2 *(aspergillus nidulans)*, and the cloned mammalian homologues: mRAD18Sc and hRAD18. The RING zinc finger domain mediates protein-protein interactions. The RAD18 protein plays a pivotal role in the post-replication repair process together with the Rad6 protein. The RAD18 has ssDNA binding ability, forms a stable complex with RAD6 in vivo and exhibits ssDNA dependent ATPase activity. The RAD18 seems to interact with both the N-and the C-termini of RAD6 via two acidic regions in the middle and the C-terminus of RAD18 (Bailly et al., 1997a). RAD6 is a ubiquitin conjugating enzyme catalyzing mono and / or polyubiquitination of target proteins. Polyubiquitinated proteins usually undergo proteasomal degradation while monoubiquitination may regulate the stabilization, folding and / or activity of proteins. The role of *RAD6* and *RAD18* in PRR is largely unknown. A possible function of RAD6 is in modifying the chromatin structure as part of DNA repair, mutagenesis, and other processes, which are affected in the *rad6* mutant. However, ubiquitination of H2B histone in yeast, required for optimal mitotic cell growth and meiosis, primarily depends on RAD6 activity. *Rad18* mutants are highly sensitive to UV and other genotoxic agents, but show no pronounced defect in sporulation, meiotic recombination, and UV induced mutagenesis. The RAD18 plays a more general role in PRR, rather than being involved only in error free repair. In particular, the RING zinc finger domain is present in proteins with ubiquitin ligating enzyme activity (E3 enzymes) (see chapter 16). The hRAD18 like RAD6 has also been shown to interact with HR6A and HR6B (Tateishi et al., 2000; Xin *et al., 2000*). In vivo, hRAD18 protein binds to hHR6 protein through a conserved ring finger motif. A human homologue *hRAD18* maps on chromosome 3p24-25, where deletions are often found in lung, breast, ovary, and testis cancers. Stable transformants with hRAD18 mutated in this motif become sensitive by UV, methyl methanesulfonate, and mitomycin C, and are defective in the replication of UV-damaged DNA (Tateishi et al., 2000).

mRAD18SC: Systematic screening of public EST databases revealed the presence of a putative murine homologue of *RAD18*, which is referred to as *mRAD18SC*. The cDNA in both directions revealed a large open reading frame of 1530-bp coding for a 509 amino acid polypeptide with a calculated molecular mass of 57.3-kDa and a pI 6.95. The 5' untranslated region (UTR) harbors the Kozak translation initiation sequence GCGGCCATGG. This ATG is first codon. In cDNA, the stop codon TAG at position +1528 is followed by a short stretch of 3' UTR containing a poly (A) tail. Surprisingly, no polyadenylation signal is found in the 3' UTR in front of the poly (A) tail. The N-terminus of the protein contains a C_3HC_4 RING-Zn^{2+}- finger

```
 -29                           GGCACGAGGCGGTCTCATCAGGTGCGGCC
   1  ATGGAGGTCCTGGCCGAGCCGCGATGCCCTCCGGGGTTGGCGGTGATGAAGACAATAGAT
   1   M  E  V  L  A  E  P  R  C  P  P  G  L  A  V  M  K  T  I  D
  61  GACTTGCTGCGCTGTGGGATTTGCTTTGAGTATTTCAACATTGCAGTGATAATCCCCCAG
  21   D  L  L  R  C  G  I  C  F  E  Y  F  N  I  A  V  I  I  P  Q
 121  TGCTCTCACAACTATTGTTCACTCTGTATAAGAAAGTTTTTATCCTATAAAACTCAGTGC
  41   C  S  H  N  Y  C  S  L  C  I  R  K  F  L  S  Y  K  T  Q  C
 181  CCAACTTGTTGCGTGGCAGTAACGGAGCCAGACCTGAGAAATAATCGCCTCTTAGATGAA
  61   P  T  C  C  V  A  V  T  E  P  D  L  R  N  N  R  L  L  D  E
 241  CTGGTAAAAAGCATGAATTTTGCACGGACTCACCTGTTGCAGTTTGCTTTAGAGTCACCA
  81   L  V  K  S  M  N  F  A  R  T  H  L  L  Q  F  A  L  E  S  P
 301  CCCATATCTCCTGTGTCTTCCACCTCAAAGAAGGTTGTTGTTAAAGTGCATAATGCTGAC
 101   P  I  S  P  V  S  S  T  S  K  K  V  V  V  K  V  H  N  A  D
 361  GCCGCCCAACACCCTGTCAAACAGGCGAACAGGTTAATGGATAAGTTCCTGATTAGAGAA
 121   A  A  Q  H  P  V  K  Q  A  N  R  L  M  D  K  F  L  I  R  E
 421  ACTGGTGACTGTGTATTTGAGTTGTTGGGAAAAGAAAATGAGAGGAAATTCAGCCCTCAA
 141   T  G  D  C  V  F  E  L  L  G  K  E  N  E  R  K  F  S  P  Q
 481  AAAGAGCTAAGCACCTCTGCTGAGATTAAAGAGACAAGTCTCCTAGGAAAGCCGGTACTG
 161   K  E  L  S  T  S  A  E  I  K  E  T  S  L  L  G  K  P  V  L
 541  GGGCTCTCGGATGCTAATGGTCCTGTGACTCCCTCTACATCCACTATGAAACTGGATACT
 181   G  L  S  D  A  N  G  P  V  T  P  S  T  S  T  M  K  L  D  T
 601  AAAGTGTCTTGTCCTGTTTGTGGGGTCAGCATTCCAGAAAATCATATCAATAAGCATTTA
 201   K  V  S  C  P  V  C  G  V  S  I  P  E  N  H  I  N  K  H  L
 661  GACAGTTGTTTATCACGTGAAGAGAAAAAGGAGAGCCTGCGAAGTTCTGCTCACAAAAGG
 221   D  S  C  L  S  R  E  E  K  K  E  S  L  R  S  S  A  H  K  R
 721  AAGCCGTTGCCCAAAACTGTATATAACTTGCTCTCTGATCGTGATTTAAAGAAAAAGCTG
 241   K  P  L  P  K  T  V  Y  N  L  L  S  D  R  D  L  K  K  K  L
 781  AAACAGTATGGCTTATCTGTTCCAGGAAACAAACAGCAGCTTATTAAAAGGCATCAGGAG
 261   K  Q  Y  G  L  S  V  P  G  N  K  Q  Q  L  I  K  R  H  Q  E
 841  TTTGTGCACATGTATAATGCCCAGTGCGATGCTTTGCATCCTAAATCAGCTGCTGAAATC
 281   F  V  H  M  Y  N  A  Q  C  D  A  L  H  P  K  S  A  A  E  I
 901  GTCCAAGAAATTGAAAGCATGGAGAAGACCAGGATGCGCCTTGAAGCAAGTAAACTCAAT
 301   V  Q  E  I  E  S  M  E  K  T  R  M  R  L  E  A  S  K  L  N
 961  GAAAATGTCATGGTTTTTACAAAGAACCAAACAGAGAAGGAAATTGAGGAAGTTCACAGT
 321   E  N  V  M  V  F  T  K  N  Q  T  E  K  E  I  E  E  V  H  S
1021  GAATATCGTAAAAAGCACCAGAATGCATTCCAGCTTCTGGTGGATCAGGCCAAAAAAGGG
 341   E  Y  R  K  K  H  Q  N  A  F  Q  L  L  V  D  Q  A  K  K  G
1081  TATAAGAAAACTGGCAGAGTTTCACAAGCTGCAGCGATGAGAACAGATGAACCTGCAGAG
 361   Y  K  K  T  G  R  V  S  Q  A  A  A  M  R  T  D  E  P  A  E
1141  ACACTGCCGTCGATGAGAACAGATGAACCTGCAGAGACACTGCCGTCGATGAGAACAGAT
 381   T  L  P  S  M  R  T  D  E  P  A  E  T  L  P  S  M  R  T  D
1201  GAACCTGCAGAGACACTGCCGTTGATGAGAGCAGATGAACCTGCAGAAACACTGCCGTCT
 401   E  P  A  E  T  L  P  L  M  R  A  D  E  P  A  E  T  L  P  S
1261  GAGTGTATCGCACAAGAAGATAATGTGAGCTTCTCAGATACTGTCTCAGTAACAAACCAC
 421   E  C  I  A  Q  E  D  N  V  S  F  S  D  T  V  S  V  T  N  H
1321  TTTCCTCAGCCCCAGCTGGACTCGCCAGGCCCGTCGGAGCCTGAGAGGCCGGATGATTCT
 441   F  P  Q  P  Q  L  D  S  P  G  P  S  E  P  E  R  P  D  D  S
1381  TCTAGTTGTACTGATATTCTTTTCTCCTCGGACTCAGACTCATGCAATAGAAATGATCAA
 461   S  S  C  T  D  I  L  F  S  S  D  S  D  S  C  N  R  N  D  Q
1441  AACAGAGAAGTCAGCCCACAACAGACTCGCCGCACTAGAGCCAGTGAATGTGTTGAGATT
 481   N  R  E  V  S  P  Q  Q  T  R  R  T  R  A  S  E  C  V  E  I
1501  GAACCAAGAAACAAGCGGAATAAGAATTAGTATGGGCTTTGTGCCAGCTTTCCGTACAGT
 501   E  P  R  N  K  R  N  K  N  *
1621  GACTTTTAAGTCTAAAn
```

Fig. 34.4. The mRAD18Sc coding sequence, 3' UTR, 5'UTR, and amino acid sequence. Black box, C_3HC_4-RING-Zn^{2+}-finger(residues 26-63); drak gray box C_2HC-Zn^{2+}-finger (residuces 204-223) (cysteines and histidines are highlighted with a black background), putative phosphorylation site (residues 469-474). Reproduced with permission from R. van der Laan et al. Genomics 69; 86-94: 2000 © Elsevier.

(RING-zinc-finger), and the middle contains a $CX_2CX_{11}HX_3$ C-type zinc finger (**Fig. 13.4**), both highly conserved between yeast RAD18 and the fungal homologue, NuvA and Uvs-2. The presence of a potential nuclear KKKLK suggests that mRAD18Sc resides in the nucleus. The size of the mRAD18Sc is comparable to that of RAD18 with significant homology over the entire length (Van der Laan et al., 2000).

The degree of sequence conservation between mRAD18SC, RAD18, and homologous sequences identified in other species is entirely consistent with the evolutionary relationship of these organisms. The mRNA expression of *mRAD18SC* was observed in thymus, spleen, brain, and ovary, but most pronounced in testis, with the highest level in pachytene stage of primary spermatocytes, suggesting that mRAD18SC plays a role in meiosis phase of spermatogenesis. The *mRAD18SC* gene was mapped on mouse chromosome 6F (Van der Laan et al., 2000). Testis from 21-day-old mice displayed a marked increase of *mRAD18S* mRNA expression, in particular the 3.1-kb transcript. In addition, spermatocytes showed a very high level of expression of this 3.1-kb transcript. In testis of 36-day-old mice, the *mRAD18Sc* mRNA level is decreased, probably due to a large number of round spermatids, which dilutes the signal from the spermatocytes (van der Laan *et al.*, 2000, c/r Baarands et al., 2001). During mammalian spermatogenesis, mRAD18Sc as E2 enzyme may be involved in the ubiquitin pathway.

13.3.3. RAD30 Gene

During post-replication DNA repair, specialized DNA polymerases participate during DNA replication, to add only few nucleotides in a growing DNA chain. One of the many DNA polymerases in *S. cerevisiae* is encoded by *RAD30*. In humans, the homologue *RAD30A* is mutated in patients with the Xeroderma pigmentosum variant, resulting in increased incidence of sunlight induced skin cancers (c/r Van der Laan et al, 2000). The expression profile of the *RAD30A* gene is not known. But second homologue (*RAD30B*) is expressed predominantly in human testis. In mouse RAD30B is mainly in round spermatids where the RAD30B repairs lesions (McDonald et al., 1999). Alternatively, use of the post-replication DNA repair pathway during gametogenesis may facilitate mutagenesis (Barrendes et al., 2001).

DinB genes: The *E. coli dinB* gene is an SOS gene known to be required for phage-untargeted mutagenesis. It exhibits a potent mutagenic activity without any exogenous treatment to damage DNA. The product DinB is structurally related to the *E. coli* Umu C protein and the *S. cerevisiae* Rev1 and Rad30 proteins, all of which are shown to be involved in bypass synthesis at a DNA lesion. Human and mouse cDNAs encoding DinB homologues produce proteins, which are highly similar to DINB1in humans and DinB1 in mouse and less similar to UmuC, Rev1 or Rad30. Both *DINB1* and *DinB1* genes are expressed abundantly in testis. Transient expression of the mouse cDNA in cultured mouse cells produce nearly 10 folds increase in the incidence of point mutations. These results suggested that a mutagenic mechanism, a so-called untargeted type, also operates in mammalian cells and thus human cells have multiple DNA polymerases for translesion synthesis which are homologous to the *S. cerevisisae*, *Rev3* and *Rad30* proteins (Ogi et al., 1999).

13.4. CELL CYCLE CHECK POINT CONTROL

13.4.1. RAD1

Freire et al., (1998) identified a human and mouse homologues of the *S. pombe* DNA damage checkpoint control gene *rad1(+)* and its *S. cerevisiae* homologue RAD17. The human gene hRAD1 is located on chromosome 5p13 and is most homologous to *S. pombe rad1 (+)*. This gene encodes a 382-amino acid residue protein that is localized mainly in the nucleus and is expressed at high levels in proliferative tissues. In addition to functioning in DNA repair checkpoints, *S. cerevisiae* RAD17 plays a role during meiosis to prevent progress through prophase 1 when recombination is interrupted. Consistent with a similar role in mammals, RAD1 protein is abundant in testis, and is associated with both synapsed and unsynapsed chromosomes during meiotic prophase I of spermatogenesis, with a staining pattern distinct from that of the recombination proteins Rad51 and Dmc1 (Freire et al., 1998).

13.4.2. ATR/ATM Gene Products

ATM and Meiotic Prophase: The *ATM* gene product has been implicated in mitogenic signal transduction, chromosome condensation, meiotic recombination, and cell cycle control. The ATR (ataxia telangiectasia and RAD3-related) protein is present on meiotic prophase chromosome cores and paired cores (SCs). Its striking characteristic is that the protein forms dense aggregates on the cores and SCs of the last chromosomes to pair at the zygotene pachytene transition. ATR-mRNA and protein are expressed in human and mouse testis (Keegan et al., 1996). In seminiferous tubules ATR shows localization to the nuclei of cells during meiosis. A nuclear association of ATR and ATM proteins with meiotic chromosomes suggests a direct role for these proteins in recognizing and responding to DNA strand interruptions that occur during meiotic recombination (Keegan et al., 1996). Several reports have described the phenotype of mice with a targeted disruption of *ATM*, the mouse homologue of the human gene *(ATM)* that is mutated in ataxia telangiectasia (AT). *ATM* deficient mice provide an excellent model for this disease. The effects of mutations in *ATM* indicated that these genes are involved in the proper functioning of a variety of mitotic; postmitotic and meiotically active cells. Most AT patients and all *ATM*-deficient mice, of both sexes are infertile due to complete absence of fertile gametes in adult gonads. The seminiferous tubules of mutant males contain spermatogonia and Sertoli cells, but no normal spermatocytes and display a complete absence of spermatids or mature sperm (Barlow et al., 1998). Using electron microscopy and immunolocalization of meiotic proteins in mutant adult spermatocytes, Barlow et al (1998) found that male gametogenesis is severely disrupted in *ATM*-deficient mice as early as leptonema of prophase I, resulting in apoptotic degeneration. A small number of mutant cells progress into later stages of meiosis, but no cells proceed beyond prophase I. ATR, a protein related to ATM, DMC1, and RAD51 were mislocalized to chromatin and had reduced intensity in developing synaptonemal complexes in spermatocytes from *ATM*-deficient mice indicating dysregulation of meiotic events. High levels of p53, and Bax proteins contribute to the severe meiotic phenotype of ATM-deficient mice. Gametogenesis is severely disrupted in ATM-deficient mice in the earliest stages of meiotic prophase I, resulting in apoptotic degeneration. ATM is required for proper assembly of Rad51 onto the chromosomal axial elements during meiosis. In addition, p53, p21 and Bax are elevated in testes from ATM-deficient mice. To determine whether these elevated protein levels are important factors in the meiotic disruption of *ATM*-deficient mice, Barlow et al., (1997b) analysed the meiotic phenotype of *ATM/p53* or *ATM/p21* double mutants. In these double mutants, meiosis progressed to later stages but was only partly rescued. Assembly of Rad51 foci on axial filaments was defective and spermatogenesis proceeded only to pachytene of prophase I. Because *ATM*-deficient mice are viable but completely infertile, studies suggested that the Rad51 and Bax represent tissue specific responses to the absence of ATM (Barlow et al., 1997a). It appeared that meiotic telomere movements occur independently of ATM signaling. Sertoli cells in the *ATM* mutants displayed numerous heterochromatin and telomere clusters. Thus, ATM appears to be removed from spermatid nuclei prior to the occurrence of DNA nicks, which emanate as a consequence of nucleoprotamine formation (Scherthan et al., 2000). Putative DNA damage checkpoint proteins, ATR, ATM and RAD1, are not associated with RAD51/DMC1 recombination sites where DNA breaks are expected to be present (Moens et al., 1999).

ATM and Telomeres: Because of the homology of the human *ATM* gene to the *TEL1* and *rad3* genes of yeast, it has been suggested that mutations in ATM could lead to defective telomere maintenance. The *ATM* gene product, which is defective in the cancer prone disorder AT, influences chromosome end associations and telomere length. The possible hypothesis explaining these results suggests that the defective telomere metabolism in AT cells is due to altered interactions between telomerase and the nuclear matrix. The *ATM* gene influences the interactions between telomeres and the nuclear matrix and that alternations in telomere chromatin could be at least partly responsible for the pleiotropic of the *ATM.* Influence of inactivation of ATM on the interaction of telomeres with nuclear matrix in somatic and germ cells has been summarized (Pandita and Dhar, 2000).

ATM and Chk 1: Checkpoint pathways prevent cell-cycle progression in the event of DNA lesions. The genetic and regulatory interaction between *Atm* and mammalian Chk1 appears to be important for integrating DNA damage repair with cell-cycle arrest. Homologues of yeast Chk1 in human and mouse Chk1 (Hu/Mo) have kinase activity and are expressed in the testis. The Chk1 accumulates in late zygotene and pachytene spermatocytes and is present along synapsed meiotic chromosomes. The Chk1 protein levels and its association with meiotic chromosome depend upon a functional ATM gene product, but Chk1 is not dependent upon p53 for meiosis I functions. Human CHK1 to human chromosomes gene is located to 11q22-23. The ATM dependent presence of *Chk1* in mouse cell and along meiotic chromosomes, and the late pachynema co-localization of *ATR* and *Chk1* on the unsynapsed axes of the paired X and Y-chromosomes, suggested that *Chk1* acts as an integrator of *ATM* and *ATR* signals (Flaggs et al., 1997) (see Chapter 19).

13.5. CRE RECOMBINASE

Specific DNA rearrangements can be generated by site-directed recombination using recombinases such as Cre from bacteriophage P1. Targeting of Cre expression to testicular germ cell would provide a way to generate mutations at predetermined loci during meiosis, and thus, to define the role of spermatogenesis of any gene of interest. A second expected outcome from such studies is the possibility to gather information on the mechanism of the chromosome breaking and joining events in genetic recombination. Finally, exchanges between predetermined sites could be used for large-scale remodeling of the gene. Transgenic mice have been generated expressing a testicular Cre-recombinase driven by promoter sequences derived from the gene encoding SCP1, expressed at an early stage of the male meiosis. Recombination at target *LoxP* sites was examined during germinal differentiation in mice harboring Scp1-Cre and a second transgene where *LoxP* sites flank either the β*geo* coding region, the *Pgk1* promoter, or a *tk-neo* cassette inserted into the *Rxr*α-locus. Another study showed transgenic line possessing Pgk-2-driven expression of the Cre-recombinase for identifying spermatogenic genes that function at or after the spermatocyte stage (Ando et al, 2000). Weber et al., (2003) have obtained a PrP-Cre-ER(T) transgenic mouse line (28.8) that selectively expresses the tamoxifen-inducible Cre- R(T) recombinase in testis under the control of a mouse Prion protein (PrP) promoter-containing genomic fragment. Cre-ER(T) is expressed in spermatogonia and spermatocytes, but not in Sertoli and Leydig cells (Weber et al., 2003).

13.6. REFERENCES

Ando H, Haruna Y, Miyazaki J-I, Okabe M, Nakanishi Y. Speramtocyte-specific gene excision by targeted expression of Cre recombinase. Biochem Biophy Res Comm 2000; 272: 125-28.

Arnheim N, Shibata D. DNA mismatch repair in mammals. Role in disease and meiosis. Curr Opin Genet Develop 1997; 7: 364-70.

Ashley T, Plug AW, Xu J, Solari AJ, Reddy G, Golub EI, Ward DC. Dynamic changes in Rad51 distribution on chromatin during meiosis in male and female vertebrates. Chromosoma 1995; 104: 19-28.

Baarends WM, van der Laan R. Grootegoed JA. DNA repair mechanisms and gametogenesis. Reproduction 2001; 121: 31-39.

Baarends WM, Wassenaar E, Hoogerbrugge JW, van Cappellen G, Roest HP, Vreeburg J, et al. Loss of HR6B ubiquitin-conjugating activity results in damaged synaptonemal complex structure and increased crossing-over frequency during the male meiotic prophase. Mol Cell Biol 2003; 23: 1151-62.

Bailly V, Parkash S, Parkash L. Domains required for dimerizaion of yeast rad6 ubiquitin-conjugaing enzme and Rad18 DNA binding protein. Mol Cell Biol 1997; 17: 4536-43.

Baker SM, Bronner CE, Zhang L, Plug AW, Robatzek M, Warren G, Elliott EA, Yu J, Ashley T, Amheim N, et al. Male mice defective in the DNA mismatch repair gene PMS2 exhibit abnormal chromosome synapsis in meiosis. Cell 1995; 82: 309-19.

Baker SM, Plug AW, Prolla TA, Bronner CE, Harris AC, et al. Involvement of mouse Mih1 in DNA mismatch repair and meiotic crossing over. Nat Genet 1996; 13: 336-42.

Barlow AL, Benson FE, SC West, Hulten MA. Distribution of the Rad51 recombinase in human and mouse spermatocytes. The EMBO Journal 1997a; 17: 5207-15.

Barlow C, Liyanage M, Moens PB, Deng CX, Ried T, Wynshaw-Boris A. Partial rescue of the prophase I defects of Atm-deficient mice by p53 and p21 null alleles. Nature Genetics 1997b; 17: 462-66.

Barlow C, Liyanage M, Moens PB, Tarsounas M, Nagashima K, Brown K. Rottinghaus S, Jackson SP, Tagle D, Ried T, Wynshaw-Boris. A Atm deficiency results in severe meiotic disruption as early as leptonema of prophase I Development 1998; 125: 4007-17.

Baumann P, Benson FE, West SC. Human Rad51 protein promotes ATP-dependent homologous pairing and strand transfer reactions in vitro. Cell 1996; 87: 757-66.

Brookman KW, Lamerdin JE, Thelen MP, Hwang M, Reardon JT, et al. ERCC4 (XPF) encodes a human nucleotide excision repair protein with eukaryotic recombination homologs. Mol Cell Biol 1996; 16: 6553-62.

Caldecott KW, McKeown CK, Tucker JD, Liungquist S, Thompson LH. An interaction between the mammalian DNA repair protein XRCC1 and DNA ligase III. Mol Cell Biol 1994; 68-76.

Cartwright R, Dunn AM, Simpson PJ, Tambini CE, Thacker J. Isolation of novel human and mouse genes of the recA/RAD51 recombination repair gene family. Nucleic Acids Res 1998a; 26: 1653-9.

Cartwright R, Tambini CE, Simpson PJ, Thacker J. The XRCC2 DNA repair gene from human and mouse encodes a novel member of the rec/RAD51 family. Nucleic Acids Res 1998b; 26: 3084-9.

de Vries SS, Baart EB, Dekker M, Siezen A, de Rooij DG, de Boer P, te Riele H. Mouse MutS-like protein Msh5 is required for proper chromosome synapsis in male and female meiosis. Genes Develop 1999; 13: 523-31.

de Wind N, Dekker M, Claij N, JansenL, van Klink Y, Radman M, Riggins G, van der Valk M, van't Wout K, te Riele H. HNPCC-like cancer predisposition in mice through simultaneous loss of Msh3 and Msh6 mismatch-repair protein functions Nature Genetics 1999; 23: 359-62.

Dosanjh MK, Collins DW, Fan W, Lennon GG, Albala JS, Shen Z, Schild D. Isolation and characterization of RAD51C, a new human member of the RAD51 family of related genes. Nucleic Acids Res 1998; 26: 1179-84.

Edelmann W, Cohen P, Kane M, Lau K, Morrow B, Benett S, Umar A. Kunkel T, Cattoretti G, Chaganti R, Pollard JW, Kolodner RD, Kucherlapati R. Meiotic pachytene arrest in MLH1-deficient mice. Cell 1996; 85: 1125-34.

Edelmann W, Cohen PE, Kneitz B, Winand N, Lia M, Heyer J, Kolodner R, Pollard JW, Kuchherlapati R. Mammalian MutS homologue 5 is required for chromosome pairing in meiosis. Nature Genet 1999; 21: 123-27.

Flaggs G, Plug AW, Dunks KM, Mundt KE, Ford JC, Quiggle Mr, Taylor EM. Atm-dependent interactions of a mammalians Chk1 homolog with meiotic chromosomes. Curr Biol 1997; 1; 977-86.

Freire R, Murguia JR, Tarsounas M, Lowndes NF, Moens PB, Jackson SP. Human and mouse homologs of Schizosaccharomyces prombe rad1 (+) and Saccharyomyces cerevisisae RAD17: linkage to checkpoint control and mammalian meiosis. Genes Dev 1998; 12: 2560-73.

Friedberg EC, Gerlach VL. Novel DNA polymerases offer clues to the molecular basis of mutagenesis. Cell 1999; 98: 413-16.

Geeta Vani R, Rao MRS. Cloning of the cDNA encoding rat homologue of the mismatch repair gene MSH2 and its expression during spermatogenesis. Gene 1997; 185: 19-26.

Geeta Vani R, Varghese CM, Rao MR. Cloning of rat MLH1 and expression anaylsis of MSH2, MSH3, MSH6, and MLH1during spermatogenesis. Genomics 1999; 62: 460-7.

Grootegoed JA, Baarends WM, Roest HP, Hoeijmakers JHJ. Knockout mouse model and gametogenic failure. Mol Cell Endocrinol 1998; 145: 161-66.

Haff T, Golub EI, Reddy G, Radding GM, Ward DC. Nuclear foci of mammalian Rad51 recombination protein in somatic cells after DNA damage and its localization in synaptonemal complexes. Proc Natl Acad Sci USA 1995; 92: 2298-302.

Habu T, Taki T, West A, Nishimune Y, Morita T. The mouse and human homologs of DMC1, the yeast meiosis-specific homologous recombination gene, have a common unique form of exon-skipped transcript in meiosis. Nucleic Acids Res 1996; 24: 470-7.

Her C, Doggett NA. Cloning, structural characterization, and chromosomal localization of the human orthologue of Saccharomyces cerevisiae MSH5 gene. Genomics 1998; 52: 50-61.

Ikeya T, Shinohara A, Sato S, Tabata S, Ogawa T. Localization of mouse Rad51 and Lim15 proteins on meiotic chromosomes at late stages of prophase 1. Genes Cells 1996; 1: 379-89.

Intano GW, McMahan CA, Walter RB, McCarrey JR, Walter CA. Mixed spermatogenic germ cell nuclear extracts exhibit high base excision repair activity. Nucleic Acids Res 2001; 29: 1366-72.

Jansen J, Olsen AK, Wiger R, Naegeli H, de Boer P, van Der Hoeven F, Hoime JA, Brunborg G, Mullenders L. Nucleotide excision repair in rat male germ cells: low level of repair in intact cells contrasts with high dual incision activity in vitro. Nucleci Acids Res 2001; 29: 1791-800.

Kawaabata M, Saeki K. Multiple alternative transcripts of the human homologue of the mouse TRAD/R51H3/RD51D gene, a member of the rec A/RAD51 gene family. Biochem Biophys Res Commun 1999; 257: 156-62.

Keegan KS, Holtzman DA, Plug AW, Christenson ER, Brainerd EE, Flaggs G, Bentley NJ, et al. The Atr and Atm protein kinases associate with different sites along meiotically pairing chromosomes. Genes Dev 1996; 10: 2423-37.

Kneitz B, Cohen PE, Avdievich E, Zhu L, Kane MF, Hou H Jr, Kolodner RD, et al. MutS homolog 4 localization to meiotic chromosomes is required for chromosome pairing during meiosis in male and female mice. Genes Develop 2000; 14: 1085-97.

Koken MH, Hoogerbrugge JW, Jasper-Dekker I, de Wet J, Willemsen R, et al. Expression of the ubiquitin-conjugating DNA repair enzymes HHR6A and B suggests a role in spermatogenesis and chromatin modification. Dev Biol 1996; 173: 119-32.

Kovalenko OV, Golub EI, Bray-Ward P, Ward DC, Radding. A novel nucleic acid binding protein that interacts with human rad51 recombinase. Nucleic Acids Res 1997; 25: 4946-53.

Leasure CS, Chandler J, Gilbert DJ, Householder DB, Stephens R, Copeland NG, Jenkins NA, Sharan SK. Sequence chromosomal location and expression analysis of the murine homologue of the RAD51I2/RAD51C. Gene 2001; 271: 59-67.

Lee BI, Sannon M, Stubbs L, Wilson DM. Expression specificity of the mouse exonuclease I (mExo1) gene. Nucleic Acids Res 1999; 27: 4114-20.

Lipkin SM, Moens PB, Wang V, Lenzi M, Shanmugarajah D, Gilgeous A, Thomas J,Cheng J,Touchman JW, Green ED, Schwartzberg P, Collins FS, Cohen PE. Meiotic arrest and aneuploidy in MLH3-deficient mice. Nat Genet 2002; 31: 385-90.

Loseva EF, Kalnin VV, IL inskaia LE. Cloning of a new murine gene coding for a protein immunologically related to RecA protein from Escherichia Coli. Genetika 2000: 36: 613-21.

Maduro MR, Casella R, Kim E, Levy N, Niederberger C, Lipshultz LI, Lamb DJ.Microsatellite instability and defects in mismatch repair proteins: a new aetiology for Sertoli cell-only syndrome. Mol Hum Reprod 2003; 9: 61-8.

Masson JY, Davies AA, Hajibagheri N, Van Dyck E, Benson FE, Stasiak AZ, Stasiak A, West SC. The meiosis specific recombinase hDmc1 forms ring structures A and interacts with hRad51. The EMBO Journal 1999; 18: 6552-60.

Masson JY, Tarsounas MC, Stasiak AZ, Stasiak A, Shah R, McIlwraith MJ, Benson FE, West SC. Identification and purification of two distinct complexes containing the five RAD51 paralogs. Genes Dev 2001; 15: 3296-307.

McDonald JP, Rapic-Otrin V, Epstein JA, Broughton BC, Wang X, Lehmann AR, Wolgemuth DJ, Woodgate

R. Novel human and mouse homologs of *Saccharomyces Cerevisiae* DNA polymerase eta. Genomics1999; 60: 20-30.

Moens B, Tarsounas M, Morita T, Habu T, Rottinghaus ST, Freire R, Jackson SP, Barlow C, Wynshaw-Boris A. The association of ATR protein with mouse meiotic chromosome cores. Chromosoma 1999; 108: 95-102.

Moens PB, Chen DJ, Shen Z, Kolas N, Tarsounas M, Heng HHQ, Spyropoulos B. Rad51 immunocytology in rat and mouse spermatocytes and oocytes. Chromosoma 1997; 106: 207-15.

Moens PB, Freire R, Tarsounas M, Spyropoulos B, Jackson SP. Expression and nuclear localization of BLM, a chromosome stability protein mutated in Bloom's syndrome, suggest a role in recombination during meiotic prophase. J Cell Sci 2000; 113 (pt4): 663-72.

Muris DF, Bezzubova O, Buerstedde JM, Vreeken K, Balajee AS, Osgood CJ, et al. Cloning of human and mouse genes homologous to RAD52, a yeast gene involved in DNA repair and recombination. Mutat Res 1994; 315: 295-305.

Nitta M, Saijo M, Kodo N, Matsuda T, Nakatsu Y, Tamai H, Tanaka K. A novel cytoplasmic GTPase XAB1 interacts with DNA repair protein XPA. Nucleic Acids Res 2000; 28: 4212-8.

Ogi T, Kato T Jr, Kato T, Ohmori H. Mutation enhancement by DINB1, a mammalian homologue of the Escherichia coli mutgenesis protein dinB. Genes Cell 1999; 4: 607-18.

Ohinata Y, Sutou S, Mitsui Y. A novel testis-specific RAG2-like protein, Peas: its expression in pachytene spermatocyte cytoplasm and meiotic chromatin. FEBS Lett 2003; 537: 1-5.

Pandita TK, Dhar S. Influence of ATM function on interactions between telomeres and nuclear matrix. Radiat Res 2000; 154: 133-9.

Paquis-flucklinger V, Santucci-Darmanin S, Paul R, Saunieres A, Turc-Carel C. Cloning and expression anaylsis of a meiosis-specific MutS homolog: the human MSH4 gene. Genomics 1997; 44: 188-94.

Pittman DL, Weinberg LR, Schimenti JC. Identification, characterization, and genetic mapping of Rad51d, a new mouse and human RAD51/RecA related gene. Genomics 1998; 49: 103-11.

Richardson LL, Pedigo C, Ann Handel M. Expression of deoxyribonucleic acid repair enzymes during spermatogenesis in mice. Biol Reprod 2000; 62: 789-96.

Roest HP, Klaveren JV, Wit JD, Van Gurp CG, et al. Inactivation of the HR6B ubiquitin-conjugating DNA repair enzyme in mice causes male sterility associated with chromatin modification. Cell 1996; 86: 799-810.

Santucci-Darmanin S, Neyton S, Lespinasse F, Saunieres A, Gaudray P, Paquis-Flucklinger V. The DNA mismatch-repair MLH3 protein interacts with MSH4 in meiotic cells, supporting a role for this MutL homolog in mammalian meiotic recombination. Hum Mol Genet 2002 ;11: 1697-706.

Santucci-Darmanin S, Vidal F, Scimeca JC, Turc-Carel C, Paquis-Flucklinger V. Family of SRY/Sox proteins is involved in the regulation of the mouse Msh4 (MutS Homolog 4) gene expression. Mol Reprod Dev 2001; 60: 172-80.

Sato S, Kobayashi T, Hotta Y, Tabata S. Characterization of a mouse recA-like gene specifically expressed in testis. DNA Res 1995a; 2: 147-50.

Sato S, Seki N, Hotta Y, Tabata S. Expression profiles of a human gene identified as a structural homologue of meiosis specific recA-like genes. DNA Res 1995b; 2: 183-6.

Scherthan H, Jerratsch M, Dhar S, Wang YA, Goff SP, Pandita TK. Meiotic telomere distribution and Sertoli cell nuclear architecture are altered in Atm- and Atm-p53-deficient mice. Mol Cell Biol 2000;20:7773-83.

Schmuttte C, Marinescu RC, Sadoff MM, Guerrette S, Overhauser J, Fishel R. Human exonuclease I interacts with the mismatch repair protein hMSH2. Cancer Res 1998; 58: 4537-42.

Shannon M, Lamerdin JE, Richardson L, McCutchen-Maloney SL, Hwang MH, Handel MA, Stubbs L, Thelen MP. Characterization of the mouse Xpf repair gene and differential expression during spermatogenesis. Genomics 1999; 62: 427-35.

Shen Z, Pardington-Purtymun PE, Comeaux JC, Moyzis RK, Chen DJ. UBL1 a human ubiquitin like protein associating with human RAD51/RAD52 proteins. Genomics 1996; 36: 271-9.

Tarsounas M, Morita T, Pearlman RE, Moens B. RAD51 and DMC1 form mixed complexes associated with mouse meiotic chromosome cores and synaptonemal complexes. J. Cell Biol 1999; 147: 207-19.

Tateishi S, Sakuraba Y, Masuyama S, Inoue H, Yamaizumi M. Dysfunction of human Rad18 results in defective postreplication repair and hypersensitivity to multiple mutagens. Proc Natl Acad Sci (U S A) 2000; 97: 7927-32.

Thompson LH, Brookman KW, Jones NJ, Allen SA, Carrano AV. Molecular cloning of the human XRCC1 gene, which corrects, defective DNA strand break repair and sister chromatid exchange. Mol Cell Biol 1990; 10: 6160-71.

Tishkoff DX, Amin NS, Viars CS, Arden KC, Kolodner RD. Identification of a human gene encoding a homologue of Saccharomyces cerevisiae EXO1, an exonuclease implicated in mismatch repair and recombination. Cancer Res 1998; 58: 5027-31.

van der Laan R, Roest HP, Hoogerbrugge JW, Smit EM, Slater R, Baarends WM, Hoeijmakers JH, Grootegoed JA. Characterization of mRAD 18Sc, a mouse homolog of the yeast postreplication repair gene RAD18. Genomics 2000; 69: 86-94.

van der spek PJ, Visser CE, Hanaoka F, Smit B, Hagemeijer A, Bootsma D, Hoeijmakers JH. Cloning, comparative mapping and RNA expression of the mouse homologues of the Saccharomyces cerevisisae nucleotide excision repair gene RAD23. Genomics 1996; 31: 20-7.

Vidal F, Sage J, Cuzin F, Rassoulzadegan M. Cre expression in primary spermatocytes : a tool for genetic engineering of the germ line. Mol Reprod Dev 1998; 51: 274-80.

Wagner K, Mincheva A, Korn B, Lichter P, Popperl H. Pbx4, a new Pbx family member on mouse chromosome 8, is expressed during spermatogenesis. Mech Dev 2001; 103: 127-33.

Walter CA, Trolian DA, McFarland MB, Street KA, Gurram GR, McCarrey JR. Xrcc-1 expression during male meiosis in the mouse. Biol Reprod 1996; 55; 630-5.

Wang TF, Kleckner N, Hunter N. Functional specificity of MutL homologs in yeast: evidence for three Mlh1-based heterocomplexes with distint roles during meiosis in recombination and mismatch correction. Proc Natl Acad Sci USA 1999; 96: 13914-19.

Weber P, Schuler M, Gerard C, Mark M, Metzger D, Chambon P. Temporally controlled site-specific mutagenesis in the germ cell lineage of the mouse testis. Biol Reprod. 2003; 68: 553-9.

Wilson TM, Rivkees SA, Deutsch WA, Kelley MR. Diferential expression of the apurinic 1 apyrimidinic endonuclease (APE/ref-1) multifunctional DNA base excision repair gene during fetal development and in adult rat brain and testis. Mutat Res 1996; 1996; 362: 237-48.

Xin H, Lin W, Sumanasekera W, Zhang Y, Wu X, Wang Z. The human RAD18 gene product interacts with HHR6A and HHR6B. Nucleic Acids Res 2000; 28: 2847-54.

Yamamoto A, Taki T, Yagi H, Habu T, Yoshida K, Yoshimura Y, Yamamoto K, Matsushiro A, Nishimune Y, Morita T. Cell cycle-dependent expression of the mouse Rad51 gene in proliferating cells. Mol Genet 1996; 25: 1-12.

Yoshida K, Kondoh G, Matsuda Y, Habu T, Nishimune Y, Morita T. The mouse RecA- like gene Dmc1 is required for homologus chromosome synapsis during meiosis. Mol Cell 1998; 1: 707-18.

Zhou ZQ, Walter CA. Cloning and characterization of the promoter of Baboon XRCC1, a gene involved in DNA strand-break repair. Somat Cell Mol Genet 1998; 24: 23-39. Arnheim N, Shibata D. DNA mismatch repair in mammals. Role in disease and meiosis. Curr Opin Genet Develop 1997; 7: 364-70.

Chapter 14

TRANSCRIPTIONAL CONTROL

14.1. TRANSCRIPTION IN EUKARYOTES

During the last two decades, considerable information has been gathered on the regulation of gene expression in eukaryotes. These genes express in specialized cells, and are variously described as luxury genes or smart genes as opposed to house keeping genes, which are constitutive in their expression. A large number of transcription factors, which take part in the formation of a transcription complex, viz., RNA polymerase and associated proteins to initiate transcription, have been identified. Specific regulatory DNA sequences, which respond to stimuli and provide sites for binding of transcription factors, have also been known. Specific signal transduction pathways (involving intermediate receptor molecules) have been discovered in many cases. In some cases, regulation of gene expression is exercised at the post-transcriptional level involving RNA processing or translation.

Britten-Davidson model or the gene battery model is one of the first models, which was proposed for the regulation of protein synthesis in eukaryotes (Davidson, 1965). The model assumes the presence of four classes of sequences: (i) *Producer gene,* comparable to a structural gene in prokaryotic operation, (ii) *Receptor* site which is comparable to operator gene in bacterial operon and one such receptor site is assumed to be present adjacent to each producer gene, (iii) *Integrator* gene which is comparable to regulator gene and is responsible for synthesis of an activator RNA that may or may not give rise to proteins before it activates the receptor site and, (iv) a *Sensor* site that regulates the activity of the integrator gene, which can be transcribed only when the sensor site is activated. These sites are recognized by agents such as hormones and proteins, which change the pattern of gene expression. For example hormone protein complex or a transcription factor may bind to a sensor site and cause the transcription of integrator. In this model, the producer gene and integrator gene are involved in RNA synthesis. On the other hand, receptor and sensor sites are those sequences, which help only in recognition without taking part in RNA synthesis. It is also proposed that receptor sites and integrator genes may be repeated a number of times, so as to control the activity of a large number of genes in the same cell. Repetition of receptor ensures that same activator recognizes all of them, and several enzymes of one pathway are simultaneously synthesized, or the transcription of same gene be needed in different developmental stages. This is achieved by multiplicity of receptor sites and integrator genes. Each producer gene may have several receptor sites, each responding to one activator so that a single activator though can recognize several genes, but different activators may activate the same gene at different times. An integrator gene may also fall in cluster with same sensor site.

In Britten-Davidson model, a set of structural genes controlled by one sensor site is called a battery. Sometimes, when major changes are needed, it is necessary to activate several sets of genes. If one sensor site is associated with several integrators, it will cause transcription of all integrators at the same time thus causing transcription of several producer genes through receptor sites. The repetition of integrator and receptor sequences is consistent with the knowledge that sufficient repeated DNA sequences occur in eukaryotes.

14.2. CHAUVINIST GENES

The gene expression in various phases of spermatogenesis is the result of multiple processes. Genes expressed during spermatogenesis encode proteins necessary both for house keeping activities as well as for processes, which are specific to germ cells. Out of several thousands of transcripts present in a cell, only a few of these genes have been identified in germ cells. Products of some of these genes are essential for structures and functions specific to spermatogenic cells, where as products of others are expressed in developmentally regulated patterns and finally are transcribed only in spermatogenic cells. The first 359 cDNA clones from 180-kb DNA sequence of human testis indicated that these sequences could be sorted into several categories with a high proportion of (67%) being novel. The pattern of gene sequences in testicular transcripts appeared to be different in comparison to similar studies on human brain (Affara et al, 1994). These characteristic genes have been introduced as chauvinist genes, because of their favored expression by spermatogenic cells with out fail. Nearly 100 chauvinist genes are known that are expressed during the meiotic phase alone (review Eddy, 1998) and several have been added during recent years. Chauvinist gene products can be classified into three general categories. i) Those homologous proteins, which are expressed only in spermatogenic cells, but closely related to proteins expressed in somatic cells. Such proteins include histone H1t, HSP70-2, calreticulin$_t$ cytochrome C and Ldh-C. ii) Another category of gene products are those, which are unique to spermatogenic cells with no significant overall similarity to any protein expressed in any other cell type. Synaptonemal complex protein1 (SCP1), Sry protein encoded on Y chromosome, transition proteins 1 and 2 (TP1 and TP2) and Prms (Prm 1 and 2) are unique to male germ cells. Such proteins are not significantly homologous to any known protein in an organism. iii) In addition, some genes in spermatogenic cells can produce specific transcripts that differ from those produced in somatic cells because of presence of alternative transcription start sites, polyadenylation signals, or they are produced from spermatogenic cell-specific exons. Examples include angiotensin converting enzyme, hexokinase-1, and gene products required in meiotic recombination (Eddy, 1998). Many of these gene products are regulated developmentally, and form unique structural components of spermatogenic cells, such as the synaptonemal complex, sperm acrosomal and flagellar proteins as well as for unique functional processes, such as meiotic recombination and transcriptional regulation. Several unique components of signal transduction pathways, as well as chaperone proteins that assist other proteins during their folding and assembly into functional complexes are the products of chauvinist genes. Moreover, genes expressed in somatic cells are often down-regulated in germ cells, whereas chauvinist genes that encode highly similar proteins with apparently comparable functions are up-regulated in germ cells. These features are important for gametogenesis, and for the survival of species through reproduction. The autosomal Pgk-2 gene product is an example of a tissue specific gene that is activated at a specific stage of spermatogenesis and coincident with the entry of spermatocytes into meiotic prophase. The X-linked structural genes are other examples of genes that are linked to X-chromosome inactivation during spermatogenesis.

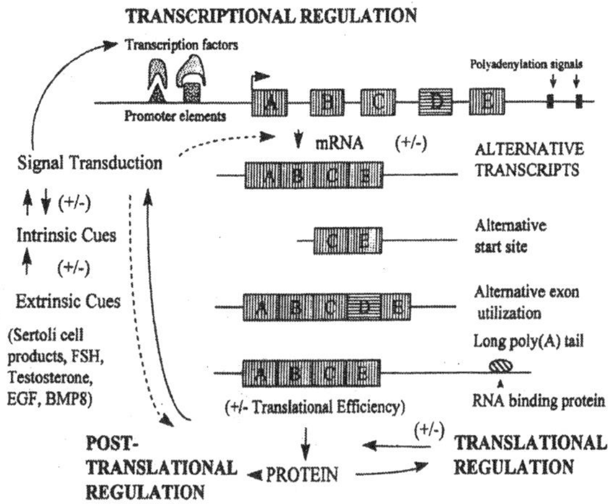

Fig.14.1. Regulation of gene expression during spermatogenesis at transcription, translation and post-translational levels. i) Transcription is regulated by intrinsic genetic program, acting through signal transduction processes. The intrinsic program is perhaps influenced by extrinsic cues. ii) Alternative transcripts are relatively common in spermatogenic cells. The methods, how these alternative transcripts are produced, are shown in the figure. iii) Translational regulation results in mRNAs storages, which are utilized for proteins, needed after transcription ends in condensing spermatids. iv) Post-translational regulation modulated through signal transduction mechanisms has a significant role in influencing these processes. Reprinted with permission from E. M. Eddy. Seminars in Cell Develop Biol 9; 451-57: 1998 © Elsevier.

14.3. REGULATORY FACTORS IN GENE EXPRESSION OF GERM CELLS

14.3. 1. Intrinsic and Extrinsic Factors

From developmental studies, the stage specific requirement of various proteins during spermatogenesis, and the conserved nature of the process, it seems that the primary regulator of spermatogenesis is an intrinsic genetic program, which is responsible for germ cell specific gene expression. The primary regulator of spermatogenesis as an intrinsic genetic program became more evident when rat spermatogonial stem cells, transplanted to the mouse testis, were able to form morphologically normal rat sperm (Clauther et al., 1996). The intrinsic regulation of gene expression in spermatogenesis occurs at three levels: transcription, translation and post-translation (see Chapters 14 -15). Transcriptional regulation is the primary determinant of gene expression. However, translational regulation has a greater role in germ cells than in other cell types. Post-translational regulation occurs through modification of proteins, perhaps in response to extrinsic and intrinsic cues, to effect transcription and translation during spermatogenesis **(Fig.14.1)**.

Besides intrinsic factors, spermatogenesis is subjected also to secondary regulation by endocrine factors transmitted through Sertoli cells, several growth factors, or by local short-loop paracrine and autocrine signals. How these extrinstic cues influence germ cell development is not understood, but they do so through their effects on converging signal transduction pathways that modulate the primary effects of the intrinsic genetic program. These pathways probably effect gene expression by causing post-translational modifications of transcription factors and other proteins that regulate alternative transcript processing mechanisms that occur in spermatogenic cells. Various signal transduction pathways have been identified in spermatogenetic cells (Chapters 2-4). In addition, parameters such as chromatin structure, association with the nuclear matrix, timing of replication during S phase, and DNA methylation have all been suggested as additional mechanisms operative within a mammalian cell to "organize" the transcriptional activity of its genome (McCarry, 1998).

14.3.2. Chromatin Structure

The first tier of control of over-expression of genetic domains utilizes chromatin structure. Hence, the structure of chromatin must undergo dramatic changes, ranging between structural conformations that enhance the ability of promoter sequences to bind transcription factors, and other conformations that inhibit access of transcription factors. The inhibitory state is accomplished by supercoiling the chromatin into a highly compact 'solenoid' like structure within a diameter of $\simeq$30 nm. Thus, prior to initiation of transcription of a tissue specific gene, the 'closed' compact structure must become unwound yielding a chromatin fibre of 10nm diameter. The series of events involved in derepression of a gene prior to actual initiation of transcription constitute a process that has been termed 'potentiation' (Kramer et al., 2000). Potentiation results in the exposure of a region in the promoter that is hypersensitive to digestion by DNase I. These regions in chromatin are known as 'DNase I hypersensitive sites' and are accessible as open configuration for interaction of transcription factors. The gene potentiation has been examined and established showing that chromatin domain functions as a discrete structural unit during differentiation. Besides potentiation, CpG islands are necessary for chromatin organization and expression of some of the genes during spermatogenesis. It is now known that changes occur in chromatin structure during the differentiation of a stem cell into a mature spermatozoan. In the spermiogenesis phase nucleosomal chromatin structure is converted into a fiber structure and ultimately into a highly condensed nucleoPrm filament. Chromatin condensation also depends on other factors viz poly-(ADP) ribosylation.

14.3.3. Role of Nuclear Matrix in Replication

The nuclear matrix is a three-dimensional structure within the eukaryotic nucleus with which many activator proteins are associated. Thus, proteins participating in DNA replication, transcription, and RNA splicing have been found associated with nuclear matrix. When a particular gene is in close proximity with the matrix, it is likely that such a gene may replicate earlier in S phase, since it is juxtaposed to replication factors and associated DNA polymerases embedded in the matrix.

14.3.4. Role of Chromosomal Proteins in Transcription

Besides DNA, the chromatin also consists of histones and non-histones. Although it is known since early 1960's that histones might be involved in repressing gene activity, the specific regulation by non-histones was demonstrated only during 1970s, when chromatin reconstitution experiments demonstrated that the mRNA, which is synthesized in vitro from reconstituted chromatin, mainly depended on the source of non-histone proteins. Certain non-histones are known to induce RNA synthesis in vitro. A large number of transcription factors are non-histone proteins. Observations also suggested that regulation of protein synthesis actually involves an interaction between histones and non-histones. Whereas histones inhibit protein synthesis non-specifically, non-histones induce RNA synthesis rather specifically. Some histone proteins may also function as transcription factors. In eukaryotes, including germ cells of testis, a number of DNA binding proteins were isolated from a variety of organisms and their corresponding binding regions on DNA were identified. Seven general transcription factors for RNA polymerase II (TFIID, TFIIA, TFIIB, TFIIE, TFIIF, TFIIJ, TFIIH), which are essential for transcription of all genes including house keeping genes and smart genes have been reported from various systems. However, for transcription of smart genes, additional transcription factors were found to form a complex to activate the essential proteins coupled to the RNA polymerase II that transcribes the gene into hnRNA. Only when all components are present, the transcription of smart genes will take place. Several hundred different transcription factors have been found to-date (Faisst and Meyer, 1992).

14.3.5. *In-situ* Modification of Nucleoproteins

Chemical alterations of proteins, such as histone by poly-(ADP)- ribosylation, acetylation and of transcriptional factors by phosphorylation, have significant influence on transcriptional activity of genes (see Sections 14.5, 14.6 and 14.7).

14.4. GENE METHYLATION AND GENE EXPRESSION IN SPERMATOGENESIS

14.4.1. DNA Methylation of Testis Specific Genes

An important parameter associated with the actively expressed genes is methylation/ demethylation state of DNA. Genomic methylation patterns originate during gametogenesis and are postulated to be involved in important developmental events, including gene regulation, embryogenesis, and genomic imprinting. In mammalian cells, only cytosines present in 5'-CpG-3' dinucleotides of DNA are found in a methylated state. It has been observed that the same cytosines in the same gene sequence can be methylated in one cell type and unmethylated in another cell type. Though the mechanism by which demethylation influences transcription is yet to be determined (see details further), evidences suggest that methylated DNA can be more readily 'packaged' into a stably repressed, closed chromatin structure. Methylated DNA binding proteins' that appear to be inhibitory to transcription have been shown to bind methylated DNA in a non-sequence specific manner and hence offer steric hindrance to transcription factors, which bind to DNA either directly, or indirectly by stabilizing an inaccessible chromatin configuration. Another possibility is that methylated DNA binding protein may control the association of DNA sequences with the nuclear matrix. Treatment of male rats with 5-azacytidine, a drug that blocks DNA methylation, resulted in abnormal embryo development when germ cells were exposed throughout spermatogenesis. Among spermatogenic cells, spermatogonia were more susceptible to the hypomethylating effects of 5-azacytidine than were spermatocytes (Doerkesen et al., 2000). Aoto et al (1995) identified a cDNA clone, named AZ1, from C3H10T1/2 cells that had been transiently exposed to 5-azacytidine. The transcript induced with 5-azacytidine treatments was highly expressed in mouse testis. The nucleotide sequence of AZ1 contained a 2841-nucleotide open reading frame, and the expected amino acid sequence had a molecular mass of 107254 Da. The protein was localized to the pre-acrosome region of round and elongated spermatids and was not detected in more advanced stage of spermatids (Aoto et al., 1995).

During meiotic phase of rat spermatogenesis, while transcriptional regulation of the TH2B gene appears to be different from that of somatic H2B gene, the sequence organization of TH2B regulatory region is remarkably similar to that of somatic H2B gene. Besides specific elements contained in their promoter, there is a strong correlation between DNA hypomethylation and the germ cell-specific expression of TH2B histone gene. Genomic sequencing revealed that all analyzed CpG sites in the promoter region of TH2B gene are methylated in both male and female somatic tissues, but not in testis, where as these CpG sites are unmethylated as early as spermatogonia type A and up to sperm during spermatogenesis. Evidencs are consistent with the hypothesis that DNA methylation inhibits gene activity by preventing the binding of transcription factors to their recognition sequences. For example, it was shown that (i) the binding of ubiquitous transcription factors to the promoter region of TH2B gene may be blocked in nuclei of liver, and (ii) DNA methylation can directly interfere with the binding of transcription factors recognizing a hexamer (ACGTCA) motif. DNA methylation in vitro showed that the expression of TH2B gene is inhibited by DNA methylation in vivo (Choi and Chae 1991). Methylated lysine 9 of histone H3 (Me9H3) is a marker of heterochromatin in divergent animal species. It localizes to both constitutive and facultative heterochromatin and replicates late in S-phase of the cell cycle. Significantly, Me9H3 is enriched in the inactive mammalian X chromosome (Xi) in female cells, as well as in the XY body during meiosis in the male, and forms a G-band pattern along the arms of the autosomes. The Me9H3 is neither necessary nor sufficient for localization of heterochromatin protein 1 (HP1) to chromosomal DNA (Cowell et al., 2002).

The genes encoding TP1, mouse Prm1 and Prm2, all of which are expressed postmeiotically, are marked by methylation during early spermatogenesis in the mouse. Whereas gene of TP1 became progressively less methylated during spermatogenesis, the Prm genes became progressively more methylated, in contrast to the methylation of β-actin gene that expressed throughout spermatogenesis. These findings provided evidence that both de novo methylation and demethylation events are occurring after the completion of DNA replication, during meiotic prophase in the testis (Choi et al.,1997; Trasler et al., 1990). A specific demethylation event has also been observed in the *Pgk-2* gene in pro-spermatogonia at about the time of birth, about 10 days before the onset of transcription, which first occurs in primary spermatocytes. *Pgk-2, ApoA1* and *Oct3-4* genes were unmethylated in adult spermatogenic cells in the testis, but were remethylated in mature spermatozoa in the vas deferens. Surprisingly, this type of remethylation is part of the process of sperm maturation, which occurs in the epididymis (Ariel et al., 1991; 1994). It is also interesting to note that the CpG islands are far more abundant in mammals. It remains to be seen if there is any correlation between the evolution of CpG islands and TP2 in mammals or it has any role in gene silencing (see review Kundo and Rao, 1999). During last few years, few more genes, which are expressed in germ cell specific manner and depend on demethylation, are added in the list.

14.4.2. CpG Islands and Spermatogenesis

CpG islands are stretches of DNA sequences in the size range of 0.5-2kb that are enriched in the $(CpG)_n$ repeat and are present in close association with housekeeping genes as well as some tissue-specific genes in the mammalian genome. CpG islands are the hot spots for chromatin remodeling, which in turn can regulate gene expression. How a CpG island can govern the chromatin organization and consequent gene expression is the subject of intensive research (Kundo and Rao, 1999). In addition to its role in modulating gene expression, methylation of CpG islands has also been implicated as one of the mechanisms underlying genomic imprinting in mammals. There seems to be an intimate relationship between the influence of CpG islands on chromatin organization and the histone acetylation of CpG island chromatin domain. CpG dinucleotides present in the human genome are most often replaced in mouse by

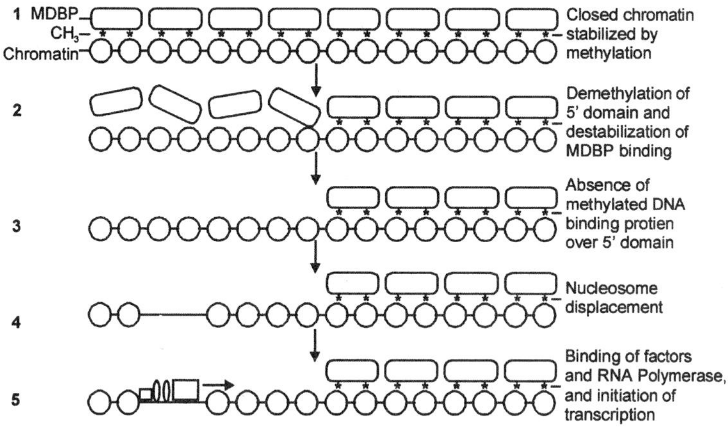

Fig.14.2. A model showing transcriptional activation of tissue specific gene. In non-expressing cells, a repressed gene is found in closed configuration characterized by: i) super-coiled chromatin, ii) hypermethylation (CH_3), and iii) association of methylated DNA-binding proteins (MDBP) (Stage 1). Prior to initiation of transcription, a part of gene gets demethylated due to dissociation of MDBP (Stage 2). In subsequent step(s), chromatin structure is unwound from solenoid structure, histone are acetylated (not shown) and nucleosome displacement occurs (Stage 3 and 4). Transcription by RNA polymerase ensues after binding of tissue specific and other ubiquitous transcription factors to their cognate sequences in enhancer and core promoter regions, respectively (Stage 5). Reprinted with permission from J.R. McCarrey. Seminars in Cell Develop Biol 9; 459-66: 1998 © Elsevier.

TpG or its complement, CpA, on account of mutation of the CpG to TpG. Although DNA methylation spread through the genome as vertebrates evolved, the promoters as well as the cis-acting regulatory elements were kept free from methylation by some unknown mechanism. The non-methylated CpG nucleotides along with an increased G+C content might have generated the present CpG islands. Methylation of genes is known to provide compact and condensed structure to chromatin. In humans, chromosomes 19 and 22 are highly enriched in CpG islands, while very few islands are present in chromosomes 18. Therefore, it is generally believed that the mammalian chromosomes are organized into domains with characteristic CpG island density. Cytosine in CpG islands is frequently (60-90%) methylated in vertebrate genomes. Methylation of DNA at cytosine in the CpG residues is used to explain genomic imprinting, which is defined by those genes whose expression is determined by their parental origin (Jaenisch, 1997). The CpG island associated genes have been shown to be hypermethylated in tumor cell lines and solid tumors. The methylation of CpG islands offers a putative mechanism for turning off those genes whose activities are not required during development of tumorigenesis. The molecular basis of transcriptional-repression by CpG-methylation is presently explained by two models. The so-called "direct mechanism" suggests that methylation of CpG islands may prevent the binding of transcriptional factors, although, several lines of evidences seem to contradict this model (**Fig.14.2**). The alternative model however, suggests that the transcriptional regulation could operate via specific binding of repressor(s) which recognize methyl CpG dinucleotides. Two methyl CpG binding repressors MeCP1 and MeCP2 are known. MeCP2 is present in all cell types except the germ cells and is found localized to the pericentromeric heterochromatin in mouse. MeCP2 has been shown to repress transcription from methylated promoters but not from unmethylated promoters (Kundo and Rao, 1999)(**Fig.14. 3**).

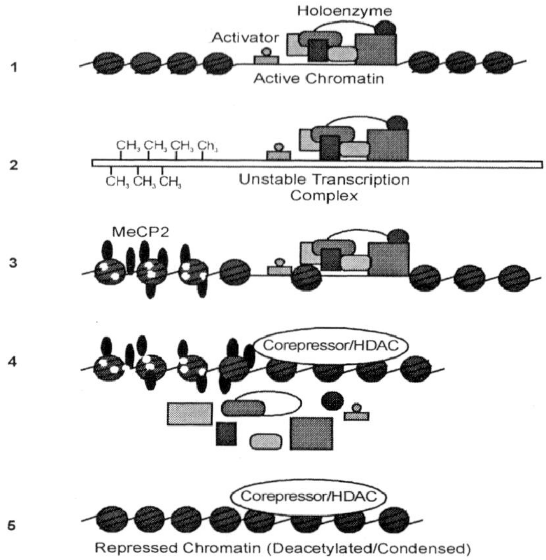

Fig.14.3. A CpG methylation dependent chromatin model showing inactivation of transcription. 1. Unmethylated CpG island containing active chromatin:activator and RNA polymerase holoenzyme are bound to the relatively nucleosome free region of the chromatin. 2. Partially methylated promoter with unstable transcription complex. 3. Upon assembly, the methylated sites of the promoter are recognized by methyl binding Cp repressor (MeCP2). 4. Binding of MeCP2 leads to recruitment of co-repressor and histone deacetylase complex (co-repressor/HDAC), displacing the activator and holoenzyme components. 5. Deacetylation of histones helps in chromatin condensation and transcription repression. Reprinted with permission from T. K. Kundu and M R S Rao. J Biochem (Japan) 125; 217-22: 1999 © The Japanese Biochemical Society.

14.4.3. Methyl-CpG-Binding Proteins

The MeCP2, MBD1, MBD2, MBD3, and MBD4 comprise a family of proteins that contain a methyl-CpG binding domain (MBD). Except for MBD4, these proteins are involved in gene silencing imposed by methylated DNA. A human gene codes for a protein that is 42% identical to MBD3 and 38% identical to MBD2 but lacks the methyl-CpG binding domain. The corresponding mouse *Mbd3L1* gene as counterparts of MBD2 and/or MBD3 was cloned from testis. During spermatogenesis, expression of MBD3L1 is observed only in round spermatids, suggesting a role for the gene product in the postmeiotic stages of male germ cell development. The MBD3L1 protein is localized to discrete areas in the nucleus and contains an N-terminal transcriptional repression domain. This repression is independent of histone deacetylase inhibition. A homologue of MBD3L1, MBD3L2, was also identified and cloned. Expression of MBD3L2 was found in germ cell tumors and some somatic tissues (Jiang et a., 2002).

14.4.4. DNA Methyltransferase

Methylation patterns are transmitted by clonal inheritance, through the strong preference of mammalian DNA (cytosine–5) methyltransferase (DNA MTase) for hemimethylated DNA. DNA MTase is present in the mouse testis and DNA MTase mRNA is most abundant in the testis, ovary, and spleen. A single DNA MTase transcript of 5.2-kb is present in all tissues, whereas an additional transcript of 6.2 kb was restricted to the testis. The predominant form of DNA MTase in mammals is Dnmt1. The mRNA for Dnmt1, the predominant maintenance and denovo DNA (cytosine – 5) methyl transferase in mammals, is present at high levels in postmitotic murine germ cells but undergoes alternative splicing of sex specific 5' exons, which control the production of enzyme during specific stages of gametogenesis. A spermatocyte specific 5' exon interferes with translation and prevents production of Dnmtl during the prolonged crossing-over stage of male meiosis (Mertineit et al., 1998; Trasler et al., 1992: Numata et al., 1994; Benoit and Trasler 1994). Dnmtl is subject to unusual transcriptional and post-transcriptional regulatory mechanisms in germ cells that depend on alternative splicing of sex specific 5' exons.

14.5. POLY-(ADP) - RIBOSYLATION

Poly-(ADP)-ribosylation of nuclear proteins has significant influence on gene expression and its regulation. Poly (ADP)-ribosylation (pADPR) of nuclear proteins increases several fold higher in the pachytene spermatocytes than in the premeiotic germ cells of the rat. Among the histones of the pachytene nucleus, histone subtypes H2A, H1 and H3 were poly-(ADP)-ribosylated. Approx. 2.5% of the solubilized pachytene chromatin is represented by the poly-(ADP)-ribosylated (PAC) domains. The DNA of pachytene PAC domains has internal strand breaks, significant length of gaps and ligatable ends, namely 5'phosphoryl and 3'-hydroxyl termini, as compared to very few gaps, in regenerating liver. A 20-kDa protein and the testis specific histone H1t were found selectively enriched in the pachytene PAC domains (Satyananarayana and Rao, 1989). Besides tissue-specific histone, H1t, core histones and three proteins (α, β and γ) with low mobility were the major acceptors of poly (ADP-ribose) in testis extracts. It shows that poly-(ADP)-ribosylation of nuclear proteins in rat testis is particularly active in the early stages of meiosis (Quesada et al., 1989). Two major enzymes participating in pADPR in spermatogenesis have been identified. They are: i) poly-(ADPR)polymerase and ii) poly(ADP)ribosyltransferase.

14.5.1. Poly-(ADPR) -Polymerase

Poly (ADPR) polymerase (PARP) is a chromatin-associated enzyme with a presumptive role in DNA repair during replication and recovery from strand breaks caused by genotoxic agents. PARP catalyzes the attachment and elongation of ADP ribose polymers (pADPR) to a variety of acceptor proteins (including PARP itself, and histones). Upon DNA damage PARP binds to DNA strand breaks and transfers ADP-ribose residues from NAD^+ to acceptor protein and to ADP-ribosyl protein adducts. This leads to branched polymers of protein-coupled poly-(ADP-ribose) (pADPR). Testis and-thymus shows highest activity of PARP. The degradation of the enzyme is also one of the classic indicators of apoptosis (Bantzer et al, 1999; Tramontano et al., 2000). Different stages of spermatogenesis revealed different subnuclear localization patterns of pADPR with significant increase in PARP. In pre-meiotic and post-meiotic cells, pADPR is localized in pattern overlapping with lamin and topoisomerase II at the nuclear rim. In primary spermatocytes pADPR is associated with three loci corresponding to the chromosomes at the nuclear periphery (Lankenau et al., 1999). The GC rich nature of its upstream gene promoter alongwith lack of TATA and CAAT boxes, a feature most common with house keeping genes is consistent with the major regulatory function played by positive transcription factor Sp1 in rat PARP gene transcription. Sp1 has been shown to interact with five distinct GC or GT boxes present in rat PARP promoter. However, lower expression of the enzyme in liver is controlled by another factor distinct from Sp1, that binds PARP promoter at the site overlapping F2Sp1 element (Laniel et al., 1997) and thus PARP belongs to CTF-NF1 family of transcription factors.

14.5.2. Poly(ADP-ribosyl)transferase (pADPRT)

Chromatin condensation also depends on other factors viz poly-(ADP) ribosylation. The enzyme participating in chromatin condensation, poly (ADP ribosyl) transferase (pADPRT) is positively correlated with degree of chromatin condensation with interchromatin spaces being virtually free of pADPRT. During spermatogenesis, poly-(ADP)ribosylation increases gradually that parallels chromatin condensation. The highest concentration was found in late stages of sperm differentiation (Mosgoeller et al., 1996). Human and murine mono (ADP) ribosyl transferases are testis specific. In human one specific gene (TART 1) and the other in murine (TART2) are testes specific (Braren et al., 1997). The predicted TART1 product contains hydrophobic N- and C- terminal signal peptides characteristic for GPI anchored surface proteins, where as TART2 contain 4 cysteine residues and also a glutamic acid conserved residue at active site (Braren et al., 1997).

14.6. ACETYLATION AND REGULATION OF PROTEIN FUNCTIONS

14.6.1. Histone Acetylation

An additional event associated with transcriptional activation of tissue specific genes is acetylation of the histones that are present in chromatin. Specifically, the lysine-rich amino-terminal tail of histone H4 becomes hyperacetylated that reduces its affinity for DNA. This further relaxes the constrained structure of chromatin to facilitate binding of transcription factors. Acetylation-dependent chromatin reorganization by a testis-specific bromodomain-containing protein (BRDT) has been demonstrated (Pivot-Pajot et al., 2003). Mouse spermatogonia and preleptotene spermatocytes contain acetylated core histones H2A, H2B and H4, whereas no acetylated histones were observed throughout meiosis in leptotene or pachytene spermatocytes. Histones remained unacetylated in most round spermatids. Acetylated forms of H2A and H2B, H3 and H4 reappeared in step 9 to 11 elongating spermatids, and disappeared later in condensing spermatids. This suggests that deacetylases are responsible for maintaining a deacetylated state of histones in these cells. TSA treatment could not induce histone acetylation in condensing spermatids, possibly suggesting that acetylated core histones are replaced by transition proteins. Therefore, the regulation of histone deacetylase activity / concentration appears to play a major role in controlling histone hyperacetylation and probably histone replacement during spermiogenesis (Hazzouri et al., 2000). The correlation between histone acetylation and increased transcription has been known for several years. The discovery of GCN5, the first nuclear acetylase and histone deacetylase-1 (HDAC1), the first deacetylase has verified that acetylation of histones is an important controlling step in transcription. However, the precise mechanism by which acetylation of histones augments transcription remains largely unknown. It appears that acetylation of histones involves a combination of events, which enchances the processivity of RNA polymerase II. Presently, we know more about acetylation of non-histone proteins than we know about histones. Hyperacetylated histones accumulate in actively transcribed chromatin, whereas hypoacetylated histones are the diagnostic features of repressed chromatin. Several initially identified

Fig.14.4. Regulation and action of acetylases (Ac). i) Regulation of acetylase activity by kinases involved in DNA repair (DNA-PK) and cell cycle progression. ii) Regulation of protein functions by acetylases. Reprinted with permission from T. Kouzarides. The EMBO Journal 19; 1176-79: 2000 © http:// www.nature.com/.

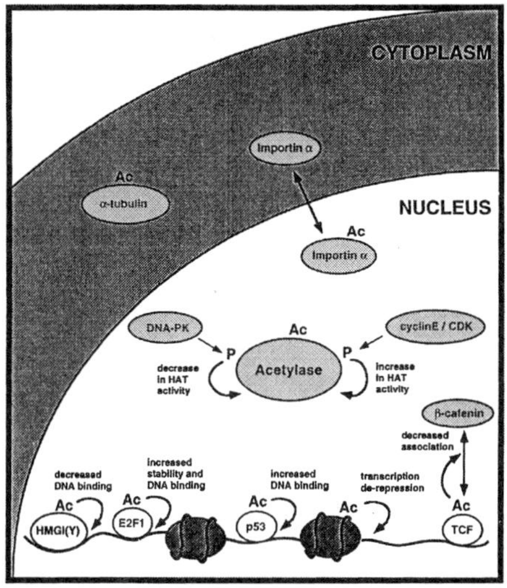

transcriptional regulators were shown to possess histone acetyltransferase (HAT) activity or seemed to be associated with histone deacetylase (HDAC). Role of binding of MeCP2 and histone deacetylation, involved in the methylation dependent alteration of chromatin structure is not known. But observations suggest that the gene regulation by histone acetylation may operate in both methylation dependent and independent manners (Kundo and Rao 1999; Struhl, 1998) (**Fig. 14.4**). Role of acetylases and deacetylases in testis germ cells and their isoform are yet to be investigated.

14.6.2. ESET Histone Methyltransferase

Covalent modifications of histone tails play important roles in gene transcription and silencing. An ERG (ets – related gene)-associated protein with a SET (suppressor of variegation, enhancer of zest and trithorax) domain (ESET) that was found to have the activity of a histone H3-specific methyltransferase has been identified. The ESET histone methyltransferase is associated with histone deacetylase-1 (HDAC1) and HDAC2, and that ESET interacts with the transcription co-repressors mSin3A and mSin3B. An N-terminal region containing a tudor domain is responsible for ESET interaction with mSin3A/B and association with HDAC1/2. The ESET represses the transcription when bound to a promoter gene. This repression by ESET is independent of its histone methyltransferase activity, but correlates with its binding to the mSin3 co-repressors. These data suggest that ESET histone methyltransferase can form a large, multi-protein complex with mSin3A/B co-repressors and HDAC1/2 that participates in multiple pathways of transcriptional repression (Yang et al., 2003).

14.6.3. Functional Significance of Acetylation

In the site-specific DNA binding transcription factors, p53, E2F1, EKLF and GATA1, the acetylation sites fall close to the DNA-binding domain and hence acetylation results in the stimulation of DNA binding. In contrast, the lysines acetylated within the HMG1 (Y), transcription factors fall within the DNA binding domain and result in disruption of DNA binding. Thus, the view that acetylaton is stimulatory for transcription (as with histones) is not true for transcription factors. Besides affecting DNA binding, acetylatin also regulates protein interactions. Acetylation also changes stability of proteins as in E2F1 (Martinez-Balbas et al., 2000; Hazzouri et al., 2000). The correlation has also been made between acetylated α-tubulin and microtubule stability (Chapter 8).

14.7. PHOSPHORYLATION / DEPHOSPHORYLATION

Proteins can be activated or inactivated by covalent phosphorylation and dephosphorylation respectively, where a protein can undergo two alternate conformational changes, i.e. between phosphorylated and non-phosphorylated states. Many protein kinases as well as protein phosphatases participate during phosphorylation and dephosphrylation reactions, in response to growth factor stimuli. Protein phosphorylation is involved in the control of various cell functions. To fulfill the different functions, it has been postulated that the human genome encodes ~2,000 different kinases, which participate in protein phosphorylation. Although several proteins kinases have been shown to be involved in spermatogenesis, only a few kinases have been characterized whose expression is restricted specifically to germ cells or to the testis. While phosphorylation by c-kit receptor tyrosine kinase is important in the survival of early spermatogonia, transcription factors, CREB and CREM are phosphorylated by a number of kinases in response to different stimulii. Phosphorylation of CREB at Ser-133 and CREM at Ser-119 is pre-requisite for transcriptional activation of several genes (see Chapter 16). Phosphorylation and dephosphorylation are important requirements for regulation of cell cycle during mitotic and meiotic phases of spermatogenesis (Chapter 11).

14.8. PROTEIN BINDING SITES IN DNA

In multicellular vertebrate species such as mammals, each individual cell possesses a copy of approximately several thousand (>10^5) genes in the genome, but expresses only a small fraction of these genes, while others remain in a stably repressed state. Many protein-encoding genes transcribed by RNA polymerase II possess a complex promoter in the 5'-region upstream from the transcribed sequence. This is typically known as a 'core promoter' region that functions to bind ubiquitous transcription factors and co-activators, which act together to attract RNA polymerase II to initiate transcription. In addition, in tissue-specific genes, promoter elements commonly known as 'enhancers' or 'repressors' are often located upstream of the core promoter and bind tissue-specific transcription factors. These are required to activate transcription and to ensure that initiation occurs at the appropriate location at the beginning of the gene. Gene transcription is regulated by the binding of single or combination of transcription factors to characteristic promoter motifs in the DNA sequence upstream of the protein coding region, thereby inducing changes in chromatin structure and modulating activity of the transcriptional machinery. Several unique transcription factors have been identified in germ cells (Chapters 15-16). Among them, the best characterized are the cAMP responsive element binding protein (CREB) and the closely related cAMP-responsive element modulator (CREM) transcription factors that are activated by cAMP/protein A kinase signaling pathway. Different CREB and CREM protein isoforms encoded by alternative transcripts suppress or activate transcription by binding to cAMP response elements (CRE) in gene promoters during specific phases of spermatogenesis (McCarry, 1998; Eddy 1998). Transcription factors bind specifically to short sequence elements within promoters and enhancers of a gene, and are able to either activate or repress transcription initiated by the RNA polymerase. Transcription factors may have separate (i) DNA binding domains and (ii) transcription activation domains.

(a) *DNA Binding Domains*. Among protein motifs, which are involved in regulation of transcription through DNA binding are: (i) steroid receptors, which are activated by corresponding steroids, leading to binding of these receptors to DNA and thus initiating transcription; (ii) zinc finger motif has a DNA binding domain which was recognized initially in transcription factor TFIIA, required for transcription of 5S rRNA by RNA polymerase III; (iii) helix turn helix, first recognized in phage λ repressors, is now known to be present in several transcription factors in *Drosophila* (homeobox) and mammals; (iv) helix loop helix motif, found in some developmental regulators, regulates expression of genes coding for some eukaryotic DNA binding proteins, and is involved in protein dimerization and DNA binding; (v) leucinc zippers consist of a stretch of amino acids with a leucine residue at every seventh position (see Section 14.9).

(b) Transcription Activation Domains: The transcription activation domains are separate from DNA binding domains and each consists of 30-100 amino acids. The transcription activation domains function through protein-protein interactions and often help in establishing contacts with components of transcription complex, which leads to activation of transcription. Three types of activation domains have been identified: a) Acidic domains, first identified in yeast transcription factors GAL4 and GCN4, have been found in glucocorticoid hormone receptor and also in AP-1/Jun transcription factors. Acidic domains show two features in common, i.e. there are regions of significant negative charge and they can form amphipathic α helical structures. The acidic domains help in association of TFIIB and TFIID. (b) Glutamine rich domains are found in SP1 and also in *Drosophila's* antennapaedia and ultra bithorax, in yeast's HAP1, HAP2, and GAL11 and in several mammalian factors (OCT-1, OCT-2, JUN, AP-2, SRF). (c) Proline rich domains have been identified in CTF/NF-1 and in mammalian factors (AP-2, JUN, OCT-2, SRF). There are several proteins in eukaryotes, which influence transcription with out binding DNA. Now it is clear that sigma factor of prokaryotes is a transcription factor similar to TFs of eukaryotes. However, our knowledge on molecular biology of testicular germ cells as compared to other systems is very meager.

14.9. FAMILIES OF DNA BINDING PROTEINS

Among other activities, DNA binding proteins are responsible for replicating the genome, and for transcribing active genes. As stated earlier, one of the largest and most diverse classes among DNA binding proteins are the transcription factors that regulate gene expression. To understand how genetic information is utilized, it is essential to understand the structure and DNA binding properties of these transcription factors. Understanding these structural details enhances our understanding of the molecular mechanisms involved in repression and activation of gene expression. A list of more than 400 transcription factors had been compiled as early as in 1992 (Faisst and Meyer, 1992). Our aim would be to focus on DNA binding domains from those transcription factors identified mainly in testis. Based on structural motifs present for recognition of DNA, many DNA binding proteins can be grouped into classes such as (a) Helix-turn-helix (HTH) proteins, (b) the homeodomains, (c) zinc finger proteins, (d) the steroid receptors, (e) leucine zipper proteins, and (f) the helix -loop- helix proteins. Two smaller families have been identified that use β-sheet for DNA binding. Inspite of the fact that there is no single

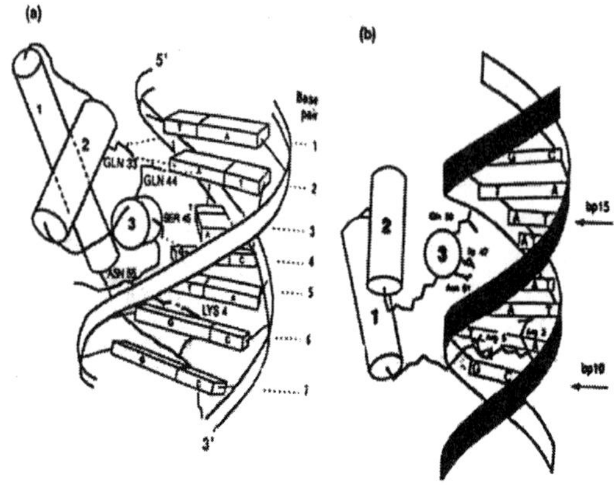

Fig.14.5. Helix-turn-helix protein-DNA interactions. (a) Sketch emphasizes the HTH unit and hydrogen bonds, showing recognition of alpha helix 3, which is inserted into major groove of DNA half site of the operator. Helices 2 and 3 constitute HTH motif. Reprinted with permission from S.R. Jordan and C.O. Pabo Science 242; 893-97:1998© AAAS, 1988). (b) Structure of an engrailed homeodomain-DNA complex showing critical contacts made by recognition helix in major groove and the contacts made by N-terminal basic region in minor groove Reprinted with permission from C.R. Kissinger et al., Cell 63;579-90: 1990 © Elsevier.

pattern or simple code, each of the motif that has been characterized, involves a simple secondary structure (usually α helix) that is complementary to the structure of B-DNA. Side chain contacts play an important role in site specific binding. It is important to realize that there are many DNA binding proteins that do not belong to any of the known families.

14.9.1. Helix-Turn-Helix (HTH) and Homeodomain

Helix-turn-Helix (HTH) was the first DNA recognition motif in the structures that have now been determined for many HTH proteins and protein DNA complexes. It occurs in a large family of prokaryotic DNA binding proteins including CAP, λ|Cro and λ Rep. The HTH motif consists of 20-residues segment with a α helix (1-7 residues), a turn (8-11 residues) and a second α helix (12-20 residues). Thus HTH motif consists of two α helices packed together at an angle of about 120° with a tight turn at the elbow between them. The structure is not stably folded of its own, but there is usually a third helix protruding from the rest of the protein, which stabilizes it. The second helix in the helix turn helix lies in the major groove of the DNA and contributes the main amino acid residues responsible for specific binding. Hence it is called the recognition helix. The whole motif is fixed in a particular orientation by hydrogen bonds to phosphates on DNA backbone (review- Pabo and Sauer, 1992)(**Fig14. 5**).

Among eukaryotic regulatory proteins, the most important domain with HTH motif is the 60 residues long homeodomain motif found in the products of a number of developmental genes in *Drosophila* and has similarity with prokaryotic HTH. The polypeptide chain representing homeodomain, folds as a 3-helix bundle, in which second and third helices form the HTH. Unlike the isolated HTH unit, the 60 residues homeodomain forms a stable, folded structure that can bind DNA by itself. The eukaryotic motif is, however, different with prokaryotic HTH in two aspects. First, the recognition helix is substantially longer, containing at its C-terminal end more residues that interact with the phosphates, leading to a somewhat different orientation in the major groove of the DNA. Second, the parent protein of the homeodomain binds as a monomer, though some proteins also contain other DNA binding regions (eg. POU-domain). Thus, the homeodomain is a DNA binding motif that has a broader role in eukaryotic gene regulation.

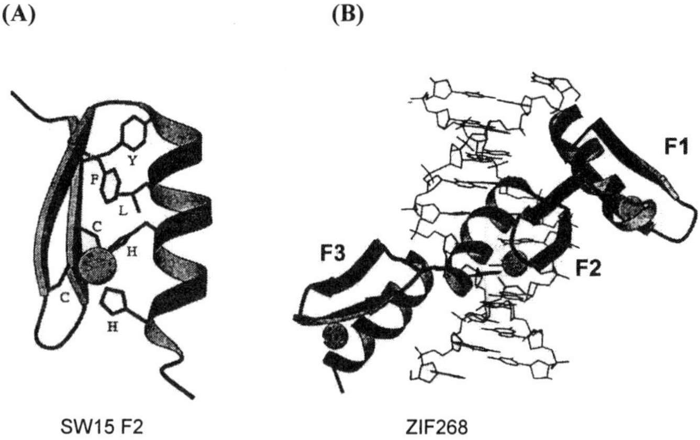

Fig.14.6. Zinc fingers of first class (TFIIIA type) and their interaction with DNA. (A) Structure of yeast protein SW15 derived from 2D-NMR in solution. The two sub-structures forming the finger, the β-sheet, and the helix are pinned together by Zn ion (shaded ball) ligated to a pair of cysteine (C) and a pair of histidine (H). The structure is further stabilized by three invariant Phe (F), Tyr (Y), and Leu (L) residues, packed together at the top of the domain (From Neuhaus et al., 1992). (B) DNA-zinc figer ZIF268 complex formed by interaction of three zinc fingers. Each zinc finger binds in a similar manner to three base pairs in major groove. Balls represent zinc ions as in (A). Reprinted with permission from N.P. Pavletich and C.O. Pabo, Science 252;80-12:1991© AAAS.

14.9.2. Zinc Finger Proteins

Zinc fingers are one of the major structural motifs involved in protein DNA interactions. The zinc finger motif was first identified by Miller et al. in the *Xenopus* transcription factor IIIA (TFIIIA), where it is repeated consecutively nine times. It consists of a 30 amino acid sequence containing two His, two Cys, and three hydrophobic amino acids, all at conserved positions. Zinc finger proteins are involved in many aspects of eukaryotic gene regulation. Homologous zinc fingers occur in: a) proteins induced by differentiation and growth signals, b) in proto-oncogene products, c) in general transcription factors, d) in proteins that regulate genes in development, and in regulatory genes of eukaryotic organisms. Proteins in this family usually contain tandem repeats of the 30-residue zinc finger motif, with each motif containing the sequence pattern Cys-$X_{2\text{ or }4}$-Cys-X_{12}-His-$X_{3\text{-}5}$His. The crystal structure of a zinc finger complex containing three fingers from zif268 shows that the zinc fingers bind in the major groove of B-DNA and wrap partway around the double helix. The zif268 fingers make a set of hydrogen bonds with bases in the major groove. Critical base contacts are made by an arginine that immediately precedes the α-helix. All these contacts involve hydrogen bonds with guanines on the G-rich strand of the consensus-binding site: 5'-GCGTGGGCG-3'.(Pabo and Sauer, 1992).

Zinc Finger Proteins (with Zn Fingers, Zn Clusters and Zn Twists) - Class I: In some proteins, there are zinc binding sites with distinct DNA binding motifs. In one such motif, a loop of amino acids protrudes out as zinc finger from zinc binding site. In other motifs, a zinc twist or a zinc cluster is found. More information is available about zinc fingers than for zinc twists and/or zinc clusters. The zinc fingers are described as Cys2/His2 (class I Zn fingers) and Cys2/Cys2 fingers (second class Zn fingers). Cys2/His2 fingers have the following consensus sequence: Cys-$X_{2\text{-}4}$-Cys-X_3-Phe-X_5-Leu-X_2-His-X_3-His. Each finger consists of about 23 amino acids, linked to another finger by 7-8 amino acids. Several zinc fingers are required for binding to DNA. On one extreme, these fingers may involve almost the entire protein as in TFIIIA (9 fingers) and on the other extreme, only a small domain is involved in forming zinc fingers, as in ADR1 (2 fingers). Since zinc fingers are found in several known transcription factors, this feature helps in recognizing proteins that function as transcription factors. TDF (testis determining factor) factors (ZFX and ZFY) are Zinc fingers coded by X and Y chromosomes (Chapter 10).

NMR studies showed that the structured region of the zinc finger motif comprises about 25 amino acids forming a compact unit with a 12 amino acid helix packed against a β-hairpin. The mode of binding to DNA was first determined from the crystallographic analysis of a three zinc finger peptide from Zif268 bound to its cognate DNA. The main interaction comes from its helical region. The three fingers bind in an equivalent manner to a segment DNA 9 base pairs long, so that each finger contacts 3 base pairs, with no gaps between the three sites. The linkers between fingers are largely extended and make no significant DNA contacts. However it does not mean that all zinc fingers bind the same way **(Fig.14.6)**. Before making generalizations, it will be necessary to solve the structures of other zinc finger DNA complexes to see if other stable docking arrangements exist, and whether zinc fingers bind to AT rich sequences in fundamentally different ways than they bind to GC-rich sequences, and also to understand how zinc finger proteins interact with RNA in the 7S particle formed by TFIIIA (Pabo and Sauer 1992).

Steroid Receptors: A Second Class of Zinc Fingers: Soon after the discovery of zinc fingers in TFIIIA, related sequence motifs were found in several other proteins or sequences of molecules that bind DNA. The steroid receptor family includes receptors for the steroid hormones, retinoids, vitamin D, thyroid hormones, and a number of other important compounds. These proteins contain separate domains for hormone binding, DNA binding, and for transcriptional activation. The DNA binding domain, which contains about 70 residues, has eight conserved cysteine residues. The DNA binding domains of such receptors include two motifs in tandem, each about 30 amino acids long. But each motif contains two pairs of Cys (Cys_2/Cys_2) rather than a pair of Cys and a pair of His (Cys_2/His_2), as in the first class. Though they do bind Zn, but the three dimensional structure of two such DNA binding domains showed that the receptor DNA binding domain is structurally distinct from the TFIIIA type of zinc finger. Each of the two motifs in each domain folds up into an irregular loop followed by a α helix. But the two together form a single structural unit with their helices crossing at right angles, so that the DNA recognition helix (from the first motif) is supported by the helix from the second motif. Hormone receptors bind to palindromic sites (response elements, RE) on the DNA as dimers and the DNA binding domains alone also form dimers. A crystal structure of the glucocorticoid receptor complex also shows that the peptide binds as a dimer, even though NMR studies had shown that the peptide exists as a monomer in solution. The first helix of each subunit fits into major groove of DNA, and side chains from helix make contacts with edges of base pairs. The second major helix provides phosphate contact with DNA and provides dimerization interface. The crystal structure also gives information about the loop regions that precede the major helices. Mutations altering the amino acids in finger regions have been shown to alter the DNA binding ability of these zinc finger proteins, suggesting that fingers are responsible for DNA binding. Rarely fingers may bind RNA rather than DNA or may not bind any nucleic acid at all.

14.9.3. Leucine Zipper

The helix-loop-helix and leucine zipper family have many similarities, which suggest that the leucine zipper and helix-loop-helix proteins have important roles in differentiation and development. The leucine zipper motif was first discovered as a conserved sequence pattern in several eukaryotic transcription factors, and now it is clear that this motif appears in a wide variety of transcription factors from fungi, plants and animals. The DNA binding domains of leucine zipper proteins generally contain 60-80 residues, and contain two distinct sub-domains: the leucine zipper region mediates dimerization, while a basic region contacts the DNA (Pabo and Sauer, 1992). Basic leucine zipper class of proteins can form homo- and heterodimers. CAAT/Enhancer binding protein (C/EBP) from SV40 and several other proteins contained a leucine zipper which is a stretch of amino acids rich in leucine, occupying every seventh position (heptad repeat) in the potential zipper. These leucine residues may form an amphipathic α helix. The leucine zippers form dimers among similar molecules (e.g. Jun-Jun homodimer) or among dissimilar molecules (e.g. Jun-Fos heterodimer). Thus the hormodimer of the proto-oncogene, c-Jun is unstable and binds DNA inefficiently, whereas the *Jun-Fos* heterodimer is stable and binds AP-1 DNA sites with higher affinity. Thus *Fos* is a positive regulator of *Jun*. The use of heterodimers thus allows a combinatorial mode of action and increases the repertoire of regulatory functions for this family of proteins. Heterodimer formation has several roles in the leucine zipper family. Heterodimers can limit the activity, such as CREB is antagonized by formation of heterodimers with CREM. In several cases, heterodimer formation may allow for different combinations of activation and/or repression domains and thus change the regulatory properties of a molecule bound at a fixed DNA site (Pabo and Sauer, 1992). Earlier it was believed that leucines of one protein may interdigitate with the leucines of the zipper of another protein in reverse orientation. However, it has now been established that zipper regions associate in parallel (Leucines overlap and line up side by side) when they form a dimer in the form of a colied-coil (two right handed helices wind around each other). The crystal structure of the GCN4 (an activator protein from yeast) leucine zipper-DNA complex was solved recently. Each of the two basic leucine zipper monomers, 56 amino acids long, forms a smoothly curved continuous α-helix; the two packed together at their C-terminal ends as a coiled-coil, which gradually diverges to allow the residues in the basic region to follow the major groove of each DNA half site. The DNA is thus grasped by a pair of splayed chopsticks. It gives an impression that the basic regions that were disordered in the free proteins become ordered into helices on binding to DNA. Amino acid sequences compared in atleast 11 regulatory leucine zipper proteins including four mammalian proteins (C/EBP, CREB, Jun, Fos), and other organisms revealed a consensus sequence with general features: (i) proline and glycine are absent or rarely found, thus permitting formation of α helix; (ii) 2-5 heptad repeats of leucine over a region of GC-40 were found to help in dimer formation for DNA binding, (iii) basic regions containing arginine and lysine (arg/lys) are found close to leucine zipper, helping in DNA binding (zippering of two molecules bring arg/lys in proper position for combining with dyad symmetric motifs in DNA); (iv) zippering leads to formation of Y shaped dimers; (v) an asparagine is always found at the same position and interrupts α helix (like proline and glycine), thus allowing bending of arms of Y shaped dimer to permit DNA motif meant for binding **(Fig. 14. 7)**.

14.9.4. The Helix-Loop-Helix (HLH)

Another group of proteins having similarities to the basic leucine zipper group displays the helix-loop-helix motif, which so far is characterized by sequence regularities. A basic region of 15 amino acids lays immediately N-terminal to a 15 amino acid segment with helical characteristics, which is separated from a second such putative helix by a region of variable length. Members of this family also function as dimers, and in almost all cases a heterodimer is the active DNA binding species. Thus the protein Myc needs the protein Max to be effective. Like the leucine zipper proteins, the HLH proteins have a basic region that contacts the DNA and a neighboring

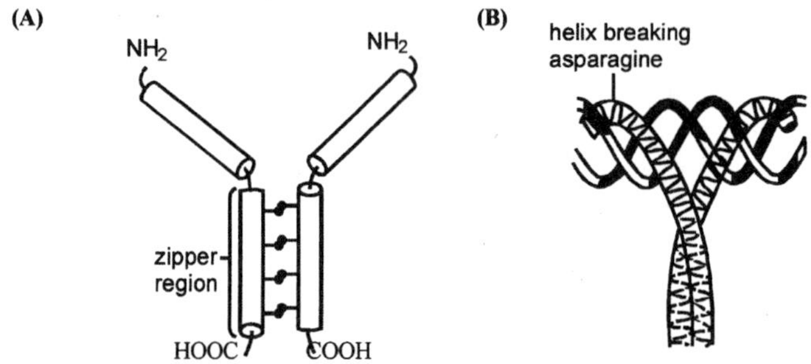

Fig.14.7. Schematic representation of leucine zipper-DNA complex showing DNA binding regions of basic-LZ protein. The protruding regions of α-helices (Fig. A) bend at asparagine residues forming a Y-shaped dimmer to establish a grip on DNA motif (Fig. B).

region that mediates dimer formation. This dimerization region forms an α-helix, a loop, and a second α-helix. Like the leucine zipper proteins, the HLH proteins play important roles in differentiation and development, and their activity is modulated by heterodimer formation. Heterodimer formation is used in many different ways in this family of proteins, mixing positive activators and negative regulators to modulate gene activity.

14.9.5. β-Sheet Motifs

Most of the major families of DNA-binding proteins that have been structurally characterized bind with an α-helix in the major groove. The MetJ, Arc, and Mnt repressors are interesting because they belong to a family of prokaryotic regulatory proteins, which interact with DNA in a different way. Here, a pair of anti-parallel β-ribbon formed between a symmetrical dimer of protein interacts with DNA in major grooves. Thus β-sheet can be used in recognition but it seems clear that a α helix offers greater intrinsic stability in more suitable element for DNA binding characteristics.

TATA Box Binding Proteins: In TFIID (transcription factor IID for RNA polymerase II) TATA box binding protein (TBP), a large β-sheet is used rather than a ribbon. TBP is required for initiation of transcription for all three eukaryotic RNA polymerases. The whole molecule looks like a saddle being formed of two identical halves related by a pseudodyad. TBP bound to DNA crystals have been analysed. The dimensions of saddle are such that rather than sitting astride the DNA helix, TBP sits almost parallel to the helix of DNA. This leads to the distortion of the minor groove so drastically that the double helix is unwound by about 110 and also bent through 80°.The bending of TATA box brings other factors together leading to a further unwinding of the helix and the separtion of two strand at the start site, required for initiation of transcription. This is a fitting example, which illustrates how TBP can exploit the conformation deformity of TATA sequences to initiate DNA transcription.

14.9.6. Other Families

There are several other families of DNA binding proteins with cysteine rich motifs (other than zinc fingers) that may serve as metal binding domains. For example, a cysteine rich motif (with the general form Cys-X-Asn-Cys-X_{17}-Cys-Asn-X-Cys), discovered in the GATA factor, is specifically expressed in erythroid cells and Sertoli cells in testis. RAD-1 and RAD18 are prototypical members of a distinctive family of DNA binding proteins that contain another cysteine rich motif. The LIM motif is a distinct cysteine rich region that occurs on the N-terminal side of the homeodomain in several regulatory proteins from elegans. The ACE-1 transcription factor of yeast has a cysteine rich region that appears to form a cluster with 6-7 copper ions. Although cysteine rich and histidine rich motifs are easily detected in sequence comparisons and have received considerable attention, several other families of DNA- binding proteins have been characterized recently. One prominent family is the set of POU proteins, which contain homeodomain. The POU specific domain also binds DNA. There also are families of transcriptional regulatory proteins related to c-ets, to the non-histone high mobility group (HMG) proteins to serum response factor, to NF-kB and to *rel* oncogene (reviewed in Pabo and Sauer, 1992). Most of these transcriptional factors have been identified in germ cells or Sertoli cells in testis. However, how far they are specific in the regulation of spermatogenesis is yet to be investigated. But, their abundance in germ cells indicates that testis offers a good source of material for physico-chemical investigations for DNA- protein interactions.

Although the diversity of known DNA binding motifs and contacts does not suggest any simple rule(s) or patterns describing site-specific recognition, Pabo and Sauer, (1992), based on analysis of several known DNA-protein complexes, made few broad generalizations:

- Site-specific recognition always involves a set of contacts with the bases and with the DNA backbone, in which hydrogen bonding is critical for recognition (although hydrophobic interactions also occur). A complex typically has of the order of 1-2 dozen hydrogen bonds at the protein/DNA interface.
- Side chains are important for site-specific recognition and make most of the critical contacts. The folding and docking of the entire protein help to control the "meaning" that any particular side chain has in site-specific recognition.
- Most of the base contacts are in the major groove. Contacts with purines (which are larger and offer more hydrogen bonding sites in the major groove) seem to offer more advantage.
- Most of the major motifs in transcription factors contain a α helical region that fits into the major groove of B-form DNA. However, there are examples of β sheets and or ribbons extended regions of polypeptide chain that play critical roles in some proteins. More examples are needed for identification of specific sites in DNA.
- Multiple DNA binding domains usually are required for site-specific recognition. The same motif may be used more than once, as occurs when the active binding species is a homodimer or heterodimer, or when a single polypeptide contains tandem recognition motifs. Different motifs (an extended arm and a HTH unit; a homeodomain and POU specific domain, etc) may also be used as complex.

14.10. REFERENCES

Affara N, Bentley E, Davey P, Pelmear A, and H. Jones MH. The identification of novel gene sequences of the human adult testis. Genomics 1994; 22: 205-10

Aoto H, Tsuchida J, Nishina Y, Nishimune Y, Asano A and Tajima S. Isolation of a novel cDNA that encodes a protein localized to the pre-acrosome region of spermatids. J. Biochem 1995; 234: 8-15.

Ariel M, Ceder H. McCarrey J. Developmental changes in methylation of spermatogenesis-specific genes include reprogramming in the epididymis, Nature Genet. 1994; 7: 59-63.

Ariel M, McCarrey J, Cedar H. Methylation patterns of testis-specific genes, Proc, Natl Acad. Sci USA 1991; 88: 2317-21.

Bantzer F, Menissier-de Murcia J, Barlow C, Wynshaw – Boris A, de Murcia G. Poly(ADP-ribose) polymerase activity is not affected in atraxia telangiectasia cells and knockout mice. Carcinogenesis 1999; 20: 177-80.

Benoit G, Trasler JM. Developmental expression of messenger ribonucleic acid, protein, and enzyme activity in the mouse testis. Biol. Reprod 1994; 50: 1312-19.

Braren R, Firner K, Balasubramanian S, Bazan F, Thiele HG, Haag F, Koch-Nolte F. Use of EST database resource to identify and clone novel mono(ADP-ribosyl) transferase gene family members. Adv Exp Med Biol 1997; 419: 163-8.

Choi YC, Chae CB. cDNA hypomethylation and germ cell-specific expression of testis-specific H2B histone. Gene 1991 characterization of MBD3L1 as a testis-specific transcriptional repressor. Genomics. 2002; 80: 621-9.

Choi YC, Aizawa A, Hecht NB. Genomic analysis of the mouse protamine 1, protamine 2, and transition protein 2 gene cluster reveals hypomethylation in expressing cells. Mamm Genome 1997; 8: 317-22.

Clouthier DE, Averbock MA, Maika SD, Hammer RE, Brinster RL. Rat spermatogenesis in mouse testis. Nature 1996; 381: 418-21.

Cowell IG, Aucott R, Mahadevaiah SK, Burgoyne PS, Huskisson N, et al. Heterochromatin, HP1 and methylation at lysine 9 of histone H3 in animals. Chromosoma. 2002; 111: 22-36.

Davidson JN. Biochemistry of Nucleic Acids. Cox and Wyman, London, 1965.

Doerkesen T, Benoit G, Trasler JM. Deoxyribonucleic acid hypomethylation of male germ cells by mitotic and meiotic exposure to 5-azacytidine is associated with altered testicular histology. Endocrinology 2000; 141: 3235-44.

Eddy EM. Regulation of gene expression during spermatogenesis. Seminars Cell Develop Biol 1998; 9: 451-57.

Faisst S, MeyerS. Compilation of vertebrate – encoded transcription factors. Nucleic Acids Res 1992; 20: 3-26.

Hazzouri M, Pivot-Pajot C, Faure AK, Usson Y, Pelletier R, Sele B, Khochbin S, Rousseaux S. Regulated hyperacetylation of core histones during mouse spermatogenesis: Involvement of histone deacelylases. Eur J Cell Biol 2000;79: 950-60.

Jiang CL, Jin SG, Lee DH, Lan ZJ, Xu X, O'Connor TR, Szabo PE, Mann JR, Cooney AJ, Pfeifer GP. MBD3L1 and MBD3L2, two new proteins homologous to the methyl-CpG-binding proteins MBD2 and MBD3: characterization of MBD3L1 as a testis specific transcriptional repressor. Genomics 2002;80:621-9.

Jordan SR, Pabo CO. Structure of the lambda complex at 2.5A resolution: Details of the repressor-operator interactions. Science 1988; 242: 893-97.

Kissinger CR, Liu B, Martin-Blanco E, Kornberg TB, Pabo CO.Crystal Structure of an engrailed homeodomain-DNA. complex at 2.8 A resolution: A framework for understanding homeodomain-DNA interactions. Cell 1990; 63: 579-90.

Kouzarides T. Acetylation: a regulatory modification to rival phosphorylation. The EMBO Journal 2000; 19: 1176-79.

Kramer JA, McCarrey JR, Djakiew D, Krawetz SA. Human spermatogenesis as a model to examine gene potentiation. Mol Reprod Dev 2000; 56(2 Supp): 254-8.

Kundu TK, Rao MRS. CpG islands in chromatin organization and gene expression. J Biochem 1999; 125: 217-22.

Laniel MA, Bergeron MJ, Poirier GG, Guerin SL. A nuclear factor other than Sp1 binds the GC-rich promoter of the gene encoding rat poly(ADP-ribose) polymerase in vitro. Biochem Cell Biol 1997; 75: 427-34.

Lankenau S, Burkle A, Lankenau DH. Detection of poly(ADP-ribose) synthesis in Drosophila testes upon gamma-irradiation. Chromosoma 1999; 108: 44-51.

Li E, Bestor TH, Jaenisch R. Targeted mutation of the DNA mehtyltransferase gene results in embryonic lethality. Cell 1992; 69: 915-26.

Martinez-Balbas M, Bauer UM, Nelsen SJ, Brehm A, Kouzarides T. Regulation of E2F activity by acetylation. The EMBO Journal 2000; 19: 662-71.

McCarrey Jr. Spermatogenesis as a model system for analysis of regulatory mechanisms associated with tissue-specific gene expression. Seminars Cell Dev Biol 1998; 9: 459-66.

Mertineit C, Yoder JA, Taketo T, Laird DW, Trasler JM, Bestor TH. Sex-specific exons control DNA methyltransferase in mammalian germ cells. Development 1998; 125: 889-97.

Mosgoeller W, Steiner M, Hozak P, Periner E, Wesieska Gadek J. Nuclear architectural distribution of poly (ADP-ribosyl) transferase, a multi functional enzyme. J Cell Sci 1996; 109 (pt2) 409-18.

Neuhaus D, Nakaseko y, Schwabe Jw R, Klug A. Solution structures of two zinc-finger domains from SW15 obtained using Two-dimensional ^{1}H Nuclear magnetic resonance spectroscopy. J Mol Biol 1992;228:637-51.

Numata M,Ono T, Isaki S. Expression and localization of the mRNA for DNA(cytosine)-methyl transferase in mouse seminiferous tubules. J Histochem Cytochem 1994;42:1271-6.

Pabo CO, Sauer RT. Transcription factors: structural families and principles of DNA recognition. Annu Rev Biochem 1992;61:1053-95.

Pavletich N P, Pabo CO. Zinc Finger-DNA Recognition: Crystal Structure of a Zif268 DNA complex at 2.1 A. Science 1991;52:09-13.

Pivot-Pajot C, Caron C, Govin J, Vion A, Rousseaux S, Khochbin S. Acetylation-dependent chromatin reorganization by BRDT, a testis-specific bromodomain-containing protein. Mol Cell Biol 2003; 23: 5354-65.

Quesada P, Farina B, Jones R. Poly(ADP-ribosylation) of nuclear proteins in rat testis correlates with active spermatogenesis. Biochim Biophys Acta 1989; 1007: 167-75.

Satyanarayana K, Rao MR. Characterization of poly(ADP-ribosyl) ated domains of rat pachytene chromatin.Biochem J 1989;261:775-86

Struhl K, Histone acetylation and transcriptional regulatory mechanisms. Gene Dev 1998; 12: 599-06.; 25: 20504-11.

Tramontano F, Malanga M, Farina B, Jones R, Quesada P. Heat stress reduces poly(ADPR) polymerase expression in rat testis. Mol Hum Reprod 2000; 6: 575-81.

Trasler JM, Alcivar AA, Hake LE, Bestor T, Hecht NB. DNA methyltransferase is developmentally expressed in replicating and non-replicating male germ cells, Nucleic Acids Res. 1992; 20: 2541-45.

Trasler JM, Hake LE, Johnson PA, Alcivar AA, Millette CF, Hecht NB. DNA methylation and demethylation events during meiotic prophase in the mouse testis. Mol Cell Biol 1990; 10: 1828-34.

Yang L, Mei Q, Zielinska-Kwiatkowska A, Matsui Y, Blackburn ML, Benedetti D, Krumm AA, Taborsky GJ Jr, Chansky HA. An ERG (ets-related gene)-associated histone methyltransferase interacts with histone deacetylases 1/2 and transcription co-repressors mSin3A/B. Biochem J 2003; 369(Pt 3): 651-7.

Chapter 15

PROTEINS IN TRANSCRIPTIONAL ACTIVITY OF SPERMATOGENIC CELLS

15.1. TRANSCRIPTIONAL ACTIVITY IN GERM CELLS

15.1.1. RNA Polymerases in Eukaryotes

In contrast to prokaryotic cells, the process of transcription is much more complex in eukaryotes. While prokaryotes require only one RNA polymerase, there are at least three different nuclear RNA polymerases in eukaryotes, whose properties can be distinguished on the basis of their sensitivity to certain inhibitors of transcription, e.g. amanitin. Each of these enzymes is a large protein (500-kDa) with two large and several smaller subunits, some of them showing homology to the subunits of *E. coli* RNA polymerase ($\alpha_2\beta\beta'\sigma$). The largest subunit of 200-kDa is homologous to β'-subunit where as another of 140-kDa has homology with β-subunit of *E.Coli* RNA polymerase. The second largest subunit of RNA polymerase II also shares the catalytic sites of ribonucleases, which are used for RNA cleavage and help in proof reading. RNA polymerase I synthesizes rRNA, whereas RNA polymerase II transcribes hnRNA, mRNA. However, RNA polymerase III synthesizes tRNA and 5S RNA. These RNA polymerases recognize different promoter sequences.

RNA polymerases bind at specific start sites in the promoter region of a gene to initiate transcription. For RNA polymerase II, the three regions have been defined as TATA box (TATAA/$_T$AA/$_T$), GC rich sequences (GGGCGG) and CAAT box (GGT/$_A$CAATCT) in β globin gene. These regions are present in many eukaryotic genes but not necessary. Eukaryotic promoters often also possess elements located 100 or 200-bp upstream, which interact with proteins other than RNA polymerase and thus regulate the activity of promoter. These elements termed enhancers, can be moved several hundred or even thousands of base pairs upstream or downstream without alteration of their activities. There are other regulatory elements known as silencers, which repress gene expression. Silencers, like enhancers, can function at great distances from genes they repress. They are sites for binding of proteins just like enhancers. In 5S RNA genes transcribed by RNA polymerase III, the promoter is located well within the transcription unit, 50-bp downstream rather than upstream from the start point. The location of downstream promoters for RNA polymerase II may vary in different genes.

15.1.2. Initiation of RNA Synthesis in Eukaryotes

The regulatory DNA sequences (promoters, enhancers, silencers, etc.) for genes transcribed by each of the three RNA polymerases in eukaryotes are not the same. A number of transcription activators are also involved in the formation of a transcription complex, required for initiation of transcription in packaged-chromatin. Although TFIID or a part of it, TATA binding protein (TBP) is required for all the three RNA polymerases, each RNA polymerase has its own set of transcription factors. Transcription factors, defined as proteins, needed for initiation of transcription, are not a part of RNA polymerase. But these factors help in binding of an RNA polymerase to DNA to form a transcription complex. Transcription occurs only after the formation of this complex. The regulatory DNA sequences and the transcription factors used by RNA polymerase II in mRNA synthesis may be either common for many genes expressed constitutively (for house keeping genes) or may be specific for tissue or stimulus specific expression of genes, whose expression is regulated. It means that for transcription of different genes, no single combination of DNA sequences or factors is adequate and essential, and that the transcription may be initiated in many different ways. The common feature of this transcription, however, involves binding of a number of transcription factors to DNA sequences upstream of start point. The number of transcription factors involved in the initiation process of transcription increases as the variety of genes transcribed by the polymerase increases (Gupta, 1999).

15.1.3. Pre-Initiation Complex with RNA Polymerase II

For constitutive expression of common genes (called house keeping genes) in all cells, a generic promoter cannot bring about regulated expression. Initiation of transcription on this generic promoter by RNA polymerase II requires the action of several transcription factors used as initiation factors (variously designated in humans, rat and yeast). In human following order is required: (i) TFIID binds at the TATA box, which permits the association of TFIIA and TFIIB. (ii) After binding of TFIIB, which forms the so called DB complex, RNA polymerase associates to promoter site and then escorted to the promoter by TFIIF to form a transcription complex. (iii) Subsequent addition of TFIIE, TFIIH and TFIIJ in a sequential way helps in the initiation of transcription. The requirement of TFIIA and TFIIE has not been found universal. The function of TFIIA can be performed by other transcription factors (Gupta, 1999). Besides the above transcription factors, hundreds of other transcription factors have been discovered, which are needed for tissue or stimulus specific transcription. Hence the order of assembly of these proteins is probably different for different genes and therefore may not allow a prescribed pathway.

15.2 RNA SYNTHETIC MACHINERY IN SPERMATOGENESIS

15.2.1. RNA in Meiosis and Post-meiotic Stages

The rate of RNA synthesis decreases from a high level in spermatogonia and spermatocytes to a low level in early spermatids and becomes absent in late spermatids and mature spermatozoa. Newly synthesized RNA in spermatogonia and spermatocytes consisted of a variety of molecular weight species, including 18S and 28S ribosomal RNAs. High molecular weight RNAs, especially ribosomal RNAs, decrease drastically in early spermatids, leading to the synthesis of only small molecular weight RNAs. Studies on RNA synthesis, in the mouse revealed maximal ^{3}H-uridine incorporation during meiosis, although RNA synthesis was low during the preleptotene stage and undetectable in leptotene, zygotene, and early pachytene spermatocytes. ^{3}H-Uridine incorporation increased rapidly to a transient peak in mid to late pachytene spermatocytes followed by a decline during remaining prophase stages and finally reaching to an undetectable level during the meiotic divisions. The high levels of RNA synthesis have been confirmed in pachytene spermatocytes from mouse, rat, hamster and human testis (Eddy and O'Brien, 1998). Nucleolar RNA synthesis also showed a peak during the zygotene stage in both mouse and human spermatocytes, providing evidence that ribosomal RNAs are synthesized during meiosis. RNA synthesis during the pachytene stage of meiosis, in both mouse and human spermatocytes, was primarily localized adjacent to the autosomes. This peri-chromosomal localization of ^{3}H-uridine suggests that RNA synthesized during meiosis is predominantly heterogeneous nuclear RNA (hnRNA), which is precursor of cytoplasmic mRNAs. Several other studies gave evidence that mRNAs transcribed during meiosis, particularly during the pachytene stage are used for protein synthesis. However, in contrast to the autosomes, the condensed X and Y-chromosomes in mouse and human pachytene spermatocytes incorporated little or no RNA precursors. Subsequent studies have provided further evidence for transcriptional inactivation of the sex chromosomes during meiosis, including the loss of mRNAs transcribed from several X-linked genes and the absence of RNA polymerase II and other components of the RNA splicing machinery from the condensed XY chromosome pair. Earlier monitoring 3H-uridine incorporation during meiosis indicated that much of the RNA synthesized during the pachytene stage accumulated in the nucleus and was then released into the cytoplasm during the division stages. A significant portion of this RNA persisted throughout spermatid development, and was ultimately discarded in the residual bodies, although spermatids are richer in RNA polymerase II and its machinery (Schmidt and Schibler, 1995). These overall results suggested that long-lived mRNAs synthesized during meiosis may be stored for translation during spermiogenesis. However, definitive evidence for translation of meiotic transcripts in spermatids is difficult to obtain, since many genes are transcribed both during meiosis and in the early stages of haploid differentiation, and since many of early studies did not consider ploidy for analysis (review Eddy and O'Brien, 1998).

15.2.2. RNA Polymerases

The transcription machinery during spermatogenesis appears to be unique. Synthetic machinery in spermatogenic cells revealed that general transcriptional factors such as TATA binding protein, transcription factor II H, splicing factors as snRNP and polyadenylation specificity factor, are expressed in meiotic and post-meiotic germ cells, whereas RNA polymerase I and II are present in all cell types, but their activities decrease from spermatocytes to mid-spermatids and finally disappear from spermatozoa. Hormone-induced polymerases showed major decrease in activity which began near the chronological mid-point of development. The reduction in polymerase II activity was accompanied by a parallel 200-fold decease in 3H-amanitin binding. The reduction in polymerase activity appears, therefore, to be due to an actual reduction in the cellular content of RNA polymerase II molecules. Thus transcription in maturing testes is regulated, at least in part, by the varying concentrations of the RNA polymerases (Gillam et al., 1979).

15.2.3. TATA Binding Protein

Among the several transcription factors (section 15.1.3), TFIID is the most important and has been purified as a multi-protein complex. A subunit of TFIID is TATA binding protein (TBP), which is actually required by all the three eukaryotic RNA polymerases (even for promoters lacking a TATA box) (Hernandez et al, 1993). However, there are other TBP associated factors (TAFs), which differ in different RNA polymerases. Therefore, it is suggested that each of the three RNA polymerases requires an essential transcription factor, which is a complex containing

TBP and TAFs; the composition of TAFs differs in the three cases. The TATA binding protein (TBP) functions in promoter recognition and initiation of transcription by the three eukaryotic RNA polymerases. Whether TBP might play a role in determining cell size dependent transcription rates, it was noticed that TBP accumulated to a much higher level in testis as compared to all somatic tissues and cell types examined. Within the testis, TBP over-expression was localized to a subset of the germ cells, which appeared to be in the early haploid stage. However, subsequent studies showed that RNA polymerase II was abundant in nuclei of round spermatids. Levels of mRNA and protein encoded by the TATA binding protein (*tbp*) gene increased dramatically during late spermatogenesis in rodents, culminating in a highly testis-enriched expression pattern. Whereas adult spleen and liver contained roughly 0.7and 2.3 molecules of TBP mRNA per haploid genome equivalent, testis contained 80-200 molecules of TBP mRNA per haploid genome equivalent. Nuclear TBP levels in testis cells were on average 8 and 11 fold higher than in liver and spleen cells respectively.

TBP mRNA levels increase substantially in meiotic and post-meiotic spermatogenic cells in rat and mouse, and are more abundant in these cells than in somatic cells. However, TBP protein levels appeared to be higher in post-meiotic cells than in meiotic cells (Schmidt and Schibler.1995; Persengiev *et, al.* 1996). These changes in protein suggest that TBP expression in pachytene spermatocytes may not relate directly to meiosis but rather to mRNA processing events that occur in post-meiotic cells, including the delayed translation of many mRNAs. In contrast to somatic tissues over-expression, testis-specific TBP mRNA arises primarily as a result of transcriptional up-regulation. This up-regulation was stage-dependent, occurring specifically in meiotic and post-meiotic cells. Transcription of the TBP gene relative to somatic tissue and pre-puberal testis is due to transcriptional induction and in large part, for the increased abundance of TBP mRNA in spermatogenic cells. Analysis suggested that translational repression is an important determining factor in the unexpectedly low ratio of top protein-mRNA in male germ cells. Thus, transcriptional induction of TBP during spermatogenesis reflects a cell-specific homeostatic mechanism that maintains TBP protein concentrations at sufficiently high levels throughout male germ cell development (Persegiev et al., 1996)

Besides TBP, two other components of the general RNA polymerase II machinery, TFIIB and RNA polymerase II, are also over-expressed in testis. TBP and RNA polymerase II are particularly rich in round spermatid nuclei. These results suggested a molecular explanation for how early spermatids are able to accumulate all of the mRNA necessary for the final week of spermiogenesis (Schmidt and Schibler, 1995). Although TBP mRNA or TBP protein have not been detected in spermatozoa, it could be conceivable that TBP mRNA in testis is required for testis specific functions (Chapter 14).

Most TBP mRNAs in testis did not initiate at the first exon used by somatic cells (here designated exon IC). Using cDNA amplification method, 5'end variants of TBP mRNA, and the corresponding cDNAs were identified and cloned from liver and testis. In liver, a single promoter/ first exon is used to generate a steady state level of roughly five molecules of TBP mRNA per diploid cell equivalent. In testis, round spermatids contain an estimated 1000 TBP mRNA molecules per haploid cell. The precise locations of the three major initiation exons were mapped on the gene. The identification of the strong testis-specific promoter/first exon was suggested for understanding spermatid-specific *tbp* gene regulation (Schmidt et al., 1997).

TATA-Binding Protein-Related Factor: The gene encoding TATA-binding protein-related factor-2 (TRF2/TLF/TLP/TRP), essential for the progress of spermiogenesis is abundantly expressed in mammalian testis. Mice lacking the TBP related factor-2 *(TRF2)* gene, which is highly expressed in the testis, have a severe defect in spermiogenesis. The expression of TRF2 is both cell- and stage-specific. TRF2 expression was first detected in the late pachytene

```
  1 MDADSDVALD ILITNVVCVF RTRCHLNLRK IALEGANVIY KRDVGKVLMK LRKPRITATI
 61 WSSGKIICTG ATSEEEAKFG ARRLARSLQK LGFQVIFTDF KVVNVLAVCN MPFEIRLPEF
121 TKNNRPHASY EPELHPAVCY RIKSLRATLQ IFSTGSITVT GPNVKAVATA VEQIYPFVFE
181 SRKEIL
```

Fig.15.1. Amino acid sequence of Tata binding protein related factor-2. Source: http//www. ncbi.nlm.nih.gov (Accession number NP_035733).

spermatocytes at stage VIII and increased during subsequent stages. After meiotic divisions, the TRF2 expression declined continuously in round spermatids from stage I to stage V (Zhang et al., 2001a). A sequence database search revealed that mouse TRF2 is encoded by two mRNAs containing the same protein-coding region and different 5'-untranslated regions. These two mRNAs are distinguished from each other by the expression patterns: ubiquitous and testis-specific expression. In transgenic animals deficient in TRF2, spermatid development fails during early spermiogenesis with increased apoptosis and failure to develop testicular spermatozoa. Infertility phenotype in homozygote males (Zhang et al., 2001b).The ubiquitously expressed form of TRF2 mRNA was present at a very low level, whereas expression of the testis-specific form was first detectable in the 14-day-old testis, and the mRNA level abundantly increased at the later stages of testicular development. The increase in TRF2 level during testicular development is consistent with the expression pattern of TRF2 mRNA. Thus, the presence of the testis-specific form of TRF2 mRNA may account for over-expression of the TRF2 gene in the testis (Sugiura et al., 2003) **(Fig.15.1)**.

15.2.4. TFIID Subunit TAF7

Transcription regulation in male germ cells can involve specialized mechanisms and testis-specific paralogues of the general transcription machinery (Pointud et al., 2003). The TAF7L, a germ-cell-specific paralogue of the TFIID subunit TAF7 has been described. TAF7L is expressed in most of the male germ-cells, but its intracellular localization dynamically changes from cytoplasm in spermatogonia and early spermatocytes to nuclear in late pachytene spermatocytes and haploid round spermatids. Import of TAF7L into the nucleus coincides with decreased expression of TAF7 and a strong increase in nuclear TBP, which suggests that TAF7L replaces TAF7 as a TFIID subunit in late pachytene spermatocytes and in haploid cells. A subpopulation of TAF7L is tightly associated with TBP in both pachytene and haploid cells and TAF7L interacts with the TFIID subunit TAF1. However, TAF3, TAF4 and TAF10 are all strongly expressed in early spermatocytes, in contrast to TBP and TAF7L, which are down-regulated in haploid cells. Hence, different subunits of the TFIID complex are regulated in distinct ways during male germ-cell differentiation (Pointud et al., 2003).

TFIIA: CREM and transcription factor IIA (TFIIA) are expressed in a spatially and temporally coordinated fashion during differentiation of germ cells. These proteins are co-localized intracellularly in spermatocyte and spermatids. CREM interacts with TFIIA, a general transcription factor that stimulates RNA polymerase II-directed transcription. The interaction is restricted to the activator isoforms of CREM and does not require Ser117. Importantly, CREM does not interact with TFIIAtau-ALF, a testis-specific TFIIA homologue (De Cesare et al., 2003).

15.2.5. Transcription Elongation Factor (S-II)

Elongating polymerases (both bacterial and eukaryotic) are associated with a series of elongating factor proteins, which decrease the likelihood that RNA polymerase would dissociate before it

```
  1 MGKEEEIARI ARRLDKMVTR KNAEGAMDLL RELKNMPITL HLLQSTRVGM SVNALRKQSS
 61 DEELIALAKS LIKSWKKLLD VSDGKSRNQG RGTPLPTSSS KDASRTTDLS CKKPDPPRTP
121 STPRITTFPQ VPITCDAVRN KCREMLTLAL QTDHDHVAVG VNCEHLSSQI EECIFLDVGN
181 TDMKYKNRVR SRISNLKDAK NPGLRRNVLC GAITPQQIAV MTSEEMASDE LKEIRKAMTK
241 EAIREHQMAR TGGTQTDLFT CNKCRKKNCT YTQVQTRSSD EPMTTYVVCN ECGNRWKFC
```

Fig.15.2. Amino acid sequence of transcription elongation factor S-II-T1, testis-specific mouse S-II. Source: http//www. ncbi.nlm.nih.gov. (Accession JC5430).

reaches the end of the gene. These factors associate RNA polymerase shortly after initiation has occurred. The S-II is a eukaryotic factor involved in the process of transcriptional elongation, and was initially found as a factor that stimulates the transcriptional activity of RNA polymerase II in vitro. Rat and mouse cDNA have been characterized for the testis-specific transcription elongation factor. S-II has been implicated in the process of transcriptional pausing, mainly through analysis using adenovirus and histone H3.3 genes as templates. S-II promotes transcription by RNA polymerase II against the inhibitory effects of elongation on these genes. Different isotypes of *S-II* have been discovered and shown to be expressed in restricted tissues (SII-T1) (Xu *et al.,* 1994; Ito *et al* 1996). S*II-T1* mRNA was not detected in spermatogonia or spermatids and appeared to be expressed exclusively in spermatocytes in the mouse (Ito *et al.*, 1996). The deduced rat S-II and SII-T1 protein sequences are highly similar, except for a unique intervening sequence of 46 residues in the SII-T1 protein (Xu et al., 1994), suggesting that these proteins are products of alternatively spliced transcripts from the same gene. Testis RNA from W/W^v mutant mice also suggested that SII-T1 is a specific transcription elongation factor essential for spermatogenesis (Ito et al., 1996). Umehara et al (1997) investigated the expression profile of tissue specific S-II-T1 during development using mouse *S-II-T1*, C-DNA. (1) Expression of *S-II-T1* is markedly reduced in the testes of adult WBBGF1-W/W mutant mice, which lack testicular germ cells, indicating that its expression is specific to testicular germ cells. (2) The onset of mouse *S-II-T1* mRNA appearance is seen about 10-14 days after birth, suggesting that *S-II-T1* is not transcribed in premeiotic and early meiotic cells such as spermatogonia, leptotene spermatocytes, or zygotene spermatocytes. (3) Mouse *S-II-T1* transcripts accumulate in meiotic pachytene spermatocytes and are detected in round and elongated spermatids. Thus the expression of mouse *S-II-T1* is restricted to testicular germ cells during and after meiosis in the course of spermatogenesis. **Figure 15.2** shows the deduced amino acid sequences of mouse S-II-T1, which is closely related to human and rat S-II-T1 with 89 and 99% sequence identity in overall primary structure, respectively.

15.3. RNA PROCESSING IN GERM CELLS

15.3.1. Alternative RNA Splicing

Gene expression during spermatogenesis is regulated also by the generation of alternative transcripts that encode different isoforms of proteins. Alternative transcripts are relatively common in this cell type and are often produced in a specific phase of spermatogenesis. Alternative transcripts can be generated by utilization of different promoters that activate transcription start sites, either up stream of the usual start site, as occur for cytosolic aspartate-aminotransferase, or down stream of the usual start site, as occurs for angiotensin-converting enzyme. They may also be produced by utilization of alternative exons, as occurs for hexokinase 1, or of alternative poly-adenylation signals for β1-4-galactosyltransferase. However, the mechanisms responsible for producing most alternative transcripts in spermatogenic cells are not known.

15. 3.2. Polyadenylation

The stored mRNAs in spermatids usually have poly (A) tails greater than 150 nt and are associated with ribosomes. They appear to undergo shortening of Poly [A^+] tract app. 30 nucleotides coincident with becoming associated with polysomes and translationally active. The poly A binding protein is a conserved protein that stabilizes mRNA and its translation. The initial RNA transcripts (hnRNA) derived from coding regions get modified at 5' end by gaining a methylated guanine (caps) and at 3' by polyadenylation (tails). Capping at 5' end occurs rapidly after start of transcription. Removal of cap leads to loss of translation activity. However, 3' end of mRNA is generated in two steps. First the transcript is cut at an appropriate location by a nuclease. In second step poly (A) is added to the newly generated end by poly (A) polymerase (PAP) utilizing ATP as substrate. On the average ~30% hnRNA and 47% mRNA are polyadenylated. In a region ~30 nucleotides upstream of the site of poly-A addition, there is a AAUAAA sequence which perhaps sends a signal for cleavage of polyadenylation. In most of the histones mRNAs, polyadenylation at 3'end is missing. Many mRNAs in male germ cells lack canonical AAUAAA but are normally polyadenylated (Wallace et al., 1999). During spermatogenesis there seem to be two distinct patterns of poly (A) size change. The poly (A) tracts of some mRNAs are longer in early haploid cells than in meiotic cells, whereas Prm mRNAs are translationally repressed with long poly (A) tracts in early haploid cells and translationally active with short poly (A) tracts in late haploid cells (Kleene et al., 1994).

15.3.3. Testis Specific Poly (A) Polymerase

A testis specific poly (A) polymerase (TPAP) gene is most abundantly expressed coincident with the additional elongation of mRNA poly(A) tails in round spermatids. The amino acid sequence of TPAP contained 642 aa residues, and shared a high degree of identity (86%) with that of a nuclear poly(A) including three catalytic Asp residues, an ATP-binding site, and an RNA-binding domain. The TPAP lacked an approximately 100-residue C-terminal sequence carrying one of two bipartite-type nuclear localization signals, and part of a Ser/Thr rich domain found in PAP II. The TPAP is recognized as a 70-kDa protein in the cytoplasm of spermatogenic cells, that may participate in the additional extension of mRNA poly(A) tails in the cytoplasm of male germ cells (Kashiwabara et al., 2000). The TPAP is stimulated by poly A binding protein II (Okamura et al., 2000). In mice, the absence of TPAP results in the arrest of spermiogenesis. TPAP-deficient mice display impaired expression of haploid-specific genes that are required for the morphogenesis of germ cells. The TPAP deficiency also causes incomplete elongation of poly(A) tails of particular transcription factor mRNAs. Although the overall cellular level of the transcription factor TAF10 is unaffected, TAF10 is insufficiently transported into the nucleus of germ cells. Thus TPAP can govern germ cell morphogenesis by modulating specific transcription factors at post-transcriptional and post-translational levels (Kashiwabara et al., 2002).

15.3.4. Poly-A Binding Protein

A 3'poly-(A) tail of about 200 bases is added post-transcriptionally in the nucleus to non-histone higher eukaryotic mRNAs. After transport of poly (A) mRNAs to the cytoplasm, the poly (A) functions in the cytoplasmic control of mRNA translation and stability. Changes in length of mRNAs are correlated with the activation and inactivation of translation of mRNAs in oocytes and early embryos. The poly (A) binding protein (PABP) that binds tightly to poly (A) plays a central role in regulating mRNA translation and its stability, which is important

```
  1 MNSSDPGCPM ASLYVGDLHP DVTEAMLYEK FSSAGPILSI RVYRDVITRR SLGYASVNFE
 61 QPADAERALD TMNFDVIKGK PVRIMWSQRD PSLRRSGVGN VFIKNLNKTI DNKALYDTFS
121 AFGNILSCKV VSDENGSKGH GFVHFETEEA AERAIEKMNG MLLNDRKVFV GRFKSQKERE
181 AELGTGTKEF TNVYIKNFGD RMDDETLNGL FGRFGQILSV KVMTDEGGKS KGFGFVSFER
241 HEDAQKAVDE MNGKELNGKH IYVGRAQKKD DRHTELKHKF EQVTQDKSIR YQGINLYVKN
301 LDDGIDDERL QKEFSPFGTI TSTKVMTEGG RSKGFGFVCF SSPEEATKAV SEMNGRIVAT
361 KPLYVALAQR KEERQAHLTN QYIQRMASVR SGPNPVNPYQ PASSSYSVAA VPQTQNCVPC
421 CPSQIAQPRP SARWIAQGSR PHPFPNVPGA IHPAAPRSSL TTVRPSSSHV QVTTAHRITN
481 TSAQITGQRP APASSATATP VHSIPQYKYA AGVQNSQQHL NAQLAQQPAV CIQGQEPWTA
541 SMLVTAPQEP KQMLGERLFP LIQAMHPTLA GKITGMLLDI DNSEPLRMLE SPVSRCSRAE
601 EAVATLQAHQ VKEAAQKAVG STSGVPTV
```

Fig.15.3. Amino acid sequence of cytoplasmic poly-A binding protein 2 from mouse testis germ cells. Source: http//www. ncbi.nlm.nih.gov (Accession NP_035163).

during meiosis. The PABP in yeast and the homologous proteins are now known from a variety of fungi, plants, and animals. The PABPs contain about 600 amino acids, which are organized into four-repeated 90 amino acid RNA binding domain proximal to the carboxyl terminus. The 90-amino acid domain is also found in a diverse family of RNP-proteins. Binding of PABP to poly (A), poly (G), and poly (U) in vitro suggests that PABP binds unidentified RNA sequences in addition to poly (A), and that combinations of the four RNA-binding domains have different specificities. The PABP is necessary for translational initiation in yeast and can protect the 3' ends of mRNAs from degradation in cell-free systems. Since, numerous mRNAs in meiotic and haploid spermatogenic cells are translationally repressed, the study of the expression of PABP mRNAs during spermatogenesis is important. The PABP mRNA levels increased as germ cells entered meiosis, reached a maximum in the early post-meiotic stages, and decreased to a nearly non-detectable level towards the end of spermatogenesis. The 70kDa PABP is wide spread in mammalian testis with maximum level in post-meiotic round spermatids. The presence of PABP in elongating spermatids, a cell type in which PABP mRNA is nearly absent, suggests that PABP is a stable protein in the later stages of male germ cell development. The high level of testicular PABP in round spermatids and in mRNPs suggests a role for PABP in the storage as well as in the subsequent translation of developmentally regulated mRNAs in the mammalian testis (Gu et al, 1995; Kleene et al., 1994).

Two PABP cDNAs with different sequences from mouse testis cDNA libraries predicted amino acid sequence of one PABP1 to be nearly identical (98.9%) to human liver PABP, while the amino acid sequence of the second, PABP2 (PABPt) showed 80% identity to mouse and human PABPs **(Fig. 15.3)**. Northern blots revealed that there is one major PABP mRNA species in liver, muscle, kidney, and brain, two in spleen, and at least four in testis. The PABP mRNA in testis are surprisingly the vast majority of all PABP mRNA size variants, which sediment more slowly than single ribosome, indicating strong translational repression. The existence of two PABP isoforms in mouse spermatogenic cells can influence cytoplasmic gene expression during spermatogenesis (Kleene et al., 1994). Although PABP mRNA is present in all cells, the levels in testis are 5- to 10-fold higher than in somatic tissues. There are multiple sizes of PABP transcripts in testis, but a 5'-untranslated region (UTR) probe for PABP2 hybridized only with a 2.7-kb transcript. Southern blot analysis indicated that three or more PABP genes exist in the mouse genome. The differences in the nucleotide sequences of the PABP1 and PABP2 cDNAs suggested that they represent mRNAs transcribed from different genes (Kleene et al., 1994). The cDNA clone isolated from mouse testis for PABP2 by Okamura et al., (2000) has an open reading frame encoding 302 amino acids, constituting a protein with a strong similarity to mouse and bovine poly (A) binding protein-2. The mPABP2 is known to bind growing poly (A) tail and stimulates poly (A) polymerase, which catalyzes the polymerization of the mRNA poly (A) tail. The mPABP2 cDNA is a single transcript of 1.2kb detectable exclusively in adult testis. The mPABP2 protein was expressed in the nucleus at specific stages from late pachytene spermatocytes to round spermatids. The *Pabp3* gene encoding mPABP2 is located near position 19.5 on mouse chromosome 14 (Okamura et al., 2000).

Cytoplasmic Polyadenylation Element Binding Protein: The cytoplasmic polyadenylation element binding (CPEB) protein regulates cytoplasmic polyadenylation of mRNAs as a trans factor in oogenesis and spermatogenesis. The CPEB protein contains two RNA recognition motifs and a Zn-finger structure. Proteins with similar structures are known from the genome database. A member of the CPEB protein family, CPEB2 is abundantly expressed in testis. Based on comparison of the amino acid sequences of CPEB2, CPEB family can be divided structurally and, perhaps, functionally into two groups: the CPEB group, and the CPEB2-KIAA0940-KIAA1673 group (Kurihara et al., 2003). The CPEB2 maps to mouse chromosome distal 5B. However, it was detected in all tissues that were examined. The CPEB2 is expressed post-meiotically in mouse spermatogenesis, suggesting a possible role in translational regulation of stored mRNAs in transcriptionally inactive haploid spermatids (Kurihara et al., 2003).

15.3.5. Cleavage Stimulation Factor - 64

Two distinct forms of the cleavage stimulation factor (CstF-64) are known: a somatic M(r) 64,000 form present in mouse male germ cells and in brain, and a variant M(r) 70,000 form specific to meiotic and post-meiotic germ cells. The gene for the somatic CstF-64 is located to the X chromosome, which is inactivated during male meiosis. The variant CstF-64 is an autosomal homologue activated during male meiosis. The variant form was named "tau CstF-64." The mRNA for the 64000 M(r) subunit of the CstF-64 is expressed at least 250-fold greater in mouse testicular RNA than in liver RNA. The mRNA for the 160000 M_r subunit of the cleavage and poly-adenylation specificity factor was similarly over-expressed, as was the mRNA for the large subunit of RNA polymerase II. The X-linked CstF-64 protein, expressed before and after but not during meiosis in the mouse (Wallace et al., 1999), suggests that over-expression of mRNA transcription and processing factors play an essential role in post-meiotic germ cell mRNA metabolism (Schmidt and Schibler, 1995; Dass et al., 2001a). The cloning and characterization of the mouse tauCstF-64 cDNA, which maps to chromosome 19, has been described (Dass et al., 2001b). The mouse tauCstF-64 protein meets the criteria of the variant CstF-64, including antibody reactivity, size and germ cell expression for tauCstF-64 from testis, and has substituted amino acids. Elevated levels of CstF-64 enhance the formation of 1-kb testis-brain RNA binding protein (TB-RBP) mRNA in male germ cells. Cst64, thus, modulates the synthesis of TB-RBP in male germ cells possibly by an alternative processing of TB-RBP pre-mRNA (Chennathukuzhi et al., 2001).

15.4. TRANSLATIONAL ACTIVITY IN GERM CELLS

The patterns of translational regulation of various mRNAs in the various testicular cell types, and models that account for the observed inhibition of the initiation step of translation have been described and translational activity of more than 40 different mRNAs in rodent testes has been analyzed by determining the proportions of inactive free-mRNPs and active polysomal mRNAs in sucrose gradients. From testicular somatic cells mRNAs sediment primarily with polysomes, indicating that they are translated efficiently, whereas the vast majority of mRNAs in late meiotic and haploid spermatogenic cells display high levels of free mRNPs, indicative of a block to the initiation of translation. Translational regulation of gene activity during post-meiotic stage of germ cells seems to be more important than in somatic cells. During this phase the nucleus condenses in step 9-12 spermatids of the mouse as the histones are shed from the chromatin and are replaced by TP 1 and TP 2 and then by the Prms. There are numerous examples of controls over the translation of individual mRNAs in meiotic and haploid cells; the proportions of various mRNAs in free mRNPs range from virtually none to virtually all, and individual mRNAs are activated at specific stages in elongated spermatids (Kleene, 1996).

15.4.1. Eukaryotic Translation Elongation Factor 1 (eEF1)

During protein synthesis, after formation of 70S initiation complex, the first amino-acyl tRNA (AA-tRNA) enters 'A' site of RNA. Factors responsible for this entry include elongation factors (EF-Tu and EF-Ts). EF-G is another elongation factor, which requires GTP. The corresponding eukaryotic elongation factors have been called eEF1 and eEF2. The somatic form of eEF-1 α (eEF-1 α-S) mRNA is virtually undetectable in male and female germ cells of the adult gonad but is very abundant in embryonic cells after the morula stage. In contrast, another form of eEF-1 α (eEF-1 αO) mRNA is highly concentrated in oogonia and in previtellogenic oocytes but is undetectable in eggs and embryos. The eEF–1 αO mRNA is also present in spermatogonia and spermatocytes of adult testis. Although germ cells contain very little eEF-1 αS retropseudogenes, several eEF-1 retropseudogenes exist in *Xenopus laevis* chromosomes. These genes are thought to arise in germ cells from reverse transcription of mRNA and subsequent integration of the cDNA copies into chromosomal DNA. It is suggested that eEF-1 α S pseudogenes are generated in primordial germ cells of the embryo before they differentiate into oogonia or spermatogonia (Abdallah et al., 1991).

15.5. m-RNA BINDING PROTEINS

The expression of many genes during early stages of spermatogenesis is regulated at the level of transcription, where as during the later phases of spermatogenesis, transcription ceases and stored mRNAs get activated to synthesize proteins as in Prms and transition proteins. Although little is known about the mechanisms inactivating and activating "paternal " mRNAs in germ cells, a number of RNA-binding proteins (RBPs) likely to be involved in mRNA translational control have been identified in the mammalian testis. Testis/brain (TB)-RBP binds to the 3'-untranslated regions (UTRs) of many testicular mRNAs and thus serves as a repressor of translation in vitro. Other proteins that bind to the 3' UTRs of testicular mRNAs include two proteins of 53-kDa and 55-kDa that bind to specific 22 and 20 nt sequences in the 3' UTRs of Prms mRNAs. Another protein with properties of a translational inhibitor of the Prms, and SPNR, is a 71-kDa protein that localizes to cytoplasmic microtubules of the testis. Multiple poly (A) binding proteins are also expressed in the testis. One of these, a 7-kDa protein, binds to both translated and stored mRNA in male germ cells. Another germ cell specific superoxide dismutase RNA-binding protein (SOD-RBP) binds to the 5'UTR of a testis-specified SOD-1 mRNA and can represses SOD-1 mRNA translation in vitro. Transcripts for several RNA-binding proteins have structural roles in modulating the conformation and organization of DNA or RNA. They are also associated with machinery of transcription, message stability, and translation. These proteins include ATP-dependent RNA helicases, poly (A)-binding protein, Y-box-binding proteins, and RNA-binding proteins. Some of these proteins may participate in RNA processing leading to the formation of unique transcripts, or may be involved in effecting translational delay. Thus, a growing number of testicular RNA-binding proteins have been identified, that are likely to facilitate the repression and activation of mRNA translation and gene transcription in male germ cells (see also Chapter 10).

15.5.1. Eukaryotic Translation Factor (eIF-4E)

Eukaryotic translation factor (eIF-4E) is a component in the regulation of translational efficiency of mRNAs and its increased expression may accelerate cell growth and division. A clone containing rat eIF-4E cDNA, isolated from a rat testis cDNA library is highly conserved in human, rat and mouse. Testis showed extraordinary elevated expression of eIF-4E, more than 50 fold, which was much greater than any tumor cell lines examined so far. At least half of the purified testicular eIF-4E protein was phosphorylated, a ratio similar to that in other rat tissues such as liver. The eIF-4E mRNA was mainly observed in post-meiotic germ cells. it seems that abundant eIF-4E in testis may play an important role in spermatogenesis through translational regulation of stage specific mRNAs during germ cell development (Asai et al., 1995).

15.5.2. RNA Helicases

PL10- A Homologue of eIF-4A as a Helicase: Two putative ATP-dependent RNA helicases have been reported that are expressed initially in pachytene spermatocytes in mouse. They share high homology to mouse translation initiation factor eIF-4A and other members of the DEAD box (Asp-Glu-Ala-Asp) family of proteins that are involved in many aspects of RNA metabolism, including splicing, translation, and ribosome assembly. One helicase, referred to as PL10, is encoded by two transcripts detected only in spermatogenic cells, at high levels in pachytene spermatocytes and lower levels in spermatids, and by a larger transcript found at low levels in the liver (Leroy et al., 1989). Southern analysis indicated that the PL10 transcripts come from a single copy gene, suggesting that spermatogenic cells and liver contain alternative transcripts. The murine PL10 expression is developmentally regulated, with high levels of transcripts being present during the meiotic and haploid stages of spermatogenesis. The deduced protein is shown to be highly homologous to the murine translation initiation factor eIF-4A and to other proteins that are also homologous to eIF-4A, including the Drosophila protein vasa. By consensus sequence conservation and comparison of secondary structure prediction, putative nucleotide binding and DNA/RNA binding domains were proposed to being shared by all these proteins. PL10 protein has helicase activity similar to that of eIF.4A and suggests its possible role in spermatogenic process **(Fig. 15.4)**.

RNA Helicase Like Protein (RHELP): Suk et al., (2001) isolated a RNA helicase like protein (RHELP) **(Fig.15.5.)** which had sequence homology to p68 RNA helicase, a prototypic member of the DEAD box protein family. The cDNA revealed that RHELP contained DEAD sequence motif and other conserved motifs of the DEAD box protein family, involved in the RNA processing, ribosome assembly, spermatogenesis, embryogenesis, and cell growth and division. RHELP showed 42% amino acid sequence identity to human p68 RNA helicase and yeast DBP2 RNA helicase, respectively among the DEAD box protein family. RHELP is expressed in most tissues including the liver, lung, tonsil, thymus, and muscle in addition to the pancreatic islets. The cDNA encoding the mouse P68 RNA helicase hybridized with an mRNA restricted to late pachytene and diplotene spermatocytes and spermatids (Lemaire and Heinlein, 1993).

Gonadotropin Regulated Testicular RNA Helicase (GRTH): The GRTH from rat Leydig cell, mouse testis, and human testis cDNA libraries is another member of DEAD box protein family,

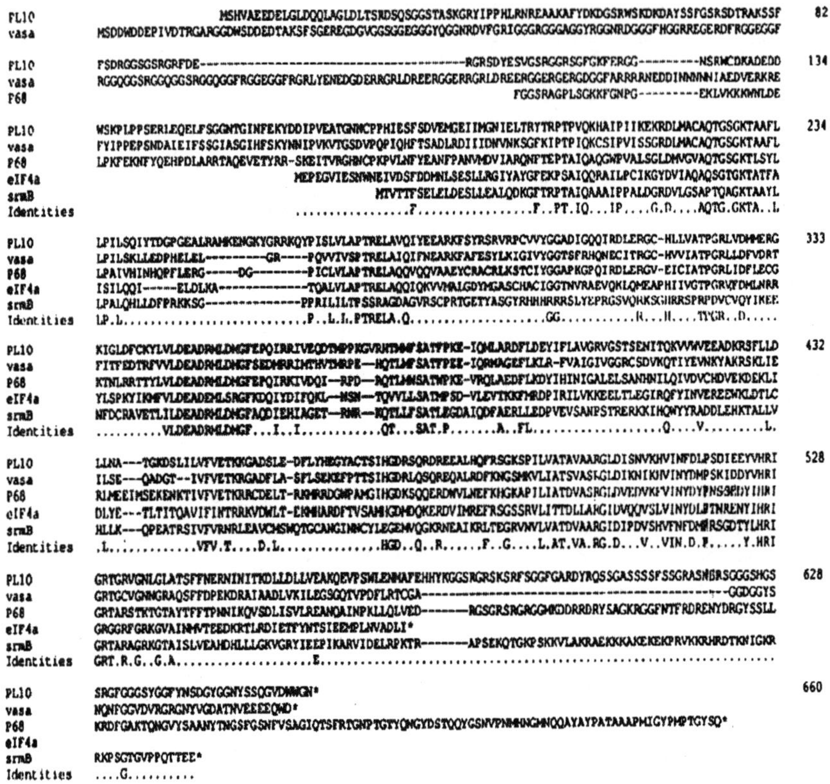

Fig.15.4. Comparison of deduced amino acid sequences of germ cell specific PL10 with P68 (human nuclear protein), eIF-4A (murine), vasa (*Drosophila*) and SrmB (E. Coli) proteins. Consensus identical residues in at least four of five proteins are given in last row (identities). Reprinted with permission from P. Leroy et al. Cell 57; 549-59: 1989 © Elsevier.

which is transcriptionally up-regulated by hCG via cyclic AMP induced androgen formation in the Leydig cell. It has ATPase and RNA helicase activities and increases translation in vitro. The GRTH gene transcription is stimulated by hCG via cyclic AMP-induced androgen formation in Leydig cells. It is produced in both somatic (Leydig cells) and germinal cells and is developmentally regulated. The GRTH is predominantly localized in the cytoplasm, may function as a translational activator. This helicase could be relevant to the control of steroidogenesis and the paracrine regulation of androgen dependent spermatogenesis in the testis (Tang et al., 1999) **(Fig. 15.6)**. Three ATGs codons with the potential for generation of multiple protein species were identified. Germ cells primarily contained major proteins of 61/56-kDa whereas Leydig cells expressed 48/43-kDa species. A 3rd ATG was weakly utilized and yielded a 33-kDa protein only in germ cells. In round spermatids, hCG caused a significant decrease of 61-kDa species and an induction 48/43 kDa species, whereas no changes were observed in pachytene spermatocytes. Studies have demonstrated a cell-specific and hormone-dependent alternative usage of ATG codons in the testis. The expression of GRTH proteins is regulated by gonadotropin/androgen at the translational level (Sheng et al., 2003). Male mice with null

```
60
RHELP           DNLEYDSDGNPIAPTKKIIDPLPPIDHSEIDYPPFEKNFYNEHEEITNLTPQQLIDLRHK
p68 (human)     GGS---AGPLSGKKFGNPGEKLVKKKWNLDELPKFEKNFYQEHPDLARRTAQEVETYRRS
p72 (human)     GG-------LPPKKFGNPGERLRKKKWDLSELPKFEKNFYVEHPEVARLTPYEVDELRRK
DBP (yeast)     YGYDQRGQGRNFYESDGPGANLVKKDWKNETLIPFQKDFYKEHENVRNRSDAEVTEYRKE
                                     *            *   ** **              *
                                                                            120
RHELP           LNLRVSG-AAPPRPGSSFAHFGFDEQLMHQIRKSEYTQPTPIQCQGVPVALSGRDMIGIA
p68 (human)     KKITVRG-HNCPKPVLNFYEANFPANVMDVIARQNFTEPTAIQAQGWPVALSGLDMVGVA
p72 (human)     KEITVRGGDVCPKPVFAFHHANFPQYVMDVLMDQHFTEPTPIQCQGFPLALSGRDMVGIA
DBP (yeast)     KEIVVHG-LNVPKPVTTFEEAGFPNYVLKEVKQLGFEAPTPIQQQAWPMAMSGRDMVGIS
                    * *    * *   *    *               ** ** *  * * ** ** *
                                                                            180
RHELP           KTGSGKTAAFIWPMLIHIMDQKELEPGDGPIAVIVCPTRELCQQIHAEJKRFGKAYNLRS
p68 (human)     QTGSGKTLSYLLPAIVHINHQPFLERGDGPICLVLAPTRELAQQVQQVAAEYCRACRLKS
p72 (human)     QTGSGKTLAYLLPAIVHINHQPYLERGDGPICLVLAPTRELAQQVQQVADDYGKCSRLKS
DBP (yeast)     ATGSGKTLSYCLPAIVHINAQPLLSPGDGPIVLVLAPTRELAVQIQQECTKFGKSSRIRN
                 ******     *   **  *  *  *****     *****  *
                                                                            240
RHELP           VAVYGGGSMWEQAKALQEGAEIVVCTPGRLIDHVKKKATNLQRVSYLVFDEADRMFDMGF
p68 (human)     TCIYGGAPKGPQIRDLERGVEICIATPGRLIDFLECGKTNLRRTTYLVLDEADRMLDMGF
p72 (human)     TCIYGGAPKGPQIRDLERGVEICIATPGRLIDFLESGKTNLRRCTYLVLDEADRMLDMGF
DBP (yeast)     TCVYGGVPRGPQIRDLIRGVEICIATPGRLLDMLDSNKTNLRRVTYLVLDEADRMLDMGF
                   ***     *   *  * **  ******        *** * ********** ****
                                                                            300
RHELP           EYQVRSIASHVRPDRQTLLFSATFRKKIEKLARDILIDPIRVVQGDIG-EANEDVTQIVE
p68 (human)     EPQIRKIVDQIRPDRQTLMWSATWPKEVRQLAEDFLKDYIHINIGALELSANHNILQIVD
p72 (human)     EPQIRKIVDQIRPDRQTLMWSATWPKEVRQLAEDFLRDYTQINVGNLELSANHNILQIVD
DBP (yeast)     EPQIRKIVDQIRPDRQTVMFSATWPKEVQRLARDYLNDYIQVTVGSLDLAASHNIKQIVE
                * * * *    ******   ***  *    ** * * *      *     *    ***
                                                                            360
RHELP           ILHSGPSKWNWLTRRLVEFTS--SGSVLLFVTKKANAEELANNLKQEGHNLGLLHGDMDQ
p68 (human)     VCHDV-EKDEKLIRLMEEIMSEKENKTIVFVETKRRCDELTRKMRRDGWPAMGIHGDKSQ
p72 (human)     VCMES-EKDHKLIQLMEEIMAEKENKTIIFVETKRRCDDLTRRMRRDGWPAMCIHGDKSQ
DBP (yeast)     VVDNA-DKRARLGKDIEEVLKDRDNKVLIFTGTKRVADDITRFLRQDGWPALAIHGDKAQ
                       *   *     *           *   *             *     ***   *
                                                                            420
RHELP           SERNKVISDFKKKDIPVLVATDVAARGLDIPSIKTVINYDVARDIDTHTHRIGRTGRAGE
p68 (human)     QERDWVLNEFKHGKAPILIATDVASRGLDVEDVKFVINYDYPNSSEDYIHRIGRTARSTK
p72 (human)     PERDWVLNEFRSGKAPILIATDVASRGLDVEDVKEVINYDYPNSSEDYVHRIGRTARSTN
DBP (yeast)     DERDWVLNEFRTGKSPIMVATDVASRGIDVKGITHVFNYDFPGNTEDYVHRIGRTGRAGA
                **  *   *     *   ***** **  *     * ***         ******  *
                                                                            480
RHELP           KGVAYTLLTPKDSNFAGDLVRNLEGANQHVSKELLDLAMQNAWFR----KSRFKGGKGKK
p68 (human)     TGTAYTFFTPNNIKQVSDLISVLREANQAINPKLLQLVEDRGSG-----RSRGRGG----
p72 (human)     KGTAYTFFTPGNLKQARELIKVLEEANQAINPKLMQLVDMRGGGGGGGGRSRYRTTS---
DBP (yeast)     KGTAYTYFTSDNAKQARELVSILSEAKQDIDPKLEEMARYSSGG---------RGG----
                 * ***  *         *   *  * *     *
```

Fig.15.5. Amino acid sequence of RHELP and other members of the DEAD box protein family. Asterisks indicate consensus sequences. The sequence motifs of RHELP conserved in the DEAD box protein family are bold-typed and underlined. Gap(-) are introduced to achieve maximum homology. Reprinted with permission from K. Suk et al. Biochim Biophys Acta 1501;63-69: 2000 © Elsevier.

mutation in GRTH gene displayed normal gonadotropin and androgen profile, but these mice were sterile, in association with azoospermia and arrest of spermatogenesis (Tsai-Morris et al, 2004).

15.5.3. Protamine m-RNA Binding Protein

The testis-specific mouse Prm genes are transcribed in haploid round spermatids, their mRNAs stored as cytoplasmic ribonucleoprotein particles and translated about 1 week later in elongating spermatids after cessation of transcription. The in vitro translational efficiencies of Prm-1

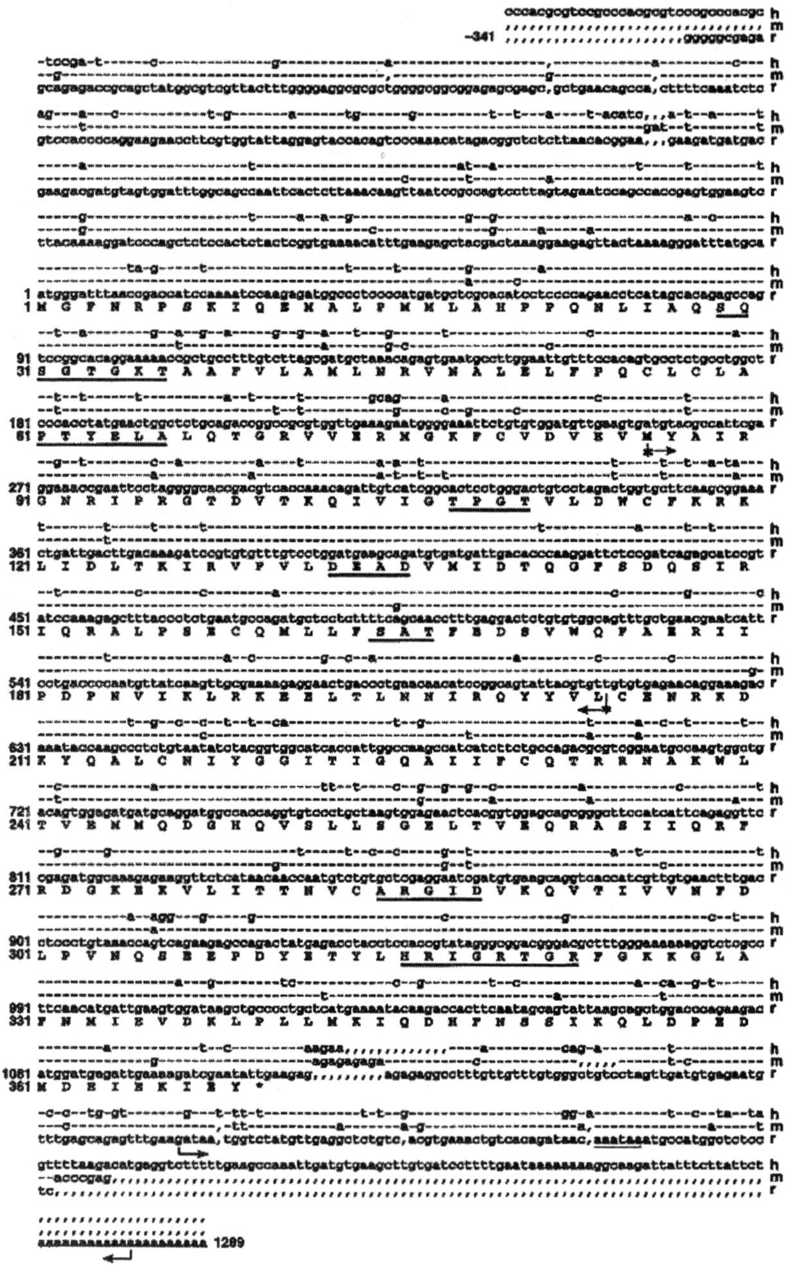

Fig.15.6. Rat, mouse, and human GRTH cDNA and predicted amino acid sequence of rat encoded protein. The ORF is shown below. The deduced amino acid sequences of GRTH contain all conserved domains of DEAD-box family of RNA helicases (underlined). Reprinted with permission from P.-Z. Tang et al. J Biol Chem 274; 37932-40: 1999 © American Society for Biochemistry and Molecular Biology.

mRNA isolated from round spermatids translates as efficiently as Prm-1 mRNA from elongating spermatids. Translation of Prm-1 mRNA is normally repressed in round spermatids. The presence of the binding activity in the cell type and sub-cellular compartment associated with Prm-1 and Prm-2 mRNA storage suggested that the activity might be actively engaged in translational repression of these mRNAs (Fajardo et al., 1994).

There are numerous reports demonstrating the pivotal role of 3' untranslated regions (UTR)' in mRNA stability, transport, localization, and translational control (Jackson, 1993). Since 3' UTR sequences are involved in many aspects of mRNA metabolism, Schumacher et al (1995) identified proteins that could bind to sequences within the 3'UTR of the spermatid-specific Prm 1 (*Prm-1*) mRNA. The mouse *Prm-1* gene encodes a male germ cell-specific mRNA that is stored as a cytoplasmic mRNP from the time it is first transcribed in post-meiotic round spermatids until its translation up to a week later in elongating spermatids. It was found subsequently that a 40-kDa Prm1 RNA binding protein (PRBP) that binds to this sequence is present in spermatids when the mRNA is repressed translationally. Lee et al., (1996) obtained two independent clones that encode PRBP, a Prm-1 RNA-binding protein (PRBP). The PRBP contains two copies of a double stranded-RNA-binding domain and binds to a portion of the *Prm*-1 3' untranslated poly region shown to be sufficient for translational repression in transgenic mice, as well as to poly (I) -poly (C). PRBP protein is present in multiple forms in cytoplasmic extracts prepared from wild type mouse testes and is absent from testes of germ cell deficient mouse mutants, suggesting that PRBP is restricted to the germ cells of the testis. Recombinant PRBP protein inhibits the translation of multiple mRNAs in a wheat germ lysate, suggesting that PRBP acts to repress translation in round spermatids. While this protein lacks complete specificity for Prm-1 containing RNAs in vitro, the properties of PRBP are consistent with it acting as a general repressor of translation (Lee et al., 1996) **(Fig. 15. 7)**.

Studies in transgenic mice have shown that sequences of the last 62 nucleotides in the 3'untranslated region (UTR) regulate translation of Prm-1 mRNA in spermatids. Other studies demonstrated that an 18-kDa phosphoprotein present in testis and brain extracts binds to YH sequence elements in the 3'UTR of Prm-2 mRNA and represses its translation in *vitro.* In addition, 48- and 50-kDa proteins bind to a conserved 20- to 22-bp element in the 3' UTR of Prm-1 and -2 mRNAs. These results indicated that multiple mRNA binding proteins interact to repress mRNA transcription. Alternatively some of these proteins have functions other than translational control, such as effecting mRNA stability and localization (Wu et al., 2000).

15.5.4 Testis Nuclear RNA Binding Protein (Tenr)

In a molecular screening of DNAs that encode Prm RNA-binding protein, Schumacher et al., (1995a) isolated seven independent clones that encode Tenr, a testis nuclear RNA-binding protein. Tenr is a 72-kDa protein that has one copy of a RNA-binding domain. Tenr binds to a variety of RNAs in vitro and it does not bind to a single stranded or double stranded DNA. The *Tenr* gene is transcribed exclusively in the testis, and its mRNA is restricted to cells from the pachytene spermatocyte stage through the round spermatid stage. It is first detected post-meiotically, demonstrating that the *Tenr* mRNA is under translational control. The Tenr shows a lattice like nuclear distribution suggesting its association with the nuclear scaffold.

15.5.5 Testis-Brain RNA Binding Protein (Translin)

An 18k-Da phosphoprotein in testis has been suggested earlier to bind to transcript C, a sequence containing two highly conserved elements, Y and H, in the 3' untranslated region (UTR) of mouse Prm 2 mRNA and represses its translation in vitro. Phosphatase studies have

```
   1 CCGGGCTTGGCCCCGGCCCTAGCTCCTCGGCGGTGTACTGGGGTGCGTGGAGGCTGTAGTCACGGTGGCGCCCGCGGGGACGGAGGAGGGA

  92 ATG AGT GAA GAG GAT CAG GGC TCC GGC ACT ACT ACA GGC TGC GGG CTG CCC AGC ATA GAG CAA ATG CTG
      M   S   E   E   D   Q   G   S   G   T   T   T   G   C   G   L   P   S   I   E   Q   M   L   23

 161 GCC GCC AAC CCG GGC AAG ACC CCG ATC AGC CTT CTT CAG GAG TAT GGG ACC AGA ATA GGA AAG ACG CCC
      A   A   N   P   G   K   T   P   I   S   L   L   Q   E   Y   G   T   R   I   G   K   T   P   46

 230 GTG TAC GAC CTT CTC AAA GCC GAG GGC CAA GCC CAT CAA CCT AAT TTC ACC TTT CGG GTC ACC GTT GGC
      V   Y   D   L   L   K   A   E   G   Q   A   H   Q   P   N   F   T   F   R   V   T   V   G   69

 299 GAC ACC AGC TGC ACT GGT CAG GGC CCC AGC AAG AAG GCA GCC AAG CAC AAG GCA GCT GAG GTG GCC CTC
      D   T   S   C   T   G   Q   G   P   S   K   K   A   A   K   H   K   A   A   E   V   A   L   92

 368 AAA CAC CTC AAA GGG GGG AGC ATG CTG GAA CCA GCC CTG GAG GAC AGC AGT TCT CTT TCT CTC CTA GAC
      K   H   L   K   G   G   S   M   L   E   P   A   L   E   D   S   S   S   L   S   L   L   D   115

 437 TCT TCA CCG CCT GAG GAC ACT CCT GTC GTT GCT GCA GAA GCT GCT GCC CCT GTT CCA TCT GCT GTA CTA
      S   S   P   P   E   D   T   P   V   V   A   A   E   A   A   A   P   V   P   S   A   V   L   138

 506 ACC AGG AGC CCT CCC ATG GAG ATG CAG CCC CCT GTC TCT CCT CAG CAG TCT GAG TGC AAC CCC GTC GGT
      T   R   S   P   P   M   E   M   Q   P   P   V   S   P   Q   Q   S   E   C   N   P   V   G   161

 575 GCT CTG CAG GAG CTG GTG GTG CAA AAA GGC TGG CGT TTG CCA GAG TAC ATG GTG ACC CAA GAG TCT GGG
      A   L   Q   E   L   V   V   Q   K   G   W   R   L   P   E   Y   M   V   T   Q   E   S   G   184

 644 CCT GCT CAC CGC AAA GAG TTC ACC ATG ACT TGC CGG GTG GAG CGT TTC ATT GAG ATT GGC AGT GGC ACT
      P   A   H   R   K   E   F   T   M   T   C   R   V   E   R   F   I   E   I   G   S   G   T   207

 713 TCC AAA AAG CTG GCA AAG CGT AAC GCA GCA GCT AAG ATG CTC CTT CGA GTG CAC ACT GTA CCT CTG GAG
      S   K   K   L   A   K   R   N   A   A   A   K   M   L   L   R   V   H   T   V   P   L   D   230

 782 GCC CGG GAT GGC AAT GAG GCA GAG CCT GAT GAC GAT CAT TTT TCC ATT GGC GTG AGC TCC CGC CTG GAT
      A   R   D   G   N   E   A   E   P   D   D   D   H   F   S   I   G   V   S   S   R   L   D   253

 851 GGA CTG AGG AAT CGT GGG CCA GGC TGC ACC TGG GAT TCC TTG CGG AAT TCT GTG GGA GAA AAG ATC CTA
      G   L   R   N   R   G   P   G   C   T   W   D   S   L   R   N   S   V   G   E   K   I   L   276

 920 TCC CTT CGC AGT TGC TCC GTG GGC TCT CTA GGG GCT CTG GGC TCT GCC TGC TGC AGT GTC CTC AGT GAG
      S   L   R   S   C   S   V   G   S   L   G   A   L   G   S   A   C   C   S   V   L   S   E   299

 989 CTC TCT GAG GAG CAG GCT TTC CAT GTC AGC TAT CTG GAT ATT GAG GAA CTG AGC CTG AGT GGG CTC TGC
      L   S   E   E   Q   A   F   H   V   S   Y   L   D   I   E   E   L   S   L   S   G   L   C   322

1058 CAG TGC CTA GTG GAA CTG TCC ACC CAG CCA GCC ACT GTG TGT TAT GGT TCT GCA ACC ACC AGG GAG GCA
      Q   C   L   V   E   L   S   T   Q   P   A   T   V   C   Y   G   S   A   T   T   R   E   A   345

1127 GCC CGA GGT GAT GCT GCT CAC CGC GCC CTA CAG TAC CTC AGG ATC ATG GCG GGT AGC AAG TAG CATCCCA
      A   R   G   D   A   A   H   R   A   L   Q   Y   L   R   I   M   A   G   S   K               365

1197 CTGCAGTGATGGATATGCATCTTTTACTTCTTGCTCCTTCTGCCCCTGGGTCCATGTATCCACCTAGCTCTGGTACCCTCCAGAGGTGCCA

1288 TCTCTACCTCTGACACAGCCTGTCTGCCTTGAGACTGAGGAAGGCACAGGCAAGCAAGGAGCCATGAACCACAGGGCCCCAGCCAGCACAG

1379 GATTTGTCCTCATTCCGTGGGTGATGGATGAATCTATTGGAGTCCTGAATAAATGCTGCTCTTTGGCTTCCAAAACCCTGCTCTCTCTGGC

1470 CTTGGGTCATGAGGGTATAGGAATAAACATGGTGAGCTTGTTCTCCTTTTG
```

Fig.15.7. Nucleotide and deduced amino acid sequence of the prbp cDNA. An upstream in frame stop codon (TAG) is indicated by a double underline. The polyadenylation signals are underlined. Reproduced with permission from K. Lee et al, Mol Cell Biol 16; 3023-34: 1996 © American Society for Microbiology.

shown that it must be phosphorylated to bind to RNA. Most importantly, a partially purified 18kDa-protein fraction selectively represses, in a cell free translation system, a human growth hormone (hGH) reporter gene linked to the Y and H elements of the 3'UTR of Prm2 mRNA. This suggests that this RNA binding protein is one of the factors controlling the temporal translation of stored mRNAs such as Prm2 in testis. In addition to testis, brain also contains a protein that binds to transcript C. Both the testis and brain proteins are found exclusively in the non-polysomal fractions of their postmitochondrial extracts. The testis and brain proteins appear to be identical according to numerous criteria: the identical mobility of their complexes with transcript C in native gels, identical optimal pH, identical lability to increased salt concentration, identical chromatographic properties, identical molecular sizes as judged from UV-cross linking, and identical peptide mapping. In addition, the phosphoprotein from testis and brain, hereafter called testis-brain RNA-binding protein (TB-RBP), also specifically binds to a similar sequence in the 3' UTR of brain Tau mRNA. TB-RBP plays a role in mRNA storage, translocation, and / or localization in brain and testis (Han et al., 1995). The TB-RBP moves from the nucleus to the cytoplasm and through intercellular bridges of male germ cells. Based on its RNA binding capabilities, a role for TB-RBP in the distribution of equal amounts of mRNAs in haploid male germ cells has been proposed (Morales et al., 1998). The TB-RBP is present primarily in the nuclei of male germ cells during meiosis and in the cytoplasm of the male cells after metaphase I of meiosis. TB-RBP suppresses translation in vitro and attaches mRNAs to microtubules by binding to conserved elements in the 3' untranslated regions of specific mRNAs.

Numerous functions have been proposed for TB-RBP and its human homologue, translin, ranging from mRNA transport and translational regulation to DNA rearrangement and repair. The TB-RBP/translin is a DNA- and RNA-binding protein with multiple functions. As an RNA-binding protein, TB-RBP binds to conserved sequence elements often present in the UTRs of specific mRNAs modulating their translation and transport. To gain insight into the likely functions of this protein, three proteins, the transitional endoplasmic reticulum ATPase (TER-ATPase), a cytoskeletal γ actin, and Trax, were specifically immunoprecipitated and confirmed that TB-RBP interacts with the TER-ATPase in vitro and in vivo and colocalizes with actin in the cytoplasm of male germ cells. This supports the hypothesis that TB-RBP serves as a link in attaching specific mRNAs to cytoskeletal structures and suggests an involvement for the ubiquitously expressed TER-ATPase in intracellular and/or intercellular mRNA transport (Wu et al., 1999). TB-RBP is a sequence-dependent RNA-binding protein that binds to conserved Y and H sequence elements present in many brain and testis mRNAs. Recombinant TB-RBP binds to microtubules assembled in vitro. An intracellular association between TB-RBP and specific target mRNAs suggests the involvement of TB-RBP in microtubule-dependent mRNA transport in the cytoplasm of cells (Wu and Hecht, 2000). Using reporter constructs and gel mobility shift assays it was possible to show that the TB-RBP associates with mRNPs in a sequence (Y element) dependent manner. Blocking the TB-RBP Y element binding site disrupts and mis-localizes mRNPs containing (α) calmodulin dependent kinase II (α-CAMKII) and ligates mRNAs. In addition, the suppression of kinesin heavy chain motor protein alters the localization of α-CaMKII mRNA (Severt et al., 1999).

The mouse TB-RBP cDNAs contains an open reading frame of 228 amino acids with a leucine zipper domain within its C-terminus, a transmembrane helix, and a group of putative phosphrylation sites. The TB-RBP shows 99% identity to the human protein, translin, a recombination hotspot-binding protein associated with chromosomal translocations (Aoki et al., 1995 c/r Wu et al., 1997). As shown for translin, TB-RBP also binds to single stranded DNAs containing a broad range of consensus sequences, many of which are similar to the Y and H RNA binding sequences. Based upon its DNA binding to target sequences in clustered breakpoint regions, it was proposed that TB-RBP might be involved in DNA recombination or DNA repair in male germ cells (Wu et al., 1997). The single copy of mouse TB-RBP gene encodes three mRNAs of 3.0-, 1.7-, and 1.0-kb which only differ in their 3' UTRs. The 1.0-kb TB-RBP mRNA predominated in testis, while somatic cells preferentially express the 3.0-kb TB-RBP mRNA. The 1-kb mRNA is translated several fold more efficiently than the 3-kb TB-RBP in rabbit reticulocyte lysates, and cells with elevated levels of the 1-kb TB-RBP mRNA express high levels of TB-RBP. Over-expression of CstF 64 increases the poly (A) site selection for the 1-kb TB-RBP mRNA. It appears that the levels of the polyadenylation factor CstF-64 modulated the levels of TB-RBP synthesis in male germ cells by an alternative processing of the TB-RBP pre-mRNA (Chennathukuzhi et al., 2001). The *Gapds* mRNA is the first mRNA shown to have a functional TB-RBP binding site in its 5' UTR (Yang et al., 2003). Selective precipitation of glyceraldehyde-3-phosphate dehydrogenase (*Gapds)* mRNA in a TB-RBP-mRNA complex, and its appearance in non-polysomal fractions suggested that TB-RBP is involved in the post-transcriptional regulation of *Gapds* gene expression during spermiogenesis. The *Gapds* mRNA translation was inhibited by recombinant TB-RBP or by a TB-RBP mutant protein in vitro (Yang et al., 2003).

15.5.6. Y- Box Proteins

The vertebrate Y box proteins represent a family of nucleic acid-binding proteins containing an 85 amino acid domain termed the cold-shock domain (CSD), which contains sequences that are over 40% identical to the small bacterial cold shock proteins and binds single stranded nucleic acids by their ribonucleoprotein (RNP)-1 and RNP-2 like motifs. The bacterial cold shock proteins act as transcription factors, a property conserved in the Y box proteins to either promote or repress transcription of a variety of genes. In addition to binding to DNA, Y box binding proteins also bind to RNAs in a sequence independent manner and facilitate the translational regulation of stored mRNAs. Hence they have duel function. All vertebrate Y-box proteins contain three domains: a varable N-terminal domain, a highly conserved nucleic acid binding domain (CSD = cold shock domain) and a C-terminal domain. The N-terminal domain is alanine and proline rich (A/P domain). The vertebrate Y box proteins contain several interspersed alternate basic or acidic (B/A) clusters in their C-terminal region that binds single stranded DNA or RNA in vitro (Kohno et al., 2003).

MSY1 and MSY2 (Xenopus Homologue FRGY1 and FRGY2): In *Xenopus* oocytes, the Y-box protein FRGY2 (mRNP3+4) is abundant and found in cytoplasmic translationally repressed maternal mRNAs. The mammalian homologue of the *Xenopus* germ cell-specific nucleic acid-binding protein FRGY2 (mRNP3+4) was designated MSY2 in 1998. The FRGY1 and FRGY2 positively regulate transcription from promoters containing a Y-box (CTGATTGGCCAA). In addition, the FRGY2 binds maternal mRNA within 60-80S mRNP storage particles and regulates translation on the developing oocyte and embryo. The binding of intact maternal mRNA by MSY2 is required for its cytoplasmic retention (Yu et al., 2003). Mammalian homologue, MSY1 mRNA accumulates over 100-fold more in testis than in other tissues and is developmentally regulated, initially appearing at pachytene stage of spermatogenesis. This is the stage of maximal transcription and translation in the spermatocyte. The MSY1 protein form a 60-80S mRNP fraction of testis, which like the frog oocyte contains stored, untranslated mRNAs, is bound to mRNA of this fraction.

Many Y-box proteins have been cloned, including mouse MSY1, which shares 85% identity with *Xenopus* FRGY1. MSY1 is most highly expressed in the testis but is also expressed in somatic tissues. Mammalian germ cell homologues, p48/p52 (MSY2), have been identified in mouse testis. Evidence indicates that a Y box-binding family of transcriptional regulators is identical to p56, a predominant protein of mRNP complexes. The p56 protein is highly enriched in oocytes and testis. FRGY2 is identical to an RNA-binding protein designated mRNP3+4 or p54/p56. They are bound to non-polysomal RNA and also interact with Y box element in Prm1 promoter (Kwon et al., 1993; Nikolajczyk et al., 1995, Tafuri et al., 1993) **(Fig.15.8)**. The 5' flanking region of the testis-specific, haploid-expressed mouse Prm gene contains an element with a 9- of -12 nucleotide identities with the previously defined Y-box consensus sequence. In the testis, MSY2 is alternatively spliced, giving rise to polypeptides that differ in the amino-terminal region upstream of the CSD. The *Msy2* undergoes alternative splicing to yield alternate N-terminal regions upstream of the cold shock domain. *Msy2* is a single copy gene. MSY2 (P48/

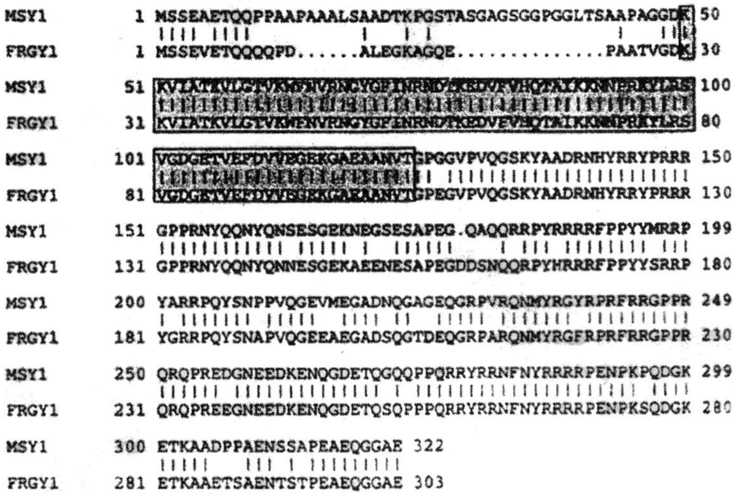

Fig.15.8. Amino acid sequence of MSY1 showing identity (85%) and similarity (90) with FRGY1. The shaded region shows highly conserved cold shock domain required for DNA interaction. Reprinted with permission from S.R Tafuri et al. J Biol Chem 268; 12213-20: 1993 © American Society for Biochemistry and Molecular Biology.

52) is first detected in early pachytene spermatocytes and progressively increases during the remainder of meiotic prophase to a post-meiotic peak in steps 1-8 round spermatids; thereafter, declined as elongated spermatids underwent nuclear condensation and elongation. The pattern of synthesis of these mRNA binding proteins together with their association with the chromatoid suggests a role as germ cell-specific mRNA stabilizing and/or storage proteins (Oko et al., 1996; Gu et al., 1998).

Contrin: A number of closely related Y-box gene products including dbpA, dbpB and contrin (dbpC) have been identified in recent years (Tekur et al., 1999 c/r Kohno et al., 2003). Contrin is closely related to YB-1 and dbpA and is specifically expressed in mouse germ cells. The Dbpc or contrin is highly homologous to germ cell specific Y-box protein, MSY2. Since YB-1 is over-expressed in cancer cells and is considered a promising target for cancer therapy, a similar role for Y-box proteins in testicular germ cells for chemical contraception needs to be investigated (Kohno et al., 2003).

15.5.7. Other mRNA-Binding Proteins in Germ Cells

Spermatid Perinuclear RNA Binding Protein: The protein product of spermatid peri-nuclear RNA-binding protein (SPNR), contained two RNA-binding domains similar to those found in the *Drosophila* staufen protein. Within the testis it is expressed exclusively in post-meiotic germ cells and that it is localized to the manchette, a spermatid specific microtubular array. Both strands of the 3-kb *SPNR* cDNA were sequenced. The entire nucleic acid- and the deduced amino acid sequences are shown in **Fig 15.9**. The *SPNR* cDNA codes for a 649-amino acid protein of 71-kDa. It has significant homology with a family of known RNA-binding proteins that includes *Xenopus* 4F-1 and 4F-2, *Drosophila* staufen, human Tar-binding, murine Prm RNA binding protein and human protein kinase. The *SPNR* protein has two copies of the RNA-

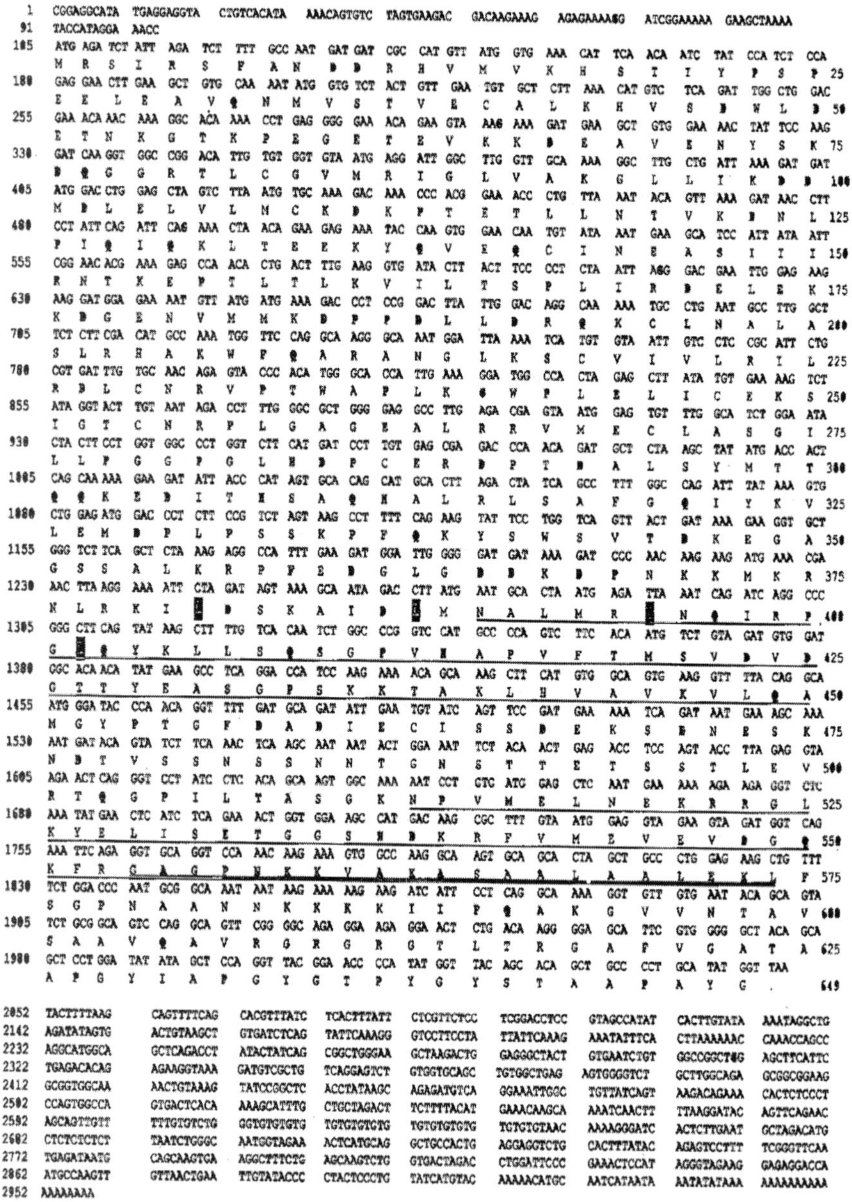

Fig.15.9. c-DNA and predicted amino acid sequences of *Spnr cDNA.* RNA binding motifs are underlined with the most highly conserved portion of the motif doubly underlined. Leucine residues within leucine zipper motif are boxed. Reprinted with permission from M. Schumacher et al, J Cell Biol 129;1023-1032:1995© The Rockefeller University Press.

binding domains both of which are located in the carboxy-terminal third of the *SPNR* protein. Just upstream of both binding motifs are stretches of conserved amino acids. This similarity is found upstream of the RNA binding domain in many of the other family members. The SPNR protein also contains a putative leucine zipper domain between amino acids 381 and 402. Just upstream of this motif is a highly charged region with a short stretch of aspartic acid residues followed by a short basic region containing several lysine and arginine residues. Leucine zippers are often found preceded by highly charged regions. The *SPNR* gene was mapped on mouse and human metaphase chromosomes. The SPNR mRNA is expressed at high levels in the testis, ovary, and brain, and is present in multiple forms in those tissues. Although the SPNR protein is expressed too late to be directly involved in the translational repression of *Prm-1* specifically, it is suggested that the SPNR protein may be involved in other aspects of spermatid RNA metabolism, such as RNA transport or translational activation (Schumacher et al., 1995b) (**Fig. 15.9**).

The SPNR's mutant mice with reduced fertility or sterility showed structurally abnormal manchettes, and that despite the highly abnormal manchettes and microtubule aggregates formed in *azh, hop-sterile*, t^{w2}, and t^{w8} mutants, SPNR remains associated with the manchettes. Localization of SPNR to the abnormal manchettes suggested that SPNR is tightly bound to the manchette. The gene encoding SPNR was cloned in a screen for RNA-binding proteins that could bind to the 3'-untranslated region (UTR) of the testis-specific Prm 1 (Prm-1) mRNA in vitro. The appearance of SPNR temporally coincides with the translational activation of the *Prm1* mRNA. SPNR and other family members that contain the double stranded RNA-binding domain bind double stranded RNA and highly structured single stranded RNAs in vitro. The SPNR interacts with the Prm1 mRNA in vivo, although in vitro RNA-binding studies with portions of the SPNR protein failed to reveal sequence specific binding to any RNAs, including the *Prm1* 3'UTR. In addition to its function as an RNA-binding protein, SPNR is also a bonafide microtubules associated protein (Schumacher et al., 1998).

RNA Binding Zinc Finger Proteins: Zinc finger protein possesses a region common to a small class of RNA binding proteins. The *Zfr* (zinc finger RNA binding) encodes a protein of 1052 amino acids with three widely spaced Cys_2His_2 zinc fingers. Outside the zinc finger, ZFR protein shares a region that is highly conserved between several RNA binding proteins containing copies of the double stranded RNA binding motif. Zfr is expressed at highest levels within the testis, ovary and brain and highly expressed during meiosis I in males and females. Zfr is also expressed in Sertoli cells in the testis and granulosa cells in the ovary where it is localized to the nucleus. An apparent homologue of ZFR exists in invertebrates (*D.melanogaster*), nematodes (*C.elegans*) and humans (*H. sapiens*) suggesting it as an ancient protein (Meagher et al., 1999).

PUF Proteins: PUF proteins are a conserved family of RNA binding proteins that regulate RNA stability and translation by binding to specific sequences in 3'-untranslated regions. The common function of PUF proteins may be to sustain mitotic proliferation of stem cells. *Drosophila* PUMILIO and *C. elegans* FBF are essential RNA binding proteins for self-renewal of germ stem cells. However, PUF-8, from *C. elegans*, related to PUMILIO, performs a different function in meiotic germ cells viz., PUF-8 is required to maintain meiosis and prevents the return of primary spermatocytes to mitosis. PUF-8 deficient primary spermatocytes complete meiotic prophase but do not undergo normal meiotic divisions. Instead, they dedifferentiate back into mitotically cycling germ cells and form rapidly growing tumors (Subramaniam and Seydoux, 2002). These findings support the view that PUF proteins regulate multiple transitions during germ-line development.

Transcriptional Control of Testis Specific Cytochrome C_t: In testes there are two cytochromes, somatic (Cyt Cs) and testis specific (Cyt Ct). Cys Cs and Cyt Ct differ in 14 out of 104 amino acids (see Chapter 28). *Cyt.* c_r gene was studied by examining DNA-protein interactions in its proximal promoter. The sequence from -136 to -127, bound by liver but not to testicular nuclear proteins, is similar to that of the binding site of a somatic *c-mos* repressor protein. Lastly, different nuclear proteins from mouse liver and testis bound to a region from -18 to + 31 that contains a putative Y box at -13 to -2. Studies established that this putative Y box binds a 52-kDa mouse testicular homologue of the *Xenopus* germ cell-specific Y box protein and a competing 50-kDa protein present in both liver and testis nuclear extracts. The testis-specific expression of the mouse *Cyt.* c_T gene during spermatogenesis may be regulated by the differential binding of tissue-specific nuclear proteins to its proximal promoter region (Yiu et al., 1997). Besides RNA binding proteins in transcriptional control of Cyt Ct in testes, there exist several RBPs, which are involved in transcriptional or translational control in germ cells of testis. Although transcriptional control of many germ cell specific genes in testis has been studied during last decade, our understanding on identification and characterization of translational components is limited (see Chapters 10 and 16).

15.6. mRNA AND RNP IN SPERMATOZOA

Studies demonstrated that human ejaculated spermatozoa contain the mRNA transcripts of β1 integrins and supported the role of these proteins in adhesive mechanisms in the process of fertilization. Subsequent reports demonstrating the presence of mRNAs for sperm-specific nucleoproteins, the Prm1, Prm2, and the transition protein TP-1 as well as β-actin, Hsp70, Hsp90, c-MYC, HLAI, HLAII, and recently observed phosphodiesterase subtypes have challenged the accepted view of the transcriptional dormancy of terminally differentiated spermatozoa. It appears that spermatozoa are a repository of information regarding meiotic and post-meiotic gene expression in the human and are likely to contain transcripts for many homologues of genes identified from animal models, which play an essential role in spermatogenesis. Further studies indicated the presence of high levels of transcribed repetitive sequences in human spermatozoa, including medium reiteration repeats (MERs) and short and long nuclear interspersed repeats (SINES and LINES). Both SINES and LINES belong to the retroposon repeat elements, which are thought to proliferate via an intermediate RNA that is converted into DNA by reverse transcriptase (RT). Evidences for the presence of reverse transcriptase in ejaculate sperm based on the detection of transcripts for the ORF2 of LINE 1 encoding such an enzyme had been obtained with murine and human sperm. Sperm cells incubated with human poliovirus RNA can take up exogenous RNA molecules and internalize them in nuclei. Direct PCR amplification of DNA extracted from RNA incubated spermatozoa indicated that poliovirus RNA is reverse transcribed in cDNA fragments (Giordano et al., 2000). The evidence has to be generated, that could prove that stored mRNAs and RT in sperm may reverse transcribe paternal genes in fertilized egg. In conclusion it suggests that: i) Ejaculate spermatozoa are useful tool in the identification of genes linked to an infertile phenotype. ii) Spermatozoa (or spermatids) may express a reverse transcriptase, the role of which is unknown. iii) RNA isolated from spermatozoa or washed semen samples may facilitate the detection of mutations and deletions in testis expressed AZF-linked genes. iv) Sperm ejaculates as a non-invasive biopsy of spermatid need to be evaluated for infertility (Miller 1997; 2000) (see Chapter 10).

15.7. REFERENCES

Abdallah B, Hourdry J, Krieg PA, Denis H, Mazabraud A. Germ cell-specific expression of a gene encoding eukaryotic translation elongation factor 1α (eEF-1 α) and generation of eEF-1α retropseudogenes in Xenopus laevis. Proc Natl Acad Sci U S A 1991; 88: 9277-81.

Asai A, Shibuya M, Fujimoto H, Kuchino Y. Abundant expression of translation initiation factor EIF-4E in post-meiotic germ cells of the rat testis. Lab Invest 1995; 73: 890-8.

Chennathukuzhi VM, Lefrancois S, Morales CR, Syed V, Hecht NB. Elevated levels of the polyadenylation factor CstF 64 enhance formation of the 1kB testis brain RNA-binding protein (TB-RBP) mRNA in male germ cells. Mol Reprod Dev 2001; 58: 460-9.

Dass B, Attaya EN, Wallace AM, MacDonald CC. Overexpression of the CstF-64 and CPSF-160 polyadenylation protein messenger RNAs in mouse male germ cells. Biol Reprod 2001; 64: 1722-9.

Dass B, McMahon KW, Jenkins NA, Gilbert DJ, Copeland NG, MacDonald CC. The gene for a variant form of the polyadenylation protein CstF-64 is on chromosome 19 and is expressed in pachytene spermatocytes in mice. J Biol Chem 2001; 276: 8044-50.

De Cesare D, Fimia GM, Brancorsini S, Parvinen M, Sassone-Corsi P. Transcriptional control in male germ cells: general factor TFIIA participates in CREM-dependent gene activation. Mol Endocrinology 2003;17: 2554-65.

Eddy EM, O'Brien DA. Gene expression during mammalian meiosis. Curr Topics Dev 1998; 141-95.

Fajardo MA, Butner KA, Lee K, Braun RE. Germ cell-specific proteins interact with the 3' untranslated regions of Prm-1 and Prm-2 mRNA. Dev Biol 1994; 166: 643-53.

Gillam S, Aline R Jr, Wylie V, Ingles CJ, Smith M. RNA synthesis and RNA polymerase activities in germ cells of developing rainbow trout testis. Biochim Biophys Acta 1979; 565: 275-92.

Giordano R, Magnano AR, Zaccagnini G, Pittoggi C, Moscufo N, Lorenzini R, Spadafora C. Reverse transriptase activity in mature spermatozoa of mouse. J Cell Biol 2000; 148: 1107-13.

Gu W, Kwon Y, Oko R, Hermo L, Hecht NB. Poly(A) binding protein is bound to both stored and polysomal mRNAs in the mammalian testis. Mol Reprod Dev 1995; 40: 273-85.

Gu W, Tekur S, Reinbold R, Eppig JJ, Choi YC, Zheng JZ, Murray MT, Hecht NB. Mammalian male and female germ cells express a germ cell-specific Y-box protein, MSY2. Biol Reprod 1998; 59: 1266-74.

Gupta PK. Genetics 1999, Rastogi Publications, Meerut, India.

Han JR, Gu W, Hecht NB. Testis brain RNA-binding protein, a testicular translational regulatory RNA-binding protein, is present in the brain and binds to the 3' untranslated regions of transported brain mRNAs. Biol Reprod 1995;53:707-17.

Hernandez N. TBP, a universal eukaryotic transcription factor? Gene Dev 1993; 7: 1291-08.

Ito T, Xu Q, Takeuchi H, Kubo T, Natori S. Spermatocyte-specific expression of the gene for mouse testis-specific transcription elongation factor S-II. FEBS Lett 1996; 385: 21-4.

Jackson RJ. Cytoplasmic regulation of mRNA function: the importance of the 3'untranslated region. Cell 1993; 74: 9-14.

Kashiwabara S, Noguchi J, Zhuang T, Ohmura K, Honda A, Sugiura S, Miyamoto K, Takahashi S, Inoue K, Ogura A, Baba T. Regulation of spermatogenesis by testis-specific, cytoplasmic poly(A) polymerase TPAP. Science 2002; 298(5600): 1999-2002.

Kashiwabara S, Zhuang T, Yamagata K, Noguchi J, Fukamizu A, Baba T. Identification of a novel isoform of poly(A) polymerase, TPAP, specifically presnt in the cytoplasm of spermatogenic cells. Dev Biol 2000; 228: 106-15.

Kleene KC Wang MY, Cutler M, Hall C, Shih D. Developmental expression of Poly(A) binding protein mRNAs during spermatogenesis in the mouse. Mol Reprod Dev 1994; 39: 355-364.

Kleene KC. Patterns of translational regulation in the mammalian testis. Mol Reprod Develop. 1996; 43: 268-281.

Kohno K, Izumi H, Uchiumi T, Ashizuka M, Kuwano M. The pleiotropic functions of the Y-box binding protein, YB-1. Bioessays 2003, 25, 691-98.

Kurihara Y, Tokuriki M, Myojin R, Hori T, Kuroiwa A, Matsuda Y, Sakurai T,Kimura M, Hecht NB, Uesugi S. CPEB2, a novel putative translational regulator in mouse haploid germ cells. Biol Reprod. 2003; 69: 261-8.

Kwon YK, Murray MT, Hecht NB. Proteins homologous to the Xenopus germ cell-specific RNA-binding proteins p54/p56 are temporally expressed in mouse male germ cells. Dev Biol 1993; 158: 90-100.

Lee K, Fajardo MA, Braun RE. A testis cytoplasmic RNA-binding protein that has the properties of a translational repressor. Mol Cell Biol, 1996; 16: 3023-34.

Lemaire L, Heinlein AO. High level expression in male germ cells of murineP68 RNA helicase mRNA. Life Sci 1993; 52: 917-926.

Leroy P, Alzari P, Sassoon D, Wolgemuth D, Fellous M. The protein encoded by a murine male germ cell-specific transcript is a putative ATP dependent RNA helicase. Cell 1989; 57: 549-59.

Meagher MJ, Schumacher JM Lee K, Holdcraft RW, Edelhoff S, Disteche C, Braun RE. Identification of ZFR, an ancient and highly conserved murine chromosome-associated zinc finger protein. Gene 1999; 228: 197-211.

Miller D. Analysis and signifcance of messenger RNA in human ejaculated spermatozoa. Mol Reprod Dev 2000; 56: 259-64

Miller D. RNA in the ejaculate spermatozoon: a window into molecular events in spermatogenesis and a record of the unusual requirements of haploid gene expression and post-meiotic equilibration. Mole Hum Reprod 1997; 3: 669-76.

Morales CR, Wu XQ, Hecht NB. The DNA/RNA-binding protein, TB-RBP, moves from the nucleus to the cytoplasm and through intercellular bridges in male germ cells. Dev Biol 1998; 201: 113-23.

Nikolajczyk BS, Murray MT, Hecht NB. A mouse homologue of the Xenopus germ cell-specific ribonucleic acid/ deoxyribonucleic acid-binding proteins p54/p56 interacts with the protamine 2 promoter. Biol Reprod 1995;52:524-30.

Okamura T, Imai Y, Kon Y, Goto M, Yamamoto M, Watanabe T. Molecular cloning and characterization of mouse testis poly(A) binding protein II encoded by the *Pabp3* gene, which transcomplements meiotic mutant sme2 of S. pombe. Biochem Genet 2000; 38: 1-11.

Oko R, Korley R, Murray MT, Hecht NB, Hermo L. Germ cell-specific DNA and RNA binding proteins p48/52 are expressed at specific stages of male germ cell development and are present in the chromatid body. Mol Reprod Dev 1996;44:1-13.

Persengiev SP, Robert S, Kilpatrick DL. Transcription of the TATA binding protein gene is highly up-regulated during spermatogenesis. Mol Endocrinol 1996; 10: 742-7.

Pointud JC, Mengus G, Brancorsini S, Monaco L, Parvinen M, Sassone-Corsi P, Davidson I. The intracellular localisation of TAF7L, a paralogue of transcription factor TFIID subunit TAF7, is developmentally regulated during male germ-cell differentiation. J Cell Sci 2003;116(Pt 9):1847-58.

Schmidt EE, Ohbayashi T, Makino Y, Tamura TA, Schibler U. Spermatid specific overexpression of the TATA-binding protein gene involves recruitment of two potent testis-specific promoters. J Biol Chem 1997; 272: 5326-34.

Schmidt EE, Schibler U. High accumulation of components of the RNA polymerase II transcription machinery in rodent spermatids. Development 1995; 121: 2373-83.

Schumacher JM, Artzt K, Braun RE. Spermatid perinclear ribonucleic acid binding protein inds microtubules in vitro and associates with abnormal manchettes in vivo in mice. Biol Reprod 1998; 59: 69-76.

Schumacher JM, Lee K, Edelhoff S, Braun RE. Distribution of Tenr, an RNA-binding protein, in a Lattice-like network within the spermatid nucleus in the mouse. Biol Reprod 1995a; 52: 1274-83.

Schumacher JM, Lee K, Edelhoff S, Braun RE. Spnr, a murine RNA-binding protein that is localized to cytoplasmic microtubules. J Cell Biol 1995b; 129: 1023-32.

Severt WL, Biber TU, Wu X, Hecht NB, DeLorenzo RJ, Jakoi ER. The suppression of testis-brain RNA binding protein and kinesin heavy chain disrupts mRNA sorting in dendrites. J Cell Sci 1999; 112(pt21): 3691-702.

Sheng Yi, Tsai-Morris, Dufau ML. Cell specific and hormone regulated expression of gonadotropin regulated testicular RNA helicase gene (GRTH/Ddx25) resulting from alternative utilization of translation initiation codons in the testis. J Biol Chem 2003; 278: 27796-803.

Subramaniam K, Seydoux G.Dedifferentiation of primary spermatocytes into germ cell tumors in C. elegans acking the pumilio-like protein PUF-8. Curr Biol. 2003; 13: 134-9.

Sugiura S, Kashiwabara S, Iwase S, Baba T. Expression of a testis-specific form of TBP-related factor 2 (TRF2) mRNA during mouse spermatogenesis. J Reprod Dev. 2003; 49: 107-11.

Suk K, Kim S. Kim YH, Oh SH, Lee MK, Kim KW, Kim HD, Seo YS. Identification of a novel human member of the DEAD box protein family. Biochem Biophy Acta 2001; 1501: 63-69.

Tafuri SR, Familari M, Wolffe AP. A mouse Y box protein, MSY1, is associated with paternal mRNA in spermatocytes. J Biol Chem 1993; 268: 12213-20.

Tang PZ, Tsai-Morris CH, Dufau ML. A novel gonadotropin-regulated testicular RNA helicase. A new member of the dead-box family. J Biol Chem 1999; 274: 37932-40.

Tsai-Morris C-, Sheng Yi, Lee E, Lei K-J, Dufau ML. Gonadotropin-regulated testicular RNA helicase (GRTH/Ddx25) is essential for spermatid development and completion of spermatogenesis. Proc Natl Acad Sci (USA) 2004;101:6373-78.

Umehara T, Kida S, Hasegawa S, Fujimoto H, Horikoshi M. Restricted expression of a member of the transcription elongation factor S-II family in testicular germ cells during and after meiosis. J Biochem 1997; 121: 598-03.

Wallace AM, Dass B, Ravnik SE, Tonk V, Jenkins NA, Gilbert DJ. Copeland NG, MaeDonald CC. Proc Natl Acad Sci (USA) 1999; 96: 6763-68.

Wu XQ, Gu W, Meng X, Hecht NB. The RNA-binding protein, TB-RBP, is the mouse homologue of translin, a reconmbination protein associated with chromosomal translocations. Proc Natl Acad Sci (U S A) 1997; 94: 5640-5.

Wu XQ, Hecht NB. Mouse testis brain ribonucleic acid-binding protein / translin colocalizes with microtubules and is immunoprecipitated with messenger ribonucleic acid encoding myelin basic protein, α calmodulin kinase II, and protamines 1 and 2. Biol Reprod 2000; 62: 720-5.

Wu XQ, Lefrancois S, Morales CR, Hecht NB. Protein-protein interactions between the testis brain RNA-binding protein and the transitional endo plasmic reticulum ATPase, a cytoskeletal gamma actin and Trax in male germ cells and the brain. Biochemistry 1999; 38: 11261-70.

Xu Q, Nakanishi T, Sekimizu K, Natori S. Cloning and identification of testis specific transcription elongation factor S-II. J Biol Chem 1994;269:3100-3.

Yang J, Chennathukuzhi V, Miki K, O'Brien DA, Hecht NB. Mouse testis brain RNA-binding protein/translin selectively binds to the messenger RNA of the fibrous sheath protein glyceraldehyde 3-phosphate dehydrogenase-S and suppresses its translation in vitro. Biol Reprod. 2003; 68: 853-9.

Yiu GK, Murray MT, Hecht NB, Deoxyribonucleic acid-protein interactions associated with transcriptional initiation of mouse testes specific cytochrome gene. Biol Reprod 1997; 56: 439-49

Yu J, Hecht NB, Schultz RM. Requirement for RNA-binding activity of MSY2 for cytoplasmic localization and retention in mouse oocytes. Dev Biol 2003; 255: 249-62.

Zhang D. Penttila TL, Morris PL,Roeder RG.Cell- and stage-specific high-level expression of TBP-related factor 2 (TRF2) during mouse spermatogenesis. Mech. Dev. 2001a; 106:203-205.

Zhang D, Penttila TL, Morris PL, Teichmann M. Roeder RG. Spermiogenesis deficiency in mice lacking the Trf2 gene Science 2001b; 292 (5519):1153-1155.

Chapter 16

TRANSCRIPTION FACTORS ASSOCIATED WITH SPERMATOGENESIS

16.1. CRE-TRANSCRIPTION FACTORS (b-ZIP CLASS PROTEINS)

General transcription factors are differentially regulated in germ cells. For example, TATA binding protein accumulates in early haploid germ cells at much higher levels than in any other somatic cell type. In addition to TBP, TFIIB and RNA polymerase II are also over-expressed in testis. These remarkable features are consistent with the potent transcriptional activity, which occurs in coordinated manner during the germ cell differentiation. But CREM transcription factor responsive to the cAMP-signaling pathway is crucial for a normal germ cell differentiation program. During the process of development a committed germ cell passes through 3 ½ cycles, driven by the alternating influences of cAMP and androgen hormones. Cyclic-AMP increases transcriptional activation through several closely related bZIP transcription factors. The CREB and CREM are known to be expressed in a variety of isoforms, many of which are inhibitory to cAMP induced transcription. Different signaling pathways converge on CREB and CREM proteins, which control cell function by modulating their phosphorylation state, which acts as a molecular switch in transcriptional machinery (De Cesare and Sassone-Corsi, 2000). CRE-binding proteins have been found to play an important role in the physiology of male gametogenesis.

16.2. CRE- MODULATOR (CREM) PROTEIN

16.2.1. Signal Transduction

The c-AMP and diacylglycerol (DAG) are two major messengers, which induce cell signals in many cells. The c-AMP pathway is characterized by a specific protein kinase A while DAG signal is activated by protein kinase C with their ultimate targets, which control cAMP-responsive element (CRE) and 12-O-tetradecanoylphorbol 13-acetate (TPA) responsive element (TRE) in DNA, respectively. There is extensive cross talk between DAG and cAMP pathways. Intracellular levels of cAMP are primarily controlled by adenylate cyclase, which is activated by extracellular stimulus mediated through interaction of receptors with G proteins (see Chapters 18 and 19). The cAMP so produced binds to two PKA regulatory subunits, thus converting the inactive form of PKA into an active form of PKA. The released active PKA catalytic subunit is translocated to the nucleus where it phosphorylates serine residues of nuclear proteins having the sequence: X-Arg-Arg-X-Ser-X. The PKA mediated phosphorylated proteins ultimately influence the transcriptional regulation of various genes

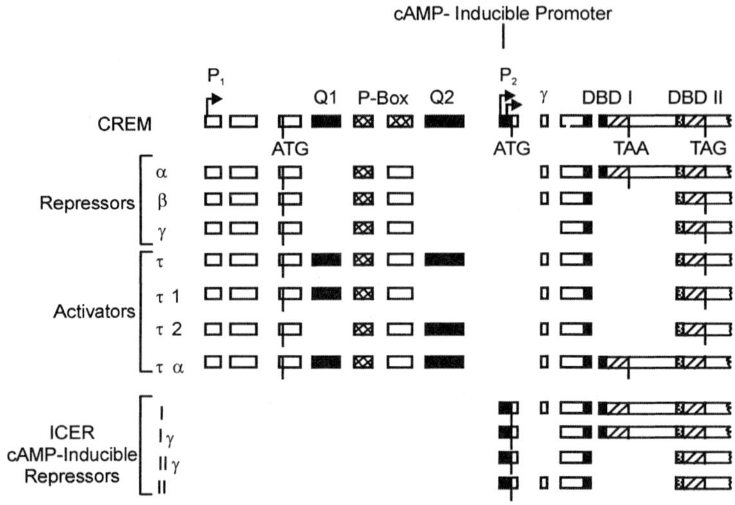

Fig.16.1. Activators and repressors from same gene. Schematic representation of CREM gene and formation of various activator and repressor CREM isoforms. The P1 promoter is GC-rich and directs non-inducible pattern of expression. The P2 promoter is highly inducible by cAMP-dependent signaling pathway. Reprinted with permission from D. De Cessare and P. Sassone-Corsi. Prog Nucl Acid Res Mol Biol 64; 343-369 © Elsevier.

through distinct cAMP-inducible promoter responsive sites. The consensus CRE site is formed of an 8 bp palindromic sequence: TGACGTCA.

***Regulation of CREM During Spermatogenesis*:** Based on high abundance of CREM in spermatogenic cells, it is considered that CREM acts as a developmental switch during spermatogenesis. Alternative splicing of the exons encoding the activation domain, produce different CREM isoforms, expressed at different times during the differentiation of germ cells. The abundant CREM transcript in the adult encodes exclusively the activator forms, whereas in pre-pubertal testis only the repressor forms are detected, albeit at low levels. Thus the CREM developmental switch also has a reverse function (Foulkes et al., 1992). The CREM in germ cells is under exquisite hormonal regulation. Since FSH does not have direct effect on spermatogenesis, the effect on CREM activation in germ cells by FSH seems to be mediated through Sertoli cells. Administration of FSH results into a rapid induction of the CREM transcript. The induction of CREM transcripts by FSH is not transcriptional. Instead, by a mechanism of alternative polyadenylation, AUUUA destabilizer elements present in the 3'-untranslated region of the gene are excluded, dramatically increasing the stability of the CREM message. CREM is the first example of a gene whose expression is directly modulated by a pituitary hormone during spermatogenesis. The implication of this observation is that hormone can regulate gene expression at the level of RNA processing (Foulkes et al., 1993; DeCesare and Sassone-Corsi, 2000).

16.2.2. Germ Cell Specific CREM Isoform

In mammals, the CREB, CREM, and activating transcription factor 1 (ATF-1) are encoded by three different genes. Each of these genes has large potential to encode several isoforms

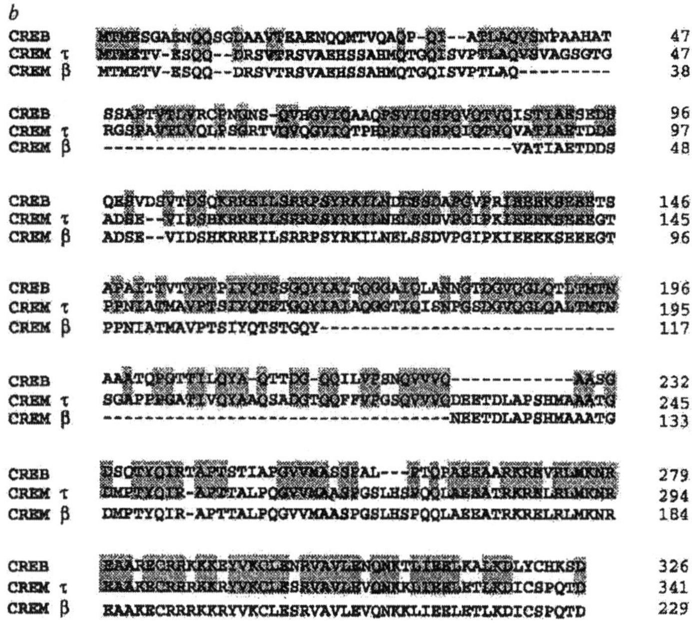

Fig.16.2. The deduced amino acid sequence of *CREMτ* cDNA. Amino acid sequence similarity between CREMτ (341 aa) and with human placental CREB (326 aa). Sequence identity is shown by shaded regions. Reprinted with permission from N. S. Foulkes et al. Nature 355; 80-84; 1996 © Macmillan Magazines Ltd http://www.nature.com/.nature

through various mechanisms. CRE binding proteins (CREB, CREM, ATF1) share a high degree of homology and belong to the basic domain leucine zipper (bZip) protein class. As stated in Chapter 14, bZip class of proteins contain a leucine zipper; an α-helical coiled coil structure, and an adjacent basic domain, which is approximately 50% rich in lysine and arginine residues, required for direct contact of DNA. Mostly, Zip proteins act as dimmers, which are formed after heterodimerization. Dimers bind to a regulatory DNA consensus DNA sequence, CRE, which is present in the regulatory region of cAMP responsive genes. The CRE consensus sequence has also been found in other distinct promoter elements, in which it apparently confers different transcriptional properties. An example is for X-box motif present in MHC class II gene.

Various RNA processing events generate a family of CREM proteins. Control seems to be exerted at three different levels: alternative splicing, alternative polyadenylation, promoter usage and alternative translational initiation. The modular structure combined with extensive differential splicing permits the CREM gene to encode a family of transcription factors with different activation properties. Additional versatility is produced by generating transcripts with different 3'ends. Different isoforms of CRE- binding proteins can act both as activators and repressors of transcription **(Fig.16.1)**. The CREM gene also encodes a transcriptional activator, CREMτ. The cDNAs have been characterized for CREMτ1 and CREMτ2, which incorporate singly N-terminal or C-terminal glutamine rich domains and still work as activators.

CREMτ differs from CREM agonists by the coordinate insertion of two-glutamate rich domains that confer transcriptional activation function. During spermatogenesis there is an abrupt switch in CREM expression. In pre-meiotic germ cells CREM is expressed at amounts in the antagonist form. Subsequently, from the pachytene spermatocyte stage onward, a splicing event generates exclusively the CREM activator, which accumulates in very high amounts. The 1,023-bp open reading frame of CREMτ encodes a protein of 341 amino acids very similar to CREB. In transfection assays, CREM in contrast to the other CREM isoforms, functions as a transcriptional activator when co-expressed with PKA. The two inserted amino acid regions in CREM are at least three times richer in glutamine residues than the rest of the protein (Foulkes et al., 1992) **(Fig 16.2**). The expression of testis specific isoforms of the catalytic and regulatory subunits of PKA (Chapter 19) implies the requirement of a specialized cAMP signal transduction pathway and the CREMτ may represent a nuclear target of this pathway.

Phosphorylation of CRE- proteins at specific sites is pre-requisite for transcriptional activation. CREMτ is efficiently phosphorylated at a serine residue at position 117 by the germ cells specific PKA, indicating that it constitutes a natural target of the adenylate cyclase pathway during spermatogenesis. Phosphorylation of serine-117 turns CREMt into a powerful activator **(Fig.16.3)**. The rise in CREMτ protein coincides with the transcriptional activation of several genes. CREMτ efficiently binds to CRE present in the promoter of these genes, suggesting that they could constitute down stream targets of CREM. The male germ cell specific RT7 promoter is cAMP inducible and activated by CREMτ and efficiently transcribed in vitro with nuclear extracts from seminiferous tubules, confirming a cascade of transcriptional events during spermatogenesis (Delmas et al., 1993).

CREM isoforms and mRNA expression have been analyzed in the cynomolgus monkey (Macaca fascicularis). Two activator isoforms (2 with and without exon) from monkey display high sequence identity to mouse and human isoforms. CREM activator expression was confined to pachytene spermatocytes and round spermatids in specific spermatogenic stages. Comparative analysis of testicular CREM expression showed several transcripts in rat, mouse, hamster, and marmoset; two transcripts in cynomolgus and rhesus monkey and one in man suggesting an evolutionary trend from multiple activator isoforms to a single activator transcript in men (Behr et al., 1999; 2000). A CREM transcript isoform from a monkey testis showed a novel 5' exon θ2 of 113-bp, which provides at its 3'-end an in-frame ATG to the downstream reading frame. Moreover, a second leader exon, θ1 of 289-bp encodes a putative open reading frame of 26 amino acids. In-vitro translation, CREM-θ1 and CREM-θ2 splice variants from human testis yielded not only full-length proteins but also shorter repressor products resulting from downstream translation initiation. While CREM-θ1 mRNA was exclusively expressed in the testis, the CREM-θ2 transcripts were abundantly expressed in the testis and at very low levels or absent in other tissues. Differential activation of promoters during mouse postnatal testicular development suggested cell- and stage-specific regulatory mechanisms for these CREM promoters (Gellersen et al., 2002).

16.2.3. CREM Activator Protein

Examination of phosphorylation state of CREM at various stages of spermatogenic cycle showed that CREM is unphosphorylated at the time when it activates post-meiotic genes. This differential pattern suggests that molecular mechanism of transcription activation by CREM in germ cells must be different from factors, which involve phosphorylation at Ser117. The answer was obtained with the search of an activator protein for CREM. Murine testis c-DNA library revealed a clone, which coded a protein that interacted with CREM with high

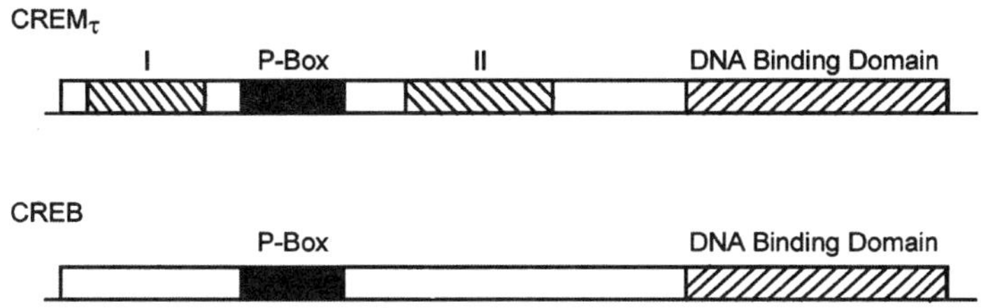

Fig.16.3. Schematic diagrams of amino acid sequences of (A) CREMτ and (B) CREB, showing DNA binding domain (DBD), cysteine rich insertion (I and II) sequences (specific to CREMτ), and kinase phosphorylation sites (P-Box).

affinity. The ORF sequence of this clone encodes a protein of 284 aa, and was named as ACT (activator of CREM in testis). The ACT shows the presence of four LIM domains and one amino terminal half LIM motif. The LIM domain is a conserved cysteine and histidine rich structure of two repeated zinc fingers (see Chapter 19 for Lim) and it functions in protein-protein interactions. The ACT belongs to the class of the LIM only proteins (LMO). While CREM and ACT are coordinately expressed in testis, the ACT is abundantly and exclusively expressed in testis. In the seminiferous epithelium, CREM transcripts accumulate in spermatocytes and spermatids, but the protein is detected only in haploid spermatids. The absence of CREM from spermatocytes reflects a strict translational control in germ cells. It is noteworthy that ACT is able to convert inactive CREM mutants into transcriptionally active molecules both in yeast and in mammalian cells. Thus, with its intrinsic capacity for activation, ACT can convert CREM into powerful transcription activator. The expression of CREM activator protein in spermatids coincides with the transcriptional activation of several genes containing CRE motifs in their promoter region. These genes encode mainly structural proteins required for spermatozoon differentiation, suggesting a role for CREM in the activation of genes required for the late phase of spermatid differentiation (Fimia et al., 1999; De Cesare and Sassone-Corsi; 2000; Sassone-Corsi, 2000).

16.2.4. CREM and Spermatogenic Genes

Genetic evidences indicated that CREM participates in testis and developmental specific regulation of post-meiotic genes during spermiogenesis. CREM acts as a master controller of haploid gene expression. The expression of several testis specific genes (already known to be regulated by CREM) was dramatically reduced or lost in the CREM mutant mice (-/-). Among these genes are Prm1, Prm2, TP1, TP2, RT7 (Blendy et al., 1996 Nantel et al., 1996 and others), calspermin, Krox-20 and Krox-24 in heterozygous and homozygous mutant males. Besides these proteins, CREM also regulates the expression of mitochondrial capsule selenoprotein (MCS), CRY-51 and ACE. These studies on inactivation of CREM gene demonstrated the importance of this gene in spermiogenesis. In contrast, other genes expressed in pre-meiotic germ cells such as that for proacrosin binding protein and *Hoxa-1* are expressed normally in the mutant mice. It is likely that CREM function must be restricted to the late phase of transcription before the compaction of the DNA.

CREM deficiency causes germ cells to cease differentiation and to undergo apoptosis (De Cesare and Sassone-Corsi, 2000; Sassone- Corsi 2000). Nevertheless, the importance of CREM in spermiogenesis is not precisely known. Several reports have addressed the physiological role of CREM using inactivation of the CREM gene by homologous recombination in mice (Blendy et al., 1996; Nantel et al., 1996). In both studies the targeting vector used to generate the mice was constructed to delete the leucine zipper domains of CREM. These leucine zipper domains are required for both dimerization of CREM and DNA binding (binding to the CRE). The homozygous CREM deficient mice (-/-) appeared healthy and seemed to have normal growth and development. Nevertheless, despite normal mating behaviour, the males were sterile (while the females were fertile). Serum testosterone and FSH levels were not reduced. Homozygous CREM mutant mice (-/-) demonstrated a complete lack of spermatozoa, while the heterozygous (+/-) mice exhibited a 46% reduction in the number of sperm. CREM inactivation blocks spermiogenesis at the early spermatid stage and inhibits morphological differentiation and development to elongated spermatid. Importantly, CREB expression is unchanged with respect to normal litter mates in CREM deficient mice. Furthermore, the expression of ATF-1, and the levels of the co-activator CREB binding protein and ATF-2 remained unchanged. This demonstrated that no redundant function can substitute for CREM in germ cells (Nantel et al., 1996). In CREM deficient male mice apoptosis of germ cell increases 10 fold. Thus the stringent requirement of CREM is manifested by the lack of maturation of the germ cells and by their entering the cell death pathway. CREM and transcription factor IIA (TFIIA) are expressed in a spatially and temporally coordinated fashion during differentiation of germ cells. These proteins are co-localized intracellularly in spermatocyte and spermatids. CREM interacts with TFIIA, a general transcription factor that stimulates RNA polymerase II-directed transcription. The interaction is restricted to the activator isoforms of CREM and does not require Ser117. Importantly, CREM does not interact with TFIIAτ-ALF, a testis-specific TFIIA homologue. These investigations contribute to the understanding of the highly specialized roles of transcriptional regulation in haploid germ cells (De Cesare et al., 2003).

CREM as an Index of Fertility in Men : In about one third of infertile men the cause of impaired spermatogenesis is not known. Indeed, several studies suggested that a switch from the expression of CREM repressors to CREM activator is present in post- meiotic germ cells in normo-spermic men where as in oligo-spermic men only CREM repressor was detected. In situ hybridization confirmed the presence of CREM activators in post-meiotic germ cells in testis specimens showing conserved spermatogenesis, but not in specimens showing maturation arrest at the spermatid stage. Thus lack of a switch in the expression of CREM gene isoforms seems to be related to impaired spermatogenesis in humans (Peri et al., 1998; Weinhauser et al., 1998).

16.2.5. Location of CRE in Promoters of Germ-Cell Specific Genes

CRE like sequences are located between position –40 and –100 in the post-mitotically expressed (Prm-1 and Prm-2, RT7, TP-1, calspermin and angiotensin converting enzyme (ACE)) genes. The CRE like sequence acts as a binding site for a common testis specific transcription factors. Several testis specific genes, however, do not show this sequence, suggesting that the use of a CRE like element is restricted to a subset of promoters, such as those which act post-meiotically. Surprisingly, only a few sequences involved in the glucocorticoid /androgen response pathways have been identified despite the profound effects of these hormones on testis and germ cell development. Hence following binding sequences have been described in order to provide understanding of the stage specific expression of several testis specific genes.

Angiotensin Converting Enzyme: In vitro transcription and transfection studies demonstrated that CREM protein is a transcriptional activator of the testis ACE promoter (Zhou et al., 1996). Gel shift experiments identified CRE like site found at position – 55 in the testis ACE promoter. Recombinant CREMτ protein and testes nuclear proteins have a similar specificity of binding to the testis ACE-CRE. ACE gene produces two mRNA species from tissue specific promoters. The transcription start site of the mRNA for the smaller testicular ACE_T is located within an intron of the larger transcription unit that encodes the pulmonary isozymes, ACE_{P}. A 298-bp DNA fragment, 5' to the rabbit ACE_T mRNA transcription initiation site, can activate the testicular expression of a transgenic reporter gene. Using the same transgenic reporter system a putative CRE like site (CRE_T) was identified within this DNA fragment at –52 and a TATA like binding site at –27, absolutely essential for transcriptional activation. Moreover, ACE mRNA was not expressed in the testes of mice homozygous for a null mutation in the transcription factor CREM. Observations indicated that ACE_T mRNA expression in the testes is regulated by the putative CRE, present 5' to the transcription start site and the corresponding transcription factor CREM (Howard et al, 1993; Kessler et al., 1998; Sadhukhan et al,1998).

Calspermin: Testis specific calspermin transcription can be produced in heterologous cells by utilization of a promoter located in an intron of the CaMK IV gene. Critical motifs within this promoter are two CRE like sequences located about -70 and -50bp upstream of the transcriptional initiation site. Both CRE motifs are foot printed by the authentic testis specific transcriptional activator CREM or by CREM present in adult testis nuclear extract. Whereas a 2.1-kb DNA fragment containing the calspermin promoter is inactive in transfection assays, activity could be restored by co-transfection of CREM and protein kinase A or CaMK IV but not CaM kinase llα (Sun et al., 1995).

Lanosterol-14-α Demethylase (CYP51): Lanosterol-14-α demethylase (CYP51) produces MAS sterols, intermediates in cholesterol biosynthesis that can reinitiate meiosis in mouse oocytes. Gene, *CYP51* is regulated by a sterol/sterol regulatory element binding protein (SREBP) dependent pathway in liver and other somatic tissue. *CREM-/-* mice are deficient in germ cell specific *CYP51* mRNAs in testis while expression of somatic *CYP51* transcripts was unaffected. The regulation by CREMτ is not characteristic for all cholesterogenic genes expressed during spermatogenesis. The –334 / +314-bp *CYP51* region can mediate both the sterol/SREBP dependent as well as the cAMP / CREMτ-dependent transcriptional activation. It appeared that two regulatory pathways mediating expression of *CYP51* (SREBP-dependent expression) in liver and other somatic cells and in haploid male germ cells controlling the operation of cholesterogenic genes are under different control (Rozman e al., 1999).

Ldh-c Gene is Regulated by Multiple Factors: The human *Ldh-c* shows an exceptionally large window of expression throughout pre- and post-meiotic stages of the male germ cell lineage. Three classes of transcriptional regulators were defined in human *Ldh-c* core promoter. 1) The Sp1 transcription factor is a testis- "enriched" protein that is absent from most somatic tissues and that appears to play a major role in determining *Ldh-c* expression in the testis. Highest levels of Sp1 during spermatogenesis correlate with maxima of *Ldh-c* expression. 2). The testis specific CREM like sequence located at position –433 might contribute to post-meiotic transduction. 3) Factors present in tissues for negative regulation of *Ldh-c* expression appear to bind both the CRE like sequence and an adjacent hormone response element. The presence of this element could be involved in regulating *Ldh-c* through the glucocorticoid/ androgen pathways at the early stages of *Ldh-c* expression (Bonny et al., 1998).

16.3. CRE-BINDING (CREB) PROTEIN

CREB is predominantly a positive modulator of cAMP responsive genes. However, in the testis, alternative exon splicing results in the expression of repressor CREB isoforms. Like CREM, dimers of CREB bind to the palindromic CRE element 5'-TGACGTCA-3' (or similar motifs) found in the control regions of many genes. CREB gene expression in Sertoli cells and germ cells of the rat testis occurs in concomitant with the FSH induced rise in cellular cAMP levels, and the transcription of the CREB gene is auto-regulated by cAMP dependent transcriptional proteins. The 5'flanking sequence of human CREB promoter gene has a high content of guanosines and cytosines and lacks canonical TATA and CCAAT boxes typically found in the promoter genes in eukaryotes. The CRE bind recombinant CREB and endogenous CREB or CREB like proteins contained in placental JEG-3 cells, and also confer cAMP- inducible transcriptional activation to a heterologous minimal promoter. The expression of the CREB gene is positively autoregulated (Meyer et al., 1993).

CREB mRNAs are located in Sertoli cells and Leydig cells and in germ cells. Rat testicular sections showed a remarkably high expression of CREB in the haploid round spermatids and to some extent in pachytene spermatocytes and Sertoli cells. Although most of the CREB antigen is detected in the nuclei, some CREB antigen is also present in the cytoplasm. The cytoplasmic CREB results from the translation of a unique alternatively spliced transcript of the CREB gene that incorporates an exon containing multiple stop codons inserted immediately up stream of the exons encoding the DNA binding domain of CREB. The RNA containing the alternatively spliced exon encodes a truncated transcriptional transactivtor protein lacking both the DNA binding domain and nuclear translocation signal of CREB. The C-terminally truncated CREB isoforms are lacking DNA binding dimerization function, and hence their physiological role remains unclear. In the Sertoli cells, a striking cyclical (12 day periodicity) increase in the levels of CREB mRNA coincides with the splicing out of the restrictive exon containing the stop codons. Because FSH-stimulated cAMP levels in Sertoli cells are also cyclical, and the CREB gene promoter contains cAMP responsive enhancers, it was suggested that the alternative RNA splicing controls a positive auto-regulation of CREB gene expression mediated by cAMP (Waeber et al., 1991).

16.3.1. CREB mRNA Isoforms

Several alternatively spliced exons in the CREB mRNA that contains in frame stop codons have been described in mouse, rat, and human CREB genes. The W, Z, and Y (or ψ) exons are testis specific and are most strongly detected in germ cells. In mouse, mRNA CREBδ and CREBα encode proteins with CRE-binding activity and identical transactivation potential. The additional CREB mRNA isoforms potentially encode CREB related proteins. From the structural organization of the mouse CREB gene it was concluded that the multiple transcripts are generated by alternative splicing. Specific CREB mRNA isoforms are expressed at a high level in the adult testis; expression of these isoforms is induced after commencement of spermatogenesis, predominantly in the primary spermatocytes. Comparison of the CREB gene with isolated CREM cDNAs illustrates that the two genes have arisen by gene duplication and have diverged to encode transcriptional activators and repressors of the cAMP signal transduction pathway (Ruppert et al., 1992) **(Fig.16.4)**. Since during rat spermatogenesis, the levels of cAMP in seminiferous tubules undergo stage dependent cyclical fluctuations, it was found that changes in cAMP levels are accompanied by alternative exon splicing of the RNA encoding the CREB expressed in both the Sertoli and germ cells.

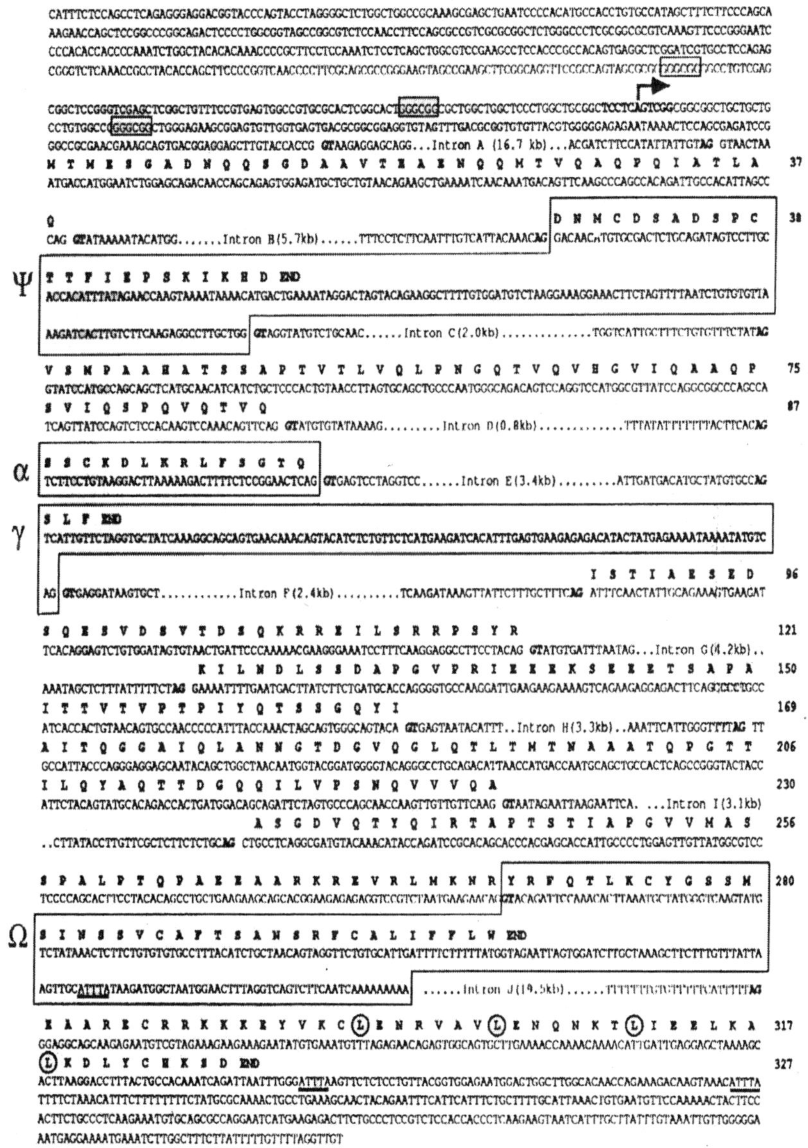

Fig.16.4. DNA and predicted amino acid sequence of mouse CREB cDNA including 5' flanking and intron sequences. Boxes denoted by ψ, α,γ, and Ω indicate inserts specific for CREB cDNA clones pmcCREBψ, pmcCREBα, pmcCREBγ and pmcCREB Ω. Intron boundaries (AT and AG) are written in bold letters. The amino acid numbers (1-327 aa residues) shown on right depict the most abundant isoform CREBχΔ. Leucines putatively involved in the formation of leucine zippers are circled. Amino acids, specific to other isoforms, are given in respective boxes and not numbered. Arrow indicates the putative transcription initiation site. In the promoter region, binding sites for SP1 (sequence 5'-GGGCGG) are boxed. Reprinted with permission from S. Ruppert et al. The EMBO Journal 11; 1503-12: 1992 © http://www.nature.com/.

Exons Y and W are expressed exclusively in the testis, and they introduce stop codons into the normal proteins coding frame of CREB. The splicing in W was shown to activate the internal translation of two novel alternative products of the CREB mRNA containing the DNA binding domain (I-CREBs). The I-CREBs act as potent inhibitors of activator isoforms of CREB. The polycistronic RNA that encodes two internally translated CREB repressor isoforms (I-CREBs) consists of the carboxy-terminal DNA binding domain, devoid of the transactivation domain. Analysis of different stages of the spermatogenic cycle revealed stage dependent splicing of both exons W and Y in the CREB transcripts confirming a cyclic variation in CREB mRNA in testis. Maximal splicing of exons W and Y occurs independently at different stages of the spermatogenic cycle, stages II-VI and IX, respectively. The distinct spermatogenic cycle dependent regulation of the splicing of exons W and Y provides further evidence in support of a functional relevance for CREB-W and Y mRNA isoforms in spermatogenesis (Daniel et al, 1998). Girardet et al (1996) reported the alternative splicing of an additional exon Z in CREB mRNA expressed in human but not in mouse or rat testis. Insertion of exon Z abolishes the synthesis of one of the two inhibitor CREBs due to the introduction of an in frame stop codon within exon Z. It was shown that Z is not spliced into mRNAs in mouse and rat testes due to the evolution of mutations in the splice signals flanking exon Z. These findings suggested that the splicing of exon Z might be part of a human specific mechanism to regulate cAMP-dependent regulatory pathways in spermatogenesis by extinguishing the expression of a CREB repressor (Girardet et al., 1996).

16.3.2. CREB Promoter

Transient transfections of primary Sertoli cells showed a conserved 300-bp region of the CREB promoter surrounding the transcription start site that is required for both basal and cAMP inducible expression of the CREB gene. This promoter region contains three Sp1-binding sites flanking the transcription initiation site and two CREs located 65 and 85-bp downstream of the transcription initiation site. It was shown that the Sp1 motifs bind Sp1 in Sertoli cell extracts and contribute to basal promoter activity. The model of FSH and cAMP mediated CREB auto-regulation of its own promoter may also explain the dramatic stage specific oscillations in Sertoli cells of CREB mRNA levels during the 12-day cycle of spermatogenesis in rat (Walker et al., 1995)

16.3.3. Hormonal Control of CREB

The expression of CREB protein mRNAs in Sertoli and germ cells during spermatogenesis is dependent on both androgen and FSH. However, the effects of androgen or FSH on the regulation of CREB protein mRNAs are different (West et al., 1994). Expression of CREB transcripts was reduced by EDS induced with-drawl but not by FSH or EDS+ testosterone. CREB is highest around the base of stage VII-VIII tubules and this was shown to be androgen dependent (West et al., 1994). Binding of FSH to Sertoli cells activates cAMP dependent pathway resulting in CREB phosphorylation, which was shown by the addition and with-drawl of forskolin to cultured primary Sertoli cells.

16.3.4. CREB Regulation by NF-κβ and Other Factors

NF-κβ transcription factors are constitutively expressed in the nuclei of Sertoli cells of rodent testis, and specifically bind to κβ enhancer motifs within the promoter of the CREB gene. NF-κβ transcription factors are potential regulators of CREB expression since the CREB promoter contains consensus NF-κβ binding motifs. The family of NF-κβ or Rel transcription factors consists of five known mammalian subunits (RelA, RelB, c-Rel, p50, and p52). Multiple combinations of homo- and heterodimers are possible, thus providing the potential to generate both transactivators and transrepressors of transcription. Tumor necrosis factor α (TNF-α) is a regulator of NF-κβ activity and therefore a regulator of CREB gene expression. In the testis TNF-α is secreted primarily by round spermatids and a 55-kDa TNF-α receptor has been detected in Sertoli cells. TNF-α increases the activity of NF-kβ in rat Sertoli cells and that NF-κβ levels in the nuclei of Sertoli cells are highest during the stages in which round spermatids are present (Delfino and Walker, 1999). Over-expression of NF-kβ proteins in primary Sertoli and fibroblast cells activates the CREB promoter in transient transfection assays. Stimulation of Sertoli cells with TNFα, produced by round spermatids stimulated the elimination of NF-kβ, the translocation of additional NF-kβ to the nucleus, and increased NF-kβ binding to CREB-promoter kβ-enhancer elements. It was concluded that NF-kβ contributes to the up-regulation of CREB expression in Sertoli cells and raises the possibility that NF-kβ may induce other Sertoli cell genes required for spermatogenesis. It is also likely that NF-kβ may be a general regulator of CREB in non-testis tissues (Delfino and Walker, 1999).

16.3.5. Activating Transcription Factor 4 (ATF4/CREB2)

Activating transcription factor 4 (ATF4/CREB2) is another member of CREB family. The male ATF4(-/-) mice are sub-fertile, despite the fact that they produce sufficient sperm and are able to fertilize wild-type eggs in vitro. Observations predicted that ATF4 is required for the normal differentiation of the lamina propria layer of the vas deferens at sexual maturation. The morphology of the ATF4(-/-) lamina propria and the constriction of the lumen are consistent with the obstruction in the vas deferens contributing to the sub-fertility of the ATF4(-/-) males (Fischer., 2004).

16.4. OTHER LEUCINE ZIPPER PROTEINS

16.4.1. RT7 : A Germ Cell Specific Protein

The rat RT7 cDNA nucleotide sequence is not homologous to any sequences present in the Gen-Bank library. RT7 RNA is expressed at very high levels in rat early spermatids, while its expression is not detectable in any other organ or tissue. The RT7 transcription start site demonstrated that RT7 encodes a putative 90 amino acid protein, of which the N-Terminus is predicted to fold as an amphipathic α helix with features resembling the leucine zipper structure found in a family of transcription factors. However, unlike the leucine zipper proteins, the RT7 α helix is not preceded by a basic region. Analysis of the RT7 promoter sequence shows that it contains a putative testis specific regulatory sequence found in Prm1 and Prm2 promoters, as well as binding sites for several other transcription factors (van der Hoorn et al., 1990; 1991; Shankar et al., 1998) (see Chapter 29).

16.4.2. Nurit Protein

A novel murine spermatid-specific gene, designated *nurit*, was identified in a two-hybrid screen for proteins that bind Nek1 kinase. Nurit protein harbors a leucine-zipper motif, and two additional coiled-coil regions. The C-terminal coiled-coil domain mediates homodimerization of the protein. Nurit homologues are found in primates, pig and rodents.

Nurit is transcribed through the elongation stage of the spermatids, but is absent from mature spermatozoa. The protein is restricted, from its first detectable appearance, to a unique spermatid organelle called the 'flower-like structure'. The function of this structure is unknown, though it may be involved in transporting proteins to be discarded via the residual bodies. Nurit is the first marker of the flower-like structure (Feige et al., 2002).

16.5. HOMEOBOX PROTEINS IN GERM CELLS

Homeodomain containing genes produce proteins that act as sequence specific transcription factors that transactivate or repress the expression of other genes. Many homologues of invertebrate homeodomain proteins have been identified in mammals, including mice and humans. The best studied vertebtrate homeobox containing genes are *Antp* like *Hox* genes, which exist in four major clusters in the mouse genome. Each cluster exhibits intriguing similarities to the complement of genes within the *Drosophila Antp* and *Bithorax* clusters, not only in homeodomain sequences but also in the temporal order of activation, anterior boundary of expression during embryogenesis and possible role in segmentation (c/r Komuro et al., 1993). In adult animals, homeobox gene appears to be expressed exclusively in glial cells of the CNS including cerebrum, midbrain, cerebellum and spinal cord, and germ cells of the testis. Several homeobox proteins and genes are expressed in male reproductive tissue, suggesting the possibility that they may regulate developmental events during male gametogenesis. For example, the POU homeobox gene *sperm-1* is expressed transiently prior to meiosis in germ cells, whereas *Hox-a-4* is expressed specifically in post-meiotic germ cells. *Hox-b-4* is expressed in both germ cells and somatic cells in the testis, whereas *Hox-1-4* is expressed by Leydig cells. Non-testicular somatic tissues also abundantly express the *Hox-b-4* and *Hox-1-4* genes.

Oct: Goldsborough et al., (1990) cloned and sequenced four POU-box containing genes expressed in mouse testis. Three of these genes code for the mouse homologues of the previously described human Oct-1, Oct-2 and rat Brn-3. A fourth called Oct-11 codes for a protein identical to Oct-1 and Oct-2 over the entire POU specific domain and is closely related over the POU-homeodomain. Using cDNA from mouse testis, the sequences for all four POU-boxes were derived from at least two PCR products: Oct-1and Oct-11. Mouse and human Oct-1 differ at a position (amino acid 113), which is potentially important for interaction with VP16 (c/r Goldsborough et al., 1990).

Pem gene: Among the cDNA isolated is a homeobox gene, *Pem*, which is distantly related to the Prd/pax homeobox gene family. The *Pem* gene is expressed at varying levels by most immortalized and malignant rodent cell lines regardless of lineage. The *Pem* gene exhibits stage and lineage specific expression during development and confined to reproductive tissues. In testis *Pem* gene is dramatically induced on day 9 in mice, when germ cells are known to enter meiotic prophase, where as mRNA levels are elevated on day 12-14 post partum. The Pem gene expression depends on androgens and gonadotrophins (Lindsey and Wilkinson 1996a, b). The promoter sequences from the *Pem* homeobox gene that directs Sertoli cell specific expression have been identified. A 0.6-kb 5'-flanking sequence directs transgene expression specifically in the testis and the epididymis but not in any other tissue tested. There are at least two regulatory regions in the *Pem* proximal promoter: one that directs androgen receptor-dependent expression specifically in Sertoli cells and another that confers stage-specific expression in neonates and adults by acting as a negative regulator (Rao et al., 2003).

MH-3: The *MH-3* gene, which contains a homeobox is expressed in testis and confined to mouse chromosome 6. This gene is expressed in male germ cells during late meiosis. In the embryo, *MH-3* transcripts are present at day 11.5 post-coitum, a stage in mouse development when gonadal differentiation has not yet occurred. The sequence of the carboxyl half of the homeobox shows homology to the helix turn helix domain of several DNA binding proteins that control developmental processes. The analysis of mouse MH-3 and its expression in the adult and in the embryo showed that this fragment contained 126 nucleotides of homeobox sequences and 99 nucleotides of sequences 3' to the homeo box. At the nucleotide level, the *MH-3* homeo box is 79% and 66% homologous to the deformed (Dfd) and Antp homeo box, respectively. For the predicted amino acid sequence, the homologues were 83% and 75% respectively. The sequence of the *MH-3* gene is homolgous to but distinct from murine boxes earlier described. No termination code was present in 3' to MH-3 homeobox. The important features of *MH-3* homeobox were that: (i) codon has been deleted in the MH-3 and (ii) the five amino acids immediately 3'of the *MH-3* homeobox are identical to those 3'of the Dfd homeobox (Rubin et al., 1986).

Gtx: Komuro et al., (1993) screened a mouse embryonic cDNA library using a *Drosophila* homeodomain probe whose sequence is divergent from the *Hox* class genes. The new gene was denoted *Gtx*, for Glial and testis specific homeobox gene. *Gtx* is located to the distal portion of mouse chromosome 7 and does not co-segregate with any other mapped homeobox genes. The *Gtx* homeodomain contains unique residues at positions predicted to contact DNA bases. It binds with high affinity to the MEF-2 motif, a binding site for the serum response factor related proteins. The GTX efficiently competed with RSRF to bind to MEF-2 element in vitro. Co-transfection of *Gtx* prevented the serum-induced activation of the MEF-2 containing reporter genes. *Gtx* is a novel cell type specific homeobox gene that has the potential to act as a transcriptional repressor for subset of serum inducible genes in germ cells (Komuro et al., 1993)**(Fig.16.5)**.

Sperm 1: The identification of Sperm-1, a testis specific POU-domain protein expressed in primary spermatocytes just prior to the onset of meiosis, suggests that this protein may be involved in the regulation of male germ cell differentiation. The Sperm-1 is most related to the transactivator Oct-3/4, which is expressed in the early embryo, primordial germ cells, and the egg. However, in contrast with Oct-3/4, rat Sperm-1 is selectively expressed during a 36 to 48 hr period immediately preceding meiosis I in male germ cells. Although the POU-domain of Sperm-1 is divergent from the POU-domains of Oct-1 and Oct-2, Sperm-1 preferentially binds to a specific variant of the classic octamer DNA- response element in which the optimal sequence differs from that preferred by Oct-1 and Pit-1 (Anderson et al., 1993; Zini et al., 1996).

16.6. ZINC FINGER PROTEINS

16.6.1. Zinc Fingers of Class 1 and 2

Zfy-1/Zfy–2: The sex testis determining genes *Zfy1* and *Zfy2* in mice encode proteins containing an acidic amino terminus and a carboxyl terminus composed of 13 zinc fingers. The zinc finger domain is conserved among human and mouse zinc finger X and Y genes. A 6-amino acid

```
   1     C CGC CGC CCA TTC AGC GCA ACA GCC GTC GGT CCT CTC GCT TTC CCG    46

  47   TAG GGG CCG TCG GCG TTC GTT TGA AAC GCG GTC CAC CCG TCC CAG CGT    94

  95   AGC CGG CGC TCT TCG GCG CCG CGC GCA AAC TTC CCG AGC CGG CGG GTG   142

 143   CGG GCG GTG GCA GCG GGG CCC GGA TGG GCG CCC GGG TCG GAG GCG GCG   190

 191   GCG CCC ATG GAC GCT AAC CGC CCG GGT GCG TTT GTG CTG AGC AGC GCG   238
   1           Met Asp Ala Asn Arg Pro Gly Ala Phe Val Leu Ser Ser Ala    14

 239   CCT TTA GCC GCG CTG CAC AAC ATG GCT GAG ATG AAG ACG TCG CTG TTC   286
  15   Pro Leu Ala Ala Leu His Asn Met Ala Glu Met Lys Thr Ser Leu Phe    30

 287   CCC TAC GCG CTG CAG GGC CCG GCG GGC TTC AAG ACA CCC GCC CTA GGC   334
  31   Pro Tyr Ala Leu Gln Gly Pro Ala Gly Phe Lys Thr Pro Ala Leu Gly    46

 335   AGC CTT GGC GCG CAG TTG CCT CTA GGC ACT CCG CAC GGC ATC AGC GAC   382
  47   Ser Leu Gly Ala Gln Leu Pro Leu Gly Thr Pro His Gly Ile Ser Asp    62

 383   ATC CTG GGA CGG CCG GTG GGC GCA GCG GGT GGC GGC CTC CTA GGA AGT   430
  63   Ile Leu Gly Arg Pro Val Gly Ala Ala Gly Gly Gly Leu Leu Gly Ser    78

 431   CTG CCC CGT CTC AAC GGG CTC GCC TCG TCT GCA GGT GTC TAC TTC GGG   478
  79   Leu Pro Arg Leu Asn Gly Leu Ala Ser Ser Ala Gly Val Tyr Phe Gly    94

 479   CCC GCA GCC GCC GTG GCT CGG GGC TAC CCC AAG CCG CTG GCG GAA CTG   526
  95   Pro Ala Ala Ala Val Ala Arg Gly Tyr Pro Lys Pro Leu Ala Glu Leu   110

 527   CCT GGG CGC CCG CCC ATC TTC TGG CCT GGG GTG GTG CAG GGC TCT CCC   574
 111   Pro Gly Arg Pro Pro Ile Phe Trp Pro Gly Val Val Gln Gly Ser Pro   126

 575   TGG AGG GAC CCG CGA CTG GCC GGC TCC GCC CAA GCC GGC GGG GTC CTG   622
 127   Trp Arg Asp Pro Arg Leu Ala Gly Ser Ala Gln Ala Gly Gly Val Leu   142

 623   GAT AAG GAT GGC AAG AAG AAA CAC TCG CGG CCG ACT TTC TCC GGC CAG   670
 143   Asp Lys Asp Gly Lys Lys Lys His Ser Arg Pro Thr Phe Ser Gly Gln   158

 671   CAG ATC TTC GCG CTG GAG AAG ACT TTC GAG CAG ACC AAG TAT TTG GCA   718
 159   Gln Ile Phe Ala Leu Glu Lys Thr Phe Glu Gln Thr Lys Tyr Leu Ala   174

 719   GGC CCA GAG CGC GCG CGG CTT GCC TAC TCT CTG GGC ATG ACC GAG AGC   766
 175   Gly Pro Glu Arg Ala Arg Leu Ala Tyr Ser Leu Gly Met Thr Glu Ser   190

 767   CAA GTG AAG GTG TGG TTC CAG AAT CGG CGG ACC AAG TGG CGC AAG CGG   814
 191   Gln Val Lys Val Trp Phe Gln Asn Arg Arg Thr Lys Trp Arg Lys Arg   206

 815   CAC GCG GCA GAG ATG GCG TCG GCT AAA AAG AAG CAA GAC TCG GAT GCC   862
 207   His Ala Ala Glu Met Ala Ser Ala Lys Lys Lys Gln Asp Ser Asp Ala   222

 863   GAG AAG CTG AAG GTG GGT GGC TCA GAC GCG GAG GAC GAT GAC GAA TAC   910
 223   Glu Lys Leu Lys Val Gly Gly Ser Asp Ala Glu Asp Asp Asp Glu Tyr   238

 911   AAC CGG CCC CTG GAC CCC AAC TCC GAT GAC GAG AAG ATC ACG CGG CTT   958
 239   Asn Arg Pro Leu Asp Pro Asn Ser Asp Asp Glu Lys Ile Thr Arg Leu   254

 959   CTC AAA AAG CAC AAA CCC TCG AAC TTG GCG CTC GTT AGC CCG TGT GGT  1006
 255   Leu Lys Lys His Lys Pro Ser Asn Leu Ala Leu Val Ser Pro Cys Gly   270

1007   GGC AGC GCG GGG GAC GCC TTG TGA GGA TGC GGC CAG GCC GGA GAA CCC  1054
 271   Gly Ser Ala Gly Asp Ala Leu                                      277

1055   GAG AAC CGG GAC TCG CGG CAT GCC CCG ACG CCA GCC GCC CAG CCG CAG  1102

1103   TGT GTA TAT ATA TTT TTA CAG AAT AAG TTA TAA AGC GGA CGT TGG CGC  1150

1151   GGC CTT GGC GTG ATG GCG GAG TAC GGG GTT TGG GCC GAT CAC TTT GTA  1198

1199   TAA TCA ATA AAT TAT TTA ACA CGT CCT CGT CGG AGC CGT GGC TCC C(A)n 1244
```

Fig.16.5. The nucleotide and deduced amino acid sequence of c-DNA of *Gtx*. Homeodomain region is boxed. Proline and glycine residues found in region preceding the homeobox are underlined with solid and dashed lines. The region following the homeodomain is rich in acidic amino acids (boxed) and basic amino acids (bold lines). The putative first polyadenylation signal is shown in bold. Reprinted with permission from E Komuro et al. The EMBO Journal 12; 1387-01: 1993 © http://www.nature.com/.

deletion in the Zfy-2 zinc finger domain by Nagamine et al., (1990) suggested that *Zfy-1* and *Zfy-2* transcription is linked to spermatogenesis and that that its transcription increased with the initiation of meiosis. The high levels of these mRNAs are found in post-meiotic round spermatid cells; Zfy-2 being slightly greater than Zfy-1. The expression of Zfy-1 in fetal testes supports the hypothesis that this gene plays a role in testis differentiation besides the postulated role in testis determination (Nagamine et al., 1990). In order to obtain a genomic clone of *Zfy-1* from a Y chromosome of Mus musculus omesticus (YDOM) origin, Kalikin et al (1989) cloned a human zinc finger Y clone. The specificity of the clone confirmed that the clone was of *Zfy-1* origin. The gene is expressed post-meiotically.

Zip 29: A cDNA from mouse testis encodes a protein (Zip-29) with 14 copies of the zinc finger (Zf) motif commonly found in transcriptional regulatory proteins. The expression of *Zip-29* is restricted to the testis in adult mice, but also occurs during embryonic development. Within the testis, Zfp-29 mRNA is enriched in round spermatids (Denny and Ashworth, 1991).

Zip 35: A novel mouse gene, *Zip-35*, encodes a protein with a block of 18 zinc finger domains and an N-terminal region rich in acidic residues. The 2.4-kb mRNA encoding this polypeptide is selectively expressed in adult testis. A considerable increase in expression of Zip35 is restricted to spermatocytes at the pachytene stage of meiotic prophase (Cunliffe et al, 1990).

TSGA: A cDNA corresponding to testis specific expression (TSGA), was isolated from testis cDNA library. The TSGA sequence was confined to cells within the seminiferous tubules and expressed only in male germ cells. The steady-state level of TSGA transcripts reaches a maximum during the meiotic and the post-meiotic stages of germ cell development, suggesting a meiotic or post-meiotic function for the encoded protein. The TSGA encodes a putative protein having 1,214 amino acids and contains a zinc finger structure (Hoog et al., 1991).

Protein E: Protein E isolated from cuttle fish sperm contains a motif-Cys-X_2-Cys-X_{23}-His-Cys-X_2-Cys-, which is likely to adopt a zinc finger conformation. The sequence of protein E does not correspond to that of any known protein, but presents some similarities with a part of Zfy protein, a putative human transcription factor specifically expressed in germinal cells and involved in spermatogenesis (Martin-Ponthieu et al., 1994).

Ctfin51: Noce et al., (1992) isolated three clones from a mouse spermatocyte cDNA library encoding proteins with C2H2-type zinc finger motifs, which are expressed by the *Drosophila* Kruppel gene. Three cDNAs were designated *Ctfin33, Ctfin51, Ctfin92.* Ctfin 51 amino acid sequence revealed seven zinc finger motif in C-terminal region. Ctfin 51mRNA expresses in spermatocytes after the pachytene stage and in early stage round spermatids of pre-puberal and adult males. Ctfin 51 was localized within nuclei of spermatocytes and spermatids. Ctfin 51 had a M_r 75,000 in extracts of the testis and the ovary. The *Ctfin51* gene encodes a DNA binding regulatory protein functionally associated with meiosis in both male and female gametogenesis (Noce et al., 1992).

Nelki et al (1990) isolated a cDNA clone coding for a zinc finger protein of the C2H2 type. The polypeptide encoded by this clone has outside the zinc finger region, an N-terminal portion of 73 amino acids of which 25% are basic. There are 12 zinc fingers in the open reading frame with 9 out of the 11 linkers having the sequence Thr-Gly-Glu-Lys-Pro, indicating it to be a member of the Kruppel family. At the C-terminus there are 9 amino acids after the last finger of which 6 are acidic and none basic. The open reading frame is bounded by stop codons at nucleotide 12 (upstream from the initiator methionine) and at nucleotide 153.

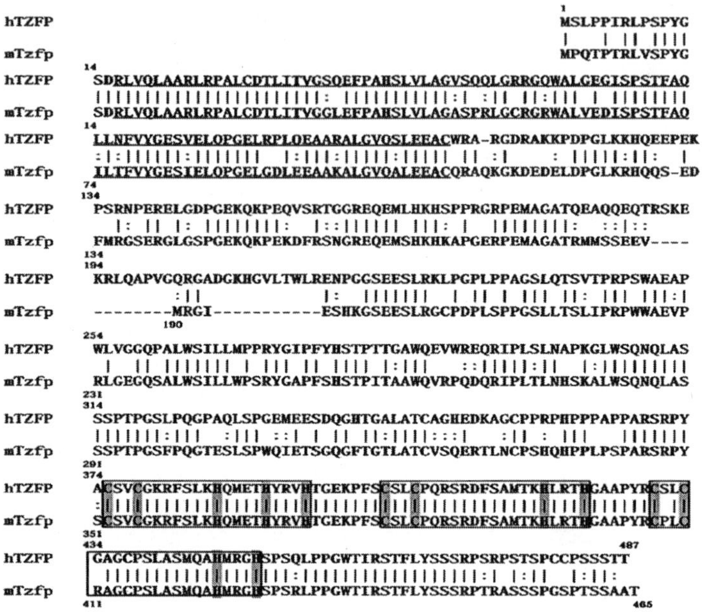

Fig.16.6. Amino acid sequences of human TZFP and homology with mouse Tzfp. Three C_2H_2 zinc fingers are boxed. The potential HTH/POZ domain is underlined. Vertical bars show identical residues and colons indicate similar residues. Reprinted with permission from C-J C Tang et al. J Biol Chem 276; 19631-39: 2001 © The American Society for Biochemistry and Molecular Biology.

Tzfp: Tang et al, (2001) described a testis zinc finger protein (Tzfp), which binds to the upstream flanking sequence of *Aie1* gene (Chapter 19). The mouse Tzfp gene is mapped to chromosome 7B2-B3, and encodes a transcription factor containing 465 aa. Tzfp has a conserved N-terminal BTR/POZ domain and three C-terminal PLZF-like C2H2 zinc fingers. The zinc finger domain of Tzfp binds to a TGTACAGTGT element (Tzfp binding site =tbs) located at the upstream flanking region of Aie1 gene. The N-terminal BTB/POZ domain of TZFP has a repressor activity. It appears that TZFP regulates genes, which carry tbs sites (Tang et al., 2001) **(Fig 16.6).**

Zfp 37: A gene designated *Zfp-37* encodes Zfp-37 transcripts at high levels in the testis, the only adult tissue in which Zfp-37 expression was observed. Zfp-37 was also expressed at lower level in the mid gestation embryo and placenta. The major testicular transcripts are of 2.3- and 2.6-kb. A 4.0-kb transcript was detected at lower levels in the testis as well as in embryo and placenta. Zfp-37 was most abundant in germ cells, which have completed meiosis and undergo the complex morphogenetic changes of spermiogenesis (Burke and Wolgemuth, 1992).

Gli: The oncogene *GLI* is expressed in some cases of human malignant glioma and undifferentiated childhood sarcoma and is the prototype for a gene family characterized by a highly conserved set of five tandem zinc fingers and a consensus cysteine/ histidine link. This zinc finger motif binds DNA with sequence specificity and mediates transcriptional regulation. Gli transcripts are present in mouse embryonic development as well as in normal adult uterus, brain, testis, and limb. Findings support a role for gli family genes in normal craniofacial and digital development in mammals (Walterhouse et al., 1993). The Gli members are differentially expressed during spermatogenesis in mice. Gli and Gli3 mRNA are present in mouse germ cells, where as Gli2 is not. Further, Gli and Gli3 exhibited stage specific expression, selectively in type A and type B spermatogonia. This shows that Gli and Gli3 participate in DNA binding during mitotic stage of spermatogeneis (Persengiev et al, 1997). In contrast to Gli3, Gli1 shifts from nucleus to cytoplasm and localized to spermatids indicating that the zinc finger transcription factor Gli 1 is the mediator of signaling of the Hedgehog (Hh) family (Kraft et al, 2001).

GLIS: A transcription factor GLI-similar 3 (GLIS3) is a 83.8-kDa nuclear protein containing five C2H2-type Kruppel-like zinc finger motifs that exhibit 93% identity with those of GLIS1. However, little homology exists outside their zinc finger domains. GLIS3 can function as a repressor and activator of transcription. Deletion mutations determined that the N- and C-termini are required for transcriptional activity. GLIS3 binds to the GLI-RE consensus sequence and is able to enhance GLI-RE-dependent transcription. The GLIS3 is expressed in specific regions in developing kidney and testis and in a highly dynamic pattern during neurulation. The temporal and spatial pattern of GLIS3 during embryonic development suggests that it may play a critical role in a variety of cellular processes during development (Kim et al., 2003).

Gfi: The Gfi-1 proto-oncogene encodes a zinc finger protein with six C2H2 type C terminal zinc finger motifs and is activated by provirus integration in T-cell lymphoma lines selected for interleukin-2 in culture and in primary retrovirus-induced thymomas. The Gfi-1 expression in adult animals is restricted to the thymus, spleen, and testis and is enhanced in mitogen-stimulated splenocytes. Gfi- is a 55-kDa nuclear protein that binds DNA in a sequence specific manner. The Gfi-1 binding site was defined by TAAATCAC(A/T)GCA. Potential Gfi-1 binding sites have been detected in a large number of eukaryotic promoter enhancers, including the enhancers of several proto-oncogenes and cytokine genes and the enhancers of the human cytomegalovirus (HCMC), major immediate early promoter, which contains two such sites. The Gfi-1 may play a role in HCMV biology and may contribute to the oncogenesis and T-Cell activation by repressing the expression of genes that inhibit these processes (Zweidler-Mckay et al., 1996). The role of Gfi in immune cells of testis by Sertoli cells is awaited.

Egr4 : Egr4 (NGFI-C, pAT133), a member of the Egr family of zinc finger transcription factors, is thought to be involved in cellular growth and differentiation. Egr4 null mice through targeted mutagenesis were found phenotypically normal with the exception that males, but not females, were infertile. Egr4 is expressed at low levels within male germ cells during meiosis and is critical for germ cell maturation during the early mid-pachytene stage. In Egr4 null male germ cells, a limited degree of spermiogenesis occurs but this is accompanied by markedly abnormal spermatozoon morphology and severe oligozoospermia. Study raises the possibility that Egr4 may contribute to some forms of human idiopathic male infertility (Tokuzawa et al., 1999).

MZF6D: The *MZF6D* is a mouse testis gene, which is selectively expressed in meiotic spermatocytes. The MZF6D protein contains an N-terminally located repressor domain, a KRAB domain, followed by at least seven successive Kruppel zinc-finger motifs. The KRAB domain of MZF6D, which consists of a KRAB A box and the newly identified KRAB C box, has been shown to interact with TIF1β, which is the common co-repressor of all KRAB zinc-finger proteins. In adult mouse testes MZF6D localizes to meiotic spermatocytes, suggesting a specific role for MZF6D in the regulation of spermatogenesis (Looman et al., 2003).

KRAB- A zinc finger protein, HZF12: Kruppel-related zinc finger proteins probably constitute the largest individual family of transcription factors in mammals. These proteins often carry a potent repressor domain called the Kruppel Associated Box (KRAB), which effectively represses transcription through interaction with transcriptional intermediary factor 1β (TIF1β). A human KRAB A zinc finger protein, HZF12 has been characterized by Looman et al., (2004). The gene encoding HZF12 is located on Chromosome (Chr) 19p13.11-p12, and a 4.4-kb transcript from this gene is expressed in a variety of adult and fetal tissues. Two additional, larger transcripts are expressed in testis only. Surprisingly, the KRAB A domain of HZF12 is followed by a 21-amino acid domain, encoded by a separate exon. This domain, which was designated KRAB C, was also identified in more than 25 additional human, mouse, and rat KRAB zinc finger proteins. This novel KRAB domain strengthens the interaction with TIF1β, thereby improving the ability of these KRAB zinc finger proteins to recruit TIF1β to specific sites (Looman et al., 2004).

Zfp94, Zfp95, Zfp96: Three novel zinc finger genes (*Zfp94, Zfp95, Zfp96*) code three protein products belonging to the LeR family (leucine-rich zinc fingers). One of them, ZFP95 contains a domain homologous to the Kruppel-associated box. Though all three genes are expressed at high levels in testis among other tissues, but testis-specific transcripts were observed for Zfp95 and Zfp96. The testis-specific transcript of Zfp95 showed highest expression in pachytene spermatocytes, while that of Zfp96 was highly expressed in pachytene spermatocytes, in round spermatids and residual bodies. In particular, the testis-specific transcripts of Zfp95 and Zfp96 were greatly reduced in heterozygous, and completely absent in homozygous testis RNA from atrichosis mutant mice (Weissig et al., 2003).

Znf230: The cDNA of mouse gene *Znf230* of 982-bp encodes a 230 aa protein with a ring finger domain at its C-terminus. Mouse c-DNA- *Znf230* and predicted protein showed 91% and 98% sequence identity with its human homologue. This cDNA is testis specific whereas the longer transcripts of 2.4 and 4.4kb are ubiquitously expressed. The expression of *Znf230* in testis is developmentally regulated and first detected at day 6 pn, and reaches adult level during round spermatids. The protein of *Znf230* exhibits DNA binding activity and its ring finger domain may function as an activator module in transcription (Qiu et al., 2003).

BTEB2 protein: Rat BTEB2 protein is a transcription factor with three zinc fingers that binds to GC box, and is expressed in the placenta, intestine, and testis. The BTEB2 gene contained 4 exons. Deletion analysis of the promoter region of the BTEB2 gene revealed that at least three regions are important for the activity. Upon investigation of cis-acting elements in the regions, the GC box, CCAAT box and NF-1 binding sites were found. As binding factors, Sp1, CBFa, and NF-1, which bind to the DNA elements were identified (Mori et al., 2003).

16.6.2. GATA Binding Proteins

The GATA-binding proteins are a family of zinc finger transcription factors that regulate gene expression, differentiation, and cell proliferation by binding to the consensus DNA sequence (A/T) GATA (A/G). Three members of the GATA-binding protein family, GATA-1, GATA-4 and GATA-6, have been shown to be expressed in the murine gonad and are thus of interest as potential regulators of testicular development and function. GATA-1 binds to the consensus WGATAR sequence through a zinc finger DNA-binding motif, which is highly conserved among the multiple members of this transcription factor family. The carboxyl finger is responsible for site-specific DNA association, while the amino finger is required for nuclear localization of GATA-1. GATA-1 mRNA is found in erythroid, megakaryocytic and mast cells and is required for functional erythropoiesis, thereby suggesting that it also contributes to the transcriptional regulation of genes specifically expressed in all these hematopoietic cell lineages. GATA-1 identical to hemotopoietic tissue is also expressed in the seminiferous tubules of mouse testis. This novel transcript utilizes a testis-specified promoter and first exon is located 8-kbp 5' to the hematopoietic cell first exon, but the remaining five exons, which code for the GATA-1 protein are common in both testis and erythroid mRNAs (reviewed in Yomogida et al., 1994). GATA-1 is expressed only in the Sertoli cell and its expression in Sertoli cells is induced concomitantly with the first wave of spermatogenesis during pre-puberal testis development, subsequently the number of GATA-1 positive cells declines, and present in tubules in stages VII, VIII and IX of spermatogenesis in the adult mouse testis. In contrast, virtually every Sertoli cell in mutant W/W^v *jsd/jsd* to cryptorchid mice (all of which lack significant numbers of germ cells) express GATA-1, showing that the expression of this transcription factor is negatively controlled by the maturing germ cells (Yomogida et al., 1994).

GATA-4 and GATA-6 transcripts are present in postnatal mouse ovary, where the exact role of these factors remains unknown. In testis, GAATA-4 mRNA appears in the fetal testis at 13.5 days postcoitum. Both GATA-4 and GATA-6 transcripts have been observed in late fetal, neonatal, juvenile, and adult Sertoli cells. In addition, GATA-4 mRNA was detected in interstitial cells throughout development and both Sertoli and Leydig cells in postnatal animals. The inhibin β promoter harboring essential GATA-binding sites can be *trans*-activated by GATA-4. It seems that GATA-4 and GAATA-6 play differing roles in the maturation and function of testicular somatic cells (Ketola et al., 1999). Secretion of Mullerian inhibiting substance (MIS) by Sertoli cells of the fetal testis and subsequent regression of the Mullerian ducts in the male embryo is a crucial event that contributes to proper sex differentiation. The GATA-4 and nuclear receptor steroidogenic factor (SF-1) are early markers of Sertoli cells that have been shown to regulate MIS transcription. GATA and SF-1 binding sites are adjacent to one another in the MIS promoter. Co-expression of both factors lead to a strong synergistic activation of the MIS promoter. (Tremblay et al., 1999) (See Chapter 3).

16.6.3. Basonuclin

For years, it has been known that zinc finger proteins are expressed in male germ cells. Where tested, these proteins have been found to be strictly nuclear. Basonuclin is a zinc finger protein that was throught to be restricted to keratinocytes of stratified squamous epithelia. In epidermis, basonuclin is associated with the nuclei of mitotically active basal cells but not in terminally differentiating keratinocytes. A novel form of basonuclin is also expressed in stratified epithelia. Unexpectedly, both forms are present in testis, where a surprising locatilization pattern was seen. While basonuclin RNA expression occurs in mitotically active germ cells, protein was not detected until the meiotic stage, where basonuclin localized to the appendage of the distal centriole of spermatocytes and spermatids. Near the end of spermiogenesis, basonuclin also accumulated in the acrosome and mitochondrial sheath surrounding the flagellum. A perfect six amino acid residue mitochondrial targeting sequence is present in basonuclin 1a (BSN1a) but not in the BSN1b form. Findings suggest a unique

```
         M R [R R P E P G] R T R G G R A R E T
hBSN1a:  ATGCGGCGGCGCCCGGAGCCGGGGCGGACGCGGGGCGGCCGGGCCCGGGAGACG
mBSN1a:  ATGCGGCGG---CCGGAGCCGGGGCGAC----GGGCAG---------GGGACGCG

         R R Q P R H  R S G R R M A E
         CGCCGGCAGCCCCGGCA-CCGCAGCGGTCGCAGGATGGCCGAG
         CGCCGGGAAGGCCGGCATCCGCAGCGGTCGCAGGATGGCCGAGgtaagc
                                           M A E ↑

hBSN1b:  GTCAGCCAGATGTGACAAGATTGTGGCGAGGCGAGTGAAGGAGATCTGATATTA
mBSN1b:  GTCAGCCAGATGTGACAAGATTGTGGCGAGGCGAGTGAAGGAGATCTGATATTA

         ATCAGCGCGTAGCTCCGAGCCAAACTTTCACAAATGCGGTGT AGAAACATGTTC
         ATCAGCGCGTAGCTCCGAGCCAAACTTTCACAAATGCGGTGT AGAAACATGTTC

         TTCTCATTTAAGGCATCTCTTTGTGGCTGCGGGGCTGCCACCGCTCCGAGTCTG
         TTCTCATTTAAGGCATCTCTTTGTGGCTGCGGGGCTGCCACCGCTCCGAGTCTG

         ACAG
         ACAGgttaaat
             ↑
```

Fig.16.7. Sequence relation between two forms of basonuclin. Nucleotide sequences of human and murine basonuclin-1a and -1b (hBSN1a/hBSN1b and mBSN1a/mBSN1b) and amino acid sequence of hBSN1a. The hBSN1b differs from hBSN1a in 5'sequences. Splice sites are marked by arrows. Reproduced with permission from Y. J. Yang et al. J Cell Biol 137; 657-69: 1997 © Rockefeller University Press.

role for basonuclin in certrosomes within the developing spermatid, and a role for one of the protein forms in germ cell mitochondrial function. Its localization with the acrosome suggested that it may also perform a function during fertilization (Yang et al., 1997).

The two basonuclin sequences (BSN1a and BSN1b) characterized from both mouse and human are provided in **Fig.16.7**. The BSN1b exon is highly conserved and is nearly identical between mouse and human. The BSN1a exon found in the originally reported sequence is less conserved between the two species, and different translation start sites are predicted. Both mouse and human forms of BSN1a but not BSN1b contains the sequence RRPEPG, which has been shown to be a mitochondrial targeting sequence for proteins (see c/r Yang et al., 1997). However, in another study, immunocytochemical staining detected basonuclin in the nuclei of spermatogonia and spermatocytes at various developmental stages. During spermiogenesis, it relocated from the nucleus to the midpiece of the flagellum of the spermatozoa. The dual presence of basonuclin in differentiated spermatozoa and oocytes suggests that it may play a role in their differentiation and the early development of an embryo (Mahoney et al., 1998). The human gene is located on chromosome 15. The basonuclin 2 homologue is present in chicken also. Although the amino acid sequence of basonuclin 2 differs extensively from that of basonuclin 1, the two proteins share essential features. Both contain three paired zinc fingers, a nuclear localization signal, and a serine stripe. The basonuclin 2 mRNA has a wider tissue distribution than the basonuclin 1 mRNA: it is particularly abundant in testis, kidney, uterus, and intestine. The extreme conservation of the basonuclin 2 amino acid sequence across vertebrates suggests that basonuclin 2 serves an important function, presumably as a regulatory protein of DNA transcription (Vanhoutteghem and Djian, 2004).

16.6.4. Ret Finger Protein

The human Rfp (ret finger protein) has a tripartite motif, consisting of two zinc fingers (the RING finger and the B-box) and a coiled-coil domain, and belongs to the B-box zinc finger protein family. The Rfp becomes oncogenic when its tripartite motif is recombined with the tyrosine kinase domain from the c-ret proto-oncogene. The mouse Rfp cDNA shared a 98.4% homology with the human sequence. The gene is mapped to human chromosome 6 and mouse chromosome 13 indicating that it was linked to several other genes encoding proteins that possess common domains. Rfp transcripts and protein are ubiquitous in day 10.5-13.5 mouse embryos, but restricted in adult mice, with the highest level of expression in pachytene spermatocytes and round spermatids of differentiating sperm. The Rfp protein was detected within cell nuclei as nuclear bodies similar to the PODs (PML oncogene domain) observed with another B box family member, PML (promyelocytic leukemia protein) (Cao et al., 1996).

16.6.5. RING Finger Proteins

Sperizin: In agreement with their nuclear localization, classical zinc finger proteins are characterized by their binding affinity to DNA. On the other hand, the RING-finger motif, which was first characterized as a cysteine rich stretch of amino acids in the human ring1 gene, is a new branch of the zinc finger family (Freemont, 1993). Some of the genes belonging to this family are known to regulate development and cell differentiation. Really interesting new gene (Ring) domain, a specialized zinc finger of 40-60 amino acids stretch that binds two atoms of zinc, is found in wide variety of proteins including several protooncogenes and products of genes involved in autosomal recessive juvenile parkinsonism, Parkin. Database search revealed a second RING-finger motif (RING-H2 family) with the conserved fourth cysteine of the motif replaced by histidine. Within this family, there are proteins that have been suggested to function in protein-protein and or protein membrane interactions. Later, a third subclass of the RING finger family was shown to consist of a second cysteine rich motif (B-box) and a coiled-coil motif on the C-terminal side of the RING finger. The RING-H2 subfamily to which sperizin belongs, consists of more than 40 proteins. The function of these proteins is distinct from the majority of RING finger proteins in which interactions with lipid or other proteins of the cytoplasm or plasma membrane rather than interactions with DNA may be involved. The cytoplasmic localization of GFP tagged sperizin suggests that sperizin may also function in the cytoplasm. Sperizin is a new member of the RING-finger family of proteins that may be involved in spermatogenesis. A ring finger protein gene in murine encodes a protein with a RING zinc finger motif, spermatid specific ring zinc finger Sperizin. *Sperizin* gene does not transcribe in pre-puberal testis, but becomes detectable at day 23, exclusively in the round spermatid. The sperizin gene is intronless, and GFP-tagged sperizin was found to be localized in the cytoplasm of ectopically expressed somatic cultured cells. The chromosomal localization of the sperizin gene is assigned to chromosome 17 (Fujii et al., 1999) **(Fig.16.8)**.

Haprin: A cDNA clone for a gene named *haprin* encodes a haploid germ cell-specific RING finger protein. This protein is a novel member of the RBCC (RING finger, B-box type zinc finger, and coiled-coil domain) motif family that plays roles in several cellular processes, such as exocytosis. It is transcribed exclusively in testicular germ cells after meiotic division. The molecular weight of Haprin protein is 82-kDa and localized in the acrosomal region of elongated spermatids and mature sperm and was not present in acrosome-reacted sperm. Inhibition of acrosome reaction by Ab against the RING finger domain of Haprin indicated that the Haprin plays a role in the fertilization (Kitamura et al., 2003).

SNURF (or RNF4): A nuclear RING finger protein, termed SNURF (or RNF4) is a co-regulator of androgen receptor-dependent transcription. Two SNURF mRNA transcripts of 3.0 and 1.6-kb in size are present in adult rat testis. Both transcripts are capable of encoding full-length SNURF protein. The 3.0-kb SNURF mRNA is expressed in Sertoli ells, whereas the expression

```
Sperizin    1'              MARFAWTRVAPVALVTFWLVLSLSPTDAQVNL--SSVDFLDLPALLGVPVD
                                   ...*.* .*  .*.....  . *.  .. .* **** .* ..
C-RZF       1" MLLSIGMLMLSATQIYTIVTVQLFAFLNLLPVEADILAYNFENGTQTFDDLPARFGYRLP

Sperizin   50' PKRARGYLLVARPADACHAIE-GPWPDNHSLDPLVLVRPLGCSWEQTGRRAQRAGATAAS
               ..  .*.*. ..*..**..*.  * .** * . .**.*.*.*... .  .***** .**
C-RZF      61" AEGLKGFLINSKPENACEPIAPPPLRDNSSTAFIVLIRRLECNFDIKVLNAQRAGYKAAI

Sperizin  109' VGPEAPGQLREF--EDLEVTVRCDQPARVLLPHAEPCPDPE------CHPVVVASWALA-
               *   ....* ..  .*.*.  . * *.  .   .. . . *        * *......*.
C-RZF     121" VHNVDSDDLISMGSNDIEILKKIDIPSVFIGEASANSLKEEFTYEKGGHVVLIPEFSLPL

Sperizin  160' ----RALALAASTLFVLRQLWPWVRGLGSRGTA----VKTQTCQKAQVRTFTRLSD--LC
                   .. . ... ..*  ..  .. . .*  *    ....  .* .*..*.. ..  .*
C-RZF     181" EYYLIPFLIIVGICLILIVIFMITKFVQDRHRARRNRLRKDQLKKLPVHKFKKGDEYDVC
                       Zinc Finger motif
Sperizin  210' AICLDDYEEGERLKILPCAHAYHCRCIDPWFSRAGASCPLCKQSVASTHDGSTDGSVGGE
               *****.**.*..*.****.*****.*.***....  .**.***.*......*.... ...
C-RZF     241" AICLDEYEDGDKLRILPCSHAYHCKCVDPWLTKTKKTCPVCKQKVVPSQGDSDSETDSSQ

Sperizin  270' EPPLPGHRPPIWAIQARLRSRRLELLARTVPCRRCSSTTSLGVAENVAQSEATSELS    326
               *   ....*.    *........ *.
C-RZF     301" EENEVSENTPLLRPLASVSTQSFGALSESHSHQNMTESSEYEEDDNDNIDSSDAESGVNE 381
```

Fig.16.8. Amino acid sequence of sperizin and its comparison with chicken ring finger protein (C-RZF). Asterisks and dots between amino acid sequences of sperizin and C-RZF represent identical and similar amino acids respectively. The Ring-finger motif is underlined with thick line. Potential phosphorylation sites for casein kinase II or PKC are shown by dotted or double underline. Reprinted with permission from T. Fujii et al. Genomics 57; 94-101: 1999 © Elsevier.

of the 1.6-kb transcript appears after day 30 of postnatal life and is restricted to step 4-11 spermatids. Increased accumulation of SNURF in step 4-11 spermatids is not due to androgen signaling. Perhaps SNURF is involved in the regulation of processes required for late steps of spermatid maturation (Yan et al, 2002).

A Double Ring-H_2 Domain in RNF32: Evidence suggests that RING finger proteins function in the ubiquitin pathway as E3 ligases. A variant of the RING domain is the RING-H2 domain, in which one of the cysteines is replaced by a histidine. A novel gene, RNF32 is located on chromosome 7q36 and contains 37-kb of genomic DNA and consists of 9 constitutive and 8 alternatively spliced exons, most of which are alternative first exons. RNF32 is expressed during spermatogenesis, most likely in spermatocytes and/or in spermatids, suggesting a possible role in sperm formation (van Baren et al., 2002). The gene produces a long and a short transcripts: the short transcript contains exon 1-4 only. This gene encodes two RING-H2 domains separated by an IQ domain of unknown function and is the first reported gene with a double RING-H2 domain. In humans, RNF32 overlaps with a processed retroposon located on the opposite strand, C7orf13. The RNF32 is specifically expressed in testis and ovary, whereas C7orf13 is testis specific, suggesting that its expression may be regulated by elements

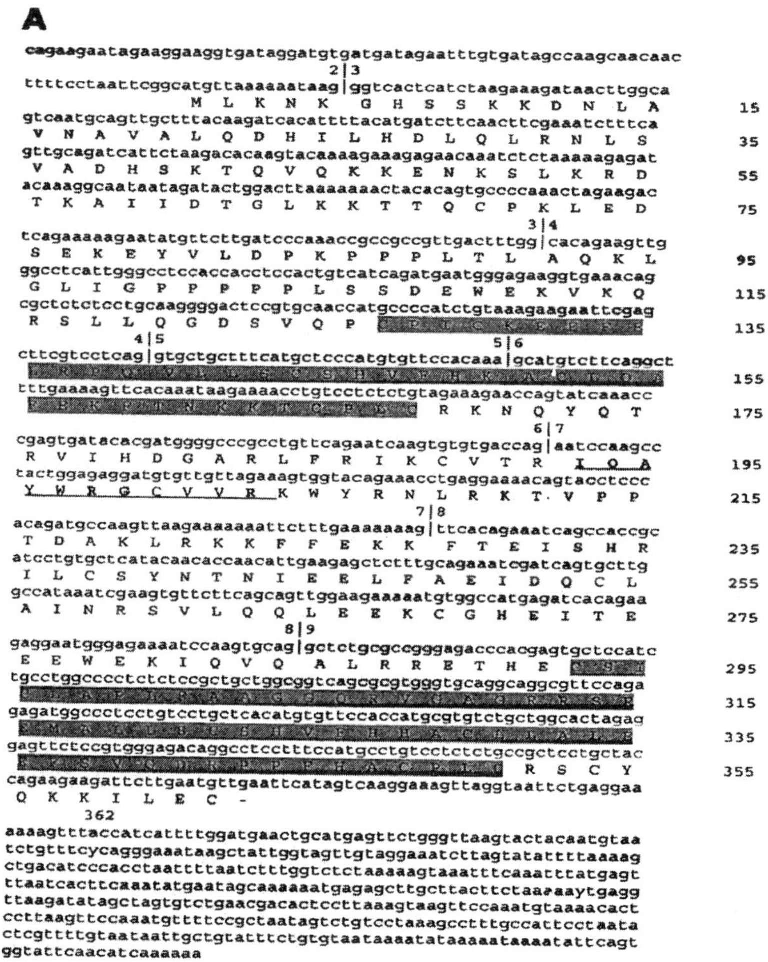

Fig.16.9. Human RNF32 protein and DNA sequence, starting with exon 2. Ring H2 domains are shaded, the IQ domain is underlined, and the consensus residues are bold. Splice sites are indicated by vertical line. Reproduced with permission from M.J.van Baren et al. Biochem Biophys Res Commun 292; 58-65: 2002 © Elsevier.

in the RNF32 promoter region **(Fig. 16.9)**.

The complex imprinted locus in chromosome 15q11-q13 encodes two genes, ZNF127 and ZNF127AS. The ZNF127 gene encodes a protein with a RING (C3HC4) zinc finger and multiple C3H zinc finger motifs. The intronless ZNF127 gene is expressed ubiquitously, but the entire coding sequence and 5' CpG island overlaps a second gene, ZNF127AS. Analyses of adult testis, sperm and fetal oocytes demonstrated a gametic methylation imprint with unmethylated paternal germ cells. ZNF127 is part of the coordinately regulated imprinted domain affected in Prader-Willi syndrome patients with imprinting mutations (Jong et al.,1999).

16.7. RNA BINDING PROTEINS AS TRANSCRIPTION FACTORS

The expression of many genes during early stages of spermatogenesis is regulated at the level of transcription, where as during the later phases of spermatogenesis, transcription ceases and stored mRNAs get activated to synthesize proteins as in Prms and transition proteins. Although little is known about the mechanisms inactivating and activating "paternal " mRNAs in germ cells, a number of RNA-binding proteins (RBPs) likely to be involved in mRNA translational control have been identified in the mammalian testis. Testis/brain (TB)-RBP, SPNR, and superoxide dismutase RNA-binding protein (SOD-RBP) bind to the 5'UTR of a testis-specified mRNA and can represses corresponding mRNA translation in vitro. Other RNA binding proteins comprise ATP-dependent RNA helicases, poly (A)-binding protein, Y-box-binding proteins and others. Some of these proteins may participate in RNA processing leading to the formation of unique transcripts, or may be involved in effecting translational delay. Thus, a growing number of testicular RNA-binding proteins have been identified, that are likely to facilitate the repression and activation of mRNA translation and gene transcription in male germ cells and have been discussed in Chapters 10 and 15.

Zinc finger RNA binding protein: The ZFR (Zinc finger RNA binding) is a murine Zinc finger Chrome associated protein, which possesses a region common to a small class of RNA binding proteins. ZFR encodes a protein of 1052 amino acids with their widely spaced Cys2 His2 zinc fingers. Out side the zinc fingers, ZFR shares a region that is highly conserved between several RNA binding proteins containing copies of double stranded RNA binding motif. The ZFR is highly expressed in testis and ovary during meiosis and in brain. It is also expressed in Sertoli cells and mapped to chromosome 15 region A. Its homologues do exist in invertebrates and humans (Meagher etal, 1999) (see Chapter 15).

Cold shock proteins: A cold shock domain (CSD) is found in many eukaryotic transcriptional factors, which specifically bind the DNA of a Y-box cis element. The CSD is present in the sequence of the *Xenopus* RNA binding proteins FRGY1 and FRGY2 (Chapter 15). Nishiyama et al, (1997) isolated the cold inducible RNA binding protein (Cirp) from a mouse testis cDNA library. In mice and humans, the Cirp is constitutively expressed in the germ cells of testes, and which is decreased at an elevated temperature. In somatic cells, Cirp is induced by mild cold stress (32^0 C) and mediates the cold induced suppression of cell growth. Human RNA binding motif protein 3 (RBM3) isolated from the Xp11.2 region of chromosome encodes a protein structurally related to Cirp and both *Cirp and RBM3* transcripts are induced by cold stress in human cells. The mouse *Rbm3 cDNA* encodes an 18-kDa protein with 94% identity in amino acid sequence to that of human RBM3; the murine *Rbm3* mRNA and protein were detected in Sertoli cells, but not in germ cells at any stage of development. In the TAMA26 mouse Sertoli cell line, *Rbm3* expression was increased or decreased after temperature shift from 37^0 to 32^{0}C or 39^{0}C respectively. The level of *Rbm3* was decreased in Sertoli cells when mouse testis was exposed to heat stress for experimental cyrptorchidism. The Rbm3 seems to play a distinct role from those played by Cirp in spermatogenesis and cyrptorchidism by regulating the gene expression in Sertoli cells (Danno et al, 2000).

16.8. TCFL5–A BASIC HELIX-LOOP-HELIX PROTEIN

A novel gene (*TCFL5* = transcription factor like 5) is expressed specifically in primary spermatocytes in the testis. The cDNA contains an open reading frame of 1356-bp, encoding a 452-amino-acid protein that includes a basic Helix-Loop-Helix (bHLH) motif. The gene, which was mapped to chromosome region 20q 13.3 qter consists of six exons and spans approximately 24-kb of genomic DNA. Immunohistochemical staining located the gene product, TCFL5 exclusively in cell nuclei of primary spermatocytes at the pachytene stage. TCFL5 protein functions as a transcription factor by regulating cell proliferation or differentiation of cells through binding to a specific DNA sequence like bHLH proteins (Maruyama et al., 1998).

16.9. OTHER TRANSCRIPTION FACTORS IN TESTIS

Heat Shock Transcription Factor: Heat shock transcription factors (HSFs) are generally known as regulators of cellular stress response. The mammalian HSF1 functions as a classical stress factor, whereas HSF2 is active during certain developmental processes, including embryogenesis and spermatogenesis. HSF2 mRNA expression in testis is specific to cell type as well as cell stage specific dependent. In adult testis, HSF2 mRNA is found at highest levels in spermatocytes and round spermatids. HSF2 protein is localized to the nuclei of spermatocytes and round spermatids. The constitutive HSF2 DNA binding activity present in testis is able to interact with promoter sequences of the hsp70.2 gene (Sarge et al., 1994). Mouse cells express not one but two distinct HSF2 isoforms and that levels of these are regulated in a tissue dependent manner. The testis expresses predominantly the 71-kDa HSF2-α isoform while the heart and brain express primarily the 69-kDa HSF2-β isoform. These isoforms are generated by alternative splicing of HSF2 pre-mRNA, which results in the inclusion of a 18 amino acid coding sequence in the HSF2-α mRNA that is skipped in the HSF2-β mRNA. HSF2 alternative splicing is also developmentally regulated, since results reveal a switch in the expression from HSF2-β mRNA isoform to the HSF2-α isoform during testis postnatal development. HSF2-α protein, the predominant isoform in testis cells, is a more potent transcriptional activator than the HSF2-β isoform (Goodson et al., 1995). Expression of the alternatively spliced HSF2 α and HSF2 β isoforms is developmentally regulated in a stage specific manner. Cellular localization and stage specific expression of the HSF2 isoforms indicated that HSF2 in addition to its functions as a nuclear transcription factor, may be involved in other cellular processes during spermatogenesis (Alastalo et al., 1998).

STAT Proteins: STAT (signal transducer and activator of transcription) proteins have been shown to be essential transcription factors which mediate biological effects of cytokines. Although most of the STATs have been shown to be widely expressed, STAT4 mRNA has been detected in only a few tissues, including the testis. The STAT4 was abundantly and exclusively expressed in male germ cells, which have completed meiosis, at the round and elongating spermatid stages. Cytolocalization showed that the level of STAT4 protein increased in parallel in both cytoplasm and nuclei. Interestingly, STAT4 was localized to the condensing perinuclear theca of spermatids (Herrada and Wolgemuth 1997),

Leukemia inhibitory factor receptor mediates STAT-3 and STAT-1: Leukemia inhibitory factor (LIF), a multifunctional member of the gp 130 cytokine family is structurally and functionally related to inteulerkin-6 (IL-6), oncostatin M, and ciliary neurotrophic factor. These related cytokines bind specific cell surface receptors that are coupled to a common gp130 signal transducing receptor component in diverse cell types and induce the homodimerization of the gp 130 protein with the IL-6 receptor chain (for IL-6) and in heterodimerization of gp 130 and the LIF receptor component (for LIF, ciliary neurotrophic factor, and oncostatin M). Jenab and Morris (1998) demonstrated that testicular LIF interacts with receptor chains in target cells and has the potential to result in sequential phosphorylation and translocation of latent to cytoplasmic STAT-3 (and STAT-1) protein to the nucleus to increase *c-fos* mRNA levels and to increase the amounts of AP -1 binding proteins. In testis, LIF exposure increases the survival of Sertoli cells and proliferating gonocytes. Together with stem cell factor, LIF treatment promotes survival and proliferation of mouse primordial germ cells while inhibiting their apoptosis. Specific DNA response elements similar to IFN-γ activation on site (TTCC NNN AA) were mapped to c-fos and Jun B promoter region (Jenab and Morris, 1998).

Transcription factor AP-1: The products of two proto-oncogenes, c-fos and c-jun, have been implicated in signal transcription pathways as regulators of gene expression. Both proto-oncogenes are members of gene families encoding closely related proteins that together make up transcription factor AP-1. AP-1 has been associated with both mitosis and differentiation. A study on expression of five AP-1 family members (c-fos, fra-1, fra-2, c-jun and junB) in European red fox, a seasonal breeder suggested a distinct although not necessarily unrelated roles for the different components of transcription factor AP-1 in the regulation of spermatogenesis (Cohen et al., 1993).

TASS-1—A Member of Ets Family: The β-galactosyltransferase-1 (β1-4GalT-1) gene is transcribed beginning in late pachytene spermatocytes from a male germ cell specific start site. A 796-bp genomic fragment that flanks the germ cell start site and contains two putative CRE like motifs, directs correct male germ cell expression of the β galactosidase reporter gene in late pachytene spermatocytes and round spermatids of transgenic mice (Shaper et al., 1994). Further more, expression of β1-4GalT-1 in male germ cells requires an essential 14-bp regulatory element (5'-GCCGGTTTCCTAGA-3') that is distinct from the two CRE like sequences. This cis element is located 16-bp upstream of the germ cell specific start site and binds a male germ protein, termed TASS-1 (transcriptional activator in late pachytene spermatocytes and round spermatids). The presence of the Ets signature binding motif 5'-GGAA-3' on the bottom strand of the TASS-1 sequence suggests that TASS-1 is a novel member of the Ets family of transcription factors. Additional transgenic analyses established that a 87-bp genomic fragment containing the TASS-1 regulatory element is sufficient for correct germ cell specific expression of the β galactosidase reporter gene (Charron et al., 1999). Ets family members have been reported to be expressed in murine testis. This group includes Elk-1, ER71, ER81, ERP, GABPα and PEA3 (c/r Charron et al., 1999). The ER71 mRNA was the only one found in testis and in embryo.

Mitochondrial transcription factor A: Mitochondrial transcription factor A (mt-TFA) is a key regulator of mammalian mitochondrial DNA transcription. A testis specific isoform of mouse mt-TFA lacks the mitochondrial targeting sequence and is present in the nucleus of spermatocytes and elongating spermatids, thus representing the first mammalian gene encoding protein isoforms targeted for the mitochondria or the nucleus. The presence of the mitochondrial activator in the nucleus raises the possibility of a role for this protein in both genetic systems. It seems that a single gene encodes the mt-TFA and a testis specific nuclear HMG-box protein (Larsson et al., 1996).

16.10. REFERENCES

Alastalo TP, Lonnstrom M, Leppa S, Kaarniranta K, Pelto-Huikko M, Sistonen L, Parvinen M. Stage-specific expression and cellular localization of the heat shock factor 2 isoforms in the rat seminiferous epithelium. Exp Cell Res 1998: 10: 16-27.

Andersen B, Pearse RV 2d, Schlegel PN, Cichon Z, Schonemann MD, Bardin CW, Rosenfeld MG. Sperm 1: a POU-domain gene transiently expressed immediately before meiosis I in the male germ cell. Proc Natl Acad Sci U S A 1993; 90: 11084-8.

Behr R, Hunt N, Lvell R, Wessels J, Weinbauer GF. Cloning and expression analysis of testis-specific cyclic 3' –5'- adenosine monophosphate-responsive element modulator activators in the nonhuman primate (Macaca fascicularis): comparison with other primate and rodent species. Biol Reprod 2000; 62:1344-51.

Behr R, Weinbauer GF. Germ cell-specific cyclic adenosine 3'–5' monophosphate response element modulator expression in roent and primate testis is maintained despite gonadotropin deficiency. Endocrinology 1999; 140: 1502-14.

Blendy JA, Kaestner KH, Winbauer GF, Nieschlag E, Schutz G, Severe. Impairment of spermatogenesis in mice lacking the CREM gene. Nature 1996; 380: 162-65.

Bonny C, Cooker LA, Goldberg E. Deoxyribonucleic acid - protein interactions and expression of the human testis-specific lactate dehydrogenase promoter: transcription factor Spl plays a major role. Biol Reprod 1998; 58: 754-59.

Burke PS, Wolgemuth DJ. Zfp-37, a new murine zinc finger encoding gene, is expressed in a developmentally regulated pattern in the male germ line. Nucleic acids Res 1992; 20: 2827-34.

Cao T, Shannon M, Handel MA, Etkin LD. Mouse ret finger protein (rfp) proto-oncogene is expressed at specific stages of mouse spermatogenesis. Dev Genet 1996; 19: 309-20.

Charron M, Shaper NL, Rajput B, Shaper JH. A novel 14-base-pair regulatory element is essential for in vivo expression of murine β4-galactosyltransferase-I in late pachytene spermatocytes and round spermatids. Mol Cell Biol 1999; 19: 5823-32.

Cohen DR, Vandermark SE, McGovern JD, Bradley MP. Transcriptional regulation in the testis: a role for transcription factor AP-1 complexes at various stages of spermatogenesis. Oncogene 1993; 8: 443-55.

Cunliffe V, Koopman P, McLaren A, Trowsdale J. A mouse zinc finger gene which is transiently expressed during spermtogenesis. The EMBO Journal 1990; 9: 197-5.

Daniel PB, Habener JF. Cyclical alternative exon splicing of transcription factor cyclic adenosine monophosphate response element-binding protein (CREB) messenger ribonucleic acid during rat spermatogenesis. Endocrinology 1998; 139: 3721-29.

Danno S, Itoh K, Matsuda T, and Fujita J. Decreased Expression of mouse Rbm3, a cold shock protein, in Sertoli cells of cryptorchid testis. Am J Pathol 2000; 156: 1685-92.

De Cesare D, Fimia GM, Brancorsini S, Parvinen M, Sassone-Corsi P. Transcriptional control in male germ cells: general factor TFIIA participates in CREM-dependent gene activation. Mol Endocrinol 2003;17: 2554-65.

DeCesare D, Sassone-Corsi P. Transcriptional regulation by cyclic AMP-responsive factors. Prog Nucleic Acid Res Mol Biol. 2000; 64: 343-79.

Delfino FJ, Walker WH. NF- κB induces cAMP-response element-binding protein gene transcription in Sertoli cells. J Biol Chem 1999; 274: 35607-13.

Delmas V, van Der HF, Mellstrom B, Jegou B, Sassone-Corsi P. Induction of CREM activator proteins in spermatids : down-stream targets and implications for haploid germ cell differentiation. Mol Endocrinol 1993; 7: 1502-14.

Denny P, Ashworth A. A zinc finger protein-encoding gene expressed in the post-meiotic phase of spermatogenesis. Gene 1991; 106: 221-7.

Feige E, Chen A, Motro B. Nurit, a novel leucine-zipper protein, expressed uniquely in the spermatid flower-like structure. Mech Dev 2002; 117: 369-77.

Fimia GM, DeCesareD, Sassone Corsi P. CBP independent activation of CREM and CREB by LIM only protein ACT. Nature 1999; 398: 165-69.

Fischer C, Johnson J, Stillwell B, Conner J, Cerovac Z et al. Activating transcription factor 4 is required for the differentiation of the lamina propria layer of the vas deferens. Biol Reprod 2004; 70: 371-8.

Foulkes NS, Mellstrom B, Benusig E, Sassone-Corsi P. Developmental switch of CREM function during spermatogenesis from antagonist to activator. Nature 1992; 355: 80-84.

Foulkes NS, Schlotter F, Pevet P, Sassone-Corsi P. Nature 1993; 362: 264-67.

Freemont PS. The RING finger: A novel protein sequence motif related to the zinc finger. Ann N Y Acad Sci 1993;684:174-92.

Fujii T, Tamura K, Copeland NG, Gilbert DJ, Jenkins NA, Yomogida K, Tanaka H, Nishimune Y, Nojima H, Abiko Y. Sperizin is a murine ring zinc-finger protein specifically expressed in haploid germ cells. Genomics 1999; 57: 94-101.

Gellersen B, Kempf R, Sandhowe R, Weinbauer GF, Behr R. Novel leader exons of the cyclic adenosine 3',5'-monophosphate response element modulator (CREM) gene, transcribed from promoters P3 and

P4, are highly testis-specific in primates. Mol Hum Reprod 2002; 8: 965-76.
Girardet C, Walker WH, Habener JF. An alternatively spliced polycistronic mRNA encoding cyclic adenosine 3'–5' monophosphate (cAMP)-responsive transcription element-binding protein) in human testis extinguishes expression of an internally translated inhibitor CREB isoform. Mol Endocrinol 1996; 10: 879-91.
Goldsborough A, Ashworth A, Willison K. Cloning and sequencing of POU-boxes expressed mouse testis. Nucleic Acids Res 1990; 18: 1634.
Goodson ML, Park-Sarge O, Sarge K. Tissue dependent expression of Heat shock factor 2 isoforms with distinct transcriptional activities. Mol Cell Biol 1995; 15: 5288-93.
Herrada G, Wolgemuth DJ. The mouse transcription factor Stat4 is expressed in haploid male germ cells and is present in the perinuclear theca of spermatozoa. J Cell Sci 1997; 110(pt-14): 1543-53.
Hoog C, Schalling M, Grunder-Brundell E, Daneholt B. Analysis of a murine male germ cell-specific transcript that encodes a putative zinc finger protein. Mol Reprod Dev 1991; 30; 173-81.
Howard T, Balogh R, Overbeek P, Bernstein KE, Sperm-specific expression of angiotensin-converting enzyme (ACE) is mediated by a 91-base-pair promoter containing a CRE-like element. Mol Cell Biol 1993; 13: 18-27.
Jenab S, Morris PL. Testicular leukemia inhibitory factor (LIF) and LIF receptor mediate phosphorylation of signal transducers and activators of transcription (STAT-3) and STAT-1 and induce c-fos transcription and activator protein-1 activation in rat Sertoli but not germ cells. Endocrinology 1998; 139: 1883-90.
Jong MT, Gray TA, Ji Y, Glenn CC, Saitoh S, Driscoll DJ, Nicholls RD. A novel imprinted gene, encoding as RING zinc-finger protein, and overlapping antisense transcript in the Prader-Willi syndrome critical region. Hum Mol Genet 1999; 8: 783-93.
Kalikin LM, Fujimoto H, Witt MP, Verga V, Erickson RP. A genomic clone of Zfy-1 from a YDOM mouse strain detects post-meiotic gene expression of Zfy in testes. Biochem Biphoys Res Commun 1989; 165: 1286-91.
Kessler SP. Rowe TM. Blendy JA, Erickson RP, Sen GC. A cylic AMP response element in the angiotensin-converting enzyme gene and the transcription factor CREM are required for transcription of the mRNA for the testicular isozymes. J Biol Chem 1998; 273: 9971-75.
Ketola I, Rahman N, Toppari J, Bielinska M, Porter-Tinge SB, Tapanainen JS, Huhtaniemi IT, Wilson DB, Heikinheimo M. Expression and regulation of transcription factors GATA-4 and GATA-6 in developing mouse testis. Endocrinology 1999; 140: 1470-80.
Kim YS, Nakanishi G, Lewandoski M, Jetten AM. GLIS3, a novel member of the GLIS subfamily of Kruppel-like zinc finger proteins with repressor and activation functions. Nucleic Acids Res 2003; 31: 5513-25.
Kitamura K, Tanaka H, Nishimune Y. Haprin, a novel haploid germ cell-specific RING finger protein involved in the acrosome reaction. J Biol Chem. 2003; 278: 44417-23.
Komuro E, Schalling M. Jahn D, Bodmer R, Jenkins NA, Copeland NG, Izumo S. Gtx : a novel murine homeobox containing gene, expressed specifically in glial cells of the brain and germ cells of testis, has a transcriptional repressor activity in vitro for serum-inducible promoter. EMBO Journal 1993; 12: 1387-01
Kraft TL, Patterson J, Won Yoon J, Doglio L, Walterhouse DO, Iannaccone PM, Goldberg E. GLI1 localization in the germinal epithelial cells alternates between cytoplasm and nucleus:upregulation in transgenic mice blocks spermatogenesis in pachytene. Biol Reprod 2001; 65: 1663-71.
Larsson NG, Garman JD, Oldfors A, Barsh GS, Clayton DA. A single mouse gene encodes the mitochondrial transcription factor A and a testis-specific nuclear HMG-box protein. Nat Genet 1996; 13: 296-02.
Lindsey JS, Wilkinson MF. Pem: a testosterone- and LH-regulated homeobox gene expressed in mouse Sertoli cell and epididymis. Dev Biol 1996b; 179: 471-84.
Lindsey S, Wilkinson F. An androgen-regulated homeobox gene expressed in rat testis and epididymis. Biol Reprod 1996a; 55: 975-83.
Looman C, Hellman L, Abrink M. A novel Kruppel-Associated Box identified in a panel of mammalian zinc finger proteins. Mamm Genome 2004 ;15: 35-40.
Looman C, Mark C, Abrink M, Hellman L. MZF6D, a novel KRAB zinc-finger gene expressed exclusively in meiotic male germ cells. DNA Cell Biol. 2003; 22: 489-96.
Mahoney MG, Tang W, Xiang MM, Moss SB, Gerton GL, Stanley JR, Tseng H. Translocation of the Zinc finger protein basonculin from the mouse germ cell nucleus to the midpiece of the spermatozoon during spermiogenesis. Biol Reprod 1998; 59: 388-94.
Martin-Ponthieu A, Wouters-Tyrou D, Pudlo B, Buisine E, Sautiere P. Isolation and characterization of a small putative zinc finger protein form cuttlefish epididymal sperm cells. Eur J Biochem 1994; 220: 463-8.
Maruyama O, Nishimori H, Katagiri T, Miki Y, Ueno A, Nakamura Y. Cloning of TCFL5 encoding a novel human basic helix-loop-helix motif protein that is specifically expressed in primary spermatocytes at

the pachytene stage. Cytogenet Cell genet 1998; 82: 41-5.

Meagher MJ, Schumacher JM Lee K, Holdcraft RW, Edelhoff S, Disteche C, Braun RE. Identification of ZFR, an ancient and highly conserved murine chromosome-associated zinc finger protein. Gene 1999; 228: 197-211.

Meyer TM, Waeber G, Lin J, Beckman W, Habener JF. The promoter of the gene encoding 3' –5' cyclic adenosine monophosphate (cAMP) response element binding protein contains cAMP response elements: evidence for positive auto regulation of gene transcription. Endocrinology 1993; 61: 1403-11.

Mori D, Okuro N, Fujii-Kuriyama Y, Sogawa K. Gene structure and promoter analysis of the rat BTEB2 gene. Gene 2003; 304: 163-70.

Nagamine CM, Chan K, Hake LE, Lau YF. The two candidate testis-determining Y genes (Zfy-1 and Zfy-2) are differentially expressed in fetal and adult mouse tissues. Genes Dev 1990; 4: 63-74.

Nantel F, Monaco Sassone L, Foulkes NS, Masquilier D, LeMeur M, Henriksen K, Dierich A, Parvinen M, Corsi P. Spermiogenesis deficiency and germ-cell apoptosis in CREM-mutant mice. Nature 1996; 380: 159-62.

Nelki D, Dudley K, Cunningham P, Akhavan M. Cloning and sequencing of a zinc finger cDNA expressed in mouse testis. Nucleic Acids Res 1990; 18: 3655.

Nishiyama H, Itoh K, Kaneko Y, Kishishita M et al A glycine-rich RNA-binding protein mediating cold inducible suppression of mammalian cell growth. J Cell Biol 1997; 137: 899-908.

Noce T, Fujiwara Y, Sezaki M, Fujimoto H, Higashinakagawa T. Expression of a mouse zinc finger protein gene in both spermatocytes and occytes during meiosis. Dev Biol 1992; 153: 356-67.

Peri A, Krausz C, Cioppi S, Forti G, Francavilla S, Serio M. Cyclic adenosine 3' 5' – monophosphate-responsive element modulator gene expression in germ cells of normo-and oligoazoospermic men. J Clin Endocrinol Metab 1998; 83: 3722-6.

Persengiev SP, kondova II, Millette CF, Kilpatrick DL. Gli family members are differentially expressed during the mitotic phase of spermatogenesis. Oncogene 1997; 14: 2259-64.

Qiu W, Zhang S, Xiao C, Xu W, Ma Y, Liu Y, Wu Q. Molecular cloning and characterization of a mouse spermatogenesis-related ring finger gene znf230. Biochem Biophys Res Commun 2003; 306: 347-53.

Rao MK, Wayne CM, Meistrich ML, Wilkinson MF. Pem homeobox gene promoter sequences that direct transcription in a Sertoli Cell-specific, stage specific, and androgen dependent manner in the testis in vivo Mol Endocrinol 2003;17:273-33.

Rozman D, Fink M, Fimia GM, Sassone-Corsi P. Waterman MR. Cyclic adenosine 3'-5' monophosphate (cAMP)/cAMP-responsive element modulator (CREM)-dependent regulation of cholesterogenic lanosterol 14α-demethylase (CYP51) in spermatids. Mol Endocrinol 1999; 13: 1951-62.

Rubin MR, Toth LE, Patel MD, D'Eustachio P Nguyen-Huu C. A mouse homeobox gene is expressed in spermatocytes and embryos. Science 1986; 237: 663-67.

Ruppert S, Cole TJ, Boshart M, Schmid E, Schutz G. Multiple mRNA isoforms of the transcription activator protein CREB: generation by alternative splicing and specific expression in primary spermatocytes. EMBO J 1992; 11: 1503-12.

Sadhukhan R, Sen GC, Ramchandran R, Sen I. The distal ectodomain of angiotensin-converting enzyme regulates its cleavage – secretion from the cell surface. Proc Natl Acad Sci (USA) 1998; 95: 138-43.

Sarge KD, Park-Sarge OK, Kirby JD, Mayo KE, Morimoto RI. Expression of heat shock factor 2 in mouse testis: potential role as a regulator of heat-shock protein gene expression during spermatogenesis. Biol Reprod 1994: 50: 1334-43.

Sassone Corsi PS. CREM: a master switch regulating the balance between differentiation and apoptosis in male germ cells. Mol Reprod Develop 2000; 56: 228-29.

Shankar S, Mohapatra B, Suri A. Cloning of a novel human testis mRNA specifically expressed in testicular haploid germ cells, having unique palindromic sequences and encoding a leucine zipper dimerization motif. Biochem Biophy Res Comm 1998; 243: 561-5.

Shaper NL, Hardulin-Lepers A, Shaper J H. Male germ cell expression of murine β1, 4-galactosyltransferase. A 796-base pair genomic region, containing two cAMP-responsive element (CRE)-like elements mediates male germ cell specific expression in transgenic mice. J Biol Chem 1994; 269:25165-171.

Sun Z, Sassone-Corsi P, Means AR. Calspermin gene transcription is regulated by two cyclic AMP response elements contained in an alternative promoter in the calmodulin kinase IV gene. Mol Cell Biol 1995; 15: 561-571.

Tang C-J C, Chuang CK, Hu HM, Tang TK. The zinc finger domain of T2fp binds to the *tbs* motif located at the upstream flanking region of Aie1 (aurora-c) kinase gene. J Biol Chem 2001;276:19631-39.

Tokuzawa Y, Kaiho E, Maruyama M, Takahashi K, Mitsui K, Maeda M, Niwa H,Tourtellotte WG, Nagarajan R, Auyeung A, Mueller C, Milbrandt J. Infertility associated with incomplete spermatogenic arrest and oligozoospermia in Egr4-deficient mice. Development 1999; 126: 5061-71.

Tremblay JJ, Viger RS. Transcription factor GATA-4 enhances Mullerian inhibiting substance gene transcription through a direct interaction with the nuclear receptor SF-1. Mol Endocrinol 1999; 13: 1388-101.
Van Baren MJ, van der Linde HC, Breedveld GJ, Baarends WM, Rizzu P, de Graaff E, Oostra BA, Heutink P. A double ring-H2 domain in RNF32, a gene expressed during sperm formation. Biochem Biophys Res Commun 2002; 292: 58-65.
Van der Hoorn FA, Tarnasky HA, Nordeen SK. A new rat gene RT7 is specifically expressed during spermatogenesis. Dev Biol 1990; 142: 147-54.
Van der Hoorn FA, Tarnasky HA. Factors involved in regulation of the RT7 promoter in a male germ cell-derived in vitro transcription system. Proc Natl Acad Sci USA 1992; 89: 703.7.
Vanhoutteghem A, Djian P. Basonuclin 2: an extremely conserved homolog of the zinc finger protein basonuclin. Proc Natl Acad Sci U S A 2004; 101: 3468-73.
Waeber G, Meyer TE, LeSieur M, Hermann HL, Gerard N, Habener JF. Developmental stage-specific expression of cyclic adenosine 3'–5' monophosphate response element-binding protein CREB during spermatogenesis involves alternative exon splicing. Mol Endocrinol 1991; 5: 1418-30.
Walker WH, Fucci L, Habener JF. Expression of the gene encoding transcription factor cyclic adenosine 3'-5' Monophosphate (cAMP) response element-binding protein (CREB): regulation by follicle stimulating hormone-induced cAMP signaling in primary rat Sertoli cells. Endocrinology 1995; 136: 3534-45.
Walterhouse D, Ahmed M, Slusarski D, Kalamaras J, Boucher D, Holmgren R. Iannaccone P. gli, a zinc finger transcription factor and oncogene, is expressed during normal mouse development. Dev Dyn 1993; 196: 91-102.
Weinhauer GF, Behr R, Bergmann M, Nieschlag E. Testicular cAMP responsive element modulator (CREM) protein is expressed in round spermatids but is absent or reduced in men with round spermatid maturation arrest. Mol Hum Reprod 1998; 4: 9-15.
West AP, Sharpe RM, Saunders PTK. Differential regulator of cyclic adenosine 3' –5' monophosphate (cAMP) response element-binding protein and cAMP response element modulator messenger ribonucleic acid transcripts by follicie-stimulating hormone and androgen in the adult rat testis. Biol Reprod. 1994; 50: 869-81.
Weissig H, Narisawa S, Sikstrom C, Olsson PG, McCarrey JR, Tsonis PA, Del Rio-Tsonis K, Millan JL. Three novel spermatogenesis-specific zinc finger genes. FEBS Lett 2003; 547: 618.
Yan W, Hirvonen-Santti SJ, Palvimo JJ, Toppari J, Janne OA. Expression of the nuclear RING finger protein SNURF/RNF4 during rat testis development suggests a role in spermatid maturation. Mech Dev 2002; 118: 247-53.
Yang CH, Gallicano GI, Yu QC, Fuchs E. An unexpected localization of basonuclin in the centrosome; mlitochondria, and acrosome of developing spermatids. J Cell Biol 1997; 137: 657-70.
Yomogida K, Ohtani H, Harigae H, Ito E, Nishimune Y, Engel JD, Yamamoto M. Developmental stage – and spermatogenic cycle-specific expression of transcription factor GATA-1 in mouse Sertoli cells. Development 1994; 120: 1759-66.
Zhou Y, Sun Z, Means AR, Sassone-Corsi P, Bernstein KE. cAMP-response element modulator τ is a positive regulator of testis angiotensin converting enzyme transcription. Proc Natl Acad Sci USA 1996; 93: 12262-66.
Zini A, Mielnik A, Schlegel PN. POU-domain gene expression during spermatogenesis. World J Urol 1996; 14: 274-7.
Zweidler-Mckay PA, Grimes HL, Flubacher MM, Tsichlis PN. Gfi-1 encodes a nuclear zinc finger protein that binds DNA and functions as a transcriptional repressor. Mol Cell Biol 1996; 16: 4024-34.

Chapter 17

PROTO-ONCOPROTEINS

17.1. INTRODUCTION

A number of genes, which have potential to cause tumor are present in viruses and in the host cells. They are called as oncogenes and designated as *v-onc* (present in viruses) and *c-onc* (present in host cells). Oncogenes, present in host cells are normally not oncogenic and hence called as proto-oncogenes. The ability of activated oncogenes to induce abnormal cell growth suggests that their normal progenitors, the proto-oncogenes, may function to regulate the growth and differentiation of normal cells. The identification of some proto-oncogenes as growth factors, hormone receptors, and transcription factors supports the function of these genes as cellular regulatory elements, but the normal physiologic functions of most proto-oncogenes remain unknown. Products of proto-oncogene fall into five functional categories that appear to constitute elements in signal transduction pathways: growth factors, tyrosine protein kinases, guanosine-triphosphate (GTP)-binding proteins, serine and threonine protein kinases, and nuclear proteins. Testis expressed proto-oncogenes can be divided into three groups: i) Proto-oncogenes that are ubiquitously expressed in many tissues and cell types and have identical transcripts in somatic and in germ cells. These include, among others, the c-*myc* and c-*fos* (Mautner, 1995). ii) The second group comprises of those proto-oncogenes that are predominantly expressed during embryonic development and during gametogenesis. These might have a single transcript as in the case of *int-1* and several transcripts as in the case of c-*kit*. iii) The third group consists of proto-oncogenes that are expressed in various adult tissues but have testis specific transcript. One example would be the c-*abl* tyrosine kinase that has testis specific transcripts in both mouse and human. Another example would be the *Pim-1* serine/threonine kinase. The *Pim-1,* among normal tissues, is predominantly expressed in hematopoietic tissues and in the testis (Domen et al, 1993, Buckley et al, 1995). The development of male germ cells in the testis provides an ideal system that is particularly well suited for analysis of gene expression within a cell lineage that represents a spectrum of developmental stages ranging from mitotically proliferating stem cells to mature spermatozoa.

17.2. C-KIT AND STEM CELL FACTOR (SCF)

The proto-oncogene, c-kit is a homologue of viral oncogene v-kit, which was initially identified as transforming gene of feline sarcoma retrovirus, HZ4-Fesv. The p145 c-kit oncoprotein is a cell surface receptor of transmembrane tyrosine kinase family. In humans, c-kit expression has been detected in several normal fetal and adult tissues such as brain, skin, adrenal gland, breast and bone marrow and also in several human malignancies. In the embryonal gonad the c-kit tyrosine kinase receptor and its ligand SCF are required for the survival and proliferation of primordial germ cells. In the adult testis, the c-kit receptor is re-expressed in differentiating spermatogonia, but not in spermatogonial stem cells, whereas SCF is expressed by Sertoli cells and is under FSH action. The SCF stimulates DNA synthesis in type A spermatogonia cultured in vitro, and injection of anti-c-kit antibodies block their proliferation in vivo. The receptor tyrosine kinase kit and its only known ligand, Kit Ligand (KL) or stem cell factor (SCF) are encoded at the white spotting (W) and steel (SL) loci in the mouse, respectively. It has been identified on mouse chromosome 5 and human chromosome 4. The signal transduction system of the c-kit receptor and its ligand (KL/SCF) is an interesting candidate for paracrine stimulation of germ cell proliferation and differentiation like macrophage colony stimulating factor.

Mutation Analysis: The kit receptor plays roles in Kit-mediated signaling events in primordial germ cells (PGCs), spermatogenesis and oogenesis, all of which follow distinct, well-studied development programs. In testis, kit expression starts at postnatal day 5 (P5) and is restricted to differentiating type A spermatogonia, type B spermatogonia, primary spermatocytes and Leydig cells. Many alleles of variable severity at both the W and SL loci have been described and characterized. Mutations either at W locus abolish or partially impair Kit receptor function whereas Kit receptor/W point mutations affect gametogenesis, melanogenesis and hematopoiesis during embryogenesis and in postnatal development to similar degrees. In contrast, mutations that affect Kit or SCF/KL expression may affect cellular targets of W and SL mutations differentially. The mutations affect steps in spermatogenesis and oogenesis including the survival and proliferation of spermatogonia and oocyte growth, causing impaired fertility. In weak W and SL alleles, effects on spermatogenesis and oogenesis are observed as well. The SCF expression mutation SL^{panda}, St and SL^{con} impair female fertility, while males are fertile. In contrast, a cytoplasmic domain mutation of the membrane growth factor SCF, SL^{17H}, does not impair female fertility, but causes male sterility. It seems that the function of c-kit is more necessary for development of primordial germ cells (PGC) than for postnatal germ cell.

Localization in Spermatogenic Cells: Among male germ cells, the expression pattern of c-kit correlates with gonial proliferation. C-kit bands of 80, 140, and 150-kDa are observed on Western blot, indicating that c-kit is a name related to several proteins sharing some common domains. Only the band of 150kDa correleated with positive staining of c-kit in tubules in differentiating hamster spermatogonia at stages VII-VIII of adult spermatogenesis and at day 13-14 during the first spermatogenic wave. It remains attached to the cell until late pachytene. This suggests that c-kit may play a role in preparing the germinal cells to enter meiosis (Vigodner et al., 2001). Nishina et al (1992) investigated its expression during differentiation of F9 cells. In differentiated cells treated with retinoic acid (RA), c-kit gene expression is induced by agents such as forskolin or theophyline, which are known to elevate cellular cAMP level. This indicated that the expression of c-kit gene is regulated by the level of intracellular cAMP in differentiated F9 cells induced by RA. In mi/mi mast cell deficient mice, the c-kit expression was deficient in mast cells but not in erythroid precursors, testicular germ cells, and neurons. Thus, the regulation of the c-kit transcription by the mi factor (a transcription factor of bHLH/ leucine zipper family) was dependent on cell types (Isozaki et al, 1994).

The c-kit mRNA was found to express at high levels in spermatogonia, and at lower levels in meiotic pachytene spermatocytes. Moreover, two novel testis-specific c-kit transcripts of 3.5 and 2.3-kb are present in post-meiotic haploid germ cells. These results suggested a role of c-kit not only during testis development in the embryo, but also throughout all stages of male germ cell development after birth. Characteristics of the receptor tyrosine kinases include an extracellular domain with 5 immunoglobulin-like domains and a kinase, which is divided into 2 subdomains by an insertion sequence of variable length. (Robinson et al, 2001; Sorrentino et al, 1991). Thus c-kit may function in gametogenesis both during early embryogenesis and in postnatal development. Although the experimental results supported the notion that the survival and /or proliferation of the type A1-A4 spermatogonia is c-kit dependent, it has been suggested that the As (Ao) are c-kit independent (Sandlow et al, 1999). Rossi et al, (1992) cloned a c-kit mRNA of 3.2-kb expressed in post-meiotic male germ cells. This transcript initiates in the genomic region immediately upstream of the exon coding for the second box of the split c-kit tyrosine kinase domain. The open reading frame (ORF) contains 12 novel amino acids in frame with the C-terminals of 190 amino acids of the c-kit protein. It lacks the upstream region in the 5.5- kb c-kit mRNA encoding the extracellular and transmembrane domain, the ATP-binding site and the kinase insert domain present in the c-kit.

The synthesis of the c-kit receptor by the spermatogonia was established by hybridization of total RNA with a specific cDNA for mouse c-kit receptor. The c-kit receptor protein expressed at the later stages of spermatogenesis as well, specifically in the acrosomal granules of the round spermatids, and the acrosomal region of testicular spermatozoa of mouse and rat (Sandlow et al, 1997; 1999). It also showed that the c-kit receptor is localized to the region of developing acrosome (Sandlow et al, 1999). During postnatal development of the rat testis, five different c-kit receptor mRNA transcripts were identified in testis and the level of expression of a particular transcript varied depending upon the developmental stage of the testis (Ali et al., 1996). Two mRNA transcripts migrated as 4.8-kb and 12-kb. The staining for c-kit receptor was observed in the cytoplasm of the isolated type A spermatogonia. The c-kit band of 160-kDa appeared to be stimulated with kit ligand, and the amount of phosphorylated protein increased significantly (Dym et al, 1995). Thus multiple forms of mRNAs exist for the c-kit receptor in the rat testis, and they are regulated differentially during post-natal development.

C-kit and Germ Cell Tumors: Besides in normal spermatogenesis, c-kit activity is also a pivotal determinant of testicular tumor development and the kinase defective mutants of c-kit are valuable for treating c-kit dependent cancer, as well as for clarifying the c-kit mediated carcinogenesis. The c-kit gene is expressed in seminomas and dysgerminomas, a subset of human germ cell tumors (GCTs). A novel missense mutation (D816H) was found in the phosphotransferase domain in tumors of seminoma/dysgerminoma differentiation. The c-kit alleles in nonneoplastic tissues from these patients were wild type, suggesting that the mutant alleles were acquired and selected for malignant transformation (Tian et al, 1999). Additionally, c-kit expression was detected in the seminoma but not in the nonseminoma subtype of human GCT. Stem cell factor was not expressed at the mRNA level in tissue from either subtype of GCT although the protein was detected in the cytoplasm of rare tumor cells (Strohmeyer et al, 1995).

17.2.1. Truncated Form of c-Kit

With the onset of meiosis, c-kit expression ceases, but a truncated c-kit product, (tr-kit) is specifically expressed in post-meiotic stages of spermatogenesis, and is accumulated in mature spermatozoa. Shorter version of c-kit protein (tr-kit) is deficient in extracellular, the transmembrane and part of the intracellular tyrosine-kinase domains. The c-kit protein synthesized by mutant Wn/Wn cultured mast cells (CMC) was reduced in size and not expressed on their surface. The c-kit product synthesized in Wn/Wn CMC mutant was truncated almost all cytoplasmic domain and was less glycosylated. In c-kit Wn-transfected cells, both glycosylation and extracellular expression of c-kit protein was also impaired, although no truncation was detected. Thus Wn mutant form of c-kit product is insufficient in maturation, which is associated with impairments in the transport to the plasma membrane (Koshimizu et al, 1994).

The 5'end of the alternative c-kit transcripts, maps within an intron of the mouse c-kit gene. This intron contains a promoter, active in nuclear extracts of round spermatids, and two discrete sequences upstream of the transcriptional start site bind spermatid-specific nuclear factors. Transgene expression is confined to haploid germ cells of seminiferous tubules, starting from spermatids at step 9, and disappearing at step 13, indicating that cryptic promoter within the 16th intron of the mouse c-kit gene is active in a short temporal window at the end of the transcriptional phase of spermatogenesis. In agreement with these data, a polypeptide of the size predicted by the open reading frame of the spermatid-specific c-kit cDNA accumulates in the later stages of spermatogenesis and in epididymal spermatozoa (Albanesi et al, 1996).

Functions of c-kit in Fertilization: Microinjection of a truncated form of the c-kit tyrosine kinase present in mouse spermatozoa (tr-kit) activates mouse eggs parthenogenetically, and tr-kit induced egg activation is inhibited competitively by preincubation with an inhibitor of phospholipase C (PLCγ1) containing the src homology (SH) domains. Tr-kit-induced egg activation is also suppressed by co-injection of antibodies raised against the PLCγ1 SH domains, but not against the PLC γ1COOH-terminal region. These studies indicated that tr-kit activates PLCγ1, and that SH3 domain of PLCγ1 is essential for tr-kit-induced egg activation (Sette et al, 1998; 2000; Rossi et al, 2000). The c-kit mAb ACK2 reacted specifically with the acrosomal region and the principal piece of fixed non-capacitated mouse sperm but did not react with the acrosome region in acrosome-reacted sperm. The mAb recognized c-kit proteins of 33, 48, and 150 kDa in mouse sperm extracts, which released 48-and 150-kDa proteins into the media and tyrosine autophosphorylated during acrosome reaction. The level of c-kit phosphorylation increased significantly on stimulation with SCF. These observations suggested that c-kit is present in mature sperm, and its binding to SCF might result in the activation of PLCγ1 and PI3-kinase, leading to receptor autophosphorylation, and ultimately play a role in capacitation and/or the acrosome reaction (Feng et al, 1998).

17.2.2. Activation of Phosphatidylinositol 3'-Kinase Pathway

How the c-kit receptor mediates different cellular responses in distinct cell populations during embryonic development and in the adult animal, is not known? Studies on bone marrow derived mast cells (BMMCs) have provided insight into the mechanism by which c-kit mediates various cellular responses including cell proliferation, adhesion, actin reorganization, membrane ruffling and secretion. The kit receptor activation leads to autophosphoryation, the phosphorylation of various substrates and the association with signaling molecules, thereby activating distinct signaling cascades. Molecules known to interact with the c-kit receptor in vivo include the p85 subunit (p85α and p85β) of class IA phosphatidylinositol 3'kinases (PI3-kinases), phospholipase Cγ-1, the Grb2 adaptor protein, the Src kinase and the tyrosine phosphatases SHP1 and SHP2, but not the adaptor protein Shc and the exchange factor Nck. But, c-kit receptor activation causes phosphorylation and activation of the Shc adaptor protein, Ras and the Vav GDP/GTP exchange. In BMMCs, mutation of the kit receptor binding site for class IA PI3-kinase adaptor proteins, kitY719, and for src, KitY567, was shown to affect cell proliferation, survival, adhesion and secretion to differing degrees (Serve et al., 1995; Vosseller et al,. 1997; Timokhna et al., 1998). Whereas kit-mediated PI3-kinase activation

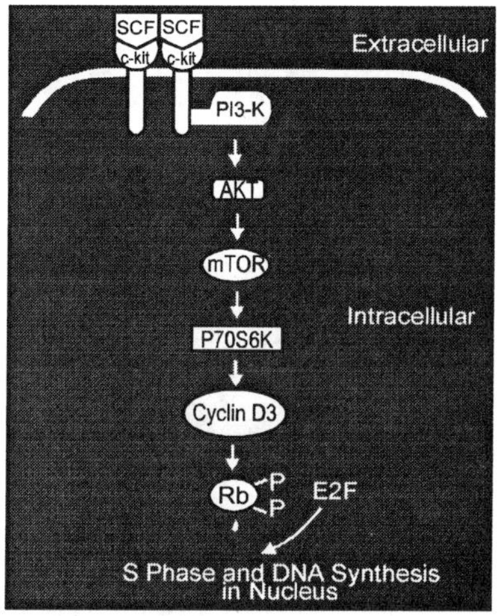

Fig. 17.1. c-kit/PI3-k signaling in spermatogonia. The model shows that c-kit/PI3-K promotes cell cycle progression via the AKT/p70S6K/ cyclin D3 pathway. Reprinted with permission from L.X. Feng et al. J Biol Chem 275; 25572-6: 2000 © The American Society for Biochemistry and Molecular Biology.

is important to the mitogenic and survival response in BMMCs, kit-mediated PI-kinase activation is critical for various responses including secretory response, cell adhesion, actin polymerization and membrane ruffling. Activation of PI3-kinase results in the rapid accumulation of phosphatidylinositol-3, 4-bisphosphate (PI3,4-P_2 or PIP2) and phosphatidyl-inositol-3, 4, 5- trisphosphate (PI3,4,5-P_3 or PIP3). The PI3,4-P_2 and PI3,4,5-P_3 are important second messengers regulating the catalytic activity of downstream signaling molecules via binding to pleckstrin homology (PH) and FYVE domains, thereby mediating activation of a wide array of downstream targets including the protein kinases PDK1, AKT, PKCδ and the small GTPase Rac 1 (reviewed in Kissel et al, 2000) (Chapter 22). Studies with type A spermatogonia indicated that SCF induces DNA synthesis and proliferation. The signaling pathway for function of SCF/c-kit involves PI3-kinase and p70S6-kinase activities. Feng et al, (2000) demonstrated that SCF activates PI3-kinase and p70S6-kinase. The SCF could induce cyclin D3 expression and phosphorylation of the retinoblastoma protein through a pathway that is sensitive to both wortmannin and rapamycin. The AKT, but not protein kinase C-zeta, is used by SCF/c-kit/PI3-kinase to activate p70S6 kinase. Thus, SCF/c-kit uses a rapamycin-sensitive PI3-kinase/p AKT/70S6-kinase/cyclin D3 pathway to promote spermatogonial cell proliferation **(Fig 17.1)**. In mast cells, the activation of PI3-kinase and phosphhatidylinositol 3'-phosphates play a critical role in mediating cell adhesion and secretion and have contributory roles in mediating cell survival and proliferation.

Although kit/SCF-R induced activation of PI3-kinase is essential for male fertility, the role of individual kit/SCF-R-induced signaling pathways in the control of development processes in the intact animal is not clear. To examine the function of SCF–induced PI3-kinase activation in vivo, Blume-Jensen et al., (2000) mutated the codon for Tyr 719, the PI3-kinase binding site in Kit/SCF-R, to Phe in the genome of mice by homologous recombination. Homozygous mutant mice were viable, but the mutation completely disrupted PI3-kinase binding to Kit/ SCF-R and reduced SCF-induced PI3-kinase-dependent activation of AKT by 90%. The mutation resulted into a gender and tissue-specific defect. Although there were no haematopoietic or pigmentation defects **(Fig.17.2)** in homozygous mutant mice, males were sterile due to a block in spermatogenesis, with initially decreased proliferation and subsequent extensive apoptosis occurring at the spermatogonial stem-cell level. In contrast, female homozygotes were fully fertile (Blume-Jensen et al, 2000). Using a knock-in strategy in p85

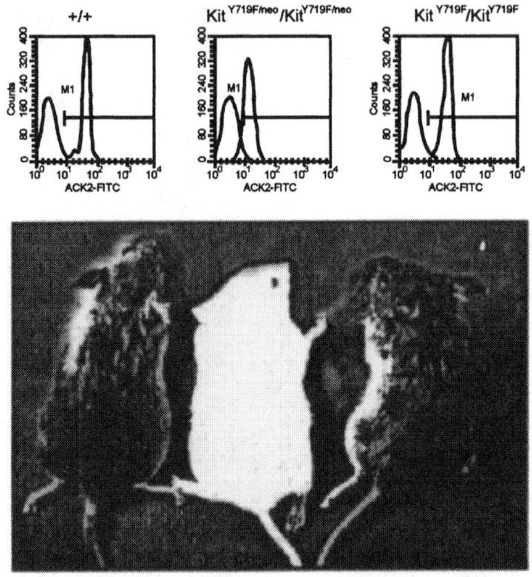

Fig.17.2. Targeted mutation of Tyr719 in the 129/Sv kit locus does not affect pigmentation in kit^{Y719F}/Kit^{Y719F} mice. Cell surface expression of kit in kit^{Y719F}/Kit^{Y719F}, $kit^{Y719neo}/kit^{Y719neo}$ and control BMMC's using FACS (top) and pigmentation phenotype in mice (below). Reprinted with permission from H. Kissel et al. The EMBO Journal 19; 1312-26: 2000 © http://www.nature.com/.

subunit of PI3-kinase in the kit gene, mutant mice did not show pigment deficiency or impairment of steady state hematopoiesis. Therefore, a block in kit receptor-mediated PI3-kinase signaling may be compensated for in hematopoiesis, melanogenesis and primordial germ cell development, but is critical in spermatogenesis and oogenesis (Blume-Jensen et al, 2000; Kissel et al 2000).

17.2.3. Stem Cell Factor (SCF) or c-Kit Ligand (KL)

In situ hybridization studies showed that stem cell factor (SCF or KL) mRNA is expressed in haematopoietic tissues, mesenchymal cells and gonads along the migratory pathway of melanoblasts and primordial germ cells. In spermatogenesis, SCF is expressed mainly by Sertoli cells and is under FSH action. The SCF stimulates DNA synthesis in type A spermatogonia cultured in vitro, and injection of anti-c-kit antibodies block their proliferation in vivo. A point mutation in the c-kit gene, which impairs SCF-mediated activation of PI3-kinase does not cause any significant reduction in PGCs number during embryonic development, or in spermatogonial stem cell population. However, males are completely sterile due to block in the initial stages of spermatogenesis, associated to abolishment of DNA-synthesis in differentiating A1-A4 spermatogonia. In the presence of SCF, more A type spermatogonia incorporate thymidine at stage XII of rat seminiferous tubules in vitro than in the absence of SCF, implying that the increased DNA synthesis might result from enhanced survival of spermatogonia. Not only spermatogonia but also spermatocytes and spermatids were protected from apoptosis in presence of SCF, which also mediated the pro-survival effect of FSH. Yan et al, (2000) confirmed that FSH pro-survival effect on germ cells is mediated partially through the SCF/c-kit pathway. The survival of late meiotic and dividing spermatocytes is also regulated by SCF through an indirect mechanism probably mediated by Sertoli cells. The SCF is important regulator of spermatid output (Packer et al, 1995).

In order to define the role of kit ligand encoded at the mouse steel (SL) locus in spermatogenesis, Tajima et al (1991) examined its production in Sertoli cells. Sertoli cells derived from +/+ and W^v/W^v but not SL^d/SL^d mutant mice produce biologically active SCF growth factor as a membrane bound form. The SCF produced by Sertoli cells acts on germ cells that have the c-kit receptor on their cell surface and stimulates their growth and differentiation. Treatment of Sertoli cells with dibutyryl cAMP, forskolin and cholera toxin increased SCF production, whereas FSH, growth factors and testosterone did not stimulate SCF production (Tajima et al, 1994). To study self-renewal and differentiation of spermatogonial stem cells, Ohta et al (2000) transplanted undifferentiated testicular germ cells of the GFP transgenic mice into seminiferous tubules of mutant mice with male sterility such as those dysfunctioned at Steel (*SL*) locus encoding the SCF or Dominant white spotting (*W*) locus encoding the receptor c-kit. In the seminiferous tubules of SL/SL^d or SL^{17H}/SL^{17H} mice, transplanted donor germ cells proliferated and formed colonies of undifferentiated c-kit (-) spermatogonia, but were unable to differentiate further. However, these undifferentiated but proliferating spermatogonia, retransplanted into *SL* (+) seminiferous tubules of *W* mutant, resumed differentiation, indicating that the transplanted donor germ cells contained spermatogonial stem cells and that stimulation of c-kit receptor by its ligand was necessary for maintenance of differentiated type A spermatogonia but not for proliferation of undifferentiated type A spermatogonia. The transplantation efficiency in the seminiferous tubules of SL^{17H}/SL^{17H} mice depended upon the stem cell niche on the basement membrane of the recipient seminiferous tubules (Ohta et al, 2000).

The SCF exists in soluble (SCFs) and membrane-bound (SCFm) forms, which are required for a complete spermatogenesis, and are generated by alternative splicing of optional exon 6, encoding sites of proteolysis. In the mouse testis, the alternative splicing of SCF is developmentally regulated. The SCFs predominates in fetal and neonatal gonads and is then replaced by SCFm in the prepubertal and adult gonads. While sequencing SCF exon6, Mauduit et al, (1999) suggested that the flanking intronic sequences perfectly follow the GT/AT rule and the basal splicing machinery might not be responsible by itself for exon 6 skipping. Moreover, freshly isolated Sertoli cells mainly express SCFm, but a switch to SCFs occurs after 48 h of culture. UV cross-linking experiments showed that nuclear Sertoli cell protein(s) bind in a sequence-specific manner to exon 6.

Gas 6-Sky signaling: It seems that Gas 6-Sky signaling plays a role in primordial germ cells (PGCs) growth, sexual differentiation, and Sertoli cell functions in vivo. Gene encoding the receptor tyrosine kinase, *Sky*, is expressed in both PGCs and their supporting cells in male genital ridges after 11.5-day postcoitum. Sky expression was not detected in female genital ridges. Gas 6, the ligand for Sky, was also expressed in interstitial cells, which surround Sky positive cells in genital ridges after birth. Sky expression in testis was restricted to Sertoli cells, and Gas 6 was detected around peritubular cells and Leydig cells. The expression of *Sky* in Sertoli cells diminished at 3 weeks of age, when haploid germ cells first appeared. On the other hand, the expression in Sertoli cells was markedly up-regulated in the testis of germ cell-deficient W/W^v and jsd/jsd mice. It suggests that signals from differentiated germ cells suppress *Sky* gene expression in Sertoli cells (Matsubara, 1996).

17.3. THE MYC FAMILY

A relatively new and largely unstudied member of the *myc* family, *c-myc,* encodes a protein that is highly homologous to the N-terminal transcriptional regulatory domain of c-Myc. The *myc* family of genes plays an important role in several cellular processes including proliferation, apoptosis, differentiation, and transformation. In early 1980, reports appeared on the identification of two amplified coding sequences with strong homology to *c-myc* in human neuroblastomas and lung carcinomas, which were dubbed the *N-myc* and *L-myc* gene. Each of the three mammalian *myc* family genes has the same characteristic three-exon structure with the major polypeptide open reading frame residing in the second and third exons. *B-myc* contains homology to *c-myc* exon2, and encodes a 188-amino-acid protein. *B-myc* polypeptide analogous to the putative protein encoded by human *L-myc* transcripts contains only first and second exon sequences. *B-myc* is expressed almost ubiquitously. Another gene, *S-myc* exhibits homology to the second and third *Myc* exons. The three bonafide members of the *myc* gene family (*c-*, *N-*, and *L-myc*) are differentially expressed in mammalian development, where as the *c-* , *L-* and *S-myc* expression has been seen in mammalian testes or testicular tumors.

17.3.1. The C-Myc

The human *c-myc* gene encodes two polypeptides (Myc-1 and Myc-2) of 439 and 453 amino acids with apparent molecular weights of 65- and 68-kDa respectively. The Myc proteins are phosphorylated at multiple sites by casein kinase II, contain α-helical domains and separate areas rich in acidic and basic amino acids, localize to the nucleus, and bind DNA. A region of human Myc, encoded in gene's third exon directs polypeptides translocation to the nucleus. This signal is defined as PAAKRVKLD specified by amino acid 320 to 328. The Myc family of oncoproteins possesses a (BR) H-L-H (LZ) chimeric architecture. The (BR) H-L-H domain of the *myc* gene family covers ~ 55 amino acids positioned about 30 residues from the proteins' carboxy termini; the remaining 30 amino acids constitute their LZ domain (Marcu et al, 1992).

c-Myc in Testis: Normal testis expresses only small amounts of p62 c-Myc. Seminomas showed increased nuclear and cytoplasmic staining. Undifferentiated teratoma showed little activity, whereas p62 c-Myc was abundant in the nuclei of differentiating epithelial structures. The c-Myc proto-oncogene is a reliable marker of the "Go-early G1" transition, and its down-regulation is believed to be necessary to obtain cellular differentiation. In murine spermatogenesis, the level of *c-Myc* transcripts does not correlate with the rate of cellular divisions in degenerating testes. To define whether *c-myc* is only expressed during somatic cell division or is also expressed during meiotic cell division, the production of c-myc mRNA and protein were investigated in the mouse testis. In somatic cells the level of *myc* transcription is not restricted to particular cell types but correlates closely with the rate of cell division. Such transcription involves the use of two active *myc* promoters and produces two mRNA species that are differentially represented among the transcripts of different tissues. In constrast to somatic cells, mitotically and meiotically dividing germ cells have very few *myc* transcripts and appear to proliferate, at least for a few divisions, in the absence of *myc* transcription (Stewart et al, 1984). It was found that c-myc mRNA and protein are expressed in a cell-cycle-dependent manner only in spermatogonia and not in spermatocytes and spermatids (Koji et al, 1988). However, the c-Myc protein is present in mature ejaculated sperm cells with a possible role in sperm cell function. Since the half-life of c-Myc protein has been shown to be short, the presence of *c-myc* mRNA in human sperm was suspected and confirmed by the presence of c-myc mRNA in human sperm cells. The *c-myc* specific DNA probe reacted with the postacrosomal mid-piece and tail regions of both noncapacitated as well as capacitated sperm (Kumar et al., 1993). One Myc 1-6E10 was shown to bind to a 62kDa protein identifiable with the *c-myc* product (p62c-Myc). Without induction, the *c-myc* gene mRNA for non-sperm-type protein was detected at a level that remained essentially constant during both activation and inactivation of the c-*myc* gene. Change in the amount of c-Myc protein alone does not appear to bring about a switch of the expression of two phosphoglycerate kinase genes during spermatogenesis in mouse testis (Koikeda et al., 1989). Transgenic female rats expressing rat *c-myc* gene were fertile, but the male transgenic rats were sterile. Over-expression of the c-myc gene induces apoptosis at the prophase meiosis of the primary spermatocytes thereby causing male sterility in the c-myc transgenic rats (Kodaira et al, 1996).

Hormonal regulation: In vivo administration of hCG on *c-myc* oncogene expression caused a transient elevation of c-myc mRNA in Ledyig cells. It seems that the growth promoting effects of hCG on Leydig cells may be mediated by the transient expression of *c-myc* protooncogene (Lin et al, 1988). Testosterone stimulates *c-myc* mRNA levels in primary Sertoli cell cultures and the induction of *c-myc* mRNA was dependent on the concentration of testosterone. It is likely that testosterone induces *c-myc* mRNA levels in the primary Sertoli cells from prepubertal rats, and then transient expression of c-myc is responsible for some of the regulatory roles of testosterone-dependent genes in the Sertoli cells. The biological significance of testosterone-dependent c-myc induction is not known (Lim et al, 1994). In Sertoli cell cultures. FSH induces c-myc mRNA levels in the primary Sertoli cells from prepubertal and early pubertal rats (Lim and Hwang, 1995).

c-Myc binding proteins: The AMY-1 is a c-Myc-binding protein, which stimulates c-Myc transcription. The AMY-1 is also associated with protein kinase-A anchor protein 84/149 (S-AKAP84/AKAP149) in the mitochondria of somatic cells and sperm. The AAT-1, a novel testis-specific AMY-1-binding protein, forms a quaternary complex with AMY-1. Three isoforms of AAT-1, AAT-1α, -β, and -γ, are derived from an alternative splicing of the transcripts of the *aat-1* gene, which was mapped to human chromosome 3q13-3q21. AAT-1 is specifically expressed in the spermatid and mature sperm, as was AMY-1. The AAT-1α, colocalized in mitochondria with AMY-1, binds to the N-terminal half of S-AKAP84/AKAP149 in a quaternary complex with AMY-1 and a regulatory subunit (RII) of PKA, in which AAT-1α is associated with RII via S-AKAP84/AKAP149 in testis and HeLa cells (Yukitake et al., 2002a). The *amap-1* gene, which was mapped at human chromosome 17q21 encodes AMAP-1. The AMY-1 binds to and colocalized with AMAP-1 in human 293T and HeLa cells. The AMAP-1 was expressed post-meiotically in the testis, as was AMY-1 (Yukitake et al., 2002b). These results suggested that AAT-1, AMAP-1 and AMY-1 play roles in spermatogenesis.

17.3. 2. Other Myc Proteins in Testis

A relatively new and largely unstudied member of the *myc* family, encodes a protein that is highly homologous to the N-terminal transcriptional regulatory domain of c-Myc. High level *B-myc* expression is restricted to primarily hormonally-controlled tissues, with the highest level of expression in the epididymis. Like other Myc family proteins, B-Myc is a short-lived nuclear protein, which is phosphorylated on residues Ser-60 and Ser-68. Rapid proteolysis of B-Myc occurs via the ubiquitin-proteasome pathway. Over-expression of B-myc significantly slows the growth of fibroblasts and COS cells suggesting that B-Myc functions as an inhibitor of cell proliferation (Gregory et al, 2000).

In contrast to expression of the closely related *c-myc* gene, *N-myc and L-myc* expression is restricted in adult tissues. Small amounts of the L-myc RNA can be detected in normal adult testis. Expression of *N-myc* gene, which is usually limited in the neoplasms derived from neuroectoderm, is detected in seminomas (80%) and in embroyonal carcinomas (20%). Expression of *N-myc* gene was detected in teratocarcinomas and normal testes (Saksela et al,

1989). A significant level of *N-myc* expression may be essential for undifferentiated tumors including seminoma and embryonal carcinoma, whereas *c-erbB-1* and possibly *c-erbB-2* may have important roles in the differentiated tumors such as immature teratoma. Rat *s-myc* gene and human *myc L2* genes are expressed in rat embryo chondrocytes and human testis, respectively. The s-Myc expression suppresses the growth activity and tumorigenicity of glioma cells, indicating that s-Myc acts as a negative regulator in tumor growth. In addition, s-Myc overexpression can effectively induce apoptotic cell death in human and rat glioma cells without serum deprivation, which is distinct from c-Myc mediated apoptosis (Kuchino et al, 1996).

17.4. THE MYB FAMILY

The *Myb* gene family currently consists of three members, named A-, B- and c-*myb*. These genes encode nuclear proteins that bind DNA in a sequence-specific manner and function as regulators of transcription. The *c-myb* encodes a transcriptional activator that is essential for the development of the hematopoietic system but appears to lack major roles in non-hematopoietic cells. The two conserved *myb*-related genes, *A-myb* and *B-myb,* have raised the possibility that these genes are functional equivalents of *c-myb* in non-hematopoietic tissues.

17.4.1. A-Myb and B-Myb in Germ Cells

In male mice, *A-myb* is expressed mainly in male germ cells. In female mice, *A-Myb* is expressed in breast ductal epithelium. Mice homozygous for a germline mutation in A-*myb* develop to term but show defects in growth after birth and male infertility due to a block in spermatogenesis. Morophological examination of the testes of A-*myb* -/- males showed that the germ cells enter meiotic prophase and arrest at pachytene stage. In homozygous null A-*myb* female mice, the breast epithelial compartment showed under-development of breast tissue following pregnancy and was unable to nurse their newborn pups. Thus A-*myb* plays a critical role in spermatogenesis and mammary gland development (Toscani et al, 1997). The *B-myb* has been implicated in the control of cell proliferation, particularly at the G1/S transition of the cell cycle. Mouse *A-myb* maps to the proximal region of chromosome 1

Abundant expression of *A-myb* mRNA occurs primarily in the testis of adult mice. The *A-myb* mRNA is detectable in spermatogenic cells during terminal differentiation. The *A-myb* and *B-myb* genes are both expressed in a cell and stage specific manner during testis development. A-myb mRNA expression however increases at postnatal day 10, when primary spermatocytes first appear. In adult testis, A-myb highly expressed in a sub-population of spermatogonia and in primary spermatocytes and was not detectable in spermatids. This expression of A-myb is consistent with the meiotic arrest that is observed in A-myb-deficient male mice. *Xenopus* A myb is related to human A-myb. In adult male *Xenopus*, virtually all A-myb expression is in the testis, particularly within the nuclei of spermatogonial cells. Expression falls dramatically during meiosis and is undetectable furing subsequent spermiogenesis (Latham et al., 1996; Sleeman, 1993; Trauth et al., 1994).

Murine A-myb cDNA gene encodes a protein of 751 amino acids with an estimated molecular weight of 83-kDa. The cDNA sequence of multiple independent cDNA clones reveals the presence of alternatively spliced mRNAs that encode smaller proteins. The A-myb RNA expression predominantly occurs in the testis and peripheral blood leucocytes transcripts, with very low levels expression in the ovaries, spleen and brain (Mettus et al,

1994; Takahashi et al, 1995). The A-Myb can activate transcription from the promoter containing Myb-binding sites in many cells. In addition to two domains (a DNA-binding domain and a transcriptional activation domain), two negative regulatory domains have been identified in A-myb. It indicated that A-Myb functions as a transcriptional activator mainly in testis and peripheral blood cells, and the regulatory mechanism of A-Myb activity is similar to that of c-Myb and v-Myb (Takahashi et al, 1995).

In normal cells, B-myb is expressed at similar levels during all stages of embroyogenesis. A tight linkage occurs between B-myb expression and proliferative activity. By contrast, the highest levels of B-myb expression are found during intermediate stages of spermatogenesis. It has also been shown that B-myb mRNA isolated from the testis differs in size from that of other tissues. B-myb plays a general role during proliferation of most cells and raises the possibility that the function of B-myb in cells undergoing meiosis may be different from its role in cells dividing mitotically (Sitzmann et al, 1996). The *B-myb* mRNA is expressed highly in gonocytes of the fetal testis and in spermatogonia and early spermatocytes in the adult mouse and decreases at day 18 post partum, coincident with the initial appearance of late pachytene spermatocytes. It is likely that B-myb plays a critical role in controlling the proliferation or differentiation of gonocytes and spermatogonia and possibly the somatic lineages as well, whereas A-myb is required for progression through the first meiotic prophase. These distinct roles for A-myb and B-myb during spermatogenesis reflect distinct transactivation potentials of two proteins during spermatogenesis and differentiation of the Sertoli and other somatic cell types of the testis (Latham et al., 1996; Sleeman, 1993; Trauth et al., 1994).

17.5. THE JUN FAMILY

The growth and differentiation of somatic cells are regulated by rapid expression of immediate early genes (IEGs), which encode proteins serving as transcriptional activators and enhance necessary cell responses. The present members of nuclear proto-oncogene *FOS* (*c-fos B, fos B2, Fra1,* and *Fra2*) and *Jun (c Jun, Jun B,* and *Jun D*) belong to IEGs. The products of two proto-oncogenes, c-fos and c-jun, have been implicated in signal transduction pathways as regulators of gene expression. The *Jun* proto-oncogene encodes a nuclear protein (JUN) that is homologous to the yeast transcription factor GCN4. The JUN protein dimerizes with the *FOS* gene products and other members of the bZIP family of transcription factors as a leucine zipper to form activator protein 1 (AP1) (a transcription factor). The expression of members of this transcription factor has been associated with cellular pathways that result in both mitosis and differentiation. The AP1 regulates the expression of genes containing the tumor promoter response element (TRE) sequence (TGACTCA) or the cyclic AMP responsive element (CRE)-like site (TGACGTCA). As an immediate early response gene, *Jun* plays a crucial role in regulation of cell proliferation and programmed cell death in the context of oxidative stress.

17.5.1. Jun-B, C-Jun, Jun-D

The members of the jun family of protooncogenes (*junB, c-jun, and junD*) share a high degree of sequence homology and function as transcriptional regulators. Alcivar et al (1990) examined the relative adundance of the *junB* and *c-jun* proto-oncogenes during development of the mouse testis. JunB and c-jun mRNA levels are low in total RNA from intact immature or mature testes. Dissociation of testicular cells, however, increases the levels of *junB* and *c-jun* mRNAs.

In 8 days old mouse, the mRNA levels for both proto-oncogenes are higher in type B spermatogonia and in the interstitial cell fraction than in type-A spermatogonia. In testis of 17-day-old mice, the highest mRNA levels for both proto-oncogenes are seen in preleptotene, spermatocytes with decreasing levels in leptotene/zygotene spermatocytes and prepuberal pachytene spermatocytes. The *junB* and *c-jun* mRNAs are nearly undetectable in pachytene spermatocytes, round spermatids, and residual bodies/cytoplasts. Comparative pattern of junD mRNA expression during spermatogenesis to that of *junB* and *c-jun* in 8-day-old mice indicated that the level of the 1.8kb *junD* mRNA is higher in type B spermatogonia than in type A spermatogonia. In 17-day-old mice, the highest *junD* mRNA levels occur in preleptotene spermatocytes. The pattern of expression of *junD* differs from that of *junB* and *c-jun* during spermatogenesis most notably in that 1) *junD* mRNA levels do not increase following dissociation of testicular cells, and that 2) in constrast to the nearly undetectable levels of *junB* and *c-jun* mRNAs in adult postmeiotic testicular cells, *junD* mRNAs are seen at high levels (Alcivar et al, 1991).

Using a seasonal breeder, the European red fox (Vulpes vulpes), Cohen et al (1993) studied the expression of five AP-1 family members (c-fos, fra 1, fra-2, c-jun and junB) with a view to elucidating their role in the regulation of spermatogenesis. Unique patterns of expression, falling into three broad categories, were observed for the five genes: continuous expression throughout the spermatogenic cycle (c-fos); expression only at times corresponding to the onset and shutdown of spermatogenesis (fra-1, fra-2 and c-jun); and (iii) expression only at the onset of the cycle (junB). Furthermore, the proteins are expressed in both pre-meiotic and post-meiotic cell types, suggesting a role in haploid, as well as diploid, gene expression in this tissue. This suggests distinct and although not necessarily unrelated roles for the different components of transcription factor AP-1 in the regulation of spermatogenesis (Cohen et al, 1993).

To evaluate the possibility that protooncogenes mediate long-term effects of LH, hCG and various growth factors, their action on the levels of *c-fos, c-jun, jun-B, and c-myc* m RNAs indicated that the regulation of protooncogene mRNAs in Leydig cells is multifactorial. They show differential responsiveness of the members of the Jun family to several factors, consistent with the hypothesis that the c-FOS and JUN families of regulatory proteins could play a role in mediating long term response to the complex array of hormones and growth factors to which Leydig cells are exposed in vivo. The transcriptional processes may mediate the effects of LH on cellular adaptive responses or steroid biosynthesis (Schultz et al., 1995). In the rat Sertoli cell, FSH rapidly inhibits c-jun gene expression, while it stimulates c-fos and jun-B as well as the expression of the more slowly responding tissue plasminogen activator (tPA). The immediate early effect of FSH is on the expression of genes encoding components of the transcription factor (Hamil et al, 1994). The mRNA levels for c-jun, as well as mRNA levels for c-myb, following retinoid stimulation supported the model that modulation of these immediate early genes is the first step in the process, which initiates a cascade of vitamin A-inducible gene expression in Sertoli cells necessary for maintaining spermatogenesis (Page et al, 1996).

17.5.2. C-FOS AND C-FOS RELATED ANTIGENS

c-Fos and Gene Methylation: The ontogenic initiation and maintenance of methylation of certain specific CpG sites in the coding region of c-fos, through testicular cells, sperm, and fetal, neonatal, and adult somatic tissues showed that 1) sperm derived methylated sites get demethylated in early development. However, unlike other genes, they remain so at least up to day 13.5 post coitum (pc); 2) de novo methylation proceeds unidirectionally in a step-wise, site-specific manner. The c-fos thus provides a novel pattern of de novo methylation in development. Also, it suggests that close proximity of CpGs may prevent methylation (Chandrasekhar and Raman, 1997).

Fra-1 and HMG Box Proteins: The functions of HMG-box transcription factors are mediated, in part, by activation of members of the AP-1 transcription factor family. The fra-1 is expressed during spermatogenesis. The promoter of the rat fra-1 gene contains several potential binding sites for HMG-box DNA-binding proteins. Hence purified SRY protein binds strongly to one of the putative fra-1 HMG-box response elements and that SRY enhances the transcription of rat fra-1 promoter constructs in co-transfection assays (Cohen et al, 1994).

c-Fos Gene and Sertoli Cell Activation by FSH: The FSH induces a transient expression of c-fos in cultured Sertoli cells. This induction is probably mediated by cAMP and likely involves an increased transcription of the c-fos gene (Hall et al, 1988). Although the action of FSH on Sertoli cells is considered to be mediated by cAMP, dibutyryl cAMP (dbcAMP), an analog of cAMP, induces much less c-fos mRNA expression than FSH, suggesting that additional cAMP independent mechanisms may mediate the effect of FSH, on c-fos. Sertoli cell differentiation of basement membrane is mediated by the c-fos protooncogene (See Chapter 1).

17.6. C-ROS TYROSINE KINASE

17.6.1. Epididymis c-Ros and Infertility

The protooncogene, *c-ros* is an orphan receptor gene, which codes an intracellular protein tyrosine kinase (PTK) domain and a long extracellular domain. Chicken c-ros is specifically expressed in certain epithelial cells of kidney, intestine, lung, bursa, thymus, and testis, and expression is regulated temporarily and spatially. It has been suggested that c-ros may play a role not only in the initial induction events in the organogenesis, but also in the mature function of those organs. Whereas the expression of c-*ros* in other organs rapidly disappears in neonatals, that in the epithelium of the proximal epididymis (caput) is up-regulated during prepubertal development, before the arrival of spermatozoa from the testis, and persists in adults. This coincides with the differentiation of the proximal caput epididymidis into the so-called initial segment. Since mammalian spermatozoa are stored in a hypertonic environment in the epididymis and would experience relative hypoosmotic conditions upon ejaculation, consideration of regulatory volume decrease (RVD) is particularly pertinent. Targeted mutations of c-ros in the mouse revealed an essential role of the gene in male fertility. The epididymal defects interfere with sperm maturation and the ability of sperm to fertilize in vivo. Interestingly, sperm isolated from c-ros -/- animals could fertilize ovum in vitro (Sonnenberg-Riethmacher et al, 1996). The receptor tyrosine kinase c-ros knockout mouse did not reveal any gross abnormality that could explain infertility in vivo. The majority of motile spermatozoa exhibited angulation in the tail. A lack of response of immature wild-type sperm and mature knockout sperm to the channel blockers suggested that there is normally a development of the volume regulatory mechanisms upon maturation that is defective in sperm from the knockout animals. The resultant flagellar angulation may account for the reduction in sperm numbers in the oviduct of mated females and the failure to fertilize in vivo. This finding provided a starting point for further research to establish the link between abnormal epididymidis and sterility. The infertility of c-ros knockout male mice could be explained by the sperm's inability to enter the oviduct, as a result of their bent tails forming the entangled sperm mass and their compromised flagellar vigor within the uterus (Yeung et al., 1998; 2000).

Transgenic male mice bearing inactive mutations of the receptor tyrosine kinase c-ros down-regulated expression of CRES, and lipocalin mouse epididymal protein 17 (MEP17), whereas other caput-enriched genes (glutathione peroxidase 5, α disintegrin and ADAM7, bone morphogenetic proteins 7 and 8α, A-raf, CCAAT/enhancer binding protein β, PEA3) were unchanged. Genes normally absent from the initial segment were expressed in the undifferentiated proximal caput of the KO. More distally located proteins were unchanged in expression. The genes of over 70 transporters, channels, and pores were detected in the caput epididymidis, but in the KO, only three were downregulated and six upregulated. The changes in these genes could affect sperm function by modifying the composition of epididymal fluid and explain the infertility of the KO males (Cooper et al., 2003) (see chapter 32).

17.6.2. Regulation of c-Ros Receptor by Protein Tyrosine Phosphatase

Male "viable motheaten" (me(v)) mice, with naturally occurring mutation in the gene of the SH2 domain protein tyrosine phosphatase SHP-1, are sterile. Ros and SHP-1 are co-expressed in epididymal epithelium. Elevated phosphorylation of Ros in the epididymis of me (v) mice suggested that Ros signaling is under control of SHP-1 in vivo. Phosphorylated Ros strongly and directly associated with SHP-1 in yeast two-hybrid assay. Strong binding of SHP-1 to Ros was selective compared to six other receptor tyrosine kinases and the interaction was mediated by the SHP-1 NH_2 terminal SH2 domain and Ros phosphotyrosine 2267. Over-expression of SHP-1 resulted in Ros dephosphorylation and effectively down-regulated Ros dependent proliferation and transformation. Therefore SHP-1 seems an important downstream regulator of Ros signaling (Keilhack et al, 2001).

c-ros promoter: The most 5' c-ros cDNA of chicken has been isolated and sequenced. Using the 5' cDNA as a probe, three genomic DNA clones containing the 5' c-ros cDNA sequence were isolated. The transcription initiation site for the c-ros mRNA in kidney and intestine were mapped. The sequence of the 1.3-kb region upstream of the initiation site contains TATA and CAAT boxes at 26 and 54 nucleotides, respectively, upstream of the initiation site. In addition, transcription factor binding sites for AP1, AP2, and Oct1 and several direct and inverted repeats are present within 1-kb upstream of the initiation site. The 1.3-kb DNA was shown to be functionally active. Serial delations of this putative c-ros promoter allowed to define a minium c-ros promoter and to identify positive and negative regulatory regions. By gel mobility shift experiments, their specific binding pattern to nuclear extracts from kidney, intestine, and thymus corresponds to the tissue specificity and temporal control of c-ros mRNA expression (Chen et al, 1995).

17.7. PIM-1 AND PIM-2

The fact that over-expression of *Pim-1* can contribute to lymphomagenesis was proven by over-expressing a *Pim-1* transgene in lymphoid cells. The mouse *Pim-1* gene encodes two cytoplasmic protein-serine/threonine kinases. The highest expression is found in haematopoietic tissues. High expression has also been noted in testis and ES cells. Expression can be induced by growth factors and mitogens. The gene is evolutionally conserved. Inactivation of both *Pim-1* alleles in ES cells or mice did not reveal any obvious abnormalities (Domen et al, 1993). Human *Pim2* gene *(hPim-2)* encodes a protein that shares 90% identity and 93% similarly at the primary structure level, with the mouse *Pim-2* gene. Structurally, like the mouse Pim-2, hPim-2 is also a serine threonine kinase. At the RNA level, two *hPim-2*

```
      1                                                       50
Hpim2 LAPPSPGSPA ALPRASTPCG LSGFSGLNLR SATSMLTKPL QGPPAPPGTP
mpim2 LAPPSPGSPA ALPRASTPCG LSGFSGLNIR STSSMLTKPL QGHPSPPVTP
hpim1 ---------- ---------- ---------- ---------- ----------
      51              I                         II       100
Hpim2 TPPPGGKDRE AFEAEYRLGP LLGKGGFGTV FAGHRLTDRL QVAIKVIPRN
Mpim2 TQPPGGKDRA AFEAEYRLGP LLGKGGFGTV FAGHRVTDRR QVAIKVISRN
Hpim1 -----GKEKE PLESQYQVGP LLGSGGFGSV YSGIRVSDNL PVAIKHVEKD

      101                          IV                V   150
Hpim2 RVLGWSPLSD SVTCPLEVAL LWKVGAGGGH PGVIRLLDWF ETQEGFMLVL
Mpim2 RVLGWSTVSD SVTCPLEVAL LWKVGEGNGH PGVIRLLDWF ETPEGFMLVL
Hpim1 RISDWGELPN GTRVPMEVVL LKKV--SSGF SGVIRLLDWF ERPDSFVLIL

      151                VIA                       VIB   200
Hpim2 ERPLPAQDLF DYITEKGPLG EGPSRCFFGQ VVAAIQHCHS RGVVHRDIKD
Mpim2 ERPMPAQDLF DYITEKGPLG ESCSRSFFTQ VVAAVQHCHA RGVVHRDIKD
Hpim1 ERPEPVQDLF DFITERGALQ EELARSFFWQ VLEAVRHCHN CGVLHRDIKD

      201       VII                 VII                 IX 250
Hpim2 ENILIDLRRG CAKLIDFGSG ALLHDEPYTD FDGTRVYSPP EWISRHQYHA
Mpim2 ENILIDLCRG SIKLIDFGSG ALLHDEPYTD FDGTRVYSPP EWISRHQYHA
Hpim1 ENILIDLNRG EIKLIDFGSG ALLKDTVYTD FDGTRVYSPP EWIRYHRYHG

      251                    X                 XI        300
Hpim2 LPATVWSLGI LLYDMVCGDI PFERDQEILE AELHFPAHVS PDCCALIRRC
Mpim2 LPATVWSLGV LLYDMVCGDI PFERDQEILE AELHFPAHVS PDCCALIRRC
Hpim1 RSAAVWSLGI LLYDMVCGDI PFEHDEEIIR GQVFFRQRVS SECQHLIRWC

      301                    325
Hpim2 IAPKPSSRPS LEEILLDPWM QTPAE
Mpim2 LAPKPCSRPS LEEILLDPWM QSPAE
Hpim1 LALRPSDRPT FEEIQNHPWM Q----
```

Fig.17.3. Comparison of amino acid sequences of Pim-2 (hpim2) deduced from human *pim-2* gene, the mouse Pim2 (mpim2) and human Pim1 (hpim1) proteins. The sub-domains characterizing protein kinase superfamily are marked with roman numeral above the first amino acid of each domain. Reprinted with permission from D. Baytel et al. Biochim Biophys Acta 1442; 274-85 : 1998. © Elsevier.

transcripts were identified. The first, 2.2-kb, is highly expressed in hematopoietic tissues and in leukemic and lymphoma cell lines. It is present at considerable high levels in testis, small intestine, colon and human colorectal adenocarcinoma cell (SW480). A second transcript, 5.0-kb in size, is detected only in spleen, thymus, small intestine and colon. Biopsies from testes of men with complete or partial spermatogenesis revealed that the gene is expressed in primary spermatocytes. In the absence of germ cells, signal could be detected over specific cells in the well developed interstitial region. These results suggest a role for hPim-2 in proliferating cells as well as during meiosis. A possible connection between hPim-2 and apoptosis has also been suggested (Baytel et al, 1998)(**Fig.17.3**).

The *pim-1* gene of the mouse has been implicated as an oncogene by its frequent activation by provirus insertion in MuLV-induced T-cell lymphomas. Direct evidence for the oncogenic potential of the *pim-1* gene in inducing T-cell lymphomas has been provided by transgenic mice bearing copies of a modified murine, *pim-1* transgene. The human *pim-1* gene has been localized to a region of chromosome 6 known to be involved in translocations in acute myeloid tumors, undifferentiated leukemias, T-cell lymphomas and malignant melanomas. Whether the *pim-1* gene can be oncogenic in humans constitutes an area of continuing investigation. The nucleotide sequences of both the cloned murine *pim-1* gene and a number of human *pim-1* cDNA have been reported. The sequence of the human *pim-1* gene together with its immediate promoter sequence(Ppim), shows a full length of 6.1-kb. Human genomic *pim-1* gene, together with its immediate 5'-upstream promoter sequence (*Ppim*) shares an overall nucleotide (nt) sequence identity of 53% with the murine *pim-1* gene. It consists of six exons and five introns and contains a protein-coding region that is identical in nt sequence to a full-length human *pim-1* cDNA. The gene codes for a predicted Pim-1 protein of 313 amino acids (aa) with an M_r of 35690 and a pI of 5.7. The deduced aa sequence of the human Pim-1 has 94% identity with the murine Pim-1 whereas the nucleotide sequences of the two genes are 88% identical. All of the conserved amino acid residues of the mouse *pim-1* gene are conserved in the predicated human protein. The human *Ppim* region is G + C-rich (69%) and shares greater than 80% identity with the murine *Ppim*. The *Ppim* has no TATA- or CAAT-box sequences but contains a number of nt sequences similar to the putative binding sites of several presumptive transcription (Reeves et al, 1990) **(Fig.17.4)**. The RNA transcript of human *pim-1* is approx. 3.0-kb and it is highly expressed in the erythroleukemia cell line K562 (Zakut-Houri et al, 1987). A 2.4-kb *pim-1* transcript, expressed in the germ cells of mouse testis is selectively expressed in haploid postmeiotic early spermatids. The evidence for a developmentally regulated expression of pim-1 in haploid spermatids suggests a possible developmental role for this protooncogene product. The 2.4-kb pim-1 transcript present in postmeiotic cells differs in size from the 2.8-kb transcript usually detected in somatic tissues. Similar testis-specific transcripts have been seen for *mos* and *abl* gene. Analysis of expression of the three members of the *ras* gene family indicated a distinct temporal regulation in the expression of the *Harvey, Kristen,* and *N-ras* genes in these germ cells (Sorrentino et al., 1988).

17.8. C-MOS FACTOR

The *c-mos* proto-oncogene is a cellular homologue of the Moloney sarcoma virus oncogene *(v-mos)*. Various *c-mos* genes from vertebrates including human, monkey, rat, mouse, chicken and *Xenopus* predicated it as a single coding exon, which encodes a serine/threonine kinase. The *c-mos* transcripts are dominantly expressed in reproductive systems including oocytes, ovary and spermatogenic cells and in the embryo, and at low levels in various other tissues. Different *c-mos* transcripts have the same 3'-end but differ at their 5'-ends. The Mos proteins differ significantly in size, according to their origins: 37-kDa and 43-kDa/v-mos transformed cells, 39-kDa/*Xenopus* oocytes, 39-kDa/mouse oocytes, 43-kDa/mouse and rat testes, 43-kDa and 75-kDa/rat muscle, 40-kDa and 37-kDa/human cells. The significance for the size difference is not known.

Specificity of Germ Cell c-Mos : The *c-mos* proto-oncogene protein serine/threonine kinase is specifically expressed in germ cells and has been found to play a key role in meiotic maturation of both *Xenopus* and mouse oocytes. The Mos is required for the initiation of meiosis as well as for progression from meiosis I to meiosis II. Experiments in which *c-mos* was disrupted by homologous recombination in mice further confirmed that *c-mos*-deficient oocytes complete meiosis 1 normally but fail to arrest at metaphase II and instead undergo parthenogenetic activation. To explore the role of pp39mos in male germ cell meiosis, Rosenberg et al, (1995) constructed transgenic mice carrying either the c-Mos or v-Mos genes. All male transgenic mice bearing the v-Mos but not the c-Mos construct were sterile due to arrest of germ cells at metaphase 1. Overexpression of pp39 v-mos during male meiosis results in alteration of various cell cycle related kinases (particularly MAP kinases) and cytostatic factor like arrest at metaphase I.

The expression of *mos* is developmentally regulated in gonadal tissue. As such the level of *mos* transcripts is low for the first 3 weeks after birth and reaches adult levels by day 30. The *mos* transcripts are present in haploid germ cells. RNA transcripts are associated with monosomes and polysomes. The *c-mos* transcripts found in testis RNA are estimated to be approximately 1.7-kb. The S1 nuclease assay detected entire mos RNA as open reading frame.

```
                                                       * *     *              * *
a   1 MLLSKINSLA HLRAAPCNDL HATKLAPGKE KEPLESQYQV GPLLGSGGFG SVYSGIRVSD NLPVAIKHVE
b     MLLSKINSLA HLRAAPCNDL HATKLAPGKE KEPLESQYQV GPLLGSGGFG SVYSGIRVSD NLPVAIKHVE
c     MLLSKINSLA HLRARACNDL HATKLAPGKE KEPLESQYQV GPLLGSGGFG SVYSGIRVSD NLPVAIKHVE
d     MLLSKINSLA HLRAKPCNDL HATKLAPGKE KEPLESQYQV GPLLGSGGFG SVYSGIRVAD NLPVAIKHVE

                         *
a  71 KDRISDWGEL PNGTRVPMEV VLLKKVSSGF SGVIRLLDWF ERPDSFVLIL ERPEPVQDLF DFITERGALQ
b     KDRISDWGEL PNGTRVPMEV VLLKKVSSGF SGVIRLLDWF ERPDSFVLIL ERPEPVQDLF DFITERGALQ
c     KDRISDWGEL PNGTRVPMEV VLLKKVSSGF SGVIRLLDWF ERPDSFVLIL ERPEPVQDLF DFITERGALQ
d     KDRISDWGEL PNGTRVPMEV VLLKKVSSDF SGVIRLLDWF ERPDSFVLIL ERPEPVQDLF DFITERGALQ

                                  *     *                *** 
a 141 EELARSFFWQ VLEAVRHCHN CGVLHRDIKD ENILIDLNRG ELKLIDFGSG ALLKDTVYTD FDGTRVYSPP
b     EELARSFFWQ VLEAVRHCHN CGVLHRDIKD ENILIDLNRG ELKLIDFGSG ALLKDTVYTD FDGTRVYSPP
c     EELARSFFWQ VLEAVRHCHN CGVLHRDIKD ENILIDLNRG ELKLIDFGSG ALLKDTVYTD FDGTRVYSPP
d     EDLARGFFWQ VLEAVRHCHN CGVLHRDIKD ENILIDLSRG EIKLIDFGSG ALLKDTVYTD FDGTRVYSPP

      *                  *                                                   *
a 211 EWIRYHRYHG RSAAVWSLGI LLYDMVCGDI PFEHDEEIIR GQVFFRQRVS SECQHLIRWC LALRPSDRPT
b     EWIRYHRYHG RSAAVWSLGI LLYDMVCGDI PFEHDEEIIR GQVFFRQRVS SECQHLIRWC LALRPSDRPT
c     EWIRYHRYHG RSAAVWSLGI LLYDMVCGDI PFEHDEEIIR GQVFFRQRVS SECQHLIRWC LALRPSDRPT
d     EWIRYHRYHG RSAAVWSLGI LLYDMVCGDI PFEHDEEIIK GQVFFRQRVS SECQHLIKWC LSLRPSDRPS

a 281 FEEIQNHPWM QDVLLPQETA EIHLHSLSPG PSK
b     FEEIQNHPWM QDVLLPQETA EIHLHSLSPG PSK
c     FEEIQNHPWM QDVLLPQETA EIHLHSLSPG PSK
d     FEEIRNHPWM QGDLLPQAAS EIHLHSLSPG SSK
```

Fig.17.4. Comparison of amino acid sequences deduced from *pim-1* gene (a, human, Reeves et al.1990; b, human, Domen et al. 1987; c, unpublished data c/r Reeves et al 1990; d, mice Selton et al, 1986). Boxes represent domains of *Ppim* that are highly conserved in most protein kinases (PKs), while asterisks show residues that are strictly conserved in all known PKs. Thick underline bold-face amino acids denote two residues coded by a single human pim-1 cDNA that differs from reported human pim-1 sequences. Differences of amino acids with mice pim-1 are also underlined. Reprinted with permission from R. Reeves et al. Gene 90; 303-07: 1990 © Elsevier.

The *c-mos* RNA transcripts from other origin appear to end at the same 3' position, and the tissue specific size variations appear to be due to use of different promoters. For example, the testicular and ovarian RNA transcripts initiate approximately 280 and approximately 70-bp, respectively, upstream from the first initiation codon, but both end at a common site downstream from the mos open reading frame. In contrast, 1.4-kb transcript is present in ovary RNA and at least two major transcripts of 2.3- and 1.3-kb in embryo RNA. The latter transcripts have in common sequences of at least 1-kb, representing most of the c-mos open reading frame (Propst and Vande Woude, 1985, Propst et al., 1987). In rat male germ cells, three *c-mos* RNA species of 5, 3.6 and 1.7-kb are found, both before and after meiosis with highest level in early spermatids (van der Hoorn et al., 1991). In murine spermatogenesis, c-mos RNA (1.7-kb) expression is exclusive to early spermatids (post-meiotic cells) or in pachytene spermatocytes besides spermatids. Testicular 43-kDa c-Mos protein was first detected in mouse testicular germ cells and a similar 43-kDa c-Mos protein is found in pachytene spermatocytes of rats. However, in the *c-mos*-deficient mice produced by gene targeting, reproduction is normal in the male, whereas parthenogenetic activation of oocytes occurs in the female. Thus, the role of the *c-mos* product in spermatogenesis is relatively unknown. This suggests that *c-mos* is expressed in both male and females and has a possible role in meiosis, germ cell development, fertilization and early embryogenesis (Goldman et al., 1987). The *c-mos* gene is reportedly expressed in epididymis. The 4A6 C-Mos mAb reactive antigen has been found on the basement membrane and luminal surface of the epithelial cells in the caput epididymis as well as in the proximal corpus epididymis, the cauda epididymis, and the vas deferens of mice (Heikinheimo et al., 1995).

A 43-kDa c-Mos protein was detected in spermatogenic cells close to the occurrence of the first meiotic division during postnatal development The c-Mos has a pI value of around 9.0-9.6 showing a hydrophobic nature, and phosphhorylarted in vitro on serine. The protein is neither ubiquitinated nor glycosylated. N-Terminal amino acid sequence of 43kDa c-Mos is: Asp-Glu-Gly-Gly-Asn-Leu-Gln- The rat testicular 43k c-Mos protein is translated from a 5'-

upstream sequence of the predicated consensus AUG start codon, probably by an unusual translational rule (Nagao, 1995). Comparison of c-Mos RNA expression in mouse and rat testes reveals species-specific variations in the regulation of gene expression. Murine *c-mos* transcripts isolated from testis have 5'UTRs of approximately 300 nt with a series of four overlapping ORFs upstream of the AUG codon that initiates the Mos ORF. Ovarian c-mos transcripts have shorter 5'UTRs (70-80 nt) and contain only 1-2 of the upstream ORFs (uORFs). The 5'UTR characteristic of testis-specific c-mos mRNA strongly represses translation relative to the translation of transcripts that contain a 5'UTR derived from beta-globin mRNA; this is mainly due to the four uORFs. Each of the four upstream AUG triplets are recognized as a start site for translation, and no single uAUG dominates the repressive effect. The uORFs repress translation by a mechanism that is not affected by the amino acid sequence in the C-terminal region of the uORF-encoded peptides. The very short uORF (AUGUGA) present in ovary-specific transcripts does not repress translation. Testis sections from transgenic mice suggested that the uORFs can dramatically change the pattern of expression in spermatogenic cells (Steel et al, 1996). A cis-acting enhancer element(s) is responsible for the cell type specific *c-mos* (rat) expression. An enhancer (300 bases) close to the repressor sequences acts on heterologous promoters in an orientation independent manner over large distances (Van der Hoorn, 1987).

c-Mos Promoter: The rat testis *c-mos* transcription start site is located 0.56-kb upstream of the coding region. A fragment containing the rat testis *c-mos* promoter directs transcription in a nuclear extract derived from rat seminiferous tubules, but not in a liver nuclear extract. DNAase I footprint and gel-retardation assays showed binding of a novel testis specific nuclear factor to rat testis *c-mos* promoter at a site homologous to the testis-specific cis-acting element identified in the promoter of the RT7 gene, which is specifically expressed in haploid male germ cells (Van der Hoorn, 1992). A negative regulatory element (NRE) upstream of the *c-mos* promoter that suppresses c-mos transcription in transfected NIH 3T3 cells has been identified (Xu and Cooper 1995). The protein was found to bind to a region of the NRE, which was shown to be required for suppression of c-Mos transcription. This factor is present in nuclear extracts of several somatic cell lines but not in male germ cells in which *c-mos* transcribed, suggesting that it is a somatic cell repressor of *c-mos* transcription. The binding site of the candidate repressor within the c-mos NRE consists of sequences related to putative NREs identified in two other male germ cell-specific genes (encoding *Prm 2 and Pgk2*). The *c-mos* repressor could bind and could be UV cross-linked to *Prm 2* and *Pgk 2* gene sequences as a protein with an apparent molecular mass of 30-kDa. The repressor-binding site is also conserved in germ cell-specific genes encoding Cyt Ct and HSP-70, suggesting that the *c-mos* repressor may be generally involved in suppressing transcription of germ cell-specific genes in somatic cells (Xu and Cooper, 1995, Higgy et al, 1995). In addition, since mitotic divisions of germ cells occurred prior to PGK-c-mos transgene expression, observations suggested that *c-mos* overexpression in spermatocytes causes an alteration in cell-cell interactions (Higgy et al, 1995).

Rat testis promoter contains a TATA box and a binding site for a testis specific transcription factor (TTF-D) as well as a region, which can act as an enhancer. The enhancer is located 2kb up stream of c-mos AUG start codon. The promoter binds three factors at sites -I, -II and III. The member, NF-1/CTF family of factors binds at site II where as a rat homologous protein of human hCut/CDF, mouse Cux/CDP and canine clox binds at site III. This protein hereafter called Cux/CDP, a 160kDa protein containing multiple repeats and a homeodomain, negatively regulates the mammalian *c-myc, gp91-phox* and *NCAM* genes. Cux repeat CR3 and the homeodomain are both required for efficient binding to enhancer site III. Transfections of

CAT constructs containing enhancer fragment linked to a minimal promoter demonstrated that Cux/CDP expresses *c-mos* enhancer activity (Higgy et al, 1997).

Van der Hoorn and coworkers identified testis specific nuclear factor binding sites in the testis specific promoter of the *c-mos* gene and the *Odf1* gene, which are 80% identical. A testis-specific nuclear factor, TTF-D, is able to complex with both binding sites and can stimulate *Odf1* promoter activity. The TTF-D is detectable in mouse testis as early as day 11 pp and contains three peptides of 22, 25 and 35-kDa in size. The TTF-D binds specifically to its cognate single and double-stranded DNA. Both double stranded and single-stranded binding site oligonucleotide DNA can specifically repress *Odf1* promoter activity. This suggests that TTF-D is involved in positive transcription regulation of a pre-meiotic and a post-meiotic gene in the testis (Oosterhuis and van der Hoorn, 1999). The role of *c-mos* repressor in other germ cell specific proteins needs to be investigated.

Methylation of c-mos gene: The methylation status of several CpG sites, present both upstream and within the coding region of the c-mos gene, has shown that *inch HpaII* and *HhaI* sites examined in the 5' half of the coding region are unmehylated in both the *c-mos* expressing and non-expressing tissues. A *HhaI* site, h3, present 380bp downstream of the transcription start site, is unmehylated in germ cells, but is partially methylated in the somatic tissues, inversely correlating with the expression status of the gene. In contrast to these tissues, in the mouse fibroblast cell line L929, all the analysed sites are completely methylated (Kanduri and Raman, 1999).

17.9. INT-1 AND INT–2 ONCOPROTEINS

Int-1: The *int-*1 proto-oncogene seems to be an extreme example of a developmentally regulated gene. During mouse embryogenesis, it is expressed only in the testis and in central nervous system among several adult mouse tissues. In embryos 11-15 days after conception, expression of the gene is restricted to the developing central nervous system in regions of the neural tube other than the teleencephalon (Jakobovits et al., 1986). This severely restricted pattern of expression in normal cells is unusual among proto-oncogenes, most of which are expressed in a wide variety of cells and tissues. The *int* –1 appears thus far to be active in only one neoplastic setting, mammary carcinomas induced by the mouse mammary tumor virus (MMTV). The int-1 is activated by adjacent insertions of proviral DNA in induced tumors and has transforming activity in certain mammary epithelial cell lines. Protein products of the gene in cells transfected or infected with retroviral vectors expressing *int-1* have been identified (Brown et al, 1987). Four protein species, of 36,000, 38,000, 40,000 and 42,000 Mr, are structurally related to each other. The int-1 RNA is detected only in post-meiotic germs cells undergoing differentiation from round spermatids to mature spermatozoa. Thus, int-1 mediates development events at these two sites (Shackleford and Varmus, 1987). The protein product of *int-*1 is modified by proteolytic cleavage and glycosylation and appears to enter a secretory pathway.

Int-2: Int-2 was discovered as a protooncogene transcriptionally activated by MMTV during mammary tumorigenesis in mouse. Sequence analysis showed int-2 to be a member of the fibroblast growth factor family of genes. Using in situ hybridization, expression was found at a number of sites during embryonic development. An analysis of the *int-2* transcripts found in embryonal carcinoma cells revealed six major classes of RNA initiating at three promoters and terminating at either of two polyadenylation sites. All size classes of RNA encompass the same open reading frame. These proteins represented glycosylated and non-glycosylated forms of the same primary product with or without the signal peptide removed. Findings suggest the potential for a dual role of int-2: an autocrine function acting at the cell nucleus, and a possible paracrine action through a secreted product (Dicksen et al, 1990).

17.10. CELLULAR-ABELSON PROTO-ONCOGENE (C-ABL) PROTEIN

The *abl* proto-oncogene was initially identified as the viral transforming gene (v-able) of Abelson murine leukemia virus (A-MuLV) from a chemically thymectomised mouse inoculated with Moloney murine leukemia virus (M-MuLV). The v–abl gene consists of M-MuLV gag sequences fused to mouse c-abl, resulting in expression of p160 fusion protein and p120 gag Abl translational products. The mammalian *c-abl* gene is ubiquitously expressed and encodes for two 145-kDa isoforms due to alternative splicing of two distinct first exons. These isoforms are designated type I and type IV c-abl proteins in the mouse and the type la and type IIb proteins in humans. The type IV/1b, but not the type I/Ia isoform is myristylated at an amino-terminal glycine residue (Laneuville, 1995). The amino-terminal half of c-Abl protein contains the Src-homology (SH) regions 1, 2, and 3 that mediate catalytic, phosphotyrosine protein binding and kinase inhibitory functions, respectively. The carboxy-terminal half of c-Abl protein codes a lysine-rich motif required for nuclear localization, a bipartite DNA-binding domain, and F-and G-actin binding domains. The c-Abl proto-oncoprotein is normally distributed both in the nucleus where it is bound to chromatin, and in the cytoplasm where it co-localizes with F-actin.

Characterization and Localization of c-Abl Protein: The appearance of the c-abl testicular variant was coincident with the entry of the germ cells into their haploid state and suggested that the regulated expression of this proto-oncogene may be important in the normal differentiation of the male germ cells. In normal mouse tissues the expression of c-cbl was highest in testis and thymus; the predominant species in testis tissue being a 3.5-kb mRNA of a 135-kDa protein (p135 c-Abl), in both normal and transformed cells. Subcellular fractionation showed that p100gag-Abl and p135c-Abl are both located in the nucleus. The translational product of a *c-abl* is a 135-kDa nuclear protein (Langdon et al, 1989). Although the testis contains high levels of a unique c-abl mRNA, along with lower amounts of two c-abl mRNAs common to somatic cells, Ponzetto et al., (1989) detected only a single polypeptide of approximately 150kDa, indistinguishable from p150c-Abl observed in murine tissues and cell lines. The p150 c-Abl was also detected in enriched populations of germ cells, including late stage spermatids, which contain the highest levels of the c-abl transcript. In mature testis, and specifically in late spermatids, p150c-abl co-precipitated with phosphoproteins of approximately 74-kDa proteins, which were labeled during kinase assay. These proteins were phosphorylated predominantly on serine and their phosphopeptide maps differed from that of p150c-abl. The p74 phosphoproteins are not found in association with p150c-Abl in germ cells at earlier developmental stages, or in other tissues or cell lines examined (Ponzetto et al, 1989). Thus mouse c-abl protein appears to be a heterogenous protein, which needs further characterization.

The *c-abl* gene encodes a protein tyrosine kinase and is transcribed from at least two promoters giving rise to transcripts of two size classes of approximately 5 and 6-kb in length. These mRNAs only differ in their most 5' exon and encode proteins of similar size but with different N-termini. An additional abundant c-abl mRNA of 4-kb is present In the mouse testis. This mRNA was shown to express in the haploid male germ cells of the adult mouse.

The 4-kb c-abl mRNA abundant in elongating spermatids and also in round spermatids, arises from alternative polyadenylatioin of an RNA transcribed from the same promoter as the 5-kb mRNA. The site of polyadenylation is unusual in this shorter-transcript as it is not preceded by the highly conserved hexanucleotide AAUAAA. The use of this polyadenylation site removes 1.2-kb of 3' sequences present in the somatic *c-abl* mRNAs, but does not affect the main open reading frame of the transcript. *C-abl* transcripts are also detected in oocytes (Meijer et al, 1987, Iwaoki et al, 1993); 4kb *c-abl* mRNA transcripts are also abundant in pachytene spermatocytes. The c-Abl interacts directly with meiotic chromosomes. In concert with a requirement for c-Abl at the pachytene stage, Khabanda et al (1998) showed that, in contrast to wild type mice, testes from Abl-/- mice exhibit defects in spermatogenesis. These findings demonstrated that c-Abl plays a functional role in meiosis.

The c-Abl mAb, against the protein tyrosine kinase domain reacted specifically with the acrosomal region of capacitated and non-capacitated human sperm. The c-Abl mAb predominantly recognized two protein bands of 145-kDa and 95-kDa in sperm and testes. The 95-kDa protein band reacted stronger than the 145-kDa band and was the only band detected in solubilized sperm preparation. The 95-kDa protein was autophosphorylated at the tyrosine residues, which was inhibited in the presence of c-Abl mAb. The tyrosine phosphorylation of sperm proteins, especially of 95-kDa protein, has been shown to have a vital role in sperm capacitation/acrosomal exocytosis and binding to zona pellucida of oocytes. Thus the c-Abl or c-Abl like proteins are present in mature human sperm cells that are tyrosine auto-phosphorylated (Naz, 1998). The c-Abl tyrosine-kinase is activated by ionizing-radiation and other DNA-damaging agents. The DNA-dependent protein kinase (DNA-PK) and the ataxia telangiectasia mutated (ATM) gene product, effectors in the DNA damage response contribute to the induction of c-Abl activity.

C-abl and Apoptosis: p53 plays a central role in the induction of apoptosis of spermatogonia in response to ionizing radiation. In p53 (-/-) testes, however, spermatogonial apoptosis still can be induced by ionizing radiation, so p53 independent apoptotic pathways must exist in spermatogonia. Hamer et al (2001) showed that p53 homologues p63 and p73 are present in testis and p73, not p63 is localized in spermatogonia. In the testis, ionizing radiation elevates cytoplasmic c-Abl that in turn interacts with p73. Although less efficient than the p53 route, this may represent an additional, cytoplasmic, apoptotic pathway.

17.11. Bcl-2 FAMILY

Bcl-2: It is well established that the *Bcl-2* proto-oncogene suppresses apoptosis induced by a multitude of stimuli in variety of cell types. The*Bcl-2* was shown to extend the survival of growth factor dependent hematopoietic progenitor cells after factor withdrawal. In subsequent studies it was shown that *Bcl-2* promotes survival of such factor dependent cells by suppressing apoptosis. However, high level of Bcl-2 expression induces apoptosis via the caspase cascade. Mutational analysis of Bcl-2 indicated that its pro-apoptotic activity is separable from its anti-apoptosis activity. In addition to the anit-apoptosis activity, recent studies have revealed that Bcl-2 exhibits other novel activities. For example, Bcl-2 has been shown to inhibit cell proliferation, which is a function separable from anti-apoptosis activity. Among the Bcl-2 homology (BH) domains previously recognized in Bcl-2 family proteins, the BH3 domain is found in both Bcl-G(L) and Bcl-G(S), but only the longer Bcl-G(L) protein possesses a BH2 domain. The Bcl-G(L) mRNA is expressed widely in adult human tissues, whereas Bcl-G(S) mRNA was found only in testis (Guo et al., 2001). The human *Bcl-2* gene is characterized by unusually long 5'- and 3'-untranslated regions (Uhlmann et al 1998). Deletion of the sequences flanking the *Bcl-2* ORF dramatically increases the level of protein expression. Both 5'- and 3'-flanking sequences contribute to the negative modulation of protein expression from the *Bcl-2* open reading frame. It appears that these sequences exert the negative regulatory effect in an orientation dependent manner.

Other Members of Bcl Family : In recent years members of Bcl-2 family including Bcl-2, Bcl-xL, Bcl and Bax have been involved in death/survival of spermatogonia, but the expression of Bcl-2 in testis has been a subject of controversy (Furuchi et al, 1996). It appears that inhibition of apoptois of spermatogonia and growth of spermatogonial stem cells by Bcl-2 is intrinsic property in a cell-intrinsic manner. It is suggested that failure of differentiation of the accumulated spermatogonia in the transgenic testes is not due to the abnormality of the Bcl-2 misexpressing spermatogonia, but may be caused by extrinsic problems including improper interaction of spermatogonia with supporting cells (Sugiyama et al, 2001). Deficiencies in Bcl-2 affect the radiation response of different cell populations in small but different ways, which may be due to variation between cells in their innate capacity for apoptosis, their dependence on different members of the Bcl-2 family gene and their cell-cycle status and p53 expression (Hoyes et al, 2000). Furuchi et al (1996) investigated the effect of misexppressing Bcl-2 in spermatogonia in transgenic mice using the human Bcl-2 cDNA under the control of the human polypeptide chain elongation factor 1α (EF-lα) promoter. It seems that spermatogonial apoptosis is part of the normal program of mammalian spermatogenesis and is regulated by a pathway affected by Bcl-2.

BAX: The BAX, a heterodimeric partner of Bcl-2, counters Bcl-2 and promotes apoptosis in gain of function experiments. A *Bax* knockout mouse proved viable but displayed lineange-specific aberrations in cell death. The BAX deficient male mice are sterile due to accumulation of atypical premeiotic germ cells but no mature haploid sperm cells (Knudson et al, 1995). The BAX has been found as Bcl-2 associated molecule that antagonizes Bcl-2 and accelerates apoptosis. Androgen withdrawal, in rat testis after EDS treatment does not lead to significant changes in the levels of clusterin, Bcl-xl, Bak, and Bad. However, the expression of Bcl-2 and Bax was upregulated at 8 days after EDS administration. The induction of Bax at this time suggests that it may play a role in germ cell apoptosis following androgen withdrawal. The concomitant elevation in Bcl-2 expression may represent a survival mechanism for remaining germ cells. There was also a decline in the expression of Fas-L and Fas-R in the pachytene spermatocytes and spermatids; Fas-R was also present in Sertoli cells, although Fas-L staining was minimal (Woolveridge et al, 1999).

To determine whether BAX or BAK activity is required for Sertoli cell death in *Bclw* mutant animals, Ross et al, (2001) analyzed survival of Sertoli cells in *Bclw/Bax* and *Bclw/Bak* double homozygous mutant mice. While mutation of *Bak* had no effect, ablation of *Bax* suppressed the loss of Sertoli cells in *Bclw* mutants. Thus, Bclw mediates survival of postmitotic Sertoli cells in the mouse by suppressing death-promoting activity of BAX. Because Bcl-2 functions as a death repressor.

Bcl-w: Bcl-w is a death-protecting member of the Bcl-2 family of apoptosis regulating proteins. Bcl-w is expressed in both germ and Sertoli cells of testis. Mice that are mutant for Bcl-w display progressive and nearly complete testicular degeneration. In Bcl-w deficient mice, a complete block occurs at the development of elongated spermatids during first wave of spermatogenesis. The Bcl-w RNA expression takes place in Sertoli cells of mouse testis.

Consequently it is suggested that the diploid germ cell death may be an indirect effect of defective Sertoli cell function. Since Bcl-w is not expressed in spermatids, the loss of this cell type most likely results from defective Sertoli cell function (Russel et al, 2001; Ross et al., 2001).

Boo: The DIVA or Boo, a new member of the Bcl-2 family, is apparently expressed only in reproductive tissues; its role, however, is controversial. Inohara et al (1998 c/r Song et al., 1999) observed this protein in spermatids, where it performs a pro-apoptotic role, while Song et al, (1999) suggested that it is in fact anti-apoptotic, and observed transcripts in epididymis but not testis. The expression of Boo was highly restricted to the epididymis implicating it in the control of sperm maturation. Boo contains the conserved BH1 and BH2 domains, but lacks the BH3 motif. Like Bcl-2, Boo possesses a hydrophobic–terminus and localizes to intracellular membranes. The N-terminal region of Boo has strong homology to the BH4 domain found to be important for the function of some anti-apoptotic Bcl-2 homologues. Boo inhibits apoptosis, homodimerizes or heterodimerizes with some death promoting and suppressing Bcl-2 family members. More importantly, Boo interacts with Apaf-1 and forms a multimeric protein complex with Apaf-1 and caspase-9. The Bak and Bik, two pro-apoptotic homologues disrupt the association of Boo and Apaf-1. Furthermore, Boo binds to three distinct regions of Apaf-1. Like Ced-9, the mammalian homologues Boo and Bcl-x_L interact with the human counterpart of Ced-4, Apaf-1, and thereby regulate apoptosis (Song et al., 1999)

17.12. OTHER PROTOONCOGENE PRODUCTS IN TESTIS

c-Met: The c-met receptor is a p190MET tyrosine kinase proto-oncoprotein that through its bindings to its ligand, hepatocyte growth factor (HGF), induces mitogenic, motogenic, and morphogenic activities in a variety of cell types. A specific band of 195-kDa corresponding to the intact c-met receptor is present in the acrosomal region of the human sperm. The c-met receptor is tyrosine phosphorylated/autophosphorylated during capacitation. Incubation of human sperm hepatocyte growth factor of mAb2 to c-met receptor enhances the degree of tyrosine phosphorylation/autophosphorylation of the c-met receptor several fold (Herness and Naz, 1999).

Raf proteins: The p69 A-raf is ubiquitously expressed, but levels of expression vary among different tissues. The proto-oncogene *c-raf-1* and the related genes *A-raf* and *B-raf* encode serine/threonine protein kinases. All three *raf* family genes are expressed in mouse testis. The c-raf-1 mRNA is ubiquitously expressed in both somatic and germ cells as a 3.1-kb transcript; the level of c-raf-1 is developmentally regulated in germ cells. Though *A-raf* is expressed predominantly in the somatic cells as two transcripts of 2.6 and 4.3-kb, its expression in Leydig cells appears to be elevated in testes undergoing spermatogenesis. In contrast, *B-raf* is expressed as two major transcripts of 2.6 and 4.0kb. While 4.0-kb transcript first appears at low levels in pachytene spermatocytes with more abundance of 2.6-kb transcript restricted to post-meiotic spermatids (Wadewitz et al, 1993). The *A-raf and c-raf-1* are also expressed in mouse epididymis. In epididymis *c-raf-1* mRNA was expressed as a 3.1-kb transcript at uniform levels throughout the length of the epididymis and in all types of epididymal epithelial cells. However, neither the germ cell-specific nor the somatic transcripts of B-raf are present in any region of the epididymis (Winer et al, 1993).

Ret finger protein (rfp): The human ret finger protein (Rfp) has a tripartite motif, consisting of two zinc fingers (the RING finger and the B-box) and a coiled-coil domain, and belongs to the B-box zinc finger protein family. The Rfp becomes oncogenic when its tripartite motif is recombined with the tyrosine kinase domain from the *c-ret* proto-oncogene (see Chapter 16).

ErbA-T: The *c-erbA* gene belongs to a multigene family that encodes transcriptional regulatory proteins including the v-erbA oncogene product, steroid receptors, and the vitamin D3 receptor. One clone *erbA-T-1* was found to encode a 490-amino acid protein (ErbA-T). The ErbA-T polypeptide shows high homology with the proteins encoded by both the chicken *c-erbA* and the human *c-erbA*-β genes, but most closely related to the chicken gene. The chicken c-*erbA* and the human *c-erbA*-β genes encode high affinity receptors for thyroid hormone, and that the erbA-T protein binds specifically to 3,5,3'-triiodo-L-thyronine (Benbrook and Pfahl, 1987).

Elk-1 and Elk-2 (ets family): The myb-ets-containing acute leukemia virus, E26, causes a mixed erythroid and myeloid leukemia in chicks. Genes (*ets-1, ets-2, and erg*) with variable relatedness to the *v-ets* oncogene of the E26 virus have been characterized. Two new members (*elk-1 and elk-2*) of the ets oncogene superfamily have now been identified. The *elk-1* cDNA clone encodes a 428-residue protein, whose predicted amino acid sequences showed 82% similarity to the 3'-region of v-ets. The elk or related sequences appeared to be transciptionally active in testis and lung. In human genome, *elk-1* and *elk-2* map to chromosome regions Xp11.2 and 14q32.3, respectively. This suggests the possibility that rearrangements of elk loci may be involved in pathogenesis of certain tumors (Rao et al., 1989).

CDC37 : The *cdc37* encodes a 50-kDa protein, CDC37 that targets intrinsically unstable oncoprotein kinases including Cdk4, Raf-1, and U-src to the molecular chaperone Hsp-90, an interaction that is thought to be important for the establishment of signaling pathways. The CDC37 co-expresses with cyclin D1, a finding consistent with a positive role in cell proliferation. The CDC37 is found to collaborate with MMTV-c-myc in the transformation of multiple tissues, including mammary and salivary glands in females and in testis in males. The CDC37 collaborates with cyclin D1 to transform the female mammary gland. It can function as an oncogene in mice and suggests that the establishment of protein kinase pathways mediated by Cdc37-Hsp90 can be a rate-limiting event in epithelial cell transformation (Stepanova et al, 2000).

MCF-2 (DBL): The MCF-2 (DBL) proto-oncogene is a prototype guanine nucleotide exchange factor (GEF) that modulates Rho GTPases such as Rho, Rac, and Cdc42. The complete mouse *MCF-2* cDNA has been studied. Mouse and human *MCF-2* (DBL) cDNAs showed 75.08% identity where as amino acid sequences showed 74.52% identity. Mouse *MCF-2* mRNA is expressed in brain, kidney, intestine, and testis. The brain-specific transcript is an alternatively spliced derivative that omits the 48bp exon 11 (Komai et al., 2003).

17.13. REFERENCES

Albanesi C, Geremia R, Giorgio M, Dolci S, Sette C, Rossi P. A cell-and developmental stage specific promoter drives the expression of a truncated c-kit protein during mouse spermatid elongaton. Development 1996; 12: 1291-02.

Alcivar AA, Hake LE, Kwon YK, Hecht NB. JunD mRNA expression differs from c-jun and junB mRNA expression during male germinal cell differentiation. Mol Reprod Dev 1991; 30: 187-93.

Ali S, Ravindranath N, Jia MC, Musto N, Tsujimura T, Kitamura Y, Dym M. Expression of multiple c-kit receptor

messenger ribonucleic acid transcripts during postnatal development of the rat testis. Biochem Biophys Res Comm 1996; 218: 104-12.
Baytel D, Shalom S, Madgar I, Weissenberg R, Don J. The human Pim-2 proto-oncogene and its testicular expression. Biochim Biophys Acta 1998; 1442: 274-85.
Benbrook D, Pfahl M. A novel thyroid hormone receptor encoded by a cDNA clone from a human testis library. Science 1987; 238(4828): 788-91.
Blume-Jensen P, Jiang G, Hyman R, Lee KF, O'Gorman S, Hunter T. Kit/stem cell factor receptor-induced activation of phosphatidylinositol 3'-kinase is essential for male fertility. Nat Genet 2000; 24: 157-62.
Brown AM, Papkoff J, Fung YK, Shackleford GM, Varmus HE. Identification of protein product encoded by the proto-oncogene int-1. Mol Cell Biol 1987; 7: 3971-7.
Buckley AR, Buckley DJ, Leff MA, Hoover DS, Magausoa. Rapid induction of *pim-1* expression by prolactin and interleukin-2 in rat Nb2 lymphoma cells. Endocrinology 1995; 136: 5252-59.
Chandrasekhar K, Raman R. De novo methylation of the proto-oncogene, c-fos, during development occurs step-wise and directionally in the laboratory mouse. Mol Reprod Dev 1997; 48: 421-32.
Chen J, Tong J, Tanaka-Sukegawa I, Wang LH. Cloning and functional characterization of the chicken c-ros promoter. Cell Growth Differ 1995; 6: 1523-30.
Cohen DR, Sinclair AH, McGovern JD. SRY protein enchances transcription of Fos-related antigen 1 promoter constructs. Proc Natl Acad Sci USA 1994; 10: 4372-6.
Cohen DR, Vandermark SE, McGovern JD, Bradley MP. Transcriptional regulation in the testis: a role for transcription factor AP-1 complexes at various stages of spermatogenesis. Oncogene 1993; 8: 443-55.
Cooper TG, Wagenfeld A, Cornwall GA, Hsia N, Chu ST et al., Gene and protein expression in the epididymis of infertile c-ros receptor tyrosine kinase-deficient mice. Biol Reprod. 2003;69:1750-62.
Dickson C, Acland P, Smith R, Dixon M, Deed R et al.etc. Characterization of int-2 a member of the fibroblast growth factor family. J Cell Sci (Suppl) 1990; 13: 87-96.
Domen J, van der Lugt NM, Laird PW, Saris CJ, Berns A. Analysis of Pim-1 function in mutant mice. Leukemia 1993; 7: 108-12.
Dym M, Jia MC, Diramai G, Price JM, Rabin SJ, Mocchetti I, Ravindranath N. Expression of c-kit receptor and its autophosphorylation in immature rat type A spermatogonia. Biol Reprod 1995; 52: 8-19.
Feng H, Sandlow JI, Sandra A. The c-kit receptor and its possible signaling transduction pathway in mouse spermatozoa. Mol Reprod Dev 1998; 49: 317-26.
Feng LX, Ravindranath N, Dym M. Stem cell factor/c-kit up-regulates cyclin D3 and promotes cell cycle progression via the phosphoinositide 3-kinase/p70 S6 kinase pathway in spermatogonia. J Biol Chem 2000; 275: 25572-6.
Furuchi T, Masuko K, Nishimune Y, Obinata M, Matsui Y. Inhibition of testicular germ cell apoptosis and differentiation in mice misexpressing Bcl-2 in spermatogonia. Development 1996; 122: 1703-9.
Goldman DS, Kiessling AA, Millette CF, Cooper GM. Expression of c-mos RNA in germ cells of male and female mice. Proc Natl Acad Sci 1987; 84: 4509-13.
Gregory MA, Xiao Q, Cornwall GA, Lutterbach B, Hann SR. B-Myc is preferentially expressed in hormonally controlled tissues and inhibits cellular proliferation. Oncogene 2000; 19: 4886-95.
Guo B, Godzik A, Reed JC. Bcl_G, a novel pro-apoptotic member of the Bcl-2 family. J Biol Chem 2001; 276: 2780-5.
Hall SH, Joseph DR, French FS, Conti M. Follicle stimulating hormone induces transient expression of the protooncogene c-fos in primary Sertoli cell cultures. Mol Endocrinol 1988; 2: 55-61.
Hamer G, Gademan IS, Kal HB, de Rooij DG. Role for c-Abal and p73 in the radiation response of male germ cells. Oncogene 2001; 20: 4298-304.
Hamil KG, Conti M, Shimasaki S, Hall SH. Follicle-stimulating hormone regulation of AP-1 : inhibition of c-jun and stimulation of jun-B gene transcription in the rat Sertoli cell. Mol Cell Endocrinol 1994; 99: 269-77.
Heikinheimo O, Dong KW, Lanzendorf SE, Gordon K, Toner JP, Gibbons WE. Primate reproductive organs reveal a novel pattern of proto-oncogene c-mos and transcription factor Oct-3 mRNA expression. Mol Reprod Dev 1995; 42: 397-06.
Herness EA, Naz RK. Presence and tyrosine phosphorylation of c-met receptor in human sperm. J Androl 1999; 20: 640-7.
Higgy NA, Tarnasky HA, Valarche I, Nepveu A, van der Hoorn FA. Cux/CDP homeodomain protein binds to an enhancer in the rat c-mos locus and represses its activity. Biochim Biophys Acta 1997; 1351: 313-24.
Higgy NA, Zackson SL, van der Hoorn FA. Cell interactions in testis development: overexpression of c-mos in spermatocytes leads to increased germ cell proliferation. Dev Genet 1995; 16: 190-200.
Hoyes KP, Cai WB, Potten CS, Hendry JH. Effect of bcl-2 deficiency on the radiation response of clonogenic cells in small and large intestine bone marrow and testis. Int J Radiat Biol 2000; 76: 1435-42.
Isozaki K, Tsujimura T, Nomura S, Morii E, Koshmizu U, Nishimune Y, Kitamura Y. Cell type-specific deficiency of c-kit gene expression in mutant mice of mi/mi genotype. Am J Pathol 1994; 145: 827-36.
Iwaoki Y, Matsuda H, Mutter GL, Watrin F, Wolgemuth DJ. Differential expression of the proto-oncogenes c-abl and c-mos in developing mouse germ cells. Exp Cell Res 1993; 206: 212-9.
Jakobovits A, Shackleford GM, Vamus HE, Martin GR. Two protooncogenes implicated in mammary carcinogenesis; int1 and int2 are independently regulated during mouse development. Proc Natl Acad Sci (USA) 1986; 83: 7806-10.

Kanduri C, Raman R. Tissue-specific characterization of DNA methylation in the gonad-specific proto-oncogene, c-mos, in the male laboratory mouse. Int J Dev Biol 1999; 43: 91-4.

Keilhack H, Muller M, Bohmer SA, Frank C, Weidner KM, Birchmeier W, Ligensa T. Negative regulation of Ros receptor tyrosine kinase signaling. An epithelial function of the SH2 domain protein tyrosine phosphatase SHP-1. J Cell Biol 2001; 22: 325-34.

Khabanda S, Pandey P, Morris PL, Whang Y, Xu Y, Sawant S, Zhu LJ, Kumar N, Yuan ZM, etc. Functional role for the c-Abl tyrosine kinase in meiosis I. Oncogene 1998; 16: 1773-7.

Kissel H, Timokhina I, Hardy MP, Rothschild G, Tajima Y, Soares V, et al. Point mutation in kit receptor tyrosine kinase reveals essential roles for kit signaling in spermatogenesis and oogenesis without affecting other kit responses. The EMBO Journal 2000; 15: 1312-26.

Knudson CM, Tung KS, Tourtellotte WG, Brown GA, Korsmeyer SJ. Bax-deficient mice with lymphoid hyperplasia and male germ cell death. Science 1995; 270 (5233): 96-9.

Kodaira K, Takahashi R, Hirabayashi M, Suzuki T, Obinata M, Ueda M. Overexpression of c-myc induces apoptosis at the prophase of meiosis of rat primary spermatocytes. Mol Reprod Dev 1996; 45; 403-10.

Koikeda S, Tamaru M, Goto M, Mori N, Takada S, Obinata M, Masamune Y, Nakanishi Y. Analysis of the expression of two phosphoglycerate kinase genes in a mouse cultured cell line during activation and inactivation of the c-myc gene. Chem Pharm Bull 1989; 37: 1103-5.

Koji T, Izumi S, Tanno M, Moriuchi T, Nakane PK. Localization in situ of c-myc mRNA and c-myc protein in adult mouse testis. Histochem 1988; 20: 551-7.

Komai K, Mukae-Sakairi N, Kitagawa M, Shiozawa S. Characterization of novel splicing variants of the mouse MCF-2 (DBL) proto-oncogene. Biochem Biophys Res Commun 2003; 309: 906-9.

Koshimizu U,Tsujimura T, Isozaki K, Nomura S, Furitsu T, Kanakura Y, Kitamura Y, Nishimune Y. W_n mutation of c-kit receptor affect its post-translational processing and extracellular expression. Oncogene 1994; 9: 157-62.

Kumar G, Patel D, Naz RK. C-MYC mRNA is present in human sperm cells. Cell Mol Biol Res 1993; 39: 111-7.

Kuchino Y, Asai A, Kitanaka C. Myc mediated apoptosis. Prog Mol Subcell Biol 1996; 16: 104-29.

Langdon WY, Hy and CD, Grumont RJ, Morse HC 3rd. The c-able proto-oncogene is preferentially expressed in thymus and testis tissue and encodes a nuclear protein. J Virol 1989; 63: 5420-4.

Latham KE, Litvin J, Orth JM, Patel B, Mettus R, Reddy EP. Temporal patterns of A-myb gene expression during testis development. Oncogene 1996; 13: 1161-8.

Li Q, Kondoh G, Inafuku S, Nishimune Y, Hakura A. Abrogation of c-kit/Steel factor-dependent tumorigenesis by kinase defective mutants of the c-kit receptor: c-kit kinase defective mutants as candidate tools for cancer gene therapy. Cancer Res 1996; 56: 4343-6.

Lim K, Hwang BD. Follicle-stimulating horomone transiently induces expression of protoonocogene c-myc in primary Sertoli cell cultures of early pubertal and prepubertal rat. Mol Cell Endocrinol 1995; 111: 51-6.

Lim K, Yoo JH, Kim KY, Kweon GR, Kawak ST, Hwang BD. Testosterone regulation of proto-oncogene c-myc expression in primary Sertoli cell cultures from prepubertal rats. J Androl 1994; 15: 543-50.

Lin T, Blaisdell J, Barbour KW, Thompson EA. Transient activation of c-myc Protooncogene expression in Leydig cells by human chorionic gonadotropin. Biochem Biophys Res Commun 1988; 157: 121-6.

Marcu KB, Bossone SA, Patel AJ. Myc function and regulation Ann Rev Biochem 1992; 61: 809-60.

Matsubara N, Takahashi Y, Nishina YM, Mukouyama Y, et al. A receptor tyrosine kinase, Sky, and its ligand Gas 6 are expressed in gonads and support primordial germ cell growth or survival in culture. Dev Biol 1996; 180: 499-510.

Mauduit C, Chatelain G, Magre S, Brun G, Benahmed M, Michel D. Regulation by pH of the alternative splicing of the stem cell factor pre-mRNA in the testis. J Biol Chem 1999a; 274: 770-5.

Mauduit C, Hamamah S, Benahmed M. Stem cell factor/c-kit sytem in spermatogenesis. Hum Reprod Update1999b; 5: 535-45.

Mautner J, Joos S, Werner T, Elic D, Bronkamm GW, Polack A. Identification of two enhancer elements down-stream of the human c-myc gene. Nucleic Acids Res 1995; 23: 72-80.

Meijer D, Hermans A, von Lindern M, van Agthoven T et al. Molecular characterization of the testis specific c-abl mRNA in mouse. EMBO J 1987: 6: 4041-8.

Mettus RV, Litivin J, Wali A, Toscani A, Latham K, Hatton K, Reddy EP. Murine A-myb: evidence for differential splicing and tissue-specific expression. Oncogene 1994; 9: 3077-86.

Nagao Y. Expression of c-mos protein in cultured rat spermatogenic cells and evidence that rat testicular 43 kDa c-mos protein has an elongated amino acid sequence at the N-terminus. Biochim Biophys Acta 1995; 1245: 130-43.

Naz RK. c-Abl oncoprotein is expressed and tyrosine phosphorylated in human sperm cell. Mol Reprod Dev 1998; 51: 210-7.

Nishina Y, Kobarai Y, Sumi T, Kosaka M, Nishikawa S, Nishimune Y. Expression of c-kit Protooncogene is stimulated buy cAMP in differentiated F9 mouse teratocarcinoma cells. Exp Cell Res 1992; 198: 352-6.

Ohta H, Yomogida K, Dohmae K, Nisdhimune Y. Regulation of proliferation and differentiation in spermatogonial stem cells : the role of c-kit and its ligand SCF. Development 2000; 127: 2125-31.

Oosterhuis JH, van der Hoorn FA. Testis-specific TTF-D binds to single-stranded DNA in the c-mos and Odf1 promoters and activates Odf1. J Biol Chem 1999; 274: 11708-12.

Packer AI, Besmer P, Bachvarova RF. Kit ligand mediates survival of typa A spermatogonia and dividing spermatocytes

in postnatal mouse testes. Mol Reprod Dev 1995; 42: 303-10.
Page KC, Heitzman DA, Chermin MI. Stimulation of c-jun and c-myb in rat Sertoli cells following exposure to retinoids. Biochem Biophys Res Commun 1996; 222: 595-602.
Ponzetto C, Wadewitz AG, Pendergast AM, Witte ON, Wolgemuth DJ. P150 c-abl is detected in mouse male germ cells by an in vitro kinase assay and is associated with stage-specific phosphoproteins in haploid cells. Oncogene 1989; 4: 685-90.
Propst F, Rosenberg MP, Iyer A, Kaul K, Vande Woude GF. C-mos proto-oncogene RNA transcripts in mouse tissues: structural features, developmental regulation, and localization in specific cell types. Mol Cell Biol 1987; 7: 1629-37.
Propst F, Vande Woude GF. Expression of c-mos proto-oncogene transcripts in mouse tissues. Nature 1985; 315(6019): 516-8.
Rao VN, Huebner K, Isobe M, ar-Rushdi A, Croce CM, Reddy ES. *elk*, tissue-specific ets-related genes on chromosomes X and 14 near trnslocation breakpoints. Science 1989; 244(4900): 66-70.
Reeves R, Spies GA, Kiefer M, Barr PJ, Power M. Primary structure of the putative human oncogene, *pim-1*. Gene 1990; 90: 303-07.
Robinson LL, Gaskell TL, Saunders PT, Anderson RA. Germ cell specific expression of c-kit in the human fetal gonad. Mol Hum Reprod 2001; 7: 845-52.
Rosenberg MP, Aversa CR, Wallace R, Propst F. Expression of the v-Mos oncogene in male meiotic germ cells of transgenic mice results in metaphase arrest. Cell Growth differ 1995; 6: 325-36.
Ross AJ, Amy SP, Mahar PL, Lindsten T, Knudson CM, Thompson CB, Korsemeyyer SJ, MacGregor GR. Bcl_w mediates survival of postmitotic Sertoli cells by regulatiin BAX activity. Dev Biol 2001; 239: 295-308.
Rossi P, Marziali G, Albanesi C, Charlesworth A, Geremia R, Sorrentino V. A novel c-kit transcript, potential encoding a truncated receptor, originates within a kit gene intron in mouse spermatids. Dev Biol 1992: 152: 203-7.
Rossi P, Sette C, Dolci S, Geremia R. Role of c-kit mammalian spermatogenesis. J Endocrinol Invest 2000; 23: 609-15.
Russel LD, Warren J, Debeljuk L, Richardson LL, et al. Spermatogenesis in Bcl_w deficient mice Biol Reprod 2001; 65: 318-32.
Saksela K, Makela TP, Alitalo K. Oncogene expression in small-cell lung cancer cell lines and a testicular germ-cell tumor: activation of the N-myc gene and decreased RB mRNA. Int J Cancer 1989; 44: 182-5.
Sandlow JI, Feng HL, Sandra A. Localization and expression of the c-kit receptor protein in human and rodent testis and sperm. Urology 1997; 49: 494-500.
Sandlow JI, Feng HL, Zheng LJ, Sandra A. Migration and ultrastructural localization of the c-kit receptor protein in spermatogenic cells and spermatozoa of the mouse. J Urol 1999; 161: 1676-80.
Schultz R, Kononen J, Pelto-Huikko M. Induction of immediate early mRNAs and proteins by hCG in interstitial cells of rat testis. Endocrinology 1995; 144: 417-24.
Serve H, Yee NS, Stella G, Sepp-Lorenzino L, Tan JC, Besmer P. Differential roles of PI3-kinase and kit tyrosine 821 in kit receptor mediated proliferation, suvival and cell adhesion in mast cells. The EMBO Journal 1995; 14: 473-83.
Sette C, Bevilacqua A, Geremia R, Rossi P. Involvement of phospholipase Cgamma1 in mouse egg activation induced by a truncated form of the C-kit tyrosine kinase present in spermatozoa. J Cell Biol 1998; 24: 1063-74.
Shackleford GM, Varmus HE. Expression of the proto-oncogene int-1 is restricted to postmeiotic male germ cells and the neural tube of mid-gestational embryos. Cell 1987; 3: 89-95.
Sitzmann J, Noben-Trauth K, Kamano H, Klempnauer KH. Expression of B-Myb during mouse embryogenesis. Oncogene 1996; 12: 1889-94.
Sleeman JP. Xenopus A-myb is expressed during early spermatogenesis. Oncogene 1993; 8: 1931-41.
Song Q, Kuang Y, Dixit VM, Vincenz C. Boo, a novel negative regulator of cell death, interacts with Apaf-1. The EMBO Journal 1999; 18: 167-78.
Sonnenberg-Riethmacher E, Walter B, Riethmacher D, Godecke S, Birchmeier C. The c-ros tyrosine kinase receptor controls regionalization and differentiation of epithelial cells in the epididymis. Genes Dev 1996; 10: 1184-93.
Sorrentino V, Giorgi M, Geremia R, Besmer P, Rossi P. Expression of the c-kit proto-oncogene in the murine male germ cells. Oncogene 1991; 6: 149-51.
Sorrentino V, McKinney MD, Giorgi M, Geremia R, Fleissner E. Expression of cellular protooncogenes in the mouse male germ line: a distinctive 2.4 kilobase *pim-1* transcript is expressed in haploid postmeiotic cells. Proc Natl Acad Sci USA 1988; 85: 2191-95.
Steel LF, Telly DL, Leonard J, Rice BA, Monks B, Sawicki JA. Element in the murine c-mos messenger RNA 5'-untranslated region repress translation of downstream coding sequences. Cell Growth Differ 1996; 7: 1415-24.
Stepanova L, Finegold M, DeMayo F, Schmidt Ev, Harper JW. The oncoprotein kinase chaperone CDC37 functions as an oncogene in mice and collaborates with both c-myc and cyclin D1 in transformation of multiple tissues. Mol Cell Biol 2000; 20: 4462-73.
Stewart TA, Bellve AR, Leder P. Transcription and promoter usage of the myc gene in normal somatic and spermatogenic cells. Science 1984; 9: 707-10.
Strohmeyer T, Reese D, Press M, Ackerman R, Hartmann M,Slamon D. Expression of the c-kit proto-oncogene and its ligand stem cell factor (SCF) in normal and malignant human testicular tissue. J Urol 1995; 153: 511-15.
Sugiyama N, Obinata M, and Matsui Y. Bcl-2 inhibits apoptosis of spermatogonia and growth of spermatogonial stem

cells in a cell-intrinsic manner. Mol Reprod and Dev 2001; 58: 30-8.
Tajima Y, Onoue H, Kitamura Y, Nishimune Y. Biologically active kit ligand growth factor is product by mouse Sertoli cells and is defective in Sld mutant mice. Development 1991;113: 1031-5.
Tajima Y, Sawada K, Morimoto T, Nishimune Y. Switching of mouse spermatogonial proliferation from the c-kit receptor independent type to the receptor-dependent type during differentiation. J Reprod Fertil 1994; 102: 117-22.
Takahashi T, Nakagoshi H, Sarai A, Nomura N, Yamamoto T, Ishii S. Human A-myb gene encodes a transcriptional activator containing the negative regulatory domains. FEBS Lett 1995; 358: 89-96.
Tian Q, Frierson HF Jr, Krystal GW, Moskaluk CA. Activating c-kit gene mutations in human germ cell tumors. Am J Pathol 1999; 154: 1643-7.
Timokhina J, Kissel H, Stella G, Besmer P. Kit signaling through PI-3 kinase and Src kinase pathways: an essential role for Rac 1 and JNK activation in mast cell proliferation. The EMBO Journal 1998; 17: 6250-62.
Toscani A, Mettus RV, Coupland R, Simpkins H, Litvin J, Orth J, Hatton KS, Reddy EP. Arrest of spermatogenesis and defective breast development in mice lacking A-myb. Nature 1997; 386(6626): 713-7.
Trauth K, Mutschler B, Jenkins NA, Gilbert DJ, Copeland NG, Klempnauer KH. Mouse A-myb encodes a trans-activator and is expressed in mitotically active cells of the developing central nervous sytem, adult testis and B lymphocytes. The EMBO Journal 1994; 13: 5994-05.
Uhlmann EJ, Subramanian T, Vater CA, Lutz R, Chinnadurai G. A potent cell death activity associated with transient high level expression of Bcl-2. J Biol Chem 1998;273:17926-32.
Van der Hoorn FA, Spiegel JE, Maylie – Pfenninger MF, Nordeen SK. A 43 kD c-mos protein is only expressed before meiosis during rat spermatogenesis. Oncogene 1991; 6: 929-32.
Van der Hoorn FA. C-mos upstream sequence exhibits species – specific enhancer activity and binds murine-specific nuclear proteins. J Mol Biol 1987; 193: 255-66.
Van der Hoorn FA. Identification of the testis c-mos promoter : specific activity in a seminiferous tubule-derived extract and binding of a testis-specific nuclear factor. Oncogene 1992; 7: 1093-9.
Vigodner M, Lewin LM, Shochat L, Golan R. Spermatogenesis in golden hamster: the role of c-kit. Mol Reprod Develop 2001; 60: 562-8.
Vosseller K, Stella G, Yee NS, Besmer P. c-kit receptor signaling through its phosphatidyl inositide 3' kinase binding site and protein kinase C: role in mast cell enhancement of degranulation, adhesion, and membrane ruffling. Mol Biol Cell 1997; 8: 909-12.
Wadewitz AG, Winer MA, Wolgemuth DJ. Developmental and cell lineage specificity of raf family gene expression in mouse testis. Oncogene 1993; 8: 1055-62.
Winer MA, Wadewitz AG, Wolgemuth DJ. Members of the raft gene family exhibit segment specific patterns of expression in mouse epididymis. Mol Reprod Dev 1993; 35: 16-23.
Wolfes H, Kogawa K, Millette CF, Cooper GM. Specific expression of nuclear proto-oncogene before entry into meiotic prophase of spermatogenesis. Science 1998; 254: 740-5.
Woolveridge I, de Boer Brouwer M, Taylor MF, Teeds KJ, Wu FC, Morris ID. Apoptosis in the rat spermatogenic epithelium following androgen withdrawl: changes in apoptosis-related genes. Biol Reprod 1999; 60: 461-70.
Xu W, Cooper GM. Identification of a candidate c-mos repressor that restricts transcription of germ cell-specific genes. Mol Cell Biol 1995; 15: 5369-75.
Yan W, Suominen J, Toppari J. Stem cell factor protects germ cells from apoptosis in vitro. J Cell Sci 2000; 113; 161-8.
Yeung CH, Sonnenberg-Riethmacher E, Cooper TG. Receptor tyrosine kinase C-ros knockout mice as a model for the study of epididymal regulation of sperm function. Reprod Fertil (Suppl) 1998; 53: 137-47.
Yeung CH, Wagenfeld A, Nieschlag E, Cooper TG. The cause of infertility of male c-ros tyrosine kinase receptor knockout mice. Biol Reprod 2000; 53: 612-8.
Yukitake H, Furusawa M, Taira T, Iguchi-Ariga SM, Ariga H. AAT-1, a novel testis-specific AMY-1-binding protein, forms a quaternary complex with AMY-1, A-kinase anchor protein 84, and a regulatory subunit of cAMP-dependent protein kinase and is phosphorylated by its kinase. J Biol Chem 2002a; 277: 45480-92.
Yukitake H, Furusawa M, Taira T, Iguchi-Ariga SM, Ariga H. AMAP-1, a novel testis-specific AMY-1-binding protein, is differentially expressed during the course of spermatogenesis. Biochim Biophys Acta 2002b; 1577: 126-32.
Zakut-Houri R, Hazum S, Givol D, Telerman A. The c-DNA sequence and gene analysis of human pim oncogene. Gene 1987; 54: 105-11.

Chapter 18

G PROTEINS AND ASSOCIATED SIGNAL TRANSDUCTION MOLECULES

18.1. CYCLIC AMP AND CA^{2+} MEDIATED SIGNAL TRANSDUCTION

Extracellular signals (effector molecules such as hormones, growth factors, cytokines and other signaling molecules and physical environmental factors), are recognized by receptors on the surface of the cytoplasmic membrane and transmitted through transducers to amplifiers, which synthesize many molecules of second messengers that reach final targets through many successive additional steps. The G protein linked receptors form the largest family of cell surface receptors, found in all eukaryotes. Despite the chemical and functional diversity of signal molecules that bind receptors, all G proteins linked receptors have similar structures. The activation of receptor results into activation of G proteins, which activate variety of effectors including adenylate cyclase (AC), phospholipases and phosphodiesterases, as well as potassium and calcium channels. The adenylate cyclase (AC) converts ATP into cyclic-AMP (c-AMP) leading to increased level of intracellular c-AMP, Ca^{2+} and other internal mediators. In most cases c-AMP activates protein kinase A (PKA), which phosphorylates several specific proteins or enzymes, including lipase for mobilization of fat and phosphorylase for breakdown of glycogen in liver. In addition to activation of several enzymes, c-AMP also causes an elevation in cytosolic Ca^{2+}, which is another second messenger. G proteins are key components in signal transduction systems, relaying information from activated membrane receptors to intracellular effector. Individual heterotrimeric G proteins are distinguished by their subunits, which are responsible for binding and hydrolyzing GTP and for regulating specific effectors. The α-subunits of G proteins can be ADP-ribosylated by specific bacterial toxins, leading to in vivo persistent activation of Gs, the uncoupling of Gi from inhibitory receptors and impeded transition of the GDP-G protein complex to its active state.

18.2. G PROTEINS

GTP-binding proteins are widely distributed in various tissues and among organisms, and the structures of individual proteins are highly conserved among distant organisms. Mechanism of reactions catalyzed by GTP-binding proteins is essentially analogous to that proposed for translational factors. The GTP- binding proteins or GTPases act as transducers, which regulate the opening and closing of the signaling pathways. GTP-binding proteins are broadly classified into two groups: (i) High molecular weight or heterotrimeric GTP-binding proteins (G proteins) and into (ii) a low molecular weight, monomeric GTP- binding proteins ("small Gs"). The low molecular-weight GTP-binding proteins include the products of *rab, rap, rho, rac, smg21,*

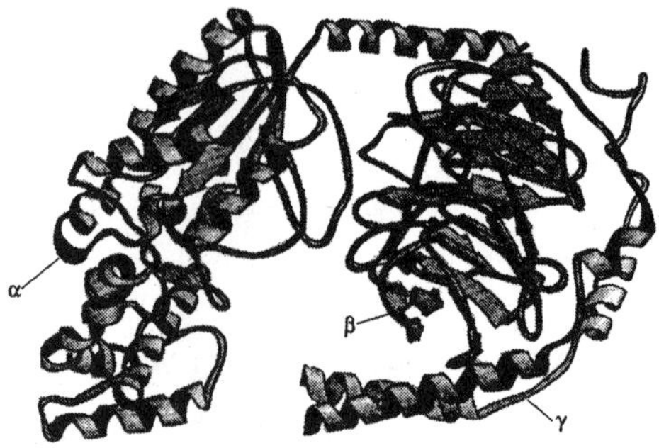

Fig18.1. The three dimensional structure of an inactive G protein, involved in visual transduction. The α subunit contains GTPase domain and binds to β subunit. The γ subunit binds to opposite side of β subunit. Adapted with permission from D.G. Lambright et al., Nature 379; 311-19: 1996. © Macmillan Magazines Ltd. http://www.nature.com/nature.

smg25, YPT, SEC4, ARF genes (small Gs). Metabolic enzymes that interact with GTP do not fall in signal category and they are not discussed here. All signal transducing GTP-binding proteins bind and hydrolyze GTP, a property, which is crucial to their function as a molecular switch for diverse cellular functions. The GTP-binding proteins undergo two alternate conformations depending on the phosphate potentials of the ligand. The GTP-bound form is an active conformation. In this form the protein can recognize and interact with its target molecules. On hydrolysis of bound-GTP to GDP and orthophosphate, the conformation as well as the reactivity of the protein is shifted to the GDP-bound form (an inactive form). An analogous mechanism is found in proteins undergoing phosphorylation and dephosphorylation reactions. Though, the mechanism of interconversion of GTP-bound to GDP-bound form with G binding proteins may vary, depending on the complexity of the reaction, the basic principle of the function of GTP-binding proteins as molecular switches is universal (Kaziro et al, 1991).

18. 2.1 Heterotrimeric G Proteins

Heterotrimeric G proteins consist of three subunits: α (39-52 kDa), β (35-36 kDa), and γ (7-10 kDa) (**Fig. 18.1**). The extent of amplification of G protein-mediated signals appears to be predetermined by the intrinsic GTPase activity of each G protein α subunit. The total number of cDNA species for mammalian Gα is 13. This includes nine genes for Gα [Gs1α, Gs2α, Gi1α, Gi 2α, Gi3α, $G_0α$, Gt1α, Gt2α, Gxα] and their additional isoforms (four genes for Gs1α and two genes for $G_0α$). The diversity for each is increasing (Kaziro et al, 1991). However, biochemical analysis has revealed a number of guanine nucleotide-binding regulatory proteins (G proteins) that mediate signal transduction in mammalian systems. Characterization of their cDNA uncovered a family of proteins with regions of highly conserved amino acid sequence. To examine the extent of diversity of the G protein family, sequences corresponding to six of the eight known G protein α subunits showed that the complexity of the G protein family is much greater than previously suspected (Strathemann et al, 1989). The GTP binding proteins reveal the presence of several regions of high degree of homology and similarities defined by Kaziro et al (1991):

(a) P region : with a consenses sequence GXXXX GKS/T. This region seems to be involved in hydrolytic process.
(b) E region : This region covering aminoacids 32-42 of Ras is effector region, the site which interacts with effector molecule.
(c) G'region : This is also involved in GTPase function in Ras proteins. The D-X X-G-Q is conserved in most GTP binding proteins.
(d) G region : It comprises NKXD as a conserved region which directly interacts with guanine ring of GTP.
(e) G" region is another region, which is closely situated to guanine ring of GTP of Ras protein. In X-ray structure Ras show the conserved motif, ETSAK.

$G_s\alpha$: Some G proteins transmit signal by regulating the production of c-AMP, which is synthesized by plasma membrane bound adenylyl cyclase. The cAMP is rapidly destroyed by one or more c-AMP phosphodiesterases that hydolyze c-AMP to 5'-AMP. All receptors that act through c-AMP are coupled to a stimulatory G protein (Gs). Gs is involved in hormonal stimulation of adenylate cyclase and opening of Ca^{2+} channels. Modification by cholera toxin of $G_s\alpha$ leads to ADP- ribosylation of Arg-201. The ribosylated form of α subunit of Gs protein does not hydrolyze GTP and leads to persistent activation of adenylate cyclase. It purifies as a mixture of 52 and 45-kDa protein and corresponds to cDNAs of Gs1α-1 and Gs1α-4 respectively. The Gsα expressed exclusively in olfactory sensory neurons is termed ($G_{olf}\alpha$) or Gs2α and also in testis, as against Gs1α expressed in all other tissues, and is therefore thought to be involved specifically in odorant signal transduction. The Gs2α gene codes a protein of 44.3 kDa and stimulates adenylate cyclase.

$G_i\alpha$: Another G protein (G_i), inhibits adenylyl cyclase, but mainly acts by regulating ion channels rather than by decreasing c-AMP content. The c-DNA cloning revealed $G_i\alpha$ to exist atleast in three forms ($G_i1\alpha$, $G_i2\alpha$, and $G_i3\alpha$). Pertussis toxin catalyzes ADP ribosylation of α subunit of Gi, preventing the subunit from interacting with the receptors. As a result, this α subunit retains its bound GDP and is unable to regulate its target proteins. The Pertussis toxin and cholera toxin are widely used to identify if a cell's response to a signal is mediated by Gs or by Gi.

G_o: G_o activates K channels as well as phospholipase C-β and inactivates Ca channels. Though the regulation of K and Ca channels is regulated by βγ complex, the action of phospholipase C-β is mediatd by α and βγ complex. Three types of **G_o**α ($G_o\alpha$–1, $G_o\alpha$–2 and $G_o\alpha$–3) have been isolated from mouse, rat and other animal tissues.

Gtα (transducin) : The cDNA have been isolated from rods and cones.

Gqα is relatively a new member. The **Gqα** proteins are thought to couple not only to adenylate cyclase but also to phospholipase C-β or K^+ channel.

The availability of complete amino acid sequences of nine G protein α- subunits allows comparison of their primary structures given elsewhere (Kaziro et al, 1991). The overall structure homology is striking. For example only one amino acid out of 394 amino acid of $G_s\alpha$ differs between human and rat. Sequence of $G_i\alpha$ from human and bovine is completely identical. Among other α subunits from different mammalian species identity is about 98%. The G protein α subunits have multiple domains, which are involved in GTP binding, GTP hydrolysis, conformation change by different guanine nucleotides, ßγ-subunits association, receptor recognition, and effector interaction and regulation. In Ras proteins, replacement of Gln-61 by Leu shows the reduction of GTP hydrolysis and leads to malignant transformation. Experiments revealed oncogenic mutations of $G_s\alpha$ in growth hormone-secreting pituitary tumors.

18.3. G Proteins in Testis Germ Cells

18.3.1 Gα

The somatic testicular cells contain G proteins sensitive to both pertussis toxin (PTX) and cholera toxin (CTX). The somatic cells (Sertoli cells, peritubular cells) contain high amounts of both Gq/11α mRNA and immunoreactive protein. In contrast, low levels of these G proteins and the corresponding mRNAs are present in the germ cells (pachytene spermatocytes and round spermatids). Thus, in the germ cells, receptor–regulated inositol phospholipid hydrolysis is not likely to be regulated via Gp/11α, but rather through the Go protein, which has been shown to be abundant in rat germ cells. Since the somatic cells are nearly devoid of Go protein, the Gpp(NH)p-stimulated phospholipase C in these cells is probably regulated by Gq and/or G11 (Haugen et al, 1993).

Both meiotic and postmeiotic germ cells (i.e. spermatocytes and round spermatids) appear to be devoid of hormone-sensitive adenyl cyclase (AC). Haploid germ cells contain a partly soluble and Mn^{2+}-dependent AC and spermatids express a unique Gsα like mRNA of chain length 0.9-kb (Paulssen et al, 1991). The 1.9-kb mRNA encoding the α subunit of Gs observed in the somatic cells of the rat testis was not seen in the spermatocytes and round spermatids as revealed by Northern blot analysis, whereas the cDNA probe for Gsα also detected a mRNA of shorter chain length (0.9 kb) exclusively found in the haploid germ cells (round and elongated spermatids). However, no ADP ribosylated protein was detected in membranes from germ cells after cholera toxin treatment. The 0.9-kb transcript may represent a haploid germ cell-specific variant of Gsα, which is functionally different from the stimulatory component of the adenylate cyclase system (Haughen et al, 1990; Paulssen et al, 1991).By using Western blot techniques $G_{i-3}\alpha$ was shown to be the only pertussis toxin (PTX) substrate present in all the testicular cells. There was lack of immunoreactive $G_{i-1}\alpha/G_{i-2}\alpha$ protein in pachytene spermatocytes and round spermatids in spite of significant levels of the corresponding mRNA. Peritubular cells and Sertoli cells contained no $G_0\alpha$ whereas high levels of both immunoreactive proteins and mRNA were found in pachytene spermatocytes. The cDNAs encoding two forms of Goα subunit were isolated from a mouse brain library. These two forms, which were called GoAα and GoBα, appeared to be the products of alternative splicing. The GoAα differs from GoBα over the C–terminus of the deduced protein sequence. Both forms are predicted to be substrates for ADP-ribosylation by pertussis toxin. GoAα transcripts are present in a variety of tissues but are most abundant in brain. The GoBα transcript is expressed at highest levels in brain and testis. It was predicted that GoAα and $G_0B\alpha$ have different functions (Strathmann et al, 1990). Haugen et al (1992) presented the evidence for the existence of a novel splice variant of α subunit of G_0 in male germ cells of rat. Northern analysis showed that $G_0\alpha$ mRNA is highly expressed in immature germ cells of rat. While $G\alpha_0 2$ mRNA is the major form in pachytene spermatocytes, a message of shorter chain length is present in large amount in haploid germ cells. This indicated the presence of a novel splice variant of $G\alpha_0$ mRNA, which may code for a $G_0\alpha$ protein important for germ cell development and that PLC activation in germ cell membrane may be mediated by $G_0\alpha$ or $G_{i-3}\alpha$ ((Haugen et al, 1992; Paulssen et al, 1991). The c-DNA clone encoding a putative G-protein coupled receptor has been isolated from a rat testis cDNA library. A corresponding 1.5-kb mRNA exclusively expressed in testis

```
   1-gggaaaaccacgctcctcggaccaagcctcgggagctaagccagatctgccagtgagcct
  61-caggctttaggaactgaagagtgtttctgaaagatctatccagcactccgatggccagca
                                                      M  A  S
 121-acaacaccgccagcatagcacaagccaggaagctggtagagcagcttaagatggaagcca
      N  N  T  A  S  I  A  Q  A  R  K  L  V  E  Q  L  K  M  E  A
 181-atatcgacaggataaaggtgtccaaggcagctgcagatttgatggcctactgtgaagcac
      N  I  D  R  I  K  V  S  K  A  A  A  D  L  M  A  Y  C  E  A
 241-atgccaaggaagacccccctcctgacccctgttccggcttcagaaacccgtttagggaga
      H  A  K  E  D  P  L  L  T  P  V  P  A  S  E  N  P  F  R  E
 301-agaagttttctgtgccatcctttaagtctttgagaggggcctgaagagcctccgggctc
      K  K  F  F  C  A  I  L
 361-ctgggacattgatgtagagttttagtgaagtgggcacctttctagtccacggcatttga
 421-agagagcgaggagaaccattctggaaactctaggctatgcatgtttaaagatctggtccc
 481-ctttatgagaatgcaagccgatccacatcctgacttaagagatctgattctgacgaactg
 541-cctggaggaggggaatatataaaaataaaattggtgtcacttcttttctgctatcccca
 601-gccccccccccaaaatcctcatgtttctgcttcatatttgaaaaataacaattaaaaca
 661-gacagctgtactgaggtaagatatgtgtgaccttcttggaatgaatattgtctttagaat
 721-accctttgataagctgagctgtcccgtgtagatgcaattcggtttaatggcattgatgta
 781-tagtcactgtgcctttctttttctttcttccttctcctctacccctccttccacccctcc
 841-ccattagagtagtgtggagataaggctggattggtctatcagattgaactccaagaatga
 901-tcacacaaaatgtttagggagatgttccccgtggtgtatcctcatggtaacaacgacaaa
 961-aaatgccggttgtctttgttctctttcactattcctaacatgtgtacatgatagctttg
1021-attctgcaagtaaaagtaaatcctgtgttgtgaaaaaaaaaaaaaa
```

Fig.18.2. Nucleotide and predicted amino acid sequence of human Gγ–2 subunit of G protein. Reprinted with permission from M.H. Modarressi et al. Biochem Biophys Res Commun 272: 610-15: 2000 © Elsevier.

is localized in spermatogonia, Leydig and Sertoli cells. The receptor-encoding mRNA and sexual maturation are correlated reaching highest levels during second and third month of age.

Karnik et al (1992) investigated the types of G proteins present in spermatogenic cells during development utilizing antibodies directed against specific regions of various G protein isotypes. The α subunits of $G_i\alpha 1$, $G_i\alpha 2$, $G_i\ \alpha 3$ and $G\alpha_0$ were detected in mouse spermatocytes and spermatids. A conserved sequence of $G\alpha_i$ subtypes localized to the proacrosomal granules of spermatocytes and the developing acrosome of spermatids. Levels of $G\alpha_0$ diminished as spermatocytes developed into spermatids such that $G\alpha_0$ was not detected in cauda epididymal sperm. Immuoreactivity using $G\alpha_0$-specific antisera did not display a distinct regionalization within any of the spermatogenic cell types. The $G\alpha_s$ was not detected in the developing spermatogenic cells or sperm. The association of $G\alpha_i$ with developing acrosome suggests a role for G proteins in the acrosome biogenesis as well as being part of a complex required later for signal transduction leading to acrosomal exocytosis (Karnik et al, 1992).

18.3.2. Gγ-Subunit

The specificity of G protein – receptor interaction is mediated mostly by the γ-subunit and hence the individual members of the γ-subunit multigene family would be expected to each have a particular expression profile. Modarressi et al (2000) identified a cDNA fragment corresponding to the γ-2 gene in human testis. The complete sequence of the Gγ-2 (GWG2) cDNA is 1066 bp long and contains an open reading frame encoding a protein of 71 amino acids. This protein is 100% homologous to the bovine, mouse, and rat G protein γ-2 subunit **(Fig. 18.2)**. The gene has two introns, one located in the 5' UTR and the other within the ORF structure, and is very similar to that of other Gγ-subunit genes. This gene is expressed in a range of foetal tissues as well as adult testis, adrenal gland, brain, white blood cells and lung but not in adult liver, muscle, sperm, prostate gland nor in the testes of two different infertile patients. There is evidence that GNG2 is expressed in malignant tissues. The human GNG2 was mapped to gene to chromosome 14q21.

18.3.3. G Proteins in Sperm

Spermatozoa from invertebrates and vertebrates contain a membrane-bound protein that is ADP-ribosylated by pertussis toxin but not by cholera toxin. The Mr of this protein is 39000 Da in invertebrate sperm, 41000Da in mammalian sperm, and 40000Da in trout spermatozoa. These studies were the first to describe a guanine nucleotide-binding coupling protein in sperm (Bentley et al, 1986). The presence of transductory G-regulatory proteins in mammalian gametes was confirmed by Garty et al (1988) using different rabbit antisera. Transductory G-proteins were confined in sperm to functionally defined regions in the head and tail, in a manner specific for each antibody. G proteins in bovine sperm tails most likely belong to a novel subtype of G protein α-subunits, whereas the putative ß-subunit could be identified as a ß2-subunit (Hinsch et al, 1992, 1995). The existence of several G-protein types in mammalian gametes suggested their possible involvement in the regulation of various effector systems, in a manner reminiscent of somatic cells.

Immunoblots of human membrane preparations demonstrated the presence of $G_i\alpha2$, $G_i\alpha3$, GαH and Gß35 where as the $G_i\alpha1$, Gαo, GαS, $G\alpha_{12}$, $G\alpha_{13}$, G16 Gα and Gß36 subunits were absent. The Gαq/11 was present on acrosome at the equatorial segment, whereas $G_i\alpha2$ was present in the acrosome, midpiece and tailpiece as against $G_i\alpha3$, which was localized in the postnuclear cap, midpiece and tailpiece. The Gß35 subunit was mostly found in mid-piece and the equatorial segment of the acrosome. The distinct pattern of distribution of G proteins suggests that they may couple to receptors or effectors, which also have discrete regions of localization in spermatozoa (Glassner et al, 1991; Merlet et al, 1999). To ascertain whether the abalone spem adenylate cyclase has associated G proteins, Kopf et al (1986) confirmed the presence of the ß-subunit common to both the stimulatory and inhibitory G heterotrimers in sperm and concluded that all the sperm tested, with the possible exception of sea urchin sperm, contain a Gi-like protein.

G_i Proteins in Acrosomal Exocytosis: Transmembrane signaling leading to changes in ionic movements that precede the acrosome reaction have established the importance of ionic effector systems in regulating the exocytotic event. These transmembrane signaling events appear to be mediated through sperm-associated G proteins of Gi class in all mammalian species. The ability of the ZP to stimulate high-affinity GTPase activity in homogenates is concomitant to ribosylation of 41kDa sperm Gi protein, which appears to be dependent on the capacitation state of the sperm (Wilde et al, 1992). Solubilized ZP and purified ZP3 stimulated both GTPγs binding and high affinity GTPase activity in a pertussis toxin (PTX) sensitive fashion; purified ZP1 and ZP2 were without effect. Immunoprecipitation with Giα subtype-specific antisera demonstrated that the egg's extracellular matrix has the ability to activate selectively sperm membrane $G_i\alpha$ subtypes, and that functional ZP3-receptor-G protein coupling is maintained in these isolated membranes, and at least one ZP3 receptor might be a G protein coupled receptor. Thus, receptor aggregation leading to signal transduction might represent a mechanism by which sperm-ZP3 binding is coupled to ZP3-induced acrosomal exocytosis.

Among reagents that promote sperm activation very efficiently, include GTPγs and aluminium fluoride complexes. The GTPγs, a poorly hydrolysable analogue of GTP, bound to these preparations in a specific and concentration dependent-fashion. Solubilized ZP also caused a significant increase in high affinity GTPase activity in the membranes and also stimulated GTPγs. ZP1 and ZP_2 did not stimulate GTPγs binding. This implies that Gi protein activation is an early event in the signal sequence leading to sperm acrosome exocytosis (Ward et al, 1992). However, it was found that ZP binding to sperm activates Gi and also activates sperm membrane AC through pertussis toxin sensitive pathway. Activation of a Gi protein in digitonin / cholate-solubilized membrane preparations of mouse sperm by the zona pellucida supported the existence of ZP receptor-G protein complex in sperm membrane. Amphiphilic peptides capable of activating G_0 and G_i also elicit the acrosome reaction in human spermatozoa. The transduction mechanism for GlcNAc- BSA- and Man-BSA- induced acrosome reaction also seems to involve G-protein of the inhibitory type. The GlcNAc- BSA- or Man-BSA-induced acrosome reaction was completely inhibited by preincubation of spermatozoa with VDCC blockers and calcium antagonists, indicating a link between the binding of sugar residues of the channel activation (Brandelli et al, 1996).

Sperm receptors and G proteins: Among many sperm proteins acting as ZP3 receptor, mammalian transient receptor potential (Trp) proteins (homologous to the *Drosophila* photoreceptor cell Trp protein), and GalTase have emerged as candidate subunits of the ion channels to mediate Ca^{2+} influx after receptor activation. It was suggested that ZP3 induces the acrosome reaction by cross-linking GalTase by activating a heterotrimeric G protein, and that the GalTase cytoplasmic domain can precipitate G_i from sperm lysates. Mouse sperm overexpresing GalTase showed accelerated G protein activation, whereas sperm from mice with a targeted deletion in GalTase had markedly less ability to bind soluble ZP3 and less ZP3-induced acrosome reaction. Thus GalTase, following aggregation of ZP3, is capable of activating pertusis toxin-sensitive G protein leading to exocytosis (Shi et al, 2001). Since ZP3 activates trimeric G proteins and phospholipase C and causes a transient Ca^{2+} influx into sperm through T-type Ca^{2+} channels, these early responses promote a second Ca^{2+}-entry pathway, there by producing sustained increase in $[Ca^{2+}]_i$ that drives acrosome reaction. It was found that Trp2 is essential for the activation of sustained Ca^{2+} influx into sperm by ZP3 (Jungnickel et al., 2001).

Mitochondrial G Proteins : The discovery of GTP-binding proteins in mitochondria is a recent event. These regulatory proteins may be participating in membrane fusion and thereby playing important roles in the physiology of the mitochondrion. So far, it has been proposed that GTP-binding protein mediated membrane fusion may be involved in protein import, steroid hormone production and mitochondrial amalgamation during spermatogenesis (Thomson, 1998).

18.4. SMALL G PROTEINS

Recent studies have clarified the functions of the small G protein superfamily, which consists of the Ras, Rho, Rab, Arf, Sar 1, and Ran families. (1) The Ras family regulates gene expression atleast through MAP kinase cascade. (2) The Rho family mainly regulates reorganization of the actin cytoskeleton. (3) The Rab, Arf, and Sar1 families regulate intracellular vesicle trafficking, and (4) the Ran family regulates nuclear transport.

18.4.1. Ras Proteins

Ras genes were first identified as the oncogenes of rat sarcoma virus and later identified as cellular genes responsible for certain human tumors. These proteins show high degree of conservation and they have essential role in cellular functions. The fact that ras proteins bind and hydrolyze GTP, as well as their association with cytoplasmic membranes through modification by lipids, suggests that Ras proteins, like G proteins, serve as transducers in transmembrane signaling system. Several differences between G proteins and ras proteins, however, also exist, viz. Ras proteins are monomeric, while G proteins are trimeric. Intrinsic GTPase activity of Ras proteins is much lower than that of G proteins. Stimulation of GTP hydrolysis is essential for Ras protein to continue its GTPase cycle. The super family of small GTPases includes >80 mammalian members (H-Ras, K-Ras and N-Ras, (referred to as Ras) which became the focus of intense scrutiny after the discovery of their frequent alterations in neoplasia. The Ras is activated upon ligand engagement of membrane receptor (most notably receptor tyrosine kinases) and mediate differentiation, proliferation, senescence and apoptosis. The functions of Ras proteins seem to depend on cell type. For example, S. *Cerevisiae* Ras1 and Ras2 stimulate adenylate cyclase as Gα does in mammalian cells. The c-raf proto-oncogene and three members of the c-ras gene family are expressed in mitotically active stem cells, throughout the prophase of meiosis and to varying extents in post-meiotic cells (Wolfes et al, 1989) (Chapter 11). Overwhelming evidence identified the MAKP (Raf-MEK-MAPK) pathway as a key effector in Ras signaling (Seita et al, 2000). The mAb against c-ras protein, p21 specifically reacts with acrosomal region of human sperm, and affects the motility parameter involved in hyperactivation of sperm. This indicated that the c-ras or c-ras-like protein is present in mature sperm cell (Naz et al, 1992).

18.4.2. Ran-GTPase and Germ Cell Specificity

Ran is a small, evolutionarily conserved, eukaryotic GTPase of the Ras superfamily that is essential for viability in every organism.The consequences of perturbing the Ran GTPase system impaired various cellular processes including DNA replication, nuclear envelope growth and structure, chromatin structure, cell cycle progression and RNA export. However, it was difficult to determine which processes are direct downstream targets of Ran and which are perturbed indirectly. In 1993, findings led to the proposal that nuclear transport is the primary function of the Ran GTPase system, and the other cellular defects observed when ran is perturbed could be attributed to a failure of nuclear trafficking. This idea became a dominant paradigm in the Ran field (Sazer and Dasso, 2000). Now it is clear that besides the direct role of Ran in nuclear cytoplasmic transport, Ran is implicated in mitotic progression, spindle assembly in *Xenopus* eggs and in actin cytoskeleton development (Sazer and Dasso, 2000).

Besides its well-characterized role in nuclear trafficking, GTP-bound Ran, but not GDP-bound Ran, stimulated polymerization of astral microtubules from centrosome assembled on *Xenopus* sperm. Moreover, a Ran allele with a mutation in the effector domain induced the formation of microtubule asters and spindle assembly, in the absence of sperm nuclei, in a γ-tubulin ring complex and XMAP 215 *Xenopus* microtubule associated protein dependent manner. Therefore, Ran could be a key-signaling molecule regulating microtubule polymerization during mitosis (Wilde and Zheng, 1999). The GTP-bound Ran induces microtubule and pseudo-spindle assembly in mitotic egg extracts in the absence of chromosome and centrosomes, and that chromosomes induce the assembly of spindle microtubules in these extracts through generation of Ran-GTP. The effects of Ran-GTP on microtubule nucleation and dynamics showed that Ran-GTP has independent effects on both the nucleation activity of centrosomes and the stability of centrosomal microtubules; and the inhibition of Ran-GTP production, even in the presence of duplicated centrosomes and kinetochores, prevents assembly of a bipolar spindle in M-phase extracts (Carazo-Salas et al, 2001). However, Hetzer et al (2000) suggested that GTP hydrolysis by Ran is required for nuclear envelope (NE) assembly in xenopus egg extracts, and is inhibited by GTPγS indicating requirement of GTPase activity. Depletion of either Ran or exchange factor for Ran (RCC1) from the assembly reaction also inhibited the step of nuclear envelope formation. Ran depletion can be complemented by addition of Ran, loaded with GTP or GDP but not GTPγS. This suggested that generation of Ran GTP by RCC 1 and GTP hydrolysis by Ran are both required for membrane fusion events that lead to nuclear envelope formation (Hetzer et al, 2000).

Germ Cell Specificity: Rat and mouse testes express both somatic and testis-specific forms of Ran-GTPase implying that the Ran-GTPase plays a significant role during sperm development. Ran-GTPase localizes in the nucleus of round spermatids and along the microtubules of the manchette in elongating spermatids. During disassembly of manchette, Ran-GTPase reactivity is visualized in the centrosome region of maturing spermatids. The circumstantial observation pointed to the role of Ran-GTPase and associated factors in microtubule nucleation as well as the potential nucleating function of spermatid centrosomes undergoing a reduction process. The localization studies, correlated with functional observations in other cell systems, suggested that Ran-GTPase may be involved both in nucleocytoplasmic transport and microtubules assembly, two critical events during the development of functional sperm (Kierszenbaum et al., 2002). Two isoforms of Ran-GTPase have been described: Ran/M1 and Ran/M2. The Ran/M2 is testis specific, whereas the Ran/M1 isoform, besides in testis is also expressed in somatic tissues. The two mRNAs, differing in 35 of the 648 nt included in the ORFs, are developmentally regulated during spermatogenesis. The expression of Ran/M1 and Ran/M2 increased in pachytene spermatocytes with progressive transcript accumulation until they reached the round spermatid stage. In the testis, the expression of both isoforms was found to be restricted to germ cells (Lopez-Casas et al., 2003). Segregation Distorter (SD) is a meiotic drive system in *Drosophila* that causes preferential transmission of the SD chromosome from SD/SD(+) males owing to induced dysfunction of SD(+) spermatids. *Sd*, the primary gene responsible for distortion encodes a mutant RanGTP, a key protein in the Ran signaling pathway required for nuclear transport and other nuclear functions. The mutant protein is enzymatically active but mislocalized to nuclei, which apparently disrupts Ran signaling by reducing intranuclear Ran-GTP level. A defect in nuclear transport may be the main cause of sperm dysfunction (Kusano et al., 2003).

18.4.3. Rap1: A Ras-like GTPase

Rap1 (Krev-1) is a small GTPase first identified as a transformation suppressor of K-ras. This GTPase is very similar to Ras, particularly in the effector region, but its function is elusive. Bos et al (1997) speculated that Rap1 plays a role in one of the specialized function in human platelets. Rap1 consists of two isoforms, Rap1a and Rap1b, which differ mainly at the C-terminus. Since most of the experimental approaches could not discriminate between the two isoforms, both were referred as Rap1. Characteristic features of Rap1 are its geranylgeranyl modification at the c-terminus, which is responsible for membrane attachment and a threonine residue at position 61. In most other GTPases this residue is a glutamine. Substitution of threonine for glutamine in Ras results in a weakly oncogenic activation. Infact, Rap1 has a 10-fold lower intrinsic GTPase activity than Ras, but is still sensitive to Rap GTPase activating proteins (RapGAPs) (Johannes et al, 1997). In mammalian testis such information is lacking.

```
  1 MEKTELIQKA KLAEQAERYD DMATCMKAVT EQGAELSNEE RNLLSVAYKN VVGGRRSAWR
 61 VISSIEQKTD TSDKKLQLIK DYREKVESEL RSICTTVLEL LDKYLIANAT NPESKVFYLK
121 MKGDYFRYLA EVACGDDRKQ TIENSQGAYQ EAFDISKKEM QPTHPIRLGL ALNFSVFYYE
181 ILNNPELACT LAKTAFDEAI AELDTLNEDS YKDSTLIMQL LRDNLTLWTS DSAGEECDAA
241 EGAEN
```

Fig.18.3. Amino acid sequence of mouse homologue of rat brain 14-3-3θ protein in male germ cells. Source: http//www. ncbi.nlm.nih.gov (Accession number AAB72023).

Rap 1 / B raf/ 14-3-3θ-Protein-complex: The 14-3-3- proteins have been implicated as potential regulators of diverse signaling pathways, in particular those involving the activity of the Raf family protein kinases. The 14-3-3q-mouse isoform is expressed almost exclusively in testis and brain. The highly conserved 14.3.3 family of proteins, found in various somatic cells and oocyes, interacts with several important signal transduction kinases, such that these 14-3-3 proteins are considered as modulators of multiple signal transduction pathways. The 14.3.3 protein is also expressed in the male germ cells. Screening of mouse spermatogenic cells revealed a cDNA clone having 735- nucleotide ORF encoding a protein of 245 amino acids (27,778 Da). The predicted protein was identical to the 14-3-3 isoform, the θ subtype from rat brain. The mRNA of 14-3-3θ protein is exclusively expressed in testis and brain. Within the testis the 14-3-3θ gene products are most abundant in meiotic prophase spermatocytes, and in differentiating spermatids. Both testicular and epididymal spermatozoa are negative (Perego and Berruti, 1997) **(Fig.18.3)**. In an effort to understand the function of 14-3-3θ in testis, Berruti (2000) sought to identify endogenous proteins that interact with 14-3-3θ in spermatogenic cells. The presence of a protein complex formed by endogenous Rap1, 93-95kDa B-Raf, and 14-3-3θ and its association with cytodifferentiating cells, and not with dividing cells, strengthens the role of Rap1/B-raf mediated signaling in cell differentiation.

18.4.4. Rab Proteins

Rab proteins are low molecular weight GTPases belonging to the ras superfamily of small GTP-binding proteins. Over 30 rab isoforms have been identified to date. They are ubiquitous proteins, which display a high degree of evolutionary conservation. Rab proteins have been implicated in intracellular vesicle traffic along the biosynthetic and endocytotic pathways. Rab proteins are thought to confer specificity to vesicles that travel in different trajectories of the exocytotic and endocytotic trafficking pathways. In the steady state they are primarily membrane – bound. The rab3 subfamily has excited much interest since these proteins appeared to be involved in the regulation of Ca^{2+}- dependent exocytosis of neural, exocrine, and endocrine cells. There are four members of this subfamily that includes Rab3a, Rab3b, Rab3c and Rab3d. The Rab3a has received the most attention due to its importance in regulated secretion in neurons and neuroendocrine cells. Although the precise role of Rab3a in the molecular events of exocytosis has not been determined, it appears to function late in exocytotic events and may serve to prepare vesicles for exocytotic release, or to stabilize the vesicle. Because of the similarities of the ZP3- induced acrosome reaction to regulated secretion in somatic cells, similar molecular mechanisms may regulate this sperm reaction. To follow along this line, the possible involvement of monomeric GTPase, in particular Rab3a has been investigated by several workers (Reviewed in Ward et al, 1999).

The membrane – cytosol cycle of Rab proteins is mediated by several regulatory proteins such as GDP dissociation inhibitor (GDI), GTPase activating protein (GAP), guanine nucleotide exchange factor including the mammalian suppressor of SEC4 and GDI displacement factor

(GDF). The biological signal generated by active rab proteins is transduced via effector proteins, such as the rab3A-specific rim and rabphillin –3A, the rab5-specific rabaptin-5 and the rab9-specific p90. A functional link has been demonstrated between rab proteins and so-called SNARE complexes, which constitute the docking and fusion machinery. Few studies suggest that Rab5, and possibly other Rab proteins, act as molecular timers rather than switches, the rate of GTP hydrolysis determining the frequency of membrane fusion events (Reviewed in Valentijn and Famieson 1998). Though, rho family has already been implicated in the regulation of the actin cytoskeleton, nonetheless, the rab actin connection is rapidly becoming an irrefutable fact, as evidenced by recent literature. In addition rab 5 has also been found associated with cytoskeletal elements. The Rab4-actin connection is far from unique. Studies have shown that rab8 colocalizes with actin in retinal photoreceptor cells where it may participate in microfilament dependent vesicle transport and morphogenesis (Valentijin and Jamieson, (1998).

Rab 3A in acrosome reaction: Inspite of the large amount of information available on acrosome reaction in several species, there is a remarkable paucity about the role of monomeric GTPase of the Rab family – well established participant in exocytosis in other cells types. One difficulty in studying the role of proteins in intact cells is the fact that they are unable to cross the cell membrane. Yunes et al (2000) established a working model of streptolysin O-permeabilized human spermatozoa, which are able to respond in a regulated way to different stimuli, such as G protein activators and calcium. Acrosome reaction is also triggered by Rab 3A. More important, recombinat Rab3A protein in the GTP-bound form caused acrosome exocytosis. Also, recombinant GDI (GDP dissociation inhibitor), a protein that releases Rab proteins from membrane, inhibited a GTPγ S-stimulated acrosome reaction. These studies indicated that 1) Rab3A is present in human spermatozoa, and 2) Rab3A or another Rab3A isoform is involved in the acrosomal exocytosis in human spermatozoa (Yunes et al, 2000). Electron microscopy provided further evidence that rab3A protein is associated with the acrosomal membrane. Synthetic peptide of the Rab3 effector domain inhibited acrosomal exocytosis triggered by calcium ionophore A23187 in a concentration dependent fashion, suggesting that Rab3A acts as an inhibitory regulator in the acrosome reaction. In view of the putative role of Rab3A protein in membrane fusion systems, these observations suggested that Rab3A could be involved in regulating acrosome reaction by controlling the membrane fusion system in sperm (Iida et al, 1999). The presence of Rab3A in membrane fractions does not change upon capacitation. However, Rab3A, localized to the head of acrosome- intact sperm, was lost during acrosomal exocytosis. Immunogold labeling and electron microscopy demonstrated a subcellular localization of Rab 3A in clusters to the peri-acrosomal membranes and cytoplasm. The presence of Rab3a in acrosomal membranes of mouse sperm suggested that Rab3A plays a role in the regulation of zona pellucida-induced acrosomal exocytosis (Ward et al, 1999). In fact, a stimulatory role for Rab3 has been proposed for the acrosome reaction of ram spermatozoa by using Rab3A peptides. On the other hand, an inhibitory role has been reported when Rab3 effector domain peptides were used concomitantly with the calcium ionophore A23187. Using α[^{32}P] GTP binding to immobilon blotted mouse sperm proteins, the presence of three or more monomeric GTP binding proteins was identified with M_r=22, 24, and 26 x 10^3. α-^{32}P-GTP binding could be competed by GTP and GDP, but not GMP, ATP or ADP (Yunes et al 2000).

Rab3A in Membrane Fusion : Very little is known about the moleculer mechanism that mediates and regulates unique fusion between acrosomal outer membrane and sperm cell membrane. The Rab3A seems to act as an inhibitory regulator in the acrosome reaction. Michaut et al

(2000) showed that membrane N-ethylmaleimide sensitive factor (NSF), a protein essential for most fusion events is present in the acrosome of several mammalian spermatozoa. The Rab3A is necessary for calcium-dependent exocytosis. The activation of Rab3A protects NSF from N-ethylmaleimide inhibition and precludes the exchange of the endogenous protein with recombinant dominant negative mutants of NSF. Furthermore, Rab3A activation of acrosomal exocytosis requires active NSF. Upon stimulation by Ca^{2+}, Rab3A switches to its active GTP-bound form, triggering the formation of a protein complex in which NSF is protected. This process is suggested to be an essential part of the molecular mechanism of membrane fusion leading to the release of the acrosomal contents. In view of the putative role of Rab3A protein in membrane fusion systems, these observations suggested that Rab3A could be involved in regulating acrosome reaction by controlling the membrane fusion system in sperm. However, it cannot be ruled out that other Rab3 isoforms may also be present in sperm. More work is required to better define the Rab3 isoforms participating in the acrosome reaction ((Iida et al, 1999; Michaut et al, 2000).

18.4.5. Rho family

Among atleast nine subfamilies that make up the superfamily of GTPases, the mammalian Rho subfamily comprises at least 14 members (Rho A, RhoB, RhoC RhoD, RhoE/Rnd3, Rnd1/Rho6 Rnd2/Rho7, Rhog, Rac1, Rac2, Rac3, Cdc42, TC10 and TTF). They are characteristically ADP-ribosylated by the exoenzyme C3 from *Clostridium botulinum.* Cells change their cytoskeletal architecture in response to environmental stimuli or during the cell cycle to make their shape to move, or to proliferate. The Rho family GDPases shuttle between the GDP-bound inactive state and the GTP–bound active state, and function as a molecular switch in some of these processes. The best-characterized function of Rho-GTPases is the regulation of the actin cytoskeleton. Functional analysis showed that Rho regulates cell adhesion, motility and cytokinesis. The Rho family members participate in gene transcription , cell cycle progression, exocytosis, and endocytosis. However, most of the studies have been carried out mainly in cultured epithelial cells and fibroblasts, and the function of Rho in specialized types of cells, remain largely unknown. Sea urchin sperm contain multiple small G proteins (28-24 kDa) whose identity and functions are unknown. The sea urchin sperm revealed the presence of 25 kDa Rho, which was identical to that Rho A from human platelets. In flagellar fractions, the 25kDa protein was ADP-ribosylated by C3. It is localized in the cytosol and in the plasma membrane whereas in sperm head, the Rho protein was detected in acrosomal and plasma membranes. Both the 25-kDa ribosylated protein and actin had same localization as bindin, characteristic of acrosome. The G-protein of 25-kDa is likely the Rho, type B. Its cellular localization suggests that Rho may participate in regulating motility and the actin polymerization that accompanies the acrosome reaction (Castellano et al, 1997).

18.4.6. Rho targets (Rhophilin and Ropporin)

Several putative Rho targets have been isolated on the basis of their selective binding to the GTP bound form of Rho and characterized. These include two splice isoforms of citron-N and citron-K, the ROCK family of serine/ theronine kinases : p160ROCK (ROCK-1) and ROK/Rho-kinase/ROCK-II, p140mDia and three molecules containing a homologus Rho – binding motif of 70 amino acid stretch, protein serine/ threonine kinase PKN, rhophilin and rhotekin (reviewed in Nakamura et al, 1999). Among them, p140mDia and ROCK family of kinases, both of which are distributed ubiquitously, regulate actin polymerization and actomyosin contractility,

```
  1 MAQTDKPTCI PPELPKMLKE FAKAAIRVQP QDLIQWAADY FEALSRGETP PVRERSERVA
 61 LCNRAELTPE LLKILHSQVA GRLIIRAEEL AQMWKVVNLP TDLFNSVMNV GRFTEEIEWL
121 KFLALACSAL GVTITKTLKI VCEVLSCDHN GGSPRIPFST FQFLYTYIAK VDGEISASHV
181 SRMLNYMEQE VIGPDGIITV NDFTQNPRVQ LE
```

Fig.18.4. Amino acid sequence of Ropporin, a sperm-specific binding protein of rhophilin. Source: http// www. ncbi.nlm.nih.gov (Accession NP060048)

respectively, and are involved in assembly of focal adhesions and stress fibers. On the other hand, rhophilin is not distributed ubiquitously but present in a limited number of tissues.

Rhophillin mRNA is highly expressed in the testis, and is induced in germ cells at the late stage of spermatogenesis. It is absent from testis of W/Wv mice deficient in germ cells. Rhophillin is localized in the principal piece of sperm flagella, indicating that rhophillin is in the peri-axonemal complex (Nakamura et al, 1999). However, the role of rhophillin in sperm functions remains unknown. Rhophilin contains the following three characteristic domains: a) The Rho binding domain is localized in N-terminal 70 amino acids. b) The amino acid sequence of the middle portion, residue 153 to 454, shows homology with YNKI proteins of *Caenorhabditis elegans, Saccharomyces cervisiae* BRO1P, Xp95 of *Xenopus oocyte* and mouse Alix. In addition, a PDZ domain is present at the C terminus. These results suggest that rhophilin acts as an adaptor protein, linking Rho to rhophilin – interacting proteins that bind to rhophilin on its PDZ domain.

One of the several positive clones for PDZ domain, from mouse testis cDNA library showed ORF of a protein of 212 amino acids designated as ropporin from a Japanese word ("oppo", the tail) **(Fig.18.4)**. The amino terminal 40 amino acid sequence of ropporin showed high homology with that of the regulatory subunit RII a of cAMP dependent protein kinase, which is involved in dimerization and binding to A-kinase anchoring proteins (Vijayaraghavan et al, 1999; Miki and Eddy, 1998). Deletion analysis indicated that the carboxy- terminal four amino acids are essential for binding of ropporin to rhophilin. The ropporin and Rho V14 coprecipitated in the presence of rhophilin in vitro. Ropporin is expressed exclusively in the testis and specifically in germ cells at the late stage of spermatogenesis. This expression pattern is similar to that of rhophilin. Moreover, ropporin is exclusively localized in the principal piece and at the end piece of sperm flagella (Fujita et al, 2000). Immunoflurorescence analysis revealed two lines of striated staining running in parallel in the principal piece of the sperm tail (Nakamura et al, 1999). Electron microscopy showed that ropporin is mostly localized in the inner surface of the fibrous sheath while rhophilin is present in the outer surface of the outer dense fiber. These results suggest that rhophillin and ropporin may form a complex in sperm flagella (Fujita et al, 2000). As rhophilin appears to be a minor component of sperm flagella compared to its major components such as A-kinase anchoring proteins (chapter 29), it is likely that rhophilin may act as a regulator of its assembly or its function. Using botullinum C3 exoenzyme, Rho was also identified in the tail membrane of bovine sperm and that the inactivation of endogenous Rho in bovine sperm decreased its motility. Since rhophilin contains a consensus sequence for a PDZ domain in its C terminus (residues 500-573) and PDZ domains have been identified in many signaling molecules, which mediate protein-protein interactions, it is likely that rhophilin may work as protein(s) through its PDZ domain. Dissection of the Rho-rhophilin pathway may help to reveal transduction mechanisms of sperm motility and may give an insight into the pathology of some cases of asthenozoospermia (Fuzita et al, 2000).

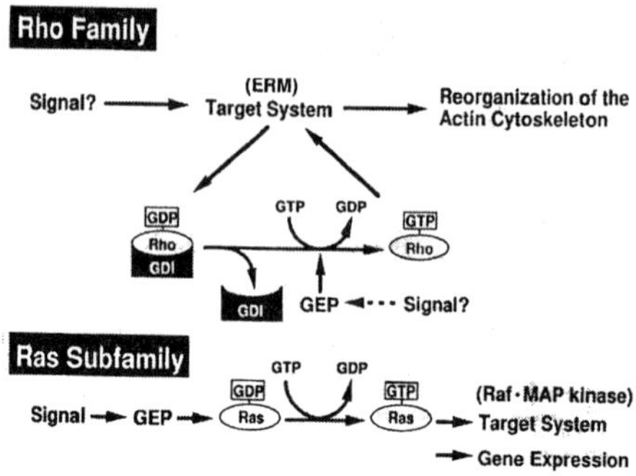

Fig.18.5. Different modes of activation and action of the Rho family and Ras subfamily systems. Reprinted with permission from T. Sasaki and Y. Takai. Biochem Biophys Res Commun 245; 641-45: 1998 © Elsevier.

18.5. REGULATORS OF RHO GTPases

There are at least three types of regulators for small G proteins: (a) GTPase activating protein (GAP), which activates conversion from GTP bound form to GDP bound form; (b) GDP dissociation inhibitor (GDI), which inhibits this reaction and (c) GDP/GTP exchange protein (GEP), which stimulates conversion from GDP bound form to GTP bound form. Of these regulators, GDI has been found for Rho and Rab families. A GDI from brain showed biochemical properties different from those regulators known for Ras GEP and Ras GAP. It was renamed RhoB GDI. Recent studies indicate that Rho GDI constitutes a family, consisting atleast three isoforms (Sasaki and Takai, 1998). Similar to all other small GTPases, GEPs, GAPs and GDIs regulate Rho-GTPases. A scenario is emerging in which GEPs, GAPs and GDIs do not regulate solely the GTP/GDP ratio of Rho-GTPases, but play additional roles in signal transduction pathways. **(Fig.18.5).**

18.5.1. GTPase–Activating Proteins

GTPase–activating proteins (GAPs) for Rho GTPases constitute a class of regulatory proteins that stimulate GTP hydrolysis. Through this catalytic function, Rho GAPs negatively regulate Rho-mediated signals. However, it has long been speculated that GAPs may also serve as effector molecules and play a role in signaling downstream of Rho and other Ras like GTPases. This idea has been substantiated in case of n-chimaerin, a Rho GAP, mainly expressed in the central nervous system. More specifically, n-chimaerin was shown to cooperate with Rac1 and Cdc42 in cytoskeletal morphology. Results support the idea that members of the chimaerin family may mediate Rho-dependent signals in various cell types, as most of them show tissue specific expression. Analysis of these Rho GAPs may therefore help in defining Rho–dependent functions in differentiated cells.

ß-Chimerin: While most of the p21 GAPs are ubiquitous, chimerins are more tissue- or even cell type-restricted. N-chimerin (α1-Chimerin) belongs to GAP family, which regulates Rho subfamily, involed in cytoskeletal organization. It contains a phorbol ester-binding domain and enriched in brain regions involved in memory and learning, and is largely present in hippocampal and cortical neurons and developmentally regulated, whereas β-chimerin is

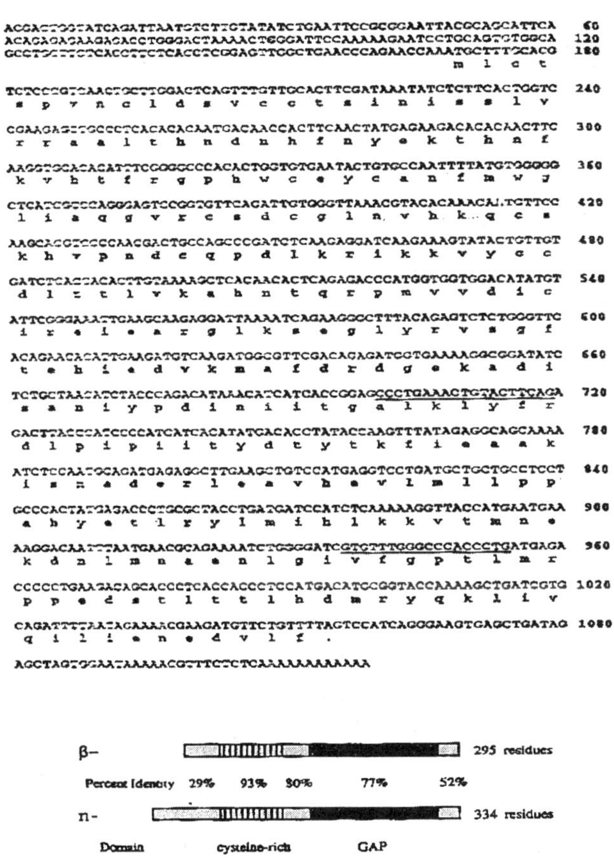

Fig.18.6. (A) Nucleotide and predicted amino acid sequence of β-chimarin from rat germ cells. Conserved regions used for degenerate primers for PCR are underlined. (B) Relationship of β-chimarin to n-chimarin. Reprinted with permission from T.Leung et al. J Biol Chem 268; 3813-16: 1993 © The American Society for Biochemistry and Molecular Biology.

present only in the late stage of sperm cell development. The presence of a highly conserved phorbol ester-binding site in β-chimerin, association of rac-GAP with the membrane, and stage specific expression provide clues to its cellular function. When round spermatids convert to mature sperm they undergo an intensive morphological transformation having drastic actin nucleation, membrane, and cytoskeletal changes. Because of the opportune appearance of chimerin during this period, these germ cells could serve as a model system for analysis of the cell specific role for chimarin / rac GAPs in signal transduction and morphological transformation. Exploiting conserved regions common to rho/rac-GAPs, Leung et al (1993) isolated a rat testis cDNA encoding a 34-kDa Rac-GAP termed ß-chimaerin, which was highly related to N-chimaerin. It contains both a GAP domain (77% identify) and the phorbol ester-binding region (93% identity) **(Fig.18.6)**. ß-chimaerin mRNA is expressed exclusively in the testis at the onset of sexual maturation, and its mRNA expression was stage-specific, paralleling acrosomal assembly at the late stage of spermatogenesis. The testis specific and stage-

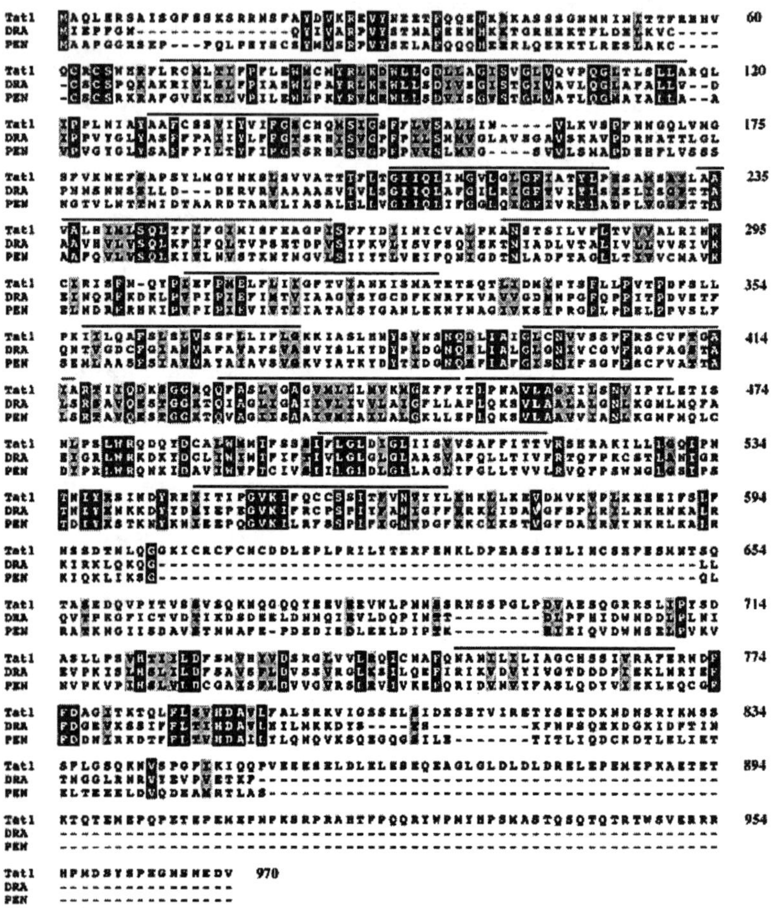

Fig.18.7. Amino acid sequence comparison of Tat1 with full-length proteins of DRA and Pendrin. Identities are shaded black, similarities are shaded gray, and predicted transmembrane domains in Tat1 are indicated by dark lines. Reprinted with permission from A Toure et al. J Biol Chem 276; 20309-15: 2001 © The American Society for Biochemistry and Molecular Biology.

dependent expression of this 30-kDa Rac-GAP chimaerin offers an alternative system for investigation of the functional role of this class of p21 GAPs, particularly in relation to cytoskeletal reorganization.

α2-Chimerin: Hall et al (1993) reported occurance of another form of chimerin, termed α2-chimerin, expressed in testes and brain. The α2-chimerin is produced by alternative splicing of human n-chimerin gene encoding an N-terminal SH2 domain in addititon to phobol ester receptor and GAP domains. Only α2-chimerin mRNA expressed in rat testes in early pachytene spermatocytes. The recombinant form of α2-chimerin had Rac-GAP activity, which was stimulated by phosphatidylserine.

MgcRac GAP and Tat-1(A sulfate Transporter): In a search for new partners of the activated form of Rac GTPase, Toure et al (1998) isolated a human cDNA encoding a new GAP for Rho family GTPases. A specific mRNA of 3.2kb was detected in low abundance in many cell types and found highly expressed in testis. A protein of 58-kDa, named MgcRacGAP was detected in human testis as well as in germ cell tumor extracts. In vitro, the GAP domain of MgcRacGAP strongly stimulates Rac1 and Cdc42 GTPase but is almost inactive on RhoA. At N-terminal to its GAP domain, MgcRacGAP contains a cysteine–rich finger like motif characteristic of the Chimaerin family of Rho GAPs. The closest homologue of MgcRacGAP is rotund RacGAP, a product of the *Drosophila rotund* locus. In situ hybridization of human testis demonstrates a specific expression of MgcRacGAP mRNA in spermatocytes similar to that of rotund RacGAP in *Drosophila* testis. The protein sequence similarity and analogous developmental and tissue specificities of gene expression support the hypothesis that Rotund RacGAP and MgcRacGAP have equivalent functions in insect and mammalian germ cells (Toure et al, 1998). A close partner of MgcRac GAP is a sulfate transporter -Tat 1(970aa with size of 109 kDa), which is related to sulfate permease family and is an integral membrane protein. Tat 1 and MgcRacGAP are co-expressed in spermotocytes and found on plasma membrane of spermatocytes and spermatids (Toure et al 2001) **(Fig. 18.7)**.

18.5.2. Rho GDP Dissociation Inhibitor (Rho GDI)

The cycling between active form and inactive form of the Rho GTPase is regulated by Rho GDP dissociation inhibitor (Rho GDI). The GDI is known to possess atleast two biochemical functions. i) It preferentially interacts with the GDP bound form and prevents it from being converted to the GTP-bound form by the action of each GEP and subsequent translocation to its target membrane. ii) After the GTP-bound form completes its function and subsequently converted to the GDP – bound form by the action of each GAP, GDI forms a complex with it and translocates it to the cytosol. The Rho GDI constitutes a family consisting of at least three isoforms, Rho GDI-α, -ß, and -γ. Rho GDI-α originally isolated from bovine brain is ubiquitous and binds all the Rho family members so far examined. The Rho GDI-ß is expressed exclusively in hematopoietic tissues and shows GDI activity on at least Rho A and Cdc42. Rho GDI-g is expressed predominantly in brain, lung, kidney, testis, and pancreas, and binds RhoA, RhoB, RhoG and Cdc42 (Togawa et al., 2000). Mice, lacking Rho GDI-α generated by homologous recombination (Rho GDIα -/- mice) were characterized by several abnormalities. i) These mice developed massive protein urea mimicking nephrotic syndrome, leading to death due to renal failure. Histologically, degeneration of tubular epithelial cells and dilatation of distal and collecting tubules were readily detected in the kidneys. ii) Rho GDIα -/- male mice were infertile and showd impaired spermatogenesis in testes. iii) Rho GDIα -/- embryos derived from Rho GDIα-/- female mice were defective in the post-implantation development. This suggests that the signaling pathway of the Rho family members regulated by Rho GDIα plays important role in maintaining the structure and physiological function of at least kidney and reproductive systems in adult mice (Togawa et al, 1999).

18.6. ADP-RIBOSYLATION FACTORS

ADP-Ribosylation Factor 4 (ARF4): The ADP–ribosylation factors (ARFs), approximately 20kDa guanine nucleotide binding proteins, involved in protein trafficking, enhance cholera toxin ADP-ribosyltransferase activity. Out of six ARF genes in mammalian tissues, only ARF 4 mRNA, considerably shorter than those present in other tissues, was detected in rat testis.

Short form of ARF4 mRNA was observed in several species in testis specific manner and appeared to involve in late stages of spermatogenesis. Rat ARF4 mRNA sequences revealed that different ARF4 mRNAs result from use of three polyadenylation signals: one AUUAAA and two AAUAAA. Sequences of 3'-untranslated regions of rat and human ARF 4mRNA are similar at similar positions. Of the ARF4 mRNAs, with sizes of 1.1, 1.3 and 0.8-kb, the 1.1-kb mRNA was predominant in adult testis, and was identified primarily in mature sperm. Shorter mRNA, thought to be more stable, may compensate for cessation of transcription at late stages of spermatogenesis (Mishima et al, 1992).

ARF–like Protein 4 (ARL4): The ARL4 is a Ras-related GTPase that has been cloned, from the 3T3-L1 pre-adipocyte cell line as an adipocyte-specific cDNA. The *Arl4* gene maps to the proximal region of mouse chromosome 12 lamed to lamb1-1, Hfhbf1 and Sos2. Compared with all other known genes of Ras related GTPase, the genomic organization of *Arl4* is unusual in its entire coding region, the 3'-untranslated region (UTR) and most of the 5'-UTR are located on a single exon. This structure suggests that *Arl4* has evolved by retroposition of an Arf (ADP-ribosylation factor) or Arf-like gene. Isolation of the 5'-UTR cDNA ends revealed heterogenous transcription initiation sites in alternative exons-1. In situ hybridization indicated that Arl4 is expressed in germ of cells of puberal and adult testis, but not in prepuberal testis suggesting that Arl4 is involved in sperm production (Jacobs et al, 1998).

ARF GTPase Activating Factor: Zhang et al (2000) identified and characterized a human ADP-ribosylation factor GTPase-activating protein (*ARFGAP1*) gene that is related to other members of the ARF GAP Family. Structurally, *ARFGAP1* encodes a polypeptide of 516 amino acids, which contained a typical GATA-1-type zinc finger motif (CXXCX(16)CXXC) with the four cytosine residues that are highly conserved among other members of the ARF GAP family. The *ARFGAP1* gene contained 16 exons ranging from 0.5 to 9.3-kb, and mapped to human chromosome 22p13.2-p13.3. The *ARFGAP1* is strongly expressed in endocrine glands and testis. The expression of ARFGAP1 in testis was about six fold higher than that in ovary, indicating a possible role in spermatogenesis. The ARFGAP1 is present in human fetal tissues and cancer cell lines (Zhang et al, 2000).

18.7. ADENYLYL CYCLASES

The adenylate cyclase (AC) converts ATP into cyclic AMP (cAMP), which is mediator of many signaling pathways. Nine isoforms of adenylyl cyclase have been isolated in mammals with a molecular weight of about 120kDa. Depending on the AC isoform considered, the AC activity can be modulated by Gαs Gαi, Gβγ Ca^{2+}/ calmodulin, or after phosphorylation by either protein kinase C (PKC), cAMP- dependent protein kinase (PKA) or calmodulin-activated kinase II thus allowing a delicate tuning of the cAMP concentration. The distribution of the isforms varies with the development and is tissue-specific.

18.7.1. Sperm Membrane Bound Adenylyl Cyclase

The modulation of sperm cAMP concentration by the ZP in mammals or by an egg jelly coat-derived factor in invertebrates could occur through intracellular ionic changes that modulates AC. Most of the stimulated AC activity is membrane associated. Mn^{2+}-ATP supports a higher AC activity than Mg^{2+}-ATP. The zona-pellucida and forskolin can stimulate AC activity in the presence of Mg^{2+} but not in presence of Mn^{2+}. The ZP-mediated stimulation of membrane-associated AC may represent a mechanism by which the ZP can affect cAMP levels and sperm functions. In mouse, the ZP increases intracellular Ca, pH, and cAMP as well as activates G protein of Gi class.The ZP also stimulates AC of mouse sperm through a pertussis toxin sensitive pathway involving Gi potein (Leclerc and Kopf, 1999). Adenylyl cyclase from sea urchin sperm does not appear to be regulated by G proteins. During sperm activation and the acrosome reaction, membrane potential changes and cAMP increases. Observations with rise in pH supported that AC in these cells is modulated by membrane potential and also by increasing $[Ca^{2+}]i$ and cAMP (Beltran et al, 1996).

In addition to a bicarbonate-regulated cytosolic soluble AC (sAC), mammalian spermatozoa, like somatic cells, appear to contain receptor/G protein-regulated AC activity that contributes to the modulation of specialized cell processes. Evidence has been provided that shows that agents, known to influence somatic membrane-associated AC (mAC) but apparently not germ cell sAC, can modulate cAMP production and functional state in mouse spermatozoa. Membrane-bound adenylyl cyclases from ram, dog, and human sperm are unresponsive to fluoride and guanlylimidodiphosphate [GMP-P(NH)P], two agents that stimulate the adenylyl cyclases of somatic cells. Ram sperm AC is also insensitive to guanine nucleotides, cholera toxin, and hormones, and appears to be devoid of the guanine nucleotide regulatory component (Ns). Sperm cell membranes also appear to lack a functional inhibitory regulatory protein of the adenylyl cyclase system (Ni), since they did not contain an ADP-ribosylatable substrate for pertussis toxin action. This suggested that the sperm cell adenylyl cyclase is unique and different from that of somatic cells. Sperm cells appear to neither contain Ns or Ni nor the ability of their adenylyl cyclase system to interact with Ns from an exogenous source. Thus membrane bound adenylyl cyclase system does not share coupling characteristics with somatic AC (Hilderbrandt et al, 1985). In intact mouse spermatozoa, the most abundant isoforms appeared to be AC2, AC3, and AC8, each with distinct distribution in the acrosomal and flagellar regions; AC1 and AC4 also appeared to be present, although less abundantly, in the midpiece and acrosomal cap regions, respectively (Baxendale and Fraser, 2003).

18.7.2. Soluble Form of Adenylyl Cyclase in Germ Cells

In mammals, a family of transmembrane (tm) adenylyl cyclases (ACs) regulated by G proteins, synthesize cAMP in response to external signals such as hormones and neurotransmitters. An additional cytosolic form of adenylyl cyclase (soluble form) (sAC) activity has recently been isolated in mammals. The expression of sAC is detectable in almost all tissues examined (Buck et al., 1999; Sinclair et al, 2000). Although both generate the second messenger cAMP, soluble AC and transmembrane tmACs define clearly distinct signaling pathways in mammals. The sAC is not predicted to possess transmembrane spanning domains, and its biochemical activity is insensitive to any of the known modulations of tmAC activity (Buck et al, 1999). A close relationship between sAC and bacterial ACs reveals a direct link between prokaryotic and eukaryotic signaling systems, and suggests that unlike transmembrane ACs, which mediate signals between cells, sACs may be functionally more similar to bacterial ACs, which control pathways important to unicellular organisms. The sAC expression is detectable in almost all tissues examined; however, northern analysis and in situ hybridization indicated that high levels of sAC message are unique to male germ cells. Elevated levels of sAC mRNA are first observed in pachytene spermatocytes and expression increases through spermiogenesis. The accumulation of high levels of message in round spermatids suggests that sAC protein plays an important role in the generation of cAMP in spermatozoa, implying possible roles in sperm maturation through the epididymis, capacitation, hypermotility and/or the acrosome

(A)

```
   1 MSARRQELQD RAIVKIAAHL PDLIVYGDFS PERPSVKCFD GVLMFVDISG FTAMTEKFST
  61 AMYMDRGAEQ LVEILNYYIS AIVEKVLIFG GDILKFAGDA LLALWKVERK QLKNIITVVI
 121 KCSLEIHGLF EAKEAEEGLD IRVKIGLAAG HITMLVFGDE TRNYFLVIGQ AVDDVRLAQN
 181 MAQMNDVILS PNCWQLCDRS MIEIERIPDQ RAVKVSFLKP PPTFNFDEFF TKCMGFMDYY
 241 PSGDHKNFLR LACMLESDPE LELSLQKYVM EIILKQIDDK QLRGYLSELR PVTIVFVNLM
 301 FKEHDKVEVI GSAIQAACVH ITSVLKVFRG QINKVFMFDK GCSFLCVFGF PGEKAPDEIT
 361 HALESAVDIF DFCSQVHKIR TVSIGVASGI VFCGIVGHTV RHEYTVIGQK VNIAARMMMY
 421 YPGIVSCDSV TYDGSNLPAY FFKELPKKVM KGVADPGPVY QCLGLNEKVM FGMAYLICNR
 481 YEGYPLLGRV REIDYFMSTM KDFLMTNCSR VLMYEGLPGY GKSQVLMEIE YLASQHENHR
 541 AVAIALTKIS FHQNFYTIQI LMANVLGLDT CKHYKERQTN LQNRVKTLLD EKFHCLLNDI
 601 FHVQFPVSRE MSRMSKIRKQ KQLEALFMKI LAQTVREERI IFIIDEAQFV DGTSWAFIEK
 661 LIRSMPIFIV MSLAPFSEVP CAAANAIMKN RNTTYITLGT MQPQEIRDKV CVDLSVSSIP
 721 RELDSYLVEG SCGIPYYCEE LLKNLDHHRV LLFQQAETEQ KTNVTWNNMF KHSVRPTDDM
 781 QLFTSISEGQ KEVCYLVSGV RLNNLSPPAS LKEISLVQLD SMSLSHQMLV RCAAIIGLTF
 841 TTELLFEILP CWNMKMMIKA LATLVESNVF NCFRSSKDLQ LALKQNVPTF EVHYRSLALK
 901 LKEGLTYGEE EELREMEGEV VECRILRFCR PIMQKTAYEL WLKDQKKVLH LKCARFLEES
 961 AHRCNHCRNV DFIPYHHFIV DIRLNTLDMD TVKRMVTSQG FKIDEEEAVF SKSELPRKYK
1021 FPENLSITEI REKILHFFDN VILKMKSSPN DIIPLESCQC KELLQIVILP LAQHFVALEE
1081 NNKALYYFLE LASAYLILGD NYNAYMYLGE GERLLKSLTN EDSWSQTFEY ATFYSLKAEV
1141 CFNMGQMVLA KKMLRKALKL LNRMFPCNLL TLTFQMHVEK NRLSHFMNQH TQEGSVPGKK
1201 LAQLYLQASC FSLLWRIYSL NFFFHYKYYG HLAAMMEMNT SLETQNDFQI IKAYLDFSLY
1261 HHLAGYQGVW FKYEILVMEQ LLNLPLKGEA IEIMAYTADT LGHIKFLMGH LDLAIELGSR
1321 AHRMWSLLRN PNKYQMVLCR LSKPLFLKSR YKHLVQVLGW LWDLSVTEED IFSKAFFYFV
1381 CLDIMLYSGF IYRTFEECLE FIHHNEDNRI LKFQSGLLLG LYSCIAVWYA RLQEWDNFNK
1441 FSDRAKHLVT RRTPTVLYYE GISRYMEGQV LHLQKQIEEQ AENAQDSGVE ILKALETLVA
1501 QNTTGPVFYP RLYHLMAYVC ILMGDGHSCD FFLNTALELS ETHGNLLEKC WLSMSKEWWY
1561 SASELTGDQW LQTVLSLPSW DKIVSGKGGQ RKRSWSWFCP PNFSMVSWSQ PQCA
```

(B)

```
   1 MSARRQELQD RAIVKIAAHL PDLIVYGDFS PERPSVKCFD GVLMFVDISG FTAMTEKFST
  61 AMYMDRGAEQ LVEILNYYIS AIVEKVLIFG GDILKFAGDA LLALWKVERK QLKNIITVVI
 121 KCSLEIHGLF EAKEVEEGLD IRVKIGLAAG HITMLVFGDE TRNYFLVIGQ AVDDVRLAQN
 181 MAQMNDVILS PNCWQLCDRS MIEIERIPDQ RAVKVSFLKP PPTFNFDEFF AKCMAFMDYY
 241 PSGDHKNFLR LACMLESDPE LELSLQKYVM EIILKQIDDK QLRGYLSELR PVTIVFVNLM
 301 FKEQDKAEVI GSAIQAACVH ITSVLKVFRG QINKVFMFDK GCSFLCVFGF PGEKAPDEIT
 361 HALESAVDIF DFCSQVHKIR TVSIGVASGI VFCGIVGHTV RHEYTVIGQK VNIAARMMMY
 421 YPGIVTCDSV TYDGSNLPAY FFKELPKKVM KGVADPGPVY QCLGLNEKVM FGMAYLICNR
 481 YEGYPLLGRV REIDYFMSTM KDFLMTNCSR VLMYEGLPGY GKSQVLMEIE YLASQHENHR
 541 AVAIALTKIS FHQNFYTIQI LMANVLGLDT CKHYKERQTN LQNRVKTLLD DKYHCLLNDI
 601 FHVQFPVSRE MSRMSKIRKQ KQLEALFMKI LEQTVREERI IFIIDEAQFV DVASWAFIEK
 661 LIRSMPIFIV MSLCPFPETP CAAANAIMKN RNTTYITLGT MQPQEIRDKV CVDLSVSSIP
 721 RELDSYLVEG SCGIPYYCEE LLKNLDHHRI LIFQQAEAEE KTNVTWNNLF KYSVKPTEDM
 781 YLYTSIAAGQ KEACYLTSGV RLKNLSPPAS LKEISLVQLD SMSLSHQMLV RCAAIIGLTF
 841 TTELLFEILP CWNMKMMIKA LATLVESNVF DCFRSSKDLQ LALKQNVTTF EVHYRSLSLK
 901 SKEGLAYSEE EQLREMEGEV IECRILRFCR PIMQKTAYEL WLKDQKKVLH LKCARFLEES
 961 AHRCNHCRNR DFIPYHHFIA DIRLNTLDMD TVKKMVKSHG FKTEDEVIFS KSEIPRKFKF
1021 PENISITETR EKILHFFDNV IIKMRTSQDD VIPLESCHCE ELLQIVILPL AQHFVALEEN
1081 NKALYYFLEL ASAYLILGDN YNAYMYLGEG ERLLKSLTNE DSWSQTFEYA TFYSLKGEIC
1141 FNMGQMVLAK KMLRKALKLL NRMFPCNLLS LTFQMHIEKN RLSHFMNQHT QEGSLPGKKL
1201 AQLFLQSSCF SLLWKIYSLN FFFHYKYYGR LAAIMQMNTS LETQNNFQII KAFLDFSLYR
1261 HLAGYEGVWF KYEILVMEQL LNLPLKGEAF EIMAYAADAL GHIKFLTGHL DLAIELGSRA
1321 HKMWSLLRNP NKYHMVLCRL SKPLFLKSRY KHLVQVLGWL WDLSVTEEHI FSKAFFYFVC
1381 LDIMLYSGFI YRTFEECLEF IHHNEDNRIL KFQSGLLLGL YSCIAVWYAR LQEWDNFYKF
1441 SNRAKTLVTR RTPTVLYYEG ISRYMEGQVL HLQKQIEEQA ENAQLSGVEL LKALETLVAQ
1501 NTTGPVFYPR LYHLMAYVCI LMGDGHSCDF FLNTALELSE TQGNLLEKCW LSMSKEWWYS
1561 APELTGDQWL QTVLSLPSWD KIVSGNVTLQ DVQKNKFLMR VNILDNPF
```

Fig.18.8. (A) Amino acid sequence of soluble form of adenylyl cyclase 10 in mouse male germ cells. Source: http//www. ncbi.nlm.nih.gov (Accession NP_766617). (B) Amino acid sequence of soluble form of rat adenylyl cyclase. Source: http//www. ncbi.nlm.nih.gov (Accession NP_067716).

reaction (Sinclair et al, 2000) **(Fig.18.8A)**. Bicarbonate directly stimulates mammalian sAC activity in vivo and in vitro in a pH-independent manner, and bicarbonate regulation of cyclase activity is conserved in early forms of life such as bacteria. The sAC is also expressed in other bicarbonate responsive tissues, which suggests that bicarbonate regulation of c-AMP signaling plays a fundamental role in many biological systems (Chen et al, 2000).

18.7.3. Olfactory Adenylyl Cyclase Type 3

During the past few years, the presence of various elements of the olfactory pathway has been reported in male germ cells. In the dog these elements are present in the late stage of spermatogenesis, on the tail midpiece of mature sperm. This observation has been extended to several mammalian species and to other components of the olfactory pathways (Vanderhaeghen et al, 1997, Weyand et al, 1994). Signal transduction of certain odorants in olfactory neurons involves the activation of the adenylyl cyclase /cAMP-gated channels, and to an influx of Ca^{2+} causing the opening of Ca^{2+} activated chloride ions channels (Defer et al, 1998). Other odorants stimulate G protein-mediated formation of InsP3 with a subsequent increase of intracellular Ca^{2+}. In each mammalian species, a restricted subset of olfactory receptors is expressed in male germ cells and displays a pattern of expression suggestive of their potential implication in control of sperm physiology. As a consequence, five olfactory receptor genes expressed in germ cells were localized in human genome. Three of these genes are located to short arm of chromosome 19 (19p13.1-19p.31.3), one to chromosome 11(11q22.1-22.3), and one to chromosome 17 (17q21-22) (Vanderhaeghen et al, 1997).

Adenylyl cyclase type 3 (AC3) was first identified specific to the olfactory transduction apparatus. Both AC3 and the olfactory $G\alpha_s$ subunit ($G\alpha_{olf}$) have been localized in the same receptor cell compartment, the distal segments of the olfactory cilia. Defer et al (1998) demonstrated that both AC3 and $G\alpha_{olf}$ are expressed within the testis and their expression is germ cell specific that occurs at the same stages of germ cell differentiation. It thus appears that male germ cells contain all the elements of the signaling cascade present in olfactory cells. In rat restis, AC3 and $G\alpha_{olf}$ are expressed in the same sub-population of germ cells, pachytene spermatocytes to spermatids, and in residual bodies. Neither AC3 nor $G\alpha_{olf}$ was found in Sertoli or in peritubular cells.

The RT-PCR confirmed that AC3 transcripts are present in germ cells, appear during the meiotic prophase, and accumulate during spermiogenesis. The presence of a predominant transcript of 7-5-kb suggested that the AC3 expressed in germ cells may be derived from a splicing variant different from the 4.5-kb transcripts expressed in somatic cells. An immunoreactive protein of 170-kDa was detected in extracts from total testis and from germ cells. The AC3 was predominantly expressed in postmeiotic germ cells from round spermatids in the cap phase to maturing elongating spermatids. This located mostly on the acrosomal membrane rather than on the plasma membrane of developing spermatids. The presence of members of the olfactory receptor family and an olfactory phosphodiesterase supported that a signal transduction system used in olfaction is present during development of male gametes (Gautier-Courteille et al, 1998) **(Fig. 18.8 B)**.

18.7.4. Pituitary Adenylate Cyclase Activating Polypeptide (PACAP)

Pituitary adenylate cyclase activating polypeptide (PACAP) is found not only in the brain, but is also abundantly expressed in the testicular germ cells. Immunoreactivity of PAC1-receptor (R)-like was seen in the cytoplasm of round spermatids, aggregated in the acrosome and coexpressed with PACAP-II. Spermatid-enriched fractions were examined for the subcellular localization of PACAP binding sites and PAC1-R-II (Li et al., 2004). Because PACAP and its receptors are coexpressed in the cytoplasm of spermatids, endogenous PACAP may directly interact with the cytosolic PAC1-R-like protein without the ligand being released into the extracellular space. However, the physiological role of testicular PACAP remains unknown (Li et al., 2004).

18.8. G-PROTEIN RECEPTORS

18.8.1. G-Protein Odorant Receptors in Germ Cells

G protein-coupled receptors (GPCRs) are involved in cell recognition and signaling. In recent years understanding of olfactory system has emerged due to identification of odor receptors in olfactory epithelium. Among the unidentified receptors ('orphan receptors'), a human genomic clone (HGMP07) was characterized by the presence of its transcripts in the testis and by its relation to a large subfamily of genes sharing extensive sequence similarities. Sequence comparison demonstrated that this gene subfamily is the human counterpart of the putative rat olfactory receptor. Another 48 members of the family were cloned. Northern blotting suggested a common receptor gene family that encodes olfactory receptors. Sperm cell receptors could be involved in chemotaxis during fertilization (Parmentier et al, 1992)**(Fig. 18.9)**. The putative odor receptors are a subfamily of seven-transmembrane-domain G protein-coupled receptors. Although their functions have not yet been definitively proven, they may act as odor receptor (GPCR). Receptor gene family has the following characteristics: 1) It is a multigene family of phylogenetically related sequences consisting of approximately 100 members in fish and up to approximately 1000 members in rodents. 2) Members of this family are expressed in the olfactory epithelium (olfactory receptors), and related genes are expressed in the gustatory epithelium of the tongue (gustatory receptors) and in sperm cell (germ-cell receptors). 3) Genes for the olfactory receptor family have been isolated from several species including fish, rat, mouse human and chick. 4) Olfactory receptor protein seems to be localized on the ciliary surface of olfactory neurons. 5) Ofactory receptors exhibit broad ligand specificity. 6) A single receptor is expressed in only 0.1%-2% of the entire olfactory neuron population (reviewed in Nef, 1993). The GPCRs consist of an extracellular N-terminus and an intracellular C-terminus separated by seven helical transmembrane domains (TM7). The extracellular region is highly glycosylated, whereas the intracellular region binds to G proteins. An epididymal GPCR, designated HE6 (for human epididymis-specific protein 6), is present in the stereocilia projecting from the apical domain of principal cells into the epididymal lumen. Kierszenbaum (2003) conceptually suggested that HE6 wears two hats: an unusually long extracellular region characteristic of cell adhesion proteins, and an intracellular region with binding site to G protein. The HE6 has another remarkable feature, which is comparable to the GPCR calcium-independent receptor of α-latrotoxin, designated CIRL. Both HE6 and CIRL are endogenously cleaved into two pieces at the GPCR proteolytic site (GPS) located adjacent to TM1, the first of the seven transmembrane helices. One fragment of the heterodimer wears the cell adhesion hat; the other retains the typical characteristics of GPCRs, thus explaining the molecular compartmentalization of cell adhesion and G protein activation functions (Kierszenbaum, 2003).

```
                                          I                                   II
HGMP07I             MMGQNQTSISDFLLLGLPIQPEQQNLCYALFLAMYLTTLLGNLLIIVLIRLDSHLHTPMYLFLSNLSFSDLCFSSVTIP  79
DTMT                MTEKNQTVVSEFVLLGLPIDPDQRDLFYALFLAMYVTTILGNLLIIVLIQLDSHLHTPMYLFLSNLSFSDLCFSSVTMP  79
OLFI15              MTEENQTVISQFLLLFLPIPSEHQHVFYALFLSMYLTTVLGNLIIIILIHLDSHLHTPMYLFLSNLSFSDLCFSSVTMP  79
OLFF3               MDSSNRTRVSEFLLLGFVENKDLQPLIYGLFLSMYLVTVIGNISIIVAIISDPCLHTPMYFFLSNLSFVDICFISTTVP  79
HGMP07J  MLLCFRFGNQSMKRENFTLITDFVFQGFSSFHEQQITLFGVFLALYILTLAGNIIIVTIIRIDLHLHTPMYFFLSMLSTSETVYTLVILP  90
                    *  * * * ****  *    *    * *** *** ** *** **  *  *  ********** **  *  *   **
                                III                                     IV
HGMP07I  KLLQNMQNQDPSIPYADCLTQMYFFLLFGDLESFLLVAMAYDRYVAICFPLHYTAIMSPMLCLALVALSWVLTTFHAMLHTLLMARLCFC 169
DTMT     KLLQNMQSQVPSIPYAGCLTQMYFFLFFGDLESFLLVAMAYDRYVAICFPLHYTTIMSPKLCFSLLVLSWVLTMFHAVLHTLLMARLCFC 169
OLFI15   KLLQNMQSQVPSIPFAGCLTQLYFYLYFADLESFLLVAMAYDRYVAICFPLHYMSIMSPKLCVSLVVLSWVLTTFHAMLHTLLMARLSFC 169
OLFF3    KMLVNIQTQNNVITYAGCITQIYFFLLFVELDNFLLTIMAYDRYVAICHPMHYTVIMNYKLCGFLVLVSWIVSVLHALFQSLMMLALPFC 169
HGMP07J  RMLSSLVGMSQPMSLAGCATQMFFFVTFGITNCFLLTAMGYDRYVAICNPLRYMVIMNKRLRIQLVLGACSIGLIVAITQVTSVFRLPFC 180
         *** *      *  * * ******* *      *** * ******** **** **   *  **    *   ** *  *  * **
                               V                                       VI
HGMP07I  ADNVIPHFFCDMSALLKLAFSDTRVNEWVIFIMGGLILVIPFLLILGSYARIVSSILKVPSSKGICKAFSTCGSHLSVVSLFYGTVIGLY 259
DTMT     A-NTIPHFFCDMSALLKLACSDTQVNELVIFIMGGLILVIPFLLIITSYARIVSSILKVPSAIGICKVFSTCGSHLSVVSLFYGTVIGLY 258
OLFI15   ADNMIPHFFCDISPLLKLSCSDTHVNELVIFVMGGLVIVIPFVLIIVSYARVVASILKVPSVRGIHKIFSTCGSHLSVVSLFYGTIIGLY 259
OLFF3    THLEIPHYFCEPNQVIQLTCSDAFLNDLVIYFTLVLLATVPLAGIFYSYFKIVSSICAISSVHGKYKAFSTCASHLSVVSLFYCTGLGVY 259
HGMP07J  ARK-VPHFFCDIRPVMKLSCIDTTVNEILTLIISVLVLVVPMGLVFISYVLIISTILKIASVEGRKKAFATCASHLTVVIVHYSCASIAY 269
             *******   ** *   *  *** *        **  **  *  **  **  *  *  *  *  * * ** *** ** * *     *
                VII
HGMP07I  LCSSANSSTLKDTVMAMMYTVVTPMLNPFIYSLRNRDMKGALSRVIHQKKTFFSL                   314
DTMT     LCPSANNSTVKETIMAMMYTVVTPMLNPFIYSLRNKDMKGALRRVICRKKITFSV                   313
OLFI15   LCPSANNSTVKETVMAMMYTVVTPMLNPFIYSLRNRDMKEALIRVLCKKKITFCL                   314
OLFF3    LSSAANNSSQASATASVMYTVVTPMVNPFIYSLRNKDVKSVLKKTLCEEVIRSPPSLLHFFLVLCHLPCFIFCY 333
HGMP07J  LKPKSENTREHDQLISVTYTVITPLLNPVVYTLRNKEVKDALCRAVGGKFS                       320
         *             * ********** ** *******  * * *
```

Fig.18.9. Amino acid sequence deduced from the HGMP07, DTMT and HGMP07J clones, compared with two rat olfactory clones. HGMP071 is a human clone corresponding to the original HGMP07. DTMT represents the major transcript present in dog sperm cells. HGMP07J is one of the genes expressed in human testis. Bars and roman numbers represent the putative transmembrane segments. Asterisks and dots indicate identical residues and conservative replacement respectively. Putative sites for N-linked glycosylation are underlined. HGMP07J has two possible initiation sites, the second of which is aligned with other putative receptors. Reproduced with permission from M. Parmentier et al, 355; 453-455:1992 © Macmillan Magazines Ltd. http//www.nature.com/nature.

18.9. ADENOSINE RECEPTORS AND LIGANDS

18.9.1. Adenosine Receptors

Since 1986 when it was proposed that adenosine (Ado) and purine nucleotides can act as neurotransmitters, a large body of data supports that extracellular Ado is both a neuromodulator and a mediator of cell-to-cell communication in physiological processes. To date, four adenosine receptor (AR) subtypes have been isolated and characterized in mammals. These include A_1 and A_3 subtypes, which inhibit adenylyl cyclase and the A_{2A} and A_{2B} subtypes, which stimulate AC. In addition to modulation of adenylate cyclase activity, activation of A_1, A_3 and A_{2B} subtypes may induce calcium mobilization through phosphoinositide hydrolysis. Functional A_1 and A_3 adenosine receptor subtypes have been characterized and localized in the mammalian testis. While A_2 adenosine receptor is present on mammalian spermatozoa, the level of this receptor subtype is either very low or absent in testicular tissue. Although, it has been proposed that in mammals the testicular adenosine recptor system regulates Sertoli cell function and the activation of the A_1-A_3 would downregulate cell activity in the seminiferous tubule, the physiological role of these receptors in the testis still remains unclear. Moreover, only inhibitory A_1 and A_3 receptor subtypes have been characterized in mammalian testis; A_1 adenosine receptors being largely present on Sertoli cells and A_3 adenosine receptors being present on spermatocytes and spermatids but not on Sertoli cells. In many cell systems, stimulatory and inhibitory signals regulate intracellular cAMP levels. In the rat Sertoli cell, AC activity is stimulated by FSH and inhibited, at least, by activation of A_1 receptors. In trout, c-

AMP may be a negative regulatr of mitotic germ cells. By contrast, in the mammalian testis, cAMP stimulates proliferation of primordial germ cells, stimulates proliferation and differentiation of Go and controls Sertoli cell production of the steel factor, which would stimulate germ cell growth and differentiation.

Adenosine and its analogues, known to stimulate adenylate cyclase activity in somatic cells via A2 receptors, can accelerate capacitation in mouse spermatozoa. Indirect evidence has suggested that adenosine can modulate mouse sperm adenylate cyclase, implicating this enzyme and cAMP in the observed functional responses. [3H]5'-N-ethylcarboxamidoadenosine (NECA), an adenosine analogue with specificity for stimulatory A2 adenosine receptors, can bind to mouse spermatozoa. This binding can be displaced by both unlabelled NECA and 2-chloroadenosine, another A2 receptor agonist, suggesting that the NECA binding is specific for A2 receptors. In addition to demonstrable binding of [3H] NECA, both NECA and 2-chloroadenosine significantly stimulated AC activity. Furthermore, the hydrolysis-resistant GTP-analogue, Gpp(NH), alone and in the presence of either NECA or 2-chloroadenosine also stimulated enzyme activity. In somatic cells, expression of response to adenosine usually requires GTP and G proteins (Fraser and Duncan, 1993). Adenosine agonist, NECA stimulates human sperm for fertilizaiton during capacitation in vitro, with specificity for A2 receptors (Fenichel et al., 1996). NECA increased the number of capacitated spermatozoa able to respond to A23187. This effect was associated with an increase in cAMP production and with an enhanced phosphorylation of a 95-kDa protein in uncapacitated sperm. This was accompanied by positive regulation of c-AMP dependent PKA suggesting a role of adenosine receptors during capacitation The effects of NECA on stimulation of human sperm motility in a dose-dependent manner and inhibition by various adenosine receptor antagonist suggest that the NECA binding sites and A2 receptors may be coupled to one or more G proteins on sperm (Shen et al, 1993). Stimulatory A2 receptors were localized primarily on the acrosomal cap region and the flagellar principal piece, the staining was much more pronounced in uncapacitated than in capacitated spermatozoa. The A1 receptors showed a very similar distribution, but the staining was markedly greater in capacitated than in uncapacitated cells. It appears that specific adenosine receptors function in a reversible manner in one or other capacitation state, resulting in regulation of cAMP (Adeoya-Osiguwa and Fraser, 2002).

Analyses of sedimentation profiles of sperm solubilized fractions in presence of agonist also showed the existence of a population of A1 receptors tightly coupled to Gi, the pertussis toxin-sensitive component of the G protein family (Minelli et al, 1997); the A1 adenosine receptor in human spermatozoa is coupled to $G\alpha_i2$ signals via InsP3, and regulates the capacitation of ejaculated spermatozoa (Allegrucci et al, 2001). Reactive oxygen species stimulated the expression of A1 receptor, which confers cytoprotection in various tissues. Increase in adenosine A1 receptor in cisplatin treated rats indicated that cisplatin enhances A1 adenosine receptor in testis, possibly through promotion of oxidative stress (Bhat et al, 1999).

18.9.2. Fertilization Promoting Peptide (FPP) and Adenosine Receptors

The FPP is a fertilization promoting peptide (pGlu-Glu-ProNH$_2$), which is structurally similar to TRH (thyrotrophin releasing hormone; pGlu-His-ProNH$_2$) (see chapter 31). Studies revealed that FPP elicits a biphasic response in mouse spermatozoa, first stimulating capacitation, secondly inhibiting the spontaneous acrosome reaction (Green et al., 1996). Adenosine, like FPP, present in seminal plasma evokes biphasic response in sperm AC. The FPP alongwith adenosine had a greater effect than the two molecules used separately, indicating involvement of two specific receptors in AC/cAMP signal transduction pathway, which is important in mammalian sperm capacitation (Green et al., 1996). Mouse spermatozoa have been shown to have adenosine receptor. In addition a protein, TCP11 coded by t-complex gene that acts as a putative receptor for FPP was identified by Fraser (1997). Antibody and FPP interactions with sperm led to the hypothesis that (1) TCP11 may be the receptor for FPP, and (2) FPP binding to TCP11 would result in modulation of the AC/cAMP signal transduction pathway. The TCP11 and adenosine receptors are very different entities and it is unlikely that FPP would interact directly with both. Studies provide a evidence, which suggest the involvement of two different types of adenosine receptors and two types of $G\alpha_s$ and $G\alpha i_o$ subunits in the regulation of capacitation, resulting in modulation of AC activity and availability of cAMP (Fraser and Adeoya-Osiguwa, 2001).

18.10. RECEPTOR GUANLYL CYCLASE

Guanlyl cyclase, responsible for the enzymatic synthysis of cGMP, was first described in 1969 and shown to be distinct from adenylyl cyclase. While adenylyl cyclase activity was found in only the particulate fractions, guanylyl cyclase was found in both soluble and particulate fractions. In the late 1970s, sodium azide, sodium nitroprusside, nitric oxide, and various other agents were shown to activate the soluble form of guanylyl cyclase in broken cell preparations; purified soluble guanylyl cyclase from bovine lung contained associated heme. Although not yet directly demonstrated, the subsequent model developed for regulation of the cytoplasmic form of guanylyl cyclase showed that nitric oxide or a similar molecule binds to the associated heme, resulting in enzyme activation (Chinkers and Garbers, 1991).

In the early 1980s, peptides obtained form sea urchin eggs that activate spermatozoa and a peptide obtained from the heart (atrial natriuretic peptide) that stimulates natriuresis, diuresis and vasodilation were isolated and shown to cause elevations of cGMP in intact cells and activation of a nonionic detergent- soluble form of particulate guanylyl cyclase in broken cell preparations. Since these initial observations, now it is known that the particulate guanylyl cyclase can serve as cell surface receptor for a variety of extracellular peptide ligands. The mechanisms by which sea urchin egg natriuretic peptides, and bacterial heat-stable enterotoxins activate the particulate form of GC and the mechanism by which NO and related molecules activate the soluble form of the enzyme are well known.

Deduced amino acid sequences of various members of the guanylyl cyclase family show a consensus sequence consisting of residues conserved among at least four of six proteins. All known membrane guanylyl cyclases contain a domain that is highly homologous to protein kinases. This domain is located between plasma membrane and catalytic domain. In all cases, this region is more closely related to protein tyrosine kinases than to protein serine kinases. The vast majority of the residues that are highly conserved among the 11 subdomains of the catalytic domains of protein kinases are also conserved in each of the membrane guanylyl-cyclase. Still a puzzling observation is the nucleotide-binding sub-domain Gly-X-Gly-X-X-Gly, which is poorly conserved in most of the membrane bound GCs and absent from GCc. The sperm plasma membrane form of guanylyl cyclase appears to be regulated by its state of phosphorylation. Dephosphorylation of the sperm enzyme appears to act as a desensitization step. The state of enzyme phosphorylation also alters kinetic behaviour of the enzyme. The membrane form of GC can be distinguished from soluble form of GC using MeGTP as substrate and measuring velocity. The plasma membrane form displays positive cooperativity whereas soluble form shows linear behavior as a function of MeGTP concentration. However, the GC followed linear kinetics after its dephosphorylation (review- Chinkers and Garbers, 1991). Both subunits of soluble GCs from rat and bovine lung have been sequenced. The carboxyl domains of both subunits of 77-kDa (α subunit) and 70 kDa (ß subunit) are homologous to the carboxy region of catalytic region of membrane GCs; but none of them contains protein kinase like domain. The α- and ß- subunits share 45% sequence similarity. The α- and ß subunits showed highest mRNA levels in lung and brain.

The sequences of cDNA clones, encoding different membrane guanylyl cyclases has been obtained. Although some of the sequences appear to encode the same protein from different species, proteins from sea urchin testis, from mammaliam tissues that appear to be natriuretic peptide receptors, and from rat small intestinal mucosa have been identified. Hydropathic analysis of the deduced amino acid sequences suggests that all of the membrane guanylyl cyclases contain a single transmembrane domain. The intracellular region, which represents about half of the protein, contains c-GMP dependent protein kinase like domain and a cyclase catalytic domain in carboxyl region to produce c-GMP. The chromosomal locations of genes encoding the human GC-A, GC-B, and ANP- C-receptor have been assigned as follows: the GC-A gene to 1q21-22, the GC-B gene to 9p12-p21, and the ANP-C-receptor gene to 5p13-14. The rat GC-A gene contains 22 exons spaced over a total of 17.5-kb pairs. The 5' flanking region has shown the presence of Sp1–binding sites and no TATA box.

18.10.1 Natriuretic Peptides of Egg as Ligand of Guanylyl-Cyclase

Several related peptides have been shown to cause vasodilation, diuresis and natriuresis, and have been designated atrial natriuretic peptide (ANP), brain natriuretic peptide (BNP), and C-type natriuretic peptide (CNP). Study on the mechanism of action of these natriuretic peptides exhibits diversity of function in reproductive system and other systems that act via cGMP pathway. Most of the biological effects of natriuretic peptides appear to be mediated by activation of membrane bound GC activity. The ANP is also a chemoattractant of human sperm. To determine whether ANP as egg-derived peptides from sea urchins can act as a chemoattractant to human spermatozoa, Anderson et al (1995) suggested that a specific sperm receptor exits for the chemoattractant activity of ANP that is associated with guanylate cyclase. The chemoattractant activity of ANPs is independent of the presence of extracellular calcium ions and is independent of the action of ANP as a stimulus of the acrosome reaction. However, factors besides ANP are also responsible for the chemoattractant activity present in follicular fluid (Anderson et al, 1995). Peptides secreted by echinoderm eggs can also activate and cause directional changes in sperm motility in a species-specific manner. The ANP markedly stimulates acrosomal exocytosis of capacitated human spermatozoa and acts through cGMP pathway. Typically, ANP exerts some of its actions via activation of the ANP receptor (ANPR-A), a particulate guanylyl cyclase-linked receptor, and subsequent formation of cGMP. The ANP-stimulated acrosome reaction was inhibited by the competitive ANPR-A antagonist anantin, indicating that acrosome reaction is a receptor-mediated process. The stimulatory effect of ANP on acrosome reaction was mimicked by cGMP analogue, 8-bromo-cGMP. Although, the sperm motility can be stimulated by 8-bromo-cGMP, indicating that the egg peptides stimulate sperm motility by virtue of their effects on cGMP, the stimulation of motility was seen at peptide concentrations that did not elevate cGMP. Down-regulation of PKC and removal of extracellular Ca^{2+} influx abolished ANP induced acrosome reaction. It suggests that ANP via Ca^{2+} influx, PKC activation, and stimulation of particulate guanylyl cyclase may play a role in the induction of acrosome reaction of human spermatozoa (Rotem et al, 1998) (see Chapters 3 and 30 for NO action).

18.11. PHOSPHODIESTERASES IN TESTIS

Signal transduction via second messengers follows receptor activation by hormones or other stimuli. For production of the cyclic nucleotides, transduction mechanisms involve the activation of adenylate cyclase or guanylate cyclase for the synthesis of the appropriate cyclic nucleotide. Conversely, attenuation or modulation of cell response can involve controlled removal of these intracellular cyclic nucleotides. This catabolic regulation is accomplished by phosphodiesterases (PDEs), which convert cyclic nucleotides to monophosphates. The PDEs represent a diverse group of enzymes that have been classified into ten families according to their individual specificities for substrates (c-AMP, c-GMP) or cofactors (Ca^{2+}/calmodulin, cGMP) viz., c-AMP specific PDEs, cGMP-specific PDEs, Ca^{2+}/ calmodulin-stimulated PDEs, cGMP-stimulated PDEs, and c-GMP inhibited PDEs. The PDEs exist in various isoforms with tissue specific expression in testis, and dictate compartmentalization of c-AMP, which may dictate the specificity of signaling response. The use of c-AMP-PDE inhibitors and c-AMP analogues, resistant to PDE hydrolysis indicated that c-AMP specific PDEs may function during spermatogenesis and epididymal maturation of spermatozoa. The administration of dibutyryl-c-AMP has been shown to inhibit mitosis in spermatogonia.

18.11.1. PDE1 and PDE2 Genes

Four genes (PDE1/IVc, PDE2/IVa, PDE3/IVd, and PDE4/IVb) encoding different PDE isoforms that specifically hydrolyze c-AMP are present in the rat. These genes are differentially expressed in the somatic and germ cells of seminiferous epithelium of the testis. Multiple mRNAs for PDE1, PDE2 and PDE4 have been identified in rat male germ cells and their developmental expression has been investigated in purified spermatogenic cell populations. The PDE1 mRNAs (4.0 and 2.8-kb) were found to be abundant in pachytene spermatocytes, decreased in round spermatids and absent from condensing spermatids/ residual bodies. However, multiple forms of PDE2 mRNA transcripts (4.0, 3.5, 3.1, 1, 2.8 and 2.4-kb) were abundant in round spermatids and lower amounts in condensing spermatids/ residual bodies of rat testis. Transcripts related to rat PDE2 were also present in mouse round spermatids. Chromatography of germ cell cytosol indicated that the rat PDE2 enzyme may function during spermiogenesis and in spermatozoa (Welch et al, 1992). However, in another study, the signals corresponding to PDE1/IVc and PDE2/IVa mRNA were localized in two layers of seminiferous epithelium. The rat PDE1/IVc mRNA was present in the region of the epithelium corresponding to the region of middle-late pachytene spermatocytes (Morena et al., 1995).

18.11.2. PDE3 Isoforms

The PDE3/IVd and PDE4/IVb mRNAs are distributed throughout the span of seminiferous epithelium, and in the Sertoli cell cytoplasm. It indicated that different c-AMP-PDE genes are active in different cells of the seminiferous tubules, and the rat PDE3/IVd gene is expressed in the Sertoli cell in a cyclical fashion (Morena et al, 1995). Sertoli cells PDE is inhibited by 3-Methoxyethanol (ME) and its metabolite, methoxy acetic acid (MAA). It seems that although pachytene spermatocytes are selectively affected by ME and MAA, changes in gene expression detected in Sertoli cells could mediate the action of ME and MAA on spermatocytes indirectly (Syed and Hecht, 1998). The rat PDE 3/ IVd gene codes for multiple forms of mRNA with divergent 5' regions in mammalian cells. At least three mRNA derived from rat PDE-3/IVd gene (rat PDE 3.1, rat PDE 3.2, rat PDE 3.3 mRNAs) are present in Sertoli cells, thyroid and brain. Rat PDE 3.1 and PDE 3.2 mRNAs were dependent for hormone stimulation, while PDE

```
A
  1 GTQAGLRSGA PSGHRQLLGV GPTLSPSSTF YRFSTSGADS SSLSPPLPAL GRTREQTLLG
 61 ARSRGRVLLP AIPDPFLWRM ENLGVGEGAE ACSRLSRSRG RHSMTRAPKH LWRQPRRPIR
121 IQQRFYSDPD KSAGCRERDL SPRPELRKSR LSWPVSSCRR FDLENGLSCG RRALDPQSSP
181 GLGRIMQAPV PHSQRRESFL YRSDSDYELS PKAMSRNSSV ASDLHGEDMI VTPFAQVLAS
241 LRTVRSNVAA LARQQCLGAA KQGPVGNPSS SNQLPPAEDT GQKLALETLD ELDWCLDQLE
301 TLQTRHSVGE MASNKFKRIL NRELTHLSET SRSGNQVSEY ISRTFLDQQT EVELPKVTAE
361 EAPQPMSRIS GLHGLCHSAS LSSATVPRFG VQTDQEEQLA KELEDTNKWG LDVFKVAELS
421 GNQPLTAIIF SIFQERDLLK TFQIPADTLA TYLLMLEGHY HANVAYHNSL HAADVAQSTH
481 VLLATPALEA VFTDLEILAA LFASAIHDVD HPGVSNQFLI NTNSELALMY NDASVLENHH
541 LAVGFKLLQA ENCDIFQNLS AKQRLSLRRM VIDMVLATDM SKHMNLLADL KTMVETKKVT
601 SLGVLLLDNY SDRIQVLQNL VHCADLSNPT KPLPLYRQWT DRIMAEFFQQ GDRERESGLD
661 ISPMCDKHTA SVEKSQVGFI DYIAHPLWET WADLVHPDAQ DLLDTLEDNR EWYQSKIPRS
721 PSDLTNPERD GPDRFQFELT LEEAEEEDEE EEEGEETAL AKEALELPDT ELLSPEAGPD
781 PGDLPLDNQR T

B
  1 MSRISGLHGL CHSASLSSAT VPRFGVQTDQ EEQLAKELED TNKWGLDVFK VAELSGNQPL
 61 TAIIFSIFQE RDLLKTFQIP ADTLATYLLM LEGHYHANVA YHNSLHAADV AQSTHVLLAT
121 PALEAVFTDL EILAALFASA IHDVDHPGVS NQFLINTNSE LALMYNDASV LENHHLAVGF
181 KLLQAENCDI FQNLSAKQRL SLRRMVIDMV LATDMSKHMN LLADLKTMVE TKKVTSLGVL
241 LLDNYSDRIQ VLQNLVHCAD LSNPTKPLPL YRQWTDRIMA EFFQQGDRER ESGLDISPMC
301 DKHTASVEKS QVGFIDYIAH PLWETWADLV HPDAQDLLDT LEDNREWYQS KIPRSPSDLT
361 NPERDGPDRF QFELTLEEAE EEDEEEEEG EETALAKEAL ELPDTELLSP EAGPDPGDLP
421 LDNQRT
```

Fig18.10. Two variants of phosphodiesterase PDE4C from human lung and testis. (A) 791 amino acid variant from lung and (B) 426 amino acids variant from testis (Source: http//www. ncbi.nlm.nih.gov. accession numbers AAB96875 and AAB96876 respectively).

3.2 and PDE 3.3 were translated with properties identical to native cAMP dependent PDEs found in Sertoli cells (Sette et al, 1994).

18.11.3. PDE4 Isoforms

Type 4 phosphodiesterases (PDE4) hydrolyze cAMP with high affinity and are regulated by several intracellular signaling pathways. Four closely related genes encoding PDE4 are present in the mammalian genome and each gene encodes several different transcriptional units originating in at least 14 different PDE4 proteins. Phosphorylation of the PDE4D3 variant, but not of the PDE4D1 and PDE4D2 variants, was demonstrated in FRTL-5 cells. Similarly PDE4A5 form is activated by GH via a SP6 kinase-mediated phosphorylation (Salanova et al, 1999). Besides, it has been proposed that the heterogeneity at the amino terminus of the different PDE4s reflects differences in subcellular localization of the variants. Observations have led to the proposal that the amino terminus of different PDE4s contains signals for compartmentalization. An additional PDE4A variant, PDE4A4, contains a polyproline region that interacts with SH3 domains in reconstituted system. The human counterpart is present in both the soluble and particulate fraction of COS cells. Although these observations are suggestive of differential targeting of different PDE4s, most of the conclusions are based on over-expression of recombinant proteins, with little information on subcellular localization of the native PDE4. Using germ cells as a model, Salanova et al (1999) compared the expression and compartmentalization of variants derived from two PDE4 genes during spermatid development in situ. A PDE4A splicing variant was found to accumulate in round spermatids, while PDE4D variant expressed in pachytene spermatocytes and during the spermatid elongation phase (Salanova et al, 1999). In addition to the presence of PDE4D variant in germ cells, the PDE4D protein is expressed in Sertoli cells of the seminiferous tubules. The expression of this protein is dependent of FSH stimulation of the Sertoli cells. With the elucidation of the structure of the PDE4D gene, it has emerged that five different splicing variants are derived from this gene. The short PDE4D1 and PDE4D2 variants are expressed in immature Sertoli cell and regulated by c-AMP. The long variants (PDE4D4, PDE4D4 and PDE4D5) are regulated by PKA dependent phosphorylation and may be targeted to different subcellular structures.

The PDE4A isoforms expressed in the rat testis were compared to the variants expressed in the brain. The mRNAs coding PDE4A are present in mature rat and mouse germ cells, with corresponding apparent masses of 98.8 and 86 kDa and different from those proteins present in the brain extracts (113 and 76 kDa). The RT-PCR of the different splicing mRNA variants expressed in testis confirmed the presence of atleast one novel PDE4A mRNA that is distinct from the PDE4A splicing variants identified in the other tissues. The expression of two isoform of PDE4A is maximal in round spermatids and spermatozoa, and absent from Sertoli and interstitial cells. The testes isoform of 98.8kDa is also present in epididymal sperm (Naro et al, 1996). Rat testis at different ages of development showed two transcripts (PDE4D1 and PDE4D2) expressed at day 10 and 15 of age and became undetectable thereafter. An additional The PDE4D transcript appeared at day 30-day and increased during testis maturation. This latter transcript codes for a long variant of the PDE4D gene and is expressed in germ cells as demonstrated in isolated pachytene spermatocytes and round spermatids. The presence of corresponding PDE4D protein with a molecular mass of 98-kDa was established by immunoprecipitation and western blots. The PDE4A transcripts were also expressed in pachytene spermatocytes and round spermatids. Two polypeptides encoded by these PDE4A transcripts were expressed in pachytene spermatocytes, reached a maximum in round spermatids and declined there after. The PPE4D was localized in the manchette and periacrosome region of spermatids. Conversely, the PDE4A was mostly soluble in the cytoplasm of round spermatids. Thus, PDE4D and PDE4A variants are expressed at different stage and localized to distinct subcellular structures of developing spermatids (Salanova et al, 1999)

Human PDE4C : Obernolte et al (1997) cloned multiple splice variants of human *PDE4C*. The two splice variants, *PDE4C-791* and *PDE4C-426*, were isolated from a fetal lung library. The longest open reading frame (ORF) of 791 aa is encoded by *PDE4C-791*, which is similar to a c-DNA described by Engels et al (1995) except that an alternative 5' end sequence upstream of the first methionine extends the *PDE4C-791* ORF by 79 amino acids. The *PDE4C-426* variant contains 3 insertions that are located 5' to the catalytic domain and encode several in-frame stop codons. The predicted 426 amino acid protein initiates at a methionine 365 aminoacid within PDE4C-791. Two additional splice variants, *PDE4C-Δ54* and *PDE4C-Δ109* were found in testis mRNA. The *PDE4C-Δ54* contained a novel 5' end region and a deletion of 162 nt; the predicted protein deletes 54 amino acids from the amino terminal region (**Fig.18.10**). The *PDE4C-Δ54* protein produced in baculovirus infected cells was enzymatically active and sensitive to PDE4-specific inhibitors. The PDE4C - Δ109 protein is similar to PDE4C- Δ54 but has an additional 55 amino acids deleted in the catalytic domain: it lacked enzymatic activity. Analysis of uncloned total mRNA from other tissues confirmed the presence of mRNAs with the two deletions and three insertions that were observed in cDNA clones. The *PDE4C- Δ54* variant was found in testis and 5' extended region of *PDE4C-791* was present only in lung and melenoma cell line G361 (Obernolte et al, 1997).

18.11.4. Other PDE Isoforms

PDE 8: Cyclic nucleotide phosphodiestease type 8 hydrolyzes cAMP with high affinity. Rat PDE8A is a protein of 823 amino acid residues, whereas PDE8B protein is predicted as an N-terminal truncated form of 760 amino acid residues. Both of the rat PDE8 proteins include REC, PAS and catalytic PDE domains. Rat PDE8A transcripts were rich in the liver and testis, and those of rat PDE8B were particularly abundant in the brain (Kobayashi et al., 2003).

PDE11: The c-DNAs encoding PDE11A were isolated by RT-PCR. The catalytic domain of PDE11A was identical to that of PDE 11A1 (490 amino acids) but had distinct N-terminus from PDE11A1. The PDA11A3 cDNA encoded a 684-amino acid protein including one complete and one incomplete GAF domain in the N-terminal region. The protein PDE11A4, which was composed of 934 amino acids including two complete GAF domains and shared 630 c-terminal amino acids with PDE11A3, had a distinct N-terminus having the putative phosphorylation sites for c-AMP and c-GMP dependent protein kinases. The PDE11A4, but not PDE11A3, was phosphorylated by both c-AMP and cGMP-dependent protein kinases in vitro. The *PDE11A3* transcripts were specifically expressed in testis, whereas *PDE11A4* transcripts were particularly abundant in prostate. This indicated that the *PDE11*A gene undergoes tissue-specific alternative splicing that generates structurally and functionally distinct products (Yuassa et al, 2000).

The Sperm PDEs: Evidence was provided for the involvement of PDE in sperm motility and capacitation. The elevation of c-AMP levels in the presence of PDE inhibitors such as methylisobutylxanthine or caffeine is accompanied by a marked stimulation of spermatozoan respiration and motility. Of the 10 families of PDE that exist in mammalian tissues, the calcium – calmodulin – dependent (type 1), the c-AMP-specific (type 4), and cGMP specific PDE (type 5) have been found in human spermatozoa. Sildenafil is a potent inhibitor of human sperm PDE (type 5). Sildenafil inhibited PDE when c-AMP and c-GMP were used as substrates. The inhibitor increased c-AMP concentration and tyrosine phosphorylation of two fibrous sheath proteins (p105/81) during capacitation. Thus sildenafil triggers human sperm motility and capacitation via its inhibitory action on PDE (type 5) with a resultant rise in c-AMP levels (Lefievre et al, 2000). The calcium-calmodulin dependent PDE type 1 (PDE 1) and the cAMP-specific PDE type 4 (PDE 4) have been implicated in sperm function. Specific PDEs regulate capacitation of bovine sperm in a manner independent of those that mediate motility (Fournier et al., 2003; Lefievre et al., 2002).

18.12. c-AMP-GEF PATHWAY

Although cAMP and PKA have long been shown to mediate specific intracellular signaling events including the transcription of specific genes via the CREB-CBP complex, recent observations have indicated that PKA does not account for all of the intracellular targets of cAMP. For example, thyroid stimulating hormone (TSH) stimulation of thyroid cell proliferation is not completely blocked by PKA inhibitors. The TSH and FSH can stimulate protein kinase B (PKB) phosphorylation by a PKA independent, PI3-K/PDK1-dependent pathway. An FSH inducible kinase, Sgk, has been shown to be a close relative of PKB. The Sgk is also a target of PI3-K-PDK1 pathway, indicating that some effects previously ascribed to PKB may be mediated by this inducible kinase. The identification of cAMP-binding proteins that exhibit guanine nucleotide exchange (GEF) activity (cAMP-GEFs; Epacs) has opened new doors for cAMP action that include activation of small GTPases such as Rap1a, Rap2, and possibly Ras. These GTPases are known activators of downstream kinase cascades, including p38MAPK and ERK1/2 as well as PI3-K. Thus, FSH and TSH activation of PKB and Sgk may occur via this alternative cAMP pathway that involves cAMP-GEFs and the activation of the PI3-K/ PDK1 pathway (Richards, 2001).

18.13. REFERENCES

Adeoya-Osiguwa SA, Fraser LR. Capacitation state-dependent changes in adenosine receptors and their regulation of adenylyl cyclase/cAMP. Mol Reprod Dev. 2002; 63: 245-55.

Allegrucci C, Liguori L, Minelli A. Stimulation by n6-cyclopentyladenosine of A1 adenosine receptors, coupled to $G\alpha_{i2}$ protein subunit, has a capacitative effect on human spermatozoa. Biol Reprod 2001; 64 1653-9.

Anderson RA Jr, Feathergill KA, Rawlins RG, Mack SR, Zaneveld LJ. Atrial natriuretic peptide: a chemoattractant of human spermatozoa by a guanylate cyclase-dependent pathway. Mol Reprod Dev 1995; 40: 371-8.

Baxendale RW, Fraser LR. Evidence for multiple distinctly localized adenylyl cyclase isoforms in mammalian spermatozoa. Mol Reprod Dev 2003; 66: 181-9.

Beltran C, Zapata O, Darszon A. Membrane potential regulates sea urchin sperm adenylylcylcyclase. Biochem 1996; 11: 7591-8.

Bentley JK, Garbers DL, Bomino SE, Noland TD, Van Dop C. Spermatozoa contain a guanine nucleotide-binding protein ADP-ribosylated by pertussis toxin. Biochem Biophys Res Commun 1986; 138: 728-34.

Berruti G. A novel rap1/B-raf/14-3-3 theta protein complex is formed in vivo during the morphogenetic differentiation of postmeiotic male germ cells. Exp Cell Res 2000; 257: 172-9.

Bhat SG, Nie Z, Ramkumar V. Cisplatin up-regulates adenosine A(1) receptors in rat testes. Eur J Pharmacol 1999; 1: 35-43.

Bos JL, Franke B, M'Rabet L, Reedquist K, Zwarkruis F. In search of a function for the Ras-like GTPase Rap1. FEBS Lett 1997; 410: 59-62.

Brandelli A, Miranda PV, Tezon JG. Voltage-dependent calcium channels and Gi regulatory proteinmediate the human sperm acrosomal exocytosis induced by N-acetylglucosaminly/mannosyl neoglycoproteins. J Androl 1996; 17: 522-9.

Buck J, Sinclair ML, Schapal L, Cann MJ, Levin LR.Cytosolic adenylyl cyclase defines a unique signaling molecule in mammals. Proc Natl Acad Sci USA 1999; 96: 79-84.

Carazo-Salas RE, Gruss OJ, Mattaj IW, Karsenti E. Ran-GTP coordinates regulation of microtubules nucleation and dynamics during mitotic spindle assembly. Nat Cell Biol 2001; 3: 228-34.

Castellano LE, Martinez-Cadena G, Lopez-Godinez J, Obregon A, Garcia-Soto J. Eur J Cell Biol 1997; 74: 329-35.

Chen Y, Cann MJ, Litvin TN, Iourgenko V, Sinclair ML, Levin LR, Buck J. Soluble adenylyl cyclase as an evolutionarily conserved bicarbonate sensor. Science 2000; 28: 625-8.

Chinkers M, Garbers DL. Signal transduction by guanylyl cyclases. Ann Rev Biochem 1991; 60: 553-73.

Defer N, Marinx O, Poyard M, Lienard MO, Jegou B, Hanoune J. The olfactory adenylyl cyclase type 3 is expressed in male germ cells. FEBS Lett 1998; 424: 216-20.

Fenichel P, Gharib A, Emiliozzi C, Donzeau M, Menezo Y. Stimulation of human sperm during capacitation in vitro by an adenosine agonist, with specificity for A2 receptors. Biol Reprod 1996; 54: 1405-11.

Fournier V, Leclerc P, Cormier N, Bailey JL. Implication of calmodulin-dependent phosphodiesterase type 1 during bovine sperm capacitation. J Androl. 2003; 24: 104-12.

Fraser LR, Adeoya-Osiguwa S. Fertilization promoting peptide-a possible regulator of sperm function in vivo. Vitam Horm 2001; 63: 1-28.

Fraser LR, Duncan AE. Adenosine analogues with specificity for A2 receptors bind to mouse spermatozoa and stimulate adenylate cyclase activity in uncapacitated suspensions. J Reprod Fertil 1993; 98: 187-94.

Fraser LR, Hosseini R, Hanyaloglu AM, Talmor A, Dudley RK. TCO-11, the product of a mouse t-complex gene, plays a role in stimulation of capacitation and inhibition of the spontaneous acrosome reaction. Mol Reprod Dev 1997; 48: 375-82.

Fujita A, Nakamura K, Kato T, Watanabe N, Ishizaki T, Kimura K, Mizoguchi A, Narumiya S. Ropporin, a sperm specific binding protein of rhophilin, that is localized in the fibrous sheath of sperm falgella. J Cell Sci 2000; 113(Pt 1): 103-12.

Gautier-Courteille C, Salanova M, Conti M. The olfactory adenylyl cyclase III is expressed in rat germ cells during spermatogenesis. Endocrinology 1998; 139:2588-99.

Glassner M, Jones J, Kligman I, Woolkalis MJ, Gerton GL, Kopf GS. Immunocytochemical and biochemical characterization of guanine nucleotide-binding regulatory proteins in mammalian spermatozoa. Dev Biol 1991; 146: 438-50.

Green CM, Cockle SM, Watson PF, Fraser LR. A possible mechanism of action for fertilization promoting peptide, a TRH releted tripeptide that promotes capacitation and fertilizing ability in the mammalian spermatozoa. Mol Reprod Dev 1996; 45: 244-52.

Hall C, Sin WC, Teo M, Michael GJ, Smih P, Dong JM,et al. α2-chimerin, an SH2-containing GTPase-activating protein for the ras-related protein p21[rac] derived by alternate splicing of the human n-chimerin gene, is selectivity expressed in brain regions and testes. Mol Cell Biol 1993;13: 4986-98.

Haugen TB, Eskild W, Hansson V. Evidence for a novel splice variant of the α subunit of Go in rat male haploid germ cells. Biochem Biophys Res Commun 1992; 168: 41-7.

Haugen TB, Paulssen Ej, Gordeladze JO, Hansson V. A unique mRNA species for the α subunit Gs is present in rat haploid germ cells. Biochem Biophys Res Commun 1990; 168: 91-8.

Haugen TB, Paulssen RH, Hansson V. Cell-specific expression of Gq/11 protein and mRNA in rat seminiferous tubules. FEBS Lett 1993; 329: 96-8.

Hetzer M, Bilbao-Cortes D, Walther TC, Gruss OJ, Mattaj IW. GTP hydrolysis by Ran is required for nuclear envelope assembly. Mol Cell 2000; 5: 1013-24.

Hinsch KD, Hinsch E, Aumuller G, Tychowiecka I, Schultz G, Schill WB. Immunological identification of G protein α-and β-subunits in tail membranes of bovine spermatozoa. Biol Reprod 1992; 47: 337-46.

Hinsch KD, Schwerdel C, Habermann B, Schill WB, Muller-Schlosser F, Hinsch E. Identification of heterotrimeric G proteins in human sperm tail membranes. Mol Reprod Dev 1995; 40: 345-54.

Iida H, Yoshinaga Y, Tanaka S, Toshimori K, Mori T. Identification of Rab3A GTPase as an acrosome-associated small GTPase-binding protein in rat sperm. Dev Biol 1999; 211: 144-55.

Jacobs S, Schurmann A, Becker W, Bockers TM, Copeland NG, Jenkins NA, Joost HG. The mouse ADP-ribosylation factor-like 4 gene: two separate promoters direct specific transcription in tissues and testicular germ cells. Biochem J 1998; 335(Pt2): 259-65.

Jungnickel MK, Marrero H, Birnbaumer L, Lemos JR, Florman HM. Trp2 regulates entry of Ca^{2+} into mouse sperm triggered by egg ZP3. Nat Cell Biol 2001; 3: 499-502.

Karnik NS, Newman S, Kopf GS, Gerton GL. Developmental expression of G protein α subunits in mouse spermatogenic cells: evidence that $G\alpha_i$ is associated with the developing acrosome. Dev Biol 1992; 152: 393-402.

Kaziro Y, Itoh H, Kozasa T, Nakafuku, Satoh T. Structure and function of signal-transducing GTP-binding proteins. Annu Rev Biochem 1991; 60: 349-400.

Kierszenbaum AL, Gil M, Rivkin E, Tres LL. Ran, a GTP-binding protein involved in nucleocytoplasmic transport and microtubule nucleation, relocates from the manchette to the centrosome region during rat spermiogenesis. Mol Reprod Dev 2002; 63: 131-40.

Kierszenbaum AL. Epididymal G protein-coupled receptor (GPCR): two hats and a two-piece suit tailored at the GPS motif. Mol Reprod Dev. 2003; 64: 1-3.

Kobayashi T, Gamanuma M, Sasaki T, Yamashita Y, Yuasa K, Kotera J, Omori K. Molecular comparison of rat cyclic nucleotide phosphodiesterase 8 family: unique expression of PDE8B in rat brain. Gene 2003; 319: 21-31.

Kopf GS, Woolkalis MJ, Gerton GL. Evidence for a guanine nucleotide binding regulatory protein in invertebrate and mammalian sperm. Identification by islet-activating protein-catalyzed ADP-ribosylation and immunochemical methods. J Biol Chem 1986; 5: 7327-31.

Kusano A, Staber C, Chan HY, Ganetzky B. Closing the (Ran)GAP on segregation distortion in Drosophila. Bioessays 2003; 25: 108-15.

Lambright DG, Sondek J, Bohm A, Skiba NP, Hamm HE, Sigler PB. The 2 A° crystal structure of a heterotrimeric G protein. Nature 1996; 379: 311-19

Leclerc P, Kopf GS. Evidence for the role of heterotrimeric guanine nucleotide-binding regulatory proteins in the regulation of the mouse sperm adenylyl cyclase by the egg's zona pellucida. J Androl 1999; 20: 126-34.

Lefevre L, de Lamirande E, Gagnon C. The cyclic GMP-specific phosphodiestrase inhibitor, sildenafil, stimulates human sperm motility and capacitation but not acrosome reation. J Androl 2000; **21: 929-37.**

Lefievre L, de Lamirande E, Gagnon C. Presence of cyclic nucleotide phosphodiesterases PDE1A, existing as a stable complex with calmodulin, and PDE3A in human spermatozoa. Biol Reprod 2002; 67: 423-30.

Leung T, How BE, Manser E, Lim L Germ cell β-chimaerin, a new GTPase-activating protein for p21rac, is specifically expressed during the acrosomal assembly stage in rat testis. J Biol Chem 1993;268: 3813-6.

Li M, Funahashi H, Mbikay M, Shioda S, Arimura A. Pituitary adenylate cyclase activating polypeptide-mediated intracrine signaling in the testicular germ cells. Endocrine 2004; 23: 59-76.

Lopez-Casas PP, Lopez-Fernandez LA, Parraga M, Krimer DB, del Mazo J. Developmental regulation of expression of Ran/M1 and Ran/M2 isoforms of Ran-GTPase in mouse testis. Int J Dev Biol 2003; 47:

307-10.
Merlet F, Weinstein LS, Goldsmith PK, Rarick T, Hall JL, Bisson JP, de Mazancourt P. Identification and localization of G protein subunits in human spermatozoa. Mol Hum Reprod 1999; 5: 38-45.
Michaut M, Tomes CN, De Blas G, Yunes R, Mayorga LS. Calcium triggered acrosomal exocytosis in human spermatozoa requires the coordinated activation of Rab3A and N-ethylmaleimide-sensitive factor. Proc Natl Acad Sci 2000; 97: 9996-100011.
Miki K, Eddy EM. Identification of tethering domains for protein kinase A type Iα regulatory subunits on sperm fibrous sheath protein FSC1. J Biol Chem 1998; 273: 34384-90.
Minelli A, Allegrucci C, Mezzasoma I. [3H]-(R) – N6-phenylisopropyladenosine agonist binding to the solubilized A1 adenosine receptor from bovine epididymal spermatozoa. Arch Biochem Biophys 1997; 337: 54-61.
Mishima K, Price SR, Nightingale MS, Kousvelari E, Moss J, Vaughan M. Regulation of ADP-ribosylation factor (ARF) expression. Cross-species conservation of the developmental and tissue-specific alternative polyadenylation of ARF 4 mRNA. J Biol Chem 1992; 267: 24109-16.
Modarressi MH, Taylor KE, Wolfe J. Cloning characterization, and mapping of the gene encoding the human G protein γ2 subunit. Biochem Biophys Res Commun 2000; 272: 610-5.
Morena Ar, Boitani C, de Grossi S, Stefanini M, Conti M. Stage and cell specific expression of the adenosine 3',5' monophosphate-phophodiestrase genes in the rat seminiferous epithelium. Endocrinology 1995; 136: 687-95.
Nakamura K, Fujita A, Murata T, Watanabe G, Mori C, Fuijta J, Watanabe N. Rhophilin, a small GTPase Rho-binding protein, is abundantly expressed in the mouse testis and localized in the principal piece of the sperm tail. FEBS Lett 1999; 445: 9-13.
Naro F, Zhang R, Conti M. Developmental regulation of unique adenosine3',5'-monophosphate-specific phosphodiestrase variants during rat spermatogenesis. Endocrinology 1996; 137: 2644-72.
Naz RK, Ahmad K, Kaplan P. Expression and function of ras proto-oncogene proteins in human sperm cells. J Cell Sci 1992; 102(Pt 31): 487-94.
Nef P. Early events in olfaction: Diversity and spatial patterns of odorant receptors. Receptor Channels 1993;1: 259-66.
Obernolte R, Ratzliff J, Baecker PA, Daniels DV, Zuppan P, Jarnagin K, Shelton ER. Multiple splice variants of phosphodiestrase PDE4C cloned from human lung and testis. Biochim Biophys Acta 1997; 1353: 287-97.
Parmentier M, Libert F, Schurmans S, Schiffmann S, Lefort A, Eggerickx D, Ledent C, Mollereau C, Gerard C, Perret J et al. Expression of members of the putative olfactory receptor gene family in mammalian germ cells. Nature 1992; 355: 453-5.
Paulssen RH, Paulssen EJ, Gordeladze JO, Hansson V and Haugen TB. Cell specific expression of guanine nucleotide binding proteins in rat testicular cells. Biol Reprod 1991; 45: 566-71.
Perego L, Berruti G. Molecular cloning and tissue-specific expression of the mouse homologue of the rat brain 14.3-3 theta protein: characterization of its cellular and developmental pattern of expression in the male germ line. Mol Reprod Dev 1997; 47: 370-9.
Richards JS. New signaling pathways for hormones and cyclic adenosine 3', 5'- monophosphate action in endocrine cells. Mol Endocrinol 2001; 15: 209-18.
Rotem R, Zamir N, Keynan N, Barkan D, Breitbart H, Naor Z. Atrial natriuretic peptide induces acrosomal exocytosis of human spermatozoa. Am J Physiol 1998; 274 (2 Pt1):E 218-23.
Salanova M, ChunSY, Iona S, Puri C, Stefanini M, Conti M. Type 4 cyclic adenosine monophosphate-specific phophodiestrases are expressed in discrete subcellular compartments during rat spermiogenesis. Endocrinology 1999; 140: 2464-72.
Sasaki T and Takai Y. The Rho small G protein family-Rho GDI system as a temporal and spatial determinant for cytoskeletel control. Biochem Biophy Res Comm 1998; 245: 641-45.
Sazer S, Dasso M. The Ran decathlon : multiple roles of Ran. J Cell Sci 2000;113:111-18.
Scita G, Tenca P, Frittoli E, Tocchetti A, Innocenti M, Giardina G, Fiore PPD. Signaling from Ras to Rac and beyond not just a matter of GEFs. The EMBO Journal 2000; 19: 2393-98.
Sette C, Vicini E, Conti M. The rat PDE3/IVd phosphodiestrase gene codes for multiple proteins differentially activated by cAMP-dependent protein kinase. J Biol Chem 1994; 269: 18271-4.
Shen MR, Linden J, Chen SS, Wu SN. Identification of adenosine receptors in human spermatozoa. Clin Exp Pharmacol Physiol 1993; 20: 527-34.
Shi X, Amindari S, Paruchuru K, Skalla D, Burkin H, Shur BD, Miller DJ. Cell surface β-1, 4-galactosyltransferase-I activates G protein-dependent exocytotic signaling. Development 2001; 128: 645-54.
Sinclair ML, Wang XY, Mattia M, Conti M, Buck J, Wolgemuth DJ, Levin LR. Specific expression of soluble adenylyl cyclase in male germ cells. Mol Reprod Dev 2000; 56: 6-11.

Strathmann M, Wilkie TM, Simon MI. Alternative splicing produces transcripts encoding two forms of the α subunit of GTP-binding protein Go. Proc Natl Sci 1990; 87: 6477-81.

Strathmann M, Wilkie TM, Simon MI. Diversity of the G-protein family sequences from live additional α subunits in the mouse. Proc Natl Acad Sci USA 1989; 86: 7407-9.

Syed V, Hecht NB. Rat pachytene spermatocytes down-regulate a polo-like kinase and upregulate a thiol-specific antioxidant protein, whereas Sertoli cells down-regulate a phosphodiestrase and up-regulate an oxidative stress protein after exposure in methoxyethanol and methoxyacetic acid. Endocrinology 1998; 139: 3503-11.

Thomson M. What are guanosine triphosphate-binding proteins doing in mitochondria. Biochim Biophys Acta 1998; 1403: 211-8.

Togawa A, Miyoshi J, Ishizaki H, Tanaka M, Takakura A et al. Progressive impairment of kidneys and reproductive organs in mice lacking Rho GD1α. Oncogene 1999; 18: 5373-80.

Toure A, Dorseuil O, Morin L, Timmons P, Jegou B, Reibel L, Gacon G. MgcRacGAP, a new human GTPase activating protein for Rac and Cdc42 similar to Drosophila rotundRacGAP gene product, is expressed in male germ cells. J Biol Chem 1998; 273: 6019-23.

Toure A, Morin L, Pineau C, Becq F, Dorseuil O, Gacon G. Tat-l, a novel sulfate transporter specifically expressed in human male germ cells and potentially linked to rho-GTPase signaling. J Biol Chem 2001; 276: 20309-15.

Valentijn JA, Jamieson JD. On the role of rab GTPases : what can be learned from the developing pancreas. Biochem Biophys Res Comm 1998; 243: 331-36.

Vijayaraghavan S, Liberty GA, Mohan J, Winfery VP, Olson GF, Carr DW. Isolation and molecular characterization of AKAP110, a novel, sperm-specific protein kinase A-Anchoring protein. Mol Endocrinol 1999; 13: 705-17.

Walensky LD, Roskams AJ, Lefkowitz RJ, Snyder SH, Ronnett GV. Odorant receptors and desensitization proteins colocalize in mammalian sperm. Mol Med 1995; 1: 130-41.

Ward CR, Faundes D, Foster JA. The monomeric GTP binding protein, rab3a, is associated with the acrosome in mouse sperm. Mol Reprod Dev 1999; 53: 413-21.

Ward CR, Storey BT, Kopf GS. Activation of a Gi protein in mouse sperm membranes by solubilized proteins of the zona pellucida, the egg's extracellular matrix. J Biol Chem 1992; 267: 14061-7.

Welch JE, Swinnen JV, O'Brien DA, Eddy EM, Conti M. Unique adenosine 3', 5'-monophosphate-phosphodiestrase messenger ribonucleic acids in rat spermatogenic cells: evidence for differential gene expression during spermatogenesis. Biol Reprod 1992; 46: 1027-33.

Weyand I, Godde M, Frings S, Weiner J, Muller F, Altenhofen W, Hatt H, Kaupp UB. Cloning and functional expression of a cyclic-nucleotide gated channel from mammalian sperm. Nature 1994; 368 (6474); 859-63.

Wilde A, Zheng Y. Stimulation of microtubule aster formation and spindle assembly by the small GTPase Ran. Science 1999; 284(5418): 1359-62.

Wilde MW, Ward CR, Kopf GS. Activation of a G protein in mouse sperm by the zona pellucida, an egg-associated extracellular matrix. Mol Reprod Dev 1992; 31: 297-306.

Wolfes H, Kogawa K, Millette CF, Cooper GM. Specific expression of nuclear proto-oncogenes before entry into meiotic prophase of spermatogenesis. Science 1989; 245 (4919): 740-3.

Yuasa K, Omori K, Yanaka N. Binding and phosphorylation of a novel male germ cell-specific cGMP-dependent protein kinase-anchoring protein by cGMP-dependent protein kinase Iα. J Biol Chem 2000; 275: 4897-95.

Yunes R, Michaut M, Tomes C, Mayorga LS. Rab3A triggers the acrosome reaction in permeabilized human spermatozoa. Biol Reprod 2000; 62: 1084-9.

Zhang C, Yu Y, Zhang S, Liu M, Xing G, Wei H, et al. Characterization, chromosomal assignment and tissue expression of a novel human gene belonging to the ARF GAP family. Genomics 2000; 63: 400-8.

Chapter 19

PROTEIN KINASES

Protein phosphorylation is an important process in the control of various cell functions. To fulfill various cell functions through phosphorylation, it has been postulated that the human genome encodes ~2,000 different kinases, representing ~2-4% of encoded genes. Although several protein kinases have been shown to be involved in spermatogenesis, only a few kinases have been characterized whose expression is restricted specifically to germ cells or to the testis. During last decade, a series of Ser/Thr protein kinases have been identified with strong expression pattern in testis. In addition to protooncogene products and cell cycle regulatory kinases already discussed (see Chapters 11 and 17), several non-specific protein kinases have been identified in spermatogenic cells. Based on their primary structures and substrate specificities, all protein kinases contain conserved residues that are important for nucleotide binding and phosphotransfer. The majority of known protein kinases fall into two mutually exclusive subgroups: Serine/threonine protein kinases and tyrosine protein kinases. Serine/threonine kinases contain residues, which are capable to discriminate different hydroxy amino acids phosphoreceptors. However, a small number of protein kinases have been identified that do not conform to these conventions. These kinases have sequence motifs that would initially suggest that they are serine /threonine specific, but they are able to autophosphorylate on both serine/threonine and tyrosine, and in some instance can also phosphorylate exogenous substrates on tyrosine. Examples of such dual specificity protein kinases include the *MCK1* gene product of *Cerevisiae*, the mammalian mitogen activated protein (MAP or ERK) kinases, which are implicated in mitotic and meiotic signaling, and the murine C1K/STγ protein kinase. Dual specificity protein kinases, therefore, appear to share an implicit involvement in signal transduction and in the control of mitosis and meiosis. Althoguh the *Cdc* gene has been generally considered as a universal regulator of cell cycle, information from *S. cervisiae* and *A nidulans* have suggested existence of alternative signal pathway involved in controlling mitosis. For example *nim A* gene of *A nidulans* works independently of Cdc2 for initiation of mitosis, for spindle formation and for nuclear envelope breakdown. *Nim A* gene products are conserved in mammalian cells as well (Letwin et al, 1992). This section would deal with those kinases which are exclusively or mostly expressed in spermatogenic cells.

I. SERINE/THREONINE KINASES

19.1. C-AMP DEPENDENT PROTEIN KINASE-A

The cAMP- dependent protein kinase (PKA) is a major enzyme in cellular signal transduction and is thought to mediate most of the physiological responses to cAMP in eukaryotic cells.

Below a cAMP threshold concentration, PKA exists as an inactive tetramer of two catalytic (C) and two regulatory (R) subunits (CR_2C). Two R subunits form a dimer with each protomer attaching to the substrate-binding site of α C subunit. Activation of adenylate cyclase by extracellular signals raises the intracellular concentration of cAMP, and at a certain threshold concentration cAMP binds to the R subunits of the PKA tetramer. Binding of cAMP to the enzyme elicits physical dissociation of the regulatory (R) and catalytic (C) subunits, resulting in phosphotransferase activity to phosphorylate its substrate. Free C subunits do not phosphorylate large number of proteins in cells. The RI subunits give rise to type I PKA, whereas RII subunits give rise to type II PKA. Sperm contain both RI and Rll isoforms of the regulatory subunits. Immunocytochemical analysis showed that Rlα and RIβ subunits are predominantly localized in the acrosomal region of the sperm whereas Rll is confined to the sperm flagellum. This differential localization implies distinct roles for the type-I and type II PKA isoforms in sperm function. Some isoforms of R also associate with binding proteins collectively termed PKA anchoring proteins. It is believed that through these interactions PKA is targeted to specific subcellular compartments.

Studies on PKAs involving cDNA cloning and sequencing have revealed an unexpected multiplicity in isoforms representing different gene products. Four different regulatory subunits and three different catalytic subunits for PKA have been identified, at least at the gene/mRNA level. These have been designated RIα, RIβ, RIIα , RIIβ, Cα, Cβ and Cγ. The $C_ß$ gene codes for atleast three isozymes (Cß1, Cß2 and Cß3) that arise from the use of three different exons, each of which contains a distinct initiator methionine codon. While Cα and $C_ß1$ are widely expressed, $C_ß2$ and $C_ß3$ are neuronal specific. AII $C_ß$ isoforms have similar activity in gene induction and interact equally well with R subunits of PKA. In addition, the c-DNA for catalytic subunit, Cγ from human testis is testis specific (Beebe et al., 1990), and demonstrated a testis – specific expression for this subunit.

19.1.1. Catalytic Sub-units of PKA

Somatic variant Cα : The catalytic subunit (C) Cα, of somatic PKA occurs in a wide variety of tissues. A number of eukaryotic proteins are known to be blocked and amino terminal is N-myristylated on glycine. In other proteins, the presence of myristate at the amino terminus helps target the protein to membranes and has been shown to be essential for the function of a number of kinases including Src, c-Abl, and cGMP-dependent protein kinase. The N-myristylation C subunit is conserved across species suggesting an essential function for this co-translation modification. However, N-myristylated cataliytic C isoforms coexist in an organism with non-myristylated isoforms. For example, the mouse Cß1 is N-myristylated, whereas the mouse Cβ2 and Cβ3 are not. In *C elegans*, the C subunit consists of at least 12 different isoforms which generate myristylated and nonmyristylated variants. Most of the Cα protein in sperm is another variant protein. The amino-terminal sequence of Cα2 is similar to that of ovine sperm Cs (San Agustin et al, 1998). Human sperm also expresses a highly related Cα2 homologue. The Cα2 subunit forms holoenzymes with either RIIα or RIα, and both activate at the same concentration of cyclic nucleotide.

Cα-s/Cs Variant of Human Sperm: The cDNA encoding a splice variant of the human Cα catalytic subunit of cAMP-dependent PKA was identified by Reinton et al (2000). This isoform differes in the N-terminal part of the deduced amino acid sequence, corresponding to the part encoded by exon 1 in the murine Cα gene, but similar to ovine Cα variant. The Cα-s mRNA was expressed exclusively in pachytene spermatocytes of human testis and the Cα-s protein was present in ejaculated human sperm. Specific antibody indicated that Cα-s was localized in midpiece region of sperm. The most of Cα-s was associated with particulate and could not be released from the sperm midpiece by cAMP treatment alone. Furthermore, RIα, RIIα and hAKAP220 anchor protein has been identified in this region of sperm that could target Cα-s (Reinton et al, 2000).

It seems that an aternative promoter, which is present in the first intron of the Cα gene, is transcriptonally active in male germ cells. Transcription from this promoter coincides with the appearance of pachytene spermatocytes and gives to a Cα2 that contains a distinctive 7 amino acid N-terminus of Cα1 (Desseyn et al, 2000). The Cα protein does not contain the myristylation signal present on Cα1 and migrates at a lower molecular weight on SDS/PAGE. Since, PKA plays a pivotal role in sperm motility and capacitation, the distinctive biochemical properties of the unmyristylated Cα2 may be essential for fertility in male.

C subunit of Ovine Sperm (Cs): Catalytic sub-unit of sperm (C_s) appears to have unusual solubility properties. It is generally accepted that, in the presence of cAMP, Cs is soluble in the cytoplasm. However, the Cs is bound to internal sperm structures by two types of bonds, one sensitive to detergent and the other one sensitive to cAMP. Mass spectrometry revealed that the mass of Cs from ram sperm was ~890 Da less than that of C from ovine striated muscle. The amino terminus of Cs was blocked, but partial sequence from internal fragments was an exact match to the sequence of bovine Cα. The Cs clearly differed in sequence from the Cβ and Cγ isoforms. An amino-terminal endoproteinase lysine-C fragment from Cs revealed that the amino-terminal 14 amino acids and amino-terminal myristate of Cα are replaced by six different amino acids and by amino-terminal acetate in Cs. The predicted mass difference due to this replacement is 899 Da. The region of homology between C_s and Cα begins at exactly the exon 1/exon 2 boundry in Cα, suggesting that the Cα gene results from an alternate exon Cα gene. The shorter amino terminus and lack of a myristate moiety may expose a hydrophobic portion of the catalytic core of Cs, allowing Cs to bind to a hydrophobic site within the flagellum and thus accounting for the unusual solubility properties of the subunit (San Agustin et al., 1998).

Cs does not contain the consensus myristylation motif but was acetylated on its amino-terminal alanine. Sequencing of cDNA clones encoding ovine Cs and Cα1 (the predominant somatic isoform) revealed that Cs is the product of an alternative transcript of the Cα gene. The Cs cDNA clones from murine and human testes indicated that Cs transcripts were detected only in testis and not in any other tissue, including ciliated tissues and ovaries. Now it is becoming clear that both Cα and Cs transcripts are present in testis; Cs is expressed specifically in spermatogenic cells and is the only C isoform detected in mature sperm. Immunohistochemistry of mouse testis using antibodies specific for Cs and Cα1 showed that Cα1 is present in somatic testicular cells, spermatogonia, and preleptotene spermatocytes but not in cells that are in later stages of spermatogenesis. In contrast, Cs is expressed only in mid-pachytene and later stage spermatocytes and in spermatids. In murine and ovine sperm, Cs is

```
  1 MGNAAAAKKG SEQESVKEFL AKAKEDFLKK WENPAQNTAH LDQFERIKTL GTGSFGRVML
 61 VKHTETGNHY AMKILDKQKV VKLKQIEHTL NEKRILQAVN FPFLVKLEFS FKDNSNLYMV
121 MEYVPGGEMF SHLRRIGRFS EPHARFYAAQ IVLTFEYLHS LDLIYRDLKP ENLLIDQQGY
181 IQVTDFGFAK RVKGRTWTLC GTPEYLAPEI ILSKGYNKAV DWWALGVLIY EMAAGYPPFF
241 ADQPIQIYEK IVSGKVRFPS HFSSDLKDLL RNLLQVDLTK RFGNLKNGVN DIKNHKWFAT
301 TDWIAIYQRK VEAPFIPKFK GPGDTSNFDD YEEEEIRVSI NEKCGKEFSE F
```

Fig.19.1. Alpha-catalytic subunit of cAMP-dependent protein kinase (C-alpha 1) from ovine sperm (Accession AAF76423). Source: http//www. ncbi.nlm.nih.gov. with permission from J.T San Agustin et al, J. Biol. Chem. 273, 24874-83: 1998 © American Society for Biochemistry and Molecular Biology.

```
  1 MGNAAAKKDT EQETVNEFLA KARGDFLYRW GNPAQNTASS DQFERLKTLG TGSYGRVMLV
 61 RHRETGNHYA MKILDKQKVV RLKQVEHTLN EKRILQAINF PFLVKLQFSF KDNSNLYLVM
121 EYVPGGEMFS HLRRVGRFSE PQACFYAAQV VLAFQYLHSL DLIHRDLKPE NLLIDQQGYL
181 QVTDFGFAKR VKGRTWTLCG TPEYLAPEI
```

Fig.19.2. Amino acid sequence of γ-catalytic subunit of cAMP-dependent protein kinase-A from monkey testis. Source:http//www.ncbi.nlm.nih.gov (Accession O62846).

located primarily in sperm tail components, including the midpiece mitochondria and the axoneme **(Fig19.1)**. Quantitative analysis of Western blots indicated that individual ovine sperm contain approximately 4 x 10^5 molecules of Cs, a seemingly large number for a protein that is active catalytically (San Agustin and Witman, 2001; San Agustin et al; 2000). The conservation of Cs throughout mammalian evolution suggests that the unique structure of Cs is important in the subunit's localization or function within the sperm. The deduced amino acid sequence of C-subunit of cAMP-dependent PKA from rainbow trout testis showed 75-80% identity to sequences previously reported in other organisms. However, the N-terminal regions of C subunit from the trout testis appeared slightly shorter than those from other tissues, suggesting its specificity to germ cells (Itoh et al., (2003).

Cβ 2 variant: The Cβ2 is a ubiquitously expressed protein with characteristic properties of a cAMP-dependent protein kinase catalytic subunit (Thullner et al, 2000). Like Cα, Cβ also is widely distributed but is expressed in lesser amounts than Cα in most tissues except the brain. Splice variants of both Cα and Cβ also are known. The β2, a 46kDa splice variant of the Cβ isoform is the largest isoform described for C subunits from PKA in mammals. It differs from Cβ in the first 15 N-terminal residues, which are replaced by a 62-residue domain with no similarity to other known proteins. The Cβ2 protein, identified in cardiac tissue by MS microsequencing, has a very low abundance of about 2% of total C subunits from bovine cardiac tissue. The abundance of the Cβ2 protein differs dramatically between tissues, with lowest level in testis. Two sites in the protein are phosphorylated: Thr-244 in the activation segment and Ser-385 close to the C-terminus.

Cγ : The Cγ appears to be expressed only in germ cells of human testis (Beebe et al, 1990). The cloning and characterization of the human and rhesus monkey genes encoding the testis-specific C-γ subunit have been described by Reinton et al (1998). The human C-γ gene is intronless with an open reading frame similar to the previously published cDNA sequence. The 3' and 5' flanking regions share high similarity with the Cα non-translated regions (82%) also outside the regions corresponding to the C-γ cDNA. The human gene is flanked by an Alu-related sequences in the 3' nontranslated region. The observation that the Cγ gene is intronless and collinear with Cα mRNA, together with the presence of remnants of a poly (A) tail and flanking direct repeats, indicates that the Cγ gene is a Cα -derived retroposon, specifically transcribed in primate testis germ cells. The 5'-flanking region of this gene has a high G/C content and a putative TATA box situated at 138 compared to the translation initiation codon. Cloning and sequencing of a partial C-γ rhesus monkey gene demonstrated conservation of the sequence in primates **(Fig 19.2)**.

19.1.2. Regulatory Subunits of PKA

ϕYen et al, (1990) documented the presence of significant mRNA levels for 5 different PKA subunits (RIα, RIβ, RIIα, RIIβ, and Cα) in germ cells and demonstrated differential expression patterns for these subunits during rat spermatogenesis. Messenger RNAs for RI (RIα and RIβ) and Cα appear to be induced at premeiotic germ cell stages, whereas mRNAs for RII (RIIα and RIIβ) are first expressed at haploid stages. The individual PKA subunits may convey specific functions in developing germ cells and mature sperm. Germ cell PKA-mRNA was smaller in size compared with somatic cell PKA-mRNA. Specific, truncated forms of RIα, RIIα, RIIβ and Cα mRNA appear to be selected in the germ cells.

In testis, there is a good correlation between the amount of mRNA and the respective immunoreactive PKA subunit proteins in Sertoli cells. In other types of cell, such as germ cells and Leydig tumour cells, this was not always the case. Furthermore, large amounts of small-sized, germ cell-specific mRNAs for RIα (1.7-kb) and RIIα (2.2-kb) were also found in the developing rat testis after 30 to 40 days of age, but the large amounts of mRNA were only partially reflected at the protein level. Pachytene spermatocytes and round spermatids were practically devoid of both RIIα and RIIβ protein. During spermatid differentiation, there was a decrease in RIα and an increase in RIIα protein. Distribution of the various PKA subunits in testicular cell types is cell specific (Landmark et al, 1993). Immunocytochemical staining revealed that RIα and RIβ subunits are both localized predominantly in the acrosomal segment of bovine sperm head. In addition RIß was observed in the midpiece of the tail while RIα was detected in the connecting piece. RIIα is prominent in the axonemal region of the flagellum, but not observed in the head region. The differential localization of different subunits suggested distinct roles for each of these isoforms in sperm functions such as motility and the acrosome reaction.

RI*α *: By using 5' RACE on rat testis cDNA, Dahle et al (2001) identified three alternatively spliced mRNAs of the RIα subunit of PKA that differed in their 5' untranslated regions. Two of these 5' regions showed similarity with the human RIα exons 1a and 1b, while the third mRNA (1c) constituted a novel mRNA splice variant. The 1c mRNA was specifically expressed in testis and only in post-meiotic germ cells. In contrast, the RIα1b and a RIα1a mRNAs were present both in pre-meiotic germ cells and somatic cells of the testis, and the expression of both RIα1a and RIα1b mRNAs were stimulated by cAMP in Sertoli cells. In sperm, the Rlα protein was expressed after meiosis, and targeted to various subcellular structures via anchoring proteins. The Rlα1c haploid-specific mRNA, therefore, may be important for the regulation of Rlα expression in sperm (Dahle et al, 2001). The carboxyl-terminal 19 amino acids of type RIα subunit of cAMP-dependent protein kinase were investigated to determine their contributions to cAMP selectivity (Kapphahn and Shabb, 1997). The parent RIα subunit contained an Ala to Thr mutation at position 334 so that it would bind both cAMP and cGMP with high affinity. Stop codons were introduced into the parent cDNA construct at positions corresponding to Val-375, Asn-372, Gln-370, and Cys-360. A transition in t ½ showed that Tyr 371 is an important cAMP-binding determinant. To test this further, Tyr-371 was mutated to Ala, Phe, and Arg in the parent construct. Activation constant (Ka) suggested that mutation of Tyr-371 enhanced B domain cGMP selectivity. The Ki for cGMP vs cAMP with isolated B domains containing Tyr 371-Arg or Tyr-371-Phe mutations showed that B domain cAMP selectivity was minimally affected by alteration of Tyr-371. These results showed that aromatic stacking is not important for determining B-domain cyclic nucleotide selectivity and that the main function of Tyr-371 is stabilization of the B-domain cAMP-binding pocket through hydrogen bonding with Glu-324.

RII*α *Subunit : A human-specific isoform of the RIIα subunit is present on the axonemal microtubule wall, whereas a different isoform of broader specificicity is present in the cytoplasm at the periphery of the coarse fibers and fibrous sheath of sperm. This isoform is also found in the mitochondria. The human-specific RIIα subunit is likely linked to microtubules by a unique binding protein of 72kDa. These findings are in agreement with the concept of a concerted mechanism involving phosphorylation of both the axonemal microtubules and the fibrous structures for the regulation of mammalian sperm motion (Pariset and Weinmàn 1994).

Promoter region of R IIα : Isolation and characterization of the 5' flanking region (1.2 kb) and exon 1 of the human RIIα gene of cAMP dependent PKA revealed the presence of a major transcriptional start site located at 208 nt upstream of start for translation. The 5'-flanking region of the RIIα gene did not contain a TATA box and was highly G/C rich. Several potential sites for transcription factors were identified in the 5' flanking sequence, which may be responsible for the germ cell-specific regulation of this gene. The human testis RIIα cDNA contains a region with little or no homology to the corresponding rat skeletal muscle cDNA. Whether this difference could arise due to organ-specific splice mechanisms or represented a species difference, Foss et al (1997) showed that the low homology region of RII α cDNA is found entirely in exon 1, and does not originate from a tissue-specific alternate splicing of this distinct region.

19.1.3. Functions of Protein Kinase A

The cAMP dependent protein kinase A plays a central role in sperm motility, capacitation and acrosome reaction (Visconti et al., 1997; Harrison et al., 2000; Skalhegg et al., 2002). Tyrosine phosphorylation in sperm is regulated either directly or indirectly by action of PKA on tyrosine kinase or phosphatases (Visconti et al., 2002).

Sperm Capacitation: Capacitation of epididymal mouse sperm in vitro is accompanied by a time-dependent increase in PKA activity. While PKA enzyme activity is found in both the soluble and insoluble fractions of the sperm, the increase in PKA activity accompanying capacitation is associated with enzyme activity found in the soluble fraction.The PKA appears to be functional in multiple compartments of sperm cell. A fraction of the insoluble PKA is presumably tethered by AKAP82, the major anchor protein of the fibrous sheath of the sperm flagellum, which anchors and compartmentalizes PKA to the cytoskeleton via the RII subunit of PKA. Binding domain of RII subunit- was mapped to a 57-amino-acid residue region at its N-terminus. This necessitates further understanding of the contribution of both the soluble and insoluble forms of PKA to sperm function (Visconti et al, 1997) (see Chapter 29).

Role in Protein Anchoring: Rat sperm flagellar contains polypeptides having molecular masses of 120, 80 and 57-kDa that are capable of binding the regulatory subunit of α type II cAMP-dependent protein kinase (RIIα). All three polypeptides are tightly associated with the fibrous sheath of sperm. The RIIα, possibly of 57-kDa and capable of phosphorylation by PKA can form tight, specific complexes with the fibrous sheath. It also demonstrated the presence of functional binding sites on RII. The N-terminal amino acids at positions 1-50 of RIIβ were the primary determinant for RII-binding protein interaction. The major 80-kDa polypeptide expressed during spermiogenesis, suggests a developmental role for RII anchoring to the sperm flagellar fibrous sheath (Macleod et al, 1994).

PKA in Cell cycle: The PKA oscillation is required for the phase following DNA replication and the PKA activity is needed from mitosis exit to the formation of the nuclear envelope. The PKA was not required for the assembly of ORC2, Cdc6 and MCM3 onto chromatin. Inhibition of PKA activity, however, blocked the release of Cdc6 from chromatin and subsequent DNA replication. It shows that PKA activation in late M phase is required for following S phase and chromosomal replication (Costanzo et al, 1999).

PKA in Cell Signaling: Tasken et al, (2001) demonstrated that Sertoli cells of the testis contain an insoluble PDE4D3 isoform, which is shown to target to centrosomes. Colocalization of PDE4D with PKA-RIIα and RIIβ in the centrosomal region and co-precipitation of RII subunits with PDE4D3 from cytoskeletal extracts indicated a physical association of the PKA and PDE proteins and with that of the centrosomal PKA-anchoring protein, AKAP450. Thus, a controsomal complex that includes PDE4D and PKA constitutes a novel signaling unit that may provide accurate spatio-temporal modulation of cAMP signals (Tasken et al, 2001).

19.1.4. PKA Inhibitor Proteins

The protein kinase inhibitor (PKI) proteins belong to a heterogeneous family of proteins. The protein kinase inhibitor family includes three genes encoding small, heat-stable inhibitors PKIα, PKIβ and PKIγ of the PKA. Each PK isoform contains a PKA inhibitory domain and a nuclear export domain, enabling PKI to both inhibit PKA and remove it from the nucleus. The PKIβ isoform, also known as testis PKI, is highly expressed in germ cells of the testis and is found at more modest levels in other tissues. In adult rat testis and cerebellum, several inhibitors with known or unknown composition are known, which are present as a monophospho form in testis, whereas no phosphorylation of these is evident in cerebellum. The PKIβ and PKIγ are the products of separate genes. The PKIβ form from rat testis is a 70-amino acid protein, but the genomic sequence suggested that an alternate form might exist, arising as a consequence of alternate translational initiation with 78 amino acids (PKIβ–78). PKIβ-78, like PKIβ-70, is a high affinity and specific inhibitor of the cAMP-dependent protein kinase. The PKIß-γ and PKIβ-X, in contrast, also inhibit the cGMP-dependent protein kinase (Kumar et al, 1997). The PKI β changes its form within germ cells with development that is initiated in testis tubules and continues as the germ cells migrate through the epididymis. In order to investigate its physiological role, Belyamani et al, (2001) generated PKIβ knockout mice by gene targeting. These mice exhibited a partial loss of PKI activity in testis but remained fertile with normal testis development and function. The PKIβ knockout females also reproduce normally. Remarkably double knockout mice were also viable and fertile with no obvious physiological defects in either males or females.

The PKIγ cDNA sequence predicted an open reading frame of 75 amino acids, showing 35% identity to PKIα and 30% identity to PKIβ. The FSH can increase the expression of both PKIα and PKIβ. Amino acid residues important for the high affinity of PKIα and PKIβ1 as well as nuclear export of the catalytic (C) subunit of cAMP-dependent protein kinase were found conserved in PKIγ. The 1.3-kb PKIγ message is widely expressed, with highest levels in heart, skeletal muscle, and testis (Collins and Uhler, 1997). The FSH- regulated expression of PKIα in the Sertoli cell likely occurs via normal route of second messenger signal transduction. In contrast, the FSH-dependent PKIβ expression must arise due to Sertoli cell –germ cell interaction (Van Patten et al., 1997).

19.2. PROTEIN KINASE-C

The protein kinase C (PKC), a multi-isozyme family is involved in synaptic transmission, memory, cellular growth, differentiation, transformation, metabolism, contraction, ion channel activity,

```
  1 MAPFLRISFN SYELGSLQAE DDASQPFCAV KMKEALTTDR GKTLVQKKPT MYPEWKSTFD
 61 AHIYEGRVIQ IVLMRAAEDP MSEVTVGVSV LAERCKKNNG KAEFWLDLQP QAKVLMCVQY
121 FLEDGDCKQS MRSEEEAMFP TMNRRGAIKQ AKIHYIKNHE FIATFFGQPT FCSVCKEFVW
181 GLNKQGYKCR QCNAAIHKKC IDKIIGRCTG TATNSRDTIF QKERFNIDMP HRFKVYNYMS
241 PTFCDHCGTL LWGLVKQGLK CEDCGMNVHH KCREKVANLC GINQKLLAEA LNQVTQKASR
301 KPETPETVGI YQGFEKKTAV SGNDIPDNNG TYGKIWEGSN RCRLENFTFQ KVLGKGSFGK
361 VLLAELKGKE RYFAIKYLKK DVVLIDDDVE CTMVEKRVLA LAWENPFLTH LICTFQTKDH
421 LFFVMEFLNG GDLMFHIQDK GRFELYRATF YAAEIICGLQ FLHGKGIIYR DLKLDNVMLD
481 KDGHIKIADF GMCKENIFGE NRASTFCGTP DYIAPEILQG LKYSFSVDWW SFGVLLYEML
541 IGQSPFHGDD EDELFESIRV DTPHYPRWIT KESKDIMEKL FERDPAKRLG VTGNIRLHPF
601 FKTINWNLLE KRKVEPPFKP KVKSPSDYSN FDPEFLNEKP QLSFSDKNLI DSMDQTAFKG
661 FSFVNPKYEQ FLE
```

Fig.19.3. Amino acid sequence of protein kinase Cδ from rat testis. Source:http//www.ncbi.nlm.nih.gov (Accession P09215).

exocytosis, and gene expression. The PKC activators are provided by various phospholipases (Chapter 22). Enhanced phosphoinositide turnover provides Ca^{2+} and early diacylglycerol (DAG) via activation of phospolipase(s) C (PLC). Activated phospholipase D provides phosphatidic acid (PA), which can be converted to DAG by a specific PA phosphohydrolase. Activated phospholipase A_2 generates arachidonic acid, which is capable of activating specific PKC isoforms. Human sperm PKC activity was found to be relatively low and was distributed in the soluble fraction (45%) and the particulate fraction (55%). Bull and ram sperm PKC was found mainly in soluble fraction.

The complex gene family of PKC encodes multiple isoforms, comprising of atleast 11 different isozymes, expressed in various mammalian tissues. This kinase family is classified into 3 groups: i) conventional (cPKC) α, ß, and γ, which require Ca^{2+} and diacylglycerol for maximal activity, ii) novel (nPKC) δ, ∈,η, and θ isoforms, which are Ca^{2+} independent but require DAG and iii) atypical (aPKC) ς, λ and τ, which are Ca^{2+} and DAG insensitive. Expression of individual PKC isoform depends on cell type and on its developmental stage. Stimulation by specific hormones or by phrobol ester leads to sites where they phosphorylate their specific target proteins. Studies have shown that activated PKC isoforms bind to anchoring proteins termed RACKs, which are believed to be positioned in close proximity to PKC substrate (Luria et al, 2000).

19.2.1. PKCδ and PKCθ Isoforms in Testis

In mice testes, there is an adundant expression of PKCδ and PKCθ, while other PKCs (α, β, γ, ε, ξ, and η) generally were not detected. However, these isoenzymes demonstrated a distinctive cellular distribution when evaluated by in situ hybridization. Previously it was shown that PKCδ gene is also present in spermatid of Wsh/Wsh mice testes and PKCθ gene was present in interstitial cells of +/+, Wsh/Wsh and W/W^v mice testes. These studies provided the evidence of selective cell distribution of the PKC isoenzymes and suggested that PKC has the functional significance in testes

Signals of PKCδ gene expression were detected specifically at the spermatid stage. However, the expression was not detected in testes of germ cell-deficient W/W^v mice even at 12 weeks. It showed that the PKCδ gene expression may be controlled by specific developmental processes that play a role in spermatogenesis. The PKCθ isoform in mouse testes displays the highest homology to PKCδ, lacks the Ca^{2+} binding C2 domain and, thus belongs to the subfamily of

Ca^{2+} independent PKC enzymes, which also include the PKCδ, ε, ξ and η isoforms. The PKCδ is localized in the spermatids whereas PKCθ was recognized only in the interstitial cells of the testes. Signals of PKCθ isoform expression were confirmed in the interstitial cells of testes in germ cell-deficient W/W^v mice (Kim et al., 1999; Kim and Shin, 1999, Shin et al, 1998).

An alternative splicing variant of mouse PKC δ-II has a 78:bp (26 amino acid) insertion at the recognition sequence in the V3 region of PKC δI. A cDNA encoding a new varient of PKC δ (PKC δ III, AF219629), which has a 83:bp insertion at the same site in the V3 region has been isolated by Ueyama et al (2000) by RT-PCR using rat testis RNA as a template **(Fig 19.3)**. In rats, the 83 bp insertion causes in-frame termination, and rat PKCδ III protein is expressed as a truncated form, having only the regulatory domain without a catalytic domain. Genomic DNA analysis revealed that the difference between mouse PKC δ-II and rat PKC δ-III is derived from the different sequence at the 5' splicing donor sites. These results suggest that PKC δ-III may show a dominant negative effects against PKC δ-I, and that the modulation of signal transduction by alternative splicing variant may play a crucial role in the physiological and/or pathological conditions, and the pathogenesis of disease.

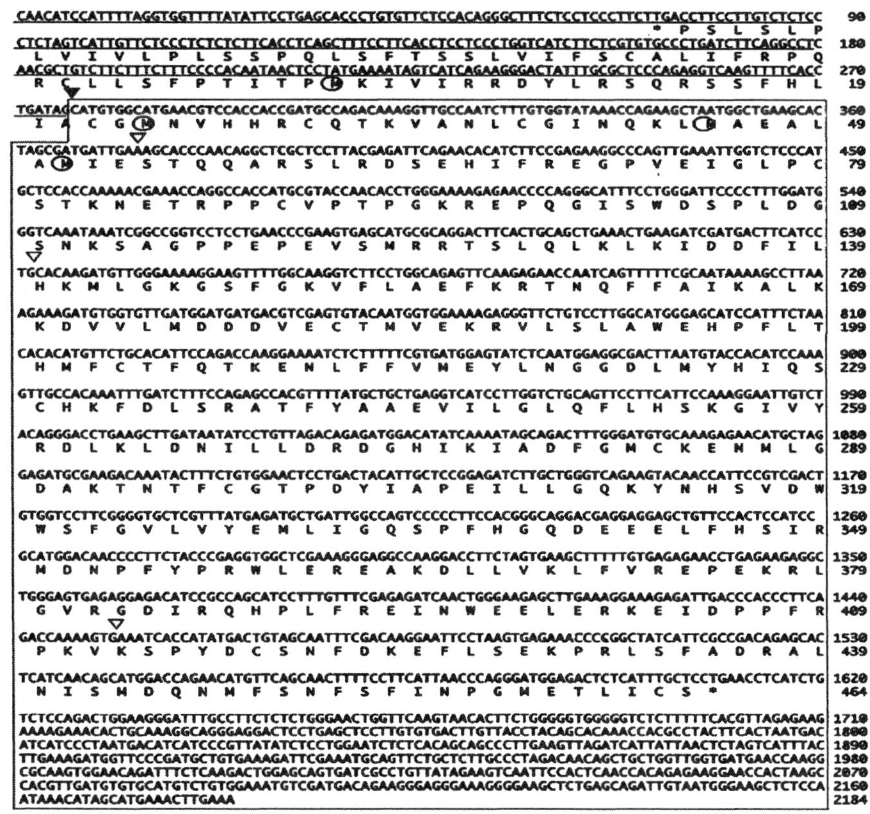

Fig.19.4. Nucleotide and deduced amino acid sequences of PKCθII. Possible initiating amino acids (M) are circled. Asterisks indicate the stop codon. The unique sequence of the PKCθII is underlined, whereas the sequence identical to PKCθI is boxed. A solid arrowhead shows the location of the θII exon/exon 8 boundary, whereas open arrowheads show the boundaries of C1, V3, C3/V4/C5, and V5 domains. Adapted from Niino et al J. Biol Chem 276;36711-7:2001

A smaller transcript of PKCθ is expressed in the mouse testis, the cDNA of which is referred to PKCθ-II as against the original PKCθ as PKCθ-I. The PKCθ-II cDNA has 2184 bp and 464 amino acids in the possible ORF, consisting of the 5' sequence of 20 amino acids and the PKCθ-1 sequence of 444 amino acids. Genomic DNA analysis revealed that transcription of PKCθ-II is initiated from the *PKCθII*-specific exon, which is located between exons 7 and 8 of the PKCθ gene, indicating that alternative splicing is the mechanism by which PKCθ-II is generated. The PKC θ-II is generated specifically in age dependent manner with sexual maturation and is expressed in the seminiferous tubules of the mouse testis. Consistent with its molecular structure lacking the C1 regulatory domain, PKCθ-II is constitutively active being independent of PKC activators, e.g. phosphatidylserine and phorbol ester (Niino et al, 2001) **(Fig.19.4)**.

19.2.2. Functions of Protein Kinase-C

Protein kinase C plays a pivotal role in cell signaling as a serine/threonine kinase, in particular for Ca^{2+} mobilizing ligands. Enhanced phosphoinositide turnover provides Ca^{2+} and early DAG via activation of phospolipase C (PLC). Activated phospholipase A_2 generates arachidonic acid, which is capable of activating specific PKC isoforms.

Sperm Motility: Activation of human sperm PKC by the phorbol ester 12-0-tetradecanoylphorbol 13-acetate (TPA) or by the cell-permeable DAG analogue 1-oleoyl-2 acetylglcycerol (OAG) results in enhanced flagellar motility. Separate signaling events elicited by Ca^{2+}- elevation or PKC activation supports the notion that PKC is involved in the regulation of flagellar motility of sperm via phosphorylation of a subunit of the outer arm dynein (Itoh et al., 2003)

PKC and Acrosome Reaction: The stimulatory effect of TPA on acrosome reaction of capacitated sperm was abolished by PKC down regulation or by the use of PKC inhibitors (Rotem et al. 1992, De Jonge et al. 1991). The potential sperm ligands ZP3 and progesterone are known to stimulate phosphoinositide turnover and acrosome reaction in mammalian sperm. The progesterone stimulated protein phosphorylation is mediated by PKC. Moreover, formation of DAG by ZP3 and progesterone is believed to activate PLA2. Information supports the notion that PKC acts upstream to phospholipase A_2 activation. Lysophosphatidic acid (LPA) is known to induce signal transduction cascades in many cell types via binding to signal cell-surface receptors. Under conditions by which LPA activates PKC, there is significant stimulation of the acrosome reaction, which is inhibited by PKC inhibitors. Since PKCα is a known regulator of phospholipase D (PLD), the possible regulatory relationships between PKCα and PLD1 was established in bovine sperm. A bilateral regulation of PKCα and PLD1 activation during the sperm acrosome reaction seems to be possible (Garbi et al, 2000).

PKC in Preoteoglycan Synthesis: Activation of PKC pathway results in increase of preoteoglycan synthesis in peritubular cells from immature rat testis. In cultured peritubular cells (PT) from rat testis, PKC was activated by phorbol 12-myristate 13-acetate (PMA), which enhances the synthesis of proteoglycans (PG) and to a lesser extent their catabolism. The PMA dramatically enhanced production of hyaluronan by PT cells (Thiebot et al., 1999). Besides these effects, activation of PKC by PMA induces qualitative increase in relative proportion of heparin sulfate PG (HSPG) in cell membrane. An inverse relationship between PG synthesis and FSH responsiveness in immature rat Sertoli cells is possible the suggestion. The PMA induced upregulation of cell membrane PG and particularly HSPG could be one of the mechanisms involved in the repression of FSH-stimulated steroidogensis induced by PKC activation (Fagen et al., 1999).

PKC and γ-Glutamyl transpeptidase: Regulation of γ-glutamyl transpeptidase (γ-GTP) activity by Ca^{2+}-and PKC-dependent pathways in Sertoli cells has been studied by Meroni et al (1997) who showed an inhibitory role of Ca^{2+} calmodulin- and PKC–dependent pathways in the regulation of basal γ-GTP activity. Similar to what has been shown for other Sertoli cell parameters, a PKC-dependent pathway can interact with the FSH-stimulated cAMP-dependent pathway. The precise steps involved in this interaction are unknown.

19.2.3. PKC Substrates in Sperm

Spermatozoa possess several substrates for PKC and other kinases. The SSeCKS is a major protein kinase C substrate, which has tumour suppressor activity in models of src– and ras-induced oncogenic transformation. The mitogenic regulatory activity of SSeCKS is likely manifested by its ability to bind key signaling proteins such as protein kinase C, PKA and calmodulin, and to control actin-based cytoskeletal architecture. SSeCKs expression is ubiquitous with highest expression in mouse gonads. In rat testis, its expression co-rrelates with the induction of puberty and in maturing mouse peripheral acrosomal membranes, in the control of cytoskeletal and tissue architecture formation of migratory processes and cell migration during embryogenesis (Gelman et al, 2000).

Scorilas et al, (2001) identified a putative serine / threonine kinase substrate (TSKs) gene that maps to chromosome 19q13.3. This gene is composed of 11 exons and 10 intervening introns and is likely the human homologue of the mouse testis-specific serine kinase substrate gene. The predicated protein-coding region of the gene is 1779 bp, encoding a 592-amino-acid polypeptide with a molecular mass of 65.1-kDa. Genomic analysis of the region 19q13.3 placed the TSKS gene close to the known genes IRF3, RRAS, and α-adaptin A. The TSKS exhibits high levels of expression exclusively in human testicular tissue, and its expression is down - regulated in cancerous testicular tissue.

The nucleotide sequence of the full length HSD-3.8 cDNA encoding a human sperm component and a substrate for kinases consists of 3818bp with a reading frame of 2778bp and a deduced polypeptide composed of 926 amino acids. The protein is present on the surface of the post-acrosomal zone of human spermatozoa and of germ cells within the seminiferous epithelium of human testis. Large pachytene primary spermatocytes stained intensely. Immunization of female rats with recombinant protein exhibited marked reduction in their fertility. Deduced HSD- 3.8 polypeptide sequence revealed the presence of tetratricopeptide repeat (TPR) motif, a P-loop sequence that acts as a binding site for ATP/GTP and phosphorylation sites for PKC, CK2 and cAMP/cGMP–dependent protein kinases. The polypeptide encoded by the 0.7-kb fragment of HSD-3.8 gene is a GTP binding protein and possesses GTPase activity and to be phosphorylated by PKC in vitro. In conclusion, HSD-3.8 is a GTP binding protein and its activity may be regulated by phosphorylation (Lin et al, 2001).

19.3. Ca^{2+}/CALMODULIN DEPENDENT PROTEIN KINASES

Various effects of Ca^{2+} in cell are mediated via its interaction with calmodulin and subsequent activation of Ca^{2+}/ calmodulin (CaM) dependent protein kinases. The CaM kinases (CaMK) play a variety of role in cellular processes including muscle contraction, neurotransmitter release, cell cycle regulation and transcriptional control in cell. While CaMK-II has been related to learning and memory, the biological function of other multifunctional CaM kinases is largely unknown.

19.3.1. CaMKI and CaMKII

The CaMK expresses in central nervous system. There are three isoforms of CaMK-I: CaMK-Iβ1, CaMK-Iβ2, and CaMK-Iγ (c/r Naito et al., 1997). The CaMK-Iβ2 is dominant in the cerebrum and cerebellum, whereas CaMK-Iβ1 is present in peripheral tissues including testis. Alternative splicing is a means for tissue–specific expression of CaMK-Iβ. Naito et al (1997) showed the existence of an isoform-specific activation mechanism of CaMK-I and alternative splicing specifically regulating CaMK-I (CaMK-Iβ2) in the CNS. The Thr-177 residue of CaMK–I is phosphorylated by not only CaMK kinase (CaMKK) but also by CaMK-Iβ for activation of the enzyme.

The involvement of CaMK II in the meiotic chromosomal pairing process has been proposed. The cDNA of another protein, which contains the association domain of α isoform of calmodulin dependent protein kinase II (CaM-KIIα), was cloned from rat skeletal muscle. This protein, called α KAP, consisted of 200 amino acid residues with a molecular weight of 22,583D. A cDNA representing a unique Ca^{2+}/ calmodulin dependent protein kinase, expressed in brain, testis and spleen, is only 32% identical to the various isoforms of CaMK-II. The sequence of the COOH-terminal 169 amino acids is identical to that of male germ cell-specific calspermin. This identity extends to the nucleic acid sequence and includes all but the first 130 nucleotides of the calspermin cDNA. In situ hybridization demonstrated that both kinase and calspermin mRNAs are expressed during spermatogenesis. The kinase mRNAs is first detected in early meiotic cells and declines in haploid cells (Means et al., 1991).

PncK: A serine/threonine kinase, Pnck encodes a 38kDa protein kinase whose catalytic domain shares 45-70% identity with other members of CaM kinase family. The gene *Pnck* localizes to mouse chromosome Xq28 that is associated with multiple distinct mental retardation syndrome. The Pnck is upregulated during intermediate and late stages of murine fetal development with highest levels of expression in developing brain, bone, and gut. The Pnck is also expressed in a tissue-specific manner in adult mice with highest levels of expression detected in brain, uterus, ovary, and testis. The chromosomal localization of Pnck and restricted spatial distribution suggest that Pnck may be involved in a variety of developmental processes including development of the central nervous system and reproductive process (Gardner et al, 2000).

19.3.2. CaMK IV in Testis

The CaMKIV is a multifunctional enzyme dependent on both calcium and calmodulin for activity. The CaMK-IV and calspermin (CaS) are expressed and encoded in testis by a common gene, *Camk4*. This gene has three promoters: the first two regulate expression of α and β isoforms of CaMKIV, while the third promoter lies within the 10th kinase intron and controls transcription of calspermin. While CaMKIV is expressed in several tissues, including brain and thymus, in addition to the testis, the calspermin is found only in the testis. Within the testis, the mRNAs encoding CaMKIV and calspermin have distinct expression patterns. The CaMK IV mRNA is expressed within pachytene spermatocytes. On the other hand, calspermin mRNA is localized to postmeiotic round spermatids. In the testis, CaMKIV has been proposed to regulate calspermin expression by phosphorylation and activation of the testis-specific transcription factor CREM, which in turn drives transcription of several germ cell genes including calspermin,

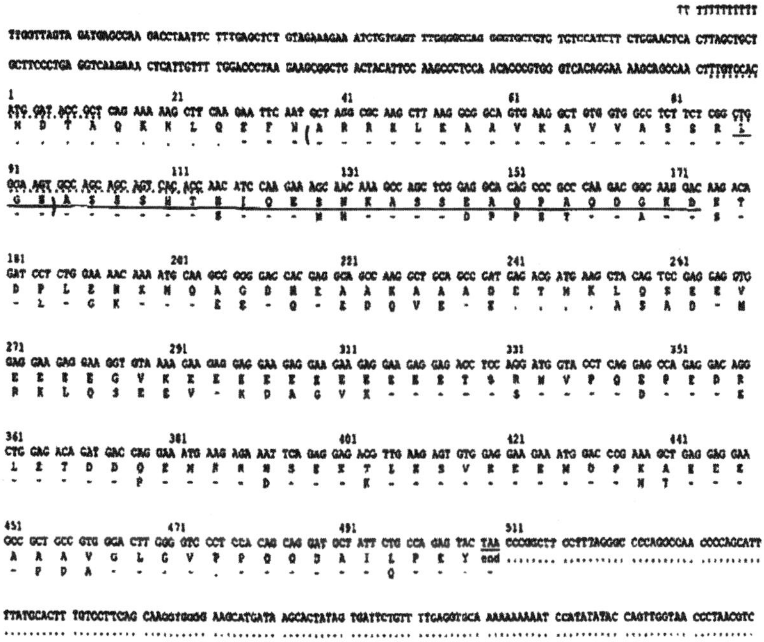

Fig.19.5 Nucleotide and deduced amino acid sequences for calspermin cDNA. The termination codon (TAA) is underlined. The deduced amino acid sequence of mouse brain ĕ1CM-1 is shown under the calspermin DNA. Dashes represent identical amino acids and dots indicate gaps. The parentheses represent the calmodulin domain. Underlined nucleotides (nt -8 to +12 and 88-114) are sequences used for screening of library. Reproduced with permission from T Ono et al. J Biol Chem 264: 2084-87; 1989 © American Society for Biochemistry and Molecular Biology.

whose promoters contain CRE elements. In mouse testis, the expression and regulation of CaMKIV and calspermin is stage-dependent and appears to be coordinately regulated. The CaMKIV is expressed in spermatogonia and spermatids but excluded from spermatocytes, while calspermin is found only in spermatids. In germ cells CaMKIV is associated with chromatin. A fraction of CaMKIV in spermatids is hyperphosphorylated and specifically localized in nuclear matrix. Calspermin does not colocalize to the nuclear matrix with CaMKIV. Thus CaMKIV may be involved in chromatin remodelling during nuclear condensation of spermatids (Wu and Means, 2000; Wu et al, 2001).

Male CaMKIV-/-mice are infertile with impairment of spermiogenesis in late elongating spermatids. The sequential deposition of sperm basic nuclear proteins of chromatin was disrupted, with a specific loss of protamine-2 and prolonged retention of transition protein –2 (TP2) in step-15 spermatids. Protamine-2 is phophorylated by CaMKIV in vitro, implicating a connection between CaMKIV signaling and exchange of basic nuclear proteins in sperm of infertile men, suggesting that these observations may have clinical implications for the understanding of human male infertility (Wu et al, 2000b). In addition to male infertility, female fertility is markedly reduced in CaMKIV-deficient mice due to impaired follicular development and ovulation. The CaMKIV is expressed in ovary where it is localized in granulosa cells. In absence of calspermin, CaMKIV expression and localization are unaffected (Wu et al., 2000; 2001).

```
  1 MHISPNNRNA IHPGDRILEI NGTPVRTLRV EEVEDAIKQT SQTLQLLIEH DPVPQRLDQL
 61 RLDARLPPHM QSTGHTLMLS TLDTKENQEG TLRRRSLRRS NSISKSPGPS SPKEPLLLSR
121 DISRSESLRC SSSYSQQIFR PCDLIHGEVL GKGFFGQAIK VTHKATGKVM VMKELIRCDE
181 ETQKTFLTEV KVMRSLDHPN VLKFIGVLYK DKKLNLLTEY IEGGTLKDFL RSVDPFPWQQ
241 KVRFAKGISS GMAYLHSMCI IHRDLNSHNC LIKLDKTVVV ADFGLSRLIV EERKRPPVEK
301 ATTKKRTLRK SDRKKRYTVV GNPYWMAPEM LNGKSYDETV DVFSFGIVLC EIIGQVYADP
361 DCLPRTLDFG LNVKLFWEKF VPTDCPPAFF PLAAICCKLE PESRPAFSKL EDSFEALSLF
421 LGELAIPLPA ELEDLDHTVS MEYGLTRDSP P
```

Fig.19.6. Amino acid sequence of testis specific lim kinase 2 (Lim2t). Source: http//www. ncbi.nlm.nih.gov (Accession JE0240).

19.3.3. Calspermin : A Germ Cell Homologue of CaMKIV with no Kinase Activity

Testis expresses a calmodulin-binding protein, calspermin (CaS), which does not have a homologue in somatic tissues. This protein of 18735 Da is most abundant calmodulin binding molecule present in adult mammalain testis. Calspermin is encoded by a 1.1-kb mRNA present predominantly in post-meiotic male germ cells and cannot be detected in any other mammalian tissue. The sequence of the NH_2-terminal half of the calmodulin-binding region (FNARRKLK) is identical to the one present in all members of the family of multisubstrate calmodulin-dependent protein kinases. Calspermin consists of unique terminal half of the COOH-terminal of 164 amino acids present in CaMKIV, including the CaM-binding domain (Tobimatsu et al., 1989) **(Fig.19.5)**. Calspermin has no kinase activity and instead may be involved in regulation of high levels of calmodulin found in germ cells. The CaMK-IV and calspermin are expressed and encoded in testis by a common gene, *Camk4*. This gene has three promoters: the first two regulate expression of α and β isoforms of CaMKIV, while the third promoter lies within the 10th kinase intron and controls transcription of calspermin. In mouse testis, the expression and regulation of CaMKIV and calspermin is stage-dependent and appears to be coordinately regulated. Calspermin is found only in spermatids. The function of calspermin within the testis is not clear, although it has been speculated to play a role in binding and sequestering calmodulin during the development of the germ cell. In the testis, CaMKIV has been proposed to regulate calspermin expression by phosphorylation and activation of the testis-specific transcription factor CREM, which in turn drives transcription of several germ cell genes including calspermin, whose promoters contain cAMP-responsive elements. Using Cre/lox technology to delete calspermin while leaving kinase expression intact, Wu et al, (2001) found that calspermin is not required for male fertility.

Calspermin gene transcription is regulated by two cyclic AMP response elements contained in an alternative promoter in the calmodulin kinase IV gene. The testis-specific calspermin transcript can be produced in heterologous cells by utilization of a promoter located in an intron of the CaMKIV gene. Critical motifs within this promoter are two CRE-like sequences located about –70 and –50 bp upstream of the transcriptional initiation site. Both CRE motifs are footprinted by the authentic testis-specific transcriptional activator CREMτ or by CREMτ present in adult testis nuclear extract. The 2.1-kb DNA fragement containing the calspermin promoter is inactive when transfected into NIH 3T3 cells, whereas activity can be restored by co-transfection of CREMτ and protein kinase A or CaMKIV but not CaMK-IIα. Restoration of activity is greatly reduced by mutation of the two CRE motifs (Sun et al 1995).

19.4. LIM KINASES

Lim proteins are defined by the presence of one or more double zinc-finger domains. The LIM motif is defined by the presence of zinc co-ordinating amino acid residues with the consensus sequence C X_2 C $X_{16\text{-}22}$ H X (C/H) X_2 C X_2 $CX_{16\text{-}23}$ CX_2 (C/H/D). The LIM kinases belong to serine/threonine family. These kinases have a 50% overall amino acid identity, and share characteristic structural features consisting two LIM domains at N-terminus, an interal PDZ like domain and C-terminal protein kinase domain. The LIM domain and the PDZ domain are thought to function in protein-protein interactions, making it likely that they are involved in the regulation of kinase activity and/or subcellular localization of LIMKs. While *Limk1* mRNA is predominantly expressed in developing neural tissues, *Limk2* mRNA is expressed in a variety of tissues of developing and adult vertebrates, suggesting that LIMK1 and LIMK 2 have some distinct functions. The *Limk1* and *Limk2* genes were mapped to human chromosome regions 7q 11.23 and 22q12, respectively. These kinases are involved in cell growth and morphogenesis. The mouse *Limk2* gene spans more than 50-kb and consists of at least 16 exons. Exon/intron boundaries of the mouse *Limk2* gene are exactly conserved with those of the mouse *Limk1* gene. The TESK1 together with TESK2 constitute another subgroup of the LIMK/TESK family of protein kinases (see Testis specific serine protein kinases).

Mizuno and coworkers identified several isoforms of *Limk2* transcripts in rat encoding proteins with distinct N-terminal and C-terminal structures. While *Limk2a* encodes a protein with a LIMK1–like structure composed of two *LIM* domains, a PDZ-like domain and a protein kinase domain, *Limk2b* codes for an N-terminally modified protein lacking zinc finger of the first *LIM* domain, and *Limk2c* codes for a c-terminally truncated protein lacking a protein kinase domain. An additional exon (exon 2b) coding the *Limk2b*-specific 5'-terminal sequence was assigned to locate between exons 2 and 3 in the mouse *Limk2* gene. RT-PCR analysis revealed tissue-specific expression of *Limk2a* and *Limk2b* transcripts (Ikebe et al, 1998). Mouse testis LIM-K2 transcript, *Limk2t* lacked LIM domains at the N-terminus due to usage of a testis-specific, alternative initiation exon. Limk2t is detected in intact adult testis, but not germ cell deficient or immature testis. The Limk2t was restricted in pachytene spermatocytes to round spermatids and expressed in early stages of spermatogenic cells and somatic cells in testis.

Ikebe et al, (1998) characterized the mouse *Limk2* gene and two *Limk2* transcripts (*Limk2a* and *Limk2b*), coding for proteins with distinct N-terminal LIM structures and two additional transcripts of the mouse *Limk2* gene. One is a 1.7-kb transcript, termed *Limk2t*, which is specifically expressed in the testis and codes for an N-terminally truncated form of LIMK2 consisting of only a part of a PDZ – like domain and a protein kinase domain. The other transcript, termed *Limk2c*, is specifically expressed in the brain and codes for a protein with a 6-amino – acid insert within the protein kinase domain. Exons specific to the 5'-terminal extra sequence of *Limk2t* and the insert sequence of *Limk2c* are located between exons 5-6 and exons 8-9 in the mouse *Limk2* gene, respectively **(Fig. 19.6)**.

The LIM-K phosphorylates and inactivates cofilin, and actin-depolymerizing factor regulating actin reorganization. Takahashi et al, (2002) generated *Limk2* gene-deficient mice in which three LIMK2 isoforms were disrupted in a Cre-mediated fashion. Impaired cofilin phosphorylation was clearly observed in Limk2 -/- fibroblasts thereby suggesting that Cdc42 or Rho-dependent LIMK activation did not occur. However, *Limk2-/-* mice did not exhibit embryonic lethality or any phenotypic abnormalities in postnatal growth and development, except for spermatogenesis in the testis, which showed degeneration of spermatogenic cells and apparent association with increased apoptosis. It seems that LIMK-2 especially the testis-

```
FHL1    MSEKFDCHYCRDPLQGKKYVQKDGRHCCLKCFDKFCANTCVDCRKPISADAKEVHYKNRY
FHL4    MSE-FKCHHCEESLQGKKYVQKDGANYCVTCFDSHCANICQECHKPIGADSKEVCYKEQF
        *** * ** * ..*********** . *. ***  *** * .*.***.**.*** **...
             I--------HL domain--------I      I---------------------

FHL1    WHDNCFRCAKCLHPLASETFVSKDGKILCNKCATREDFPRCKGCFKAIVAGDQNVEYKGT
FHL4    WHNTCFQCTKCSQLLATETFVAWDKNILCNKCATRVTFPKCKGCLKDIEEGDHNVEYKGS
        **. **.*.** . **.****. * . ********  **.****.* *  **.******.
        ---------LIM domain 1-----------I       I------------LIM----

FHL1    VWHKDCFTCSNCKQVIGTGSFFPKGEDFYCVTCHETKFAKHCVKCNKAITSGGITYQDQP
FHL4    IWHKNCFVCTNCKDIIGTKNFFPKDEGFYCVTCYDALFTKHCMKCKKPITSGGVSYQDQP
        .***.** *.***..*** .****.* ****** .. *.***.**.* *****..*****
        ---domain 2---------------------I       I------------LIM----

FHL1    WHAECFVCVTCSKKLAGQRFTAVEDQYYCVDCYKNFVAKKCAGCKNPITGFGKGSSVVAY
FHL4    WHSECFVCVSCSKELSGQRFTAMDDQYFCVDCYKNYIAKKCAGCKNPITGFGKGANVVAH
        **.******.*** *.******..***.*******..*****************..***
        ---domain 3--------------------I      I-------------LIM----

FHL1    EGQSWHDYCFHCKKCSVNLANKRFVFHNEQVYCPDCAKKL
FHL4    EQNSWHDYCFNCKTCSVNLANKHFVFHDEQVYCPDCARNL
        * .*******.** ********.**** *********..*
        ---domain 4-------------------------I
```

Fig.19.7. Clustal alignment of deduced amino acid sequence of mouse FHL1 and FHL4. Identical and similar amino acids are marked by asterisks and dots respectively. Lim domains are shown. Adapted with permission from M. J Morgan and A J Madgwick. Biochim Biophys Acta 255; 251-55: 1999 © Elsevier.

specific isoform LIMK2t plays an important role in proper progression of spermatogenesis by regulation of cofilin activity and/or localization in germ cells.

The FHL LIM protein family (formerly the SLIM protein family) members were defined by their expression in skeletal muscle and by their particular secondary structural arrangement of LIM domains. The FHL proteins consist of four LIM domains and an N-terminal single zinc finger domain that meets the consensus of the C-terminal half of the LIM domain motif (hence four–and–a Half–LIM proteins). Each LIM domain is separated by eight amino acid residues. The FHL1 (SLIM1) and FHL3 (SLIM2) are expressed at relatively high levels in skeletal muscle, whereas FHL2 (DRAL, SLIM3) mRNA is expressed at high levels in the heart. A fourth member of the FHL-LIM protein family, FHL4 was identified by Morgan and Madgwick (1999). The sequence of a LIM protein, FHL4 was found to contain the defining secondary structure of the FHL family of LIM proteins. The expression of FHL4 mRNA was restricted in the seminiferous tubules of the testis (**Fig.19.7**).

19.5. TESTIS SPECIFIC SERINE PROTEIN KINASES

19.5.1. TESK-1

The cDNA clones encoding a novel protein serine/threonine kinase, termed TESK1 and TESK2 (testis-specific protein kinase 1 and 2) different from tssk have been identified (Toshima et al, 1998). The genomic organization and chromosomal localization of the mouse TESK1 showed that TESK1 has a unique structure composed of an N-terminal protein kinase domain and C-terminal proline rich domain. RNAs from rat tissues revealed that the major and relatively broad band of *Tesk1* mRNA of about 3.6-kb was almost exclusively expressed in the testis, whereas a 2.5-kb band was faintly detectable in other tissues and cell lines. The *Tesk1* mRNA was detected in germ cell-enriched preparations from spermatocytes and round spermatids of mouse

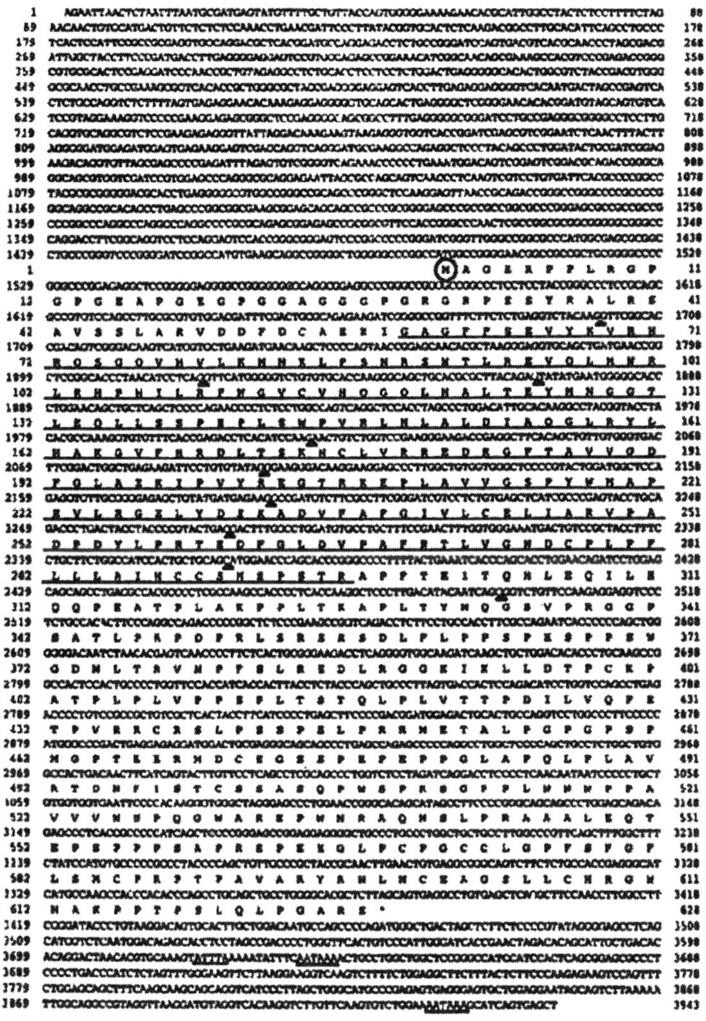

Fig.19.8. The cDNA and predicted amino acid sequences of mouse TESK1. The protein kinase domain is underlined. The polyadenylation signals and an mRNA instability signal in 3'-UTR are underlined. Exon/intron boundaries are marked by wedges. Reprinted with permission from J. Toshima et al. Gene 206; 237-45: 1998 © Elsevier.

testes, after the 18-20th day of postnatal development, when mouse spermatocytes begin to undergo meiosis and generate haploid round spermatids on 18-20th postnatal days. Based on these findings, it is likely that the *Tesk1* gene product may play a role at and/or after the meiotic stages of spermatogenesis. Expression of *Tesk1* mRNA and protein in the rat testes is developmentally regulated and increases after 20-22 postnatal days. Prominent expression of *Tesk1* mRNA and protein was detected at stages of late pachytene spermatocytes to round spermatids, but not in Sertoli and Leydig cells (Toshima et al, 1998) **(Fig.19.8)**. The expression of a lacZ reporter gene linked to a 9.0-kb 5' flanking region of *Tesk1* gene has been examined in transgenic mice (Toshima et al, 2001b). A high level of lac Z expression was observed in testicular germ cells after pachytene spermatocytes with expression patterns being similar to those of *Tesk1* mRNA in rat testis. Expression of lacZ was also detected in renal proximal tubules, cardiac myocytes, and specific neurons in the central nervous system in adult transgenic mice and in neural tissues of embryonic mice. This suggests diverse and specific physiological functions of TESK1 including those in spermatogenesis and neural development (Toshima et al, 2001b).

Characterization: Rat TESK1 contains 628 amino acid residues, composed of an N-terminal protein kinase consensus sequence followed by a C-terminal proline-rich region. Human TESK1 contains 626 amino acids, sharing 92% amino acid identity with its rat counterpart. Chromosomal localization of human *Tesk1* gene was assigned to 9p13. Anti-TESK1 antibody raised against the C-terminal peptide of TESK1 recognized two polypeptides of 68 and 80-kDa in cell lysates of COS cells transfected with human TESK1 (Toshima et al, 1995). The kinase domain of TESK1 contains in the catalytic loop of subdomain VIB, an unusual DLTSKN sequence, which is not related to the consensus sequence of either serine/threonine kinases or tyrosine kinases. Toshima et al (1999) suggested that TESK1 has kinase activity with dual specificity of both serine/threonine and tyrosine residues. In an in vitro kinase reaction, the kinase domain of TESK1 underwent autophosphorylation on serine and tyrosine residues and catalyzed phosphorylation of histone H3 and myelin basic protein on serine, threonine, and tyrosine residues. Site-directed mutagenesis analysis revealed Ser-215 within the "activation loop" of the kinase domain as the site of serine autophosphorylation of TESK1. Replacement of Ser-215 by alanine almost completely abolished serine autophosphorylation but retained histone H3 kinase activity. Results suggested that autophosphorylation of Ser-215 is an important step to regulate the kinase activity of TESK1 (Toshima et al., 1999).

Gene Structure: The mouse *Tesk1* gene spans 6.1-kb and contains 10 exons and 9 introns. The protein kinase domain is located in exons 1-9, while the proline-rich domain is in exons 9 and 10. The deduced 627 amino acid sequence of mouse TESK1 shows 97% and 94% identity with the rat and human TESK1, respectively. Sequence of the 5'-flanking and –untranslated region is devoid of a TATA box, but does contain several potential binding sites for transcription factors, including Sp1, AP-1 c-Myc, SRY and CREM. Since, CREM is implicated in the activation of several male germ cell-specific genes, it is suggested that the expression of the *Tesk1* gene is under the control of CREM transcriptional activity. The *Tesk1* gene was mapped on mouse chromosome 4A5-C1 by in situ hybridization (Toshima et al, 1998b). Sequencing of cDNA clones revealed that the centre divider gene (cdi) in *Drosophila encodes* an 1140-amino acid protein that is an ortholog of the mammalian testis specific TESK1 protein kinase. This kinase is different from other protein kinases because of sequence differences in the residues conferring substrate specificity. The unique sequence is conserved in Cdi, suggesting that Cdi/TESK1 represents a novel class of signaling proteins (Mathews and Crews, 1999).

Relation to Lim Kinases: The kinase domain of TESK1 is phylogenetically related to those of LIM-kinase 1 and LIM-kinase 2, with about 50% sequence identity. Like LIM-kinase, TESK1 phosphorylated cofilin specifically at Ser-3 both in vitro and in vivo. When expressed in Hela cells, TESK1 stimulates the formation of actin stress fibers and focal adhesions. In contrast to LIM-kinases, the kinase activity of TESK1 is not enhanced by Rho-associated kinase (ROCK) or p21-activated kinase, indicating that TESK1 is not their downstream effector. The kinase activity of TESK1 and the level of cofilin phosphorylation are increased by plating cells on fibronectin. Y27632, a specific inhibitor of ROCK, inhibited LIM-kinase-induced cofilin phosphorylation but did not affect fibronectin-induced or TESK1 induced cofilin phosphorylation in Hela cells. Expression of a kinase negative TESK1 suppressed cofilin phosphorylation and formation of stress fibers and focal adhesions induced in cells plated on fibronectin. Thus, TESK1 functions downstream of integrins and plays a key role in integrin-mediated actin reorganization, presumably through phosphorylating and inactivating cofilin but are regulated in different ways and play distinct roles in actin reorganization in living cells (Toshima et al, 2001a).

19.5.2. TESK 2

The open reading frame of TESK2 encodes a putative 555-amino-acid protein, including a protein kinase consensus sequence in the N-terminal half. The protein kinase domain of TESK2 is structurally similar to TESK1 with 48% over all identity in amino acids and 64% identity in the kinase domain, and to those of the LIMK1 and LIMK2 kinases (42 and 39% identity, respectively). The TESK2, together with TESK1, constitutes a second subgroup of the LIMK/TESK family of protein kinases. Chromosomal localization of human TESK2 was assigned to 1p32. Expression analysis revealed a single mRNA species of 3.0-kb predominantly expressed in testis and prostate and low expression in most other tissues. Rat testicles expressed a single species of TESK2 mRNA of approximately 3.5-kb. However, the transcript was first detectable in rat testis after day 30 of postnatal development and was predominantly expressed in round spermatids (Rosok et al, 1999). Like TESK1, TESK2 phosphrylated cofilin specifically at Ser-3 and induced formation of actin stress fibers and focal adhesions. In contrast to TESK1, which is predominantly expressed in testicular germ cell, TESK2 is expressed predominantly in Sertoli cells. Thus TESK1 and TESK2 seem to play distinct roles in spermatogenesis. A TESK2 mutant lacking the C-terminal noncatalytic region had about a 10-fold higher kinase activity in vitro and, when expressed in HeLa cells, induced punctate actin aggregates in the cytoplasm and unusual condensation and fragmentation of nuclei, followed by apoptosis. It has been proposed that the C-terminal region plays important roles in regulating the kinase activity and cellular functions of TESK2 (Toshima et al., 2001c).

19.5.3. Testis Specific Serine Kinase(s)

Kueng et al (1997) isolated a murine testis-specific serine kinase (TSSK)-1 and a closely related family member, TSSK2. The TSSK2 is probably the orthologue of the human DGS-G gene. These serine kinases are expressed only in testis of sexually mature males. Both kinases are localized in the cytoplasm of late spermatids and to the structures resembling residual bodies. The TSSK1 and TSSK2 were not present in sperm and did not reveal autophosphorylation but they led to serine phosphorylation of a coprecipitating protein of ~65-kDa. A search for interacting proteins from mouse testis, led to the isolation of a cDNA, interacting specifically with both TSSK1 and Tssk 2, and encoding the coprecipitated 65kDa protein phosphoryated by these. Results represent the demonstration of the involvement of a distinct kinase family, together with a substrate in the cytodifferentiation of late spermatids to sperms (Kueng et al, 1997) In contrast to present study, the TSSK, present in human testis was localized to the equatorial segment of ejaculated human sperm (Hao et al., 2004).

The ORF of *TSSK1* and *TSSK2* encode proteins of 367 and 358 amino acids respectively, with conserved kinase domains typical of superfamily of Ser/Thr kinases. Both genes were intronless and mapped to chromosomes 5 and 22 respectively. The human and mouse homologues of TSSK1 and TSSK2, together with TSSK 3 and SSTK/FKSG82, constitute a

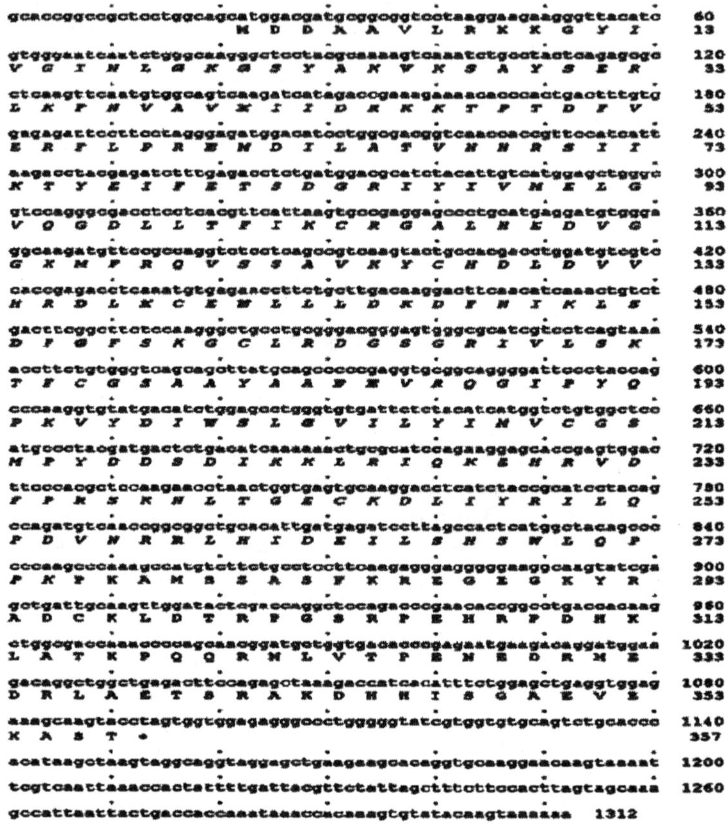

Fig.19.9. Nucleotide sequence and deduced amino acid sequence of tssk 2. The kinase domain is presented in italics and the conserved amino acids typical for a serine/threonine kinase domain are indicated in bold. Reproduced with permission from Keung et al. J Cell Biol 139; 1851-59: 1997 © Rockefeller University Press.

kinase subfamily closely related to the calmodulin kinases and SNF/nim 1 kinase subfamilies **(Fig .19.9)**. Human TSSK1 and -2 mRNAs are expressed exclusively in the testis. But mRNAs for these kinases are also detected in other tissues. In addition, TSKS, the putative substrate of TSSKs was also identified. The TSKS had an ORF of 592 amino acids and was expressed exclusively in the testis (Hao et al., 2004).

19.6. CASEIN KINASES

19.6.1. Casein Kinase 1

Casein kinase 1 (CK1) is known to exist in multiple isoforms in mammals involved in growth and morphogenesis in eukaryotic cells. Using a partial cDNA fragment corresponding to an isoform termed CK1γ, three full-length rat testis cDNAs were cloned that defined three separate members

```
CK1γ1 MDHSNREKDD RQRTTKTMAQ RNTHCSRP-- SGTSTSSGVL MVGPNFRVGK KIGCGNFGEL 58
CKIγ2 MDFDKKGGKG ELEEGRRMSK TGTNRSNHGV RNSGTSSGVL MVGPNFRVGK KIGCGNFGDV 60
CKIγ3 MDNKKKDKD- --KSDDRMAR PSGRSGHSTR GTGSSSSGVL MVGPNFRVGK KIGCGNFGEL 57
CK1α  ---------- ---------- ---------M ASSSGSKAEF IVGGKYKLVR KIGSGSFGDI 31

CK1γ1 RLGKNLYTNE YVAIKLEPIK SRAPQLHLEY RFYKQLGSAG EGLPQVYYFG PCGKYNAMVL 118
CKIγ2 RLGKNLYTNE YVAIKLEPIK SRAPQLHLEY RFYKQL-S-T TGVPQVYYFG PCGKYNAMVL 118
CKIγ3 RLGKNLYTNE YVAIKLEPMK SRAPQLHLEY RFYKQLGS-G DGIPQVYYFG PCGKYNAMVL 116
CK1α  YLAINITNGE EVAVKLESQK ARHPQLLYES KLYKILQG-G VGIPHIRWYG QEKDYNVLVM 90

CK1γ1 ELLGPSLEDL FDLCDRTFTL KTVLMIAIQL LSRMEYVHSK NLIYRDVKPE NFLIGRQGNK 178
CKIγ2 ELLGPSLEDL FDLCDRTFTL KTVLMIAIQL ITRMEYVHTK SLIYRDVKPE NFLVGRPGSK 178
CKIγ3 ELLGPSLEDL FDLCDRTFSL KTVLMIAIQL ISRMEYVHSK NLIYRDVKPE NFLIGRPGNK 176
CK1α  DLLGPSLEDL FNFCSRRFTM KTVLMLADQM ISRIEYVHTK NFIHRDIKPD NFLMGI--GR 148

CK1γ1 KEHVIHIIDF GLAKEYIDPE TKKHIPYREH KSLTGTARYM SINTHLGKEQ SRRDDLEALG 238
CKIγ2 RQHSIHIIDF GLAKEYIDPE TKKHIPYREH KSLTGTARYM SINTHLGKEQ SRRDDLEALG 238
CKIγ3 AQQVIHIIDF GLAKEYIDPE TKKHIPYREH KSLTGTARYM SINTHLGKEQ SRRDDLEALG 236
CK1α  HCNKLFLIDF GLAKKYRDNR TRQHIPYRED KNLTGTARYA SINAHLGIEQ SRRDDMESLG 208

CK1γ1 HMFMYFLRGS LPWQGLKADT LKERYQKIGD TKRSTPIEAL CENFPEEMMT YLRYVRRLDF 298
CKIγ2 HMFMYFLRGS LPWQGLKADT LKERYQKIGD TKRATPIEVL CESFPEEMAT YLRYVRRLDF 298
CKIγ3 HMFMYFLRGS LPWQGLKADT LKERYQKIGD TKRATPIEVL CENFPEEMAT YLRYVRRLDF 296
CK1α  YVLMYFNRTS LPWQGLKAAT KKQKYEKISE KKMSTPVEVL CKGFPAEFAM YLNYCRGLRF 268

CK1γ1 FEKPDYEYLR NLFTDLFERK GYTFDYAYDW VGRPIPTPVG SVHVDSGASA ITRESHTHRD 358
CKIγ2 FEKPDYDYLR KLFTDLFDRS GYVFDYEYDW AGKPLPTPIG TVHPDVPSQP PHRDKAQLHT 358
CKIγ3 FEKPDYDYLR KLFTDLFDRK GYMFDYEYDW IGKQLPTPVG AVQQDPALSS NREAHQHRDK 356
CK1α  EEAPDYMYLR QLFRILFRTL NHQYDYTFDW TMLKQKAAQQ AASSSGQGQQ AQTPTGF--- 325

CK1γ1 RPSQQQPLRN QPRSLTAEWF VLAPLSHPPA PT-------- ---------- ---------- 390
CKIγ2 KNQA------ ---------- ---------- ---------- LNSTNGELNT DDPTAGHSNA 382
CKIγ3 IQQSKNQSAD HRAAWDSQQA NPHHLRAHLA ADRHGGSVQV VSSTNGELNT DDPTAGRSNA 416
CK1α  ---------- ---------- ---------- ---------- ---------- ---------- 325

CKIγ2 PIAAPAEVEV ADETKCCCFF KRRKRKSLQR HK 414
CKIγ3 PITAPTEVEV MDETKCCCFF KRRKRKTIQR HK 448
CK1α  ---------- ---------- ---------- -- 325
```

Fig.19.10. Alignment of predicted protein sequences of CK1α, CK1γ1, CK1γ2, and CK1γ3. Reproduced with permission from L. Zhai et al. J Biol Chem 270: 12717-24; 1995 © American Society for Biochemistry and Molecular Biology.

of CK1 subfamily. The isoforms, designated CK1γ 1, CK1γ2, and CK1γ3, have predicted molecular masses of 43-, 45.5-, and 49.7kDa. The CK1 γ3 may also exist in an alternatively spliced form. These proteins are more than 90% identical to each other within the protein kinase domain but only 51-59% identical to other casein kinase 1 isoforms within this region. Messages for CK1γ1 (2-kb), CK1γ2 (1.5 and 2.4-kb), and CK1γ3 (2.8-kb) were detected by hybridization of testis RNA. The CK1γ3 is ubiquitously present in all tissues but CK1γ1 and CK1γ2 messages were restricted to testis. All three CK1γs are autophosphorylated in presence of ATP and Mg^{2+}. The *YCK1* and *YCK2* genes in *S. cerevisiae* encode casein kinase 1 homologues, defects in which lead to aberrant morphology and growth arrest. Expression of mammalian CK1γ1 or CK1γ3 restored growth and normal morphology to a yeast mutant carrying a disruption of YCK1 and a temperature-sensitive allele of YCK2, suggesting an overlap of function between the yeast YCK proteins and these CK1 isoforms (Zhai et al, 1995) **(Fig.19.10)**.

A clone of immature cDNA for human CK1γ2 on comparison with the sequence of rat CK1γ2 showed complete identity with rat CK1 γ2 as well as intron-like sequence of 1006 bp, part of which was homologous to the Alu sequence. The predicated amino acid sequence consisted of 16 residues, 94% of which were identical to that of the rat homologue. Although there are two Src homology-3 (SH3) domain-binding motifs (Pro-X-X-Pro consensus), Pro-Lys-Val-Pro and- Pro-Ser-Glu-Pro in the C-terminal region of rat CK1 γ 2, the latter part only was conserved in the human counterpart. The latter motif is important for binding to the signal transduction adaptor protein Nek (NEK). The human *CK1γ2* gene was mapped to chromosome 19p13.3 (Kitabayashi et al, 1997).

Kasuda et al (2000) isolated two types of cDNA for human CK1γ1 (hCK1γ1). One of them (hCK1γ1S) was found to encode a polypeptide of 393 amino acids, which is highly homologus with rat CK1γ1 (rCK1γ1). The other type of cDNA (hCK1γ1L) encodes a polypeptide consisting of 422 amino acids, which is quite identical in the kinase domain, but different in the C-terminal sequence from hCK1γ1S. For example, hCK1 γ1L has a characteristic sequence of 50 amino acids at the C-terminal end; this motif was shown to be shared by the casein kinase γ2 and γ3 from rat and human, suggesting that it is a signature sequence of the γ-isoforms. In this sense, newly isolated hCK1γ1L might be the original form of CK1γ1. The rCK1γ1 and hCK1γ1S. RT-PCR analysis showed that hCK1γ1S mRNA is predominatly present in the testis, whereas the adundance of hCK1γ1L mRNA was nearly the same in all tissues examined. Thus hCK1 γ1L may have a unique functional role different from that of hCK1γ1S and rCK1γ1. The human *hCK1γ1* gene *(CK1γ1)* was mapped to chromosome 15q22.1.>q22.31 (Kusuda et al, 2000).

A cDNA clone encoding the enzyme, designated casein kinase-1δ, contained an open reading frame of 1284 nucleotides that predicts a polypeptide of 428 amino acids with a Mr of 49,121. The predicted amino acid sequence shares 76% identity with casein kinase 1α, a 37-kDa form bovine brain. Northern analysis of rat or rabbit RNA revealed three hybridizing species of 3.5-4.1-kb, 2.2 and, 1.9 kb. The largest messge was detected in all tissues examined, whereas the 1.9- and 2.2-kb species were found predominantly in testis. Expression of the casein kinase 1δ cDNA in E.coli resulted in active enzyme that phosphorylated casein, phosvitin, and the peptide substrate DDDDVASLPGLRRR. Casin kinase 1δ, therefore, represents a separate member of the casein kinase 1 family distinguished by its larger size and unique kinetic behavior with respect to heparin (Graves et al., 1993).

19.6.2. Casein Kinase II

The protein kinase, CK II (CK2) is a cyclic-AMP and calcium-independent serine-threonine kinase that is composed of two catalytic α-subunits and two regulatory β-subunits (α2 β2). The CK2 is not casein kinase in vivo, but over 100 substrates are known in vitro.The CSII is a probable physiolgical substrate for $p34^{cdc2}$ and its functional properties can be affected in a cell cycle dependent manner (Litchfield et al., 1992). The CK2 has been implicated in DNA replication, regulation of basal and inducible transcription, translation and control of metabolism. The mammalian enzyme is an oligomeric protein with a molecular mass of 130-140 kDa. The larger α–subunit (Mr 34-44 kDa) possesses the catalytic activity; the smaller ß-subunit (24-28 kDa) is phosphorylated when the holoenzyme undergoes autophosphorylation in vitro. The CK2 phosphorylates acidic proteins such as casein and phosvitin using either ATP or GTP as phosphroyl donor. The enzyme can be activated by polyamines and is sensitive to inhibition by heparin. Phosphorylation by casein CK2 modulates the activity of its substrates and in other cases the enzyme works synergistically with other protein kinases by rendering its substrates more susceptible to additional phosphorylation by the other enzymes. Although a large number of enzymes and other proteins are phosphorylated by CK2 in other cells, most are not present in immature spermatozoa. However, proteins such as tubulin, calmodulin, ornithine decarboxylase, calcium channel proteins, and regulatory subunit of PKA, which are substrates of CK2 in other cells, are also present in mature spermatozoa and hence, are potential substrates of sperm (Chaudhary et al, 1991). Casein kinase II from bovine sperm has a molecular mass of

130 kDa by gel filtration and displays three polypeptide bands with molecular masses of 26, 33, and 36 kDa by SDS-PAGE. Three sperm polypeptides cross-react with Abs raised against calf thymus casein kinase II. Incubation of the holoenzyme with either ^{32}P-ATP or ^{32}P-GTP results in phosphorylation of 26kDa subunit. The enzymatic activity with casein as substrate is strongly inhibited by heparin and greatly stimulated by spermine. Low concentration of spermine strongly enhanced the phosphorylation of 92 and 106-kDa cytosolic proteins, which may be endogenous substrates of the kinase. Thus, casein kinase II plays a role in mediating the phosphorylation state of sperm proteins (Chaudhary et al, 1991).

Casein kinase is present in goat caudal-epididymal sperm plasma membrane. The kinase is a strongly basic protein with pI of 9.5 with a molecular mass of 310-kDa and affinity for protein substrates in the order membrane protein > phosvitin > protamine. ATP, rather than GTP or PP is the donor of phosphate for the phosphorylation reaction. Heparin and polyamines at a relatively high concentration inhibited the enzyme activity. The isolated enzyme is an ecto-protein kinase whose catalytic site in sperm is oriented primarily to the surface of viable sperm cell to cause phosphorylation of the endogenous outer cell-surface phosphoproteins (Mitra et al, 1994). The specific activity of the ectokinase has been found to increase markedly during forward progression of spermatozoa, implicating thereby that the sperm outer membrane kinase and its protein substrate may have an important role in the regulation of flagellar motility.

The *Csnk2α2* encodes the *Csnk2α2* catalytic subunit of CK2α1 that is predominantly expressed in testis. The CK2α1 and *Csnk2α2* isoforms (products of the gene *Csnk2α1* and *Csnk2α2*, respectively) are highly homologous, but the reason for their redundancy and evolutionary conservation is unknown. The *Csnk2α2* is preferentially expressed in late stages of spermatogenesis. Male mice in which *Csnk2α2* had been disrupted were infertile, with oligospermia and globozoospermia (round headed sperm). The primary spermatogenic defect in *Csnk2α2-/-* testis was a specific abnormality of anterior head shaping of elongating spermatids. Few sperm cells in epididymis exhibited head abnormalities similar to those described in human globozoospermia. The *Csnk2α2* null testes revealed extensive germ cell degenerative processes at all stages of spermatogenesis, including the first spermatogenesis wave. CK2α1 deficiency could impair the phosphorylation of nuclear proteins of male germ cells leading to a particular cell-death pathway characterized by NE protrusions and an unusual pattern of chromatin modifications in spermatids. Thus *Csnk2α2* may be a candidate gene for these inherited syndromes (Xu et al, 1999; Escalier et al., 2002)

19.7. AURORA LIKE KINASES

Aurora kinases are involved in mitotic events that control chromosome segregation. Members of this kinase subfamily possess two distinct domains, a highly conserved catalytic domain and an N-terminal non-catalytic extension that varies in size and sequence. Mutations in IPL1 and aurora cause abnormal chromosome segregation and centrosome separation.

19.7.1. Aurora Like Kinase 3 (ALK3)

The cDNAs encoding human protein kinases ALK1 and ALK2 sharing high amino acid sequences with *Drosophila Aurora* and *S. cerevisiae* IpL1 kinases whose mutations cause abnormal chromosome segregation were isolated (Kimura et al, 1999). A third human cDNA (ALK3) highly homologous to aurora/IPL1 cDNA encodes 309 amino acids with a predicted molecular mass of 35.9 kDa. The C-terminal domain of ALK3 protein shares high amino acid sequence identities with those of Aurora/IPL1 family protein kinases including human ALK, human ALK2, Xenopus aurora and others, whereas the N-terminal domain of ALK3 protein shares little homology with any other Aurora/IPL1 family members. The ALK3 gene was assigned to human chromosome 19q13.43. and its expression was limited to testis where it is localized to centrosome during mitosis from anaphase to cytokinesis. The expression levels of AIK3 in several cancer cell lines were elevated several fold compared with normal fibroblasts. It is suggested that ALK3 might play a role(s) in centrosome function at later stages of mitosis (Kimura et al, 1999; Yang et al., 2000). To investigate the role of variable noncatalytic region, Giet and Prigent (2001) over-expressed and purified *Xenopus laevis* aurora (pEg2) histidine-tagged N-terminal peptide from bacterial cells. Unlike the full-length protein, the N-terminal domain localizes to the centrosomes in a microtubule dependent manner. The centrosome localization of the N-terminal peptide was disrupted by nocodazole whereas localization of the catalytic domain was not, suggesting that in order to efficiently localize to the centrosome, pEg2 kinase required the non-catalytic N-terminal domain and the presence of microtubule (Giet and Prigent, 2001).

19.7.2. AIE 1 and AIE/2

Two protein kinases, designated AIE1 (mouse) and AIE2 (human) share high amino acid identities with the serine/threonine kinase domain of yeast IPL1, fly aurora, and frog pEg2. Both AIE1 and AIE2 contain a typical S/Tkinase domain (251 aa) flanked by a short polypeptide at both ends. Two other AIE-related kinases (STK-1 and IAK1/Ayk1) are also present in mature mouse oocytes. The central kinase domain of AIE1 revealed 77.6% and 66.3% identity with STK-1 and IAK1/Ayk1. While AIE1 and AIE2 are specifically expressed in testis, STK-1 and IAK1/Ayk1 are expressed in many tissues rich in proliferating cells. The AIE1 can phophorylate casein, AIE1 itself, and an uncharacterized cellular protein (p16). The kinase activity of AIE1 is heat labile (Tseng et al, 1998).

The mouse *Aie1* gene spans approximately 14-kb and contains seven exons. The sequences of the exon-intron boundaries of the *Aie1* gene conform to the consensus sequences (GT/AG) of the splicing donor and acceptor sites of most eukaryotic genes. The gene structure is highly conserved between mouse *Aie1* and human *AIE2*, with less homology in the sequences outside the kinase-coding domain. The *Aie*1 locus was mapped to mouse chromosome 7A2-A3. The *Aie1* mRNA likely is expressed at a low level on day 14 and reaches its plateau on day 21 in postnatal testis. The expression of the *Aie1* transcript was restricted to meiotically active germ cells, with the highest levels detected in spermatocytes at the late pachytene stage (Hu et al., 2000). The testis zinc finger protein (Tzfp) binds to the upstream flanking sequence of the *Aie1* gene (Chapter 16).

19.8. MICROTUBULE ASSOCIATED PROTEIN KINASES

19.8.1. Polo Like Kinase

The polo gene of *D. melanogaster* is the founding member of the polo-like kinase family, which is conserved among eukaryotes. The POLO has been implicated in the organization and function of the mitotic apparatus and has been shown to be required for normal spermatogenesis. The POLO localizes to centrosomes, the centromere / kinetochore and spindle midzone. Observations of mutant testes indicated cytological abnormalities in all meiotic cell types, including variable

DNA content and multipolar spindles. Primary sprmatocytes in *polo* mutant testes contain abnormal DNA content and chromosome segregation. A-tubulin shows several abnormalities of the meiotic spindle, including a reduced central spindle, suggesting that polo has multiple functions during spermatogenesis (Herrmann et al, 1998). One cDNA cloned from murine embryonic germ (EG) cells encodes a nonreceptor type serine / threonine kinase and is predominantly expressed in the testis, ovary, and spleen of adult mouse. The entire coding region shows that this clone, designated PLK1 (polo like kinase 1), is identical with STPK13 present in murine erythro-leukemia cells. The protein encoded by PLK1 is closely related to the product of *Drosophila* polo that plays a role in mitosis and meiosis. The pattern of expression of the PLK1 gene suggests that the gene product is involved in completion of meiotic division, and like the *Drosophila* polo protein, is a maternal factor active in embryos at the early cleavage stage. Furthermore, rat pachytene spermatocytes down-regulate a polo-like kinase after exposure to methoxyethanol and methoxyacetic acid (Syed and Hecht, 1998).

19.8.2. p56 KKIAMRE

A 56 kDa protein kinase (p56 KKIAMRE) shares homology with p42 KKIALRE and is related to the proline-directed protein kinase group of signal transducing enzymes. The p56KKIAMRE and p42 KKIALRE protein kinases exhibit mutually exclusive expression in reproductive tissues viz., p56 KKIAMRE in testis and p42 KKIALRE in ovary, whose functions are unknown. The p56 KKIAMRE and p42 KKIALRE are activated by treatment of cells with epidermal growth factor (EGF) . Although p56 KKIAMRE and p42 KKIALRE contain the MAPK dual phosphorylation motif Thr-Xaa-Tyr (Thr-Asp-Tyr), phosphorylation on Thr and Tyr within this motif is not required for EGF-stimulated protein kinase activity (Taglienti et al, 1996).

19.8.3. MAST205 with Kinase Activity

The association of c-mos with tubulin and of type II cAMP-dependent kinase with MAP2 provides a means for rapid phosphorylation of specific microtubule proteins that could conceivably affect transient microtubule organization. Purified anti-MAP recognized 205-kDa protein, termed MAST205, which colocalizes with the spermatid manchette. Sequencing of full-length cDNA clones encoding MAST205 revealed it to be a novel serine/threonine kinase with a catalytic domain related to those of the A and C families. The testis-specific MAST205 RNA increases in abundance during prepuberal testis development peaking at the spermatid stage. Binding of MAST205 to microtubules requires interaction with other MAPs, since it does not bind to MAP-Free tubulin. A 75-kDa substrate protein from testis microtubules may form part of the MAST205 protein complex, which binds microtubules (Walden and Cowen, 1993). The MAST205 contains a stretch of 89 amino acids with homology to a dlg homology region (DHR) domain. The DHR domain was thought to be involved in signal transduction at tight, septate, and synaptic junctions. Interestingly, a 29-kDa region of MAST205 involved in microtubule binding contains the DHR domain. While MAST205 RNA level were similar in pachytene spermatocytes, round spermatids, and residual bodies, MAST205 protein could be detected only in round spermatids and residual bodies. Kinase activity was low in pachytene spermatocytes, high in round spermatids, and maximal in residual bodies, indicating that MAST205-associated kinase is modified during spermatid maturation. The MAST205 protein and associated kinase activity were not present in spermatozoa. The MAST205 mRNA, detected in human testis is indicative of conservation of MAST 205 in other mammalian species (Walden and Millette, 1996) (see Chapter 8).

19.9. CELL CYCLE CHECK POINT KINASES

Check Point Kinase 1: Checkpoint pathways prevent cell-cycle progression in the event of DNA lesions. Checkpoints are well defined in mitosis, where lesions can be the result of extrinsic damage, and they are critical in meiosis, where DNA breaks are a programmed step in meiotic recombination. The genetic and regulatory interaction between Atm and mammalian Chk1 appears to be important for integrating DNA-damage repair with cell-cycle arrest structural homologues of yeast Chk1 in human and mouse. The Chk1 (Hu/Mo) has protein kinase activity and is expressed in the testis. Chk1 accumulates in late zygotene and pachytene spermatocytes and is present along synapsed meiotic chromosomes, and localizes along the unsynapsed axes of X- and Y-chromosomes in pachytene spermatocytes. The association of Chk1 with meiotic chromosomes and levels of Chk1 protein depend upon a functional Atm gene product, but Chk1 is not dependent upon p53 for meiosis-I functions. The Chk1 is located at 11q22-23, a region marked by frequent deletions and los of heterozygosity in human tumors. Atm dependent intractions of mammalian Chk1 with meiotic chromosomes suggest that Chk1 acts as an integrator for Atm and Atr signals and may be involved in meiotic recombination (Flaggs et al., 1997).

Check Point Kinase 2: Check point kinase 2 is emerging as a critical mediator of genotoxic stress cellular responses. The Chk2 is a transducer of DNA damage signals and a tumour suppressor whose germ-line mutations predispose to diverse tumor types. Unlike its downstream targets such as the p53 tumour suppressor, the expression patterns of Chk2 in tissues and tumours remain unknown. Localization of the Chk2 protein was examined during normal development of human testes, and at various stages of germ cell tumour (GCT) pathogenesis (Bartkova et al, 2001). The Chk2 is abundant in foetal germ cells and adult spermatogonia, yet only weakly expressed during the meiotic and later stage of spermatogenesis. High levels of Chk2 detected in majority of GCTs suggest that down modulation or lack of Chk2 is not simply attributable to quiescence or differentiation, but they suggest a role for Chk2 in mitotic rather than meiotic divisions, and support the concept of foetal origin of GCTs. Upon ionizing radiation, Chk2 is activated to phosphorylate Cdc25C, leading to G2 phase arrest. The p53 is a substrate of Chk2, which phosphorylates and stabilizes p53 in response to ionizing radiation. The p53-deficient p53(-/-) mice demonstrated that Chk2 expression is under control of p53 in a tissue-specific manner (Chin and Li, 2003).

ATR/ATM: The *Atr* mRNA and protein are expressed in human and mouse testis. Analysis of specific cells in seminiferous tubules shows localization of ATR to the nuclei of cells in the process of meiosis I. ATR and ATM proteins of approximately 300 and 350-kDa relative molecular mass, respectively, have associated protein kinase activity. ATR and ATM interact directly with meiotic chrosomsomes. ATR is found at sites along unpaired or asynapsed chromosomal axes whereas ATM is found along synapsed chrosomoal axes (Keegan et al,1996) (Chapter 13).

Mps: A vertebrate homologue of yeast Mps1p is shown to be a kinetochore- associated kinase whose activity is necessary to establish and maintain the checkpoint. Since high levels of Mad2 overcome checkpoint loss in Mps1-depleted extracts, Mps1 acts upstream of Mad2-mediated inhibition of APC/C. Mps1 is essential for check point, as it is required for retention of active CENP-E at kinetochores, which in turn is necessary for kinetochore assocation of Mad1 and Mad2 (Abrieu et al., 2001).

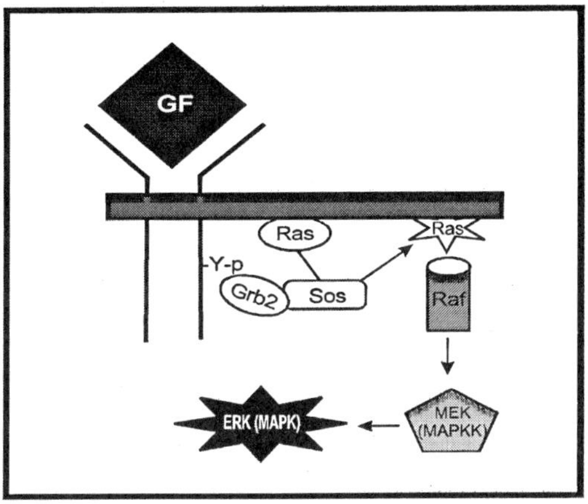

Fig. 19.11. Activation of MAPK pathway. Growth Factor (G.F.) binds to its receptor and causes tyrosine autophosphorylation, which causes Grb2 binding with Sos. Ras is activated by Sos mediated nuclotide exchange. This triggers a kinase cascade by activating MAPKKK, which phosphorylates MAPKK, which in turn phosphorylates MAPK (ERK).

19.10. MITOGEN ACTIVATED PROTEIN KINASES

The growth, division and differentiation of eukaryotic cells are influenced by a wide range of extracellular signals. Perception enquiry of these signals is mediated by the activation of MAPK (mitogen activated protein kinase). The MAPKs, also known as extracellular signal-regulated kinases (ERKs) are serine / threonine kinases. But some of them show dual specificity kinase activity. Two major routes have been described that lead to the activation of MAPK. One is initiated by the activation of receptor tyrosine kinases (RTKs), while the other begins with the activation of G-proteins coupled hepatahelical receptors. In the case of RTKs, intermediary proteins relay signals resulting in the activation of GTP binding protein, ras or PKC. Activated ras then leads to the activation / phosphorylation of MEK (MAPK kinases) and consequently, MAPKK at threonine and tyrosine in a Thr-X-Tyr (TXY) motif. Either raf or MEKK (MEK kinase) is capable of serving as the intermediary between ras and MEK. Once MAPK is activated, it is competent to phosphorylate a veriety of substrates, including transmembrane receptors and elements of cytoskeleton **(Fig.19.11)** Cell cycle during mitosis and meiosis is dependent on activation of MAPK pathway (see Chapter 11). In *C.elegans*, Church et al (1995) found that exit from pachytene is dependent on induction of three members of the MAP kinase signaling. One of these, *mek-2*, is a *C. elegans* MEK (MAP kinase). The other two genes, *mpk-1/sur-1* (MAP kinase) and *let-60 ras* have been shown to act in combination with mek-2 to permit exit from pachytene. It appeared that the expression of *mpk-1 / sur-1* is required within the germline to permit exit from pachytene. The MAPK pathway has been identified in spermatozoa, where ERKs play a role in capacitation (de Lamirande and Gagnon, 2002) and acrosome reaction (Luconi et al., 1998; du Plessis et al., 2001). The MAPK isoform, ERK2, adapter protein Shc (Morter et al., 1998) and ras have been localized to sperm head indicating that these are required for phosphorylation of sperm head. In spermatozoa centrosomal proteins may be candidate substrtes for MAPK pathway (Urnar and Sakkas, 2003).

19.10.1. ERK 1/ ERK2/ERK7 and Other MAP-Kinases

Two forms of MAP kinases, p42ERK2 and p44ERK1, are expressed in spermatogenic cells at different stages. The ERK1 and ERK2 were phosphorylated (activated) in early spermatogenic cells from primitive spermatogonia to zygotene primary spermatocytes, while only a small quantity of phosphorylated MAP kinase could be detected in pachytene primary spermatocytes and spermatids. ERK1 and ERK2 were also present in epididymal spermatozoa, and their phosphorylation was increased while spermatozoa passed through epididymis and vas deferens during maturation. It appears that MAP kinase activation may contribute to the mitotic proliferation of primative spermatogonia, an early phase of spermatogenic meiosis, and later sperm motility acquirement (Lu et al, 1999a).

In cellular signaling pathways, a new 61-kDa member of the MAPK family, termed extracellular signal regulated kinase 7 (ERK7) was characterized. Althought it has the signature TEY activation motif of ERK1 and ERK2, ERK7 is not activated by extracellular stimuli that typically activate ERK1 and ERK2 or by common activators of c-Jun N-terminal kinase (JNK) and p38 kinase. Instead, ERK7 has appreciable constitutive activity in serum-starved cells that is dependent on the presence of its C-terminal domain. The C-terminal tail, not the kinase domain, of ERK7 regulates its nuclear localization and inhibition of growth. These observations elucidate a new type of MAP kinase whereby interactions via its C-terminal tail, rather than extracellular signal-mediated activation cascades regulate its activity, and function (Abe et al, 1999).

P38-δ MAPK : Hu et al (1999) cloned murine c-DNA clones encoding p38–δ MAPK. The P38δ gene is localized to mouse chromosome 17A3B and human chromosone 6p21.3. The P38-δ is expressed primarily in the lung, testis, kidney, and gut epithelium. Interestingly, p38 δ and p38-α were similar in serine/threonine kinases but differed in substrate specificity. Overall, p38-β resembles p38-γ, whereas p38-β resembles p38-α. Moreover, p38-δ is activated by environmental stress, extracellular stimulantes, and MAPK kinase3, -4 and –7, suggesting that p38-δ is a unique stress-responsive protein kinase (Hu et al, 1999).

Mixed Lineage Kinase 2 (MLK2) : The MLK2 is a mammalian protein kinase that activates stress-activated protein kinases/c-jun N-terminal kinases (SAPK/JNKs) through direct phosphorylation of their upstream activator, SEK1/JNKK. In rat testis MLK2 mRNA expression was first evident at a very low level on day 25 after birth and present on day 35 till adulthood. The MLK2 RNA transcript was detected in primary spermatocytes and round spermatids, but not in Leydig or Sertoli cells or in the testis after induced cryptorchidism. The SEK1/JNKK transcripts, on the other hand, were present at all stages of testicular development and in all cell types tested, but was absent from spermatogonia. These findings indicated that the JNK pathway is most likely ubiquitous in rodent testicular cells, while the cell-specific pattern of MLK2 expression suggested that it might be involved in the regulation of processes specific to post-mitotic germ cells and in the regulation of nuclear events (Phelan et al, 1999).

T-LAK Cell-Originated Protein Kinase (TOPK): A MAPKK: A protein kinase, TOPK (T-LAK cell-originated protein kinase) was purified from a lymphokine-activated killer T (T-LAK) cells subtraction cDNA fragment library. The open reading frame of the TOPK gene encodes a protein of 322 amino acids, possessing a protein kinase domain profile. The cap site analysis of the 5'-end of TOPK mRNA revealed two forms, a major full-length form and a minor spliced form at the 5'-site, both encoding the same protein. The TOPK is related to dual specific mitogen-activated protein kinase kinase (MAPKK) and phosphorylated p38 MAPK but not

ERK1/2 or SAPK / JNK. The TOPK protein can be associated with p38 in vitro. Tissue distribution of TOPK mRNA expression was specific for the testis, T-LAK cells, activated lymphoid cells, and lymphoid tumours. On the other hand, deactivated T-LAK cells, did not show TOPK mRNA expression. This shows that TOPK is a member of a MEK3/66-related MAPKK that may be enrolled in the activation of lymphoid cells and support testicular functions (Abe et al, 2000).

19.10.2. p21-Activated Kinases (PAKs)

P21-activated Kinase-6 (PAK 6): Transcription activation by steroid hormone receptors is mediated through interaction with cofactors. Recently a androgen receptor interacting protein, provisionally termed P21-activated kinase-6 (PAK6) has been identified. The PAK6 shares a high degree of sequence similarity with p21-activated kinase (PAK). PAK6 is a 75-kDa protein that contains a putative amino-terminal Cdc42/Rac interactive binding motif and a carboxyl terminal kinase domain. A ligand-dependent interaction between androgen receptor (AR) and PAK6 was confirmed in vivo and in vitro. PAK6 is highly expressed in testis and prostate tissues. PAK6 co-translocates into the nucleus with androgen receptor in response to androgen. Transient transfection experiments showed that PAK6 specifically represses AR- mediated transcription. A new function for a PAK homologous protein suggests a signal transduction pathway which may cross-talk with androgen receptor pathway to regulate AR function in normal and malignant prostate cells (Yang et al, 2001).

Pak interacting exchange factor (Pix-d): βPix, a *Pak*-interacting guanine nucleotide exchange factor is known to be involved in the regulation of Cdc42/Rac GTPases and Pak kinase activity. Three β Pix isoforms, β Pix-a, -b, and –c already are known. Manser et al (1998) cloned a new class of Rho-p21 guanine nucleotide exchange factor binding tightly through its N-terminal SH3 domain to a conserved proline-rich PAK sequence with a binding constant (k_d) of 24 nM. This PAK –interacting exchange factror (Pix), which is widely expressed and enriched in *Cdc42-* and *rac1*-driven focal complexes, is required for PAK recruitment to these sites. Pix participates in Cdc42 to Rac1 signalling involving the Pix/PAK complex. The cDNA of a novel Pix splice variant named β pix-d was isolated from a mouse brain cDNA library. The cloned βPix isoform, βPix-d, lacks leucine zipper domain that is present in other Pix isoforms, and has a 11 amino acid addition at carboxyl terminus and distinct 3' UTR. The βPix-d message was present mainly at brain and testis but at lower level in other tissues. Specific probes showed that β Pix-d isoform is expressed mainly in CNS of rat embryo (Kim and Park, 2001).

BRDT gene: The RING3 gene encodes a 90-kDa mitogen–activated nuclear protein. In proliferating cells, RING3 has serine–threonine kinase and autophosphorylation activities. A cDNA identified encodes a protein of 947 amino acids with extensive homology to RING3, D26362, and fsh. Similar to these proteins, it possesses two bromodomain motifs and a PEST sequence. Northern analysis shows transcripts of 3.5 and 4.0kb expressed specifically in testis. The gene has been named BRDT gene. Radiation hybrids localize the gene to chromosome 1p between markers WI-7719 and WI-3099 (DIS2154) (Jones et al 1997).

19.11. RAC-PROTEIN KINASES

The cDNA of RAC protein kinases (RAC-PK) α and β were cloned from mammalian tissues. The predicted ORF(s) encode 480 and 481 amino acids of RAC-PKα and -β respectively, and

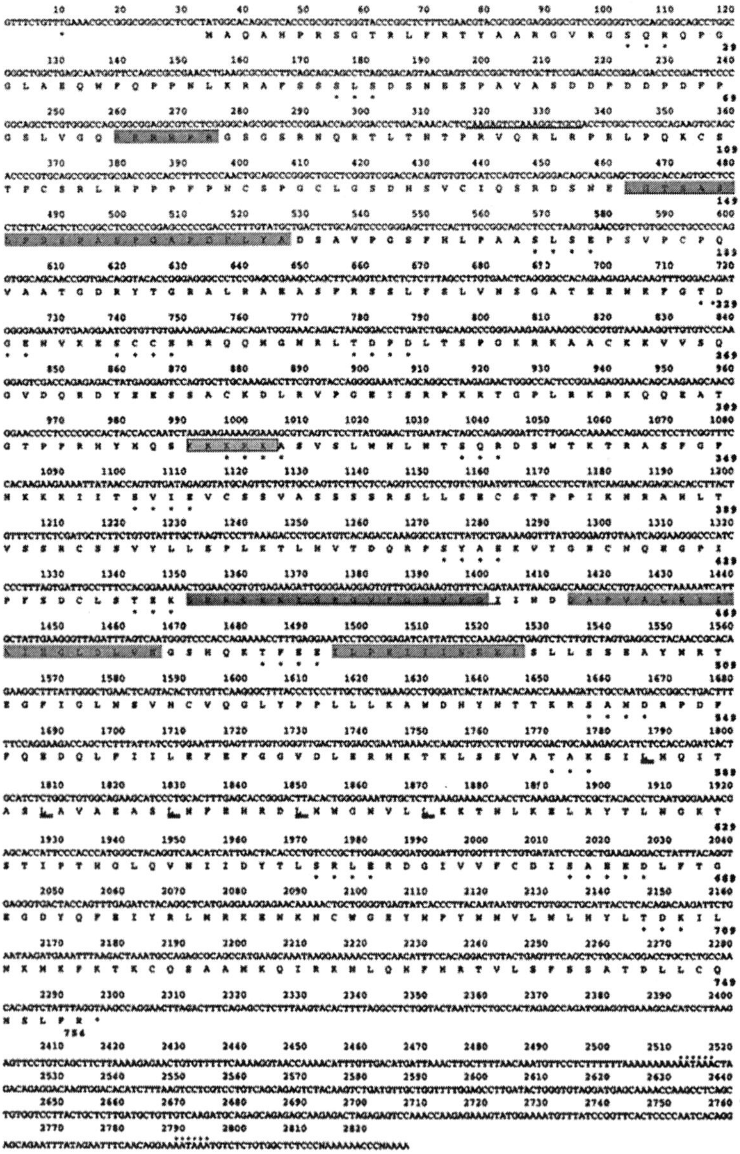

Fig.19.12. Nucleotide and deduced amino acid sequences of haspin cDNA. Underline indicates the primer sequence for 5'-RACE (at nt 317-335) and a deleted region (aa 447-453) of deletion mutant haspin. The putative polyadenylation signals are indicated by dots. Shaded regions indicate a basic amino acid sequence (aa 76-81), a region homologous to MEF2B (aa 144-162), and parts of protein kinase consensus sequences (residues 440-498). The gray box indicates a putative nuclear localization signal (aa 320-324). The leucine zipper motif (aa 585-613) is double underlined. Each series of asterisks indicates a potential target site for some protein kinase, and one asterisk indicates a stop codon in the haspin ORF. Reproduced with permission from H.Tanaka et al. J Biol Chem 274: 17049-57; 1999 © American Society for Biochemistry and Molecular Biology.

the rat RAC-PKα and β have sequences conserved among different mammalian species such as the pleckstrin homology domain at their amino-terminal region and the protein-serine/threonine kinase catalytic domain at their carboxyl-terminal region. Two RAC-PKs are widely distributed in rat tissues and associate with protein kinase Cξ through the pleckstrin homology domain in vitro, suggesting the interaction between RAC-PK and protein kinase C. The c-DNA encoding third member of RAC protein kinase family called RAC-PKγ was obtained from rat brain library. The amino acid sequence of RAC-PKγ was highly related to RAC-PKα and -β, which have a pleckstrin homology domain and a protein-serine/threonine kinase catalytic domain at the amino and carboxyl-terminal regions, respectively. The RAC-PKγ expresses abundantly in brain and testis. The pleckstrin homology domains of the three subtypes of RAC-PK associate with both protein kinase C- subspecies, and -β, γ-subunits of G proteins. It was further demonstrated that the pleckstrin homology domains of RAC protein kinases could associate more than one protein to regulate the activity and/or intracellular distribution of this enzyme family by different ways (Konishi et al, 1995).

19.12. HASPIN

Haspin belongs to the Ser/Thr protein kinase family and plays a role in cell cycle control. It catalyses phosphorylation of proteins in presence of Mg^{+2} and is expressed in germ cells within the testis of adults and of embryos from day 24 onwards. But it is also present in adult thymus and weakly expressed in spleen, lung and whole embryo. It can also autophosphorylate on both serine and threonine residues. Tanaka et al (1999) cloned the coding region of a mouse germ cell-specific cDNA encoding a unique protein kinase whose catalytic domain contains only three consensus subdomains (I-III) instead of normal 12 (**Fig 19.12**). The protein possesses intrinsic Ser/Thr kinase activity and is exclusively expressed in haploid germ cells, nuclei, and was named Haspin (for haploid germ cell-specific nuclear protein kinase). It has a M_r 83,000. Ectopically expressed Haspin is detected exclusively in the nuclei of cultured somatic cells. Even in the absence of kinase activity, however, Haspin causes cell cycle arrest at G_1, resulting in growth arrest of the trsansfected somatic cells.

19.13. MALE GERM CELL ASSOCIATED KINASE (MAK)

The gene, designated *mak* (male germ cell-associated) encodes a protein kinase distantly related to Cdc2 kinase. Its transcripts are expressed exclusively in testicular germ cells at and after meiosis. *Mak* products identified as 66- and 60-kDa proteins specifically appeared in rat testes after puberty; and most abundant in the fraction of the late pachytene stage, followed by a dramatic decrease in postmeiotic haploid cells. The *Mak* products were associated with a 210-kDa cellular protein, which was efficiently phosphorylated on serine and threonine residues. It seems that MAK complex plays a role in meiosis during spermatogenesis and 210-kDa protein is one of the physiological substrates for this protein kinase (Jinno et al., 1993). Sequence analysis of the cDNA corresponding to the 2.6-kb transcript revealed that the predicted product of rat Mak consisted of 622 amino acids and contained protein kinase consensus motifs in its amino-terminal region. Comparison of the deduced amino acid sequence of Mak in the kinase domain with those of other protein kinase genes demonstrated that *Mak* was approximately 40% identical to the *Cdc2-CDC28* gene family in *S pombe, S cerevisiae*, and humans but less identical to most other protein kinase gene products. Expression of Mak is highly tissue specific, and detected almost exclusively in testicular cells entering after meiosis but hardly detectable in ovarian cells. Mak mRNA was mainly localized in pachytene and diplotene spermatocytes (Matsushime et al., 1990).

19.14. OTHER SERINE / THREONINE KINASES

Serine/Arginine (SR) Protein Kinase (SRPK-1 And -2)

Arginine/serine protein kinases constitute a novel group of enzymes that can modify arginine/ serine (RS) dipeptide motifs. The SR splicing factors that are essential for pre-mRNA splicing are among the best-characterized proteins that contain RS domains. Two SR protein-specific kinases, SRPK1 and SRPK2, have been considered as highly specific for the phosphorylation of these proteins, thereby contributing to splicing regulation. However, despite the fact that SR proteins are more or less conserved among metazoa and have a rather ubiquitous tissue distribution, SRPK1 is predominantly expressed in testis. The SRPK1 mRNA is abundant in all germinal cells but not in mature spermatozoa. The SR- protein specific kinase 1 (SRPK1) is identified as a specific kinase for SR splicing factors. The SRPK clones were isolated from mouse brain. The SRPK1 cDNA clone encodes 106-kDa protein with 648 amino acids (92% identical to human SRPK1) and a 120-kDa protein with 681 amino acids (58% identical to human SRPK1). They were designated mSRPK1 and mSRPK2, respectively. The mSRPK1 expresses in all tissues examined but the expression of mSRPK2 is restricted to testis, lung, and brain. Both kinases phosphorylated SF2/ASF, a member of SR proteins in vitro with identical phosphopeptide mappings, indicating that these kinases phosphorylate the same site of SF2/ ASF. Over-expression of mSRPK2 caused disassembly of cotransfected SF2/ASF and endogenous SC35. Thus, SRPK family may regulate the disassembly of the SR proteins in a tissue-specific manner (Kuroyanagi et al, 1998). The RS kinase activity was found primarily in the cytosol. Testis-specific substrates of SRPK1 include human protamine 1 as well as a cytoplasmic pool of SR proteins present in the testis. SRPK1 may have a role not only in pre-mRNA splicing, but also in the condensation of sperm chromatin (Papoutsopoulou et al., 1999).

PKN and PKC: A protein kinase (PKN) with catalytic domain homologous to PKC present in every tissue is adundant in testis, spleen and brain. The soluble fraction of rat brain enzyme migrates as a band of apparent mass of 120kDa. Using the pseudosubstrate sequence of PKC-δ as phosphate acceptors, the kinase activity of PKN is stimulated by unsaturated fatty acids or by detergents such as sodium deoxycholate and SDS. The echancement of kinase activity of PKN in the presence of arachidonic acid, for phosphate acceptors and inhibition of autophosphorylation activity by pretreatment with alkaline phosphatase, can distinguish PKN from many known protein kinases (Kitagawa et al, 1995).

Tousled Like Kinase (TLk): Tousled is associated with various differentiative processes including differentiation of the reproductive organs. Tousled like kinase sequences have been reported in *C. elegans*, murine, and in the human. Positive hybridization signals suggest evolutionary conservation of TLk throughout the phylogentic ladder. Shalom and Don (1999), using the yeast IME2 meiotic gene (serine / threonine kinase) as a probe identifed a putative serine / threonine kinase, which did not exhibit significant homology to IME2, but did show significant sequence homology to the Tousled kinase in *Arabidopsis*. Four distinct TLk transcripts were detected in mouse testis, at least one of which is testis–specific. In normal testis, TLk is expressed in pachytene spermatocytes and in round spermatids. Transcripts differ from one another in their 3'untranslated region, and in the length of their 5' region. Within the coding region, three of the putative peptides share the kinase and C-terminal domains but differ in their N-terminal domain, suggesting that the latter may be involved in the regulation of TLk's function. Although TLk might have an essential role in all tissues, these kinases are likely to take part in the complex array of phosphorylations during spermatogenesis (Shalom and Don, 1999).

Spermatogenesis-Related Protein Kinase (SPK): A cDNA clone denoted by SPK (spermatogenesis-related protein kinase) is present in human testis, whose deduced protein is 99% identical to human PDZ-binding kinase (pbk), which is a mitotic kinase. Its isoform, which is about 300 bp longer at 3' end, was also isolated. The *SPK* gene was assigned to human chromosome 8p21.2 by radiation hybrid. The SPK mRNA was predominantly expressed in testis and localized to the outer cell layer of seminiferous tubules (Zhao et al, 2001).

SN related kinase (SNRK): The SNRK is related to SNF1 family of protein kinases in catalytic domain. The noncatalytic domain comprises several intriguing structural features, including a glycine-rich region, two PEST sequences, and a bipartite nuclear localization signal, which is preceded by a stretch of ten consecutive acidic residues. In addition, a high degree of sequence similarity is detected with other SNF1-related proteinases in a small region (30-35 amino acids) flanking the C-terminus of the catalytic domain. This domain (designated the SNH domain) appears to difine the subfamily of SNF-1 related protein kinases and might represent a new type of regulatory domain of protein kinases (Becker et al, 1996a). The SNRK – mRNA of 5kb is present in testis as well as in several other tissues of rat.

Fatty Acid Stimulated Protein Kinase: The goat testis possesses a protein kinase (PKx), which is stimulated by a number of unsaturated fatty acids (FA) of which arachidonic acid is the best activator in absence of cAMP or Ca^{2+}. Phosphatidylcholine, phosphatidylethanolamine, phosphatidylserine and diacylglycerol have no effect either alone or in combination. The PKx is a cytosolic protein and migrates as a band with molecular mass of 116-kDa. It phosphorylates histone H1, histone IIIs and protamine–SO_4 but not casein. However, the best phosphorylation was obtained with a substrate based on PKC pseudosubstrate sequence (RFARKGSLROKNV). The protein kinase, PKx phosphorylates two endogenous cytosolic proteins of 60 and 68-kDa at Ser and Thr residues. The Ab-cross reactivity suggests that PKx is not related to any known PKC family (Roy et al, 1999).

II. PROTEIN TYROSINE KINASES

First group of activity of protein tyrosine kinases (PTK) comprises those, which were detected with viral transforming proteins, particularly with those of oncogenic retroviruses. There are at least six cellular genes of this kind, which all appear to encode proteins with PTK activity. The second type of PTKs activity was detected in association with growth factor receptors. Receptor PTKs are stimulated by their respective ligands.

It is now clear, that all these proteins do indeed have a catalytic domain capable of transferring phosphate to proteins. A stretch of about 260 amino acids within each of these proteins has homology with sequence of the catalytic subunit of the serine-specific cyclic AMP-dependent protein kinase (PKA). Despite this homology the PTKs show strict specificity for tyrosine phosphorylation in protein. While there is little doubt that the enzymes in question do phosphorylate proteins on tyrosine, there are suggestions that they may also phosphorylate nonprotein substrates, such as phospatidylinositol. If these activities prove to be intrinsic, the true function of PTKs will have to be reassessed. While PI3-Kinases transfer phosphate group to phospholipid, phosphatidylinositol, these enzymes elude easy classification since they are also related to protein kinases (Keith and Schreiber, 1995). Initially it appeared that one could classify PTKs on the basis of distinction between genes, which are part of retroviruses and those associated with growth factor receptors. However, there is overlap between these two types. An example of each type is pp60v^{src} and the epidermal growth factor receptor (EGFR) respectively.

ProteinTyrosine Kinases of Retroviruses: The pp60v^{src}, the 60kDa product with 526 amino acids, of the Rous sarcoma virus (RSV) src gene, is the prototypic retroviral PTK, while the EGF receptor is by far the best understood growth factor receptor PTK. The products of the v-src, v-yes, v-fps, v-fes, v-abl, and v-ros retroviral oncogenes all have PTK activities. The v-src, v-yes, v-fps, and v-ros genes are the oncogenes of different chiken sarcoma viruses, while the v-fgr and v-fes genes are oncogene of cat sarcoma virus and v-abl gene is oncogene of mouse lymphoma virus. Despite the frequency of oncogenes encoding PTKs, it should be stressed that not all retroviral oncogene products have this activity. The v-sis, v-ras, v-myc, v-myb, v-fos, v-ski, v-rel, v-erb-A, and v-ets oncogenes have little or no relationship to those of the PTK family. The v-mil (or v-mht), v-raf, v-fms, v-mos, and v-erb-B have sequence homology with the catalytic domain of the protein kinases, and yet these proteins do not have detectable PTK activity.

Receptor Protein Tyrosine Kinases: Several growth factor receptors in testis are associated with PTK activity. An epidermal growth factor (EGF) - activated protein kinase activity was first detected in membranes made from A431 cells. After the discovery of EGF receptor PTK activity, other growth factor receptors were examined for a similar ligand-stimulated property. The platelet derived growth factor (PDGF), insulin, and insulin-like growth factor 1 (IGF-1) receptors all prove to have ligand stimulated autophosphorylating activities specific for tyrosine. In recent years such activities have been found in spermatogenic cells (see Chapter 3). These receptors also can phosphorylate exogenous substrates in vitro. The complete amino acid sequence of the human EGF receptor has been deduced from the nucleotide sequence of a series of over-lapping cDNA clones. Sasaki et al., (2003) identified 13 receptor TK genes, one of which appeared to be novel. The full-length sequence of this cDNA indicated that it encoded a novel TK receptor of the Eph family. The putative amino-acid sequence of this TK was 63.0% identical to that of human EphA1. This TK was designated as *ephA9*. The *ephA9* RNA transcripts were present in the kidney, lung, testis, and thymus but not in the spleen, brain, or liver (Sasaki et al., 2003).

19.15. TYROSINE KINASES IN TESTIS

19.15.1. Src Kinase

Src family of tyrosine kinases play important role in cell signal transduction in various cells. A Src-related protein-tyrosine kinase, named Xyk, may act upstream of the calcium release in fertilization of the *Xenopus* egg (Sato et al, 2000). Src tyrosine kinase activity in Leydig cells is related to luteinizing hormone responsiveness. Src tyrosine kinase has an important role in regulating steroid secretion in MA10 Leydig cells. The regulation may in part be due to src modulation of phosphodiesterase activity, although other components of the LH-signaling pathway may also be involved (Taylor et al, 1996). The membrane extract of rat testes revealed the existence of pp60^{-src} kinase activity. The expression of pp60^{-src} kinase activity in germ cells is associated with the development of rat testis. The peak period coincided with the timing, when the spermatogenesis by meiosis just began. The pp60c^{-src} is most abundantly expressed in the spermatids and is a developmentally regulated gene (Nishio et al, 1995).

19.15.2. Testis Specific Tyrosine Kinase FER (FerT)

The FER is expressed in a variety of tissues but includes a particular mRNA species detectable only in testes. Compared to tyrosine phosphorylation events on membranes in mammalian cells, nuclear tyrosine phosphorylation events are relatively rare. Yet several nuclear tyrosine kinases have been identified and characterized. These can be divided into two main groups: src-related and non-related nuclear tyrosine kinases. The p94^{fer} is an evolutionarily conserved tyrosine kinase encoded by the *FER* locus in mouse, rat and human. The p94^{fer} accumulates in both cytoplasm and nucleus of cells, and has been found in most mammalian cell lines analyzed. The p51^{fer} and p94 fer differ in their N-termini but they do share common SH2 and kinase domains.

The testes-specific transcript of FER, termed ferT and the first meiosis specific tyrosine kinase has been cloned from a mouse testis cDNA library (Fishman et al 1990; Hazen et al, 1993 c/r Schwark et at, 1998). The FerT is generated by an alterative splicing mode and likely also through utilization of a different promoter. The 453 amino acid encoded protein (p51 fer T) is a truncated version of the FER encoded protein p94 fer. The p51 ferT contains a unique N-terminal domain (the first 44 amino acids) that is absent from p51^{fer}. The ferT mRNA is undetectable in RNA from testes of newborn mice at times preceding week 3 of postnatal development. These findings suggested that expression of fertT within the testis is restricted to certain stages of germ cell development (Keshet et al, 1990). Keshet et al (1990) showed that ferT is transiently expressed during spermatogenesis and the expression was restricted to spermatocytes at the pachytene stage of meiotic prophase. Transfection experiments in CHO cells confirmed the nuclear localization of p51^{ferT} in eukaryotic cells (Hazan et al, 1993). Ectopic expression of p94 fer and p51^{ferT} led to tyrosine phosphorylation of 66-kDa and 120-kDa protein in CHO cells. These proteins may mediate the cellular functions of the FER tyrosine kinases. The TMF, which was shown to bind the TATA element in RNA polymerase II promoters, impaired the functioning of p94fer and p51^{ferT} in a cell free transcription system. Both p94fer and p51^{ferT} phosphorylated the TMF in vitro and in vivo at the carboxy-terminal. The p94^{fer} and p51^{ferT} may thus modulate the suppressive activity of TMF during cellular growth in a defined differentiation process (Schwartz et al., 1998).

19.15.3. G protein-Coupled Receptor Kinase 4 (GRK4)

The GRK4 presents some peculiar characteristics that make it a unique member within the GRK multigene family. For example, this is the only GRK for which four splice variants (GRK4-α, -β, -γ, -δ) have been identified. The GRK4 also mediates the homologous desensitization of type 1 metabotropic glutamate (mGlu1) receptors and phosphorylates agonist occupied form of G protein coupled receptor leading to homologous mode of desensitization. The c-DNAs clones have been identified in rat (Virlon et al, 1998). The GRK4α, but not GRK4β, -γ, and -δ, was able to phosphorylate rhodopsin in an agonist-dependent manner. The GRK4 mRNA is poorly expressed in testis and most rat tissues but is heterogeneously distributed in the kidney. The differential tissue distribution of GRK4β suggests that individual GRK4 variants may serve distinct physiological functions. Expression of GRK4 mRNA has been shown in testis and in the spermatogonia cell line GC-1 spg. Specific GRK4 immunoreactivity was found on sperm membranes, the acrosomal membranes, and to the outer mitochondrial membranes. The GRK4-γ was the only detectable isoform in human sperm. Thus, the association of GRK4 with specialized sperm organelles, which are essential for fertilization, indicates that this kinase is involved in sperm formation (Sallese et al, 1997).

19.15.4. Tyro-Receptors

Tyro 3, Axl, and Mer: Tyro-3 family receptors are essential regulators of mammalian spermatogenesis. Lu et al (1999) generated and analysed null mutations in the mouse genes encoding three structurally related receptors with tyrosine kinase activity: Tyro 3, Axl, and Mer. Mice lacking any single receptor, or any combination of two receptors, are viable and fertile, but male animals that lack all three receptors produce no mature sperm, owing to the progressive death of differentiating germ cells. This degenerative phenotype appears to result from a failure of the tropic support that is normally provided by Sertoli cells. Tyro 3, Axl and Mer are all normally expressed by Sertoil cells during postnatal development, whereas their ligands Gas6 and protein S, are produced by Leydig cells before sexual maturity, and by both Leydig and Sertoli cells thereafter. It was found that the concerted activation of Tyro 3, Axl and Mer in Sertoli cells is critical to the role that these cells play as nurtures of developing germ cells. These receptors may also be essential for the tropic maintenance of diverse cell types in the mature nervous, immune and reproductive systems (Lu et al, 1999). The FLt3 gene encodes the sixth member of tyrosine receptor family and possesses an open reading frame of 3000 nt that appears to code for a protein of 1000 amino acids. The deduced structure of the FLT3 protein presents all the characteristics of a receptor-type kinase of this family. The gene is expressed in placenta, in various adult tissues including gonads and brain, and in hematopoietic cells. The FLt3 transcript is 3.7-kb long, except in the testis, where two shorter post-meiotic transcripts are detected (Rosnet et al., 1991).

Erb Type 1 Tyrosine Kinase Receptors: The erb type 1 tyrosine kinase receptors are important for mediating proliferation and differentiation. Their aberrant activation may contribute to tumor development and progression. They comprise epidermal growth factor receptor (EGFR), c-erbB-2, c-erbB-3, and c-erbB-4. There are overlaps and differences in the cellular distribution of the four receptors, which is of significance since they can form heterodimers that give differing responses to different ligands. The different distributions could be of particular importance in cancer if therapeutic modalities are identified to inhibit or stimulate the different ligands (Walker, 1998).

Glutamate Receptor: RT-PCR analysis showed the presence of mGlu-1, -4 and -5 (but not -2 or -3) receptor mRNA in the rat testis. In the rat testis, both mGlu-1a and Glu-5 receptors were highly expressed in the germinal cell line. It is likely that these receptors are functional, since the agonist, (IS, 3R)-1-aminocyclopentane 1-3 dicarboxylic acid, was able to stimulate inositol phospholipid hydrolysis in testes. While human testes showed a high expression of mGlu-5 receptors inside the seminiferous tubuli, mGlu-1 reactivity was restricted to intertubular spaces. The mGlu-5 receptors are also present in mature spermatozoa where they are localized in the mid-piece and tail. The localization of mGlu-5 coincided with that of β-arrestin, a protein that is critically involved in the homologous desenitisation and internalization of G protein-coupled receptors (Storto et al, 2001).

19.16. PHOSPHORYLATION OF SPERM TYROSINE

Sperm Motility: The initiation and maintenance of motility of spermatozoa involves phosphorylation of sperm proteins, and motility is highly regulated by a cascade of phosphorylation-dephosphorylation events affecting the activities of protein kinase substrates. In sperm of most of the species, flagellum appears to be the principal site, which is rich for tyrosine phosphorylation. Tyrosine phosphorylated proteins have been localized in sperm flagella (Urner and Sakkas, 2003). Western blot and immunocytochemical analyses demonstrated that epididymal maturation in vivo is associated with a progressive loss of phosphotyrosine residues from the sperm head. As cells pass from the caput to the cauda epididymis, tyrosine phosphorylation becomes confined to a narrow band at the posterior margin of the acrosomal vesicle. Epididymal maturation of rat spermatozoa is also associated with an acquired competence to respond to high levels of intracellular cAMP by phosphorylating tyrosine residues on the sperm tail. Immature caput spermatozoa are incapable of exhibiting this response, despite the apparent availability of cAMP and protein kinase A (Lewis and Aitken, 2001). It is also suggested that glycogen synthase kinase-3 (GSK-3) regulated by phosphorylation, could be a key element underlying motility initiation in the epididymis and regulation of mature sperm function. It follows that tyrosine phosphorylation situation would be opposite to that observed in somatic cells where tyrosine phosphorylation has been shown to increase rather than decrease GSK-3 activity. To explore the mechanism underlying hyaluronan-induced sperm motility, Ranganathan et al, (1995) demonstrated the enhanced cellular protein phosphorylation in hyluronic acid stimulated cauda spermatozoa. Labeling of spermatozoa with orthophosphoric acid yielded a wide range of labeled phosphoproteins in presence of hyaluronan. In addition, the role of 34-kDa hyaluronan binding protein in hyaluronan-induced spermatozoa is supported by elevated production of inostiol triphosphate accompanied by increased phosphorylation. This suggests the importance of hyaluronan binding protein phosporylation in the transduction of signals resulting from the interaction of hyaluronate with the sperm surface.

Sperm Capacitation: Protein tyrosine phosphorylation is associated with sperm capacitation and the acrosome reaction in several mammalian species (Leyton et al., 1992). The phosphorylation in mid-piece is the pre-requisite for capacitation. However, non-phosphorylated sperm can bind zona-pelluda. During capacitation, human spermatozoa undergo tyrosine phosphorylation of a characteristic set of proteins, only one of which has thus far been cloned and localized. Phosphorylation of tyrosine residues on a cohort of specific proteins in mouse, bovine, and human sperm indicated that activation of sperm tyrosine kinase occurs during the capacitation process. Although distinct sets of high-molecular weight tyrosine phosphorylated proteins have been recognized during capacitiation in human (80-105kDa), mouse (40-120kDa), and bovine (40-120kDa) spermatozoa, little is known about the structure of these substrates of tyrosine phosphorylation, particularly in case of human spermatozoa. Only one capacitating-induced tyrosine-phosphorylated protein substrate of human spermatozoa has been reported and it is a fibrous sheath associated A- kinase anchor protein (AKAP). The PKA anchoring proteins have been demonstrated in fibrous sheath of both mouse and human spermatozoa. The AKAPs in other cell types have been implicated in signal transduction by forming complexes with protein kinases and phosphatases within specific cellular compartments. (see chapter 29).

Acrosome Reaction: Tyrsoine phosporylation in sperm appears to be important during various steps involved in acrosome reaction, it was evidenced that fertilization inhibition of PTK activity prevents acrosomal exocytosis and blocks fertilization (see Chapter 21).

```
  1 MKPITKQQGE LVGSRISHVW QSAGISKELL EEVGQNGSRA RISVQVHNAT CTVRIAAVTK
 61 GGVGPFSDPV CKYYTGGNTG YFCANVSRMS THRSFKLNNT LHIPCRGRPQ PNVTCRDLKR
121 CNVSDEVQRG MPGNVTPCTR LGRLCPLFNS GAWQRRSCAH HLWLLLWIIL IGLVLYISLA
181 IRKRVQETKF GNAFTEEDSE LVVNYIAKKS FCRRAIELTH SLGVSEELQN KLEDVVIDRN
241 LLILGKILGE GEKGTVYEGL WNIPEGKEVK IPVAIKTLKL DTMANKEILD EASVMKGFGN
301 PHVVRLLGIC MTSTIYVITE YCLLVYRRNK DKAEQHRSNC AELNPPLQTL LKFMVDIALG
361 MEYLSNRNFL HRDLAARNCM LRDDMTVCVA DFGLSKKIYS GDYYRQGRIA KMPVKWIAIE
421 SLADRVYTKS DVWAFGVTMW EIATTLRGMT PYPGVQNHEM YDYLLHGHRL KQPRTAWNCT
481 EIRIRLLKLP ILGSRTMRPM TIFSMATRLS SPKTAWMNCM KKCTLAGEPI PKTGPTFSVL
541 RLQLEKLLES LPDVRNQADV IYVNTQLLES EGLARVHPCS TGLEHHPCSE HRPRPHLYNC
```

Fig.19.13. Amino acid sequence of tyrosine kinase from human sperm with activity for zona pellucida interaction. Reproduced with permission from D.J. Burks et al. Science 269(5220); 83-86: 1995 © AAAS.

19.17. TYROSINE KINASES IN SPERM

Non-receptor protein tyrosine kinases may also be present in sperm head (Leclerc and Goupil, 2002) and in cell cytoplasm (Beruti and Birgonova, 1996). Sperm head possesses c-yes, a member of non-receptor Src family of tyrosine kinases, which depends on c-AMP for tyrosine phosphorylation. The C-yes is inhibited by Ca^{2+} ions (Leclerc and Goupil, 2002).

19.17.1. Zona Receptor Kinase

A 95-kDa mouse and human protein with characteristics of a protein tyrosine kinase (PTK) has been identified as a receptor for ZP3 of egg's extracellular matrix **(Fig.19.13)**. Hence, the 95-kDa protein is termed as zona receptor kinase (ZRK). Demonstrated similarity between mouse and human ZP3 suggests the existence of a ZRK homologue in human sperm. The testis-specific c-DNA encodes a protein similar to receptor tyrosine kinase and appears to be expressed only in testicular germ cells. Antibodies against a synthetic peptide from the intracellular domain recognized human sperm protein that contains phosphotyrosine. Synthetic peptides corresponding to regions of the predicted extracellular domain inhibited sperm binding to human zona pellucida (Leyton et al., 1992; Burks et al, 1995). Since, tyrsoine phosporylation in sperm appears to be important during various steps involved in acrosome reaction, it was evidenced that fertilization inhibition of PTK activity prevents acrosomal exocytosis and blocks fertilization. The PTK activity is dependent on exposure to isolated receptor. Aggregation of ZP3 receptor is an initial signal, which leads to acrosomal exocytosis..

Hu9 cDNA: Burks et al (1995) isolated a cDNA clone termed *Hu9*, encoding a transmembrane PTK (ZRK), and reactive with antibodies prepared against synthetic peptide of 15 residues in the Hu9 intracellular domain (residues 539 to 553). *Hu9* 2.2-kb insert revealed an open reading frame of 1800 nt predicting a protein of 600 amino acids. The predicted amino acid sequence of the protein coded by *Hu9* contains features found in all PTK's catalytic-domain, similar (55% identity) to that of c-Eyk, a receptor-like PTK, coded by the proto-oncogene of *v-eyk*. Binding sites for several src-containing proteins are present in the intracellular region of *Hu9* including growth factor receptor bound proteins (Geb2), Shc and Nck. The putative Hu9 intracellular domain contains several potential glycosylation sites and is cysteine rich (Burks et al; 1995) **(Fig.19.14)**. However, another study (Bork, 1996) revealed that *hu* cDNA encodes for human protooncogene c-mer instead ZRK, bringing into question the existence of human homologue of mouse ZRK. Nonetheless, studies indirectly indicate that a receptor protein tyrosine kinase is functional in sperm (Urnar and Sakkas, 2003). Protein tyrosine kinase mechanisms involving ZRK appear conserved within the family Felidae and are regulated by a cAMP/PKA pathway

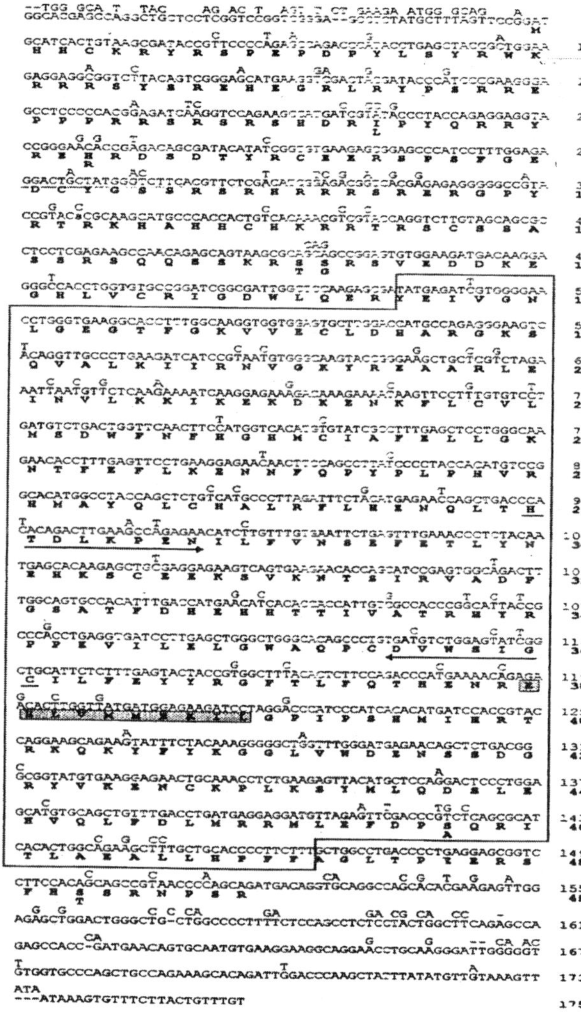

Fig.19.14. Amino acid sequence of rat Cdc2-like kinase (Clk3). With permission from W. Becker et al. Biochim Biophys Acta 1312; 63-67: 1996 © Elsevier.

(Pukazhenthi et al, 1998).

19.17.2 . Sp 42

Sp42, the boar sperm tyrosine kinase, is a male germ cell-specific product with highly conserved tissue expression extending to other mammalian species. The protein is effectively expressed in the germ cells and is not a sperm-associated protein secreted by the epididymal epithelium or male accessory glands. The Sp42 is not glycosylated and deglycosylation did not result in the reduction of molecular mass of Sp42. The observation was extended to human, mouse and rat, which share highly conserved tissue expression of Sp42. This suggests a possible role played by the cytoplasmic tyrosine kinase in the cell-signaling network specific to haploid male germ cells (Beruti and Borgonova et al, 1996).

III. DUAL SPECIFICITY PROTEIN KINASES

19.18. CDC2-LIKE KINASE-3 (CLK3): A LAMMER KINASE

A new family of protein kinases has been defined by structural criteria including a characteristic motif (LAMMER) in subdomain X of the catalytic domain. These 'LAMMER' kinases appear to play a fundamental role in cellular regulation, since the family comprises members from mammals, Drosophila, plants, and yeast. The first cloned mammalian member of the LAMMER kinase was Cdc2-like kinase1 (CLK1) (also called STY). This enzyme has been identified as a 'dual specificity' protein kinase because the recombinant protein is able to phosphorylate serine/ theronine as well as tyrosine residues. The CLK family of kinases has been implicated in mRNA splicing. A cDNA clone of protein kinase with high similarity to the CLK subfamily was isolated from a rat brain library and characterized. Its deduced amino acid sequence exhibited a 99% identity with human CLK3 and was therefore designated rat CLK3. In addition to the protein kinase domain, the sequence (490 amino acids) comprises an N-terminal domain with a strikingly high protein of basic amino acids. A fusion protein of CLK3 catalyzed autophosphorylation of the kinase but not phosphorylation of the exogenous substrate, histone or casein. The mRNA of CLK3 was detected predominately in testis, suggesting that this kinase regulates a testicular function (Becker et al, 1996b) **(Fig.19.14)**. The CLK3 is abundantly expressed in the reproductive system of male mice. Specifically, high level of CLK3 protein expression is found in mature spermatozoa in the testis and epididymis. The majority of the CLK3 in the testis is a full-length kinase-containing protein. Within the mature spermatozoa CLK3 is localized to the acrosome and sperm tail. CLK3 is expelled from the sperm following the acrosome reaction and gets inactivated, or degraded by the proteases released by the sperm during the acrosome reaction. The CLK family of kinases has been implicated in mRNA splicing; however, the bulk of the CLK3 protein in these cells is located in the cytoplasm, suggesting that CLK3 may have additional roles in the cell (Menegay et al, 1999).

19.19. NIMA LIKE KINASES

Dual specificity protein kinases appear to share an implicit involvement in signal transduction and in the control of mitosis and meiosis. Althoguh the Cdc gene has been generally considered as a universal regulator of cell cycle, information from *S. cervisiae* and *A. nidulans* have suggested existence of an alternative signal pathway involved in controlling mitosis. For example *nim A* gene of *A. nidulans* works independently of Cdc2 for initiation of mitosis, for spindle formation and for nuclear envelope breakdown. *Nim A* gene products are conserved in mammalian cells as well (Letwin et al., 1992).

19.19.1. Nek1 and Nek2

Nek1 / STK2: Screeing of mouse cDNA resulted in isolation of cDNAs that encode a mammalian protein kinase of 774 amino acids termed Nek1. The Nek1 is a mammalian relative of the fungal

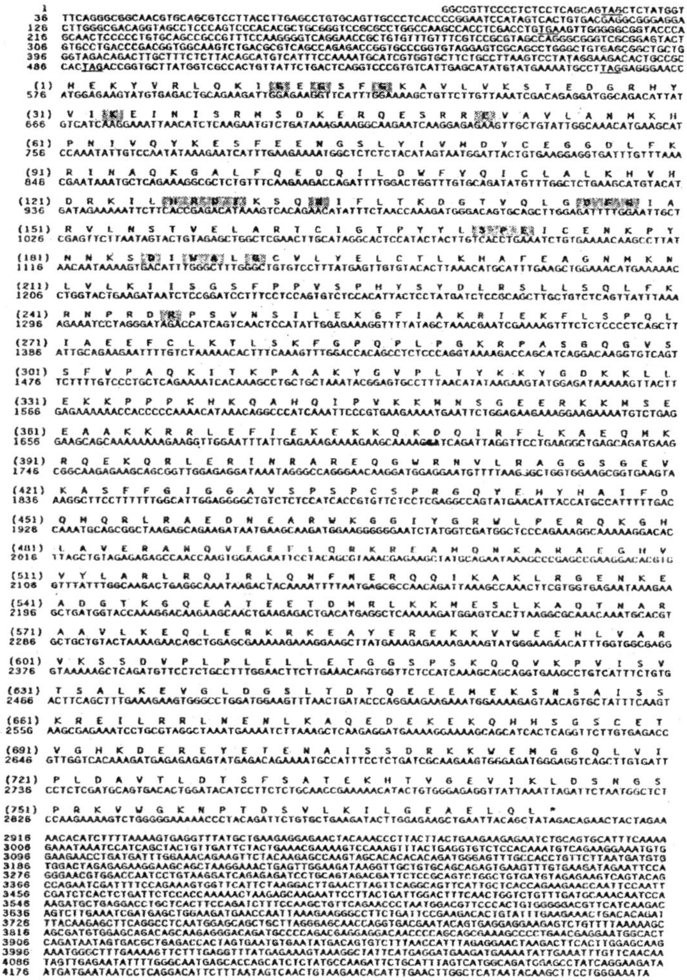

Fig.19.15. The cDNA and predicted amino acid sequence of nek1. The conserved protein kinase catalytic domain amino acids are highlighted. Reprinted with permission from K.Letwin et al. The EMBO Journal 11; 3521-31:1992 © http://www.nature.com.

NIMA cell cycle regulator. The Nek1 contains an N-terminal protein kinase domain, which is most similar (42% identity) to the catalytic domain of a protein kinase, which controls initiation of mitosis in *A. nidulans*. In addition, both Nek1 and NIMA have a long, basic C-terminal extension, and are therefore similar in overall structure. Despite its identification with anti-phosphotyrosine antibodies, Nek1 contains sequence motifs characteristic of protein serine/threonine kinases. The Nek1 kinase domain phosphorylates exogenous substrates primarily on serine/threonine, but also on tyrosine, indicating that Nek1 is dual specificity kinase. Like NIMA, Nek1 preferentially phosphorylated ß-casein in vitro. In situ RNA analysis of *nek1* expression in mouse gonads revealed a high level of expression in both male and female germ cells, with a distribution consistent with its role in meiosis (Letwin et al, 1992) **(Fig 19.15)**. Hayashi et al, (1999) isolated a murine serine/threonine kinase 2 (mSTK2) cDNA that is homologous to Nek1 serine / threonine kinase. The kinase domain of mSTK2 is highly similar to NIMA/Nek family but the C-terminal region is not similar to any protein except for human STK2 (hSTK2). Similar to Nek1, mSTK2 is expressed ubiquitously among various organs and is upregulated in the testis. The expression and localization of mSTK2 are not associated with the cell cycle progression of mitogen-activated lymphocyte and DNA-transfected fibroblast. The substrate specificity of mSTK2 is similar to NIMA, but the phosphorylation is observed exclusively on threonine residues rather than serine (Hayashi et al, 1999).

Nek2: In search of a function(s) for Nek 2 kinase during development, Uto et al (1999) isolated cDNAs encoding a Xenopus homologue of mammalian Nek2 and showd that Xenopus Nek2 has two structural variants (Nek2A and Nek2B); Nek2A, most likely a C-terminally spliced form, corresponds to the human and mouse Nek2 whereas NeK2B is probably a terminally unspliced form of Nek2. As a consequence to this (probable) alternative splicing, Nek2B is deficient in the C-terminal 70-amino acid sequence of Nek2A, which contains a PEST sequence, a motif for rapid degradation. The Nek2A is expressed predominantly in the testis (presumably in spermatocytes) and very weakly in the stomach and, during development, only after the merula stage. By contrast Nek 2B is expressed mainly in ovary and in both primary and secondary oocytes and early embryos up to the merula stage. Thus Nek2A and Nek2B may play meiotic and mitotic roles in spatially and temporally manner both during Xenopus development, and Nek2B rather than NeK2A may play an important role in early development (Uto et al, 1999).

A cDNA coding for a mouse ortholog of human Nek2 analyzed for its expression in different organs, showed high expression in testis and low level of transcripts in mitotically active organs, such as intestine, thymus, and skin and 10.5-day embryos brain. In situ hybridization analysis revealed that transcription of NeK2 occurs in stage specific manner during spermatogenesis. The strongest signals were seen in cells undergoing meiosis, besides presence of Nek2 transcripts in haploid cells (stage 1 and 11 spermatids). In addition to actively dividing follicle cells surrounding the oocytes and in the oviduct, this kinase is highly expressed in spermatocytes and, to a lesser extent, in early spermatids (Tanaka et al, 1997). Similar observation came from Rhee and Wolgemuth (1997) who also suggested that the presence of Nek2 protein was not unique to male germ cell meiosis, as it was found in meiotic pachytene stage oocytes as well. Furthermore, in an in vitro experimental setting in which meiotic chromosome condensation was induced with okadaic acid, a concomitant induction of Nek2 kinase activity was also observed. The expression of Nek2 in meiotic prophase is consistent with the hypothesis that in vivo, Nek2 is involved in G2/M phase transition of cell cycle with the evidence that in vivo, mouse Nek2 is involved in events of meiosis, including but not limited to chromosomal condensation (Rhee and Wolgemuth, 1997).

19.19.2. Murine Nek3 and Nek4

Chen et al (1999) cloned the mouse *Nek3* and *Nek4* genes. Mouse and human Nek3 are probably homologous, whereas murine *Nek4* cDNA is probably the orthologue of human STK2. The Nek4 is highly conserved between mouse and human, whereas Nek3 is somewhat less conserved (96.5 and 88% identity in the kinase domains, respectively). Preferential expression of Nek3 is seen in mitotically active tissue, whereas Nek4 is highly abundant in the testis. Within the developing testicular germ cells, nek1, 2 and 4 exhibit differential patterns of expression suggesting overlapping but not identical function. The *Nek1* is located between Cpe and

D8Mit8 on chromosome 8 at around 32 CcM. While *Nek2* is mapped to the distal region of chromosome 1, *Nek3* is mapped to the most centromeric region of chromosome.

19.19.3. Human Nek6 and Nek7

Phylogenetic analysis suggests that NEK6, NEK7, and *C. elegans* F19H6.1 constitute a subfamily within the NIMA family of protein kinases. Kimura and Okano (2001) characterized human NEK7, a gene structuralluy related to NIMA. Its open reading frame encodes a 302-amino acid protein and is 77% identical to human NEK6 protein. Tissue distribution of NEK7 is restricted to a subset of tissues contaning lung, muscle, testis, brain, heart, liver, leukocyte, and spleen but NEK6 transcripts were detected in all tissues examined. The human NeK7 gene is assigned to human chromosome 1 by radiation hybrid mapping.

19.20. DYRK-1B

The DYRK-1A is a dual specificity protein kinase whose action depends on phosphorylation of tyrosines in activation loop. The DYRK-2, DYRK-3, and DYRK-4 are predominantly expressed in testis, whereas DYRK-1 is ubiquitously expressed in rat tissues (Leder et al., 1999). The DYRK-1A gene on human chromosome 21 encodes a protein kinase presumed to be involved in the pathogenesis of mental retardation in Down's syndrome. Leder et al (1999) described a highly similar homologue, DYRK-1B, which, in contrast to DYRK-1A, is predominately expressed in muscle and testis. The human DYRK-1B gene was mapped to chromosome 19 (19q12-13.11) by radiation hybrid analysis. The amino acid sequences of DYRK-1A and DYRK-1B are 84% identical in the N-terminus and the catalytic domain but show no extended sequence similarity in the C-terminal region. DYRK-1B contains all motifs characteristic for the DYRK family of protein kinases. In addition a sequence comprising bipartite nuclear localization motif is also present. The protein was localized mainly in the nucleus of COS-7 transfected cells. The DYRK-1B is testis and muscle specific and involved in the regulation of nuclear functions (Leder et al, 1999).

19.21. C-GMP DEPENDENT PROTEIN KINASE

The *c*GMP-dependent protein kinase (cGKs) is a member of a family of cyclic nucleotide-dependent protein kinases that also includes cAMP-dependent protein kinases (cAKs). The cGK is a major cellular receptor of cGMP and plays important roles in cGMP-dependents signal transduction pathways. There exist two forms of cGK (I and II) that are encoded by distinct genes and two different isoforms of cGK-I (designated Iα and Iβ) that are produced by alternative splicing. Endogenous cGK-I is co-precipitated with anti-GKAP42 (42-kDa cGMP-dependent protein kinase anchoring protein) antibody from mouse testis suggesting that cGK-I interacts with GKAP42. Kinase-deficient mutant of cGK-Iα stably associated with GKAP42 and that binding of cGMP to cGK-Iα facilitated their release from GKAP42 (Yuasa et al 2000). The generation of nitric oxide (NO) in penile erectile tissue and the subsequent elevation of cGMP levels are important for normal penile erection. Hedlund et al (2000) reported that mice lacking cGMP-dependent kinase-1 (cGK-1) have a low ability to reproduce. Elevation of cAMP by forskolin, however, induces similar relaxation in normal and cGK-1 null corpus cavernosum. In addition, sperm derived from cGK1–null mice is normal and can undergo acrosomal reactions, and can efficiently fertilize eggs. These data identify cGK-1 as the downstream target of cGMP

```
   1 MGILLPPVSQ VTNNHINHTQ HHRNRSLDSA LQRIPEVEVS SPNAESENTF CSPSILGATM
  61 CSKIATTAAT VTSALSPPGT PTPALMSTAC TPTIGAVKGN VQHPVPKHEQ SISLAAACRS
 121 SAVVESCPTL RMRPKSSESV AANGQEMIGV IVSSTGNGSS SSNNSCSSAG STKKREDLTS
 181 LGSDDSGIIC GSESDQISLN RICHSHESLD SGEMDADADA EEECVDLMDT TSIDEEYNIM
 241 QPDHLCLYPP QAASTELDCR TLRPSRKQKR QQQQVLERKN LPSQNQIESE DSVDAGVDAR
 301 TRYRVMNMSQ AAINVELGTA SNEIDPDVNY RESHQDPIAD APSLSTPSVA STASATPSVT
 361 PTPIATPTPT ESTKNDQVLF KNFFGATKNA IFRTAQSIIE NHEKKNAAKQ KEQPDHSGTV
 421 SLVGSPNGGS SVPQDLVKSP TDISKKKEFF SLLTSSSSIK AAAAAAANSA ASTPPATPTE
 481 ALAANSNSKL QPLSDPSADL KVKERTLNLR LPGDSEGSEG ALTKCSSINS IHKLKLPVPG
 541 VVKYFISEKA PVSDVLQKPE KGQSGLLRFF ESPVFNIHFA VHYLFYSKEP GVLSFIGNKI
 601 FSFPDQEVDL YIPQLVVMYI QMDELAEVLD PYLTIRCRKS VDFSLKCLWL LEAYNYQVDS
 661 LGNSHNSSRK SKLALMKEIF SKREHKQTQN DLKSAGTGGL HVGEPRSVVA LAKKTHHRSQ
 721 SDATVLLADF RSPHTLSISH RMYQQPPYTT QTLPTTPAKL CLGDLTSGHA FDNGCTCFET
 781 VRGQVNGLLG QRTLCSCGAP KTSPQKEFMK ALMNVGKNLT SWPSKAEKTS ALRMSLNLIN
 841 KNLPARVWLP LYSDIPHHVV RITEEKTAVL NSKDKTPYII YVEVVEIPDI YTSPLIPKMM
 901 PSLRHTKSEE HLDGSCLNSH QHLSCSRTSS CSSSQQGTSC RRKKGADVDG GCGDAGPHNS
 961 CSNVHSLGGL QFQEDDVWSQ EEDEITAQYL NMRKVSERDT ISQISLDSCD SRDQGPPVVF
1021 NIGDVRLRHC SNLSCENTKS FSNDPEDPSA AALKEPWHEK EKLIRESSPY GHLSNWRLLS
1081 AIVKCGDDLR QELMATQLLQ MFKIIWQEEQ VDLWVRPYKI VCLSNDSGLI EPILNTVSLH
1141 QIKKNSNKSL KEYFIDEYGS PSGESFRRAQ KNFVQSCAAY CLISYLLQVK DRHNGNILFH
1201 SDGHIIHIDF GFILSISPKN LGFEQSPFKL TPEFVEVMGG TSSEHWREFN KLLLVGMMSA
1261 RKHMDRIINF VEIMRSNAHL PCFKNGCSGT VQNLRKRFHM NLTEQEMERK VEQLVQDSLK
1321 SLSTKLYDGY QYYTNGIL
```

Fig.19.16. Amino acid sequence of phosphatidylinositol 4-kinase â-isoform [*D. melanogaster*]. Source: http//www. ncbi.nlm.nih.gov (Accession AAK27793).

in erectile tissue and provide evidence that cAMP signaling cannot compensate for the absence of the cGMP/cGK-1 signaling cascade in vivo (Hedlund et al, 2000).

19.22. PHOSPHOLIPID RELATED KINASES

Phosphoinositide dependent kinases (PIKs) comprise a group of enzymes, which transfer phosphate groups to the phospholipid, phosphatidylinositol (PI). They are easily distinguished. Their subtle amino acid sequence difference reliably indicates their preference for the 3- or 4-position of the inositol ring of PI. The PIKs elude easy classification although they are more related to lipids than to protein kinases (Keith and Schreiber, 1995).

19.22.1. Phoshoinositide-Dependent Protein Kinase–1 (PDK1)

The PDK1 is a serine/threonine kinase that phosphorylates and activates Akt and p70 (S6K), two down stream kinases of phosphatidylinositol 3-kinase (PI3-kinase). A cDNA encoding the PDK1 predicted mouse PDK1 (mPKD1) protein to contain 559 amino acids and a COOH-terminal pleckstrin homology domain. A 7-kb mPDK1 mRNA is broadly expressed in mouse tissues and in embryonic cells. In the testis, a high level expression of a tissue-specific 2-kb transcript was also detected. Anti-mPDK1 antibody recognized multiple proteins in mouse tissues with molecular masses ranging from 60 to 180-kDa. The mPDK1 phosphorylated the conserved threonine residue (Thr402) in the activation loop of protein kinase $C\xi$ and activated the enzyme in vitro and in cells. Findings suggested that there may be different isoforms of mPDK1 and that the protein is an upstream kinase that activates divergent pathways down stream of PI 3-kinase (Dong et al., 1999).

19.22.2. Phosphatidylinositol 3-Kinase

The PI 3-kinase is another kinase and operates downstream of tyrosine phosphorylation in the signal transduction cascade, which leads to the human sperm acrosome reaction (Fisher et al, 1998). The 3-phosphoinositide dependent protein kinase 1 (PDK1) and PKC ξ and a tyrosine kinase are involved in the regulation of ERK pathway. The ERK pathway is an important route in sperm capacitation and acrosome reaction (Du Plesis et al., 2001; De Lamirande et al., 2002). A cDNA encoding a human homolgue of mouse p55 PI3-kinase, a regulatory subunit of phosphatidylinositol 3-kinase (hp55γ) was isolated. The hp55γ protein interacts with the activated IGFIR but not with the kinase–negative mutant recetpor. The putative hp55γ protein is composed of a unique amino terminal region followed by a proline-rich motif and two Src homology 2 (SH2) domains, which are highly homologous to those in mouse p55 PI3-kinase, rat p55γ, human p85 α and bovine p85γ; it contains no SH3 domain. Hp55γ mRNAs are expressed in most of the human fetal and adult tissues with particulary high abundance in adult testis. Splice variant(s) of hp55γ, one of which has a deletion of 36 amino acids at the amino terminus and another which has an insertion of 59 amino acids at position 256 between the SH2 domains, were also identified. The hp55γ has the ability to interact with the IGFIR and IR and may be involved in PI 3-kinase activation by these receptors (Dey et al., 1998).

19.22. 3. Phosphatidylinositol 4-Kinase

A remarkable variation of the universal process of cytokinesis occurs during gametogenesis in many organisms: instead of separating, the daughter cells develop as a syncytium, with clonally related cells connected by inter-cellular bridges, called ring canals. The end game of cytokinesis can follow one of two pathways depending on developmental context: resolution into separate cells or formation of a stable inter-cellular bridge. It was shown that the four wheel drive (fwd) gene of *D. melanogaster* is required for intercellular bridge formation during cytokinesis in male meiosis. In *fwd* mutant males, contractile rings form and constrict in dividing spermatocytes, but cleavage furrows are unstable and daughter cells fuse together, producing multinucleate spermatids. The *fwd* is shown to encode a phosphatidylinositol-4 kinase (PI-4 kinase), a member of a family of proteins that perform the first step in the synthesis of the key regulatory membrane phospholipid PIP2. Wild-type activity of the fwd PI-4 kinase is required for tyrosine phosphorylation in the cleavage furrow and for normal organization of actin filaments in the constricting contractile ring, thus, suggesting a critical role for PI-4 kinase and phosphatidylinositol derivative during the final stages of cytokinesis (Brill et al, 2000) **(Fig.19.16)**.

19.22.4. Diacylglycerol Kinase

Diacylglycerol kinase plays a central role in the metabolism of diacylglycerol by converting diacylglycerol (DAG) into phosphatidic acid thus initiating re-synthesis of phosphatidylinositols. The DAG is a second messenger reversibly activating PKC. In addition, DAG is a potential precursor for polyunsaturated fatty acids. Frolov et al (2001) described a type III *Drosophila* diacylglycerol kinase isoform, (DGKepsilon). *Drosophila* DGKepsilon is mapped to the cytological position 49C1-3. DGKepsilon mRNA is 1.9-kb in length and is broadly distributed throughout developing testes. Comparison of the *Dropophila* DGKepsilon with the human homologue revealed that the first zinc finger-like motif is specific for the type III isoform. Although the testis-specific DIG kinase activity is dependent upon the dose of DGKepsilon gene, the deletion of DGKepsilon does not modulate the total cellular diacylglycerol level (Frolov et al, 2001).

19.23. OTHER PROTEIN KINASES

19.23.1. 5'-AMP-Activated Protein Kinase (AMPK)

The AMPK cascade plays an important role in the regulation of energy homeostasis within the cell. The mammalian 5'-AMP-activated protein kinase (AMPK) is a heterotrimeric protein consisting of α -, β- and γ-subunits. The α-subunit is the catalytic subunit. The non-catalytic subunits AMPK-β and AMPK-γ form, together with the catalytic AMPK-α, the active kinase complex in mammals and its homologue in yeast. The gene for AMPKγ-1 has been designated as *Prkag1*. Mouse *Prkag1* cDNA from testis (1623 nt) codes a protein of 330 amino acids and a ubiquitously expressed 1.8-kb transcript. A comparison between mouse, rat and human Prkag1 cDNA and protein sequences shows that the gene is highly conserved among these species with a homology of 96% at the protein level. Southern blot analysis indicates that there is more than one gene for *Prkag* in the mouse genome. The *Prkag1* contains 12 exons with short introns. Analysis of 50 interspecific backcross mice mapped the mouse gene to the distal region of chromosome 15 (Shamsadin et al., 2001). Cheung et al (2000) isolated and characterized two isoforms of the γ subunit, termed γ2 and γ3. Both γ2 (569 amino acids) and γ3 (492 amino acids) have a long N-terminal domain, which is not present in the γ1 isoform. As with γ1, mRNA encoding γ2 is widely expressed in human tissues, whereas significant expression of γ3 mRNA was only detected in skeletal muscle. Using isoforms-specific antibodies, AMPK activity was found associated with different γ isoforms in different rat tissues (Cheung et al., 2000).

19.23.2. Glycogen Synthase Kinase-3

Glycogen synthase kinase-3 (GSK3) is a multisubstrate, proline-directed kinase that phosphorylates τ, β-amyloid and neurofilaments. The expression levels of the two GSK-3 isoforms, α and β mRNA and proteins in different human tissues demonstrated that GSK-3 α is encoded by a 2.6-kb mRNA and GSK-3β by 8.3- and 2.8-kb mRNA. The two GSK-3β mRNA species were variably expressed in different tissues. Both GSK-3α and GSK-3β mRNA were prominently expressed in testis, thymus, prostate and ovary but were low in adult lung and kidney. The 51kDa GSK-3α-protein is highly expressed in lung, ovary, kidney and testis, whereas 46kDa GSK-3β-protein is highly expressed in lung, kidney and brain. Glycogen synthase kinase-3β (GSK-3β) regulates cell metabolism, cell cycle, and cell fate through the phosphorylation of a diverse array of substrates. The GSK-3 also supports a role in mammalian meiosis and spermatogenesis. The differential expression of GSK-3α and GSK-3β mRNA and proteins and the lack of relationship between transcription and translation in some tissues revealed that GSK-3α and GSK-3β are subject to different means of regulation (Lau et al, 1999), whereas the GSK-3α appears to regulate bovine sperm motility (Vijayaraghavan et al., 1996; 2000). While GSK-3α is ubiquitous in the seminiferous tubules, GSK-3β is expressed in premeiotic type B spermatogonia, in both meiotic preleptotene and leptotene spermatocytes, as well as in Sertoli cells in both the mouse and rat in addition to its activity in spermatids. This provides the basis for future studies of GSK-3β signaling in mammalian meiosis and spermatogenesis (Guo et al., 2003).

19.23.3. Other Non-specific Protein Kinases in Germ Cells: These include ribosomal protein S6 kinase (Zhang et al., Genomics, 61, 314-8, 1999), STE20 like kinase in guinea pigs (Itoh et al, ABB 340, 210-7, 1997), Shf a Shb-like adapter protein involved in regulation of apoptosis (Lindholm et al., BBRC 278, 537-43, 2000). Krct is another Ser/Thr protein kinase widely expressed in murine tissues (Stairs et al Hum Mol Genet 7, 2157-66, 1998). A ubiquitious protein kinase with multiple intracellular locations has been described by Brede et al, (Genomics 70, 82-92, 2000). A murine heme regulated inhibitor (HRI), the heme regulated eukaryotic initiation factor 2a (eIF2a) kinase has been shown to be present in testis (Berlanga et al., JBC 273, 32340-6, 1998). A CSK type tyrosine kinase (ctk) transcript similar to Src family kinases has been identified in testis and brain and seems to be involved in male germ cells differentiation (Kaneko et al, Oncogene 10, 945-52, 1995). A serine/threonine kinase gene designated *Gek1* is present in mouse primordial germ cell-derived embryonic germ cell. Gek1 mRNA is also detected in several other tissues of adult mice and central nervous system in embryos. The deduced amino acid sequence of Gek1 in the kinase domain is related to those encoded by the *S. cerevisiae STE20, CDC15*, and *D. melanogaster ninaC*. Gek1 is located on chromosome 11, near the wr locus, showing neuronal and reproductive defects (Yanagisawa et al., 1996). The specific involvement of only few kinases in the cytodifferentiation occurring during spermiogenesis has been described.

Phosphorylase kinase (Phk) is a regulatory enzyme of glycogen metabolism. The Phk is clinically and genetically heterogeneous protein and is responsible for glycogon storage disease (1 in 100000 births). Two isoforms of Phk, encoded by separate genes are known for the subunits α (muscle [α M] and liver [α L] isoforms), and γ (muscle [γ M] and testis [γ T] isoforms), wherease only one gene appears to exist for the β-subunit. The subunit δ is calmodulin; identical calmodulins are expressed from three different human genes. Maichele et al (1996) reported that autosomal liver-specific Phk deficiency is associated with mutations in the gene encoding the testis/liver isoform of the catalytic γ subunit (Phkγ2). The *Phkγ2* gene product is the predominant isoform of the catalytic γ subunit of Phk not only in testis but also in liver, erythrocytes and, possibly, other-muscle tissues.

19.24. REFERENCES

Abe MK, Kuo WL, Hershenson MB, Rosner MR. Extracellular signal-regulated kinase 7 (ERK7), a novel ERK with a C-terminal domain that regulates its activity, its cellular localization, and cell growth. Mol Cell Biol 1999; 19: 1301-12.

Abe Y, Matsumoto S, Kito K, Ueda N. Cloning and expression of a novel MAPKK-like protein kinase, lymphokine-activated killer T-cell-originated protein kinase, specifically expressed in the testis and activated lymphoid cells. J Biol Chem 2000; 275: 21525-31.

Abrieu A, Magnaghi-Jaulin L, Kahana JA, Peter M, Castro A, Vigneron S, Lorca T, Cleveand DW, Labbe JC. Mps1 is a kinetochore-associated kinase essential for the vertebrate mitotic checkpoint. Cell 2001; 106: 83-93.

Bartkova J, Falck J, Rajpert De Meyts E, Skakkebaek NE, Lukas J, Bartek J. Chk2 tumour suppressor protein in human spermatogenesis and testicular germ-cell tumours. Oncogene 2001; 20: 5897-902.

Becker W, Heukelbach J, Kentrup H, Joost HG. Molecular cloning and characterization of a novel mammalian protein kinase harboring a homology domain that defines a subfamily of serine / threonine kinases. Eur J Biochem 1996a; 235: 736-43.

Becker W, Kentrup H, Heukelback J, Joost HG. cDNA cloning and characterization of rat a Lammer kinase predominately expressed in testis. Biochim Biophy Acta 1996b; 1312: 63-7.

Beebe JS, ϕyen O, Sandberg M, Frϕysa A, Hanson V, Jahnsen T. Molecular cloning of a tissue specific protein kinase (Cγ) from human testis – representing a third isoform for the catalytic subunit of cAMP dependent protein kinase. Mol Endocrinol 1990; 4: 465-75.

Belyamani M, Gangolli EA, Idzerda RL. Reproductive function in protein kinase inhibitor deficient mice. Mol Cell Biol 2001; 21: 3959-63.

Berruti G, Borgonovo B. Sp42 the boar sperm tyrosine kinase, is a male germ cell-specific product with a highly conserved tissue expression extending to other mammalian species. J Cell Sci 1996; 109(pt4): 851-8.

Bork P. Sperm-egg binding protein or proto-oncogene. Science 1996; 27: 11431-32.

Brill JA, Hime GR, Scharer Schuksz M, Fuller MT. A phospholipids kinase regulates actin organization and

intercellular bridge formation during germline cytokinesis. Development 2000; 127: 3855-64.
Burks DJ, Carballada R, Moore HDH, Saling PM. Interaction of a tyrosine kinase from human sperm with the zona pellucida at fertilization. Science 1995; 269(5220): 83-86.
Chaudhary PS, Nanez R, Casillas ER. Purification and characterization of polyamine stimulated protein kinase (Casein kinase II) from bovine spermatozoa. Arch Biochem Biophys 1991; 288: 337-42.
Chen A, Yanai A, Arama E, Kilfin G, Motro B. NIMA-related kinases: isolation and characterization of murine nek3 and nek4 cDNAs and chromosomal localization of nek1, nek2 and nek3. Gene 1999; 234: 127-37.
Cheung PC, Salt IP, Davies SP, Hardie DG, Carling D. Characterization of AMP-activated protein kinase γ-subunit isoforms and their role in AMP binding. Biochem J 2000; 346pt3: 659-69.
Chin MY, Li G.Tissue-specific regulation of checkpoint kinase 2 expression by p53. Exp Mol Pathol 2003; 75: 131-6.
Church JL, Guan KL, Lambie EJ, Three genes of MAP kinase cascade *mek2*, (*mpk-1*) and *Let-60* ras are required for meiotic cell cycle progression in *Caenorhabditis elegans*. Development for 1995; 121: 2525-35.
Collins SP, Uhler MD. Characterization of PKIγ, a novel isoform of the protein kinase inhibitor of cAMP-dependent protein kinase. J Biol Chem 1997; 272: 18169-78.
Costanzo V, Avvedimento EV, Gottesman ME, Gautier J, Grieco D. Protein kinase A is required for chromosomal DNA replication. Curr Biol 1999; 9: 961-6.
Dahle MK, Reinton N, Orstavik S, Tasken KA, Tasken K. Novel alternatively spliced mRNA (1c) of the protein kinase A R1&α; subunit is implicated in haploid germ cell specific expression. Mol Reprod Dev 2001; 59: 11-6.
De Jonge CJ, Han JL, Mack SR, Zaneveld LID. Effect of phorbol diesters, synthetic diacyglycerols and a protein kinase C inhibitor on the human sperm acrosome reaction. J Androl 1991; 12: 62-70.
De Lamirande E, Gagnon C. The extracellular signal-regulated kinase (ERK) pathway is involved in human sperm function and modulated by the superoxide anion Mol Hum Reprod 2002; 8: 124-35.
Desseyn JL, Burton KA, McKnight GS. Expression of a nonmyristylated variant of the catalytic subunit of protein kinase A during male germ-cell development. Proc Natl Acad Sci USA 2000; 97: 6433-8.
Dey BR, Furlanetto RW, Nissley SP. Cloning of human p55γ, a regulatory subunit of phosphatidylinositol 3-kinase, by a yeast two-hybrid library screen with the insulin-like growth factor-1 receptor. Gene 1998; 209: 175-83.
Dong LQ, Zhang RB, Langlais P, He H, Clark M, Zhu L, Liu F. Primary structure, tissue distribution and expression of mouse phosphoinositide dependent protein kinase-1, a protein kinase that phosphorylates and activates protein kinase Cξ. J Biol Chem 1999; 274: 8117-22.
Du Plessis SS, Page C, Franken DR. The zona pellucida-induced acrosome reaction of human spermatozoa involves extracellular signal regulated kinase activation. Andrologia 2001; 33: 337-42.
Escalier D, Silvius D, Xu X. Spermatogenesis of mice lacking CK2alpha': failure of germ cell survival and characteristic modifications of the spermatid nucleus. Mol Reprod Dev 2003; 66: 190-201.
Fagnen G, Phamantu NT, Bocquet J, Bonnemay PJ. Activation of protein kinase C increases proteoglycans synthesis in immature rat Sertoli cells. Biochim Biophys Acta 1999; 1472: 250-61.
Fisher HM, Brewis IA, Barratt CL, Cooke ID, Moore HD. Phosphoinositide 3-kinase is involved in the induction of the human sperm acrosome reaction downstream of tyrosine phosphorylation. Mol Hum Reprod 1998; 4: 849-55.
Fishman K, Edman JC, Shackleford GM, Turner JA, Rutter WJ and Nir U. A murine fer testis-specific transcript (ferT) encodes a truncated Fer protein. Mol. Cell Biol 1990; 10: 146-53.
Flaggs G, Plug AW, Dunks KM, Mundt KE, ford JC, Quiggle MR, et al. Atm-dependent interactions of a mammalian chk1 homolog with meiotic chromosomes. Curr Biol 1997; 7: 977-86.
Flesch FM, Colenbrander B, van Golde LM, Gadella BM. Capacitation induces tyrosine phosphorylation of proteins in the boar sperm plasma membrane Biochem Biophys Res Comm 1999; 262: 787-92.
Foss KB, Solberg R, Simard J, Myklebust F, Hansson V, Jahnsen T, Tasken K. Molecular cloning, upstream sequence and promoter studies of the human gene for the regulatory subunit RIIα of cAMP-dependent protein kinase. Biohchim Biophys Acta 1997; 1350: 98-108.
Frolov MV, Benevolenskaya EV, Birchler JA. Molecular analysis of a novel Drosophila diacylglycerol kinase, DGKε. Biochim Biophys Acta 2001; 1538: 339-52.
Garbi M, Rubinstein S, Lax Y, Breitbart H. Activation of protein kinase Cα in the lysophosphatidic acid induced bovine sperm acrosome reaction and phospholipase D1 regulation. Biol Reprod 2000; 63: 1271-77.
Gardner HP, Rajan JV, Ha SI, Copeland NG, Gilbert DJ, Jenkins NA, Marquis ST, Chodosh LA. Cloning characterization and chromosomal localization of Pnck, a Ca^{2+} calmodulin dependent protein kinase. Genomics 2000; 63: 279-88.
Gelman IH, Tombler E, Vargas J Jr. A role for SSeCKS, a major protein kinase C substrate with tumour

suppressor activity, in cytoskeletal architecture, formation of migratory processes, and cell migration during embryogenesis. Histochem J 2000; 32: 13-26.

Giet R, Prigent C. The non-catalytic domain of the Xenopus laevis aurora A kinase localizes the protein to the centrosome. J Cell Sci 2001; 114(pt11): 2095-104.

Graves PR, Haas DW, Hagedom CH, Dpaoli-Roach PJ. Molecular cloning, expression, and characterization of a 49-kilodalton casein kinase I isoform from rat testis. J Biol Chem 1993; 25: 6394-401.

Guo TB, Chan KC, Hakovirta H, Xiao Y, Toppari J, Mitchell AP, Salameh WA. Evidence for a role of glycogen synthase kinase-3 beta in rodent spermatogenesis. J Androl 2003; 24: 332-42.

Hao Z, Jha KN, Kim YH, Vemuganti S, Westbrook VA, Chertihin O, Markgraf K, et al., Expression analysis of the human testis-specific serine/threonine kinase (TSSK) homologues. A TSSK member is present in the equatorial segment of human sperm. Mol Hum Reprod 2004 Mar 25

Harrison RA, Miller NG. cAMP-dependent protein kinase control of plasma membrane lipid architecture in boar sperm. Mol Reprod Dev 2000; 55: 220-8.

Hayashi K, Igarashi H, Ogawa M, Sakaguchi N. Activity and substrate specificity of the murine STK2 Serine/ Threonine kinase that is structurally related to the mitotic regulator protein NIMA of Aspergillus nidulans. Biochem Biophys Res Commun 1999; 264: 449-56.

Hedlund P, Aszodi A, Pfeifer A, Alm P, Hofmann F, Ahmad M, Fassler R, Andersson KE. Erectile dysfunction in cyclic GMP dependent kinase I-deficient mice. Proc Natl Acad Sci USA 2000' 97: 2349-54.

Herrmann S, Amorim I, Sunkel CE. The POLO kinase is required at multiple stages during spermatogenesis in Drosophila melanogaster. Chromosoma 1998; 107: 440-51.

Hu HM, Chuang CK, Lee MJ, Tseng TC, Tang TK. Genomic organization expression, and chromosome localization of a third aurora – related kinase gene, Aie1. DNA Cell Biol 2000; 19: 679-88.

Hu MC, Wang YP, Mikhail A, Qiu WR, Tan TH. Murine p38-δ mitogen-activated protein kinase, a developmentally regulated protein kinase that is activated by stress and proinflammatory cytokines. J Biol Chem 1999; 274: 7095-102.

Ikebe C, Ohashi K, Mizuno K. Identification of testis-specific (Limk2t) and brain specific (Limk2c) isoforms of mouse LIM-kinase 2 gene transcripts. Biochem Biophys Res Commun 1998; 246: 307-12.

Itoh A, Inaba K, Ohtake H, Fujinoki M, Morisawa M. Characterization of a cAMP-dependent protein kinase catalytic subunit from rainbow trout spermatozoa. Biochem Biophys Res Commun 2003; 305: 855-61.

Jinno A, Tanaka K, Matsushime H, Haneji T, Shibuya M.; Testis-specific Mak protein kinase is expressed specifically in the meiotic phase in spermatogenesis and is associated with a 210-Kilodalton cellular phosphoprotein. Mol Cell Biol 1993; 13: 4146-56

Jones MH, Numata M, Shimane M. Identification and characterization of BRDT: a testis-specific gene related to the bromodomain gene RING3 and Drosophila fsh. Genomics 1997; 45: 529-34.

Kapphahn MA, Shabb JB. Contribution of the carboxyl-terminal regional of the cAMP-dependent protein kinase typ1α regulatory subunit to cyclic nucleotide interactions. Arch Biochem Biophys 1997; 348: 347-56.

Keegan KS, Holtzman DA, Plug AW, Christenson ER, Brainerd EE, Flaggs G, Bentley. The Atr and Atm protein kinases associate with different sites along meiotically pairing chromosomes. Genes Dev. 1996; 10: 2423-37.

Keith CT, Schreiber S, PIK-related kinases: DNA repair, recombination and cell cycle checkpoints. Science 1995; 270: 50-51.

Keshet E, Itin A, Fischman K, Nir U. The testis-specific transcripts (fer T) of the tyrosine kinase Fer is expressed during spermatogenesis in stage-specific manner. Mol Cell Biol 1990; 10:5021-25.

Kim HM, Shin T. Expression and localization of protein kinase Cθ isoform in mouse testis. Life Sci 1999; 65: 805-12.

Kim T, Park D. Molecular cloning and characterization of a novel mouse βPix isoform. Mol Cells 2001; 28: 89-94.

Kimura M, Matsuda Y, Yoshika T, Okano Y. Cell cycle-dependent expression and centrosome localization of a third human aurora/Ipl1-related protein kinase, AIK3. J Biol Chem 1999; 274: 7334-40.

Kimura M, Okano Y. Identification and assignment of the human NIMA related protein kinase 7 gene (NEK7) to human chromosome 1q31.3. Cytogenet Cell Genet 2001; 94: 33-8.

Kitabayashi AN, Kusuda J, Hirai M, Hashimoto K. Cloning and chromosomal mapping of human casein kinase 1γ2 (CSNK1G2). Genomics 1997; 46: 133-7.

Kitagawa M, Mukai H, Shibata H, Ono Y. Purification and characterization of a fatty acid-activated protein kinase (PKN) from rat testis. Biochem J 1995; 310(pt2): 657-64.

Konishi H, Kuroda S, Tanaka M, Matsuzaki H, Ono Y, Kameyama K, Haga T, Kikkawa U. Molecular cloning and characterization of a new member of the RAC protein kinase family: association of the pleckstrin homology domain of three types of RAC protein kinase with protein C subspecies and βγ subunits of G proteins. Biochem Biophys Res Commun 1995; 216: 526-34.

Kueng P, Nikolova Z, Djonov V, Hemphill A, Rohrback V, Boehlen D, Zuercher G, Andres AC, Ziemiecki A. A novel family of serine/threonine kinases participating in spermatogenesis. J Cell Biol 1997; 139: 1851-9.

Kumar P, Van Patten SM, Walsh DA. Multiplicity of the β form of the cAMP-dependent protein kinase inhibitor protein generated by post-translational modification and alternate translational initiation. J Biol Chem 1997; 272: 20011-20.

Kuroyanagi N, Onogi H, Wakabayashi T, Hagiwara M. Novel SR-protein-specific kinase, SRPK2, disassembles nuclear speckles. Biochem Biophys Res Commun 1998; 242: 357-64.

Kusuda J, Hirai M, Tanuma R, Hashimoto K. Cloning expression analysis and chromosome mapping of human casein kinase 1 γ1 (CSNK1G1): identification of two types of cDNA encoding the kinase. Cytogenet Cell genet 2000; 90: 298-302.

Landmark BF, Yen O, Skalhegg BS, Fauske H, Jahnsen T, Hansson V. Cellular location and age-dependent changes of the regulatory subunits of cAMP-dependent protein kinase in rat testis. J Reprod Fertil 1993; 99: 323-34.

Lau KF, Miller CC, Anderton BH, Shaw PC. Expression analysis of glycogen synthase kinase-3 in human tissues. J Pept Res 1999; 54: 85-91.

Leclerc P, Goupil S. Regulation of the human sperm tyrosine kinase c-yes. Activation by cyclic adenosine 3'-5'-monophosphate and inhibition by Ca^{2+}. Biol Reprod 2002; 67: 301-37.

Leder S, Weber Y, Altafaj X, Estivill X, Joost HG, Becker W. Cloning and characterization of DYRK1B, a novel member of the DRYK family of protein kinases. Biochem Biophys Res Commun 1999; 254: 474-9.

Letwin K, Mizzen L, Motro B, Ben-David Y, Bernstein A, Pawson T. A mammalian dual specificity protein kinase, Nek1, is related to the NIMA cell cycle regulator and highly expressed in meiotic germ cells. The EMBO Journal 1992; 11: 3521-31.

Lewis B, Aitken RJ. Impact of epididymal maturation on the tyrosine phosphorylation patterns exhibited by rat spermatozoa. Biol Reprod 2001; 64: 1545-56.

Leyton L, LeGuen P, Bunch D, Saling P. Regulation of mouse gamete interaction by a sperm tyrosine kinase. Proc Nat Acad Sci USA 1992; 89: 11692-95.

Lin W, Zhou X, Zhang M, Li Y, Miao S, Wang L, Zong S, Koide SS. Expression and function of the HSD: 3.8 gene encoding a testis-specific protein. Mol Hum Reprod 2001; 7: 811-8.

Litchfield DW, Luscher B, Lozeman FJ, Eisenman RN, Krebs EG. Phosphorylation of casein kinase II by p34cdc2 in vitro and at mitosis. J Biol Chem 1992; 267: 13943-51.

Lu Q, Gore M, Zhang Q, Camenisch T, Boast S, Casagranda F, Lai C, Skinner MK, Klein R, Matsushima GK, Earp HS. Goff SP, Lemke G. Tyro-3 family receptors are essential regulators of mammalian spermatogenesis. Nature 1999; 398: 723-8.

Lu Q, Sun QY, Breitbart H, Chen DY. Expression and phosphorylation of mitogen activated protein kinases during spermatogenesis and epididymal sperm maturation in mice. Arch Androl 1999; 43: 55-66.

Luconi M, Krausz C, Barni T, Vanelli GB, Forti G, Baldi E. Progesterone stimulates p42 extracellular signal-regulated kinase ($p42^{erk}$) in human spermatozoa Mol Hum Reprod 1998; 4: 254-58.

Luria A, Tennenbaum T, Sun QY, Rubinstein S, Breitbart H. Differential localization of conventional protein kinase C isoforms during mouse oocyte development. Biol Reprod 2000; 62: 1564-70.

Macleod J, Mei X, Erlichman J, Orr GA. Association of the regulatory subunit of a type II cAMP-dependent protein kinase and its binding proteins with the fibrous sheath of rat sperm flagellum. Eur J Biochem 1994; 225: 107-14.

Maichele AJ, Burwinkel B, Maire I, Sovik O, Kilimann MW. Mutations in the testis/liver isoform of the phosphorylase kinase γ subunit (PHKG2) cause autosomal liver glycogenosis in the gsd rat and in humans. Nat Genet 1996; 14: 337-40.

Mandal A, Naaby-Hansen S, Wolkowicz MJ, Klotz K, Shetty J, Retief JD, Coonrod SA, Kinter M, Sherman N, Cesar F, Flickinger CJ, Herr JC. FSP5, A testis specific 95-kilodalton fibrous sheath antigen that undergoes tyrosine phosphorylation in capacited human spermatozoa. Biol Reprod 1999; 61: 1184-97.

Manser E, Loo TH, Koh CG, Zhao ZS, Chen XQ, Tan L, Tan I, Leung T, Lim L. PAK kinases are directly coupled to the PIX family of nucleotide exchange factors. Mol Cell 1998; 1: 183-92.

Matsushime H, Jinno A, Takagi N, Shibuya M. A novel mammalian protein kinase gene (mak) is highly expressed in testicular germ cells at and after meiosis. Mol Cell Biol 1990; 10: 2261-8.

Matthews BB, Crews ST. Drosophila center divider gene is expressed in CNS midline cells and encodes a developmentally regulated protein kinase orthologous to human TESK1. DNA Cell Biol 1999; 18: 435-48.

Means AR, Cruzalegui F, LeMagueresse B, Needleman DS, Slaughter GR, Ono T. A novel Ca^{2+} / calmodulin-dependent protein kinase and a male germ cells-specific calmodulin-binding protein are derived from the same gene. Mol Cell Biol 1991; 11: 3960-71.

Menegay H, Moeslein F, Landreth G. The dual specifically protein kinase CLC3 is abundantly expressed in mature mouse spermatozoa. Exp Cell Res 1999; 15: 463-73.
Meroni S, Canepa D, Pellizzari E, Schteingart H, Cigorraga S. Regulation of γ-glutamyl transpeptidase activity by Ca^{2+}- and protein kinase C dependent pathways in Sertoli cells. Int J Androl 1997; 20: 189-94.
Miller MA, Steele RE. Lemon encodes an unusual receptor protein-tyrosine kinase expressed during gametogenesis in Hydra. Dev Biol 2000; 224: 286-98.
Mitra S, Nath D, Majumder GC. Purification and characterization of a protein kinase from goat sperm plasma membrane. Biochem Cell Biol 1994; 72: 218-26
Morgan MJ, Madgwick AJA. The fourth member of the FHL family of Lim proteins is expressed exclusively in the testis. Biochem Biophys Res Commun 1999; 255; 251-55.
Morter C, Iborra A, Martinez P. Phosphorylation of Shc proteins in human sperm in response to capacitation and progesterone treatment Mol Reprod Dev 1998; 50: 113-20.
Naito Y, Watanabe Y, Yokokura H, Sugita R, Nishio M, Hidaka H. Isoform-specific activation and structural diversity of calmodulin kinase I. J Biol Chem 1997; 272: 32704-8.
Niino YS, Irie T, Takaishi M, Hosono T, Huh N, Tachikawa T, Kuroki T. PKCθ-II, a new isoform of protein kinase C theta specifically expressed in the seminiferous tubules of mouse testis. J Biol Chem 2001; 276: 36711-7.
Nishio H, Tokuda M, Itano T, Matsui H, Takeuchi Y, Hatase O. pp60c-src expression in rat spermatogenesis. Biochem Biophys Res Commun 1995; 206: 502-10.
Ono T, Slaughter GR, Cook RG, Means AR. Molecular cloning and distribution of rat calspermin, a high affinity calmodulin binding protein. J Biol Chem 1989; 246: 2081-87.
Pangilinan F, Li Q, Weaver T, Lewis BC, Dang CV, Spencer F. Mammalian BUB1 protein kinases: map positions and in vivo expression. Genomics 1997; 46: 379-88.
Papoutsopoulou S, Nikolakaki E, Chalepakis G, Kruft V, Chevaillier P, Giannakouros T. SR protein-specific kinase 1 is highly expressed in testis and phosphorylates protamine 1. Nucleic Acids Res 1999; 27: 2972-80.
Pariset C, Weinman S. Differential localization of two isoforms of the regulatory subunit RIIα of cAMP-dependent protein kinase in human sperm: biochemical and cytochemical study. Mol Reprod Dev 1994; 39: 415-22.
Paupard MC, MacLeod J, Wasco W, Orr GA. Major 56,000-dalton, soluble phosphoprotein present in bovine sperm is the regulatory subunit of a type II cAMP-dependent protein kinase. J Cell Biochem 1988; 37: 161-75.
Phelan DR, Loveland KL, Devereux L, Dorow DS. Expression of mixed lineage kinase 2 in germ cells of the testis. Mol Reprod Dev 1999; 52: 135-40.
Pukazhenthi BS, Long JA, Wildt DE, Ottinger M, Armstrong DL, Howard J. Regulation of sperm function by protein tyrosine phosphorylation in diverse wild field species. J Androl 1998; 19: 675-85.
Ranganathan S, Bharadwaj A, Datta K. Hyaluronan mediates sperm motility by enhancing phosphorylation of proteins including hyaluronan binding protein. Cell Mol Biol Res 1995; 41: 467-76.
Reinton N, Haugen TB, Orstavik S, Skalhegg BS, Hansson V, Jahnsen T, Tasken K. The gene encoding the Cγ catalytic subunit of cAMP-dependent protein kinase is a transcribed retroposon. Genomics 1998; 49: 290-7.
Reinton N, Orstavik S, Haugen TB, Jahnsen T, Tasken K, Skalhegg BS. A novel isoform of human cyclic 3', 5'- adenosine monophosphate-dependent protein kinase, cα-s, localizes to sperm midpiece. Biol Reprod 2000; 63: 607-11.
Rhee K, Wolgemuth DJ. The NIMA-related kinase 2, Nek2, is expressed in specific stages of the meiotic cell cycle and associates with meiotic chromosomes. Development 1997; 124; 2167-77.
Rosnet O, Marchetto S, DeLapeyriere O, Birnbaum D. Murine Flt3, a gene encoding a novel tyrosine kinase receptor of the PDGFR/CSF1R family Oncogene 1991; 6: 1641-50.
Rosok O, Pedeutour F, Ree AH, Aasheim HC. Identification and characterization of TESK2, a novel member of the LIMK/TESK family of protein kinases, predominantly expressed in testis. Genomics 1999; 61: 44-54.
Rotem R, Paz GF, Homonnai ZT et al. Calcium-independent induction of acrosome reaction by protein kinase C in human sperm. Endocrinology 1992; 131: 2235-43.
Roy K, Mandal AK, Sikdar R, Majumdar S, Ono Y, Sen PC. Unsaturated fatty acid-activated protein kinase (PKx) from goat testis cytosol. Biochim Biophys Acta 1999; 1434: 161-9.
Sakkas D, Leppens-Luisier G, Lucas H, Chardonnens D, Campana A, Franken D, Urner F. Localization of tyrosine phosphorylated proteins in human sperm and relation to capacitation and zona pellucida binding Biol Reprod 2003; 68: (in press)
Sallese M, Mariggio S, Collodel G, Moretti E, Piomboni P, Baccetti B, De Blasi A. G protein-coupled receptor

kinase GRK4. Molecular analysis of the four isoforms and ultrastructural localization in spermatozoa and germinal cells. J Biol Chem 1997; 272: 10188-95.

San Agustin JT, Leszyk JD, Nuwaysir LM, and Witman GB. The catalytic subunit of the cAMP-dependent protein kinase of ovine sperm flagella has a unique amino-terminal sequence. J Biol Chem 1998; 273: 24874-83.

San Agustin JT, Wilkerson CG, Witman GB. The unique catalytic subunit of sperm cAMP-dependent protein kinase is the product of an alternative Cα mRNA expressed specifically in spermatogenic cells. Mol Biol Cell 2000; 11: 3031-44.

San Agustin JT, Witman GB. Differential expression of the Cs and Cα1 isoforms of the catalytic subunit of cyclic 3', 5'- adenosine monophosphate-dependent protein kinase testicular cells. Biol Reprod 2001; 65: 151-64.

Sasaki E, Hikono H, Kaku Y, Kuwana T, Naito M, Sakurai M. ephA9, a novel avian receptor tyrosine kinase gene. Gene 2003; 316: 103-10.

Sato K, Tokmakov AA, Iwasaki T, Fukami Y. Tyrosine kinase-dependent activation of phospholipase Cγ is required for calcium transient in Xenopus egg fertilization. Dev Biol 2000; 224: 453-69.

Schwartz Y, Ben-Dor I, Navon A, Motro B, Nir U. The phosphorylation of the TATA element FER nuclear tyrosine kinases. FEBS Lett 1998; 339-45.

Scorilas A, Yousef GM, Jung K, Rajpert-De Meyts E, Carsten S, Diamandis EP. Identification and characterization of a novel human testis-specific kinase substrate gene which is down regulated in testicular tumors. Biiochem Biophys Res Commun 2001; 285: 400-8.

Shalom S, Don J. Tlk, a novel evolutionarily conserved murine serine threonine kinase, encodes multiple testis transcripts. Mol Reprod Dev 199; 52: 392-405.

Shamsadin R, Jantsan K, Adhamn I, Engel W. Cloning, organization, chromosomal localization and expression analysis of the mouse Prkagl gene. Cytogenet Cell Genet 2001; 92: 134-8.

Shin T, Jin J, Kim J, Kim H, Lee C. Immunohistochemical study of protein kinase C in the testes of cattle and pigs. J Vet Med Sci 1998; 60: 631-3.

Skalhegg BS, Huang Y, Su T, Idzerda RL, McNight GS, Burton KA. Mutation of the Cα subunit of PKA leads to growth retardation and sperm dysfunction Mol Endocrinol 2002; 16: 630-39.

Storto M, Sallese M, Salvatore L, Poulet R, Condorelli DF, Dell-Albani P, Marcello MF, Romeo R, Piomboni P, Barone N, Nicoletti F, Nicoletti F, De Blasi A. Expression of metabotropic glutamate receptors in the rat and human testis. J Endocrinol 2001; 170: 71-8.

Sun Z, Sassone-Corsi P, Means AR. Calspermin gene transcription is regulated by two cyclic AMP response elements contained in an alternative promoter in the calmodulin kinase IV gene. Mol Cell Biol 1995; 15: 561-71.

Syed V, Hecht NB. Rat pachytene spermatocytes down-regulate a polo-like kinase and up-regulate a thiol-specific antioxidant protein, whereas Sertoli cells down-regulate a phosphodiesterase and up-regulate an oxidative stress protein after exposure to methoxyethanol and methoxyacetic acid. Endocrinology 1998; 139: 3503-11.

Taglienti CA, Wysk M, Davis RJ. Molecular cloning of the epididymal growth factor-stimulated protein kinase p56 KKIAMRE. Oncogene 1996; 13: 2563-74.

Takahashi H, Koshimizu U, Miyazaki J, Nakamura T. Impaired spermatogenic ability of testicular germ cells in mice deficient in the LIM-Kinase 2 gene. Dev Biol 2002; 241: 259-72.

Takahashi H, Koshimizu U, Nakamura T. A novel transcript encoding truncated LIM kinase 2 is specifically expressed in male germ cells undergoing meiosis. Biochem Biophys Res Commun 1998; 249: 138-45.

Tanaka H, Yoshimura Y, Nozaki M, Yomogida K, Tsuchida J, Tosaka Y, Habu T, Nakanishi T, Okada M, Nojima H, Nishimune Y. Identification and characterization of a haploid germ cell-specific nuclear protein kinase (Haspin) in spermatid nuclei and its effects on somatic cells. J Biol Chem 1999; 274: 17049-57.

Tanaka K, Parvinen M, Nigg EA. The in vivo expression pattern of mouse Nek2, a NIMA-related kinase, indicates a role in both mitosis and meisois. Exp Cell Res 1997; 237: 264-74.

Tasken KA, Collas P, Kemmner WA, Witczak O, Conti M, Tasken K. Phosphodiesterase 4D protein kinase a type II constitute a signaling unit in the centrosomal area. J Biol Chem 2001; 276: 21999-2002.

Taylor CC, Limback D, Terranova PF. Src tyrosine kinase activity is related to luteinizing hormone responsiveness : genetic manipulations using mouse MA10 Leydig cells. Endocrinology 1996; 137: 5735-8.

Thiebot B, Langris M, Bonnamy PJ, Bocquet J. Activation of protein kinase C pathway by phorbol ester results in a proteoglycans synthesis increase in peritubular cells from immature rat testis. Biochim Biophys Acta 1999; 1426: 151-67.

Thullner S, Gesellchen F, Wiemann S, Pyerin W, Kinzel V, Bossemeyer D. The protein kinase A catalytic subunit Cβ2 : molecular characterization and distribution of the splice variant. Biochem J 2000; 351(pt1): 123-32.

Toshima J, Nakagawara K, Mori M, Noda T, Mizuno K. Structural organization and chromosomal localization of the mosue tesk1 (testis-specific protein kinase 1) gene. Gene 1998; 206: 237-45.
Toshima J, Ohashi K, Okano I, Nunoue K, Kihshioka M, Kuma K, Miyata T, Hirai M, Baba T, Mizuno K. Identification and characterization of a novel protein kinase, TESK1, specifically expressed in testicular germ cells. J Biol Chem 1995; 270: 31331-7.
Toshima J, Tanaka T, Mizuno K. Dual specificity protein kinase activity of testis-specific protein kinase 1 and its regulation by autophosphorylation of serine-215 within the activation loop. J Biol Chem 1999; 274: 12171-6.
Toshima J, Toshima JY, Amano T, Yang N, Narumiya s, Mizuno K. Cofilin phosphorylation by protein kinase testicular protein kinase 1 and its role in integrin-mediated actin reorganization and focal adhesion formation. Mol Biol Cell 2001a; 12: 1131-45.
Toshima J, Toshima JY, Suzuki M, Noda T, Mizuno K. Cell-type specific expression of a TESK1 promoter linked lacZ gene in transgenic mice. Biochem Biophys Res Commun 2001b; 286: 566-73.
Toshima J, Toshima JY, Takeuchi K, Mori R, Mizuno K. Cofilin phosphorylation and actin reorganization activities of testicular protein kinase2 and its predominant expression in testicular Sertoli cells. J Biol Chem 2001c; 276: 31449-58.
Tseng TC, Chen SH, Hsu YP, Tang TK. Protein kinase profile of sperm and eggs: cloning and characterization of two novel testis-specific kinases (AIE1, AIE2) related to yeast and fly chromosome segregation regulators. DNA Cell Biol 1998; 17: 823-33.
Ueyama T, Ren Y, Ohmori S, Sakai K, Tamaki N, Saito N. cDNA cloning of an alternative splicing variant of protein kinase C δ (PKCδIII), a new truncated form of PKCδ, in rats. Biochem Biophys Res Commun 2000; 269: 557-63.
Urnar F, Sakkas D. Protein phosphorylation in mammalian spermatozoa. Reproduction 2003; 125: 17-26.
Uto K, Nakajo N, Sagata N. Two structural variants of Nek2 kinase, termed Nek2A and Nek2B, are differentially expressed in *Xenopus* tissues and development. Dev Biol 1999; 208: 456-64.
Van Patten SM, Donaldson LF, McGuinness MP, Kumar P, Alizadeh A, Griswold MD, Walsh DA. Specific testicular cellular localization and hormonal regulation of the PKIα and PKIβ isoforms of the inhibitor protein of the cAMP-dependent protein kinase. J Biol Chem 1997; 272: 20021-9.
Vijayaraghavan S, Mohan J, Gray H, Khatra B, Carr DW. A role for phosphorylation of glycogen synthase kinase -3α in bovine sperm motility regulation. Biol Reprod 2000; 62: 1647-54.
Vijayaraghavan S, Stephens DT, Trautman K, Smith GD, Khatra B, Edgar F, Silva De Ce, Greengard P. Sperm motility development in the epididymis is associated with decreased glycogen synthase kinase-3 and protein phosphatase 1 activity. Biol Reprod 1996; 54: 709-18.
Virlon B, Firsov D, Cheval L, Reiter E, Troispoux C, Guillou F, Elalouf JM. Rat G protein-coupled receptor kinase GRK4: identification, functional expression, and differential tissue distribution of two splice variants. Endocrinology 1998 139: 2784-95.
Visconti PE, Johnson LR, Cyaski M, Fornes M, Moss SB, Gerton GL, Kopf GS. Regulation localization and anchoring of protein kinase A subunits during mouse sperm capacitation. Dev Biol 1997; 192: 351-63.
Visconti PE, Westbrook VA, Chertihin O, Demarco I, Sleight S, Diekman AB. Novel Signaling pathways involved in sperm acquisition of fertilizing capacity. J Reprod Immunol 2002; 53: 133-50.
Walden PD, Cowan NJ. A novel 205-kilodalton testis-specific serine/threonine protein kinase associated with microtubules of the spermatid manchette. Mol Cell Biol 1993; 13: 7625-35.
Walden PD, Millette CF. Increased activity associated with the MAST205 protein kinase complex during mammalian spermatogenesis. Biol Reprod 1996; 55: 1039-44.
Walker RA. The erbB/HER type 1 tyrosine kinase receptor family. J Pathol 1998; 185: 234-5.
Wu JY, Gonzalez-Robayna IJ, Richards JS, Means AR. Female fertility is reduced in mice lacking Ca^{2+} calmodulin-dependent protein kinase IV. Endocrinology 2000a; 141: 4777-83.
Wu JY, Means AR. Ca^{2+}/Calmodulin-dependent protein kinase IV is expressed in spermatids and targeted to chromatin and the nuclear matrix. J Biol Chem 2000b; 275: 7994-99.
Wu JY, Ribar TJ, Cummings DE, Burton KA, McKnight GS, Means AR. Spermiogenesis and exchange of basic nuclear proteins are impaired in male germ cells lacking Camk4. Nat Genet 2000b; 275: 448-52.
Wu JY, Ribar TJ, Means AR. Spermatogenesis and the regulation of Ca^{2+} calmodulin-dependent protein kinase IV localization are not dependent on calspermin. Mol Cell Biol 2001; 21: 6066-70.
Xu X, Toselli PA, Russell LD, Seldin DC. Globozoospermia in mice lacking the casein kinase IIα catalytic subunit. Nat Genet 1999; 23: 118-21.
Yanagisawa M, Mukouyama Y, Watanabe T, Obinata M, Matsui Y. A novel serine/threonine kinase gene, Gek1, is expressed in meiotic testicular germ cells and primoridial germ cells. Mol Reprod Dev 1996; 45: 411-20.
Yanai A, Arama E, Kilfin G, Motro B. *Ayk1*, a novel mammalian gene related to *Drosophila aurora* centrosome separation kinase, is specifically expressed during meiosis. Oncogene 1997; 14: 2943-50.

Yang F, Li X, Sharma M, Zarnegar M, Lim B, Sun Z. Androgen receptor specifically interacts with a novel p21-activated kinase, PAK6. J Biol Chem 2001; 276: 15345-53.

Yang SC, Huang CH, Chen NJ, Chou CK, Lin CH. Functional implication of human serine/threonine kinase, hAIK, in cell cycle progression. J Biomed Sci 2000; 7: 484-93.

ϕyen O, Myklebust F, Scott JD, Cadd GG, McKnight GS, Hansson V, Jahnsen T. Subunits of cyclic adenosine 3', 5'-Monophosphate-dependent protein kinase show differential and distinct expression patterns during germ cell differentiation alternative polyadenylation in germ cells gives rise to unique smaller-sized mRNA species. Biol Reprod 1990; 43: 46-54.

Yuasa K, Omori K, Yanaka N. Binding and phosphorylation of a novel male germ cell-specific cGMP-dependent protein kinase-anchoring protein by cGMP-dependent protein kinase Ia. J Biol Chem 2000; 275: 4897-95.

Zhai L, Graves PR, Robinson LC, Italiano M, Culbertson MR, Rowles J, Cobb MH, DePaoli-Roach AA, Roach PJ. Casein kinase Iγ subfamily. Molecular cloning, expression and characterization of three mammalian isforms and complementation of defects in the *Saccharomyces cervisiae YCK* genes. J Biol Chem 1995; 270: 2717-24.

Zhao S, Dai J, Zhao W, Xia F, Zhou Z, Wang W, Gu S, Ying K, Xie Y, Mao Y. PDZ-binding kinase participates in spermatogenesis. Int J Biochem Cell Biol 2001; 33: 631-6.

Chapter 20

PROTEIN PHOSPHATASES

20.1. INTRODUCTION

In comparison to protein kinases, comparatively little is known about the protein phosphatases (PPs) of mammalian sperm. Protein phosphatases have been detected in the sperm of cattle, goats, pigs, dogs, humans and roosters (Ahmad et al., 1995; Ashizawa et al, 1997). Smith et al (1996) demonstrated that the primate spermatozoa contain serine-threonine- specific protein phosphatase 1, which is sensitive to heat and acid stable inhibitors 1 and 2. Protein tyrosine phosphatases have also been identified in mouse testis and implicated in spermatogenesis, sperm velocity, and capacitation (Leclerc et al., 1996, 1997; Emiliozzi and Fenichel, 1997; Galantino-Homer et al., 1997). The studies on the role of protein phosphatases in epididymal maturation or motility are limited. The membrane–associated protein phosphatase of epididymal spermatozoa of hamster was identified as a protein tyrosine phosphatase (PTPase). The membrane-associated PTPase may not be essential for acquisition of motility. However, it appears that the activity is essential for the sustenance of motility since sodium orthovanadate, which specifically inhibits PTPase activity, also inhibits motility of sperm (Uma Devi et al, 1999). A 520-kDa phosphoprotein phosphatase that dephosphorylates serine and threonine residues of histones was isolated from the goat cauda-epididymal sperm plasma membrane (Barua et al, 1999). The enzyme is a glycoprotein and a specific phosphatase as it displayed higher affinity for dephosphorylation of large molecular weight phosphate esters. Two nuclear phosphoprotein phosphatases (PP1 and -2) that cause dephosphorylation of ^{32P}histone, are present in goat testis. In addition to histones, both the isozymes can cause dephosphorylation of protamines, casein, and testicular nuclear proteins. The membrane-bound PP1 is distinct from the cytosolic PP1. The PPM-1 is an ecto-enzyme whose function in sperm is unknown (Barua et al, 1999; Devi et al, 1999).

I. SERINE-THREONINE PHOSPHATASES

20.2. CLASSIFICATION

Classification of serine/threonine PPs is based on substrate specificity and response to a defined set of inhibitors and activators. Type 1 PP (PP1) is sensitive to the heat- and acid-stable inhibitors 1 and 2 (I_1 and I_2) (Cohen, 1989), whereas type 2 PPs (PP2s) are insensitive to inhibitors 1 and 2. Inactivation of PP1 by I_1 (PP1-I_1) is dependent on I_1 phosphorylation by cAMP- dependent protein kinase, whereas I_2 does not require phosphorylation for PP1 inhibition. The inactive PP-I_2 complex is activated by dissociation of PP1 and I_2, a reaction controlled by glycogen synthase kinase-3 (GSK 3) phosphorylation of I_2. Type 2 PPs can be sub-classified into PP2A, PP2B, and PP2C on the basis of cation requirements. The PP1 and PP2A do not require divalent cations for activity whereas PP2B and PP2C require Ca^{2+} / calmodulin and Mg^+, respectively for activity. Several isoforms are known to exit for each PP type. For example, the catalytic-subunit of PP1 is encoded by three different genes, termed *PP1α*, *PP1ß*, and *PP1γ*. The *PP1γ* can undergo alternative splicing generating two proteins that differ at their extreme C-terminal ends. *PP1γ*1 is ubiquitously expressed in most tissues, whereas PP1γ2 is expressed solely in the testis (Smith et al, 1996).

20.3. PROTEIN PHOSPHATASE TYPE-1

Protein phosphatases Type-1 (PP1) are involved in diverse cellular activities, ranging from glycogen metabolism to chromatin structure modification, mitosis, and meiosis. The holoenzymes are composed of two or more subunits, including a catalytic subunit (PP1c) and one or more regulatory subunits. Many eukaryotes possess several catalytic subunit genes, which encode highly conserved isoforms. There are at least 4 isotypes of the catalytic subunit of type 1 protein phosphatase (PP1), viz, PP1-α, PP1-γ1, PP1-γ2, and PP1-δ in rat as shown by cDNA cloning. In rodents, one of these isoforms, PP1c γ2 and its mRNA, appear to be expressed predominantly in testes. All isoforms of PP1 (α, γ1, γ2, and δ) are phosphorylated by Cdc2-cyclin B in vitro. The amino acid sequences of these PP1s differ in C-terminal regions. Phosphorylation by Cdc2-cyclin B is specific and involves a C-terminal threonine most likely at position 320 in PP1α. Threonine-320 is conserved in γ1, δ, and in the testis-specific PPγ2, and is the only Thr that fits the Cdc2- consensus sequence in the C-terminal region. It appears that all the phosphatase–1 isoforms are potentially regulated at M-phase (Puntoni and Villa–Moruzzi, 1997).

20.3.1. Protein Phosphatase-1γ2

The PP1γ2 participates in rat spermatogenesis. The antibody against a synthetic 21 mer amino acid sequence of the C-terminal region of the PP1γ2 interacted specifically with a rat testicular protein of 39-kDa. The PP1γ2 was located in the nuclei of late spermatocytes and of early spermatids (Shima et al., 1993). The purified PP1γ2 in native form had an apparent Mr of 170 kDa. The holoenzyme from nondenaturing gel revealed it to be composed of a catalytic subunit and two noncatalytic subunits of 78 kDa and 55 kDa. Partial amino acid sequence analysis of 78-kDa-subunit suggested it to be a 78-kDa glucose-regulated protein (Grp78), which is a member of the 70kDa heat shock protein family. The 78kDa-protein possibly functions as a chaperone by confining substrate specificity of PP1γ2 (Chun et al, 1994). The Grp 78 subunit is a regulatory subunit of PP1γ2 (Chun et al, 1999). Two cDNAs sequences (1320bp and 1180 bp) of the 55-kDa subunit of PP1γ2 from rat testis were cloned. These represented similar structure up to 1180bp, suggesting them to be generated by alternative splicing. The 1320 bp –cDNA is a homologue of the human sds22 α1 (thus, named rat sds22 α1). The 1180 bp-cDNA was a new splice-variant since its sequence at the 3'-end had not been identified in human sds22α genes (named rat sds22α3). The 1320bp cDNA is ubiquitously expressed in various tissues including the immature testis. However, the expression of 1180bp-cDNA might encode the 55-kDa-subunit to associate with PP1γ2 in rat testis and could be involved in spermatogenesis by controlling PP1γ2 activity (Chun et al, 2000). The serine/threonine phosphatase PP1γ2 is a testis-specific protein phosphatase isoform in spermatozoa. This enzyme appears to play a key role in motility initiation and stimulation. In general, protein phosphatases are regulated by their binding proteins. Its activity decreases during sperm maturation in the epididymis. Inhibition of PP1γ2 initiates and stimulates sperm motility. The purified PP1γ2 was complexed to a protein of 43kDa in size, which is a mammalian homologue of sds22, that binds to a yeast PP1 binding protein. Phosphatase activity is catalytically inactive when PP1γ2 is complexed to sds22. Both PP1γ2 and sds22 proteins are present in sperm tail and at distinct locations in the head (Huang Z, et al. 2002). Further studies suggested that dissociation of sds22 from p17 or some other post-translational modification of sds22 is required for its binding with and inactivation of PP1γ2 (Mishra et al., 2003).

Whether PP1cγ2 performs a testis-specific function, the PP1cγ gene was disrupted by targeted insertion in murine embryonic stem cells. Mice derived from these cells were viable, and homozygous females were fertile. However, males homozygous for the targeted insertion were infertile. The defects in meiosis were inferred from revealing severe impairment of spermiogenesis beginning at the round spermatid stage in addition to the presence of polyploid spermatids (Varmuza et al, 1999). The major fraction of axonemal PP1 is located in the central pair apparatus associated with the C1 microtubule (Mitchell and sale. 1999). As predicted, axoneme contains two microcystin-sensitive protein phosphatases: protein phosphatases type-1 (PP1) and protein phosphatase type 2A (PP2A). Catalytic subunit of PP1 is anchored at the central pair of the microtubule apparatus that controls flagellar motility. The axonemal PP2A is likely anchored on the outer doublet microtubules, possibly in a position to directly control phosphorylation of inner arm dynein. This system of control, which involves the central pair, radial spoke, and inner arm dynein I1, is likely to regulate flagellar wave. The discovery that PP1 and PP2A are anchored in the wild–type 9+2 axoneme, but not anchored in specific *Chlamydomonas* mutants, provides a new opportunity to define subunits that anchor phosphatases in the cell. Yang et al (2000) cloned the PP1c, a *Chlamydomonas* flagellar axeonemal microcystin sensitive 35-kDa protein, which is found anchored in the axoneme. Analysis of flagella and axoneme from mutant strains revealed that PP1c is primarily but not exclusively, anchored in the central pair apparatus, and associated with the C1 microtubule. The mRNA of two PP-1 (dis 2m1 and dis2m2) proteins were not identical and were specificaly expressed (Kitagawa et al, 1990).

20.4. PROTEIN PHOSPHATASE TYPE-2

20.4.1. Protein Phosphatase-2A

The PP2A selectivity hydrolyzes phosphate groups that are esterified to serines 716 and 730 (Han et al., 2001). Kitagawa et al, (1990) examined the expression of mRNAs for catalytic subunits of serine/threonine protein phosphatases of two PP-1 (dis2m1 and dis2m2), and two isotypes of PP-2A (PP-2Aα and (PP-2Aβ). The mRNAs of all the four catalytic subunits expressed in two different sizes in most of the tissues. The smaller sized mRNAs of dis2m1 and PP-2Aβ, of 1.8-kb and 1.4-kb in length, respectively, were specifically highly expressed in testis. Both these smaller sized mRNAs began to be expressed at the age when meiosis started and were detected in at the pachytene stage of meiotic prophase.

Hatano et al (1994) isolated a rat cDNA encoding part of the β-isotype of the B regulatory subunit (BR-β) of PPαA. The isolated cDNA encoded the region corresponding to amino acid positions 8(R) to 177 (N) of human BR. The homology between nucleotide and amino acid sequences of the rat and human BR-β were 95.7%, respectively. The BR-β mRNA was specifically expressed in rat brain and testis. The length of mRNA in these two organs was different. The

BR-β mRNA was expressed specifically in elongated spermatids, while mRNA of the α-isotype (BR-α) was expressed equally in all spermatogenic cells. The BR-α seemed to regulate the activity of the PP2A catalytic subunit in spermatids, and to involve in spermatogenic maturation, especially spermatid elongation (Hatano et al, 1993).

A murine protein α4, known to be present in lymphocytes binds to PP2A. A similar α4-related gene, named α4-b is present selectively in the brain and testis and absent from most of the vital organs. The α4-b putative protein is highly homologous (66% identity in amino-acid sequence) to the α4-molecule. The α4-b protein is associated with the catalytic subunit of PP2A (PP2A), involved in the regulation of phosphatase activity in neuronal cells and in spermatogenic cells (Maeda et al., 1999). In sperm axoneme also, besides microcystin-sensitive PP2A and PP1, two other axonemal proteins of 62 and 37-kDa were also isolated using microcystin-Sepharose affinity. Based on direct peptide sequence, these proteins are the A- and C-subunit of PP2A. It was proposed that PP2A is anchored on the doublet microtubules, possibly in a position to directly control inner arm dynein activity ((Mitchell and sale. 1999; Yang et al, 2000).

20.4.2. Ca^{2+} /Calmodulin Dependent Protein Phosphatase 2B (Calcineurin)

Calcineurin (CN), a Ca^{2+}/calmodulin regulated protein phosphatase 2B (CaM – PP 2B) plays an important role in many biological processes including T-cell signal transduction. Calcineurin is a major calmodulin (CaM) binding protein in brain and is the only PP that is directly regulated by Ca^{2+}/CaM. Calcineurin is a heterodimer of one catalytic subunit (CNA) and one regulatory subunit (CNB or CNβ). The catalytic subunit, CNA has high homology with other PPs The regulatory subunit CNB belongs to EF-hand Ca^{2+} binding protein family. Though it is similar to calmodulin in many ways, its tertiary structure is entirely different from calmodulin (Guerini, 1997; Shibasaki et al., 2001). The CN is a Thr/Ser protein phosphatase and stimulated by Mn^{2+} In vitro, the phosphatase activity of calcineurin is inhibited by hydrogen peroxide, superoxides and glutathione in a dose-dependent manner whereas antioxidants stimulated phosphatase activity. This suggested that the phosphatase activity of calcineurin is modulated by redox factors (Sommer et al., 2000).

Calcineurin A to Calcineurin B: Limited proteolysis of calcineurin, with clostripam defines four functional domains in calcineurin A (61-kDa), which in presence of CaM is converted to a 42kDa polypeptide that contains the calcineurin B binding domain and a 14kDa fragment. The 14-kDa fragment binds calmodulin in a Ca^{2+} dependent manner with high affinity. In the absence of CaM, clostripam rapidly degrades both CaM and inhibitory domain. The preserved catalytic domain in proteolytic product is increased in presence and absence of Ca^{+2}. The calcineurin B binding domain and calcineurin B appear unaffected by proteolysis both in the presence and in the absence of calmodulin. These results show that calcineurin A is organized in functionally distinct domains connected by proteolytically sensitive hinge regions. The catalytic, inhibitory, and calmodulin-binding domains are readily removed from the protease-resistant core, which contains the calcineurin B binding domain. Calmodulin stimulation of calcineurin is dependent on intact inhibitory and calmodulin-binding domains (Hubbard and Klee, 1989).

Catalytic subunit (CNA): The murine cDNA of the catalytic subunit of CaM-PP showed sequence of 1964-bp containing an open reading frame that encodes a protein of 513 amino acids (Mr 58,706), the predicted isoelectric point of which (pI 7.1) was much more basic than those of brain isoforms (pI 5.6-5.8). The deduced amino acid sequence was 77-81% identical to

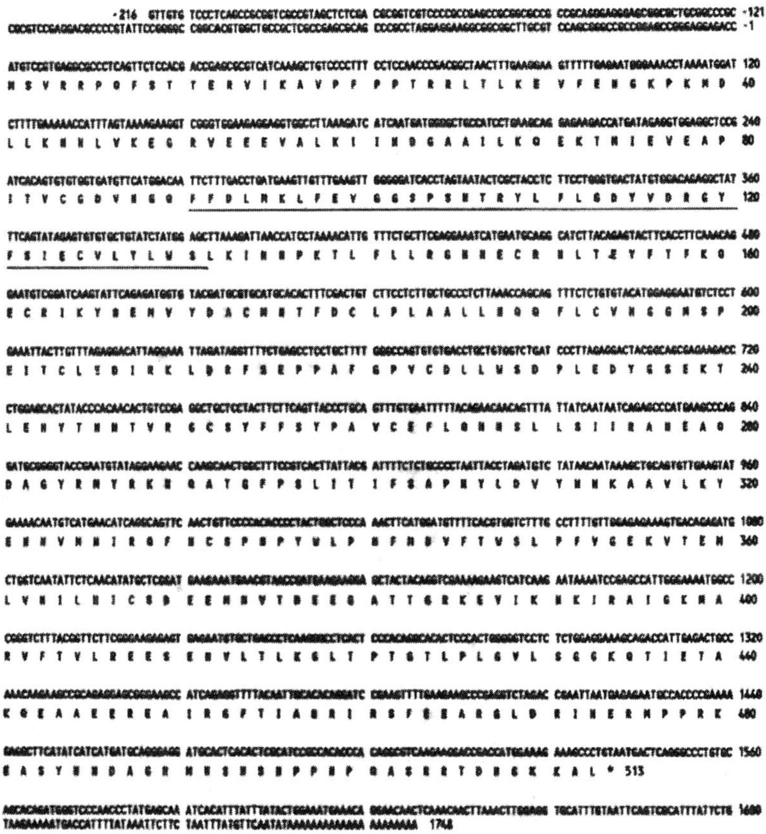

Fig.20.1. Nucleotide and deduced amino acid sequences for MTα-1, a cDNA for catalytic subunit of CaM-PrP from mouse testes. The underlined region (bp 271-392) represents the segment that was amplified for PCR. Reprinted with permission from T. Muramatsu et al. Proc Natl Acad Sci (USA) 89; 529-33: 1992 © National Academy of Sciences (USA).

two other murine CaM-PP genes and displayed a distinct Southern blot hybridization pattern, indicating that it was derived from a separate gene (type 3). High amounts of a 2800-nt mRNA transcript were observed in testis, where as mRNA species were not detectable in brain **(Fig 20.1)**. It seems that this CaM-PP represents a non-neural isoenzyme. Measurements of CaM-PP mRNA during testicular development showed a dramatic increase in expression during week 4-6, correlating with the stages of spermatogenesis (Muramatsu et al, 1992a). A cDNA for alternatively spliced variant of the testis-specific subunit of CaM-PP from a human testis showed that the nucleotide sequence of 2134-bp encodes a protein of 502 amino acids (Mr 57, 132) and pI 7.0. The cDNA sequence differs from the murine form of this gene by a 30-bp deletion in the coding region, the position of which matches those in the two other genes for the catalytic subunit. The deduced sequence of the human proteins is only 88% identical to the homologous murine form, in striking contrast to the other two CaM-PP catalytic subunits, which are highly conserved between mouse and man (Muramatsu and Kincaid, 1992b).

Ragulatory subunit (CNB): The murine cDNA sequence of the regulatory subunit of CaM-PP from brain-specific isoform (β1) encodes a 170 amino acid protein (M 19.3 kDa), whereas that of the testis isoform (β2) contains 179 residues (M_r 20.7-kDa); these two sequences show approximately 80% amino acid identity. An oligonucleotide probe for the brain isoform (β1) hybridized to a single mRNA of 3.6-kb in many tissues, whereas using the β2 probe, two mRNAs of 1.8-kb and 0.8-kb were detected only in testis. The mRNA for the testis-specific isoform increases markedly during development, its pattern being virtually identical to that of mRNA for a testicular form of the catalytic subunit (α3). These results are consistent with the biological co-regulation of catalytic and regulatory subunits of a testis-specific isoenzyme during germ cell maturation (Ueki et al, 1992).

The calcineurin B (CNB) clone from rat testis differs from rat brain cDNA by an addition of 138 bp in the coding region. The clone corresponding to the initiation codon of the clone from a brain is not ATG but AAG. The 5'- noncoding regions of these cDNAs are also different. The addition in the coding region results in the gain of 46 amino acids at the N-Terminus. These findings suggest that two distinct isoforms of CNBα are derived from the same gene through a process involving alternative utilization of two promoters. RT-PCR followed by Southern blot analysis suggested that CNB α2 (testis from) is specifically expressed in the testis, and its expression is developmentally regulated (Chang et al, 1994).

A rat testis calcineurin B-like protein (CBLP) is composed of 176 amino acids and contains four putative Ca^{2+} binding domains. The overall predicted amino acid sequence homology between CBLP and calcineurin B is 82%. The CBLP transcripts are specifically and highly expressed in the testis (Mukai et al, 1991). The deduced amino acid sequence is similar to that of human brain calcineurin β subunit with respect to four putative calcium binding sites. However, the rat testis cloned cDNA had six amino acids polypeptide tail at carboxy-terminal, which was absent in human brain calcineurin β subunit. This amino acids tail makes the carboy-terminal highly hydrophilic, in contrast to the human brain β subunit, which is hydrophobic at carboxy –terminus, and eleven amino acids at the N terminus of the cloned cDNA were completely different from the corresponding region of the brain calcineurin β subunit (Sugimoto et al., 1991).

The two subunits of scallop testis calcineurin deduced amino acid sequences of the catalytic subunit (calcineurin A=486 amino acid residues, M_r 55,005.91), and the regulatory subunit (calcineurin B=170 residues, M_r 19,237.67) showed high similarity to those of mammalian calcineurins, especially to the brain-type rather than to the testis-specific isoforms. Northern blot analysis showed that only a single species for each subunit was expressed in testis. The results were consistent with the one as to the expression of the testis-specific isoform of calcineurin A in mouse, which occurs immediately after meiosis. Its sequence similarity to the mammalian brain type isoforms may indicate that the mammalian testis–specific isoforms appeared in evolution after the divergence of mammals from the mollusks and then diverged rapidly for specific functions in testis (Uryu et al, 2000).

Ontogeny: The CN has a wide tissue distribution. The CN is not evenly distributed but concentrated in specific cells, especially in brain, kidney and testis. The common CNB subunit transcrips and protein are present in all tissues where as testis specific CNB subunit transcripts are found in brain, lung, thymus and heart, while protein is detected only in testis. This indicates that testis specific CNB subunit gene is regulated at both transcriptional and post-transcriptional levels (Su et al, 1995). In the mouse testis CaM is unusually rich in spermatogenic stages from mid-pachytene to elongating spermatids. The antibodies raised against calcineurin from scallop testis reacted with subunit B, but not subunit A of calcineurin isoforms from mouse brain and testis. In nuclei of round or elongating spermatids calcineurin started to accumulate in nuclei from the acrosomal cap phase, peaked at the initial stage of nuclear elongation, and decreased thereafter. There was almost no signal in the spermatogenic cells at other stages, including spermatogonia, spermatocytes, mature sperm, and other somatic cells in the seminiferous tubules. It suggested a role for calcineurin in remodeling of the nuclear chromatin in metamorphosing spermatids (Moriya et al, 1995). Testis-specific isoform of CNβ expressed only 3 weeks after birth, when meiosis begins. The mature sperm is most likely to have only the testis-specific isoform of CNβ, whereas the brain type isoform was not detected in rat sperm. Testes from different rodent species showed that all rodent species examined had their own homologues corresponding to a testis-specific isoform of CNβ. Each homologue is specifically expressed in post-meiotic phase of spermatogenesis, as was seen in rats. Two isoforms of calcineurin β subunit (β1 and β2) exist in rat testis. Both β1 and β2 are recovered in calmodulin binding protein fraction. The β2 showed same Mr as that of brain calcineurin β and found in wide variety of tissues. The β1, with additional six amino acid polypepeptide sequence showed higher molecular weight than brain β and was specific for testis (Nishio et al, 1992a,b). The expression of testis-specific calcineurin B subunit isoform β1 was examined in developing rat testis. Two mRNA signals of 4.0-kb and 0.9-kb present in testis, are expressed synchronously with the start of meiosis at 3 weeks after birth, and increased depending on the maturation of spermatogenesis. The β1mRNA was specifically localized to spermatocytes where meiosis occurs but its activity in spermatogonia, spermatids, Sertoli or Leydig cells was hardly detectable (Miyamoto et al, 1994).

20.4.3. Protein Phosphatase 2C

Type 2C protein phosphatases (PP2C), a class of ubiquitous and evolutionally conserved serine/threonine protein phosphatases, are encoded by at least four distinct genes and implicated in the regulation of various cellular functions. Travis and Welch (1997) cloned a cDNA of PP2C from human skeletal muscle, which encodes unique acidic domain. This PP is 34% identical to mammalian PP2Cα and PP2Cβ and named as PP2Cγ. The PP2Cγ is widely expressed, and is most abundant in testis, skeletal muscle, and heart. Seventy five percent of the 54 amino acid residues are glutamate or aspartate. Like other PP2Cs, recombinant PP2Cγ requires Mg^{2+} and Mn^{2+} for activity. The PP2Cβ gene is tissue specific and developmentally regulated. The presence of five distinct isoforms of protein phosphatase 2Cβ (PP2Cβ1-5) is known. Kato et al (1996) demonstrated that the mRNA levels of PP2Cβ-3, -4 and –5 and PP2Cβ protein level increase during the course of the first wave of spermatogenesis in neonatal mouse testis. These isozymes were expressed predominantly in pachytene spermatocytes and in more highly differentiated germ cells. The substrate specificity of PP2Cβ-4 determined with artificial substrates differed from those of PP2Cβ-3, and –5, suggesting that the difference in the structure of PP2Cβ-3, -4 and –5 reflect their unique physiological function in testicular germ cells.

Two isoforms of PP2C, PP2Cβ-3 and PP2Cβ-4 were sequenced from mouse testis. A deletion of 48 nt of PP2Cβ-4 cDNA in comparison with PP2Cβ-3 cDNA resulted in different COOH-terminal sequences of 12 and 15 amino acids, respectively. These COOH-terminal sequences of PP2Cβ-3 and PP2Cβ-4 were further found to be different from those of isoforms MPPβ1 and MPPβ2 of mouse PP2Cβ (Terasawa, et al, 1993 c/r Kato et al 1996). The common sequence of 378 amino acids from these four isoforms of mouse PP2Cβ exhibited 95% identity with the corresponding sequence of rat PP2Cβ. The mRNA of approximately 2.0-kb for PP2Cβ3 and PP2Cβ4 were expressed only in testis, while the mRNA of 3.3-kb and 8.5-kb for MPPβ1 and

MPPβ2, respectively, were found in somatic tissues (Hou et al, 1994).

To understand the regulatory mechanism of expression of PP2Cβ, the comparison of mouse genomic sequence with PP2Cβ cDNA sequences provided information on the structure and localization of intron/exon boundaries and showed that PP2Cβ isoforms with different 5'-termini were generated by alternative splicing of its pre-mRNA. The 5'-flanking region of exon 1 had features characteristic of a housekeeping gene: it was GC-rich, lacked TATA boxes and CAAT boxes in the standard positions, and contained potential binding sites for the transcription factor Sp1. In the 5'-flanking region of exon 2, several consensus sequences were found, such as a TATA-like sequence and negative regulatory element box-1, -2 and –3'. These regions act as distinct promoters. Analysis of PP2Cβ transcripts showed that exon-1 transcripts were expressed ubiquitously in several tissues, whereas exon-2 transcripts were predominantly expressed in the testis, intestine and liver. These results suggested that the alternative usage of two promoters within the PP2Cβ gene regulates tissue-specific expression of PP2Cβ mRNA (Ohnishi et al, 1999).

Wip1 gene product: The human wild type p53-induced phosphatase 1 (*Wip 1*) gene encodes a type 2C protein phosphatase (PP2C) that is induced by ionizing radiation in a p53 dependent manner. The mouse *Wip1* gene is composed of six exons and spans over 36-kb of DNA and mapped near the p53 gene on mouse chromosome11. The mouse cDNA sequence predicts a 589-amino-acid protein with a molecular mass of 66-kDa. Comparison of human and mouse Wip1 sequences revealed 83% overall identity at the amino acid level. The 5'-flanking region of exon 1 had promoter elements characteristic of a house keeping gene. The *Wip1* coding sequences share conserved functional regions with other PP2Cs from a diverse array of species. Expression of *Wip1* mRNA was detected ubiquitously in adult and embryonic tissues, though expression in the testis was much higher than in other tissue (Choi et al, 2000).

Mg dependent PPβ 4 (Type-2Cβ 4): A full-length cDNA clone (pTK-3) encoding an isoform of Mg^{2+}-dependent protein phosphatase-β (MgPPβ-4) has been isolated from a mouse melanocyte cDNA library. The mRNA corresponding to the pTK–3 insert was a splicing variant of a single pre-mRNA that also encodes MgPPβ1 and β2 (Terasawa et al, 1993). Sequence of MgPPβ-4 differed from those of MgPPβ-1 and -2 at the carboxyl terminal end. The β-4 mRNA was expressed only in testis and intestine and not in any other tissues. Specific expression of mRNA signals of two other isoforms of MgPPβ, MPPβ-3 and –5 was also demonstrated in testis and intestine. The carboxyl terminal end of MgPP β-5 was found to have a chimera structure composed of part of MPPβ-1 and part of MPPβ-3. The mRNA expression levels of MgPPβ-3, –4, and –5 alter according to the maturation of mouse testis. It suggests that the complex structure of MgPPβ isoforms and their tissue and developmental stage-specific expression reflect the variety of their physiological functions (Kato et al, 1995).

20.5. PROTEIN PHOSPHATASE-4 or PPx

The protein phosphatase-4 (PP4) also called PPx is a serine/threonine protein phosphatase that predominantly localizes to centrosomes and plays a role in microtubule organization at centrosomes. The PP4 catalytic subunit from porcine testis is inhibited by the antitumour drug, fostriecin and also by several naturally occurring tumour promoters and toxins. The gene for human PP4 catalytic subunit localizes to 16p11.2 (Hu et al., 2001). While the nucleotide sequences of murine and human PP4 are distinct, their amino acid sequences are identical. The PP4 is

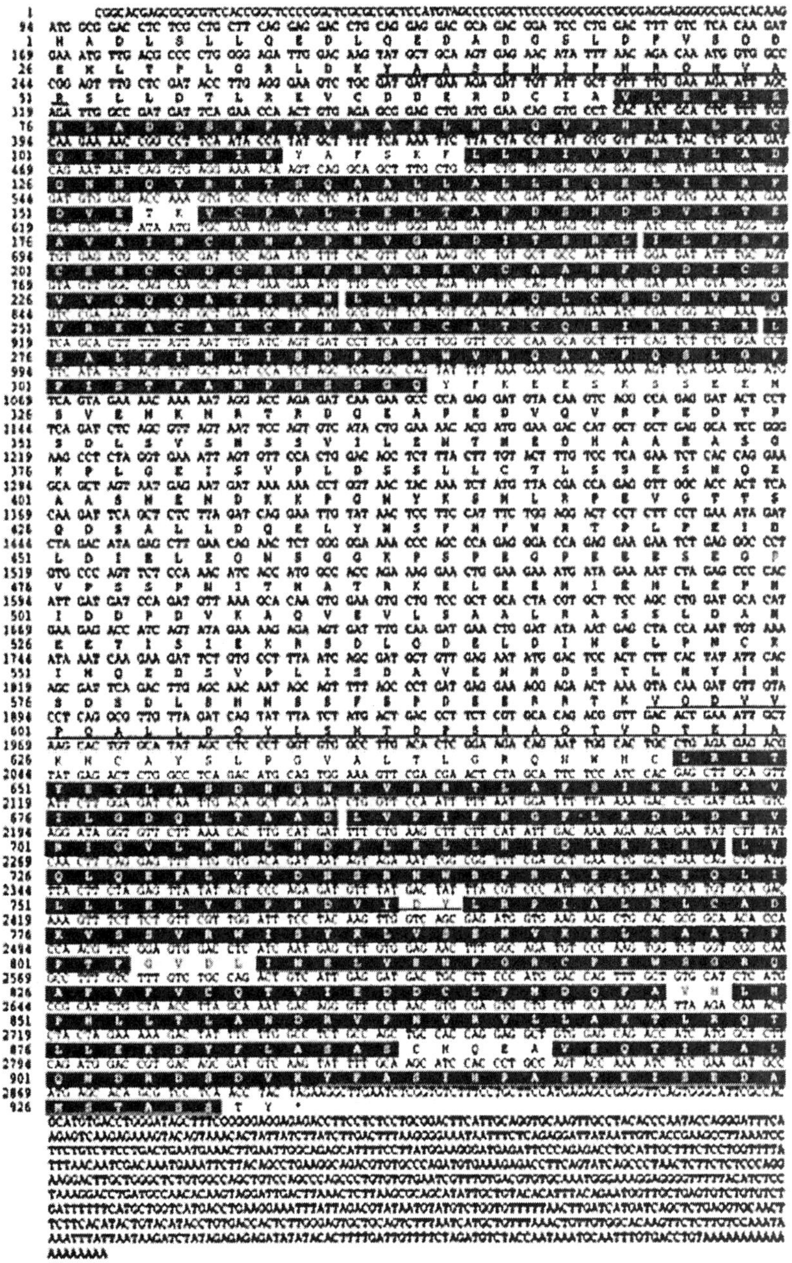

Fig.20.2. Sequence of PP4R1. The peptides obtained from purified bovine PP4R1 are underlined. Regions homologous to the repeats found in the A subunit of PP2A are printed in white type on black. Reproduced with permission from S.Kloeker and B.E. Wadzinski. J Biol Chem 274: 5339-47; 1999 © American Society for Biochemistry and Molecular Biology.

conserved across species. In adult tissues, PP4 is expressed at high levels in the testis, kidney, liver, and lung, and at lower levels in virtually all tissues. The PP4 is a developmentally regulated protein phosphatase (Hu et al., 2001).

The catalytic subunit of PP4 has more than 65% amino acid identity to the catalytic subunit of PP2A. Despite this high homology, PP4 does not appear to associate with known PP2A regulatory subunits. Bovine testis PP4 existed in two complexes of approximately 270-300 and 400-450 kDa as determined by gel filtration. The smaller PP4 complex contained two major proteins: the PP4 catalytic subunit plus a protein that migrated as a doublet of 120 –125 kDa on SDS-PAGE. The associated protein, termed PP4R1 and PP4C also bound to microcystin-Sepharose, and resolved into two major peaks, at 35 (PP4C) and 105-kDa (PP4R1) (Kloeker and Wadzinski, 1999) **(Fig.20.2)**. The PP4 forms a complex with 50-kDa protein and the complex is located at microtubule organization contents. The regulatory subunit (termed PP4R2) of PP4 comprises 453 amino acids, and a molecular mass of 50.4-kDa. Immunocytological detection of PP4R2 at centrosomes suggested that it might target PP4C to this location. Native 450-kDa and 600 PPF4 complexes are inactive, but can be activated by basic proteins, suggesting that PP4R2 may also regulate the activity of PP4C at centrosomal organizing centers (Hastie and Cohen, 1998; Hastie et al, 2000).

20.6. Cdc25 PROTEIN PHOSPHATASES

The *Cdc25A* and *Cdc25C*, which encode threonine /tyrosine phosphatases involved in cell cycle regulation, are found to be differentially localized in human and rodent testicular germ cells during spermatogenesis. Northern blot analysis identified two *Cdc25A* transcripts in rat testis. A transcript of 3.8-kb was expressed in the early germ cells including spermatocytes and The round spermatids but not in Sertoli cells. Cdc25C expressed two transcripts of 2.1-kb and 1.9-kb, which were present in spermatocytes and round spermatids respectively. The Cdc25A may participate during mitosis of spermatogonia (Mizoguchi and Kim, 1997). Murine *Cdc25C* transcripts of 2.1-kb were detected in midgestation embryos and in several adult tissues, including testis and ovary. The highest levels of *Cdc25C* transcripts were found in the testis where its expression was localized in germ cells, specifically in late, pachatene–diplotene spermatocytes and round spermatids, whereas *Cdc25B* expression was most readily detected in the somatic cells (Wu and Wolgemuth, 1995). The *Cdc25C* (-/-) mice are viable and do not display any obvious abnormalities. Among adult tissues, its transcripts are most abundant in testis, followed by thymus, ovary, spleen, and intestine. However, mice lacking Cdc25C were fertile, indicating that Cdc25C does not play an essential function during spermatogenesis or oogenesis in the mouse (see Chapter 11).

20.7. INHIBITORS OF PROTEIN PHOSPHATASES

In vitro treatment of spermatozoa with inhibitors of protein phosphatases may be of great value in understanding some cases of explained infertility (Hortas et al, 2001). The sperm PP1, being mostly cytosolic is inversely related to the motility of monkey epididymal sperm (Smith et al, 1999). Okadaic acid (OA) and caliculin A (CA) are specific PP inhibitors. Protein phosphatase-1 (PP1) and PP2A are completely inhibited by CA at 1.0 nM, whereas PP1 is 100 fold less sensitive to OA inhibition while PP2A has similar sensitivity to OA and CA. On the other hand PP2ß and PP2C are essentially insensitive to OA and CA. Human and rhesus monkey sperm contain PP1 and its regulators. The PP has been involved in Ca^{2+} dependent regulation of sperm flagellar motility. Caliculin A treatment of human sperm resulted in an increase in motility and in the acceleration of mean path velocity (Smith et al, 1996; 1999).

Testicular germ cells also contain protein inhibitors of PPs. The c-DNAs for three isotypes of protenous inhibitor–2 (I_2), ($I_2\alpha1$, $I_2\alpha2$ and $I_2\beta$) have been characterized. The $I_2\alpha2$ is an alternatively spliced form of $I_2\alpha1$, encoding a protein with substitution of three amino acids for 14 amino acids of the $I_2\alpha1$ protein at the C-terminus. The latter $I_2\beta$ is derived from a different gene and encodes a protein of 126-aminoacid residues with conserved regions of $I_2\alpha1$ protein at positions 22-47 and 111-126. The $I_2\alpha2$ and $I_2\beta$ are expressed exclusively in the testis, and the expression of three forms of I_2 coincides with sperm maturation (Osawa et al., 1996).

II. PROTEIN TYROSINE PHOSPHATASES

Protein tyrosine phosphatases (PTPase) have been classified into two subfamilies, receptor-type PTPs and nonreceptor-type PTPases based on the presence or absence of the extracellular and transmembrane portions. Nonreceptor type PTPase ordinarily comprises a single PTPase domain and a variable length of noncatalytic segment. Some nonreceptor-type PTPases show dual specificity for both phosphotyrosine and phosphoserine or phosphothreonine. The PTPase domain consists of approximately 240 amino acids carrying several consensus motifs including a sequence (I/V) HCXAGXXR (S/T)G (X can be any amino acid). The cysteine residue within the motif is involved in the formation of the catalytic core of the enzyme.

20.8. PROTEIN TYROSINE PHOSPHATASE-1

The protein tyrosine phosphatase-1 (PTPase-1) from mouse, rat and human are homologous proteins. The cDNA encoding an intracellular PTPase from a mouse testis was found to contain an open reading frame of 1,296 nt, as well as 5' (83 nt) and 3' (289 nt) non-coding regions. The deduced sequence of 432 amino acids of mouse PTPase-1 exhibited 93% and 83% identity to rat PTPase-1 and human PTPase-1B, respectively. Murine PTPase-1 mRNAs were most abundant in testis, and were detected in size of 4.4-kb, 2.4-kb, 2.2-kb, and 2.0-kb. The PTP-1 transcripts of 4.4-kb and 2.0-kb, but not 2.4-kb and 2.2-kb, were also present in other tissues. Genomic blot analysis showed that a single copy of the PTP-1 gene is contained in the mouse genome and that introns are present in mammalian PTPase-1 genes (Miyasaka and Li, 1992).

20.8.1. Tesis SpecificCytoplasmic PTPase

Kaneko et al (1993) cloned five cDNAs encoding putative PTPases from the murine testis. Two of them were identical to the mouse cytoplasmic PTPase and PTPase-1B. The remaining three were likely to represent the murine counterparts of human PTPaseδ, human PTPase ε and rat striatum-enriched PTPase. Northern blot hybridization revealed that the transcripts of PTPase-1B, mouse cytoplasmic PTPase and the murine homologue of human PTPase δ were quantitatively and/or structurally regulated during germ cell development. Testis specific PTPase (Typ) of cytoplasmic origin may play an important role(s) in spermatogenesis and/or meiosis. The Typ is specifically expressed in testicular germ cells that undergo meiosis. Developmentally, Typ is detected between 2 and 3 weeks after birth, in parallel with the onset of meiosis. The ORF of cDNAs of Typ codes a protein of 426 amino acids. The predicted Typ protein contained a

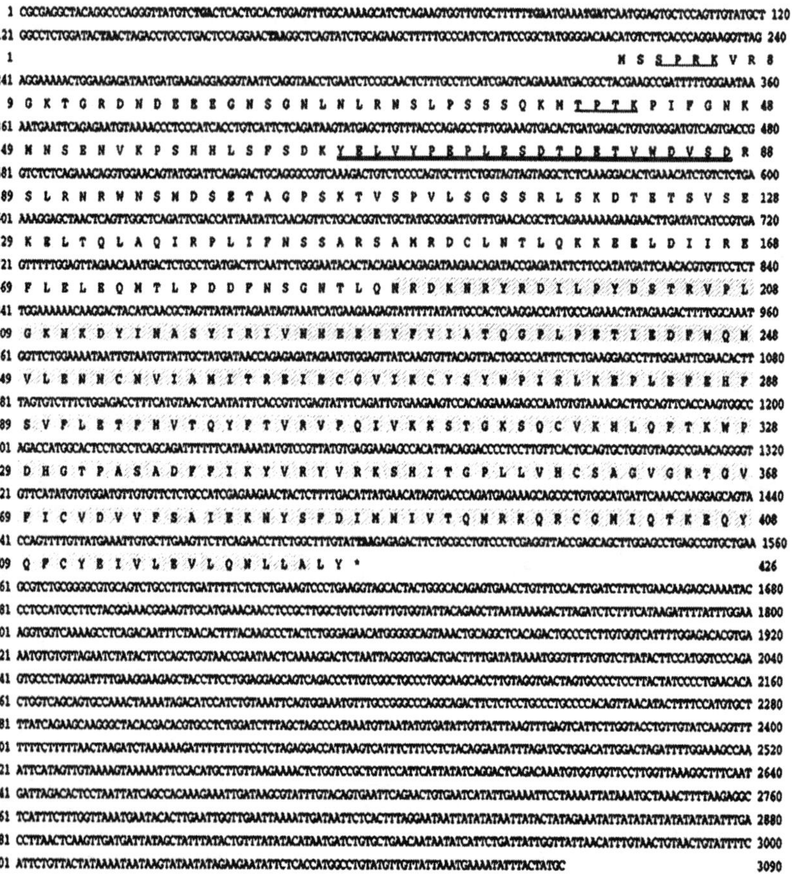

Fig.20.3. The composite nucleotide sequence and the deduced amino acid sequence of mouse tyrosine phosphatase (typ). In-frame stop codons preceding and terminating the open reading frame are shown in bold. The shaded portion defines the conserved catalytic domain that is homologous to the other PTPs. The potential phosphorylation sites for p34cdc2 kinase (S/TPXK) are underlined by thin lines and the putative PEST sequence, by a thick line. Reproduced with permission from M. Ohsugi et al. J Biol Chem 272: 33092-99; 1997 © American Society for Biochemistry and Molecular Biology.

single catalytic domain at the carboxyl-terminal half. No hydrophobic stretch for a possible transmembrane sequence or signal sequence was found, suggesting that Typ is a cytoplasmic protein-tyrosine phosphatase. The amino-terminal half of Typ did not share significant homologies with the other known proteins but contained a region rich in PEST residues (Ohsugi et al, 1997) **(Fig 20.3)**.

Park et al (2000) identified a novel mouse PTPase cDNA, which contains a single open reading frame of the predicted 926 amino acids. The predicted protein showed significant identity with human megakaryocyte PTPase (91% in nucleotide sequences and 94% in amino acid sequences). The expression of this PTPase is highly enriched in the testis in mouse and human and has been termed as a testis-enriched phosphatase (TEP). This PTPase transcribes two mRNA species of 3.7 and 3.2-kb in mouse testis, and specifically expressed in spermatocytes and spermatids within seminiferous tubules (Park et al, 2000).

Cytoplasmic PTPase–RL10: The cytoplasmic-type PTPase-RL10/PTPase D1/PTPase 2E contains an ezrin-like domain and associates with the c-Src protein tyrosine kinase. Testes of wild-type mice and W/W^v mutant mice lacking germ cells were analyzed for the expression of PTPase-RL10, its isoform PTPase-RL10b and c-Src. The PTPase-RL10 transcripts of 5.7kb and 2.9kb in size are detected in the testis, and that the 5.7kb transcripts are expressed in pachytene spermatocytes and somatic cells including Sertoli cells, in which c-Src transcripts were identified. The 2.9kb transcript encoded an isoform, PTPase-RL10b that has the catalytic domain but not the domains that associate with c-Src. Thus, PTPase-RL10b is expressed mainly in the testis Sertoli cells (Tokuchi et al., 1999).

20.8.2. Transmembrane PTPase with Tensin Homology (TPTE)

A 2.5-kb mRNA is strongly and exclusively expressed in the testis. The corresponding full-length cDNA encodes a predicted polypeptide of 551 amino acids with at least two potential transmembrane domains and a tyrosine phosphatase motif. The cDNA has sequence homology to chicken tensin, bovine auxilin and rat cyclin-G associated kinase (GAK). The entire polypeptide sequence also has significant homology to tumor suppressor PTEN/MMAC1 protein TPTE. This novel gene/protein has been termed transmembrane phosphatase with tensin homology. This gene has highly homologous copies on HC13, 15, 22, and Y, in addition to its HC21 copy or copies. The estimated minimum number of copies of TPTE gene in the haploid human genome is 7 in male and 6 in female. The TPTE is conserved between humans and other species. Its expression pattern, sequence homologies, and the presence of a potential tyrosine phosphatase domain suggest that it may be involved in signal transduction pathways of the endocrine or spermatogenetic function of the testis. The TPTE is, possibly the gene located closest to the human centromeric sequences on human chromosome, 21 and 13 (Chen et al, 1999, Guipponi et al 2000). The *TPTE* gene consists of 24 exons and spans approximately 87-kb. The position of this copy of *TPTE* was confirmed on the short arm of chromosome 21 (Guipponi et al, 2000).

20.8.3. Osteotesticular Protein Tyrosine Phosphatase (OST-PTP)

Osteotesticular tyrosine phosphatase (OST-PTP) is a class III receptor-type tyrosine phosphatase (RPTPase). The expression of the rat OST-PTP gene, *Esp*, is restricted to osteoblasts and Sertoli cells, and the transcript level in osteoblasts is highly upregulated by parathyroid hormone and cAMP. Mauro et al, (1994) have isolated the cDNA for a receptor-like PTPase, OST-PTP. The deduced 1711-residue protein possesses an extracellular domain with 10-fibronectin type-III repeats and a cytoplasmic region with two catalytic phosphatase domains in cytoplasmic region. In primary rat-osteoblasts, the 5.8-kb OST-PTP transcript is upregulated in differentiating cultures and down regulated in late stage mineralizing cultures. In addition, a presumed alternate transcript of 4.8-5.0 kb, which may lack PTPase domains, is present in proliferating osteoblasts, but not detectable at other stages. The OST-PTP may function in signaling pathways during bone remodeling, as well as serve a broader role in cell interactions associated with differentiation in bone and testis. The rat *Esp* gene sequence possesses two potential binding sites for Osf2/Cbfa1, an osteoblast-specific transcription factor in the promoter region. The *Esp* is composed of 35 exons, but spans merely 20-kb making it the most compact RPTP gene identified. Each FN-III repeat is encoded by a single exon flanked with phase 1 introns. Two phosphatase domains are encoded by 16 exons in a genomic organization similar to those in *RPRPα, RPTPγ*, and *Ptprc* genes. The *Esp* was mapped to rat chromosome 13q1 (Lathrop et al, 1999). The cytoplasmic domain (CD) CD1 of OST-PTP is enzymatically active and appears to be influenced by the catalytically inactive second domain CD2. The activity of CD1 is specific to phosphorylated tyrosine. Full-length OST-PTP protein expressed in COS cells has a molecular mass of 185-kDa, and shows strong tyrosine phosphatase activity (Chengalvala et al, 2001).

III. DUAL SPECIFICITY PROTEIN PHOSPHATASES

20.9. DSP FAMILY

Dual-specificity protein phosphatases (DSPs) comprise a new family of PTPases characterized by the ability to dephosphorylate phospho-seryl/threonyl residues in addition to phosphotyrosyl residues. A single mutation converts a novel phosphotyrosine-binding domain into a dual-specific phosphatase (Wishart et al, 1995). Increasing number of DSPs has been identified by cDNA. Most of them are reported to dephosphorylate both phospho-threonyl and tyrosyl residues in the T-X-Y motif of the mitogen-activated protein kinase family, and to cause inactivation of their kinase activity. The DSP family includes mitogen-activated protein kinase phosphatase (MKP)-1, human vaccinia HI homologous phosphatase-2 (hVH-2)/MKP-2/ threonine-tyrosine phosphatase-1 (TYP-1), hVH 3/B23 hVH-5/M3-6, MKP-3/rVH-6/PYST1, MKP-X/PYST2, MKP-4 and phosphatase of activated cell-1 (PAC-1) (Nakamura et al., 1999).

20.9.1. Testis-and Skeletal-Muscle-Specific DSP (TMDP)

A target disruption of the protein phosphatease PP1γ gene, which encodes an isoform of type1 serine/threonine protein phosphatase, induces meiotic defect in the testis identified a novel DSP, termed TMDP (testis-and skeletal-muscle-specific DSP). Nucleotide sequence analysis of TMDP cDNA indicated that the open reading frame of 597-bp encodes a protein of 198 amino acid residues with a predicted molecular mass of 22.5- kDa. The deduced amino acid sequence contains a motif for a conserved catalytic domain of DSPs and shows highest similarity to human Vaccinia HI-related phosphatase (45.5% identity) but low homology to the mitogen-activated protein kinase phosphatase and CDC25 subfamilies of DSPs. The N-terminal region of TMDP shows no similarity with other proteins in the database whereas the C- terminal half of TMDP is fairly conserved in other DSPs. Three amino acid residues conserved in all PTPase and thought essential for the enzyme activity are also conserved at position Asp-106, Cys-138, and Agr-144. The TMDP is most abundantly expressed in the testis and the expression is characterized as follows: (i) TMDP mRNA first appeared with the onset of meiosis; (ii) TMDP mRNA was abundantly expressed in spermatocytes and round spermatids. Nek1, a dual-specificity kinase, is most abundantly expressed in round haploid spermatids. The expression pattern of Nek1 is almost parallel to those with TMDP.

20.9.2. Low Molecular Weight Dual Specificity Phosphatases

Five putative dual specificity protein phosphatases (DSPs) designated LMW-DSP1, -DSP4, - DSP6, - DSP10, and –DSP11, were cloned by Aoki el al, (2001) and LMW-DSP2 by Aoyama et al (2001). The MKP3 is a cytosolic protein-tyrosine phosphatase that reverses the mitogen-activated protein kinase family by dephosphorylating critical Tyr and Thr residues. Sequence analysis revealed that LMW-DSPs lacked Cdc25 homology domain that is conserved in most known DSPs/MAP kinase phosphatases (MKPs). While LMW-DSP1 showed maximum similarity to plant DSPs, LMW-DSP4 exhibited the highest similarity to human YVH1 and rat GKAP, though its C-terminal region was shorter than that of the human and rat clones. The LMW-DSP6 was found identical to TMDP, whereas LMW-DSP11 seemed to be a mouse orthologue of human VHR. The LMW-DSP10 was found to have a DSP catalytic-like domain, but the critical cysteine residue for catalytic activity was missing. The phosphatase activity of the recombinant LMW-DSPs was inhibited by orthovanadate but not sodium fluoride. However, none of the DSPs could dephosphorylate MAP kinases such as ERK1, P38, and SAPK/JNK in transiently transfected COS7 cells under the conditions used. Analysis revealed that LMW-DSP1, -DSP6, -DSP10, and –DSP11 were specifically expressed in testis, while LMW-DSP4 expressed broadly. The testis-specific expression and apparent absence of dephosphorylation action on MAP kinases suggest that LMW-DSP1, -DSP6, -DSP10, and –DSP11 play specific role in testis. It was conceivable that a distinct class of low molecular mass DSPs is present and plays a role in dephosphorylating unknown molecules other than MAP kinases. Two LMW-DSPs, specific to testis were described by Hood et al., (2002). They are not altered in any of several disease states examined. These proteins, expressed in the nucleus and the cytoplasm, are able to dephosphorylate the phosphotyrosine analogue pNPP in vitro and are inhibited by sodium orthovanadate. These LMW-DSPs might belong to a new subclass of testis-specific proteins that act independently of the MAPK signal transduction cascade and do not depend on N-terminal docking regions for substrate binding (Hood et al., 2002).

The LMW-DSP2 open reading frame of 194 amino acids contained a single DSP catalytic domain but lacked the Cdc25 homology domain, which is conserved in most known DSPs. The LMW-DSP2 was specifically expressed in testis. Recmobinant LMW-DSP2 phosphatase activity was inhibited completely by sodium orthovanadate but not sodium fluoride, pyrophosphate, and okadaic acid. The substitution of critical amino acid residues, aspartic acid and cysteine resulted in a dramatic reduction of phosphatase activity. The LMW-DSP2 dephosphorylated and deactivated p38, to a higher extent, and stress-activated protein kinase (SAPK)/c-Jun N-terminal kinase (JNK), but not extracellular signal-regulated kinase 1/2 mitogen-activated protein kinases. A mutation in a conserved docking motif of p38 and SAPK/JNK as well as in a cluster of aspartic acids of LMV-DSP2 did not affect the deactivation of the mitogen-activated protein kinases by LMV-DSP2 (Aoyama et al., 2001) **(Fig.20.4)**. Among the DSPs lacking the Cdc25 homology domain, LMV-DSP2 is the first one that dephosphorylates and deactivates SAPK/JNK. The MKP-3 is a cytosolic protein-tyrosine phosphatase that exemplifies a new class of DSP that reverses the mitogen-activated protein kinase family by dephosphorylating critical Tyr and Thr residues. The MKP-3 is 36% identical to MKP-1 but not expressed in testis (Muda et al, 1996).

20.10. PTPase IN GERM CELLS-SERTOLI CELLS INTERACTIONS

Inter-Sertoli cell tight junctions (TJ) in the rat testis permit the passage of preleptotene and leptotene spermatocytes to cross the blood-testis barrier from the basal compartment to the

(A)

```
   1 CGAGCGCGGACGCGACGCGGCGCGGCCATGGGGAGTGGGATGAGCCAGATCCTGCCGGGC
                                 M  G  S  G  M  S  Q  I  L  P  G    11
  61 CTGTACATTGGCAACTTCAAAGACGCAAGAGATGCAGAACAGTTGAGCAGGAACAAGGTG
      L  Y  I  G  N  F  K  D  A  R  D  A  E  Q  L  S  R  N  K  V    31
 121 ACACACATTCTTTCTGTGCACGATACTGCCAGGCCCATGTTGGAGGGAGTTAAATACCTG
      T  H  I  L  S  V  H  D  T  A  R  P  M  L  E  G  V  K  Y  L    51
 181 TGTATTCCAGCGGCAGACACACCATCTCAAAACCTGACAAGACATTTCAAAGAAAGCATT
      C  I  P  A  A  D  T  P  S  Q  N  L  T  R  H  F  K  E  S  I    71
 241 AAATTCATTCATGAGTGCCGACTCCAGGGTGAGAGCTGTCTTGTACATTGCCTGGCTGGG
      K  F  I  H  E  C  R  L  Q  G  E  S  C  L  V  H  C  L  A  G    91
 301 GTCTCCAGGAGTGTGACATTGGTGATCGCATACATCATGACTGTCACCGACTTTGGCTGG
      V  S  R  S  V  T  L  V  I  A  Y  I  M  T  V  T  D  F  G  W   111
 361 GAAGATGCCTTGCACACTGTTCGTGCGGGGAGGTCCTGTGCCAACCCCAACCTGGGCTTT
      E  D  A  L  H  T  V  R  A  G  R  S  C  A  N  P  N  L  G  F   131
 421 CAAAGGCAGCTGCAGGAGTTTGAGAAACATGAAGTGCACCAGTATCGGCAATGGCTGAGA
      Q  R  Q  L  Q  E  F  E  K  H  E  V  H  Q  Y  R  Q  W  L  R   151
 481 GAAGAGTATGGAGAGAACCCTTTGCGGGATGCAGAAGAAGCCAAAAATATTCTGGCTGCC
      E  E  Y  G  E  N  P  L  R  D  A  E  E  A  K  N  I  L  A  A   171
 541 CCGGGAATTCTGAAGTACTGGGCCTTTCTCAGAAGACTGTAATGTACCTGAAGTTTCTGA
      P  G  I  L  K  Y  W  A  F  L  R  R  L  *                     184
 601 AATATTGCAAAGTTCAGGCTGGTGCTGCCAAAAAGAAAAGTGATGTAAAGTTTATTTTTA
 661 AGAATCCAATAGTGATTTGTATACTTGTTTTTTTTTCATTTTAAACCAAATGCATGTATA
 721 ATCATGTTGGAATATGTTAAGATCTATGGATATTCTGTAGCAAGAGAAAATATCTTTGCC
 781 TTAACTCCACTGCTGTGGTTGTTCCTTGGACCTGACCGATGCTCATACAATAATCTCAAG
 841 AGCCCTGTCTGTTTCGTAATAGTAACTACTTCTCATGAACACTACCCAAGGAGGAAGCCT
 901 GCACCTGGGAAGTGTGCAGTGTGAGCTCTGCCCTCCTGTTAAGTTCTCCAGCTCTAGACA
 961 TGTCTCTTGGTGTGTGTTTTATCTACTGGTGTTATTCTACATGGTAGAATTACCAAAAGC
1021 TATTCAGATTTCTTAATAAAGGGCAAATCAACCAAAAAAAAAAAAAAA
```

(B)

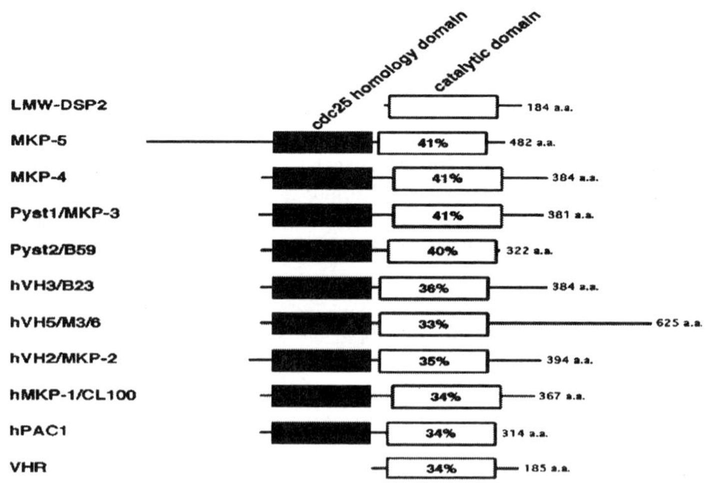

Fig.20.4. cDNA cloning of LMW-DSP2. (A) the nucleotide and deduced amino acid sequences of the cDNA encoding LMW-DSP2. Nucleotide and amino acid residues are numbered on the right and left, respectively. The amino acid consensus sequence DX26(V/L)X(V/I)HCXAG(I/V)SRSXT(I/V)XXAY(L/I)M (where X is any amino acid) found in all DSPs is underlined. Starting at ATG, catalytically essential amino acid resides aspartic acid 57 and cysteine 88 are indicated by bold letters. The polyadenylation signal sequence is boxed. An asterisk indicates the stop codon. (B) Schematic presentation of LMW-DSP2 amino acid (a.a.) identity to other DSPs/MKPs. Reproduced with permission from K Aoyama et al. J Biol Chem 276: 27575- 83; 2001 © American Society for Biochemistry and Molecular Biology.

adluminal compartment of the seminiferous epithelium so that these cells can continue their further development into spermatids. However, the mechanism by which this event is regulated remains a mystery. During understanding the biology of this event and its regulation, Li et al (2001b) studied transepithelial electrical resistance (TER) across the Sertoli cell epithelia tight junctions in vitro and assessed the effects of different inhibitors of phosphatases and kinases on the inter-Sertoli TJ barrier. It was shown that inhibitors of PTPs and inhibitors of protein Ser/Thr PPase (PP1) could perturb the assembly and maintenance of the inter-Sertoli tight junction permeability barrier. The inhibitory effect of PTP inhibitors (PTPi) was abolished by pre-treating Sertoli cells with protein tyrosine kinase inhibitor (PTKi), which illustrated the specificity of the PTPi treatment. A c-AMP dependent (PKA) activator and inhibitor of calcium – diacylglycerol–dependent PKC also perturbed the inter-Sertoli TJ permeability barrier, which suggested that opening and closing of the inter-Sertoli tight junctions during spermatogenesis is likely a regulated process, at least in part, by the PKA/PKC pathways. Thus the interplay of protein kinases and phosphatases, which regulate the intracellular phosphoprotein content of Sertoli cells possibly via PKA and PKC signal transduction pathways, plays a crucial role in modulating the assembly and maintenance of inter-Sertoli tight junctions in the testis (Li et al, 2001b).

Myotubularin protein, which is a member of the PTP family, is associated with X-linked myotubular myopathy. The MTM may be a useful marker monitoring the events of cell-cell interactions in the testis.The full-length cDNA encoding the entire ORF of rat myotubularin (rMTM) was isolated. Among the three clones of approximately 2.9-kb cDNAs that were sequenced, one clone was different from the other, and contained seven extra amino acids. Its sequence was found between Gln-22 and Phe-422 within the SET (Suvar 3-9, Enhancer –of – zeste, Trithorax) interacting domain (SID) of rMTM. The rMTM ORF had 1713-bp encoding for a 571 amino acid polypeptide of M_r of 65.8-kDa. The amino acid sequence revealed a 53.1 % identity with human myotubularin protein (hMTM1). Immunoblot analysis using a Ab against a 22 amino acid synthetic peptide NH_2-TKVNERYELCDTYPALLAVPAN based on sequence of rMTM, detected a 66-kDa protein in both Sertoli and germ cell cytosol. The rMTM mRNA was found in various tissues but was predominantly expressed in the testis, ovary, and skeletal muscle. Sertoli cell rMTM expression was stimulated by germ cells and enhanced when inter – Sertoli junctions were being assembled in vitro (Li et al., 2000). Studies on Sertoli-germ cell co-cultures and -Sertoli cells have shown that the high levels of testicular rMTM expression in the testis might be maintained by germ cells. These results illustrated that germ cells play a very active role in regulation of testicular function by altering the phosphoprotein content (Li et al, 2001a).

20.11. REFERENCES

Ahmad K, Bracho GE, Wolf DP, Tash JS. Regulation of human sperm motility and hyperactivation component by Ca^{2+}-calmodulin and phosphoprotein phosphatases. Arch Androl 1995; 35: 187-208.

Aoki N, Aoyama K, Nagata M, Matsuda T. A growing family of dual specificity phosphatases with low molecular masses. J Biochem 2001; 130: 133-40.

Aoyama K, Nagata M, Ohsima K, Matsuda T, Aoki N. Molecular cloning and characterization of a novel dual specificity phosphatase, LMW-DSP2, that lacks the cdc25 homology domain. J Biol Chem 2001; 276: 27575-83.

Ashizawa L. Hashimoto K, Tsuzuki Y. Regulation of fowl sperm flagellar motility by protein phosphatase type 1 and its relationship with dephosphorylation of axonemal and/or accessory cytoskeletal proteins. Biochem Biophys Res Commun 1997; 235: 108-112.

Barua M, Ghosh AK, Majumder GC. Partial purification and characterization of a phosphorprotein phosphatase from sperm plasma membrane. Reprod Fertil Dev 1999; 11: 379-86.

Chang CD, Mukai H, Kuno T, Tanaka C. cDNA cloning of an alternatively spliced isoform of the regulatory subunit of Ca^{2+}/calmodulin-dependent protein phosphatase (calcineurin Bα2). Biochim Biophys Acta 1994; 1217: 174-80.

Chen H, Rossier C, Morris MA, Scott HS, Gos A, Bairoch A, Antonarakis SE. A testis-specific gene, TPTE, encodes a putative transmembrane tyrosine phosphtase and maps to the pericentromeric region of human chromosomes 21 and 13, and to chromosomes 15, 22, and Y. Hum Genet 1999; 105: 399-409.

Chen MS, Hurov J, White LS, Woodford-Thomas T, Piwnica-Worms H. Absence of apparent phenotype in mice lacking Cdc25C protein phosphatase. Mol Cell Biol 2001; 21: 3853-61.

Chengalvala MV, Bapat AR, Hurlburt WW, Kostek B, Gonder DS, Mastroeni RA, Frail DE. Biochemical characterization of osteo-testicular protein tyrosine phosphatase and its functional significance in rat primary osteoblasts. Biochemistery 2001; 40: 814-21.

Choi J, Appella E, Donehower LA. The structure and expression of the murine wildtype p53-induced phosphatase 1 (WP1) gene. Genomics 2000; 64: 298-306.

Chun YS, Park JW, Kim GT, Shima H, Nagao M, Kim MS, Chung MH. A sds22 homolog that is associated with the testis-specific serine/threonine protein phosphatase 1γ2 in rat testis. Biochem Biophys Res Common 2000; 14: 972-6.

Chun YS, Park JW, Kim MS, Shima H, Nagao M, Lee SH, Park SW, Chung MH. Role of the 78-kDa glucose-regulated as an activity modulator of protein phosphatase γ2. Biochem Biophys Res Common 1999; 259: 300-4.

Chun YS, Shima H, Nagasaki K, Sugimura T, Nagao M. PP1γ2, a testis-specific protein-serine/threonine-phosphatase type 1 catalytic subunit, is associated with a protein having high sequence homology with the 78-kDa glucose-regulated protein, a member of the 70-kDa heat shock protein family. Proc Natl Sci USA 1994; 91: 3319-23.

Cohen P. The structure and regulation of protein phosphatases. Ann Rev Biochem 1989;58:453-509.

Emiliozzi C, Fenichel P. Protein tyrosine phosphorylation is associated with capacitation of human sperm in vitro but is not sufficient for its completion. Biol Reprod 1997; 56: 674-79.

Galantino-Homer HL, Visconti PE, Kopf GS. Regulation of protein tyrosine phosphorylation during bovine sperm capacitation by a cyclic adenosine 3'5'-monophosphate – dependent pathways. Biol Reprod 1997; 56: 707-19.

Guerini D. Calcineurin not just a simple protein phosphatase. Biochem Biophys Res Commun 1997; 235: 271-75.

Guipponi M, Yaspo ML, Riesselman L, Chen H, De Sario A, Roizes G, Antonarakis SE. Genomic structure of a copy of the human TPTE gene which encompasses 87 kb on the short arm of chromosome 21. Hum Genet 2000; 107: 127-31.

Han MH, Han DK, Aebersold RH, Glomset JA. Effects of protein kinase CK2, extracellular signal-regulated kinase 2, and protein phosphatase 2A on a phosphatidic acid-preferring phospholipase A1. J Biol Chem 2001; 276: 27698-708.

Hastie CJ, Carnegie GK, Morrica N, Cohen PT. A novel 50 kDa protein forms complexes with protein phosphatase 4 and is located at centrosomal microtubules organizing centers. Biochem J 2000, 347 : 845-55

Hastie CJ, Cohen PT. Purification of protein phosphatase 4 catalytic subunit: inhibition by the anti-tumor drug fostriecin and other tumor supressors and promoter. FEBS Lett 1998; 24: 357-61.

Hatano Y, Shima H, Haneji T, Miura AB, Sugimura T, Nagao M. Expression of PP2A B regulatory subunit β isotype in rat testis. FEBS Lett 1993; 324: 71-5.

Hood KL, Tobin JF, Yoon C. Identification and characterization of two novel low-molecular-weight dual specificity phosphatases. Biochem Biophys Res Commun 2002; 298: 545-51.

Hortas ML, Castilla JA, Gil MT, Samaniego F, Morell M, Redondo M. Alternations in sperm protein phosphorylation in male infertility. Androlgia 2001; 33: 282-6.

Hou EW, Kawai Y, Miyasaka H, Li SS. Molecular cloning and epxresson of cDNAs encoding two isoforms of protein phosphatase 2Cβ from mouse testis. Biochem Mol Biol Int 1994; 32: 273-80.

Hu MC, Shui JW, Mihindukulasuriya KA, Tan TH. Genomic structure of the mose PP4 gene: a developmentally regulated protein photophatase. Gene 2001; 31: 89-99.

Huang Z, Khatra B, Bollen M, Carr DW, Vijayaraghavan S. Sperm PP1γ2 is regulated by a homologue of the yeast protein phosphatase binding protein sds22. Biol Reprod 2002; 67: 1936-42.

Hubbard MJ, Klee CB. Functional domain structure of calcineurin A: mapping by limited proteolysis. Biochemistry 1989; 28: 1868-74.

Kaneko Y, Takano S, Okumura K, Takenawa J, Higashituji H, Fukumoto M, Nakayama H, Fujita J. Identification of protein tyrosine phosphatase expressed in murine male germ cells. Biochem Biophys Res Commun 1993; 15: 625-31.

Kato S, Kobayashi T, Kusuda K, Nishina Y, Nishimune Y, Yomogida K, Yamamoto M, et al. Differentiation-dependent enhanced expression of protein phosphatase 2Cβ in germ cells of mouse seminiferous

tubules. FEBS Lett 1996; 396 : 293-7.
Kato S, Terasawa T, Kobayashi T, Ohnishi M, Sasahara Y, Kusuda K, Yanagawa Y, Hiraga A, Matsui Y, Tamura S. Molecular cloning and expression of mouse Mg^{2+} dependent protein phosphatase β-4 (type 2Cβ4). Arch Biochem Biophys 1995; 318: 387-93.
Kitagawa Y, Sasaki K, Shima H, Shiuya M, Sugimura T, Nagao M. Protein phosphatases possibly involved in rat spermatogenesis. Biochem Biophys Res Commun 1990; 171: 230-5.
Kloeker S, Wadzinski BE. Purification and identification of a novel subunit of protein serine/threonine phosphatase 4. J Biol Chem 1999; 274: 5339-47.
Leclerc P, de Lamirande E, Gagnon C. Cyclic adenosine 3'5' monophosphate dependent regulation of protein tyrosine phosphorylation in relation to human sperm capacitation and motility. Biol Reprod 1996; 55: 684-92.
Leclerc P, de Lamirande E, Gagnon C. Regulation of protein tyrosine phosphorylation and human sperm capacitation by reactive oxygen derivatives. Free Radic Biol Med. 1997; 22: 643-56.
Li JC, Lee TW, Mruk TD, Cheng CY. Regulation of Sertoli cell myotubularin (rMTM) expression by germ cells in vitro. J Androl 2001; 22: 266-77.
Li JC, Mruk D, Cheng CY. The inter-Sertoli tight junction permeability barrier is regulated by the interplay of protein phosphatases and kinases: an in vitro study. J Androl 2001; 22: 847-56.
Li JC, Samy ET, Grima J, Chung SS, Mruk D, Lee MW, Silvestrini B, Cheng CY. Rat testicular myotubularin, a protein tyrosine phosphatase expressed by Sertoli and germ cells, is a potential marker for studying cell-cell interactions in the rat testis. J Cell Physiol 2000; 185: 366-85.
Lthrop W, Jordan J, Eustice D, Chen D. Rat osteotesticular phosphatase gene (Esp) : Genomic structure and chorosome location. Mamm Genome 1999; 10: 366-70.
Maeda K, Inui S, Tanaka H, Sakaguchi N. A new member of the α4-related molecule (α-4) that binds to the protein phosphatase 2A is expressed selectively in the brain and testis. Eur J Biochem 1999 264: 702-6.
Mauro LJ, Olmsted EA, Skrobacz BM, Mourey RJ, Davis AR, Dixon JE. Identification of a hormonally regulated protein tyrosine phosphatase associated with bone and testicular differentiation. J Biol Chem 1994; 269: 30659-67.
Mishra S, Somanath PR, Huang Z, Vijayaraghavan S. Binding and inactivation of the germ cell specific protein phosphatase PP1γ2 by sds22 during epididymal sperm maturation. Biol Reprod 2003; 69: 1572-9.
Mitchell D, Sale WS. Characterization of a Chlamydomonas insertional mutant that disrupts flagellar central pair microtubule-associates structures. J Cell Biol 1999; 144: 293-304.
Miyamoto K, Matsui H, Tomizawa K, Kuwata Y, Itano T, Tokuda M, Hatase O. In situ localization of rat testis-specific calcineurin B subunit isoform beta 1 in the developing rat testis. Biochem Biophys Res Common 1994; 203: 1275-83.
Miyasaka H, Li SS. The cDNA cloning, nucleotide sequence and expression of an intracellular protein tyrosine phosphatase from mouse testis. Biochem Biophys Res Commun 1992; 20: 818-25.
Mizoguchi S, Kim KH. Expression of cdc25 phosphatases in the grem cells of the rat testis. Biol Reprod 1997; 56: 1474-81.
Moriya M, Fujinaga K, Yazawa M, Katagiri C. Immunohistochmical localization of the calcium/calmodulin-dependent protein phosphatase, calcineurin, in the muse testis: its unique accumulation in spermatid nuclei. Cell Tissue Res 1995; 281: 273-81.
Muda M, Boschert U, Dickinson R, Martinou JC, Martinou I, Camps M, Schlegel W, Arkinstall S. MKP-3 a novel cytosolic protein-tyrosine phosphatase that exemplifies a new class of mitogen-activated protein kinase phosphatase. J Biol Chem 1996; 271: 4319-26.
Mukai H, Chang CD, Tanaka H, Ito A, Kuno T, Tanaka C. cDNA cloning of a novel testis-specific calcineurin B-like protein. Biochem Biophys Res Commun 1991; 179: 1325-30.
Muramatsu T, Giri PR, Higuchi S, Kincaid RL. Molecular cloning of a calmodulin-dependent phosphatase from murine testis: identification of a developmentally expressed nonneural isoenzyme. Proc Natl Acad Sci USA 1992; 95: 529-33.
Muramatsu T, Kincaid RL. Molecular cloning and chromosomal mapping of the human gene for the testis-specific catalytic subunit of calmodulin-dependent protein phosphatase. Biochem Biophys Res Common 1992; 188: 265-71.
Nakamura K, Shima H, Watanabe M, Haneji T, and Kikuchi K. Molecular cloning and characterization of a novel dual-specificity protein phosphatase possibly involved in spermatogenesis. Biochem J 1999; 344: 819-25.
Naz K, Ahmad K, Kumar R. Role of membrane phosphotyrosine proteins in human spermatozoal function. J Cell Sci 1991; 99: 157-65.
Nishio H, Matsui H, Etoh S, Moia LJ, Tokuda MJ, Itano T, Hatase O. Identification of testis calcineurin beta

subunit isoform by a monoclonal antibody and detection of a specific six amino acid sequence. Biochem Biophys Res Common 1992; 182 : 24-8.
Nishio H, Matsui H, Moia LJ, Taketa S, Miyamoto K, Tokuda M, Itano T, Nakahara S, Hatase O. The evidence for post-meiotic expression of a testis-specific isoform of a regulatory subunit of calcineurin using a monoclonal antibody. Biochem Biophys Res Commun 1992; 187: 828-31.
Ohnishi M, Chida N, Kobayashi T, Wang H, Ikeda S, Hanaba M, Yanagawa Y, Katsura K, Hiraga A, Tamura S. Alternative promoters direct tissue-specific expression of the mouse protein phosphatase 2Cβ gene. Eur J Biochem 1999; 263: 736-45.
Ohsugi M, Kuramochi S, Matsuda S, Yamamoto T. Molecular cloning and characterization of a novel cytoplasmic protein-tyrosine phosphatase that is specifically expressed in spermatocytes. J Biol Chem 1997; 272: 33092-9.
Osawa Y, Nakagama H, Shima H, Sugimura T, Nagao M. Identification and characterization of three isotypes of protein phosphatase inhibitor-2 and their expression profiles during testis maturation in rats. Eur J Biochem 1996; 15: 793-8.
Park KW, Lee EJ, Lee S, Lee JE, Choi E, Kim BJ, Hwang R, Park KA, Baik J. Molecular cloning and characterization of a protein tyrosine phosphatase enriched in testis, a putative homologue of human PTPMEG. Gene 2000; 17: 45-55.
Puntoni F, Villa-Moruzzi E. Phosphorylation of protein phosphatase–1 isoforms by cdc2-cyclin B in vitro. Mol Cell Biochem 1997; 171: 115-20.
Shibasaki F, Hallin U, Uchino H. Caleineurin as a multifunctional regulator. J Biochem 2002; 131: 1-15.
Shima H, Haneji T, Hatano Y, Kasugai I,m Sugimura T, Nagao M. Protein phosphatase 1 γ2 is associated with nuclei of meiotic cells in rat testis. Biochem Biophys Res Common 1993; 194: 930-7.
Smith GD, Wolf DP, Trautman KC, Cruz e silva EFD. Greengard P, Vijayaraghavan S. Primate sperm contain protein phosphatase 1, a biochemical mediator of motility. Biol Reprod 1996; 54: 719-27.
Smith GD, Wolf DP, Trautman KC, Vijayaraghavan S. Motility potential of macaque epididymal sperm: the role of protein phosphatase and glycogen synthase kinase – 3 activities. J Androl 1999; 20"47-53.
Sommer D, Fakata KL, Swanson SA, Stemmer PM. Modulation of the phosphatase activity of calcineurin by oxidants and antioxidants in vitro. Eur J Biochem 2000; 267: 2312-22.
Su Q, Zhao M, Weber E, Eugster HP, Ryffel B. Distribution and activity of calcineurin in rat tissues. Evidence for post-transcriptional regulation of testis-specific calcineurin B. Eur J Biochem 1995; 230: 469-74.
Sugimoto M, Matsui H, Etoh S, Shimizu T, Nishio H, Moia LJ, Tokuda M, Itano T, Takenaka I, Hatase O. regulatory subunit (calcineurin B). Biochem Biophys Res Commun 1992; 187: 537-43.
Uma Devi K, Basheer Ahmad M, Shivaji S. Acrosome membrane associated protein kinase activity in hamster spermatozoa. FEBS Lett 1996; 396: 123-26.
Uma Devi K, Jha K, Shivaji S. Plasma membrane – associated protein tyrosine phosphatase activity in hamster spermatozoa. Mol Reprod Dev 1999; 53: 42-5.
Uma Devi, Basheer Ahmad M, Shivaji S. A maturation related differential phosphorylation of the plasma membrane proteins of the epididymal spermatozoa of hamster by endogenous protein kinases. Mol Reprod Dev 1997; 47: 341-50.
Uryu M, Nakatomi A, Watanabe M, Hatsuse R, Yazawa M. Molecular cloning of cDNA encoding two subunits of calcineurin from scallop testis: demonstration of stage-specific expression during maturation of the testis. J Biochem 2000; 127: 739-46.
Varmuza S, Jurisicova A, Okano K, Hudson J, Boekelheide K, Shipp EB. Spermiogenesis is impaired in mice bearing a targeted mutation in the protein phosphatase 1Cγ gene. Dev Biol 1999; 1: 98-110.
Wishart MD, Denu JM, Williams JA, Dixon JE. A single mutation converts a novel phosphotyrosine binding domain into a dual-specificity phosphatase. J Biol Chem 1995; 270: 26782-5.
Wu S, Wolgemuth DJ. The distinct and developmentally regulated patterns of expression of members of the mouse Cdc25 gene family suggest differential functions during gametogenesis. Dev Biol 1995; 170;195-206.
Yang P, Fox L, Colbran RJ, Sale WS. Protein phosphatases PP1 and PP2A are located in distinct positions in the *Chlamydomonas* flagellar axoneme. J Cell Sci 2000; 113: 91-102.

Chapter 21

ION CHANNELS AND AQUAPORINS

21.1. ION CHANNELS IN SPERM PHYSIOLOGY

Ion channels are key elements in the interaction between sperm and the egg. Components from the outer layer of the egg induce ion permeability changes in sperm that regulate sperm motility, chemotaxis, and the acrosome reaction (Darszon et al., 1999). It has been shown that sperm motility is associated with a complex ionic event. If immotile spermatozoa were re-suspended in a sodium-containing medium, they initiated or regained forward motility. Later, it was also found that an efflux of K^+ was associated with sodium-induced motility. Calcium ions (Ca^{2+}) have also been shown to be important in sperm motility initiation. How Ca^{2+} reaches the axoneme in the core of the flagellum to switch on hyperactivation is unknown (Ho and Suarez, 2001). These observations have led to the speculation that ion channels are the mediators of the ionic events involved in sperm motility initiation. Ion fluxes are also fundamental in sperm in response to egg. Acrosome reaction is shown to be mediated by elevation of intracellular Ca^+. In addition to Ca^{2+} influx, a sequence of ionic changes such as Na^+ influx and efflux of H^+ and K^+ has also been observed in association with the acrosome reaction. The variety in the types of ion channels observed in rat and human spermatozoa plasma membrane suggests that ion channels may play different roles in sperm physiology and gamete interaction.

21.2. POTASSIUM CHANNELS

High concentrations of potassium ions (K^+) improve the rate of acrosome reaction in human sperm in vitro. Incubation of capacitated sperm with different concentrations of potassium chloride results in dose-dependent increase in $[Ca^{2+}]_i$ similar to that observed with progesterone (P). The increase in $[Ca^{2+}]_i$ by K^+ and progestrone is blocked by EGTA. The combined treatment of K^+ caused an additive effect on the increase in $[Ca^{2+}]_i$. It would appear that human sperm plasma membrane possesses different types of K^+ channels in addition to Ca^{2+} channels.

21.2.1. Ca^{2+} Activated K^+ Channel

Both Ca^{2+} and K^+ ions have been implicated in various sperm functions. Wu et al (1998b) suggested that sperm may possess a Ca^{2+} activated K^+ channel, which has been implicated in the process of sperm activation and gamete interaction. Another study investigated *Xenopus oocytes* to express sperm ion channel by injection of RNAs extracted from the rat testis. The RNA-injected oocytes expressed an outwardly rectifying current, which was dependent on K^+

```
  1 MDNKETNGEL EQSDEADPSG QNLDDGETDS KQEENLINVS PPKTPPGPPP PLKNGGRGQK
 61 PPKIPICHQN GKLPKEVEWT EDRGEDRKDS LTLQSKLDHG AYTDEKQDLL TYLDRHGINS
121 PVKLTPDETG GSSALDILGI IEERDTGALG SDPSSTMQAM AKPVGFLQRQ LWTVLQPSDN
181 RLSMKLFGSK KGLQKEKYRL RKAGVLIIHP CSHFRFYWDL LMLCLIMANV ILLPVVITFF
241 HNKDMSTGWL IFNCFSDTFF ILDLICNFRT GIMNPKSAEQ VILNPRQIAY HYLRSWFIID
301 LVSSIPMDYI FLLAGGQNRH FLEVSRALKI LRFAKLLSLL RLLRLSRLMR FVSQWEQAFN
361 VANAVIRICN LVCMMLLIGH WNGCLQYLVP MLQEYPDQSW VAINGLEHAH WWEQYTWALF
421 KALSHMLCIG YGKFPPQSIT DVWLTIVSMV SGATCFALFI GHATNLIQSM DSSSRQYREK
481 LKQVEEYMQY RKLPSHLRNK ILDYYEYRYR GKMFDERHIF REVSESIRQD VANYNCRDLV
541 ASVPFFVGAD SNFVTRVVTL LEFEVFQPAD YVIQEGTFGD RMFFIQQGIV DIIMSDGVIA
601 TSLSDGSYFG EICLLTRERR VASVKCETYC TLFSLSVQHF NQVLDEFPAM RKTMEEIAVR
661 RLTRIGKESS KLKSRLESPT IRDTAPLFPI PPDTPSFVTD IEKNRFFGDD TDDVHIRTRV
721 DVERGSHENV IAIMDGSLSD LRMENEIQAR KSSSGKRRKF QQQTTEL
```

Fig.21.1. Hyperpolarization-activated channel [Strongylocentrotus purpuratus](Accession NP_999729) Source: http//www. ncbi.nlm.nih.gov. (with permission from R.Gauss et al. Nature 393: 583-7; 1998 © http://www.nature.com/Nature).

concentration and inhabitable by K^+ channel blockers. The Ca^{2+} ionophore-stimulated current activation with similar current characteristics in the RNA-injected oocytes suggested the expression of a Ca^{2+} activated K^+ channel. Predominant Ca^{2+} activated K^+ channel reactivity is also associated with spermatogenic cells. Using the same *Xenopus oocytes* expression system, Wu et al, (1998a) investigated the specific purinoceptor subtype P_{2u} involved in mediating the effect of extracellular ATP and other nucleotides. The current responses to different nucleotides using the two – electrode voltage clamp technique showed an order of relative effectiveness: UTP>ATP>ADP> adenosine, consistent with pharmacological classification of P_{2u} receptor. The ATP also showed an inhibitory effect on the Ca^{2+} activated K^+ current, elicited by the Ca^{2+} ionophore ionomycin. A similar inhibitory effect was observed with UTP. RT-PCR confirmed the expression of P_{2U} receptor mRNA in isolated spermatogenic cells. Thus, the expression of P_{2U} receptor and its dual role, both stimulatory and inhibitory, in the modulation of the Ca^{2+}-activated K^+ channel in spermatogenic cells appeared possible (Wu et al., 1998a).

21.2.2. Cyclic Nucleotides and K^+ Channels

Several lines of evidence suggest that cyclic nucleotides and Ca^{2+} play important roles in chemotaxis of invertebrate sperm and fertilization, whereas their mechanisms of action in vertebrate sperm are not well known. Cyclic nucleotide-gated (CNG) channels are key elements of cGMP– and cAMP–signaling pathways in vertebrate photoreceptor cells and in olfactory sensory neurons, respectively. Labarca et al, (1996) described cAMP regulated K^+ selective channel from sea urchin sperm plasma membrane that may have role during sea urchin sperm chemotaxis and/or the acrosome reaction.

SPIH: Gauss et al, (1998) cloned, from sea urchin testis, the cDNA encoding a channel polypeptide, SPIH. Functional expression of SPIH gives rise to weakly K^+-selective hyper-polarization-activated channels and is under the dual control of voltage and cAMP. The SPIH channel is confined to the sperm flagellum and may be involved in flagellar beating. SPIH currents exhibit all the similarities of hyper-polarization-activated currents (Ih), which participate in the rhythmic firing of central neurons, control pace-making in the heart, and curtail saturation by bright light in retinal photoreceptors. Because of their sequence and functional properties, SPIH channel forms a class of its own within the superfamily of voltage-gated and cyclic-nucleotide-gated channels (**Fig.21.1**).

```
  1 MAKISTQYSH PTRTHPSVRT MDRDLDCIEN GLSRTHLPCE ETSSELQEGI AMETRGLAES
 61 RQSSFTSQGP TRLSRLIISL RAWSARHLHQ EDQRPDSFLE RFRGAELQEV SSRESHVQFN
121 VGSQEPPDRG RSAWPLARNN TNTCNNSEKD DKAKKEEKEK KEEKKENPKK EEKKKDSVVM
181 DPSSNMYYHW LTVIAVPVFY NWCLLVCRAC FDELQSEHLM LWLVLDYSAD ILYGMDMLVR
241 ARTGFLEQGL MVMDASRLWK HYTQTLHFKL DVLSLVPTDL AYFKLGMNYP ELRFNRLLKL
301 ARLFEFFDRT ETRTNYPNMF RIGNLVLYIL IIIHWNACIY FAISKFIGFG TDSWVYPNVS
361 NPEYGRLSRK YIYSLYWSTL TLTTIGETPP PVKDEEYLFV VIDFLVGVLI FATIVGNVGS
421 MISNMNASRA EFQAKIDSIK QYMQFRKVTK DLETRVIRWF DYLWANKKTV DEKEVLKSLP
481 DKLKAEIAIN VHLDTLRKVR IFQDCEAGLL VELVLKLRPA VFSPGDYICK KGDIGREMYI
541 IKEGKLAVVA EDGITQFVVL GDGSYFGEIS ILNIKGSKSG NRRTANIRSI GYSDLFCLSK
601 DDLMEALTEY PEAKKALEEK GRQILMKDNL IDEELAKAGA DPKDIEEKVE HLETSLDSLQ
661 TRFARLLAEY NATQMKVKQR LSQLESQVKM GLPPDGDAPQ TEASQP
```

Fig.21.2. Amino acid sequence of cyclic nucleotide gated channel l from bovine testis (Accession 2010407A) Source: http//www. ncbi.nlm.nih.gov. (with permission from I. Weyand et al. Nature 368: 859-63; 1994 © http://www.nature.com/Nature).

CNG Channels in Bovine Testis: Expression of a cyclic nucleaotide gated channel (CNG) has been studied from bovine testis (Weyand et al, 1994). The functional polypeptide is present in sperm but is also specifically expressed in cone photoreceptor cells. These channels might be involved in chemotaxis of sperm by controlling Ca^{2+} entry through a cyclic-nucleotide signaling pathway. The CNG channels form heterooligomeric complex composed of at least two distinct subunits (α and β). The α subunit of cone photoreceptors is also present in mammalian sperm. Several transcripts of β subunits in testis have been identified. The α and β subunits are expressed in a characteristic temporal and spatial pattern and precursor cells. In mature sperm, the α subunit is observed along the entire flagellum, whereas the short β subunit is restricted to the principal piece of the flagellum. These findings suggest that different forms of CNG channel coexist in the flagellum. The CNG channels serve as a Ca^{2+} entry pathway that responds more sensitively to cGMP than to cAMP **(Fig.21.2)**. Assuming that CNG channel subtypes differ in their Ca^{2+} permeability, dissimilar localization of α and β subunits may give rise to a pattern of Ca^{2+} micro-domains along the flagellum, thereby providing the structural basis for control of flagellar bending waves (Wiesner et al., 1998).

Sperm K^+ Channel Activation by Egg Speract: Transduction by sperm of the instructive signal provided by the egg peptide, speract from sea urchin eggs involves rapid, complex changes in internal ion and cyclic nucleotide content. Hypotonically swollen sperm provide insight into the underlying processes and in identifying K^+ channel activation as an initial ionic event in gamete recognition. A sustained hyper-polarization of swollen sperm is promoted by less than 2.5pM speract and is followed by transient re-polarization followed by depolarization at high concentration of speract that is dependent on external Ca^{2+}. Monophasic increase in pH is produced only by still higher speract, indicating that hyper-polarization may not directly promote alkalinization. Increased K^+– selective membrane permeability is found after speract at greater than 2.5pM, suggesting that hyper-polarization results from persistent activation of K^+ channels (Babcock et al., 1992). Speract–induced accumulation of cGMP or AMP is selectively enhanced by the phosphodiesterase inhibitors, IBMX or papaverine, respectively. The earliest known ionic event in speract signal transduction results in rise of both cGMP content and K^+ permeability, which are maximal within 15 sec of speract stimulation followed by a decline of both after intracellular rise of pH (pH_i) in response to hyper-polarization. The study further suggested that elevation of pHi initiates an inactivation of guanylylcyclase that leads to K channel closure. Regardless of the mechanism that mediates $[Ca^{2+}]_i$ action, sperm K channels are downstream targets of cGMP and are implicated in a feedback loop that both terminates guanylyl cyclase activity and leads to their own inactivation (Cook and Babcock, 1993). Other studies strengthened the view that speract transiently stimulates a membrane guanylyl cyclase and activates a K^+-selective channel that hyperpolarizes sperm. Speract hyperpolarizes and increases the cGMP content of flagellar vesicles. Galindo et al, (2000) showed that intra-testicular GTPγS and GTP but not GDPβS, enhance this hyper-polarization. The increase in vesicular cGMP content produced by speract may directly mediate opening of the channel that hyperpolarizes sperm membrane vesicles. Similar mechanisms presumably operate in intact sperm (Galindo et al., 2000). Sanchez et al (2001) gave first direct evidence that cation selective channels upwardly regulated by cAMP operate in sea urchin sperm. Due to their poor selectivity among monovalent cations, channel activation in sea water could contribute to sperm membrane re-polarization during speract response.

21.2.3. SLO3

The Slo channel comprises a multi-gene family, defined by a combination of sensitivity to voltage and variety of intracellular factors. Among multi-gene gated channels, Slo1 provides negative feed back for entry of Ca^{2+} into cells. Many ion channels respond to membrane depolarization or intracellular ligands. In the Slo1 channel, these features are combined in a channel that opens in response to both depolarization and increase in intracellular $[Ca^{2+}]_i$. Slo1 channels cloned from mouse and humans show strong conservation of sequence and functional properties. Schreiber et al, (1998) described the cloning and expression of Slo3, a novel potassium channel abundantly expressed in mammalian spermatocytes. Slo3 represents a unique type of potassium channel regulated by both intracellular pH and membrane voltage. The Slo3 is primarily expressed in testis of mouse and human. Because of its sensitivity to both pH and voltage, Slo3 could be involved in sperm capacitation and /or the acrosome reaction. The protein sequence of mSlo3 (the mouse Slo3 homologue) is similar to Slo1, the large conductance calcium-and voltage-gated potassium channel. A Slo3 homologue is also present in human and conserved with regard to sequence transcript size, and tissue distribution. Because of its high testis-specific expression, Slo3 channels may be useful target in the control or enhancement of fertility (Schreiber et al., 1998) **(Fig.21.3)**.

21.2.4. Rectifier K^+ Channel

It is thought that the membrane potential of sperm hyperpolarizes during capacitation, possibly due to the opening of K^+ channels. Munoz-Garay et al, (2001) documented the presence of a K^+–selective inwardly rectifying current. Macroscopic current activated at membrane potentials below the equilibrium potential for K^+ and its magnitude was found dependent on the external K^+ concentration accompanied by the change of resting membrane potential of mouse sperm from -52 ± 6 to -66 ± 9 mV during capacitation in vitro. A mechanism is proposed whereby opening of inwardly rectifying K^+ channels may produce hyper-polarization under physiological conditions and contribute to the cellular changes that give rise to the capacitated state in mature sperm. Immunofluorescent detection of sperm–bound Ca^{2+}– activated, delayed rectifier K^+ channels, indicated that these ion channels are uniformly distributed over the surface of both heads and tails of rat spermatozoa. The K^+ channel transcripts are present in the cytoplasm of primary spermatocytes and post–meiotic elongating spermatids. Among various multiple K^+ channel transcripts in rat tissues identified, some are observed only in testis. An attempt to obtain a full length rat testis K^+ channel cDNA sequence gave an assembled sequence of 3693-

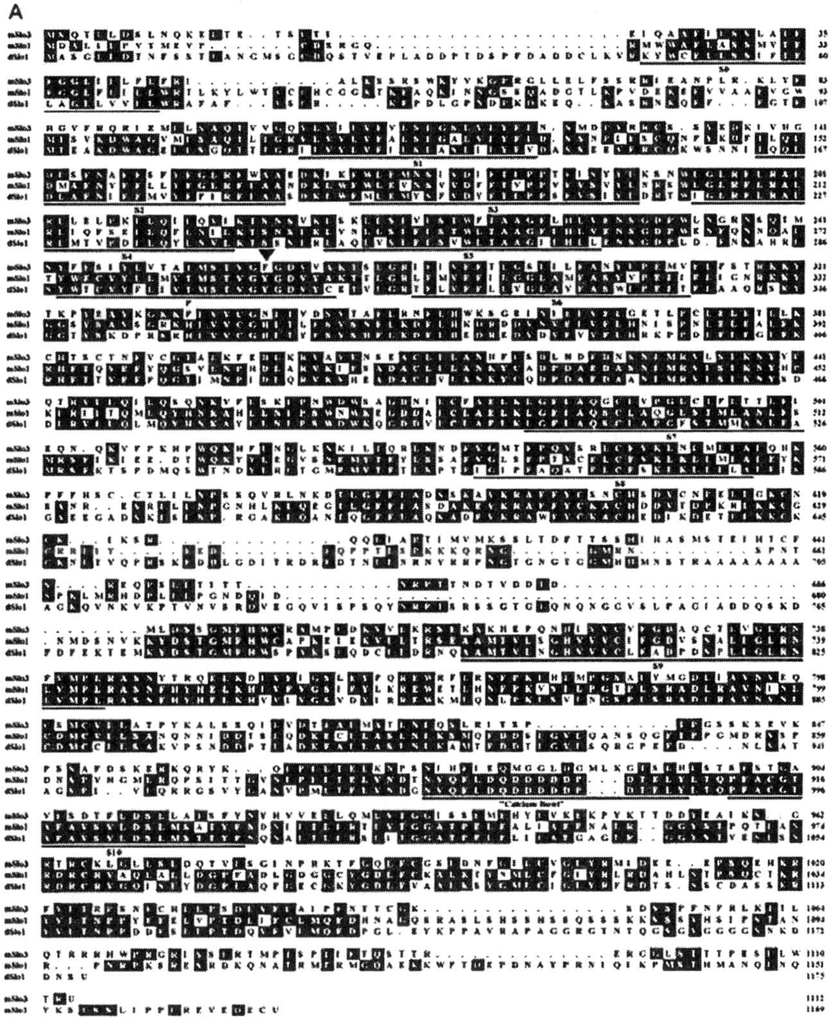

Fig.21.3. Primary sequence of a K^+ ion channel: mSlo3. Alignment of primary sequence of mSlo3 with BK Ca^{2+}-activated K^+ channels, mSlo1 (mouse) and dSlo1 (Drosophila). Hydrophobic segments are designated S0-S10. An arrowhead indicates a phenylalanine residue (F) in the pore region critical for ion selectivity. The region designated "Calcium Bowl" has been implicated in the regulation of mSlo1 by calcium. The overall corc and tail organization of Slo1 has been conserved in Slo3. mSlo1 (mbr5) and dSlo1 (splice variant A2C2E2G5I0) sequences are shown. Reproduced with permission from M. Schreiber et al. J Biol Chem 273: 3509-16; 1998 © American Society for Biochemistry and Molecular Biology.

bp with >90% homology to a delayed rectifier K^+ channel. DNA sequencing suggested that multiple related K^+ channels, which differed at their 5'-ends are amplified in rat testis (Jacob et al., 2000b).

Kir 5.1 is a member of the inward rectifier potassium channel super-family, which does not form functional channels when expressed by itself in *Xenopus laevis* oocytes with high levels of Kir 5.1 mRNA expression in testis. To determine the cell- specific expression of this channel in the testis, a polyclonal antibody against an external epitope of Kir 5.1 was tested for its specificity in *Xenopus* oocytes, expressing several cloned Kir subunits. Strong immunoreactivity for Kir 5.1 was found in rat testis and, particularly, in spermatogonia, primary and secondary spermatocytes, spermatids and in the head and body of spermatozoa. The intensity of Kir immunofluorescence increased with age at every stage in the development of sperm during spermatogenesis in rats (Salvatore et al., 1999).

21.2.5. Volume Regulatory K^+ Channel

Effect of the ion-channel blocker, quinine on human sperm volume, kinematics and mucus penetration, and the involvement to potassium channels showed that volume regulation in human spermatozoa and the linear trajectory of their motion may depend on largely calcium–independent, potassium channels, and possibly volume- sensitive organic anion channels (Yeung and Cooper, 2001). It has been proposed that swelling–activated potassium channels are involved in regulatory volume decrease in boar and bull sperm. However, boar spermatozoa may contain fewer swelling – activated chloride channels than do bull spermatozoa (Petrukina et al., 2001).

21.3. SODIUM CHANNELS

Ionic changes triggered by progesterone (P) in human spermatozoa show a rapid increase in the cytoplasmic Ca^{2+} concentration, which is obliterated by chelation of extracellular Ca^{2+}. Ca^{2+} fluxes were insensitive to verapamil and pertussis toxin (PTx), thus suggesting that they did not occur via voltage-gated channels and did not involve a pertussis toxin-sensitive G protein. Progesterone also caused a depolarization of the plasma membrane in Na^+ containing as well as in choline – or methyl-glucamine-containing saline; depolarization was larger in the absence of extracellular Ca^{2+}, suggesting that Na^+ fluxes occurred through the same channel. It was suggested that progesterone activates a membrane ion channel that is permeable to monovalent cations as well as to Ca^{2+}. In human sperm progesterone induces depolarization of plasma membrane, Ca^{2+} influx and K^+ efflux. Na^+ is partly responsible for progesterone induced depolarizing effect but was not required for Ca^{2+} influx. Patrat et al (2000) demonstrated that progesterone stimulates a Na^+ influx that could be involved in the acrosome reaction completion. Progesterone initiated human sperm acrosome reaction is dependent on the presence of extracellular Na^+ (Na^+ depletion decreases cytosolic pHi), suggesting involvement of a Na dependent pHi regulatory mechanism during P-initiated acrosome reaction. The decreased pHi resulting from Na^+ depletion is reversible and mediated by a Na^+/H^+ exchange (NHE) mechanism (Garcia and Meizel, 1999). Sperm activation seems to involve extracellular ATP suggesting that human sperm expresses an ATP- gated Na^+ channel that might have an important role in sperm activation before egg fertilization (Foresta et al., 1996). These studies implicated presence of Na^+ channel in the plasma membrane of sperm.

Several sodium channels from somatic cells belong to DEG/ENaC (Epithelia type Na Channel) superfamily, whose members have similar sequences and the same predicted structure. They function as amiloride sensitive sodium channel to transport Na in epithelia of kidney, colon and lung or play a role in taste perception and also important for mechanotransduction. Brain type of Na channel (BNaC) has been cloned from human testis, which was named hTNaC1 (for human testis sodium channel 1). The hTNaC1 has 532 amino acid residues with two

```
  1 MKPTSGPEEA RRPASDIRVF ASNCSMHGLG HVFGPGSLSL RRGMWAAAVV LSVATFLYQV
 61 AERVRYYREF HHQTALDERE SHRLIFPAVT LCNINPLRRS RLTPNDLHWA GSALLGLDPA
121 EHAAFLRALG RPPAPPGFMP SPTFDMAQLY ARAGHSLDDM LLDCRFRGQP CGPENFTTIF
181 TRMGKCYTFN SGADGAELLT TTRGGMGNGL DIMLDVQQEE YLPVWRDNEE TPFEVGIRVQ
241 IHSQEEPPII DQLGLGVSPG YQTFVSCQQQ QLSFLPPPWG DCSSASLNPN YEPEPSDPLG
301 SPSPSPSPPY TLMGCRLACE TRYVARKCGC RMVYMPGDVP VCSPQQYKNC AHPAIDAMLR
361 KDSCACPNPC ASTRYAKELS MVRIPSRAAA RFLARKLNRS EAYIAENVLA LDIFFEALNY
421 ETVEQKKAYE MSELLGDIGG QMGLFIGASL LTILEILDYL CEVFRDKVLG YFWNRQHSQR
481 HSSTNLLQEG LGSHRTQVPH LSLGPRPPTP PCAVTKTLSA SHRTCYLVTQ L
```

Fig21.4. Amino acid sequence of DEG/ENaC channel from human testis (Accession number NP_004760). Source: http//www. ncbi.nlm.nih.gov.

hydrophobic transmembrane domains. The longest sequence identified in human testis libraries contains 209-bp of 5'-untranslated region, 1596-bp of coding sequence, and 104-bp of 3'-untranslated region (excluding the poly A) tail. The 3'–untranslated region contains a consensus polyadenylation signal with poly (A) tail. It encodes a protein of 532 amino acids, a calculated molecular mass of 59,152 Da. The hydropathy analysis predicts two transmembrane regions with N-terminus and C-terminus localized in the cytosol similar to other DEG/ENaC family members. A potential N-linked glycosylation site (N-X-S/T) is present near the first third of extracellular loop. Four potential protein kinase phosphorylation sites are present at the cytosolic domains. There is no consensus protein A phosphorylation site in the cytosolic domains. Near the C-terminus, there is an SH3 (Src homology 3) – binding motif of proline-rich sequence (PRPPTPP) as found in αENaC. However, hTNa-C1 lacks a proline-rich PY motif, which is found in ENaCs. The hTNaC1 shares 82% sequence identity with rat sensory neuron Na channel (DRASIC), 50% identity with human brain Na channel (BNaC2), 29% identity with kidney Na channel (α- ENaC) and 26% identity with MEC-4 **(Fig. 21.4.)**. The Northern blot analysis with poly (A) RNA of human tissues revealed the selective expression of hTNaC1 in testis at 7 kb and absent in other tissues. The amino acid identity of DEG/ENaC superfamily between human and rat is usually very high; 99% identity of BNaC1/MDEG and 92% identity of BNaC2/ASIC. The role of hTNaC in testis as a H^+-gated Na channel is unclear. However, testis also expresses α ENaC (Waldmann et al., 1995). Spermatozoa have been shown to have a Na selective channel (Ishibashi and Marumo, 1998).

Amiloride-sensitive sodium channels have been identified in reproductive and in early developmental processes of several species. Darboux et al., (1998) identified a gene called *dGNaC1*that is specifically expressed in testis, ovary and early embryos of *D. melanogaster.* The corresponding protein belongs to the superfamily of cationic channels blocked by amiloride.

```
  1 MTISDSELDS SKGIDLTFRR RRKAGSVACE GFITTYHEYC RNTSIHGVQY LGEQERPFRE
 61 RIFWLFVFLI SIYGCSTLIQ SAYTKWTETP VIVSFAEKST PVWNIPFPAV TVCSETKRVL
121 KQKGKETTYA DLYSQFSEDM RASRVFRPEN VSALEMEEFR TLLHVCNTQI IEEDIPLIAG
181 DDLDYFDVLQ RMLPQFDRYF FYCRWLSRFG ECETFFRKTL TEEGICYTFN GLRATEIYRD
241 DTYQYQHSGE PLEMENISSQ HTAWTLETGY ALDSDVETFP ARVLSAGARS GIFLALQSFK
301 QEVDYACRGP VQGFKVLLHA PDDVPQVSKQ FVRIPMGKEV LIAVKPNMIT MSSGIAEYHP
361 VRRQCFLSHE RSLRFFKVYT ESNCQLECLA NFTLTKCGCV KFSMPRNVDM PVCGEDKIHC
421 YDRAERELLV REFKRVKALN AGRENSRSVE SACNCMPACT SLVYNTEISQ ANFDLEEMLV
481 AEGDTEFLKE YPGSQMSRLS IYFKQSQFIT SKRSELYGMT EFLANCGGIF GLFMGFSILS
541 LVEMIYHFTL RLFTNLKRLV KG
```

Fig.21.5. Protein sequence of gonad-specific amiloride-sensitive sodium channel 1 [*D melanogaster*] (Accession CAA76129) Source: http//www. ncbi.nlm.nih.gov (Darboux et al, 1998).

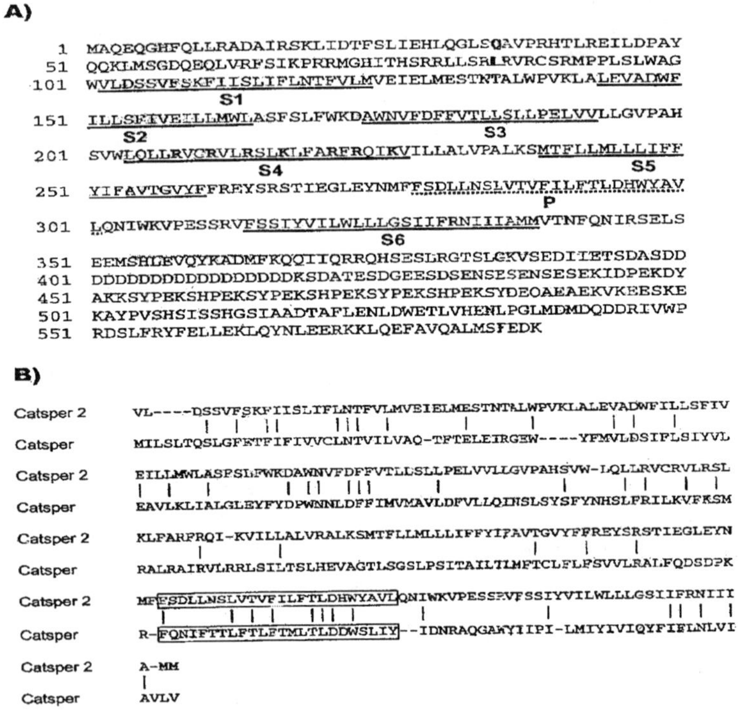

Fig.21.6. (A) Amino acid sequence of Catsper2 from sperm. The transmembrane segments (S1-S6) and ion selectivity P region are underlined. (B) Comparison of amino acid alignment of Catsper with Catsper 2. Reprinted with permission from T.A. Quill et al. Proc Natl Acad Sci (USA) 98; 12527-31: 2001 © National Academy of Sciences (USA).

Expression of *dGNaCl* in *Xenopus* oocytes generates constitutive current that does not discriminate between Na^+ and Li^+, but is selective for Na^+over K^+ **(Fig.21.5).**

21.4. CATION CHANNELS

In boar sperm plasma membrane, the most prominent was a nonselective cation channel, which conducted K, Na, Cs, Ca, and Ba. Channel opening did not show a strict dependence on voltage but was partially blocked by verapamil, nitredipine, and ruthenium red. A channel with these characteristics was observed in the plasma membrane overlying the sperm head (Cox et al., 1991).

21.4.1. Voltage Gated Ion Channels

Quill et al, (2001) reported a voltage-gated ion channel (CatSper2) that is expressed in male germ cells but not in other cells **(Fig.21.6)**. The putative channel contains 6 transmembrane segments, making it more similar to the voltage-gated potassium channel, but the ion selectivity pore domain sequence resembles that of a Ca (v) channel. The mRNA is expressed during the meiotic or post-meiotic stage of spermatogenesis, and the protein is localized to the sperm flagellum, suggesting a role in the regulation of sperm motility. The mRNA for the channel is present in mouse, rat and human sperm cells, and the gene is found on chromosome 2 E5-F1 in the mouse and 15q13 in the human. Another sperm voltage-gated cation channel (CatSper) that has features similar to the one reported is expressed within the flagellum and is required for normal fertility of mice. However, expression of CatSper2 alone or co-expression with CatSper in cultured cells, or attempts to co-immunoprecipitate the two proteins from germ cells led to demonstrate that these two unique but similar α-like subunits form either a homo-or heterotetramer.

CatSper is located specifically in the principal piece of the sperm tail. Targeted disruption of the gene results in male sterility in otherwise normal mice. Sperm motility is decreased markedly in CatSper-/- mice, and CatSper -/- sperm are unable to fertilize intact eggs. In addition, cyclic – AMP – induced Ca^{2+} influx is abolished in the sperm of mutant mice. CatSper is thus vital to c-AMP-mediated Ca^{2+} influx in sperm motility and fertilization (Ren et al., 2001). It is possible, therefore, that two independent α subunits, different from other known voltage-gated channel, regulate sperm motility. Although disruption of the gene for the sperm-specific CatSper fails to significantly alter sperm production, protein tyrosine phosphorylation that is associated with capacitation, induction of the acrosome reaction, forward velocity, or percentage of motility, yet CatSper2-/- males are completely infertile. The defect that was identified in the null sperm cells was a failure to acquire hyperactivated motility, which seemed to render spermatozoa incapable of generating "power" needed for penetration of the extra-cellular matrix of the egg. Thus, CatSper2 is responsible for driving hyper-activated motility (Quill et al., 2003). Although, molecular mechanisms for sperm to follow directional cues are unknown, Gao et al., (2003) demonstrated that the *Drosophila* PKD2 cation channel operates in sperm for directional movement inside the female reproductive tract.

21.5. CALCIUM CHANNELS

Calcium influx through voltage-sensitive channels controls a variety of cellular processes including secretion and muscular contraction. Modulation has been attributed to alterations in the phosphorylation state of channel subunit proteins to the direct interaction of G protein subunits with ion channel constituents and to voltage-dependent processes. Two classes of voltage sensitive Ca^{2+} channels have been described in somatic cells on the basis of their biophysical characteristics. 1) High voltage activated (HVA) and 2) low voltage activated (LVA) channels. The HVA Channels require large depolarization to > -20 mV for activation and channel opening, conduct maximal currents at >+10 mV, and may account for the L-, N-, P-, Q- and R-type currents. These channels have a heteromeric protein composition consisting of a pore forming $\alpha 1$ subunit and auxiliary regulatory α,γ and δ subunits. Voltage-operated Ca^{2+} channel structure and the relationship between channel subunit expression and the characteristics of consequent Ca^{2+} currents has been reviewed (Publicover and Barratt, 1999; Jagannathan et al., 2002). 2) The low voltage activated (LVA) channel that has a voltage threshold of approximately – 60 mV, conducts maximal current at –30 to –20 mV and accounts for the T-type current. There is presently no consensus regarding the molecular components of T channels. Pharmacological evidences indicate that there may be a diversity of T channels. Electrophysiological studies on rodent spermatogenic cells have revealed the presence of a T type voltage-operated Ca^{2+} current. This current has pharmacological attributes consistent with those of the putative channel responsible for Ca^{2+} influx mediating the acrosome reaction. Ni, mibefradil and amiloride have been used to discriminate T type channels from HVA channels (Jagannathan et al., 2002). Recombinant channel proteins have shown that most LVA channels can function without associating with any auxiliary subunits.

21.5.1. Presence of Two Ca^{2+} Channels in Sperm

Entry of Ca^{2+} through Ca^{2+} channels triggers the acrosome reaction of spermatozoa during fertilization. These Ca^{2+} channels share some properties with brain, cardiac, and skeletal muscle tubule Ca^{2+} channels, and may be involved in increasing $[Ca^{2+}]_i$ before the acrosome reaction (Tiwari-Woodruff and Cox, 1995; O 'Toole et al, 2000). Florman et al, (1992) provided evidence that sperm possess voltage dependent calcium channel (VDCC) and that activation of sperm VDCC is sufficient to induce sperm acrosomal exocytosis. The entry of Ca^{2+} requires two Ca^{2+} permeable channels. The glycoprotein of the zona pellucida (ZP3) depolarizes sperm membranes by activating a pertussis toxin-insensitive mechanism with the characteristics of a poorly selective cation channel. The ZP3 also activates a pertussis toxin-sensitive pathway that produces a transient rise in internal pH. The concerted effects of depolarization and alkalinization open voltage-sensitive Ca^{2+} channels. Studies indicated that the presumed G-protein target of pertussis toxin probably produces a required but indirect activation of the putative sperm VDCC. The intervening events may include alteration of the voltage sensitivity of the VDCC, membrane depolarization, or both. Thus depolarization–induced acrosome reaction may provide a useful system to investigate subsequent events in the exocytosis process. These observations suggest that mammalian sperm utilize membrane potential-dependent signal transduction mechanisms and that a depolarization pathway is an upstream transducing element coupling adhesion to secretion during fertilization (Arnoult et al., 1996; Jungnickel et al, 1999).

The role of internal calcium stores in human uncapacitated spermatozoa was studied by determining the effects of two inhibitors of Ca^{2+} ATPase, thapsigargin and cyclopiazonic acid on Ca^{2+} in plasma membrane potential and acrosome reaction. Study provided evidence that internal Ca^{2+} store depletion can evoke the opening of Ca^{2+-} channels on sperm plasma membrane, thus showing the existence of "capacitative" Ca^{2+} entry into these specialized cells. The addition of thapsigargin to human spermatozoa induced a dose-dependent increase in acrosome reaction, but only when Ca^{2+} was present in external medium. These inhibitors induced a depolarization dependent Ca^{2+} influx from external medium, and this was preceded by a transient hyper-polarization, caused by activation of Ca^{2+}-dependent K^+ channels. It demonstrates that human spermatozoa possess internal Ca^{2+} stores and the capacitative Ca^{2+} entry pathway present in these cells regulates events that are fundamental for the acrosome reaction (Gonzalez-Martinz et al., 2001; Rossato et al., 2001; O'Toole et al., 2000).

A linear fucose sulfate polymer (FSP > 106-Da), a major component of sea urchin egg jelly, induces the sperm acrosome reaction. Two Ca^{2+} channels are activated during acrosome reaction induction. The first opens 1 sec after FSP addition, and the second opens 5 sec after the first. Mild acid hydrolysis of FSP results in a linear decrease in polymer size. Hydrolyzed FSP of ~60-kDa blocks intact FSP from inducing the acrosome reaction. Thus hydrolysed FSP opens second Ca^{2+} channel, not the first and does not induce acrosome reaction. Therefore, full size of intact FSP is required to open both Ca^{2+} channel involved in triggering the acrosome reaction (Hirohashi and Vacquier, 2001). Thus the entry of Ca^{2+} requires two Ca^{2+} permeable channels, which have been defined as VDCCs.

21.5.2. T-Type Low Voltage Activated VDCC

A functional low voltage-activated T type Ca^{2+} channel is expressed during the meiotic and post–meiotic stages of mammalian spermatogenesis, and can be observed readily in mouse spermatogenic cells, which lack detectable levels of other voltage-sensitive Ca^{2+}currents (Lievano et al., 1996). A T-type Ca^{2+} channel is expressed during differentiation of the male germ lineage in the mouse and is retained in sperm, where it is activated by contact with the egg's extracellular matrix and controls sperm acrosomal exocytosis. Arnoult et al, (1997) examined the regulation of this Ca^{2+} channel in dissociated spermatogenic cells from the mouse using the whole-cell patch-clamp technique. T currents were enhanced, or facilitated, after strong depolarization or high frequency stimulation. The same facilitation was produced by antagonists of protein tyrosine kinase activity. Conversely, antagonists of tyrosine phosphatase activity blocked voltage-dependent facilitation of the current (VDCC). These data were consistent with the presence of a two – state model, in which T channels are maintained in a low (or zero) conductance state by tonic tyrosine phosphorylation and can be activated to a high conductance state by a tyrosine phosphatase activity. The positive and negative modulations of these channels by the tyrosine phosphorylation state provides a plausible mechanism for the control of sperm activity during the early stage of mammalian fertilization (Arnoult et al., 1997). Progesterone also induces the acrosome reaction in non-capacitated cryo-preserved bovine spermatozoa through intracellular mechanisms dependent on protein kinase C and the voltage-dependent calcium channel (Cordoba and Beconi, 2001). Detection of LVA channels has been difficult. Jacob and Benoff (2000) detected only transcripts encoding for domain IV and C-terminus in rat testis RNA, whereas PCR products could not be generated with cDNA from human spermatozoa (Jacob and Benoff, 2000). Espinosa et al., (1999), obtained PCR products from mouse spermatogenic cell cDNA. Son et al., (2000) used degenerate primers on cDNA from human testicular biopsies to obtain a 489 bp fragment of α1H. The full-length sequences of both α1G and α1H LVA channels from human testicular cDNA have been obtained by Jagannathan et al, (2002).

A detailed analysis of Ca^{2+} currents expressed in acutely dissociated mouse primary spermatocytes using patch–clamp recordings demonstrated that the only voltage–gated Ca^{2+} channels present belongs to the family of T-type Ca^{2+} currents. Accordingly, Ni^{2+} and amiloride reduced current amplitude by 75 and 62%, respectively. Unexpectedly, nifedipine also reduced peak currents at concentrations as low as 2 μM. Because these channels represent the primary pathway for voltage–gated Ca^{2+} entry in mouse spermatocytes, they may also participate in regulating meiotic cell division and sperm differentiation (Santi et al., 1996;1998). To identify the members of the voltage-dependent Ca^{2+} channel family present in sperm, Lievano et al, (1996) looked for the expression of the *α1A*, *α1B*, *α1C*, *α1D,* and *α1E* genes in mouse testis and in purified spermatogenic cells. All 5 genes are expressed in mouse testis, and in contrast only *α1E*, and to a minor extent *α1A*, are expressed in spermatogenic cells. In agreement with these findings, only T-tpye Ca^{2+} channels sensitive to the dihydropyridine nifedipine were observed in patch-clamp recordings of pachytene spermatocytes. These results suggested that low-threshold Ca^{2+} channels are the dihydropyridine-sensitive channels involved in the sperm acrosome reaction (Arnoult et al, 1996, Lievano et al, 1996). However, evidence for the presence of α1C, α1G, and α1H in mouse pachytene spermatocytes and in round and condensing spermatids has also been presented (Espinosa et al., 1999).

Westenbroek and Babcock (1999) probed rodent sperm with anti-peptide antibodies directed against cytosolic domains of cloned rat brain α1A, α1C, and α1E Ca channel subunits. Each Ab recognized a 200 to 245-kDa band on immunoblots of rat with a smaller (110-kDa) α1C band. Mouse sperm showed characteristic patterns of punctuate α1A-, α1C-, and α1E-immunoreactivity. For α1A, the puncta were larger, less numerous, and more variable in distribution than for α1C and α1E. They were absent from the acrosomal crescent, but present over the sperm head, often at the apical tip and equatorial segment. These subunits were also found at irregular intervals along both the mid-piece and the principal piece of the flagellum. It was suggested that multiple Ca channel proteins are present in mature sperm and are regionally localized in ways that may give them different regulatory roles (Westerbroek and Badcock, 1999; Wennemuth et al., 2000). Analysis of PCR products showed that only α1H subunits are expressed in human testes. The expression of the α1H subunit may be tissue specific since its mRNA was not detected in the human ovary. This suggested that the acrosome reaction of human spermatozoa is highly associated with T-type Ca^{2+} channels and was mainly mediated by calcium influx through α1H T-type Ca^{2+} channels (Linares-Hernandez et al, 1998; Son et al, 2000).

21.5.3. Neoglycoproteins and Ca^{2+} Channels

Hexose (mannose) or two amino sugars (glucosaminyl or galactosaminyl residues) when covalently conjugated to a protein backbone (neoglycoproteins, NGP) mimick the mouse ZP3 glycoprotein action and induce the acrosome reaction in capacitated mouse spermatozoa (Loeser et al, 1999). To elucidate the mechanism underlying sperm-neoglycoprotein interaction and the induction of the acrosome reaction Loeser et al (1999) demonstrated that two known L-type Ca^{2+} channel blockers prevented the induction of the acrosome reaction by three neoglycoproteins (Man-BSA, Glc NAc-BSA, Gal NAc–BSA). The fact that the L-type Ca^{2+} channel blockers (verapamil, diltiazem) had no inhibitory effect on sperm surface galactosyltransferase or α-D-mannosidase, suggests that the reagents block the acrosome reaction by a mechanism other than binding to the active site of the enzymes. The neoglycoproteins are known to rapidly increase intracellular free Ca^{2+} in human spermatozoa. It was also suggested that mannose-BSA, but not progesterone, activates T-type Ca^{2+} channels in human spermatozoa for the following reasons: (i) the capacity of Man-BSA to increase $[Ca^{2+}]_i$ was inhibited by the specific T-type Ca^{2+} channel blocker, mibefradil while progesterone was not inhibited by mibefradil; (ii) the effect of Man-BSA to elevate $[Ca^{2+}]_i$ was inhibited more potently by Ni^{2+} This supported the idea that Man-BSA and progesterone were activating distinct Ca^{2+} channels, one of which was a T-type Ca^{2+} channel activated by mannose-BSA whereas the other Ca^{2+} channel that was activated by progesterone had yet to be defined at the molecular level (Blackmore and Eisoldt, 1999). However, progesterone ac celerated the onset of capacitation in mouse sperm via T-type calcium channels and this could be associated with calcium influx and tyrosine phosphorylation (Senuma et al, 2001).

Results on pretreatment of spermatozoa with pertussis toxin showed that the transduction mechanism for NGPs involves G proteins of the inhibitory type (Gi). An increase in the acrosome reaction of capacitated spermatozoa was observed in presence of potassium ions only when the pH was raised to 8.5; and acrosome reaction was inhibited by verapamil, Ni^{2+}, Cd^{3+}, La^{3+} indicating the participation of VDCC activated by membrane depolarization. The GlcNAc-BSA- or Man-BSA- induced acrosome reaction was completely inhibited by pre-incubation of spermatozoa with VDCC blockers and calcium antagonists, indicating a link between the binding sugar residues of the NGPs and channel activation (Brandelli et al., 1996). Lopez-Gonzales et al, (2001) analyzed the effects of CaM antagonists (W7 and trifluoroperazine) on voltage-dependent T-type Ca^{2+} currents in mouse spermatogenic cells and on the zona pellucida-induced acrosome reaction in sperm. It was found that CaM antagonists decreased T- currents suggesting a functional interaction between CaM and the sperm T-type Ca^{2+} channel and consistent with the involvement of T-channels in the acrosome reaction. Patch-clamp experiments demonstrated that serum albumin induced an increase in Ca^{2+} T current density and significant shifts in the voltage dependence of both steady-state activation and inactivation of the channels. In contrast, β-estradiol significantly inhibited Ca^{2+} channel activity in a concentration-dependent and essentially voltage-independent fashion (Espinosa et al, 2000; Krasznai et al, 2001). In conclusion, the induction of the acrosome reaction in human spermatozoa by sNGPs involves VDCC and G-like regulatory proteins similar to the induction described for ZP in other mammalian species.

21.5.4. L-Type Hight Voltage Dependent VDCC in Testis

The mechanisms of modulation of sperm functions have been examined primarily in high voltage-activated Ca^{2+} channels. For example, L-type currents are positively modulated by protein kinases A and C, as well as by G_s, and are suppressed by G_0. Similar mechanisms can regulate other high voltage-activated Ca^2 channels. The entry of Ca^{2+} through Ca^{2+} channels, with characteristics similar to those of L-type voltage – sensitive Ca^{2+} channels found in cardiac and skeletal muscle is crucial step in the sequence of events leading to exocytosis in progesterone-stimulated human spermatozoa. An influx of Na^+ also may play a role, but at a point prior to the opening of Ca^{2+} channels (O Toole et al., 1996).

Whether the pharmacological activity of dihydropyridines, which block calcium influx through voltage-dependent calcium channels, contributes to acrosome loss and in the production of an infertile state, Goodwin et al, (1997) obtained a protein, which is expressed in rat and human spermatozoa that is related both antigenically and by cDNA sequence to the α1 subunit of the rat cardiac muscle VDCC, which forms the pore of the channel. Using RT-PCR, Goodwin et al, (1997) isolated a clone of nt 3908 – 6077 from rat testis mRNA whose sequence was largely identical to that of the α1 subunit of the rat cardiac muscle calcium channel, but had an 84 base change, attributable to splicing and alternate exon usage. This base change inserts a peptide cassette encoding an amphipathic membrane-spanning helix that constitutes part of the ionic pore of the skeletal muscle calcium channel and may serve as an intra-membrane dihydropyridine binding site. These data unequivocally demonstrated that an L-type VDCC is expressed in rat testis and that VDCC isoforms from rat testis and heart differ in deduced amino acid composition in and around potential bindings sites for calcium channel blocking drugs such as the dihydropyridine (Goodwin et al., 1997; 1998).

α1 Subunits of VDCC: The main structural constituent of all VDCCs is the single pore forming α1 subunit, which is primary determinant for any chemical or pharmacological variation of VDCCs. The VDCC is composed of four homologous domains, which are linked by cytoplasmic linker regions. Each domain is made up of six transmembrane helices (S1-S6)**(Fig.21.7)**. Between S5 and S6 segments in each domain, there is non-helical P loop. Ten α1 subunits have been identified by molecular cloning. Of these α1A, α1B, α1C, α1D, α1E, α1F and α1s all appear as HVA channel. The α1G, α1H and α1I have been recently discovered (Jagannathan et al., 2002). In HVA, α1 subunit is associated with an auxiliary subunit such as β, δ and possibly γ subunit. Transcripts for a number of HVA channels encoding α1A, α1B, α1C, α1D and α1E subunits have been detected in mouse spermatogenic cells (Espinosa et al., 1999), and in rat and human testis (Goodwin et al., 1997; 1998). In addition, α1C mRNA was detected from human sperm (Goodwin, 2000).

In addition, the α1A gene product mediates both P- and Q-type high voltage activated currents, possibly in association with distinct auxiliary subunits. Similarly, recombinant α1E

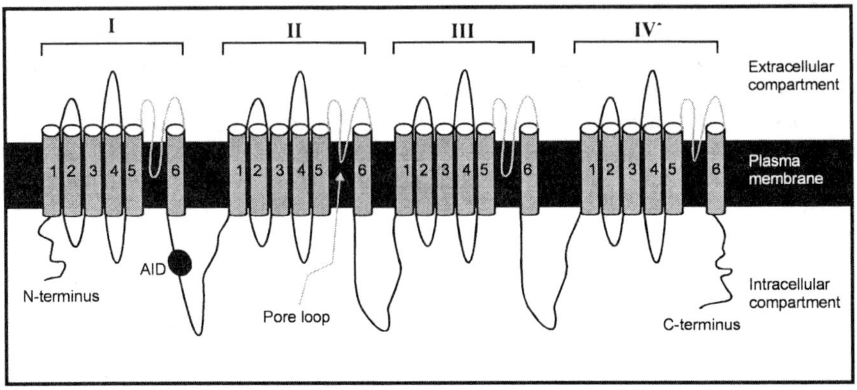

Fig. 21.7. Putative structure of voltage-operated Ca^{2+} channels showing four repeating domains (I-IV). Each domain is composed of six trans-membrane α helical regions (1-6), interspersed by pore loop linkers.

protein that was initially described to produce a low voltage–activated Ca^{2+} current, now known to produce a high voltage activated class and may be the R-type current. While transcription of these Ca^{2+} channel genes occurs during spermatogenesis, the associated high voltage–activated current components are present either at low copy number or in an inactive form (Florman et al., 1998). Changes in phospholipids and removal of cholesterol from membranes, resulting in increased membrane fluidity (Visconti and Kopf, 1998; Baldi 2000; Flesch and Gadella, 2000) are factors, which can change VDCC currents. Thus, sperm T channels represent both an essential step in the ZP3 signal transduction pathway and a site of regulation during capacitation. But the mechanisms by which K^+ permeability and ZP3 regulated cation channel conductance are regulated during capacitation are not understood.

Expression of α1C mRNA: Expression of the mRNA for testis specific VDCC and its co-ordinated translation is initiated early in rat male germ line and continues throughout spermatogenesis. Goodwin et al (1999) reported the complete mRNA and deduced amino acid sequence of the α1C pore-forming subunit of the rat testis-specific L-type calcium channel. This subunit is transcribed from the α1C gene, which is also expressed in brain and cardiac muscle. Sequence analysis of nucleotide bases 3908-6077 of the subunit expressed in rat testis differed from cardiac sequences only in a 84 base pair region corresponding to exons 31/32 encoding a putative dihydropyridine binding region. Sequence analysis of bases 3048-3936 identified a second difference between the rat testis and rat cardiac α1 sequence in a 60 base pair region corresponding to exons 21/22 and encoding another putative dihydropyridine binding site. Rat testis and cardiac isoforms are products of same gene. Testis specific exon 32 of VDCC α1 subunit was confined to seminiferous tubules and was associated with the germ cell lineage from type A spermatogonia to mature spermatozoa. Western blot analysis of rat protein extracts detected primarily 175-220-kDa proteins in the size range of VDCC. The cardiac- and testis-specific isoforms of the α1C subunit are produced by alternate splicing of the same primary transcript. The testis-specific isoform differs from that of cardiac tissue at its amino terminus and in transmembrane segments IS6, IIIS2 and IVS3, which are also dihydropyridine binding sites. In somatic tissues, segments S2 and S3 regulate channel activation while the amino terminus and segment IS6 contribute to channel inactivation kinetics. The amino terminus and IS6 segment of the testis-specific α1C subunit are also expressed respectively, in the brain and in smooth muscle from lung where they alter the electrophysiological characteristics of the subunit to produce relatively slow inactivation kinetics. Further results identified RNA transcripts of a highly conserved region within the α-1C (pore-forming) subunit of L-type voltage dependent calcium channel from human donors. Data were consistent with results from in-situ RT-PCR of rat testis sections indicating that the testis-specific calcium channel of that species was expressed uniformly in all stages of the germinal epithelium, including mature spermatozoa (Goodwin et al, 2000).

Auxiliary Subunits of VDCC: The evidence for the presence of the Ca^{2+} channel β auxiliary subunits in spermatogenic cells and sperm has been also provided. Serrano et al, (1999) demonstrated the expression of all four known genes encoding the β subunits proteins in spermatogenic cells and sperm. Further more in spermatogenic cells both α1 and β subunits are diffusely distributed throughout the cytoplasm while they appear to be regionally localized in sperm (Serrano et al., 1999). Examination of depolarization-evoked Ca^{2+} entry indicated that mature sperm possess a larger palette of voltage-gated Ca^{2+} channels than previously thought. Such diversity may permit specific responses to multiple cues encountered on the path to fertilization. Wennemuth et al, (2000) showed the involvement of a previously undetected Ca(V) 2.2 (N type) channel, which was substantiated by immunodetection of Ca^{2+} channel α (1B) subunits in sperm. Along with this, the participation of Ca(V)2.3 (R-Type) channel specified by α(1E) subunits was also substantiated. However, to investigate whether α(1) 2.3, specified by α(1E) subunit of the voltage-dependent Ca^{2+} channel codes for the LVA current, experiments were performed in pachytene spermatocytes from Ca(v)2.3+/+ and Ca(v)2.3-/-mice, which showed a typical LVA Ca^{2+} currents in pachytene spermatocytes. It was suggested that the Ca(V)2.3 channel makes no detectable contribution to the LVA Ca^{2+} current in the pachytene spermatocyte. Instead, Ca(v)3 family such as Ca(V) 3.1 may be likely candidates responsible for LVA currents in pachytene spermatocytes (Sakata et al., 2001).

In nutshell, T-type low voltage-activated Ca^{2+} channel genes are synthesized during mouse spermatogenesis, as revealed by direct examination using patch clamp methods. No other functional channels can be detected during spermatogenesis, although transcription of other channel genes has been reported. T channels are retained in sperm following terminal differentiation and play an essential role in the initiation of the acrosome reaction during zona pellucida contact. Since conductance through T channels is regulated by membrane potential, question arises how eggs depolarize sperm membrane potential?

21.6. AMINO ACID NEUROTRANSMITTER RECEPTOR/Cl CHANNELS

Two different amino acid neurotransmitter receptor/Cl channels, a γ-aminobutyric acid (GABA) receptor–like / Cl channel and a glycine receptor (GlyR) / Cl channel have been found in mammalian sperm and shown to be involved in the sperm acrosome reaction. The identity of the receptor / cl channel involved in the acrosome reaction depends upon the stimulatory ligand. For example, the acrosome reaction initiated by the egg zona-pellucida requires the participation of the sperm glycine receptor/Cl channel, whereas the acrosome reaction initiated by progesterone involves the sperm GABA receptor–like/Cl-channel.

21.6.1. $GABA_A$ Receptor/Cl^- Channel

Progesterone exerts its effects on capacitated human sperm at least partially by interacting with a unique sperm steroid receptor/Cl^- channel complex resembling a neuronal $GABA_A$ receptor/Cl^- channel complex (GBRC) but apparently possessing a different sensitivity to progestins. This interaction results in an increased Cl^- efflux essential to acrosome reaction initiation (Wistrom and Meizel, 1993). Lavendustin A, a potent, specific inhibitor of tyrosine kinase activity, strongly inhibits the P-initiated human acrosome reaction and the essential P-mediated Cl^- efflux. Lavendustin B, a weak tyrosine kinase inhibitor had no significant effect. This suggests that, as part of acrosome reaction initiation, progesterone mediates tyrosine phosphorylation of the sperm GABAA$^-$ like receptor/Cl^- channel (Meizel and Turner, 1996). Activation of sperm amino acid neurotransmitter receptor/Cl^--channels may result in plasma membrane depolarization and there by exert at least a partial control over voltage sensitive sperm Ca^{2+} channels and Ca^{2+} influx important to the acrosome reaction (Meizel, 1997; Meizel and Lurns, 1996; Espinosa et al. 1998).

21.6.2. Glycine Receptor (GlyR) / Cl Channel

Using a monoclonal antibody specific for the α (48-kDa) and β (58-kDa) subunits of the rat spinal cord GlyR (mAb Gly R4a), Melendrez and Meizel (1996) provided the direct evidence for GlyR in mammalian sperm. The GlyR4a was detected on the porcine sperm periacrosomal plasma membrane, a site supporting GlyR involvement in the acrosome reaction. This was one of the first direct demonstration of both α and β subunits of GlyR in a non-neuronal cell. Studies on porcine and human sperm also suggested a role for sperm glycine receptor/Cl channel in the zona pellucida-initiated acrosome reaction, suggesting that a glycine receptor / Cl channel is involved in the zona-initiated mammalian sperm acrosome reaction (Melendrez and Meizel 1996; Sato et al (2000a,b).

21.7. ANION CHANNELS

21.7.1. Voltage-Dependent Anion Selective Channels (VDACs)

Voltage-dependent anion selective channels (VDACs) also known as mitochondrial porins are small channel proteins involved in the translocation of metabolites across the mitochondrial outer membrane. The VDACs or eukaryotic porins were first identified in the mitochondrial outer membrane that are able to form hydrophilic pore structures in membranes. A single channel-forming protein is found in yeast, whereas higher eukaryotes express multiple VDACs, with humans and mice each harboring three distinct channels (VDAC1-, 2-, 3) encoded by separate genes. Axonemal defects may be caused by associated non-axonemal components such as VDAC and illustrate that normal mitochondrial function is required for stability of the axoneme. Sampson et al (2001) gave evidence suggesting that infertility in mice is due to loss of mitochondrial VDAC-3 in sperm. Mice lacking VDAC3 are healthy, but males are infertile. Although the sperm count was normal, sperm exhibited markedly reduced motility. Two thirds of epididymal axonemes showed abnormal structure, with loss of a single microtubule doublet at a conserved position within the axoneme. Defect in testicular sperm was rarely observed, suggesting that instability of a normally formed axoneme occurs with sperm maturation.

Sequencing of voltage-dependent anion channel 2 (VDAC2, porin-2) cDNA from bovine testis has been reported (Hinsch et al., 2001). The murine, rabbit, and human subtypes show

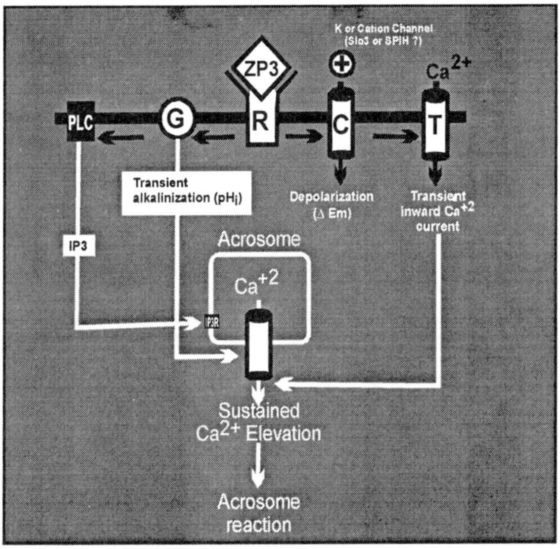

Fig.21.8. Model for acrosome reaction based on ion channel activation by zona-pellucida. Abbreviations used: R-Receptor, G-Protein, C-Cation, T-T Currents, PLC-Phospholipase, ZP3-Zona-pellucida Protein 3, IP3- Inositol triphosphate, IP3R-Inositol triphosphate receptor (IP3R is present in acrosomal cap).

high homology at both the nucleotide and amino acid levels. Messenger RNA analysis revealed expression of VDAC2 in bovine testis, with high content of VDAC2 protein found in late spermatocytes, spermatids, and spermatozoa. In contrast, VDAC1 (porin-1) is exclusively localized in Sertoli cells. Testicular VDAC2 plays an important role in providing energy metabolites and in germ cell apoptosis (Hinsch et al, 2001). One of the most abundant proteins in ODF is a 30-32-kDa polypeptide. Peptide sequences of this protein matched with VDAC2 and VDAC3. While VDAC1, 2, and 3 were detected in bovine sperm, only VDAC2 and 3 were solubilized from bovine ODFs (Hinsch et al., 2004) (Chapter 29).

21.7.2. Close Cell-Cell Chloride Channel (CLC CL^- Channel)

The functions of some CLC CL^- channels are evident in human diseases, which are caused due to their mutations. The role of the broadly expressed CLC CL^- channel in testis has been described by Bosl et al, (2001). Several important functions have been attributed to CLC-2, but contrary to these expectations CLC-2 deficient mice lacked overt abnormalities except for a severe degeneration of the testes, which led to selective male infertility. Seminiferous tubules did not develop lumina, and germ cells failed to complete meiosis. Beginning around puberty there was a massive death of primary spermatocytes and later also of spermatogonia. Tubules were filled with abnormal Sertoli cells, which normally express CLC-2 in patches adjacent to germ cells. Thus, CLC-2 disruption entails the death of cell types which depend on supporting cells that form the blood-testes and retina barrier (Bosl et al., 2001).

21.8. VDCC AND ACROSOME REACTION

Elevation of intracellular Ca^{2+} is prerequisite in the control of acrosome reaction. A model for the mechanisms by which zona pellucida adhesion regulates sperm Ca^{2+} level is presented in **Fig. 21.8)** (Florman et al., 1998). A minor increase in $(Ca^{2+})i$ does not initiate acrosome reaction whereas addition of ZP3 produces an additional increase to 300-500 nM within minutes, at which level $[Ca^{2+}]_i$ values either are stabilized or slowly decline. Peak rates of $[Ca^{2+}]_i$ elevation during this response are ~150 nM/min, which are associated with the initiation of acrosome reaction. The evidence shows that T currents in spermatogenic cells are inhibited by several drugs and ions and the elevation of $[Ca^{2+}]_i$ and the acrosome reaction induced by ZP3 are inhibited by these compounds. This indicated that ZP3 stimulation of sperm evokes a T-type Ca^{2+} current that is an essential component of the signal transduction regulating acrosome reaction (Florman et al., 1998).

Capacitation of Sperm is Pre-requisite for Aerosome Reaction: Spermatozoa maintain a $[Ca^{2+}]_i$ of 50-100 nM that increases to 125-175 mM during capacitation. During capacitation, an increase in $[Ca^{2+}]i$ occurs uniformly throughout the head and flagellar regions of the sperm. The mechanisms for this physiological alteration in the sperm $[Ca^{2+}]_i$ have not been ascertained. During capacitation of spermatozoa, the plasma membrane gets hyperpolarized from $\simeq$-30mv to $\simeq$-60mv. Binding of zona pellucida to its receptor on sperm depolarizes the membrane potential to $\simeq$ -25mv. Depolarization following ZP binding is caused on account of opening of an unidentified voltage insensitive cation channel, which is permeable to Na^+ and bivalent cations (viz, Ca^{2+}, Mn^{2+}, Ba^{2+}).

An important requirement for induction of acrosome reaction via VDCC is that channels should remain closed to avoid premature Ca^{2+} influx and acrosome reaction. Hence capacitation of sperm is prerequisite state before onset of acrosome reaction by ZP3. It means that ZP3 transducing mechanisms are regulated during capacitation. It implies that sperm T channels are modulated in several ways during capacitation, viz (1) conductance state, which is controlled by membrane potential. During capacitation membrane potential hyperpolarizes due to enhanced permeability of K^+ and shift of Em to Ek. Hyperpolarization may act to prime T channels for subsequent activation by ZP3 (Zeng et al., 1995; Arnoult et al., 1999). The increase in pHi that occurs during capacitation (Baldi, 2000) may activate K^+ rectifier channel and increase K^+ permeability that leads to hyperpolarization (Gonzalez-Martinez et al., 2001). Besides above, Slo3 present in mouse and human spermatocytes is also sensitive to pHi. (2) Another way of T channel activation during capacitation may be the sensitivity of cation channels, which may produce an inward current carried by Ca^{2+} and T channel activation. (3) Finally sperm T channels may also be modulated during capacitation by tyrosine phosphorylation. It has been found that tyrosine phosphorylation of either a T channel or its regulator decreases current through the channels. Conversely, dephosphorylation by tyrosine phosphatase enhances T current. This type of modulation has not been observed in somatic cell T channel, and thus may represent a novel regulatory mechanism (Dolphin, 1998). Alternatively, tyrosine phosphorylation- dependent modulation may be more easily detectable in the germ cell model system, where other Ca^{2+} currents are not present (Florman et al., 1998). Thus, capacitation can be considered to be associated with a wave of protein tyrosine phosphorylation during which the T channel, or a channel regulator, is one substrate for a capacitation–dependent tyrosine kinase. The resultant negative modulation of T currents would reduce Ca^{2+}current and could minimize spontaneous acrosome reaction. This implies that ZP3 activates a tyrosine phosphatase activity during the initiation of acrosome reaction. Phosphorylation of sperm increases with capacitation and localizes mainly to the principal piece of human sperm. Following binding to the zona pellucida, the percentage of sperm with phosphotyrosine residues localized to both the neck and the principal piece was significantly higher in bound sperm than in capacitated sperm in suspension. Evidences also support that different compartments of human spermatozoa undergo a specific sequence of phosphorylation during both capacitation and upon binding to the zona pellucida. Tyrosine phosphorylation in the principal and neck piece may be considered a prerequisite for fertilization in humans (Sakkas et al., 2003).

Role of Ca-ATPase: An increase in the concentration of intracellular free Ca^{2+} and in the phosphotyrosine content of specific proteins characterizes human sperm capacitation. Whether tyrosine phosphorylation regulates the intracellular free Ca^{2+} concentration through modulation of Ca^{2+}-ATPase activity or the phosphotyrosine content is under Ca^{2+} regulation was investigated using Ca^{2+}-ATPase modulators and tyrosine kinase inhibitors. The presence of the Ca^{2+}-ATPase-inhibitor thapsigargin during human sperm capacitation caused an incease in the cytoplasmic free Ca^{2+} that was associated with an increase in the phosphotyrosine content of specific sperm proteins. Conversely, a decrease in protein tyrosine phosphorylation was observed, when a Ca^{2+}-ATPase activator, was present during the incubation period. On the other hand, thapsigargin had no effect on the phosphotyrosine content or the cytoplasmic Ca^{2+} concentration. These findings suggested that Ca^{2+}-ATPases are involved in the filling of internal Ca^{2+} stores, such as the acrosome, and are inhibited later during capacitation. Their inhibition allows an increase in cytoplasmic free Ca^{2+}, which is involved in the subsequent increase in the phosphotyrosine content of specific sperm proteins (Dorval et al., 2003).

Depending on the increase in intracellular Ca^{2+}, there were two sperm subpopulations, called LR (low responsive) and HR (high responsive) in response to thapsigargin. In addition to their high increase in intracellular Ca^{2+}, sperm from the HR population showed a higher protein phosphotyrosine content in association with higher proportion of acrosome reacted sperm, as compared with LR sperm. The thapsigargin-mediated Ca^{2+} increase and acrosomal exocytosis suggested that, during the acrosome reaction, the signaling pathway mediated by src-related tyrosine kinases is involved upstream of the capacitative Ca^{2+} entry. The calcium-induced increase in tyrosine phosphorylation has a functional correlate in sperm exocytosis. Results showed that both tyrosine kinases and phosphatases played a central role in sperm exocytosis (Dorval et al., 2003; Tomes et al., 2004).

In conclusion, it now appears that a transient Ca^{2+} influx through T type channel in patch clamped spermatogenic cells is an essential early event in response to zona binding. The initial spilt of $[Ca^{2+}]_i$ induces a second sustained Ca^{2+} influx, which is mediated primarily by a store operated channel activated after depletion of small Ca^{2+} store probably in acrosome (c/r Jagannathan et al., 2002). The importance of sustained Ca^{2+} influx in acrosome reaction has been demonstrated. Inspite of presence of T type channel, mature spermatozoa also possess a number of HVA VDCC. Although mouse model has been studied in detail, it is not known how far this works in humans. However, phosphorylation of sperm is pre-requisite for induction of acrosome reaction by ZP3.

21.9. AQUAPORINS (WATER CHANNELS)

The discovery of water channels (Aquaporins) was a major breakthrough in research on water transport. Aquaporins are a family of major intrinsic membrane proteins (MIP) that function as

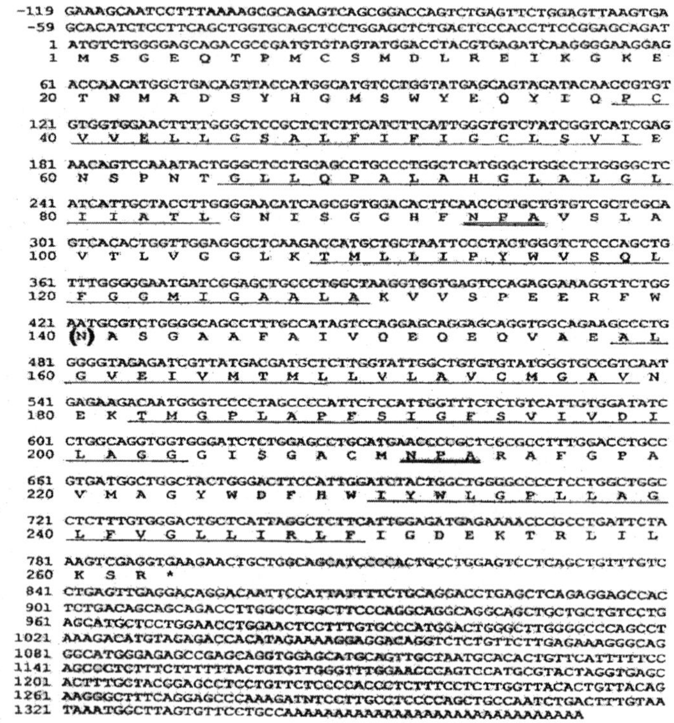

Fig21.9. Nucleotide and predicted amino acid sequence ofaquaporin -8 from rat testis. Transmembrane domains are underlined. NPA boxes are double underlined. A glycosylation site is given in parenthesis. Reprinted with permission from K. Ishibashi et al., Biochem Biophys Res Commun 237: 714: 1997 © Elsevier.

water-selective channels (except-aquaporin-3 and aquaporin-7, which are permeable to urea and glycerol as well) in the plasma membranes of many cells. They have six putative transmembrane domains of25-30kDa molecular weight. More than a dozen of aquaporins have been discovered in rat and humans. Aquaporin 0 (MIP26) maintains fluid balance in the lens. Aquaporin-1(AQP-1 or AQP-CHIP) is asymmetric homotetramer, and is involved in water reabsorption in the kidney's proximal tubules and descending thin limbs of Henle, in the formation of aqueous humor in eye, cerebrospinal fluid formation in brain, and airway hydration in lung. Aquaporin-2 is the water channel that is activated by vasopressin to enhance water reabsorption in kidney collecting duct. While major function of aquaporin 3 is to reabsorb water in kidney, it also contributes in maintaining transparency of cornea. Aquaporin-4 is involved in cerebrospinal fluid transport in brain, water transport in the kidney collecting duct, aqueous humor transport in the eye, and airway hydration in the lung. Aquaporin-5 apparently is coupled to fluid secretion in exocrine tissue. Although the function of aquaporin–6 is not known, its presence in the kidney suggests a potential role in water transport (review: Dibas et al., 1998). The AQP1 has been shown to be present at testis efferent ductules and seminal vesicles, it is not present in sperm (Liu et al., 1995). While AQP2 mRNA is absent from testis, AQP3 and AQP4 proteins have not been detected there in.

21.9.1. Aquaporins in Male Reproductive Tract

Though aquaporin 2 (AQP2), the vasopressin-regulated water channel, was originally identified in renal collecting duct principal cells, recent observation indicated that this water channel may have extra-renal functions, possibly related to sperm concentration in the male reproductive tract. The amino acid sequence of vas deferens AQP2 showed 100% identity to the renal protein and expressed in the distal portion (ampulla) of the vas deferens but not in the proximal portion, nearest the epididymis. The AQP2 is a constitutive apical membrane protein, not vasopression regulated in vas deferens (Stevens et al., 2000).

Ram and human spermatozoa have a high coefficient of osmotic water permeability (pf) with a low activation energy (Ea) suggesting the presence of water channels in plasma membrane of sperm. Sperm membranes have been examined for the presence of two known water channel proteins; CHIP28 and glucose transporters belonging to the GLUT family of proteins. Data from ram and human spermatozoa showed that CHIP28 is not present in sperm membranes but that glucose transporters may have secondary water channel function (Curry et al., 1995).

21.9.2. Aquaporin 7 and 8

The report on high water permeability of human and ram sperm is intriguing, but its physiological significance remains unknown (Curry et al., 1995). It is reported that the high water permeability of human sperm is mercury-resistant and not mediated by AQP1. In search for its explanation Ishibashi et al (1997b) reported an aquaporin-7 (AQP7) abundantly expressed in testis. However, the expression of AQP7 is transient and limited to late phase of spermatogenesis. The mature spermatozoa express only small amount of AQP7.

In search of major water channel, a new member of water channel aquaporin-8 (AQP-8) was identified in rat testis (Ishibashi et al., 1997a). The expression of AQP8 in Xenopus oocytes stimulated the osmotic water permeability (pf) 8.5 folds. The increase of pf was inhibited by mercury chloride and reversed by mercaptoethanol. The Ea for the water permeability was low (5.1 Kcal/mol). The AQP8 did not facilitate glycerol transport. A 1.5-kb transcript of AQP8 is abundantly present in testis and slightly in liver **(Fig.21.9)**. In situ hybridization of testis revealed the expression of AQP8 mRNA in all stages of spermatogenesis from primary spermatocytes to spermatids. The results suggested that AQP8 is also present in spermatozoa (Suzuki-Tayota et al., 1999; Ishibashi et al., 1997b; Calamita et al., 2001; Kageyama et al., 2001).

Aquaporin-7 differs from aquaporin-8 in many ways. Aquaporin-7 appears to play a role in the cryo-preservation of the sperm as well as in spermatogenesis, whereas aquaporin-8 is responsible for the secretion of pancreatic juice and spermatogenesis. While aquaporin 7-8 are most abundantly present in rat testis, aquaporin-7 plays an important role in epididymis for sperm maturation. However, structural and functional differences between AQP7 and AQP8 are noteworthy. AQP8 has consensus glycosylation and phosphorylation site but AQP7 does not have a consensus glycosylation and a phosphorylation site. AQP8 mRNA was found in testis but not in epididymis where as AQP7 was present in both. The AQP8 shows a band of 25 kDa and a diffused component of 32 to 40 kDa corresponding to glycosylated protein. AQP7 appears as 23-24 kDa protein and found in both testis and epididymis, The AQP8 is absent in epididymis. The AQP8 is present intracellularly as well as on plasma membrane of germ cells through out spermatogenesis, whereas AQP7 was localized in spermatids and spermatozoa only and was present predominantly on plasma membrane (Calamita et al., 2001). Functionally, both AQP7 and AQP8 stimulate osmotic water permeability when expressed in *Xenopus oocytes*.

The water permeability of AQP7 was mercury resistant while AQP8 was mercury sensitive. The AQP7 transports both water and glycerol, while AQP8 transports only water. There is also some difference in the distribution pattern of two aquaporins in seminiferous tubules. The AQP7 mRNA expression varies among seminiferous tubules probably depending on the different maturation stages of spermatogenesis indicating its transient nature of expression. On the other hand, AQP8 mRNA was distributed uniformly in seminiferous tubules suggesting that the protein may be constantly expressed. These structural, functional, and distribution differences of AQP7 and AQP8 may reflect their different roles in spermatogenesis (Kageyama et al., 2001).

AQP7 encodes a 269-amino acid protein that contained the conserved NPA motifs of MIP family proteins. It has the amino acid sequence homology (30%) with other apuaporins and highest with AQP3 (48%) suggesting that both AQP3 and AQP7 belong to a subfamily in the MIP family. Injection of AQP7-RNA into *Xenopus* oocytes expressed a 26-kDa protein. AQP7 also facilitated glycerol and urea transport by 5-and 9-fold, respectively. Northern blot analysis revealed a 1.5 kb pair transcript expressed abundantly in testis. The Ea for stimulated water was low. (2.1-kcals/mol). Immunoreactivity of AQP7 is restricted within the elongated spermatids, testicular spermatozoa, and residual bodies remaining in the seminiferous epithelium. The whole head and distal tail, where the elongated spermatid had only a limited amount of cytoplasm did not show immunoreactivity throughout spermiogenesis. After spermiation, the immunoreactivity of AQP7 remained at the middle piece and in the cytoplasmic droplet in the testicular spermatozoon. Since this water channel protein is localized on the plasma membrane covering the condensing cytoplasmic mass of elongated spermatid, and since the seminiferous tubule fluid is hypertonic, it is suggested that AQP7 contributes to the volume reduction of spermatids (Suzuki-Toyota et al., 1999). Cell volume reduction is one of the most distinct morphological change during spermiogenesis that may be largely attributable to water efflux from the cell. A strong candidate for a water efflux is AQP7 channel, in the rat testis.

The gene of aquaporin-8 encodes a 263-amino-acid protein that contains the conserved NPA motifs of MIP family proteins **(Fig. 21.9)**. The AQP8 has amino acid sequence identity with other aquaporins (~35%) and highest with a plant water channel, AQP-TIP (39%), suggesting that AQP8 is a unique member in mammalian aquaporins. While the AQP8 protein of a 25-kDa is absent from epididymis, AQP7 appeared as a 23-24 kDa band both in testis and epididymis. The AQP8 may be involved in the cytoplasmic condensation occurring during differentiation of spermatids into spermatozoa and in the generation of seminiferous tubule fluid.

21.9.3 Aquaporin 9

AQP9 is permeable to water and to some solutes. In the male reproductive tract, AQP9 is enriched on the apical (but not basolateral) membrane of nonciliated cells in the efferent duct and principal cells of the epididymal (rat and human) and vas deferens, where it could play a role in fluid reabsorption. A 30- kDa protein has been isolated from brush-border membrane vesicles from the epididymis and from epithelial cells of the prostate and coagulating gland, where fluid transport across the epithelium is important for secretory activity. It shows that AQP9 is a constitutively inserted apical membrane protein whose cell surface expression is not acutely regulated by vesicular trafficking. It seems that AQP9 can provide a route via which apical fluid and solute transport occurs in several regions of the male reproductive tract (Pastor-Saler et al, 2001).

21.10. REFERENCES

Arnoult C, Kazam IG, Visconti PE, Kopf GS, Florman HM. Control of the low voltage-activated calcium channel of mouse sperm by egg ZP3 and by membrane hyperpolarization during capacitation Proc Nat Acad Sci USA 1999; 96: 6757-62.

Arnoult C, Lemos JR, Florman HM. Voltage-dependent modulation of T-type calcium channels by protein tyrosine phosphorylation. EMBO J 1997; 16: 1593-9.

Arnoult C, Zeng Y, Florman HM. ZP3-dependent activation of sperm cation channels regulates acrosomal secretion during mammalian fertilization. J Cell Biol 1996; 134: 637-45.

Babcock DF, Bosma MM, Battaglia DE, Darszon A. Early persistent activation of sperm K^+ channels by the egg peptide speract. Proc Natl Acad Sci USA 1992; 89: 6001-5.

Baldi E, Luconi M, Bonaccorsi L, Muratori M, Forti G. Intracellular events and signaling pathways involved in sperm acquisition of fertilizing capacity and acrosome reaction. Front Biosci 2000; 5: E110-23.

Benoff S. Receptors and channels regulating acrosome reactions. Human Fertil (Camb) 1999; 2: 42-55.

Blackmore PF, Eisoldt S. The neoglycoprotein mannose bovine serum albumin but not progesterone activates T type calcium channels in human spermatozoa. Mol Hum Reprod 1999; 5: 498-506.

Bosl MR, Stein V, Hubner C, Zdebik AA, Jordt SE, Mukhopadhyay AK, Davidoff MS, Holstein AF, Jentsch TJ. Male germ cells and photoreceptors, both dependent on close cell-cell interactions, degenerate upon ClC-2 Cl^- channel disruption. EMBO J 2001: 20: 1289-99.

Brandelli A, Miranda PV, Tenzon JG. Voltage-dependent calcium channels and Gi regulatory protein mediate the human sperm acrosomal exocytosis induced by N-acetylglucosaminyl/mannosyl neoglycoproteins. J Androl 1996; 17: 522-9.

Calamita G, Mazzone A, Cho YS, Valenti G, Svelto M. Expression and localization of the aquaporin-8 water channel in rat testis. Biol Reprod 2001; 64: 1660-6.

Chan HC, Zhou TS, Fu WO, Wang WP, Shi YL, Wong PY. Cation and anion channels in rat and human spermatozoa. Biochim Biophys Acta 1997; 1323: 117-29.

Cook SP, Babcock DF. Selective modulation by cGMP of the K^+ channel activated by speract. J Biol Chem 1993; 268: 22402-7.

Cordoba M, Beconi MT. Progesterone effect mediated by the voltage-dependent calcium channel and protein kinase C on noncapacitated cryopreserved bovine spermatozoa. Andrology 2001; 33: 105-12.

Cox T, Campbell P, Peterson RN. Ion channels in boar sperm plasma membranes: characterization of a cation selective channel. Mol Reprod Dev 1991; 30: 135-47.

Curry MR, Millar JD, Watson PF. The presence of water channel proteins in ram and human sperm membranes. J Reprod Fertil 1995; 104: 297-303.

Darboux I, Lingueglia E, Champign Y, Coscoy S, Barbry P, Lazdunski M. dGNaC1, a gonad specific amiloride-sensitive Na+ channel. J Biol Chem 1998;273: 9424-29.

Darszon A, Labarca P, Nishigaki T, Espinosa F. Ion channels in sperm physiology. Physiol Rev 1999; 79: 481-510.

Dibas AI, Mia AJ, Yorio T. Aquaporins (water channels) role in vasopressin-activated water transport. Proc Soc Exp. Biol Med. 1998; 219: 183-99.

Dolphin AC. Mechanisms of modulation of voltage-dependent calcium channels by G proteins. J Physiol (London) 1998; 506: 3-11.

Dorval V, Dufour M, Leclerc P. Role of protein tyrosine phosphorylation in the thapsigargin-induced intracellular Ca(2+) store depletion during human sperm acrosome reaction. Mol Hum Reprod. 2003;9:125-31.

Espinosa F, de La Vega-Beltran JL, Lopez-Gonzalez I et al. Mouse sperm patch clamp recording reveal single CL^- channels sensitive to nifumic acid, a blocker of sperm acrosome reaction. FEBS Lett 1998; 426:47-51.

Espinosa F, Lopez-Gonzalez I, Munoz-Garay C, Felix R, De le Vega-Beltran JL, Kopf GS, Visconti PE, Darszon A. Dual regulation of the T-type Ca^{2+} current by serum albumin and β-estradiol in mammalian spermatogenic cells. FEBS Lett 2000; 475: 251-6.

Espinosa F, Lopez-Gonzalez I, Serrano CJ, Gasque G, de la Vega-Beltran JL, Trevino CL, Darszon A. Anion channel blockers differently affect T-type Ca^{2+} currents of mouse spermatogenic cells, α1E currents expressed in Xenopus oocytes and the sperm acrosome reaction. Dev Genet 1999; 25: 103-14.

Flesch FM, Gadella BM. Dynamics of the mammalian sperm plasma membrane in the process of fertilization. Biochim Biophys Acta 2000; 1469: 197-235.

Florman HM, Corron ME, Kim TD, Babcock DF. Activation of voltage-dependent calcium channels of mammalian sperm is required for zona pellucida – induced acrosomal exocytosis. Dev Biol 1992; 152: 304-14.

Florman MH, Arnoult C, Kazam IG, Li C, O' Toole CMB. A perspective on the control of mammalian

fertilization by egg-activated Ion channels in sperm: a tale of two channels. Biol Reprod 1998; 59: 12-16.
Foresta C, Rossato M, Chiozzi P, Dio Virgilio F. Mechanism of human sperm activation by extracellular ATP. Am J Physiol 1996; 270: (6ptl): 1709-14.
Galindo BE, Beltran C, Cragoe EJ Jr, Darszon A. Participation of a K^+ channel modulated directly by cGMP in the speract-induced signaling cascade of stronglylocentrotus purpuratus sea urchin sperm. Dev Biol 2000; 221: 285-94.
Gao Z, Ruden DM, Lu X. PKD2 cation channel is required for directional sperm movement and male fertility. Curr Biol. 2003;13:2175-8.
Garcia MA, Meizel S. Regulation of intracellular pH in capacitated human spermatozoa by a Na^+/H^+ exchanger. Mol Reprod Dev 1999; 52: 189-95.
Gauss R, Seifert R, Kaupp UB. Molecualr identification of a hyperpolarization – activated channel in sea urchin sperm. Nature 1998; 393 (6685): 583-7.
Gonzalez-Martinez MT, Galindo BE, De la Torre L, Guerrero A, Morales E, Zapata O, Rodriguez E, Florman HM, Darszon A. A sustained increase intracellular Ca^{2-} is required for the acrosome reaction in sea urchin sperm Dev Bio 1997; 236: 193-202.
Goodwin LO, Karabinus DS, Pergolizzi RG, Benoff S. L-type voltage-dependent calcium channel α1C subunit mRNA is present in ejaculated human spermatozoa. Mol Hum Reprod 2000; 6: 127-36.
Goodwin LO, Leeds NB, Guzowski D, Hurley IR, Pergolizzi RG, Benoff S. Identification of structural elements of the testis-specific voltage dependent calcium channel that potentially regulate its biophysical properties. Mol Hum Reprod 1999; 5: 311-22.
Goodwin LO, Leeds NB, Hurley I, Cooper GW, Pergolizzi RG, Benoff S. Alternative splicing of exons in the α1 subunit of the rat testis L-type voltage dependent calcium channel generates germ line-specific dihydropyridine binding sites. Mol Hum Reprod 1998; 4: 215-26.
Goodwin LO, Leeds NB, Hurley I, Mandel FS, Pergolizzi RG, Benoff S. Isolation and characterization of the primary structure of testis-specific L-type calcium channel: implications for contraception. Mol Hum Reprod 1997; 3: 255-68.
Hinsch KD, Asmarinah, Hinsch E, Konrad L. VDAC2 (porin – 2) expression pattern and localization in the bovine testis. Biochim Biophys Acta 2001; 1518: 329-33.
Hinsch KD, De Pinto V, Aires VA, Schneider X, Messina A, Hinsch E. Voltage-dependent anion-selective channels VDAC2 and VDAC3 are abundant proteins in bovine outer dense fibers, a cytoskeletal component of the sperm flagellum. J Biol Chem 2004 Jan 22
Hirohashi ND, Vacquier VD. High molecular mass egg fucose sulfate polymer is required for opening both Ca^{2+} channels involved in triggering the sea urchin sperm acrosome reaction. J Biol Chem 2001; Nov. 7.
Ho HC, Suarez SS. An inositol 1,4,5-trisphophate receptor-gated intracellular Ca^{2+} store is involved in regulating sperm hyperactivated motility. Biol Reprod 2001; 65: 606-15.
Ishibashi K, Kuwahara M, Gu Y, Kageyama Y, Tohsaka A, Suzuki F, Marumo F, Sasaki S. Cloning and functional expression of a new water channel abundantly expressed in the testis permeable to water, glycerol, and urea. J Biol Chem 1997a; 272: 20782-86.
Ishibashi K, Kuwahara M, Kageyama Y, Tohsaka A, Marumo F, Sasaki S. Cloning and functional expression of a second new aquaporin abundantly expressed in testis. Biochem Biophys Res Commun 1997b; 237: 714-18.
Ishibashi K, Marumo F. Molecular cloning of a DEG/ENaC sodium channel cDNA from human testis. Biochem Biophy Res Comm 1998; 245: 589-93.
Jacob A, Benoff S. Full length low voltage-activated ('T-type') calcium (Ca^{2+}) channel α1G mRNA is not detected in mammalian testis and sperm. J Androl 2000a (Supp); 56: 48.
Jacob A, Hurley IR, Goodwin LO, Cooper GW, Benoff S. Molecular characterization of a voltage-gated potassium channel expressed in rat testis. Mol Hum Reprod 2000b; 6: 303-13.
Jagannathan S, Barratt CLR, Publicover J. Characterization of α1H, a T-type calcium ion channel from human germ cells. Euro J Neurosci 2000; 12: 383.
Jagannathan S, Publicover SJ, Barratt CLR. Voltage-operated calcium channels in male germ cells. Reproduction 2002; 123: 203-15.
Jungnickel MK, Marrero H, Birmbaumer L, Lemos JR, Florman HM. Trp2 regulated Ca^{2+} entry into mouse sperm triggered by egg ZP3. Nature Cell Biol 2001; 3: 499-502.
Kageyama Y, Ishibashi K, Hayashi T, Xia G, Sasaki S, Kihara K. Expression of aquaporins 7 and 8 in the developing rat testis. Andrologia 2001; 33: 165-9.
Krasznai Z, Marian T, Izumi H, Damjanovich S, Balkay L, Tron L, Morisawa M. Membrane hyperpolarization removes inactivation of Ca^{2+} channels, leading to Ca^{2+} influx and subsequent initiation of sperm motility in the common carp. Proc Natl Acad Sci USA 2000; 97: 2052-7.
Labarca P, Santi C, Zapata O, Morales E, Beltran C, Li'evano A, Darszon A. A cAMP regulated K^+ selective

channel from the sea urchin sperm plasma membrane. Dev Biol 1996; 174: 271-80.
Lievano A, Santi CM, Serrano CJ, Trevino CL, Bellve AR, Hernandez-Cruz A. Darszon A. T-type Ca^{2+} channels and α1E expression in spermatogenic cells and their possible releance to the sperm acrosome reaction. FEBS Lett 1996; 388; 150-4.
Linares-Hernandez L, Guzman-Grenfell AM, Hicks-Gomez JJ, Gonzalez-Martinez M. Voltage dependent calcium influx in human sperm assessed by simultaneous optical detection of intracellular calcium and membrane potential. Biochim Biophys Acta 1998; 1372: 1-12.
Liu C, Goa D, Pretson GM, McGann LE, Benson CT, Citser ES, Citser JK. High water permeability of human spermatozoa is mercury resistant and not mediated by CHIP28. Biol Reprod 1995; 52: 913-19.
Loeser CR, Lynch C 2nd, Tulsiani DR. Characterization of the pharmacological sensitivity profile of neoglycoprotein-induced acrosome reaction in mouse spermatozoa. Biol Reprod 1999; 61: 629-34.
Lopez-Gonzalez I, De La Vega-Beltran JL, Santi CM, Florman HM, Felix R, Darszon A. Calmodulin antagonists inhibit T-type Ca^{2+} currents in mouse spermatogenic cells and the zona pellucida induced sperm acrosome reaction. Dev Biol 2001; 236: 210-9.
Meizel S, Turner KO. Chloride efflux during the progesterone-initiated human sperm acrosome reaction is inhibited by lavendustin A, tyrosine kinase inhibitor. J Androl 1996; 17: 327-30.
Meizel S. Amino acid neurotransmitter receptor/chloride channels of mammalian sperm and the acrosome reaction. Biol Reprod 1997; 56: 569-74.
Melendrez CS, Meizal S. Immunochemical identification of the glycine receptor/channel in porcine sperm. Biochem Biophys Res Common 1996; 223: 675-8.
Melendrez CS, Meizel S. Studies of porcine and human sperm suggesting a role for a sperm glycine receptor/ Cl- channel in the zona pellucida – initiated acrosome reaction. Biol Reprod 1995; 53: 676-83.
Munoz-Garay C, De la Vega-Beltran JL, Delgado R, Labarca P, Felix R, Darszon A. Inwardly rectifying K^+ channels in spermatogenic cells: functional expression and implication in sperm capacitation. Dev Biol 2001; 234: 261-74.
O'Toole CM, Arnoult C, Darszon A, Steinhardt RA, Florman HM. Ca^{2+} entry through store-operated channels in mouse sperm is initiated by egg ZP3 and drives the acrosome reaction. Mol Biol Cell 2000; 11: 571-84.
O'Toole CM, Roldan ER, Fraser LR. Role for Ca^{2+} channels in the signal transduction pathway leading to acrosomal exocytosis in human spermatozoa. Mol Reprod Dev 1996; 45: 204-11.
Pastor-Saler N, Bagnis C, Sabolic I, Tyskowski R, Mckee M, Van Hoek A, Breton S, Brown D. Aquaporin 9 expression along the male reproductive tract. Biol Reprod 2001; 65: 384-93.
Patrat C, Serres C, Jouannet P. Induction of a sodium ion influx by progresterone in human spermatozoa. Biol Reprod 2000; 62: 1380-6.
Petrunkina AM, Harrison RA, Hebel M, Weitze KF, Topfer Petersen E. Role of quinine sensitive ion channels in volume regulation in boar and bull spermatozoa. Reproduction 2001; 122: 327-36.
Publicover SJ, Barratt CL. Voltage-operated Ca^{2+} channels and the acrosome reaction: which channels are present and what do they do? Human Reprod 1999; 14: 873-9.
Quill TA, Ren D, Clapham DE, Garbers DL. A voltage-gated ion channel expressed specifically in spermatozoa. Proc Natl Acad Sci USA 2001; 98: 12527-31.
Quill TA, Sugden SA, Rossi KL, Doolittle LK, Hammer RE, Garbers DL. Hyperactivated sperm motility driven by CatSper2 is required for fertilization. 1: Proc Natl Acad Sci USA 2003 D; 100: 14869-74.
Ren D, Navarro B, Perez G, Jackson AC, Hsu S, Shi Q, Tilly JL, Clapham DE. A sperm ion channel required for sperm motility and male fertility. Nature 2001; 413(6856): 581-2.
Rossato M, Di Virgilio F, Rizzuto R, Galeazzi C, Foresta C. Intracellular calcium store depletion and acrosome reaction in human spermatozoa role of calcium and plasma membrane potential. Mol Hum Reprod 2001; 7: 119-28.
Sakata Y, Saegusa H, Zong S, Osanai M, Murakoshi T, Shimizu Y, Noda T, Aso T, Tanaba T. Analysis of Ca^{2+} currents in spermatocytes from mice lacking Ca(v) 2.3 (α(1E)Ca^{2+} channel. Biochem Biophys Res Commun 2001; 288: 1032-6.
Sakkas D, Leppens-Luisier G, Lucas H, Chardonnens D, Campana A, Franken DR, Urner F. Localization of tyrosine phosphorylated proteins in human sperm and relation to capacitation and zona pellucida binding. Biol Reprod. 2003;68:1463-9.
Salvatore L, D'Adamo MC, Polishchuk R, Salmona M, Pessia M. Localization and age-dependent expression of the inward rectifier K^+ channel subunit Kir 5.1 in a mammalian reproductive system. FEBS Lett 1999; 449: 146-52.
Sampson MJ, Decker WK, Beaudet AL, Ruitenbeek W, Armstrong D, Hicks MJ, Craigen WJ. Immotile sperm and infertility in mice lacking mitochondrial voltage-dependent anion channel type 3. J Biol Chem 2001; 276: 39206-12.
Sanchez D, Labarca P, Darszon A. Sea urchin sperm cation-selective channels directly modulated by cAMP.

FEBS Lett 2001; 503: 111-5.
Santi CM, Darszon A, Hernandez-Cruz A. A dihydropyridine-sensitive T-type Ca2+ current is the main Ca^{2+} current carrier in mouse primary spermatocytes. Am J Physiol 1996; 271(5pt1): 1583-93.
Santi CM, Santos T, Hernandez-Cruz A, Darszon A. Properties of a novel pH-dependent Ca^{2+} permeation pathway present in male germ cells with possible roles in spermatogenesis and mature sperm function. J Gen Physiol 1998; 112: 33-53.
Sato Y, Son JH, Meizel S. The mouse sperm glycine receptor/chloride channel: cellular localization and involvement in the acrosome reaction initiated by glycine. J Androl 2000a; 21: 99-106.
Sato Y, Son JH, Tucker RP, Meizel S. The zona pellucida – initiated acrosome reaction: defect due to mutations in the sperm glycine receptor / CL^- channel. Dev Biol 2000b; 227: 211-8.
Schreiber M, Wei A, Yuan A, Gaut J, Salto M, Salkoff L. SLo3, a novel pH-sensitive K^+ channel from mammalian spermatocytes. J Biol Chem 1998; 273: 3509-16.
Senuma M, Yamano S, Nakagawa K, Irahara M, Kamada M, Aono T. Progesterone accelerates the onset of capacitation in mouse sperm via T-type calcium channels. Arch Androl 2001; 47: 127-34.
Serrano CJ, Trevino CL, Felix R, Darszon A. Voltage-dependent Ca^{2+} channel subunit expression and immunolocalization in mouse spermatogenic cells and sperm. FEBS Lett 1999; 462: 171-6.
Son WY, Lee JH, Lee JH, Han CT. Acrosome reaction of human spermatozoa is mainly mediated by α1H T-type calcium channels. Mol Hum Reprod 2000; 6: 893-7.
Stevens AL, Breton S, Gustafson CE, Bouley R, Nelson RD, Kohan DE, Brown D. Aquaporin 2 is a vasopresin-independent, constitutive apical membrane protein in rat vas deferens. Am J Physiol Cell Physiol 2000; 278: 791-802.
Suzuki-Toyota F, Ishibashi K, Yuasa S. Immunohistochemical localization of a water channel, aquaporin 7 (AQP7), in the rat testis. Cell Tissue Res 1999; 295; 279-85.
Tiwari-Woodruff SK, Cox TC. Boar sperm plasma membrane Ca^{2+} selective channels in planer lipid bilayers. Am J Physiol 1995; 268 (5ptI1): 1284-94.
Tomes CN, Roggero CM, De Blas G, Saling PM, Mayorga LS. Requirement of protein tyrosine kinase and phosphatase activities for human sperm exocytosis. Dev Biol. 2004;265:399-415.
Visconti PE, Kopf GS. Regulation of protein phosphorylation during sperm capacitation. Biol Reprod 1998; 59: 1-6.
Waldmann R, Champigny G, Bassilana F, Voilley N, Lazdunski M. J Biol Chem 1995; 270: 27411-14.
Wennemuth G, Westenbroek RE, Xu T, Hille B, Babcock DF. CaV2.2 and CaV2.3 (N-and R-type) Ca^{2+} channels in depolarization-evoked entry of Ca^{2+} into mouse sperm. J Biol Chem 2000; 275: 21210-7.
Westenbroek RE, Babcock DF. Discrete regional distributions suggest diverse functional roles of calcium channel α1 subunits in sperm. Dev Biol 1999; 15: 457-69.
Weyand I, Godde M, Frings S, Weiner J, Muller F, Altenhofen W, Hatt H, Kaupp UB. Cloning and functional expression of a cyclic-nucleotide gated channel from mammalian sperm. Nature 1994; 368(6474); 859-63.
Wiesner B, Weiner J, Middendorff R, Hagen V, Kaupp UB, Weyand I. Cyclic nucleotide-gated channels on the flagellum control Ca^{2+} entry into sperm. J Cell Biol 1998; 142: 473-84.
Wistrom CA, Meizel S. Evidence suggesting involvement of a unique human sperm steroid receptor/Cl-channel complex in the progesterone-initiated acrosome reaction. Dev Biol 1993; 159: 679-90.
Wu WL, So SC, Sun YP, Chung YW, Grima J, Wong PYD, Yan YC, Chan HC. Functional expression of P_{2U} receptors in rat spermatogenic cells: dual modulation of a Ca^{2+}-Activated K^+ channel. Biochem Biophys Res Commun 1998a; 248: 728-32.
Wu WL, So SC, Sun YP, Zhou TS, Yu Y, Chung YW, Wang XF, Bao YD, Yan YC, Chan HC. Functional expression of a Ca^{2+} activated K^+ channel in Xenopus oocytes injected with RNAs from the rat testis. Biochim Biophys Acta 1998b; 1373: 360-5.
Yeung CH, Cooper TG. Effects of the ion-channel blocker quinine on human sperm volume, kinematics and mucus penetration, and the involvement of potassium channels. Mol Hum Reprod 2001; 7: 819-28.
Zeng Y, Clark EN, Floramn HM. Sperm membrane potential hyperpolarization during capacitation regulates zona pellucida dependent acrosomal secretion. Dev Biol 1995; 171: 554-63.

Chapter 22

ACTION OF PHOSPHOLIPASES

22.1. INOSITOL TRIPHOSPHATE MEDIATED SIGNAL TRANSDUCTION

In response to extra- or intracellular signals, a number of phospholipases (PLS) including phosphoinositide specific PLC, phosphatidyl choline specific PLC, PLD, phosphatidic acid (PA) phosphohydrolase, PLA2 and sphingomyelinase can be activated. These enzymes produce second messenger that functions in the regulation of cell activity. Phosphatidic acid (PA) is one messenger that has received increasing attention. The PA is produced by PLD catalyzed degradation of phosphatidylinositol-4,5-biphosphate [PI(4,5)P2] or [PIP2] to diacylglycerol (DAG) and subsequent diacylglycerol kinase catalyzed phosphorylation of diacylglycerol. Some external signals stimulate cells through G-linked receptors, which activate inositol-phospholipid signaling pathway. In this action G proteins activate plasma membrane bound phopholipase C-β (PLCβ), which stimulates the turn over of a membrane bound lipid and change in intracellular concentration of Ca^{2+}. The PLC acts on PI(4,5)P2, which is a minor component of plasma membrane bilayer. The receptor that operates through PI(4,5)P2 pathway mainly activates G_q form of G protein, which regulates PLC in same way as G_s activates adenylyl cyclase (Chapter 18). The activated phospholipase C cleaves PI(4,5)P2 to form two products: inositol-1,4,5-triphosphate (InsP3/IP3) and diacylgycerol. The InsP3 and DAG produced by PI(4,5)P2 following PLC action take two different routes for cell signaling **(Fig.22.1)**.

In 1980s, the activation of receptors, linked through hydrolysis of phosphatidyl inositide and mobilization of Ca^{2+} was demonstrated in many tissues including in germ cells, in response to many stimuli. Inositol-1,4,5- trisphosphate is a water-soluble molecule that leaves the plasma membrane and diffuses rapidly though the cytosol. On reaching the endoplasmic reticulum (ER), it binds to and opens InsP3-gated Ca^{2+} release channels in the ER membranes and quickly mobilizes and raises the concentration of Ca^{2+} by 10 to 20 fold in the cytosol. The Ca^{2+} acts as a ubiquitous intracellular messenger to propagate the signal further depending on the cell function. The initial Ca^{2+} response can be terminated by one or more mechanisms. For example InsP3 can be rapidly dephosphorylated to form IP_2 or phosphorylated to IP_4 (which may function as another intracellular mediator). Alternatively, the Ca^{2+} that enters the cytosol can also be rapidly pumped out of the cell to terminate the Ca^{2+} signal. At the same time the $InsP_3$ produced by the hydrolysis of $PI(4,5)P_2$ also raises the concentration of Ca^{2+} in the cytosol. The other cleavage product of PI(4,5)P2- diacylglycerol- embedded in the membrane, has two potential signaling roles, such as, it can be further cleaved to release arachidonic acid, which can act as a messenger by itself. An important function of diacylglycerol is to activate a serine/threonine

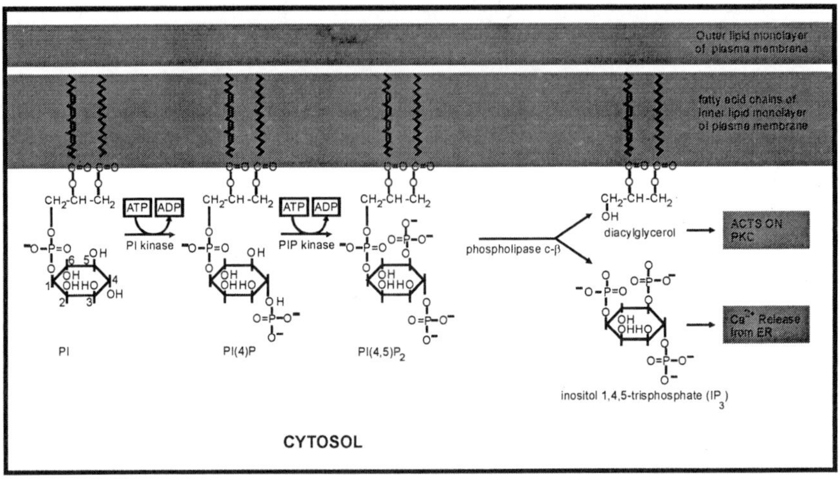

Fig.22.1. Activation of phospholipase C (PLC) path way and production of PI(4) phosphate [PI(4)P] and PI(4,5) biphosphate [PI(4,5)P2] by phosphorylation of phosphatidylinositol (PI) and PI(4)P respectively. During signaling response, PI(4,5)P2, present in plasma membrane generates two intracellular mediators by hydrolytic action of phospholipase Cβ (PLCβ): IP3 and diacylglycerol. During cell signaling, PLCβ is activated by G protein linked receptors. Other types of PLCs are also activated by other receptor mechanisms as is seen in sperm acrosome reaction and sperm capacitation. Among two intracellular mediators generated by PLC, IP3 diffuses through cytosol and releases Ca^{2+} from endoplasmic reticulum, while diacylglycerol activates protein kinase C.

protein kinase C (PKC), which is activated by the combined action of initial rise of Ca^{2+}, diacyl glycerol, and negatively charged membrane phosphatidylserine. Activated PKC phosphorylates specific proteins, which are required for cell functions. The two signaling pathways of inositol phospholipid can be mimicked by pharmacological agents such as Ca^{2+} ionophore or by phorbol esters. Many of the effector mechanisms involving phosphatidic acid have been noted in sperm functioning and spermatogenesis.

22.2. INOSITOL TRIPHOSPHATE RECEPTORS IN GERM CELLS

Many extracellular signals induce an increase in cytosolic Ca^{2+} level, not just those that work via G proteins. For example, sperm during fertilization of egg suddenly triggers rise of cytosolic Ca^{2+} in the egg that results into embryonic development. The binding of several hormones to the cell surface receptors lead to the activation of phospholipase C that generates the InsP3. The intracellular receptor for InsP3 is a calcium channel, which permits efflux of Ca^{2+} sequestered in intracellular stores, out into the cytoplasm. The importance and complexity of change in Ca^{2+} concentration in transmitting information implies that the calcium signaling is strictly regulated. The inositol-1,4,5-triphosphate receptor (IP3R) occupies a central position among the multitude of molecules involved in intracellular calcium metabolism and possesses regulatory properties that may account for the spatio-temporal complexity of calcium signals.

During spermatogenesis, the activity of $[Ca^{2+}]_i$ release channels play an important role in different specific cellular functions. Accordingly, mRNAs for the three IP3R subtypes are found throughout spermatogenesis, showing distinct distribution patterns of the mature IP3Rs during sperm differentiation. Multiple forms of IP3 receptors reported are generated either by alternative splicing or by the expression of multiple encoding IP3 receptors. These receptors have been also reported in mammalian sperm cells (Alba et al., 2000; Walensky and Snyder, 1995) where they are functionally associated with acrosome. At early stages, IP3Rs are distributed throughout the cytoplasm, and as differentiation proceeds they become selectively localized to the Golgi complex. The distribution of IP3Rs and the larger Ca^{2+} release responses suggested that IP3Rs may be involved in proliferation of spermatogonia. Characterization of bovine IP3R has shown that the functional size of the IP3R binding domain is a protein of 66-kDa in agreement with other IP3R receptors. In contrast, bovine cerebellum IP3Rs have one affinity state and a relatively high density. They are functional and release internal calcium upon the binding of the second messenger. The A1 adenosine receptor agonist R-PIA elicits almost the same effect as InsP3, which might be of some help in understanding the physiological role of these inhibitory adenosine receptors in mammalian spermatozoa (Minneli et al., 2000). The three genes encoding ryanodine receptor proteins (RyRs) are expressed at all stages of spermatogenesis. However, immunocytochemical studies detected types 1 and 3 in spermatogenic cells and only type 3 in mature sperm. In contrast to IP3Rs, RyRs remain scattered in the cytoplasm throughout differentiation (Trevino et al, 1998).

22. 3. PHOSPHOLIPASES

Phospholipases (PLs) that catalyze the hydrolysis of membrane phospholipids, are classified according to the bond cleaved in a phospholipid into: PLA1 (EC 3.1.1.3), PLA2 (EC 3.1.1.4), PLB (EC 3.1.1.5), PLC (EC 3.1.4.3), and PLD (EC 3.1.4.4). The PLA2 and PLC are involved in several biological phenomena, such as, signal transduction, photoreception, biosynthesis of lung surfactant, sperm motility, fertilization and many inflammatory reactions.

22.3.1. Phospholipase A2

Phospholipases A2 (PLA2) are ubiquitous enzymes that are present in most cells and tissues. Hydrolysis of cellular phospholipids by PLA2s causes release of free fatty acids with concomitant formation of lysophospholipids. The newly released free fatty acids may affect cellular functions, since they activate protein kinase and MAP kinase, and the concomitantly generated lysophospholipids may cause damage to the cellular membranes. The products of PLA2 catalyzed phospholipids hydrolysis may include arachidonic acid and lyso-PAF, which are the precursors of a wide spectrum of pro-inflammatory mediators (including prostaglandins, thromboxans, leukotrienes and PAF). Notably, cytosolic (c) PLA2 becomes catalytically active in the presence of free Ca^{2+} at concentrations present in stimulated cells and preferentially cleaves arachidonic acid-containing phospholipids. A variety of agonists, growth factors and cytokines, as well as stressful stimuli activate cPLA2 to hydrolyze cellular phospholipids thereby liberating fatty acids and lysophospholipids and providing the precursor substrates for the biosynthesis of eicosanoids and platelet-activating factor. Once formed these lipid mediators may act as intracellular messengers acting on protein targets in the cell where they have been formed or, alternatively, they may leave the cell surface of the parent and/or neighbouring cells. The release of arachidonic acid, formation of the intermediate metabolite (endoperoxide or hydroperoxide) and conversion to the final eicosanoid can take place in two saparate cell types. Many of the newly generated lipids derived from products of PLA2 action have profound biological properties. They promote inflammatory reactions and participate in processes that lead to tissue injury. PLA2s are therefore attractive targets for the development of inhibitors that may be useful novel therapeutic agent for the treatment of inflammatory and degenerative diseases. Using functional criteria for classification, one can distinguish at least four different subfamilies of PLA2 enzymes. First, secretory PLA2s (sPLA2s) have a molecular mass of 14- kDa and characterized by a catalytic requirement for Ca^{2+} and a rigid three dimensional structure maintained by disulphide bridges. Second, an 85-kDa Ca^{2+}-sensitive cytosolic PLA2 (cPLA2) has been purified and characterized. Third group comprises iPLA2s with molecular masses ranging from 29- to 85-kDa. Fourth, the important features that differentiate the PAF acetylhydrolases (PAF-Ahs) from the above described PLA2s is their remarkable specificity for short and/or oxidized acyl groups at the sn-2 position of phospholipids and their Ca^{2+} independence. The cPLA2 is the only known PLA2 that exhibits functional properties indicative of a receptor-regulated PLA2 and is thus likely to be involved in receptor-mediated eicosanoid production and intracellular signal transduction. A wide range of extra-cellular stimuli activate cPLA2. These stimuli induce growth factors, mitogens, vasoactive peptides, cytokines and interferons. However, non-receptor mediated stimulation of cells by stressful stimuli, including oxidation, hyperglycemia, UV light and shear stress, may also promote cPLA2 activation. A wide variety of extra-cellular stimuli were found to cause rapid phosphorylation of cPLA2 suggesting that phosphorylation may also play an important role in the regulation of cPLA2 activation. Studies with cultured cells and in-vitro kinase assays showed that cPLA2 is a substrate for the p42 MPA kinase (also referred as ERK2) (Chapter 19). Many subsequent reports have proposed that phosphorylation of cPLA2 by ERK kinases is a critical step in the sequence of events leading to the mobilization of arachidonic acid in stimulated cells. Many different cell types contain cPLA2.

The cPLA2 macrophage cDNA comprised a total of 2880 nt, including about 200 nt for the 5'-untranslated region and about 500nt for the 3'-untranslated region. While the complete chromosomal DNA for cPLA2 has not been sequenced, there appear to be at least 7 introns, and some features of the genetic control mechanisms of the cPLA2 gene were revealed by analysis of its 5'-flanking sequence. It was suggested that there may be coordinate regulation of cPLA2 prostaglandin synthase 2. Upstream of the 5'-untranslated segment of the cDNA cloned earlier, a 5' exon is separated by an intron of about 6kb. The 5' flanking region or the first exon contains exact or modified copies of sequence elements that may contribute to transcriptional regulation. There are no perfect TATA or CCAAT elements, but close homologues have been noted, along with a homologue of the Sp1 element. In addition there are consensus site for NFκβ, NF-IL-6, AP-1, AP-2, PEA3, OCT, C/EBP and GRE. The cPLA2 cDNA encodes a 749 amino acid protein with a predicted molecular mass of 85.2-kDa. The inferred sequence of murine cPLA2 is more than 95% homologous to the human cPLA2 sequence indicating great structural similarity between cPLA2 from different species (Review-Kramer and Sharp, 1997). A low molecular weight PLA2 mRNA has been identified in mouse testis (PLA2 Group IIC). This protein is specific to cells undergoing meiosis. A single 1.6 kb transcript is present only in testis. This gene is transcribed mainly in pachytene spermatocytes, secondary spermatocytes and round spermatids. Expression of the gene is seen at all stages of seminiferous epithelium, especially in stages VI-VII (Chen et al., 1997). Presence of germ cell specific PLA in testis warrant further investigations.

22.3.2. PLA2 in Sperm

Characteristics: Calcium-dependent phospholipase A2 of mouse sperm shows maximum activity at pH 8.0 and almost totally membrane-associated. The activity is stimulated during the ionophore-induced acrosome reaction and is equally distributed between plasma/outer acrosomal and inner acrosomal membrane fractions. The membrane-associated PLA2 had an absolute requirement for low concentrations of Ca^{2+}; Sr^{2+}, Mg^{2+}; other divalent and monovalent cations can not substitute for Ca^{2+}. Mouse sperm did not show any activity of PLA1 (Thakkar et al, 1983). Human sperm phospholipase A is optimally active at pH 7.5, calcium-dependent, and exclusively catalyzed the release of fatty acid from the 2-position of phospholipids. Hydrolysis of dipalmitoyl phosphatidylcholine was specific to the sn-2 position by human spermatozoa PLA. Phospholipase A2 activity is inhibited by Zn^{2+} and Mn^{2+} (Thakkar et al, 1984). Human sperm phospholipase A2 has an estimated molecular mass of 16.7 kDa with N-terminal sequence ofYNYQFGLMIVITKGHFAMV. Glutamine-4, phenylalanine-5, methionine-8, and isoleucine-9 appeared to be highly conserved throughout evolution (Langlais et al, 1992). PLA2 activity was inhibited in a dose-dependent manner by an oligomer of prostaglandin in vitro and in situ. Sperm phospholipase A2 and its modulators may contribute to membrane fusion events in mammalian fertilization (Fry et al, 1992).

In human spermatozoa, PLA2 exists as a zymogen (Antaki et al, 1989). Precursor form of PLA2 can be activated by endogenous proteases (acrosin) as well as exogenous proteases of seminal plasma and follicular fluid (Plasmin, Kallikrein). The interrelationship of proteases and pro-phospholipase A2 could activate a dormant fusogenic system: the resulting effect would lead to membrane fusion by lysolipids, key components in the acrosome reaction (Guerette et al 1988). Oxidative reactions of cis-unsaturated fatty acids relieve their natural inhibitory activity, and polymerization of an inactive fatty acid metabolite yields a potent inhibitor of phospholipase A2 activity in vitro and in situ (Franson et al, 1990). Fresh spermatozoa possess more than one form of PLA2 activity as judged by the biphasic nature of the curve obtained during enzyme inactivation. The behaviour of PLA2, during the effect of temperature and in radiation inactivation experiments indicated that the low molecular weight component in the seminal plasma as well as in spermatozoa is temperature resistant. However, in fresh spermatozoa, a second form of PLA2 was sensitive to changes in temperature (Antaki et al, 1988). Cold shock disrupts sperm on three levels; membrane molecular organization, intracellular Ca^{2+} regulation, and gross morphology/motility. Two molecular forms of PLA2 were also kinetically distinguishable, one with an apparent Michaelis constant of 3.0μM and the other with Km of 630μM. Both forms of the enzyme were Ca^{2+} dependent and heat stable; however, the low-Km activity was less resistant to 60°C (Anderson et al, 1990).

In bull sperm, PLA2 is having overlapping distribution with calmodulin. Bull seminal plasma contains a Ca^{2+} independent (form 1) and Ca^{2+} dependent PLA2. Both enzymes bind to spermatozoa (Ronkko et al, 1991). Bovine seminal PLA2 has a Mr of 100 kDa, identified by gel filtration and subunit of 60-kDa band. The enzyme was acid-labile and did not display affinity for heparin (Soubeyrand et al, 1997a). It is suggested that sperm PLA2 and one of its modulators, the LPC, may contribute to membrane-fusion events in mammalian fertilization (Riffo and Parraga, 1997). The BSP proteins may act as spermatozoa stabilizing agents by preventing premature lipolysis of the sperm surface (Soubeyrand and Manjunath, 1997b). In hamster sperm, a PLA2 is preferentially located in the acrosome and mitochondria. On the other hand, the presence of another PLA2 in the plasma membrane covering the acrosome was also suggested (Riffo et al, 1992). Phospholipase and lysophospholipase activities are present in goat spermatozoa. These activities decrease substantially during transit of spermatozoa from the caput to the cauda epididymidis. Goat sperm phospholipase hydrolyses phosphatidyl-ethanolamine, -choline and -inositol and phosphatidic acid (Atreza and Anand, 1985).

Localization: PLA2 is localized in the epithelial cells of the human prostate and bovine seminal vesicles as well as in the fibrous connective tissue of bovine Cowper's gland. Ejaculated human spermatozoa revealed an immune-reaction, which was not uniform, and the reaction was restricted to the middle piece and the acrosomal and post-acrosomal regions. Human seminal plasma, spermatozoa and prostate PLA2s were immunochemically related to those from bovine Cowper's gland, seminal vesicles and seminal fluid (Ronkko, 1995).

Functions of PLA2: It seems possible that PLA2 and one of its reaction products might contribute to membrane-fusion events during mammalian fertilization (Riffo and Parraga, 1996). A rab3-peptide stimulates exocytosis of ram sperm acrosome via interaction of with c-AMP and PLA2 metabolites. A functional coupling between rab 3 protein, PLA metabolites and PKA in membrane fusion during acrosome reaction has been suggested (Garde and Rolden, 1996). In the presence of Ca^{2+} exposure of boar spermatozoa to bicarbonate lead to a partial activation of phospholipase A2. Bicarbonate primed spermatozoa underwent subsequent activation upon stimulation with progesterone and rendered spermatozoa capable of undergoing exocytosis in response to this steroid. Capacitation was completed in a relatively short period of time, suggesting a role of PLA2 and carbonate in the process (Roldan and Vazquez, 1996).

PLA2 in Acrosome Reaction: G-proteins, calcium, and PLA2 have all been implicated in the cascade of signaling events leading to the acrosome reaction in human spermatozoa. GTP-binding proteins require Ca^{2+} and PLA2 to accomplish their stimulatory effect, and that Ca^{2+} is also required when the acrosome reaction-bypassing the action of PLA2- is stimulated by arachidonic acid (AA). On the other hand, a massive influx of Ca^{2+} was completely unable to induce the acrosome reaction if PLA2 was inhibited, suggesting that both, an increase of $[Ca^{2+}]i$ and PLA2 activation are required for the acrosome reaction to occur (Dominguez et al., 1996; 1999). The sequence leading to exocytosis of the sperm acrosome involves aleast three Ca^{2+}- requiring processes, the first one represented by breakdown of the polyphosphoinositides and the final one by membrane fusion. Besides the intermediate Ca^{2+} requiring event by stimulating ram spermatozoa, it seems that another as yet unidentified Ca^{2+} dependent event may occur before PLA2 activation (Roldan and Fragio, 1993a,b). In ram spermatozoa treatment with Ca^{2+} and A23187 or ionomycin stimulated the release of arachidonic acid (20:4) and exocytosis of the acrosome in a time- and concentration-dependent manner. Diacylglycerol did not appear to be the source of 20:4. On the other hand, generation of 20:4 was significantly correlated with breakdown of phosphatidylcholine, phosphatidylserine, and phosphatidylethanolamine under a variety of conditions, thus indicating that 20:4 release was due to phospholipase A2 activity. Phospholipase A2 plays a fundamental role in the exocytosis of the acrosome elicited by Ca^{2+} and ionophore stimulation. It is possible that activation of this enzyme constitutes an essential Ca^{2+} dependent event underlying exocytosis in response to physiological stimuli (Roldan and Fragio, 1993). In spermatozoa the messenger role of DAG is related to the activation of phospholipase A2, which can in turn generate an array of metabolites directly or indirectly involved in bringing about exocytosis of the acrosome (Roldan and Fragio, 1994). Since ram sperm lacks PKC and since both diacyl- and alkylacylglycerols increased PLA2 activity and exocytosis, stimulation of PLA2 activity by these diglycerides may take place independently from protein kinase C activation (Roldan et al, 1994; Roldan and Mollinedo, 1991).

22.3.3. Phosphatidic Acid Preferring Phospholipase A1

Experiments in several laboratories have provided evidence that phosphatidic acid functions in cell signaling. However, the mechanisms that regulate cellular phosphatidic acid levels

```
  1 MNYPGHGSPR SSERNGGRGG DGAAWELGSD TEPAFGGSVC RFDHLPVGEP GDDEVPLALL
 61 RGEPGLHLAP GAEDHNHHLA LDPCLSDDNY DFSSAESGSS LRYYSEGESG GGGSSSSLHP
121 PQQPLVPSNS GGGGAAGGGP GERKRTRPGG AAARHRYEVV TELGPEEVRW FYKEDKKTWK
181 PFIGYDSLRI ELAFRTLLQA TGARARAQDP DGDHVCGPAS PAGPASSSVE DEDEDRVCGF
241 CPRIAGHGRE MEELVNIERV CVRGGLYEVD VTQGECYPVY WNQSDKIPVM RGQWFIDGTW
301 QPLEEEESNL IEQEHLSRFR GQQMQESFDI EVSKPIDGKD AIHSFKLSRN HVDWHSVDEV
361 YLYSDATTSK IARTVTQKLG FSKASSSGTR LHRGYVEEAT LEDKPSQTTH IVFVVHGIGQ
421 KMDQGRIIKN TAMMREAARK IEERHFSNHA THVEFLPVEW RSKLTLDGDT VDSITPDKVR
481 GLRDMLNSSA MDIMYYTSPL YRDELVKGLQ QELNRLYSLF CSRNPNFEEK GGKVSIVSHS
541 LGCVITYDIM TGWNPVRLYE QLLQKEEELP DERWMSYEER HLLDELYITK RRLREIEERL
601 HGLKASSMTQ TPALKFKVEN FFCMGSPLAV FLALRGIRPG NTGSQDHILP REICNRLLNI
661 FHPTDPVAYR LEPLILKHYS NISPVQIHWY NTSNPLPYEY MKPSFLHPAK DPTSISENEG
721 ISTIPSPVTS PVLSRRHYGE SITNIGKASI LGAASIGKGL GGMLFSRFGR SSASQPSETS
781 RDSIEDEKKP VASPPMTTVA TQTLPHSSSG FLDSALELDH RIDFELREGL VESRYWSAVT
841 SHTAYWSSLD VALFLLTFMY KHEHDNNVKP SLDPV
```

Fig.22.2. Amino acid sequence of phosphatidic acid-preferring phospholipase A1 from bovine testis (Accession AAC03019) Source: http//www. ncbi.nlm.nih.gov.

remain obscure. In response to extracellular or intracellular signals, a number of phospholipases can be activated including phosphatidic acid (PA) phosphohydrolase. Bovine testis contains a soluble phospholipase A1 (PLA1) that preferentially catalyzes the hydrolysis of phosphatidic acid (PA) in mixed micelle assay system. Moreover, the enzyme hydrolyzes phosphatidic acid molecular species containing two unsaturated fatty acids in preference to those containing a combination of saturated and unsaturated fatty acyl groups. Under certain conditions, the enzyme also displays lysophospholipase activity toward lysophosphatidic acid. The phospholipase A1 is not likely to be a lysosomal enzyme because its optimum pH is 7.5-8.5. High levels of activity are found in mature testis and brain but no measurable activity is seen in liver, spleen, or heart. The fact that the activity of the phospholipase A1 in mature bovine testis is more than 10-fold higher than that in newborn calf testis, raises the possibility that the enzyme plays a regulatory role in spermatogenesis or sperm function (Higgs and Glomset, 1994). The native enzyme of 110-kDa is a homotetramer in solution and preferentially catalyses the hydrolysis of PA. The activity of this PA-preferring PLA1 (PA-PLA1) was present in high speed supernatant fractions from mature testis and brain, but not from newborn testis and other tissues. This restricted localization implied that the PLA1 may be taking part in a signaling pathway in brain and testis. In addition, PA appeared to activate the PA-PLA1 reaction by promoting the enzyme's binding to micelles. The open reading frame of PA-PLA1 encoded an 875-amino acids protein with a calculated molecular mass of 97576-Da and a pI of 5.61. The sequence includes a region similar to a lipase consensus sequence containing the putative active site serine and also includes a potential, coiled-coil-forming region. Mutation of the putative active site serine (amino acid 540) demonstrated that it was essential for enzyme activity. Northern blot analysis revealed at least five different messages with the highest overall message level in mature testis. Two possible alternatively splice regions in the open reading frame also were identified. A search of the data base identified six related proteins: one in *Caenorhabditis elegans*, two putative lipases in yeast, and three proteins separately encoded by the *Drosophila* retinal degeneration B gene and its mouse and human homologues (Higgs et al., 1998). The soluble phosphatidic acid-preferring phospholipase A1 expressed in mature bovine testes but not in newborn calf testes, seems to contribute to the formation of the function of sperm **(Fig.22.2)**. A murine phosphatidic acid (PA)-PLA1β is specifically expressed in human testis. The sequence of mPAPLA1β encodes a 460-amino acid protein containing a lipase domain with significant homology to mPA-PLA1α. The mPA-PLA1β contains a short lid and deleted β9 loop, which are characteristics of PLA1 molecules in the lipase family. Both mPA-PLA1β and mPA-PLA1α recombinant proteins exhibited PA-specific PLA1 activity and were vanadate-sensitive. When mPAPLA1β-expressing cells were treated with bacterial phospholipase D, the cells produced lysophosphatidic acid (LPA) (Hiramatsu et al., 2003).

22.3.4. Phopholipase B

Guinea pig intestinal phospholipase B is a calcium-independent phospholipase hydrolyzing sequentially the acyl ester bonds at sn-2 and sn-1 positions of glycerophospholipids. Trypsin treatment of phospholipase B from epididymal membranes reduced its size to 140 kDa, coinciding with the appearance of a significant phospholipase A2 activity (Delagebeaudeuf et al, 1998).

22.3.5. Phospholipase - C

Suh et al., (2001) classified 11 mammaliann PLC Isoforms on the degree of amino acid homology into four classes: PLC-β1-4, PLC-γ1-2, PLC-δ1-4, and PLC-ε. Several isozymes of PLC have been found in sperm and oocytes (Fukami et al, 2001). All PLC Isoforms contain two catalytic domains, X and Y. The PLC-β Isoforms have a C-terminal of 400 aa residues downstream Y domain, which is associated with membranes and also regulates enzyme activity. The PLC-β is regulated by G-protein–coupled receptors, which are activated by a large number of extracellular signals. The PLC-β Isoforms are also regulated by PDZ-containing proteins (Suh et al., 2001). The phosphoinositide-specific phospholipase C is activated by calcium and inhibited by EGTA, although endogenous calcium supported a half-maximal activity. Both bull and rabbit spermatozoa contain a phosphatidylcholine-specific phospholipase C (PC-PLC) and a phosphatidylinositol-specific phospholipase C (PI-PLC). Both enzymes had optimum at pH 7.5. Activity of PC-PLC remains unaffected by varying concentrations of Ca^{2+}, whereas PI-PLC activity is significantly increased. The bulk of PI-PLC was found to be associated with inner acrosomal membrane of bull and rabbit sperm (Hinkovska-Galchev and Srivastava 1992) just as 80% of particulate phospholipase C activity is located in sperm head (Ribbes et al, 1987).

Sperm Capacitation and PLC: It was shown that two intracellular events, which occur during capacitation of bovine sperm are the formation of actin filaments on the plasma and outer acrosomal membranes and the attachment of a PI-PLC to this membrane bound F-actin. The role of this PI-PLC using a PI-PLC of bacterial origin was elucidated. The bacterial PLC is different from the endogenous sperm PI-PLC in that it is calcium independent and not inhibited by neomycin. Using bovine sperm it was shown that bacterial PLC can restore actin release from extracted membranes as well as membrane fusion in a cell-free assay when the endogenous PI-PLC is inhibited by neomycin. The role of membrane bound F-actin in regulating membrane fusion showed that cortical F-actin has two roles in regulating sperm exocytosis. One is to form a scaffolding to hold phospholipase C at the membrane. It also functions as a physical barrier to membrane fusion, which is removed by the increase in intracellular calcium and pH, which precede fusion (Spungin et al, 1995). Sperm plasma membranes possess a thapsigargin insensitive calcium pump and calcium channels, which are opened following phosphorylation by PKC. The acrosomal membranes possess a calcium pump, which is inhibited by thapsigargin, and calcium channels, which are opened by cAMP. These observations suggest a model of acrosomal exocytosis, which involves a calcium rise that occurs in two stages resulting from calcium mobilization from internal stores followed by influx of extracellular calcium (Spungin and Breitbart, 1996). A recombinant PLC2 on incubation with CK2, ERK2, PP2A showed that: a) CK2 phosphorylated the phospholipase on serines 93, 105, and 716; b) ERK2 phosphorylated the enzyme on serine 730; c) PP2A selectively hydrolyzed phosphate groups that were esterified to serines 716 and 730; d) CK2α formed a stable, MgATP/MgGTP-dependent complex with the phospholipase. These observations may be relevant to the process of capacitation in which PLC loses activity after phosphorylation.

PLC, DAG and Acrosome reaction: Choi et al (2001) evaluated the biological significance of phospholipase Cβ1 gene mutation in mouse sperm in the acrosome reaction, fertilization, and embryo development and concluded that mutation of the PLCβ1 gene in the mouse sperm reduces the acrosome reaction rate, fertilization rate, and embryo development rate, which may be the etiologic factors responsible for the low reproductive rate of PLC β 1-/- mouse. During acrosome reaction, calcium entry into the cells activates both a phospholipase A2 and a phospholipase C, leading to the production of substances, like lysophospholipid, diacylglycerol or phosphatidic acid (Bennet et al, 1987). Treatment with ionophore and Ca^{2+} lead to rise in total diacylglycerol and exocytosis of ram spermatozoa. Diacylglycerols are generated via PLC. Increase in diacylglycerols is paralleled by rise in monoacyl- or monoalkylglycerols 1 implying that unlike somatic cells, spermatozoa catabolize diacylglycerols via a 2-diglyceride lipase. Activation of PLC appears to be affected by phosphoinositide-derived diacylglycerol. The 1,3-diacylglycerol and alkyl acylglycerol also stimulated PLC activity, suggesting that the effect is unlikely to be mediated via protein kinase C. Since diacylglycerols are known to be essential in the molecular sequence leading to membrane fusion in mammalian spermatozoa, results suggested that their generation via PLC constitutes a fundamental event during acrosomal exocytosis in response to physiological agonists (Roldan and Murase, 1994). Human spermatozoa undergoing acrosomal exocytosis after treatments with progesterone and Ca^{2+} ionphore separately resulted in rapid and considerable generation of DAG, followed by a limited rise in phosphatidic acid (PA). The PLC activity appeared to be important in the generation of DAG, but phospholipase D activity was not. In addition, polyphosphoinositide-specific phosphoinositidase C activation and hydrolysis of phosphatidylinositol 4,5-bisphosphate appeared to be a necessary prerequisite for activation of the PLC pathway. Finally the DAG formed appeared to be important in acrosomal exocytosis. It suggested that DAG plays a key role in events leading to membrane fusion during human sperm acrosomal exocytosis stimulated by natural agonists (O'Toole et al.,1996). Phosphoproteins and phospholipases were also found to be involved in the membrane fusion step of sperm exocytosis. During this action the cortical F-actin forms a scaffolding to hold phospholipase C at the membrane. F-actin also functions as a physical barrier to membrane fusion, and which is removed by the increase in intracellular calcium and pH, which precede fusion (Spungin et al, 1995).

Phospholipase Cδ4 and Acrosome Reaction: Splicing isoforms of PLCδ4 are predominantly expressed in testis. Male mice in which the PLCδ4 gene had been disrupted either produced few small litters or were sterile. In vitro fertilization studies showed that insemination with PLCδ4 -/- sperm resulted in significantly fewer eggs becoming activated and that the calcium transients associated with fertilization were absent or delayed. The PLCδ4-/- sperm were unable to initiate the acrosome reaction required for fertilization and induced by the zona pellucida suggesting that PLCδ4 functions in the acrosome reaction that is induced by the zona pellucida during mammalian fertilization (Fukami et al., 2001).

Phospholipase D and Acrosome reaction: Induction of the acrosome reaction and activation of sperm PKCα can be caused by lysophosphatidic acid (LPA), which is known to induce signal transduction cascades in many cell types via binding to specific cell-surface receptors.

Under certain conditions by which LPA activates PKCα, there is a significant stimulation of the acrosome reaction, which is inhibited by PKC inhibitors. Protein kinase Cα belongs to the Ca^{2+} dependent classical PKC family of isoforms, and its activation depends upon the presence of Ca^{2+} in the medium. Protein kinase Cα in emergency can be sortened as a regulator of phospholipase D (PLD). The possible regulatory relationship between PKCα and PLD1 was ascribed to their co-existence in bovine sperm, and to their co-immunoprecipitation. The PKCα-PLD1 complex decomposes after treatment of the cells with LPA resulting in the translocation of PKCα to the plasma membrane and translocation of PLD1 to the particulate fraction. This suggested a bilateral regulation of PKCα and PLD1 activation during the sperm acrosome reaction (Garbi et al, 2000). On the other hand, Ca^{2+} entry triggers a late activation of PLD, which is not involved in the early generation of diglycerides. Hence PLD does not make a substantial contribution in events leading to exocytosis of the sperm acrosome. Therefore, generation of diglycerides may take place primarily via phospholipase C (Roldan and Dawes, 1993).

22.3.6. Sperm PLC in Egg Activation

Upon fertilisation by sperm, mammalian eggs are activated by a series of intracellular Ca^{2+} oscillations that are essential for embryo development. The mechanism by which sperm induces this complex signalling phenomenon is unknown. One proposal is that the sperm introduces an exclusive cytosolic factor into the egg that elicits serial Ca^{2+} release. The 'sperm factor' hypothesis has not been ratified because a sperm-specific protein that generates repetitive Ca^{2+} transients and egg activation has not been found. In echinoderms, egg activation appears to involve an egg PLCγ. Injections of sperm extracts trigger Ca^{2+} oscillations in mammalian egg. The addition of sperm extracts to sea urchin egg homogenates causes Ca^{2+} release and InsP3 production. Further, depleting homogenates of phosphatidylinositol lipids using a phosphatidylinositol-specific phospholipase C blocked the sperm extract from causing InsP3 production and a Ca^{2+} rise. This indicated that sperm extracts contained an InsP3-generating phospholipase C, which may play a role in Ca^{2+} release at fertilization (Jones et al, 1998).

The boar sperm PLC activity is Ca^{2+} dependent. The PLC activity of a single boar sperm in a mammalian egg is enough to generate 400 nM InsP3 in 1 min that may be sufficient to account for the observed Ca^{2+} changes in an egg at fertilization. The ability of boar sperm extract to generate InsP3 was examined using sea urchin egg homogenate fractions. The sperm PLC activity triggered InsP3 production from a PIP2-enriched non-microsomal egg compartment that contained yolk platelets. This sperm PLC activity, which is active at nanomolar Ca^{2+} levels and hydrolyzes PIP2 from intracellular membranes, could be involved in the Ca^{2+} changes observed at fertilization (Rice et al, 2000). Thus, the PLC from boar sperm generates InsP3 and hence causes Ca^{2+} release when added to sea urchin egg homogenate. This PLC activity is also associated with the ability of sperm extracts to cause Ca^{2+} oscillations in mammalian eggs following fractionation. Consistent with the sperm PLC acting on egg InsP3, the ability of sperm extracts to release Ca^{2+} was blocked by pre-incubation with the PLC inhibitor U73122 or by the addition of neomycin to the homogenate. The Ca^{2+} releasing activity was detectable in sperm from other species also as well as in whole testis extracts. However, such activity was not present in extracts of other tissues. A sperm-specific phospholipase C, PLC ζ, was found to trigger Ca^{2+} oscillations in mouse eggs indistinguishable from those at fertilization. Absence of PLCζ from sperm extracts abolished Ca^{2+} release in eggs. Moreover, the PLCζ content of a single sperm was sufficient to produce Ca^{2+} oscillations as well as normal embryo development to blastocyst (Saunders et al., 2002). The PLCζ homologues from human and cynomolgus

(A)

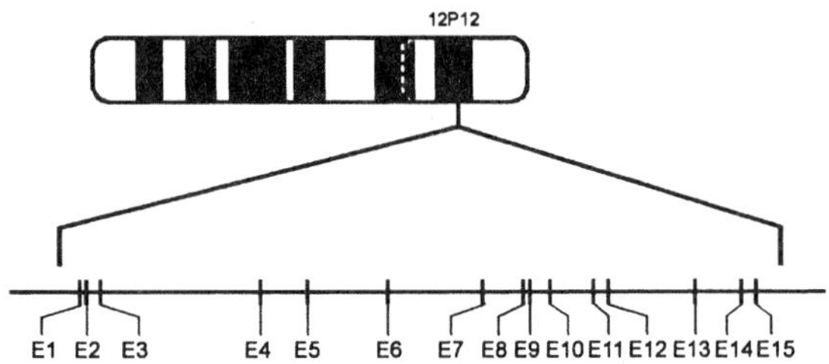

(B)

1 MEMRWFLSKI QDDFRGGKIN LEKTQRLLEK LDIRCSYIHV KQIFKDNDRL KQGRITIEEF
61 RAIYRIITHR EEIIEIFNTY SENRKILLAS NLAQFLTQEQ YAAEMSKAIA FEIIQKYEPI
121 EEVRKAHQMS LEGFTRYMDS RECLLFKNEC RKVYQDMTHP LNDYFISSSH NTYLVSDQLL
181 GPSDLWGYVS ALVKGCRCLE IDCWDGAQNE PVVYHGYTLT SKLLFKTVIQ AIHKYAFMTS
241 DYPVVLSLEN HCSTAQQEVM ADNLQATFGE SLLSDMLDDF PDTLPSPEAL KFKILVKNKK
301 IGTLKETHER KGSDKRGDNQ DKETGVKKLP GVMLFKKKKT RKLKIALALS DLVIYTKAEK
361 FKSFQHSRLY QQFNENNSIG ETQARKLSKL RVHEFIFHTR KFITRIYPKA TRADSSNFNP
421 QEFWNIGCQM VALNFQTPGL PMDLQNGKFL DNGGSGYILK PHFLRESKSY FNPSNIKEGM
481 PITLTIRLIS GIQLPLTHSS SNKGDSLVII EVFGVPNDQM KQQTRVIKKN AFSPRWNETF
541 TFIIHVPELA LIRFVVEGQG LIAGNEFLGQ YTLPLLCMNK GYRRIPLFSR MGESLEPASL
601 FVYVWYVR

Fig.22.3. (A) Schematic representation of human phospholipase-cæ (PLCζ) gene. The fifteen exons (E1 to E15) identified within 179456 bp contig are aligned to a 54.8 kb region on the chromosome 12 (Cox et al, 2002). (B) Amino acid sequence of human PLCζ (Accession NP_149114). Source: http//www. ncbi.nlm.nih.gov.

monkey testes have been isolated. A shorter X-Y linker region in human PLCζ in comparison to mouse protein predicts a different isoelectric point. Microinjection of complementary RNA for both human and cynomolgus monkey PLCζ triggers Ca^{2+} oscillations in mouse oocytes, similar to those seen during fertilization in mice. Human PLC ζ elicited greater potency than PLC ζ from monkeys and mice. Thus sperm PLCζ is the molecular trigger for egg activation during fertilization (Cox et al., 2002) **(Fig.22.3)**. However, another study suggested that a Src-related protein-tyrosine kinase (PTK), named Xyk, may act upstream of the calcium release in fertilization of the *Xenopus* egg. Whether PTK is related to the activation of PLCγ in the fertilization-induced calcium signaling, studies showed that *Xenopus* egg PLCγ is consistently tyrosine phosphorylated and activated within a few minutes after fertilization but not after A23187-induced egg activation. The fertilization-induced activation of PLCγ by Xyk activity was blocked by PTK inhibitor (PP1) when microinjected into the egg. In addition, a PLC inhibitor inhibited sperm-induced InsP3 production and the calcium transient and subsequent calcium-dependent events, which were also inhibited by PP1. These results supported the idea that *Xenopus* egg fertilization requires Src-family PTK-dependent PLCγ activity that acts upstream of the calcium-dependent signaling pathway (Sato et al, 2000). Thus it is likely that sperm specific PLC isozyme is responsible for Ca^{2+} release either directly or through Tyr phosphorylation, during egg fertilization. In contrast, recombinant PLCβ1, -γ1, -γ2, -δ1, all of which had higher specific activities than boar sperm extracts, were not able to release Ca^{2+} in the sea urchin egg homogenate. In addition, these PLCs were not able to cause Ca^{2+} oscillations following microinjection into mouse eggs. These results imply that the sperm PLC possesses distinct properties that allow it to hydrolyze InsP3 in eggs (Jones et al, 2000). Whether phosphatidylinositol 3-kinase could have a role in this process, Mehlmann et al found that several inhibitors of PI3K-mediated signaling had no effect on Ca^{2+} release at fertilization (Mehlmann et al, 2001).

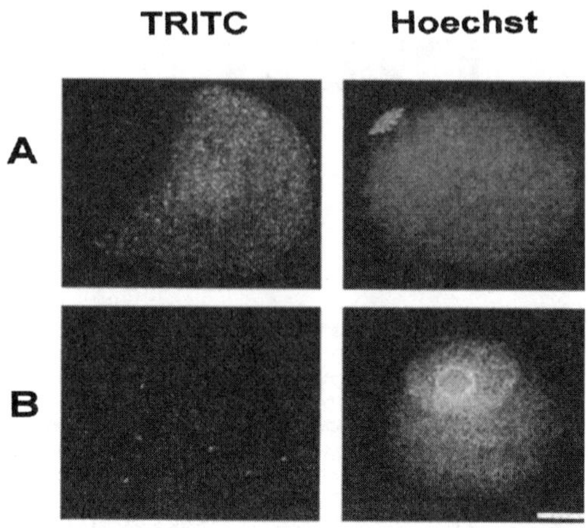

Fig.22.4 The SH3 domain, and not the SH2 domains, of PLC1 inhibits tr-kit-induced cortical granules exocytosis and pronuclear formation in mouse eggs. Eggs were co-injected with cell extracts containing recombinant tr-kit and either of GST-PLC1-SH3 (A) or GST-PLC1-SH2SH2 (B) and fixed 4 h after injection. Tr-kit-induced cortical granule reaction was inhibited by co-injection of GST-PLC1-SH3, but not by GST-PLC1-SH2SH2*, with a similar rate as for pronuclear formation. Bar, 30 µm. (* 500 µg/ml giving a final concentration in the egg: ~10 µg/ml). Reproduced with permission from C. Sette et al. J Cell Biol 142; 1063-74: 1998 © Rockefeller University Press.

22.3.7. Role of Egg PLC in Egg Activation

Although inositol triphosphate (InsP3) functions in releasing Ca^{2+} in eggs at fertilization, it is not known how fertilization activates the phospholipase C that produces InsP3.

Mouse and human: Dupont et al (1996) demonstrated that mouse oocytes possess mRNAs coding for PLCβ1, PLCβ3 and PLCγ isoenzymes. Immunodetection of PLCγ1 revealed that PLCγ1 protein is present in mature mouse oocytes, ruling out the possibility that mRNA was being transcribed but not expressed. The presence of PLCβ protein was not detectable, but the presence of this isoform could be inferred from functional studies. The soluble tyrosine kinase inhibitors, which could affect signaling through PLCγ, hindered but never completely inhibited Ca^{2+} spiking in response to fertilization. Thus the activation of PLC to generate InsP3 was evident in fertilization (Dupont et al, 1996). Microinjection of a truncated form of the c-kit tyrosine kinase present in mouse spermatozoa (tr-kit) activated mouse eggs parthenogenetically, and tr-kit-induced egg activation was inhibited by pre-incubation with an inhibitor of PLC (Sette, et al, Development 1997; 124: 2267-74). Co-injection of GST fusion proteins containing the src-homology (SH) domains of the PLCγ1 competitively inhibited tr-kit- induced egg activation. Tr-kit-induced egg activation was also suppressed by co-injection of antibodies raised against the PLCγ1 SH3 domains, but not against the PLCγ1 COOH-terminal region. In transfected COS cells, co-expression of PLCγ1 and tr-kit increased, DAG and inositol phosphate production, and the phosphotyrosine content of PLCγ1 with respect to cells expressing PLCγ1 alone. These data indicated that tr-kit activates PLCγ1, and that the SH3 domain of PLCγ1 is essential for tr-kit-induced egg activation (Sette et al, 1998) **(Fig.22.4)**.

Starfish Egg: In starfish eggs injected with a PLCγ SH2 domain fusion protein that inhibits activation of PLCγ, Ca^{2+} release at fertilization was delayed or completely inhibited at high protein : sperm ratio. The PLCγSH2 protein is a specific inhibitor of PLCγ in the egg, since it did not inhibit PLCβ activation of Ca^{2+} release. Furthermore, injection of a PLCγ SH2 domain protein mutated at its phosphotyrosine binding site, or the SH2 domains of another protein (the phosphatase SHP2), did not inhibit Ca^{2+} release at fertilization. These results again indicated that during fertilization of starfish eggs, activation of PLCγ by an SH2 domain-mediated process stimulates the production of InsP3 that causes intracellular Ca^{2+} release (Carroll et al, 1997). Microinjection of the tandem SH2 domains of PLCγ produced a concentration-dependent inhibition of Ca^{2+} release and also inhibited subsequent Ca^{2+} dependent events. However, the Ca^{2+} action potential, sperm entry, and sperm pronuclear formation were not prevented by injection of the PLCγSH2 domain protein. Taken together, Carroll et al suggested that PLCγ plays a key role in several fertilization events in sea urchin eggs, including Ca^{2+} release and DNA synthesis. However, the action potential, sperm entry, and male pronuclear formation can occur in the absence of PLCγ activation or Ca^{2+} increase (Carroll et al, 1999; Heyers et al, 2000). Coward et al., (2003) reported evidence for the existence of a similar sperm-derived factor in a commercially important species of teleost fish, the Nile tilapia Oreochromis niloticus (L). Protein extracts obtained from tilapia spermatozoa exhibit PLC activity similar to that seen in mammalian sperm extracts, and also induced calcium release when added directly to the homogenate (Coward et al., 2003).

Sea urchin Eggs: In sea urchin eggs, the first calcium signal is followed by other calcium transients that allow progression through mitotic control points of the cell cycle of the early embryo. How sperm induces these calcium transients is not clearly understood. In sea urchin eggs, both InsP3 and ryanodine receptors contribute to generate the fertilization calcium transient, while the InsP3 receptor generates the subsequent mitotic calcium transients. It seemed that PLCγ might be the origin of the peaks of InsP3 production that punctuate the first mitotic cell cycles of the fertilized sea urchin egg. It was found that several fragments of sea urchin PLCγ containing the two SH2 domains share similarities with SH2 domains of PLCγ from mammals. Microinjection of these fragments into embryos interferes with mitosis. A related construct made from bovine PLCγ also delayed or prevented entry into mitosis and blocked or prolonged metaphase. This indicated that PLCγ plays a key role during fertilization and early development (Shearer et al, 1999).

Xenopus Eggs: The γ isoform of PLC is present in both the unfertilized and the fertilized egg and contributes to the initial phase of PLC activation. Fertilization also results in translocation of a significant fraction of PLC-γ from the cytosol to the membrane compartment of the egg (Carroll et al, 1999; Rongish et al, 1999). However, it was shown that injection of an inhibitory antibody against the type1 InsP3 receptor reduces Ca^{2+} release at fertilization, indicating that the Ca^{2+} release requires InsP3. *Xenopus* eggs were injected with specific inhibitors of PLCγ and PLCβ. The Src-homology 2 (SH2) domains of PLCγ were used to inhibit SH2-mediated activation of PLCγ, and an antibody against G_q family G-proteins was used to inhibit G_q-mediated activation of PLCβ. Absence of inhibition of Ca^{2+} rise at fertilization egg indicated that Ca^{2+} release at fertilization of *Xenopus* eggs requires type1 InsP3 gated Ca^{2+} channels, but not SH2 domain-mediated activation of PLCγ or G_q-mediated activation of PLCβ (Runft et al, 1999).

22.4. HORMONE-SENSITIVE LIPASE

Hormone-sensitive lipase (HSL) is a triaclglycerol lipase and a cholesterol esterase expressed at high levels in adipocytes, testes, and adrenals. In adipocytes, HSL catalyzes the rate-limiting step in the hydrolysis of triglycerides into fatty acids and glycerol. The HSL activation is mediated through phosphorylation by the cAMP-dependent protein kinase. In rat testis, HSL mRNA and protein are expressed in the seminiferous tubuli with a stage-dependent pattern corresponding to the appearance of haploid germ cells. Several isoforms of HSL produced by a single gene have been characterized than from adipocytes. Human adipose tissue expresses a 2.8-kb mRNA that encodes an 88-kDa protein. The mRNA and protein species expressed in testis are larger 3.9-kb and 120-kDa, respectively. Human adipocyte and testis (HSL_{tes}) HSLs are 775 and 1076 amino acids long, respectively than from adipocytes. The HSL_{tes} differs from the adipocyte form by a unique NH_2-terminal region. Elucidation of the HSL gene organization showed that nine coding exons are common to both forms. The additional sequence in HSL_{tes} is encoded by a 1.2-kb long testis specific exon. When a gene is expressed in somatic tissues and in germ cells, tissue-specific expression often results from alternate promoter use. The promoter of the adipocyte form of HSL is located 13 kb downstream of the HSL_{tes} 5-flanking region suggesting that the expression of the different forms of HSL is controlled by several tissue-specific promoters. In mouse HSL_{tes} mRNA appears during spermiogenesis in round spermatids. Two constructs containing 1.4 and 0.5-kb of the human HSL_{tes} gene 5'-flanking region, upstream of the chloramphenicol acetyl transferase were cloned. Enzyme activity in male and female transgenic mice showed that 0.5-kb of the HSL_{tes} promoter was sufficient to direct expression only in testis. The CREMτ did not transactivate the HSL_{tes} promoter. Gel retardation assays showed presence of four testis-specific binding regions (TSBR) that were identified using testis and liver nuclear extracts. Mutation of the AACAAAG motif in TSBR4 abolished the binding. Moreover, binding of the high mobility group domain of Sox5 induced a bend within TSBR4. These results showed that 0-5 kb of the human HSL_{tes} promoter bind Sox proteins and contain cis-acting elements essential for the testis specificity of HSL (Blaise et al., 1999). In seasonal breeder, mink, two HSL immunoreactive bands of 104- and 108-kDa were detected in the interstitial tissue (ITf)- and seminiferous tubule (STf)-enriched fractions. The HSL was localized to interstitial macrophages, Sertoli cells, spermatids, and the equatorial segment of spermatozoa. Epididymal spermatozoa expressed a 104-kDa HSL isoform. These studies suggested that modulation of cholesterol metabolism in testicular compartments might be regulated by HSL isoforms expressed by distinct cells (Kabbaj et al., 2003).

22.5. REFERENCES

Alba M, Cinzia A, Roberto R, Isabella M. Molecular and binding characteristics of IP3 receptors in bovine spermatozoa. Mol Reprod Dev 2000; 56: 527-33.

Anderson RA Jr, Johnson SK, Bielfeld P, Feathergill KA, Zaneveld LJ. Characterization and inhibitor sensitivity of human sperm phospholipase A2: evidence against pivotal involvement of phospholipase A2 in the acrosome reaction. Mol Reprod Dev 1990; 27: 305-25.

Antaki P, Guerette P, Chapdelaine A, Roberts KD. Detection of prophospholipase A2 in human spermatozoa. Biol Reprod 1989; 41: 241-6.

Antaki P, Langlais J, Ross P, Guerette P, Roberts KD. Evidence for two forms of phospholipase A2 in human semen. Gamete Res 1988; 19: 305-14.

Atreja SK, Anand SR. Phospholipase and lysophospholipase activities of goat spermatozoa in transit from the caput to the cauda epididymis. J Reprod Fertil 1985; 74: 687-91.

Bennet PJ, Moatti JP, Mansat A, Ribbes H, Cayrac JC, Pontonnier F, Chap H, Bouste-Blazy L. Evidence for the activation of phospholipases during acrosome reaction of human sperm elicited by calcium ionophore A23187. Biochim Biophys Acta 1987; 919: 255-65.

Benoff S. Modelling human sperm-egg interactions in vitro: signal transduction pathways regulating the acrosome reaction.

Mol Hum Reprod 1998; 4: 453-71.
Blaise R, Grober J, Route P, Tavernier G, Daegelen D, Langin D. Testis expression of hormone-sensitive lipase is conferred by a specific promoter that contains four regions binding testicular nuclear proteins. J Biol Chem 1999; 274: 9327 – 34.
Carroll DJ, Albay DT, Terasaki M, Jaffe LA, Foltz KR. Identification of PLCγ-dependent and – independent events during fertilization of sea urchin eggs. Dev Biol 1999; 206: 232-47.
Carroll DJ, Ramarao CS, Mehlmann LM, Roche S, Terasaki M, Jaffe LA. Calcium release at fertilization in starfish eggs is mediated by phospholipase Cγ. J Cell Biol 1997; 138: 1303-11.
Chen J, Shao C, Lazar V, Srivastava CH, Lee WH, Tischfield JA. Localization of group IIc molecular weight phospholipase A2 mRNA to meiotic cells in the mouse. J Cell Biochem 1997; 64: 369-75.
Choi D, Lee E, Hwang S, Jun K, Kim D, Yoon BK, Shin HS, Lee JH. The biological significance of phospholipase Cβ1 gene mutation in mouse sperm in the acrosome reaction, fertilization, and embryo development. J Assist Reprod Genet 2001; 18: 305-10.
Coward K, Campos-Mendoza A, Larman M, Hibbitt O, McAndrew B, Bromage N, Parrington J. Teleost fish spermatozoa contain a cytosolic protein factor that induces calcium release in sea urchin egg homogenates and triggers calcium oscillations when injected into mouse oocytes. Biochem Biophys Res Commun. 2003; 305: 299-304.
Cox LJ, Larman MG, Saunders CM, Hashimoto K, Swann K, Lai FA. Sperm phospholipase C zeta from humans and cynomolgus monkeys triggers Ca^{2+} oscillations, activation and development of mouse oocytes. Reproduction 2002; 124: 611-23.
Delagebeaudeuf C, Gassama–Diagne A, Nauze M, Ragab A, Li RY, Cadevielle J, Ferrara P, Fauvel J, Chap H. Ectopic epididymal expression of guinea pig intestinal phospholipase B. Possible role in sperm maturation and activation by limited Proteolytic digestion. J Biol Chem 1998; 273: 13407-14.
Dominguez L, Yunes RM, Fornes MW, Burgos M, Mayorga LS. Calcium and phospholipase A2 are both required for the acrosome reaction mediated by G-proteins stimulation in human spermatozoa. Mol Reprod Dev 1999; 52: 297-302.
Dominguez L, Yunes RM, Fornes MW, Mayorga LS. Acrosome reaction stimulated by the GTP non-hydrolizable analogue GTP γS is blocked by phospholipase A2 inhibitors in human spermatozoa. Int J Androl 1996; 19: 248-52.
Dupont G, McGuinness OM, Johnson MH, Berridge MJ, Borgese F. Phospholipase C in mouse oocytes: characterization of β and γ isoforms and their possible involvement in sperm-induced Ca^{2+} spiking. Biochem J 1996; 316(Pt2): 583-91.
Franson R, Raghupathi R, Fry M,Saal J, Vishwanath B, Ghosh SS, Rosenthal MD. Inhibition of human phospholipases A2 by cis-unsaturated fatty acids and oligomers of prostaglandin B1. Adv Exp Med Biol 1990; 279: 219-30.
Fry MR, Ghosh SS, East JM, Franson RC. Role of human sperm phospholipase A2 in fertilization : effects of a novel inhibitor of phospholipase A2 activity on membrane perturbations and oocyte penetration. Biol Reprod 1992; 47: 751-9.
Fukami K, Nakao K, Inoue T, Kataoka Y, Kurokawa M, Fissore RA, Nakamura K, Atsuki M, Mikoshiba K, Yoshida N, Takenawa T. Requirement of phospholipase Cδ4 for the zona pellucida induced acrosome reaction. Science 2001; 292(5518): 920-3.
Garbi M, Rubinstein S, Lax Y, Breitbart H. Activation of protein kinase Cα in the lysophosphatidic acid-induced bovine sperm acrosome reaction and phospholipase D1 regulation. Biol Reprod 2000; 63: 1271-7.
Garde J, Roldan ERS. Rab3-peptide stimulates exocytosis of the ram sperm acrosome via interaction with cyclic AMP and phospholipase A2 metabolites. FEBS Lett 1996; 361: 263-68.
Guerette P, Langlais J, Antaki P, Chapdelaine A, Roberts KD. Activation of phospholipase A2 of human spermatozoa by proteases. Gamete Res 1988; 19: 203-14.
Han MH, Han DK, Aebersold RH, Glomset JA. Effect of protein kinase CK2, extracellular signal-regulated kinase 2 and protein phosphatase 2A on a phosphatidic acid-preferring phospholipase A1. J Biol Chem 2001; 276: 27698-708.
Heyers S, Sousa M, Cangir O, Schmoll F, Schellander K, van der ven H, Montag M. Activation of mouse oocytes requires multiple sperm factors but not sperm PLCγ1. Mol Cell Endocrinol 2000; 166: 51-7.
Higgs HN, Glomset JA. Identification of a phosphatidic acid-preferring phospholipase A1 from bovine brain and testis. Proc Natl Acad Sci USA 1994; 91: 9574-8.
Higgs HN, Han MH, Johnson GE, Glomset JA. Cloning of a phosphatidic acid-preferring phospholipase A_1 from bovine testis. J Biol Chem 1998; 273: 5468-77.
Hinkovaska-Galchev V, Srivastava PN. Phosphatidylcholine and phosphatidylinositol-specific phospholipase C of bull and rabbit spermatozoa. Mol Reprod Dev 1992; 33: 281-6.
Hinkovska VT, Momchiliva AB, Petkova DH, Koumanov KS. Phospholipase A2 activity in ram spermatozoa plasma membranes. Int J Biochem 1987; 19: 569-72.
Hiramatsu T, Sonoda H, Takanezawa Y, Morikawa R, Ishida M, Kasahara K, Sanai Y, Taguchi R, Aoki J, Arai H. Biochemical and molecular characterization of two phosphatidic acid-selective phospholipase A1s, mPA-PLA1α and mPA-PLA1β. J Biol Chem 2003; 278: 49438-47.
Jones KT, Cruttwell C, Parrington J, Swann K. A mammalian sperm cytosolic phospholipase C activity generates inositol trisphosphate and causes Ca^{2+} release in sea urchin egg homogenates. FEBS Lett 1998; 437: 297-300.
Jones KT, Matsuda M, Parrington J, Katan M, Swann K. Different Ca^{2+} releasing abilities of sperm extracts compared with tissue extracts and phospholipase C isoforms in sea urchin egg homogenate and mouse eggs. Biochem J 2000; 346 Pt3: 743-9.
Kabbaj O, Yoon SR, Holm C, Rose J, Vitale ML, Pelletier RM. Relationship of the hormone-sensitive lipase-mediated modulation of cholesterol metabolism in individual compartments of the testis to serum pituitary hormone and testosterone concentrations in a seasonal breeder, the mink (Mustela vison). Biol Reprod 2003; 68: 722-34.
Kramer M, Sharp JD. Structure function and regulation of Ca^{2+} sensitive cytosolic phospholipase A2 (cPLA2). FEBS Lett 1997; 410: 49-53.

Langlais J, Chafouleas JG, Ingraham R, Vigneault N, Roberts KD. The phospholipase of human spermatozoa: purification and partial sequence. Biochim Biophys Res Commun 1992; 182: 208-14.

Mehlmann LM, Chattopadhyay A, Carpenter G, Jaffe LA. Evidence that phospholipase C from the sperm is not responsible for initiating Ca^{2+} release at fertilization in mouse eggs. Dev Biol 2001; 15: 492-501.

Minelli A, Allegrucci C, Rosati R, Mezzasoma I. Molecular binding characteristics of IP3 receptors in bovine spermatozoa. Mol Reprod Dev 2000; 56: 527-33.

O'Toole CM, Roldan ER, Hampton P, Fraser LR. A role for diacylglycerol in human sperm acrosomal exocytosis. Mol Hum Reprod 1996; 2: 317-26.

Ribbs H, Plantavid M, Bennet PJ, Chap H, Douste-Bilazy L. Phospholipase C from human sperm specific for phosphoinositides. Biochim Biophys Acta 1987; 919: 245-54.

Rice A, Parrinogton J, Jones KT, Swann K. Mammalian sperm contain a Ca^{2+} sensitive phospholipase C activity that can generate InsP(3) from PIP(2) associated with intracellular organelles. Dev Biol 2000; 228: 125-.35.

Riffo M, Gomez Lahoz E, Esponda P, Medicina FD, Chile UD. Immunocytochemical localization of phospholipase A2 in hamster spermatozoa. Histochem 1992; 97: 25-31.

Riffo MS, Parraga M. Study of the acrosome reaction and the fertilizing ability of hamster epididymal cauda spermatozoa treated with antibodies against phosphatases A2 and/or lysophosphatidylcholine. J Exp Zool 1996; 275: 459-68.

Roldan ER, Fragio C. Phospholipase A2 activity and exocytosis of the ram sperm acrosome: regulation by bivalent cations. Biochim Biophys Acta 1993a; 1168: 108-14.

Roldan ER, Fragio C. Phospholipase A_2 activation and subsequent exocytosis in the Ca^{2+} / ionphore induced acrosome reaction of ram spermatozoa. J Biol Chem 1993b; 268: 13692-70.

Roldan ER, Fragio C. phospholipase A2 and subsequent exocytosis in ram spermatozoa. Evidence that the effect is not mediated via protein kinase C. Biochem J 1994; 297 (Pt1): 225-32.

Roldan ER, Martinez-Dalmau R, Mollinedo F. Diacylglycerol and alkylacylglycerol stimulate ram sperm phospholipase A2. Int J Biochem 1994; 26: 951-8.

Roldan ER, Mollinedo F. Diacylglycerol Stimulates the Ca^{2+} dependent phospholipase A2 of ram spermatozoa. Biochem Biophys Res Commun 1991; 176: 294-300.

Roldan ER, Murase T. Polyphosphoinositide-derived diacylglycerol stimulates the hydrolysis of phosphatidylcholine by phospholipase C during exocytosis of the ram sperm acrosome. Effect is not mediated by protein kinase C. J Biol Chem 1994; 269: 23583-9.

Roldan ER, Vazquez JM. Bicarbonate/CO2 induces rapid activation of phospholipase A2 and renders boar spermatozoa capable of undergoing acrosomal exocytosis in response to progesterone. FEBS Lett 1996; 396: 227-32.

Rolden ER, Dawes EN. Phospholipase D and exocytosis of the ram sperm acrosome. Biochim Biophys Acta 1993;1210:48-54.

Rongish BJ, Wu W, Kinsey WH. Fertilization-induced activation of phospholipase C in the sea urchin egg. Dev Biol 1999; 215: 147-54.

Ronkko S, Lahtinen R, Vanha-Perttula T. Phospholipases A2 in the reproductive system of the bull. Int J Biochem 1991; 23: 595-603.

Ronkko S. Immunohistochemical localization of phospholipase A2 in human and bovine male reproductive organs. Comp Biochem Physiol B Biochem Mol Biol 1995; 110: 503-9.

Runft LL, Watras J, Jaffe LA. Calcium release at fertilization of Xenopus eggs requires type 1 PIP(3) receptors but not SH2 domain-mediated activation of PLCγ or G(q)-mediated activation of PLCβ. Dev Biol 1999; 214: 399-11.

Sato K, Tokmakov AA, Iwasaki T, Fukami Y. Tyrosine kinase-dependent activation of phospholipase Cγ is required for calcium transient in Xenopus egg fertilization. Dev Biol 2000; 224: 453-69.

Saunders CM, Larman MG, Parrington J, Cox LJ, Royse J, Blayney LM, Swann K, Lai FA. PLCζ: a sperm-specific trigger of Ca(2+) oscillations in eggs and embryo development. Development 2002; 129: 3533-44.

Sette C, Bevilacqua A, Geremia R, Rossi P. Involvement of phospholipase Cγl in mouse egg activation induced by a truncated form of the C-kit tyrosine kinase present in spermatozoa. J Cell Biol 1998; 142: 1063-74.

Shearer J, De Nadai C, Emily-Fenouil F, Gache C, Whitaker M, Ciapa B. Role of phospholipase Cγ at fertilization and during mitosis in sea urchin eggs and embryos. Development 1999; 126: 2273-84.

Soubeyrand S, Khadir A, Brindle Y, Manjunath P. Purification of a novel phospholipase A2 from bovine seminal plasma. J Biol Chem 1997a; 272: 222-7.

Soubeyrand S, Manjunath P. Novel seminal phospholipase A2 is inhibited by the major proteins of bovine seminal plasma. Biochim Biophys Acta 1997b; 1341: 183-8.

Spungin B, Breitbart H. Calcium mobilization and influx during sperm exocytosis. J Cell Sci 1996; 109 (Pt7): 1947-55.

Spungin B, Margalit I, Breitbart H. Sperm exocytosis reconstructed in a cell-free system: evidence for the involvement of phospholipase C and actin filaments in membrane fusion. J Cell Sci 1995; 108 (Pt6): 2525-35.

Suh P-G, Hwang J-I k, Ryu SH, Donowitz M, Kim JH. The role of PDZ-containing proteins in PLC-β-mediated signaling. Biochem Biophys Res Commun 2001; 288: 1-7.

Thakkar JK, East J, Franson RC. Modulation of phospholipase A2 activity associated with human sperm membranes by divalent cations and calcium antagonists. Biol Reprod 1984; 30: 679-86.

Thakkar JK, East J, Seyler D, Franson RC. Surface-active phospholipase A2 in mouse spermatozoa. Biochim Biophys Acta 1983; 754: 44-50.

Trevino CL, Santi CM, Beltran C, Hernandez-Cruz A, Darszon A, Lomeli H. Localization of inositol trisphosphate and ryanodine receptors during mouse spermatogenesis: Possible functional implications. Zygote 1998; 6: 159-72.

Walensky LD, Snyder SH. Inositol 1,4,5-triphosphate receptors selectively localized to the acrosomes of mammalian sperm. J Cell Biol 1995; 130: 857-69.

Chapter 23

ACROSOMAL ENZYMES

23.1. INTRODUCTION

Sperm-egg interaction is a species-specific event, which is initiated by the recognition and binding of complementary molecule(s) present on sperm surface (receptor) and the surface of the zona-pellucida (ZP ligand). This is a carbohydrate-mediated signal transduction event, which is believed to be a pre-requisite that enables the acrosome reacted spermatozoa to penetrate the ZP before the egg is fertilized. The hydrolytic enzymes released at the site of sperm-egg binding, alongwith the enhanced thrust generated by the hyperactivation of the bound spermatozoa are important factors that regulate the penetration of zona pellucida and fusion of the gametes. Evidences strongly suggest that sperm-egg binding leading to the acrosomal exocytosis is a complex event that likely reflects interaction between multiple sperm surface receptors and multivalent ZP. The ZP in a species is composed of three glycoproteins namely, ZP1, ZP2, and ZP3. These glycoproteins at least in the mouse are synthesized and secreted during oogenesis. It is generally accepted that hydrolytic enzymes, such as arylsulfatase, β-glucuronidase, N-acetylglucosaminidase and hyaluronidase, and perhaps proteases aid in the process of dispersion of the cumulus mass by hydrolyzing hyaluronic acid and/or polypeptide substrates. The protease group of enzymes includes serine proteases, proacrosin, acrosin, metalloproteases, ADAM proteins, dipeptidylpeptidase II, aspartic protease, cathepsin D, cysteine proteases such as calpain, CRES, cathepsins B, L, and S-like. Many of these enzymes are released during acrosome reaction when the outer acrosomal membrane and the overlapping sperm plasma membrane fuse, allowing the release of the acrosomal contents. In addition to the glycohydrolases and , proteinases the sperm acrosome contains esterases, sulfatases, phosphatases, and phospholipases. Some of these enzymes have been described in sufficient details elsewhere (Zaneveld and de Jonge, 1991), whereas phospholipases, metalloproteases, and ADAM proteins have been discussed in chapters 22, 25 and 26 respectively.

23.2. ACROSOME BIOGENESIS

The acrosome of spermatozoan is formed during spermiogenesis phase of spermatogenesis that lasts for about 20 days in rat. The Golgi apparatus is actively involved in the formation of sperm acrosome during Golgi phase and cap phase. During acrosome phase and maturation phase, the acrosomic system undergoes structural transformation and condensation, but without participation of Golgi apparatus. During Golgi phase, the proacrosomic granules appear as electron dense material and coalesce into larger proacrosomic granules (step 1-2).

Each granule shows a less dense outer zone and a more dense inner zone. In the last step of Golgi phase (step 3), the proacrosomic granules fuse to form a single acrosomic granule, which forms a contact with the nucleus. The dense inner zone of the granule remains in contact with the inner acrosomal membrane and faces towards nuclear envelope. The acrosomal granule however, continues to grow in size with the help of secreted material of Golgi apparatus and forms a hemispherical cap like structure over the surface of the nucleus (step 4-7). In step 8 of spermiogenesis, the Golgi apparatus separates from the acrosomic system, which continues to change its shape with the change of nucleus, to which it is attached and till it becomes falciform (Clermont et al., 1990). During this phase of spermiogenesis (steps 1-7 spermatids), various Golgi secreted glycoproteins and lysosomal enzymes are delivered to the growing acrosomic system. Thus, the juxtaposition of the Golgi apparatus to the developing acrosome of spermatids has been taken as the evidence for direct contribution of Golgi apparatus to the formation of acrosome system of spermatozoa.

Acrosome assembly occurs throughout spermiogenesis and continues during post-testicular sperm maturation in the epididymis, resulting in a structurally polarized membrane-bounded organelle that contains assorted hydrolases and a stable infrastructure termed the acrosomal matrix. Specific post-translational modifications of the major acrosomal matrix proteins are evident in late, step 16. Spermatids and matrix protein processing continues within specific acrosomal subdomains of caput epididymal spermatozoa. Many of the molecules thus arranged gradually become compartmentalized during sperm passage through the epididymis. Some of them are further modified during the fertilization process (Olson et al., 2003; Yoshinaga and Toshimori, 2003). Sperm surface proteins/glycoproteins are processed by (i) elimination or masking of the proteins; (ii) modification of the existing glycoproteins and by (iii) selective proteolysis to convert enzymatically inactive precursor forms to an enzymatically active mature form. Testicular spermatozoa are non-motile and unable to fertilize an egg. They acquire progressive forward motility and fertilizing capacity during passage through epididymis, which provides environments for biochemical modifications necessary to mature sperm (see Chapter 34). Functionally, the anterior acrosome is involved in the acrosome reaction or sperm-zona pellucida interaction, while the equatorial segment (posterior acrosome) is involved in sperm-egg fusion.

23.3. GLYCOSYLTRANSFERASES AND GLYCOSIDASES

The terminal sugar residues are suggested to be recognized by the male and female gametes during fertilization. The terminal sugar residues considered to be recognized by the spermatozoa include mannosyl, sialyl, glucosaminyl, α-galactosyl, glucosaminyl, α-galactosyl, and β-galactosyl. Mammalian spermatozoa contain a host of acid glycohydrolases with catalytic properties similar to enzymes present in lysosomes. These enzymes, contained in acrosome, are glycoproteins in which glycan portion contributes about 10% of molecular mass. The glycan portion is mainly N-linked (asparagine linked oligosaccharides). Lysosomal hydrolases are synthesised on polysomes of ER where they are modified and further processed in Golgi apparatus (Tulsiani et al., 1998). A single glycohydrolase contains multiple glycan chains of mannose type. Sperm glycosidases and proteases play relevant roles in sperm-egg binding and penetration in different species. Among mammals, glycosyltransferases and glycosidases have been involved in non-catalytic carbohydrate binding activities (Miller et al., 1992; Hunnicat et al., 1996 a,b).Two sets of glycan modifying enzymes, namely glycosyltransferases, which add sugar residues from the nucleotide sugar (donor) to glycolipids from the existing glycoconjugates are present in sperm acrosome and are thought to have an important role. Lectin binding studies provide evidence that glycan moieties of sperm plasma membrane glycoproteins undergo extensive modification during epididymal transit (Tulisiani et al., 1998). Alterations of sperm plasma membrane are not limited to the effects of glycan modifying enzymes alone, since several other sperm surface molecules are modified by proteolytic processing in epididymis. For examples, rat sperm plasma membrane mannosidase, guinea-pig sperm plasma membrane proteins PH20 and PH 30, mouse sperm antigen M42, rat sperm glycoprotein CE9, and antigen 2B1 undergo proteolysis during sperm maturation. In addition to the modification of sperm plasma membrane proteins/glycoproteins, spermatozoa undergo intra-acrosomal modifications, acrosin and acrogranin. Evidence presented demonstrate that at least one sperm surface glycoprotein (Mr 86-kDa) is fucosylated in vitro when caput spermatozoa are incubated with GDP fucose (Tulsiani et al., 1998).

23.4. β1, 4-GALACTOSYLTRANSFERASE

The nature of the sperm receptor for ZP3-oligosaccharides has been the subject of active debate, with numerous sperm proteins being implicated as ZP3 receptors. One of them is β1,4-galactosyltransferase (GalTase). In somatic cells, GalTase exists in two subcellular pools performing two distinct functions. The bulk of GalTase is localized in the Golgi complex, where it participates in glycoconjugate biosynthesis. A minor pool of GalTase is expressed on the cell surface, possibly as a result of an alternative cytoplasmic domain, where it functions as a signal transducing receptor for extracellular glycoside ligands. The GalTase's dual subcellular distribution results, at least in part, from the synthesis of two GalTase isoforms with different cytoplasmic domains. The shorter isoform, which contains an additional 13-amino acid sequence over-rides the Golgi retention signal and targets some GalTase molecules to the cell surface. The longer GalTase isoform mediates the binding of sperm to ZP3 oligosaccharides in the zona pellucida and is the only isoform found in mature sperm. The binding of sperm to ZP3 induces the acrosome reaction. The acrosome reaction appears to be induced by multivalent oligosaccharides on ZP3 that aggregate sperm-bound receptors, thus eliciting intracellular signals. On the sperm surface, it binds to terminal GlcNAc residues of O-linked oligosaccharides of ZP3 perturbing GalTase or its ZP3 ligand affects sperm-zona binding (Gong et al., 1995). It seems that GalTase serves as a generalized gamete receptor in mammals. The PNA-positive glycoprotein of 135-150 kDa present on plasma membrane of sperm from the caput (but not cauda epididymis) is degalactosylated by digestion with purified luminal fluid β-D-galactosidase. This suggested a role for glycoprotein modifying enzymes in the modification of sperm surface glycoproteins during epididymal maturation.

Characterization: The organization of the 5' end of the murine GalTase-1 gene is unusual since its three transcriptional start sites are contained within an ~725-bp contiguous piece of DNA. In somatic cells and tissues, the GalTase-1 gene specifies two transcripts of 4.1 and 3.9-kb. These two transcripts arise as a consequence of initiation at two different start sites separated by ~200 bp. Since the two start sites are positioned either upstream of the first two in-frame ATGs (4.1-kb) or between these two in-frame ATGs (3.9-kb), translation of each mRNA results in the synthesis of two catalytically identical, structurally related protein isoforms that differ only in the NH_2 – terminal cytoplasmic domains. The third, most distal transcriptional start site is used exclusively during the later stages of spermatogenesis. In spermatogonia, only the 4.1-kb transcript can be detected and it is identical in structure to the 4.1-kb mRNA found in somatic cells. As spermatogonia develop into early pachytene spermatocytes, transcription of the GalTase-1 gene is reduced to barely detectable levels. Continued differentiation to late pachytene spermatocytes and haploid round spermatids is coincident with renewed GalTase-1 expression to levels comparable to that observed in spermatogonia. However, the 4.1-kb GalTase-1 transcript is replaced with two truncated germ cell-specific transcripts of 2.9- and 3.1-kb. These two transcripts encode the same open reading frame as the 4.1-kb transcripts. However, their respective 3'-untranslated regions are only ~0.8 or ~1.0-kb in length.

Interaction of GalTase with ZP3: The GalTase has been localized on the surface of all mammalian sperm including that of human, bovine, porcine, equine, rat, mouse, guinea pig, and rabbit but its functional significance remains untested except in mouse and pig (Larson and Miller, 1997). Guinea pig, mouse, and rat sperm had higher level of GalTase than bovine, porcine, and rabbit sperm. Bull sperm have a protein of molecular weight similar to mouse at ~60 kDa. The GalTase epitopes are confined to the anterior cap of fresh or capacitated bull sperm. The GalTase expresses as an integral membrane protein on dorsal anterior surface of sperm, specifically interacts with sperm-binding oligosaccharides of ZP3 to mediate gamete recognition. Sperm exposed to anti-GalTase antiserum show reduced fertilization in vitro. If sperm GalTase is blocked by competitive inhibitors, or if its binding site on ZP3 is selectively destroyed, sperm-oocyte binding is inhibited. The GalTase-null sperm are unable to bind ZP3 in radiolabelled binding assays, fail to undergo ZP3-induced acrosome reactions, and are relatively unable to penetrate the zona pellucida (Lu and Shur, 1997). Nevertheless, GalTase-null spermatozoa still bind to the zona pellucida, albeit without undergoing an acrosome reaction. This implies that other sperm proteins cooperate in some way with GalTase to facilitate adhesion to the zona pellucida. A large scaffolding protein has been identified that associates with GalTase cytoplamic domain and may be responsible for orchestrating its signal transduction mechanisms that result into acrosome reaction (Nixon et al., 2001).

However, in mouse sperm GalTase specifically recognizes those ZP3 oligosaccharides that have sperm-binding activity, but does not interact with other zona pellucida glycoproteins. In constrast, all ZP glycoproteins are recognized by non-sperm GalTase, demonstrating a more stringent substrate specificity for the sperm enzyme. After the release of the sperm acrosome, the transferase relocalizes to a new membrane domain where it can no longer bind to ZP3, which is consistent with the inability of acrosome-reacted sperm to bind ZP3 or to initiate binding to the ZP (Miller et al., 1992). Binding of ZP3 to sperm induces the acrosome reaction through aggregation of GalTase and activation of a heterotrimeric G-protein cascade. The cytoplasmic domain of sperm surface GalTase is bound to activated heterotrimeric G protein complex that contains the $G_i\alpha$ subunit. Aggregation of GalTase by multivalent ligands elicited G protein activation. Sperm from transgenic mice that overexpressed GalTase had higher rates of G protein activation than did wild-type sperm, which rendered transgenic sperm hypersensitive to their ZP3 ligand. Thus, the cytoplasmic domain of cell surface GalTase appears to enable it to function as a signal-transducing receptor for extracellular oligosaccharide ligands (Gong et al., 1995; Tengowski et al., 2001). It appears that, although GalTase is able to bind porcine zona proteins, its function in porcine sperm-zona binding is not necessary or sufficient for sperm-zona binding. This supports the contention that porcine sperm-zona binding requires redundant gamete receptors (Rebeiz and Miller, 1999).

GalTase in Germ Cell-Sertoli Cell Adhesion: Since GalTase is present on the surface of early spermatogenic cells long before it is required for sperm-egg recognition. Pratt et al,

(1993) determined whether GalTase on germ cells functions during adhesions to Sertoli cells consistent with such a function. Therefore, in support of adhesive action, GalTase was localized to areas of putative germ cell-Sertoli cell contact. More directly, anti-GalTase IgA and Fab fragments inhibited the initial adhesion of spermatocytes to Sertoli cell monolayers. Direct enzyme assays showed that Sertoli cells also express surface GalTase; most of which was confined to the basal cell surface where it was inaccessible to germ cell, but where it may function in adhesion to the underlying basal lamina.

Galactose Receptor on Sperm Surface: Galactosyl receptor with an apparent molecular mass of 54-kDa has been shown to be a cell surface molecule present in sperm. It belongs, together with hepatocyte asialoglycoprotein receptor to the family of C- type lectins requiring Ca^{2+} for carbohydrate binding specificity mediated by a galactose recognition domain (Drickamer and Taylor, 1993 c/r Rivkin et al., 2000). Rat testis galactsoyl receptor gene generates two mRNA species: one species, designated liver-type, is identical to rat hepatic lectin receptor 2/3 and has 301 amino acid residues, (RHL-2/3); the other designated testis-type, contains one unspliced intron (86 nt), which alters the reading frame and has 166 amino acid terminus that changes the amino acid sequence of the carboxyl terminus. As a result the carbohydrate recognition domain (CRD) (glutamine-proline-aspartic acid/QPD) and flanked Ca^{2+} binding amino acid sequence were not present in the testis-type protein. A transcript with unspliced intron is present in rat sperm but not liver. It appears that sperm galactosyl receptor plays a role in sperm egg binding (Rivkin et al., 2000).

23.5. OTHER GLYCOSYL TRANSFERASES

β1, 2-N-Acetyl Glucosaminyl Tranferase

The α3-D-mannoside-β1, 2-N-acetylglucosaminyltransferase I (GnT I) (EC 2.4.1.101) catalyzes the conversion of oligomannose to complex or hybrid N-glycans of glycoproteins. Its activity is present in rat testicular and cauda epididymal fluids. The enzyme activity of testicular fluid had a pH optimum of 6.0, whereas that of the cauda epididymal fluid was optimal at pH 7.0. The enzyme in testicular fluid had an absolute requirement for either Co^{2+}, or for Mn^{2+}, Mg^{2+} and Ca^{2+}, and being stimulated by these cations, whereas that of cauda epididymal fluid had an absolute requirement for Mn^{2+} or Ca^{2+}, with Co^{2+} and Mg^{2+} being ineffective. The apparent Km value for α1-3 α 1-6mannopentose of GnT I in the testicular and epididymal fluids was 0.57 and 0.38 mM, respectively. This suggested that two forms of GnT I exist in the testicular and epididymal fluids (Nozaki et al., 2003).

GALβ1, 3GALNAC/GALβ1, 4 GlcNAc α2, 3-SIALYL TRANSFERASE

The mRNA expression of sialyltransferase genes is regulated in a cell specific manner. The mRNAs of human Galβ1,3GalNAc/Galβ1,4GlcNAcα2,3 sialyltransferase gene (hST3Gal IV) consist of six isoforms, type A1, A2, B1, B2, B3, and BX. These mRNAs are transcribed from different promoters, pA, pB1, pB2, pB3, and pBX, respectively. Type B mRNAs are expressed in several cells, whereas type A mRNAs are specifically expressed in testis, ovary, and placenta, suggesting that pA promoter activity is especially high in these tissues. Germ-cell specific transcriptional regulation of the hST3Gal IV pA promoter has been demonstrated. Using a luciferase assay, pA promoter activity was shown to be high in testis (Taniguchi et al., 2003).

23.6. HYALURONIDASE AND PH-20

Hyaluronidases (HDases) comprise a family of the glycosidases that has been exceedingly difficult to purify and characterize, because of their low concentration and structural instability. A number of mammalian hyaluronidases have been discovered, including ubiquitous and tissue – specific forms (Gupta and Goldberg 1981; Csoka et al, 1999). At least three types of HDases hydrolyse hyaluronan via different mechanisms. The first group comprises endo-β-N-acetyl-D-hexosaminidases, which degrade the high-molecular weight substrate to the tetrasaccharide as the main end product. Testicular enzyme also catalyzes transglycosylation reaction resulting also in the formation of hexa-, di- and octasaccharides during hydrolysis of hyaluronic acid (HA). A second type of HDase represented from leech, is a β-endoglucuronidase. Enzymes of this type may be present in hookworms. Lastly, the bacterial lyases that act via β-elimination yield disaccharides as the main end products. The HDase activity has been detected in the venom of snakes, fish, bee, wasp, scorpion, spider etc. These enzymes supposedly serve to degrade hyaluronan thereby facilitating the diffusion of other venom. The gene coding this form, called *HYAL1* gene maps to human chromosome 3p21.3. A second HDase gene *HYAL2* was identified that has homology to PH20. The *HYAL1* and *HYAL2* genes encode two distinct forms of lysosomal HDases. Mutation in *HYAL1* gene causes a newly described lysosomal disorder, mucopolysaccharidosis IX (Triggs-Raine 1999). Hyaluronidase in sperm acrosome was implicated in the mechanism of sperm–oocyte interaction (Austin, 1960). This protein is present on the posterior head and the acrosomal membrane of mammalian sperm. It has been shown that PH–20 plays an essential role in fertilization. Guinea pigs immunized with their own PH–20 protein become sterile (Primakoff et al., 1997). Upon fertilization, recombinant PH-20 protein can digest outer layer of oocyte. Inhibitors of PH-20 could thus block fertilization and thereby conceivably act as contraceptives.

23.6.1. Characterization

The enzyme from bee venom has been investigated in detail. The mature HDase is composed of 249 amino acids including four cysteine residues potential sites for N-glycosylation are present in the sequence (Gmachl and Kreil, 1993). The most widely recognized tissue – specific source of hyaluronidase is the mammalian testis, identified originally as a source of "spreading factor". Although the mammalian hyaluronidases are derived from different genes, it is reasonable that these enzymes share certain structural and biological characteristics with one another and with other glycosidases. Active site of HDase is not known. However, after analysis of bee HDase (Gmachl and Kreil, 1993), it was shown that significant homology existed between insect enzyme and a protein termed PH–20, which had earlier been detected with monoclonal antibodies.

A comparison of the sequences of bee venom HDase and the guinea pig PH–20 protein revealed a 36% identity in a region encompassing about 300 amino acids. This raised the possibility that the PH-20 protein possesses HDase activity **(Fig 23.1)**. This notion was corroborated for the human, monkey and mouse PH–20 proteins. The PH–20 mRNA could only be detected in testis (Kreil, 1995). It is noteworthy that the mRNA encoding the bee venom enzyme could also be detected in testes from drones but not in other bee tissues. In vitro mutagenesis of PH–20 showed that two acidic amino acids are critical for enzyme activity (Arming et al., 1997) and the region of amino acid sequence 142-172 is essential for HDase activity at neutral pH, whereas the region of amino acid sequence 277-297 may be more important at a lower pH. It is likely that these two regions are the acid / base catalyst site and the nucleopathic site, respectively, of PH–20 HDase (Yudin et al., 2001).

```
...CGGTCGGATCGTCGATTCATGTCTCGGCCTCTCGTGATCACGGAAGGGATGATGATT
                          M  S  R  P  L  V  I  T  E  G  M  M  I
GGAGTGTTGCTAATGCTAGCCCCGATAAACGCGTTATTACTCGGCTTCGTACAGAGCACC
 G  V  L  L  M  L  A  P  I  N  A  L  L  L  G  F  V  Q  S  T

CCCGACAACAACAAAACCGTACGGGAGTTCAACGTTTACTGGAACGTGCCCACCTTTATG
 P  D  N  N  K  T  V  R  E  F  N  V  Y  W  N  V  P  T  F  M
TGCCATAAATACGGGCTACGGTTCGAAGAAGTATCGGAGAAATATGGTATTCTACAGAAC
 C  H  K  Y  G  L  R  F  E  E  V  S  E  K  Y  G  I  L  Q  N
TGGATGGATAAGTTTCGGGGCGAGGAGATCGCGATCCTTTACGACCCTGGAATGTTCCCG
 W  M  D  K  F  R  G  E  E  I  A  I  L  Y  D  P  G  M  F  P
GCGTTGCTGAAAGACCCGAATGGGAACGTGGTGGCGAGGAACGGCGGTGTCCCGCAACTG
 A  L  L  K  D  P  N  G  N  V  V  A  R  N  G  G  V  P  Q  L
GGCAATCTCACCAAGCATCTGCAAGTATTTCGGGACCACTTGATCAATCAGATCCCGGAC
 G  N  L  T  K  H  L  Q  V  F  R  D  H  L  I  N  Q  I  P  D
AAGTCGTTTCCCGGCGTGGGGGTGATCGATTTCGAAAGTTGGAGGCCGATATTCAGACAG
 K  S  F  P  G  V  G  V  I  D  F  E  S  W  R  P  I  F  R  Q
AACTGGGCCTCCCTCCAGCCTTACAAGAAACTGTCCGTAGAGGTGGTTCGCCGTGAGCAT
 N  W  A  S  L  Q  P  Y  K  K  L  S  V  E  V  V  R  R  E  H
CCGTTCTGGGACGATCAGAGGGTGGAGCAGGAGGCGAAACGAAGGTTCGAGAAATACGGG
 P  F  W  D  D  Q  R  V  E  Q  E  A  K  R  R  F  E  K  Y  G
CAGCTTTTCATGGAGGAGACGTTGAAAGCGGCGAAACGGATGAGGCCGGCCGCCAATTGG
 Q  L  F  M  E  E  T  L  K  A  A  K  R  M  R  P  A  A  N  W
GGATACTACGCCTACCCTTATTGCTACAATCTGACGCCGAATCAGCCGAGCGCCCAATGC
 G  Y  Y  A  Y  P  Y  C  Y  N  L  T  P  N  Q  P  S  A  Q  C
GAAGCGACCACCATGCAGGAGAACGATAAAATGTCGTGGCTGTTCGAGTCGGAAGACGTC
 E  A  T  T  M  Q  E  N  D  K  M  S  W  L  F  E  S  E  D  V
CTCCTTCCGTCCGTTTACTTGAGATGGAATCTGACGAGCGGCGAAAGAGTGGGCCTGGTC
 L  L  P  S  V  Y  L  R  W  N  L  T  S  G  E  R  V  G  L  V
GGTGGCCGCGTGAAGGAGGCGTTGAGAATAGCGAGGCAAATGACGACCAGCAGGAAGAAG
 G  G  R  V  K  E  A  L  R  I  A  R  Q  M  T  T  S  R  K  K
GTTCTACCATATTACTGGTACAAATATCAGGATCGAAGGGACACGGATTTGAGCAGGGCT
 V  L  P  Y  Y  W  Y  K  Y  Q  D  R  R  D  T  D  L  S  R  A
GACCTCGAGGCAACTTTACGAAAAATCACGGACCTCGGCGCCGACGGGTTCATCATTTGG
 D  L  E  A  T  L  R  K  I  T  D  L  G  A  D  G  F  I  I  W
GGAAGTTCCGACGATATAAACACGAAGGCGAAGTGCCTACAATTCAGGGAATACCTGAAC
 G  S  S  D  D  I  N  T  K  A  K  C  L  Q  F  R  E  Y  L  N
AACGAGTTGGGCCCTGCCGTTAAACGAATCGCGTTGAACAACAACGCGAACGATCGACTG
 N  E  L  G  P  A  V  K  R  I  A  L  N  N  N  A  N  D  R  L
ACGGTGGACGTGAGCGTGGACCAAGTGTGACCGTAGCCACCGAGCTCGGGGAAGCTGGAG
 T  V  D  V  S  V  D  Q  V  /
```

23.1. Nucleotide sequence of cloned hyaluronidase cDNAs and the deduced amino acid sequence of preproenzyme. The sequence of synthetic oligonucleotide used for RACE protocol and AATAAA polyadenylation signal are underlined. Arrows indicate site for signal peptidase. Reprinted with permission from M. Gmachl and G. Kreil. Proc Natl Acad Sci (USA) 90; 3569-73: 1993 © National Academy of Sciences (USA).

Deduced amino acid sequence of human PH – 20 has 509 residues and is 59% identical with guinea pig PH – 20, suggesting that they may have a conserved function and immunogenticity. Southern blots showed that there is a single PH – 20 gene in human genome and that human PH–20 is testis-specific. The mouse PH-20 (Spam 1) is uniformly distributed over the surface of the caput sperm, while in caudal sperm it is localized to the anterior and posterior regions. The hyaluronidase activity of Spam 1 in acrosome intact caput sperm was significantly lower than that of caudal sperm. The increase in enzymatic activity in caudal sperm was accompanied by a reduction in the molecular weight. The protein structure from Spam 1 DNA sequence revealed four putative N-linked glycosylation sites. It was suggested that the activation of Spam1 during epididymal maturation is regulated by deglycosylation (Deng et al., 1999). Bee venom hyaluronidase is related to the amino – terminal part of mature guinea pig PH–20, in a region encompassing most of the bee enzyme and probably also the bulk of the 41 to 48-kDa segment of PH–20 (aa 17-307), 104 of 290 amino acids (36% including four cysteins are identical. These findings indicated that 41-48kDa segment of PH-20 is a hyaluronidase (Gmachl and Kreil, 1993)**(Fig.23.2)**. The amino-terminal region of 41-48 kDa linked via disulfide bonds to the 27-kDa carboxyl-terminal segment, which also contains the glycosyl phosphatidylinositol membrane anchor. The role of sperm hyaluronidase in cumulus penetration became clearer with the discovery that PH–20 genes had significant DNA homology with bee venom hyaluronidase. Subsequently, the hyaluronidase activity of PH–20 was shown to be involved in sperm penetration of the mouse and macaque cumulus (Myles and Primakoff, 1997); the

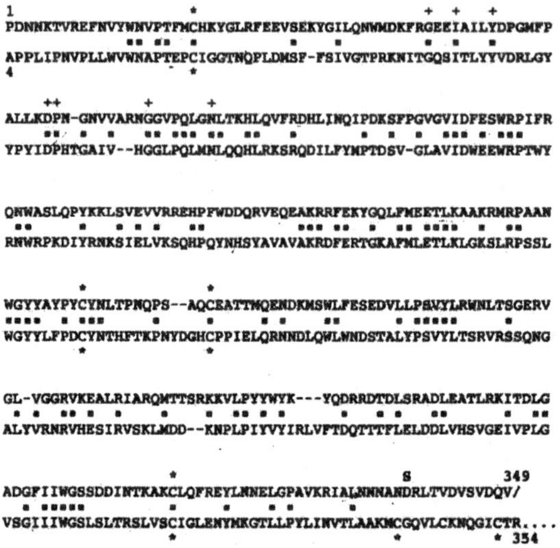

Fig. 23.2. Comparison of the amino acid sequence of bee venom hyalurondiase and PH-20 from guinea pig sperm. Identities betweens the two sequences are marked by a black square, and the cysteines are emphasizsed by asterisks. Gaps introduced to maximize homology are indicated by dashes. Residues also present in a region considered to form part of the ligand-binding pocket in β-integrin are indicated by a plus sign. PH-20 contains a total of 495 residues. Reprinted with permission from M. Gmachl and G. Kreil. Proc Natl Acad Sci (USA) 90; 3569-73: 1993 © National Academy of Sciences (USA).

plasma membrane-bound protein rather than a soluble enzyme appears to be responsible for this activity (Li et al., 1997a). The PH–20 and CD-46 were demonstrated on the inner acrosomal membrane of sperm induced to acrosome react by ionophore treatment and by zona binding. Ultrastructure of oocytes showed that anti-PH-20 lgG prevented sperm penetration of the macaque zona pellucida by interfering with secondary sperm-zona binding, rather than primary sperm-zona binding or the zona-induced acrosome reaction. It indicated that thus acrosin did not appear to be critical for sperm penetration of the macaque zona pellucida (Yudin et al., 1999).

Gmachl and Kreil (1993) predicted that mammalian sperm would contain a GPI-linked sperm surface hyalurondiase. The GPI-linked mouse sperm hyaluronidase is inhibited by the flavonoid apigenin, which is a competitive inhibitor of bovine testicular hyaluronidase as well as the recombinant form of PH–20. The biochemical characteristics of hyaluronidase activity, molecular mass, GPI linkage, and its presence on both acrosome – intact and acrosome-reacted sperm suggest that HDase is, in fact GPI – linked (Lin et al., 1994). Thaler and Cardulla (1995) quantified the activity of the soluble acrosomal HDase as well as membrane – bound HDase, on both acrosome-intact and acrosome- reacted mouse sperm. The membrane – bound hyaluronidase was specifically released by PI-PLC, indicating that it is GPI-linked. This showed the existence of two isoforms of hyaluronidase: a soluble form within the acrosomal vesicle, which is released during acrosomal exocytosis and a GPI-linked form, which is present on the surface of both acrosome–intact and acrosome–reacted sperm. Thus, hyaluronidase joins a growing list of GPI-linked cell surface enzymes including alkaline phosphatase, 5'-nucleotidase, acetylcholinesterase, alkaline phosphodiesterase I, trehalase, and p63 proteinase. It is important to note that mouse sperm have several prominent GPI-linked proteins, two of which appear exclusively in acrosome-intact sperm; several additional GPI-linked components appear in acrosome-reacted sperm, concomitant with the exposure of the inner acrosomal membrane following acrosomal exocytosis.

Guinea pig: The PH-20 is anchored on both the sperm plasma and inner acrosomal membranes. The derived amino acid sequence from cDNA shows a mature protein of 468 amino acids containing six N-linked glycosylation sites and twelve cysteines, eight of which are tightly clustered near the COOH terminus. The sequence indicates no relationship to the mouse sperm adhesion protein galactosyl transferase. Genomic DNA indicated PH20 as a single copy gene with a message of approximately 2.2 kb. Guineapig gene is a homologue of the PH-20 gene in mouse, rat, hamster, rabbit, bovine, monkey, and human genomic DNA, showing that the PH-20 gene is conserved among mammals (Lathrop, 1990). The PH-20 protein is localized to the posterior head plasma membrane of the mature sperm cell. Following the exocytotic acrosome reaction, it moves into the newly incorporated inner acrosomal membrane (IAM), placing it in a position favorable for a role in binding of sperm to the egg zona pellucida The PH-20 protein was observed to move against a concentration gradient in the posterior head plasma membrane and suggested that an active process served to concentrate PH-20 protein toward the boundary separating the posterior head and IAM regions. The sperm PH-20 acts during two different stages of fertilization. On acrosome-intact sperm, PH-20 has a hyaluronidase activity that is required for sperm penetration through the cumulus cell layer that surrounds the oocyte. On acrosome-reacted sperm, PH-20 has a required function in sperm-zona binding (secondary binding). An anti-PH-20 monoclonal antibody that inhibited sperm-zona binding had no effect on hyaluronidase activity. Conversely, apigenin, a hyaluronidase inhibitor, blocked PH-20 hyaluronidase activity by 93% without inhibiting sperm-zona binding. Similarly, anti-PH-20 monoclonal antibody that inhibited hyaluronidase activity by 95% inhibited sperm-zona binding only partially. Thus PH-20 is bifunctional and has two activities: a hyaluronidase activity and a second, separate activity required for secondary sperm-zona binding (Hunnicut et al., 1996a). A structural relationship exists between PH-20 and the soluble hyaluronidase released during the acrosome reaction. During the acrosome reaction, PH-20 undergoes endoproteolytic cleavage into two disulfide-linked fragments whereas the released sPH-20 is not cleaved, suggesting the possible activity of a membrane-bound endoprotease on PH-20. The sPH-20 may arise from the enzymatic release of PH-20 from its membrane anchor, possibly at the time of acrosome reaction (Hunnicut et al., 1996b).

Murine: Acrosome intact mouse sperm incubated with anti-PH-20 antibodies can not pass through the cumulus layer and thus can not reach the zona pellucida indicating that PH-20 enables acrosome intact sperm to penetrate the cumulus barrier. The murine PH-20 (SPAM1) gene was mapped to the proximal region of chromosome 6 (MMU 6). Based on the map location and physiological characteristics of its encoded sperm antigen, the gene is an attractive candidate for the sperm dysfunction seen in Robertsonian (Rb) (6.16) translocation heterozygotes and the reduced fertility of homozygotes. The expression of SPAM1 mRNA was found to be tissue specific and developmentally regulated, with a haploid expression (Zheng et al., 1997).

SPAM1 gene, which covers approximately 10.5-kb of genomic DNA, is encoded by four exons, and the splice site consensus sequence is maintained in all intron-exon junctions, similar to that reported for the human homologue. Two transcription initiation sites were detected. One was assigned to the residue C and the other (a minor site) to the residue G, at positions 1 and – 56 respectively. These are at 313 and 369 nt upstream of the translation initiation codon, ATG. In about 770-bp upstream region of *SPAM1* that has been cloned and sequenced, multiple transcription factor binding sites including a CRE were found. The eight-nucleotide CRE sequence (TGATGTCA) of the promoter region can bind to the transcription factor CREM in gel mobility shift assays using mouse testis nuclear extract, and the binding could be inhibited by a 28-bp oligonucleotide containing the CRE sequence. Similar binding and inhibition assays using rat nuclear extract suggested the existence of a rat CRE sequence and the involvement of CREM in rat SPAM1 expression. The CRE is necessary for the transcriptional activity of the murine promoter. The SPAM1 transcripts are absent in CREM-knockout mice suggesting that the murine *SPAM1* expression is under the control of CREM, and that this transcriptional regulator for SPAM1 might be conserved in other mammals, at least in the rat (Zheng et al., 1999).

In mice bearing the 6.15 Robertsonian translocation (Rb) sperm dysfunction associated with the Rbs has been shown to lead to transmission ratio (TRDs) in heterozygotes. The severity of the TRDs is directly related to the severity in the alteration of expression of the gene *SPAM1,* which maps to proximal mouse Chromosome 6 (Chr 6) near the translocation junction and encodes a sperm antigen with HDase activity. Catalytic kinetics studies indicated that reduced *SPAM1* (PH-20) HDase activity in the Rb(6.15) mice results from a qualitative defect, while for Rb (6.16) with the greater TRD, both a qualitative and a quantitative deficiency of *SPAM1* exist. Entrapment of spontaneous SPAM1 mutations, owing to recombination suppression near the Rb junctions, is proposed as the major underlying defect of the sperm dysfunction (Zheng et al., 2001).

The Spam1 or PH-20 is also synthesized by the epididymal epithelium, preferentially in the distal region, and is released into the luminal fluid. Whereas testicular and epididymal Spam1 have hyaluronidase activity at neutral pH, they are under different transcriptional regulation. The poly (A) tails are significantly shorter in the epididymis than in the testis. Epididymal Spam1 exists in three isoforms with the isoelectric point (pl) ranging from 7.3 to 9.0, whereas testicular Spam1 shows four Isoforms with pl ranging from 6.6 to 9.0. Two isoforms with a pl ranging from 7.6 to 9.0 were observed for caudal sperm. Lectin blotting recognized a 67-kDa band in the epididymis and caudal sperm, but not in the testis. Enzymatic deglycosylation studies confirmed the presence of an O'linked glycan in all the three cell types. Epididymal Spam1 is secreted predominantly as insoluble particles, which when treated with phosphatidylinositol-specific phospholipase C or Triton X-100, revealed that the majority of epididymal Spam1 is released with its lipid anchor, a form in which it can bind to sperm (Zhang et al., 2003).

Rat: Rat sperm 2B1 antigen, the orthologue of guinea pig PH–20 is a plasma membrane – bound glycoprotein that is endoproteolytically cleaved during passage through the epididymis and subsequently migrates from the tail to the acrosomal domain during capacitation. Unlike guinea pig PH20, however, sperm surface 2B1 is insensitive to phosphatidylinositol activity. It contains an internal sequence motif for attachment of a glycosyl phosphatidylinositol (GPI) anchor, and the cleavage from a single into a two-chain molecule causes a significant shift in the optimum pH of HDase activity. Functionally rat sperm 2B1 glycoprotein is attached to the plasma membrane via a GPI anchor and this is an important factor in its ability to migrate from the tail to the acrosomal domain during capacitation. Endoproteolytic cleavage of 2B1 serves to optimise its HDase activity immediately before fertilization, thereby facilitating penetration of spermatozoa through the cumulus oophorus (Seaton et al., 2000).

Rat sperm 2B1 is encoded by a 2.2-kb RNA transcript that is abundantly expressed in the testis. The deducted protein sequence contains 512 amino-acid residues with a strong candidate signal sequence and C-terminal transmembrane domain and shows a high degree of sequence similarity to guinea pig, rabbit, monkey, and human PH20 and a lower degree of similarity to honey bee and whiteface hornet venom HDase. Rat 2B1 antigen possesses HDase activity, suggesting that it is a bifunctional protein with putative roles in the dispersion of cumulus oophorus cells as well as zona adhesion. However, while it would appear that 2B1 is the rat homologue of the guinea pig PH20 antigen, they differ in a number of important biochemical respects (Including their mode of attachment to the sperm membrane and distribution between soluble and membrane-bound fractions), as well as in their localization on the sperm membrane (Hou et al., 1996).

Bull: In order to assess if the soluble HDase from bull testes is related to the PH-20 polypeptide, the sequence of the membrane-bound PH-20 hyaluronidase from bovine sperm was determined by cDNA cloning, and compared with the sequence of commercial bovine enzyme. It showed that the soluble 60-kDa HDase from bovine testes is a glycoprotein derived from the sperm PH-20 enzyme. As compared to the primary translation product of the PH-20 mRNA, it lacks the signal peptide at the amino terminus and 56 amino acids at the carboxyl end. A bull sperm protein of 80-kDa, which has many homologies with PH-20, has been identified (Lalancette et al; 2001). The 80-kDa bull protein possesses HDase activity.

Stallion: Stallion sperm expresses a surface-associated HDase localized to the posterior sperm head region in ejaculated sperm. The inner acrosomal membrane (IAM) displays intense HDase suggesting that the IAM and HDase plays significant role in zona penetration by sperm. The differential expression of PH-20 in ejaculated and epididymal sperm could be involved in cumulus penetration, sperm-egg recognition, and oolemmal fusion in this species (Meyers, 2001).

Fox: The fox PH20 shows high levels of homology to PH20 proteins of other species. Unlike other PH20 proteins the fox protein does not appear to be membrane associated through a GPI-linkage nor does it show the presence of a transmembrane domain at the C-terminus of the protein. The PH20 protein in fox sperm is located on the inner acrosomal membrane and its transcription is seasonally regulated, with the mRNA expressed during those months when spermatogenesis is at its peak (ten Have et al., 1998).

Monkey: The PH-20 is uniformly distributed over most of the monkey sperm head. However, after acrosome reaction, most spermatozoa labeled intensely over the anterior sperm head, whereas labeling of the posterior sperm head was greatly reduced. Transmission electron miscroscopy revealed gold particles distributed uniformly on the plasma membrane overlying the acrosome, the equitorial segment, and most of the post-acrosomal region (Overstreet et al., 1995). The soluble component released during the acrosome reaction included both the 64-kDa form and a 53-kDA form of PH-20. The 64-kDa polypeptide had a broad range, with the majority of the activity at neutral pH (pH 7). The 53-kDa polypeptide in sperm extracts only exhibited activity at acid pH (pH 4). The hyaluronidase activities of both enzymes were inhibited by apigenin (Cherr et al., 1999). Plasma membrane PH-20 had also an apparent molecular weight of 64-kDa and the optimum pH for its hyaluronidase activity was 6.5. However, the PH-20 associated with denuded sperm heads was localized by immunogold label to the persistent inner acrosomal membrane (IAM). Soluble form of hyaluronidase, which is released at the time of the acrosome reaction, is derived from the IAM. This soluble hyaluronidase is composed of both the 64-kDa form and 53-kDa form of PH-20. The 53kDa form appears to be processed from the 64-kDa form at the time of the acrosome reaction (Li et al., 1997). Exposure of sperm to mono-specific, bivalent-polyclonal antibodies to PH-20 causes a rapid clustering of PH-20. The predominant morphological consequence of PH-20 redistribution is its aggregation along the lateral edge of the sperm head. Anti-PH-20 IgG prevented sperm penetration of macaque zona pullucida by interferring with secondary zona binding or zona induced acrosome reaction. Since acrosin was not detected on inner acrosomal membrane of sperm, it was suggested that PH-20 but not acrosin is involved in sperm penetration of the macaque zona pellucida (Yudin et al., 1999).

Human: Nucleotide sequence for 1919-bp of human SPAM-1 (PH–20) from a series of overlapping cDNA clones confirmed the sequence identity within a 1527-bp ORF to be 71-74% to the guinea pig gene and the similarity to be 60% for the predicated protein of 509 amino acids. Southern blot analysis of human genomic DNA and DNA from somatic cell hybrids indicated that the gene *(SPAM1)* is unique and does not form part of larger family and that it maps to chromosome 7. The 1919-bp of the gene that has been cloned covers approximately 11-kb of genomic DNA and is encoded by at least 4 exons. Northern analysis of poly (A)+ mRNA from a range of 16 human tissues demonstrated that the expression of the gene as a single 2.4-kb transcript is strictly limited to the testis (Jones et al., 1995). Like macaque sperm PH-20, two forms of PH-20 are observed in human sperm. Non-capacitated sperm extracts, capacitated sperm extracts, and the acrosome-reaction supernatant had hyaluronidase activity at neutral pH (pH 7) and acid pH (pH4). The 64-kDa form had hyaluronidase activity at both neutral and acid pH, but the 53-kDa form was only active at acid pH. Human PH-20 had similar localization to that of macaque sperm (Li et al., 1997). Expression of PH – 20 is elevated in primary laryngeal cancer tissue and seems to be even higher in metastatic lesions compared with normal laryngeal tissue. PH-20 is a useful tumor marker and prognostic tool for laryngeal cancer (Godin et al., 2000).

23.6.2. Post-Testicular Modifications

In caput sperm, SPAM1 is uniformly distributed over the surface of the head while in caudal sperm, it is localized to the anterior and posterior regions. The hyaluronidase activity of SPAM1 in acrosome-intact caput sperm was significantly lower than that in caudal sperm. The increase in enzymatic activity in caudal sperm is accompanied by a reduction in the molecular weight. Caput sperm had a major band at 74-kDa and a minor band at 67-kDa; while for the caudal sperm showed a major band of approximately 67-kDa and minor bands of approximately 70 and – 56kDa. Additionally, the bands from caput sperm were 4.9 to 7.7-fold were less intense than from caudal sperm. Protein structure from *SPAM1* cDNA sequence revealed four putative N-linked glycosylation sites, and all sites were functional. After endoglycosidation, caput and caudal sperm showed the major band at 56-kDa, the size of the membrane-anchored polypeptide. Thus the activation of SPAM1 during epididymal maturation is regulated by deglycosylation (Deng et al., 1999).

The two common isoforms of SPAM1 observed in mice sperm are also present in the caput, corpus, and cauda epididymides, with the highest expression detected in the corpus. The endogenous production of enzymatically active (hyaluronidase) SPAM1 by epididymal cells is supported by the detection of steady-state SPAM1 epididymal mRNA in both wild type and germ cell-deficient mice. In situ hybridization showed the transcript to be localized to the principal cells of the epithelium, suggesting a mechanism for transportation of SPAM1 from the epididymal epithelium to sperm during their transit and storage in the cauda (Deng et al., 2000). In the epididymis SPAM1 is found in membranous vesicles of the principal cells of the epithelium in all three regions. Epididymal SPAM1 may be a secretory protein and released in the luminal fluid where it may effect post-testicular maturation and function of sperm. The *Spam1* transcripts from epididymal and testicular showed that *Spam1* were under different transcriptional regulation (Zhang and Martin-De Leon, 2001).

The cellular location of equine PH-20 is modified during epididymal transit. Western blots revealed significant differences in electrophoretic migration of PH-20 proteins from caput and cauda epididymal sperm where as no effect was seen after deglycosylation of PH-20. However, N-deglycosylation resulted in the loss of hyaluronidase activity of sperm from both epididymal regions, whereas O-deglycosylation or trypsinization did not affect hyaluronidase activity. In caput epididymal sperm, the PH-20 protein is distributed over the entire sperm head whereas in caudal sperm it is restricted to postacrosomal region (Rutllant and Meyers, 2001).

23.6.3. Contraceptive Effects

The immunization of male guinea pigs with PH-20 reproducibly results in infertility with a durations of 6-12 months or longer. The 100% infertility can reproducibly be induced in male guinea pigs immunized with purified PH-20. The males are extremely responsive to PH-20 immunization since infertility could be induced with a single injection of only 5 μg or a lower dose of PH-20. Surprisingly lower weight animals regained fertility (93%) whereas in heavy weight animals, fertility was not regained (Primakoff et al., 1997). To identify immunodominant regions on gpPH-20 that may be related to this contraceptive effect, Chan et al (1999)screened a set of overlapping peptides that cover the entire 494-residue sequence. Multiple clusters of peptide sequences showed specific reactivity. Octameric synthetic peptides indicated two regions (residues 94-119 and residues 424-444) to be highly immunogenic and both were surface accessible when native gpPH-20 is in solution or anchored on sperm surface. Both anti-peptide antibodies are specific for gpPH-20 and one of them inhibited hyaluronidase activity partially (Chan et al., 1999). Fully effective contraception has been achieved by immunizing either male or female guinea pigs with purified guinea pig PH-20. Unfortunately, PH-20 immunized males showed two separate effects. In almost all infertile, PH-20 immunized males, the cauda epididymidis were empty and contained no sperm, or contained only abnormal sperm. A second effect was the induction of experimental autoimmune orchitis (EAO), representing a report of EAO induced by a purified testis/sperm molecule of known functions (Tung et al, 1997).

Substrates for HDase

The substrate for HDase is haluronan (hyaluronic acid, HA), which is a ubiquitous component of the extracellular matrix (ECM) of vertebrates. This linear polysaccharide is composed of a repeating unit with the structure [D-glucuronic acid (1-β-3)-N-acetyl-D-glucosamine (1-β-4)]. The HA can form highly viscous solution and thereby influence the properties of ECM. The HA rich extracellular matrix of the cumulus oophorus of egg is known to facilitate fertilization. It has been suggested that HA may enhance fertilization in a number of species, and in macaque sperm. The HA has been shown to increase the acrosome reaction that follows sperm binding to the zona pellucida. It seems that HA increases intracellular Ca^{2+} in macaque sperm through interaction with plasma membrane PH-20. Perhaps, HA binding to plasma membrane PH-20 induces receptors that results in intracellular signaling (Cherr et al., 1999). It supports the hypothesis that HA in the cumulus matrix may act to prime the fertilizing sperm for induction of the acrosome reaction by constituents of the cumulus and/or zona pellucida (Cherr et al., 1999; Sabeur et al., 1998). A region of the PH-20 molecule, termed peptide 2 (amino acids 205-235), has amino acid charge homology with other HA binding proteins. The Peptide 2 region of PH-20 is involved in binding HA, which results in the cell signaling events related to the elevation of Ca^{2+} during sperm penetration of the cumulus (Vines et al., 2001). Among mannose as another ligand for hyaluronic acid binding protein1(HABP1), only N-linked mannosylated zona-glycoproteins bound to sperm HABP1. Labeled HABP1 interacts with ZP of intact oocyte of Bubalus bubalis, which can be competed with unlabeled HABP1 or excess d-mannosylated albumin. This suggests that sperm surface HABP1 may act as mannose binding sites for zona recognition (Ghosh and Datta, 2003).

23.7. GLYCOSIDASES

23.7.1. N-Acetyl-β-D-Glucosaminidase

The activity of N-acetyl-β-D glucosaminidase (EC:3.2.1.52) (β-NAG) in the mammalian male reproductive tract has been well documented. The epididymis shows the highest activity of this enzyme (Hall et al., 1996). The enzyme has been studied extensively in human tissues, where it occurs in two major isozymic forms, A and B, with molecular composition consisting of αβ and ββ subunits, respectively. The two isoforms have similar kinetic properties but can be differentiated on the basis of their reactivity towards natural substrates such as trisaccharides derived from hyaluronic acid and glycopeptides. In addition to N-acetyl-β-D-glucosaminidase activity, the native enzyme exhibits N-acetyl-β-D-galactosaminidase activity. Since these two enzymatic activities have not been separated biochemically, the enzyme is commonly known as β-hexosaminidase (Hex). The regionalization of its activity in mammalian epididymis, suggests a physiological role for the enzyme in the maturation of sperm as they pass through the epididymis and/or in the fertilization process (Kapoor and Gupta, 1986).

The highest specific activity of N-acetyl-β-D-glucosaminidase (β-NAG) has been found in the different parts of the epididymis, where the activity seemed to be partly in secretory and partly in non-secretory tissue bound form. Epididymal spermatozoa also possess moderate β-NAG activity. The major secretory forms of β-NAG in caput and cauda epididymis are distinct in elution profiles. The β-NAG activity derived from bull testis and caput epididymidis had smaller weights than the secretory enzymes in seminal plasma, seminal vesicle secretion and cauda epididymis. Disruption of the Hex gene encoding the β subunit of Hex has led to the generation of a mouse model of human Sandhoff disease. The morphological appearance and topographical arrangement of the testis cell Hex-deficient ($HexB^{-/-}$) and wild – type ($HexB^{+/+}$) mice appeared to be unaffected. Study revealed an increase in the size and number of lysosomes in all epithelial cell types lining the efferent ducts and entire epididymis as well as myoid cells and macrophages of the testis (Adam et al., 1999; Miranda et al., 2000). Brandelli et al (1994) reported the effect of different neoglycoproteins on the human sperm acrosome reaction. The BSA-Glc-Nac-induced acrosome reaction was inhibited by N-acetylglucosamine (GlcNAc), p-nitrophenyl-GlcNAc and purified soluble β-NAG. The induction of the acrosome reaction with BSA-Man could be inhibited by mannose, while soluble α-mannosidase was only partially effective. This suggested that binding sites for GlcNAc and mannose may be involved in the induction of acrosome reaction in human sperm. The characteristics of the BSA-GlcNAc induction suggest that the β-NAG may be the mediator of this effect (Brandelli et al., 1994). N-acetyl-β-D glucosaminidase is the major glycosidase activity present in vitelline envelopes of *Bufo arenearum* sperm. Optimum pH of β-NAG is 3.5 and inhibited at pH 7.5. The β-NAG is the only glycosidase that binds to vitelline envelopes in vitro under conditions that resemble natural fertilization media. The enzyme migrates as a single band with Mr of 45 kDa. The

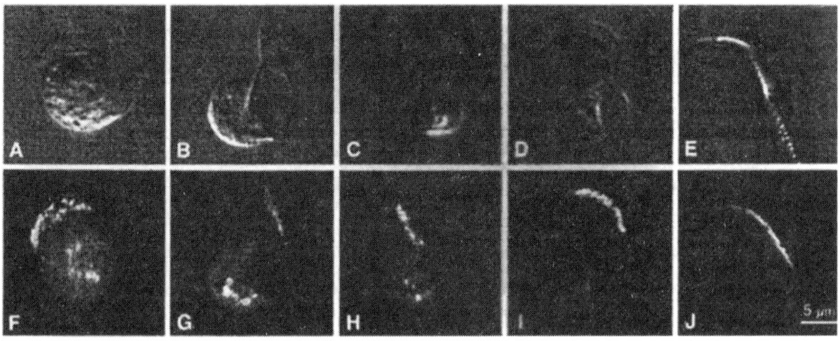

Fig.23.3. Immunolocalization of beta-glucuronidase in acrosome during elongation phase of spermiogenesis in rat. Panel A-E: various phases of acrosome formation; panel F-G: beta-glucuronidase activity during various phases of acrosome formation. Reprinted with permission from A Abou-Haila and DRP Tulsiani. Arch Biochem Biophys 379; 173-82: 2000 © Elsevier.

inhibition of the enzyme results in the inhibition of fertilization suggesting that β-NAG plays an important role in toad fertilization (Martinez et al., 2000).

23.7.2. β-D-Glucuronidase

β-D-glucuronidase (EC 3.2.1.31) an exoglycosidase, is one of the most extensively studied enzymes. It has been reported in all mammalian tissues including male reproductive tissues and spermatozoa. The sperm β-glucuroinidase is thought to be an acrosomal enzyme which, along with arylsulfatase, N-acetyl-β-D-glucosaminidase, and hyaluronidase has an important role in digestion of cumulus cells surrounding the ovulated oocyte. The lysosomal β-glucuronidase purified from the rat, mouse liver and human semen is homotetrameric glycoprotein of an apperent molecular mass of280-kDa (Gupta and Singh, 1984). The microsomal β-glucuronidase in mouse tissues consists of four identical subunits, which correspond to noncovalent complexes formed between the pro-enzyme (X-form) and one to four molecules of egasyn, 64-kDa glycoprotein with carboxyl esterase activity (Abou-Haila et al., 1999). Although β-glucuronidase has been extensively studied, understanding of the expression, properties and localization of various molecular forms of the enzyme in spermatogenic cells and spermatozoa is very limited. β-D-glucuronidase from testicular germ cells and spermatozoa has been characterized (Gupta and Singh 1984; Abou- Haila et al., 1999). Germ cells (spermatocytes, round spermatids, and condensed/elongated spermatids) were shown to contain only the lysosomal form of the enzyme that was localized in the Golgi apparatus, Golgi vesicles, lysosome like multivesicular bodies of spermatocytes, spermatids, and the forming/formed acrosome of spermatids/spermatozoa. **Fig.23.3** shows the distribution of β-D-glucuronidase during successive formation of the sperm acrosome in the elongation and maturation phases and also in the fully developed acrosome of a testicular spermatozoon.

23.7.3. β-D-Galactosidase

Acid β-D-galactosidase (EC3.2.1.23) is an exo-glycosidase that cleaves β-linked terminal galactosyl residues from a variety of natural and artificial substrates. The enzyme has been reported in all tissues examined, including male reproductive tissues such as testis and epididymis. When cultured in vitro, rat epididymal principal cells synthesize the enzyme in precursor forms that are processed to mature form before being secreted in the medium. The cDNA for human placental and testicular β-D-galactosidase has been cloned and sequenced. In spermatozoa, β-D-galactosidase is present in the cytopolasmic droplet and the acrosome. Rat sperm contains a different β-D- galactosidase, which has also been reported in sperm membranes. Acrosomal enzyme is readily released during sperm disruption. It has been suggested that sperm associated β-D- galactosidase has a role either as a lectin or as a hydrolase during fertilization. Only 10-14% of total β-D- galactosidase activity of epididymal fluid was associated with spermatozoa. The remaining enzyme activity is present in the luminal fluid. Enzyme showed a progressive increase in spermatozoa from the caput to the corpus or proximal cauda followed by a sharp decline in the distal cauda epididymis sperm. However, the changes in β-D- galactosidase activity during sperm maturation in the epididymis were accompanied by changes in the molecular form of the enzyme. The sperm associated β-D-galactosidase did not appear to be due to adsorption and / or binding of the luminal fluid β-D-galactosidase 97-kDa form in the fluid from the caput, and two forms of 97-kDa and 84-kDa from the fluid of corpus and cauda regions. The observed difference in the molecular forms of the luminal fluid was found to be due to differential glycosylation, since deglycosylation of various forms of β-D- galactosidase generated a single immunoreactive form of 70 kDa (Skudlarek et al., 1993; Chyko et al., 2000).

The Luminal Fluid β-D-Galactosidase: Whereas the enzyme activity of β-D-galactosidase from epididymal luminal fluid was optimum at acidic pH (3.5) with a synthetic PNP substrate and as expected for an acid hydrolase, a neutral pH optimum (6.8) occurred with a [^{3}H] galactose labelled glycoprotein substrate. Thus, this enzyme may be active in the neutral environment of the epididymal lumen. Predominant form of epididymal spem β-D-galactosidase gradually shifts from 82- to 80-kDa as spermatozoa migrate from the caput to the cauda epididymis. When the luminal fluid or sperm forms of β-D- galactosidase were de-glycosylated, the resultant protein was resolved at the position of 70-kDa. This suggested that the luminal fluid and acrosomal forms of β-D-galactosidase are differentially glycosylated.

The enzyme was localized within acrosomal cap of spermatids and in acrosome of epididymal spermatozoa and within cytoplasmic droplets, cap of spermatids and germ cells. Spermatids produce two forms of β-D- galactosidase, 90 and 88-kDa and a 56kDa protein. Treatment of β-D- galactosidase with N-glycannase or Endo-H revealed that both 90- and 88-kDa forms become a 70-kDa polypeptide showing the presence of an extensive N-linked high mannose / hybrid-type glycans on these proteins. Treatment of the 56-kDa form of β-D- galactosidase with Endo H or N-glycanase resulted in the appearance of 52 and 50-kDa forms, respectively. It seems that high molecular weight β-D-glycosidase is associated with acrosome, while 50-kDa is present in cytoplasmic droplets (Chyko et al., 2000). In another study (Skudlarek et al., 2000), the germ cells from rat testis were found to contain β-galactosidase and four other glycohydrolases (β-D-galactosidase, α-D-mannondam, α-L-fucosidase, and β-N-galactosidase). With the exception of α-L-fucosidase, the other enzymes demonstrated a two- to three fold higher activity per cell in spermatocytes than in round spermatids indicating several molecular forms of β- galactosidase. One of these forms (62-kDa) was seen only in round spermatids. The net result was the formation of predominantly 64- and 62-kDa forms in spermatocytes and round spermatids respectively (Skudlarek et al., 2000).

23.7.4. Arylsulfatase A

Arylsulfatase A co-purified with β- galactosidase from rabbit testes indicated that both enzymes aggregate into macromolecular complex at pH 4.0, while both dissociated at pH 8.0. β-galactosidase/arylsulfatase A are co-localized on the sperm surface and in the acrosome and post-acrosomal regions of sperm. Throughout the zona induced acrosome reaction both enzymes remain associated with the detached acrosomal cap and post-acrosomal region of acrosome reacted spermatozoa. Because the acrosome is an acidic subcellular compartment, internal β-galactosidase and arylsulfatase A are probably aggregated in acrosome intact spermatozoa and dissociate as they are exposed to increase in pH during the acrosome reaction (Nikolajezyk and O'Rand, 1992).

23.7.5. Sperm α-L-Fucosidase

Sperm acrosome contains α-L-fucosidase, an exo-glycohydrolase that cleaves terminal α-fucosyl residues from glycoproteins and glycolipids. Although the biological significance of acrosomal exo-glycohydrolases remains elusive, one study reported abnormal spermiogenesis and sperm maturation in dogs suffering from fucosidosis disorder caused by severe deficiency in lysosomal α-fucosidase (Veermachaaneni et al., 1998). It may also be of significance to note that male English springer spaniels, genetically deficient in α-fucosidase are sterile. Although testis specific glycosidases appear to exist in testis of certain species, it is not clear whether acrosomal glycosidases are testis-specific isoenzymes or not. α-fucosidase activity has been detected in mammalian sperm, seminal plasma, and reproductive organs of the bull, and has been localized to the acrosome of bull spermatozoa. Isoelectric focusing of NaCI solubilized extracts of rat epididymal sperm indicated the presence of one major isoform with a pI near 7.2±0.1 (Form A) and a minor acidic isoform with a pI near 5.2±0.1 (Form B). The NaCI-solubilized α-L-fucosidase from epididymal sperm contained one or two closely spaced subunits of about 54 and 50 kDa. The relationship of the plasma-membrane associated sperm α-L-fucosidase to the lysosomal α-L-fucosidase is not completely understood but both enzymes have two non-identical subunits and a multimeric structure with at least some antigenic epitopes in common. The two fucosidase isoforms present in sperm from the distal cauda had similar pH-activity curves (with optima near pH 7) and comparable K_M values for 4-methylumbelliferyl α-L-fucopyranoside. Isoform A is considerably more thermostable than isoform B. Neuraminidase treatment leading to disappearance of isoform A and appearance of form B suggests that isoform A is derived by sialylation of isoform B near the end of epididymal maturation (Abascal et al., 1998). Hancock et al., (1993) showed the presence of the 54-kDa form of α-fucosidase in testicular germ cells and the existence of a 52 kDa mature polypeptide in epididymal sperm. Acquisition and processing of sperm α-fucosidase takes place prior to or concomitant with the arrival of sperm in the epididymis. Hancock et al, (1993) demonstrated the synthesis and N-glycosylation of α-fucosidase in rat pachytene spermatocytes and round spermatids and provided evidence for the processing of the enzyme during transit through the ex-current duct system. It was suggested that in addition to the enzyme synthesized by testicular germ cells, epididymal sperm may acquire and process a soluble form of α-fucosidase from luminal fluid. Comparative characterization of human seminal fluid and sperm α-L-fucosidases showed that seminal fluid α-L-fucosidase has a broad pH optimum curve with a number of near equal maxima between pH 4.8 and 7.0, while sperm fucosidase has a major optimum between pH 3.4 and 4.0. Seminal fluid contains three to six isoforms of α-L-fucosidase with isoelectric points (pI) of 5-7, compared to two distinct isoforms of fucosidases with pI at 5.2±0.2 and 7.0±0.2. It appears that seminal fluid fucosidase corresponds to a protein band with a M_r of 56 kDa while sperm possess a fucosidase of 51kDa. The overall results indicated the presence of a low-abundance plasma membrane associated human sperm α-L-fucosidase, which is different in its properties from human seminal fluid α-L-fucosidases, and whose function is not well known (Alhadeff et al., 1999; Khunsook et al, 2003). A mammalian plasma membrane-associated α-L-fucosidase of 51-kDa has been purified (Khunsook et al., 2002). Isoelectric focusing indicated the presence of two major α-L-fucosidase isoforms (pIs 6.5 and 6.7) and a possible minor isoform (pI 6.3). The unusual properties of human sperm α-L-fucosidase argue in support of a potentially important, but presently unknown, role for this enzyme in human reproduction.

23.7.6. Mannosidase

Mannosidase is mainly a sperm membrane protein. However, due to its role in fertilization, it shall be pertinent to discuss mannosidase with group of glycohydrolases already discussed. Sperm surface mannosidase has been shown to be present in rete testis and is modified during spermatogenesis and sperm maturation. Most of the mannosidase activity present in the germ cell extract is antigenically similar to the enzyme present on the cauda spermatozoa. Cell fractionation techniques indicated that the germ cell associated mannosidase activity is an integral component of the plasma membrane, and sperm surface mannosidase is first expressed on the testicular germ cells and localized on the periacrosomal region of the sperm head. In addition, membrane-bound α-D-mannosidase undergoes proteolytic processing during maturation of sperm. This was revealed by the presence of three specific immunoreactive bands (apparent Mr = 135, 125 and 115-kDa) in the membranes from testis, caput, and corpus spermatozoa. However, the cauda sperm plasma membranes showed only one immunoreactive band of apparent Mr 115-kDa. The involvement of the sperm enzyme in sperm egg interaction is suggested by several studies such as; i) the enzyme is an integral membrane component, ii) the enzyme is optimally active at neutral pH, and its catalytic site is oriented towards the sperm surface, iii) the mouse ZP2 and ZP3 contain high mannose/hybrid-type glycan chains. These haptens are believed to be the ligand for the sperm mannosidase. The inclusion of the sperm mannosidase inhibitors or competitive sperm egg binding assays in vitro inhibited the number of sperm bound per egg in dose-dependent manner in the mouse and rat. Moreover, the increase in sperm-associated mannosidase during maturation correlated with the fertilizing ability of rat spermatozoa. Results are consistent with the receptor-like role for the sperm mannosidase. A Polyclonal antibody raised against the enzyme purified from rat epididymal luminal fluid cross – reacted with the α-D-mannosidase activity present on cauda spermatozoa and plasma membranes prepared from rat testis in an enzymatically inactive precursor form that undergoes lytic processing in the testis and epididymis to generate an enzymatically active mature form in the epididymal cauda spermatozoa (reviewed in Pereira et al., 1998). α-D-mannosidase involved in N-glycan processing showed cell specificity and distinct sub-compartmentalization within the Golgi apparatus of cells in the testis and epididymis (Igdoura et al., 1999). The cell-specific expression and distinct Golgi sub-compartmental localization suggests that α-mannosidases play different roles during N-glycan maturation.

The cDNA encoding a mouse homolgue of porcine epididymis – specific 135-kDa α-D-mannosidase (MAN2B2, D28521) codes a protein of 1018 amino acids in its open reading frame, with 62% amino acid identity to that of porcine MAN2B2 **(Fig.23.4)**. Adult mouse testis contained higher amounts of mRNA encoding the MAN2B2 homologue than the epididymis, though porcine MAN2B2 was mainly expressed in the narrow region between the caput and corpus epididymis. The mRNA of the mouse MAN2B2 homologue was exclusively expressed in type A spermatogonia at stages IX-XI of spermatogenesis until the cell developed into type B spermatogonia. It seems that the expression of the MAN2B2 homologue can serve as a good marker for the late stages of type A spermatogonia and may have an important

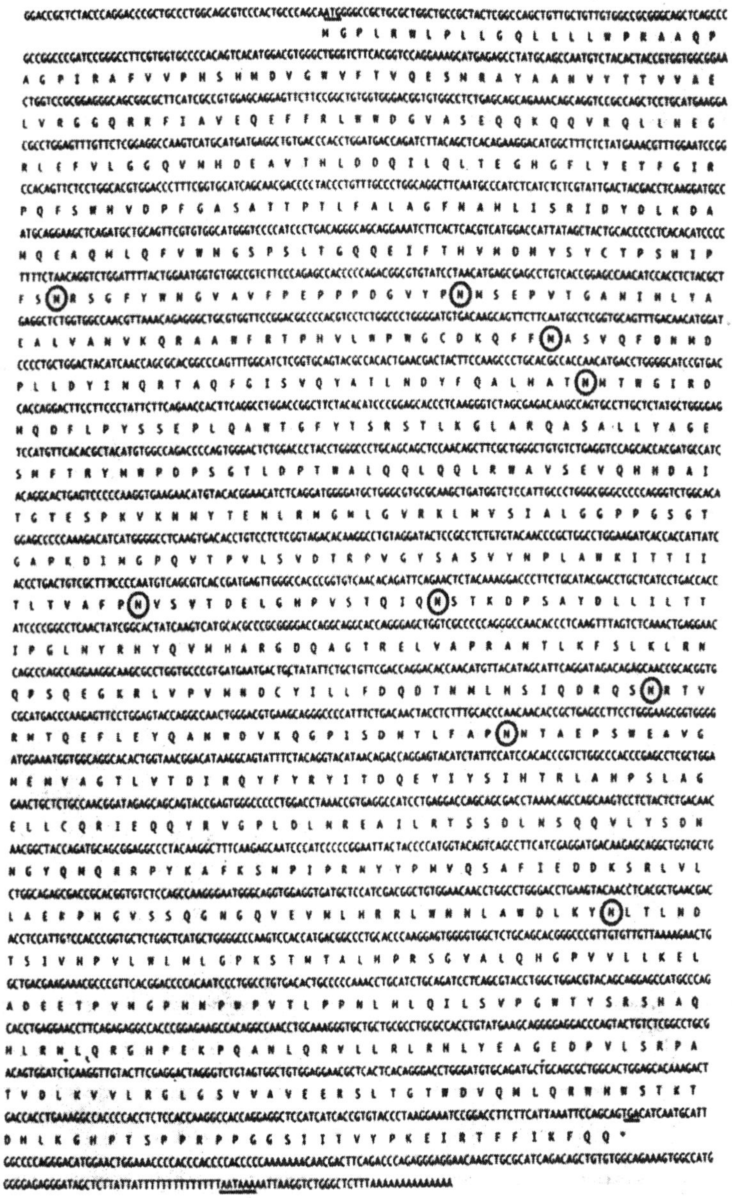

Fig. 23.4. The nucleotide and predicted amino acid sequence of mouse mannosidase (MAN2B2) cDNA. The first potential translation initiation codon ATG is preceded by 52 bp containing Kozak's conserved sequence. An open reading frame is formed by 3054 bp that code 1018 amino acids, followed by 198 bp of 3' untranslated region. The polyadenylation signal AATAAA is located 20 bp upstream of the polyadenylation site. Reprinted with permission from S. Hiramoto et al.. Biochem Biophys Res Commun 241; 439-45: 1997 © Elsevier.

role to play in the early steps of spermatogenesis in mice (Hiramoto et al., 1997; Tascou et al., 2000).

A135-kDa protein from porcine cauda epididymal fluid has been purified. This protein is secreted by epididymal epithelial cells and then binds to the equatorial segment of sperm head as a 17-kDa fragment. During epididymal maturation, the fragment migrates on the sperm surface and is localized to a crescent-shaped area just behind the acrosome and to the apical rim of the sperm head. Okamura et al, (1995) isolated cDNA encoding the 135-kDa protein in order to elucidate the mechanism of binding and localization on the sperm surface and to investigate its role in sperm maturation. The overall amino acid sequence of 135-kDa protein was significantly homologous (25.7%) to the α-mannosidase precursor of *Dictiostelium discoideum* (P34098). The 135-kDa-protein could digest both p-nitro-phenyl-α-D-mannoside and high mannose oligosaccharide (Man_8 Glc NAc_2), which suggested that it is a α-mannosidase homologue **(Fig. 23.4)**. Rat sperm plasma membrane α-D-mannosidase is very similar to the porcine epididymal enzyme (MAN2B2) in properties such as molecular mass, optimum pH and substrate specificity. It has been strongly suggested that the two enzymes from two species are closely related.

23.8. PROTEINASES

In addition to glycohydrolases, a number of proteinases have been identified in sperm acrosome. These include serine specific proteinase (acrosin / proacrosin), cysteine proteinases (cathepsins), metalloproteinases, a dipeptidyl carboxypeptidase and calpain II.

23.8.1. β Acrosin

Ontogeny: β-Acrosin, a serine proteinase present in the acrosome of the sperm, has been implicated in recognition and binding of the sperm to the zona pellucida, the ovum and the sperm penetration through the pellucida. The enzyme is synthesized in a zymogen proacrosin form, which is activated to the mature enzyme during acrosome reaction. The biosynthesis of proacrosin occurs in early haploid spermatids and its primary sequence has been elucidated in boar, bull, rat mouse, guinea pig and human (Tranter et al., 2000). Using Northern blot analysis of RNAs from various mouse tissues, the acrosin gene transcript was found to be present only in testis. The 1,800-base acrosin message was first detectable in 18-day-old testis. In situ hybridisation analysis suggested that the acrosin gene is expressed only in early stages of the round spermatid in many species (Kashiwabara et al., 1990). In the mouse, proacrosin is transcribed in spermatocytes, but protein synthesis starts in haploid spermatids and is restricted to the emerging acrosome. In another study, four lines of transgenic mice harboring 2.3-kb of 5'-untranslated region of the rat proacrosin gene fused to the CAT-gene was found to be present in 10-100 copies per genome in different strains. The 5'untranslated region of rat proacrosin gene could properly direct CAT gene expression to spermatocytes and CAT-mRNA translation to round spermatids as in mouse. It showed that cis-acting sequences are conserved in mouse and rat. Acrosin is involved in dispersal of the acrosomal matrix after acrosome reaction. Thus, the control of the ZP glycan chains over proacrosin activation may regulate both sperm penetration rate and limited proteolysis of zona pellucida proteins (Moreno et al., 1999). It seems that β-Acrosin binds non-enzymically to zona pellucida through interaction of basic residues located on the surface of β-acrosin, and forms strong ionic bonds with polysulphate groups of the carbohydrate moiety of the zona glycoproteins.

A

RAM 1 DNTTCDGPCG VRFRQNRQGG VR
BOAR 1 RDNATCDGPCG LRFRQKLESG MR

B

1 VVGGMSAEPG AWPWMVSLQI FMYHNNRRYH TCGGILLNSH WVLTAAHCFK NKKKVTDWRL
61 IFGANEVVWG SNKPVKPPLQ ERFVEEIIIH EKYVSGLEIN DIALIKITPP VPCGPFIGPG
121 CLPQFKAGPP RAPQTCWVTG WGYLKEKGPR TSPTLQEARV ALIDLELCNS TRWYNGRIRS
181 TNVCAGYPRG KIDTCQGDSG GPLMCRDRAE NTFVVVGITS WGVGCARAKR PGVYTSTWPY
241 LNWIASKIGS NALQMVQLGT PPR

C

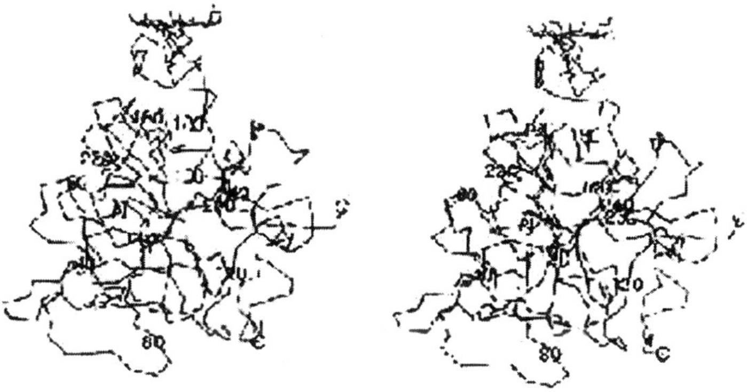

Fig.23.5. (A) Amino acid sequence of L chains of beta acrosin from ram and boar sperm (Accession 1FIW_L and Accession 1FIZ_L), and (B) A chain of beta acrosin from boar spermatozoa (Accession 1FIZ_A). Source: http//www. ncbi.nlm.nih.gov. (C) Structure and stereo diagrams showing Cβ traces of the structures of ram beta acrosin. Reprinted with permission from R. Tranter et al. Structure 8; 1179-88: 2000 © Elsevier.

Characterization: β-acrosin from boar and ram comprises two polypeptides: a heavy (H) chain and a light (L) chain. The protein has six disulphide bonds: four intra-chain bonds within the heavy chain and two inter-chain linkages between the heavy and the light chains. The interchain disulfide bonds are linked between L-5 and H-114, and L-9 and H-122. While heavy chain of both β-acrosins exhibits a structural homology to trypsin, the light chain is covalently associated in a manner, similar to blood coagulation enzymes. Nonetheless, the unique arrangement and strict conservation of the two disulphides in all forms of β-acrosin suggests that the light chain might play a functional role beyond activation from the proenzyme.

In boar, single chain glycoprotein of proacrosin has an estimated molecular mass of 55-kDa. During its activation, the two chains glycoprotein, α-acrosin is formed with a light chain and a heavy chain held together by two disulfide bridges. The light chain consists of 23 amino acid residues and one carbohydrate attachment site. The 37-kDa heavy chain with approximately 320 amino acids was found to contain the proteolytically active serine centre. Without loss of enzymatic activity, smaller isoforms of acrosin called β- and γ-acrosin can be found after proacrosin activation. The c-DNA clones for the boar pre-proacrosin showed the nucleotide sequence of 1418 bp and includes an insert of 46-bp of 5'-untranslated region. The deduced amino acid sequence includes the active-site residues histidine, asparagine and serine of the catalytic triad of the serine proteinase super-family and is collinear with the boar acrosin light chain and of a small region of the NH_2-terminal sequence of the heavy chain, determined by amino acid sequencing. The pre-proacrosin cDNA contains at the 3' end a 381-bp sequence, which codes for an amino acid sequence not yet found in any other serine proteinase. This amino acids sequence is rich in proline (42 out of 127 amino acids) and is suggested to be involved in the recognition and binding of the spermatozoa to the zona pellucida of the ovum. The mRNA for preproacrosin is synthesized as an approximately 1.6-kb-long molecule only during postmeiotic stages of boar and bull spermatogenesis (Adham et al., 1989)(**Fig.23.5a**).

The deduced amino acid sequence from cDNA of mouse acrosin indicated that mouse acrosin is initially synthesized as a single chain polypeptide with 16-residue signal peptide followed by a 23-residue light chain and then a 394-residue heavy chain. Mouse proacrosin contains 417 amino acid residues with a calculated molecular mass of 47 kDa. The cDNA-derived sequence of mouse proacrosin shows a high degree of similarity with human and porcine proacrosins and major portion of bovine trypsin, including the active site residues, the recognition site for substrate, the location of 12 cysteine residues, and two potential N-glycosylation sites. The sequence homology suggests that mouse proacrosin is converted to a mature acrosin, which consists of the light and heavy chains with a combined molecular mass of 35587 Da, by cleavage of the peptide bond between Arg^{23} and Ile^{24} and sequential removal of 23-, 26- and 50-residue COOH-terminal segments (Kashiwabara et al., 1990).

The histidine, aspartic acid and serine are located at positions 70, 124, and 222, respectively, in the mouse proacrosin sequence. It was suggested that aspartic acid at position 216 acts as recognition residue for specific hydrolysis of peptide bonds containing arginine and lysine. These sequence similarities also indicated that the NH_2-terminal amino acid is Lys in mouse proacrosin. Human and porcine proacrosins have a proline-rich segment in the COOH-terminal region. In mouse proacrosin, most of the proline residues in the corresponding region are substituted by other amino acids; this sequence is rich in serine, leucine, and threonine residues. However, the sequence around the - NH_2 and COOH-terminal cleavage sites of porcine proacrosin during its maturation are well conserved among mouse, porcine, and human proacrosins.

Gene structure: The acrosin gene consists of five exons separated by four introns. Organization of this gene is very similar to those of the genes for other typical serine proteases, except for the phase class of the first intron. The start site of transcription initiation in the acrosin gene is heterogenous, including three major sites. Thus, the structure and organization of the mouse acrosin gene are different from those of the human gene in two respects: the number of transcription initiation sites and the phase class of the third intron. The putative promoter regions of the mouse and human acrosin genes lack typical sequences of TATA, CAAT, and GC boxes, but contain a consensus, GGGTGGG, known to be specific for the phosphoglycerate kinase-2 gene, and the protamine-1 and -2 gene that are uniquely expressed during spermatogenesis (Watanabe et al., 1991). A comparison of the 5'flanking sequences between the mouse and human acrosin genes shows two highly homologues regions. It is suggested that the GGGTGGG motif is a candidate for the cis-acting elements, which play important roles in the regulation of acrosin gene transcription.

The proacrosin gene is first transcribed in late pachytene spermatocytes and translated in spermatids, showing that the mRNA for proacrosin is under translational control. A highly conserved 17bp (F) element: 5'-AACTTCAAATGGCTCC/T-3') is located in the proacrosin promoter. By using this DNA element as a target in yeast one hybrid assay, a cDNA fragment coding for the C-terminal part of the transcription factor YY1 was isolated, and its binding to this F1 element was confirmed in EMSA (see Chapter 7).

3D Structure: The crystal structures of ram and boar β-acrosins have been solved at 2.1 Å and 2.9 Å resolution, respectively. The heavy chain, for which 260 residues in ram acrosin and 263 residues in boar acrosin are visible, adopts a folded structure similar to that observed in trypsin and other serine proteases such as blood coagulating factor Xa (Fxa), tissue plasminogen activator (tPA) and urokinase (uPA). This fold comprises a bilobal structure with each sub-domain formed from anti-parallel β barrel structure. The sub-domains are related by a pseudo-2 fold axis, with the catalytic triad (His-57, Asp102 and Ser-195) located close to the bilobal interface. All the core features of the trypsin fold were also present in β-acrosin, with significant changes being restricted to surface loops. Many of the inserted loop regions in β-acrosin are similar to those present in urokinase (uPA). Two features, which appear unique to β-acrosin include an insertion of five residues in the 74-79 loop and an additional small helix at the carboxyl terminus formed by residues 242-252 **(Fig.23.5c)** (Tranter et al, 2000).

In crystals of boar β-acrosin, the carboxyl terminus of the heavy chain is inserted into the active site of the neighbouring molecule. The enzyme structure shows distinctive positively charged surface "patches" close to the active site, in association with carbohydrate from adjacent site, and capable to bind sulfate ions. Two separate effector sites, a site for proteolytic activity, and a positively charged region on the surface in the structure that may act as a receptor for ZP glycoproteins were evident in β acrosin. The spatial proximity of these two effector sites suggests synergy between them. Though the overall structures of ram and boar β-acrosin heavy chains are very similar, yet they differ in the carboxy-terminus, which shows five additional residues beyond the terminal helix in boar β-acrosin (Tranter et al, 2000).

The light chain in both structures is located on the opposite face of the heavy chain from the active site, similar to the accessory chains found in the blood coagulation enzymes. However, the composition and conformation of the light chains in all these serine proteases differ significantly. In both forms of β-acrosin, the six residues segment between the two cysteins where the peptide chain is anchored to the heavy chain adopts a similar, extended conformation without having any secondary structure. From relative positions of light chain carboxy terminus and the amino terminus of the heavy chain, it was apparent that considerable distortion could be necessary to form the pro-enzyme in a single polypeptide chain. Apart from the disulphide linkages, the light chain makes very limited contacts with heavy chain (Tranter et al, 2000).

***The Active Site of* β*-acrosin*:** The conformation of the active site residues in ram and boar β-acrosin bears considerable homology to the similar region of trypsin and other trypsin-like enzymes. The S1 pocket, which regulates specificity for Arg and Lys substrates, is highly similar in crystal structures of β-acrosin and in trypsin like enzymes. In all cases the pocket is surrounded by the segments 213-220, and 225-29, with the conserved Trp-215 at the entrance to the packet and Asp-189 and Ser-195 at its space. The S2 and S4 pockets confer inhibitor specificity for proteinases, whereas the S2 pocket is marked by a tyrosine at position 99, leading to restricted accommodation for side chains in the P2 position and hence a preference for substrates with Gly at this location. Significantly many of the reported auto-cleavage sites in the carboxyl terminus of the heavy chain in β-acrosin have proline, leucine, or valine in the P2 position. Within proteinases, β-acrosin is unique in contributing two arginines into S4, a feature, which can help in the design of specific β-acrosin inhibitors. Although the active site of boar β-acrosin is similar to that of ram β-acrosin, the molecular packing in these crystals is entirely different (Tranter et al 2000).

β-Acrosin as a Receptor for ZP Glycoproteins: Both proacrosin and β-acrosin have been considered to act as secondary binding molecules for receptors on zona pellucida glycorproteins. The presence of two regions of concentrated positively charged amino acids on the periphery of the active site is a consistent feature in β-acrosin. These regions differ from those described for thrombin and are formed primarily by residues from the 35-40, 60-65, and 87-94 loop regions (exosite I) and the 145-150, 170-178, 186-188, and 220-225 loops (exosite II). Many of the positively charged residues in these regions are highly conserved in β-acrosin. The two patches straddle the catalytic active site described above, forming a semi-continuous ring of positive surface charge at its periphery. β-acrosin has also strong affinity for sulfated polymers. The affinity for anions involves residues 60-65, 87-94 and 250-253. The features of the ram and boar crystal structures demonstrate the ability of two exosite regions to act as generic anion receptors. There are two conserved N-linked glycosylation sites in mammalian acrosin sequences: one close to the N terminus of the light chain at Asn-2 and the other in the heavy chain at Asn-169. In ram β-acrosin five sugar residues are located on the periphery of the face of the enzyme containing the active site. The overall results reveal that β-acrosin has many features common with trypsin like proteases such as active site, heterodimer and anchoring similarity (Tranter et al., 2000).

Functions of* β*-Acrosin: During fertilization, proacrosin auto-activates to form β-acrosin, in which there is a "light" chain cross-linked to a "heavy" chain by two disulphide bonds. β-acrosin is thought to be multifunctional with its role in acrosomal exocytosis as receptor for zona pellucida proteins, and as a protease to facilitate penetration of spermatozoa into the egg. There are conflicting views on the role of β-acrosin during fertilization. Traditionally, it has been known that acrosin helps in the penetration of spermatozoa through zona pellucida, where it acts as a lysin. However, doubts have been raised on numerous occasions notably in knock out mice in which sperm lacking acrosin gene are still capable of penetrating the zona. Besides, the dispersal of acrosomal contents following acrosome reaction, there are evidences, which suggest its role as a secondary binding protein during sperm-egg interaction (Wasserman, 1999).

Although Acr-/- mouse sperm lacking the acrosin protease activity still penetrated zona pellucida and fertilized the egg, the mutant sperm exhibited a delay in penetration of the zona pellucida solely at the early stages after insemination. The ability of sperm from Acr+/+, Acr+/- and Acr-/- mice to adhere and bind to the zona pellucida, did not change in vitro, and the release of acrosomal proteins from Acr-/- mouse sperm was significantly delayed during acrosome reaction despite normal membrane. It indicated that the delayed sperm penetration of the zona pellucida in the Acr-/- mouse resulted from the altered rate of protein dispersal from the acrosome and provided the evidence that the major role of acrosin is to accelerate the dispersal of acrosomal components during acrosome reaction. These results reflected the difference of the serine protease system for the sperm penetration through the egg zona, possibly explaining why the acrosin-deficient mouse sperm were capable of penetrating the zona pellucida (Yamagata et al., 1998). In another report, homozygous Acr+/+ mice were fertile and yielded litters comparable in number and size to those of Acr-/- mice, sperm homozygous to Acr-/- were able to penetrate the zona pellucida, fertilize the ovum, and produce viable offspring. However, spermatozoa lacking acrosin showed a delayed fertilization. Thus Acr-/- sperm had a selective disadvantage when they were in competition with $Acr_{+/+}$ sperm (Adham et al., 1997).

Sp32 - As Acrosin Binding Protein: The SP32 is a 32-kDa protein that acts as a binding protein specific for 55-, 53-, and 49-kDa forms of pro-acrosin, but not capable of binding to a 43-kDa acrosin intermediate , or to the 35-kDa mature acrosin. Sp32 significantly accelerated

A

```
  1 MWGSRAQQSG PDRGGACLLA AFLLCFSLLH AQDYTPSQTP PPTSNTSLKP RGRVQKELCG
 61 KTKFQGKIYG GQIAKAERWP WQASLIFRGR HICGAVLIDK TWLLSAAHCF QRSLTPSDYR
121 ILLGYNQLSN PSNYSRQMTV NKVILHEDYS KLSRLEKNIV LIQLHHPVIY STHIFPACVP
181 DGTTKVSPNN LCWISGWGML SADKFLQAPF PLLDAEVSLI DEEECTTFFQ TPEVSITEYD
241 VIKDDVLCAG DLTNQKSSCR GDSGGPLVCF LNSFWYVVGL ANWNGACLEP IHSPNIFTKV
301 SYFSDWIKQK KANTPAADVS SAPLEEMASS LRGWGNYSAG ITLKPRISTT LLSSQALLLQ
361 SIWLRIL
```

B

```
  1 MCGVRAKKSG LSGYGAGLLA ALLGVSFLSQ HAQTAEPTNV TNAANNTTIQ IMKSTLSLSE
 61 VCGKTKFQGK IYGGQIAGAE RWPWQASLRL YGRHICGAVL IDKNWVLGAA HCFQRSQEPS
121 DYHVMLGYTD LNSPTRYSRT MSVQKVIVHK DYNRFHTQGS DIVLLQLRSS VEYSSHILPA
181 CVPEENIKIP KEKACWASGW GYLREDVRIP LPNELYEAEL IIMSNDQCKG FFPPPVPGSS
241 RSYYIYDDMV CAADYDMSKS ICAGDSGGPL VCLLEGSWYV VGLTSWSSTC EEPIVSPSVF
301 ARVSYFDKWI KDNKKSSSNS KPGESPHHPG SPENENPEGN NKNQGTVIKP VCTALLLSQT
361 LLQQLI
```

Fig.23.6. Amino acid sequence of mouse testicular serine protease 1 (A) and serine protease 2 (B). Source: http//www. ncbi.nlm.nih.gov. Accession NP_033381 (A) and Accession NP_033382 (B).

auto-activation of proacrosin at a basic pH in vitro and affected the maturation pathway of proacrosin. In the presence of Sp32, the 55-and 53-kDa proacrosins are converted to 49-kDa, instead of the 43-kDa intermediate. The Sp32 is initially synthesized as a 61-kDa precursor protein with a putative signal peptide at the amino terminus. The carboxyl-terminal half of the precursor molecule corresponds to the mature Sp32. The binding of Sp32 to proacrosin may be involved in packaging the acrosin zymogen into the acrosomal matrix (Baba et al., 1994).

23.8.2. Cathepsins

Cathepsin A is a multifunctional lysosomal carboxypeptidase that also functions as a protective and an activator protein for neuraminidase and β-galactosidase. In normal rats, cathepsin A expression is noted in lysosomes of Sertoli and Leydig cells but not in germ cells of the testis as well as non-ciliated cells of the efferent ducts. A cell and region-specific expression of cathepsin A has been seen in testis and epididymis, but its expression is not regulated by testicular or pituitary factors (Luedtke et al., 2000). Cathepsin D, a lysosomal acid proteinase, participates in the catabolism of proteins, peptides and proteoglycans in various tissues. Testicular cathepsin D exhibits an apparent molecular weight of 42-kDa with two identical subunits of 20-kDa. Purified cathepsin D catalyzes the conversion of pro-acrosin to acrosin (Srivastava and Ninjoor, 1982).

23.8.3. Testicular Serine Protease -1 and -2 (TESP1 and TESP2)

Complementary DNA clones encoding each of two similar but different serine proteases, TESP1 and TESP2 present in testis and sperm have been identified. The nucleotide sequence of these clones initially synthesized a pre-proproteins of 367 and 366 amino acids respectively. Comparison of these TESPs with typical serine proteases suggests that each TESP zymogen is converted into mature enzyme consisting of a light and a heavy chain covalently linked by a single pre-existing disulfide bond. The conversion may be accomplished by another protein with a trypsin like cleavage specificity since it is unlikely that the mature TESP1 and TESP2 are capable of splitting the Lys-Ile bond between the light and heavy chains. The TESP1 and TESP2 genes are expressed only in the testis and the transcripts are abundantly present in the haploid round spermatids. Both of these TESPs are localized in the sperm acrosome and are released during the acrosome reaction (Kohno et al., 1998) **(Fig.23. 6)**.

23.8.4. Testisin

A cDNA encoding a human serine proteinase, testisin is abundantly expressed in the testis and is lost in testicular tumors. The testisin cDNA is 1073 nt long, including 942 nucleotides of open reading frame and a 113-nucleotide 3' untranslated sequence range of normal human with a predicted protein of 314 amino acids. Human testisin consists of a 19-amino acid (aa) signal peptide, a 22-aa proregion, and a 273-aa catalytic domain, including a unique 17-aa COOH-terminal hydrophobic extension that is predicted to function as a membrane anchor. Deduced amino acid sequence of testisin shows 44% identity to protestisin and contains features of serine proteinases. Testisin with molecular mass of 35-39-kDa was expressed in the cytoplasm and on the plasma membrane of premeiotic germ cells and was not detected in germ cells derived from testicular tumors. The testisin gene was localized to the short arm of human chromosome 16 (16p 13.3) associated with heterozygosity in sporadic testicular tumors. These findings demonstrated a new cell surface serine proteinase, loss of which may have a direct or indirect role in the progression of testicular tumors of germ cells origin (Hooper et al., 1999).

The esp-1/testisin cDNA from rat testes consisted of 1099 nt with a single open reading frame encoding 328 amino acids and an expected molecular mass of 36.6- kDa. The deduced amino acid sequence of rat Esp-1/Testisin had 89% and 62% identity with its murine and human counterparts, respectively, and appeared to be a trypsin-type serine protease with a hydrophobic region at the C-terminus. Rat esp-1/testisin mRNA was predominantly expressed in testis, as in human and mouse. However, its immunohistochemical distribution was predominantly in the elongated spermatids at steps 12 to 19, and not in the primary spermatocytes and round spermatids. This different distribution profile suggests that Esp-1/Testisin plays a role in species-specific proteolytic events during spermatogenesis and fertilization (Nakamura et al., 2003).

23.9. OTHER ENZYMES IN ACROSOME

Membrane associated protein kinase has been identified in hamster acrosomal membranes. The kinase activity was a cAMP independent type. The use of specific stimulators and inhibitors indicated that the activity was not due to casein kinase, protein kinase A or protein kinase C but due to predominant residue in the proteins. A specific isoform of protein disulfide isomerase of liver has been localized in the developing acrosome of the spermatids. In addition to the acrosome, the protein appeared in the nucleus of spermatids during maturation phase, and was localized in the nucleus of epididymal spermatozoa. Almost all of the protein disulfide isomerase in the testis was extractable by an isotonic buffer. The isoform of protein disulfide isomerase is an intra-acrosomal soluble protein that begins to enter the nucleus of mature spermatids in the testis and tightly binds to the nuclear components in epididymal spermatozoa (Ohtani et al., 1993). Presence of ADAM proteins and metalloproteinases and other proteins has been discussed in subsequent chapters 25-27, whereas cysteine rich proteinases have been described during sperm maturation in epididymis (Chapter 34), and non-ezymatic acrosomal proteins in Chapter 24.

23.10. REFERENCES

Abascal J, Skalaban SR, Grimm KM, Aviles M, Martinez ML, Castells MT, Ballesta J, Alhadeff JA.Alteration of the isoform composition of plasma-membrane-associated rat sperm α–L-fucosidase during late epididymal maturation : comparative characterization of the acidic and neutral isoforms. Biochem J 1998; 333: 201-7.

Abou-Haila A, Fouquet JP, Tulsiani RP. Characterisation and immunolocalization of β–D-Glucuronidase in mouse testicular germ cells and spermatozoa. Exp Cell Res 1999; 247 :48-60.

Abou-Haila A, Tulsiani DR. Mammalian sperm acrosome: formation, contents and function. Arch Biochem Biophys 2000; 379: 173-82.

Adam HI, Somani IH, Huang JQ, Mahuran D, Gravel RA, Trasler JM, Hermo L. Abnormalities in cells of the testis, efferent ducts, and epididymis in juvenile and adult mice with β-hexosaminidase A and B deficiency. J Androl 1999; 20: 779-802.

Adham IM, Klemm U, Maier WM, Hoyer-Fender S, Tsaousidou S, Engel W. Molecular cloning of preproacrosin and analysis of its expression pattern in spermatogenesis. J Biochem. 1989; 182: 563-68.

Adham IM, Nayernia K, Engel W. Spermatozoa lacking acrosin protein show delayed fertilization. Mol Reprod Dev 1997; 46: 370-76.

Alhadeff JA, Khunsook S, Choowongkomon K, Baney T, Heredia V, Tweedle A, Bean B: Characterisation of human semen α–L-fucosidases. Mol Hum Reprod 1999; 5:809-15.

Arming S, Strobl B, Wechselberger C, Kreil G. In vitro mutagenesis of PH-20 hyaluronidase from human sperm. Eur J Biochem 1997; 247: 810-14.

Austin CR. Capacitation and the release of hyaluronidase from the spermatozoa. J Reprod Fertil 1960; 3: 310-11.

Baba T, Niida Y, Michikawa Y, Kashiwabara S, Kodaira K, Takenaka M, Kohno N, Gerton G L, Arai Y: An acrosomal protein sp32, in mammalian sperm is a binding protein specific for two proacrosins and an acrosin intermediate. J Biol Chem 1994; 269: 10133-40.

Brandelli A, Miranda PV, Tezon JG: Participation of glycosylated residues in the human sperm acrosome reaction: possible role of N-Acetylglucosaminidase. Biochim Biophys Acta 1994; 1220: 299-304.

Cardullo RA, Wolf DE: Distribution and dynamics of mouse sperm surface galactosyltransferase: implications for mammalian fertilization. Biochemistry 1995; 34: 10027-35.

Chan CP, Gupta S, Mark G E. Identification of linear surface epitopes on the guinea pig sperm membrane protein PH-20. Life Sci 1999; 64: 1989-2000.

Chayko CA, Orgebin-Crist MC, Skudlarek MD, Tulsiani DR. Biosynthesis, processing and subcellular localization of rat sperm β–D-Galactosidase. Biol Reprod 2000; 63: 688-96.

Cherr GN, Yudin AI, Li MW, Vines CA, Overstreet JW. Hyaluronic acid and the cumulus extracellular matrix induce increases in intracellular calcium in macaque sperm via the plasma membrane protein PH-20. Zygote 1999; 7: 211-22.

Clermont Y, Hermo L, Rambourg A, Thorne-Tjomsland G. Structural and Cytochemical characteristics of the Golgi apparatus during the formation of acrosome system in spermatids, In: Cellular and molecular events in spermiogenesis (Hamilton DW, Waites GMH eds), Cambridge University Press, Cambridge, pp 17-39,1990.

Csoka TB, Scherer SW, Stern R. Expression analysis of six paralogous human hyaluronidase genes clustered on chromosomes 3p21 and 7q31. Genomics 1999; 60: 356-61.

Deng X, Czymmek K, Martin-DeLeon PA. Biochemical maturation of Spam1 (PH-20) during epididymal transit of mouse sperm involves modifications of N-linked oligosaccharides. Mol Reprod Dev 1999; 52: 196-06.

Deng X, Czymmek K, Martin-DeLeon PA: Biochemical maturation of spam1 (PH-20) during epididymal transit of mouse sperm involves modifications of N-linked oligosaccharides. Mol Reprod Dev 1999; 52: 196-06.

Deng X, He Y, Martin-Deleon PA. Mouse Spam1 (PH-20): evidence for its expression in the epididymis and for a new category of spermatogenic-expressed genes. J Androl 2000; 21: 822-32.

Ghosh I, Datta K.Sperm surface hyaluronan binding protein (HABP1) interacts with zona pellucida of water buffalo (Bubalus bubalis) through its clustered mannose residues. Mol Reprod Dev 2003; 64: 235-44.

Gmachl M, Kreil G. Bee venom hyaluronidase is homologous to a membrane protein of mammalian sperm. Proc Natl Acad Sci USA 1993; 90: 3569-73.

Godin DA, Fitzpatrick PC, Scandurro AB, Belafsky PC, Woodworth BA, Amedee RG, Beech DJ, Beckman BS. PH20: a novel tumor marker for laryngeal cancer. Arch Otolaryngol Head Neck Surg 2000; 126: 402-4.

Gong X, Dubois DH, Miller DJ , Shur BD :Activation of a G protein complex by aggregation of on the surface of sperm. Science 1995; 269: 1718-21.

Gupta GS, Goldberg E. Isolation, properties, immunological specificity and localization of mouse testicular

hyaluronidase. Biochim Biophys Acta 1981; 757: 364-73.
Gupta GS, Singh GP. Isolation and characterization of major form of β-glucuronidase from human seminal plasma. Biochim Biophys Acta 1983; 748; 398-404.
Hall JC, Perez FM, Kochins JG, Pettersen CA, Li Y, Tubbs CE, LaMarche MD.Quantification and localization of N-Acetyl-βD-Hexosaminidase in the adult rat testis and epididymis. Biol Reprod 1996; 54: 914-29.
Hancock LW, Raab LS, Arnonson Jr NN: Synthesis and processing of rat sperm-associated α–L-Fucosidase . Biol Reprod 1993; 48, 1228-38.
Hiramoto S, Tamba M, Kiuchi S, Jin, Yin-Zhe, Bannai S, Sugita Y, Dacheux F, Dacheux JL, Yoshida M, Okamura N. Stage-specific expression of a mouse homologue of the porcine 135kDa α-D-Mannosidase (Man2B2) in Type A spermatogonia. Biochem Biophys Res Commun 1997; 241: 439-45.
Hooper JD, Nicol DL. Dickinson JL, Eyre HJ etal Jestisin, a new serine protease expressed by premeiotic testicular germ cells and lost in germ cell tumors. Cancer Res 1999; 59: 3199-205.
Hou ST, Ma A, Jones R, Hall L. Molecular cloning and characterization of rat sperm antigen 2B1, a glycoprotein implicated in sperm-zona binding. Mol Reprod Dev 1996; 45: 193-203.
Hunnicutt GR, Mahan K, Lathrop WF, Ramarao CS, Myles DG, Primakoff P. Structural relationship of sperm soluble hyaluronidase to the sperm membrane protein PH-20. Biol Reprod 1996a; 54: 1343-9.
Hunnicutt GR, Primakoff P, Myles DG. Sperm surface protein PH-20 is bifunctional: one acrtivity is a hyaluronidase and a second, distinct activity is required in secondary sperm-zona binding. Biol Reprod 1996b; 55: 80-6.
Igdoura S A, Herscovics A, Lal A, Moremen K W, Morales C R, Hermo L. α-Mannosidases involved in N-glycan processing show cell specificity and distinct subcompartmentalisation within the golgi apparatus of cells in the testis and epididymis. Eur J Cell Biol 1999; 78: 441-52.
Jones MH, Davey PM, Aplin H, Affara NA. Expression analysis genomic structure, and mapping to 7q31 of the human sperm adhesion molecule gene SPAM1. Genomics 1995; 29: 796-800.
Kapoor DK, Gupta GS. Purification, biochemical properties and active sites of N-acetyl-β-D-hexosaminidases from human seminal plasma. Biochem J 1986; 236: 103-09.
Kashiwabara SI, Baba T, Takada M, Watanabe K, Yano Y, Arai Y. Primary structure of mouse proacrosin deduced from the cDNA sequence and its gene expression during spermatogenesis . J Biochem 1990; 108: 785-91.
Khunsook S, Bean BS, McGowan SR, Alhadeff JA. Purification and characterization of plasma membrane-associated human sperm alpha-L-fucosidase. Biol Reprod 2002; 68:109-16.
Kohno N, Yamagata K, Yamada S, Kashiwabara S. Two novel testicular serine proteases, TESP1 and TESP2, are present in the mouse sperm acrosome. Biochim Biophys Res Commun1998; 245: 658-65.
Kreil G. Hyaluronidases- a group of neglected enzymes. Protein Science 1995; 4: 1666-69.
Lalancette C, Dorval V, Leblanc V, Leclerc P. Characterization of an 80-kilodalton bull sperm protein identified as PH-20. Biol Reprod 2001; 65: 628-36.
Larson JL, Miller DJ: Sperm from a variety of mammalian species express β1,4-Galactosyltransferases on their surface. Biol Reprod 1997; 57, 442-53.
Lathrop WF, Carmichael EP, Myles DG, Primakoff P. cDNA cloning reveals the molecular structure of a sperm surface protein, PH-20, involved in sperm-egg adhesion and the wide distribution of its gene among mammals. J Cell Biol 1990; 111 (6pt2): 2939-49.
Li MW, Cherr GN, Yudin AI, Overstreet JW. Biochemcial characterization of the PH-20 protein on the plasma membrane and inner acrosomal membrane of cynomolgus macaque spermatozoa. Mol Reprod Dev 1997; 48: 356-66.
Lin Y, Kimmel LH, Myles DG, Primakoff P. Molecular cloning of the human and monkey sperm surface protein PH-20. Proc Natl Acad Sci USA 1993; 90: 10071-5.
Lin Y, Mahan K, Lathrop WF, Myles DG, Primakoff P. A hyaluronidase activity of the sperm plasma membrane protein PH-20 enables sperm to penetrate the cumulus cell layer surrounding the egg. J Cell Biol 1994; 125: 1157-63.
Lu Q, Shur BD. Sperm from β1,4-Galactosyltransferases null mice are refractory to ZP3-induced acrosome reactions and penetrate the zona pellucida poorly. Development 1997; 124: 4121-31.
Luedtke CC, Andonian S, Igdoura S, Hermo L. Cathepsin A is expressed in a cell-specific manner in the testis and epididymis and is not regulated by testicular or pituitary factors. J Histochem Cytochem 2000; 48: 1131-46.
Martinez ML, Marteloto L, Cabada MO. Purification and Biological Characterization of N-Acetyl –βD Glucosaminidase from *Bufo arenarium* Spermatozoa. Mol Reprod Dev 2000; 57 : 194-203.
Meyers SA, Equine sperm-oocyte interaction: the role of sperm surface hyaluronidase. Anim Reprod Sci 2001; 68: 291-303.
Miller DJ, Macek MB, Shur BD: Complementarity between sperm surface β1,4-Galactosyltransferases and egg-coat ZP3 mediates sperm–egg binding. 1992; 357: 589-92.
Miranda P V, Gonzalez-Echeverria F, Blaquier J A, Mahuran D J, Tezon J G : Evidence for the participation

of β-hexosaminidase in human sperm-zona pellucida interaction in vitro. Mol Hum Reprod 2000; 6: 699-706.

Moreno RD, Sepulveda MS, de loannes A, Barros C. The polysulphate binding domain of human proacrosn/acrosin is involved in both the enzyme activation and spermatozoa-zona pellucida interaction. Zygote 1998; 6: 75-83.

Myles DG, Primakoff P. Why did the sperm cross the cumulus? To get to the oocyte. Functions of the sperm surface proteins PH-20 and fertilin in arriving at and fusing with the egg. Biol Reprod 1997; 56: 320- 7.

Nakamura Y, Inoue M, Okumura Y, Shiota M, Nishikawa M, Arase S, Kido H. Cloning, expression analysis, and tissue distribution of esp-1/testisin, a membrane-type serine protease from the rat. J Med Invest 2003; 50: 78-86.

Nikolajczyk BS, O'Rand MG: Characterisation of rabbit testis β-galactosidase and arylsulfatase: purification and localization in spermatozoa during the acrosome reaction Biol Reprod. 1992; 46:366-78.

Nixon B, Lu Q, Wassler MJ, Foote CI, Ensslin MA, Shur BD. Galactosyltransferase function during mammalian fertilization. Cells Tissues Organs 2001; 168: 46-57.

Nozaki H, Matsuzawa T, Nakamura T, Arai I, Urashima T. Are there two forms of beta 2-N-acetylglucosaminyltransferase I in rat testicular and epididymal fluids? Biochim Biophys Acta. 2003; 1649: 140-5.

Ohtani H, Wakui H, Ishino T, Komatsuda A, Miura AB. An isoform of protein disulfide Isomerase is expressed in the developing acrosome of spermatids during rat spermiogenesis. Histochemistry 1993;100:423-29.

Okamura N, Tamba M, Liao H, Onoe S, Sugita Y, Dacheux F, Dacheux J. Cloning of complementary DNA encoding a 135 kilodalton protein secreted from porcine corpus epididymis and its identification as an epididymis-specific α-mannosidase. Mol Reprod Dev 1995; 42: 141-48.

Overstreet JW, Lin Y, Yudin AI, Meyers Sas, Primakoff P, Myles DG, Katz DF, Vandervoort CA. Location of the PH20 protein on acrosome-intact and acrosome-reacted spermatozoa of cynomolgus macaques. Biol Reprod 1995; 52: 105-14.

Pereira BMJ, Abou-Haila A, Tulsiani DRP: Rat sperm surface mannosidase is first expressed on the plasma membrane of testicular germ cells. Biol Reprod 1998; 59: 1288- 95.

Pratt SA, Scully NF, Shur BD: Cell surface β1,4Galactosyltransferase on primary spermatocytes facilities. Their initial adhesion to sertoli cells in vitro. Biol Reprod 1993; 49 : 470-82.

Primakoff P, Woolman-Gamer l, Tung KS, Myles DG. Reversible contraceptive effect of PH-20 immunization in male guinea pigs. Biol Reprod 1997; 56; 1142-6.

Rebeiz M, Miller DJ: Porcine sperm surface β1,4-Galactosyltransferases binds to the zona pellucida but is not sufficient to mediate sperm–zona pellucida binding. Mol Reprod Dev 1999; 54:379-87.

Rivkin E, Tres LL, Kaplan-Kraicer R, Shalgi R, Kierszenbaum AL: Molecular cloning of rat sperm galactosyl receptor, a C-type lectin with in vitro egg binding activity. Mol Reprod Dev 2000; 56;401-11.

Rutllant J, Meyers SA. Post-translational processing of PH-20 during epididymal sperm maturation in the horse. Biol Reprod 2001; 65: 1324-31.

Sabeur K, Cherr GN, Yudin AI, Overstreet JW. Hyaluronic acid enhances induction of the acrosome reaction of human sperm through interaction with the PH-20 protein. Zygote 1998; 6: 103-11.

Seaton GJ, Hall L, Jones R. Rat sperm 2β1 glycoprotein (PH-20) contains a C-terminal sequence motif for attachment of a glycosylphosphatidylinositol anchor. Effects of endoproteolytic cleavage on hyaluronidase activity. Biol Reprod 2000; 62: 1667-76.

Shur BD. Is sperm galactosyltransferase a signaling subunit of a multimeric gamete receptor? Biochem Biophys Res Commun 1998; 250: 537-43.

Skudlarek M D, Tulsiani D R, Nagdas S K, Orgebin-Crist M C : β-D-Galactosidase of rat spermatozoa: Subcellular distribution, substrate specificity, and molecular changes during epididymal maturation. Biol Reprod 1993; 49: 204-13.

Skudlarek MD, Abou-Haila A, Tulsiani DR: Rat spermatogenic cell β-D-galactosidase: characterization, biosynthesis, and immunolocalisation. Exp Cell Res 2000; 26: 139-49.

Srivastava PN, Ninjoor V. Isolation of rabbit testicular cathepsin D and its role role in the activation of proacrosin. Biochim Biophys Res Commun 1982; 109: 63-69.

Taniguchi A, Hioki M, Matsumoto K.Transcriptional regulation of human Galbeta1, 3GalNAc/Galbeta1, 4GlcNAc alpha2, 3-sialyltransferase (hST3Gal IV) gene in testis and ovary cell lines. Biochem Biophys Res Commun 2003; 301: 764-8.

Tascou S, Nayernia K, Engle W, Burfeind P. Refinement of the expression pattern of a mouse homologue of the porcine 135-kDa α-D-mannosidase (MAN2B).Biochem Biophys Res Commun 2000; 272: 951-52.

ten Have J, Beaton S, Bradley MP. Cloning and characterization of the cDNA encoding the PH20 protein in the European red fox Vulpes vulpes. Reprod Fertil Dev 1998; 10: 165-72.

Tengowski MW, Wassler MJ, Shur BD, Schatten G: Subcellular localization of β1,4-Galactosyltransferases onbull sperm and its function during sperm-egg interactions Mol Reprod Dev 2001; 58: 236-44.

Thaler CD, Cardullo RA. Biochemical characterization of a glycosyl phosphatidylinositol-linked hyaluronidase on mouse sperm. Biochemistry 1995; 34: 7788-95.
Tranter R, Read JA, Jones R: Effector sites in the three-dimensional structure of mammalian sperm β-Acrosin. Structure 2000; 8: 1179-88.
Triggs-Raine B, Salo TJ, Zhang H, Wicklow BA, Natowicz MR. Mutations in *HYAL1*, a member of a tandemly distributed multigene family encoding disparate hyaluronidase activities, causes a newly described lysosomal disorder, mucopolysaccharidosis IX. Proc Natl Acad Sci USA 1999; 96: 6296-300.
Tulsiani DR, Abou-Haila A, Loeser C, Pereira MJ. The biological and functional significance of the sperm acrosome and acrosomal enzymes in mammalian fertilization. Expt Cell Res 1998; 240: 151-64.
Tulsiani DRP, Orgebin-Crist MC, Skudlarek MD. Role of luminal fluid glycosyltransferases and glycosidases in the modification of rat sperm plasma membrane glycoproteins during epididymal maturation. J Reprod Fert (Suppl). 1998; 53,85-97.
Tung KS, Primakoff P, Woolman-Gamer L, Myles DG. Mechanism of infertility in male guinea pigs immunized with sperm PH-20. Biol Reprod 1997; 56: 1133-41.
Veeramachaneni DN, Smith MO, Ellinwood NM. Deficiency of fucosidase results in acrosomal dysgenesis and impaired sperm maturation. J Androl 1998; 19: 444-49.
Vines CA, Li MW, Deng X, Yudin AI, Cherr GN, Overstreet JW. Identification of a hyaluronic acid (HA) binding domain in the PH-20 protein that may function in cell signaling. Mol Reprod Dev 2001; 60: 542-52.
Wasserman PM. Mammalian fertilization: molecular aspects of gamete adhesion, exocytosis, and fusion.. Cell 1999; 96: 175-83.
Watanabe K, Baba T, Kashiwabara S I, Okamoto A, Arai Y. Structure and organization of the mouse Acrosin gene. J Biochem 1991; 109: 828-33.
Yamagata K, Honda A, Kashiwabara S I, Baba T : Difference of acrosomal serine protease system between mouse and other rodent sperm. Dev Genet 25(2) : 115-22 (1999).
Yamagata K, Murayama K, Okabe M, Kiyotaka T, Nakanishi T, Kashiwabara S I, Baba T. Acrosin accelerates the Dispersal of sperm acrosomal proteins during acrosome reaction. J Biochem Mol Biol 1998; 273; 10470-74.
Yoshinaga K, Toshimori K. Organization and modifications of sperm acrosomal molecules during spermatogenesis and epididymal maturation. Microsc Res Tech 2003; 61:39-45.
Yudin AI, Li M-W, Robertson KR, Cherr GN, Overstreet JW. Characterization of the active site of monkey sperm hyaluronidase. Reproduction 2001; 121: 735-43.
Yudin AI, Vandevoort CA, Li MW, Overstreet JW. PH-20 but not acrosin is involved in sperm penetration of the macaque zona pellucida. Mol Reprod Dev 1999; 53: 350-62.
Zhang H, Martin-DeLeon PA. Mouse epididymal Spam1 (PH-20) is released in vivo and in vitro, and Spam1 is differentially regulated in testis and epididymis. Biol Reprod 2001; 65: 1586-93.
Zhang H, Martin-Deleon PA.Mouse epididymal Spam1 (pH-20) is released in the luminal fluid with its lipid anchor. J Androl 2003; 24: 51-8.
Zheng Y, Deng X, Zhao Y, Zhang H, Martin-DeLeon PA. Spam1 (PH-20) mutations and sperm dysfunction in mice with the Rb(6.16) or Rb(6.15) translocation. Mamm Genome 2001; 12: 822-9.
Zheng Y, Martin-Deleon PA. Characterization of the genomic structure of the murine Spam1 gene and its promoter: evidence for transcriptional regulation by a cAMP-responsive element. Mol Reprod Dev 1999; 54: 8-16.
Zheng Y, Martin-Deleon PA. The murine Spam1 gene: RNA expression pattern and lower steady-state levels associated with Rb (6.160 translocation. Mol Reprod Dev 1997; 46: 252-57.

Chapter 24

ACROSOMAL PROTEINS (NON-ENZYMATIC)

24.1. ACROSOMAL MATRIX

Compartmentalization: The interior of the mammalian sperm acrosome contains a structural frame work, the acrosomal matrix (AM) that may regulate the distribution of hydrolases within the acrosome and their release during acrosome reaction, a receptor-regulated exocytotic process involving fusion between the peri-acrosomal plasma membrane and the outer acrosomal membrane. In several mammalian species, the acrosomal contents are segregated into spatially distinct domains of differing ultrastructural appearance. Specific interaction between acrosomal matrix polypeptides and hydrolases represent a mechanism to sequester hydrolases within the acrosome and to regulate their release during acrosome reaction. During acrosome reaction, specific hydrolases exhibit different temporal ordered release patterns. For example, dipeptidyl peptidase is rapidly released during the acrosome reaction and this enzyme is sequestered to a specific domain along the dorsal surface of the acrosomal apical segment. In contrast, a considerable amount of pro-acrosin is associated with a particulate sperm fraction, and that is released more slowly during the acrosome reaction. This particulate nature and temporal release of different hydrolases may reflect a sequential requirement for specific hydrolases as sperm encounter different elements of the egg vestments during fertilization. The difference in solubility properties and release rates between acrosomal proteins appears to reflect their compartmentalization within the acrosome and their association with, or assembly into stable acrosomal matrix assemblies (Olson et al., 1998b). To test the hypothesis that compartmentalization affects the release of acrosomal components during the course of secretion in guinea pig sperm, the recovery of acrosomal constituents in the medium surrounding the sperm induced to undergo exocytosis provided strong support to the hypothesis that compartmentalization plays a significant role in the release of proteins during acrosome reaction (Kim et al., 2001b). The mechanism regulating hydrolase release during the mammalian acrosome reaction showed that specific domains of acrosomal matrix of bovine spermatozoa function to maintain a particulate proacrosin pool and to regulate proacrosin/ acrosin release to sonicated sperm suspensions. The acrosomal contents in structural appearance and specific elements of the acrosomal matrix remain intact and associated with the hybrid membrane complex after the acrosome reaction. The acrosomal matrix elements also resist solubilization by various extraction regimens, emphasizing their structural stability.

Acrosomal Lamina Complex (ALM): A stable acrosomal assembly from cauda epididymal hamster spermatozoa comprises two distinct acrosomal matrix domains and the detergent-insoluble membrane skeleton of the outer acrosomal membrane that binds proacrosin and N-acetylglucosaminidase. The acrosomal structure was termed the acrosomal lamina-matrix (ALM) complex. The ALM represents a stable infrastructure that may segregate hydrolases within the acrosome and regulate their release during the acrosome reaction. It may also represent a cytoskeleton-like framework that affects the vesiculation of the outer acrosomal membrane during the acrosome reaction and that maintains the integrity of the shed acrosomal cap (Olson, et al., 1998b).

Outer Acrosomal Membrane Matrix Complex: To determine whether outer acrosomal membrane associated matrix complex (OMC) is composed of a unique set of acrosomal proteins and to define its fate during both capacitation and acrosome reaction, a purified OMC fraction of 32-kDa was isolated from sperm. The OMC32 was restricted to the stable matrix assembly but was not associated with the inner acrosomal membrane of the equatorial segment. The OMC32 antigen was related to a family of polypeptides between 38 and 19-kDa and its N-terminal sequence was similar to the SP-10 family of human and baboon spermatozoa (Olson et al.,1996).

24.2. ACROSOMAL MATRIX PROTEINS

Acrosomal matrix contains a protein of 50-kDa that is restricted to the ventral-most region of the apical segment of guinea pig cauda epididymal sperm. Native AM50 is resistant to solubilization. However, during the acrosome reaction, AM50 was converted into a 42-43kDa double protein (AM50AR) and released into the medium. The AM50AR remained associated with proteases after the acrosome reaction (Wesbrook-Case et al., 1994). It is possible that the role of these matrix elements is not limited to regulating protein distribution and release during the acrosome reaction but they could also function in the membrane fusion events or in sperm-zona interactions. The sperm acrosomal matrix contains a member of the pentaxin family of calcium-dependent binding proteins. Noland et al (1994) isolated guinea pig sperm acrosomal apical segment and mapped it by 2D-PAGE. Although complex, the two dimentional PAGE was dominated by two 50-kDa polypeptides (p50 and proacrosin) and the other proteins of 67-kDa (p67) and a 32-kDa polypeptide (sp32). When acrosomal matrix was purified following Triton X-100 extraction, p50 was the major component having p67 proacrosin and sp32 as less prominent constituents. Molecular cloning demonstrated that p50 was a unique, testis-specific member of the pentaxin family of calcium-dependent binding proteins. Olson et al (1998a) defined the temporal expression, processing and localization of major matrix proteins of 29-kDa (AM29) and 22-kDa (AM22) during spermatogenesis and post-testicular sperm maturation in the epididymis. The AM29 and AM22 of mature spermatozoa are structurally related and appear to arise from a common 40-kDa precursor protein expressed in round spermatids. The precursor protein, present in the acrosome of round spermatids, undergoes size processing during the terminal stages of spermiogenesis so that the mature matrix polypeptides are evident in epididymal spermatozoa. The processing of major proteins of the acrosomal matrix occurs in a temporally regulated fashion after their transport to the acrosome and the processed product can assemble into ultrastructurally distinct matrix elements.

24.2.1. Mouse Sp 56

The mouse sperm protein Sp56 was identified on the basis of its specificity, and high affinity for ZP3 protein recognized by sperm. The Sp56 polypeptide is present in testes and epididymus,

```
-32 MITWSFIDLW RTSHSTLFQM TLATVLMAPV LGDCGPPPLL PFASPTNQLY ESTTFPSGTV
 28 LKYTCHHGFK RVNSSHLSCD ENGSWVYSTF CARKRCKNPG ELVNGKVEIP SDLLVGSIIE
 88 FSCSKGYLLI GSATSRCEVQ GKGVDWSDSL PECVIATCEP PPPISNGKHS GRDDDLYTFG
148 SVVIYNCDPT FTLLGNASIV CTVVNRTVGV WRPHPPACQK IVCHRPQIPK GYLAPGFRQF
208 YAYRDALEIR CKKGFILRGS SVIHCEANGE WFPSIPTCEP NGCTNIPDIS YASWEGYKFP
268 LRNFEVFEIG AKLKYQCKPG YRASLNDPQT VTCQENLTWS STNGCERICC PTPDMEKIKI
328 VSERRDFTGT CIYAYGDYVF YICNEGSYPM STDGRSSCQA DGKWDPAIPS CQADSGLQNR
388 LALFTFPNIS ETNVTNKTYL FGHEENSTEH AMKGVCLKPM VINGNLSVER VIYAELENIT
448 IQCDPGYTIV GSPNIICSNR TWYPEVPSCQ MEVLEDCRIV SRGAQLLHCL SSPEDVHRAL
508 KVYKLFLEIE RLEHQKEKWI QLHRKPQSMK INRSFRLCN
```

Fig.24.1. Amino acid sequence of mouse zona pellucida 3 receptor pro:ein, Sp56 (Accession number NP_033607) Source: http//www. ncbi.nlm.nih.gov. (Bookbinder et al. Science 269; 86-89: 1995).

but not in other mouse tissues. The Sp56 is a homomultimeric peripheral membrane protein confined to the plasma membrane overlying the sperm's acrosome, and it has a specific affinity for ZP3's functional domain oligosaccharide (FD-oligo). Purified Sp56, which binds to the ZP of unfertilized mouse eggs but not to fertilized embryos, blocks sperm-egg recognition (Bookbinder, et al, 1995) **(Fig. 24.1)**. The Sp56 mRNA was detected with Sp56 specific antisense RNA probe in the nuclei and cytoplasm of round spermatids, and in no other cells. The detection of Sp56 mRNA in round spermatids may indicate that transcription activation of the Sp56 gene occurs in a haploid cells. Mouse and hamster sperm Sp56 has special affinity for ZP3 and is responsible for sperm-egg interaction (Cheng et al., 1994). However, Sp56 was not detected in guinea pig sperm or human sperm, which do not bind to mouse egg ZP.

Mouse Sp56 has been shown an acrosomal component (Foster et al, 1997; Kim et al, 2001a; Cohen and Wasserman, 2001). The Sp56 first appears in late meiotic cells and accumulated during spermiogenesis. The form of Sp56 in pachytene spermatocytes and spermatids has high molecular weight than was found in sperm. The size difference could be due to alteration in carbohydrate side chains. Recombinant mouse sperm Sp56 has shown contraception potential after immunization in female mice (Hardy and Mobbs, 1999). Sequencing of a 1-kb PCR product confirmed that this polynucleotide encoded Sp56, as well as all six, independently identified stretches of Sp56 amino acid sequence. The Sp56 cDNA encodes a 547-amino acid ORF. A presumptive 32-amino acid signal peptide is located between the first in-frame methionine and the experimentally determined NH_2-terminus of Sp56, as one would predict from the extracellular location of sp56. The protein sequence contains no obvious transmembrane-domain. The predicted polypeptide molecular mass of sperm Sp56 from the cDNA sequence is 62-kDa. The discrepancy between the molecular mass of sperm sp56 (56-kDa) and a calculated molecular mass of 62-kDa appears to be due to posttranslational truncation of Sp56 in spermatogenic cells. The Sp56 amino acid sequence derived from the cDNA does not indicate the presence of regions having significant homology to carbohydrate recognition domains (CRDs) of previously characterized lectins. The presence of CRDs within the Sp56 polypeptide sequence is expected, because Sp56 is a lectin that has an affinity for α-galactose and a specific affinity for ZP3 (Bookbinder et al., 1995).

Comparison of the Sp56 amino acid sequence to proteins in databases indicated that Sp56 is a member of a superfamily of protein receptors that contain multiple consensus repeats of approximately 60 amino acids in length, termed Sushi domains. The Sp56 open reading frame contains six contiguous Sushi domains, followed by 44 amino acids specific to Sp56, a seventh Sushi domain, and a highly acid COOH-terminal domain of 70 amino acids. The presence of seven Sushi repeats in Sp56, each of which contains two disulfides, is consistent with protein electrophoretic studies that showed that Sp56 contains numerous intramolecular disulfides. The Sp56 is most closely related to one member of the superfamily, the alpha subunit of complement 4B-binding protein (C4BPα).

24.2.2. Guinea pig AM 67: An Orthologue of Sp56

Guinea pig sperm acrosomal matrix glycoprotein AM 67 is most closely related to mouse sperm Sp56. Based on homology between guinea pig AM 67 and mouse Sp56, AM67 was localized in acrosomal matrix. The AM-67 is transcribed as a 1.9-kb testis specific mRNA. Using a PCR, two cDNA clones that encode the AM67 peptides, were isolated. AM67 is transcribed as a 1.9-kb testis-specific mRNA. The complete AM67 sequence encodes a pre-pro-polypeptide of 533 amino acids with a mass of 59,768 D. Following cleavage of a probable signal sequence, the polypeptide was predicted to have a Mr of 56,851 and seven consensus sites for asparagine-linked glycosylation. The deduced amino acid sequence of AM67 is most similar to those of the mouse sperm protein Sp56 and the α-subunits of complement component 4-binding protein from various mammalian species. This shows that Sp56 and AM67 are orthologues and suggests that Sp56 may function in acrosomal matrix-zona pellucida interactions during and immediately following the acrosome reaction in the mouse (Foster et al., 1997; Kim et al., 2001b).

24.2.3. Acrins

Acrin1 (MN7) is a 90-kDa intra-acrosomal antigen, which is restricted to the anterior acrosome of the mouse, rat, hamster, and human spermatozoa and follows a trafficking pathway during rat spermiogenesis. Acrin1 first appears in the proacrosomic vesicles of the early Golgi phase spermatids, and then localized in the electron-lucent matrix region of the acrosomic vesicles of the late Golgi phase spermatids (Yoshinaga et al., 1998). Acrin1 is not proteolytically modified during epididymal sperm maturation. The intra-acrosomal Acrin1 (MN7) shows not only a specific localization in the apical segment of the guinea pig sperm acrosome, but also a distal alteration during maturation.

The localization of another intra-acrosomal sperm Acrin2 demonstrated its presence in the membrane of the endoplasmic reticulum of early stage spermatids, but was not detectable in the developing acrosome until spermatids reached the maturation phase. In the final stage of spermiogenesis, Acrin2 became localized in the outer acrosomal membrane (OAM) matrix associated materials, both in the small region posterior to the dorsal matrix and along the ventral margin of the acrosomal apical segment (Yoshinaga et al., 1998; 2000; 2001). The acrosomal location of acrin2 in caput epididymidal sperm was almost identical to that observed in the final step spermatids. But during maturation it was more restricted in area, until in distal cauda epididymidal sperm it remained only in the dorsal region. Acrin2 was recognized as a 165-kDa protein in the mature sperm and showed a reduction in molecular weight during sperm passage through the epididymis.

24.2.4. SP17: A Zona-Binding Protein

To better understand sperm-zona pellucida binding, O'Rand and associates analysed a zona binding protein in sperm, designated as SP17 due to its molecular size of 16891 Da (Richardson et al., 1994). The SP17 mRNA is present in rabbit, mouse and human testes but was not detected initially in any somatic tissues. Recent reports showed that although SP17 expression is highest in the testis, it is present in all the mouse somatic tissues examined and is highly conserved throughout all mammalian species. The SP17's central domain, which is necessary for heparin binding, exhibits the greatest sequence divergence of all three domains present in SP17. In the rabbit, SP17 is the 17-kDa member of the rabbit sperm autoantigen (RSA) family of sperm specific auto-antigens and is encoded by two mRNAs of 0.9 and 1.1-kb. Each mRNA has a unique 5'untranslated region but both have identical coding regions. The deduced amino acid sequence of SP17 zona binding protein showed several interesting features, including a similarity to the N-terminus of human testis c-AMP-dependent protein kinase. In the rabbit as well as other similar species in which the corona radiata (granulosa) cells adhere tightly to the zona pellucida and sythesize zona glycoproteins, the fertilizing spermatozoon may have been already begun acrosome reaction within the cumulus oophorus. During final phase of cumulus passage SP17 would be available for initial zona binding. The SP17 has been proposed to belong to the subfamily of lectins with specificity of galactose (Richardson et al., 1994) (see Benoff, 1997). Detailed observations of Richardson et al, (1994) led to propose that during the acrosome reaction, SP17 becomes exposed on the rabbit sperm surface and moves posteriorly on the sperm head while aggregating into punctuate foci; finally SP17 remains associated with both the acrosomal ghost and with the equatorial region of the head of acrosome reacted sperm. These characteristics suggested that SP17 is involved in secondary binding of the ZP (Richardson et al, 1994). The anti-RSA antibodies inhibited sperm-egg interaction in vivo and in vitro.

Characterization: The SP17 is a 22- to 24-kDa triplet of proteins in rabbit spermatozoa and is unaffected by capacitation. However, during the acrosome reaction, SP17 is processed from a 22- to 24-kDa triplet of proteins to 17-19 kDa by removal of amino acids from the C-terminal end. Sequence analysis of SP17 revealed that 52% amino acids of the sequence 108-37 were 52% identical to the calmodulin-binding domain of neuromodulin and contained an IQ motif found in other calmodulin binding proteins (Wen et al., 1999). Three cDNA clones encoding the sperm specific SP17 from the baboon (Papiopapio) differing in the lengths at their 3' untranslated regions encoded mRNA transcripts of 0.8-1.35-kb. The ORF encodes 163 amino acids with a predicted molecular mass of 18.8-kDa. The baboon SP17 protein sequence is 97% identical to human Sp17 but differs by the addition of 12 amino acids at the C-terminal end, providing an additional site for protein kinase C phosphorylation. The baboon mRNA was specific to the baboon testes. The native baboon sperm SP17 protein exists as a doublet with an apparent Mr of 26.5 and 27.2 kDa.

The mouse mRNA for SP17 encodes a 149-amino acid protein with a predicted molecular weight of 17,296. The mouse SP17 (mSP17) cDNA sequence is 82% identical to the rabbit SP17 cDNA, while the murine protein is 74% identical to the rabbit protein. The native SP17 has an apparent molecular mass of 24-kDa. In the absence of proteolytic inhibitors, part of the C-terminal in native mSP17 is cleaved, giving rise to an 18-kDa band. The SP17's central domain (amino acids 61-117) spanning exon 3 is critical for heparin binding. The SP17 has two additional functional domains, an N-terminal domain similar to the dimer-interaction site in the cAMP-dependent protein kinase IIα regulatory subunit and a C-terminal calmodulin – binding domain. The SP17 is present in spermatocytes and spermatids in the testis. In spermatozoa, SP17 is not available to bind antibody on the surface of live, acrosome-intact spermatozoa. In fixed spermatozoa, staining is observed along the length of the principal piece, weekly along the mid-piece and over the acrosomal region of the head (Kong et al., 1995).

The mouse gene for SP17 is 6.5-kb and contains four exons. The *Sp17* gene is induced in metastatic cells and during mucosal immune responses, and the protein appears to play an important role in cell migration and/or adhesion in somatic cells, as well as in male germ cells (Wen et al., 2001). Additionally a second gene (Sp17-2), which showed the complete absence of introns, is also present in testis. This Sp17-2 gene has likely arisen by reverse transcription (RT) of a spliced Sp17-1 mRNA with subsequent integration into the human genome. Its ORF is interrupted by stop codons, giving rise to a pseudogene. The possibility of additional Sp17 species within the human genome has been suggested (Buchli et al., 2002). Lea et al, (1996) sequenced two testis specific cDNAs (1.36-kb and 1.6-kb) of human hSP17. Each cDNA gave rise to identical protein sequences and differed only in the 5'untranslated region. The predicted amino-acid sequence revealed a protein of 17.5-kDa, which exhibited a high degree of homology with both rabbit and mouse SP17.

SP17 as Cancer-Testis Antigen: Although SP17 was originally described as a testis-specific antigen,emerging evidence indicates that it may be more ubiquitously expressed than was previously thought. With the use of a specific antiserum, SP17 was found on the surface of malignant lymphoid cells, including B- and T-lymphoid cell lines, and on the surface of primary cells of B-lymphoid tumors.The SP17 promotes heparin sulfate-mediated cell aggregation and thereby plays a role in regulating adhesion and migration of normal and malignant lymphocytes (Lacy and Sanderson, 2001). The SP17 is detectable in tumor cells from myeloma patients. Since a high proportion of normal individuals develop antibodies against SP17 following vasectomy, Sp17 is likely to be a highly immunogenic protein in vivo. The SP17 is therefore, a member of the cancer testis antigen family and could be an ideal target for immuno-therapy of multiple myeloma (Lim et al., 2001). Based on this, it was possible to generate donor-derived Sp17-specific CTL for administration following allogeneic stem cell transplant to augment graft versus myeloma (GVM) effect without inducing a global GVHD. These CTLs were able to lyse autologous Epstein-Barr virus transformed lymphoblastoid cells in a SP17 dependent manner. Analysis of the CTL indicated that they were predominantly CD8 in phenotype and they produced IFN-γ and very little IL-4. Results suggested the potential for the generation and administration of donor – derived Sp17-specific CTL that augment GVM without inducing GVHD following alllogeneic stem cell transplant for multiple myeloma (Chinva–Internati et al., 2001).

24.3. ACROSOMAL VESICLE PROTEIN 1 (SP-10)

The *ACRV1* gene in humans encodes a testis-specific, differentiation antigen, acrosomal vesicle protein 1 (SP-10) that arises within the acrosomal vesicle during spermatogenesis, and is associated with the acrosomal membranes and matrix of mature sperm. This gene consists of 4 exons and its alternative splicing generates 11 distinct transcripts, which encode protein isoforms ranging from 81 to 265 amino acids. The longest transcript is the most abundant comprising 53-72% of the total acrosomal vesicle protein 1messages; the second largest transcript comprises 15-32%; the third and the fourth largest transcripts account for 3.4-8.3% and 8.7-12.5%, respectively; and the remaining 7 transcripts combined account for < 1% of the total acrosomal vesicle protein 1 message. It is suggested that phenomena of cryptic splicing and exon skipping occur within this gene. The acrosomal vesicle protein 1 may be involved in sperm-zona binding or penetration, and it is a potential contraceptive vaccine immunogen for humans. Some of the transcripts are of special interest: i) The variant (8) lacks a 210 nt fragment within the exon II and the entire exon III consisting of 120 nt, as compared to variant 1. Isoform **h** encoded by this variant is 110 aa shorter than isoform **a** encoded by variant 1. ii) Transcript variant 9 lacks two fragments of -210 and 135 nt in exon II, as compared to variant 1. Isoform i encoded by this variant is 80 aa shorter than isoform **a** encoded by variant **1**. iii)

```
Isoform e (variant 5) (176aa) accession 64494

  1 MNRFLLLMSL YLLGSARGTS SQPNELSGSI DHQTSVQQLP GEQPSGEQPS GEHLSGEQPL
 61 SELESGEQPS DEQPSGEHGS GEQPSGEQAS GEQPSGTILN CYTCAYMNDQ GKCLRGEGTC
121 ITQNSQQCML KKIFEGGKLQ FMVQGCENMC PSMNLFSHGT RMQIICCRNQ SFCNKI

Isoform f (variant 6) (170aa) accession 64496

  1  MNRFLLLMSL YLLGSARGTS SQPNELSGSI DHQTSVQQLP GEQPSDEQPS GEHGSGEQPS
 61  GEQASGEQPS  GEHASGEQAS GAPISSTSTG TILNCYTCAY MNDQGKCLRG EGTCITQNSQ
121 QCMLKKIFEG  GKLQFMVQGC ENMCPSMNLF SHGTRMQIIC CRNQSFCNKI

Isoform h (variant 8) (155aa) accession 64498

  1 MNRFLLLMSL YLLGSARGTS SQPNELSGSI DHQTSVQQLP GEQPSGEQPS GEHLSGEQPL
 61 SELESGEQPS DEQPSGEHGS GEQPSGEQAS GEQPSGEHAS GEQASGAPIS STSTGGKLQF
121 MVQGCENMCP SMNLFSHGTR MQIICCRNQS FCNKI

 Isoform i (variant 9) (150aa) accession 64499

  1 MNRFLLLMSL YLLGSARGTS SQPNELSGSI DHQTSVQQLP GEQASGEQPS GEHASGEQAS
 61 GAPISSTSTG TILNCYTCAY MNDQGKCLRG EGTCITQNSQ QCMLKKIFEG GKLQFMVQGC
121 ENMCPSMNLF SHGTRMQIIC CRNQSFCNKI

Isoform k (Variant 11) (81aa) accession 64500

  1 MNRFLLLMSL YLLGSARGTS SQPNELSGSI DHQTSVQQLP GGKLQFMVQG CENMCPSMNL
 61 FSHGTRMQII CCRNQSFCNK I
```

Fig.24.2. Amino acid sequences of various isoforms of acrosomal vesicle protein 1(Accession numbers are indicated with each isoform). Source: http//www. ncbi.nlm.nih.gov.

Transcript variant 11 lacks a 552 nt fragment including part of the exon II and the entire exon III, as compared to variant **1**. Isoform **k** encoded by this variant is 184 aa shorter than isoform a encoded by variant **1**. Five acrosomal vesicle protein **1** isoforms representing SP10 in humans are shown in **Fig.24.2** .

The SP-10 was first identified in human sperm, and subsequently detected in several mammalian species including baboon, macaques, pig and fox. The SP-10 cDNAs have been cloned and sequenced in human, macaques and baboons. A SP-10 homologue, MSA-63, has been identified in mouse. Extracts from whole human testis and epididymal ejaculated, and capacitated sperm showed that a full-length 45-kDa SP-10 precursor protein is present in the testis. The SP-10 peptides of 32, 30, 28 and 26-kDa were obtained from proteolytic processing of the SP-10 precursor protein in the testis and/or alternative splicing. In addition, SP-10 peptides of 25-18 kDa were detected in extracts of caput epididymal sperm that probably resulted from the proteolytic processing of the 45- and 32-26kDa SP-10 peptides in the initial segment or caput epididymis. After a follicular fluid induced acrosome reaction, SP-10 was found on the inner acrosomal membrane in the equatorial segment and was associated with hybrid vesicles. This localization is consistent with the hypothesis that SP-10 may be involved in sperm-zona binding /penetration. In situ hybridization demonstrated SP-10 mRNA primarily in round spermatids found in stages IV, V and VI than in previous stages, and rarely found in spermatogonia or spermatocytes. SP-10 mRNA declined when spermatids underwent polarization, nuclear condensation and elongation. The appearance of SP-10 transcription or SP-10 mRNA stability or both occur as spermatids develop from the Golgi phase to the cap phase (Foster et al., 1994; Kurth et al., 1993). The SP-10 proximal promoter, which activates transcription in spermatids, functions as an insulator in somatic cells. Insulator activity mapped to the -186/-135 region and mutation of two ACACAC motifs compromised the insulator function. In conclusion, the evolutionarily conserved SP-10 insulator regulates transcription of a germ cell differentiation marker (Reddi et al., 2003).

Characterization: Three overlapping SP-10 specific-cDNAs were isolated from a human testes cDNA expression library. These cDNAs hybridised to a 1.35-kb RNA that was present only in testes. Sequencing of these cDNAs (SP-10-5, SP-10-8, and SP-10-10) produced a 1117-bp sequence containing a 265-amino acid coding region for the SP-10. Hydrophobicity plots showed a very hydrophobic amino terminus characteristic of a signal peptide. The amino acid sequence for SP-10 deduced from the cDNA sequence predicted a protein of 28.3 kDa. Thus three different repeating amino acid motifs were identified in SP-10. The first motif (Ser-Gly-Glu-Gln-Pro/Ala) occurs seven times. There were two additional variants of the first repeat (Val-Gly-Glu-Gln-Pro) and (Ser-Asp-Glu-Gln-Pro), which differs by only one amino acid. The second motif (Ser-Glu-His-Gly or Ala) was repeated three times, and the third motif (Ser-Gly-Glu-His) was repeated four times. These three motifs comprised 76 of the 108 amino acids between amino acids 66 and 174 (Wright et al., 1990). The sequence (Ser-[Asp or Glu]-X-X-Pro), which occurs at residue 140, has also been suggested as a possible target site for O-linked glycosylation. Thus three different amino acid repeats occur a total of 16 times in the central third of the SP-10 protein. Interestingly, cDNA SP-10-10 has an internal 57-base pair (19 amino acids) inframe deletion that is not present in SP-10-5 suggesting that alternative splicing generates more than one SP-10 mRNA. The sequence analysis also identified a consensus polyadenylation sequence at position 1094, 236bp 3' of the TAG termination codon, and a putative eukaryotic RNA degradation sequence 71 bp 3' of the stop codon. The 5' sequence (CCAG) that flanked the initiator methionine is similar to a consensus sequence found 5' to eukaryotic start codons (Wright et al., 1990).

The SP-10 is encoded by a single gene. Mapping and sequencing of the 8-kb SP-10 gene shows that the SP-10 mRNA consists of four exons of 119, 487,113, and 390-bp (E1, E2, E3 and E4 respectively) with exon coding for a distinct structural domain within the SP-10 protein. An in-frame alternatively spliced form of the SP-10 mRNA, identified during SP-10 cDNA characterization, employs the same 3'splice site as the 487-bp exon and unique 5' splice site within the 487-bp exon. Comparisons of human, baboon, and macaque SP-10 cDNA showed that a 60-bp deletion may have occurred during the evolution of the human SP-10 gene by homologous recombination. This deletion may be responsible for the reduced level of alternatively spliced SP-10 message within the human testis. The exons, E1, E2, E3 and E4 of SP-10 gene code for distinct structural domains that have been identified within the SP-10 protein sequence. The E1 codes for the 18 amino acids hydrophobic SP-10 signal peptide sequence, E2 codes for a 167 amino acid central domain consisting of repeated hydrophilic peptide motifs, whereas E4 codes for a relatively hydrophobic 43 amino acid carboxyl terminal domain that is the most highly conserved region when human, baboon, macaque, and mouse (MSA-63) SP-10 cDNAs were compared. The four exons are separated by three introns: Intron-1 approximately of 2350-bp, Intron-2 of 1312-bp, and Intron-3 of 3200- bp. All the exon/intron border sequences agree with the mammalian consensus splice sequence motif AG/GTNAG——CAG/G (Wright et al., 1993).

The core promoter of mouse mSP-10 lacked a TATA box but contained a canonical initiator (Inr) element surrounding the transcription start site. In transgenic mice, the -108 to +28 bp or the – 266 to +28 bp SP-10 5' flanking region is sufficient to direct round spermatid-specific expression of a green fluorescent protein reporter gene. On the other hand the –91 to +28-bp mSP-10 gene fragment lacked promoter activity in vivo.

SP-10 : as a Immuno-contraceptive: From the standpoint of development of SP-10 as contraceptive vaccine, the question of tissue specificity is of paramount importance because a testis-specific vaccine immunogen reduces the likelihood that autoimmune diseases will arise in females administered with such an immunogen. Northern blots showed SP-10 is expressed exclusively in the testis (Freeman et al., 1994). It is a strong immunogen in female mice and baboons. However, being localized within the acrosomal compartment, it becomes accessible to antibody only after the acrosome reaction is initiated. To be effective as contraceptive immunogen, such intra-acrosomal antigens must evoke antibodies that are concentrated in the oviduct at sufficient levels. It was found that IgG and IgA antibodies were produced against recombinant SP-10 in primate oviductal fluids after systemic immunization. The SP-10 was recognized by primate oviductal fluid IgG and IgA of the endogenous contraceptive target on both human and macaque sperm (Kurth et al., 1997).

Mouse Sperm Antigen - 63 and SP-10: Liu et al (1992) showed that mouse sperm antigen-63 (MSA-63) preparation comprises a group of proteins ranging from 24-84 -kDa with isoelectric points (pIs) ranging from 4.0 to 6.0. A high degree of homology was observed between MSA-63 and human sperm SP-10 at DNA/protein levels (Liu et al., 1992). The cDNA encoding the putative mouse homologue for human acrosomal protein SP-10 was cloned and sequenced. The *Acrv1* was localized in the proximal portion on mouse chromosome 9, in a region that exhibits synteny with human 11q 23, for *ACRVI* (Reddi et al., 1995). The MSA-63 was not expressed until post-meiotic stages of spermatogenesis. MSA-63 is conserved among different mammalian species. Only one specific mRNA 1.5kb in size was identified in testis among different adult mouse tissues. The mouse SP-10 gene, like its human counterpart exhibited a germ cell type-and stage dependent pattern of gene expression. Transcription of the mSP-10 gene was abundant in postmeiotic round spermatids at stage II of the cycle of seminiferous epithelium, and the mRNA signal rapidly declined in spermatids at stage VII. Unlike other testis-specific gene lacking TATA boxes, such as acrosin and Ldh3 that initiate from multiple sites, the SP-10 gene utilized only one major transcription start site embedded within the Inr sequence. This indicated that core promoter elements other than TATA may play a role in accurate initiation of SP-10 transcription (Reddi et al., 1999).

Fox Sperm Protein FSA-1: Beaton et al., (1995) isolated and characterized a cDNA, *cFSA Acr 1*, encoding a testis-specific fox sperm antigen. The antigen located on the inner acrosomal compartment expressed during spermatogenesis on the developing acrosome of round and elongating spermatids. The deduced amino acid sequence of cFSA-Acr 1 revealed that the clone has high homology to both human and baboon sperm protein SP-10, and the mouse MSA-63. The region of highest homology is within the carboxyl terminus. In the middle of the ORF, the fox sequence showed absence of a unique sequence from human, baboon SP-10, and mouse MSA-63 protein.

24.4. HAMSTER P26H AND ITS HUMAN ORTHOLOGUE

During epididymal transit, the male gamete is subjected to major surface modifications such as changes in lipid composition and acquisition of new epididymal protein as well as posttranslational modifications of existing sperm proteins (see Chapter 32). Using the hamster as a model, a 26-kDa protein, termed P26h has been described, that shows immunocontraceptive

```
  1 MKLNFTGLRA LVTGAGRGIG RGTAKALHAS GAKVVAVSLI NEDLVSLAKE CPGIEPVCVD
 61 LGDWEATEKA LGRIGPVDLL VNNAAVALVQ PFIQSTKEVF DRSFNVNVRS VLQVSQMVAK
121 GMINRGVAGS IVNISSMVAY VTFPGLATYS STKGAITMLT KAMAMELGPY KIRVNSVNPT
181 VVLTDMGKKV SADPEFAKKL KERHPLRKFA EVEDVVNSIL FLLSDSSAST SGSGILVDAG
241 YLAS
```

Fig.24.3. Amino acid sequence of surface protein from hamster sperm P26h deduced from cDNA (Accession AAD03695) Source: http//www. ncbi.nlm.nih.gov.

properties in active immunization of male hamsters (Gaudreault et al., 2002). This protein is localized on the sperm acrosome and is acquired during epididymal transit. The P26h plays a role in sperm egg interactions as shown by the ability of anti-P26h IgG and the corresponding Fab fragment to inhibit sperm-zona pellucida binding in vivo and invitro. The human homologue of hamster P26h, P34H, has been proposed as a marker of male fertility. The P26h is phosphatidylinositol anchored to sperm plasma membrane during epididymal transit, and prostasome-like particles may be involved in this transfer (Legare et al., 1999). The P26h is exclusively located on the surface covering the acrosomal cap of the mature spermatozoa. The P26h was released when live spermatozoa were treated with a solution of phospholipase C specific to phosphatidylinositol. In contrast, the P26h remained associated to the sperm surface following treatment with trypsin. P26h was also associated to epididymal prostasomes. The hamster sperm P26h is strikingly homologous with mouse lung carbonyl reductase and is highly expressed in the testis, but its physiological functions in the testis are unknown. A testicular cDNA was screened to clone the full-length P26h cDNA. The longest transcript obtained from the library revealed a cDNA of 1081-bp coding for a 244-amino acid protein and was in agreement with the behaviour of the predicted size of the translational product.

The predicted hamster P26h amino acid sequence shows 85% identity with mouse AP27 protein and porcine carbonyl reductase, members of the short-chain dehydrogenase/reductase (SDR) family of proteins, and hence P26h appears a member of the SDR family of proteins involved in the process of mammalian gamete interactions. A major transcript of 1-kb is present in testis. The P26h gene was predominantly transcribed in seminiferous tubules of the testis and at a lower level in the corpus epididymidis and not in other somatic tissues (Gaudreault et al., 1999) **(Fig.24.3)**. The P26h transcripts appear at 3 week of age, within the first wave of spermatogenesis in the hamster and expressed in spermatocytes and round spermatids, whereas the protein P26h was found in the cytoplasm of round spermatids and elongated spermatids. It was undetectable in testicular spermatozoa (Gaudreault et al., 2001).

Recombinant P26h resembles NADP(H)-dependent lung carbonyl reductase in the tetrameric structure, broad substrate specificity, inhibitor sensitivity, and activation by arachidonic acid, differs in a preference for NAD(H) and high efficiency for the oxidoreduction between 5α-androstane-3α,17β-diol and 5α-dihydrotestosterone. The replacement of Ser 38-Leu39-Ile40 in P26h with the corresponding sequence (Thr38-Arg39-Thr40) of MLCR led to a switch in favour of NADP (H) specificity, suggesting the key role of the residues in the coenzyme specificity. While the P26h mRNA was detected only in the testis of the mature hamster, its enzyme activity was found mainly in the testis mitochondrial fraction and in the nuclear fraction of the epididymis. In testicular cells, P26h mainly exists as a tetrameric dehydrogenase in mitochondria and plays a role in controlling the intracellular concentration of a potent androgen, 5α-dihydrotestosterone, during spermatogenesis, in which it may be incorporated in mitochondrial sheaths of spermatozoa (Ishikura et al., 2001). A 34-kDa (P34H) human epididymal sperm protein that shows antigenic and functional homologies with the hamster P26h is localized on the acrosomal cap of human spermatozoa and has been proposed to be involved in the interaction with the zona pellucida (Boue et al., 1996). In human testis p34 was not detected on those spermatozoa found within the seminiferous tubules or in the vasa defferentia (see Chapter 34).

24.5. PERIACROSOMAL PLASMA MEMBRANE PROTEIN (PM52)

Cowan and Myles, (1993) examined the expression and localization of several guinea pig sperm surface antigens and identified three distinct surface domains on testicular sperm that were established at specific stages of spermiogenesis. Two antigenically related, size-variant integral membrane proteins of 52 kDa (PM52) and 35 kDa that are localized to the periacrosomal domain of cauda epididymal guinea pig spermatozoa were characterized (Westbrook-Case et al., 1994). The PM52 is first expressed in acrosome phase spermatids and localizes exclusively to the cytoplasmic lobe. Both cytoplasmic vesicles and the plasma membrane of the cytoplasmic lobe were found labelled with PM-2. During early stages of expression, PM52 appeared to be absent from the head region, but significant PM52 accumulation over the spermatids head was noted in late acrosomal phase spermatids. Throughout spermiogenesis, PM52 extended posteriorly to the annulus, which represents a barrier preventing PM52 diffusion over the posterior tail. Following the migration of the annulus to the midpiece- principal piece junction, PM52 began to disappear from the flagellar region, and at the completion of spermiogenesis most of the PM52 was restricted to the acrosomal segment. Spermatids and epididymal sperm PM52 exhibited identical sizes indicating that they are not proteolytically modified during epididymal maturation. Screening a guinea pig testis cDNA library and sequence determination of full-length PM52 clones demonstrated identity of a sperm membrane protein termed "Sperad" (Quill and Garber, 1996; Olson et al., 1998). The Sperad is closely related to a large family of putative cell adhesion molecules and is a protein which contains two extracellular Ig-like domains, a transmembrane segment, and an intracellular proline rich domain. The full-length molecule comprises a Mr of 55kDa. The role of Sperad in cell adhesion has not been confirmed, but it appears to be involved in heterotypic interactions prior to interaction of sperm with egg plasma membrane (Quill and Garbers, 1996).

24.6. ACTIN AND ACTIN BINDING PROTEINS IN ACROSOME

Inspite of several decades of investigations, the complex nature of the mature acrosome is not fully understood. Biochemical and ultrastructural studies have provided evidence for the involvement of cytosketetal domains such as actin, calmodulin and α-spectrin-like antigens in the organization of the acrosome. In addition, the organelle contains filamentous structures primarily associated with outer acrosomal membrane. However, the functional significance of the filamentous structures, if any, is not yet known (Abou-Halia and Tulsiani, 2000). In both prokaryotes and eukaryotes, cell shape is determined largely by the internal cytoskeleton, of which actin is a major component. Restructuring of the actin-cytoskeleton has been shown to be important for polarized cell growth and formation of mating process. In mammals, activation of T cells and neutrophils by chemotactic agents leads to a repositioning of cytoskeletal elements and organelles. These changes in the disposition of actin are controlled by actin binding proteins, which include actin-capping proteins, β-thymosins, cofilin/destrin and profilin. Actin binding proteins have specific effects on the actin cytoskeleton. Leung et al, (1993) isolated a rat testis cDNA encoding a 34-kDa rac-GAP termed β-chimerin, as it was highly related to n-chimerin, containing both α GAP domain (77% identity) and the phorbol ester binding region (93% identity). β-chimerin RNA is exclusively expressed in testes at onset of sexual maturation. It's expression is stage specific and parallels acrosomal assembly at the late stage of spermatogenesis. A corresponding testis specific 30-kDa rac-GAP was also detected (see Chapter 18).

Fodrin: Fodrin, a spectrin like protein is present in gametes, zygotes and embryos from sea urchins and mice. Mammalian fodrin comprises two polypeptides with molecular weights of approximately 240-kDa (α) and 235-kDa (β). Immunologically α subunit from sea urchin eggs and mammals are cross-reactive. Fodrin–specific fluorescence was localized to the acrosome of the sperm and was distributed over the entire egg of sea urchin near the surface in a punctate pattern similar to the distribution of polymeric actin. During sperm incorporation, the fodrin-specific fluorescence is found at the site of sperm incorporation, in the fertilization cone. After fertilization, the intensity of fodrin fluorescence increases. Study suggests that fodrin participates in the actin-mediated events at the cell surface during fertilization and early development in both mice and sea urchins (Schatten et al., 1986).

Scinderin: Scinderin is an actin filament-severing protein of 80-kDa found in chromaffin cells, fetal and adult tubules, interstitial cells, spermatozoa, aorta, and vena cava. Scinderin levels are higher in epididymal than in ejaculated spermatozoa. During sperm movement from epididymis to ejaculate sperm, scinderin showed reallocation in acrosome region. In Sertoli cells, scinderin was detected near the cell surface and within the cytoplasm. In the epididymis, scinderin was localized next to the surface of the cells; in the tail, it collected near the base of the principal cells. In Sertoli cells and epididymal cells, scinderin may contribute to the regulation of tight junctional permeability and to the release of the elongated spermatids by controlling the state of perijunctional actin. In germ cells, scinderin may assist in the shaping of the developing acrosome and influence the fertility of the spermatozoa (Pelletier et al., 1999).

24.7. FUSION PROTEINS IN SPERM

A hallmark of fertilization is a high degree of species specificity, implying gamete-specific recognition signals. A large number of sperm protein signal molecules, which might participate in cell-cell adhesion leading to fertiliation have been discussed in chapter 23, 26, 27 and 32. Fusion proteins form a distinct class of molecules that result in sperm–egg fusion.

Equatorin: Equatorin from sperm head is possibly involved in sperm-oocyte fusion. This protein contained in posterior acrosome is detectable only after acrosome reaction, but not in intact spermatozoa (Manandhar and Toshimori 2001).

Synaptosomal-Associated Protein of 25-kDa: The SNAP-25 is a palmitoylated integral membrane protein expressed in neuronal and neuroendocrine tissues. The SNAP-25 forms a complex with vesicle associated membrane protein (VAMP) and syntaxin, which is thought to regulate the fusion of plasma and vesicle membranes during exocytosis. The SNAP-25 is epressed in sea urchin sperm and shares greater identity with mammalian SNAP-25 than with SNAP-23, which is expressed ubiquitously in non-neural tissues. Findings support the involvement of SNAP-25 in invertebrate sperm acrosome reaction, through increased association with VAMP and syntaxin driving the fusion of plasma and acrosomal membranes (Schulz et al., 1997; 1998).

SNARE (Soluble N-ethylmaleimide-sensitive Factor Attachment Protein Receptor): The SNARE proteins are key players in membrane fusion during regulated exocytosis in nerve terminals and secretory cells. Similar proteins have been found to exist in acrosome. Vesicle-associated membrane protein/synaptobrevin, a SNARE on the membrane of a vesicular carrier, and syntaxin 1, a SNARE on the target membrane, as well as the Ca^{2+} sensor synaptotagmin 1 are present in mammalian sperm acrosome. Sperm SNAREs are sloughed off during acrosome reaction (Ramalho-Santos et al., 2000). Various isoforms of syntaxin 2 and syntaxin 4A of SNARE proteins have been identified in rodent testes and spermatozoa. Syntaxin 2 is colocalized with acrin 1, over acrosomal region. The syntaxin 2 may be involved in acrosome reaction and membrane fusion during mammalian fertilization (Katafuchi et al., 2000).

Sp18: The 18-kDa protein is known to be released from abalone sperm during acrosome reaction. It coats the acrosomal process where it is thought to mediate fusion between sperm and egg cell membranes. The Sp18 acrosomal protein aggregates negatively charged (but not neutral) large unilamellar liposomes. These proteins mediate the fusion of lipid bilayers. The secondary structure of Sp18 shows the presence of a strong amphipathic alpha helix that appears to be active during perturbation of phospholipid bilayers. The crystal structure of this fusagenic sperm protein shows extreme surface properties, which may mediate sperm-egg fusion during fertilization (Kresge et al., 2001). Sperm adhesions are a new family of secretory proteins expressed in male genital tract of pig, horse and bull. They are major products of seminal plasma and are associated to sperm surface during sperm maturation. Part of these secretory proteins have been discussed in chaptor 34.

24.8. BINDIN

The bindin is a major protein component of the acrosome granule of sea uchrin sperm, which mediates the species-specific adhesion of sperm to the egg during fertilization. Bindin from *Arbacia punctulata purpuratus* sperm demonstrates a distinct adhesive preference for eggs of the same species and among cross-reactive -species. Bindin cDNA clones from these genera show substantial sequence similarity in both the mature bindin domain and the pro-bindin precursor region. The most striking identity is a region of 42 conserved amino acids in the central part of the mature bindins. This conserved element on N- and C- terminus is more divergent suggesting that they are responsible for the species-specific properties of bindin. The mature region contains a putative transmembrane segment between residues 431-457 that is absent from *S. purpuratus* bindin. This structural element may account for the observation that isolated A. *punctulata* bindin uniquely forms multilamillar structures reminiscent of lipid bilayers and binds significant amounts of phospholipid and detergent (Glabe and Clark, 1991). Another study showed that a few specific regions of the bindin molecule are involved in the sperm-egg contact and that certain of these regions mediate the species specificity of the interaction in a sequence-specific manner (Minor et al., 1993).

The functional domain structure of bindin has been analyzed using a series of carboxyl- and amino terminal deletion analogs of recombinant bindin. S. *franciscanus* sperm bindin agglutinates both *S. franciscanus* and *S. purpuratus* eggs whereas *S. purpuratus* sperm bindin displays a much more restricted specificity for homologous *S. purpuratus* eggs. Consistent with this result, cross-species spem adhesion experiments demonstrated that *S. franciscanus* sperm also bind more efficiently to *S. purpuratus* eggs in comparision to *S.*

```
  1 MSCPLRKKAL PLTMLLLLLS FHVLITPVSK ANKETNRSVH FIPTVEFAVN TFNQESQDEY
 61 AYRMEHIMSS WREKVNFPTV YSMRLQLRRT ICKKFEESLD ICPFQESHGL NNTFTCLFTV
121 GTYPWITKFK LFRSVCS
```

Fig.24.4. Amino acid sequence of mouse testatin: a cystatin-related gene expressed product during early testis development (Accession number CAA77090) Source: http//www. ncbi.nlm.nih.gov.

purpuratus sperm attachment to *S. franciscanus* eggs. In this function, residues 1-74 or residues 122-236 can be deleted without a loss of the species-specific agglutination properties of bindin. Comparison of two sequences reveals that all of the active deletion analogs have a central segment in common to residues 75 through 121 of *S. pupuratus* bindin, except for Arg 77, which is an Ala in *S. franciscanus*. However, substitution of Arg 77 for Ala does not alter the species specificity of recombinant bindin, suggesting that the more divergent amino-and carboxy-terminal sequences flanking the conserved central domain determine the species specificity of bindin and either end is sufficient to impart specificity (Lopez et al., 1993).

24.9. CYSTATIN-RELATED EPIDIDYMAL SPERMATOGENIC PROTEIN

A possible process, by which acrosomal enzymes are regulated, is through the binding of specific inhibitors. The CRES (Cystatin-related epididymal spermatogenic) is a member of the cystatin super-family of cysteine protease inhibitors, and shows substrate specificities against papain like cysteine proteases, such as cathepsins B, S, H, and L (Freije et al., 1991). The in vivo function of these proteins is not well understood. The CRES gene represents a new subgroup in the family of cystatins. The CRES exhibits highly restricted expression in the reproductive tract suggesting its role in reproduction. CRES protein is present in elongating spermatids in the testis and is synthesized and secreted by the proximal caput epididymal epithelium. The presence of CRES protein in developing germ cells and in the luminal fluid surrounding maturing spermatozoa prompted Syntin and Cornwall (1999) to examine the presence of CRES in spermatozoa. The CRES protein is highly expressed in elongating spermatids in the testis and localized in sperm acrosome and is released during the acrosome reaction. The human ortholog of mouse CRES (Testatin1) is highly expressed in the human testis, specifically within clusters of round spermatids. Human CRES is predominantly 19-kDa protein and a minor 14-kDa protein. However, in contrast to the acrosomal localization of CRES in mouse spermatozoa, human CRES was strictly localized to the equatorial segment (Tohonen et al., 1998; Wassler et al., 2002) **(Fig.24.4)**. While the 19 and 14-kDa CRES proteins were present in testicular and proximal caput epididymal spermatozoa, the 14-kDa CRES protein was the predominant form present in mid-caput to cauda epididymal spermatozoa. Testatin 1, expressed during fetal gonads and adult testis could play a role in tissue reorganization during early testis development (Tohnen et al, 1998).

24.10. OTHER NON-ENZYMATIC ACROSOMAL PROTEINS

Vitronectin: Vitrpnectin (Vn) is an intrinsic protein of spermatozoa The Vn has been extracted from fresh human sperm and detected on the surface of living, capacitated sperm. The Vn is transcribed exclusively in germ cells at the spermatcyte and round spermatid stages, and localized within the acrosomal region of ejaculated sperm. The Vn appears to be a specific product of intra-testicular spermatozoa that is released from sequestered location within sperm following acrosome reaction. It seems that Vn is positioned to play a strategic role in gamete interactions leading to fertilization (Bronson et al., 2000; Fusi et al., 1994)(see chapter 25).

Sperm–Oocyte Binding Protein (SOB1): Several human sperm proteins are involved in sperm-oocyte interactions. The mABs of SOBs inhibit human sperm binding in hamster egg binding assay (Lefevre et al., 1997; 1999). The SOB-1 is a glycoprotein of testicular origin that is localized on subequatorial region of epididymal and ejaculated spermatozoa. The SOB1 from human macaque and rodent sperm has an apparent molecular weight of 100-kDa, which is close to 94.8-kDa predicted by cDNA (The difference may be due to glycosylation). Northern blot analysis shows that SOB-1 is conserved among mouse, rat, rabbit, dog and rat. The SOB-1 shows 32% homology with mouse sperm fibrous sheath components (Lefevre et al, 1999). The c-DNA sequence indicated that the ORF encodes a 853 amino acids protein, with a molecular mass of 94.7-kDa. This testis-specific cDNA is 27, 32 and 34.4% homologous to three sperm proteins: HI, FSC1 and AKAP82 respectively. A single 3-kb transcript was demonstrated only in the testis. It is a single copy gene, well conserved among mammals and located on human chromosome 12 at band p13. Further studies are required to elucidate the function of SOB1. Using a probe derived from *Sob1*, a human testis-specific protein 2P1 was identified from mouse. A database search revealed that 2P1 was 91% identical to ORF1 of E3-3, a rat gene probably involved in the regulation of alternative splicing. Sequencing showed that 2P1 has a destabilization motif in its 3'-untranslated region. 2P1 shows strong expression in testis and weak expression in epididymis. The 2P1 mRNA was absent in spermatogonia but expressed in spermatocytes (Hammami-Hamza et al, 2003).

Decapacitation Factor: Epididymal mouse spermatozoa have surface-associated decapacitation factor (DF) that can be removed precociously by centrifugation resulting in acceleration of capacitation and increased fertilizing ability. The DF has the potential of binding to a GPI anchored receptor via fucose receptor. Fucose binding sites are in the same region where Ca^{2+} ATPase, the enzyme regulated by DF was localized. The DF modulates capacitation by regulating enzyme activity and hence the intracellular Ca^{2+} concentration (Fraser, 1998).

AZ1: Aoto et al (1995) identified a cDNA clone, named AZ1 that produced the transcript, which was highly expressed in mouse testis. As the mutant mouse jsd/jsd, which has a defect in germ cell maturation barely expressed the transcript. The message was expressed specifically in spermatocytes. The presence of mRNA in the testes of 16-day old mice suggested that its expression starts at the pachytene spermatocytes stage. The elucidated nucleotide sequence contained a 2841 nucleotide open reading frame, and the expected amino acid sequence had a molecular mass of 107,254 Da. The protein was localized to the pre-acrosome region of round and elongated spermatids. However, it was not detected at a more advanced stage of spermatids, i.e. just before their release from Sertoli cells.

Caltrin Proteins: Rat caltrin, guinea pig caltrin 1, and the mouse seminal vesicle trypsin inhibitor protein P12, which also inhibits Ca^{2+} uptake into epididymal spermatozoa (mouse caltrin 1) bind specifically to the sperm head on the acrosomal region. Caltrin 1 plays an important role in the control of sperm functions such as Ca^{2+} influx in the proper time to ensure successful fertilization (Winnica et al., 2000).

Protein C inhibitor (PCI): The PCI is a plasma glycoprotein belonging to the serpin superfamily of serine protease inhibitors, of which protease inhibitor is the prototype. Sperm acrosin is inhibited by PCI. Induction of the acrosome reaction in ejaculated human spermatozoa leads to the disappearance of PCI from the plasma membrane overlying the acrosomal head and the appearance of a strict distribution at the equatorial segment of human spermatozoa. The activity of acrosin in sperm extracts could be effectively inhibited by PCI, which also blocks sperm-egg binding (see Chapter 26).

Tep22: Neesen et al (2002) isolated a cDNA from testis that was named *Tep22*. The gene encoding Tep22 consists of three exons and is localized in the telomeric region of mouse chromosome 12. The Tep22 is predominantly expressed in spermatocytes and spermatids of the murine testis. Four Tep22 transcripts ranging from 647 to 1122 nt were detected in testes of 15-day-old mice. While 5'-UTRs of *Tep22* were variable, the ORFs of all transcripts of Tep22 had a length of 567bp. The Tep22 is translationally repressed for several days and appears first on day 8 as a 22-kDa band. Tep22 was localized in the acrosomal region of early elongating spermatids, while the surrounding cytoplasm was barely labeled. During further germ cell development, the intensity of the staining in the acrosomal region decreased and was no longer detectable in late stages of elongating spermatids, whereas the intensity of the Tep22 protein increased in the cytoplasm. Finally, Tep22 was incorporated into the midpiece of spermatids and was also present in the mitochondrial sheath of mature spermatozoa. It suggested that Tep22 is involved in the biogenesis of the acrosome as well as in the function of the midpiece of murine spermatozoa (Neesen et al., 2002).

24.11. REFERENCES

Aoto H, Tsuchida J, Nishina Y, Nishimune Y, Asano A, Tajima S: Isolation of a novel cDNA that encodes a protein localized to the pre-acrosome region of spermatids. Eur J Biochem 1995; 234: 8-15.

Beaton S, Ten Have J, Cleary A, Bradley MP: Cloning and characterization of the cDNA encoding the fox sperm protein FSA-Acr.1 with similarities to the SP-10 antigen. Mol Reprod Dev 1995; 40: 242-52.

Benoff S. Carbohydrates and fertilization: An over view. Mol Hum Reprod 1997; 3: 599-637.

Bookbinder LH, Cheng A, Bleil JD. Tissue- and species-specific expression of SP56, a mouse sperm fertilizing protein Science 1995; 269: 86-89.

Boue F, Blais J, Sullivan R. Surface localization of P34H, an epididymal protein, during maturation, capacitation, and acrosome reaction of human spermatozoa. Biol Reprod 1996; 54: 1009-17.

Bronson R, Peresient T, Golightly M, Preissner K. Vitronectin is sequestered within human spermatozoa and liberated following acrosome reaction. Mol Human Reprod 2000; 6: 977-82.

Buchli R, De Jong A, Robbins DL. Genomic organization of an intron-containing sperm protein 17 gene (Sp17-1) and an intronless pseudogene (Sp17-2) in humans: a new model. Biochim Biophys Acta 2002; 1578 :29-42.

Cheng A, Le T, Palacios M, Bookbinder LH, Wasserman PM, Suzuki F, Bleil JD. Sperm-egg recognition in the mouse: characterization of sp56, a sperm protein having specific affinity for ZP3. J Cell Biol 1994; 125: 867-78.

Chiniva-internati M, Wang Z, Xue Y, Bumm K, Hahn AB, Lim SH. Sperm protein 17 (Sp17) in multiple myeloma. opportunity for myeloma-specific donor T cell infusion to enhance graft-versus-myeloma effect without increasing graft-versus-host disease risk. Eur J Immunol 2001; 31: 2277-83

Cohen N, Wassarman PM. Association of egg zona pellucida glycoprotein sp56 during fertilization in mice. Int J Dev Biol 2001; 45: 569-76.

Cowan AE, Myles DG, Koppel DE. Migration of the guinea pig sperm membrane protein PH-20 from one localized surface domain to another does not occur by a simple diffusion-trapping mechanism. Dev Biol 1991; 144: 189-98.

Foster JA, Friday BB, Maulit MT, Blobel C, Winfrey VP et al. AM67 a secretory component of the guinea pig sperm acrosomal matrix is related to mouse sperm protein sp56 and the complent component 4-binding proteins. J Biol Chem 1997; 272: 12714-22.

Foster JA, Klotz KL, Flickinger CJ, Thomas TS, Wright RM, Castillo JR, Herr JC. Human SP-10: Acromal distribution, processing and fate after the acrosome reaction. Biol Reprod 1994; 51: 1222-31.

Fraser LR. Interactions between a decapacitation factor and mouse spermatozoa appear to involve fucose residues and a GPI-anchored receptor. Mol Reprod Dev 1998; 51: 193-02.
Freemann AJ, Wright RM, Flickinger CJ, Herr JC. Tissue specificity of the acrosomal protein SP-10: A contraceptive vaccine candidate molecule. Biol Reprod 1994; 50, 615-21.
Fusi FM, Lorenzetti I, Mangili F, Herr JC, Freeman AJ, Gailit J, Bronson RA: Vitronectin is an intrinsic protein of human spermatozoa released during the acrosome reaction. Mol Reprod Dev 1994; 39: 337-43.
Gaudrealt C, El Alfy M, Legare C, Sullivan R. Expression of hamster sperm protein P26h during spermtogenesis. Biol Reprod 2001; 65: 79-86.
Gaudreault C, Le gare C, Berube B, Sullivan R. Hamster sperm protein, p26h: a member of the short-chain dehydrogenase/reductase superfamily. Biol Reprod 1999; 61: 264-73.
Gaudreault C, Montfort L, and Sullivan R. Effect of immunization of hamsters against recombinant P26h on fertility rates. Reproduction 2002; 123: 307-13.
Glabe CG, Clark D. The sequence of the Arbacia punctulata bindin cDNA and implications for the structural basis of species specific sperm adhesion and fertilization. Dev Biol 1991; 143: 282-8.
Hammami-Hamza S, Doussau M, Allemand I, Segretain D, Gasc JM, Finaz C. 2P1, a novel male mouse cDNA specifically expressed during meiosis. Int J Dev Biol 2003; 47: 71-6.
Hardy CM, Mobbs KJ. Expression of recombinant mouse sperm protein sp56 and assessment of its potetial for use as an antigen in an immunocontraceptive vaccine. MolReprod Dev 1999; 52: 216-24.
Ishikura S, Usami N, Kitahara K, Isaji T, Oda K, Nakagawa J, Hara A. Enzymatic characteristics and subcellular distribution of a short-chain dehydrogenase/reductase family protein, P26h, in hamster testis and epididymis. Biochemistry 2001; 40: 214-24.
Katafuchi K, Mori T, Toshimori K; Iida H. Localization of syntaxin isoform, syntaxin 2, to the acrosome region of rodent spermatozoa. Mol Reprod Dev 2000;57:375-83.
Kim KS, Cha MC, Gerton GL. Mouse sperm protein sp56 is a component of the acrosomal matrix. Biol Reprod 2001a; 64: 36-43.
Kim KS, Foster JA, Gerton GL. Differential release of guinea pig sperm acrosomal components during exocytosis. Biol Reprod 2001b; 64: 148-56.
Kong M, Richarson RT, Widgren EE, O'Rand MG. Sequence and localization of the mouse sperm autoantigenic protein, Sp17. Biol Reprod 1995; 53: 579-90.
Kresge N, Vacquier VD, Stout CD. The crystal structure of fusagenic sperm protein reveals extreme surface properties. Biochemistry 2001; 40: 5407-13.
Kurth BE, Weston C, Reddi P, Bryant D, Bhattacharya R, Flickinger CJ, Herr JC: Oviductal antibody response to a defined recominant sperm antigen in macques. Biol Reprod 1997; 57: 981-89.
Kurth BE, Wright RM, Flickinger CJ, Herr JC: Stage specific detection of mRNA for the sperm antigen SP-10 in human testes. Anat Record 1993; 236: 619-25.
Lacy HM, Sanderson RD. Sperm protein 17 is expressed on normal and malignant lymphocytes and promotes heparan sulfate-mediated cell-cell adhesion. Blood 2001; 98: 2160-65
Lea IA, Richardson RT, Widgren EE, O'Rand MG. Cloning and sequencing of cDNAs encoding the human sperm protein, Sp17. Biochim Biophys Acta 1996; 1307: 263-6
Lefevre A, Duquenne C, Rousseau-Merck MF, Rogier E, Finaz C. Cloning and characterization of SOB1, a new testis specific cDNA encoding a human sperm protein probably involved in oocyte recognition. Biochem Biophys Res Commun 1999; 259: 60-66.
Lefevre A, Martin Ruiz C, Chokomian S, Duquenne C, Finaz C. Characterizationof SOB2, a human sperm protein with a potential role in oocyte membrane binding. Mol Hum Reprod 1997; 3: 506-16.
Legare C, Berube B, Boue F, Lefievre L, Morales CR, El-Alfy M, Sullivan R. Hamster sperm antigen P26h is a phosphatidylinositol-anchored protein. Mol Reprod Dev 1999; 52: 225-33.
Leung T, How BE, Manser E, Lim L. Germ cell β-chimaerin, a new GTPase-activating protein for p21rac, is specifically expressed during the acrosomal assembly stage in rat testis. J Biol Chem 1993; 268: 3813-6.
Lim SH, Wang Z, Chriva-internati M, Xue Y. Sperm protein 17 is a novel cancer-testis antigen in multiple myeloma. Blood 2001; 97: 1508-10
Liu MS, Aebersold R, Fann CH, Lee CY: Molecular and developmental studies of a sperm acrosome antigen recognized by HS-63 monoclonal antibody. Biol Reprod 1992;46: 937-48.
Lopez A, Miraglia SJ, Glabe CG. Structure/function analysis of the sea urchin sperm adhesive protein bindin. Dev Biol 1993; 156: 24-33.
Manandhar G, Toshimori K. Exposure of sperm head equatorin after acrosome reaction and its fate after fertilization in mice. Biol Reprod 2001; 65: 1425-36.
Minor JE, Britten RJ, Davidson EH. Specific inhibition of fertilization by a peptide derived from the sperm protein bindin. Mol Biol Cell 1993; 4: 375-87.
Neesen J, Hartwich T, Brandhorst G, Aumuller G, Glaser B, Burfeind P,Mendoza-Lujambio I. Tep22, a novel

testicular expressed gene, is involved in the biogenesis of the acrosome and the midpiece of the sperm tail.. Biochem Biophys Res Commun. 2002;297:737-48.

Noland TD, Friday BB, Maulit MT, Gerton GL: Sperm acrosomal matrixcontains a novel member of the pentaxin family of calcium-dependent binding proteins. J Biol Chem 1994; 269: 32607-14.

Olson GE, Winerey VP. Westbrook VA: Targeting of the domain-specific integral Membrane protein PM52 to the periacrosomal plasma membrane during guinea pig spermatogenesis. Mol Reprod Dev 1998a; 50: 103-12.

Olson GE, Winfrey VP, Nagdas SK. Acrosome biogenesis in the Hamster: Ultrastructurally distinct Matrix regions are assembled from a common precursor polypeptide. Biol Reprod 1998b; 58: 361-70.

Olson GE, Winfrey VP, Nagdas SK. Structural modification of the hamster sperm acrosome during posttesticular development in the epididymis. Microsc Res Tech 2003; 61: 46-55.

Pelletier R, Trifaro JM, Carbajal ME, Okawara Y, Vitale ML. Calcium-dependent actin filament-severing protein scinderin levels and localization in bovine testis, epididymis, and spermatozoa. Biol Reprod 1999; 60: 1128-36.

Quill TA, Garbers DL. Sperad is a novel sperm-specific plasma membrane protein homologous to a family of cell adhesion proteins. J Biol Chem 1996; 271: 33509-14.

Ramalho-Santos J, Moreno RD, Sutovsky P, Chan et al. SNAREs in mammalian sperm: possible implications for fertilization. Dev Biol 2000; 223: 54-69.

Reddi P P, Naaby-Hansen S, Aguolnik Irena, Tsai JY, Silver LM, Flickinger CJ, Herr JC. Complementary deoxyribonucleic acid cloning and characterization of mSP-10 : The mouse homologue of human acrosomal protein SP-10. Biol Reprod 1995; 53, 873-81.

Reddi PP, Flickinger CJ, Herr JC. Round spermatid-specific transcription of the mouse SP-I0 gene is mediated by a 294-base pair proximal promotor. Biol Reprod 1999; 61: 1256-66.

Reddi PP, Shore AN, Shapiro JA, Anderson A, Stoler MH, Acharya KK. Spermatid-specific promoter of the SP-10 gene functions as an insulator in somatic cells. Dev Biol 2003; 262: 173-82.

Richardson RT, Yamasaki N, O'Rand MG. Sequence of a rabbit sperm zona pellucida binding protein and localization during the acrosome reaction. Dev Biol 1994; 165: 688-701.

Schatten H, Cheney R, Balczon R, Willard M, Cline C, Simerly C, Schatten G. Localization of fodrin during fertilization and early development of sea urchins and mice. Dev Biol 1986; 118: 457-66.

Schulz JR, Sanchez R, Vacquier VD. Incresed association of synaptosome-associated protein of 25kDa with syntaxin and vesicle-associated membrane protein follwing acrosomal exocytosis of sea urchin sperm. J Biol Chem 1998; 273: 24355-9.

Schulz JR, Wessel GM, Vacquier VD. The exocytosis regulatory proteins syntaxin and VAMP are shed from sea urchin sperm during acrosome reaction. Dev Biol 1997; 191: 80-7.

Syntin P, Cornwall GA. Immunolocalization of CRES (Cystatin-Related Epididymal Speratogenic) protein in the acrosomes of mouse spermatozoa. Biol Reprod 1999; 60: 1542-52.

Tohonen V, Osterlund C, Nordqvist K. Testatin: a cystatin-related gene expressed during early testis development. Proc Natl Acad Sci (USA) 1998; 95: 14208-13

Wassler M, Syntin P, Sutton-Walsh HG, Hsia N, Hardy DM, Cornwall GA. Identification and characterization of cystatin-related epididymal spermatogenic protein in human spermatozoa: localization in the equatorial segment. Biol Reprod. 2002; 67: 795-803.

Wen Y, Richardson RT, O'Rand MG. Processing of the sperm protein Sp17 during the acrosome reaction and characterization as a calmoudlin binding protein. Dev Biol 1999; 206 : 113-32

Wen Y, Richardson RT, Widgren EE, o'Rand MG. Characterization of Sp17: a ubiquitous three domain protein that binds heparin. Biochem J 2001; 257 (Pt1): 25-31.

Westbrook-Case VA, Winfrey VP, Oison GE. Characterizaiton of two antigenically related integral membrane proteins of the guinea pig sperm periacrosomal plasma membrane. Mol Reprod Dev 1994; 39: 309-21.

Winnica DE,Novella ML,Dematteis A, Coronel CE: Trypsin/acrosin inhibitor activity of rat and guinea pig caltrin proteins. Structural and functional studies. Biol Reprod 2000; 63:42- 48.

Wright R M, John E, Klotz K, Flickinger, Johnn C H : Cloning and sequencing of cDNAs coding for the human intra-acrosomal antigen SP-10. Biol Reprod 1990;42 : 693-701.

Wright R M, Suri A K , Kornreich B, Flickinger C J, Herr J C : Cloning and characterization of the human acrosomal protein SP-10. Biol Reprod 1993;49: 316-325.

Yoshinaga K, Tanii I, Oh-oka T, Toshimori K. Changes in distribution and molecular weight of the acrosomal protein acrin2 (MC41) during guinea pig spermiogenesis and epididymal maturation. Cell Tissue Res 2001; 303: 253-61.

Yoshinaga K, Tanii I, Oh-oka T, Toshimori K. Transport and rearrangement of the intra-acrosomal protein acrin1 (MN7) during spermiogenesis in the guinea pig testis. Anat Rec 2000; 259: 131-40.

Yoshinaga K, Tanii I, Saxena DK, Toshimori K. Immunocytochemical alterations in the intra-acrosomal antigen MN7 during epididymal maturation of guinea pig spermatozoa. Cell Tissue Res 1998; 292: 427-33.

Chapter 25

ACTINS AND MYOSINS

25.1. INTRODUCTION

Actin occurs in two forms, G-actin (globular actin) and F-actin (filamentous actin) - a polymer of G-actin. The monomeric G-actin, which is stable in distilled water, undergoes polymerization under certain condtions to form F-actin. Each molecule of G-actin binds Ca^{2+} ions and one molecule of ATP or ADP independently. G-actin polymerizes to F-actin in presence of ATP. Actins are believed to play major roles in cell division, cell shape changes, secretory processes, phagocytosis, cell and organelle motility and muscle contraction. At least six different vertebrate actins have been identified; each is the product of different gene. Amino acid sequencing studies have shown that the multiple actin isotypes have evolved from two major classes, the cytoplasmic and the muscle actins. In mammals, two forms of cytoplasmic actins, called β- and γ-actins, are generally found. The expression patterns of different actin isomers during spermatogenesis and their role and regulatory mechanisms in spermatogenesis have remained poorly understood. In rat testis actin immunofluorescence was demonstrated in vascular smooth muscle cells, interstitial macrophages and most intensely in peritubular cells. Inside the seminiferous tubules the Sertoli cell junctions and the ectoplasmic specializations of the Sertoli cells that follow the outer contour of spermatid heads displayed distinct actin fluorescence. In addition to the locations mentioned, actin-like immunoreactivity was visualized at the ultrastructural level in the chromatid body and the subacrosomal space of spermatids as well as on the outer dense fibers of the sperm tail (Aumuller and Seitz, 1988).

25.1.1. β- and γ-Actins in Spermatogenic Cells

Of the six vertebrate actins, only two cytoplasmic actins have been detected in mouse spermatogenic cells. Two dimensional gel electrophoresis of *in vivo*-labeled mouse testis proteins has demonstrated a high level of radiolabeled γ-actin in meiotic pachytene spermatocytes, followed by a decrease in haploid cell types, i.e. round and elongating spermatids. In contrast, the amount of radiolabeled β-actin increased between pachytene spermatocytes and the subsequent post-meiosis spermatids. Thus, the rate of total cytoplasmic actin labeling did not appear to vary greatly during spermatogenesis, where as the production of β and γ-actin synthesis changed. It was suggested that the differential expression of β and γ-actin during spermatogenesis plays a role in the changing cell shapes or in the movement of mitochondria from around the nucleus to the plasma membrane and the flagellar axoneme, which occurs during this developmental process (Waters et al., 1985). However in a recent study (Ventela et al, 2000), β actin was found highly expressed in mouse spermatogenia, with almost no expression found at early stages of meiosis (leptotene spermatocytes). A gradual increase in the translation of β-actin during later stages of meiosis and early spermatogenesis, with a maximum in elongating spermatids seems to play a role in the nuclear elongation of spermatids. Expression of β-actin was regulated by FSH in a stage-specific fashion (Ventela et al., 2000).

Using several actin isotype-specific cDNA probes, Waters et al (1985) found actin mRNA of two sizes 2.1- and 1.5-kb in extracts of polyadenylated and nonadenylated RNA from sexually mature mouse testes. The 2.1-kb mRNA codes for cytoplasmic β- and γ-actin through out spermatogenesis, while the post-meiotic actin is encoded by the 1.5-kb mRNA as a smooth muscle γ-actin (SMGA). Although the 2.1-kb sequence was present in both meiotic and postmeiotic testicular cell types, it decreased many folds in the late haploid cells. The 1.5-kb actin was not detectable in meiotic pachytene spermatocytes, but was present in round and elongating spermatids and residual bodies. To differentiate between the β- and γ-actin mRNAs, Waters et al (1985) isolated a cDNA, pMGA, containing the 3'untranslated region of a mouse cytoplasmic actin that has homology to the 3'untranslated region of a human γ-actin cDNA but not to the 3'untranslated region of human α-, β-, or cardiac actins. The pMGA detected high levels of presumptive γ-actin mRNA in pachytene spermatocytes and round spermatids, with lower amounts found in elongating spermatids. Hybridization studies revealed that round spermatids in comparison to pachytene spermatocytes or residual bodies contained higher levels of β-actin mRNA. Both probes (pMGA and β-actin probe) hybridized to the 2.1-kb actin mRNA but failed to hybridize to the 1.5-kb mRNA. Thus the level of β-actin mRNA was highest in isolated round spermatids when compared with the levels in pachytene spermatocytes or residual bodies. In round spermatids of mouse testis, actin immunolabeling was detected in the subacrosomal space; labeling increased during the elongation phase, and then decreased and completely disappeared before spermiation (Prigent and Dadoune, 1993; Ventela et al, 2000; Waters et al, 1985).

The amino acid sequence deduced from the postmeiotic actin cDNA sequence was nearly identical to that of a chicken gizzard SMGA, with one amino acid replacement at amino acid 359, where glutamine was substituted for proline. The nucleotide sequence of the untranslated region of the SMGA differed substantially from those of other isotypes of mammalian actins. Testicular SMGA mRNA was present in the polysome fractions, indicating that it was translated (Kim et al, 1989). The cellular location(s) of the SMGA protein, probed with two different mAb specific for muscle actin, suggested that there are muscle isoforms of actins within the cytoplasm of developing spermatids and within apical processes of Sertoli cells (Oko et al., 1991).

However, mouse testicular RNAs revealed an additional actin mRNAs of 1.4-kb. While the 2.1-kb mRNA encodes the cytoplasmic β and γ actins, the two faster-migrating actin mRNAs encode γ enteric actin. The 1.5-kb γ enteric actin mRNA is primarily found in the nonpolysomal fraction, whereas the 1.4-kb γ enteric actin is polysomal. When the poly (A) tails were removed, the non-polysomal and polysomal γ enteric actins mRNAs both migrated at 1.3-kb, indicating that the difference in electrophoretic mobilities of the two γ enteric actin mRNAs was caused by poly (A) length differences. Sequence comparison of the 3'untranslated region of the mouse γ enteric actin to the 3' untranslated region of other testicular mRNAs that undergo partial deadenylation revealed three highly conserved sequence elements. This demonstrates that the poly (A) shortening of polysomal mRNAs seen only with testis specific mRNAs, stored as mRNPs also occurs with mRNAs of widely expressed genes in postmeiotic male germ cells (Gu et al, 1996). α-smooth muscle actin, present in peritubular cells, increased with advancing age (Steger and Wrobel, 1994). In order to explain the absence of α-actin in

```
  1 MATKNSPSPK PMGTAQGDPG EAGTLPAPEA AGIRDTGSTQ LKTKPKKIRK IKALVIDLGS
 61 QYCKCGYAGE PRPTYFISST VGKRRPEMAA DAGDNFKETY VGHELLNMEA SLKLVNPLKH
121 GVVVDWDCIQ NIWEYIFHTA MKIMPEEHAV LVSDPPLSPT SNREKYAELL FETFGIPAMH
181 VTSQALLSIY SYGKTSGLVV ESGHGVSHVV PISEGDLLPG LPSRVDYAGC DLTNYLMQLL
241 NEAGHKFSDD HLHIIEHIKK KCCYAALLPE EEMSLGLDEL HVDYELPDGK IITIGQERFR
301 CSEMLFKPSL VGCTQPGLPE LTATCLARCQ GTGFKEEMAA NVLLCGGCTM LDGFPERFQR
361 ELSLLCPGDS PTVAAAPERK TSVWTGGSIL ASLQAFQQLW VSKEEFEERG CAAIYSKC
```

Fig.25.1. Amino acid sequence of T-actin 1 from mouse testis Source : http//www.ncbi.nlm.nih.gov. (Accession P_079547).

germ cells, Warnecke and Clark (1999) made detailed analysis of the methylation state of a tissue-specific gene through early development and differentiation and mapped the methylation profile of the tissue-specific mouse skeletal α-actin promoter at all stages of development. The α-actin promoter is fully methylated in the sperm and essentially unmethylated in the oocytes and undergoes a general demethylation from morula to blastocyst stages.

25.1.2. F-Actin

Specific fluorescence emitted along the dorsal curvature of the head and as an inverted V-shaped structure in what appears to be the anterior aspect of the post-acrosomal region suggested that the filamentous (F) actin occurs in the heads of rat spermatozoa. Moreover, this F actin is concentrated in two regions of the perinuclear theca: in the subacrosomal space along the dorsal curvature of the nucleus and in the subacrosomal region in an area termed the ventral spur (Vogl et al, 1993). The distribution pattern of actin in the ovine testis, during postnatal development was investigated by Steger and Wrobel (1994). Structural F-actin was localized in peritubular myoid cells and Sertoli cells of the adult testis, Sertoli-Sertoli junctions and at contact sites of Sertoli cells with primary spermatocytes during the acrosome phase of spermiogenesis.

25.1.3. Actin-Like Proteins (T-ACTINS)

Actin-like proteins, T-ACTIN-1 and T-ACTIN-2, specifically expressed in the mouse testis have been described by Tanaka et al. (2003). The mRNA sizes and deduced molecular masses of t-actin 1/mACTl7b and t-actin 2/mACTl7a were 2.2-kb and 1.8-kb, and Mr 43.1 x 10^3 and Mr 47.2 x 10^3, respectively. The two deduced amino acid sequences had 60% homology, and had 40% homology with other actins. The T-ACTINs contained some of the conserved regions found in other actins. Although the cellular locations of these two proteins are quite different (T-ACTIN-1 in the cytoplasm and T-ACTIN-2 in the nucleus), the expression of their proteins and mRNAs is limited during spermiogenesis. In contrast, only T-ACTIN-2 was present in sperm heads and tails. The *Tact1* and *Tact2* genes are exclusively expressed and translated in haploid germ cells in testis. *Tact1* and *Tact2* were single copy genes contained on a common fragment in a head-to-head orientation, and the distance between these genes was less than 2-kb **(Fig.25.1)**. Comparison of the nucleotide sequences of genomic DNA and cDNA demonstrated that *Tact1* and *Tact2* lack introns, although all known actin or actin-related genes in mammals contain introns. Human *Tact* orthologues also lack introns and are located within 6.4-kb in a head-to-head orientation. Thus these genes arose by retroposition of a spliced mRNA transcribed from an actin progenitor gene prior to the divergence of rodents and primates. Comparison of the murine *Tact* genes with their human orthologues showed a high level of identity between the two species in the 5'-upstream and non-coding sequences as well as in the coding region, indicating that conserved elements in these regions may be involved in the regulation of haploid germ cell-specific expression. The promoter region contains no TATA-, CCAAT- or GC-boxes, although there are potential CRE-like motifs in the 5'-upstream region and the 5'-untranslated region in *Tact1* and *Tact2*, respectively. Transient promoter analyses indicated that CREMτ may activate *Tact1* and *Tact2* expression in germ cells (Hisano et al., 2003).

Bull sperm calyx has two major acidic components, which are novel members of the subfamily of actin-related proteins (Arps). These proteins have been termed Arp-T1 and Arp-T2 in humans. They are synthesized specifically in the testis, late in spermatid differentiation and localized in the calyx. Two Arps as major components in a cytoskeletal, nonmotile structure of mammalian spermatozoa suggested that certain members of this family of proteins might serve functions other than nucleation of actin filaments (Heid et al., 2002).

25.2. ACTINS IN SPERMATOZOA

The mammalian spermatozoa are endowed with a unique cytoskeleton, which consists both of ubiquitous and specific proteins, some of them arising from gene haploid transcription, that is divided into subacrosomal layer and post-acrosomal calyx. In the head, the dense perinuclear layer is made of major basic proteins (calicin, cylicin etc) associated with calmodulin and actin remnants. In the flagellum, the axonemal microtubules are mainly composed of glutamylated tubulin isoforms; the periaxonemal outer dense fibers and fibrous sheath are considered as related cytoskeletal structures on the basis of some common polypeptides (Fouquet and Kann 1994; von Bulow et al., 1997). Actin has been characterized and localized in sperm cells of many mammals (Camatini et al., 1986; Stokes and DeRosier, 1987; Fouquet et al., 1989). Nevertheless, the reported localizations obtained by different methods and/or antibodies varied from species to species and even for the same species. Actin is localized in spermatids and spermatozoa of rabbit, mouse, rat, monkey, and human. In these species, F-actin was detected between the nucleus and the acrosome of round and elongating spermatids. In sperm, actin specific labeling was localized a) around the connecting piece in the neck region of sperm from human, bull, rabbit and hamster species, although a species-specific pattern was evident on the external surface of the fibrous sheath of human sperm; b) in the perinuclear space underlying the postacrosomal sheath of bull and rabbit sperm; c) between the plasma membrane and outer acrosomal membrane along the concave margin of the hamster sperm head (Dlaherty et al., 1988). Species-specific change occurred in maturing spermatids. Localization at different stages of maturation suggests that there is a redistribution of actin in late spermatids and spermatozoa, which is a species-specific process (Fouquet and Kann, 1992). In ram spermatozoa the monoclonal anti-actin antibody recognizes single band at 43,000D. In all spermatozoa, intense actin staining was observed in the whole length of the flagellum and, in the neck and postacrosomal region of the head. Ram sperm actin is a monomeric, intracellular, membrane-associated protein. Gelsolin is also present in ram spermatozoa and precisely colocalized with actin. This suggests a physical association of actin to the plasma membrane, most likely by its intracellular side (de las Heras et al., 1997). The localization of the cytoskeletal proteins (α-tubulin, β-tubulin, actin, spectrin, tropomyosin, vimentin and cytokeratin) in human and boar spermatozoa differs remarkably. In human spermatozoa, the distribution of actin and spectrin changes after acrosome reaction. In constrast, in boar spermatozoa, the distribution was concerned with α-tubulin, β-tubulin, actin and spectrin. Redistribution of cytoskeleton proteins suggests that these proteins participate in the process of acrosome reaction of mammalian spermatozoa (Palecek et al., 1999). A sperm cytoskeletal protein (MSP) that signals oocyte meiotic maturation and ovulation in *C elegans,* has been reported by Miller et al, (2001).

Participation of Actin in Acrosome Reaction: Phosphoproteins and phospholipases are known to play a significant role in membrane fusion of sperm exocytosis, during which cortical F-actin has two roles in regulating sperm exocytosis. One is to form a scaffolding to hold phospholipase C at the membrane. It also functions as a physical barrier to membrane fusion, which is removed by the increase in intracellular calcium and pH that precede fusion (Spungin et al, 1995). The presence and absence of cytochalasin B and cytochalasin D (CB and CD) during in vitro fertilization in boar and human showed that actin polymerization plays a role in fertilization. Presence of CD (an inhibitor of the polymerization of actin) markedly decreases the fertilizing capacity of the boar spermatozoa. Fertilization capacity further decreased when CD was present during both capacitation and fertilization processes. The effect of inhibitors of actin polymerization and depolymerization in calcium ionophore-induced acrosome reacted spermatozoa indicated that actin staining was confined to the equatorial segment, post-acrosomal region and tail of human sperm. Cytochalasins B and D significantly inhibited ZP-induced acrosome reaction in a dose-dependent manner. Cytochalasin B or D had no effect on total percentage motile spermatozoa but decreased sperm velocity and hyperactivation. This suggests that actin polymerizes during capacitation and the acrosome reaction and that this polymerization is essential to the fertilization process (Castellani-Ceresa et al., 1993; Liu et al., 1999). However, in hamster sperm, G-actin originating in the head of late spermatids was redistributed to the flagellum of epididymal spermatozoa. Further changes were not seen after capacitation and acrosome reaction thus indicating no apparent effect on actin polymerization and distribution (Fouquet et al, 1991). The role of actin in acrosome reaction was confirmed by an anti-actin monoclonal antibody, which inhibits the zona pellucida-induced acrosome reaction and hyperactivated motility of human sperm. The sperm plasma membrane became permeable to the anti-actin mAb during capacitation and initiation of acrosome reaction (Liu et al, 2002).

Actin Binding Proteins in Acrosome Reaction: Actin-capping proteins are ubiquitous components of mammalian cells. They are known to regulate the polymerization state of actin and hence indirectly control the activity of the cytoskeleton and cell shape (Hernandez-Gonzalez et al., 2000). Howes et al (2001) examined the distribution of actin regulatory proteins, thymosin β 10, destrin, and actin capping protein (ACP) involved in controlling the balance between actin monomers (G-actin) and actin filaments (F-actin), and polymerization status of actin in bull spermatozoa following acrosomal exocytosis. In permeabilized testicular spermatozoa all 3 regulatory proteins were localized primarily to the acrosomal domain but during epididymal maturation they remained confined to the equatorial segment. Following ejaculation, however, they extended back into the acrosomal region. In spermatozoa induced to undergo an acrosome reaction, further rearrangement occurs with actin binding proteins appearing in the postacrosomal domain. This redistribution of actin and actin-regulatory proteins, coupled with changing levels of actin polymerization suggested a continuing role for actin in both post-testicular sperm maturation and acrosomal exocytosis (Howes et al., 2001). Confocal and electron microscopy also demonstrated visible changes of actin, tubulin and spectrin containing structures after the acrosome reaction, and proves the role of cytoskeletal proteins in this process (Dvorakova et al., 2001). Presence of actin in the acrosome and in the entire tail as well as its presence in the equatorial and in the postacrosomal regions suggested a correlation with calmodulin. Since guinea pig spermatozoa showed a change on calmodulin location during the acrosome reaction, it seemed that actin participates in calmodulin translocation to the postacrosomal region during acrosome reaction, in maintaining the acrosome structure, and perhaps also in sperm motility. The calmodulin present in the Ca^{2+} rich postacrosomal lamina could be involved in the regulation of egg activation (Fouquet et al., 1991; Moreno-Fierros et al., 1992).

25.3. ACTIN BINDING PROTEINS

Actin-capping proteins are ubiquitous components of mammalian cells. They are known to regulate the polymerization state of actin and hence indirectly control the activity of the cytoskeleton and cell shape.

25.3.1. Basic Proteins of Cyclicin Group

In the mammalian sperm head the nucleus is tightly associated with a large and dense nonfilamentous cytoskeletal structure, the calyx, major proteins of which are basic. The calyx is a large cytoskeletal component of the perinuclear theca of the mammalian sperm head. It displays remarkable morphological interspecies differences and is characterized by resistance to high ionic strength and detergents. It has a special protein composition, including the basic proteins calicin, cylicin I and II, and two major actin-capping proteins. Two basic proteins, i.e. the 60-kDa calicin and a group of very basic (pI>10) polypeptides ranging in size from 58 to 100-kDa ("multiple band protein "MBPs) are the major proteins of calyx. Calicin is a polypeptide of 588 amino acids (Mr of 66,889; pI 8.1), present in the bull and man. It is encoded by a 2.2-kb mRNA that has been detected only in testis but not in any other tissue. Calicin is homologous to the kelch protein of the ring canal structure of *Drosophila* ovaries. In particular, it contains three consecutive repeating units of 48 amino acids each, which are homologous to the so-called "β-strand folds" occurring in proteins of the kelch family, including the actin cross-linking protein, scruin of *Limulus* sperm and a series of other eukaryotic, bacterial, and viral proteins and contains BTB/POZ domain. The amino terminal domain of calicin contains a region of about 100 amino acids homologous to an extended motif shared by the kelch protein as well as various zinc finger and poxvirus proteins. It has a possible role as a morphogenic cytoskeletal element in spermiogenic differentiation as demonstrated by its absence or altered arrangement in frequent forms of human teratozoospermia such as "round-headed" or other "postacrosomal sheath defect" sperm malformations (von Bulow et al., 1997). Calicin is bound to actin with high affinity (Kd $\simeq$ 5 nM), and a stoichiometry of approximately one calicin per 12 actin monomers. Calicin forms homomultimers and is present with actin in the acrosomal region of round spermatids and is mainly localized in the postacrosomal region of late spermatids and spermatozoa. The affinity of calicin to F-actin allows targeting of calicin at the subacrosomal space of round spermatids, and its ability to form homomultimers contributes to the formation of a rigid calyx (Lecuyer et al., 2000).

A bovine and a human cDNA clone encoding one of these proteins, is termed "cyclicin" (cup or beaker). Bovine cylicin I of a calculated molecular weight of 74,788 contains a high proportion (29%) of positively charged amino acids (pI, 10.55) with numerous KKD tripeptides, and characterized by an organization of the central part of the molecule in nine repeating units of maximally 41 amino acids, each of which according to prediction should tend to form an alpha helix. The 2.4-kb cyclicin I mRNA is present only in testis. The cyclicin is an unusual cytoskeletal protein and plays a possible architectural role during spermiogenesis (Hess et

```
  1 MSLPRFQRVN FGPYDNYIPV SELSKKSWNQ QHFALLFPKP QRPGTKRRSK PSQIRDNTVS
 61 IIDEEQLRGD RRQPLWMYRS LMRISERPSV YLAARRQPLK PTRTVEVDSK AAEIGKKGED
121 KTTQKDTTDS ESELKQGKKD SKKGKDIEKG KEEKLDAKKD SKKGKKDAEK GKDSATESED
181 EKGGAKKDNK KDKKDSNKGK DSATESEGEK GGTEKDSKKG KKDSKKGKDS AIELQAVKAD
241 EKKDEDGKKD ANKGDESKDA KKDAKEIKKG KKDKKKPSST DSDSKDDVKK ESKKDATKDA
301 KKVAKKDTEK ESADSKKDAK KNAKKDAKKD AKKNAKKDEK KDAKKKGK
```

Fig.25.2. Amino acid sequence of cylicin 2 from human sperm. Source : http//www.ncbi.nlm.nih.gov. (Accession NP_001331)

al., 1997). Boar sperm contain a 66kDa protein that binds F-actin in actin pelleting assays. Sequence studies and immunological characterization with antibodies specific for human cylicin II identified the 66kDa protein as the homologue of bovine and human cylicin II, a protein of the sperm perinuclear cytoskeleton (Hess et al., 1995; Rousseaux-Prevost et al., 2003) **(Fig.25. 2).**

25.3.2. Actin Capping Proteins

Inspite of several decades of investigations, the complex nature of the mature acrosome is not fully understood. Biochemical and ultrastructural studies have provided evidence for the involvement of cytosketetal domains such as actin, calmodulin and α-spectrin-like antigens in the organization of the acrosome. In addition, the organelle contains filamentous structures primarily associated with outer acrosomal membrane. However, the functional significance of the filamentous structures, if any, is not yet known. In both prokaryotes and eukaryotes, cell shape is determined largely by the internal cytoskeleton, of which actin is a major component. Restructuring of the actin-cytoskeleton has been shown to be important for polarized cell growth and formation of mating process. In mammals, activation of T cells and neutrophils by chemotactic agents leads to a repositioning of cytoskeletal elements and organelles. These changes in the disposition of actin are controlled by actin binding proteins, which include actin-capping proteins, β-thymosins, cofilin/destrin and profilin. Individual actin binding proteins have specific effects on the actin cytoskeleton. This coordinated activity of actin binding proteins enables the actin cytoskeleton to respond to various external agents and/or localized internal signals by promoting changes in cell shape (Hurst et al., 1998). The actin may stabilize the position and/or shape of the acrosome with respect to the nucleus. To elucidate the role of actin during spermiogenesis, it is necessary to examine the role of actin-binding proteins. Thymosin β10 transcript expressed only in adult testis is a member of the β-thymosin actin-sequestering protein family and binds to monomeric actin, thus helping to regulate filament elongation. The expression of testis–specific form of a GTPase activating protein for p21 rac, β-chimaerin is expressed in post-meiotic spermatids. The timing of expression of these two actin binding proteins with the onset of sperm differentiation suggests that they may be involved in the restructuring of the actin cytoskeleton during the shaping of the sperm head. Reports suggest that the structures of actin fibres affect cell morphology, division and mobility, which in turn are capable of regulating actin polymerisation.

As a part of the investigation into the molecular mechanisms that direct differentiation of a round spermatid into an elongating spermatozoa, Hurst et al, (1998) reported a testis – specific 1.7-kb transcript from rat testis with sequence similarities to the β subunit of actin – capping proteins (ACPs) from somatic cells. The transcript contains a putative cAMP-responsive motif (CREM) upstream of the initiation codon in the DNA sequence and is expressed postmeiotically first appearing between 20 and 30 days of postnatal development.

The primary amino acid sequence is 90% identical to that of previously identified testis specific mouse protein, gsg3, both showing approximately 40% homology to the α subunit of somatic ACPs. The rat transcript is identified as a 32-kDa protein **(Fig.25. 3)**. The ACP is an intracellular protein and accumulates asymmetrically in the cytoplasm of round spermatids coincident with the position of developing acrosome. This spatial expression parallels the distribution of F-actin during sperm differentiation, supporting the hypothesis that testis-specific ACPs have an important role in determining the final shape of mature sperm heads. A disturbance in the expression of these ACPs may underlie many of the abnormalities in sperm morphology observed in infertile men (Hurst et al., 1998).

Sequencing of the ACP cDNA revealed the presence of an open reading frame of 893 bp that translated into a peptide containing 299 amino acid residues. A database search revealed a homology between the rat cDNA clone and a mouse testis-specific germ cell transcript gsg3 that also has a nucleotide sequence containing an open frame of 852 bp giving a predicted peptide length of 284 amino acids with more than 90% identity. Comparison of amino acids with the α-subunits of somatic actin capping protein revealed overall homology of approximately 40% with high levels of conservation in discrete regions of the sequence toward the C terminus.

Nishimune and coworkers (Yoshimura et al, 1999) isolated various cDNA clones specifically expressed in germ cells. One of the clones, *gsg3* encoding a testis-specific homologue of mouse actin capping protein α (ACPα) was specifically expressed in testicular germ cells. The gsg3 showed approx. 40% homology to the α subunit of somatic ACPs. Yoshimura et al (1999) obtained a mouse *gsg3* genomic clone using cDNA. Sequencing data showed that the *gsg3* gene was not interrupted by introns. The transcription initiation site for the gene was preceded not by a TATA box or GC rich promoter motifs, but by two consensus CRE motifs at the putative position. Southern blotting analysis showed that *gsg3* is a single copy gene in the mouse, and conserved in mammals. Phylogenetic analysis showed that gsg3 is a novel ACPα specific for haploid germ cells.

25.3.3. β3 Capping Protein

von Bulow et al, (1997) discovered another calyx constituent, which by amino acid sequencing and cDNA cloning was recognized as a novel isoform of the widespread β subunit of the heterodimeric actin binding "capping protein" (CP). This polypeptide, CP β3, of sperm calices, is identical with the β2 subunit present in diverse somatic cell types, except that it shows an amino-terminal extension of 29 amino acids and its mRNA is detected only in testis and, albeit in trace amount, in brain. This CP β3 mRNA contains the additional sequence, encoded by exon 1 of the gene, which is missing in β2 mRNAs. In spermiogenesis the transcription of the gene encoding the β1, β2, and β3 CP subunits is regulated specifically to include exon 1 and to give rise to the testis isoform CP β3, which is integrated into the calyx structure of the forming sperm head. This finding of an actin-binding protein isoform in an insoluble cytoskeletal structure is important in relation to the demonstrated roles of actin and certain actin-binding proteins, such as Limulus α-scruin, in spermiogenesis and spermatozoa (von Bulow et al., 1997).

25.3.4. Gelsolin

Gelsolin is a bivalent Ca^{2+} modulated actin-binding protein that severs, nucleates, and caps actin filaments. In order to gain a better understanding of the capping mechanism, Weber et al

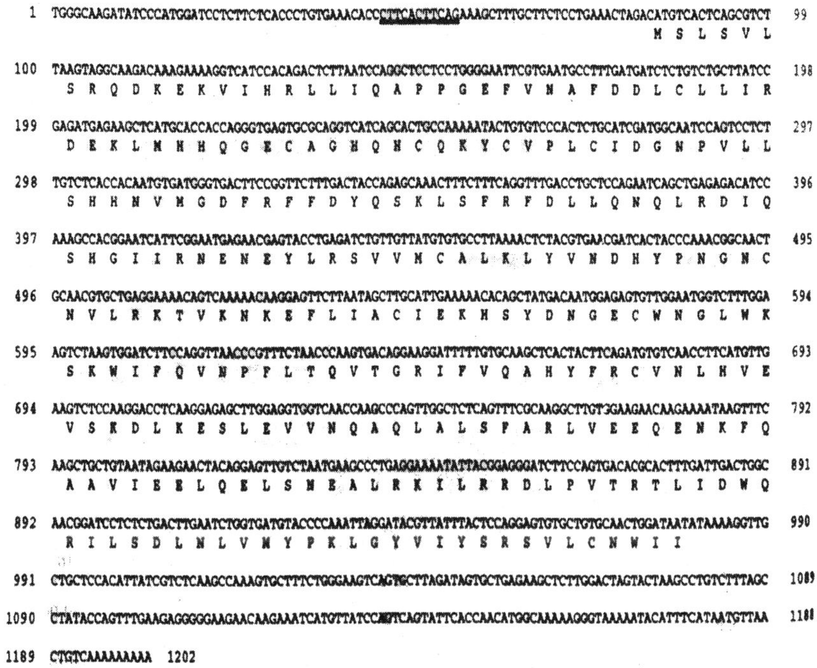

Fig.25.3. Nucleotide and predicted amino acid sequence of rat testis α-subunit of actin capping protein. Putatative c-AMP response element-τ is shown. Reprinted with permission from S. Hurst et al. Mol Reprod Develop 49; 81-91:1998 © John Wiley & Sons Inc. http://www3.interscience.wiley.com/cgi-bin/jabout/37692/ ProductInformation.html.

(1991) studied N- and C-terminal (T) gelsolin fragments, (14NT and 41CT), each of which contains a single functional actin-binding site. The very tight binding between gelsolin and the barbed filament end requires gelsolin to greatly decrease the dissociation rate constant of the terminal actin from this end. Gelsolin links two actin monomers so that they dissociate more slowly as a dimer. Gelsolin is concentrated in these intercellular adhesion complexes (ectoplasmic specialization), in addition to PIP2 and phosphoinositide-specific phospholipase C found in these structures. Treatment of isolated spermatid/junction complexes with exogenous phosphoinositide specific phospholipase C, results in release of gelsolin and loss of actin from the adhesion complexes. It is speculated that the hydrolysis of PIP2 may result in a local Ca^{2+} surge via the action of IP3 on junctional endoplasmic reticulum. This Ca^{2+} surge may facilitate the actin severing function of gelsolin within the adhesion complex (Guttman et al., 2002) (see Chapter 2).

80-kDa Protein: An actin-binding protein of 80-kDa, belonging to gelsolin family was identified from the body wall muscle of *ascidian, Halocynthia roretzi.* In the presence of Ca^{2+}, the 80-kDa protein accelerated the initial phase of actin polymerization, namely the nucleation process, decreased the level of polymerization at the steady state, caused marked reduction in viscosity of an F-actin solution, and fragmented F-actin filaments; while in the absence of Ca^{2+} it remained associated with F-actin without severing the filaments. The interaction of the 80-

kDa protein with actin was inhibited by PIP2. The 80kDa protein inhibited the growth of actin filaments at the barbed end but not at the pointed end, suggesting that 80-kDa protein acts as a barbed end capping protein. The deduced peptide sequence from cDNAs showed about 44% homology in amino acid residues with the human gelsolin sequence. In addition, 6 repeating segments were present in the sequence of the 80-kDa protein as described in the gelsolin sequence (Ohtsuka et al., 1998).

25.3.5. Thymosin β 10

β-thymosins are a family or related peptides, initially isolated from calf thymus, but now known to be present in a wide variety of mammalian and other vertebrate cells. Thymosin β_4 was the first member of this family to be characterized. Although it was initially proposed to be a thymic hormone acting at early stages of T cell maturation, the high concentration of the protein and its mRNA in a number of different tissues and in cells suggested that thymosin β_4 has a general function in many cell types. Thymosin β_4 forms a 1:1 complex with G-actin in blood platelets suggesting that it may be one of the most abundant actin-sequestering proteins in many cell types. G actin-binding proteins are one of the actin binding proteins that regulate the various forms of filamentous actin, which is essential to cell motility.

Thymosin β_{10} is another abundant member of the β-thymosin family related in structure to thymosin β_4. Thymosin β_{10} possesses the entire putative actin-binding domain found in thymosin β_4 (amino acid 16-25). One of the functions of β_{10}, like that of thymosin β_4 is to modulate actin polymerisation. The DNA encoding thymosin β_{10} has been isolated. The mRNA is present at high concentration in a number of developing tissues, as well as in adult spleen, lung, thymus and testis. In adult rat testis an additional thymosin β_{10} mRNA of higher molecular weight was identified. Nucleotide sequencing of cDNA clone complementary to the testis specific thymosin mRNA indicated that this mRNA differs from the ubiquitous thymosin β_{10} mRNA in its 5' untranslated region, beginning 14 nucleotides upstream of the translation initiation codon. The two thymosin β_{10} mRNAs are transcribed from the same gene through a combination of differential promoter utilization and alternative splicing. Both mRNAs were present in pachytene spermatocytes and only testis specific mRNA was detected in postmeiotic haploid spermatids. The thymosin β_{10} protein from adult testis was identical in size to that synthesized in brain. Presence of the protein in differentiating spermatids, suggested that the testis-specific thymosin β_{10} mRNA is translated in haploid male germ cells (Lin et al., 1991; Lin and Morrison-Bogorad, 1991). Testis specific thymosin β_{10} cDNA is 539 bp in length as compared to 446 bp that of the thymosin β_{10} cDNA from other tissues. The testis specific thymosin β_{10} cDNA has an open reading frame and 3'UTR identical to those of the ubiquitous thymosin $\beta_{10.}$ The novel cDNA has 149 nucleotides in the 5'-UTR and the ubiquitous thymosin β_{10} has only 56 nucleotides in this region. The 5'-UTRs of the two cDNAs diverge from nucleotide – 14 upstream to their 5'-ends. The 5'-UTR of the testis-specific cDNA has a high content of G and C nucleotides (66%) and contains a pair of 9-bp direct repeats (GGGAACGCC) separated by six nucleotides (Lin et al., 1991) **(Fig. 25.4).**

25.3.6. Other Actin Associating Proteins

The spectrin and ankyrin-like proteins express during the differentiation and maturation of spermatozoa in mammalian species, which express actin and calmodulin. Neither spectrin nor ankyrin were detected in the F-actin rich subacrosomal layer of spermatids in any species. In hamster and mouse maturing spermatids and spermatozoa, spectrin was mainly evidenced

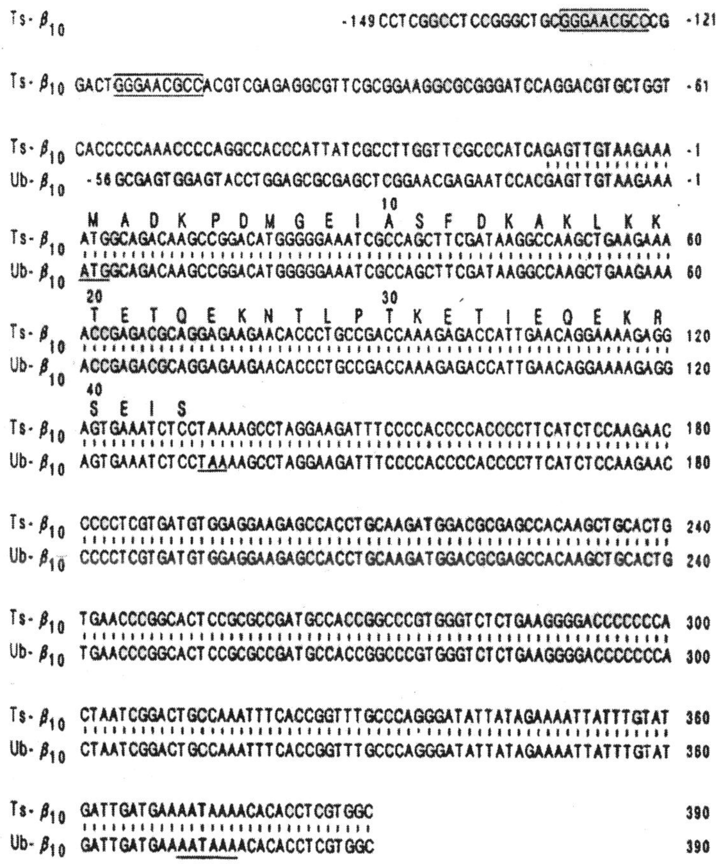

Fig.25.4. Nucleotide and predicted amino acid sequence of rat testis-specific thymosin β_{10} cDNA (Ts-β_{10}) and comparison to the ubiquitous thymosin β_{10} cDNA sequence (Ub-β_{10}).Two direct repeats of 9 nt in 5'-UTR are boxed. Initiation and termination codons and polyadenylation signal are underlined. Identical sequences are shown by vertical lines. Printed with permission from S.C. Lin et al. J Biol Chem 266; 23347-53: 1991 © American Society for Biochemistry and Molecular Biology.

around the fibrous sheath of the flagellum whereas ankyrin was detected only in the neck. In rabbit spermatozoa, the patterns of spectrin and ankyin are different from hamster and mouse. In rat, monkey and human sperm cells, these proteins were not demonstrated. Hence there was no uniform pattern of distribution of spectrin and ankyrin among the 6 species (Kann et al., 1991; D'Andrea et al., 1991). α-actinin, and tropomyosin were consistently localized in the neck of the spermatozoa alongwith actin. Actin and tropomyosin present in the postacrosomal area could be removed by sonication, whereas α-actinin in the basal plate appeared to be resistant to the treatment. α-actinin was seen over the acrosomal area, whereas in the washed sperm it appeared as a narrow cap at the margin of head. In the majority of preparations, tropomyosin could be localized in the principal piece of the tail (Fouquet and Kann, 1992; Yagi and Paranko, 1992). Yagi and Paranko (1995) reinvestigated the subcellular distribution of actin and actin-binding proteins in bovine spermatozoa subjected to various extractions.

Labeling for α-actinin and spectrin was localized in the acrosin-positive acrosmal lamina, neck, and principal piece, the latter containing also relatively extraction resistant oligomeric or polymerized actin. Spectrin reactivity is enhanced by MgCl2 in head, neck, and principal piece. Thus α-actinin, and spectrin may be involved in acrosomal vesicle formation (Yagi and Paranko, 1995). A 41kDa macrophage capping protein (MCP) is capable of forming complexes with actin in addition to capping the barbed ends of actin filaments. The protein is calcium activated in a fully reversible manner and weakly nucleates actin polymerization. The actin nucleus that survives MCP inactivation contains a minimum number of five actin molecules (Young et al., 1990).

Nectin-2 is a component of cell-cell adherens junctions and interacts with 1-afadin, an F-actin-binding protein. Disruption of both alleles of the murine nectin-2 gene resulted in morphologically aberrant spermatozoa with defects in nuclear and cytoskeletal morphology and mitochondrial localization. Homozygous null males are sterile, while homozygous null females, as well heterozygous males and females are fertile. Consistent with such a role, Nectin-2, expressed in the testes only during the later stages of spermatogenesis suggests its role in organization and reorganization of the cytoskeleton during spermiogenesis in mice.

25.4. ACTIN IN LIMULUS SPERM

Formation of actin filaments has been extensively studied in Limulus sperm and suggests a model for study in mammalian sperm. The three-dimensional structure of an actin filament bundle from the sperm of Limulus has been studied (Bullitt et al, 1988; Schmid et al, 1994 and others). The bundle is a motile structure, which by changing its twist, converts from a coiled to an extended form. The bundle is composed of actin plus two auxiliary proteins of molecular masses 50 and 60 kDa. Fraying the bundle with potassium thiocyanate created three classes of filaments: actin, actin plus the 60-kD a protein, and actin plus both the auxiliary proteins. Under electron microscopy the actin subunit appears to be bilobed with dimensions 70 X 40 X 35 Å. The inner lobe of the actin subunit, located at 20 Å radius, is a prolate ellipsoid, 50 X 25Å; the outer actin lobe, at 30 Å radius, is a 35- Å-diamspheroid. Attached to the inner lobe of actin is the 60-kDa protein, an oblate spheroid, 55 X 40 Å, at 50 Å radius. The arm like 50-kDa protein, at 55 Å radius, links the 60-kDa protein on one of actin's twin strands to the outer lobe of the actin subunit on the opposite strand. It is speculated that the 60-kDa protein may be a bundling protein and that the 50-kD protein may be responsible for the change in twist of the filaments which causes extension of the bundle.

The acrosmoal process of Limulus sperm contains a bundle of filaments composed of actin and a 102 kDa protein in a 1:1 molar ratio. Each segment of the bundle is found to obey the symmetry of space group P1, with $a = b = 147$ Å, $c = 762$ Å, $\alpha = 90^0$, $\beta = 90.6^0$, $\gamma = 120^0$. A unit cell contains a helical repeat of the filament with a selection rule following that of an actin filament. A 24 Å projection map based on the h01 view was reconstructed after averaging 5300 unit cells from six electron images. Filaments in this projection are well separated and clearly display a 21 screw symmetry. Structural analysis led to the proposal that the assembly of a stable bundle with a defined maximum diameter can be controlled by the crystallographic packing of the twisted filaments (Schmid et al., 1991c/r).

25.4.1. β-Scruin

α-Scruin is a actin bindling protein found in the acrosomal process of Limulus polyhemus sperm. Way et al, (1995a) cloned and sequenced a second scruin isoform from Limulus, β-

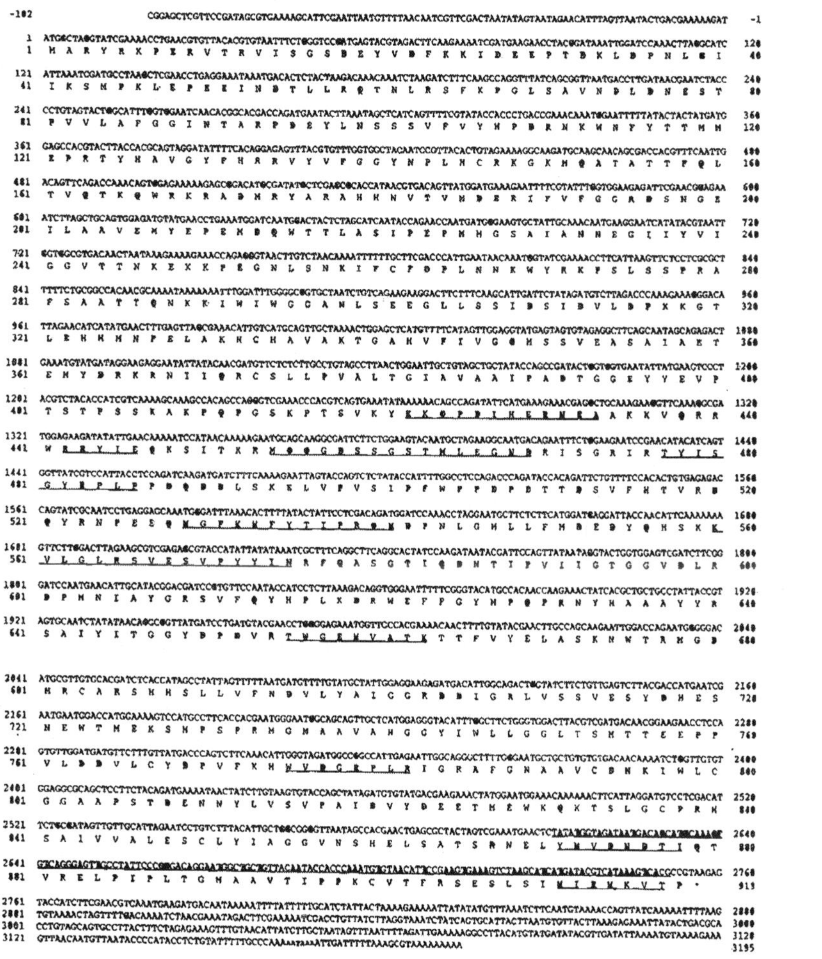

Fig. 25.5. The cDNA and deduced amino acid sequence of scruin cDNA clone (L1). Amino acid sequence obtained from NH2-terminal sequencing of scruin peptides and the sequence at the extreme COOH terminus corresponding to the probe are underlined. The polyadenylation signal is shown bold. Reproduced from M.Way et al J. Cell Biol 128; 51-60:1995© The Rockefeller University Press.

scruin, that is 67% identical to α-scruin. β-scruin and α-scruin are encoded by distinct genes. The sequence of β-scruin, like α-scruin, is organized into N- and C-terminal superbarrel domains that are characterized by a six-fold repeat of a 50 residues motif. β-scruin, like α-scruin, is found in Limulus sperm but not in blood or muscle **(Fig25.5)**. β-scruin is localized within the acrosomal vesicle at the anterior of sperm but not in the acrosomal process. In the acrosomal process of Limulus sperm, the β-propeller protein, scruin cross-links actin into a crystalline bundle. To confirm that scruin has the topology of a β-propeller protein and to understand how scruin binds actin, Sun et al., (1997) compared solvent accessibility of cysteine residues in scruin and the acrosomal process by chemical modification with (1,5-IAEDANS). In soluble scruin are C837 and C900, whereas C146, C333, and C683 are moderately reactive. This pattern of reactivity agrees with the topology of typical β-propeller protein. The chemical reactivity of cysteine in the acrosomal process implicates C837 at an actin-binding site. In constrast to soluble scruin, in the acrosomal process, C837 is completely unreative while the other cysteines become less reactive. Peptides corresponding to residues flanking C837 bind actin and narrow a possible actin-binding region to a KQK sequence. On the basis of these studies, results suggest that an actin-binding site lies in the C-terminal domain of scruin and involves a putative loop defined by C837 (Sun et al., 1997).

The deduced amino acid sequence of scruin reflects the domain organization of scruin: it consists of a tandem pair of homologous domains joined by a linker region. The domain organization of scruin is confirmed by limited proteolysis of the purified acrosomal process. Although the protein sequence of scruin has no homology to any known actin-binding protein, it has similarities to several proteins, including four open reading frames of unknown function in poxviruses, as well as kelch, a *Drosophila* protein localized to actin-rich ring canals. All proteins that show homologies to scruin are characterized by the presence of an approximately 50-amino acid residue motif that is repeated between two and seven times. Results suggest that the two domains of scruin seen in EM are superbarrel folds and they present the possibility that other members of this family may also bind actin (Way et al., 1995a).

The two domains of scruin contact two actin subunits in different strands of the same actin filaments. A correlation of Holmes' actin filament model to the density in acrosomal filament map showed that actin sub-domains 1, 2, and 3 match the model density closely. However, actin subdomain 4 matches rather poorly, suggesting that interactions with scruin may have altered actin conformation. Scruin makes extensive interactions with helix-loop-β motifs in subdomain 3 of one actin subunit and in subdomain 1 of a consecutive actin subunit along the genetic filament helix. These two actin subdomains are structurally homologous and are closely spaced along the actin filament (Schmid et al., 1994). Three-dimensional structure of acrosomal process by electron crystallographic reconstruction reveals an actin-scruin helix that is azimuthally modulated by the influence of the interactions of a filament with its neighbors. There are a variety of density connections with neighboring filaments involving scruin. Scruin commonly contacts one neighbor, but Sherman et al (1999) observed up to three interfilament connections involving both domains of the 28 scruin molecules in the unit cell. This structure indicates that promiscuous scruin-scruin contacts are the major determinants of bundle stability in true discharge.

25.5. MYOSINS

Motor proteins moving along actin filaments and microtubules are the candidates to provide the motive force for asymmetric sorting of cell contents. The role of myosins in such processes has been suggested, but few examples of their involvements are known. The first motor protein identified was skeletal muscle myosin, which generates force for muscle contraction. This myosin II is an elongated protein that is formed from two heavy chains and two copies of each of two light chains. Each heavy chain has a globular head domain at its N-terminus that contains the force generating machinery, followed by a very long sequence that forms an extended coiled-coil that mediates heavy chain dimerization. The two light chains bind close to N-terminal head domain, while the long coiled-coil tail bundles itself with the tails of other

```
Mouse X     (1774)   KLAATSEAGD  APWKFYFKLY  CFLDTDSMPK  DGVEFAFM--  ------FEQA
Hum. XV     (1374)   INPNFYGYQD  APWKIFLRKE  VFY-----PK  DSYS---HPV  QLDLL-FRQI
Hum.VIIA    (1934)   WIKKARPIKD  GIVPSL--TY  QVFFMKKLWT  -TTVPGKDPM  ADSIFHYYQE
Talin       (173)    WLDHGRTLRE  QGV--EEHEI  LLLRRKFFYS  DQNVDSRDPV  QLNLL-YVQA
Merlin      (74)     WLKMDKKVLD  HDVSKEEPVI  FHFLAKFYPE  NAEEELVQEI  TQHLF-FLQV
Ezrin       (57)     WLKLDKKVSA  QEVRKENPLQ  FKFRAKFYPE  DVAEELIQDI  TQKLF-FLQV

Mouse X     (1816)   HEAVIHGHHP  --APEESLQV  LAALRLQYLQ  GDYTPH-TSI  PP--LEEVYS
Hum. XV     (1415)   LHDTLSEACL  RISEDERLRM  KALFA-QNQL  DTQKPLVTES  ---VKRAVVS
Hum.VIIA    (1981)   LPKYLRGYHK  C-TREEVLQL  GALIY-RVKF  EEDKSYFPSI  P-KLLRELVP
Talin       (220)    RDDILNGSHP  -VSFDKACE-  FAGFQCQIQF  GPHNEQKHKA  GFLDLKDFLP
Merlin      (123)    KKQILDEKIY  C-PPEASVL   LASYAVQAKY  GDYDPSVHKR  GFLAQEELLP
Ezrin       (106)    KEGILSDEIY  C-PPETAVL   LGSYAVQAKF  GDYNKEVHKS  GYLSSERLIP

Mouse X     (1861)   VQRLRARISQ  STKTFTPYER  LEKRRTSFLE  GTLRRSFRTG  SVVRQKAEEE
Hum. XV     (1461)   TARDTWEVYF  SRIFP-----  ----------  -------ATG  SVGT------
Hum.VIIA    (2028)   -QDLIRQVSP  ----------  ----------  ----------  ----------
Talian      (268)    -KEYVKQKGE  RK--------  ----------  ----------  ----------
Merlin      (171)    -KRVINLYQM  TP--------  ----------  ----------  ----------
Ezrin       (154)    -QRVMDQHKL  TR--------  ----------  ----------  ----------

Mouse X     (1911)   QMLDMWIKEE  VCSARTSILD  KWKKLQGMTQ  EQRMAKYMAL  IKEWPGYGST
Hum. XV     (1483)   ----------  ----------  ------GVQL  LAVSHVGIKL  LRMVKGGQEA
Hum.VIIA    (2037)   ----DDW-KR  ------SIVA  YFNKHAGKSK  EEAKLAFLKL  IFKWPTFGSA
Talian      (279)    ----------  -------IFQ  AHKNCGQMSE  IEAKVRYVKL  ARSLKTYGVS
Merlin      (182)    -EMW-EER    -------ITA  WYAEHRGRAR  DEAEMEYLKT  AQDLEMYGVN
Ezrin       (165)    -DQW-EDR    -------IQV  WHAEHRGMLK  DNAMLEYLKI  AQDLEMYGIN

Mouse X     (1961)   LFDVECKEGG  FPQELWLGVS  A---EAVSVY  KRGEGKPLEV  FQYEHI----
Hum. XV     (1507)   GGQLRVLRAY  SF-ADILFVT  MPSQNMLEFN  LASEKV----  ----------
Hum.VIIA    (2076)   FFE-QTTEPN  FP-EILLI--  --AINKYGVS  LIDPKTKDIL  TTHPPIKTSM
Talian      (312)    FFLVKEKMKG  ---KNKLVPR  LLGITKECVM  RVDEKTKEVI  QEWSLINIKR
Merlin      (221)    YRAIRNKKG-  --TELLLGVD  ALGLH--IYD  PENRLTPKIS  FPWN-EIRM
Ezrin       (204)    YFEIKNKKG-  --TDLWLGVD  ALGLN--IYE  KDDKLTPKIG  FPWS-EIRM

Mouse X     (2004)   LSFGAPLANT  YKIVVDER-E  LLFETSE-VV  DVAKLMK---  -AYISMIVKK RY
Hum. XV     (1542)   ILFSA-RAHQ  VKTLVDDFIL  ELKKDSDYVV  AVRNFLPE-D  PALL
Hum.VIIA    (2120)   WSSGNTYFHI  TIGNLVRGSK  LLCETSLGY-  KMDDLLT---  -SYISQML-T AM
Talian      (359)    WAASPKSFTL  DFGD-YQDGY  YSVQTTEG-E  QIAQLIA---  -GYIDIILKK KK
Merlin      (264)    ISYSDKEFTI  KPLDKKI-DV  FKFNSSKL--  RVNKLILQLC  IGNHDLFMRR RK
Ezrin       (247)    ISFNDKKFVI  KPIDKKAPDF  VFYAP-RL--  RINKRILQLC  MGNHELYMRR RK
```

Fig.25.6. Amino acid sequence of mouse myosin X and its alignment with talin-like domains in human myosins XV and VIIa, merlin, and ezrin with corresponding N-terminal region of talin. The numbers in parentheses indicate the ordering numbers in protein sequences of first residues. Gaps are shown by dashes. The talin-like domain of myosin X has an insertion of about 50 aa. Reproduced with permission from S. Yonezawa et al. Biochem Biophy Res Commun 271; 526-33: 2000 © Elsevier.

myosin molecules. Thus all myosin proteins consist of three distinct regions: the head or motor domain, which contains the ATP and actin binding sites, the neck or regulatory domain that interacts with calmodulin and/or myosin light chains, and the tail domain, which is the site of interaction with other cellular components. Myosins, the actin-based molecular motors, are involved in various cellular events such as contraction, organelle transport, cellular morphology, and signal transduction. Myosins are classified on the basis of structural variation in the relatively conserved head region. In mammals, there exist several dozens of myosin genes belonging to eight classes, I, II, V, VI, VII, IX, X, and XV, and in at least five families mutations of genes are responsible for human diseases and/or mutant mice. According to the calf data on which structure function relationship is available, myosin X is characterized by the presence of three IQ motifs in the neck region, and a coiled-coil domain, a cluster of pleckstrin homology (PH) domains, a myosin tail homology 4 (MyTH4) domain and a talin-like domain in tail region. Myosin X functions as a dimer via the coiled-coil domain, and is involved in signal transduction via the PH domains. MyTH4 and talin-like domains, both of which have also been identified in myosins VIIa and XV, can also be considered as functional domain in myosin X.

25.5.1. Myosin X in Mouse Testis

The structure of the coding region of mouse myosin X cDNA predicted protein sequence of an approximately 240-kDa molecular mass with 2062 amino acids **(Fig.25.6).** When aligned with various proteins, the structure predicted for calf myosin X completely conserved pleckstrin homology domains and a myosin tail homology 4 domain, which were apparent in the tail region, suggesting their importance for myosin X's function. A myosin X mRNA of 8.7-kb in size is ubiquitous in various mouse tissues, while a similar size of human type X mRNA was recognized mainly in the testis. In addition to the adult-type transcripts in mice, a smaller embryo-specific mRNA, 4.8-kb in size, was identified in early to late embryonic stages, suggesting the presence of a shorter myosin X isoform in mouse embryos. *In situ* hybridization with mouse testis revealed that myosin X mRNA was restricted to Sertoli cells at stages VIII-X of spermatogenic cycle, which suggested that myosin X is involved in the supporting cells during the spermatid morphogenesis (Yonezawa et al., 2000). Amino acid alignment indicated that some of myosin X-specific structures previously identified in the tail region are highly conserved. Similar split PH subdomains are present in phospholipase C. The PH domain has now been identified in over one hundred proteins. Its primary function is thought to be targeting signaling molecules by binding to membrane phosphoinositides. From the PH domain, it is highly probable that myosin X functions, in an actin-based manner, to transport target molecules to the plasma membrane and/or to hold them on or near the membrane (Yonezawa et al., 2000).

25.5.2. Functions of Myosin in Reproduction

Smooth muscle heavy myosin and actin have been detected in mouse and rat meiotic chromosomes. Both contractile proteins are detectable in the nuclei of meiotic cells during the first prophase. The appearance and disappearance time of myosin and actin, however, is not synchronous. The two contractile proteins are associated with the synaptonemal complex (SC). Myosin seems to be associated with the central region of the SC, while actin is present in its basal knob, which is in connection with the nuclear membrane. The presence of contractile proteins in the nuclei of primary spermatocytes seems to suggest that they might play a role in the process of pairing of homologous chromosomes (De Martino et al., 1980). In normal mice, some spermatocytes fail to undergo cytokinesis after meiotic-I or -II nuclear division, and to form syncytial secondary spermatocytes and spermatids. Abnormal cytokinetic cells develop sparse and dispersed midzone spindles during the early stage. However, during late stages, single and compact midzone spindles are formed as in normal but localize asymmetrically and attach to the cortex. Myosin and F-actin were observed in midzone spindle and midbody regions of clearing cells implying that cytokinetic failure is unlikely to be due to defect in myosin or actin assembly. Depolymerization of microtubules by nocodazole resulted in the loss of the midbody-associated F-actin and myosin. These observations suggest that actin-myosin localization in the midbody could be a microtubule-dependent process that may not play a direct role in cytokinetic furrowing (Manandhar et al., 2000). Bi et al (1997) provided direct evidence, which indicated that in intact living cells, kinesin and myosin motors may mediate two sequential transport steps that recruit vesicles to the release sites of Ca^{2+} regulated exocytosis, although the identity of the responsible myosin isoform is not yet known. In addition, results provide in vivo evidence for the cargo-binding function of the kinesin heavy chain tail domain (Bi et al., 1997).

To explore the role of non-muscle myosin II isoforms during mouse oogenesis, fertilization, and early development, microinjection studies, using monospecific antibodies to myosin IIA and IIB isotypes each of which recognizes a 205-kDa protein in oocytes, but not mature sperm. Myosin IIA and IIB demonstrated differential expression during meiotic maturation and following fertilization since only the IIA isoform detects metaphase spindles or accumulates in the mitotic cleavage furrow. Microinjection of a nonphosphorylatable 20-kDa regulatory myosin light chain specifically blocks sperm incorporation cone disassembly and impedes cell cycle progression, suggesting that interference with myosin II phosphorylation influences fertilization. Although murine spermatozoa do not express myosin II, different myosin II isotypes may have distinct roles during early embryonic development (Simerly et al., 1998).

25.5.3. Myosins in Spermatogenic Cells of Non-Mammalian Species

i) Minor myosin, a myosin isoform from *Drosophila* testis is encoded by the muscle myosin heavy chain gene. Class VI unconventional myosin is required in spermatogenesis in *Drosophila*. During spermatogenesis the spermatids remain interconnected by cytoplasmic bridges until individualization. The process of individualization involves the formation of a complex of cytoskeletal proteins and membrane, the individualization complex (IC), around the spermatid nuclei. The 95F myosin is a component of the IC whose function is essential for individualization. In wild-type testes, 95F myosin localizes to leading edge of the IC. Two independent mutations in 95F myosin reduce the amount of 95F myosin in only a subset of tissues, including the testis. This reduction of 95F myosin causes male sterility as a result of defects in spermatids individualization. The IC movement was aberrant in these 95F myosin mutants, indicating a critical role for 95F myosin. The identification of a component of the IC other than actin suggested that 95F myosin is a motor that participates in membrane reorganization during individualization (Hicks et al., 1999; Miedema et al., 1995).

ii) Analysis of a *C. elegans* class VI myosin deletion mutant revealed a role for this motor protein in the segregation of cell components during spermatogenesis. Mutant spermatocytes cannot efficiently deliver mitochondria and endoplasmic reticulum/Golgi-derived fibrous-body membranous organelle complexes to budding spermatids. The segregation defects were not due to a global sorting failure, as nuclear inheritance was unaffected. It suggested that *C. elegans* myosin VI has an important role in the unequal partitioning of two organelles and cytoskeletal components, a role for this class of motor protein (Kelleher et al., 2000).

iii) Both actin and myosin have been found on ascidian sperm mitochondria and sperm tail. *Boltenia* and *Cnemidocarpa* sperm have additional actin at the tip of the head and additional myosin at the base of the head. The presence of actin and myosin on the mitochondrion and in the tail supports a means by which the force for mitochondrial movement is generated (Lambert and Lambert, 1984). Contractile proteins (actin, myosin) and tubulin have been revealed within DNA-containing nucleocytoplasm of mature spermatozoa of *Libinia emarginata* L. Antisera to actin or myosin, revealed staining at different nucleocytoplasmic sites and in radial processes of the spermatozoa. Myosin was present at the base of each of three radial extensions, whereas actin appeared throughout the nucleocytoplasmic compartment and in radial extensions (Perez et al., 1986).

iv) Myosin is present in the head portion of spermatozoa of the starfish, *Asterias amurensis*. The sperm has a specific Ca^{2+} activated ATPase activity and resembles egg myosin in forming thick filaments and in getting attached to actin filaments, subunit composition, and serological properties (Mabuchi, 1976).

v) *Fowl sperm*: The presence of myosin light chain kinase (MLCK) in fowl sperm indicates the presence of myosin involved in fowl sperm motility (Ashizawa et al., 1995a; 1995b). The motility of both intact and demembranated fowl spermatozoa was vigorous at 30°C but decreased markedly following the addition of ML-9, a specific inhibitor of myosin light chain kinase. Furthermore, the presence of a MLCK substrate peptide also inhibited the motility of demembranated spermatozoa at 30° C. In contrast, the addition of H-8 or HA1004 specific inhibitors of cAMP- dependent protein kinase, did not appreciably affect the motility of either intact or demembranated spermatozoa. This suggested that the phosphorylation of axonemal protein(s) by MLCK or MLCK-like protein, rather than by cAMP-dependent protein kinase, may be involved in the maintenance of fowl sperm motility at 30° C.

vi) Cranefly: Myosin is involved in cranefly primary spermatocyte division as revealed by myosin inhibitor (BDM) on chromosome movement and cytokinesis. Myosin ATPase inhibitor reversibly perturbed the movements of all autosomes. BDM added before anaphase onset altered chromosome movement less than BDM added during anaphase. Study confirmed the role of BDM on inhibition of myosin ATPase and indicated that the myosin is activated near the start of prometaphase. The effects of BDM on myosin indicated that the myosin used for cytokinesis of spermatocytes is activated starting from autosomal anaphase and continuing throughout cytokinesis (Silverman-Gravila and Forer, 2001).

25.6. REFERENCES

Ashizawa K, Wishart GJ, Hashimoto K, Tsuzuki Y. Dephosphorylation of a 30-kDa protein of fowl spermatozoa by the addition of myosin light chain kinase substrate peptide inhibits the flagellar motility. Biochem Biophys Res Commun 1995(a); 215: 706-12.

Ashizawa K, Wishart GJ, Nishinakama K, Sakamoto T, Tsuzuki Y. Regulatory mechanisms of fowl sperm motility: possible role of endogenous myosin light chain kinase-like protein. J Reprod Fertil 1995(b); 104: 141-8.

Aumuller G, Seitz J. Immunocytochemical localization of actin and tubulin in rat testis and spermatozoa. Hisochem 1998; 89, 261-7.

Bi GQ, Morris RL, Liao G, Alderton JM, Scholey JM, Steinhardt RA. Kinesin-and myosin-driven steps of vesicle recruitment for Ca^{2+} regulated exocytosis. J Cell Biol 1997; 138: 999-1008.

Bullitt ES, DeRosier DJ, Begg DA, Bonder EM. Three dimensional reconstruction of an actin bundle. J Cell Biol 1988;107: 597-611.

Camatini M. Anelli G, Casale A. Identification of actin in boar spermatids and spermatozoa by immunoelectron microscopy. Eur J Cell Biol 1986; 42: 311-8.

Castellani-Ceresa L, Mattioli M, Radaelli G, Barboni B, Brivio MF. Actin polymerization in boar spermatozoa: fertilization is reduced with use of cytochalasin D. Mol Reprod Dev 1993; 36: 203-11.

D'Andrea L, Fishkind DJ, Begg DA, Bonder EM. Isolation and localization of a spectrin-like protein from echinodern sperm. Cell Motil Cytoskeleton 1991; 19: 49-61.

De las Heras MA, Valcarcel A, Perez LJ, Moses DF. Actin localization in ram spermatozoa: effect of freezing/thawing, capacitation and calcium ionophore-induced acrosomal exocytosis. Tissue Cell 1997; 29: 47-53.

De Martino C, Capanna E, Nicotra MR, Natali PG. Immunochemical localization of contractile proteins in mammalian meiotic chromosomes. J. Cell Tissue Res 1980; 213: 159-78.

Dlaherty SP, Winfrey VP, Olson GE. Localization of actin in human, bull, rabbit, and hamster sperm by immunoelectron microscopy. Anat Rec. 1988; 221: 599-610.

Dvorakova K, Palecek J, Peknicova J. Changes in immunocytochemical localization of cytoskeletal proteins in boar spermatozoa after acrosome reaction induced by specific cytoskeletal inhibitors. Folia Biol (Praha) 2001; 47: 18-27.

Fouquet JP, Fraile B, Kann ML. Sperm actin and calmodulin during fertilization in the hamster: an immune electron microscopic study. Anat Rec 1991; 231: 316-23.

Fouquet JP, Kann ML, Dadoune JP. Immunogold distribution of actin during spermiogenesis in the rat, hamster, monkey, and human. Anat Rec 1989; 223: 35-42.

Fouquet JP, Kann ML. Species-specific localization of actin in mammalian spermatozoa: fact or artifact. Micros Res Tech 1992; 20: 251-8.

Fouquet JP, Kann ML. The cytoskeleton of mammalian spermatozoa. Biol Cell 1994; 81; 89-93.

Gu W, Kwon YK, Hecht NB. In postmeiotic male germ cells poly (A) shortening accompanies translation of mRNA encoding γ-enteric actin but not cytoplasmic β and γ actin mRNAs. Mol Reprod Dev 1996;

44: 141-5.

Guttman JA, Janmey P, Vogl AW. Gelsolin evidence for a role in turnover of junction-related actin filaments in Sertoli cells. J Cell Sci 2002; 115(Pt 3): 499-505.

Heid H, Figge U, Winter S, Kuhn C, Zimbelmann R, Franke W. Novel actin-related proteins Arp-T1 and Arp-T2 as components of the cytoskeletal calyx of the mammalian sperm head. Exp Cell Res. 2002; 279: 177-87.

Hernandez-Gonzalez EO, Lecona-Valera AN, Esobar-Herrera J, Mujica A. Involvement of an F-actin skeleton on the acrosome reaction in guinea pig spermatozoa. Cell Motil Cytoskeleton 2000; 46: 43-58.

Hess H., Heid H, Zimbelmann R, Franke W W. The protein complexity of the cytoskeleton of bovine and human sperm heads: the identification and characterization of cylicin II. Exp. Cell Res.1995; 218: 174-82

Hess H, Heid H, Franke WW. Molecular characterization of mammalian cylicin, a basic protein of the sperm head cytoskeleton. J Cell Biol 1997; 122: 1043-52.s

Hicks JL, Deng WM, Rogat AD, Miller KG, Bownes M. Class VI unconventional myosin is required for spermatogenesis in Drosophila. Mol Biol Cell 1999; 10: 4341-53.

Hisano M, Yamada S, Tanaka H, Nishimune Y, Nozaki M. Genomic structure and promoter activity of the testis haploid germ cell-specific intronless genes, Tact1 and Tact2. Mol Reprod Dev. 2003;65:148-56.

Howes EA, Hurst SM, Jones R. Actin and actin-binding proteins in bovine spermatozoa: potential role in membrane remodeling and intracellular signaling during epididymal maturation and the acrosome reaction. Andrology 2001; 22: 62-72.

Hurst S, Howes EA, Coadwell J, Jones R. Expression of a testis-specific putative actin-capping protein associated with the developing acrosome during rat spermiogenesis. Mol Reprod Dev 1998; 49: 81-91.

Kann ML, Feinberg J, Rainteau D, Dadoune JP, Weinman S, Fouquet JP. Localization of calmodulin in perinuclear structures of spermatids and spermatozoa: a comparison of six mammalian species: Nature 1991; 352 (6336): 543-5.

Kelleher JF, Mandeel MA, Moulder G, Hill KL, L'Hernault SW, Barshead R, Titus MA. Myosin VI is required for asymmetric segregation of cellular components during *C. elegans* spermatogenesis. Curr Biol 2000; 10: 1489-96.

Kim E, Waters SH, Hake LE, Hecht NB. Identification and developmental expression of a smooth-muscle g-actin in postmeiotic male germ cells of mice. Mole Biol 1989; 9: 1875-81.

Lambert CC, Lambert G. The role of actin and myosin in ascidian sperm mitochondrial translocation. Dev Biol 1984; 106: 307-14.

Lecuyer C, Dacheux JL. Hermand E, Mazeman E, Rousseaux J, Rousseaux-Prevost R. Actin-binding properties and colocalization with actin during spermatogenesis of mammalian sperm calicin. Biol Repord 2000; 63: 1801-10.

Lin SC, Morrison-Bogorad M. Cloning and characterization of a testis specific thymosin β10 cDNA. Expression in post meiotic male germ cells. J Biol Chem 1991; 266: 23347-53.

Liu DY, Martic M, Clarke GN, Dunlop ME, Baker HW. An important role of actin polymerization in the human zona pellucida-induced acrosome reaction. Mole Hum Reprod 1999; 5: 941-9.

Liu DY, Martic M, Clarke GN, Grkovic I, Garrett C, Dunlop ME, Baker HW. An anti-actin monoclonal antibody inhibits the zona pellucida induced acrosome reaction and hyperactivated motility of human sperm. Mol Hum Reprod 2002; 8: 37-47.

Mabuchi J. Isolation of myosin from starfish sperm heads. J Biochem (Tokyo) 1976; 80: 413-5.

Manandhar G, Moreno RD, Simerly C, Toshimori K, Schatten G. Contractile apparatus of the normal and abortive cytokinetic cells during mouse male meiosis. J. Cell Sci 2000; 113 (Pt 23): 4275-86.

Miedema K, Hanske M, Akhmanova A, Bindels P, Hennig W. Minor-myosin a novel myosin isoform synthesized preferentially in Drosophila testis is encoded by the muscle myosin heavy chain gene. Mech Dev 1995; 51: 67-81.

Miller MA, Nguyen VQ, Lee MH, Kosinski M. Schedl T, Caprioli RM, Greenstein D. A sperm cytoskeletal protein that signals oocyte meiotic maturation and ovulation. Science 2001; 291(5511):2144-7. Erratum: Science 2001; 292(5517): 639 and Science 2001; 292(5514): 53.

Moreno-Fierros L, Hernandez EO, Salgado ZO, Mujica A. F-actin in guinea pig spermatozoa: its role in calmodulin translocation during acrosome reaction. Mol Hum Reprod 1992; 33: 172-81.

Ohtsuka Y, Nakae H, Abe H, Obinata T. Functional characteristics and the complete primary structure of ascidian gelsolin: Biochim Biophys Acta 1998; 1383: 219-31.

Oko R, Hermo L, Hecht NB. Distribution of actin isoforms within cells of the seminiferous epithelium of the rat testis: evidence for a muscle form of actin in spermatids. Anat Rec 1991; 231: 63-81.

Palecek J, Peknicova J, Vitu M. Changes in immunochemical localization of cytoskeletal proteins in human and boar spermatozoa before and after acrosome reaction. Folia Biol (Praha) 1999; 45: 13-20.

Perez RA, Langford GM, Eckberg WR, Anderson WA. DNA-containing nucleocytoplasm in mature

spermatozoa of *Libinia emarginata L.* J Submicrosc Cytol 1986;18: 471-80.
Prigent Y, Dadoune JP. Immunoelectron microsopic distribution of actin in the spermatids and epididymal spermatozoa of the mouse. Acta Anat (Basel) 1993; 147: 133-9.
Rousseaux-Prevost R, Lecuyer C, Drobecq H, Sergheraert C, Dacheux JL, Rousseaux J. Characterization of boar sperm cytoskeletal cylicin II as an actin-binding protein. Biochem Biophys Res Commun. 2003; 303:182-9.
Schmid MF, Agris JM, Jakana J, Matsudaira P, Chiu W. Three dimensional structure of a single filament in the Limulus acrosomal bundle. Scruin binds to homologous helix-loop-β motifs in actin. J Cell Biol 1994; 124: 341-50.
Sherman MB, Jakana J, Sun S, Matsudaira P, Chiu W, Schmid MF. The three-dimensional structure of the Limulus acrosomal process, a dynamic actin bundle. J Mol Biol 1999; 294: 139-49.
Silverman-Gavrila RV, Forer A. Effects of anti-myosin drugs on anaphase chromosome movement and cytokinesis in crane fly primary spermatocytes. Cell Motil Cytoskeleton 2001; 50: 180-97.
Spungin B, Margalit I, Breitbart H. Sperm exocytosis reconstructed in a cell-free system: evidence for the involvement of phospholipase C and actin filaments in membrane fusion. J. Cell Sci 1995; 108 (Pt 6): 2525-35.
Steger K, Wrobel KH. Immunohistochemical demonstration of cytoskeletal proteins in the ovine testis during postnatal development. Anat Embryol (Berl) 1994; 189: 521-30.
Stokes DI, De Rosier DJ. The variable twist of actin and its modulation by actin-binding proteins. J Cell Biol 1987; 104: 1005-17.
Sun S, Footer M, Matsudaira P. Modification on Cys-837 identifies an actin-binding site in the β-propeller protein scruin. Mol Biol Cell 1997; 8: 421-30.
Tanaka H, Iguchi N, Egydio de Carvalho C, Tadokoro Y, Yomogida K, Nishimune Y. Novel actin-like proteins T-ACTIN 1 and T-ACTIN 2 are differentially expressed in the cytoplasm and nucleus of mouse haploid germ cells. Biol Reprod. 2003;69:475-82.
Ventela S, Okabe M, Tanaka H, Nishimune Y, Toppari J, Parvinen M. Expression of green fluorescent protein under β-actin promoter in living spermatogenic cells of the mouse: stage-specific regulation by FSH. Int J Androl 2000; 23: 236-42.
Vogl AW, Genereux K, Pfeiffer DC.Filamentous actin detected in rat spermatozoa.Tissue Cell 1993;25:33-48.
von Bulow M, Heid H, Hess H, Franke WW. Molecular nature of calicin, a major basic protein of the mammalian sperm head cytoskeleton. Exp Cell Res 1995; 219: 407-13.
von Bulow M, Rackwitz HR, Zimbelmann R, Franke WW. CP β3, a novel isoform of an actin-binding protein, is a component of the cytosketetal calyx of the mammalian sperm head. Exp Cell Res 1997;233:216-24.
Warnecke PM, Clark SJ. DNA methylation profile of the mouse skeletal α-actin promoter during development and differentiation. Mol Cell Biol 1999; 19: 164-72.
Waters SH, Distel RJ, Hecht NB. Mouse testes contain size classes of actin mRNA that are differentially expressed during spermatogenesis. Mol Cell Biol 1985; 5: 1649-54.
Way M, Sanders M, Chafel M, Tu YH, Knight A, Matsudaira P. β-Scruin a homologue of the actin-crosslinking protein scruin is localized to the acrosomal vescile of Limulus sperm. J Cell Sci 1995a; 108 (Pt 10): 3155-62.
Way M, Sanders M, Garcia C, Sakai J, Matsudaira P. Sequence and domain organization of scruin, an actin-cross-linking protein in the acrosomal process of Limulus sperm.. J Cell Biol 1995b; 128: 51-60.
Weber A, Pring, M, Lin SL, Bryan J. Role of the N and C-terminal actin-binding domains of gelsolin in barbed filament end capping. Biochemistry 1991; 30: 9327-34.
Yagi A, Paranko J. Actin, α-actinin, and spectrin with specific associations with the postacrosomal and acrosomal domains of bovine spermatozoa. Anat Rec 1995; 241: 77-87.
Yagi A, Paranko J. Localization of actin, α-actinin, and tropomyosin in bovine spermatozoa and epididymal epithelium. Anat Rec 1992; 233: 61-74.
Yonezawa S, Kimura A, Koshiba S, Masaki S, Ono T, Hanai A, Sonta, Kageyama T, Takahashi T, Moriyama A. Mouse Myosin X: Molecular architecture and tissue expression as revealed by northern blot and in situ hybridization analysis. Biochem Biophys Res Commun 2000; 271: 526-33.
Yoshimura Y, Tanaka M, Nozaki M, Yomogida K, Shimamura K, Yasunaga, Nishimune Y. Genomic analysis of male germ cell specific actin capping protein α. Gene 1999; 237: 193-9.
Young CL, Southwick FS, Weber A. Kinetics of the inteaction of a 41-kilodalton macrophage capping protein with actin: promotion of nucleation during prolongation of the lag period. Biochemistry 1990; 29: 2232-40.

Chapter 26

CELL ADHESION PROTEINS

26.1. CELL ADHESIONS

Metazoans possess the property of cell adhesion to hold the cells together, but adhesion does much more than what has been stated. Adhesion receptors make transmembrane connections, linking extracellular matrix and adjacent cells to the intracellular cytoskeleton. They also serve as signal transducers. Prior to fertilization, mammalian sperm must first bind to the zona pellucida (ZP), a glycoprotein matrix surrounding the egg. The ZP3 of ZP functions as both an adhesion molecule and as a secret agogue for acrosomal exocytosis. Affinity parameters substantiated the assumption that the initial adhesion event between sperm and the zona pellucida is a high affinity event, which is sufficient to tether a sperm to the extracellular matrix of ZP prior to the induction of acrosomal exocytosis. In testis adhesion also plays important role in germ-germ cells and germ cell-Sertoli cells and interactions. The proposed candidates for adhesion/signaling molecules have been listed (**Table.26.1**). In addition to these adhesions, there are some other molecules, which act during gamete interaction. These molecules are present in diverse somatic tissues and mediate cell-cell and cell-matrix interactions. These are the extra-cellular matrix proteins (ECM) (vitronectin, fibronectin, laminin, collagen, von-Williband factor etc.), NCAM and the cell adhesion molecules (CAMs) acting like receptors either for other CAMs or for ECM. Fortunately, most of the adhesion receptors fall into a relatively small number of families, the major ones of which have been shown in **Fig. 26.1**. The receptor nature of these molecules has been classified into (a) integrins, (b) cadherins (c) Ig superfamily molecule, (d) selectins (e) and mucins (Hynes, 1999).

26.2. INTEGRINS

The major family of adhesion receptors comprises of integrins (ITGs). Integrins are indispensable and represent the best-known transmembrane glycoprotein receptors. Unlike all others, they are non-covalently bound heterodimers with cytoplasmic and an extracellular domain composed of a unique specific smaller β-subunit and a distinct larger α-subunit. These heterodimers act as receptors for ECM components and membrane anchored cell surface co-receptors. Since they are involved in integrating the extracellular matrix proteins with the intracellular cytoskeleton proteins, hence they are called intergins. Integrins can be activated to bind extracellular ligands by intracellular ligands. This type of signaling is thought to involve change in conformation by various factors. In mammals, there are genes for eighteen α and eight β integrins; many αβ combinations fail to occur but at least two dozens are well defined. Most integrins are predominantly or exclusively receptors for ECM proteins such as fibronectins,

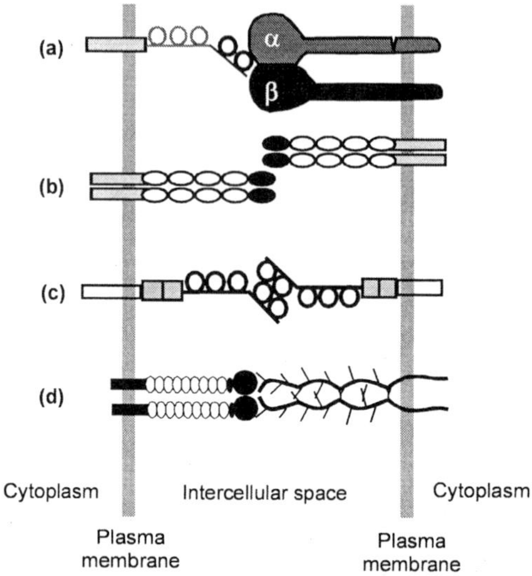

Fig26.1. Major class of cell adhesion receptors. (a) Integarins, (b) Cadherins (c) Ig superfamily (d) Selectins. Adapted with permission from R.O.Hynes. TIBS 24; M33-37: 1999© Elsevier Science

laminins and collagens but a few also play important roles in heterotypic cell adhesion. Integrins play a central role in cell adhesion to basement membranes, in the polarization of cells induced by that adhesion and in cell migration upon and through ECM.

The β chain of ITGs at N-terminus consists of a conserved region possessing ligand-binding activity, cation-binding site and is also involved in heterodimer formation. All β-subunits of ITGs possess a cyseine rich domain consisting of 40 consecutive cysteine residues. This cysteine rich domain undergoes extensive interchain disulphide binding to provide correct conformation required for its biological function. In several cell-cell interactions, integrins act as co-receptors after activation and are able to transduce signals through cytoskeletal proteins and adaptor kinases. Intergins transduce signals through the membrane after ligand binding, including the regulation of intracellular pH or intracellular free calcium, phosphorylation of tyrosine or serine/threonine residues and inositol lipid turnover. Very often ITG receptors are associated with other receptors in a multi-molecular complex, which contributes to cell activation. This molecular complex involves not only surface molecules but also cytoskeletal proteins. The cytoplasmic domain of the β-subunit binds talin, α-actin and vinculin and help in cytoskeletal reorganization. The clustering of ITGs and co-receptors leads to the activation of several signaling kinases, tyrosine kinases and mitogen activated protein kinases. Integrin receptors are expressed on sea urchin, mouse, hamster and human unfertilized oocytes. Available information suggests that they may act in similar way during fertilization and may participate in the initiation and/or propagation of the calcium signal via stimulation of phospholipases and production of inositol triphosphate (Fenichel and Durand-Clement, 1998). The integrin α6β1 is one candidate, which may serve as a sperm receptor, mediating sperm-egg binding. Integrins play particularly important roles in fertilization and embryogenesis, including the process of implantation. A disruption of the integrin expression is associated with certain types of infertility in women. However role of integrins as a receptor in egg-sperm adhesion molecules has been contradicted (Ji et al, 1998; Miller et al, 2000). Agents that affect growth and differentiation can

Table 26.1 Classification and characerstics of integrins (Shrimali and Reddy, 2000)

Family	β sub unit	α sub unit	αβ complex	Receptor name	I domain	Ligand	RGD Recognition	Distribution
VLA Proteins	β_1	α_1	$\alpha_1\beta_1$	VLA-1	+		?	Ubiquitous
		α_2	$\alpha_2\beta_1$	VLA-2	+		?	Ubiquitous
		α_3	$\alpha_3\beta_1$	VLA-3	-	Co,Fn,entacin	?	Ubiquitous
		α_4	$\alpha_4\beta_1$	VLA-4	-	Fn, VCAM-1	-	Ubiquitous
		α_5	$\alpha_5\beta_1$	VLA-5	-	Fn	+	Ubiquitous
		α_6	$\alpha_6\beta_1$	VLA-6	-	L	RGD	Ubiquitous
		α_7	$\alpha_7\beta_1$					
		α_8	$\alpha_8\beta_1$			Fn,Vn,Tn		
		α_9	$\alpha_9\beta_1$			Thr		Epith muscle
		α_V	$\alpha_V\beta_1$			Fn,Vn?	+	Cell lines
Leucocyte integrins	β_2	α_L	$\alpha_L\beta_2$	LFA-1	+	ICAM1,-2,-3	-	Leucocytes
		α_M	$\alpha_M\beta_2$	MAC-1	+	C3b,Fb ICAM-1	-	Neutrophils, Monocytes,LGL
		α_X	$\alpha_X\beta_2$	P150, P95	+	C3b, LPS	-	M, monocyte
		α_D	$\alpha_D\beta_2$		+	ICAM-3	-	
	β_7	α_4	$\alpha_4\beta_7$					
		α_H	$\alpha_H\beta_7$	LPAM-1		Fn, VCAM-1		Lymphocytes
		α_E	$\alpha_E\beta_7$		+	E-Cadherin		Lymphocytes
Cyto-adhesin	β_3	α_V	$\alpha_V\beta_3$	VitronectinR	-	Vn,Fb,Fn, VWF,Tn,Thr	+	Ubiquitous
		α_{IIb}	$\alpha_{IIb}\beta_3$	gpIIb/IIIa		Fb,Fn,VWF,Vn	+	Platelets
Others	β_5	α_V	$\alpha_V\beta_5$			Vn,Fn	+	Ubiquitous
	β_6	α_V	$\alpha_V\beta_6$			Fn,Tn	?	Lung epithelium
	β_8	α_V	$\alpha_V\beta_8$			Vn	+	Cell line
	β_4	α_6	$\alpha_V\beta_4$			Ln	+	Epithelial cells

Co = Collagen, Fb= Fibrinogen, Fn= Fibronectin, LFA= Leucocyte function assoicated antigen, Ln= Laminin, LGL= Large Granular Lymphocytes, M= Macrophage, Tn= Tenascin, Thr= Thromobospondin, Vn= Vitronectin VWF = Van Williband Factor, VLA = Very late antigen,

modulate ITG expression. Factors like TGF-1β, TNF-α, interferon-γ and NGF upregulate the synthesis of very late antigens (VLA) αv, β1, β3, β5 and β6 ITG subunits. Integrins are classified according to the mode of adhesion in which they are involved: i) adhesion to ECM for which ligands are; fibronectin, laminins, collagens, vitronectin, von Williband factors ii) cell-cell adhesion, and iii) platlet adhesion.

The mRNAs for integrin subunits α5, α6, αv, β1, β3, β5, α6 β1 and αv β3 have been detected on the plasma membrane of unfertilized mouse eggs. The GoH3, an anti-α6, indicated the role for the integrin α6β1 as a cell-cell adhesion receptor that mediates sperm-egg binding. Chen and Sampson (1999a) suggested that fertilin-β binds directly to α6β1 integrin on egg surface and mediates sperm egg fusion, and that different states of α6β1 could interact with an ECM ligand (laminin) or a membrane-anchored cell surface ligand (fertilin-β) (Almeida et al., 1993; Chen et al., 1999a,b,c).

26.2.1. Integrins and Their Ligands in Testis and Sperm

The reactivity of integrin β and α3, α5 and α6 chains has been found in the basement membrane of the human seminiferous tubules, spermatocytes, spermatids and testicular spermatozoa, whereas, the α1, α2 and α4 chains could not be detected on spermatogenic cells. Very late antigens (VLA) subunits were localized on endothelial cells, leukocytes and basement membranes. Matrix proteins such as laminin, collagen IV, and fibronectin were detectable as components of basement membranes in human testis. Germinal cells except spermatogonia express fibronectin only (Glander et al., 1998). The presence of α3, α5, α6 ITG subunits and fibronectin in human testis suggests their role in Sertoli-Sertoli cells and Sertoli-germ cell adhesion. The evidence of ligands of β1 ITGs, fibronectin, and collagen on the surface of spermatozoa of different species, gives the first hint of involvement of adhesion molecules in fertilization.

On human sperm surface, Collagen IV, α1(CD49a) and α2(CD49b) chains do not express, but laminin was detected at the acrosomal membrane, while fibronectin and β4 integrin chains were located mainly at the equatorial membrane. The fibronectin receptors α3, α4 and α5 chains of the β1 integrins were mainly located on acrosomal and equatorial membrane areas, whereas laminin receptor α6 chain was located post-acrosomal and less frequently acrosomal. While β2 chain and vitronectin receptors αV and β3 chains showed mainly post-acrosomal localization, the LFA-3 (CD58) was found constantly on post-acrosomal membrane regions. The localization of fibronectin appeared to be influenced by the presence of integrins, since the typical equatorial fibronectin band disappeared in case of an equatorial localization of integrins (Trubner et al., 1997). The β1 ITGs (and α6 ITG) subunit has been detected on human spermatozoa (Klentzeris et al., 1995). Low expression of α6, β3 and β4 ITG subunits on sperm has been assigned the cause of oligoasthenozoospermia (Reddy et al., 1998).

Eggs are rich in three cytoskeletal proteins known to be linked to the β chain of integrins: talin, vinculin and α-actinin (de Nadai et al., 1996). The β1 and αv integrin subunits were present consistently at the surface of pig oocytes, whereas the remaining α integrin subunits were not routinely detected. Sperm plasma membrane proteins of 137- and 93-kDa appeared to be the ligands for the β1 integrin subunit. These results are consistent with a αv and β1 pig oocyte integrin interacting with a ligand on the sperm plasma membrane during fertilization. The α2 and α5 integrin chains may be common mediators in adhesion-fusion mechanisms triggered by fertilization (Linfor and Berger, 2000).

There is accumulating evidence, which suggests that the RGD sequence is involved in sperm adhesion to the oolemma. In this interaction ITGs might be the oolemmal receptors responsible for recognition of RGD ligands on sperm. Many ITG subunits (α2, α6β1, αvβ3, α3, α5, αv, αm and β1) have been identified on hamster, mouse and human eggs (Shrimali and Reddy, 2000). Involvement of fibronectin in the gamete interaction was made clear by the fact that fibronectin is localized in the region where a spermatozoon is fused first with the egg

```
A.  T1  cadherin  from  mouse  testis(Accession  AAB87707)

  1 PEAKREDSKL RRDVSPETIF QIRRTVPLWE NIDVQDFIHR RLKENDSTAP PYDSLATYAY
 61 EGNDSVANSL SSLESLTADC NQDYDYLSDW GPRFKKLAEM YGGNDSDLN

B.  T2  cadherin  from  Mouse  testis  (Accession  AAB87708)

  1 DDVDRRSFSF SLAAVNPNFT VQDNEDNTAR ILARKNGFNR HEISTYLLPV VISDNDYPIQ
 61 SSTGTLTIRV CACDSQGNMQ SCSAEALLLP WPHTGALIAI LLCIIILLVI VVLFAALKEP
121 LILSKEDIRD NIVSYNDEGG GEEDTQAFDI GTLRNPAAIE EKKLRRDIIP ETLFIPRTPT
181 APGNTDVRDF INERLKEHDL DPTAPPYDSL ATYAYEGNDS VAESLSSLES GTTEGDQNYD
241 YLREWGPRFK KLAEMYGGGE SDKDS
```

Fig.26.2. Amino acid sequence of T1-cadherin (A) and T2-cadherin (B) present in mouse spermatogenic cells. Source: http//www. ncbi.nlm.nih.gov.

plasma membrane during fertilization antibodies (Hoshi et al., 1994).

26.3. CADHERINS

Cadherins are critical components of adhesive interactions between cells. Through homotypic interaction of their extracellular domains, cadherins physically link cells and provide spatial cues, which modulate cell behaviour. Cadherins convey this information by regulating intracellular signaling events via their cytoplasmic domains. Classic cardherins and protocadherins are two subfamilies within the cadherin superfamily that depends on Ca^{2+} for cadherin functioning. The extracellular domains of cadherins contain five characteristic cadherin repeats, each comprising a sandwich of β sheets. Structural and functional analyse suggested that the functional unit is a dimmer as shown (**Fig. 26.1**). Clustering of cadherins is an important aspect for their functions, and multiple dimer-dimer interactions are believed to provide sufficient local avidity to mediate cell-cell adhesion. Both sub-families of cadherins share the typical cadherin repeats in their extracellular portion but differ in their cytoplasmic regions. The cytoplasmic domain of classic cadherins (type I and II) contains conserved sequences that interact with proteins termed catenins and actin cytoskeleton. Protocadherins have six cadherin repeats. Recently, three families within the protocadherin subfamily have been described: protocadherin α, β, and γ (PCDHα, β and γ). The PCDHα and γ (and likely β) families are distinguished by unique cytoplasmic domain. Within each PCDH family, individual members differ in their extracellular domains but have identical cytoplasmic domains. Therefore, each protocadherin family has the potential to mediate numerous events.

26.3.1. Cadherins in Male Germ Cells

Studies on cadherin function in testis have been limited to classic cadherins. Seven classic cadherins expressed in mice testis are E, N, P, K, T1, T2 and OB (Munro and Blaschuk, 1996). Rat testis expresses at least 24 cadherins. Similar to brain, testis expresses a large number of cadherin superfamily members: 7 classic cadherins of (types I and II), 14 protocadherins, 2 protocadherin-related cadherins, and 1 cadherin related receptor like protein. All three protocadherin families (α, β and γ) were found in testis. The results on mRNA expression during a postnatal developmental time course and following ablation of specific testis cell types imply that cadherin expression is regulated by paracrine interactions within testis. Three members of the cadherin family of calcium dependent cell adhesion molecules (neural-cadherin [N], epithelial cadherin [E], and placental-cadherin [P]) have been shown to be present in mouse, rat and human testes. These CAMs are known to be capable of regulating the formation of intercellular junctions. They have been implicated as modulators of cell polarity, cell sorting, and cell migration. Diverse expression patterns and the large number and variety of cadherin superfamily members found in testis support a critical function for cadherin mediated cell adhesion in spermatogenesis. Ziv et al., (2002) proposed a specific cadherin organization at the fusogenic domains of both gametes.

Developmental Regulation: On the basis of PCR analysis, Munro and Blaschuk (1996) reported seven cadherins (E-cadherins, N-cadherin, P-cadherin, K-cadherin OB cadherin as well as T1-cadherin, and T2-cadherin, expressed at various stages during mouse testicular development, which was ascertained by mRNA levels of each cadherin. N-cadherin mRNA is expressed at all stages of testicular development from fetal life till adulthood, with maximal levels being present in the testes of 21 day old mice. Furthermore, E-, P-, K-, OB-, and T2-cadherin mRNAs are all expressed in the fetal gonad, followed by a dramatic decline after birth. Conversely, T1-cadherin mRNA was not detected in the fetal, newborn, and 7 day old testes but was present in 21 days old testis, being 10-fold higher in the testes of adult mice, compared to the levels found in the testes of 21 day old mice. The cytoplasmic domains of classic cadherins interact with catenins. Cadherin-catenin complex has been associated in cell transduction pathway. The high levels of cadherin expression inhibit B-catenin signaling activity. Therefore, it seems likely that N-cadherin mediated cell adhesion in testis is linked in some way to these signaling events and testicular development (Munro and Blaschuk, 1996) **(Fig.26.2)**.

In another study, analysis of rat testicular RNA indicated the presence of N- (4.3 and 3.5-kb) and P- (3.5-kb) cadherin transcripts, whereas E-Cadherin message was not detected. During postnatal rat testicular development, relative P-Cadherin mRNA levels were highest at 7 days of age and decreased to almost half by day 14. This contrasted with the developmental pattern observed for the 4.3-kb N-Cadherin transcript, which was low early in testicular development but increased to peak levels on day 42, coincident with the shedding of the first sperm. N-Cadherin mRNA concentrations decreased from 42 to 56 days and then remained constant until 91 days. The developmental patterns of P and N cadherin mRNA suggested a role for P cadherin in early testicular development, while N cadherin appeared to play a role in later stages of spermatogenesis (Cyr et al., 1992). However, other studies suggested N-cadherin to be produced by the Sertoli cells and spermatocytes. Antibodies directed against N-cadherin block Sertoli cell-spermatogenic cell adhesion in vitro. The expression of a 135-kDa N-cadherin in the human seminiferous epithelium was demonstrated on the surface of spermatogonia and primary spermatocytes, and possibly also around some early spermatids, whereas late spermatids were always negative. Expression of E-cadherin could not be detected in the human testis (Andersson et al., 1994).

The rat testicular N-cadherin was predicted a polypeptide of a 883-amino acids that displayed a 98.6% identity with the mouse homologue, and was found to be expressed by Sertoli and germ cells in the rat testis; the N-cadherin expression increased with time. The primary amino acid sequence of N-cadherin in aa-1 to aa-272 is highly conserved between the chicken and the mouse. This stretch of sequence, the known cell binding site of which may be necessary for interactions between two cadherin molecules is also present in rat N-cadherin. Since N-cadherin plays a crucial role in facilitating invasive capacity of metastatic tumor cells, the observation of germ cell released factor(s) in affecting Sertoli cell N-cadherin expression may suggest its possible role in facilitating germ cell migration during spermatogenesis (Chung et al., 1998). Rat and mouse N-cadherins share 98% homology.

Hormonal Regulation of N-Cadherin: The production of N-cadherin by Sertoli cells was stimulated by testosterone in the presence of FSH alone. The FSH can also induce an increase in Sertoli cell N-cadherin mRNA *in vitro*, which can be further enhanced by estradiol. The action of FSH is mediated by specific receptors that are functionally coupled via membrane-associated G proteins to the adenyl cyclase cAMP pathway. The cAMP analogs, on N-cadherin expression, further enhance the action of estradiol, while estradiol, testosterone, and dihydrotestosterone alone failed to stimulate Sertoli cell N-cadherin mRNA levels. In contrast, these steroids are potent regulators of testicular N-cadherin mRNA level *in vivo* (MacCalman et al., 1994).

Both the production of N-cadherin by Sertoli cells and also the binding of round spermatids to Sertoli cells are stimulated in a synergistic manner by testosterone and FSH. Thus N-Cadherin may be one of the factors that subserve the androgen dependent process of round to elongated spermatid maturation. N-cadherin inversely correlated with the expression of E-cadherin (Chung et al., 1998; Perryman et al., 1996).

Cadherins in Sperm: Fusion between hamster oolemma and spermatozoa is prevented in the presence of trypsin in Ca^{2+} free media, as is oocyte activation, implicating involvement of a cadherin like adhesion. The presence of an anti-cadherin antibody reactive peptide has been detected in human spermatozoa and testis. N-cadherin sequences from PCR products of testis, ovary and spermatozoa were identical except for occasional base changes (Goodwin et al., 2000). Immunoblotting with the pan-cadherin antibody revealed a single band of approximately 120-kDa in spermatozoa and oocyte extracts. Oocytes presented three classical (E-, P- and N) cadherins with the appropriate molecular weights of 120-130 kDa. However, in the sperm the presence of E- cadherin was positive, while N-cadherin was negative (Rufas et al., 2000 c/r Ziv et al, 2002).

26.4. IMMUNOGLOBULIN SUPERFAMILY

The third major class of adhesion receptors comprises the immunoglobulin superfamily (Ig-SF), characterized by the presence of varying numbers of Ig-related domains. Like cadherin domains, these are sandwiches of two β sheets held together by hydrophobic interactions **(Fig.26.1)**. This is a stable structure that occurs also in another domain common among adhesion molecules: fibronectin type III (Fn3) domains (boxes), which frequently occur in tandem with Ig domains (circles) in cell adhesion receptors. Fn3 domains also occur in adhesive proteins of the ECM such as fibronectin and tenascin and in the ligand binding domains of cytokine receptors. The Ig superfamily is diverse, numbering well over 100 members in vertebrates. Like the cadherin and the integrin superfamily, Ig superfamily plays important roles in development by regulating cellular differentiation, adhesion and migration. Spermatogenic immunoglobulin superfamily (Sg-IgSF), a mouse protein belonging to the Ig superfamily expressed on spermatogenic cells binds to some membrane molecules on Sertoli cells in a heterophilic manner and thereby may play diverse roles in the spermatogenesis (Wakayama et al., 2003).

Besign: Basigin (Bsg) is a highly glycosylated transmembrane protein belonging to the Ig superfamily with two Ig domains. A number of studies, including gene targeting, have demonstrated that Bsg plays pivotal role in spermatogenesis, implantation, neural network formation and tumor progression. The Bsg forms homo-oligomers in a cis-dependent manner on the plasma membrane. If the disulfide bond of the more N-terminally located Ig-like domain was destroyed by mutations, Bsg could not form oligomers. Mouse and human Bsgs exhibit high homology in the transmembrane and intracellular domains but low homology in the extracellular domain. The association of Bsg heterologous species was weak as compared with that within the same species, suggesting the importance of the extracellular domain in the association. Results indicated that the N-terminal lg-like domain is necessary and sufficient for oligomer formation by Bsg on the plasma membrane (Yoshida et al., 2000).

Nectin 2: Nectin-2 is a cell adhesion molecule encoded by a member of the poliovirus receptor gene family, which belongs to the members of Ig gene superfamily. Nectin-2 is a component of cell-cell adherens junctions and interacts with 1-afadin, an F-actin-binding protein. Disruption of both alleles of the murine nectin-2 gene resulted in morphologically aberrant spermatozoa with defects in nuclear and cytoskeletal morphology and mitochondrial localization. Homozygous null males were sterile, while homozygous null females, as well as heterozygous males and females were fertile. Nectin-2 functions at a late stage of germ cell development and expressed only during the later stages of spermatogenesis (Bouchard et al., 2000). Ozaki-Kuroda et al., (2002) have shown that the nectin-afadin system plays essential role in coupling cell-cell adhesion and the cortical actin scaffold at SspJs and in subsequent sperm morphogenesis. Mueller et al., (2003) described the generation of a mouse-line lacking a functional nectin-2 gene (nectin-2LacZ/LacZ). Although nectin-2LacZ/LacZ males produced normal number of motile spermatozoa, SEM revealed severe malformations of the sperm head and midpiece. Functional analyses indicated that the infertility phenotype of nectin-2-deficient male mice is caused by a combination of reduced migration to the oviduct, spermatozoa-zona binding, and sperm-oocyte fusion (Mueller et al., 2003). Nectin-2 (-/-) mice exhibit defective sperm morphogenesis and the male-specific infertility (Bouchard et al., 2000).

Sperad: The Sperad, containing two extracellular Ig like domains, a transmembrane segment, and an intracellular proline-rich domain closely related to a large family (biliary glycoproteins) of putative cell adhesion molecules has been identified in testis. Spread is first expressed by the haploid spermatid and is localized to the plasma membrane overlying the acrosome, supportive of its role in cell adhesion/signaling. The open reading frame of two cDNA clones predicts protein of either 32.2 or 33.3-kDa. Antibody produced to Spread recognized three sperm plasma membrane proteins. The M_r 55,000 sperm plasma membrane represents the full-length protein encoded by the clone. The acrosome reaction does not appear to alter the molecular weight of Sperad but does result in its loss from the sperm cells (Quill and Garbers, 1996).

26.5. SELECTINS

Another well-studied group of cell adhesion receptors comprises the selectins and their counter-receptors. The figure 26.1 shows a heterophilic interaction between a selectin (P-selectin) and its counter-receptor, a heavily glycosylated protein, P-selectin glycoprotein ligand-1 (PSGL-1). Binding is through the C-type lectin domain in the selectin, which recognizes specific carbohydrate moieties, particularly sialyl-Lewis (x), in the counter receptor/ligand (Hynes, 1999). P-selectin is stored in α-granules and dense granules of platelets as well as in Weibel-Palade bodies of endothelial cells, and it is rapidly redistributed to the cell surface after activation. While zona free human and hamster oocytes did not react with mAb directed against P-selectin, oocytes from both species displayed reactivity with this antibody following their contact with human spermatozoa. Ca^{2+} ionophore induced activation of zona-intact human oocytes and the expression on the oolemma of P selectin moiety as well as P-selectin also appeared to be expressed on the sperm surface following the acrosome reaction. This suggested that P-selectin might be involved in gamete interactions (Fusi et al., 1996). O-glycan structural similarities between oligosaccharides from human leukocyte, PSGL-1 and from zona pellucida glycoproteins of porcine oocytes also indicated the possible existence of a P-selectin ligand in the zona pellucida. Therefore, a search for a specific receptor for this ligand resulted to the identification of P-selectin on the acrosomal membrane of porcine sperm cells. In vitro binding of porcine acrosome reacted sperm cells to oocytes was found Ca^{2+} dependent and inhabitable with either P-selectin, P-selectin receptor-globulin, or relating antibodies. Moreover, porcine sperm cells were found capable of binding to human promyeloid cell line HL-60. Taken together, these findings further implicated a potential role for the sperm P-selectin in porcine sperm-egg interactions (Geng et al., 1997).

26.6. ADAM PROTEINS THE FAMILY OF METALLO PROTEINASE DISINTEGRINS

Disintegrins (DITGs) are soluble ligands for integrins (ITG), found in snake venoms. These are short peptides of low molecular weight involved in cell-cell and cell-ECM interactions. Disintegrins molecules are functionally just the opposite of ITGs and hence the name disintegrins. Metalloproteinase-disintegrins are a family of zinc binding metalloproteinases that were originally isolated from the venom of snakes. They are closely related to the matrix metalloproteinases. Metalloproteinase-disintegrins or ADAMs (a disintegrin and a metalloproteinase) have a unique domain structure composed of a signal sequence, pro-domain with a Cys switch, catalytic domain with a zinc-binding motif, disintegrin domain, cysteine-rich domain, a transmembrane domain, and a cytoplasmic domain. All members of this family are predicated to be membrane anchored. Thus as a group they possess four potential functions: metalloprotease activity, cell adhesion activity, cell fusion activity, and signaling activity. Till date more than thirty similar proteins have been identified which contain DITG like domain and grouped together into a family of: ***A Disintegrin and A Metalloprotease/Metalloprotease Disintegrin Cysteine (ADAM/MDC)***. Thus, ADAM proteins are type 1 transmembrane proteins expressed on the cell surface. Protein kinase C is known to stimulate ADAMs' protease activity. The ADAMs have been isolated from mammalian species, *Caenorhabditis, Xenopus, and Drosophila.* Approximately half of the ADAMs do not contain the zinc-binding motif HEXXHXXGXXHD, which is required for enzymatic activity. However, all ADAMs contain the disintegrin domain, which comprises approximately 80 amino acids in length with 15 highly conserved Cys residues. In some members this region has been found to bind integrins, although the role of this domain for the majority of the family members is unknown. Among two and a half dozen ADAMs identified, only in a few, their biological roles elucidated. It is believed that through their proteolytic, adhesive and fusogenic functions, ADAMS are implicated in a variety of important physiological and pathological functions (Evans et al., 1998; Yuan et al., 1997; Lum and Blobel, 1997). Some ADAMs are ubiquitously expressed (ADAM9, ADAM10, ADAM15, and ADAM17) and may have pleotropic effects, as has been found for ADAM15 and ADAM17. Many of the other ADAMs, however, show tissue-specific expression: ADAM12 and ADAM19 in muscle, ADAM22 in brain, and ADAM23 in brain and heart. The largest group of ADAMs (1,2,3,4,5,6,7,18,20,21,24,25,26,27,29,30, and 31) is predominately expressed in testis and is thought to be involved in spermatogenesis and fertilization. The first mammalian ADAMs discovered, ADAM1 and ADAM2 are required for sperm-egg fusion (Cerretti et al., 1999). The cloning and sequence analysis of cDNAs encoding several rat orthologues of ADAM protein family, some of which are expressed exclusively in the male reproductive tract, others exhibiting a broader tissue distribution have been reported (Frayne et al, 1997).

Atleast six ADAM mRNAs are known to express in spermatogenic cells at different times of spermatogenesis. RNAs encoding ADAM 2 and 3 are found only in testis and spermatozoa, but RNA encoding ADAMs 1, 4 and 5 are present in other tissues as well. The ADAM2, which is found on sperm surface as a heterodimer, is the only ADAM heterodimeric ADAM found on sperm. Both α and β subunits of ADAM 2 are proteolytically processed during different stages of sperm development. Proteolytic activity of metalloprotease domain is functional during spermatogenesis and brings about proteolytaic cleavage of α and β, so that mature forms of ADAMs are present on ejaculated spermatozoa. Cysteine rich domains have been found to contain a fusion peptide, which is thought to help in sperm-egg membrane fusion (Shrimali and Reddy, 2000; Finaz and Hammami-Hamza, 2000). Several ADAM transcripts are expressed in fetal, neonatal, and prepubertal testes. Cyritestin (ADAM3), ADAM5, ADAM6, and ADAM15 are expressed on day 17 in fetal testes. In contrast, no expression of fertilin α (ADAM1) and fertilin β (ADAM 2) was detected in fetal testes. Fertilinβ gene expression starts after postnatal day 2, subsequent to the expression of fertilin α, which occurs on postnatal day 1. After postnatal day 2, all the indicated ADAMs, including the fertilin α and fertilin β, continued to be expressed (Rosselot et al., 2003).

Three closely related isoforms of monkey testis MDC are expressed not only in the testis, but also in the liver, albeit at a lower level. While some members of the MDC family were specific to the reproductive tract, suggesting functions peculiar to those tissues, others had a broader tissue distribution and might therefore play a more general role in integrin-mediated cell-cell recognition, adhesion or signaling (Perry et al, 1994, 1995; Barker et al, 1994). The distribution of seven of macaque MDC transcripts, including the description of a testis-derived tMDC III, their localization in spermatogenic cells in testis sections, and their processing on the sperm surface during epididymal transit has been described in different tissues (Frayne et al., 1998; Jurry et al., 1999). Unlike many MDC proteins expressed in the reproductive tract, epididymal eMDC II possesses the extended 'catalytic centre' consensus sequence characteristic of a reprolysin-like metalloproteinase. This suggested that eMDC II had proteolytic activity (Jurry et al, 1999). Full-length clone from a Macaca fascicular corresponds to a 2.65-kb mRNA in testis. The encoded putative 82-kDa transmembrane protein (tMDC1) showed striking sequence similarity to other members of the MDC family including rat and monkey EAP-I, guinea-pig PH-30 and human MDC protein, as well as a number of snake venom components (Baker et al., 1994; Perry et al., 1994).

Shilling et al (1997) identified the disintegrin domain of five distinct MDC proteins from *Xenopus laevis* testis cDNA. Four of these sequence tags (*Xmdc9, Xmdc11.1, X mdc11.2, and Xmdc13*) showed strong similarity to known mammalian MDC proteins, and expressed in other tissues, whereas the fifth (XMDC16) represents a novel family member whose mRNA was only expressed in testis. The XMDC16 protein sequence contains a metalloprotease domain with the active-site sequence HEXXH, a disintegrin domain, a cysteine-rich region, an EGF repeat, a transmembrane domain, and a short cytoplasmic tail (Shilling et al, 1997).

26.6.1. Fertilin α and β (ADAM1 and ADAM2)

The DITGs molecule fertilin αβ is an integral membrane glycoprotein on the posterior head of the guinea pig sperm (previously pH-30). Fertilin consists of two tightly associated but immunologically distinct subunits: α and β both of which are synthesized as large precursors in the testis. Fertilin α is processed in the testis whereas fertilin β is processed in the epididymis. The α-subunit contains α potential fusion peptide. The subunit β contains a DITG domain, which is a high affinity ITG ligand. The β-subunit exhibits a hairpin loop that presents the binding sequence RGD at its tip. Both α and β subunits of fertilin have been shown to participate in sperm-egg binding and the subsequent membrane fusion. The sequence similarity between PH-30 α, PH-30 β, HRIB, and disintegrins indicated a common ancestry. The PH-30 α and β are unique, however, in that they contain transmembrane anchors. Sequence analysis of primate fertilin α genes showed that fertilin αI and fertilin αII isoforms possess a common central region and different, but related 5' and 3' regions with nucleotide identities (Jury et al., 1998).

Since fertilin α and fertilin β are sperm plasma membrane proteins, which are implicated in binding to- and fusion with the egg plasma membrane, they represent the prototypes of a gene family, renamed as ADAM in honour of its dual origin in the fields of snakes and fertility, a widely expressed and developmentally regulated family of proteins containing a disintegrin and a metalloprotease domain (Wolfsberg et al., 1995). In situ hybridization demonstrated that RNA encoding these ADAMs is expressed only in spermatogenic cells and that this expression is developmentally regulated. Mouse ADMAs display different patterns of tissue distribution. The developmental expression of fertilin α and β in pre-pubertal and mature rat testes, showed that fertilin α mRNA was present at all stages of development, suggesting that it is not exclusively expressed in post-meiotic germ cells. In contrast, fertilin β mRNA was first identified in day 19 day old testes, coincident with the presence of pachytene spermatocytes. Fertilin α has the consensus active site sequence for a zinc-dependent metalloprotease in their metalloprotease-like domain. Guinea pig fertilin α and β are made as full-length precursors, which are proteolytically processed during spermatogenesis, resulting in removal of the pro- and metalloprotease-like domains and retention of all or part of the disintegrin domain, cysteine-rich domain, epidermal growth factor (EGF) like repeat, transmembrane domain, and cytoplasmic tail. During interactions with the egg plasma membrane, the most immediate relevant roles of fertilin α and β are postulated in cell-cell adhesive and fusion functions. The guinea pig fertilin α/β complex shares molecular and biochemical features with certain viral adhesion/fusion proteins. All fertilin α subunits (guinea pig, mouse, bovine, and monkey) contain a hydrophobic sequence in their ectodomains with similarity to viral fusion peptides. Although suggestive, none of these observations constitutes proof that fertilin α is a bonafide membrane fusion protein. The bovine homologues of fertilin α and β have the same domain organization as guinea pig fertilin α and -β. In contrast to the reported sequence of guinea pig fertilin α, the mature proteolytically processed bovine fertilin α subunit contains a complete disintegrin-like domain, and a potential amphipathic β strand, which may be involved in sperm-egg fusion. Gradient sedimentation experiments suggest that the fertilin α/β heterodimer may be present on the sperm surface as a higher-order oligomer (Chen et al., 1998; Waters and White, 1997). In the mouse, it has been proposed that the disintegrin domain of fertilin β binds to the egg integrin α6 β1. Mouse genes coding fertilin α and fertilin β with asigned locus symbols Ftna and Ftnb respectively, were mapped to chromosomes 5 and 14 (Cho et al., 1996).

The nucleotide sequence of mature fertilin α contains an ORF encoding 289 amino acids with one predicted transmembrane domain. The glycosylated protein has molecular mass of 60,000 D on reducing SDS-PAGE where as the predicted size of unmodified mature fertilin α from the cDNA sequence is 29,700. Although N-linked glycosylation accounts for part of this difference, additional post-translational modifications of the molecule must also predict its electrophoretic mobility. Fertilin α contains a region (residues 90-111) that fulfills all the criteria as an internal fusion peptide. Potential fusion peptide of fertilin α overlaps a region (residues 82-102) that is similar in sequence to a potential fusion peptide of the E2 glycoprotein of rubella virus.

The nucleotide sequence of mature fertilin β contains one ORF encoding 353 amino acids, with two potential sites for N-linked glycosylation and one predicted transmembrane domain. The calculated size of unmodified fertilin β is 39,000 Da, in good agreement with the observed size of glycosylated fertilin β on reducing SDS gels. The N-terminus of 90 amino acids of mature fertilin β contains a putative integrin-binding 'disintegrin' domain that could function in the membrane-binding step that precedes sperm-egg fusion. Disintegrins (such as bitistatin, barbourin, kistrin and echistatin comprise a family of short amino acids), are soluble and highly conserved platelet aggregation inhibitors from sanke venoms which act by competitive inhibition of fibrinogen binding to the integrin gpllb/IIIa. The similarity of sperm fertilin β to these integrin ligands suggests that it binds to a receptor, probably an integrin, on the egg post membrane. Fertilin α and -β contain an epidermal growth factor (EGF)-like repeat (Zhu et al., 2000).

Human fertilin β or ADAM 2 encodes a protein of 735 amino acids (2205 nucleotide) open reading frame. Its deduced amino acid sequence contains pro-metalloproteases like (172-382), DITG like (383-475), an EGF like repeat (616-644), trans-membrane (697-707) and cytoplasmic (708-735) domain. The human fertilin β shares 90%, 56% and 55% amino acid identity with monkey, guinea pig, and mouse fertilin β homologues, respectively. A phenylalanine-glutamate-glutamate (FEE) binding tripeptide within the disintegrin like domain of human fertilin β, homologous to other fertilin β RGD-like (arginine-glycine-aspartic acid) tripeptides, could compete for recognition by integrins and other receptors. Human fertilin β maps to chromosome 8, band p11.2 (Vidaeus et al., 1997). Rabbit fertilin α and β are predicted to encode proteins of 919 and 751 amino acids, respectively, with significant homology to fertilin subunits from other species. Analysis of the predicated protein sequences from fertilin α but not fertilin β reveals presence of 21 direct repeats of the hexameric sequence A/PPPPEA at the extreme carboxy terminus, similar to a fertilin α gene isoform in the monkey. Polyclonal antibodies against recombinant subunits detected the presence of these components in rabbit sperm (Hardy and Holland, 1996) **(Fig.26.3 a, b)**. Of particular interest is the mature guinea pig fertilin β, which contains an amino-terminal 90 amino acid disintegrin domain. It has been suggested that the integrin recognition sequence (TDE) of the guinea pig fertilin β disintegrin domain mediates sperm-egg binding. The amino acid sequence at this position in human fertilin β differs from the mouse, guinea pig and monkey sequence (Gupta et al., 1996). The *Caenorhabditis elegans adm-1* gene related to PH-30 has a disintegrin-and metalloprotease-like domain and shows a domain organization identical to PH-30. ADAM-1 expression in the hypodermis, pharynx, vulva, and mature sperm is consistent with a putative role in somatic and gamete cell fusions (Podbilewicz, 1996).

Processing of Fertilin α and Fertilinβ: Fertilin α and β are made as larger precursors that are processed proteolytically at different stages of sperm maturation in the testis and epididymis. Fertilin α is processed first, most likely by a pro-protein convertase in the secretory pathway of testicular cells. Fertilin β is processed later, while spermatozoa are in transit through the epididymis. The processing of fertilin β in the epididymis correlates with the acquisition of fertilization competence of sperm, which expose an epitope that has a role in sperm-egg interactions, and triggers the relocalization of fertilin from the whole sperm head to the posterior head. The proteolytic processing of fertilin and perhaps also other sperm proteins is an important aspect of sperm maturation and activation in the epididymis (see Chapter 34). Blobel (2000) highlighted the mechanism of fertilin processing in the guineapig, and about the identity of the proteases responsible for the proteolytic processing.

Fertilin-β is processed in epididymis in two steps. In step-1, fertilin-β is converted to pro-fertilinβ*. In the second step pro-fertilinβ* is further processed to yield mature fertilin β. In distal cauda the conversion of pro-fertilinβ* to mature fertilin is complete. On testicular sperm

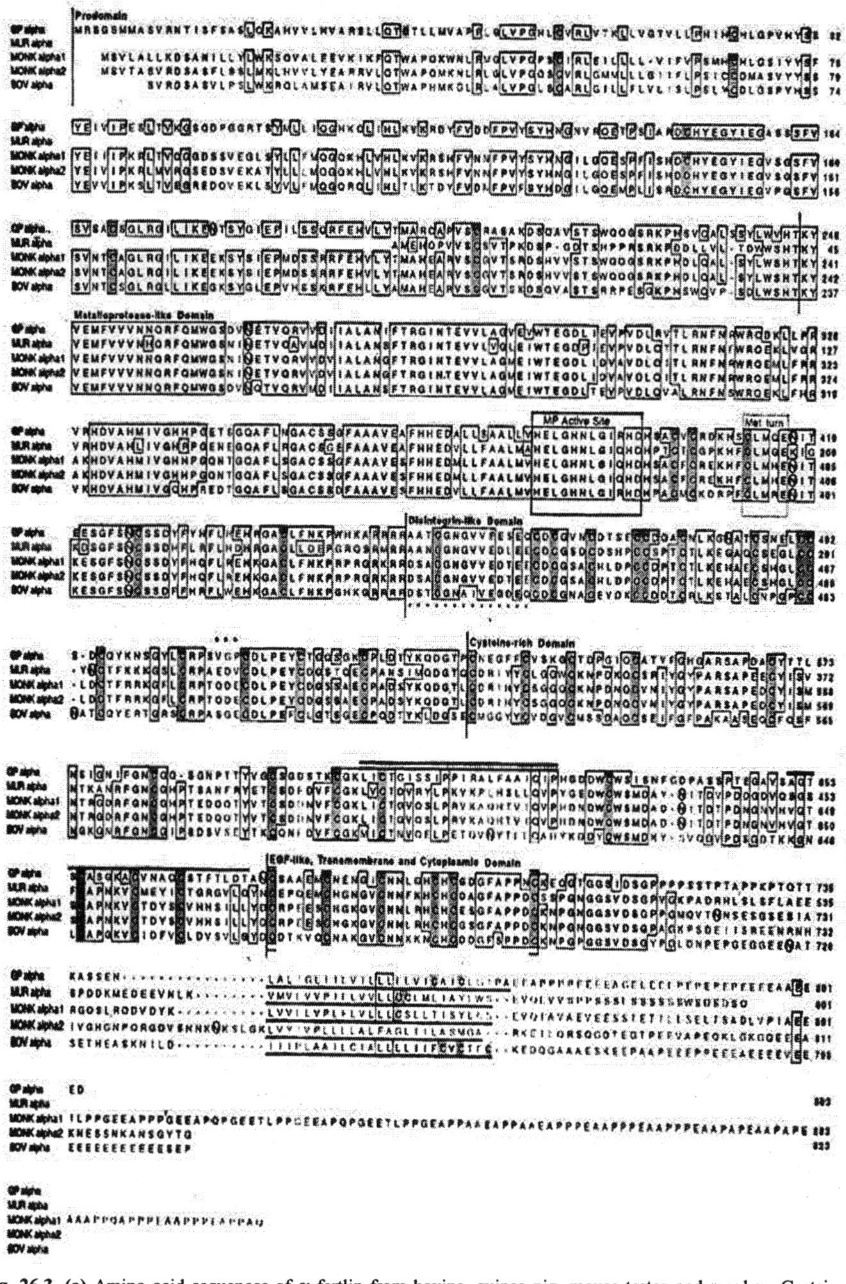

Fig. 26.3. (a) Amino acid sequences of α fertlin from bovine, guinea pig, mouse testes and monkey. Cysteine residues are shaded. Sites for N-linked glycosylaton are circled. For further details see captions of Figure 26.3(b). Reproduced with permission from S.I. waters and J.M. White. Biol Reprod 56. 1245-1254:1997 © Society for Study of Reproduction.

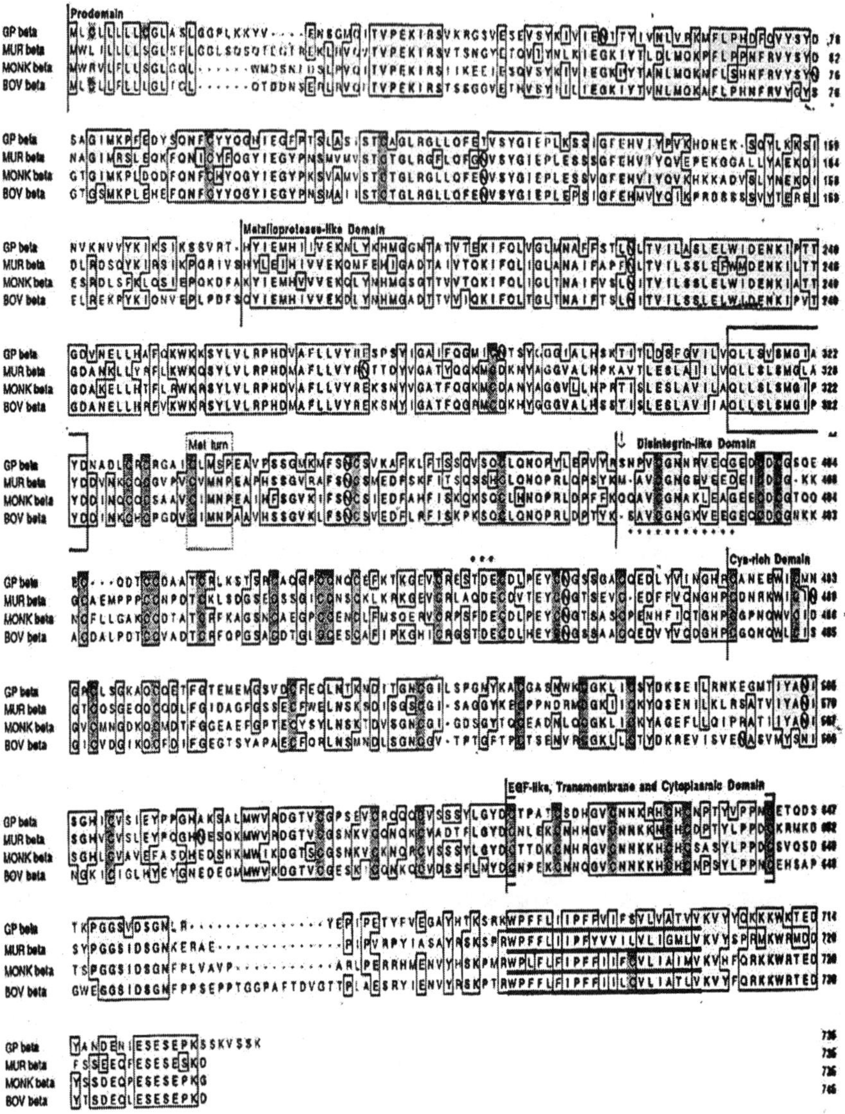

Fig. 26.3. (b) Amino acid sequences of β fertlin from bovine, guinea pig, mouse, and monkey testes. Cysteine residues are shaded. Sites for N-linked glycosylaton are circled. Amino Acids are boxed at positions of at least 75% identity. Prodomain: The first amino acid shown for each complete sequence is that of the first in-frame methionine. Metalloprotease-like domain: The location of the zinc-dependent metalloprotease (MP) active site sequence is boxed, and the met turn is in a dashed box. Disintegrin-like domain: Filled circles are above the location of the snake venom RGD sequence. Asterisks mark the N-terminal sequence determined for mature guinea pig fertilin. Cysteine-rich domain: A heavy double line is positioned over the predicted α-helical fusion peptide region of guinea pig, mouse, and monkey α. A lighter double line is positioned over the potential fusogenic β-strand region in bovine α. EGF like, transmembrane, and cytoplasmic domains: The EGF-like repeat is bracketed. The potential transmembrane domains are underlined. Reproduced with permission from S.I. Waters and J.M. White, Biol Reprod 56; 1245-54:1997 © Society for the Study of Reproduction.

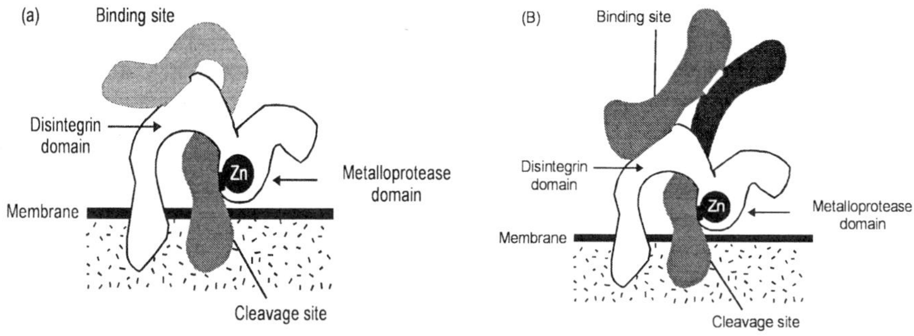

Fig.26.4 a, b. The beloved liberator model of ADAM action. The model presents two versions (a and b) of ADAM-substrate interactions (see text). Reprinted with permission from P.Primakoff and D. Myles. TIG 16; 83-87: 2000© Elsevier Science

of guinea pig fertilin is distributed on the plasma membrane over the entire sperm head, but is found only on the posterior head after sperm passed through epididymis. Fertilin from seven regions of the epididymis showed a temporal correlation between the beginning of fertilin's migration to the posterior head and the proteolytic processing of the full-length fertilin β precursor (85-kDa) to 75-kDa intermediate pro-fertilin β*. Further migration led to further cleavage of pro-fertilin β* to mature form (25-28kDa). This indicaded that further cleavage of pro-fertilin β to pro-fertilin β* may initiate fertilin's migration into the posterior head domain and, after localization to that membrane domain, proβ* is cleaved to mature β (Hunnicutt et al., 1997). The processing of mouse fertilin-β during epididymal maturation involves changes in the cytoplasmic tail domain as well as the N-terminal domains. Although Yuan et al (1997) and Chan et al (1999) have reported that mature fertilin-β is 55-57kDa, Cho et al (2000) showed that 55-kDa is an unrelated protein in the sperm extract, which cross-reacts with an antibody that recognizes precursor, but not mature fertilin β. Fertilin β knockout sperm revealed that authentic, mature fertilin β is 45-kDa and that mouse fertilin α and β exist as a heterodimer. Polyclonal antisera against a 28-residue peptide and two recombinant fusion proteins identified a 90-kDa protein in testicular sperm extracts and a 60-kDa protein in caput and cauda epididymal sperm extracts, the predicted sizes for rat fertilin β precursor and mature protein respectively (McLaughlin et al, 1997). In addition, mice lacking fertilin-β showed absence of fertilin α from mature sperm.

The Beloved Liberator Model of ADAM Action: The mature, processed fertilin is a heterodimeric complex of two membrane anchored glycoproteins ,which both contain an N-terminal disintegrin domain raising the possibility that one or both subunits can interact with an integrin on the egg. Myles and colleagues provided the first evidence that peptides containing RGD mimic a predicted integrin binding sequence in guinea-pig fertilin β, and is found in the snake venom integrin binding sequence (Primakoff and Myles, 2000). In describing a mechanism of action of ADAMs, " The beloved liberator" model has been proposed. In this model, the ADAM uses its disintegrin domain to bind to the substrate and uses its zinc containing metalloprotease active site to cleave the substrate. After cleavage, the substrate undergoes conformational change, dissociates from the disintegrin domain and large fragment of substrate is released from the cell surface. In another version, the ADAM uses its disintegrin domain to bind to the substrate-associated protein and uses its zinc containing metalloprotease active site to cleave the substrate. After cleavage, the substrate undergoes conformational change, dissociates from the substrate associated protein and the large extracellular fragment of substrate is released from the cell surface (Primakoff and Myles, 2000) **(Fig.26.4 a, b)**. While the substrates and function of the fertilin α metalloprotease domain remain to be elucidated, the sub-cellular localization of fertilin α processing is quite well defined. Fertilin contains a tetrabasic sequence RRRR between metalloprotease domain and the disintegrin domain, a consensus cleavage site for (RRRR) pro-protein convertases. The pro-protein convertase PC4 is highly expressed in the testis and also has an essential role in fertilization (Mbikay et al., 1997).

26.6.2. Fertilin as a Co-receptor for Egg Integrins

ADAM 1 contains 22 amino acid hydrophobic domain that shares significant homology to viral fusion peptides, and thus suggests a role for ADAM1 in membrane fusion during sperm-egg interaction. Sperm from mice lacking fertilin-β were shown to be deficient in sperm-egg membrane adhesion, sperm-egg fusion, migration from the uterus into the oviduct, and binding to the egg zona pellucida. Egg activation was unaffected (Cho et al., 1998). However, while the β subunit is apparently testis-specific, the finding of low levels of fertilin-α in non-reproductive tissues has cast some doubt on a unique role in fertilization. Moreover, the absence of a functional fertilin-α gene in the human would imply that this gene product is not absolutely essential for fertilization, although it could play a facilitatory role. The organization and sequence of the fertilin α genes in a range of primates, including the great apes, indicated that the gorilla gene, like that of the human, is non-functional (Jury et al, 1998; 1999).

To date, five different ADAMs (fertilin β, ADAM9, ADAM12, ADAM15, ADAM23) have been described to interact with integrins (specifically α6β1, αv β3, α9 β1, αvβ5, and/or α5β1). Best sperm-egg binding is achieved through the interaction of the disintegrin domain of fertilin -β with an integrin on the egg surface. Several evidences support that sperm fertilin is the co-receptor for egg α6β1 integrin. However, it is yet to be ascertained whether different species of mammalian eggs use RGD dependent (hamster) or RGD independent (mouse) ITGs as sperm receptors. Since there are several ADAM homologues to fertilin, it remains to be determined which sperm fertilin(s) binds to which ITGs on eggs from different species. As is true for sperm-egg binding, it is likely that other proteins in addition to ADAMs may participate in binding and fusion at the egg plasma membrane. These also may include new members of ADAM family, of which fertilin α and β are the prototypes. Other participating sperm proteins may include antigens containing sequences TDE in guinea pig, QDE in mouse, ECD in monkey, and FEE in human. The ECD containing sequences are conserved. In the context of recent knockout studies, which show that eggs lacking the α6 integrin subunit can be fertilized and eggs lacking the integrin-associated tetrapanin protein CD9 fail to fertilize (Evans, 2001). It has also been hypothesized that fertilin β and α subunits mediate interactions between sperm and egg plasma membranes namely binding and fusion, respectively. It has been suggested that fertilin β binds to an oolemmal integrin, and that the tripeptide FEE (Phe-Glu-Glu) is the integrin recognition sequence in human fertilin β (see above). The effects of a linear octapeptide SFEECDLP containing the FEE sequence, and a scrambled octapeptide with the same amino acids, SFPCEDEL, were studied on the interaction of human spermatozoa with human zona-free eggs. The effects of G4120, a potent RGD-containing (Arg-Gly-Asp) thioether-bridged cyclic peptide, which blocks both fibronectin and vitronectin receptor interactions on sperm-egg relationship between FEE- and RGD-receptor interactions on sperm-egg interactions were also studied. The FEE containing peptide inhibited sperm adhesion to oocytes and their penetration. The inhibition induced by SFEECDLP was reversible that occurred only in the presence of peptide itself, where as the G4120 containing RGD exhibited 10-fold less inhibitory effects on sperm adhesion and penetration than did SFEECDLP. Combined action of SFEECDLP and G4120 exhibited strong inhibition of both adhesion and penetration at concentrations that individually had been ineffective, suggesting co-operative interaction between the two receptor-ligand interactions during fertilization (Bronson et al., 1999).

An RGD-containing peptide (Arg-Gly-Asp-Val, RGDV) inhibits both oolemmal binding and penetration of zona-free hamster eggs by human spermatozoa in vitro. The RGD sequence is found in fibronectin, vitronectin, and other extracellular matrix proteins. The oolemmas of eggs from human and several other mammalian species contain receptors capable of binding to RGD ligands, and the integrin subunits are expressed by oocytes. These results suggest that RGD containing proteins may play a role in sperm-oolemmal interactions required for fertilization (Fusi et al., 1993). Receptors of the very late antigens (VLA) subfamily of integrins and heterodimeric receptors as $\alpha5\beta1$ amd αv $\beta1$ recognize fibronectin RGD. Fibronectin derived, RGD-containing peptides competitively inhibit sperm-oolemmal adhesion and penetration in both heterologous (human-hamster) and homolgous (hamster-hamster) gamete interactions. But the majority of fresh spermatozoa do not display fibronectin on their plasma membrane. Sperm surface expresses fibronectin following capacitation (Fusi et al, 1996). To investigate the role of fertilin-β in fertilization, Evans et al (1997a,b) expressed the putative extracellular domain of mouse fertilin-β (fertilin β-EC) in bacteria and found that recombinant fertilin β-EC binds to the region of the plasma membrane of the egg to which spermatozoa bind, thus providing evidence that fertilin β has adhesive properties. This binding is reduced in absence of Ca^{2+}, Mg^{2+} or Mn^{2+}. Peptides corresponding to the disintegrin domain of fertilin-β reduced its binding to eggs, suggesting that this domain is at least partially involved in the recognition of fertilin β. These experiments led to conclude that β1 containing integrin participates in the binding of recombinant fertilin β-EC to mouse eggs and that fertilin-α can function as a cell adhesion molecule during fertilization mediating the binding of sperm and egg plasma membrane (Evans 2001).

Subsequently, Evans et al (1998) examined the roles of the recombinant forms of disintegrin domains of fertilin α and β (αDCE and βDCE) or truncated forms that lacked the disintegrin domains (αCE and βCE) and tested their abilities to bind to eggs and to inhibit sperm-egg binding. Fertilin βDCE was able to inhibit sperm-egg binding while fertilin βCE was relatively ineffective suggesting that disintegrin domain of fertilin β is required for interactions with egg binding sites and/or for proper protein folding. Other domains of the fertilin α extracellular region (cysteine-rich and/or EGF-like repeat) also have the ability to block sperm binding suggesting that these domains of fertilin α may participate in sperm-egg adhesion (Evans et al., 1998).

Evans and associates confirmed that recombinant forms of fertilin α corresponding to either the disintegrin-like domain or the cysteine-rich domain and the EGF-like repeat can perturb sperm-egg binding, and suggested that both of these domains can participate in fertilin α-mediated adhesion events. In further examination, Wong et al (2001) suggested that a sub-domain of disintegrin-like domain with the sequence DLEECDCG outside the putative disintegrin loop but with homology to the fertilin β disintegrin loop can inhibit the binding of both sperm and recombinant fertilin α to eggs, suggesting that this is an adhesion-mediating motif of the fertilin α disintegrin-like domain. This sequence also inhibits the binding of recombinant fertilin β to eggs and thus is another peptide sequence found to block two different sperm ligands (Wong et al., 2001). It appears that fertilin β utilizes an ECD sequence within its disintegrin domain to interact with the egg plasma membrane; the Asp is especially critical. Based on what is known about different integrin subfamilies and their ligands, Zhu and Evans (2002) sought to characterize fertilin β binding sites on mouse eggs, focusing on integrin subfamilies that recognize short peptide sequences that include an Asp residue, which was speculated to have implications for the role of CD9 in the strengthening of fertilin β-mediated cell adhesion but not in initial ligand binding.

Since fertilin β contains a highly conserved motif (D/E)ECD in the disintegrin domain, it was suggested that (D/E)ECD could be the consensus sequence for recognition of disintegrins by α6β1 integrins. To determine whether a four amino acid peptide sequence with two adjacent acidic residues improved inhibition, a series of linear and cyclic peptides were synthesized, in which either one or both adjacent acidic residues in the sequence DECD were mutated to their corresponding amides (N or Q). Results showed that only one acidic residue and a reduced one are required for inhibition of fertilization (Gupta and Sampson, 2000). A widely proposed model for sperm-egg fusion suggests that fertilin α is the sperm component that promotes membrane fusion by undergoing a conformational change that exposes a virus-like, hydrophobic fusion peptide (Cho et al, 2000).

26.6.3. Role of CD9 in Sperm-Egg Interactions

The mRNAs for integrin subunits α5, α6, αv, β1, β3, β5, α6β1 and αvβ3 have been detected on the plasma membrane of unfertilized mouse eggs. The GoH3, an anti-α6, indicated the role for the integrin α6 β1 as a cell-cell adhesion receptor that mediates sperm-egg binding. Chen and Sampson (1999b) re-examined the role of egg α6β1and suggested that fertilin β binds directly to α6β1 integrin on egg surface and mediates sperm egg fusion, but suggested that different states of α6β1 can interact with an ECM ligand (laminin) or a membrane-anchored cell surface ligand (fertilin β) (Chen et al., 1999b). However, using eggs from cultured ovaries of mice, lacking the α6 integrin subunit, it was observed that the fertilization rate, fertilization index, and sperm binding were not impaired compared with wild type or heterozygous controls. A re-examination of antibody inhibition, revealed no inhibition of fusion by the GoH3 mAb. But an anti-CD9 mAb completely blocked sperm fusion with either wild type eggs or eggs lacking α6β1. Based on these results, it was concluded that the α6β1 integrin is not essential for sperm-egg fusion. Earlier, Ji et al (1998) investigated the potential role of β1 integrin in the gamete fusion process in humans and suggested that human gamete fusion can bypass the β1 requirement. β1 integrin certainly participates in human gamete fusion by acting in co-operation with multiple integrin / disintegrin couples or another cofactor, not yet identified. Status on egg receptor becomes more complicated when using α3 integrin null eggs, He et al., (2003) found that the α3β1 integrin was not essential for sperm-egg binding and fusion. These results indicated that none of the integrins known to be present on mouse eggs or to be ADAM receptors are essential for sperm-egg binding/fusion, and thus, egg integrins may not play the role in gamete fusion, previously attributed to them (He et al., 2003).

Therefore, the model in which CD9 acts by itself, or interacts with egg protein(s) other than α6β1, to function in sperm-egg fusion was proposed (Miller et al., 2000). The CD9 belongs to the tetraspan superfamily (TM4SF) of integral plasma membrane proteins. The tetraspanin CD9 has been implicated in cell motility, metastasis, and sperm-egg fusion. These proteins possess four transmembrane domains, two extracellular loops, and three short cytoplasmic segments. Tetraspan proteins have been proposed to act as "molecular facilitators" by bringing together and stabilizing molecular complexes. Tetraspan proteins have been shown to associate physically with members of the integrin family, with MHC class II glycoproteins as well as with each other. The CD9 is known to associate with several β1 integrins including α6β1. Because α6 β1 is present on murine eggs and interacts with the sperm-surface glycoprotein ADAM 2 (fertilin β), the anti-CD9 mAb, potentially inhibits sperm-egg binding and fusion in vitro. The CD9 is prominently present on the epithelium in the region between the uterus and oviduct and persists apically in oviduct. The integrin α6 subunit is found in similar apical patches in the region between the uterus and oviduct, but is confined to the basal aspect of the epithelium in the uterus and oviduct. Hence, α6β1 and CD9 both are expressed on the apical epithelial surface at the uterine-oviduct junction. Findings correlate with the observation that fertilin β "knockout" sperm traversed the uterus but did not progress into the oviduct, contributing to the infertility of fertilin β -/- male mice. This suggested that high-avidity binding between fertilin β (ADAM 2) and α6β1 requires cooperation between α6β1 and CD9. Such cooperation might assist sperm passage into the oviduct as well as sperm egg interactions. Although the egg surface tetraspanin CD9 is essential for gamete fusion (Kaji, et al, 2000) and CD9 is known to associate with integrins, recent models of gamete fusion have posited that egg CD9 acts in association with α6 β1 in fusion (Chen, et al 1999c; Kaji et al 2000, Le Naour, et al., Miyado et al 2000). Gutierrez-Lopez et al., (2003) characterized the first CD9 conformation-dependent epitope whose expression depends on changes in the activation state of associated β1 integrins. The expression of the PAINS-13 epitope depends on CD9 association with α6β1 integrin and the reactivity was mapped to the CD9 region comprising residues 112-154 in the NH_2 half of the large extracellular loop (Gutierrez-Lopez et al., 2003). Takahashi et al., (2001) presented evidence that in addition to the tetraspanin CD9, two other β1-integrin-associated proteins, the tetraspanin CD81 as well as the single pass transmembrane protein CD98 are expressed on murine eggs. Antibodies to CD9 and CD98 inhibited in vitro fertilization and binding of the ADAM 3 disintegrin domain was proposed that an egg surface "tetraspan web" facilitates fertilization and that it may do so by fostering ADAM-integrin interactions.

26.7. CYRITESTIN (ADAM 3)

The amino terminal 90 amino acids of ADAM 3 contains a putative ITG binding domain that shares 70% sequence homology with snake venom DITG, that are known to bind to ITGs on cell surface.

Mouse: Cyritestin has been identified in male germ cells and proposed to participate in the binding and/or fusion between sperm and the egg plasma membrane. Cyritestin is the product of the *Cyrn* gene on mouse chromosome 8. It is a membrane-anchored protein belonging to the ADAM family of proteins, and localized in the acrosomal region of spermatids and spermatozoa. Cyritestin is transported to the forming acrosomal vesicle through the Golgi apparatus to become part of the acrosomal membrane. Differential staining with antibodies recognizing either the metalloprotease-like domain or the cytoplasmic domain of cyritestin indicates that processing of the molecule leading to the loss of the pro-and metalloproteinase-like domains begins during germ cell stage 6 and is completed before stage 15 (Forsbach and Heinlein, 1998). The *Cyrn* mRNA is present in germ cell 4 days prior to translation. It undergoes posttranslational modification after incorporation into the acrosomal membrane and cyritestin has an apparent molecular weight of 110-kDa but is subject to processing during epididymal sperm transport, resulting in a shorter molecule lacking approximately 55-kDa from the terminal half. The N-terminal part including the metalloprotease-like domain is removed during spermiogenesis by processing of the precursor protein to a major form. Mature cyritestin contains a domain with homology to a family of integrin ligands, disintegrin. Cyritestin becomes exposed on the sperm surface after successful acrosome reaction and thus may play a role in sperm function rather than in testicular germ cell maturation (Linder et al., 1995).

Cyritestin plays a role in sperm function rather in testicular germ cell maturation. Support for this hypothesis came from in vitro fertilization experiments with synthetic peptides derived from the disintegrin domain of cyritestin. Such peptides interferred with successful fertilization at lower concentrations than did peptides corresponding to fertilin β, thus identifying cyritestin as one of the major candidates involved in sperm egg interaction events at the level of egg plasma membrane. In an effort to elucidate the physiological function of cyritestin, Shamsadin et al, (1999) disrupted its locus by homologous recombination. Homozygous null mutants were infertile, even though spermatogenesis, mating, and migration of sperm from the uterus into oviduct were normal. In vitro experiments showed that infertility is due to the inability of the cyritestin-deficient sperm to bind to the zona pellucida. However, after removal of the zona pellucida, sperm-egg membrane fusion did not reveal any differences from the wild type situation. These results demonstrated that cyritestin is crucial in the fertilization process at the level of the sperm-zona pellucida interaction. Male mice lacking the sperm cyritestin are infertile. Similar to fertilin β, cyritestin null sperm were drastically deficient in adhesion to the egg zona-pellucida and to the egg plasma membrane. Thus deletion of either of ADAM genes can result in the loss of multiple gene products. This loss of multiple gene products appears to result from a developmental mechanism during sperm differentiation (Cho et al., 1998; Nishimura et al., 2001).

Peptide mimetics, from the predicted binding sites in the disintegrin domains of the five testis-expressed ADAMs in a sperm-egg plasma membrane, showed that the active site peptide from cyritestin strongly inhibited (80-90%) sperm adhesion and fusion and was a more potent inhibitor than the fertilin β active site peptide. Antibodies against the active site region of either cyritestin or fertilin β also strongly inhibited both sperm-egg adhesion and fusion. Cyritestin, present on mature sperm and live, acrosome-reacted sperm showed staining of the equatorial region that participates in the early steps of membrane fusion (Yuan et al., 1997).

Human: The human cyritestin is a polymorphic protein that could include membrane-anchored and soluble forms. Sequencing of eight human cyritestin cDNA clones revealed that they are identical at their 5' and 3' ends but differ from each other in the length of an internal deletion, suggesting that the human cyritestin mRNA is alternatively spliced. Internal deletions that are present in some cDNA isoforms do not cause a frameshift in the C-terminal coding region. Characterization of human cyritestin genomic fragments revealed that the human genome contains two copies of the cyritestin assigned to the region p12-21 of chromosome 8 and q12 of chromosome 16, respectively. Both human genes are expressed in the testis. Amino acid sequence comparisons between cyritestin and other members of the metalloprotease-disintegrin family of proteins suggested that human and mouse cyritestin and monkey tMDC1 are homologous (Adham et al., 1998).

26.8. OTHER ADAM PROTEINS

ADAM 4 and ADAM5: Mouse genes coding ADAM4 and ADAM5 have been given the locus symbols *Adam4* (ADAM 4) and *Adam5* (ADAM5). They have been mapped to chromosomes 9 and 8 respectively as against Fertilin α and Fertilin β being localized to chromosomes 5 and 14, thus revealing a dispersed localization of *ADAM* genes. Because all four of these ADAM genes are expressed in testis and fertilin α and β have been found to be important for fertilization, their chromosomal locations were compared with known mouse mutations affecting

```
  1 MVLAEGQVML LLLGLWVLLD PGQCSPGRPS WRYISSEVVI PRKELHQGRG VQVAGWLYYS
 61 LHFGGQRHVI CLQSKKLIWA RHLLMMTQDD QGALQMDYPY IPTDCYYLGH LEDIPLSTVT
121 IDTCYGGLEG IMKLDDLTYE IKPLKDSNTF EHIVSQIVAN RNATGPMYRL GHEGDFDPFF
181 SKVNSSVAPK LSSFNFMYHR AQMKGQVQIA HEVYTVLNNI SKCIQFAVNM FSITDSFLRS
241 LGFGHYIFLL NIYNQREPIV LNDFRVPGGA IHAYYKATFH DLYQPYPSTL ITKNAPNDDQ
301 EPHRYGCCGH FNLIIISSRG RHYLLLAILA AHKIGRQIGL EYDGSTCVCQ RRATCLMRRF
361 PEITDSFSNC SFVHTQHIVS NRYISKCYYL PGRLYLNKTL LQTRCGNSLV EEREQCDCGS
421 FKHCYANACC QSDCRFTPGS ICDKQQCCTN CTYSPTSTLC RPVMNICDLP EYCGGSTYTC
481 PDNFYLQDGT PCTEDGYCYR GNCTDRTMHC KEIFGETAHN GPGDCYAINL NTFRFGHCKR
541 VQTQNVYQAC AAADKECGRL QCINVTHLPQ LQDHVSFHQS VYNEFTCFGL DEHRGTGSTD
601 AGHVRDGTPC GEGLYCFASR CNMTMNNLHY DCFPEKCNFR GICNNNKNCH CHVGWDPPLC
661 LSPGAGGSSQ SGPPPRRMRT VTDSMEPIVY LRVVFARVYC FIFALLFGVA TNVRTIKTTI
721 VQEQTATEPQ
```

Fig.26.5. Amino acid sequence of cellular disintegrin ADAM 6e from rabbit testis (Accession AAC09476) (Hardy and Holland, 1998). Source: http//www. ncbi.nlm.nih.gov.

spermatogenesis and fertility (Cho et al., 1996).

ADAM6: Disintegrin genes (ADAM 6d and ADAM 6e) were characterized from a rabbit testicular cDNA library. The cDNAs have open reading frames encoding proteins of 731 and 730 amino acids, respectively. They share an amino-acid homology of greater than 89% and a nucleotide base matching of 94% in both the coding and non-coding regions. Polyclonal anti-recombinant ADAM 6e antibodies recognized two bands with molecular masses of 42 kDa and 46 kDa (Hardy and Holland, 1998)(**Fig.26.5**).

ADAM7 and GP83: The mouse *ADAM7* gene is expressed in the caput region of the epididymis and in the anterior pituitary gonadotropes with no detectable expression in other tissues. The ADAM7 mRNA exhibited an apical localization within the proximal caput epididymal epithelium, to which no known function has been ascribed. ADAM7 gene expression requires androgens as well as testicular factors for its expression, in a region-dependent manner. The apical localization of ADAM7 mRNA is dependent upon an intact testis, because in situ hybridization analyses of the proximal caput epididymidis from a testosterone maintained castrated mouse did not show the apical localization. The ADAM7 gene maps to the central region of mouse chromosome 14, approximately 4-5 cM distal from the fertilin β locus (Cornwall and Hsia, 1997).

A human homologue of ADAM7 is GP-83, a glycoprotein secreted by the epididymis, and conjugated to mature sperm. A 2.1-kb GP-83 expressing insert has been isolated from a cDNA library of human epididymis. The 5' end RACE and 3'-RACE of the 2.1-kb insert elucidated two isoforms of GP-83 encoding cDNA sequences: an α-form of 3451-bp and β-form of 2643-bp. Both forms exhibit the same open reading frame of 2262-bp predicting a peptide of 754 amino acid residues. Deduced amino acid sequence revealed the presence of a signal sequence, prodomain, metalloproteinase, disintegrin, cysteine-rich, epidermal growth factor-like, transmembrane, and cytoplasmic domains. Human ADAM7 was located at chromosome 8p22. In vivo expression confirmed that h-ADAM7 cDNA did encode GP-83 and two transcripts of 4 kb and 3 kb in the epididymis, but not in testis or other major tissues. The functions of GP-83-encoding associated with ADAM7 gene (*h-ADMA7*) in epididymis need to be investigated (Lin et al, 2001).

ADAM12 and ADAM15: Eto et al, (2000) showed that integrin α9β1 specifically interacts with the recombinant ADAM-12 and –15 disintegrin domains in an RGD-independent manner, and suggested that interaction between ADAM-12 or – 15 and α9β1 supports cell-cell interaction. This observation is different from other reports that suggested ADAM-2/α6β1 interaction during sperm/egg fusion. Eto et al suggested that α9β1 may be a major receptor for ADAMs

```
        1       Signal Sequence       |          Pro Domain                                                                   100
ADAM21  .MAVDGTLVY IRVTLLLLWL GVFLSISGYC QAGPSQHFTS PEVVIPLKVI SRGRSAKAPG WLSYSLRFGG QKHVVHMRVK KLLVSRHLPV FTYTDERALL
ADAM20  .MAVGEPLVH IRVTLLLLWF GMFLSISGHS QARPSQYFTS PEVVIPLKVI SRGRGAKAPG WLSYSLRFGG QRYIVHMRVN KLLFAAHLPV FTYTEQHALL
ADAM29  .......... MKMLLLLHCL GVFLSCSGHI QDEHPQYHSP PDVVIPVRIT GTTRGMTPPG WLSYILPFGG QKHIIHIKVK KLLFSKHLPV FTYTDQGAIL
ADAM30 .MRSVQIFLSQ CRLLLLLVPT MLLKSLGEDV IFHPEGEFDS YEVTIPEKLS FRGEVQGVVS PVSYLLQLKG KKHVLHLWPK RLLLPRHLRV FSFTEHGELL

        101                                                                                                           200
ADAM21  EDQLFIPDDC YYHGYVEGAP ESLVVFSACF GGFRGVLKIS GLTYEIEPIR HSATFEHLVY KVNSNETQFP AMRCGLTEKE VARQQLEFEE AENSALEPKS
ADAM20  QDQPFIQDDC YYHGYVEGVP ESLVALSTCS GGFLGMLQIN DLVYEIKPIS VSATFEHLVY KIDSDDTQFP PMRCGLTEEK IA.HQMELQL SYNFTLKQSS
ADAM29  EDQPFVQNNC YYHGYVEGDP ESLVSLSTCF GGFQGILQIN DFAYEIKPLA FSTTFEHLVY KMDSEEKQFS TMRSGFMQNE ITC.RMEFEE IDNSTQKQSS
ADAM30  EDHPYIPKDC NYMGSVKESL DSKATISTCM GGLRGVFNID AKHYQIEPLK ASPSFEHVVY LLKKE..QFG NQVCGLSDDE IEWQMAPYE. ..NKA.RLRD

        201|  Catalytic Domain                                                                                     300
ADAM21  AGDWWTHAWF LELVVVVNHD FFIYSQSNIS KVQEDVFLVV NIVDSMYQQL GTYIILIGIE IWNQGNVFPM .T.SIEQVLN DFSQWKQISL .SQLQHDAAH
ADAM20  FVGWWTHQRF VELVVVVDNI RYLFSQSNAT TVQHEVFNVV NIVDSFYHPL EVDVILTGID IWTASNPLPT .SGDLDNVLE DFSIWKNYNL NNRLQHDVAH
ADAM29  YVGWWIHFRI VEIVVVIDNY LYIRYERNDS KLLEDLYVIV NIVDSILDVI GVKVLLFGLE IWTNKNLI.V .VDDVRKSVH LYCKWKSENI TPRMQHDTSH
ADAM30  FPGSYKHPKY LELILLFDQS RYRFVNNNLS QVIHDAILLT GIMDTYFQDV RMRIHLKALE VWTDFNKIRV GYPELAEVLG RFVIYKKSVL NARLSSDWAH

        301                                            Zn-Binding Motif                                               400
ADAM21  MFIKNSLISI LGLAYVAGIC RPPIDCGVDN FQGDTWSLFA NTVAHELGHT LGMQHDEEFC FC.GERGCIM NTFRVPAEKF TNCSYADFMK TTLNQGSCLH
ADAM20  LFIKDTQGMK LGVAYVKGIC QNPFNTGVDV FEDNRLVVFA ITLGHELGHN LGMQHDTQWC VC.ELQWCIM HAYRKVTTKF SNCSYAQYWD STISSGLCIQ
ADAM29  LFTTLGLRGL SGIGAFRGMC TPHRSCAIVT FMNKTLGTFS IAVAHHLGHN LGMNHDEDTC RC.SQPRCIM HEGNPPITKF SNCSYGDFWE YTVERTTCLL
ADAM30  LYLQRKYNDA LAWSFGK.VC SLEYAGSVST LLDTNILAPA TWSAHELGHA VGMSHDEQYC QCRGRPNCIM GSGR...TGF SNCSYISFFK HISSGATCLN

        401|  Disintegrin Domain                                                                                   500
ADAM21  NPPRLGEIFM LKRCGNGVVE REEQCDCGSV QQCEQDACCL LNCTLRPGAA CAFGLCCKDC KFMPSGELCR QEVNECDLPE WCNGTSHQCP EDRYVQDGIP
ADAM20  PPPYPGNIFR LKYCGNLVVE EGEECDCGTI RQCAKDPCCL LNCTLHPGAA CAFGICCKDC KFLPSGTLCR QQVGECDLPE WCNGTSHQCP DDVYVQDGIS
ADAM29  ETVHTKDIFN VKRCGNGVVE EGEECDCGPL KHCAKDPCCL SNCTLTDGST CAFGLCCKDC KFLPSGKVCR KEVNECDLPE WCNGTSHKCP DDFYVEDGIP
ADAM30  NIPGLG..YV LKRCGNKIVE DNEECDCGST EECQKDRCCQ SNCKLQPGAN CSIGLCCHDC RFRPSGYVCR QEGNECDLAE YCDGNSSSCP NDVYKQDGTP

        501  |  Cys-Rich Domain                                                                                    600
ADAM21  CSDSAYCYQK RCNNHDQHCR EIFGKDAKSA SQNCYKEINS QGNRFGHCGI NG.TTYLKCH ISDVFCGRVQ CENVRDIPLL QDHFTLQHTH I..NGVTCWG
ADAM20  CNVNAFCYEK TCNNHDIQCK EIFGQDARSA SQSCYQEINT QGNRFGHCGI VG.TTYVKCW TPDIMCGRVQ CENVGVIPNL IEHSTVQQFH L..NDTTCWG
ADAM29  CKERGYCYEK SCHDRNEQCR RIFGAGANTA SETCYKELNT LGDRVGHCGI KN.ATYIKCN ISDVQCGRIQ CENVTEIPNM SDHTTVHWAR F..NDIMCWS
ADAM30  CKYEGRCFRK GCRSRYMQCQ SIFGPDAMEA PSECYDAVNL IGDQFGNCEI TGIRNFKKCE SANSICGRLQ CINVETIPDL PEHTTIISTH LQAENLMCWG

        601                                                                                                          |700
ADAM21  IDYHLRMN.. .ISDIGEVKD GTVCGPGKIC IHKKCVSLSV LSHVCLPETC NMKGICNNKH HCHCGYGWSP PYCQHRGYGG SIDSGPASAK R.......GV
ADAM20  TDYHLGMA.. .IPDIGEVKD GTVCGPEKIC IRKKCASMVH LSQACQPKTC NMRGICNNKQ HCHCNHEWAP PYCKDKGYGG SADSGPPPKN NME.....GL
ADAM29  TDYHLGMK.. .GPDIGEVKD GTECGIDHIC IHRHCVHITI LNSNCSPAFC NKRGICNNKH HCHCNYLWDP PNCLIKGYGG SVDSGPPPKR KKKKKFC YL
ADAM30  TGYHLSMKPM GIPDLGMIND GTSCGEGRVC FKKNCVNSSV LQFDCLPEKC NTRGVCNNRK NCHCMYGWAP PFCEEVGYGG SIDSGPPGLL R.......GA

        701 Transmembrane Domain  |  Cytoplasmic Domain                                                             800
ADAM21  FLPLIVIPSL SVLTFLFTVG LLMYL....R QCSGPKETKA HSSG*..... .......... .......... .......... .......... ..........
ADAM20  NVMGKLRYLS LLCLLPLVAF LLFCLHVLFK KRTK...SKE DEEG*..... .......... .......... .......... .......... ..........
ADAM29  CILLLIVLFI LLCCLYRLCK KSKPI....K KQQDVQTPSA KEEEKIQRRP HELPPQSQPW VMPSQSQPPV TPSQSHPRVM PSQSQPPVMP SQSHPQLTPS
ADAM29γ                                                                                   QVM PSQSQPPQNL FLFSFSISDC
ADAM30  IPSSIWVVSI IMFRLILLIL SVVFVFF..R QVIGNHLKPK QEKMPLSKAK TEQEESKTKT VQEESKTKTG QEESEAKTGQ EESKAKTGQE ESKANIESKR

        801
ADAM29  QSQPPVMPSQ SHPQLTPSQS QPPVTPSQRQ PQLMPSQSQP PVTPS*
ADAM29γ VLNFRLLYLQ AT*...... .......... .......... ......
ADAM30  PKAKSVKKQK K*........ .......... .......... ......
```

Fig.26.6. Amino acid sequences of ADAM29 and ADAM30 from human testis library, and their comparison with other ADAMs. Reproduced with permission fromD.P. Cerretti et al. Biochem Biophy Res Commun 263; 810-15 (1999) © Elsevier.

that lack RGD motifs. Considering a wide distribution of ADAMs and α9β1, this interaction may be of potential biological and pathological significance.

ADAMs 20-23: The ADAM20 and ADAM 21 cloned from a human testis cDNA library predicted translation products with 50% sequence identity with each other. The best similarity was seen to sperm fertilin-α and -β, and meltrin-γ (ADAM9), which is ubiquitously expressed. Both ADAM-20 and -21 mRNAs are exclusively expressed in testis, presumably, on mature spermatocytes. Both cDNAs were found to be tightly linked to the same marker on chromosome 14q24.1. This region is not syntenic with the loci of mouse sperm-specific ADAMs 1-5. The *ADAM 20*, but not *ADAM-21* encodes a consensus Zn^{2+} binding site of active ADAM lysin metzincin metalloproteases. But both encode putative cell-fusion peptides, required for sperm-egg fusion. It is possible that ADAM-20 and /or -21 are functional equivalent of fertilin α (Hooft van Huijsduijnen, 1998).

Based on a conserved sequence within the disintegrin domain, GE(E/Q)CDCG, seven genes have been reported in a human genomic library. Two of these genes, *ADAM20* and *ADAM21* lack introns and show testis-specific expression, while the other two genes contain introns (*ADAM22 and ADAM 23*) and are expressed predominantly in the brain. In addition, three pseudogenes were isolated; one of which evolved from ADAM21. Human chromosomal mapping indicated that ADAM22 and ADAM23 mapped to chromosome 7q21 and 2q33, respectively, while the three pseudognes 1-2, 3-3, and 1-32 mapped to chromosome 14q24.1, 8p23, and 14q24.1, respectively. An ancestral analysis of all known ADAMs indicated that the zinc-binding motif in the catalytic domain arose once in a common ancestor and was lost by those members lacking this motif (Poindexter et al, 1999).

ADAM 24 – ADAM27: Four murine *ADAM cDNAs (ADAM 24, ADAM 25, ADAM 26 and ADAM 27)* present in testis were cloned and sequenced (Zhu et al., 1999). The amino acid sequences showed that all four ADAMs contain the complete domain organization common to ADAM family members. Messenger RNA for each of the four ADAMs was found only in the testis. The conserved zinc-dependent metalloprotease active site HEXGHXXGXXHD was found in the metalloprotease domain of three of these ADAMs, suggesting that they are testis-specific proteases, which were assigned alternative names: testase 1 (ADAM24); testase 2 (ADAM25), and testase 3 (ADAM26). Adult level of these ADAMs reached by day 20 (ADAM 27), day 25 (ADAMs 24 and 25) and in the range day 25-50 (ADAM 26). Each ADAM among ADAM24 - ADAM27 is transcribed in spermatogenic cells in a regulated pattern at a specific developmental stage (Zhu et al., 1999). The testase 1 (ADAM 24) is the first example of a plasma membrane-anchored protease on mature sperm (Zhu et al., 2001). Unlike other sperm ADAMs (fertilin α and β, cyritestin) whose metalloprotease domains are removed during sperm development, testase 1 retains an active metalloprotease domain, suggesting that it acts as a protease on mature sperm. Testase 1 is a glycoprotein (88 kDa), localized to the equatorial region of the plasma membrane of cauda epididymal sperm. The pro-domain of the testase 1 precursor (108 kDa) is proteolytically removed in caput epididymis to produce processed (mature) testase 1 (88 kDa). Testase 1 is unique among all other ADAMs in that its proteolytic processing occurs on the sperm plasma membrane instead of at an intracellular site (the Golgi). The cytoplasmic tail of testase 1 could be phosphorylated in vitro by PKC. Thus testase 1 apparently has a cytoplasmic PKC phosphorylation site(s). Protein kinase C is known to stimulate other ADAMs' protease activity. Since the acrosome reaction includes PKC activation, it was speculated that testase 1 protease functions in sperm penetration of the zona pellucida after sperm PKC is activated during the acrosome reaction (Zhu et al, 2001).

ADAM29–ADAM30: Human *ADAM29* gene codes a putative fusion peptide, which is highly expressed in the testis. Testis-specific expression shows three forms of *ADAM29* that encode proteins of 820, 786 and 767 amino acids. All of the amino acid differences are located in the cytoplasmic domain. Two forms of *ADAM30* genes encode proteins of 790 and 781 amino acids, with the difference in the coding region occurring in the cytoplasmic domain. The *ADAM29 and ADAM30* are mapped to human chromosome 4q34 and 1p11-13, respectively. An ancestral analysis of known mammalian ADAMs indicated that the zinc-binding motif in the catalytic domain arose once in a common ancestor and was subsequently lost by those members lacking this motif (Cerretti et al, 1999, Xu et al 1999) **(Fig.28.6)**.

ADAM31: Mouse ADAM 31 predicts a transmembrane protein with metalloproteinases, disintegrin, cysteine-rich, and cytoplasmic domains. Messenger RNA encoding ADAM 31 was most abundant in testes, but was also detected in many other tissues. The protein has a unique and restricted expression pattern. The ADAM 31 is expressed in Leydig cells of the testes, but unlike many other ADAMs, it is not found on developing sperm. ADAM 31 is highly expressed on four types of specialized epithelia: the cauda epididymidis, the vas deferens, the convoluted tubules of the kidney, and the parietal cells of the stomach (Liu and Smith, 2000).

26.9. ZONA ADHESINS

26.9.1 Mouse

Hardy and Garbers (1995) purified and cloned a sperm membrane protein, designated zonaadhesin (zonadhesin) that binds in a species-specific manner to the extracellular matrix (zona pellucida) of the egg. Mouse zonaadhesin mRNA is evident only within the testis, and the protein is found exclusively on the apical region of the sperm head. The cDNA encodes a protein with a single transmembrane segment separating a 36 amino acid, highly basic intracellular C-terminus from a 2418 amino acid extracellular region. The extracellular sequence specifies a mosaic protein comprising a unique N-terminal domain, a mucin-like domain, and five tandem domains proximal to the membrane that are homologous to pre-pro von Willebrand factor. The N-terminal and mucin-like domains were absent from zonaadhesin which suggested that processing occurs during sperm maturation and/or capacitation. Within testis, it expressed primarily in haploid spermatids. The unique domain structure of zonaadhesin suggests multiple functions, one of which is to mediate sperm adhesion to the zona pellucida. The cDNA for mouse zonaadhesin (16.4-kb) shows a large species variation in the numbers and arrangements of domains. There are 20 partial D-domains, found as tandem repeats, inserted between two of the four full D-domains and an additional partial D-domain. Theses domains are homologous to the D-domains of von Willebrand factor and α-tectorin. A region at the N terminus of the mouse cDNA contains three tandem repeats homologous to MAM domains. These domains comprising of about 160 amino acids, are present in transmembrane proteins such as receptor protein tyrosine phosphatases, where they appear to function in cell/cell interactions. In addition, mouse zonaadhesin contains a mucin-like domain and a domain homologous to EGF. A putative single transmembrane segment separates a short carboxyl tail from the extracellular region. The existence of these (MAM, mucin, D-, and EGF) domains suggested that mouse zonaadhesin functions in multiple cell adhesion processes, including binding to the extracellular matrix of the egg as one of the functions (Gao and Garbers, 1998; Hardy and Garbers, 1995).

26.9.2. Pig

The pig zonadhesin precursor is a 267-kDa mosaic protein with a Type I membrane topology and a large extracellular region comprising meprin/A5 antigen/mu receptor tyrosine phosphatase, mucin and five tandem von Willebrand D (VWD) domains. Multiple mature forms of zonadhesin in the sperm head differ in their avidities for the ZP. The predominant active polypeptides of processed zonadhesin were M(r) 300-kDa, 105-kDa, and 45-kDa. The 45-kDa (p45) and 105-kDa (p105) polypeptide proteins comprised primarily the D1, D2-D3 domains respectively, and were N-glycosylated. The pig zonadhesin polypeptide chains of p105 and p45 are covalently associated. Although p105/45 form is in minority forms in sperm membrane fractions, it was the most dominant form capable of binding to the pig zona-pellucida. Like mouse sperm, pig zonaadhesin is present at the apical head of pig sperm. It was suggested that a heterogeneous combination of specific proteolysis and intermolecular disulfide bond formation in the sperm head (as evidenced by zona adhesion) generates multiple forms of zonaadhesin with differing avidities for the zona-pellucida (Hickox et al., 2001). The p300 was heavily O-glycosylated, and spanned the meprin/A5 antigen/mu receptor tyrosine phosphatase, mucin and D0 domains. Hydrolysis of the precursor polypeptide occurred in the testis, and N-terminal sequencing of p45 and p105 identified Asp806-Pro and Asp1191-Pro in the D1 and D2 domains respectively as bonds cleaved in the protein's functional maturation. Zonadhesin localized to the perimeter of the acrosome in intact ejaculated spermatozoa and to the leading edge of acrosomal matrix overlying cells with disrupted acrosomal membranes. The zonadhesin precursor is specifically proteolysed, glycosylated and assembled into particulate structures in the distal parts of the acrosome where it may mediate specific adhesion to the ZP during the initial stages of acrosomal exocytosis (Bi et al., 2003).

26.9.3. Tektorins and Zonaadhesins

The cDNA for α-tectorin predicts a protein of 239034-Da with 33 potential N-glycosylation sites, and of β-tectorin, a smaller protein of 36074-Da with 4 consensus N-glycosylation sites. The α- and β-tectorins are single copy genes only expressed in the inner ear. Both sequences terminate with a hydrophobic carboxyl end preceded by endoproteinase cleavage site suggesting that the tectorins are synthesized as glycosylphosphatidylinositol linked, membrane bound precursors, targeted to the apical surface of the inner ear epithelia by the lipid and proteolytically released into the extracellular compartment. The mouse β-tectorin contains a single zona-pellucida domain, whereas α-tectorin is composed of three distinct modules: an NH-2 terminal region similar to the part of entactin G1 domain, a large central segment with three full and two partial von Willebrand factor type D repeats, and a carboxyl-terminal region. The central, high molecular mass region of α-tectorin containing the von Willebrand factor type D repeats has homology with sperm zonaadhesin. The two major non-collagenous proteins of the tectorial membrane are similar to the components of the sperm-egg adhesion system, and as such may interact in the same manner (Legan et al., 1997).

26.10. ADHESION COMPONENTS OF IMMUNE SYSTEM

26.10.1. Membrane Cofactor Protein (MCP)

Human membrane cofactor protein (MCP) is a complement regulatory protein, which serves as a cofactor for I-mediated inactivation of complement proteins C3b and C4b, and thus plays a major role in protection of host cells from complement. The MCP also acts as a measles virus receptor and is thus down-regulated by measles virus. Studies in guinea pigs suggested that MCP is expressed predominantly in the testis, whereas in pigs, monkeys and humans MCP is ubiquitously expressed. Murine MCP is developmentally expressed in the testis, paralleling the formation of spermatids (Tsujimura et al., 1998). The molecule consists of four short consensus repeats (SCR), a serine/threonine-rich domain (ST), 13 amino acids homology of unknown significance, a 23-amino acid transmembrane domain and a short cytoplasmic tail to yield a 45-70kDa type 1 glycoprotein. The SCR1, -2, and -4 are N-glyocosylated and the ST domain is O-glycosylated. The functions of MCP are mapped on different sets of SCRs. The protein expressed in CHO cells was 47-kDa, ~6 kDa larger than the murine testis MCP. It served as a cofactor for factor I-mediated inactivation complement protein C3b in a homologous system and, to a lesser extent, in a human system. The major message of murine MCP 1.5 kb was expressed predominantly in the testis. It was not detected in mice defective in spermatogenesis with immature germ cell. The murine MCP message is expressed selectively in spermatids during germ-cell differentiation (Tsujimura et al., 1998). Human sperm have abundant MCP in a 42-kDa molecule with no O-linked sugars. The MCP is localized in the inner acrosomal membrane of spermatozoa and becomes expressed on the surface of sperm through acrosome reaction. It has been proposed that the unusual structure and distribution of sperm MCP reflect it as a protein associated with fertilization. This hypothesis was reinforced by the findings that monoclonal antibodies against human MCP can partially block fusion of human sperm to hamster eggs, and that human oocytes may have a counter receptor. In some idiopathic male infertility syndromes, MCP may be aberrant in a sperm-specific manner. Human MCP has YXXL, RRKKK and FTSL sequences in its cytoplasmic tail. A cDNA encoding the murine homologue of MCP showed 45% identity in deduced protein sequence and 62% identity in nucleotide sequence with human MCP. The protective function of MCP of host cells from complement resides in the complement control protein repeats (CCPs), with CCPs 2-4 essential for regulation. Four isoforms of MCP arise via alternative splicing. On human spermatozoa, MCP is expressed on the inner acrosomal membrane, and the Abs to CCP1 inhibit sperm-egg interactions. The MCP on human sperm has a unique M(r) pattern. Human MCP expression in the tg mice mimics the human pattern in that it is located on the inner acrosomal membrane and has a faster M(r) than MCP expressed elsewhere. Human spermatozoa express MCP bearing cytoplasmic tail, which is also utilized in most other tissues and contains several signaling motifs (Riley et al., 2002a). Though, in somatic tissues, New World monkeys express an alternatively spliced form of MCP lacking CCP1, but sperm express MCP bearing CCP1. The germ cell-specific expression pattern of this domain in New World monkeys strongly suggests an evolutionarily conserved role for MCP in fertilization (Riley et al., 2002b).

MCP is an isoform of CD46: The MCP is a unique variant of CD46. The cDNA cloning and distribution analysis of murine and guinea-pig CD46 revealed the predominant expression of these rodent CD46 homologues in the testis, especially in mature testicular germ cells. The MCP does not support the established functions of human CD46 but supports the hypothesis that CD46 on sperm serves as a fertilization related adhesion molecule towards eggs. A unique CD46 cDNA (STc/CY4) from the human testis predicted amino acid sequence, which suggested the presence of an isoform of CD46. The testis, isoform possessed a short (11 amino acids) transmembrane section (TM) and an unidentified cytoplasmic tail (CY). When expressed in CHO cells, this CD46 isoform underwent no O-glycosylation and was mostly retained in the endoplasmic reticulum. This unusual behavior of the new isoform was due in part to the short TM and the unusual sequences of the CY. The Mr of this isoform was 42,000Da, which was 20,000Da smaller than conventional CD46. These properties of the STc/CY4 isoform were similar to those of sperm CD46. The only difference between sperm CD46 and the STc/CY4 isoform expressed in CHO cells was that only the latter possessed N-linked sugars of high mannose types. Studies on sperm-egg interaction infer that O-glycosylation perturbs CD46-mediated sperm-binding to eggs and thus sperm CD46 lacking O-linked sugars can serve as an adhesion molecule and its possible role in fertilization (Hara et al, 1998). The CD46 expression is regulated by 3' untranslated region in transgenic mice. While the expression of MCP in mice carrying MCP (BC2) cDNA with 125 bp of 3'untranslated region (3'UT) was minimal, in mice carrying MCP cDNA without total 3' UT was evident in many organs. The expression was regulated at the post-transcriptional stage. In transgenic mice also the first 125 bp down-regulated the expression of MCP molecules along with β-actin and SRα promoter. Also, this region down-regulated the expression of decay accelerating factor (DAF: CD55) molecules when it was inserted into cDNA of DAF. Moreover, the first 32 bp of the 3' UT revealed the same down regulation effect as 125 bp on MCP molecules. Thus the first 125 bp (and the first 32 bp in particular) of 3' UT regulate the expression of MCP molecules in transgenic mice (Miyagawa et al., 1997).

MCP and C3b: The complement component C3b and its receptors play significant role in sperm-oocyte interaction. Damaged or dead sperm activate the alternative pathway of complement and bind complement C3 catabolites. In one study, MCP on sperm that have undergone the acrosome reaction, specifically bind dimeric C3b and that human sperm acrosomal proteases released during the acrosome reaction directly cleaved C3, facilitating its binding to MCP. Human and hamster oocytes also activate the alternative pathway of complement and can also bind human C3 fragments. Specific complement-binding molecules may play a role in the attachment of C3 catabolites to oocytes. It was found that subsaturating concentration of dimeric C3b promoted penetration of hamster oocytes by human sperm, whereas saturating doses inhibited this process. In addition, antibodies to both MCP and C3 significantly inhibited penetration of hamster oocytes by human sperm. The study provided evidence that gamete-induced generation of C3 fragments, and their binding by selectively expressed receptors on sperm and oocytes may be an initial step in gamete interaction, leading to membrane fusion and fertilization (Anderson et al., 1993).

Complementary Binding Protein: The mRNAs of the complementary binding protein α chain (C4BPα) are expressed at significant levels in the guinea pig and mouse epididymis in an androgen-dependent manner. The epididymal and liver C4BPα mRNAs are generated from a single-copy gene while the epididymal C4BPα mRNAs are transcribed from a novel transcription start sites located 100-bp downstream from those used in the liver. The C4BP is localized in the stereocilia and Golgi apparatus of the epididymal epithelial cells and the surface of spermatozoa in the lumen. At cell level C4BP is localized preferentially on the head region of the spermatozoa, and on the plasma membrane and the outer acrosomal membrane. The epididymal C4BP is synthesized in a region-restricted manner and is taken up to the sperm membranes on passage through the epididymis (Nonaka et al., 2003).

26.10.2. Protectin (CD59)

Protectin (CD59) is a complement regulatory protein, which blocks the membrane attack complex during complement activation. The CD59 is present on the human sperm surface found on the whole plasma membrane including the head and tail of fresh ejaculated, capacitated and acrosome reacted sperm. This protein, which is also present on intraluminal germ cells, has a molecular weight of 20-kDa and comparable to that of CD59 expressed on peripheral blood cells. It is bound to the membrane through a glycophospholipid tail, which could be released after treatment with phosphatidylinositol-specific phospholipase C. In association with MCP (CD46) and decay accelerating factor (CD55) located in the acrosomal membranes, CD59 may participate to the protection of male gametes against complement-mediated damage as they travel through the female genital tract. Since CD59 is known as an adhesion molecule involved in lymphocyte rosettes, and since anti-protectin antibody H19 inhibited sperm binding and penetration rate, it is likely that it participates in cell-to-cell gamete interaction (Fenichel et al., 1994).

26.10.3. Other Components of Immune System in Germ Cells

C1q: C1q is a component of the classical pathway that can react with the Fc-fragment of immunoglobulins and with other proteins, such as fibronectin, laminin, and a specific C1q receptor present on several cell types. Given its role in many adhesion systems, mainly related to phagocytosis, the effects of C1q on the interaction between human spermatozoa and zona free hamster eggs resulted in promotion of sperm-oolemma adhesion and in inhibition of penetration. The addition of C1q to the medium resulted in sperm agglutination, which varied between sperm donors. The ability of C1q to promote sperm agglutination was shown to be dependent upon capacitation, suggesting the increased expression of C1q receptors during this process (Fusi et al., 1991; Bronson et al., 1998).

CD11b/CD18 Complex: The pathogenesis of antisperm antibody (ASA)-mediated infertility is related in part to complement (C)-dependent neutrophil mediated injury to sperm in the female genital tract. Sperm bound IgG activates human C and deposits C3 fragments on motile sperm. Furthermore IgG and C3-bound motile sperm adhered to human neutrophils in vitro. As a result, this adhesion potentiates the localized release of oxygen radicals at the contact site of neutrophil/sperm membranes. To identify the neutrophil surface receptor(s) involved in neutrophil/sperm adhesion and to evaluate their relevance to the pathogenesis of neutrophil-mediated immune injury to sperm, findings implicated the CD11b/CD18 glycoprotein complex (CR3) in the adhesive events involved in ASA-and C-mediated immune destruction of motile sperm by neutrophils (D'Cruz and Haas, 1995). In addition, CD15 antigen was also expressed on human sperm cells that had undergone acrosomal loss. The mAb to CD15 inhibited sperm binding and penetration of zona-free hamster eggs and penetration of human pellucida. Hence it appeared that sperm-egg interaction might be also mediated in part by the CD15 antigen (D' Cruz et al., 1997).

26.11. REFERENCES

Adham IM, Kim Y, Shamsadin R, Heinlein UA, Von Beust G, Mattei MG, Engel W. Molecular cloning, chromosomal localization, and expression analysis of CYRN1 and CYRN2, two human genes coding for cyritestin, a sperm protein involved in gamete interaction. DNA Cell Biol 1998; 17: 161-8.

Almeida EA, Huovila AP, Sutherland AE, Stephens LE, Calarco PG, Shaw LM, Mercurio AM, Sonnenberg A, Anderson DJ, Abbott AF, Jack RM. The role of complement component C3b and its receptors in sperm-oocyte interaction. Proc Natl Acad Sci (U S A) 1993; 90: 10051-5.

Andersson AM, Edvardsen K, Skakkebaek NE. Expression and localization of N- and E-cadherin in the human testis and epididymis. Int J Androl 1994; 17: 174-80.

Barker HL, Perry AC, Jones R, Hall L. Sequence and expression of a monkey testicular transcript encoding tMDC I, a novel member of the metalloproteinases-like, disintegrin-like, cysteine-rich (MDC) protein family. Biochim Biophys Acta 1994; 1218: 429-31.

Almeida EA, Huovila AP, Sutherland AE, Stephens LE, Calarco PG, Shaw LM, Mercurio AM, Sonnenberg A, Primakoff P, Myles DG, et al, Mouse egg integrin α6β1 functions as a sperm receptor. Cell 1995; 81: 1095-104

Bi M, Hickox JR, Winfrey VP, Olson GE, Hardy DM. Processing, localization and binding activity of zonadhesin suggest a function in sperm adhesion to the zona pellucida during exocytosis of the acrosome. Biochem J 2003; 375(Pt 2): 477-88.

Blobel CP. Functional processing of fertilin : evidence for critical role of proteolysis in sperm maturation and activation Rev Reprod 2000; 5: 75-83.

Bouchard MJ, Dong Y, McDermott BM Jr, Lam DH, Brown KR, Shelanski M, Bellve AR, Racaniello VR. Defects in nuclear and cytoskeletal morphology and mitochondrial localization in spermatozoa of mice lacking nectin-2, a component of cell-cell adherens junctions. Mol Cell Biol 2000; 20: 2865-73.
Bronson R, Bronson S, Oula L, Zhang W, Ghebrehiwet B. Detection of complement C1q receptors on human spermatozoa. J Reprod Immunol 1998; 38; 1-14.
Bronson RA, Fusi FM, Calzi F, Doldi N, Ferrari A. Evidence that a functional fertilin-like ADAM plays a role in human sperm-oolemmal interactions. Mol Hum Reprod 1999; 5: 433-40.
Byers SW, Sujarit S, Jegou B, Butz S, Hoschutzky H et al. Cadherins and cadherin associated molecules in the developing and maturing rat testis. Endocrinology 1994; 134: 630-39.
Cerretti DP, DuBose CF, Black RA, Nelsons N. Isolation of two novel metalloproteinases-Disintegrin (ADAM) cDNAs that show testis-specific gene expression. Biochem Biophys Res Commun 1999; 263; 810-15.
Chen H, Pyluck AL, Janik M, Sampson NS. Peptides corresponding to the epidermal growth factor-like domain of mouse fertilin: synthesis and biological activity. Biopolymers 1998; 47: 299-307.
Chen H, Sampson NS. Mediation of sperm-egg fusion: evidence that mouse egg α6β1 integrin is the receptor for sperm fertilin β. Chem Biol 1999a; 6: 1-10.
Chen MS, Almedia EA, Huovila AP, Takahashi Y, Shaw LM, Mercurio AM, White JM. Evidence that distinct states of the integrin α6β1 interact with laminin and an ADAM. J. Cell Biol 1999b; 144: 549-61.
Chen MS, Tung KS, Coonrod SA, Takahashi Y, Bigler D, Chang A, Yamashita Y, Kincade PW, Herr JC, White JM. Role of the integrin-associated protein CD9 in binding between sperm ADAM 2 and the egg integrin α6β1: implications for murine fertilization. Proc Natl Acad Sci (U S A) 1999c; 96: 11830-5.
Cho C, Bunch DO, Faure JE, Goulding EH, Eddy EM, Primakoff P, Myles DG. Fertilization defects in sperm from mice lacking fertilin β. Science 1998; 281(5384): 1857-9.
Cho C, Ge H, Branciforte D, Primakoff P, Myles DG. Analysis of mouse fertilin in wild type and fertilin β(-/-) sperm: evidence for C-terminal modification, α/β dimerization, and lack of essential role of fertilins α in sperm-egg fusion. Dev Biol 2000; 222: 289-95.
Cho C, Primakoff P, White JM, Myles DG. Chromosomal assignment of four testis-expressed mouse genes from a new family of transmembrane proteins (ADAMs) involved in cell-cell adhesion and fusion. Genomics 1996; 34: 413-7.
Chung SS, Mo MY, Silvestrini B, Lee WM, Cheng CY. Rat testicular N-cadherin: its complementary deoxyribonucleic acid cloning and regulation. Endocrinology 1998; 139: 1853-62.
Cornwall GA, Hsia N. ADAM7, a member of the ADAM (a disintegrin and metalloprotease) gene family is specifically expressed in the mouse anterior pituitary and epididymis. Endocrinology 1997; 138: 4262-72.
Cyr DG, Blaschuk OW, Robaire B. Identification and developmental regulations of cadherin messenger ribonucleic acids in the rat testis. Endocrinology 1992; 131: 139-45.
D'Cruz OJ, Haas GG Jr. Beta 2-integrin (CD11b/CD18) is the primary adhesive glycoprotein complex involved in neutrophil-mediated immune injury to human sperm. Biol Reprod 1995; 53: 1118-30.
D'Cruz OJ, lambert H, Haas GG Jr. Expression of CD15 (Lewis) antigen on human sperm and its role in sperm-egg interaction. Am J Reprod Immunol 1997; 37: 172-83.
De Nadai C, Fenichel P, Donzeau M, Epel D, Ciapa B. Characterization and role of integrins during gametic interaction and egg activation. Zygote 1996; 4: 31-40.
Eto K, Puzon-McLaughlin W, Sheppard D, Sehara-Fujisawa A, Zhang XP, Takada Y. RGD-independent binding of integrin α9β1 to the ADAM-12 and – 15 disintegrin domains mediates cell-cell interaction. J Biol Chem 2000; 275: 34922-30.
Evans JP, Kopf GS, Schultz RM. Characterization of the binding of recombinant mouse sperm fertilin β subunit to mouse eggs: evidence for adhesive activity via an egg β1 integrin-mediated interaction. Dev Biol 1997a; 187: 79-93.
Evans JP, Schultz RM, Kopf GS. Characterization of the binding recombinant mouse sperm fertilin α subunit to mouse eggs: evidence for function as a cell adhesion molecule in sperm-egg binding. Dev Biol 1997b; 187: 94-106.
Evans JP, Schultz RM, Kopf GS. Roles of the disintegrin domains of mouse fertilins α and β in fertilization. Biol Reprod 1998; 59: 145-52.
Evans JP. Fertilin β and other ADAMS as integrin ligands: insights into cell adhesion and fertilization. Bioessays 2001; 23: 628-39.
Fenichel P, Cervoni F, Hofmann P, Deckert M, Emiliozi C, His BL, Rossi B. Expression of the complement regulatory protein CD59 on human spermatozoa: characterization and role in gametic interaction. Mol Reprod Dev 1994; 38: 338-46.
Fenichel P, Durand – Clement M. Role of integrins during fertilization in mammals. Hum Reprod 1998; 120: 31-46.
Finaz C, Hammami-Hamza S. Adhesion proteins expressed on human gamete surfaces and egg activation.

Biol Cell 2000; 92: 235-44.
Forsbach A, Heinlein UA. Intratesticular distribution of cyritestin, a protein involved in gamete interaction. J Exp Biol 1998; 201 pt6: 861-7.
Frayne J, Hall L. The gene for the human tMDC 1 sperm surface protein is non-functional: implications for its proposed role in mammalian sperm-egg recognition. Biochem J 1998; 334 (PEI): 171-6.
Frayne J, Jury JA, Barker HL, Hall L. Rat MDC family of proteins: sequence analysis, tissue distribution, and expression in prepubertal and adult rat testis. Mol Reprod Dev 1997; 48: 159-67.
Frayne J, Jury JA, Barker HL, Perry AC, Jones R, Hall L. Macaque MDC family of proteins: sequence analysis, tissue distribution and processing in the male reproductive tract. Mol Hum Reprod 1998; 4: 429-37.
Fusi F, Bronson RA, Hong Y, Ghebrehiwet B. Complement component C1q and its receptor are involved in the interaction of human sperm with zona – free hamster eggs. Mol Reprod Dev 1991; 29: 180-8.
Fusi FM, Montesano M, Bernocchi N, Panzeri C, Ferrara F, Villa A, Bronson RA. P-selectin is expressed on the oolemma of human and hamster oocytes following sperm adhesion and is also detected on the equatorial region of acrosome-reacted human spermatozoa. Mol Hum Reprod 1996; 2: 341-7.
Fusi FM, Vignali M, Gailit J, Bronson RA. Mammalian oocytes exhibit specific recognition of the RGD (Arg-Gly-Asp) tripeptide and express oolemmal integrins. Mol Reprod Dev 1993; 36: 212-9.
gamete interaction. DNA Cell Biol 1998; 17: 161-8.
Gao Z, Garbers DL. Species diversity in the structure of zonadhesin, a sperm-specific membrane protein containing multiple cell adhesion molecule-like domains. J Biol Chem 1998; 273: 3415-21.
Geng JG, Raub TJ, Baker CA, Sawada GA, Ma L, Elhammer AP. Expression of a P-selectin ligand in zona pellucida of porcine oocytes and P-selectin on acrosomal membrane of porcine sperm cells. Potential implications for their involvement in sperm-egg interactions. J Cell Biol 1997; 137: 743-54.
Glander HJ, Schaller J, Rohwedder A, Henkel R. Adhesion molecules and matrix proteins on human spermatozoa. Andrologia 1998; 30: 289-96.
Goodwin LO, Karabinus DS, Perglizzi RG. Presence of N-cadherin transcripts in mature spermatozoa. Mol Hum Reprod 2000; 64: 487-97.
Gupta S, Li H, Sampson NS. Characterization of fertilin β-disintegrin binding specificity in sperm-egg adhesion. Bioorg Med Chem 2000; 8: 723-9.
Gupta SK, Alves K, Oniell PL, Mark GE, Hollis GF. Molecular cloning of the human fertilin β subunit. Biochem Biophys Res Commun 1996; 43: 318-24.
Gutierrez-Lopez MD, Ovalle S, Yanez-Mo M, Sanchez-Sanchez N, Rubinstein E, Olmo N, et al. A functionally relevant conformational epitope on the CD9 tetraspanin depends on the association with activated β1 integrin. J Biol Chem. 2003; 278: 208-18.
Hara T, Suzuki Y, Nakazawa T, Nishimura H, Nagasawa S, Nishiguchi M, Matsumoto M, Hatanaka M, Kitamura M, Seya T. Post-translational modification and intracellular localization of splice product of CD46 cloned from human testis: Role of the intracellular domains in O-glycosylation. Immunology 1998; 93: 546-55.
Hardy CM, Holland MK. Characterization of two highly conserved but non-allelic cellular disintegrins from rabbit testis. Gene 1998; 206: 127-35.
Hardy CM, Holland MK. Cloning and expression of recombinant rabbit fertilin. Mol Reprod Dev 1996; 45: 107-16.
Hardy DM, Garbers DL. A sperm membrane protein that binds in a species specific manner to the egg extracellular matrix is homologous to von Willebrand factor. J Biol Chem 1995; 270: 26025-8.
He ZY, Brakebusch C, Fassler R, Kreidberg JA, Primakoff P, Myles DG. None of the integrins known to be present on the mouse egg or to be ADAM receptors are essential for sperm-egg binding and fusion.Dev Biol 2003; 254: 226-37.
Hickox JR, Bi M, Hardy DM. Heterogeneous processing and zona pellucida binding activity of pig zonadhesin. J Biol Chem 2001; 276: 1502-9.
Hooft van Huijsduijnen R. ADAM 20 and 21: two novel human testis-specific membrane metalloproteases with similarity to fertilin-α. Gene 1998; 206: 273-82.
Hunnicutt GR, Koppel DE, Myles DG. Analysis of the process of localization of fertilin to the sperm posterior head plasma membrane domain during sperm maturation in the epididymis. Dev Biol 1997; 191: 146-59.
Hynes RO. Cell adhesion : Old and new questions. TIBS 1999; 24: M33-37.
Ji YZ. Wolf JP, Jouannet P, Bomsel M. Human gemete fusion can bypass β1 intergin requirement. Hum Reprod 1998; 13: 862-9.
Jury JA, Frayne J, Hall L. Sequence analysis of a variety of primate fertilin α genes: evidence for non-functional genes in the gorilla and man. Mol Reprod Dev 1998; 51: 92-7.
Jury JA, Perry AC, Hall L. Identification, sequence analysis and expression of transcripts encoding a putative metalloproteinase, eMDC II, in human and macaque epididymis. Mol Hum Reprod 1999; 5: 1127-34.

Kaji K, Oda S, Shikano T, Ohnuki T, et al. The gamete fusion process is defective in eggs of CD9-deficient mice. Nat Genet 2000; 24: 279-82.

Klentzeris LD, Fishel S, McDermott H, Dowell K, Hall J, Green S. A positive correlation between expression of β1 integrin cell adhesion molecules and fertilizing ability of human spermatozoa in vitro. Mol Hum Reprod 1995; 10: 728-36.

Le Naour F, Rubenstein E, Jasmin C, Prenant M, Boucheix C. Severely reduced female fertility in CD9-deficient mice. Science 2000;287: 319-21.

Legan PK, Rau A, Keen JN, Richardson GP. The mouse-tectorins. Modular matrix proteins of the inner ear homologous to components of the sperm-egg adhesion system. J Biol Chem 1997; 272: 8791-801.

Lin YC, Sun GH, Lee YM, Guo YW, Liu HW. Cloning and characterization of a complementary DNA encoding a human epididymis-associted disintegrin and metalloprotease 7 protein. Biol Reprod 2001; 65: 944-50.

Linder B, Bammer S, Heinlein UA. Delayed translation and posttranslational processing of cyritestin, an integral transmembrane protein of the mouse acrosome. Exp Cell Res 1995; 221: 66-72.

Linfor J, Berger T. Potential role of αv and β1 integrins as oocyte adhesion molecules during fertilization in pigs. J Reprod Fertil 2000; 120: 65-72.

Liu L, Smith JW. Identification of ADAM31: a protein expressed in Leydig cells and specialized epithelia. Endocrinology 2000; 141: 2033-42.

Lum L, Blobel CP. Evidence for distinct serine protease activities with a potential role in processing the sperm protein fertilin. Dev Biol 1997; 191: 131-45.

Mbikay M, Tadros H, Ishida N et al. Impaired fertility in mice deficient for the testicular germ-cell protease PC4. Proc Natl Acad Sci USA 1997; 94: 6842-46.

MacCalman CD, Blaschuk OW. Gonadal steroids regulate N-cadherin mRNA levels in the mouse testis. Endocr J 1994; 2: 157- 63.

Mbikay M, Tadros H, Ishida N et al. Impaired fertility in mice deficient for the testicular germ-cell protease PC4. Proc Natl Acad Sci USA 1997; 94: 6842-46.

McLaughlin EA, Frayne J, Barker HL, Jury JA, Jones R, Ford WC, Hall L. Cloning and sequence analysis of rat fertilin α and β developmental expression, processing and immunolocalization. Mol Hum Reprod 1997; 3: 801-9.

Miller BJ, Georges-Labouesse E, Primakoff P, Myles DG. Normal fertilization occurs with eggs lacking the integrin α6β1 and is CD9-depedent. J Cell Biol 2000; 149: 1289-95.

Miyado K, Yamada G, Yamada H, Hasuwa Y, Nakamura F, et al. Requirement of CD9 on the egg plasma membrane for fertilization. Science 2000; 287: 321-24.

Miyagawa S, Mikata S, Tanaka H, Ikawa M, Kominami K, Seya T, Nishimune Y, Shirakura R, Okabe M. The regulation of membrane cofactor protein (CD46) expression by the 3'untranslated region in transgenic mice. Biochem Biophys Res Commun 1997; 233: 829-33.

Mueller S, Rosenquist TA, Takai Y, Bronson RA, Wimmer E. Loss of nectin-2 at Sertoli-spermatid junctions leads to male infertility and correlates with severe spermatozoan head and midpiece malformation, impaired binding to the zona pellucida, and oocyte penetration. Biol Reprod 2003; 69:1330-40.

Munro SB, Blaschuk OW. A comprehensive survey of the cadherins expressed in the testes of fetal immature, and adult mice utilizing the polymerase chain reaction. Biol Reprod 1996; 55: 822-27.

Nishimura H, Cho C, Branciforte DR, Myles DG, Primakoff P. Analysis of loss of adhesive function in sperm lacking cyritestin or fertilin β. Dev Biol 2001; 233: 204-13.

Nonaka MI, Hishikawa Y, Moriyama N, Koji T, Ogata RT, Kudo A, Kawakami H, Nonaka M. Complement C4b-binding protein as a novel murine epididymal secretory protein. Biol Reprod. 2003; 69:1931-9.

Ozaki-Kuroda K, Nakanishi H, Ohta H, Tanaka H, Kurihara H, Mueller S, Irie K, Ikeda W, Sakai T, Wimmer E, Nishimune Y, Takai Y. Nectin couples cell-cell adhesion and the actin scaffold at heterotypic testicular junctions. Curr Biol 2002; 12: 1145-50.

Perry AC, Barker HL, Jones R, Hall L. Genetic evidence for an additional member of the metalloproteinase-like, disintegrin-like, cysteine-rich (MDC) family of mammalian proteins and its abundant expression in the testis. Biochim Biophys Acta 1994: 1207: 134-7.

Perry AC, Jones R, Hall L. Analysis of transcripts encoding novel members of the mammalian metalloprotease-like, disintegrin-like, cysteine-rich (MDC) protein family and their expression in reproductive and non-reproductive monkey tissues. Biochem J 1995; 312: 239-44.

Perryman KJ, Stanton PG, Loveland KL, McLachlan RI, Robertson DM. Hormonal dependency of neural cadherin in the binding of round spermatids to Sertoli cells in vitro. Endocrinology 1996; 137: 3877-83.

Podbilewicz B. ADM-1, a protein with metalloprotease-and disintegrin-like domains, is expressed in syncytial organs, sperm, and sheath cells of sensory organs in *Caenorhabditis elegans*. Mol Biol Cell 1996; 7: 1877-93.

Poindexter K, Nelson N, DuBose RF, Black RA, Cerretti DP. The identification of seven metalloproteinase-disintegrin (ADAM) genes from genomic libraries. Gene 1999; 237; 61-70.

Primakoff P, Myles DG. The adam gene family. Surface proteins with adhesion and protease activity. TIG 2000; 16: 83-87.
Quill TA, Garbers DL. Sperad is a novel sperm-specific plasma membrane protein homologous to a family of cell adhesion proteins. J Biol Chem 1996; 271: 33509-14.
Reddy KV, Meherji PK, Shahani SK. Integrin cell adhesion molecules on human spermatozoa. Indian J Exp Biol 1998; 136: 456-63.
Riley RC, Kemper C, Leung M, Atkinson JP. Characterization of human membrane cofactor protein (MCP; CD46) on spermatozoa. Mol Reprod Dev. 2002a; 62: 534-46.
Riley RC, Tannenbaum PL, Abbott DH, Atkinson JP. Cutting edge: inhibiting measles virus infection but promoting reproduction: an explanation for splicing and tissue-specific expression of CD46. J Immunol. 2002b; 169: 5405-9.
Rosselot C, Kierszenbaum AL, Rivkin E, Tres LL. Chronological gene expression of ADAMs during testicular development: prespermatogonia (gonocytes) express fertilin beta (ADAM2). Dev Dyn. 2003; 227: 458-67.
Shamsadin R, Adham IM, Nayernia K, Heinlein UA, Oberwinkler H and Engel W. Male mice deficient for germ-cell cyritestin are infertile. Biol Repord. 1999; 61: 1445-51.
Shilling FM, Kratzschmar J, Cai H, Weskamp G, Gayko U, Leibow J, Myles DG, Nuccitelli R, Blobel CP. Identification of metalloprotease/disintegrins in Xenopus laevis testis with a potential role in fertilization. Dev Biol 1997; 186; 155-64.
Shrimali RK, Reddy KVR. Integrins and disintegrins: The candidate molecular players in sperm-egg interaction. Ind J Expt Biol 2000; 38: 415-24.
Takahashi Y, Bigler D, Ito Y, White JM. Sequence-specific interaction between the disintegrin domain of mouse ADAM3 and murine eggs: role of β1 integrin-associated proteins CD9, CD81, and CD98. Mol Biol Cell 2001; 12: 809-20.
Trubner M, Glander HJ, Schaller J. Localization of adhesion molecules on human spermatozoa by fluorescence microscopy. Andrologia 1997; 29: 253-60.
Tsujimura A, Shida K, Kitamura M, Nomura M, Takeda J, et al. Molecular cloning of murine homologue of membrane cofactor proteins (CD46): preferential expression in testicular germ cells. Biochem J 1998; 330 (pt1) 163-8.
Vidaeus CM, von Kapp-Herr C, Golden WL. Eddy RL, Shows TB, Herr JC. Human fertilin β : identification, characterization, and chromosomal mapping of an ADAM gene family member. Mol Reprod Dev 1997; 46: 363-9.
Wakayama T, Koami H, Ariga H, Kobayashi D, Sai Y, Tsuji A, Yamamoto M, Iseki S. Expression and functional characterization of the adhesion molecule spermatogenic immunoglobulin superfamily in the mouse testis. Biol Reprod. 2003; 68: 1755-63.
Waters SI, White JM. Biochemical and molecular characterization of bovine fertilin α and β (ADAM1 and ADAM2): A candidate sperm-egg binding/fusion complex. Biol Reprod. 1997; 56: 1245-54.
Wolfsberg TG, Straight PD, Gerena RL, Huovila AP, Primakoff P, Myles DG, White JM. ADAM, a widely distributed and developmentally regulated gene family encoding membrane proteins with a disintegrin and metalloprotease domain. Dev Biol 1995; 169: 378-83.
Wong GE, Zhu X, Prater CE, Oh E, Evans JP. Analysis of fertilin α (ADAM1) mediated sperm-egg cell adhesion during fertilization and identification of an adhesion-mediating sequence in the disintegrin-like domain. J Biol Chem 2001; 276: 24937-45.
Xu R, Cai J, Xu T, Zhou W, Ying B, Deng K, Zhao S, Li C. Molecular cloning and mapping of a novel ADAM gene (ADAM 29) to human chromosome 4. Genomics 1999; 62: 537-9.
Yoshida S, Shibata M, Yamamoto S, Hagihara M, Asai N, Takahashi M, Mizutani S, Muramatsu T, and Kenji Kadomatsu. Homo-oligomer formation by basigin, an immunoglobulin superfamily member, via its N-terminal immunoglobulin domain. J Biochem. 2000; 267: 4372-80.
Yuan R, Primakoff sP, Myles DG. A role for the disintegrin domain of cyritestin, a sperm surface protein belonging to the ADAM family, in mouse sperm-egg plasma membrane adhesion and fusion. J Cell Biol 1997; 137: 105-12.
Zhu GZ, Lin Y, Myles DG, Primakoff P. Identification of four novel ADAMs with potential roles in spermatogenesis and fertilization. Gene 1999; 234: 227-37.
Zhu GZ, Myles DG, Primakoff P. Testase 1(ADAM24) a plasma membrane-anchored sperm protease implicated in sperm function during epididymal maturation or fertilization. J Cell Sci 2001; 114: 1787-94.
Zhu X, Bansal NP, Evans JP. Identification of key functional amino acids of the mouse fertilin β (ADAM2) disintegrin loop for cell-cell adhesion during fertilization. J Biol Chem 2000; 275: 7677-83.
Zhu X, Evans JP. Analysis of the roles of RGD-binding integrins, α(4)/α(9) Integrins, α(6) integrins, and CD9 in the interaction of the Fertilin β (ADAM2) Disintegrin domain with the mouse egg membrane. Biol Reprod 2002; 66: 1193-202.
Ziv S, Rufas O, Shalgi R. Cadherins expression during gamete maturation and fertilization in the rat. Mol Reprod Dev. 2002; 62: 547-56.

Chapter 27

METALLOPROTEASES AND METALLOPROTEASE INHIBITORS

27.1. MATRIX METALLOPROTEASES

The extra-cellular matrix (ECM) or its components regulate Sertoli cell differentiation, testicular cord formation, germ cell development and the rate and the pattern of Sertoli cell migration. Coordinated activity of proteolytic enzymes and endogenous inhibitors dictates the rate and extent of ECM turnover. Proteinases such as the plasminogen activator and matrix metalloproteinases (MMP) are believed to play a role in these developmental changes. Furthermore these enzymes are produced by testicular cells in response to hormones demonstrating coordinated interactions of proteolytic enzyme expression with cell and tissue development in the testis. The family of tissue inhibitors of metalloproteinases (TIMPs) represents one of the major groups of extra-cellular matrix-degrading proteases and comprises of the collagenases, stromelysins, and gelatinases. The coordinate regulation of TIMPs and MMPs has been documented in many cell types in response to variety of stimuli. The loss of coordinated TIMP and MMP regulation resulting in an imbalance in favour of proteolysis has been proposed to underlie pathological conditions characterized by uncontrolled ECM degradation. The presence of proteases in the epididymis requires the presence of their inhibitors to maintain homeostasis as a part of regulatory mechanism for regional maturation and processing of spermatozoa.

27.1.1. MMP-1, MMP-2, and MMP-9

The production and localization of MMP and tissue inhibitors of metalloproteinases (TIMP) in the developing human gonad during mid-gestation, using zymographic techniques, indicated the presence of MMP-2, MMP-9 and all four TIMPs in both testis and ovary, with the predominant gelatinase produced by both the organs. In the testis, MMP-1, MMP-2, MMP-9 and all TIMP family members are localized to the interstitium and to varying degrees within the tubules. The MMP-9 and TIMP-4 were abundant in both Sertoli cells and gonocytes, where as MMP-1 and TIMP-1 were localized in particular to Sertoli cells (Robinson et al., 2001). Membrane-type 1 (MT1)-MMP, physiological activator of proMMP-2 under TIMP–2 control, is present within the testis together with MMP-2 and TIMP-2. In the prepuberal testis MT1-MMP immune-reactivity was uniformly distributed, whereas in the adult it was confined to the apical compartment of the tubules, where meiosis and spermiogenesis occur. The two cell lineages (somatic and germinal) express MT1-MMP and TIMP-1, while MMP-2 was of somatic origin. Follicle-stimulating hormone enhanced the expression of MMP-2 and TIMP-2 but not of MT1-

```
  1 MVAGVSLLLR ALPLLLWGCQ DAQPTQHGLP ELRQEAEAFL EKYGYLSEQG SKAPASAQFR
 61 NAIREFQWIS QLPLSGVLDQ ATLRQMTRPR CGVADTDSHA TWTERISTLL AGHRAKMRRK
121 KRFAKPGNKW YKQHLSYRLV NWPERLPEPA VRGAVRAAFQ LWSNVSALEF WEAPATGPAD
181 IRLTFFQGDH NDGLANAFDG PGGALAHAFL PRRGEAHFDG DERWSLSRRR GRNLFVVLAH
241 EIGHTLGLTH SPAPRALMAP YYKKLGRDAL LSWDDVLAVQ SLYGKPLGRS VATQLPGKVF
301 TDFEAWDPHN SQSRRPETRG PKYCHSSFDA ITVGSYFWEV TVDGNVSEPR PLQKRWPGLP
361 PGIEAAAVSL EDGDFYFFKG NRCWRFQGTK SVWGFAQLCR AGGLPRHPDA ALFFPPLRRL
421 VLFKGSRYYV LAQGGMQVEP YYPRSLRDWA GVPEEVSGAL PRPDGSIIFF RDDHYWHLDQ
481 AKLRVTSSGR WATELSWMGC WNANSGGALF
```

Fig.27.1. Amino acid sequence of mouse matrix metalloproteinase 28 (epilysin). Source: http//www.ncbi.nlm.nih.gov. (Accession NP_536701).

MMP, and promoted the activation of proMMP-2. In co-cultures, a tremendous elevation and activation of MMP-2 was observed, which might relate to the processed MT1-MMP form solely detected in germ cells. It indicates that MMP-2 synthesis and activation are under local (germ cells) and hormonal control. The FSH regulation emphasizes the importance of MMPs in testicular physiology (Longin et al, 2001).

27.1.2. Epilysin (MMP-28)

Epilysin (MMP-28) is a recently cloned member of the matrix metalloproteinase family (Lohi et al., 2001). It is expressed at highest levels in the skin by basal and supra-basal keratinocytes and in testis by developing germ cells. A 3.0-kb fragment of human genomic DNA containing 5'-flanking sequence of the epilysin gene, and a corresponding 660bp fragment is present in mouse testes. The 5'-flanking sequences contained no typical TATA-box or CCAAT sequences close to the translation initiation sites. RNAse protection assay showed that two transcription start sites are situated 210 and 230-bp upstream from the translation start site in human epilysin gene. The promoter contains a GT-box, situated 300-bp upstream from the translation start site, with homology to the consensus binding site for transcription factors of the Sp family. This site is conserved between the human and mouse promoters. Reporter gene assays indicated that deletion or mutation of the GT-box dramatically reduced the transcriptional activity both in keratinocytes and in spermatogonia. Several nuclear proteins bind specifically to this sequence. Sp1 and Sp3 are components of protein/DNA complexes as possible regulators of the epilysin gene. This indicated that epilysin promoter have distinctive structural and functional features (Illman et al., 2001)**(Fig.27.1)**.

27.1.3. Collagenase IV Metalloproteinases

Testicular peri-tubular myoid cells secrete a variety of metalloproteinases when maintained in culture in a defined medium. The predominant metalloproteinases secreted were identified as latent type IV procollagenases having molecular masses of 72-kDa and 75-kDa. The metalloproteinase of 72-kDa present in rat Sertoli cell and Sertoli-peritubular cell co-culture medium is specifically activated by p-amino phenylmercuric acetate (p-APMA) to a lower molecular mass form, indicating that it is likely to be a latent collagenase. Since, Sertoli and peri-tubular cells in co-culture do produce type IV pro-collagenase, study suggested that the secretion and activation of this enzyme and other metalloproteases is regulated by the ratio of mesenchymal cells to epithelial cells and time in culture. The production of collagenase is influenced by RGD containing peptides. Laminin stimulated Sertoli cell type IV collagenase mRNA levels. However, three peptides corresponding to different domains of the laminin molecule (CSRAKQAASIKVASADR, FALRGDNP, CLQDGDVRV) did not influence type IV collagenase mRNA. Neither soluble laminin nor any of the peptides influenced 72-kDa type IV collagenase protein levels, where as peptide FALRGDNP resulted in a selective increase in two high molecular weight metalloproteinases (83-kDa and 110-kDa) and in the activation of the 72-kDa rat type IV collagenase. Collagenase activation is not influenced by interleukin-1, phorbol ester, testosterone, and FSH. The addition of soluble laminin resulted in a redistribution of type IV collagenase from intracellular vesicles to the cell-substrate regions. Plasma membrane preparations demonstrated triton-soluble gelatinases of 76-kDa, 83-kDa, and 110-kDa and a triton-insoluble gelatinase of 225-kDa. Selective studies indicated that testicular cell type IV collagenase mRNA levels, enzyme activation, and distribution are influenced by laminin and RGD-containing peptides (Sang et al., 1990; 1991).

Peritubular cells are stimulated by dibutyryl cyclic AMP, forskolin or cholera toxin to secrete increased amounts of type IV procollagenase, though type IV collagenase having lower molecular mass of 66-kDa was not stimulated under these conditions. The cytochalasin D acts on mesenchymal-type peritubular cells, but not on epithelial type Sertoli cells, to enhance the conversion of latent type IV procollagenase into active type IV collagenase. It seems likely that conversion of IV procollagenase into type IV collagenase by peritubular cells is inhibited by factors secreted by Sertoli cells. It shows that the interactions between Sertoli cells and peritubular cells modulate net proteinase activities in discrete regions of the testis (Ailenberg et al., 1991).

27.1.4. Other Metalloproteinaes

MMP-18: Another predicted protein, which displays all the structural features characteristic of the MMP family and has closest identity with MMP-1, -3, -10, and 11, is designated as MMP-18. The MMP-18 mRNA is expressed in a wide variety of normal human tissues, including mammary gland, placenta, lung, pancreas, ovary, small intestine, spleen, thymus, prostate, testis, colon, and heart, but not in brain, skeletal muscle, kidney, liver, or peripheral blood leucocytes (Cossins et al, 1996).

MMP-23: The MMP-23 is predominantly expressed in ovary, testis, and prostate, suggesting that this MMP may play a specialized role in reproductive processes (Velasco et al, 1999).The MMP-23 cloned from an ovary cDNA library exhibits sequence similarity with other MMPs but displays a different domain structure. In this respect MMP-23 lacks a recognizable signal sequence and has a short prodomain, although it contains a single cysteine residue that can be part of the cysteine-switch mechanism operating for maintaining enzyme latency. The C-terminal domain is considerably shortened and shows no sequence similarity to hemopexin, whereas all human MMPs with the exception of natrilysin contain four hemopexin-like repeats. Furthermore, MMP-23 is devoid of structural features distinctive of the diverse MMP subclasses, including the specific residues located close to the zinc-binding site in collagenase, the transmembrane domain of membrane-type MMPs, or the fibronectin-like domain of gelatinases. The human MMP-23 gene maps to 1p36, a location which differs from location of other MMP genes (Velasco et al, 1999).

Insulin Degrading Enzyme: The expression of insulin degrading enzyme (IDE) gene in testis is correlated with sexual maturation. The IDE from rat as well as homologues from human and *Drosophila* contain the carboxyl-terminal consensus sequence A/S-K-L, for peroxisome targeting. A stretch of 43-bp surrounding an alternatively used polyadenylation site is highly conserved between rat and human, suggesting that it may contain important regulatory information. Normally the cDNA produces two IDE transcripts of 3.7 and 5.5-kb in various tissues, where as testis produces three different RNAs of entirely different sizes (3.7-, 4.1- and 6.1-kb) at relatively high abundance (Baumeister et al., 1993).

27.2. METALLOPEPTIDASES

An emerging family of metalloproteases/metallopeptidases has recently been shown to be involved in the maturation and inactivation of various neuropeptides. Because of their roles in controlling the activity of several bio-active peptides, members of the neprilysin family of zinc metallopeptidases have been identified as putative targets for the design of therapeutic agents. It is classified as belonging to clan MA whose most representative member is the bacterial metalloprotease thermolysin, and forms the rapidly growing M13 subfamily. At present it comprises six zinc-dependent metalloproteases, i.e. neutral endo-peptidase (NEP, EC 3.4.24.11) (neprilysin), endothelin-converting enzyme-1 (ECE-1, EC 3.4.24.71), ECE-2, Kell, blood group proteins, PHEX (product of the phosphate-regulating gene with homologies to endopeptidase on X chromosome), and X-converting enzyme (XCE). Neprilysin is widely distributed in brain and many peripheral tissues, particularly in brush-border membranes of the kidney and intestine. Members of the neutral end peptidase (NEP, also known as MME for membrane metallo-endopeptidase) family play significant roles in pain perception, arterial pressure regulation, phosphate metabolism, and homeostasis. The NEP appears to be responsible for the inactivation of enkephalins, substance P, and neurotensin, and possibly of atrial natriuretic peptides. The ECE-1, which is responsible for the production of the vasoconstrictive peptide endothelin by the cleavage of big-endothelin, is coded by a single gene, which is alternatively spliced to yield four isoforms with distinct subcellular localizations. The ECE-2 also cleaves big-endothelin, albeit at a lower pH, probably within secretion granules. The last three members of the family are still "orphan enzymes", i.e. without an established function. These metalloproteases share structural characteristics beyond HEXXH zinc-binding consensus sequence characteristic of the M13 family. They are highly glycosylated type II integral membrane proteins with a cluster of four conserved cysteine residues following the transmembrane domain.

27.2.1. Neutral End Peptidases (Neprilysins)

The NEPs or MMEs (neprilysins) are involved in perception of pain, arterial pressure regulation and phosphate metabolism homeostasis. Based on conserved sequences of neprilysin ECE-1, and PHEX, Ghaddar et al (2000) reported the cloning from mouse testis of a neprilysin-like peptidase called NL1. NL1 is a glycoprotein that shows strongest sequence identity with neprilysin. In constrast with neprilysin and other members of the family NL1 was secreted when expressed in cultured mammalian cells, possibly due to cleavage at a furin-like site located 22 amino acid residues in the C-terminus of the transmembrane domain. The recombinant enzyme was inhibited by phosphoramidon and thiorphan, two inhibitors of neprilysin. The NL1 mRNA was found predominantly in testis, specifically in round and elongated spermatids, suggesting its role in sperm formation or other processes related to fertility. Using a similar approach, Tanja et al., (2000) identified another member of this family in rat tissues. It is a glycosylated, type II integral membrane protein of 774 amino acids, containing a zinc-binding consensus motif, highly homologous to NEP and, therefore, designated NEPII. Multiple splice variants of NEPII mRNA with distinct expression patterns in brain regions, pituitary and testis have been characterized **(Fig.27.2)**. In situ hybridization of testis, where levels of the NEPII gene transcript are the highest, reveals localization within round spermatids.

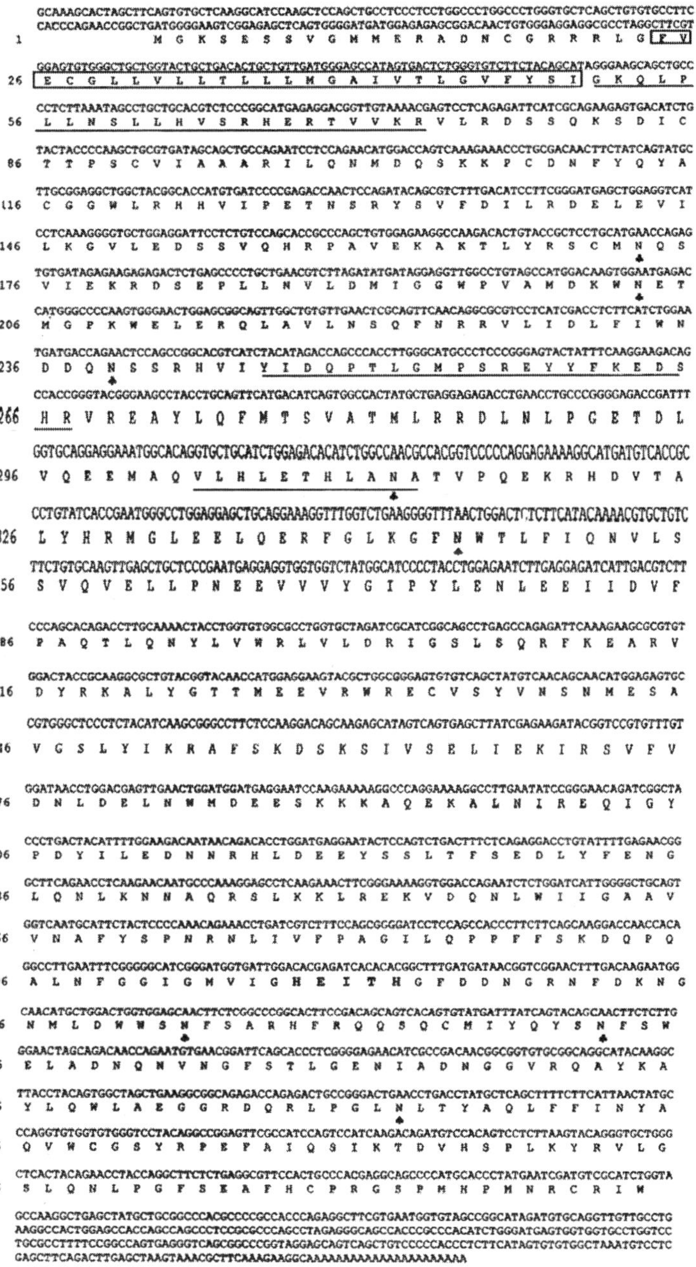

Fig.27.2. Nucleotide and deduced amino acid sequence of cDNA of neprilysin II present in testis and brain. Transmembrane domain is boxed. N-glycosylation sites are marked (♣). Bold letters show zinc binding domain. Reprinted with permission from O. Tanja et al. Biochem Biophy Res Commun 271, 565-70: 2000 © Elsevier.

Bonvouloir et al, (2001) reported the cloning of a new human NEP named for membrane metallo-endopeptidase-like-2 (MMEL-2). The MMEL2 protein has the structural characteristics of type II transmembrane proteins, although the presence of a furin-like cleavage site in the ectodomain suggests that it may be released into the medium following proteolytic cleavage. The MMEL2 protein contains the zinc-binding consensus sequence HEXXH and all the residues known to be essential for the enzymatic activity of other members of the family. The MMEL2 mRNA was detected predominantly in testis, but weak expression also was observed in brain, kidney, and heart. The human MMEL2 gene was mapped to 1p36 by in situ hybridization (Bonvouloir et al., 2001).

27.2.2. Endopeptidase 24.15

Endopeptidase-24.15 (E.C.3.4.24.15) (EP-24.15) cleaves several substrates found in the hypothalamic/pituitary/gonadal axis, including gonadotropin-releasing (GnRH) and the opioid peptides of the dynorphin family. Testicular specific activity of EP-24.15 increased linearly with age, while ovarian EP-24.15 activity increased immediately prior to puberty. Endopeptidase-24.15 with an Mr of about 70,000 from rat testes cleaves preferentially bonds on the carboxyl side of hydrophobic amino acids. Secondary enzyme-substrate interactions at sites removed from the scissile bond are indicated by the finding that a hydrophobic or bulky residue in the P3' position greatly contributes to substrate binding and catalytic efficiency. The enzyme activity is inhibited by metal chelators and by thiols but restored by low concentrations of Zn^{2+} and Co^{2+} ions. Biologically active peptides such as LHRH, bradykinin and neurotensin are cleaved at sites consistent with the specificity of the enzyme. The testis enzyme is catalytically and immunologically closely related to brain enzyme (Orlowsky et al., 1989). The recombinant rat testes EP-24.15 and the rabbit brain endooligopeptidase A showed similarity in physico-chemical features. Both enzymes are activated by dithiothreitol and irreversibly inhibited by a SH-affinity labeling dynorphin-related compound, but not inhibited by a SH-affinity labeling dynorphin-related compound or by EDTA. It is unlikely that the recombinant rat testes EP-24.15 and the rabbit brain endo-oligopeptidase A are the same molecule although they might belong to the same family of oligopeptidases (Hayashi et al., 1996).

A metallo-endopeptidase that selectively cleaves doublets of basic amino acids on the amino-terminal side of arginine residues is present in rat testes. It is present in two catalytically active forms with apparent Mr of 110,000 and 140,000-Da, respectively, and inhibited by metal chelators and divalent cations. Its activity lost by EDTA was restored by low concentrations of Zn^{2+} and Mn^{2+}. This endopeptidase is inhibited by amastatin, an amino peptidase inhibitor. A substrate specificity study indicated that cleavage occurred selectively at the amino-terminal side of an arginine residue, independent of the nature of the basic doublet. The enzyme produced such a cleavage at the Arg-Lys doublet of somatostatin 28, at the Arg-Arg doublet of dynorphin A and atrial natriuretic factor, and at the Lys-Arg doublet of preproneurotensin- (154-170). Moreover, cleavage efficiency was found to be higher for the larger substrates (Chesneau et al., 1994).

27.2.3. Endothelin-1 Converting Enzyme

The potent smooth muscle agonist endothelin-1 (ET-1) is involved in the local control of seminiferous tubule contractility, which results in the forward propulsion of tubular fluid and spermatozoa, through its action on peritubular myoid cells. The ET-1, known to be produced in the seminiferous epithelium by Sertoli cells, is derived from the inactive intermediate big endothelin-1 (big ET-1) through a specific cleavage operated by the endothelin-converting enzyme (ECE), a membrane-bound metalloprotease with ectoenzymatic activity. Endothelin–1 and its receptors are present in the human testis and epididymis. The ET-1 mRNA and protein are readily detectable in the epithelial compartment of the human epididymis, where as ET-converting enzyme-1, that converts the precursor pro-ET-1 into the active peptide ET-1, is expressed in the epididymis thus indicating an active processing of the pro-hormone. Thus ET-1 can be responsible for sperm progression through this organ, showing a paracrine mode of action (Peri et al., 1997). The expression of ECE by Sertoli cells and myoid cell cultures showed that ECE-1 (and not ECE-2) is specifically expressed in Sertoli cells. Sertoli cell monolayers were capable of cleaving big ET-1, an activity inhibited by the ECE inhibitor phosphoramidon. Myoid cells did not respond to big endothelin, nor to Sertoli cell plain medium, but to the medium conditioned by Sertoli cells in the presence of big ET-1. In situ hybridization analysis showed regional differences in ECE expression, indicating that pulsatile production of endothelin by Sertoli cells may regulate the cyclicity of tubular contraction. This suggested that endothelin-mediated spatio-temporal control of rhythmic tubular contractility might be operated by Sertoli cells through the cyclic expression of ECE-1, which is dependent upon the timing of spermatogenesis (Tripiciano et al., 1999).

27.2.4. NRD Convertase (Nardilysin)

Activation of pro-proteins during their transit through the secretory pathway requires limited proteolysis at specific sites that most commonly comprise basic residues organized as singlets or doublets. Seven mammalian pro-protein convertases involved in this type of post-translational processing have been characterized. These enzymes cleave the C-terminus of basic amino acids and the exposed basic residues are subsequently removed by carboxy-peptidase E or carboxypeptidase D. However, NRD covertases cleave various peptide substrates such as somatostatin 28, dynorphin A and atrial natriuretic factor at the N-terminus of Arg(R) residues in dibasic sites and was thus named NRD convertase (Nardilysin, EC 3.4.24.61). N-Arg dibasic convertase (NRDc) is a metallo-endopeptidase from rat brain cortex and testis and cleaves peptide substrates on the N-terminus of Arg residues in dibasic stretches. The search for the enzyme (s) responsible for the processing of somatostatin-14 led Cohen and associates to purify and characterize a Zn^{2+}- metallo-endopeptidase and to clone its cDNA. Initially isolated from rat brain cortex, the enzyme was purified and cloned from testis, where it is expressed abundantly in germ cells. The predicted 1161-residues rat NRD convertase of 133-kDa is characterized by the presence of a putative signal peptide, a highly acidic stretch of 71 residues (79% Glu and Asp) and a zinc binding motif, HXXEH. The presence of this motif, together with an overall 35% similarity to *E-coli* protease III pitrilysin (EC 3,4,25,55) and 48% similarity to rat or human insulinase (Insulysin, EC 3.4.24.56) suggested that NRD convertase is a member of the insulinase family metalloendoproteases implicated in the limited proteolysis of proteins or peptides, such as the mitochondrial matrix-processing protease. In addition a Cys residue that may be responsible for thiol sensitivity of the insulinase and N-Arg dibasic convertase was proposed. It has been hypothesized that this metallo-endopeptidase may be a member of a distinct class of processing enzymes (Pierotti et al., 1994). Although the substrates in vivo of NRD convertase have not been yet identified, its cleavage specificity and its membership of the insulinase family suggest its involvement in pro-protein processing events. Interestingly, the acidic stretch, unique to NRD convertase, is inserted 35 residues upstream of the Zn^{2+}-binding motif, within the most conserved region of the insulinase family members. The NRD convertase expression is restricted to germ cells and is maximal in elongated spermatids. In this

```
  1 MRPSQTALWL GLVLSLALLA VGWASARPPI YVSSWAVRVT KGYQEAERLA RKFGFVNLGQ
 61 IFPDDQYFHL RHRGVAQQSL TPHWGHRLRL KKEPKVRWFE QQTLRRRVKR SLVVPTDPWF
121 SKQWYMNKEI EQDLNILKVW NQGLTGRGVV VSILDDGIEK DHPDLWANYD PLASYDFNDY
181 DPDPQPRYTP NDENRHGTRC AGEVSATANN GFCGAGVAFN ARIGGVRMLD GAITDIVEAQ
241 SLSLQPQHIH IYSASWGPED DGRTVDGPGL LTQEAFRRGV TKGRQGLGTL FIWASGNGGL
301 HYDNCNCDGY TNSIHTLSVG STTRQGRVPW YSEACASTFT TTFSSGVVTD PQIVTTDLHH
361 QCTDKHTGTS ASAPLAAGMI ALALEANPLL TWRDLQHLVV RASRPAQLQA EDWRINGVGR
421 QVSHHYGYGL LDAGLLVDLA RVWLPTKPQK KCTIRVVHTP TPILPRMLVP KNVTVCCDGS
481 RRRLIRSLEH VQVQLSLSYS RRGDLEIFLT SPMGTRSTLV AIRPLDISGQ GYNNWIFMST
541 HYWDEDPQGL WTLGLENKGY YYNTGTLYYC TLLLYGTAED MTARPQTPQV TSCAHACAEG
601 HRGAVPGKSL SPLHCGRTLP HLQQAVVVAL QPHTAASDQG TGQLSPSYHT CSAA
```

Fig.27.3. Amino acid sequence of rat testicular proprotein convertase (PC4). Source: http//www.ncbi.nlm.nih.gov (Accession NP_598243).

cell type the enzyme is present in the cytoplasm, where it is strongly associated with two microtubular structures: the manchette and the flagellum. In addition, NRD convertase is still present in the flagella of mature spermatozoa. These results suggested its possible participation in the processing events associated with morphological transformation of the spermatid and/or in spermatozoon motility (Chesneau et al., 1996; Hospital et al., 1997). Fractionation studies argue in favour of a primary cytosolic localization of both peptidases, whereas biotinylation experiments showed the presence of both isoforms at the cell surface. Thus NRD convertase may also fulfill a similar function in the cytoplasm and/or at the cell surface (Hospital et al., 2000).

Rat testis NRD convertase is a Zn^{2+} dependent endopeptidase of M16 family. The cloning of two human testis cDNA species encoding NRD convertase isoforms, hNRD1 and hNRD2 has been reported (Hospital et al, 1997). Whereas the hNRD1 transcript (3.7-kb) is equivalent to the previously characterized rat cDNA (rNRD1), hNRD2 and rNRD2 are 3.9-kb forms containing a nucleotide insertion encoding a 68-residue segment. This motif, which is inserted at N-terminus of the Zn^{2+} binding site, HXXEH, is contained within the most conserved region among the insulinase family members. Sequencing revealed 92% identity between rat and human orthologues. The human gene was localized to chromosome 1p32.1-p32.2. Whereas NRD convertase is mostly expressed in testis and in 24 cell lines, low mRNA levels were detected in most of the 27 other tissues examined (Hospital et al., 1997) **(Fig.27.3)**. The NRD1 and NRD2 convertase isoforms differ by the absence (isoform 1) or presence (isoform 2) of a 68-amino acid insertion close to the active site. Both isoforms over-expressed either by vacccinia virus infection of BSC40 cells or transfection of COS-7 cells exhibited similar biochemical properties with N-terminally processed NRD convertase activity. The rat and human promoters of NRD convertase are highly conserved, containing a number of motifs, which may correspond to transcription-factor binding sites. Functionality of the rat promoter was observed with 5' deletions to 411-bp upstream of the transcriptional start site in spermatid, prostate and pituitary cell lines. Further deletion to 10-bp caused a complete loss of activity in spermatid and prostate lines. By contrast, GH3 pituitary cells displayed no reduction in promoter activity. The region 411-101-bp contained a number of transcription-factor binding sites, although differences in binding between the cell lines were not apparent (Winter and Pierotti, 2000).

Polyamine binding site: Mouse arginine-specific dibasic cleaving enzyme (dynorphin converting enzyme) (Csuhai et al., 1995; Chesneau et al., 1994) is the homologue of N-arginine dibasic convertase isolated from rat testis. A mouse NRD cDNA exhibited 98% amino acid identity with the rat cDNA. However, within a 74 residue acidic stretch, this identity drops to

82% and the corresponding acidic stretch of human NRDc is only 73% identical with that of rat NRD. Although the rat, human, and mouse NRD's behaved similarly, Tris had a pronounced effect on the kinetics of peptide hydrolysis. However, other amines including polyamines such as putrescine, spermine, and spermidine had significant influence on NRD activity suggesting that amines bind to the acidic stretch found in NRD, and that quantitative differences in the sensitivity to amines between the rat, mouse, and human enzymes can be at least partially accounted for by differences in their acidic stretch. The role of polyamines as physiological modulators of N-arginine dibasic convertase is suggestive (Csuhai et al., 1998).

27.3. ENDOPROTEASES

27.3.1. Proprotein Convertase

A common essential step for maturation of many polypeptides is caused by limited endoproteolysis of larger inactive precursors at sites marked by paired or multiple basic amino acids. Although endoproteases have been characterized, their functions are not well known in testis. A mouse testis cDNA that represents a member of a growing class of mammalian endoproteases involved in the processing of precursor proteins is present in testis. This cDNA encodes a 655-residue protein, designated proprotein convertase (PC4), containing a bacterial subtilisin-like catalytic domain closely related to those of characterized precursor-processing endoproteases, furin, PC1/PC3, PC2, and Kex2. Within this domain, the amino acid sequence of PC4 was 70, 58, 55, and 45% identical with those of mouse furin, mouse PC1/PC3, mouse PC2, and yeast Kex2, respectively. The PC4 mRNA was detectable only in the testes after the 20th day of postnatal development mainly expressed in the round spermatids. It seems that PC4 represents a prime candidate for a precursor-processing endoprotease in the testicular germ cells (Nakayama et al., 1992). The carboxy terminus of protein encoded by major PC4 mRNA differs between rat and mouse. To explain the divergence among species, Seidah et al.,(1992) cloned cDNA representing 3' region of PC4 mRNA and identified PC4-B and PC4-C mRNAs with yet different terminus. The structural gene of mouse PC4 is approximately 9.5-kb long. It contains 15 exons and 14 introns. The exon-intron organization is very similar to that of the genes for the related convertases furin, PC1, and PC2. The upstream region carries several GGGCGG and three CCAAT but no TATAA motifs. Analysis of the 5' end of PC4 mRNA in the testis has led to the identification of two novel 5' splice variants that might encode a non-secretory enzyme. The multiple forms of PC4 mRNA can all be explained by alternate splicing of primary transcripts of a single gene (Mbikay et al., 1994).

The rat cDNA sequence of PC4 (rPC4) demonstrated the presence of at least three rPC4 mRNAs resulting in the production of rPC4-A (654 amino acids), rPC4-B (619 amino acids), and rPC4-C (609 amino acids) with different C terminal sequences. Analogous to rat PC4, three cDNAs were also found for the mouse PC4. The molecular diversity of PC4 mRNA possibly results from the differential splicing and/or exon skipping of the parent gene. The PC4 mRNA, with a major form at 2.8-kb, was highly abundant in the rat testis but not detected in any other tissues. Testicular cell analysis indicated high expression levels of PC4 in germ cells but not in Leydig, Sertoli, or peritubular cells. The site of PC4 gene expression was located at the pachytene spermatocytes and the round spermatids but not in the elongating spermatids and co-localized with proenkephalin. The PC4 mRNA was first expressed postnatally between days 19 and 22, coinciding with the first stages of spermiogenesis. In order to define the function of PC4, Mbikay et al (1997) showed that mice deficient in germ cell protease PC4 were infertile. Though, spermatozoa were morphologically normal, but they failed to fuse with egg.

Role of PC4 in sperm fertilization and in early embryonic development has been demonstrated through several studies including those with PC4 nulll mice. Physiological substrates postulated or identified for PC4 include growth factors IGF-1 and IGF-2, hormonal polypeptide proPACAP, and a number of surface proteins of ADAM family such as ADAM-1, ADAM-2, ADAM-3 (procyritestin), and ADAM-5. A comparative kinetic analysis of fluorogenic substrates against PC4 and PC7 revealed that the mutant variants of human proPACAP and mouse ADAM-5 derived peptides are most efficiently and selectively cleaved by PC4. The most likely sequence motif for recognition by PC4 is KXKXXR or KXXR where X= any amino acid other than Cys and that it prefers Pro at P3, P5 and/or P2 positions. It was also suggested that PC4 is a good candidate processing enzyme for growth factors IGF-1 and 2, neuropeptide proPACAP and several ADAM proteins such as ADAM-1, 2, 3, and 5 (Basak et al., 2004).

27.3.2. Calpain

In 1964, Guroff and Guroff first described a unique proteinase in a soluble brain fraction. This protease after many twists and turns is now called "calpain". Calpain (EC 3.4.22.17) is a Ca^{2+} requiring cysteine protease. In animals such as vertebrates and nematodes, calpain forms a large gene family comprising more than 10 members. The best characterized members of the caplain superfamily include μ-calpain and m-calpain, which are now called "conventional" and "classical" calpains. "Calpain" should mean a papain-like cysteine protease that requires Ca^{2+} for its activity. The μ-and m-calpains consist of two distinct subunits, a larger 80-kDa catalytic subunit and a smaller 30-kDa regulatory subunit, forming a heterodimer structure. The 3D structures of human and rat m-calpains in the absence of Ca^{2+} have recently been solved (c/r Sorimachi and Suzuki, 2001). Calpain mediates specific Ca^{2+} dependent reactions including cell fusion. Because spermatozoa have an absolute Ca^{2+} requirement for penetration of oocytes, Rojas et al (1999) showed that whole sperm homogenate and cell fractions contain calpain activity mostly particulate in nature with optimum pH 9.0. Using specific anti-calpain I and anti-calpain II antibodies, a protein of 67-kDa for calpain II and a 75-kDa for calpain I was obtained. Also spermatozoa contain the endogenous calpain inhibitor, calpastatin, which is located in the cytosolic fraction, suggesting that calpain-calpastatin system exists in mammalian spermatozoa. The mRNA expression of calpain genes, *capn5, capn6,* and *capn11* during development suggested that *capn11* is exclusively restricted to spermatocytes during late stages of meiosis (Dear and Boehin, 1999).

27.4. TISSUE INHIBITORS OF METALLOPROTEINASES

27.4.1. Testicular Inhibitors of Metalloproteinases (TIMPs)

In addition to the metallo-proteinases, endogenous inhibitors of these enzymes may also regulate the maturation of male gonads. For example, tissue inhibitors of metalloproteinases (TIMPs), the endogenous inhibitors of matrix metalloproteinases, modulate ECM turnover and testicular development. The TIMPs are important regulators of testicular development and maturation not only in tissue remodeling but also can act as paracrine / autocrine regulator of steroidogenesis and spermatogenesis. A role for TIMP-3-sensitive-Zn^{2+}-dependent mettalloprotease in mammalian gamete membrane fusion has been proposed by Correa et al, (2000). Of the four identified TIMPs (referred to as TIMP-1-TIMP-4), TIMP-1 and TIMP-2 were produced by testicular cells (Grima et al., 1996), although TIMP-3 and TIMP-4 have not been thoroughly explored in the testis. The TIMP family members are localized to the interstitium and to a varying degree within tubules of testis. The TIMP-4 is abundant in Sertoli cells and gonocytes. Membrane type-1 MMP is under the control of TIMP-2 (Robinson et al., 2001). Members of the TIMP family range in size from 21- to 29-kDa and differ in their mode of action and selectivity for metalloproteinases. For example in tissues other than testis TIMP-1, TIMP-2, and TIMP-4 are secreted and found in the extracellular space whereas TIMP-3 is secreted and bound to the ECM. In addition to their known role as proteinase inhibitors, TIMP-1 and TIMP-2 exhibit growth factor activity and a TIMP-1 like protein to stimulate steroidogenesis by cultured Leydig cells. The role of TIMPs in testis was studied in mice deficient to express the *TIMP-1* gene (Nothnick et al., 1998). The TIMP–1 mRNA is expressed only in normal animals at highest level on day 18-27 of age. While the TIMP-2 expression was identical between genotypes with increasing pattern, the pattern of TIMP-3 was similar between genotypes and decreased on day 18-41 of age, while TIMP-4 expression was not detected. TIMP-3 expression continued to increase between genotypes. The TIMP-1 mutant mice had higher testis weight and lower serum testosterone. This suggested that each TIMP displays its own unique pattern of expression during the pre-pubertal period and may have specific roles in testicular development. Overall, TIMP-1 appears to have little effect on testosterone production in mice lacking *TIMP-1* gene (Nothnick et al., 1998).

Both Sertoli and germ cells express TIMP-1 (Mruk et al., 2003). A polypeptide from primary Sertoli cell conditioned culture medium showed two molecular variants of 31- and 29-kDa. The N-terminal amino acid sequence analysis of these two proteins revealed a sequence of NH(2)-IKMAKMLKGFDAVGNATG, which is homologous to tissue inhibitor of metalloproteases-1 (TIMP-1). The FSH stimulates Sertoli cell *TIMP-1* and *TIMP-2* activity, and also TIMP-1, TIMP-2 protein and mRNA levels. These effects were stimulated by the cAMP analogue, 8-bromo-cAMP and the phosphodiesterase inhibitor, 3-isobutyl-1-methylzanthine. The ability of FSH to stimulate Sertoli cell TIMP activity is the central role for this hormone in the control of ECM turnover during testicular development (Ulisse et al., 1994). Sertoli cells secretion of TIMP-1 and TIMP-2 seems to be coordinately up-regulated by FSH through a cAMP-PKA dependent pathway. A convergence of phorbol maristrate acetate (TPA), FSH, and cAMP mediated signals in prepuberal Sertoli cells may occur with the induction of specific AP-1 site binding complexes containing jun and fos proteins. Alterations in the coordinate TIMP-1/MMP expression have been hypothesized to involve changes in the production of a *ets* and *jun/fos* containing transacting factors. Dimeric activation of protein complexes comprising fos and jun proteins (AP-1) recognize the consensus sequences, 5'-TGA C/G TCA-3'. This sequence represents the AP-1 site. TIMP-1 gene contains AP1 sites conserved in rodents and humans. It seems that FSH stimulation of TIMP-2 expression may be regulated independently to that of TIMP-1 (Ulisse et al., 1994).

27.4.2. Epididymal Protease Inhibitor (EPPIN)

The tissue inhibitors of metalloproteinases are one class of inhibitors. Another class comprises new serine protease inhibitors, EPPIN. The *Eppin,* a gene located on human chromosome 20 expresses three mRNAs encoding two isoforms of a cystine-rich protein containing both Kunitz-type and WAP-type four disulfide core protease inhibitor consensus sequences. Analysis of *Eppin's* genomic sequence from chromosome 20q12-13.2 predicts the existence of all three splices variants of *Eppin* and that all the exons conform to the AG/GT splicing rule.The *Eppin* is a single copy gene. The TATA box transcription initiation sites are present for both of the different *Eppin* 5' UTRs; examination of the promoter region at 1800 bp upstream of the

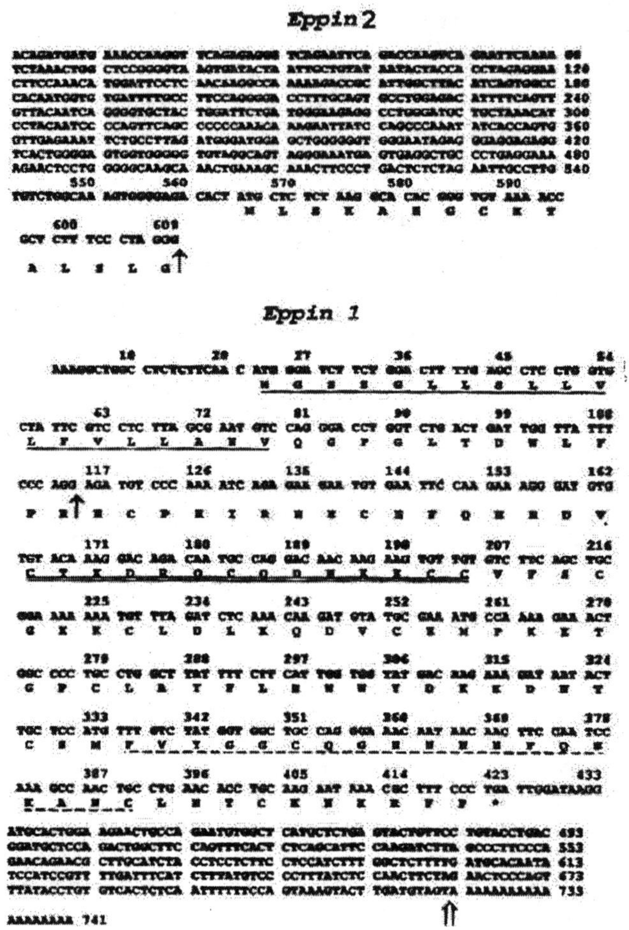

Fig.27.4. Nucleotide and deduced amino acid sequences of human *Eppin* 1 and 2. Secretory signal is single underlined; Kumtz-type sequence double underlined; and four-disulfide core (WAP-type) sequence is dash underlined. Reprinted with permission from T.R. Richardson et al. Gene 270; 93-103: 2001 © Elsevier.

start condon revealed a number of putative transcription enhancer binding sites, typical of genes expressed in the epididymis or testis. The *Eppin-1* expresses only in the testis and epididymis. The *Eppin-2* is expressed only in the epididymis, and *Eppin-3* only in the testis. Immunohistochemistry indicated a strong pattern of expression by the ciliated cells of the efferent ducts and strong staining of ejaculated spermatozoa. The Eppin represents the first member of the family of protease inhibitors characterized by dual inhibitor consensus sequences, both WAP-type and Kunitz-type. A second family member is predicted to exist on chomosome 20 approximately 4 kb downstream from *Eppin's* exon1, which has two WAP-type sequences and one Kunitz-type consensus sequence (Richardson et al., 2001) **(Fig.27.4)**.

27.4.3. Protein C Inhibitor

The protein C inhibitor (PCI) is a plasma glycoprotein belonging to the serpin superfamily of serine protease inhibitors, of which PCI is the prototype. The concentration of PCI in seminal plasma is 40 times higher than in the blood plasma. Hermans et al. (1994) described a very potent inhibitor of the sperm acrosin by PCI. The PCI is very rapidly inactivated in seminal plasma by proteinases in the reproductive tract such as prostate-specific antigen (PSA) and tissue kallikrein. The PCI has a broad protease specificity, inhabiting several proteases in homeostasis and fibrinolysis by acting as a suicide substrate. The PCI is present on the acrosomal cap of human sperm, which demonstrates the early presence of PCI in the male reproductive tract. Induction of the acrosome reaction in ejaculated human spermatozoa leads to the disappearance of PCI from the plasma membrane overlying the acrosomal head and the appearance of a strict distribution at the equatorial segment of human spermatozoa. Although a direct relation between PCI and physiological events during fertilization has not yet between established, it seems that PCI may protect spermatozoa against premature acrosome reaction and degradation, thereby modulating the acrosin activity.

27.5. REFERENCES

Ailenberg M, Stelter-Stevenson WG, Fritz IB. Secretion of latent type IV procollagenase and active type IV collagenase by testicular cells in culture. Biochem J 1991; 279: 75-80.

Basak S, Chretien M, Mbikay M, Basak A. In vitro elucidation of substrate specificity and bioassay of proprotein convertase 4 using intramolecularly quenched fluorogenic peptides. Biochem J. 2004;Pt.

Baumeister H, Muller D, Rehbein M, Richter D. The rat insulin-degrading enzyme. Molecular cloning and chararterization of tissue-specific transcripts. FEBS Lett 1993; 317: 250-4.

Bonvouloir N, Lemieux N, Crine P, Bioeau G, DesGroseillers L. Molecular cloning, tissue distribution, and chromosomal localization of MMEL2, a gene coding for a novel human member of the neutral endopeptidase. DNA Cell Biol 2001; 20: 493-8.

Chesneau V, Pierotti AR, Barre N, Creminon C, Topugard C, Cohen P. Isolation and characterization of a dibasic selective metallo-endopeptidase from rat testes that cleaves at the amino terminus of arginine residues. J Biol Chem 1994; 269: 2056-61.

Chesneau V, Prat A, Segretain D, Hospital V, Dupaix A, Foulon T, Jegou B, Cohen P. NRD convertase: a putative processing endoprotease associated with the axoneme and the manchette in late spermatids. J Cell Sci 1996 109 (pt11): 2737-45.

Correa LM, Cho C, Myles DG, Primakoff P. A role for a TIMP-3-sensitive, Zn^{+2} dependent metalloprotease in mammalian gamete membrane fusion. Dev Biol 2000; 225: 124-34.

Cossins J, Dudgeon TJ, Catlin G, Gearing AJ, Clements JM. Identification of MMP-18, a putative novel human matrix metalloproteinase. Biochem Biophys Res Commun 1996; 228: 494-8.

Csuhai E, Chen G, Hersh LB. Regulation of N-arginine dibasic convertase activity by amines: putative role of a novel acidic domain as an amine binding site. Biochemistry 1998; 37: 3787-94.

Dear TN, Boehim T. Diverse mRNA expression of the mouse calpain genes Capn5,Capn6 and Capn11 during development. Mech Dev 1999; 89: 2019.

Ghaddar G, Ruchon AF, Carpentier M, Marcinkiewicz M, Seidah NG, Crine P, Desgroseillers L, Boileau G. Molecular cloning and biochemical characterization of a new mouse testis soluble-zinc-metallopeptidase of the neprilysin family. Biochem J 2000; 347: 419-29.

Grima J, Calcagno K, Yan Cheng C, Purification, cDNA cloning, and developmental changes in the steady-state mRNA level of rat testicular tissue inhibitor of metalloproteases-2 (TIMP-2). J Androl 1996; 17: 263-275.

Hayashi MA, Gomes MD, Reboucas NA, Fernandes BL, Ferro ES, de Camargo AC. Species specificity of thimet oligopeptidase (EC 3.4.24.15). Biol Chem Hoppe Seyler 1996; 377: 283-91.

Hermans JM, Jones R, Stone SR. Rapid inhibition of the sperm protease acrosin by protein C inhibitor. Biochemistry 1994; 33:5440-44.

Hospital V, Chesneau V, Balogh A, Joulie C, Seidah NG, Cohen P, Prat A. N-arginine dibasic convertase isoforms are soluble dibasic-specific metalloendopeptidases that localize in the cytoplasm and at the cell surface. Biochem J 2000; 349 (Pt 2): 587-97.

Hospital V, Prat A, Joulie C, Cherif D, Day R, Cohen P. Human and rat testis express two mRNA species encoding variants of NRD convertase a metalloendopeptidase of the insulinase family. Biochem J. 1997; 327: 773-9.
Illman SA, Keski-Oja J, Lohi J. Promoter characterization of the human and mouse epilysin (MMP-28) genes. Gene 2001 275: 185-94.
Longin J, Guillaumot P, Chauvin MA, Morera AM, Le Maguerresse-Battistoni B. MT1-MMP in rat testicular development and the control of Sertoli cell proMMP-2 activation. J Cell Sci 2001; 114: 2125-34.
Mbikay M, Laure M, Sanson R, Tadros H, Sirosi F, Seidah NG, Chretien M. Structure of the gene for the testis specific proprotein convertase 4 and of its alternate messenger RNA isoforms. Genomics 1994; 20: 231-37.
Mbikay M, Tadros H, Ishida N et al. Impaired fertility in mice deficient for the testicular germ-cell protease PC4. Proc Natl Acad Sci USA 1997; 94: 6842-46.
Mruk DD, Siu MK, Conway AM, Lee NP, Lau AS, Cheng CY. Role of tissue inhibitor of metalloproteases-1 in junction dynamics in the testis. J Androl 2003; 24: 510-23.
Nakayama K, Kim WS, Torii S, Hosaka M, Nakagawa T, Ikemizu J, Baba T, and Murakami K. Identification of the fourth member of the mammalian endoprotease family homologous to the yeast Kex2 protease. J Biol Chem 1992; 267: 5897-5900.
Nothnick WB, Soloway PD, curry TE Jr. Pattern of messenger ribonucleic acid expression of tissue inhibitors of metalloproteinases (TIMPs) during testicular maturation in male mice lacking a functional TIMP-1 gene. Biol Reprod 1998; 59: 364-70.
Orlowski M, Reznik S, Ayala J, Pierotti AR. Endopeptidase 24.15 from rat testes. Isolation of the enzyme and its specificity toward synthetic and natural peptides, including enkephalin-containing peptides. Biochem J 1989; 261: 951-8.
Pierotti AR, Prat A, Chesneau V, Gaudoux F, Leseney AM, Foulon T, Cohen P. N-arginine dibasic convertase, a metallo-endopeptidase as a prototype of a class of processing enzymes. Proc Natl Acad Sci (U S A) 1994; 91: 6078-82.
Richardson TR, Sivashanmugam P, Hall SH, Hamil KG, Moore PA, Ruben SM, French FS, Rand M. Cloning and sequencing of human Eppin : A novel family of protease inhibitor expressed in the epididymis and testis. Gene 2001; 270: 93-102.
Robinson LL, Sznajder NA, Riley SC, Anderson RA. Matrix metalloproteinases and tissue inhibitors of metalloproteinases in human fetal testis and ovary. Mol Hum Reprod 2001; 7: 641-8.
Rojas FJ, Brush M, Moretti-Rojas I. Calpain-calpastatin: a novel, complete calcium-dependent protease system in human spermatozoa. Mol Hum Reprod 1999; 5: 520-6.
Sang QX, Stetler-Stevenson WG, Liotta LA, Byers SW. Identification of type IV collagenase in rat testicular cell culture: influence of peritubular-Sertoli cell interactions. Biol Reprod 1990; 43: 956-64.
Sang QX, Thompson EW, Grant D, Stetler-Stevenson WG, Byers SW. Soluble laminin and arginine-glycine-aspartic acid containing peptides differentially regulate type IV collagenase messenger RNA, activation, and localization in testicular cell culture. Biol Reprod 1991; 45: 387-94.
Seidah NG, Day R, Hamelin J, Gaspar A, Collard MW, Chretien M. Testicular expression of PC4 in the Rat: Molecular Diversity of a novel germ cell-specific Kex2/Subtilisin like proprotein convertase. Mol Endocrinol 1992; 6: 1559-70.
Sorimachi H, Suzuki K. The structure of calpain. J Biochem 2001; 129: 653-64.
Tanja O, Facchinetti P, Rose C, Bonhomme MC, Gros C, Schwartz JC. Neprilysin II: A putative novel metalloprotease and its isoforms in CNS and testis. Biochem Biophys Res Commun 2000; 271: 565-70.
Tripiciano A, Peluso C, Morena AR, Palombi F, Stefanini M, Ziparo E, Yanagisawa M, Filippini A. Cyclic expression of endorthelin-converting enzyme-1 mediates the functional regulation of seminiferous tubule contraction. J Cell Biol 1999; 145: 1027-38.
Ulisse S, Farina AR, Piersanti D, Tiberio A, Cappabianca L, D' Orazi G, Jannini EA et al. Follicle-stimulating hormone increases the expression of tissue inhibitors of metalloproteinases TIMP-1 and TIMP-2 and induces TIMP-1 AP-1 site binding complex(es) in prepubertal rat Sertoli cells. Endocrinology 1994; 135: 2479-2487.
Velasco G, Pendas AM, Fueyo A, Knauper V, Murphy G, Lopez-Otin C. Cloning and characterization of human MMP-23, a new matrix metalloproteinase predominantly expressed in reproductive tissues and lacking conserved domains in other family members. J Biol Chem 1999; 274: 4570-6.
Winter AG, Pierotti AR. Gene expression of the dibasic-pair cleaving enzyme NRO convertase (N-arginine dibasic convertase) is differentially regulated in the GHB pituitary and Mat-Lu prostate cell lines. Biochem J 2000; 351 (Pt 3): 755-64.

Chapter28

ISOENZYMES IN ENERGY PATHWAYS

28.1. INTRODUCTION

Testis expresses variety of isoforms of enzymes associated with energy producing pathways such as glycolysis and pentose shunt pathway. Glycolytic enzymes have also been reported to associate with tubulin and microtubules in sperm flagellum (Srere and Knull, 1998). The presence of glycolytic enzymes, glyceraldehyde-3-phosphate dehydrogenase (GAPD), aldolase (Volker and Knull, 1997), and 1-phosphofructokinase (Lehotzky et al., 1993; Vertessy et al., 1997) in microtubules have led to the suggestion that binding of these enzymes to tubulin may act as a mechanism for controlling the activity of the glycolytic pathway (Srere and Knull, 1998) possibly in sperm. The molecular bases for the testis-specific expression are diverse. Cognate genes expressed uniquely in testicular germ cells for glycolysis are PGK-2 and lactate dehydrogenase-C (Ldh3). In addition, other enzymes, which show cell specific features, include GAPD, hexokinase, phosphogluco-isomerase, aldolase, phosphoglycerate mutase and enolase.

28.2. HEXOKINASE

Hexokinase (Hk; EC 2.7.1.1) is the first essential enzyme in the main pathway of glycolysis, where it phosphorylates glucose to form glucose 6-phosphate. In mammals there are four hexokinase isozymes, which vary in their tissue distribution and kinetic properties. The hexokinase isozymes type 1-3, each consists of a single polypeptide chain of approximately 100kDa, which is inhibited by the glucose-6-phosphate product. The type 4 hexokinase (glucokinase) is similar to yeast hexokinase, being a single polypeptide of approximately 50-kDa and insensitive to inhibition by glucose-6-phosphate. Expression of glucokinase is limited to hepatocytes and pancreatic B cells and is regulated differently in these two tissues. Sequence comparisons between bovine, rat, human, and mouse Hk1 indicated that the N-terminal 15 amino acids are 100% conserved. This hydrophobic domain is the putative outer mitochondrial membrane-binding domain of somatic Hk1, which is normally cytoplasmic or bound to the outer mitochondrial membrane via contact sites with a voltage dependent anion channel (porin) through porin biding domain (PBD). The porin is the pore forming protein involved in the movement of adenine nucleotides across the outer mitochondrial membrane. Hexokinase interacts with porin in a manner such that hexokinase is provided with preferred access to the ATP, generated in the mitochondrion.

Characterization and Localization: Preliminary analysis of mouse spermatogenic cell hexokinase cDNA (*Hk1-sa*) showed it to be 99% identical to the somatic *Hk1* cDNA sequence

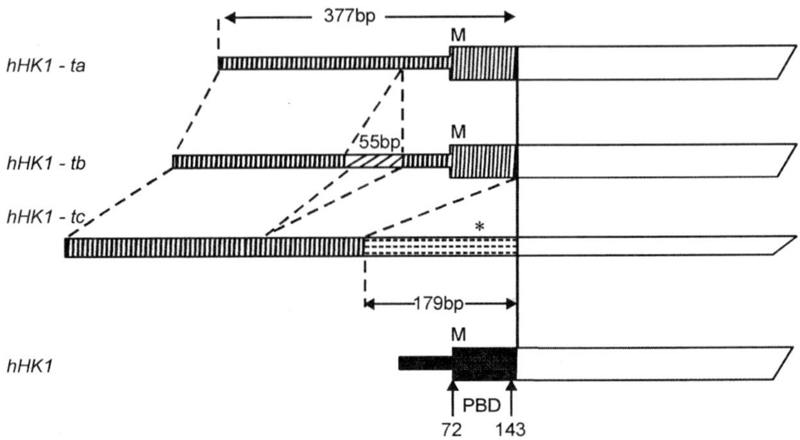

Fig.28.1. Comparison of structural features of 5'ends representing mRNAs for hexokinase-1 from human testis (*hHK1-ta, hHK1-tb*, and *hHK1-tc*) and human somatic cells (*hHK1*). Sequence differences in human *HK1* are shown by hatched box (377bp). Additional boxes of 55bp in *hHK1-tb* and 179 bp in *hHK1-tc* within or outside the hatched box respectively are indicated. Porin binding domain in *hHK1* is shown by solid box. The predicted coding regions are indicated by wide boxes, while untranslated regions are indicated by narrow boxes (Mori et al., 1996).

through most of its lengths but contained a unique sequence at the 5'-end. The expression of mRNA of this hexokinase was specific to spermatogenic cells. Analysis of two additional hexokinase cDNA clones further demonstrated three cDNAs, which contained a spermatogenic cell-specific sequence flanked on the 5'-end by sequences unique to each clone. These unique hexokinase cDNA clones were referred to as *Hk1-sb* and *Hk1-sc*. The three spermatogenic cell-specific hexokinase cDNAs lack the porin binding domain present at the 5'-end of the coding region of somatic *Hk1* genes. Cloning of mouse Hk1-s (mHk1-s) cDNA showed that the porin binding domain (PBD) in the Hk1of somatic cells is replaced by a spermatogenic cell-specific region (SSR) (Mori et al., 1993; 1996). Although the PBD binds to the outer membrane of a mitochondrion, the SSR of mHk1-s targets some specific organelle. Nakamura et al., (2003) revealed that PBD is concentrated around the mitochondrion, whereas the SSR could not be ascribed to the mitochondrion, ER, or nuclear co-localization.

The cloning of cDNAs representing three unique human type 1 hexokinase mRNAs (hHk1-ta, hHk1-tb, and hHk1-tc) expressed exclusively in testis also testified a testis-specific sequence not present in somatic Hk1, but lacked the sequence for the PBD. The hHk1-tb and hHk1-tc mRNAs each contained an additional unique sequence, similar to the spermatogenic cell-specific sequence of the mouse mRNAs. Northern analysis of RNA from mouse, hamster, guinea pig, rabbit, ram, human, and rat demonstrated absence of PBD in hexokinase1 from the testes of these mammals (Mori et al., 1996) **(Fig 28.1)**. Antisera against unique regions of Hk1-sa and Hk1-sb showed that Hk1-sa and Hk1-sb are not translated during spermatogenesis, whereas only germ cell-specific Hk1-sc was identified in round spermatids, condensing spermatids, and mature sperm. Whole of the sperm Hk1-sc is tyrosine phosphorylated, and associated with the mitochondria and with the fibrous sheath of the flagellum and was found in discrete clusters in the region of the membranes of the sperm head. The somatic Hk1 isoform was not present. The unusual distribution of Hk1-sc in sperm suggested novel functions such as extra-mitochondrial energy production (Visconti et al., 1996; Travis et al., 1998).

Hk1-sc is a Processed Form of Hk1-sa: The mechanism by which one protein is targeted to multiple sites within highly polarized sperm cell poses important questions of protein targeting. Mutation analysis of constructs, (containing a non-hydrophobic, germ cell-specific domain, present at the amino terminus of the Hk1-sc) and targeted to the endoplasmic reticulum and the plasma membrane demonstrated the presence of a complex motif, *PKIRPPLTE* (with essential residues) that represents a novel endoplasmic reticulum targeting motif in Hk1-sc. Constructs based on another germ cell specific hexokinase transcript, Hk1-sa, demonstrated the specific proteolytic removal of the amino terminal domain, resulting in a protein product identical to Hk1-sc. Such processing might constitute a regulatory mechanism governing the spatial and/ or temporal expression of the protein, Hk1-sc (Travis et al., 1999).

Phosphotyrosine Containing Hex (pY-mHk1): The mouse sperm possess at least one Hk1 isoform that is present on the sperm head and behaves as an integral membrane protein (Visconti et al., 1996). Mouse sperm also possess a phosphotyrosine containing hexokinase type 1 (Hk1) that is associated with the plasma membrane fraction of these cells (Kalab et al., J. Biol Chem 1994; 269:3810-17). The plasma membrane association of Hk1 appears unique, since somatic Hk1 is normally cytoplasmic or bound to the outer mitochondrial membrane via contact sites with porin. This suggested that germ cell hexokinases (Hk1-sa, Hk1-sb, and Hk1-sc) might not be localized to the outer mitochondiral membrane and could have alternative functions in germ cells and/or sperm. It seems that at least one of the Hk1 isoforms is tyro-phosphorylated during mouse spermatogenesis. Treatment of sperm membrane fractions to dissociate the phosphotyrosine containing Hk1 (pY-mHK1) demonstrated that pY-mHK1 has properties of an integral membrane, specifically staining both in the head and tail regions of sperm. The pY-mHk1 contains an extracellular domain (Visconti et al., 1996).

Hex as a Immunocontraceptive: Based on the sperm specific nature of an isozyme of hexokinase and association of a tyrosine-phosphorylated form of Hk1 with sperm, having properties consistent with an integral plasma membrane protein, hexokinase has been suggested as an immunogen for an immuno-contraception. The tyrosine phosphorylated form of Hk1 has an extracellular domain that is localized to both the head and flagellum of non-permeabilized cells (Visconti et al., 1996). Mature sperm from sterile tw32/tw5 mice (mutant sperm) had defects in motility and sperm-egg interaction. Characterization of Hk1 from mutant sperm indicated that at least a portion of the Hk1 present in these cells is an integral membrane protein with an extra-cellular domain located on the sperm head and the flagellum. Anti-phoshotyrosine antibodies demonstrated that Hk1 in mutant sperm is not tyrosine phosphorylated. This suggested a relation between sterility and lack of tyrosine phosphorylation (Olds-Clarke et al., 1996). In another report, however, Naz et al (1996) suggested that the hexokinase present in acrosome, mid piece, and tail regions of human sperm has M_r of 116-kDa. The 116-kDa sperm hexokinase is a glycoprotein that is different from the 95-kDa phosphotyrosine protein, and that was not phosphorylated at tyrosine residues. However, its antibodies could cause agglutination and an inhibition of fertilizing ability of human sperm (Naz et al, 1996).

28.3. GLUCOSE-6-PHOSPHATASE

Glucose-6-phosphatase (EC 3.1.3.9) has traditionally been thought to be a liver enzyme, which plays a major role in the regulation of blood glucose levels. In response to stress or low blood glucose levels, it releases glucose from glucose-6-phosphate in blood for use of other tissues. The microsomal glucose- 6-phosphatase system has been unequivocally demonstrated to be also in pancreatic islets and human fetal adrenal. Low levels have been found in intestine and gall bladder mucosa. There have also been several reports on glucose-6-phosphate hydrolysis in the adult male reproductive tract. Unfortunately, these studies were carried out before it was recognized that glucose-6-phosphatase was a multi-component system. In addition, it was not clear whether the hydrolysis of glucose-6-phosphate seen in testis was due to the action of the glucose-6-phosphatase enzyme or the action of other nonspecific phosphatases. Burchell et al., (1996), using monospecific antibodies to the liver enzyme, and specific DNA probes examined whether glucose-6-phosphatase is present in human and rat testis. Microsomal glucose-6-phosphatase activity in human fetal testis (weeks 15-20 of gestation) was approximately 25% corresponding to liver. Kinetic analysis of glucose-6-phosphatase in intact and disrupted microsomes and Southern blot analysis indicated that the specific glucose-6-phosphatase enzyme-system was also present in rat and human testis. The predominant site of expression is the Leydig cell in fetal testis. Since testicular glucose – 6-phosphatase does not play a major role in the regulation of blood glucose levels, it is proposed that it produces glucose (and/or other hexoses) at a maintained or accelerated rate for local use only. It is likely that as in adrenals, this is required for the generation of NADPH via the hexose monophosphate activity required for steroidogenesis (Burchell et al., 1996).

28.4 Glucose-phosphate Isomerase and Sperm Antigen-36

The glycolytic enzyme, glucose phosphate isomerase (GPI), also known as phosphohexose isomerase catalyzes the interconversion of glucose 6-phosphate and fructose 6-phosphate. The 50-kb gene for human GPI, located on the long arm of chromosome 19, has been cloned. Surprisingly, GPI mRNA shares high sequence homology to mRNA that codes for neuroleukin (NLK). The NLK is a neurotrophic mediator for spinal and sensory neurons. It is also secreted by activated T lymphocytes and also by tumor cells. The mouse sperm agglutinating monoclonal antibody-36 (mAb A36) was shown to induce extensive, "tangled" sperm agglutination and hence was used to isolate cDNAs encoding its cognate antigen. Sequencing of these cDNAs yielded the complete nucleotide sequence of a 3-kb cDNA that encodes the mAb-related polypeptide, designated sperm antigen-36 (SA-36), composed of 558 deduced amino acids **(Fig.28.2)**. The SA-36 cDNA contained a 5'-untranslated region of 234 nt, an open reading frame of 1674 nt, and a 3'-untranslated region of 1138 nt. The SA-36 cDNA displayed 99% homology to glucose phosphate isomerase /neuroleukin mRNA. The mAb A36 reacted specifically with GPI obtained from rabbit muscle and from baker's yeast. Taken together, it was suggested that mAb A36 cognate sperm surface antigen, encoded by SA-36 cDNA, is a glucose phosphate isomerase/NLK-like protein involved in sperm agglutination (Yakirevich and Naot, 2000). Buehr and McLaren (1981) had, earlier demonstrated that a unique isoform of glucose phosphate isomerase, absent from somatic tissues, exists in mouse sperm cells and mouse testes, but only after puberty. Evidently, more studies are needed to elucidate the relationship between sperm isoform of glucose phosphate isomerase /NLK and SA-36.

28.5. FRUCTOSE-6-PHOSPHATE, 2 KINASE / FRUCTOSE-2, 6-BIS PHOSPHATASE

Fructose-2,6-bisphosphate (Fru-2,6-P_2) is a potent allosteric regulator of 6-phosphofructo-1-kinase (Fru-6-P,2-kinase). The levels of Fru-2, 6-P_2 are maintained by a family of bifunctional enzymes, 6-phosphofructo-2-kinase/fructose-2, 6-bisphosphatase (Fru-6-P,2-kinase / Fru-2,6-Pase), which have both kinase and phosphatase activities. Each member of the enzyme family is characterized by their phosphatase : kinase activity ratio (K:B) and their tissue-specific

```
  1  MAALTRDPQF QKLQQWYREH RSELNLRRLF DANKDRFNHF SLTLNTNHGH
 51  ILVDYSKNLV TEDVMRMLVD LAKSRGVEAA RERMFNGEKI NYTEGRAVLH
101  VALRNRSNTP ILVDGKDVMP EVNKVLDKMK SFCQRVRSGD WKGYTGKTIT
151  DVINIGIGGS DLGPLMVTEA LKPYSSGGPR VWYVSNIDGT HIAKTLAQLN
201  PESSLFIIAS KTFTTQETIT NAETAKEWFL QAAKDPSAVA KHFVALSTNT
251  TKVKEFGIDP QNMFEFWDWV GGRYSLWSAI GLSIALHVGF DNFEQLLSGA
301  HWMDQHFRTT PLEKNAPVLL ALLGIWYINC FGCETHAMLP YDQYLHRFAA
351  YFQQGDMESN GKYITKSGTR VDHQTGPIVW GEPGTNGQHA FYQLIHQGTK
401  MIPCDFLIPV QTQHPIRKGL HHKILVANFL AQTEAVMRGK STEEARKELQ
451  AAGKSPEDLE RLLPHKVFEG NRPTNSIVFT KLTPFMLGAL VAMYEHKIFV
501  QGIIWDINSF DQWGVELGKQ LAKKIEPELD GSAQVTSHDA STNGLINFIK
551  QQREARVQ
```

Fig.28.2. Amino acid sequence of sperm agglutinating antigen (SA-36) deduced from cDNAs isolated from human testis (Yakirevich and Naot, 2000).

expression. The bifunctional enzyme Fru-6-P,2-kinase / Fru-2,6-Pase catalyses the synthesis and degradation of fructose-2,6-bisphosphate and participates in glucose homeostasis by regulating the intracellular concentration of fructose-2,6-bisphosphate (Fru-2,6-P_2). The Fru-2,6-P_2 is both a potent physiological activator of 6-phosphofructokinase, and an *in vitro* inhibitor of fructose-1,6 bisphosphate. The Fru-2,6-P_2 concentration is governed by a tissue–specific nature of bifunctional isozyme, which differs in the relative activities of the kinase and the phosphatase. One of the phospho-fructo kinase-2 isozyme genes, *PFKFB-3*, was induced by hypoxia through the hypoxia-inducible factor-1 (HIF-1) pathway. The basal and hypoxic expression of three members of this family in different organs of mice indicated that all four isozymes (PFKFB-1-4) are responsive to hypoxia; their basal level of expression and hypoxia responsiveness depends on organ studied. However, the highest hypoxia responses are found only in liver and testis. Results demonstrated that Fru-6-P,2-kinase genes are responsive to hypoxia in vivo, indicating a physiological role in the adaptation of the organism to environmental or localized hypoxia/ischemia (Minchenko et al., 2003).

Characterization: Several isozymes of Fru-6-P,2-kinase/ Fru-2,6-Pase from mammalian tissues have been characterized. The major isozymes from liver, heart, and testis are homodimers, consisting of subunit M_r ranging from 54,000 to 60,000Da. A comparison of the amino acid sequences reveals that the catalytic domains of these isozymes are highly conserved, but their N- and C- termini, which are regulatory for enzyme activity are completely different. In addition to differences in the primary structures, the terminal peptides of the liver and the heart isozymes contain phosphorylation sites and play important roles in regulation of the bifunctional enzyme activity. For example, phosphorylation of the N-terminus of the liver enzyme results in the inhibition of Fru-6-P, 2 kinase activation of Fru 2, 6-Pase. In contrast, the phophorylation of the heart enzyme in the C-terminus results in the activation of Fru 6-P, 2-kinase. These effects of phosphorylation suggest the importance of the terminal peptides in the regulation of the enzyme activities, which probably involve the conformational changes induced by interaction of the terminal peptides with the catalytic domain in a specific manner. Based on isoforms, these studies serve as a model for modification of other proteins structures.

On the basis of significant amino acid homology, the phosphatase activity domain of bifunctional enzyme is structurally related to the phosphoglycerate mutase and acid phosphatase families. The enzyme structure of testis origin shows the location of the C-terminus but does not reveal the N-terminal 36 amino acid residues, suggesting a mobile nature for this terminal tail. To gain insight into the roles of the terminal peptides, 24 and 30 amino acids from the N-terminus of the testis enzyme were deleted. These deletions caused a large increase in $K_{Fru\ 6\text{-}P}$, a decrease in the kinase activity, and an increase in the phsphatase activity (Tominaga et al., 1993). To determine the role of phosphorylation, a consensus amino acid sequence (RRAS) was introduced for protein kinase A in the N-terminus of the testis enzyme (which lacks any phosphorylation site). The phosphorylation of this site resulted in inhibition of the kinase and activation of the phosphatase. However, the phosphoenzyme was susceptible to thermal inactivation than the dephospho forms. These results suggested that the N-terminus is essential in stabilizing the dimmer structure, probably by strengthening the subunit interaction. Deletion or phosphorylation of the N-terminus weakens the interaction of N-terminus with the catalytic domain, which results in with change of kinetic properties of the enzyme in general (reviewed in Yuen et al., 1999).

The rat testis enzyme also referred as (RT2K) lacks the regulatory domain, but the rat liver and the bovine heart enzymes contain phophorylation site(s) in the N- and the C-termini, respectively. Tominaga et al., (1997) constructed mutant enzymes in which the N- or the C-terminal tail of the testis enzyme was replaced with that of either liver or heart enzyme. The substitution with the N-terminus of the liver enzyme (RLN-RT2K) and c-terminus of heart resulted in a small change in the kinetic properties of Fru 6-P, 2 kinase, but substitution with the N-terminus of heart enzyme increased the $K_{Fru\ 6\text{-}P}$ 18 fold without affecting the V_{max}. The phosphorylation of RLN-RT2K increased $K_{Fru\ 6\text{-}P}$ fivefold as in the liver enzyme but did not affect the Fru 2,6-Pase, unlike the liver enzyme. Mutant enzymes were more thermally labile than the wild type testis enzyme. The RLN-RT2K was more sensitive to the denaturant.

Conformation: The crystal structure of the rat testis Fru 6-P, 2 kinase/ Fru-2, 6 Pase shows that each enzyme monomer consists of independent kinase and phosphatase domains and the two subunits in the homodimmer are arranged in a head to head manner (Fig.31.2). The kinase domains are in close contact, forming an extended hydrophobic core between them, while the phosphatase domains are essentially independent of one another. The kinase domain for kinase activity is related to the superfamily of mononucleotide binding proteins with a close relationship to the adenylate kinases and the nucleotide binding portion of the G-proteins. A mutant rat testis isozyme with an alanine replacement for the catalytic histidine (H256A) in the Fru–2,6-Pase domain retains 17% of the wild type activity (Mizuguchi, et al., 1998). The crystal structure of H256A was solved at resolution of 2.4 Å by molecular replacement. The Fru-6-P was found at the Fru-2, 6 Pase active site, revealing the interaction in substrate/product binding at the active site of the enzyme. A view of the Fru-2, 6 P_2 bound enzyme complex identified the residues responsible for catalysis. The analysis disclosed distinct catalytic mechanisms for the wild type and mutant proteins. The wild type protein leads to an inefficient transfer of a proton to the leaving group Fru-6-P, which is consistent with a view of this event being rate limiting and responsible for the extremely slow turnover (0.032 s^{-1}) of the Fru-2,6-Pase in all Fru-6P, kinase/Fru-2, 6-Pase isozymes (Yuen et al., 1999) **(Fig.28.3)**.

28.6. PHOSPHOGLYCERATE KINASE-2

Phosphoglycerate kinase (ATP-3-phospho-D-glycerate1-phosphotransferase, EC 2.7.2.3) catalyzes the reversible conversion of 1,3-diphosphoglycerate to 3-phosphoglycerate, a reaction that generates one molecule of ATP in the glycolytic pathway. An X-linked phosphoglycerate kinase gene is expressed in all mammalian somatic cells and pre-meiotic cells and undergoes dosage compensation. In addition to this X-encoded phosphoglycerate kinase (PGK-1 or PGK-A) isozyme, PGK-2 (PGK-B) has been identified which is expressed in some mammalian testis and which is transcribed post-meiotically. Comparison of murine PGK-1 and PGK-2 isozymes suggested that they evolved from a common ancestral sequence. In contrast to X-linked *Pgk-1*, the *Pgk-2* gene is located on an autosome in both mouse and kangaroo and has

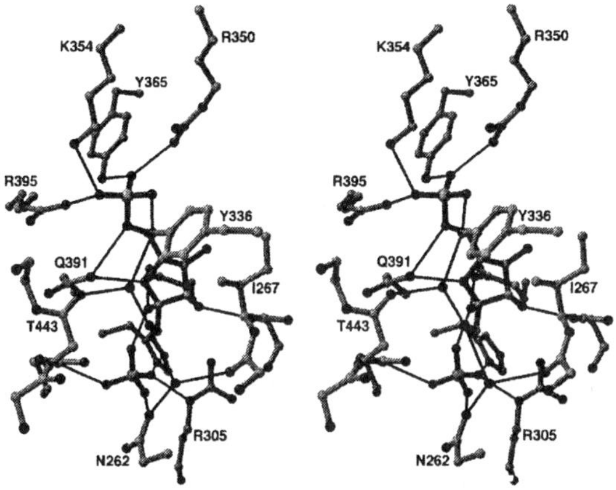

Fig 28.3. Fru-6-P interactions in the Fru-2, 6-Pase active site. A stereo view of the Fru-2, 6-Pase active with various interactions between protein and Fru-6-P are shown. The Fru-6-P 6-phosphate is bound by interactions with Arg350, Lys354, Tyr365, and Arg395. The fructose portion of Fru-6-P is bound by both hydrogen bonds and a hydrophobic stacking interaction, where Ile267 can be seen to stack with the fructose ring and Fru-6-P-O-2. The two water molecules that form hydrogen bonds between Fru-6-P-O-1 and the protein are shown as isolated spheres. One of these waters interacts with Fru-6-P, Tyr336, and the main chain carbonyl of Thr443. The other water interacts with Asn262 and the main chain carbonyl of position 266. The free phosphate (analogous to the 2-phosphate of Fru-2, 6-P2) is shown at the bottom, bound by Arg255, Asn262, Arg305, and His390. The identities of the side chains are indicated with the one-letter amino acid designation and position in the RT2K protein sequence. Hydrogen bonds between the Fru-6-P, phosphate, and protein are drawn as black lines. Printed with permission from M.H. Yuen et al, J Biol Chem 274; 2176- 84; 1999 © American Society for Biochemistry and Molecular Biology.

been mapped to the vicinity of MHC gene on chromosome 17 in mouse. However, using subcloned fragments of the cDNA, the presence of four independent phosphoglycerate kinase genes was identified. These genes have been mapped to both the human X chromosome (band q13) and chromosome 6 (p12-21.1). The genomic distribution of phosphoglycerate kinase sequences is conserved in man and mouse, not only for the X-chromosome, but also for linkage to the respective major histocompatibility complexes. Cloning of X-linked phosphoglycerate kinase sequences led to the identification of a intronless phosphoglycerate kinase pseudogene, which is localized proximal to the active gene on the X chromosome (Michelson et al., 1985).

Characterization of Pgk-2 Gene: The nucleotide sequence for the autosomal genomic clone of *Pgk-2* shows an open reading frame of 1,251 nt capable of coding a full length PGK protein of 417 amino acids. This coding sequence shows 85% homology with that of the functional human X-linked *Pgk-1* gene at the nucleotide level, and its predicted amino acid sequence is 87% homologues with that of the functional *Pgk-1* derived enzyme. The predicted sequence of the PGK-2 polypeptide is identical to the functional PGK-1 sequence at 364/417 positions (87% homology). The human *Pgk-1* gene consists of 11 exons and 10 introns encompassing a region ~ 23-kb in length. The *Pgk-2* has revealed that this autosomal sequence is completely deficient in all ten introns and contains characteristics of a processed gene or retroposon including a 165bp remnants of a poly $(A)^+$ tail and bounding direct repeats of which 52% bases are adenines. In the human *Pgk-2* gene, direct repeats of seven base pairs are located just 3' to the remnant of the poly $(A)^+$ tail and at 939-945-bp upstream from the start of the coding sequence. Such

```
PGK-Y PR    1   MSLSSKLSVQDLDLKDKRVFIRVDFNVPLDGKKITSNQRIVAALPTIKYVLEHHPRYVVL   60
PGK-1 PR    1   ....N..TLDK..V.G...VM.......MKNNQ..N....K..V.S..FC.DNGAKS...   60
PGK-2 PR    1   ....K..TLDK..VRG...IM.......MKKNQ..N....K.SI,S...C.DNGAKA...   60
ψPGK-1 PR   1   ....N..TLDKP.V.GQ..IM.......VXNNQ..N....K.TVLS..FC.DNGAKS...   60

PGK-Y PR   61   ASHLGRPNGERNE-KYSLAPVAKELQSLLGKDVTFLNDCVGPEVEAAVKASAPGSVILLE  119
PGK-1 PR   61   M......D.VPMPD....E...V..K.......L..K.........K.CANP.A.......  120
PGK-2 PR   61   M......D.VPMPD........V..K.......L..K....A...K.CANP.........  120
ψPGK-1 PR  61   M....Q.D.VPMPD?...E.F.V.FK.......L..K.........K.CANP.A.......  120

PGK-Y PR  120   NLRYHIEEEGSRK-VDGQKVKASKEDVQKFRHELSSLADVYINDAFGTAHRAHSSMVGFD  178
PGK-1 PR  121   ...F.V....KG.DAS.N....EPAKIEA..AS..K.G...V.................VN  180
PGK-2 PR  121   ...F.V....KGQDPS.K.I..EPDKIEA..AS..K.G...V.................VN  180
ψPGK-1 PR 121   ..HF.V....KG.-?S.N...?EPAKIEA..AS..K.GN.HV.G...............VS  179

PGK-Y PR  179   LPQRAAGFLLEKELKYFGKALENPTRPFLAILGGAKVADKIQLIDNLLDKVDSIIIGGGM  238
PGK-1 PR  181   ...K.G...MK...N..A....S.E.......................N.M....NEM......  240
PGK-2 PR  181   ..HK.S...MK...D..A......V.......................K.M....NEM......  240
ψPGK-1 PR 180   ...K.G...MK...N..V.V..S.X............F....H..NIM....NEM.....T  239

PGK-Y PR  239   AFTFKKVLENTEIGDSIFDKAGAEIVPKLMEKAKAKGVEVVLPVDFIIADAFSADANTKT  298
PGK-1 PR  241   ....L...N.M...T.L..EE..K..KD..S..EKN..KIT.....VT..K.DEN.K.GQ  300
PGK-2 PR  241   .Y..L...N.M...A.L..EE..K..KDI.A..QKN..RITF....VTG.K.DEN.QVGK  300
ψPGK-1 PR 240   ....L...N.M.T.T.L..EE..K..KD..SQ.EKN..KIT..I..VT.EK.DEN.R.GQ  299

PGK-Y PR  299   VTDKEGIPAGWQGLDNGPESRKLFAATVAKAKTIVWNGPPGVFEFEKFAAGTKALLDEVV  358
PGK-1 PR  301   A.VAS.......M...C....S.KY.EA.TR..Q.......V....W.A..R.....M....  360
PGK-2 PR  301   A.VAS..SP..M...C....N.NH.QV..Q.RL.......L....WDA..K......M..I.  360
ψPGK-1 PR 300   A.VAS.....XM...C....S.KY.E..TW..Q.......V....W.A..Q.....M....  359

PGK-Y PR  359   KSSAAGNTVIIGGGDTATVAKKYGVTDKISHVSTGGGASLELLEGKELPGVAFLSEKKX  416
PGK-1 PR  361   .ATSR.CIT.........CCA.WNTE..V....................V....DA..NI-X  417
PGK-2 PR  361   .ATSK.CITV........CCA.WNTE..V....................I....EA..NM-X  417
ψPGK-1 PR 360   .ATSR.CIT.....N...CCA.WNTE...R....T.........DS.V....DA.GNI-X  416
                              *                  *
```

Fig.28.4. Amino acid sequence of human PGK-2 (PGK-2 PR) and its comparison with yeast PGK protein (PGK-Y PR), human X-linked pseudogene PGK-1 protein (PGK-1 PR), and human X-linked pseudogene protein (ψPGK-1 PR). PGK-2 shows 87% and 62% homology with PGK-1 and PGK-Y respectively. Amino acids critical in yeast enzyme function are boxed and conserved in PGK-1 and PGK-2, with differences in ψPGK-1 (*). Two inframe termination codons in ψPGK-1 pseudogene are indicated by X. Reprinted with permission from J.R. McCarrey and K. Thomas, Nature 326 (6112), 501-05: 1987 © Macmillan Magzines Ltd. http://www.nature.com/Nature.

processed sequences form non-functional pseudogenes that have evolved multiple genetic lesions, which preclude translation of any transcript into a functional polypeptide. For example an X-linked processed pseudogene of *Pgk-1 (ψPgk-1)* in humans has shown to contain premature termination codons in all reading frames (McCarrey and Thomas, 1987; McCarrey et al., 1996) **(Fig.28.4)**.

Pgk-2 gene is a Transcribed Retroposon of Pgk-1: The human *Pgk2* and *ψhPgk1* pseudogenes are retroposons of *Pgk1*. The *Pgk-2* retroposon gene arose by reverse transcriptase-mediated processing of a transcript from the *Pgk-1* gene. The nucleotide sequence of the *Pgk-2* gene suggests that it arose from X-linked *Pgk-1* more than 100 million years ago by RNA-mediated gene duplication. Most retroposons form nonfunctional "processed pseudogenes" because they are incomplete copies of their progenitors representing only the transcribed sequences present in the mRNA and deficient in the flanking regulatory (promoter) sequences necessary for direct transcription. There are at least two pseudogenes of the *Pgk-1* gene in the human genome (Michelson et al. 1985) and six pseudogenes of *Pgk-1* in the murine genome (Adra et al. 1988). However, the human *Pgk-2* gene was the first reported case of a functional gene that evolved in this manner (McCarrey and Thomas 1987). Few other such functional genes have since been reported: *Zfa* and *G6pd-2* in the mouse (Ashworth et al., 1990) and *Pdha-1* in the human (Dahl et al., 1990; Hendriksen et al., 1997). Certain X-chromosomal genes (*Pgk-1* and the E1α-subunit of *Pdha-1*) remain transcriptionally silent in spermatids (after completion of the meiotic divisions). In contrast, other X-chromosomal genes such as *Ubelx* and *MHR6A* (encoding ubiquitin activating and ubiquitin conjugating enzymes, respectively) transcribed by retroposition are post-meiotically transcribed genes. The inactivation of *Pgk-1* and *Pdha-1* in spermatocytes and spermatids is compensated by testis-specific expression of the autosomal genes *Pgk-2* and *Pdha-2,* which encode isoenzymes PGK-2 and Pdha-2 respectively. The *G6pd-2* is one of the very few known expressed retroposons encoding a functional protein, and the presence of this gene is probably related to X chromosome inactivation during spermatogenesis (Hendriksen et al., 1997). Zhang and Kleene (1999) suggested that reverse transcriptase cDNA copies of mRNAs are present in meiotic and haploid spermatogenic cells, but these cDNAs are not integrated into genomic DNA. Thus, the *Pgk-2* retroposon was formed from an aberrant transcript of *Pgk-1* that initiated from an upstream start site, such that the 5' flanking sequence of the *Pgk-1* gene was duplicated along with the coding and untranslated sequence as depicted in **Fig.28.4**.

Expression of Pgk genes in metatherian mammals: Although the *Pgk-1* gene is expressed ubiquitously in mammals, expression of the *Pgk-2* gene differs widely in different mammalian species, especially when metatherians and eutherians are compared. The expression of the PGK-B (PGK-2) isozyme is restricted to testicular tissue in the vast majority of eutherians examined. Out of 22 species, only dog and the fox showed some evidence of the PGK-B isozyme in somatic tissues. However, many marsupial species showed at least some *Pgk-2* expression in somatic tissues, in addition to a high level expression in the testis. None of the marsupials showed the strict testis specific pattern of *Pgk-2* expression seen in the mouse and human (McCarrey, 1994).

Promoter Region of Pkg-2: Despite various differences between the *Pgk-1* and *Pgk-2* promoters, there are certain similarities that form the basis for the hypothesis that the *Pgk-2* promoter is derived from the *Pgk-1* promoter. These similarities include the absence of a TATA – box sequence in either of the promoter genes and the presence of CAAT/GC box pairs in both the genes. The absence of a TATA box is a common feature of many housekeeping genes, but its absence from tissue specific promoter sequence is rare (Broceno et al, 1999). The TATA box is typically the key element in the core promoter of tissue specific genes and acts to localize ubiquitous transcription factors, such as the TATA binding protein (TBP) and other associated proteins. These transcription factors in association with RNA polymerase II bind the transcription start point for proper initiation of transcription (see Chapter 15). The mechanism underlying the transcription switch from *Pgk-1* to *Pgk-2* during spermatogenesis remains obscure.

Previous results indicated that *Pgk2* expression is regulated at the transcriptional level by core promoter sequences that bind ubiquitous transcription factors and by sequences in a 40-bp upstream enhancer region (E1/E4) that binds tissue specific transcription factors. Transgenic mice carrying different *Pgk2* promoter sequences linked to CAT reporter gene (one containing only the 40-bp E1/E4 enhancer sequence plus the core promoter and two containing 515-bp of *Pgk2* promoter but with either the E1/E4 enhancer region or the Sp1-binding site in the core promoter disrupted by in vitro mutagenesis) all showed levels of expression reduced to less than half that of the wild type 515 *Pgk2/CAT* transgenic. The single disruption of any one of these binding activities did not abolish the transgene expression, but only reduced it. These results indicated multiple factor binding regions normally regulating the initiation of transcription from the *Pgk-2* promoter, and involvement of multiple transcription factors , which interact themselves to form an enhanceosome like complex with the promoter (Zhang et al., 1999).

Transgenic mice carrying the various upstream regions including the transcription promoter of the testis specific mouse *Pgk-2* gene revealed that the 1.4-kb DNA region is sufficient for determining the organ-specific and stage specific transcription of the mouse *Pgk-2* gene during spermatogenesis. When the region between -684 and +61 was used to generate transgenic mice β-galactosidase mRNA was detectable not only in testis, but also in other organs such as brain and lung. However, the timing and cell-type specificity of testicular expression of β-galactosidase mRNA were retained in these mice. Because the region between -1404 and -685 repressed the *Pgk-2* promoter in somatic cell derived cell lines, it has been suggested that the organ specificity of *Pgk-2* transcription is achieved at least partly by negative regulation. Thus the 1.4-kb region upstream of the mouse *Pgk-2* coding sequence is sufficient for organ, developmental and spermatogenic stage specific transcription of the *Pgk2* gene (Ando et al, 2000). Although the negative element has been identified but an upstream region longer than 684-bp is needed for transcriptional regulation of the *Pgk-2* gene and that negative regulation is necessary for restricting mouse *Pgk-2* transcription in the testis. It is thus possible that transcription of the mouse *Pgk-2* gene is regulated in both positive and negative manners.

Regulation of Pgk-2 Gene Transcription: The transcriptional regulation of tissue specific gene expression in higher eukaryotes involves multiple levels of control, and the *Pgk-2* gene is no exception. The PGK-A and its *Pgk1* mRNA are expressed ubiquitously in all cells possibly on transcriptionally active X chromosomes. Transcriptional activity of *Pgk-2* gene is present only in spermatogenic cells and not in any other somatic cells. In testis, the transcriptional activity of *Pgk2* initiates with the onset of meiosis and continues through all meiotic stages in spermatocytes and in post-meiotic stages in round spermatids. The *Pgk2* transcription ceases in late spermatids when entire genome becomes inactive. Thus *Pgk2* gene expression is highly cell specific in most of eutherian mammals.

Regulatory Sequences in the 5' Flanking Region: Sperm, like other metabolically active cells, can activate autosomal *Pgk2* gene to provide necessary PGK in the form of PGK2 during glycolysis. Transcription of the *Pgk2* gene is controlled by regulatory sequences located in the 5' flanking region of the gene. This region includes both core promoter sequences and tissue-specific enhancer sequences. Core promoter function has been observed in the first 188-bp upstream from the translational start site in the human *Pgk2* gene, including a 70-bp 5'-untranslated region. This core promoter contains a CAAT box and a GC box upstream from the single transcriptional start site, but lacks a TATA box. It has been known that CAAT and GC-boxes act as binding sites for the ubiquitous transcription factors, CTF-1 and Sp-1, respectively. (Zhang et al., 1999; Gebara et al., 1992). Robinson et al. (1989) described a specific region of the *Pgk-2* promoter that is required to direct appropriate tissue-specific expression of a reporter gene in transgenic mice. The enhancer activity was located within a 327 bp region of the 5' flanking region of the *Pgk-2* gene, immediately upstream from the core or minimal promoter, which is necessary to direct expression of this gene in spermatogenic cells. Later, tissue specific protein–DNA interactions in this region were also found associated that control *Pgk-2* transcription in spermatogenic cells and possibly also with the repression of *Pgk-2* transcription in somatic cells. Studies on the appearance and disappearance of DNase I-hypersensitive (DH) sites in *Pgk-1* and *Pgk-2* each gene and their correlation with transcriptional activity suggested that the occurrence of DH sites is related to periods of active transcription. Results with the *Pgk-2* gene indicated that DH sites appear coincident with, or just prior to, transcriptional activation of this gene (Kumari et al., 1996).

Demethylation Events in Pgk-2 Gene: Many ubiquitously expressed housekeeping genes bear constitutively hypomethylated CpG islands, defined as sequences of 100-bp with relatively high GC content (>50%), typical under representation of CpG dinucleotides as compared to other regions of the mammalian genome. A tissue specific correlation has been demonstrated between hypomethylation and transcription for many tissue specific genes. In the developmental context, demethylation of tissue-specific genes typically occurs prior to the initiation of transcription. In the mouse, the 5'-portion of the endogenous *Pgk-2* coding sequence undergoes a specific demethylation event that precedes transcriptional activation by 10-12 days (Ariel et al., 1991; Kafri et al., 1991) (see Chapter 14)). Demethylation depicts the transition of condensed

```
  1 MSRRDVVLTN VTVVQLRRDR CPCPCPCPCP CPVIRPPPPK VEDPPPTVEE QPPPPPPPPP
 61 PPPPPPPPPP PQIEPDKFEE APPPPPPPPP PPPPPPPPLQ KPARELTVGI NGFGRIGRLV
121 LRVCMEKGIR VVAVNDPFID PEYMVYMFKY DSTHGRYKGN VEHKNGQLVV DNLEINTYQC
181 KDPKEIPWSS IGNPYVVECT GVYLSIEAAS AHISSGARRV VVTAPSPDAP MFVMGVNEKD
241 YNPGSMTIVS NASCTTNCLA PLAKVIHENF GIVEGLMTTV HSYTATQKTV DGPSKKDWRG
301 GRGAHQNIIP SSTGAAKAVG KVIPELKGKL TGMAFRVPTP NVSVVDLTCR LAKPASYSAI
361 TEAVKAAAKG PLAGILAYTE DQVVSTDFNG NPHSSIFDAK AGIALNDNFV KLVAWYDNEY
421 GYSNRVVDLL RYMFSREK
```

Fig.28.5. Amino acid sequence of mouse testicular glyceraldehydes 3-phosphate dehydrogenase. Source: http//www.ncbi.nlm.nih.gov (Accession AAA53033).

chromatin to an open chromatin configuration in the *Pgk-2* promoter required for binding of transcription factors and onset of transcription. Remethylation of the *Pgk-2* gene in epididymal spermatozoa is indication of the return to a closed chromatin configuration. McCarrey and colleagues analyzed the methylation pattern of the mouse *Pgk-2* gene and have shown that the coding region, as well as the gene flanking region, are highly methylated in *Pgk-2* non-expressing cell types, but under-methylated in pachytene spermatocytes and spermatids where the *Pgk-2* gene is transcribed. Transgenes consisting of the *Pgk-2* core promoter indicated that specific promoter sequences can influence the pattern of tissue-specific demethylation within different genes and that a signal for spermatogenic cell specific demethylation resides within the core promoter of the mammalian *Pgk-2* gene. The tissue specific protein-DNA interactions and demethylation events distinguish the *Pgk-2* gene from the *Pgk-1* gene in mice and humans (McCarrey, 1994; Zhang et al., 1998).

28.7. GLYCERALDEHYDE-3 PHOSPHATE DEHYDROGENASE

Glyceraldehyde-3-phosphate dehydrogenase (GAPD) in addition to being a classic glycolytic enzyme is a multifunctional protein relevant in cell functions such as DNA replication, DNA repair, translational control of gene expression and apoptosis. A mouse gene *Gapd-S* has been identified that encodes the unique GAPD-S, which is expressed only in spermatids, which are morphologically differentiated to sperm and hence the enzyme is called sperm isozyme as GAPD-S. Polyclonal antibodies against GAPD-S showed a high specificity for mammalian sperm enzyme, localized along the fibrous sheath in sperm of boar, bull, rat, stallion and man. Lactate is the preferred energy substrate for round spermatids in the rat, mouse, hamster and presumably, other mammals (Mori et al, 1992; Welch et al., 1995). In contrast to round spermatids, sperm have a high rate of glycolysis and are able to utilize glucose effectively. In the mouse, glucose is required for hyperactivated motility of sperm and for fertilization in vitro. The GAPD-S appears to be the target of several reproductive toxicants that adversely affect male fertility, including α-chlorohydrin, epichlorohydrin, 6-chloro-6-deoxyglucose, and ornidazole. These compounds are converted to the common metabolite, 3-chloroacetaldehyde, a competitive substrate inhibitor of the GAPD isoform in sperm that reduces fertility in several mammalian species. The mouse *Gapd-s* gene encodes a protein of 438 amino acids, whereas the *Gapd* gene expressed in somatic cells encodes a protein of 333 amino acids. The additional 105 amino acids of GAPD-S are on the N-terminus, 49.5% of these residues are proline. The overall homology between GAPD-S and somatic GAPD is 71% **(Fig.28.5)**. The cysteine required for substrate binding is conserved, as are seven of eight amino acids that form the NAD^+ binding pocket. Although the multifunctional nature of GAPDH suggests versatility, the mechanisms regulating its expression have been reported by Mezquita et al, (1998). These workers detected alternative initiation to TATA box and alternative splicing in the 5'-regions of the pre-mRNA, resulting in at least six different types of mRNAs. The amount and the polyadenylation of the GAPDH transcripts increased in mature testis in relation to immature testis and further increased when cell suspensions from mature testis were exposed to heat shock. Thus alternative initiation, alternative splicing, and polyadenylation could provide the necessary versatility to the regulation of the expression of this multifunctional protein during spermatogenesis.

In situ hybridization studies have shown that *Gapd-s* mRNA is present only in round and condensing spermatids, first appearing during the cap phase of spermiogenesis (Welch et al., 1995). Although the steady-state level of *Gapd-s* mRNA is maximal at step 9 of mouse spermatogenesis, GAPD-S protein was not detected until steps 12-13 suggesting that *Gapd-s* is translationally regulated. Though, a major immunoreactive protein migrating with a molecular weight M_r of 69,200 is observed in condensing spermatids and cauda sperm, additional minor proteins that migrated at M_r 55200, 32500 and 27500 were also detected in sperm. The molecular weight of GAPD-S is higher than the predicted molecular weight of 47445 apparently due to a proline rich 105 amino acid domain at the N-terminus. Recombinant GAPD-S protein lacking the proline rich region migrated at M_r 38250 comparable to somatic GAPD, which also lacks the proline rich domain. However, the molecular mass of 41.5- and 238-kDa of subunit and native enzyme, respectively suggests that native GAPD-S enzyme is a hexamer.

In sperm, GAPD-S is restricted to the principal piece of the sperm flagellum where enzyme is tightly associated suggesting its potential role in regulating sperm motility. Evidence has been suggesting that GAPD-S is covalently linked to the fibrous sheath of axoneme (Bunch et al, 1998; Westhoff and Kamp 1997). It is likely that proline rich region of GAPD-S serves as an anchor to fibrous sheath in sperm. Mouse TB-RNA-binding protein/translin selectively binds to the m-RNA of the fibrous sheath protein GAPD-S and suppresses its translation in vitro (Yang et al, 2003). During sperm maturation, most of the cytoplasm of the sperm mid-piece is removed as droplets during the passage through the epididymis, which suggests that GAPD-S is completely removed from the mid-piece during sperm maturation in the epididymis (Westhoff and Kamp 1997).

28.8. PHOSPHOGLYCERATE MUTASE

Phosphoglycerate mutase (PGAM) is a glycolytic enezyme, which occurs in three dimer isoforms of two distinct 30kDa sub-units of M and B types. Both M and B types subunits are present in testis. Phosphoglycerate mutase gene in skeletal muscle *(Pgam-m)* encodes the skeletal muscle specific subunit (M) of the phosphoglycerate mutase (PGAM-M) (Tsujino et al., 1989). This subunit is also expressed in heart and testis. The *Pgam-m* gene encoding the muscle-specific subunit is transcriptionally activated during spermatogenesis. The muscle-specific *Pgam-m* gene is a meiotic gene that is included in the growing list of stage-specific genes co-ordinately expressed during spermatogenesis (Broceno et al., 1995). The PGAM-M type isoform is present in testes from several mammals including humans. The cDNAs for the A and B subunits from human and rat have been obtained (Ruiz-Lozano et al., 1994) and the transcription elements involved in muscle specific gene expression of PGAM-M have also been documented (Broceno et al., 1995). The mRNA corresponding to the muscle-specific phosphoglycerate mutase present in testis shows a longer poly (A) tail and begins to be expressed at postnatal day 22, when germ cells start to enter into meiosis. In contrast to what happens during skeletal-muscle differentiation, *Pgam-m* gene expression during spermatogenesis is not coupled to constitutive phosphoglycerate mutase *(Pgam-b)* gene repression.

Among muscle specific isoforms of PGAM, the muscle-specific PGAM-B subunit was not detected in testes of newborn mice, in which only the PGAM-A isozyme was observed. In adult males two isozymes are equally distributed. The PGAM-B was first observed between Day 14 and Day 16 of postnatal development and is localized exclusively in germ cells. Among germ cells, PGAM-B is detected in pachytene spermatocytes and in spermatids, but not in earlier stages of spermatogenesis (Fundele et al 1987). Rat skeletal muscle specific gene for *Pgam-m* is also specifically expressed in meiotic and haploid male germ cells of testis. To analyse the promoter elements that regulate the transcription of the *Pgam-m* gene during spermatogenesis, transgenic mice were developed containing 1.3 kb from the *Pgam-m* promoter linked to the *E. coli LacZ* gene. RNA analysis showed transcriptional activity in the testis with a pattern during testis development that was identical to the endogenous gene. The transgene was also active in skeletal muscle but not in the adult heart. These studies demonstrated that the 1.3 kb *Pgam-m* promoter contains sufficient sequences to specify temporarily regulated testis-specific expression as well as skeletal muscle expression (Broceno et al, 1999).

In comparison to two main extended products of 100 and 102 nt detected in skeletal muscle, coincident with the transcription start in testis, two different extended products of 135 and 139 nt have been mapped upstream of skeletal muscle transcription start point and of the putative TATA box. This suggested a different transcription start point in testis than skeletal muscle for rat *Pgam-m* gene and demonstrated that the transgenic construct contained the endogenous transcription start point for testis and skeletal muscle. The transcription start point for the endogenous *Pgam-m* gene in testis is located upstream from the one in skeletal (25-30 nt) muscle (25-30 nt) and suggests that *Pgam-m* testis transcription is driven by a TATA less promoter, which is often found when the same gene is selectively transcribed in somatic tissues and during spermatogenesis, although a cryptic TATA box cannot be excluded. Sequence alignment of regulatory elements between human and rat promoter regions revealed that the homology is essentially reduced to the –400-bp region of the *Pgam-m* genes (Broceno et al., 1999).

28.9. ENOLASE

Enolase is a member of a superfamily of enzymes, which are related by their ability to catalyze the abstraction of the α-proton of a carboxylic acid to form an enolic intermediate. The enzyme enolase exists as a dimmer in all eukaryotes and many prokaryotes, and catalyzes the conversion of 2-phoshoglycerate in a dehydration step to yield phosphoenolpyruvate. In higher eukaryotes, enolase exists as three distinct isoforms (α, β, and γ), which are encoded by separate genes. The β isoform appears to be restricted to muscle, while γ enolase has been described as neuron specific. The α isoform appears to be ubiquitously expressed in all tissues. There have been several reports of glycolytic enzymes associated with microtubules (Gitlits, 2000). Volker and Knull (1997) have defined a 43-amino-acid sequence in the carboxyl terminal region of α tubulin, which can bind and inhibit aldolase and hence termed tubulin glycolytic enzyme binding domain (T-GEBD-43 mer). Enolase is also associated with centrosomes. These reports led to the suggestion that glycolytic enzymes may not be entirely soluble in the cell, but they have transient interactions with structural components of the cell as well (Srere and Knull, 1998; Gitlits et al., 2000). Enolase is most abundant in mature spermatozoa and in residual bodies detached from elongating spermatids with little or no activity in meiotic primary spermatocytes and round spermatids. In mature sperm enolase is localized mostly to the tail. RT-PCR analysis of spermatogenic cells detected only the α isoform of enolase and appeared to be associated with microtubules and sperm flagellar motility. In presence of nucleotides, the association of microtubules is disrupted (Gitlits et al., 2000).

28.10. LACTATE DEHYDROGENASE-C

Lactate dehydrogenase (LDH) (L-lactate: NAD^+ oxidoreductase EC 1.1.1.2.7) is the terminal enzyme of glycolytic pathway and converts pyruvate to lactate in a reversible way. The LDH exists in multi-molecular tetrameric forms in a single cell. The LDH-A and LDH-B subunits and corresponding genes (*Ldh-a* and *Ldh-b*) are present in all somatic cells of mammalian origin. However, another subunit of LDH designated as LDH-C, a product of *Ldh-c (Ldh-3)* gene is exclusively present in spermatogenic cells of testis. The tetrameric form (LDH-C_4) and the subunit of LDH-C along with its *Ldh-c* gene is the best-characterized enzyme among testis or sperm specific enzymes so far described (review Gupta, 1999).

Primary Structure LDH-C: Subunits of lactate dehydrogenase, including LDH-C, have been shown to have a molecular weight of about 35-kDa in all vertebrates, but the amino acid composition differs for each of the subunits and for each species. The exceptions are D-LDHs, which are dimers of 70-kDa, in some organisms and tetramers in others (Goldberg, 1977). Pan et al. (1983) studied the 330 amino acid sequence of LDH-C_4 and compared it with LDH-A_4 and with LDH-B_4 from mouse and rat tissues. The results showed a high variation in amino acid sequence and composition among A, B, and C subunits. Li et al. (1983b) compared the amino acid sequence variation among seven LDH isozymes from different species with respect to whole dehydrogenase polypeptide chains and their four functional domains. The co-enzyme binding domain is more conserved than the substrate-binding domain. The sequence of the loop and helix D region of testicular LDH-C_4 isozyme is different from those of somatic isozymes, while NH_2-terminal arm was extremely variable. The characteristic catalytic properties of LDH-C_4 could be attributed to the marked difference in amino acid sequence of loop helix region and coenzyme and substrate binding site of LDH-C_4. Twenty residues in amino terminal of LDH isozymes are extremely variable, and their primary function is to stabilize the quaternary structure of tetrameric LDH through their interaction with carboxy terminal region of other subunits. Immunological properties of rat and mouse LDH-C_4 were compared and related to those variable 32 amino acid residues. Amino acid differences were also related to the structural-functional relationships of LDH-C_4 with somatic LDH isozymes (Li et al., 1983a).

Ldh-c Gene: Mouse and human sperm specific *Ldh-c* genes were cloned and sequenced. The mouse *Ldh-c* cDNA insert of 1236-bp consists of the protein coding sequence (999 bp), the 5' (54-bp) and 3' (113-bp) non-coding region, and the poly (A) tail (70-bp). The *Ldh-c* gene is expressed in testis but not in liver or any other tissue, and its mRNA is approximately 1400 nucleotides in length. The nucleotide and amino acid sequence of *Ldh-c* cDNA of mouse showed 73 and 72% homologies, respectively, with those of the mouse *Ldh-a*. The Southern blot analysis of genomic DNAs from mouse liver and human placenta indicated the presence of Ldh-c gene-related sequences. LDH-c has a 70% homology in amino acid sequence with B isozymes. The *Ldh-c* is a single copy gene. However, Wu et al, (1987) reported slight modifications in *Ldh-c* cDNA such as size comprising 1135 base pairs in length that can code 332 amino acids polypeptide and can express an mRNA of 1.5- kb in size. On the other hand, human *Ldh-c* comprising 1000-bp sub-fragment includes 66-bp of non-coding region with remaining base pairs that encode LDH-C protein. These structural studies revealed great similarities among the LDH isozymes and confirmed that LDH-C_4, LDH-B_4 and LDH-C_4 belong to a homologous gene family. The expression of *Ldh-c* gene is restricted to germinal epithelium and is developmentally regulated. The murine *Ldh-c* mRNA is first detected in preleptotene spermatocyte and persists post-meiotically to the round spermatid stage (Thomas et al., 1990).

```
5'-CGCCTCAACTGTCGTTGGTGTATTTTTCTGGTGTCACTTCTGTGCCTTCCTTCAAAGGTTCTCCAA ATG TCA ACT     75
                                                                      Met Ser Thr      2

GTC AAG GAG CAG CTA ATT GAG AAG CTA ATT GAG GAT GAT GAA AAC TCC CAG TGT AAA ATT ACT   138
Val Lys Glu Gln Leu Ile Glu Lys Leu Ile Glu Asp Asp Glu Asn Ser Gln Cys Lys Ile Thr    23

ATT GTT GGA ACT GGT GCC GTA GGC ATG GCT TGT GCT ATT AGT ATC TTA CTG AAG GAT TTG GCT   201
Ile Val Gly Thr Gly Ala Val Gly Met Ala Cys Ala Ile Ser Ile Leu Leu Lys Asp Leu Ala    44

GAT GAA CTT GCC CTT GTT GAT GTT GCA TTG GAC AAA CTG AAG GGA GAA ATG ATG GAT CTT CAG   264
Asp Glu Leu Ala Leu Val Asp Val Ala Leu Asp Lys Leu Lys Gly Glu Met Met Asp Leu Gln    65

CAT GGC AGT CTT TTC TTT AGT ACT TCA AAG GTT ACT TCT GGA AAA GAT TAC AGT GTA TCT GCA   327
His Gly Ser Leu Phe Phe Ser Thr Ser Lys Val Thr Ser Gly Lys Asp Tyr Ser Val Ser Ala    86

AAC TCC AGA ATA GTT ATT GTC ACA GCA GGT GCA AGG CAG CAG GAG GGA GAA ACT CGC CTT GCC   390
Asn Ser Arg Ile Val Ile Val Thr Ala Gly Ala Arg Gln Gln Glu Gly Glu Thr Arg Leu Ala   107

CTG GTC CAA CGT AAT GTG GCT ATA ATG AAA ATA ATC ATT CCT GCC ATA GTC CAT TAT AGT CCT   453
Leu Val Gln Arg Asn Val Ala Ile Met Lys Ile Ile Ile Pro Ala Ile Val His Tyr Ser Pro   128

GAT TGT AAA ATT CTT GTT GTT TCA AAT CCA GTG GAT ATT TTG ACA TAT ATA GTC TGG AAG ATA   516
Asp Cys Lys Ile Leu Val Val Ser Asn Pro Val Asp Ile Leu Thr Tyr Ile Val Trp Lys Ile   149

AGT GGC TTA CCT GTA ACT CGT GTA ATT GGA AGT GGT TGT AAT CTA GAC TCT GCC CGT TTC CGT   579
Ser Gly Leu Pro Val Thr Arg Val Ile Gly Ser Gly Cys Asn Leu Asp Ser Ala Arg Phe Arg   170

TAC CTA ATT GGA GAA AAG TTG GGT GTC CAC CCC ACA AGC TGC CAT GGT TGG ATT ATT GGA GAA   642
Tyr Leu Ile Gly Glu Lys Leu Gly Val His Pro Thr Ser Cys His Gly Trp Ile Ile Gly Glu   191

CAT GGT GAT TCT AGT GTG CCC TTA TGG AGT GGG GTG AAT GTT GCT GGT GTT GCT CTG AAG ACT   705
His Gly Asp Ser Ser Val Pro Leu Trp Ser Gly Val Asn Val Ala Gly Val Ala Leu Lys Thr   212

CTG GAC CCT AAA TTA GGA ACG GAT TCA GAT AAG GAA CAC TGG AAA AAT ATC CAT AAA CAA GTT   768
Leu Asp Pro Lys Leu Gly Thr Asp Ser Asp Lys Glu His Trp Lys Asn Ile His Lys Gln Val   233

ATT CAA AGT GCC TAT GAA ATT ATC AAG CTG AAG GGG TAT ACC TCT TGG GCT ATT GGA CTG TCT   831
Ile Gln Ser Ala Tyr Glu Ile Ile Lys Leu Lys Gly Tyr Thr Ser Trp Ala Ile Gly Leu Ser   254

GTG ATG GAT CTG GTA GGA TCC ATT TTG AAA AAT CTT AGG AGA GTG CAC CCA GTT TCC ACC ATG   894
Val Met Asp Leu Val Gly Ser Ile Leu Lys Asn Leu Arg Arg Val His Pro Val Ser Thr Met   275

GTT AAG GGA TTA TAT GGA ATA AAA GAA GAA CTC TTT CTC AGT ATC CCT TGT GTC TTG GGG CGG   957
Val Lys Gly Leu Tyr Gly Ile Lys Glu Glu Leu Phe Leu Ser Ile Pro Cys Val Leu Gly Arg   296
                                          ▼
AAT GGT GTC TCA GAT GTT GTG AAA ATT AAC TTG AAT TCT GAG GAG GAG GCC CTT TTC AAG AAG  1020
Asn Gly Val Ser Asp Val Val Lys Ile Asn Leu Asn Ser Glu Glu Glu Ala Leu Phe Lys Lys   317

AGT GCA GAA ACA CTT TGG AAT ATT CAA AAG GAT CTA ATA TTT TAA ATTAAAGCCTTCTAATGTTCCAC  1088
Ser Ala Glu Thr Leu Trp Asn Ile Gln Lys Asp Leu Ile Phe ***                           331

TGTTTGGAGAACAGAAGATAGCAGGCTGTGTATTTTAAATTTTGAAAGTATTTTCATTGATCTTAAAAAATAAAAACAAATTG  1171
GAGACCTGAAAAAAAAAAAAAAAA-3'                                                            1195
```

Fig.28.6. Nucleotide and amino acid sequence of human LDH-C_4 cDNA.The polyadenylylation signal is underlined. Arrowhead indicares internal EcoR1 site. Reproduced with permission from J. L. Millan et al, Proc Natl Acad Sci USA 1984;5311-15:1987© National Academy of Sciences (USA).

The Ldh-c gene has been sequenced from different species (Millan et al., 1987) **(Fig.28.6)**. The nucleotide and predicted aminoacid sequences of mouse somatic and testis germ cells showed 73 and 72% homologies, where as human cDNA and predicted amino acid sequence were 75 and 74% identical (Fig.28. 6). The mouse Ldh-c gene contains a TATA box, whereas the human gene contains six potential Sp1 binding sites in the same region. The Sp1 binds in the promoter region and appears to play a role in *Ldh-c* regulation at transcription level. The human *Ldh-c* gene is interrupted by six introns at positions homologous to those *Ldh-a* and *Ldh-b* genes (Takano and Li, 1989). The promoter of human *Ldh-c* has been cloned and studied in cell co-culture system. In human, testis-specific expression is driven by a 180bp fragment of promoter region (Cooker et al., 1993). Zhou and Goldberg (1996) have identified a 65-kDa protein designated as transcription initiation responder (TIR-1) in liver nuclear extract as a possible candidate for repressor, and a 103-kDa protein designated as testis initiation factor (TIF-1) in testis nuclear extract with possible role in transcription initiation.

Regulation of Ldh-c Gene: An in vitro transcription system was used to characterize murine *Ldh-c* promoter (Zhou and Goldberg, 1994; 1996). Murine *Ldh-c*, which is less than 25-kb away from *Ldh-a* on chromosome-7, arose by tandem duplication of *Ldh-a*. The *Ldh-c* gene expression starts early in leptotene/zygotene spermatocytes and continues through the elongated spermatid stage of spermatogenesis. Sequence analysis of a 720-bp fragment containing promoter region revealed several ubiquitous cis-regulatory elements, including one TATA box, one GC box, and two putative GCAAT elements. Deletion analysis demonstrated that a 60-bp fragment containing a 30-bp palindromic sequence was sufficient to direct transcription in testes nuclear

(TN) extract and that was transcriptionally inactive with liver nuclear (LN) extract. Gel retardation and Southwestern assays detected different protein binding activities to this palindromic sequence in TN and LN. The TATA box and 30-bp palindromic sequence were both required for promoter activity. Therefore, palindromic sequence contains both a negative element for repression in somatic tissues and a positive element for activation of *Ldh-c* gene in testis (Zhou and Goldberg, 1996). The human *Ldh-c* gene is a single copy gene with two alternative 5'-noncoding exons comprising an approximately 40-kb locus. The proximal region, relative to the transcription start site binds proteins extracts from human testis but not liver. The upstream region contains three putative regulatory elements: a glucocorticoid or hormone response element (HRE), a CRE, and an octamer factor binding site (OCT). The proximal promoter contains a small CpG island that is hypermethylated in testis (Bonny and Goldberg, 1995).

Cloning analysis of the m*Ldh-c* promoter revealed that a 100-bp core promoter was able to regulate testis-specific transcription in vitro and in transgenic mice. While the expression of the reporter in transgenic testes was limited to pachytene spermatocytes, the native LDH-C_4 was detected in pachytene and all subsequent germ cells. Kroft et al., (2003) found that transcription factor YY1 binds to the *mLdh-c* promoter, and that the *mLdh-c* 3' untranslated sequence does not permit a post-meiotic expression of a β-galactosidase reporter in transgenic mice whereas native *mLdh-c* mRNA is predominately meiotic, with only a low level of post-meiotic distribution. This suggested that the high level of LDH-C_4 in post-meiotic cells results from mRNA and protein stability (Kroft et al., 2003).

Hypermethylation is indicative of transcriptional repression. To address the relationship between the methylation and testicular gene expression DNA methylation patterns and mRNAs expression for LDH-A were compared with LDH-C in preparations from testes of prepubertal and sexually mature mice. At specific sites, *Ldh-a* gene was less methylated in adult testis than in spleen DNA; the decreased methylation in the testicular DNA occurred as early as type A spermatogonia. In contrast, DNA methylation of *Ldh-c* from type A and B spermatogonia from 8-day-old mice did not differ with DNA of spleen. The *Ldh-a* transcripts were low in early germ cells (6-12 day old mice) whereas *Ldh-c* transcripts appeared on day 12 and increased from 16-45 day old testis. These observations indicated that whereas DNA methylation of *Ldh-a* and *Ldh-c* at the 5'-CCGG-3' sites did not change markedly during testis development, both these genes are temporally expressed (Alcivar et al., 1991).

Of all the *Ldh* genes, promoter activity has been rigorously demonstrated only for the rat *Ldh-a*, which has been shown to be cyclic AMP-dependent in an *in vitro* transcription system (Short et al., 1991). The mouse *Ldh-c* message initiates at a single transcription start site, whereas human message has multiple initiation sites (Cooker et al., 1993). There is evidence that there are regulatory differences between the *Ldh-c* genes of mouse and rat, between mouse and baboon at the post-transcriptional level (Salehi-Ashtiani and Goldberg, 1993). These workers demonstrated that the *Ldh-c* gene message level is nearly nine-fold greater in mouse testes and remains high post-meiotically. In contrast, rat *Ldh-c* mRNA was highest in primary spermatocytes and reduced in spermatids. It was concluded that nuclear post-transcriptional events contribute to the differences in *Ldh-c* message levels.

3-D Structure of LDH-C: Temperature dependence and relative stability of LDH in the presence of dioxane H_2O indicated that hydrophobic forces may not be primary forces of importance in holding the subunits together, rather coulombic attractions prevail in the stabilization of tetrameric forms. The unit cell dimensions of mouse–C subunit are similar to the subunits of other LDH isozymes. At 7.5 Å resolution, the structure of LDH-C could not be distinguished from dogfish LDH-A. When the analysis was extended to 2.9Å resolution, the similarity to other LDH isozymes was even more apparent. The overall folding of apo-LDH-C polypeptide chain backbone

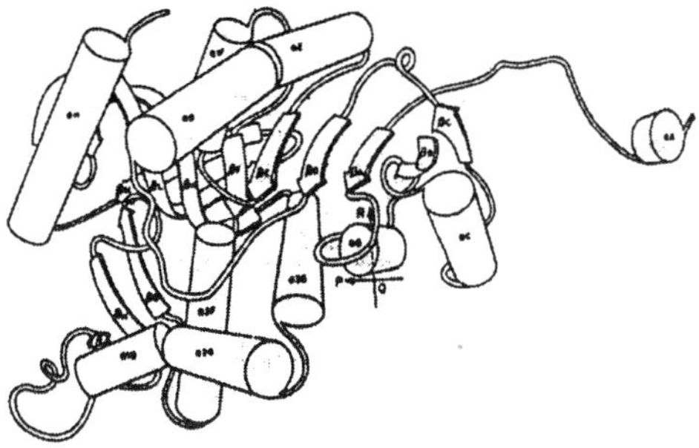

Fig.28.7. Three dimensional structure of mouse LDH-C subunit. Printed with permission from H.H. Hogrefe et al, J Biol Chem 262; 13155-64; 1987 © American Society for Biochemistry and Molecular Biology.

was found to be similar to the structures of somatic LDH. The same α-helices and β-pleated sheets observed in other LDH isozymes were apparent in LDH-C. The unique feature of mouse LDH-C structure is in co-enzyme binding loop. All other isozymes crystallized with the loop, corresponding to residues 99-112, in an open conformation. In a ternary complex with bound NAD^+ this loop moves 13 Å to cover the co-enzyme. Although mouse LDH-C_4 is isolated as the apoenzyme, it crystallizes with the loop closed. If this crystal structure reflects the conformation of LDH-C_4 in solution, it could explain the low turnover number of this isozyme. It other words if the loop is in equilibrium between open and closed conformations in solution, the efficiency of coenzyme binding would be reduced (Musick and Rossmann, 1979). Triclinic crystals of apo-LDH-C_4 in 40% saturated ammonium sulfate at 4°C showed tetrameric molecule in the unit cell with the dimensions: 84.8 Å, b = 76.6 Å, c = 63.9 Å; α=109.7°, β=89.5°, γ= 96.5° (Hogrefe et al., 1987) **(Fig.28.7)**.

LDH-C4 as a Immuno-contraceptive: The strategy behind sperm-based contraceptives is to induce antibodies in the female reproductive tract against sperm antigen at a sufficient level to block fertilization. It is envisioned that antibodies developed by this vaccine will act before fertilization events, thus terming it as a 'pre-fertilization contraceptive. Being a sperm-specific isozyme in man and animals, LDH-C4, and partially localized on the plasma membrane of sperm, it was considered and immunogen of choice (Gupta and Kinsky, 1994; Goldberg and Herr, 1999). Though the contraceptive effects were obvious in allogenic and heterologous systems, the effect was reversed when LDH-C4 was used in isogenic mice. Isogenic LDH-C4 immunization yielded protective effects on pregnancy (Gupta and Kinsky, 1994). However, due to presence of cross-reactive epitopes in LDH-C the development of an immuno-contraceptive based on LDH-C subunit does not seem to be feasible (Gupta, 1999). Nonetheless, identification of non-reactive antigenic epitopes from LDH-C can offer a potential application for development of a contraceptive vaccine (Goldberg and Herr, 1999).

28.11. PYRUVATE DEHYDROGENASE-E1α

The pyruvate dehydrogenase complex (PDH) plays a central role in aerobic energy metabolism. Seven subunits of this nuclear encoded enzyme complex are transported into the mitochondrion where they catalyze the conversion of pyruvate to acetyl CoA via an oxidative decarboxylation step for entry into the citric acid cycle. The isozyme of PDH-E1 is a heterotetramer of two α and

```
  1 MLAAFISRVL RRVAQKSARR VLVASRNSSN DATFEIKKCD LYLLEEGPPV TTVLTRAEGL
 61 KYYRMMLTVR RMELKADQLY KQKFIRGFCH LCDGQEACCV GLEAGINPSD HVITSYRAHG
121 VCYTRGLSVR SILAELTGRR GGCAKGKGGS MHMYTKNFYG GNGIVGAQGP LGAGIALACK
181 YKGNDEICLT LYGDGAANQG QIAEAFNMAA LWKLPCVFIC ENNLYGMGTS TERAAASPDY
241 YKRGNFIPGL KVDGMDVLCV REATKFAANY CRSGKGPILM ELQTYRYHGH SMSDPGVSYR
301 TREEIQEVRS KRDPIIILQD RMVNSKLATV EELKEIGAEV RKEIDDAAQF ATTDPEPHLE
361 ELGHHIYSSD SSFEVRGANP WIKFKSVS
```

Fig.28.8. Amino acid sequences of human testis specific pyruvate dehydrogenase E1α. Source: http// www.ncbi.nlm.nih.gov (Accession NP_005381 XP_352955).

two β subunits designated PDH-E1α and PDH-E1β. The PDH activity is dependent on phosphorylation and dephosphorylation state of three serine residues of E1α subunit. In human and murine somatic cells, *Pdha-1* codes for a somatic isoform and maps to the X-chromosome. The inactivation of *Pdha-1* in spermatocytes and spermatids is compensated by testis-specific expression of the autosomal gene *Pdha-2,* which encodes corresponding isozyme.

Pdha-E1α Genes: Mouse *Pdha-1* and *Pdha-2* genes coding two subunits of E1α of PDH show a high degree of homology with each other and with the human PDH genes, *PDH-A1* and *PDHA-2*. Conserved regions include mitochondrial import sequences, phosphorylation sites and a putative TPP binding site. The *Pdha-2* is a intronless retroposon that originated from reverse transcription of mRNA molecule. The *Pdha-2,* located on an autosome is only expressed in spermatogenic cells. The somatic variant is not expressed during spermatogenesis at a time when X-chromosome inactivation occurs; instead, its expression in spermatogenic cells has been linked to an autosomal, intronless gene, which encodes the testis specific isoform. Examination of various mouse tissues demonstrated that transcriptional expression of the mouse autosomal *Pdha-2* gene is tightly regulated. The *Pdha-2* expression occurs specifically in testis and that its expression is initiated at the meiotic prophase stage of spermatogenesis following birth. The human PDH genes have an analogous chromosomal arrangement to PGK genes in that two isoforms code for a functionally and structurally similar product. Comparison of human and mouse PDH and PGK genes sequences shows that the somatic sequences are more conserved relative to the testis specific isoforms and that the mouse *PdhE1α* genes show a faster rate of DNA change compared to their human counterparts (Fitzgerald et al., 1992). The human *PDH-E1α* revealed the presence of two loci: 1) *PDH-A-1* mapped to the X-chromosome (Xp22.1-22.2) is expressed in somatic tissues. This gene contains 10 introns and spans approx. 17 kb. 2) The second locus, *PDHA-2,* is localized to chromosome 4 (4q22 4q23) with expression limited to the testis. The autosomal testis specific isoform completely lacks introns and possesses characteristics of a functional processed gene. Chromosome mapping in mouse has also revealed the presence of two E1α loci, one maps to the X-chromosome and the other to chromosome 19 (Fitzgerald et al., 1992) **(Fig.28.8)**.

A cDNA clone (2.2-kb) encoding rat testis *PDH-E1α* has been shown to contain an open reading frame (from nt 974 to 2149) identical to that of liver reported earlier (Cullingford et al., BBA 1993; 1216 : 149-53) but contained a long 5'untranslated region, which had little identity to the other clone. Northern blot confirmed testis specific expression of this isoform. This clone was a gene product distinct from its X-linked somatic counterpart. The recombinant rat testis PDH E1 (containing both E1α and E1β subunits) was enzymatically active and phosphorylated in vitro by purified PDH-kinase p48 or p45, similar to the recombinant human liver enzyme (Jeng et al., 1998).

Localization of Pdh-E1α Genes: A considerable increase in the mRNA level of the testis specific *Pdh-E1α* gene was observed in spermatocytes at the pachytene stage and gradually decreased in spermatids as spermiogenesis progressed and it was not detectable in residual bodies. Transcripts of *Pdh-E1α* gene were not identified in Leydig and Sertoli cells. In contrast, somatic form of the *Pdh* gene was detected in spermatogonia, Leydig cells and Sertoli cells at a low level. Transcripts of the somatic form of the *Pdh* gene were not identified in other types of germ cells in adult mouse testis. The PDH-E1α protein is dramatically increased in primary spermatocytes and exists abundantly in pachytene spermatocytes. It was suggested that (1) the transcription switch from the somatic form of the *Pdh-E1α* gene to the testis specific *Pdh-E1α* gene occurs during the first meiotic prophase of spermatogenesis in adult mouse testis, and (2) PDH-E1α protein coded for by the testis specific *Pdh-E1α* gene is involved in the development of spermatogenic cells especially at stages after first meiotic prophase until the end of spermiogenesis in the testis (Takakubo and Dahl, 1992). Expression of E1α subunit gene *(Pdha-2)* during spermatogenesis has shown that a 2.0-kb *Pdha-2* mRNA is initially transcribed in meiotic prophase and precedes that of *Pgk-2* and corresponds to the appearance of *Ldh-3* mRNA. A second *Pdha-2* of 1.7-kb transcript is present in post-meiotic round spermatids. Polysomal analysis demonstrated that the 2.0-kb mRNA species is translated in diploid, pachytene spermatocytes, whereas the 1.7-kb mRNA species is translated in round spermatids, although a large proportion is present on the non-polysomal fraction that may act as a store for use in later stages of spermiogenesis (Fitzgerald et al., 1994).

Regulation of Pdha-2 Gene: Although *Pdha-2* is strongly up-regulated in the testis, its transcription is repressed in somatic tissues. Proximal region of the *Pdha-2* promoter between nt –187 to +22 harbors a transcriptionally active core. Transgenic mice harboring a construct containing only 187 bp of promoter and upstream sequence (core promoter) is sufficient for directing the testis specific expression of a CAT reporter gene. Like the endogenous *Pdha-2*, the CAT gene is expressed in testis in a stage specific manner (Iannello et al., 1997, Datta et al, 1999). The *"core"* promoter directs high levels of CAT reporter gene transcription. DNase I foot printing of the proximal promoter revealed four level of protection. One of these contains the consensus sequence for the Sp1 binding site and another the ATF/CREB binding site. The cis-sequences of the remaining two protected regions (designated MEP-2 and MEP-3; mouse E1α Promoter site) showed no apparent homology with cis-elements of other known transcription factors. The ATF/CREB and MEP binding can interact in a characteristic and specific manner with factors present in the nuclei of both testis and somatic cells. The factor, which recognizes the MEP-3 motif appears to be ubiquitous, whereas the MEP-2-protein complexes were tissue specific. Formation of a complex involving MEP-2 and a putative testis specific binding factor (τMEP-2BF) was first observed in testis of 2 weeks old mice and correlated with expression of *Pdha-2*. In contrast, formation of complexes between MEP-2BF (σMEP-2BP) decreased in the testis as spermatogenesis proceeds. Results suggested that 1) the MEP-2 binding factors are temporarily regulated during spermatogenesis, and 2) interactions involving these factors with the MEP-2 cis-element may be important for modulating *Pdha-2* expression (Iannello et al., 1993).

Tissue specific gene expression may be modulated by other mechanisms in addition to specific transcription factor availability and co-operativity. The methylation may be one of the mechanism by which repression of the testis-specific *Pdha-2* gene is established in somatic tissues. Methylation dependent repression of the *Pdha-2* core promoter is mediated regionally through a consensus activating transcription factor/CRE binding site located between nucleotides -54 and -62 upstream of the major transcriptional start site. Targeting of the CpG dinucleotide within this cis-element significantly disrupts the ability of this basal promoter to

```
  1 MAEQVTLSRT QVCGILREEL YQNDAFHQAD THIFIIMGAS GDLAKKKIYP TIWWLFRDGL
 61 LPKETFIVGY ARSQLTVDDI QKQSEPFFKA TPEERPKLEE FFTRNSYVVG QYDDPASYKH
121 LNSYINALHQ GMQANHLFYL ALPPTVYEAV TKNIQETCMS QTGFNRIIVE KPFGRDLQSS
181 NQLSNHISSL FREDQIYRID HYLDKEMVQN LMVLRFANRI FGPIWNGDNI VCVILTFKEP
241 FGTEGRGGYF DEFGIIRDVM QSHLLQMLCL VAMEKPATTD SDDVRNEKVK VLKCISEVET
301 DNVILGQYVG NPNGEGEAAN GYLDDPTVPR GSTTATFAAA VLYVKNERWD GVPFILRCGK
361 ALNERKAEVR LQFRDIPGDI FHQKCKRNEL VIRMQPNEAV YTTMMTKKPG MFFNPEESEL
421 DLTYGNKYKN VKLPGAYERL ILDVFCGCQM HFVRTDELRE GWRIFTPLLH KIEREKPQPF
481 PYVYGSRGPT EADELMRRVG FQYKGTYKGT HKH
```

Fig.28.9. Predicted amino acid sequence of glucose-6-phosphate dehydrogenase isozyme-2 from mouse testis (Hendriksen et a, 1997). (Source: http//www. ncbi.nlm.nih.gov. Accession NP_062341).

activate gene expression *in vitro* and completely abolishes promoter activity *in vivo.* DNase I foot printing experiment indicated that availability of the nuclear factor(s) binding this element was limited in sexually immature mouse testis, and as such, these factors may play an important role in the coordinate activation of early spermatogenic gene expression. The CpG dinucleotides associated with the hypersensitive region flanking the activating transcription factor / CRE binding site appeared to confer some conformational structure on the promoter since mutations at these specific CpG dinucleotides resulted in elevated basal levels of transcription. This raised the possibility of a bifunctional role for CpG dinucleotides in either methylation dependent or independent processes. This supports the notion that hypomethylation and transcription factor recruitment are necessary events that precede gene activation at the early stages of spermatogenesis (Iannello et al., 2000).

28.12. GLUCOSE-6-PHOSPHATE DEHYDROGENASE-2

It is well known that glucose metabolism through glycolysis provides energy to spermatozoa. However, the existence and the role of the pentose phosphate pathway (PPP) have not been unequivocally demonstrated in sperm. In characterizing the glucose metabolism of mouse sperm, it has been shown that mouse epididymal spermatozoa have a functional pentose phosphate pathway, implying that they produce NADPH, which is required for reducing reactions, and ribose phosphate, which is required for nucleic acid synthesis; and that sperm are able to fuse with the oocyte when NADPH is substituted for glucose, suggesting that sperm need to produce NADPH via the PPP in order to achieve fertilization (Urner and Sakkas, 1999).

Certain X-chromosomal genes remain transcriptionally silent in spermatids. In contrast, other X-chromosomal genes such as *Ubelx* and *MHR6A* are post-meiotically transcribed genes (Hendriksen et al., 1995). The inactivation of *Pgk-1* and *Pdha-1* in spermatocytes and spermatids is compensated by testis-specific expression of the autosomal genes *Pgk-2* and *Pdha-2,* which encode corresponding isozymes. The *Pgk-2* and *Pdha-2* are intronless retroposons that originated from reverse transcription of mRNA molecules. The autosomal gene *Zfa*, which encodes a protein of unknown function, represents another retroposon that shows testis-specific expression and that thus far has been found only in the mouse. The X-chromsomal gene glucose-6-phosphate dehydrogenase *(G6pd)* is expressed in most cell types of mammalian species. In the mouse, the chromosomal *G6pd-2* encoding a G6PD isoenzyme also does not contain introns and appears to represent a retroposed gene **(Fig.28.9)**. This gene is transcribed in postmeiotic spermatogenic cells in which the X-encoded *G6pd* sequence in a bacterial system showed that the encoded product is an active enzyme. Recombinant G6PD-2 but not recombinant G6PD-1, formed tetramers under reducing conditions. Under same conditions G6PD tetramers were also found in extracts of spermatid and spermatozoa. The G6PD-2 is one

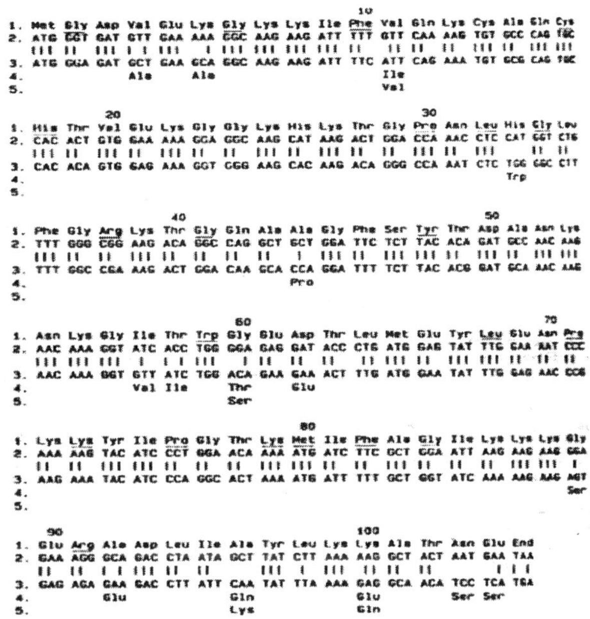

Fig.28.10. Coding region and amino acid sequences of rodent Cyt c. Line-1 indicates complete amino acid sequence of rat and mouse somatic Cyt c_s. Line-2 and -3 depict nucleotide sequence of coding regions of somatic *Cyt* c_s and testicular *Cyt* c_t genes respectively. Amino acid differences in rat and mouse Cyt c_t from Cyt c_s are given in line 4 and 5. Invariant residues are underlined. Printed with permission from J.V. Virbasius and R.C. Scarpulla, J Biol Chem 263; 6791-96: 1988 © American Society for Biochemistry and Molecular Biology.

of the very few known expressed retroposons encoding a functional protein (Hendriksen et al., 1997).

28.13. TESTICULAR CYTOCHROME C_t

The mammalian testis contains two distinct isoforms of cytochrome c, cytochrome c_s and cytochrome ct (Cyt Ct). Cytochrome c_s is a highly conserved protein encoded by a multi-pseudogene family in mammals and ubiquitously expressed in all somatic tissues in the mouse. Rat testis specific cytochrome c_t variant displays a number of notable differences with its somatic counterpart: a) In contrast to the multipseudogene family from mammalian somatic cytochrome c genes, the rat testis gene is a single copy in genomic DNA with no detectable pseudogenes. b) The rat testis specific gene is approximately 7-kb long with three introns totaling nearly 6.5-kb whereas the two introns dividing the 2.1-kb somatic gene occupy only 0.9-kb. Introns differ in position as well as size. c) The testicular variant has a longer 5'-untranslated leader (230 versus 70-bp for the somatic gene) with an upstream ORF of 129-bp beginning with an AUG in a favorable translational context. d) A single polyadenylation site in the testicular mRNA (approximately 900 nt) contrasts with the three functionally equivalent sites observed in rat somatic messages. e) In the mouse and rat the amino acid sequence of cytochrome c_s differs from that of cytochrome c_t in 14 and 15 of 104 amino acids respectively.

Although the amino acid sequences of the rat and mouse cytochrome c_s proteins are identical but cytochrome c_t proteins differ in four amino acids between the two closely related species (Virbasius et al., 1988; Morales et al., 1993) **(Fig.28.10)**.

Transcriptional Regulation: In testis low levels of cytochrome c_s mRNAs are detected in Leydig cells, myoepithelial cells, Sertoli cells, all types of spermatogonia, and during meiotic prophase. As spermatogenesis proceeds the cytochrome c_s is replaced by the testis specific cytochrome c_t. Cytochrome c_s is encoded by four different sizes of mRNA of about 1.3, 1.1, and 0.5-0.7-kb, which are widely expressed in somatic cells and in the testis germ cells ranging from type A spermatogonia to round spermatids. The 1.7-kb cytochrome mRNA appears to be predominantly expressed in the testis at late post-meiotic germ cells with most abundant in round spermatids, step 1 to 9. Cytochrome C_s was lost from the mitochondria as spermatogenesis advanced, while a relative increase occurred in cytochrome c_t during the zygotene to pachytene transition. This with other studies suggested activation of the cytochrome c_t gene during prophase of the first meiotic division. Cytochrome c_t is highly concentrated in mitochondria that are being degraded within cytoplasmic lobes of spermatids and in residual bodies. Two isoforms of cytochromes co-exist within the same mitochondrion during the transition from cytochrome c_s to cytochrome c_t. In addition, presence of both Cyt c_s and Cyt c_t in the chormatid bodies of spermatocytes and round spermatids suggested that the chromatid body is involved in the storage of these isozymes and possibly in their differential expression within germ cell mitochondria (Hess et al., 1993). Cytochrome c_t mRNAs is present in RNA preparation from isolated populations of meiotic and post-meiotic mouse germ cells. However by in situ hybridization, low levels of cytochrome c_t mRNAs, initially detected in zygotene spermatocytes, reach maximal levels in round spermatids. The absence of cytochrome c_s or c_t mRNAs in steps 10-19 spermatids suggested that the cytochrome c_t protein does not turn over rapidly in late stage of male germ cells (Morales et al., 1993).

The translation of a specific mRNA among others within a population is controlled by changes in its stability, primary structure, polyadenylation, as well as by protein-mRNA interactions. The fact that polysomal Cyt c_t mRNAs have longer poly $(A)^+$ tails than non-polysomal ones contributes to these differences in lengths. This suggests that translation of mRNA can be influenced by minor variation in their structures, alterations in 5'-UTR and 3'-UTRs as well as poly A^+ tail lengths that can lead to a differential polysomal distribution of transcripts. Cytochrome c_t RNAs sediment in two broad size classes: non-polysomal mRNAs are about 0.6 to 0.75-kb whereas polysomal mRNAs range from 0.7 to 0.9-kb. Both classes of mRNAs shorten to about 0.5-kb following deadenylation. Oligonucleotide–directed cleavage of size heterogeneity of cytochrome c_t mRNAs resides in the 5'-UTRs. Ribonuclease protection assays revealed that multiple cytochrome c_t mRNAs are transcribed from six different transcriptional start sites spanning a region of 59 nucleotides in the 5'UTR from +1 to +59. Each of the longer mRNAs has an upstream open reading frame, which starts at +8 and ends at +136 in the 5'UTR of the cytochrome c_t transcript. It has been suggested that heterogeneity in 5' untranslated region of mouse Cyt c_t mRNA leads to altered translational status of the mRNAs (Yiu et al., 1994).

Regulation of Cyt c_s and Cyt c_t Genes: Oligonucleotide RNase H cleavage of *Cyt c_s* mRNA revealed that the 1.7-kb mRNA contains over 1-kb of 5'-untranslated region, which is not present in four shorter Cyt c_s mRNAs. RNase protection assays indicated that this additional sequence arises from the utilization of an alternative transcription initiation site of the functional *Cyt c_s* gene, which is 1085-bp upstream of the initiation site for the four shorter *Cyt c_s* mRNAs. Sequence comparison of the putative promoter region with promoters of other post-meiotically expressed genes revealed several conserved regions. Utilization of this alternative transcription initiation site of Cyt c_s may be involved in the down-regulation of Cyt c_s during spermatogenesis (Hake and Hecht, 1993). The Cyt c_t-null testes undergo early atrophy equivalent to that, which occurs during aging as a consequence of a reduction in oxidative phosphorylation (Narisawa et al., 2002).

28.14. ENZYMES WITH LIMITED SPECIFICITY

Cytochrome C Oxidase: Cytochrome C oxidase II mRNA transcript is expressed in pachytene spermatocytes at high levels. Saunders et al, (1993) identified clones that encode all or part of the second subunit of cytochrome C oxidase (COX II), the terminal enzyme in the electron transport chains located in mitochondria. The COX II mRNA is strongly expressed in pachytene spermatocytes with the highest levels of expression at stages VII-VIII of the spermatogenic cycle. Although this is the androgen dependent stage, Northern analysis of total testicular RNA showed that expression of COX II mRNA was not altered detectably by EDS-induced androgen withdrawal nor the expression was altered detectably by treatment with high doses of FSH. In the absence of pachytene spermatocytes, COX II mRNA was found in preleptotene spematocytes and in Sertoli cell cytoplasm. Cytochrome C oxidase has thirteen subunits, three of which (I-III) are encoded on the mitochondrial genome. The physiological significance of the high levels of expression of COX II mRNA in pachytene spermatocytes and its stage dependent changes presumably reflect requirements for alteration in energy demand as these cells enter the final stages of meiosis (Saunders et al., 1993).

Creatine Kinase: Creatine kinase (CK, ATP creatine phosphotransferase, EC 2.7.3.2) participates in ATP regeneration, which is the primary source of energy in living organisms. The CK from herring spermatozoa has high activity and has a different electrophoretic mobility from isoenzymes present in skeletal muscle. Treatment of herring spermatozoa with tributyltin (TBT) caused a time-dependent decrease of viability after 6 h exposure. It was suggested that CK could be a good biomarker of sperm cell membranes degradation in the case when lactate dehydrogenase release from permeabilized cells is not possible for rapid determination of the effect of TBT (Grzyb et al., 2003).

Aldo-keto Reductase: The mouse aldo-keto reductase (AKR) AKR7A5 is similar to rat aflatoxin aldehyde reductase (AKR7A1) and to human brain succinic semialdehyde reductase (AKR7A2). The mouse enzyme is present in tissues including liver, kidney, testis and brain, and is able to reduce several carbonyl compounds, including succinic semialdehyde, carboxybenzaldehyde, 4-nitrobenzaldehyde and 9,10phenanthrenequinone (Hinshelwood et al., 2003). A human testis specific protein (HTSP) was identified initially by the search of the Expressed Sequence Tag database, followed by the screening of human testis cDNA library. The HTSP transcripts were detected only in human testis and not in other reproductive organs. Four isoforms, HTSP-1, -2, -3 and -4 were identified; these isoforms were generated by alternative splicing of a single gene. The HTSP gene is localized to chromosome 10. The HTSP4, was composed of 307 amino acids and shared 56% identity to mouse vas deferens protein as well as human aldose-reductase at amino acid levels, and showed activity towards 9, 10-phenanthrenequinone among the putative substrates. Accordingly, HTSP is a new member of the aldo-keto reductase superfamily with as yet unidentified function (Nishinaka et al., 2003).

Aldehyde Dehydrogenase: Mouse testis expresses a tissue-specific aldehyde dehydrogenase. Using the technique of isoelectric focusing, the expression of four to five banded "testis-specific aldehyde dehydrogenase" was confirmed in the mouse testis. However, the activity was found not only in the testis but also in the uterus and in embryonic tissues. The tissue distribution as well as substrate specificity and isoelectric points indicated that the "testis-specific aldehyde dehydrogenase" corresponds to mouse type 2 retinaldehyde dehydrogenase (Maly et al., 2003).

28.15. REFERENCES

Adra CN, Ellis NA, McBurney MW. The family of mouse phosphoglycerate kinase genes and pseudogenes. Som Cell and Mol Gen 1988; 1988: 69-81.

Alciver AA, Trasler JM, Hake LE, Salchi Ashtiani K, Goldberg E, Hecht NB. DNA methylation and expression of the genes coding for lactate dehydrogenases A and C rodent spermatogenesis. Biol Reprod 1991; 44: 527-535.

Ando H, Haruna Y, Suzuki M, Yamada S, Okabe M, Nakanishi Y. Ectopic activation of the transcription promoter for the testis-specific mouse pgk-2 gene on elimination of a cis-acting upstream DNA region. Growth Differ 2000; 42: 385-93.

Ariel M, McCarrey J, Cedar H. Methylation patterns of tesits specific genes. Proc Natl Acad Sci USA 1991; 88: 2317-21.

Ashworth AB, Skene B, Swift S, Lovell-Badge R. *Zfa* is an expressed retroposon derived from an alternative transcript of the *Zfx* gene EMBO Journal 1990; 9: 1529-34.

Bonny C, Goldberg E. CpG-rich promoter of human LDH-C is differently methylated in expressing and non-expressing tissues. Dev Genet 1995; 16: 210-217.

Broceno C, Ruiz P, Reina M, Vilaro S, Pons, G. The muscle specific phosphoglycerate mutase gene is specifically expressed in testis during spermatogenesis. Eur J. Biochem 1995;. 227: 629-635.

Broceno C, Walsh K, Pons G. A 1.3-kb upstream 5' region of the rat phosphoglycerate mutase M gene confers testis and skeletal muscle-specific expression in transgenic mice. Biochem Biophys Res Commun 1999; 263: 244-50.

Buehr M, McLaren A. An electrophoretically detectable modification of glucose phosphate isomerase in mouse spermatozoa. J Reprod Fertil 1981; 63: 169-73.

Bunch DO, Welch JE, Magyar PL, Eddy EM, O'Brien DA. Glyceraldehyde-3-phosphate dehydrogenase-S protein distribution during mouse spermatogenesis. Biol Reprod 1998; 58: 834-41.

Burchell A, Watkins L, Hume R. Human fetal testis endoplasmic reticulum glucose -6- phosphatase enzyme protein. Biol Reprod 1996; 55: 298-303.

Cooker LA, Brooke CD, Kumari M, Hofmann MC, Millian JL, Goldberg E. Genomic structure of promoter activity of the human testis lactate dehydrogenase gene. Biol Reprod 1993; 48: 1309-19.

Dahl HHM, Brown RM, Hutchison WM, Maragos C, Brown GK. A testis specific form of the human pyruvate dehydrogenase E1 alpha subunit is coded for by an intronless gene on chromosome 4. Genomics 1990; 8 : 225-32.

Datta U, Wexler ID, Kerr DS, Raz I, Patel MS.Characterization of the regulatory region of the human testis-specific form of the pyruvate dehydrogenase alpha-subunit (PDHA-2) gene. Biochim Biophys Acta 1999; 1447, 236-43.

Fitzgerald J, Dahl HHM, Iannello RC. Differential expression of two testis-specific transcripts of the mouse *Pdha-2* gene during spermatogenesis. DNA Cell Biol 1994; 13: 531-7.

Fitzgerald J, Hutchison WM, Dahl HHM. Isolation and characterization of the mouse pyruvate dehydrogenase E1α genes. Biochim Biophys Acta 1992; 1134: 83-9.

Fundele R, Winking H, Illmensee K, Jagerbauer EM. Developmental activation of phosphoglycerate mutase-2 in the testis of the mouse. Dev Biol 1987; 124: 562-6.

Gebara MM, McCarrey JR. Protein-DNA interactions associated with the onset of testis-specific expression of the mammalian *Pgk-2* gene. Mol Cell Biol 1992; 12: 1422-31.

Gitlits VM, Toh BH, Loveland KL, Sentry JW. The glycolytic enzyme enolase is present in sperm tail and displays nucleotide-dependent association with microtubules. Eur J Cell Biol 2000; 79: 104-11.

Goldberg E, Herr JC. LDH-C_4 as a Contraceptive vaccine. In: Reproductive Immunology (Gupta SK, Ed), Narosa Publishing House, New Delhi,1999, pp 309-15.

Goldberg E, Isozymes in testes and spermatozoa, In: Isozymes: Current Topics Biol Med Res (Ratazzi MS, Scandalios JG, and Whitt GS, Eds). New York, Alan R Liss, Vol. I, pp. 79-124, 1977.

Grzyb K, Rychlowski M, Biegniewska A, Skorkowski EF.Quantitative determination of creatine kinase release from herring (Clupea harengus) spermatozoa induced by tributyltin. Comp Biochem Physiol C Toxicol Pharmacol.2003;134:207-13.

Gupta GS, Kinsky RG. Effects of Immunization with sperm specific lactate dehydrogenase with and without muramyl dipeptide as adjuvant. Ind J Med Res 1994; 100: 98-105.

Gupta GS. LDH-C_4: A unique target of mammalian spermatozoa. Crit Rev Biochem Mol Biol 1999; 34: 361-85.

Hake E, Hecht NB. Utilization of an alternative transcription initiation site of somatic cytochrome C in mouse produces a testis-specific cytochrome c mRNA. J Biol Chem 1993; 268: 4788-97.

Hendriksen PJM, Hoogerbrugge JW, Baarends WM, de Boer P, Vreeburg JTM, Vos EA, van der Lende T, Grootegoed JA. Testis specific expression of functional retroposon encoding glucose-6 phosphate dehydrogenase in mouse. Genomics

1997; 41: 350-359.

Hendriksen PJM, Hoogerbrugge JW, Themmen APN, et al. Postmeiotic transcription X and Y chromosomal genes during spermatogenesis in mouse. Dev Biol 1995; 170: 730-33.

Hess RA, Miller LA, Kirdy JD, Margoliash E, Goldberg E. Immunoelectron microscopic localization of testicular and somatic cytochromes in the seminiferous epithelium of the rat. Biol Reprod 1993; 48: 1299-8.

Hogrefe HH, Griffith JP, Rossmann, MG, Goldberg E. Characterization of the antigenic sites on the refined 3 A resolution. Structure of Mouse testicular LDH-C_4. J Biol Chem 1987: 262: 13155-64.

Iannello R, Kola I, Dahl HHM. Temporal and tissue-specific interactions involving novel transcription factors and the proximal promoter of the mouse *Pdha-2* gene. J Biol chem. 1993; 268: 22581-590.

Iannello RC, Gould JA, Young JC, Giudice A, Medcalf R, Kola I. Methylation-dependent silencing of the testis-specific *Pdha-2* basal promoter occurs through selective targeting of an activating transcription factor/cAMP-responsive element-binding site. J Biol Chem 2000; 275: 19603-8.

Iannello RC, Young J, Sumarsono S, Tymms MJ. Dahl HHM, Gould J, Hedger M, Kola I. Regulation of *Pdha-2* expression is mediated by proximal promoter sequence and CpG methylation. Mol Cell Biol 1997; 17: 612-619.

Jeng J, Kallarakal AT, Kim SF, Popov KM, Song BJ. Pyruvate dehydrogenase E1 α isoform in rat testis : cDNA cloning, characterization, and biochemical comparison of the recombinant testis and liver enzymes. Comp Biochem Physiol B Biochem Mol Biol 1998; 120: 205-16.

Kafri T, Ariel M, Brandeis M, Shemar R, Urven L et al., Developmental pattern of gene specific DNA methylation in the mouse embryo and germ line. Genes Dev 1992; 6: 705-14.

Kroft TL, Li S, Doglio L, Goldberg E. A transgenic analysis of mouse lactate dehydrogenase C promoter activity in the testis. J Androl 2003; 24: 843-52.

Kumari M, Stroud JC, Anji A, McCarrey JR. Differential appearance of DNase I-hypersensitive sites correlates with differential transcription of *Pgk* genes during spermatogenesis in the mouse. J Biol Chem 1996; 271: 14390-97.

Lehotzky A, Telegdi M, Liliom K, Ovadi J. Interaction of phosphofructokinase with tubuli and microtubules. Quantitative evaluaion of the mutual effects. J Biol Chem 1993; 268: 10888-94.

Li SS-L, Feldmann RJ, Okabe M, and Pan Y-CE. Molecular features and immunological properties of lactate dehydrogenase C_4 isozymes from mouse and rat testes. J Biol Chem 1983a; 258: 7017-28.

Li SS-L, Fitch WM, Pan Y-CE, Sharief FS. Evolutionary relationships of vertebrate lactate dehydrogenase isozymes A_4 (muscle), B_4 (Heart) and C_4 (testis). J Biol Chem 1983 b; 258: 7029-32

Maly IP, Crotet V, Toranelli M.The so-called "testis-specific aldehyde dehydrogenase" corresponds to type 2 retinaldehyde dehydrogenase in the mouse. Histochem Cell Biol 2003; 119: 169-74.

McCarrey JR, Thomas K, Human testis-specific PGK gene lacks introns and possesses characteristics of a processed gene. Nature (Lond) 1987; 326: 501-505.

McCarrey JR. Evolution of tissue-specific gene expression in mammals. How a new phosphoglycerate kinase gene was formed and refined. BioScience 1994; 44: 20-27.

McCarrey JR. Kumari M, Aivaliotis MJ, Wang Z, Zhang P. Marshall F, VandeBerg JL. Analysis of the cDNA and encoded protein of the human tetis-specific PGK2 gene. Dev Genet 1996; 19: 321-332.

Mezquita J, Pau M, Mezquita C. Several novel transcripts of glyceraldehyde-3-phosphate dehydrogenase expressed in adult chicken testis. J Cell Biochem 1998; 71: 127-39.

Michelson AM. Bruns GAP. Morton CC. Orkin SH. The human phosphoglycerate kinase multigene family. HLA associated sequences and an X-linked locus containing a processed pseudogene and its functional counterpart. J Biol Chem 1985: 260: 6982-6992.

Millan JL, Driscoll CE, LeVan KM, Goldberg E. Epitopes of human testis specific lactate dehydrogenase from cDNA sequence. Proc Natl Acad Sci USA. 1987; 84: 5311-15.

Minchenko O, Opentanova I, Caro J. Hypoxic regulation of the 6-phosphofructo-2-kinase/fructose-2,6-bisphosphatase gene family (PFKFB-1-4) expression in vivo. FEBS Lett 2003; 554: 264-70.

Mizuguchi H, Cook PF, Tai C-H, Hasemann CA, Uyeda K. Reaction mechanism of fructose -2,6- biphosphatase. A mutation of nucleophilic catalyst, histidine 256, induces an alteration in the reaction pathway. J Biol Chem 1998; 274: 2166-75.

Morales R. Hake E, Hecht NB. Cellular localization of the mRNAs of the somatic and testis-specific cytochrome C during spermatogenesis in the rat. Mol Reprod Develop 1993; 34: 196-205.

Mori C, Nakamura N, Welch JE, Shiota K, Eddy EM. Testis specific expression of mRNAs for a unique human type 1 hexokinase lacking the porin-binding domain. Mol Reprod Dev 1996; 44: 14-22.

Mori C, Welch JE, Fulcher KD, O'Brien DA, Eddy EM. Unique hexokinase messenger ribonucleic acids lacking the porin-binding domain are developmentally expressed in mouse spermatogenic cells. Biol Reprod 1993; 49: 191-203.

Mori C, Welch JE, Sakai Y, Eddy EM. In situ localization of spermatogenic cell-specific glyceraldehyde 3-phosphate dehydrogenase (Gapd-s) messenger ribonucleic acid in mice. Biol Reprod 1992; 46: 859-868.

Musick WDL, Rossman MG. The structure of mouse testicular lactate dehydrogenase- C_4 at 2.9Å resolution. J Biol Chem 1979: 254: 7611-20.

Nakamura N, Komiyama M, Fujioka M, Mori C. Sorting specificity of spermatogenic cell specific region of mouse hexokinase-s (mHk1-s). Mol Reprod Dev 2003; 64: 113-9.

Narisawa S, Hecht NB, Goldberg E, Boatright KM, Reed JC, Millan JL.Testis-specific cytochrome c-null mice produce functional sperm but undergo early testicular atrophy. Mol Cell Biol. 2002; 22: 5554-62.

Naz RK, Morte C, Ahmad K, Martinez P. Hexokinase present in human sperm is not tyrosine phosphorylated but its antibodies affect fertilizing capacity. J Androl 1996; 17: 143-50.

Nishinaka T, Azuma Y, Ushijima S, Miki T, Yabe-Nishimura C. Human testis specific protein: a new member of aldo-keto

reductase superfamily. Chem Biol Interact. 2003; 143-144: 299-305.
Olds-Clarke P, Pilder SH, Visconti PE, Moss SB, Orth JM, Kopf GS. Sperm from mice carrying two t haplotypes do not possess a tyrosine phosphorylated form of hexokinase. Mol Reprod Dev 1996; 43: 94-104.
Robinson MO, McCarrey JR, Simon MI, Transcriptional regulatory regions of testis-specific PGK2 defined in transgenic mice. Proc Natl Acad Sci USA 1989; 86: 8437-41.
Ruiz-Lazano P, deLecea L, Buesa C, Perez de la Osa et al. The gene encoding rat phosphoglycerate mutase subunit M: cloning and promoter analysis in skeletal muscle cells. Gene 1994; 147: 243-48.
Salehi-Ashtiani K, Goldberg E. Differences in regulation of testis specific lactate dehydrogenase in rat and mouse occur at multiple Levels. Mol Reprod Dev 1993; 35; 1-7.
Saunders PT, Millar MR, West AP, Sharpe RM. Mitochondrial cytochrome C oxidase II messenger ribonucleic acid is expressed in pachytene spermatocytes at high levels and in a stage-dependent manner during spermatogenesis in the rat. Biol Reprod 1993; 58: 57-67.
Short S, Short ML, Milkowski DM, Jungmann RA. Functional analysis of cis-and trans regulatory elements of lactate dehyedrogenease A sub-unit promoter by in vitro transcription. J Biol Chem 1991; 266: 22164-72.
Srere PA, Knull HR. Location, Location, Location. Trends Biochem Sci 1998; 23: 319-20.
Takakubo F, Dahl HH. The expression pattern of pyruvate dehydrogenase E1α subunit genes during spermatogenesis in adult mouse. Exp. Cell Res 1992; 199: 39-49.
Takano T, Li SS-L. Human testicular lactate dehydrogenase-C gene interrupted by six introns at positions homologus to those of *Ldh-a* (muscle) and *Ldh-b* (heart) genes. Biochem Biophys Res Commune 1989; 159: 579-83.
Thomas K, Del Mazo J, Eversole P, Bellve A, Hiraoka Y, Li SS-L, Simon M. Developmental regulation of expression of the lactate dehydrogenase (LDH) multigene framily during mouse spermatogenesis. Development 1990; 109: 483-93.
Tominaga N, Minami Y, Sakakibara R, Uyeda K. Significance of the amino terminus of rat testis fructose 6 phosphate, 2 kinase:fructose-2,6-biphosphatase. J Biol Chem 1993; 268, 15951-57.
Tominaga N, Tsujikawa T, Minami Y, Watanabe F, Sakakibara R, Kosaku U. Effect of replacement of the amino and the carboxyl termini of rat testis fructose 6-phosphate, 2-kinase: fructose 2, 6-bisphosphatase with those of the liver and heart isozymes. Arch Biochem Biophys 1997; 247: 275-81.
Travis AJ, Foster JA, Rosnbaym MA, Visconti BE, Gerton GL, Kopf GS, Moss SB. Targeting of a germ cell-specific type1 hexokinase lacking a porin-binding domain to the mitochondria as well as to the head and fibrous sheath of murine spermatozoa. Mol Biol Cell 1998; 9: 263-76.
Travis AJ, Riedel D, Hofinann R, Moss B. Wilson E, Kop G. A novel NH_2 terminal, non-hydrophobic motif targets a male germ cell-specific hexokinase to the endoplasmic reticulum and plasma membrane. J Biol Chem 1999; 274: 34467-75.
Urner F, Sakkas D. A possible role for the pentose phosphate pathway of spermatozoa in gamete fusion in the mouse. Biol Reprod 1999; 60: 733-39.
Vertessy BG, Orosz F, Kovacs J, Ovadi J. Alternative binding of two sequential glycolytic enzymes to microtubules. J. Biol. Chem 1997; 272: 2552-46.
Virbasius JB, Scarpulla RC. Structural and expression of rodent genes encoding the testis specific cytochrome C. J Biol Chem 1988; 263: 6791-6.
Visconti PE, Olds-Clarke P, Moss SB, Kalab P, Travis AJ, de las Heras M, Kopf GS. Properties and localization of a tyrosine phosphorylated form of hexokinase in mouse sperm. Mol Reprod Dev 1996; 43: 82-93.
Volker KW, Knull HR. A glycolytic enzyme binding domain on tubulin. Arch. Biochem Biophys 1997; 338: 237-43.
Welch JE, Brown PR, O'Brien Da, Eddy EM, Genomic organization of a mouse glyceraldehyde 3-phosphate dehydrogenase gene (Gapds) expressed in post-meiotic spermatogenic cells. Dev Genet 1995; 16:179-89.
Westhoff D, Kamp G. Glyceraldehyde -3- phosphate dehydrogenase is bound to fibrous sheath of mammalian spermatozoa. J Cell Sci 1997; 110 (Pt 15):1821-9.
Wu KC, Chan K, Lee CYG, Lau YF. Molecular isolation and sequence determination of the cDNA for the mouse sperm specific lactate dehydrogenase – X gene. Biochem Biophys Res Commun 1987; 146: 964-70.
Yang J , Chennathukuzhi V, Miki K, O'Brien DA, Hecht NB.Mouse testis brain RNA-binding protein/translin selectively binds to the messenger RNA of the fibrous sheath protein glyceraldehyde 3-phosphate dehydrogenase-S and suppresses its translation in vitro. Biol Reprod 2003; 68: 853-59.
Yakirevich E, Naot Y. Cloning of a glucose phosphate isomerase/neuroleukin-like sperm antigen involved in sperm agglutination. Biol Reprod 2000; 62: 1016-1023.
Yiu GK, Gu W, Hecht NB. Heterogeneity in the 5' untranslated region of mouse Cytochrome C_T mRNAs leads to altered translational status of the mRNAs. Nucleic Acids Res. 1994; 22: 4599-4606.
Yuen MH, Lee MY, Cookl F, Uyeda K, Hasemann A. Crystal structure of the H256A mutant of rat testis fructose-6-phosphate,2-kinase/fructose-2, 6-bisphosphatase. J Biol Chem 1999; 274: 2176-84.
Zhang LP, Stroud JC, Eddy CA, Walter CA, McCarrey JR. Multiple elements influence transcriptional regulation from the human testis-specific *Pgk2* promoter in transgenic mice. Biol Reprod 1999; 60: 1329-37.
Zhang LP, Stroud JC, Walter CA, Adrian GS, McCarrey JR. A gene specific promoter in transgenic mice directs testis-specific demethylation prior to transcriptional activation in vivo. Biol Reprod 1998; 59: 284-92.
Zhang X, Kleene KC. cDNA copies of the testis-specific lactate dehydrogenase (LDH-C) mRNA are present in spermatogenic cells in mice, but processed pseudogenes are not derived from mRNAs that are expressed in haploid and late meiotic spermatogenic cells. Mamm Genome 1999; 10: 6-12.
Zhou W, Goldberg E. A dual function palindromic sequence regulates testis-specific transcription of the mouse lactate dehydrogenase C gene in vitro. Biol Reprod 1996; 54: 84-90.
Zhou W, XuJ, Goldberg E. A60bp core promoter sequence of murine lactate dehydrogenase – C is sufficient to direct testis-specific transcription in vitro. Biol Reprod 1994; 51: 425-32.

Chapter 29

FIBROUS SHEATH, DENSE FIBERS, AND PLASMA MEMBRANE OF SPERM

The mammalian sperm tail presents a complex organization in which number of additional structures, namely outer dense fibers and fibrous sheath surround the central axoneme and are thought to regulate flagellar motility. The development of sperm tail with propelling activity occurs during spermiogenesis. Three keratins – Sak57, Odfl and Odf2 and other proteins (the 26S proteasome and the Odfl-binding protein Spag4) are temporarily stored in the manchette before being sorted to the developing sperm tail, and hence the manchette provides a transient storage to both structural and signaling proteins (Kierszenbaum, 2001). The major protein components and their isoforms defining axoneme have been discussed in Chapters 8 and 9. In this section, we shall concentrate on iso-proteins discovered in plasma membrane, outer dense fiber and fibrous sheath encasing outer dense fibers of sperm flagellum.

29.1. OUTER DENSE FIBER PROTEINS

The flagellum of the mammalian sperm possesses a characteristic complex of nine outer dense fibers around the axoneme, which generates a 9+9+2 cross-sectional pattern. Each fiber, joined anteriorly to one of the striated columns of the connecting piece, courses longitudinally in close proximity to the corresponding doublet of the axoneme. In some species, such as the rat and hamster, they are thick structures and extend the full length of the principal piece of the sperm tail, while in others they terminate halfway along the principal piece. In various species, these fibers in contrary to the identical appearance of each doublet of the axoneme, they differ from each other in cross sectional shape and size. The biochemical nature of the accessory structures of the sperm tail, namely the proteins of ODFs of the mid-piece and the principal piece is the subject of recent investigations. Though the precise function(s) of the ODFs remains to be determined, but they are thought to impart rigidity and directionality to tail movement. The ODF proteins are synthesized by the spermatids during the latter half of spermiogenesis and are assembled around the axoneme in a proximal to distal direction. At least six distinct protein bands, for example, have been identified in rat ODFs by PAGE with apparent molecular weights of 87, 30-32, 25-26, 19-20, 13, and 11.5-12-kDa most of which are phosphorylated at serine residues (Vera et al., 1984; O'Bryan et al., 1998). To date the gene sequences encoded for about a dozen ODF proteins are known.

```
        (A) ODF1 FROM HUMAN SPERM (ACCESSION   NP_077721)

  1 MAALSCLLDS VRRDIKKVDR ELRQLRCIDE FSTRCLCDLY MHPYCCCDLH PYPYCLCYSK
 61 RSRSCGLCDL YPCCLCDYKL YCLRPSLRSL ERKAIRAIED EKRELAKLRR TTNRILASSC
121 CSSNILGSVN VCGFEPDQVK VRVKDGKVCV SAERENRYDC LGSKKYSYMN ICKEFSLPPC
181 VDEKDVTYSY GLGSCVKIES PCYPCTSPCS PCSPCNPCNP CSPCNPCSPY DPCNPCYPCG
241 SRFSCRKMIL

        (B) ODF1 FROM MOUSE SPERM (ACCESSION   CAA55965)

  1 MAALSCLLDS VRRDIKKVDR ELRQLRCIDE ISSRCLCDLY MHPYCCCDLH PYPYCLCYSK
 61 RSRSCGLCDL YYPCCLCDYK LYCLRPSLRS LERLRRTTNR ILASSCCSSN ILGSVNVCGF
121 EPDQVKVRVK DGKVCVSAER ENRYDCLGSK KYSYMNICKE FSLPPCVDEK DVTYSYGLGS
181 CVKIESPCYP CTSPCNPCNP CSPCSPCAPC ACGPCGPCGP CGPCGPCDPC NPCYPCGSRF
241 SCRKMIL

        (C) ODF1 FROM RAT SPERM (ACCESSION  P21769)

  1 MAALSCLLDS VRRDIKKVDR ELRQLRCIDE ISSRCLCDLY MHPYCCCDLH PYPYCLCYSK
 61 RSRSCGLCDL YYPCCLCDYK LYCLRPSLRS LERLRRTTNR ILASSCCSSN ILGSVNVCGF
121 EPDQVKVRVK DGKVCVSAER ENRYDCLGSK KYSYMNICKE FSLPPCVDEK DVTYSYGLGS
181 CVKIESPCYP CTSPCNPCNP CSPCSPCGPC GPCGPCGPCG PCGPCDPCNP CYPCGSRFSC
241 RKMIL
```

Fig 29.1. Amino acid sequences of outer dense fiber protein-1(Odf1) of sperm tail from human (A), mouse (B), and rat (C) (source:http///www.ncbi.nlm.nih.giv).

29.1.1. ODF27/Odf1

By searching for the rat and human homologue of the *Drosophila* germ cell product Mst(3)gl-9, a cDNA was found that encoded a 244-aminoacid polypeptide with a total amino acid composition similar to that of a 26-32-kDa protein of the ODF. This gene has been called *rts5/1*, *rt7*, and *odf27* depending on the species from which it was isolated (Henkel et al., 1992). The cDNA of *odf27* encodes the major 27-kDa protein of rat sperm ODF with a high proportion of a repetitive motif, Cys-Gly-Pro, at the carboxy-terminal end, reminiscent of the testis specific Mst(3)CGP proteins of *Drosophila melanogaster*. Nick translation probes of the *odf27* cDNA recognized two complementary mRNAs, first transcribed in round spermatids. Haploid expression of these transcripts reach a peak in steps 8-10 of spermiogensis at the time when transcription ceases and remains at high levels from steps 11 to 15 but diminish in steps 16-18 at the time when ODF protein synthesis and assembly are maximum (Morales et al., 1994). Using the *Odf27* leucine zipper as bait in a yeast two-hybrid system, a testis-specific protein was isolated, whose interaction with ODF27 depends critically on the ODF27 leucine zipper. The gene of 111-450 nt encodes a product that localizes to ODF and that the product is identical to ODF84, which contains two C-terminal leucine zippers. All leucine residues in the upstream leucine zipper are required for interaction with ODF27. The putative proteins encoded by *Odf1* in mouse, rat and man are similar **(Fig.29.1)**.

A total of 15 amino acids in the C-terminal region were deleted in the mouse protein, compared with the rat protein. Through *in situ* hybridization to metaphase chromosomes, the *Odf1* gene was localized to mouse chromosome 15 region B2-C. The chromosomal localization of the *Odf1* gene extends the hitherto known linkage group consisting of MYC (Myc), PVT1 (Pvt1), GPT (Gpt), and TG (Tg) common to human chromosome 8 and mouse chromosome 15 in the proximal direction of both chromosomes. The linkage group now extends from band q24 to band q22 of human chromosome 8 and from region D2-E to region B2-C of mouse chromosome 15 (Hoyer-Fender et al., 1995).

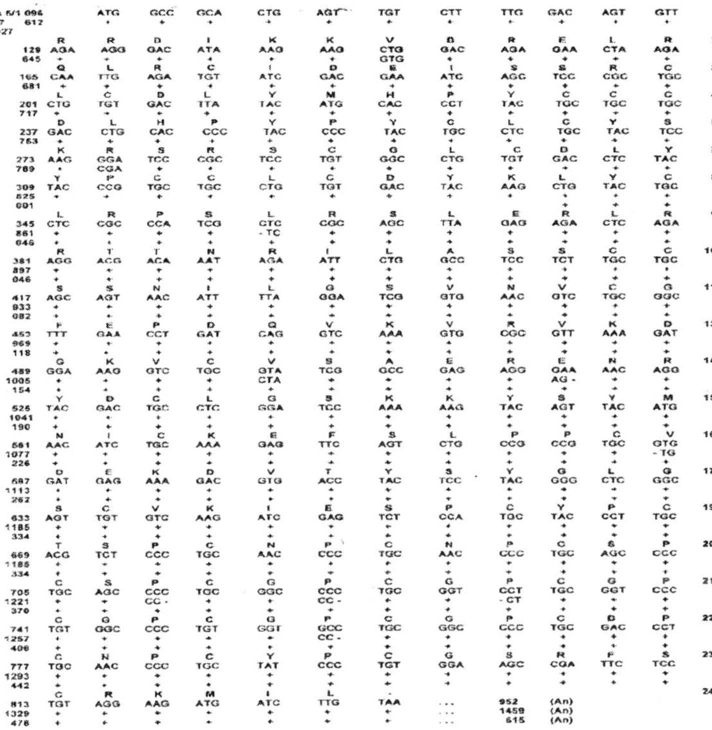

Fig. 29.2. Largest open reading frame (ORF) of ODF-27 cDNA (OD 27) and its predicted amino acid sequence. Comparison of ORF of ODF-27 with rts 5/1 cDNA and RT7 cDNA and their amino acid sequences. Adapted from C.R. Morales et al, Mol Reprod Dev 37;229-40:1994© John Wiley and Sons Inc. http://www3.interscience.wiley.com/cgi-bin/jabout/37692/ProductInformation.html.

RT7: The *RT7* gene is expressed as an abundant RNA in round spermatids and encodes a protein of 26-27-kDa. The mAbs raised against a peptide from the predicted N-terminal amphipathic α-helix of the rat RT7 protein recognizes RT7 protein or N-terminal parts of it. Several mAbs recognize RT7 protein in elongating spermatids in nonrandom fashion. RT7 can form stable complexes with itself that are associated through a region located in the N-terminal half of RT7. The RT7 protein appears as a major sperm tail component, bears homology with 27kDa-ODF polypeptide, and might be a structural component of sperm tail outer dense fibers (Higgy et al., 1994; Morales et al., 1994) **(Fig 29.2)**.

A human gene, homologous to the *Mst(3)CGP* gene family of *Drosophila melanogaster* encoding an ODF protein of 241 amino acids was described by Gastmann et al. (1993). The transcribed region has a size of approximately 1-kb and contains two exons of 416-bp and 406-bp, respectively, excluding the 3' untranslated region. The gene is expressed in testis but not in other human tissues and resembles the *Drosophila Mst(3)CGP* gene family in the male germ line. The encoded sequence corresponds to approximately 32-kDa in the extract of human sperm flagella and localized to band q22 of chromosome 8.

29.1.2: Odfl Interacting Proteins

SPAG2: Two isoforms of sperm associated antigen 2 (SPAG2), SPAG2-1 and SPAG2-2, were identified in testis and placental libraries, respectively. In Southern analysis of human genomic DNA with a probe common to the two SPAG2 isoforms indicated a single SPAG2 gene. Therefore, alternative splicing is likely the mechanism for production of variant mRNAs. In human testis SPAG2 expresses in primary spermatocytes, with decreased or arrested expression in post-meiotic cells. SPAG2 is an intracellular component of the sperm flagellum localized to ODF. SPAG2 does not exhibit the molecular characteristics of keratin like proteins, nor does SPAG2 demonstrate sequence homology with other characterized ODF proteins (Diekman et al., 1998).

Spag4: The cloning and characterization of another Odfl-interacting protein, Spag4 suggested that Spag4 mRNA is spermatid specific, and the 49-kDa protein complexs specifically with Odfl, but not Odf2, mediated by a leucine zipper. It also has the property of self-association. In contrast to Odfl and Odf2, Spag4 protein localizes to two microtubule containing spermatid structures. Spag4 is detectable in the transient manchette and it is associated with the axoneme in elongating spermatids and epididymal sperm. Data suggest a role for Spag4 in protein localization to two major sperm tail structures and the role for leucine zippers in molecular interactions during sperm tail morphogenesis (Shao et al., 1999).

Spag5: *Spag5* is another gene, which encodes a 200-kDa testicular protein that interacts strongly with Odfl. The *Spag5* is transcribed and translated in pachytene spermatocytes and spermatids. It bears 73% similarity with the mitotic zippers in the C-terminal part of the Spag5 protein, the downstream of which is involved in interaction with Odfl, thus strengthening the importance of the leucine zipper in sperm tail protein interactions (Shao et al., 1999; 2001).

OIP1: Zarsky et al, (2003) discovered a novel RING finger protein OIP1 that binds to conserved amino acid repeats in sperm tail protein Odfl.

29.1.3. ODF84/ Odf2

The *odf84* is a translationally delayed gene that codes for an 84-kDa protein. Schalles et al, (1998) reported the isolation of a cDNA (111-450 nt) encoding a 84-kDa ODF. Riboprobes generated from the clones recognized four testicular-specific transcripts of 1.6, 2.2, 2.4, and 2.8-kb in both rat and bull, of which the product of the 2.4-kb mRNA co-migrates with ODF 84 protein. The 2.2-and 2.4-kb mRNAs are first transcribed during meiotic prophase while the other two species are first expressed in rounds spermatids. The anti-84 reactive ODF proteins are synthesized and assembled in the cytoplasm of elongated spermatids (steps 9-18) with peak activity occurring in step 16 of spermiogenesis. Labeling was selective to the assembling ODF and connecting piece of the tail and to granulated bodies of the cytoplasmic lobe. Immunogold labeling provided evidence that the 84-kDa ODF protein is localized to both the cortex and medulla of the ODF in contrast to the sole medullary localization of the major 27-kDa ODF protein. Thus the 84-kDa ODF protein, encoded by the 2.4 transcript, is translationally regulated, packaged after synthesis into granulated bodies, assembled in a proximal to distal direction along the axoneme and may interact by means of leucine zippers specifically with the 27-kDa ODF protein during assembly. Its localization to both the cortex and medulla of the ODF, as opposed to exclusive medullary localization to the 27-kDa ODF protein, and the

```
  1 MSASSSGGSP RFPSCGKNGV TSLTEKKVLR TPCGAPSVTV TKSHKRGMKG DTVNVRRSVR
 61 VKTKNPPHCL EITPPSSEKL VSVMRLSDLS TEDDDSGHCK MNKYDKKIDS LMNAVGCLKS
121 EVKMQKGERQ MAKRFLEERK EELEEVAHEL AETEHENTVL RHNIERIKEE KDFTMLQKKH
181 LQQEKECLMS KLVEAEMDGA AAAKQVMALK DTIGKLKTEK QMTCTDINTL TRQKELLLQK
241 LSTFEETNRT LRDLLREQHC KEDSERLMEQ QGALLKRLAE ADSEKARLLL LLQDKDKEVE
301 ELLQEIQCEK AQAKTASELS KSMESMRGHL QAQLRCKEAE NSRLCMQIKN LERSGNQHKA
361 EVEAIMEQLK ELKQKGDRDK ETLKKAIRAQ KERAEKSEEY AEQLHVQLAD KDLYVAEALS
421 TLESWRSRYN QVVKDKGDLE LEIIVLNDRV TDLVNQQQSL EEKMREDRDS LVERLHRQTA
481 EYSAFKLENE RLKASFPPME DKLNQAHLEV QQLKASVKNY EGMIDNYKSQ VMKTRLEADE
541 VAAQLERCDK ENKMLKDEMN KEIEAARRQF QSQLADLQQL PDILKITEAK LAECQDQLQG
601 YERKNIDLTA IISDLRSRVR DWQKGSHELA RAGARLPR
```

Fig.29.3. Amino acid sequence of rat sperm outer dense fiber tail protein 2 (Source: http//www. ncbi.nlm.nih.gov. Accession NP_058909).

presence of two leucine zippers, only one of which interacts with the 27-kDa ODF, suggest that it could act as a link between proteins of the two regions of the ODF (Shalles et al., 1998). The *Odf2* gene is localized to human chromosome 9q34 (Shao et al., 1997; 1998).

The human homologue of rat *Odf2* gene has been isolàted. In rat *Odf2* or the 84-kDa ODF protein interacts with ODF1 of 27kDa; the interaction is mediated by leucine-zipper during ODF assembly along sperm axoneme. The cDNAs encoding Odf2 protein of mouse have shown variability in amino acid sequences of N-terminal part where as C-terminal is highly conserved among different species. The *Odf2* RNA transcripts were restricted to germ cells. The transcription of Odf2 starts at step5 spermatids of tubular stage V (Hoyer-Fender et al., 1998) (**Fig.29.3**). Following the screening of a rat testis expression library with an ODF-specific antiserum, a third ODF gene, *odf2*, was identified and found to encode a variety of transcripts that translated into proteins of 65-70-kDa.

Transcription of Odf2 starts in step 6 spermatids, which coincides with transcription of the major outer dense fiber protein Odfl and with the formation of the sperm tail. Computer analyses of the structure of the encoded Odf2 protein revealed an overall a-helical structure with two regions identical to the dimerization region of the leucine zipper motif (Brohman et al., 1997).

KTT4: An Orthologue of Odf2: Of several cloned cDNAs isolated from an adult rat testis, two identical clones encoding a 2.2-kb cDNA *(KTT4)* were found to contain an open reading frame of 578 amino acids including two leucine zipper motifs. KTT4 mRNA was abundant in round spermatids, and homologous mRNAs were present in testes from mice and marmosets. The *KTT4* gene is conserved in humans, monkey, mice, dogs, and cattle, and its mRNA was first detectable in pachytene spermatocytes and thereafter was abundant in round and elongating spermatids. Expression of KTT4 was not altered by EDS induced androgen withdrawal, but in rats treated with methoxyacetic acid, a marked reduction in KTT4 was noted associated with the depletion of round spermatids. Database searches showed it to be homologous to published sequences for an ODF protein of the sperm tail (Odf2/Odf84) (Turner et al., 1997)

MPM2 Reactive Sperm Protein (MSP): One component that may be involved in the regulation of sperm centrosomal activity in the oocyte is a phosphorylated protein complex (MPM-2-reactive sperm protein; MSP) with a molecular mass of 77-85-kDa. The MSPs isolated from boar sperm are identical to rat/bull 85-kDa ODF/CP85 complex of the outer dense fibers and connecting piece. It is likely that any involvement of these proteins in the regulation of sperm centrosomal activity is through the process of connecting piece disassembly (Long et al., 1997).

29.1.4. Cysteine Rich Proteins in ODF

TPX-1: Isolation of the cDNA sequence of a protein likely to be an ODF component and a homologous relationship with the so-called male sperm cell-specific gene *tpx-1* has been reported by O'Bryan et al (1998). This gene was characterized earlier from murine and human sources (Kasahara et al, 1988 c/r O'Bryan et al., 1998) (Chapters 2 and 34). The *tpx-1*-like gene encodes a 1.6-kb mRNA and a 243-amino-acid protein that has significant homology with members of the cysteine rich secretary protein (CRISP) family and partial homology with several venom/allergen proteins from both plants and insects. The protein Tpx-1 in sperm is formed during rat spermatogenesis. The expression of *tpx-1* increased dramatically to a maximum in step 11-12 spermatids. Tpx-1 mRNA is first expressed in the late pachytene spermatocytes, but the production of these tpx-1 proteins is translationally delayed for 4-5 days before being incorporated into the developing sperm acrosome. Concurrent with sperm head formation, tpx-1 protein was incorporated into the developing sperm tail, and specifically the ODFs. The tpx-1 protein was seen within structures resembling granulated bodies in the cytoplasmic lobe of elongating spermatids and was incorporated subsequently into the growing tail in a manner consistent with ODF development. In addition, tpx-1 protein was localized at the connecting piece of the neck and longitudinal columns of the fibrous sheath, suggesting common protein components in these cytoskeletal structures. As such, tpx-1 may have functional significance in the processes of sperm head development and tail formation (O'Bryan et al., 2001). Synthesis and application of peptide immunogens related to protein tpx-1 represent promising candidates for investigators into the role of tpx-1 in the immunoregulation of sperm function in mammalian models (Keah et al., 2001).

Autoantigen 1- A Tpx homologue: *A*utoantigen 1 (AA1) is the major autoantigen of the guinea pig sperm acrosome and was first detected in pro-acrosomal vesicles of pachytene spermatocyes. The ~25kDa protein was packaged into the acrosome during spermiogenesis. The cDNAs encoding autoantigen 1 predicts a precursor protein of 244 amino acids including a 21 amino acid hydrophobic, secretory signal sequence to have a molecular mass of 24kDa. The AA1 is the guinea pig homologue of the testis-specific protein tpx-1 in mice and humans and shows significant amino acid sequence homology with other CRISP's rat and mouse acidic epididymal glycoproteins (AEG also known as proteins D/E in rats) (see chapters 1 and 34). Auto-antigen 1-RNA from guinea pig spermatogenic cells shows a 1.5-kb message and is first detected in pachytene spermatocytes and strongest in round spermatids, with low activity in condensing spermatids (Foster and Gerton, 1996).

Sak57: The SAK57 is a 57-kDa acidic intermediate filament keratin that first appeared in association with the manchette in step 13 rat spermatids. However, later in development, this protein appears to become associated with the ODF. The Sak57 is present in rat spermatocytes, spermatids, and sperm, and has the characteristics of a keratin intermediate filament present during meiotic and post-meiotic stages of spermatogenesis. The 57-kDa Sak57 (Sak57=for spermatogenic cell/sperm-associated keratin) from outer dense fibers of rat sperm tails (Kierszenbaum et al, 1996) yielded two 15-mer and 10-mer fragments, which showed 70-100% homology to human, rat, and mouse keratins and corresponding to the 1Å and 2Å regions of the α-helical rod domain of keratins. Rat testis gave two proteins: i) a soluble 83-kDa protein with pI range 5.9-6.3, regarded as a precursor, and ii) both detergent-insoluble and soluble 57-

kDa protein with pI range 5.0-5.9, corresponding to the mature form Sak57. The testicular soluble form was phosphorylated. Sperm tail samples displayed only the Sak57 detergent-insoluble form with pI of 4.7-4.8 and rod shape pattern in electron microscopy. The three-dimensional arrangement of Sak57 was analysed in pachytene spermatocytes using. confocal laser scanning microsopy. Isolated outer dense fiber and sperm tails displayed an immunoreactive product in the form of linear clusters. In elongating spermatids (steps 10-11), Sak57 reactivity was predominant in the head region whereas pachytene spermatocytes showed a cortical cytoplasmic distribution. (Kierszenbaum et al., 1996; Tres and Kierszenbaum, 1996). During spermatogenic steps 8-12 Sak57 is associated with the microtubular mantle of the manchette, the transient microtubular structure largely regarded as formed by tubulin and microtubule-associated proteins and which encircles the spermatid nucleus during shaping and chromatin condensation. At later stages, Sak57 immunoreactive sites in the spermatid head region disappeared. It appears that, during early spermiogenesis, microtubular Sak57 scaffolding is associated with the spermatid nucleus during shaping and chromatin condensation. During late spermiogenesis, the dispersion of the manchette coincides with the progressive visualization of Sak57 in the peraxonemal outer dense fibers and longitudinal column of the fibrous sheath in developing spermatid tail (Tres and Kierszenbaum, 1996).

Bovine Cysteine Rich Protein: Kim et al (1995) compared the bovine and porcine cDNA of an outer dense fiber protein with the human and rat cDNA sequences and found that the coding sequences and the 5' and 3'-untranslated regions of the ODF cDNAs are highly conserved. The bovine, porcine, human and rat ODF protein sequences revealed that the protein displays a high degree of similarity, ranging from 87% to 98%. The protein is rich in cysteine and contains the CXP repeat at the C-terminus, which is different in number among mammalian species. All the 27cysteine residues in the ODF sequences except those CXP repeat are conserved in four species. The transcription start site in the bovine ODF gene is localized 98-bp upstream of the translation start site. Alignment of the 5' flanking region of bovine ODF with the rat revealed 83% similarity in the first 130 nucleotides upstream of the transcription start site. This conserved region contains a TATA-like box (TTTAAA) and binding sites for AFT/CREB transcription factors (Kim et al., 1995). One Odf gene in bovine is located to chromosome 4p11-p14 (Musilova et al., 1997).

MAK 248: The 48-kDa protein is a dimer of the 24-kDa protein, which was referred as MAK248. Structural analysis suggested that MAK248 is a novel CRISP protein and a member of the CAP (CRISP, Ag 5, PR-1) family of proteins. Based on amino acid sequence homology, it is possible that MAK248 functions as a protease inhibitor (Yudin et al., 2002).

29.2. FIBROUS SHEATH PROTEINS

The fibrous sheath (FS) is a unique cytoskeletal structure located in the principal piece of the sperm flagellum and consists of two longitudinal columns connected by closely spaced circumferential ribs that assemble from distal to proximal throughout spermiogenesis. The longitudinal columns connect to the axoneme, replacing ODFs 3 and 8 in the principal piece. The fibrous sheath is believed to influence the degree of flexibility, plane of flagellar motion, and the shape of the flagellar beat. The FS encases the ODFs and microtubules in the principal piece and extends distally beyond the termination of the ODFs to end at the principal piece/ end piece junction. The axoneme, devoid of ODFs and the FS, continues into the end-piece, completing the flagellum. The FS has been reported to be a relatively insoluble structure with extensive disulfide crosslinks. Viewed in longitudinal section, the FS displays a segmented rib like appearance. It appears that the fibrous sheath functions as a scaffold for proteins in signaling pathways that might be involved in regulating sperm maturation, motility, capacitation, hyper-activation, and/or acrosome reaction and for enzymes in the glycolytic pathway that provide energy for the hyper-activated motility of sperm allowing them to penetrate through the zonapellucida (Eddy et al., 2003).

29.2.1. Electrophoretic Studies

Biochemical analysis of FS shows that FS is composed mainly of 80-kDa protein in rat and 74-kDa in mouse. The presence of very high disulphide linkages in FS suggests that like ODF the FS may serve a passive elastic function in motility. Proteins that make up the FS in rat sperm are assembled in a distal proximal direction and are synthesized throughout stages 8-17 of spermiogenesis. In rat the FS consists of up to 17 polypeptides ranging form 75 to 14.4-kDa and also includes a major protein of 80-kDa. Human fibrous sheath revealed at least 14 protein bands of which the most intensely stained were of molecular weight 84, 72, 66.2, 57, 32 and 28.5-kDa. The rabbit fibrous sheath revealed at least 10 protein bands, of which the most intensely stained were 35.2, 32.7 and 28.5-kDa. The amino acid composition of the purified fibrous sheath from human, rat, and rabbit spermatozoa was similar, being high in aspartic acid and/or asparagine and glutamic acid and/or glutamine, serine, alanine, leucine, lysine and glycine, but low in histidine, tyrosine and isolecucine. This composition being similar in different species suggests that mammalian sperm tail fibrous sheaths are composed of similar types of proteins, although there are apparent differences in protein components between species (Kim et al., 1992).

Immunoreactive peptides of 68, 53, and 45-kDa were recognized by mAbs on immunoblots of non-reduced human sperm extracts, while a 68-kDa band and a strongly immunoreactive triplet from 54 to 51-kDa were recognized in reduced sperm extract (Beecher et al., 1993). Based on mAb 4F7 a doublet of polypeptides of about 95-kDa was identified throughout the FS of human sperm. This FS protein co-localizes with actin irrespective of the location of their abnormal assembly, suggesting that the assembly of the FS and actin could be under the control of some common morphogenetical factor (s) (Escalier et al., 1997).

29.2.2. Thioredoxins

Thioredoxin-1: Thioredoxins are proteins that participate in different cellular processes via redox-mediated reactions as general protein disulphide reductases and regulate several cellular processes such as transcription factor, DNA binding activity, apoptosis and DNA synthesis. In mammalian organisms, thioredoxins are generally ubiquitously expressed in all tissues, with the exception of Sptrx-1, which is specifically expressed in sperm cells. Spermatid-specific thioredoxin-1 (Sptrx-1) is the first member of the thioredoxin family of proteins with a tissue-specific expression pattern, found exclusively in the tail of elongating spermatids and spermatozoa. Human Sptrx-1 eluates as a 400kDa protein in gel permeation. Analysis of CD spectra of fragments 1-360 and 361-469 and comparison to spectra of full-length Sptrx-1 supports a two-domain organization with a largely unstructured N-terminal domain and a folded thioredoxin-like C-terminal domain. The Sptrx-1 behaves as an oxidant in vitro when using selenite, but not oxidized glutathione, as electron acceptor. This oxidizing enzymatic activity suggests that Sptrx-1 might govern the stabilization (by disulfide cross-linking) of the different structures in the developing tail of spermatids and spermatozoa (Jimenez et al.,

2002a). The mouse *Sptrx-1* gene and protein are similar to those described for the human orthologue. The mouse *Sptrx-1* ORF encodes for a protein of 462 aa composed of an N-terminal repetitive domain of a 15 residue motif followed by a C-terminal domain typical of thioredoxins. The mouse Sptrx-1 gene is located at 17E1.2-1.3 and mostly in the round spermatids (Jimenez et al., 2002b).

The developmental analysis of Sptrx-1 in rat seminiferous tubule showed that its expression begins at step 9, gets progressively stronger until steps 14-16, and then diminishes in steps 17 and 18 until disappears in step 19 spermatid. The SEM analysis showed its localization to the assembling longitudinal columns of the fibrous sheath, whereas the forming ribs of the fibrous sheath were unlabeled. It specifically associates to the fibrous sheath during development but does not become a permanent structural component. The pattern of Sptrx-1 during rat spermiogenesis suggests that it could be part of a nucleation center for the formation of the longitudinal columns and transverse ribs that bridge the latter (Yu et al., 2002).

Thioredoxin 2: Sadek et al (2001) reported the identification and characterization of another member of the thioredoxin family, with a tissue- specific distribution in human sperm, termed Sptrx-2. The Sptrx-2 ORF encodes for a protein of 588 amino acids with two different domains: an N-terminal thioredoxin domain encompassing the first 105 residues and a C-terminal domain composed of three repeats of a NDP kinase domain. The *Sptrx-2* gene spans about 51-kb organized in 17 exons and maps at locus 7p13.14. Sptrx-2 mRNA is exclusively expressed in human testis, mainly in primary spermatocytes, while Sptrx-2 protein expression is detected from the pachytene spermatocytes stage onwards, peaking at round spermatids stage. The sperm specific expression of Sptrx-2 together with its chromosomal assignment suggests it a potential locus for flagellar anomalies and male infertility phenotypes (Sadek et al., 2001). The cloning, developmental expression, and location of mouse and rat SPTRX-2 proteins display a high homology to their human orthologue in the thioredoxin and NDP kinase domains; the coding genes are located at syntenic positions. However, recombinant full length Sptrx-2 expressed in bacteria displayed neither thioredoxin nor NDP kinase enzymatic activity. The testis-specific expression of murine Sptrx-2 mRNA occurs mostly in round spermatids. The SPTRX-2 expression becomes prominent in the cytoplasmic lobe of step 15-18 spermatids and diminishes in step 19 just before spermiation. However, in the spermatid tail, SPTRX-2 was confined to the principal piece. The SPTRX-2 incorporation into the FS lags well behind FS assembly, suggesting it is required during the final stages of sperm tail maturation in the testis and/or epididymis, where extensive disulfide bonding of FS proteins occurs (Miranda-Vizuete et al., 2003) **(Fig.29.4)**.

Thioredoxin/Glutathione Reductase System: The study of Jurado et al., (2003) provides the first absolute expression patterns of genes coding for all known components of both thioredoxin (Trx) and glutaredoxin (Grx) systems in mouse: Trx1, Trx2, Grx1, Grx2, TrxR1, TrxR2, thioredoxin/glutathione reductase, and glutathione reductase. The study showed: i) unique expression profiles for each transcript; ii) continuous expression during embryogenesis with outstanding up-regulations of Trx1 and TrxR1 mRNAs; iii) drastic differences in mRNA stability; iv) testis-specific differences in the amounts (relative to total isoforms) of transcripts yielding the mitochondrial Grx2a and 67-kDa TrxR1 variants; and v) coordinated up-regulation of TrxR1 and glutathione reductase mRNAs in response to superoxide stress in an organ-specific manner (Jurado et al., 2003). Two thioredoxin proteins, DHD (dead head) and Trx-2 have been characterized in *D melanogaster*. The TrxT associates with the Y chromosome lampbrush

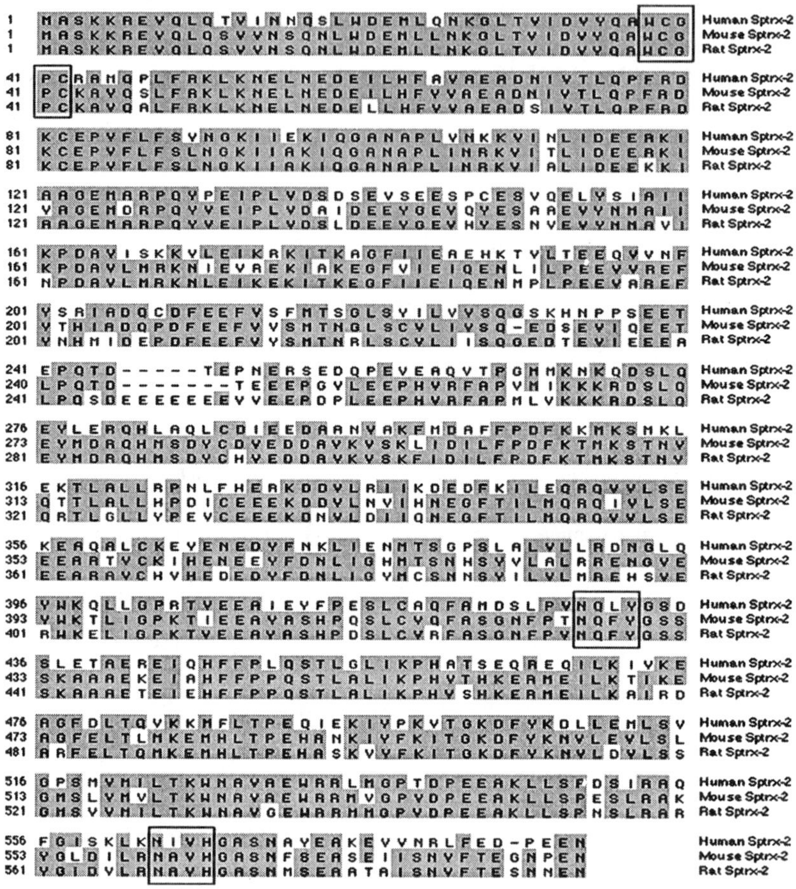

Fig.29.4. Amino acid sequence comparison of thioredoxin 2 (Sptrx 2) from human, mouse and rat sperm. Printed with permission from A. Miranda-Vizuete et al. J Biol Chem 278; 44874-85: 2003 © American Society for Biochemistry and Molecular Biology.

loops, ks-1 and kl-5 in primary spermatocytes. However, the association is lost in the absence of the Y chromosome. Nuclear thioredoxins seemed to have regulatory functions in the germ line (Svensson et al., 2003).

29.3. ANCHOR PROTEINS IN FIBROUS SHEATH

In recent years, a great emphasis has been placed on identifying and characterizing proteins capable of organizing the appropriate regulatory proteins and enzymes involved in various signaling cascades. Anchoring proteins have been identified for PKA, PKC, mitogen-activated protein kinases, protein phosphatases and calmodulin. Some A-kinase-anchoring proteins (AKAP) can bind several enzymes, regulatory proteins and compounds on distinct domains of the protein, thereby forming kinase-phosphatase complexes. For example AKAP 75/79 and calmodulin-binding protein also binds PKC, calcineurin and phospatidylinositol 4, 5-

bisphosphate (Erlichman et al, 1999) (see section 29.5). The highly polarized sperm cell is well suited for investigating the structure and function of various- anchoring proteins.

A-kinase Anchor Proteins: cAMP-dependent protein kinase is known to play a central role in mediating cellular responses to alterations in cAMP levels (Chapter 19). Compartmentalization of the sperm cell makes it an excellent model system by which one can study the role of targeting and anchoring of cAMP-dependent protein kinase (PKA) and other enzymes in regulating cell function. The anchoring of PKA is accomplished at specific sub-cellular sites by the binding of subunit R to an amphipathic helix-binding motif located within AKAPs. Synthetic peptides containing an amphipathic helix domain are able to competitively disrupt PKA binding to AKAPs. The AKAPs contain a common structural motif that binds with nanomolar affinity to the R subunit of PKA. Other sperm anchoring proteins are capable of binding with both RI and RII and have been labeled dual D-AKAPs. Since PKA has broad substrate specificity, presumably one of the primary functions of PKA anchoring is to spatially restrict its action, thus ensuring specificity of the function. In addition to PKA, some AKAPs also simultaneously bind other signal transduction molecules such as calmodulin and other protein kinases. Thus some AKAPs could serve as scaffolding proteins that coordinate the action of several signal transduction enzymes. Scaffolding proteins are responsible for forming the framework of multi-protein complex that could ensure precise regulation of the signaling cascade and allow the cell to respond specifically to a extra-cellular stress (Moss and Gerton, 2001).

Spermatozoa possess different types of AKAPs (Moss and Gerton, 2001). The AKAP82 (Carrera et al., 1994; Johnson et al., 1997) and FSP-95 (Mandal et al., 1999) are located at the fibrous sheath of principal piece and anchor RII subunit of PKA. The AKAP110 is bound to the principal piece and acrosomal region in addition to its interaction with ropporin (Vijayaragharan et al., 1999; Carr et al., 2001; Urnar and Sakkas, 2003). It has been suggested that the interaction of the regulatory subunit of PKA with sperm AKAPs is a key regulator of sperm motility (Vijayaragharan et al., 1997).

29.3.1. AKAP220

Human AKAP-220: A cDNA of 9923-bp comprising the ORF and encoding a 1901-amino–acid AKAP with an apparent mobility of 220-kDa, named human AKAP220 (hAKAP 220) was identified during human spermatogenesis. The hAKAP220 amino acid sequence has high similarity to rat AKAP220 in the 1167 C-terminal residues, but contained 727 residues in the N-terminus not present in the rat AKAP220 sequence (Lester et al., 1997). The hAKAP220 mRNA is expressed at high levels in human testis and in human pachytene spermatocytes and round spermatids. The hAKAP220 protein was present in male germ cells and mature sperm, and localized in the cytoplasm of pre-meiotic pachytene spermatocytes and in the centrosome of developing post-meiotic germ cells, while a midpiece/ centrosome localization was found in elongating spermatocytes and mature sperm. The hAKAP220 protein along with a fraction of PKA type I and II, and protein phosphatase I, was associated with cytoskeletal structures. The hAKAP220 interacts with both classes of regulatory subunits of PKA, either through separate or through a common binding motif (Reinton et al., 2000).

Lester et al., (1996) isolated cDNA encoding 1129-amino acid protein from rat testis that contains both a PKA binding region and a per-oxisome targeting motif. The mRNAs of 9.7 and 7.3-kb are present in several rat tissues with the highest levels present in the brain and testis. The protein of approximately 220-kDa co-purifies with type II PKA holoenzyme and co-localizes with pro-protein of RII in microbodies that appear to be a subset of per-oxisomes. These results suggested that AKAP 220 might play a role in targeting type II PKA for cAMP-responsive peroxisomal events.

29.3.2. AKAP4

The most abundant protein in the FS is AKAP4. Nearly half of the protein in fibrous sheaths isolated from mouse sperm is AKAP4. Although several other fibrous sheath proteins have been identified, but how the fibrous sheath assembles is not understood. The AKAP4 and two others, AKAP3 and TAKAP-80 have anchoring sites for PKA. AKAP3 also anchors ropporin, a spermatogenic cell-specific protein that is linked through rhophilin to the small Rho GTPase. The AKAP4-binding protein and its binding regions had been identified by Brown et al, (2003), who found that AKAP4 binds AKAP3 and two novel spermatogenic cell-specific fibrous sheath interacting proteins 1 and 2 (FSIP1, FSIP2).

Transcription of *Akap4, Akap3*, and *Fsip1* genes begins in early spermatid development, whereas transcription of *Fsip2* begins in late spermatocyte development. AKAP3 is synthesized in round spermatids and incorporated into the fibrous sheath concurrently with formation of the rib precursors. However, AKAP4 is synthesized and incorporated into the nascent fibrous sheath late in spermatid development. The AKAP4 precursor is processed in the flagellum and only the mature form of AKAP4 appears to bind AKAP3. Thus AKAP3 appears to be involved in organizing the basic structure of the fibrous sheath, whereas AKAP4 has a major role in completing fibrous sheath assembly (Brown et al., 2003).

29.3.3. Rat Testis AKAP80

A RII-binding overlay procedure documented that RII could specifically associate with fibrous sheath polypeptides of 120 and 80-kDa. Mei et al, (1997) cloned the rat testis-specific RII-binding protein TAKAP-80. A 1.2-kb cDNA clone from rat testis expression library hybridized to a 1.8-kb mRNA transcript present exclusively in testis at 30 days after birth, at a time corresponding to the initiation of spermiogenesis. The complete nucleotide sequence of TAKAP-80 cDNA contained a continuous ORF of 502 amino acids. The deduced amino acid sequence showed a clear demarcation of charged and hydrophobic amino acid residues. Amino acids 1-147 of the protein contained 45% charged residues, with predominance of lysine and arginine. Similarly, amino acids 268-502 also contained a high percentage of charged amino acids (35%). In contrast, amino acids 148-267 were mostly hydrophobic and contained clusters of a repeating PXXP motif where X was predominantly valine and alanine or sometimes proline. High affinity interaction site for RII was contained within amino acids 258-78 of TAKAP-80. Level of the 80-kDa immunoreactive protein was significantly higher in mature (60 days old) rat testis and correlated with the mRNA transcription levels (Mei et al., 1997). Presence of TAKAP80 in sperm suggests its role in sperm motility.

AKAP121: Feliciells et al, (1998) cloned a cDNA for rat AKAP121 and showed that its expression is regulated by TSH (thyroid stimulating hormone) and cAMP. Differentiated thyroid cells (TL5) accumulate AKAP121 upon incubation with TSH or a cAMP analogue. Levels of total and newly synthesized AKAP121 mRNA also increased after treatment. The AKAP121 is directly controlled by cAMP and PKA.

29.3.4. Sperm AKAP 82

The type II cAMP-dependent protein kinase is localized to specific sub-cellular environments through interaction of regulatory subunit (RII) dimer to RII-anchoring proteins. The RII-anchoring protein sequences possess common regions of approximately 14 residues, which display high probabilities of forming amphipathic helices. The potential amphipathic helix region of Ht31 (Leu-Ile-Glu-Glu-Ala-Ala-Ser-Arg-Ile-Val-Asp-Ala-Val-Ile) lies between residues 495 and 507. Anchoring proteins interact with RIIα via an amphipathic helix binding motif (Carrera et al, 1994). Later studies indicated that recombinant AKAP82 bound RII, and binding could be competed for with a synthetic peptide RII-binding site of AKAP82. Adult mouse, however, contains a homologue of rat TAKAP80 indicating that there are two 80-kDa AKAPs associated with the fibrous sheath of mouse sperm (Carrera et al., 1996; Visconti et al., 1997).

Mouse AKAP82 is synthesized in the cell body of condensing spermatids as a Mr 97,000 precursor (pro-mAKAP82). This precursor polypeptide is transported down the flagellum to the principal piece where it is processed by the proteolytic cleavage of the amino-terminal 179 amino acids to produce mAKAP82 and a free 179 amino acid pro-domain. Following cleavage, mAKAP82 is assembled into the FS. Mouse AKAP82 can tether PKA close to the axoneme and other components of the flagellum that are involved in sperm motility (Turner et al., 1998). The pro-AKAP82 is detected in condensing spermatids but not in epididymal sperm. In addition, two other immunoreactive proteins of Mr 109000 (p109) and 26000 (p26 representing the "pro" domain of the precursor) were present in epididymal sperm. Pro-AKAP82 was localized to the entire length of the principal piece in testicular sperm, while in epididymal sperm p109 and p26 were present only in the proximal portion of the principal piece. Pro-AKAP82 was solubilized from germ cells with Triton X-100. Similar to pro-AKAP82, the RII subunit of PKA was present in the Triton X-100 soluble fraction cf developing germ cells. In sperm, much of the RII also became particulate, consistent with the hypothesis that AKAP82 anchors RII in the flagellum. This indicated that pro-AKAP82 is synthesized in the cell body, transported down the axoneme to its site of assembly in the fibrous sheath, and then proteolytically clipped to form mature AKAP82 (Johnson et al., 1997).

In humans, both hAKAP82 (82kDa) and its predicted precursor protein, pro-hAKAP82 (97kDa) are A-kinase anchor protein polypeptides that sequester protein kinase A to FS. These proteins are able to bind the RII subunit of PKA and tyrosine-phosphorylated in a medium supporting capacitation. The cDNA sequence of human sperm fibrous sheath pro-hAKAP82 is highly homologous to the mouse pro-mAKAP82 sequence, and the functional domains of the pro-hAKAP82 protein, the protein kinase A binding, and the pro-hAKAP82/hAKAP82 cleavage sites were identical to those of the mouse protein. The genomic organization of mouse *pro-AKAP82* indicated occurrence of alternative splicing in both the mouse and human *pro-AKAP82* genes that gave at least two distinct transcripts and possibly two different proteins. Compared with pro-mAKAP82, pro-hAKAP82 was considerably less processed in human sperm. Although sperm pro-mAKAP localizes only to the proximal portion of the principal piece of the flagellum, pro-hAKAP82 localized to the entire length of the principal piece. The pro-hAKAP82 gene mapped to human chromosome Xp11.2, indicating that defects in this gene are maternally inherited (Turner et al., 1998).

The rat sperm RII 80-kDa protein is associated with fibrous sheath protein, TAKAP80, which is not similar in amino acid sequence with mouse AKAP82. Adult mouse, however, contains a homologue of rat TAKAP80, which revealed that there are two 80-kDa AKAPs associated with the fibrous sheath of mouse sperm (Carrera et al., 1996; Visconti et al., 1997). The primary structure of bovine bAKAP82 is highly conserved in amino acid sequence

```
  1 MADRVDWLQS QSGVCKVGVY SPGDNQHQDW KMDTSTDPVR VLSWLRKDLE KSTAGFQDSR
 61 FKPGESSFVE EVAYPVDQRK GFCVDYYNTT NKGSPGRLHF EMSHKENPSQ GLISHVGNGG
121 SIDEVSFYAN RLTNLVIAMA RKEINEKIHG AENKCVHQSL YMGDEPTPHK SLSTVASELV
181 NETVTACSKN ISSDKAPGSG DRASGSSQAP GLRYMSTLKI KESTKEGKCP DDKPGTKKSF
241 FYKEVFESRN AGDAKEGGRS LPGDQKLFRT SPDNRPDDFS NSISQGIMTY ANSVVSDMMV
301 SIMKTLKIQV KDTTIATILL KKVLMKHAKE VVSDLIDSFM KNLHGVTGSL MTDTDFVSAV
361 KRSFFSHGSQ KATDIMDAML GKLYNVMFAK KFPENIRRAR DKSESYSLIS TKSRAGDPKL
421 SNLNFAMKSE SKLKENLFST CKLEKEKTCA ETLGEHIIKE GLHMWHKSQQ KSPGLERAAK
481 LGNAPQEVSF ECPDPCEANP PHQPQPPENF ANFMCDSDSW AKDLIVSALL LIQYHLAQGG
541 KMDAQSFLEA AASTNFPTNK PPPPSPVVQD ECKLKSPPHK ICDQEQTEKK DLMSVIFNFI
601 RNLLSETIFK SSRNCESNVH EQNTQEEEIH PCERPKTPCE RPITPPAPKF CEDEEATGGA
661 LSGLTKMVAN QLDNCMNGQM VEHLMDSVMK LCLIIAKSCD SPLSELGEEK CGDASRPNSA
721 FPDNLYECLP VKGTGTAEAL LQNAYLTIHN ELRGLSGQPP EGCEIPKVIV SNHNLADTVQ
781 NKQLQAVLQW VAASELNVPI LYFAGDDEGI QEKLLQLSAT AVEKGRSVGE VLQSVLRYEK
841 ERQLDEAVGN VTRLQLLDWL MANL
```

Fig.29.5. Amino acid sequence of A-kinase anchor protein-3 from mouse sperm. Source: http// www.ncbi.nlm.nih.gov (Accession NP_033780).

corresponding to the region of mAKAP82 responsible for binding to regulatory subunit of protein kinase A. Bovine AKAP82 was present in both epididymal and ejaculated sperm and was localized to the entire principal piece of eh flagellum, the region in which the fibrous sheath is located. Finally, bAKAP82 bound the regulatory subunit of protein kinase A (Moss et al., 1999).

29.3.5. FSP95

Mandal et al, reported a fibrous sheath protein of 95-kDa (FSP95) that undergoes phosphorylation during capacitation of human sperm and has similarity to sperm A-kinase anchor proteins. The 853-residue containing FSP95 protein has a Mr of 94.6-kDa and a pI of 6.0. It has multiple phosphorylation sites for PKC and casein kinase II as well as one phosphrylation site for tyrosine at 435. The FSP95 is 32-34% identical to mouse and human AKAP82 in amino acids. The *FSP95* gene has 5 exons separated by 4 introns. The FSP95 gene codes a single transcript of 3.0-kb and located on chromosome 12 at locus p13.3. The human protein is highly specific and expressed only in testis. In sperm the protein is localized to the ribs of fibrous sheath in principal piece of sperm tail (Mandal et al, 1999).

29.3.6. AKAP110

Bovine, human, monkey, and mouse sperm all contain one predominant AKAP with a relative mol wt of approximately 110kDa (AKAP110) **(Fig.29.5)**. The AKAP110 cDNA has been sequenced from mouse, bovine and human libraries. Alignment of the RII-binding domain on AKPA110 to those from other AKAPs revealed that AKAPs contain eight functionally conserved positions within an amphipathic helix structure that are responsible for RII interaction. In situ hybridization analysis detected AKAP110 only in round spermatids and among male germ cells. In spermatozoa, RI regulatory subunit of PKA is located primarily in the acrosomal region of the head, and RII is located exclusively in the entire length of the tail. AKAP110 is localized to acrosome region of sperm head and along the entire piece of principal piece. This suggests that AKAP110 shares compartments with both RI and RII isoforms of PKA and may function as a regulator of both motility - and head-associated functions such as capacitation and the acrosome reaction (Vijayaragharan et al., 1999).

Carr et al (2001) have identified two sperm-specific human proteins that interact with the amphipathic helix region of AKAP110. These proteins, ropporin (also found to interact with Rho pathway) and AKAP-associated sperm protein are 39% identical to each other and share a strong sequence similarity with the conversed domain on the N-terminus of RII that is involved in dimerization and AKAP binding. Mutation of conserved residues in RII prevents binding to AKAP 110. This suggests that sperm contains several proteins that bind to AKAPs in a manner similar to RII and imply that AKAPs may have additional and perhaps unique functions in spermatozoa.

29.3.7. Mitochondrial S-AKAP84

Accumulation of S-AKAP84 and its cognate mRNA are developmentally regulated during sperm development. The anchor protein is expressed de novo during late spermiogenesis, and this is coincident with the maximal expression and subsequent anchoring of RIIα and RII subunits. The developmentally regulated sperm AKAP of 84-kDa (S-AKAP84) is localized to the germ cell mitochondria. S-AKAP84 is a splicing variant of AKAP149 expressed in the testis. S-AKAP84 comprises 593 amino acids and contains a centrally located domain that avidly binds regulatory subunits (RIIα and RIIβ) of PKAIIα and PKAIIβ. The 3.2-kb S-AKAP84 mRNA and the cognate S-AKAP84 RII binding protein are expressed mainly in the male germ cell lineage and the expression of S-AKAP84 is developmentally regulated. The protein accumulates as spermatids undergo nuclear condensation and tail elongation. The timing of S-AKAP84 expression is correlated with the de novo accumulation of RIIα and RIIβ subunits and the migration of mitochondria from the cytoplasm (round spermatids) to the cytoskeleton (mid-piece in elongating spermatids). Residues 1-30 at the NH2 terminus of S-AKAP84 constitute a putative signal/anchor sequence that may target the protein to the outer mitochondrial membrane (Lin et al., 1995).

sA c-Myc binding protein, AMY-1 (associate of Myc-1), stimulates the transcription activity of c-Myc. The AMY-1 was found to bind in vitro and in vivo to the regulatory subunit II binding region of AKAP149 and S-AKAP84, a splicing variant of AKAP149 expressed in the testis. The AMY-1 was co-expressed with S-AKAP84 post-meiotically in the testis. Furthermore, AMY-1, S-AKAP84 and regulatory subunit II, formed a ternary complex in cells. The AMY-1 was localized in the mitochondria of HeLa and sperm in association with AKAP149 and S-AKAP84 (Furusawa et al., 2001).

29.3.8. Human Testis *hi* gene

Mohapatra et al, (1998) cloned a human testis specific gene *(hi)*, which encodes for a protein having regional homologies to the domain of A kinase anchoring proteins. The ORF of *hi* encodes a protein of 860 amino acids with a predicted molecular weight of 95.8-kDa The *hi* product contains 32 cysteine residues and 34 potential phosphorylation sites. *Hi* transcript is detected as a single transcript of approximately 3.0-kb in testis and not in any other somatic tissues. Expression of *hi* transcript occurs only in round spermatids indicating post-meiotic haploid gene expression. The cDNA sequence analysis revealed a total of 2897 bases with an ORF of 2583 bases from 128 to 2707. The initiator methionine lies in a nucleotide sequence homologous to Kozak consensus sequence. The mRNA degradation consensus sequence ATTTC was identified at position 2800, 93bp downstream of TGA terminator codon preceded by the polyadenylation signal (AATAAA) at position 2859, 152bp downstream of TGA terminator codon. Sequence analysis showed one direct repeat at positions 2145 and 2302 and one inverted repeat at positions 376 and 464 within the ORF **(Fig.29.6)**.

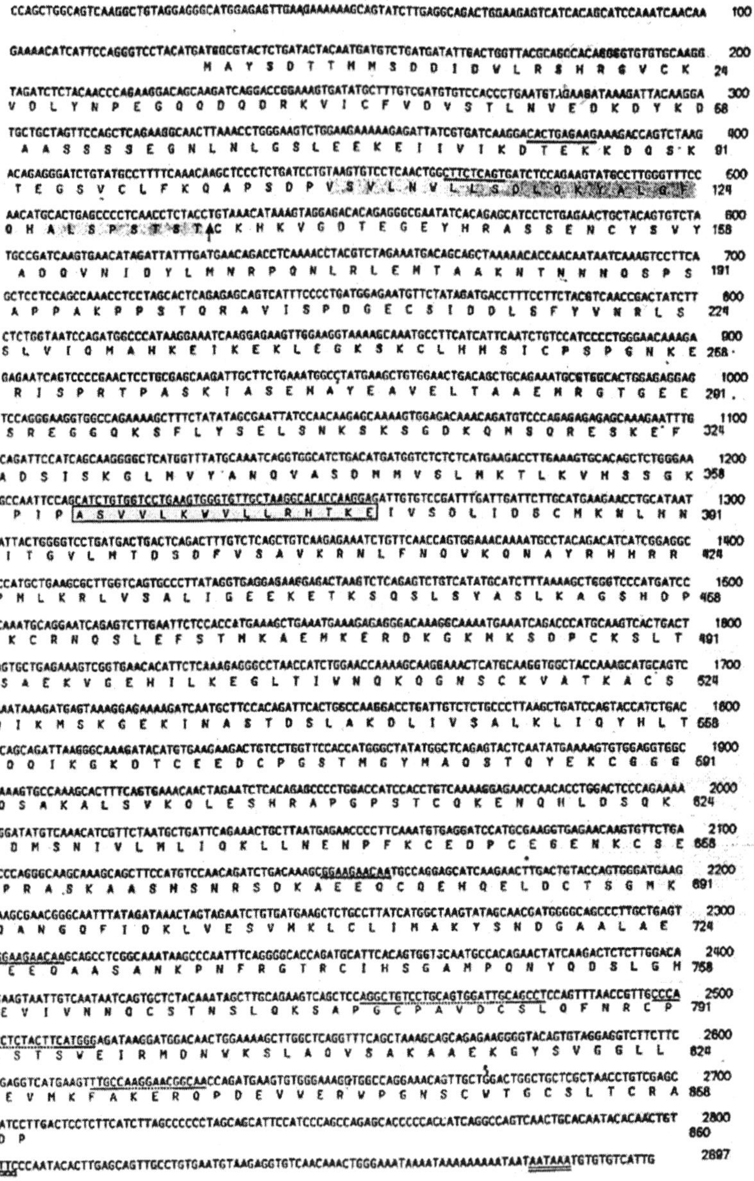

Fig.29.6..cDNA and deduced amino acid sequences for hi cDNA. One direct repeat and one indirect repeat are underscored by dotted and single solid line respectively. Hi domain having structural homology with AKAP protein is boxed. The three hairpin loops are marked by bold dotted lines. A mRNA degradation sequence (ATTTC) and polyadenylation signal (AATAAA) is shown by double dotted and solid underline respectively. Reprinted with permission from B. Mohapatra et al.. Biochem Biophys Res Commun 244; 540-45: 1998 © Elsevier.

```
gaattccggaccaaagttttcactctacacgtggtcccgaataagagcacacagtctaaacacagtgaaacagagttcac
tggctgagcctgtctctccatctaaaagacatgagactccagcatcttctgtaactgaggctcttgacaggaggttgggg
gattctagctcagccccactgcttgtgactcaatcagaaggaactgacctggtagtagaactcagaaccctcagaaccac
ttgctgctttcccaggaaggccacagtgcccgttctctccatcgtgaggtggccagaacaggaagtcatcagattcccac
cgactcccaggactcctcctccagacttgcagtaggcagtagaaacagttgctcttctccactgagagaactgtcagaaa
aacta

406     ATGAAAAGTATAGAACAAGATGCAGTGAATACTTTTACCAAATATATATCTCCAGATGCTGCTAAGCCAATA
         M  K  S  I  E  Q  D  A  V  N  T  F  T  K  Y  I  S  P  D  A  A  K  P  I   -24

478  CCAATTACAGAAGCCATGAGAAACGACATCATCGCAAAGATTTGTGGAGAAGATGGACAGGTGGATCCCAACTGT
      P  I  T  E  A  M  R  N  D  I  I  A  K  I  C  G  E  D  G  Q  V  D  P  N  C   -49

553  TTCGTTCTGGACACGGCTGTAGTCTTTAGTGCAATGGAGCAAGAGCACTTTAGTGAGTTTCTGCGAAGTCACCAT
      F  V  L  D  T  A  V  V  F  S  A  M  E  Q  E  H  F  S  E  F  L  R  S  H  H   -74

628  TTCTGTAAATACCAGATTGAAGTGCTGACCAGTGGGACTGTTTACCTGGCTGATATCCTCTTCTGTGAGTCAGCC
      F  C  K  Y  Q  I  E  V  L  T 'S  G  T  V  Y  L  A  D  I  L  F  C  E  S  A   -99

703  CTCTTTTATTTTTCTGAGTACATGGAAAAAGAAGATGCAGTGAATATCTTACAATTCTGGTTAGCAGCGGATAAT
      L  F  Y  F  S  E  Y  M  E  K  E  D  A  V  N  I  L  Q  F  W  L  A  A  D  N   -124

778  TTCCAGTCTCAGCTTGCTGCCAAAAAGGGCCAGTATGATGGACAGGAGGCCCAGAATGATGCCATGATTTTATAT
      F  Q  S  Q  L  A  A  K  K  G  Q  Y  D  G  Q  E  A  Q  N  D  A  M  I  L  Y   -149

853  GACAAGTACTTTTCCCTCCAAGCCACACACCCCCTTGGATTTGATGATGTTGTACGATTAGAAATTGAATCTAAT
      D  K  Y  F  S  L  Q  A  T  H  P  L  G  F  D  D  V  V  R  L  E  I  E  S  N   -174

928  ATCTGCAGGGAAGGTGGACCACTTCCTAATTGTTTCACAACTCCATTACGTCAGGCCTGGACAACCATGGAGAAG
      I  C  R  E  G  G  P  L  P  N  C  F  T  T  P  L  R  Q  A  W  T  T  M  E  K   -199

1003 GTCTTTTTGCCTGGTTTTCTGTCCAGCAATCTTTATTACAAATATTTGAATGATCTCATCCATTCAGTTCGAGGA
      V  F  L  P  G  F  L  S  S  N  L  Y  Y  K  Y  L  N  D  L  I  H  S  V  R  G   -224

1078 GATGAATTTCTTGGAGGGAATGTTTCCCTGGCTGCTCACGGCTCTGTCTGCCTTCCTGAGGAGTCTCACTCAGGT
      D  E  F  L  G  G  N  V  S  L  A  A  H  G  S  V  C  L  P  E  E  S  H  S  G   -249

1153 GGTTCCGATGGCTCCACTGCTCAGTCTAGTGTGAAAAAAGCCAGTATTAAAATTCTGAAAAATTTTGATGAAGCA
      G  S  D  G  S  T  A  Q  S  S  V  K  K  A  S  I  K  I  L  K  N  F  D  E  A   -274

1228 ATAATTGTGGATGCTGCAAGTCTGGACCCAGAATCTTTATATCAACGGACATATGCAGGGAAGATGTCCTTTGGG
      I  I  V  D  A  A  S  L  D  P  E  S  L  Y  Q  R  T  Y  A  G  K  M  S  F  G   -299

1303 AGAGTTAGTGATTTGGGGCAGTTCATCCGAGAGTCTGAGCCTGAACCTGATGTGAAGAAATCAAAAGGATTCATG
      R  V  S  D  L  G  Q  F  I  R  E  S  E  P  E  P  D  V  K  K  S  K  G  F  M   -324

1378 TTCTCACAAGCTATGAAGAAGTGGGTGCAAGGAAATACTGACGAGGCCCAAGAAGAGCTAGCTTGGAAGATTGCA
      F  S  Q  A  M  K  K  W  V  Q  G  N  T  D  E  A  Q  E  E  L  A  W  K  I  A   -349

1453 AAAATGATAGTGAGTGATGTTATGCAGCAGGCACACCATGATCAACCACTAGAGAAGTC"ACAAAGCTATGACTC
      K  M  I  V  S  D  V  M  Q  Q  A  H  H  D  Q  P  L  E  K  S  T  K  L  *     -372

gaaaccagagataaggaaatctgattttgaaaaaaataagaactttcccctttggttggattcttcaacacagccgatga
aaacggagaactgcatttcttttcactgctagatcactgtgttccaggaaagagtgggagacagtccgagagtggttgtt
aaatgtag                                                                      -1695
```

Fig.29.7. cDNA and deduced amino acid sequences for D-AKAP2 core. Numbers on the left are for cDNA sequence, and numbers on the right indicate amino acid position. RPP8 sequence is underlined. Printed with permission from L J Huang et al. Proc Natl Acad Sci USA 94; 11184-9: 1997 © National Academy of Sciences (USA).

29.3.9. Dual Specificity AKAPs

Using a two-hybrid screen, Huang et al., (1997) isolated a dual specificity AKAP. Because it interacts with both the type-I and type II regulatory subunit, it was defined as a dual specific AKAP or DAKAP1. Another member of dual specificity is D-AKAP2 that also binds both types of regulatory subunits. A message of 5-kb pairs was detected for D-AKAP2 in all embryonic stages and in all adult tissues tested. In brain, skeletal muscle, kidney, and testis, a 10-kb mRNA was identified. In testis, several small mRNAs were observed. The D-AKAP2 represents a protein, which is composed of 372 amino acids and includes R binding fragment (residues 333-372) at its C-terminus **(Fig.29.7)**. The R binding domain of D-AKAP2 interacts with the N-terminal dimerization domain of RIα and RIIα. A putative RGS domain was present

```
   1  CCC CAG TTC TGA GCC ATG GCC TCT GCA GTC CTC AGC TCT GTG CTC ACA ACT GCC TCT CGC    60
   1                      Met Ala Ser Ala Val Leu Ser Ser Val Leu Thr Thr Ala Ser Arg    15

  61  TTT GCC CTG CTA CAA GTG GAC AGC GGC AGT GGC TCT GAT TCC GAG CCT GGG AAA GGC AAA   120
  16  Phe Ala Leu Leu Gln Val Asp Ser Gly Ser Gly Ser Asp Ser Glu Pro Gly Lys Gly Lys    35

 121  GGC CGG AGT AAT GGG AAG TCT CAA ACT CTG GGA AAC AAA GCA ACC TCA AAT GAG AAG AAA   180
  36  Gly Arg Ser Asn Gly Lys Ser Gln Thr Leu Gly Asn Lys Ala Thr Ser Asn Glu Lys Lys    55

 181  CGG GAG AAG AGA AGA AAA AAG AAG GAA CAG CAG CAG AGT GAA GCA AAC GAG CTC AGG AAT   240
  56  Arg Glu Lys Arg Arg Lys Lys Lys Glu Gln Gln Gln Ser Glu Ala Asn Glu Leu Arg Asn    75

 241  CTT GCT TTC AAG AAA ATT CCC CAG AAA TCC TCC CAT TCC ATT TGC AAT GTT CAA CAT GAG   300
  76  Leu Ala Phe Lys Lys Ile Pro Gln Lys Ser Ser His Ser Ile Cys Asn Val Gln His Glu    95

 301  CTT TCA TCG CCA AAC CCA GCA CAG AAG GAG TCA CGG GAA GAA AAC TGG CAA GAG TGG AGA   360
  96  Leu Ser Ser Pro Asn Pro Ala Gln Lys Glu Ser Arg Glu Glu Asn Trp Gln Glu Trp Arg   115

 361  CAG AAA GAT GAA CAG CTG ACC TCT GAA ATG TTT GAA GCT GAC CTT GAG AAG GCG TTG CTG   420
 116  Gln Lys Asp Glu Gln Leu Thr Ser Glu Met Phe Glu Ala Asp Leu Glu Lys Ala Leu Leu   135

 421  TTG AGT AAG CTG GAG TAT GAA GAG CAC AAA CAG GAT TAT GAA AAT GCT GAG ACT GCT TCA   480
 136  Leu Ser Lys Leu Glu Tyr Glu Glu His Lys Gln Asp Tyr Glu Asn Ala Glu Thr Ala Ser   155

 481  ACT CAG ACA AAA GGT ATA AAT AAA AAA GAT AAA AGG AAG AAC CAC CAG GGC AAG GAC AAA   540
 156  Thr Gln Thr Lys Gly Ile Asn Lys Lys Asp Lys Arg Lys Asn His Gln Gly Lys Asp Lys   175

 541  CCT GTC ACG GTG TCA CTC AAA GAC TTC CAG TGT GAA GAT CAT ATT AGT AAA AAG GCA GAG   600
 176  Pro Val Thr Val Ser Leu Lys Asp Phe Gln Cys Glu Asp His Ile Ser Lys Lys Ala Glu   195

 601  GAA TCG AAT TCT GCT CAG ACT TTA TCA CAC GAT GGG GGA TTC TTC AAT AGA CTG GAA GAT   660
 196  Glu Ser Asn Ser Ala Gln Thr Leu Ser His Asp Gly Gly Phe Phe Asn Arg Leu Glu Asp   215

 661  GAT GTT CAT AAG ATT CTG ATT AGA GAA AAA AGG AGA GAA CAA CTC ACA GAG CAC AAC GGA   720
 216  Asp Val His Lys Ile Leu Ile Arg Glu Lys Arg Arg Glu Gln Leu Thr Glu His Asn Gly   235

 721  ACA GAT AAC TCT CCC GCT CCA GAG CAC AAC CAG GAA GTA GGT CTA AAA GAT GGA AGA ATT   780
 236  Thr Asp Asn Ser Pro Ala Pro Glu His Asn Gln Glu Val Gly Leu Lys Asp Gly Arg Ile   255

 781  GAA AGA CTA AAG TTA GAA CTT GAA AGG AAA GAC GCA GAA ATT CAG AAG CTC AAA GCT GTG   840
 256  Glu Arg Leu Lys Leu Glu Leu Glu Arg Lys Asp Ala Glu Ile Gln Lys Leu Lys Ala Val   275

 841  ATC ACT CAG TGG GAG GCA AAG TAT AAA GAA GTA AAA GCA AGA AAT GGA CAA TTA CTG AAA   900
 276  Ile Thr Gln Trp Glu Ala Lys Tyr Lys Glu Val Lys Ala Arg Asn Gly Gln Leu Leu Lys   295

 901  ATG CTT CAG GAG GGA GAG ATG AAA GAT AAG GCC GAG ATA CTT CTG CAA GTG GAT GAG TCT   960
 296  Met Leu Gln Glu Gly Glu Met Lys Asp Lys Ala Glu Ile Leu Leu Gln Val Asp Glu Ser   315

 961  CAG AGC ATC AAG AAC GAG CTG ACG GTC CAG GTG TCT TCA CTT CAT GCC GCA CTG GAA CAA  1020
 316  Gln Ser Ile Lys Asn Glu Leu Thr Val Gln Val Ser Ser Leu His Ala Ala Leu Glu Gln   335

1021  GAA AGA TCT AAA GTG AAA GTA TTA CAG GCA GAA TTA GCC AAA TAC CAG GGA GGC AGG AAG  1080
 336  Glu Arg Ser Lys Val Lys Val Leu Gln Ala Glu Leu Ala Lys Tyr Gln Gly Gly Arg Lys   355

1081  GGG AAA AGG AAC TTC GAA CCT GAC CAT TGC AGG TGA TCG TGG CCT CGG AGG CCA GCA CAG  1140
 356  Gly Lys Arg Asn Phe Glu Pro Asp His Cys Arg ***                                  366

1141  AAT TCT TTC ACA CTT TTG CAG AAA TGT GTA TAT TTA ATG CTG TGC GAC TGC TAA ACT GCA  1200
1201  GGT TTT TGT TGA AGG AAC TAA AAG CAA CTA GTT CAC TCA TAG TCT ACA GTT TTA TGT TCT  1260
1261  TTT GGC TTA AAG TTA AAA GAA AAT ACA GAA TAC CAG CAG AAC TTG ATT ATT GTT TTC TGT  1320
1321  CCA AAC TTC TTA TGA CTT TAG TTT TTC ATC AGT CAA TAA AAT TAA TTT ACT CTT CCA AAA  1380
1381  AAA AAA AAA AAA                                                                  1392
```

Fig.29.8.cDNA and deduced amino acid sequences of cyclic GMP dependent protein kinase anchor protein. Printed with permission from K. Yuassa et al. J. Biol Chem 275: 4887-95 : 2000 © American Society for Biochemistry and Molecular Biology.

near the N-terminal region of D-AKAP2. The presence of this domain raises the possibility that D-AKAP2 may interact with a Gα protein, thus providing a link between the signaling machinery at the plasma membrane and the downstream kinase (Huang et al., 1997).

29.3.10. Fibrous Sheath Component 1

FSC1 is the major structural protein of fibrous sheath. When FSC1 was used as the bait to screen a mouse testis cDNA library, two clones were isolated encoding the RIα of cAMP-dependent protein kinase. In vitro binding assay demonstrated that two RIα and one RIIα tethering domains are present on FSC1. A domain located at residues 219-232 (termed domain A) corresponds to the reported tethering domain for RII of PKA, indicating that this binding domain has dual specificity to RIα and RIIα. Another RIα tethering site (termed domain B) at residues 335-344 shows specific binding of RIα and had no significant sequence homology with known RII tethering domains. While RIα and RIIα bound to domain A, only RIα bound to domain B. This suggests that the amphipathic helix formed by domain B has a secondary structure distinct from that required for RII binding. However, domain B is likely to form an amphipathic helix, which is the secondary structure of RII of protein AKAP (Miki and Eddy, 1998). Taken together it suggests that tethering of RIα to FSC1 may provide a mechanism for the sub-cellular localization of PKA to a region of the flagellum important for sperm motility.

29.3.11. c-GMP Dependent Protein Kinase Anchor Protein

cGMP–dependent protein kinase (cGKs) is a member of a family of cyclic nucleotide-dependent protein kinases that also includes PKA. The cGK is a major cellular receptor of cGMP and plays important roles in cGMP-dependents signal transduction pathways. There exist two forms of cGK (I and II) that are encoded by distinct genes and two different isoforms of cGK-I (designated Iα and Iβ) that are produced by alternative splicing. Information on anchor proteins for c-GMP dependent protein kinases is very limited. It has been suggested that GKAP42 functions as an anchoring protein for cGK-Iα, which may participate in germ cell development through phosphorylation of Golgi–associated proteins such as GKAP42 (Yuasa et al 2000) (see Chapter 19) **(Fig.29. 8)**.

29.4. OTHER PROTEINS OF FIBROUS SHEATH

Ropporin, a sperm specific binding protein of rhophilin is present in the inner surface of fibrous sheath of sperm flagella where as rhophilin is present in outer dense fiber (Fujita et al, 2000). The cDNA of PDZ domain of ropporin has been cloned and discussed in Chapter 18. Other proteins associated with the fibrous sheath include two enzymes in the glycolytic pathway. Glyceraldehyde 3-phosphate dehydrogenase-s (GAPDS) is the product of a gene expressed only in spermatogenic cells, while hexokinase type 1-s (HK1-S) is derived from alternative transcripts present only in spermatogenic cells. Most of the other glycolytic enzymes in sperm have unique structural or functional properties (Chapter 28). The fibrous sheath also contains a spermatogenic cell-specific member of the mu-class glutathione S-transferase family (GSTM5) and an intermediate filament-like protein (FS39) (Chapter 31). Ohuchi et al (2001) isolated a cDNA designated F77 from mouse testis. The full-length F77 mRNA was 3.4-kb and showed no significant matching with proteins in the databases. F77 was mapped at a proximal position between D8Mit212 and D8Mit138 on mouse chromosome 8. F77 mRNA was detected only in germ cells of normal adult testis, mainly at round spermatids and at the tail of elongated spermatids. A cDNA clone specifically expressed during spermatogenesis consisted of 1085 nt and had an ORF of 870 nucleotides encoding a putative protein of290 aa residues with MW of33-kDa. The corresponding mRNA of 1.2-kb is exclusively expressed in the testis of adult mice and first detected late pachytene stage. The protein was localized to occupy the flagella from the connecting piece through the principal piece. This gene was named as *oppo 1* whose product plays an important role in sperm tail structure and/ or sperm movement (Nakamura et al., 2002). Modarressi et al. (2001; 2004) identified and characterized a novel gene, TSGA10 in human testis. TSGA10 encodes a 65-kDa protein of 697 amino acids that is processed to the 27-kDa fibrous sheath protein.

Clone 72/SSeCKS was identified as a PKC –binding protein. It is a high-molecular-mass heat-stable protein present in rat testis and rat brain and several cell lines, with high expression in rat testis. Down regulation of clone 72 mRNA in transformed cells suggests a role for the protein in growth regulation. There is a strong correlation between phosphorylation of the protein and actin-based cytoskeletal rearrangement. Clone 72 is involved in cytoskeletal remodeling (Erlichman et al 1999; Mei et al, 1997). Amino acids 1495-1524 of SSeCKS had the ability to interact with RII. The recombinant protein was recognized as a 280/290-kDa doublet and a 240-kDa protein on Western blots of rat testis cyosolic extracts and was not detected in mature sperm. A dramatic increase in hybridizing mRNA on day 30 after birth coincided with high expression in elongating spermatids. The expression of 72/SSeCKS like that of TAKAP80 during late-stage of spermatogenesis and the presence of a AKAP domain in clone 72/SSeCKS indicated it to be a scaffolding protein, which participates during late stage of spermatogenesis (Erlichman et al., 1999).

29.5. SPERM PLASMA MEMBRANE PROTEINS

29.5.1. Na/K-ATPase

The Na/K-ATPase or Na-pump includes a family of isozymes, all of which hydrolyze ATP in a ouabain-sensitive and or Na/K dependent fashion. Structurally, the Na/K-ATPase isozymes consist of distinct forms of two major polypeptides, the α- and the β subunits. Three structural variants of the α (α_1, α_2, and α_3) and β (β1, β2, and β3) subunits have been identified in vertebrates. Depending on the cell type, developmental stage, hormonal stimulation, and pathological state of the tissue, all Na/K-ATPase α and β polypeptides are expressed in a highly regulated manner. Association of different α and β subunits in cells results into the formation of different Na/K-ATPase isozymes with unique kinetic properties. The Na/K-ATPase is responsible for translocating Na^+ out of the cell and K^+ into the cell using the energy of hydrolysis of ATP. The electrochemical gradient it generates is necessary for many cellular functions, including establishment of the plasma membrane potential and transport of sugars and ions in and out of the cell. Families of isoforms of α and β subunits with specific functions have been identified in recent years, suggesting that the Na/K ATPases have specific roles. The evidence for the existence of an additional α subunit isoform (α_4) was presented in 1987.The cDNA, designated α_4, codes for a polypeptide consisting of 1028 amino acids with an identity of 76-78% to the other α isoforms of the Na pump. In rats, the α_4 mRNA is selectively expressed in the testis. Because testicular germ cells express the Na/K-ATPase α_1 subunit and two different isoforms of the β polypetide β1 and β3, the possibility exists for α_4 to form functionally active complexes with multiple β subunits. Where as the α_4 isoform has properties consistent with that of a Na/K-ATPase catalytic subunit, it is not certainthat the subunit can function as an ATPase and transport ions.

When α4 and the Na pump β1 subunit are co-expressed in SF9 cells, Na/K-ATPase activity is induced. The activity of α_4 is inhibited by the P-type ATPase blocker vanadate but not by compounds that inhibit the sarcoplasmic resticulm Ca-ATPase or the gastric H/K-ATPase. The enzymatic properties of the $\alpha_4\beta_1$ and $\alpha_4\beta_3$ isozymes are, however, distinct from the other Na pump isozymes. In rat testis Na/K-ATPase activity represents approximately 55% of the total enzyme of the gonad and that the α_4 polypeptide is a functional isoform of the Na/K-ATPase both in vitro and in the native tissue (Blanco et al., 1999). The α_4-subunit is expressed in the germ cells of rat testes. The highest amount of the isoform is found in spermatozoa, where it constitutes two third of the total Na/K-ATPase activity of the gametes. The characteristic localization of α_4 in the testis is supported by the drastic reduction of the polypeptide in mice, which are infertile as a consequence of arrest in maturation of the germ cells. In addition, GC-1spg cells, a murine cells line derived from testis spermatogonia, also contained the Na/K-ATPase α_4 polypeptide albeit at lower level than in the sperm. The particular pattern of expression of the Na/K-ATPase α_4-isoform and the specific enzymatic properties of the polypeptide suggests its importance for ionic homeostasis of testicular germ cells (Blanco et al., 2000).

Effect of Glucose on Sperm Na/K ATPase: A transient hyper-polarization of the plasma membrane of human sperm occurs when cells deprived of glucose are restored in glucose medium. The hyper-polarization of sperm membrane depends on glycolysis. The hyper-polarization induced by glucose in the medium containing a high concentration of KCI, was insensitive to the K^+ channel blocker TEA and the Cl^- channel blocker niflumic acid, but it was blocked by ouabain. It is likely that glucose addition causes an increase in the ATP concentration that in turn increases the Na^+/K^+ ATPase activity. Since this pump is electrogenic ($2K^+/3Na^+$) the plasma membrane gets hyperpolarized. On the other hand, CCCP, a proton ionophore, inhibited the hyper-polarization induced by glucose, whereas CCCP caused hyper-polarization of sperm was reversed in a medium without Ca^{2+} (Guzman-Grenfell et al., 2000; Lingrel et al., 2003).

Sperm α_4 Na/K ATPase Gene: The cDNA clone encoding the murine Na/K-ATPase α subunit isoform (α_4) predicted a polypeptide of 1032 amino acids, and exhibits 75% amino acid sequence identity to the rat α_1, α_2, and α_3 subunits. Within the first extracellular loop, the α_4-subunit is highly divergent from other Na/K-ATPase α subunits. The α_4-subunit gene (*Atp 1α_4*) is localized to the distal portion of mouse chromosome 1, in very close proximity to the murine Na/K-ATPase α_4 subunit gene. The α_4-subunit is expressed almost exclusively in testis, with low level of expression in epididymis. The close similarities in the organization and expression pattern of the murine and human α_4-subunit genes suggest that these two genes are orthologous (Underhill et al., 1999)

29.5.2. Periacrosomal Plasma Membrane Protein (PM52)

Two antigenically related, size-variant integral membrane proteins of 52-kDa (PM52) and 35-kDa that are localized to the periacrosomal domain of cauda epididymal guinea pig spermatozoa were characterized (Westbrook-case et al., 1994). The PM52 is first expressed in acrosome phase spermatids and localizes exclusively to the cytoplasmic lobe. During early stages of expression, PM52 appeared to be absent from the head region, but significant PM52 accumulation over the spermatids head was noted in late acrosomal phase spermatids. Throughout spermiogenesis, PM52 extended posteriorly to the annulus, which represents a barrier preventing PM52 diffusion over the posterior tail. Following the migration of the annulus to the mid-piece-principal piece junction, PM52 began to disappear from the flagellar region, and at the completion of spermiogenesis most of the PM52 was restricted to the acrosomal segment. Sperm PM52 is not proteolytically modified during epididymal maturation. Sequence determination of full-length PM52 clones demonstrated identity to a sperm membrane protein termed "spread" (Quill and Garber, 1996; Olson et al., 1998).

29.5.3. Calcium-Binding Tyrosine-Phosphorylation Regulated Protein (CABYR)

Human calcium-binding tyrosine-phosphorylation regulated protein (CABYR) is a polymorphic, testis-specific, calcium binding protein that is tyrosine phosphorylated during in vitro capacitation. Presence of a PKA regulatory subunit type IIα (RII-α) domain in the N-terminus, phosphorylation dependent Ca^{2+} binding isoforms, and localization to the principal piece of the human sperm tail suggested that CABYR might be involved in sperm motility. Human and

```
  1 MISSKPRLVV PYGLKTLLEG ISRAVLKTNP SNINQFAAAY FQELTMYRGN TTMDIKDLVK
 61 QFHQIKVEKW SEGTTPQKKL ECLKEPGKTS VESKVPTQME KSTDTDEDNV TRTEYSDKTT
121 QFPSVYAVPG TEQTEAVGGL SSKPATPKTT TPPSSPPPTA VSPEFAYVPA DPAQLAAQML
181 GKVSSIHSDQ SDVLMVDVAT SMPVVIKEVP SSEAAEDVMV AAPLVCSGKV LEVQVVNQTS
241 VHVDLGSQPK ENEAEPSTAS SVPLQDEQEP PAYDQAPEVT LQADIEVMST VHISSVYNDV
301 PVTEGVVYIE QLPEQIVIPF TDQVACLKEN EQSKENEQSP RVSPKSVVEK TTSGMSKKSV
361 ESVKLAQLEE NAKYSSVYME AEATALLSDT SLKGQPEVPA QLLDAEGAIK IGSEKSLHLE
421 VEVTSIVSDN TGQEESGENS VPQEMEGRPV LSGEAAEAVH SGTSVKSSSG PFPPAPEGLT
481 APEIEPEGES TAE
```

Fig.29.9. Amino acid sequence of calcium binding tyrosine phosphorylation regulated protein (CABYR) of human sperm fibrous sheath (Source: http//www. ncbi.nlm.nih.gov. Accession AAC35373).

mouse CABYR conserve potential functional motifs. The mCABYR is arranged into six exons spanning about 14-kb of DNA and was similar to human CABYR, in matter of its localization to the principal piece, identity in coding regions, and presence of two testis-specific transcripts of 1.4- and 2.4-kb. The mouse CABYR showed 81% nucleotide identity with human CABYR (Sen et al., 2003) **(Fig.29.9).**

29.5.4. Human Sperm Membrance Protein-I

A cDNA fragment (HSD-1) coding for part of a human sperm membrane protein (hSMP-1) was isolated from a human testis cDNA expression library using serum of a infertile patient as a probe. The complete cDNA of 2482-bp showed an ORF of 1572-bp that encodes 523 amino acid residues with a computed molecular mass of 55.08-kDa and represents a sperm specific antigen. The poly (A) mRNA was detected as a band of approximately 2.5-kb in both species. The hSMP-1 mRNA was present in human testis and was localized to the head of human sperm acrosome. The 3'-untranslated region without poly (A) tail was 886-bp long. A putative polyadenylation signal AATAAA was found 16-bp upstream from poly (A) tail. The hydropathy profile showed two regions of predominantly hydrophobic amino acids at positions 43-72 and 441-461 that may function as transmembrane domains. The absence of a signal peptide and the presence of the internal hydrophobic domain suggested that this protein is a type II transmembrane protein (Liu et al., 1996).

29.5.5. Other Protein Components of Sperm Membrane

Cleavage Signal-1 Protein: The cleavage signal-1 protein (CS-1), a doublet antigen comprised of approximately14-kDa and 18-kDa proteins has been shown to be present on the surface of sperm of various mammalian species including humans. The cDNA is 1828-bp long; the start codon assigned to the first ATG (at 98-100 nt) encodes a protein with 249 amino acid residues terminating at TAA (at 845-847 nt). The cDNA has a 97-bp 5-' and a 984-bp 3'-untranslated region. The potential polyadenylation signal (5'-AATAAA) is at 1803-1808 nt. The doublet antigen did not show any extensive homology with any known sequence, indicating that CS-1 is a unique protein (Javed and Naz, 1992).

A 15-kDa protein is present as a major component of the residual protein fraction of mouse spermatozoal heads. It is present in epididymal/vas spermatozoa as a series of five differentially phosphorylated molecules with pI 6.0-7.0 protein to a circumscribed region of the demembranated mouse sperm head. The 15-kDa protein was shown to be sperm and species-specific. The 15-kDa protein of mouse spermatozoa is synthesized during the elongation phase of spermiogenesis (stage 12-16) and is phosphorylated in the terminal period of that phase and/

or after excursion of spermatozoa from the seminiferous tubules (Pruslin and Rodman, 1985).

Protein Carboxyl-methylase: Protein carboxyl-methylase (PCM), an enzyme from rat and rabbit sperm, is known to be involved in exocytotic secretion and chemotaxis. Its substrate methyl acceptor protein(s) was demonstrated in the supernate after solubilization of the sperm cell membrane on rat and rabbit spermatozoa. A protein methylesterase that hydrolyzes methyl ester bonds created by PCM was demonstrated in rabbit but not in rat spermatozoa. Methyl acceptor protein substrates were low in testis but increased in parallel with sperm maturation in the epididymis. Fractionation of heads and tails of rat spermatozoa indicated that PCM was found exclusively in the tail fraction, where as MAP was detected both in head and tail fractions (Bouchard et al., 1980).

MC31/CE9: MC31/CE9, a member of the immunoglobulin superfamily molecules, was localized on the rat sperm flagellar plasma membrane. During spermatogenesis, MC31/CE9 mRNA first appeared in type B spermatogonia. Intensity increased progressively to pachytene spermatocytes and remained at constant level throughout the subsequent phases of spermatocytes and round spermatids, and then decreased gradually from step-11 spermatids to disappear in step-15 spermatids. The most intense immunoreactivity was found on the flagella of step-8 to step-19 elongated spermatids. The production of MC31/CE9 is post-transcriptionally regulated during spermiogenesis (Wakayama et al., 2000)

29.6. REFERENCES

Blanco G, Melton RJ, Sanchez G, Mercer RW. Functional characterization of a tetstes-specific α-subunit isoform of the sodium/potassium adenosine triphosphatase. Biochemistry 1999; 38: 13661-9.

Blannco G, Sanchez G, Melton RJ, Tourtellotte WG, Mercer RW. The α4 isoform of the Na, K-ATPase is expressed in the germ cells of the testes. J. Histochem Cytochem 2000; 48: 1023-32.

Bouchard P, Gagnon C, Philips DM, Bardin CW. The localization of protein carboxyl-methylase in sperm tails. J Cell Biol 1980; 86: 417-23.

Brohmann H, Pinnecke S, Hoyer-Fender S. Identification and characterization of new cDNAs encoding outer dense fiber proteins of rat sperm. J. Boil Chem 1997; 272: 10327-32.

Brown PR, Miki K, Harper DB, Eddy EM. A-kinase anchoring protein 4 binding proteins in the fibrous sheath of the sperm flagellum. Biol Reprod 2003; 68: 2241-8.

Carr DW, Fujita A, Stentz CL, Liberty GA, Olson GE, Narumiya S. Identification of sperm-specific proteins that interact with A-kinase anchoring proteins in a manner similar to the type II regulatory subunit of PKA. J Biol Chem 2001; 276: 17332-8.

Carrera A, Gerton GL, Moss SB. The major fibrous sheath polypeptide of mouse sperm: structural and functional similarities to the A-kinase anchoring proteins. Dev Biol 1994: 165: 272-84.

Carrera A, Moos |J, Ning XP, Gerton GL, Tesarik J, Kopf GS, Moss SB. Regulation of protein tyrosine phosphorylation in human sperm by a calcium/calmodulin-dependent mechanism: identification of A kinase anchor proteins as major substrates for tyrosine phosphorylation. Dev Biol 1996; 180: 284-96.

Diekman AB, Olson G, Goldberg E. Expression of the human antigen SPAG2 in the testis and localization to the outer dense fibers in spermatozoa. Mol Reprod Dev 1998; 50: 284-93.

Eddy EM, Toshimori K, O'Brien DA. Fibrous sheath of mammalian spermatozoa. Microsc Res Tech 2003; 61: 103-15.

Erlichman J, Gutierrez-Juarez R, Zucker S, Mei X, Orr GA. Development expression of the protein kinase C substrate binding protein (clone 72/ SSe CKS) in rat testis. Identification as a scaffolding protein containing an A-kinase-anchoring domain which is expressed during late-stage spermatogenesis. Eur J Biochem 1999; 263: 797-805.

Escalier D, Gallo JM, Schrevel J. Immunochemical characterization of a human sperm fibrous sheath protein, its developmental expression pattern, and morphogenetic relationships with actin. J Histochem Cytochem 1997; 45: 909-22.

Feliciello A, Rubin CS, Avvedimento EV, Gottesman ME. Expression of A- kinase anchor protein 121 is regulated by hormones in thyroid and testicular germ cells. J Biol Chem 1998; 273: 23361-66.

Foster JA, Gerton GL. Autoantigen 1 of the guinea pig sperm acrosome is homologue of Tpx-1 and human Tpx-1 and is a member of the cysteine secretary protein (CRISP) if any. Mol Reprod Dev 1996; 44: 221-29.
Fujita A, Nakamura K, Kato T, Watanabe N. Ishizaki T, Kimura K, Mizoguchi A, Narumiya S. Ropporin, A sperm specific binding protein of rhophilin, that is localized in the fibrous sheath of sperm flagella. J. Cell Sci 2000; 113 (Pt. 1) : 103-12.
Furusawa M, Ohnishi T, Taira T, Iguchi-Ariga SM, Ariga H. AMY-1, a c-Myc-binding protein is localized in the mitochondria of sperm by association with S-AKAP84, an anchor protein of cAMP-dependent potein kinase. J Biol Chem 2001; 276: 36647-51.
Gastmann O, Burfeind P, Gunther E, Hameister H, Szpirer C, Hoyer-Fender S. Sequence, expression, and chromosomal assignment of a human sperm outer dense fiber gene. Mol Reprod Dev 1993; 36: 407-18.
Guzman-Grenfell AM, Bonilla-Hernandez MA, Gonzatlez-Martinez MT. Glucose induces a Na^+/K^+ATPase-dependent transient hyperpolarization in human sperm I. Induction of changes in plasma membrane potential by the proton ionophore CCCPBiochim Biophys Acta 2000; 1464: 188-98.
Henkel R, Staff T, Miska W. Isolation and partial characterization of the outer dense fiber proteins from human spermatozoa Biol Chem Hoppe Seyler 1992; 373: 685-9.
Higgy NA, Pastoor T, Renz C, Tarnasky HA, van dder Hoorn FA. Testis-specific RT7 protein localizes to the sperm tail and associates with itself. Biol Reprod 1994; 50: 1357-66.
Hoyer-Fender S, Burfeind P, Hameister H. Sequence of mouse Odfl cDNA and its chromosomal localizaton: extension of the linkage group between human chromosome 8 and mouse chromosome 15. Cytogenet Cell Genet 1995; 70: 200-4.
Hoyer-Fender S, Petersen C, Brohmann H, Rhee K, Wolgemuth DJ. Mouse Odf2 cDNAs consist of evolutionary conserved as well as highly variable sequences and encode outer dense fiber proteins of the sperm tail. Mol Reprod Dev 1998; 51: 167-75.
Huang LJ, Durick K, Weiner JA, Chun J, Taylor SS. D-AKAP2, a novel protein kinase A anchoring protein with a putative RGS domain. Proc Natl Acad Sci USA 1997; 94: 11184-9.
Javed AA, Naz RK. Human cleavage signal-1 protein; cDNA cloning, transcription and immunological analysis. Gene 1992; 112: 205-11.
Jimenez A, Johansson C, Ljung J, Sagemark J, Berndt KD, Ren B, Tibbelin G, Ladenstein R, Kieselbach T, Holmgren A, Gustafsson JA, Miranda-Vizuete A. Human spermatid-specific thioredoxin-1 (Sptrx-1) is a two-domain protein with oxidizing activity. FEBS Lett 2002 a; 530: 79-84.
Jimenez A, Oko R, Gustafsson JA, Spyrou G, Pelto-Huikko M, Miranda-Vizuete A. Cloning, expression and characterization of mouse spermatid specific thioredoxin-1 gene and protein. Mol Hum Reprod 2002b; 8: 710-8.
Johnson LR, Foster JA, Haig-Ladewig, VanScoy H, Rubin CS, Moss SB, Gerton GL. Assembly of AKAP82, a protein kinase A anchor protein, into the fibrous sheath of mouse sperm. Dev Biol 1997; 192: 340-50.
Jurado J, Prieto-Alamo MJ, Madrid-Risquez J, Pueyo C. Absolute gene expression patterns of thioredoxin and glutaredoxin redox systems in mouse. J Biol Chem. 2003; 278: 45546-54.
Keah HH, O'Bryan MK, de Kretser DM, Hearn MT. Synthesis and application of peptide immunogens related to the sperm tail protein tpx-1, a member of the CRISP superfamily of proteins: J Pept Res 2001; 57: 1-10.
Kierszenbaum AL, Rivkin E, Ferfer-Sadler S, Mertz JR, Tres LL. Purification partial characterization, and localization of Sak57, an acidic intermediate filament keratin present in rat spermatocytes, spermatids, and sperm. Mol Repord Dev 1996; 44: 382-94.
Kierszenbaum AL. Spermatid manchette: plugging proteins to zero into the sperm tail. Mol Reprod Dev 2001; 59: 347-9.
Kim Y, Adham IM, Haack T, Kremling H, Engel W. Molecular cloning and characterization of the bovine and porcine outer dense fiber cDNA and organization of the bovine gene. Biol Chem Hoppe Seyler 1995; 376: 471.
Kim YH, de Kretser DM, Temple-Smith PD, Hearn MT, McFarlane JR. Isolation and partial characterization of the outer dense fiber proteins from human spermatozoa. Biol Chem Hoppe Seyler 1992; 373: 685-9.
Lester LB, Coghlan VM, Nauert B, Scott JD. Cloning and characterization of a novel A-kinase anchoring protein, AKAP220: association with testicular peroxisomes. J Biol Chem 1996; 271: 9460-5.
Lin RY, Moss SB, Rubin CS. Characterization of S-AKAP84, a novel development regulated A-kinase anchor protein of male germ cells. J Biol Chem 1995; 270: 27804-11.
Lin X, Tombler E, Nelson PJ, Ross M, Gelman IH.A novel src- and ras-suppressed protein kinase C substrate associated with cytoskeletal architecture. J Biol Chem 1996; 271: 28430-38.
Lingrel J, Moseley A, Dostanic I, Cougnon M, He S, James P, Woo A, O'Connor K, Neumann J. Functional

roles of the alpha isoforms of the Na,K-ATPase. Ann N Y Acad Sci. 2003; 986:354-9.

Liu QY, Wang LF, Miao SY, Catterall JF. Expression and characterization of a novel human sperm membrane protein. Biol Reprod. 1996; 54: 323-30.

Long CR, Duncan RP. Robl JM. Isolation and characterization of MPM-2-reactive sperm proteins: homology to components of the outer dense fibers and segmented columns. Biol Reprod 1997; 57: 246-54.

Mandal A, Naaby-Hansen S, Wolkowicz MJ et al. ESP95, a testis-specific 95-kilodalton fibrous sheath antigen that undergoes tyrosine phosphorylation in capacitated human spermatozoa. Biol Reprod 1999; 61: 1184-97.

Mei X, Singh IS, Erlichman J, Orr GA. Cloning and characterization of a testis-specific, developmentally regulated A-kinase anchoring protein (TAKAP-80) present on the fibrous sheath of rat sperm. Eur J Biochem 1997; 246: 425-32.

Miki K, Eddy EM. Identification of tethering domains for protein kinase A type Iα regulatory subunits on sperm fibrous sheath protein FSC1. J Biol Chem 1998; 273: 34384-90.

Miranda-Vizuete A, Tsang K, Yu Y, Jimenez A, Pelto-Huikko M, Flickinger CJ, Sutovsky P, Oko R. Cloning and developmental analysis of murid spermatid-specific thioredoxin-2 (SPTRX-2), a novel sperm fibrous sheath protein and autoantigen. J Biol Chem. 2003; 278: 44874-85.

Modarressi M H , Cameron J , Taylor K E, Wolfe J. Identification and characterisation of a novel gene, TSGA10, expressed in testis. Gene 2001; 262:249-255.

Modarressi MH, Behnam B, Cheng M, Taylor K E, Wolfe J, van der Hoorn FA.Tsga10 encodes a 65-kilodalton protein that is processed to the 27-kilodalton fibrous sheath protein. Biol Reprod 2004; 70, 608-615.

Mohapatra B, Verma A, Shankar S, Suri A. Molecular cloning of human testis mRNA specifically expressed in haploid germ cells, having structural homology with the A-kinase anchoring proteins. Biochem Biophys Res Comm 1998; 244: 540-45.

Morales CR, Oko R, Clermont Y. Molecular cloning and developmental expression of an mRNA encoding the 27 kDa outer dense fibre protein of rat spermatozoa. Mol Reprod Dev. 1994; 37: 229-40.

Moss SB, Gerton Gl. A-kinase anchor proteins in endocrine systems and reproduction.Trends Endocrinol Metabol 2001; 12: 434-40.

Moss SB, Turner RM, Burkert KL, VanScoy Butt H, Gerton GL. Conservation and function of a bovine sperm A-kinase anchor protein homolgous to mouse AKAP82. Biol Reprod 1999; 61: 335-42.

Musilova P, Stratil A, Jurakova M, Rubes J, Cepica S. Assignment of the gene for outer dense fiber of sperm tails (ODF) to porcine chromosome 4p11-p14. Mamm Genome 1997; 8: 628.

Nakamura Y, Tanaka H, Koga M, Miyagawa Y, Iguchi N, Egydio de Carvalho C,Yomogida K, Nozaki M, Nojima H, Matsumiya K, Okuyama A, Nishimune Y. Molecular cloning and characterization of oppo 1: a haploid germ cell-specific complementary DNA encoding sperm tail protein. Biol Reprod 2002; 67:1-7.

O'Bryan MK, Sebire K, Meinhardt A, Edgear K, Keah HH, Hearn MT, De Kretser DM. Tpx-1 is a component of the outer dense fibers and acrosome of rat spermatozoa. Mol Reprod Dev 2001; 58; 116-25.

O'Bryan MK. Loveland KL, Herszfeld D, McFarlane JR, Hearn MT, de Kretser DM. Identification of a rat testis-specific gene encoding a potential rat outer dense fibre protein. Mol Reprod Dev 1998; 59: 313-22.

Ohuchi J, Arai T, Kon Y, Asano A, Yamauchi H, Watanabe T. Sperm tail associated protein (Stap), in mouse post-meiotic testicular germ cells. Mol Reprod Dev 2001; 59: 350-8.

Olson GE, Winfrey VP, Westbrook VA, Melner MH. Targeting of the domain-specific integral protein PM52 to the periacrosomal plasma membrane during guinea pig spermiogenesis. Mole Reprod Dev 1998; 50: 103-12.

Pruslin FH, Rodman TC. Proteins of demembraned mouse sperm heads. Characterization of a major sperm-unique components. J Biol Chem 1985; 260: 5654-9.

Quill TA, Garbers DL. Sperad is a novel sperm-specific plasma membrane protein homologous to a family of cell adhesion proteins. J Biol Chem 1996; 271: 33509-14.

Reinton N, Collas P, Haugen TB, Skalhegg BS, Hansson V, Jahnsen T, Tasken K. Localization of a novel human A-kinase-anchoring protein, hAKAP220, during spermatogenesis. Dev Biol 2000; 223: 194-204.

Sadek CM, Damdimopoulos AE, Pelto-Huikko M, Gustafsson JA J, Spyrou G, Miranda-Vizuete A. Sptrx-2, a fusion protein composed of one thioredoxin and three tandemly repeated NDP-kinase domains is expressed in human testis germ cells. Genes Cells 2001; 6: 1077-1090.

Schalles U, Shao X, van der Hoorn FA, Oko R. Developmental expression of the 84-kDa ODF sperm protein: localization to both the cortex and medulla of outer dense fibers and to the connecting piece. Dev Biol 1998; 199: 250-60.

Sen B, Mandal A, Wolkowicz MJ, Kim YH, Reddi PP, Shetty J, Bush LA, Flickinger CJ, Herr JC. Splicing in murine CABYR and its genomic structure. Gene 2003; 310: 67-78.
Shao X, Murthy S, Demetrick DJ, van der Hoorn FA. Human outer dense fiber gene, ODF2, Localizes to chromosome 9q34. Cytogenet Cell Genet 1998; 83: 221-3.
Shao X, Tarnasky HA, Lee JP, Oko R, van der Hoorn FA. Spag4, a novel sperm protein, binds outer dense-fiber protein Odfl and localizes to microtubules of manchette and axoneme. Dev Biol 1999; 211; 109-23.
Shao X, Tarnasky HA, Schalles U, Oko R, van der Hoorn FA. Interactional cloning of the 84-kDa major outer dense fiber protein Odf84. Leucine zippers mediate associations of Odf84 and Odf27. J Biol Chem 1997; 272: 6105-13.
Shao X, Xue J, van der Hoorn FA. Testicular protein Spag5 has similarity to mitotic spindle protein Deepest and binds outer dense fiber protein Odfl: Mol Reprod Dev. 2001; 59: 410-6.
Si Y, Okuno M, Regulation of Microtubule sliding by a 36-kDa phosphorportein in Hamster Sperm Flagella. Mol Reprod Develop 1999; 52: 328-334.
Svensson MJ, Chen JD, Pirrotta V, Larsson J. The ThioredoxinT and deadhead gene pair encode testis- and ovary-specific thioredoxins in Drosophila melanogaster. Chromosoma 2003 Sep 17
Tres LL, Kierszenbaum AL. Sak57, an acidic keratin initially present in the spermatid manchette before becoming a component of paraaxonemal structures of the developing tail. Mol Reprod Dev 1996; 44: 395-407.
Turner KI, Sharpe RM, Gaughan J, Millar MR. Expression cloning of a rat testicular transcript abundant in germ cells, which contains two leucine zipper motifs. Biol Reprod 1997; 57: 1223-32.
Turner RM, Johnson LR, Haig-Ladewig L, Gerton GL, Moss SB. An X-linked gene encodes a major human sperm fibrous sheath protein, hAKAP82. J Biol Chem 1998; 273: 32135 – 41.
Underhill DA, Canfield VA, Dahl JP, Gross P, Levenson R. The Na, K-ATPase α4 gene *Atp1* (a4) encodes a ouabain-resistant α subunit and is tightly linked to the α2 gene (*Atp1a2*) on mouse chromosome 1. Biochemistry 1999; 38: 14746-51.
Vera JC, Brito M, Zuvic T, Burzio LO. Polypeptide composition of rat sperm Outer dense fibers. A simple procedure to isolate the fibrillar complex. J Biol Chem 1984; 259: 5970-77.
Vijayaraghavan S, Goueli SA, Davey MP, Carr DW. Protein kinase A-anchoring inhibitor peptides arrest mammalian sperm motility. J Biol Chem 1997; 272: 4747-52.
Vijayaraghavan S, Liberty GA, Mohan J, Winfery VP, Olson GF, Carr DW. Isolation and molecular characterization of AKAP110, a novel, sperm-specific protein kinase A-Anchoring protein. Mol Endocrinol 1999; 13: 705-17.
Visconti PE, Johnson LR, Oyaski M, Fornes M, Moss SB, Gerton GL, Kopf GS. Regulation, localization and anchoring of protein kinase A subunits during mouse sperm capacitation. Dev Biol 1997;192:351-63.
Wakayama T, Nagata K, Ohashi K, Mizuno K, Tani I, Yoshinaga K, Oh-Oka T. Toshimori K. The expression and cellular localization of the sperm flagellar protein MC31/CE9 in the rat testis: possible post-transcriptional regulation during rat spermiogenesis. Arch Histol Cytol 2000; 63: 33-41.
Westbrook-Case VA, Winfrey VP, Olson GE. A domain specific 50-kilodalton structural protein of the acrosomal matrix is processed and released during acrosome reaction in the guinea pig. Biol Reprod 1994; 51: 1-13.
Yu Y, Oko R, Miranda-Vizuete A.Developmental expression of spermatid-specific thioredoxin-1 protein: transient association to the longitudinal columns of the fibrous sheath during sperm tail formation. Biol Reprod 2002; 67: 1546-54.
Yuasa K, Omori K, Yanaka N. Binding and phosphorylation of a novel male germ cell-specific cGMP-dependent protein kinase-anchoring protein by cGMP-dependent protein kinase Iα. J Biol Chem 2000; 275: 4887-95.
Yudin AI, Li MW, Robertson KR, Tollner T, Cherr GN, Overstreet JW. Identification of a novel GPI-anchored CRISP glycoprotein, MAK248, located on the posterior head and equatorial segment of cynomolgus macaque sperm. Mol Reprod Dev 2002; 63: 488-99.
Zarsky H A , Cheng M, Van Der Hoorn F A. Novel RING finger protein OIP1 binds to conserved amino acid repeats in sperm tail protein ODF1. Biol Reprod 2003; 68: 543-52.

Chapter 30

PROTEINS IN ANTIPEROXIDATION

30.1. REACTIVE OXYGEN SPECIES

Although formation of reactive oxygen species (ROS) is normally associated with phagocytic leucocytes, spermatozoa were the first cell type with which this activity was observed. Now such activity has been found with variety of cells. The ROS initiate peroxidation of the unsaturated fatty acids in the sperm plasma membrane, which plays a key role in the etiology of male infertility (Aitken, 1999; de Lamirande et al, 1997). Both spermatozoa and leukocytes generate ROS, although leukocytes produce much higher levels. The clinical significance of leukocyte presence in semen is controversial. Although seminal plasma confers some protection to sperm against ROS damage, a variety of defense mechanisms comprising a number of anti-oxidants can be employed to reduce or overcome oxidative stress caused by excessive ROS. Determination of etiology of male infertility is important in order to develop effective therapies to overcome the excess production of ROS.

However, it is becoming clear that ROS can have both beneficial and detrimental effects on the spermatozoa. Hence the balancing act between the levels of ROS produced and the amounts scavenged at any moment will determine the survival of normal sperm (Aitken, 1999, Sharma and Agarwal 1996, de Lamirande 1997). Infact, white blood cells in semen and, to a lesser extent, the mature spermatozoa generate excess ROS of hydrogen peroxide and nitric oxide. But due to the paucity of antixidant mechanisms available at the site, spermatozoa have little defence against these ROS, which damage their membrane and the genetic apparatus. As a consequence, sperm acrosome reaction and fusogenic capacity of sperm are inhibited, causing reduced rate of conception. Damage to sperm DNA may result in mutagenesis whereby the genetic disturbance may occur in the offspring if the repair mechanisms in spermatozoa are insufficient.

30.1.1. Types of ROS

Human spermatozoa generate the superoxide anion ($O_2^{\bullet-}$), which spontaneously or enzymatically dismutates to hydrogen peroxide (H_2O_2). There is growing evidence that superoxide anion and H_2O_2 also participate in biological events such as capacitation and acrosome reaction. Owing to its low reactivity and short half–life (1ms), $O_2^{\bullet-}$ is not very harmful, although reaction with its targets can produce more toxic species, such as thiyl radicals (RS·). But, H_2O_2 being relatively stable and being uncharged, can freely cross cell membranes. The very low concentrations of ferric ions present in any solution/cytosol are sufficient to catalyse the formation of the hydroxyl radical ($OH^{\bullet}$) from $O_2^{\bullet-}$ and H_2O_2. These ROS can react with virtually any cell component and may produce toxic effects during their very short half-life. However, seminal plasma contains a large number of ROS scavengers, including enzymes such as superoxide dismutase (SOD), catalase and the glutathione peroxidase/reductase system. The balance between ROS generation by sperm and scavenging capacity of seminal plasma as well the site at which spermatozoa come into contact with ROS will determine the final outcome. H_2O_2 is the primary ROS responsible for the loss of sperm functions since catalase, which selectively degrades ROS, is the only scavenger to confer complete protection to spermatozoa (de Lamirande et al, 1997).

30.1.2. ROS and Sperm Function

Although, ROS are known for their damaging effects on sperm functions, there is now growing evidence to suggest that very low and controlled concentrations of ROS participate in signal transduction in sperm capacitation and acrosome reaction. The first experimental evidence for involvement of ROS in human sperm physiology came from the observation that spermatozoa exposed to $O_2^{\bullet-}$ show higher hyperactivation and capacitation than to those treated with fetal cord serum and that superoxide dismutase (SOD) prevents these effects (de Lamirande and Gagnon, 1993a, b). Furthermore, the rate of sperm capacitation induced by biological fluids is inversely correlated with the potential of these fluids to scavenge $O_2^{\bullet-}$. Human spermatozoa incubated under capacitating conditions produce higher concentration of intra- and extracellular $O_2^{\bullet-}$ than spermatozoa in medium alone. The increase in intracellular $O_2^{\bullet-}$ is not prevented by SOD and therefore may reflect an increased metabolic activity rather than a direct involvement in sperm capacitation (de Lamirande and Gagnon 1993b). The involvement of ROS in the acquisition of fertilizing ability may not be limited to $O_2^{\bullet-}$ or to human spermatozoa. Addition of any factor such as catalase that reduces the required level of ROS can affect human sperm hyperactivation and the acrosome reaction. Furthermore, addition of H_2O_2 either directly or through enzymatic generation by the combination of glucose and glucose oxidase, stimulates the acrosome reaction. The requirements of ROS for sperm capacitation and acrosome reaction were supported by O'Flaherty et al, (1999). But excessive ROS may show negative effect on acrosome reaction (Ichikawa et al, 1999).

Relation of ROS with sperm capacitation is associated with tyrosine (Tyr) phosphorylation through a cAMP-PKA-mediated pathway. Since human sperm capacitation is associated with an increased production of superoxide anion ($O_2^{*\cdot-}$), De Lamirande et al (1998) investigated whether the $O_2^{*\cdot}$ generation by spermatozoa is correlated with level of capacitation and increased tyrosine phosphorylation of two sperm proteins (p105/p81). It was found that progesterone and ultrafiltrates of human fetal cord serum, follicular fluid, and seminal plasma individually promoted sperm generation of $O_2^{*\cdot}$, tyrosine phosphorylation of p105/p81, and sperm capacitation. The production of $O_2^{*\cdot}$ by spermatozoa was rapid and transient. A typical electron paramagnetic resonance spectrum for $O_2^{*\cdot}$ spin adduct exhibited only by capacitated spermatozoa but not by fallopian tubal fluids, nor any factor in seminal plasma and noncapacitated spermatozoa, supports the role of free radicals in capacitation. It seems possible to utilize the inhibitory effect of SOD on sperm capacitation to regulate fertilization (Zhang and Zheng, 1996). Gagnon and associates evidenced the double phosphorylation of the threonine-glutamine-Tyr motif (P-Thr-Glu-Tyr-P) in human sperm proteins of 80 and 105-kDa during capacitation. The P-Thr-Glu-Tyr-P motif was located to the principal piece of spermatozoa. The regulation of P-Thr-Glu-Tyr-P is specific to nitric oxide and not to superoxide anion or hydrogen peroxide and appeared to involve MEK pathway (Thundathil et al., 2003). The study of antiperoxidative mechanisms in male reproductive tract is quite relevant, since epididymal spermatoxoa must be properly protected against reactive oxygen species, which can impair the complex maturation process.

30.1.3. Oxidative Stress and DNA damage

Reactive oxidative metabolites when produced in low level, stimulate DNA compaction; but when produced in excessive amounts, it leads to DNA fragmentation. The excessive generation of ROS associated with defective sperm function does not only attack the fluidity and integrity of the sperm plsma membrane, but also attacks the DNA in the sperm nucleus. DNA fragmentation is commonly observed in the spermatozoa of infertile patients and there is strong evidence that this damage is free radical mediated and is induced by oxidative stress (Kodoma et al., 1997; Lops et al, 1998). The amount of DNA fragmentation in human spermatozoa is negatively related with semen quality rate of *in vitro* fertilization. The fertilization failure in this context is presumably due to peroxidative damage to the sperm plasma membrane, and damaged genome, which cannot participate in the normal process of fertilization. Consequences of fertilization involving DNA damaged spermatozoa may lead to oncogenesis in the offsprings and infertility in males. For example heavy smoking induces a state of oxidative stress, which is accompanied with free radical mediated damage to sperm DNA and decrease in the antioxidant capacity of seminal plasma. This DNA damage has consequences for the health of the offspring, who show a particularly high incidence of childhood cancer. In addition to childhood cancer, another possible consequence of free radical-mediated damage to DNA in the male germ line is infertility in the offspring. This possibility relates specifically to forms of male infertility involving deletions on the long arm (q) of the Y chromosome. In Y chromosome, three regions have been identified that contain genes , which regulate spermatogenesis and fertilizing ability of sperm; these loci have been designated AZF (azoospermia factor) a, b and c (see Chapter 10) (Roberts, 1998, Aitken, 1999). The mechanism responsible for infertility and the induction of mutations associated with neoplastic disease need to be elucidated. Excessive free radical generation by the germ cells could be induced by the redox cycling of xenobiotics, increased testicular temperatures, excessive NADPH oxidase activity or the increased availability of transition metals. Oxidative stress could also be associated with impaired antioxidant protection, possibly related to dietary deficiencies, age or genetic factors. At the same time when generated in low levels, ROS are thought to enhance sperm function by stimulating DNA compaction and promoting redox regulated c-AMP mediated pathway which plays a central role in sperm capacitation and acrosome reaction (Aitken, 1999).

30.2. PROTEINS IN ANTIPEROXIDATION

30.2.1. Glutathione Peroxidase in Reproductive Tract

The family of glutathione peroxidases comprises four distinct mammalian selenoproteins. (a) The classical enzyme (cGPx) is unbiquitously distributed. Its primary function is to counteract oxidative attack. It is dispensible in unstressed animals, and accordingly ranks low in the hierarchy of glutathione peroxidases. (b) The gastrointestinal isoenzyme (GI-GPx) is most related to cGPx and is exclusively expressed in the gastrointestinal tract. (c) Plasma GPx (pGPx) behaves similar to cGPx in selenium deficiency. It is directed to extracellular compartments and is expressed in various tissues in contact with body fluid, e.g., kidney, ciliary body, and maternal/fetal interfaces as well as in reproductive tract. (d) Phospholipid hydroperoxide glutathione peroxidase (PHGPx), originally presumed to be a universal antioxidant enzyme protecting membrane lipids, appears to have adopted a variety of specific roles like silencing lipoxygenases and becoming an enzymatically inactive structural component of the mitochondrial capsule during sperm maturation. Thus, all individual isoenzymes are efficient peroxidases in principle, but beyond their mere antioxidant potential may exert cell- and tissue-specific roles in metabolic regulation, as is evident for PHGPx (Brigelius-Flohe, 1999).

30.2.2. Classical Glutathione Peroxidase in Male Accessary Sex Organs

Glutathione peroxidase did not attract any particular attention until it had been characterized as the first example of mammalian selenoprotein in 1973. Later it became obvious that selenium deficiency cannot simply be interpreted as a deficiency of GPx since several isozymes of GPx such as plasma GPx, gastro-intestinal GPx, phospholipid hydroperoxide GPx as well as proteins structurally and functionally unrelated to GPx, like type 15'- deiodinase, selenoprotein P, protein W, and thioredoxin reductase were shown to contain selenium, and more selenoproteins, likely to be discovered in future. The cumulated knowledge on GPx offers the unique opportunity to understand how enzymatic selenium catalysis operates in detail. Although GPx in testis has been identified, its characterization has yet to be elaborated (Schwaab et al, 1998a, b).

The mammalian epididymis is the site of expression and secretion of an abundant, tissue-specific, androgen-regulated, selenium-independent, glutathione peroxidase isoenzyme (GPx5), which has been proposed to play a role in protecting the membranes of spermatozoa from the damaging effects of lipid peroxidation and /or preventing premature acrosome reaction. Two glutathione peroxidase genes (gpx5 and gpx3) are known to be expressed in the mouse epididymis. The GPx5 was shown to be epididymis specific and restricted to the caput epididymidis, while GPx3 is expressed in a wide array of tissues including the caput, corpus and cauda epididymis. Both, single copy genes, are regulated by androgens as well as developmentally regulated during postnatal ontogenesis of the epididymis. The mechanisms by which these genes are regulated in the mouse epididymis and the putative roles of their enzymes in the sperm maturation have been discussed (Schwaab et al, 1998a). The study of mRNA expression of GPx isoforms in the rat male reproductive tract and its expression in testis revealed a 0.8-kb transcript in liver, testis, prostate, seminal vesicle, vas deferens, and epididymis using the cDNA probe for classic cellular GPx (Zini and Schlegel, 1997). Williams et al (1998) described the developmental expression of GPx5 transcripts and proteins in the rat epididymis and characterized the association of rat GPx5 with the sperm plasma membrane. The epididymis-specific GPx from the porcine cauda epididymal fluid consists of four identical 23kDa subunits. The c-DNA encoding the 23kDa subunit from porcine caput epididymis was cloned (Okamura et al, 1997). Although the selenocysteine codon (TGA) is contained in the cDNA of the other cytosolic type of glutathione peroxidases, it is replaced by cysteine codon (TGT) in the 23-kDa subunit cDNA, similarly to the results obtained for cDNAs encoding the epididymis-specific form of the secreted glutathione peroxidases of mouse, rat and monkey. The purified protein was proved to contain no selenium atom in the molecule. Within the epididymis and the kidney, GPx3 mRNA and protein are temporarily regulated in a tissue-specific manner. Androgen withdrawal by castration down-regulates the expression of the GPx3 gene both in the epididymis and vas deferens while GPx3 expression in the kidney was found to be androgen-independent. Epididymis GPx3 distribution is quite peculiar suggesting the existence in this organ of complex transductional and/or transcriptional regulatory processes (Schwaab et al, 1998a). The cDNA cloning of human GPx5 showed that the majority of the transcripts contained a 118 nt frame-shifting deletion, arising most likely from inappropriate excision of exon 3 during processing. Antisera raised against recombinant human GPx5 cross-reacted with rat and macaque (Macaca fascicularis) epididymal proteins of full-length GPx5 (Hall et al, 1998). Lahti et al (2001) evaluated the 5.0-kb long GPx5 and 3.8-kb-long CRISP-1 gene 5'-flanking regions. It was evident that the 5.0-kb 5'-flanking region of GPx5 promoter is suitable for directing the expression of structural genes of interest into the caput epididymis in transgenic mice.

Vernet et al (1999) characterized and cloned a secreted sperm-bound selenium-independent GPx5, the expression of which was found to be restricted to the mouse caput epididymidis. Because of the lack of selenium in the active site of this enzyme, unlike the other GPx5 characterized, it was suspected that GPx5 does not function in the epididymis as a true glutathione peroxidase in vivo. Therefore, following dietary selenium deprivation, which is known to reduce antioxidant defenses and favor oxidative stress in relation with depressed Se-dependent GPx activities, Vernet et al (1999) showed that epididymis is still efficiently protected against increasing peroxidative conditions. These data strongly suggested that the selenium-independent GPx5 could function as a back-up system for Se-dependent GPxs. This epididymal GPx protein was found to bind to the acrosomal region of the epididymal sperm and to disappear during the acrosome reaction (Okamura et al, 1997).

Gpx-5 Modulator Protein: The tissue-restricted polyoma enhancer activator protein (PEA3) of the Ets oncogene family of DNA-binding proteins is a putative modulator of the epididymis-specific gpx-5 gene. The PEA3 factor is spatially and temporarily expressed within the mouse epididymis in a manner consistent with *gpx5* characteristics of expression. The transcriptional activity of the *gpx5* promoter is modulated by the presence of the PEA3 protein. DNA sequences located within the 5' flanking region of the *gpx5* gene have the ability to bind specifically to the PEA3 protein (Drevet et al, 1998).

30.2.3. Phospholipid Hydroperoxide Glutathione Peroxidase

Phospholipid hydroperoxide glutathione peroxidase (PHGPx) also known as GPx4 is an antioxidant selenoenzyme that belongs to the superfamily of five selenoproteins, which are capable of reducing hydroperoxy lipids to the corresponding alcohols. While the "classical" glutathione peroxidase (GPx) and its variants are tetrameric and reactive on the hydroperoxides of free fatty acids, PHGPx is monomeric and interacts directly with peroxidized phospholipids and cholesterol. The PHGPx is a 19kDa monormeric enzyme that protects cells from lipid peroxide-mediated damage by catalyzing the reduction of lipid peroxides. It utilizes thiols such as glutathione to specifically scavenge phospholipid hydroperoxides, and thus protects membrane phospholipids against peroxidation. It down regulates leukotriene biosynthesis, interleukin 1-induced signal transduction, and apoptosis. The PHGPx accounts for almost the entire selenium content of mammalian testis. PHGPx has been shown to increase in rat testis after puberty and depends on gonadotropin stimulation in hypophysectomized rats (Roveri et al, 1992; Giannattasio et al, 1997). Exposure of decapsulated whole testis, however, failed to reveal any transcriptional activation or inhibition of the *PHGPx* gene by testosterone, human chorionic gonadotropin, or forskolin. Nevertheless, the specific activity of PHGPx in testis, but not of cGPx, correlated with sexual maturation. Leydig cell destruction in vivo by ethane dimethane sulfonate (EDS) resulted in a delayed decrease in PHGPx activity and mRNA that could be completely prevented by testosterone substitution. In situ hybridization studies demonstrated an uncharacteristic low level of cGPx transcription in testis, wheras PHGPx mRNA was abundantly and preferentially expressed in round spermatids, The age or gonadotropin dependent expression of PHGPx is not related to direct transcriptional gene activation by teststerone but is due to differentiation stage specific expression, in late

```
CAGGGGCCTCGCGTCTTAGCGCAGGCGCATCTGACCAATAAGAGACGTCAGTGGGCGTGC   60

CCGAGGGCGGGCAAGCCATACTCGGCCTCGCGTCCATTGGTCGGCTGCGTGAGGGGAG   120

GAGCCGCTGGCTCCGGCCGCCGAGATGAGCTGGGGCCGTCTGAGCCGCTTACTTAAGCCA  180
                        M S W G R L S R L L K P        12

GCACTGCTGTGCGGGGCTCTGGCTGCGCCTGGTCTGGCAGGCACCATGTGTGCATCCCGC  240
A L L C G A L A A P G L A G T M C A S R                  32
                         S1
GATGATTGGCGCTGTGCCGCCTCCATGCACGAATTCTCAGCCAAGGACATCGACGGGCAC  300
D D W R C A A S M H E F S A K D I D G H                  52
                              S2
ATGGTCTGCCTGGATAAGTACAGGGGTTTCGTGTGCATCGTCACCAACGTGGCCTCGCAA  360
M V C L D K Y R G F V C I V T N V A S Q                  72

TGAGGCAAAACTGACGTAAACTACACTCAGCTAGTCGATCTGCATGCCCGATATGCTGAG  420
SecG K T D V N Y T Q L V D L H A R Y A E                 92

TGTGGTTTACGAATCCTGGCCTTCCCCTGCAACCAGTTTGGGAGGCAGGAGCCAGGAAGT  480
C G L R I L A F P C N Q F G R Q E P G S                 112

AATCAAGAAATCAAGGAGTTTGCAGCCGGCTACAACGTCAAGTTTGACATGTACAGCAAG  540
N Q E I K E F A A G Y N V K F D M Y S K                 132

ATCTGTGTAAATGGGGACGATGCCCACCCACTGTGGAAATGGATGAAAGTCCAGCCCAAG  600
I C V N G D D A H P L W K W M K V Q P K                 152

GGCAGGGGCATGCTGGGAAATGCCATCAAATGGAACTTTACCAAGTTTCTCATTGATAAG  660
G R G M L G N A I K W N F T K F L I D K                 172

AACGGCTGCGTGGTGAAGCGCTATGGTCCCATGGAGGAGCCCCAGGTGATAGAGAAGGAC  720
N G C V V K R Y G P M E E P Q V I E K D                 192
               As1
CTGCCGTGCTATCTCTAGCTAGCCCTACAAGTGTGTGCCCCTACACCGAGCCCCCCTGCC  780
L P C Y L *                                             197
   As2
CTGTGACCCCTGGAGCCTTCCACCCCGGCACTCATGAAGGTCTGCCTGAAAACCAGCCTG  840

CTGGTGGGGCAGTCCTGAGGACCTGGCGTGCATCCCTGCCGGAGGAAGGTCCAGAGGCCT  900

GTGGCCCTGGGCTCGAGCTTCACCCTGGCTGCCTTGTGGGAATAAAATGTAGAAATGTAA  960

AAAAAAAAAAAAAA                                              974
```

Fig.30.1. c-DNA and deduced amino acid sequence of mouse PHGPx. Sequences (S) for RACE are indicated with arrows. Potential start sites (ATG), stop codon (*), and polyadenylation signal are marked/underlined. Boxed sequences (residues 1-27) represent potential mitochondrial targeting sequence in testis. Amino acid, Gln[108] and Trp[163] at active site may interact with selenocysteine. Sequences in 3'-UTR with double underline may incorporate selenocysteine. Printed with permission from S.Nam et al. Gene 198; 245-49: 1997 © Elsevier.

spermatids, which are under the control of Leydig cell-derived testosterone. The burst of PHGPx expression at the transition of round to elongated spermatids suggests the involvement of this selenoprotein in sperm maturation (Maiorino et at, 1998). Dependence of PHGPx activity on gondatropin indicates that enzyme has more than general function of an antioxidant.

PHGx mRNA in the mouse testes is developmentally regulated during spermatogenesis. It first appeared in pachytene spermatocytes of stage X, gradually increased in round spermatids during early spermiogenesis, and reached a peak in step 10-11 elongating spermatids. After step 12, PHGPx mRNA began to show a progessive decline in the spermatids and was weakly detected in step 16 spermatids. However, the signal was not detected in spermatogonia or early spermatocytes. According to in situ RT-PCR, *PHGPx* mRNA also expresses in Leydig cells. These findings confirm that PHGPx in testes is closely involved in spermatogenesis as well as having a general antioxidant function (Nam et al, 1998a). In rat testicular germ cells, PHGPx is distributed in mitochondria as well as in nuclei. In nuclei, PHGPx is bound to chromatin (Godeas et al 1994, 1996; Roveri et al., 1994; Zini and Schlegel, 1997).

Charaterization: A full-length cDNA clone encoding the PHGPx encodes a polypeptide of 197 amino acids that initiates the translation at ATG (145-147) and contains an inframe TGA selenocysteine codon. It also has selenocysteine insertion sequences in the 3'-UTR. Moreover, the mouse PHGPx contains the active-site residues Gln 108 and Trp 163 that interact with selenocysteine, and the N-terminal 27-amino acid residues that may act as a potential mitochondrial targeting signal. Mouse PHGPx shares a high level of amino acid identity with pig (93.4%), human (92.9%), and rat (98%) PHGPxs (Roveri et al., 1994; Nam et al, 1997) **(Fig.30.1)**. PHGPx is synthesized in two forms: i) as a 194- amino acid peptide that predominates in testis and ii) as a 170-amino acid protein that predominates in most somatic tissues and localizes in the cytoplasm (Knopp et al., 1999). Immunochemical evidence revealed that PHGPx is present in sperm head and tail midpiece mitochondria. The binding of the enzyme to spermatozoa, head, and mitochondria was barely affected by ionic strength or thiols or detergents, as compared to the detachment of PHGPx obtained from testis nuclei (Godeas et al, 1997).

The PHGPx has been implicated in antioxidative defense, but its high expression level in testicular tissue suggests a more specific function during sperm maturation. The cDNA sequences from mouse heart and testis were identical except that the testis cDNA contained an additional 109-bp at its 5'-end. With a partial cDNA with complete homology to both the testis and myocardial PHGPx cDNAs, highest level of PHGPx mRNA expression was found in the testis, followed by the kidney, heart and skeletal muscle, liver, brain, lung, and spleen. The longer PHGPx transcript was present only in the testis. The murine PHGPx gene *(Gpx4)* was mapped to a region of murine chromosome 10, located 43 cM from the centromere, that is syntenic with the human locus, which is located at the terminus of the short arm of human chromosome 19 (Pushpa-Rekha et al, 1995).

Sperm PHGPx: Mammalian spermatozoa are unusually rich in polyunsaturated fatty acids, a property that predisposes them to the deleterious effects of oxygen free radicals. Expression of PHGPx activity in testis is regulated by dietary Se and vitamin E differently from that in liver (Lei et al, 1997). Failure of expression of mitochondrial PHGPx in spermatozoa may be one of the causes of oligoasthenozoospermia in infertile men (Imai et al, 2001; Foresta et al., 2002). The presence and structure of the PHGPx mRNA was analysed in rat sperm after feeding Se deficient diet. The subtype of the PHGPx in the rat sperm was of the mitochondrial-type mRNA, which included a region corresponding to the mitochondrial transfer leader sequence. This suggests that the intracellular localization of PHGPx is regulated at the transcription level, and the analysis of the PHGPx mRNA in the sperm is useful tool for investigating the dysfunction caused by the structure of PHGPx in the sperm (Mizuno et al., 2000). Infertile sperm are associated with abnormal mitochondria with loss of mitochondrial membrane potential. The selenoprotein PHGPx changes its physical characteristics and biological function during sperm maturation. PHGPx exists as a soluble peroxidase in spermatids but persists in mature spermatozoa as an enzymatically inactive cross-linked, insoluble protein. In the midpiece of mature spermatozoa, PHGPx protein represents at least 50 percent of the capsule material that embeds the helix of mitochondria. The role of PHGPx as a structural protein may explain the mechanical instability of the mitochondrial midpiece that is observed in selenium deficiency (Ursini et al, 1999).

30.2.4. Sperm Nucleus Glutathione Peroxidase (snGPx)

A sperm nucleus glutathione peroxidase (snGPx), which is closely related to the phospholipid hydroperoxide glutathione peroxidase (PHGPx), has been identified in late spermatids. Though both isoforms originate from a joint *ph/snGPx* gene, their N-terminal peptides are encoded by alternative first exons. The expression of two enzymes is spatially and differentially regulated in various cells. While PHGPx is expressed at low levels in many organs, snGPx was only detected in testis, kidney, and in the human embryonic kidney cell line HEK293 (Borchert et al., 2003). The PHGPx can be expressed as a mitochondrial or cytosolic isoform. The *sn/phGPx* genes have been cloned from various mammalian species. The 5'-flanking region of the murine *sn/phGPx* gene has been investigated (Ufer et al., 2003). The basic promoter activity is localized in a 200-bp region immediately upstream of the translational initiation site of the cytosolic isoform (3'-ATG). It showed the presence of five distinct protein-binding regions such as for stimulating protein 1 (SP1), nuclear factor Y (NF-Y) and members of the SMAD family. The basic *phGPx* promoter constitutes a 200-bp oligonucleotide, which is localized immediately upstream of the 3'-ATG and involves functional SP1/SP3, NF-Y and SMAD binding sites. The corresponding trans-regulatory proteins may contribute to differential expression regulation of the mitochondrial and cytosolic PHGPx isoforms (Ufer et al., 2003).

30.2.5. Glutathione-S-Transferases

Glutathione S-transferases (GST) are dimeric proteins comprising a large family of dimeric enzymes that catalyse the nucleophilic addition of the tripeptide glutathione to a wide spectrum of compounds that have electrophilic functional groups. These isozymes are found in most organisms and function primarily as phase II detoxification enzymes protecting cells against both endogenous and xenobiotic alkylating agents and can catalyse reactions towards a large number of diverse substrates. GSTs may also function as intracellular binding / carrier proteins for various lipophilic ligands viz, bilirubin, heam derivatives and thyroid hormones and play an additional protective role by binding and immobilizing certain reactive electerophilic molecules and inactivating these compounds (Hayes and Pulford 1995; Tam et al, 1998). At least 16 cytosolic GSTs have been identified. On the basis of physical, immunological, structural and enzymic properties, these isozymes are grouped into five classes Alpha, Mu, Pi, Sigma and Theta based on the degree of amino and homology of their subunits. Multiple subunit types from within each of these classes usually assemble in homo- or heterodimeric combinations. However, with the exception of GST Pi, which is usually the product of a single gene, cross-correlations among the multiple Mu-or Alpha-class GSTs in different species have not been possible. Under in vitro condition, Alpha and Mu-class GSTs can act as substrates for phosphorylation and methylation, respectively. The modified proteins have a lower enzymic activity. Post-translational modifications include N-terminal processing or C-terminal truncations. Mammalian GSTs are expressed in discrete tissue-specific forms. For example livers of some species are rich in Alpha and some Mu-class GSTs but do not contain Pi, yet some Mu subunits not present in liver are major components in certain extrahepatic tissues, whereas tissue-specific expression of rGSTM5 has been observed in rat testis. Rowe et al (1998) established that a labile and poorly characterized rat subunit previously designated as Yo or GST11 and mouse mGSTM5 are similar to the human hGSTM3 subunit in terms of structure, catalytic mechanisms, and other characteristic properties. Some of the used nomenclature for rat GSTs is shown in **Table 30.1.** In view of the report (Rowe et al, 1998) on understanding of nomenclature of GST subunits, we have followed nomenclature given by individual authors in original articles.

Table 30.1. Subunit Nomenclature of GST in rat (Rowe et al, 1998)

Rowe et al, 1998	Y system	Numerical system
GSTM1	Y_{b1}	3
GSTM2	Y_{b2}	4
GSTM3	Y_{b3}	6
GSTM5	Y_o	11
GSTM6	Y_{n2}	9
GSTP1	Y_p	7
GSTA1	Y_{n1}	1
GSTA3	Y_{c1}	2
GSTA4	Y_k	8

Identification of GST Sub-units: The GST comprises subunits of Mr 25500 (Ya), 26500 (Yn), 27000 (Yb1 and Yb2) and 28500 (Yc). In rat testis, the major GST isoenzymes expressed are: subunits A3, M1, M2, M3, M5 and M6. Subunits A1, A4 and P1 are expressed in lesser amounts. There was no evidence of post-translational modifications in any GSTs with known cDNA sequence. The molecular masses of subunits M5* and M6* of class-Mu GSTs, were determined to be 25495 and 26538 Da respectively. An N-terminally modified protein from rat testis with molecular mass 25737 Da and designated rGSTA6 was sequenced. This GST is a novel class-Alpha GST that was not previously reported (Hsieh et al, 1997). The majority of rat testicular GSTs are of Yb size (60%) with molecular weight of 27-kDa. The most predominant subunits, however, are GST Yn2 (27%), followed by GST Yc (24%) and GST Yn1 (20%). Testis cytosol contains the $Y_{b2}Y_{b2}$-homodimer glutathione S-transferase D (GST-D) in addition to the glutathione S-transferases A ($Y_{b1}Y_{b1}$) and C ($Y_{b1}Y_{b2}$). Treatment of rats with phenobarbital induces the level of GST-D in testis with no increase in the activities of GST-A and GST-C. Thus a specific induction of the Y_{b2} subunit in contrast with other forms of GSTs occurs in testis (Sheehan et al, 1984). Testicular GSTs possess very high Leukotriene C4 (LTC4) synthase activity with 5, 6-Leukotriene A4Me (LTA4Me) as the substrate and prostaglandin D (PGD) synthase activity with prostaglandin H2 (PGH 2) as the substrate. Majority of rat testicular GSTs of Yb sized are involved in the synthesis of eicosanoids like LTC4 and PGD2 (Anuradha et al, 2000). In most organs of the rat the predominant forms of GSTs have alkaline pI values. In contrast, in the cytosol from rat testes almost 50% of the transferase actively is due to isoenzymes with acidic (less than 7.0) pI values. Three acidic forms of GSTs were purified from rat testis cytosol (Boyer and Kenny, 1985). One form accounted for more than 90% of the enzymic activity in the acidic fraction. This major form was a homodimer of a subunit, termed Yt. This subunit had an electrophoretic mobility that was different from the subunits that form the alkaline transferases. In addition, the two minor acidic enzymes of rat testis appeared to be heterodimers of the Yt subunit with an electrophoretic mobility identical with that of the Yb subunit present in some alkaline enzymes (Boyer and Kenney, 1985).

Hayes (1984) isolated a YnYn dimer of transferase N from rat testis and made structural and functional comparisons among Yb1, Yb2 and Yn monomers. Association-dissociation experiments between the YnYn and Yb2Yb2 homodimers demonstrated that Yn monomers can hybridize with both Yb1 and Yb2 monomers. Reversible dissociation of transferases N and C (Yb1Yb2) showed that both Yb1 and Yb2 monomers can hybridize with Yn monomer under competitive conditions. The hydridization data suggested that transferase S represents the Yb2Yn subunit combination (Hayes, 1984).

Localization of Various Subunits of GST: The testis microsomes contain mainly cytosolic form of GST, which is a member of Mu family. GST of the Mu family is present in cytoplasmic droplets of step 19 spermatids. After spermiation, cytoplasmic droplets of spermatozoa within the proximal region of the epididymis remain intensely stained. This demonstrates that, at the end of spermiogenesis, the GST Yo subunit is expressed at high levels in late spermatids. Furthermore, the presence of GST in late spermatids and cytoplasmic droplets of spermatozoa suggests that GST plays a vital role in protecting these cells from electrophilic attack. Also, of significance is the correlation between the loss of reactivity in cytoplasmic droplets of spermatozoa in the distal region of the epididymis and the concomitant increase of reactivity in principal cells of this region (Very et al, 1994). The Pi class of GSTs (GST-Pi) contains homodimers of the Yf subunit, also known as Yp or rat subunit 7, which is found in high concentrations in the testis and epididymis. In the testis, Yf was localized exclusively to Sertoli and Leydig cells. A dramatic change in the immunostaining pattern for the Yf subunit of GST-Pi occurs during postnatal development in both principal and basal cells along the epididymis. Thus, different factors seem to play a role in the regulation of the expression of the different subunits of GST, not only in different epididymal regions, but also in different cell types during postnatal development (Hermo et al, 1994, Very et al, 1994).

The distribution of the Yc and Yo subunits from the Alpha family, as well as the Yb1 and Yb2 subunits from the Mu gene family was examined in the adult rat testis and epididymis (Papp et al, 1995) and the results were compared with two other subunits, the Yf and Yo proteins (Very et al., 1993; 1994). Leydig cells were intensely strained for all six subunits, whereas, Sertoli cells were reactive for Ya, Yb1 and Yf subunits. Among germ cells, all spermatogonia, spermatocytes and step 1-15 spermatids were virtually unreactive for each of the six GSTs. However, moderate to intense staining was seen over steps 16-19 spermatids with the anti-Y_o, anti-Y_a, anti-Y_{b1} and anti Y_{b2} antibodies. Y_f, Y_o Y_{b1}, and Y_{b2} subunits were intensely reactive over the epithelial cells of rat testis. In the efferent ducts the Yc, Yb1 and Yf proteins were reactive over ciliated cells, whereas only the Yc protein was reactive over nonciliated cells. In the epididymis, immunoreactivity varied among the principal and basal cells of a given epididymal region and in intensity of reaction of the Y_c, Y_{b1}, Y_a and Y_o proteins increased from the proximal to distal segments. In contrast, the Y_{b2} was intensely expressed only in the distal caput. The distribution of Y_a and Y_c, Y_{b1} and Y_o and the Y_f were re-examined in cauda epididymidis. The principal cells showed high levels of expression of Y_a, Y_c, Y_c, and Y_o subunits, while Y_f was maintained at low levels in principal cells of all cauda regions. Thus both principal and basal cells showed varying degrees of GST expression in the different regions of the cauda epididymis, suggesting that these cells are subjected to a complex, changing environment of substrates (Andonian and Hermo, 1999).

Glutathione-S-Transferase in Apoptosis: Mueller et al, (1998) studied the localization of various subunits of GST (Y_a, Y_c, Y_{b1}, Y_{b2} Y_o and Y_f) in rat testis and epididymis at various ages (3-24 months) and indicated that only selective changes occur in the expression of GSTs at 24 months in principal cells of epididymis having both a normal and a vacuolated appearance. It appeared that aging affects region-specific changes in GST expression in the epididymis and Leydig cell distribution in the testis (Mueller et al., 1998). The GST also plays a protective role in germ cell apoptosis in testes. Male germ cells are susceptible to H_2O_2-induced stress and, upon exposure to H_2O_2 in vitro, demonstrate a typical apoptotic phenotype that includes DNA fragmentation and formation of DNA ladders. Associated with the increase of GST, other changes include considerable accumulation of products of lipid peroxidation in the germ cells after exposure to H_2O_2. If the increase in GST activity is inhibited with suitable inhibitors, the formation of products of lipid peroxidation is augmented, resulting in germ cell apoptosis. Thus, GSTs form a part of adaptive response of germ cells to oxidative stress and are important constituents in detoxifying the products of lipid peroxidation (Rao and Shaha, 2000).

30.2.6. Mu Class of Glutathione-S-Transferase in Testis

The major human Mu-class GSTs purified from hepatic, testicular and skeletal muscle tissue comprise three distinct subunits (M1, M2 and M3), which may combine to form both homodimeric and heterodimeric proteins. Two distinct subunits, M1a and M1b, which represent allelic charge variants, and polymorphic forms encoded at the GST M2 and M3 loci have been observed. A human isoenzyme of Mu-GST with an isoelectric point of 5.2, which was a major GST in testis and also present in cerebral cortex but not detected in liver, was identified and purified by Campbell (1990). Sequence analysis of peptides from testicular Mu-class GST revealed distinct features of primary structure not found in other Mu-class GSTs. These unique features in testis Mu class GSTincluded a blocked and extended amino terminus and 3 additional residues (Pro-Val-Cys) at the carboxy end. In the coding region the mRNA of the brain-testis Mu-class GST was 75% homologous with that of the liver form, and its 3'-untranslated sequence was mostly divergent, indicating that it is the product of a separate gene. Distinct catalytic and structural properties of the testis-brain Mu-class GSTs suggested that brain-testis Mu-class GSTs may be uniquely involved in blood-barrier functions common to both organs (Campbell, et al, 1990). Three GST isozymes (M1a-1a, M1a-b, M1-b) from liver and four GSTs (GST-M1a-2, M1b-2, M2-2 and M2-3) from muscle were isolated and compared with the M3-3 homodimer purified from human testis. The M1a, M1b, M2 and M3 have molecular masses of 26.7, 26.6, 26.0 and 26.3-kDa, respectively. The M1, M2 subunits isolated from either liver, skeletal muscle or testis, are catalytically distinct. The hMu class subunits are immunochemically different. The M3 subunit has a blocked N-terminus (Hussey and Hayes, 1993). Thus the testicular GST M3-3 is likely to correspond to the brain/testis Mu-class GST cDNA described by Campbell et al, (1990). A GSTM4, the fourth member of Mu class of GSTs, identified by Comstock et al (1993), was not testis specific. The deduced amino acid sequence of GSTM4 is 87% (GSTM1), 83% (GSTM2), and 70% (GSTM3) identical to the previously described human Mu class GSTs. By far the tissue richest in unique hGSTM3 subunit was testis although brain also had significant role (Listowsky et al., 1998).

In the study of Rowe et al (1998), cytosolic GST subunits were resolved and individual components was assigned to Alpha, Mu and Pi classes on the basis of their immunoreactivities. Acetylation of N-terminal residues in the GSTα1, GSTα2, GSTM3 and GSTM4 subunits were the only natural post-translational modifications detected. The unique structure of GSTM3, with N- and C-terminal peptide extensions was confirmed from cDNA sequences. Only testis and brain seemed to be rich sources of GSTM3 subunits, and this tissue specificity in GST expression was evident even in organs from different individuals. GSTM4 and GSTM5 subunits were minor components, with GSTM5 found only in brain, lung and testis of humans (Rowe et al, 1997). The rat testis-Mu-class GST cross-reacted with the specific antisera prepared against the human hGSTM3 subunit.

The physiological signifiance of Se-independent GPx activity of GSTs associated with human GSTA1.1 and hGSTA2.2 is not known. A major portion of GPx activity of human liver and testis phosphatidylcholine hydroperoxide (PC-OOH) is contributed by α class of GSTs. The hGSTA2 transfection does not affect the function of antioxidant enzymes including GPx activity toward H_2O_2 suggesting that the α class GSTs play an important role in the regulation of the intracellular concentrations of products of the of lipid peroxidation, that may be involved in the signaling mechanisms of apoptosis (Yang et al, 2001).

Rat Mu-GST (GSTM5): The rat testicular-Mu-class GST cross-reacted with the specific antisera against the human hGSTM3 subunit. Electrospray ionization MS showed that this rat GST subunit has a significantly greater molecular mass (26,541 Da) than the other rat GST subunits. The rGSTM5 peptide sequence of the protein is highly homologous to the hGSTM3 and murine subunits. All three GSTs of this subclass have N- and C-terminal extensions with terminal cysteine residues, but the two penultimate amino acids near the C terminus are divergent in the three species. The proteins of Mu subfamily, which have similar catalytic specificities and mechanisms, are cysteine rich and found mainly in testis and show properties that distinguish them from other GSTs. Moreover, the rGSTM5 subunit was not found in heterodimeric combination with other common Mu-class GST subunits. As rGST M5, mGSTM5 and h GSTM3 are closely structurally related to each other than they are to other Mu GSTs, it has been proposed that they be considered a functionally distinct and separate subfamily within class Mu (Rowe et al, 1998) **(Fig. 30. 2)**.

The rGSTM5 subunit has exceptional properties. The rGSTM5 subunit is found primarily in testis and is labile under oxidative conditions. These forms contain 7-8 cysteine residues per subunit as opposed to 3 for most other mammalian Mu-class GSTs. In addition the protein undergoes S-thiolation to form a mixed disulfide. The high cysteine contents can account for their lability. Despite their lesser catalytic efficiencies, their K_m values for GSH are lower than or comparable with those of other GSTs. A stricking feature of this class of GSTs is their tissue distribution. In all three (mouse, rat, human) species these forms are found primarily in testes and to a lesser extent in brain. Other tissues are virtually devoid of this GST isoform. The rodent and human germ cells are rich in this GST, which suggests that this class of GST may have specific or unique functions in male reproductive processes, besides these could serve in protecting germ cells from reactive electrophilic compounds and hydroperoxides (Rowe et al, 1998)

The major isoforms present in rat seminiferous tubular fluid (STF) in vivo share extensive N-terminal identity with rat (r) rGSTM1, rGSTM2, rGSTM3 and rGST-alpha. Molecular masses of rGSTM2, rGSTM3 and rGSTM alpha from liver and testis are similar, unlike STF-GSTM1 which was larger by 325 Da than its liver counter part isoform. However, peptide digest showed structural similarity between liver and STF-GST isoforms. Active synthesis and secretion of GSTs by the seminferous tubules (STs) were evident from media of ST cultures. GST-Alpha was not secreted by the STs in vitro, whereas of GST-Pi secretion was induced. Detection of GST- Mu in Sertoli cells in vitro, coupled with the recovery of GST from Sertoli cell culture media, provided evidence for Sertoli cells as prime secretor of GST, whereas, germ

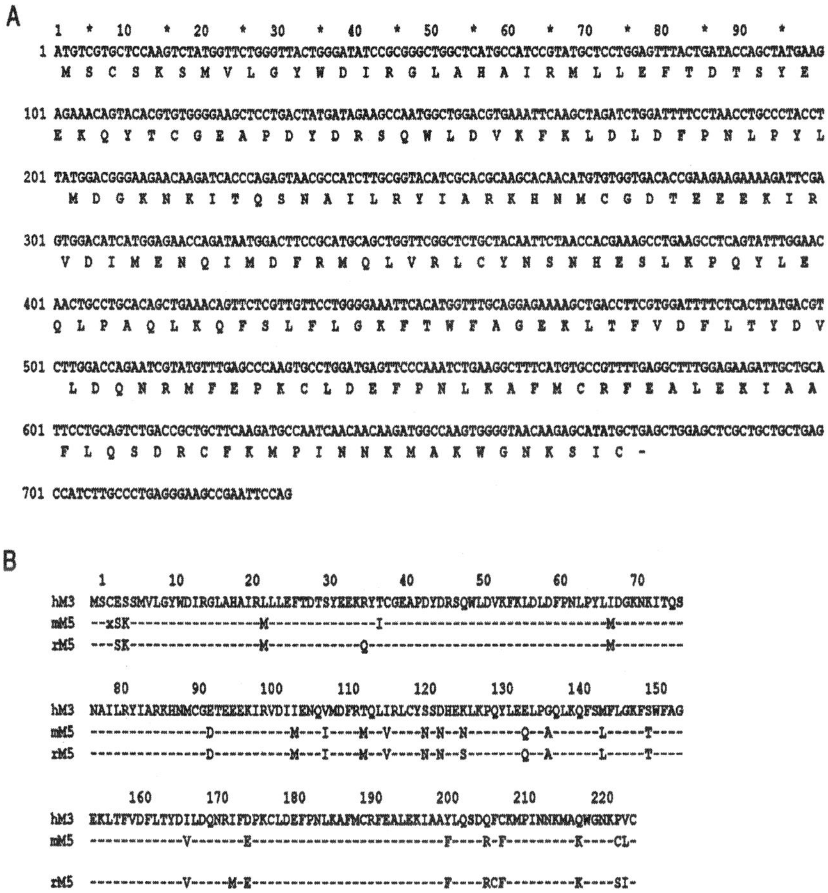

Fig.30.2, (A) Nucleotide and deduced amino acid sequences of rGSTM5. The putative ATG initiation codon and the termination TGA codon are labeled. (B) Alignment of human hGSTM3 (hM3), mouse mGSTM5 (mM5), and rat rGSTM5 (rM5) subunit sequences. Sequence identities are indicated by dashes. The X at position 2 of mM5 reflects a missing codon. Printed with permission from J.D. Rowe et al. J Biol Chem J Biol Chem 273; 9593-601: 1998. © American Society for Biochemistry and Molecular Biology.

cells synthesize but do not release GSTs. Functionally, STF-GSTM1 appeared to serve as a steroid-binding protein by its ability to bind to testosterone and oestradiol that are essential for spermatogenesis (Mukherjee et al, 1999).

Murine Mu Class GST: Two peptide sequences from a digest of mouse fibrous sheath proteins exhibit a high homology with Mu-class GSTs. A 1.1 kb cDNA clone for fibrous sheath component 2 (Fsc2) showed 84% nucleic acid and 89% amino acid sequence identity with a reported Mu-class human GST gene (hGSTM3) (Campbell et al., 1990). Sequences corresponding to those of the two fibrous sheath peptides were present in the protein encoded by the Fsc2 cDNA. The full length Fsc2 cDNA transcribed 1.1 kb mRNA in 12 of 15 somatic

```
                 1                                                     50
mGSTM3 (62%)       mp.t.. ..nt...t.s .........s ........m. ...n......
mGSTM1 (65%)       mp.... ..n....t.p .........s .......tm. .. .......
mGSTM2 (69%)       mp.t.. .......... .......... .....k.tm. ..........
hGSTM3 (89%)     ..ce.s.... .......... .......... .......t.. ..........
hGSTM5*          MSSKS-MVLG YWDIRGLAHA IRMLLEFTDT SYEEKRYICG EAPDYDRSQW

                51                                                    100
mGSTM3           .se..n.g.. .......i.. sh........ ...lg..... .......r..
mGSTM1           .ne....g.. .......i.. sh........ ...l...h. d.......r..
mGSTM2           .se....g.. .......i.. sh........ ...l...... .......r..
hGSTM3           .......... .......... .......... .......... ..........
mGSTM5*          LDVKFKLDLD FPNLPYLMDG KNKITQSNAI LRYIARKHNM CGDTEEEKIR
                            |-----------|
               101                                                    150
mGSTM3           ..t....... t.i..miv.. spdf.kq..e ..kai.ek.. l..e....rp
mGSTM1           a..v...... t.....m... .pdf.kq..e ..kti.ek.. l..e....rp
mGSTM2           .......a.. t.i..amv.. spdf.kk..e ...g..ek.. l..e....qp
hGSTM3           ...i...... ..t....... s.d..k.... ...e..g... ......w..s
mGSTM5*          VDIMENQIMD FRMQLVRLCY NSNHENLKPQ YLEQLPAQLK QFSLFLGKFT

               151                                                    200
mGSTM3           ......v... ...a...... y.m....... a....rd..a ...g.k..s.
mGSTM1           ......v... ...a...... y.m....... a....rd..a ...g.k..s.
mGSTM2           ....n.v... ...v...... h......... a.....d..g ...g.k..sd
hGSTM3           .......... .......... .......... .......... ..........
mGSTM5*          WFAGEKLTFV DFLTYDVLDQ NRIFEPKCLD EFPNLKAFMC RFEALEKIAA
                            |---------------|

               201                     225
mGSTM3           ..k.s...pr ..ft.i.q.. td
mGSTM1           ..k.s..iat ..fs...h.s ..
mGSTM2           ..k.s...sk ..fa...f.n p.
hGSTM3           .....q.c.. ........q.. ..pv.
mGSTM5*          FLQSDRFFKM PINNKMAKWG NKCLC
```

Fig.30.3. Amino acid sequences of mGSTM5 (Accession P48774) and its comparison with other murine and human Ì class GSTs. Dashes indicate homologous sequences. Differences are marked by lower case letters. Sequences from FS proteins are marked by horizontal line. Reprinted with permission from K.D. Fulcher et al,Mol Reprod Dev 42;415-25:1995© John Wiley & Sons, Inc. http:/// www3.interscience.wiley.com/cgi-bin/jabout/37692/ProductInformation.html.

tissues examined, including spermatogenic cells. However, 5'(nt 96 to 12) or 3'(nt 637 to 808) Fsc2 probes, containing mostly noncoding sequences, detected a 1.1 kb mRNA abundant in testis and spermatogenic cells, but absent or present at low levels in somatic tissues **(Fig.30.3)**. This transcript is first present during the meiotic phase of germ cell development. Thus, a mu-class GST gene *(mGSTM5)* is expressed at a specific time during the development of spermatogenic cells in the mouse. The mu-class GST protein is associated with the fibrous sheath, suggesting that it is an integral part of the mouse sperm cytoskeleton (Fulcher et al, 1995).

Goat Sperm GST: Goat sperm consist of both GST-Pi and GST-Mu. On sperm, GSTs are located at the strategic position of sperm head, since normal functioning of sperm following inhibition of GST activity, is impaired. Interference with normal motility, acrosome reaction and fertilizing ability of the goat sperm as a consequence of inhibition of GST activity is due to the functional impairment of sperm membrane, which was evident from the alteration in the lipid peroxidation status of these cells after GST inhibition. Gopalakrishanan et al, (1998a) resolved GSTPi into three major components: comprising molecular mass of 25.5-kDa and 26.5-kDa. N-terminal sequences of the first two peaks appeared to be blocked and showed the presence of Mu-GST-reactive sites. The third component showed 80% N-terminal similarity with human and rGSTP1-1. Goat sperm showed the presence of both Mu and Pi-GST on sperm surface at distinct cellular domains. Although both Pi- and Mu-GSTs seem to be involved in fertilization, the Mu-GST sites, essential for fertilization, were exposed only after 3 h of capacitation (Gopalakrishanan et al., 1998 b).

30.2.7. Superoxide Dismutase

Superoxide dismutases (SODs) protect cells from oxygen radicals by scavenging $O_2^{\bullet-}$ and by converting it into H_2O_2 which in turn is broken down into H_2O in the cytoplasm by glutathione peroxidase and in peroxisomes by catalases. Depending on the transition metal ion found at their active site, SODs can be of three types: copper/zinc (Cu/Zn SOD/SOD1), manganese (Mn SOD/SOD2), and iron (Fe SOD/SOD3). Cu/Zn-SOD is found primarily in cytosol of wide variety of cells, Mn-SOD in prokaryotic and in mitochondria of eukaryotes; and Fe-SOD in prokaryotes. An extracellular form of Cu/Zn SOD (SOD_{EX}), which is distinct from the cytosolic form in terms of molecular weight and amino acid composition, is also found in eukaryotes. Different types of testicular cells display successfully to a high oxidative stress. In testis, Sertoli and peritubular cell have elevated SOD and GSH-dependent enzyme activities associated with a high GSH content. Compared with the somatic cells, pachytene spermatocytes (PS) and round spermatids (RS) presented a different antixidant system characterized by higher SOD activity and GSH content associated with very low GSH-dependent enzyme activity. Spermatozoa exhibited the same enzymatic system as PS and RS but were devoid of GSH. Interstiial tissue has high GPx and GSH content with moderate SOD and GSH related enzymes (Bauche et al, 1994).

SOD Transcripts in Testis Germ Cells: In mammals, the SOD-1 gene is highly conserved, consisting of five exons with similar splicing sites and encoding a protein of about 16-kDa. Moreover, the 225 nucleotides flanking the transcriptional start site of the SOD-1 gene are also strongly conserved, showing similarities of 84% between rat and mouse, 56% between rat and human, and 54% between mouse and human. In order to understand the potential for the testis to be protected from reactive from reactive oxygen, the copper-SOD showed two mRNA transcripts of 0.77-kb and 0.94-kb. The two SOD-mRNA transcripts followed different patterns of expression during development. The accumulation of the SOD mRNA in the seminiferous tubules was stage specific (Jow et al 1993). In mouse testis, three different sizes of SOD-1 mRNAs of about 0.73, 0.80, and 0.93-kb are detected. The 0.73-kb mRNA is found in early stages of male germ cells and is ubiquitously expressed in all somatic tissues. The mRNAs of 0.80 and 0.93-kb are exclusively detected in post-meiofic germ cells. 0.93-kb mRNA is non–polysomal while 0.80 and 0.73-kb SOD –1 mRNAs are mostly polysome associated. The three SOD-1 mRNAs are derived from two transcripts, a ubiquitously expressed transcript and a post-meiotic transcript, which differ by 114-120 nucleotides; the additional nucleotides present in the post-meiotic mRNA are located in the 5'-untranslated region. The 0.93–kb SOD-1 mRNA originates from an alternative upstream promoter contiguous with the somatic SOD-1 promoter. In cell free system, 0.73-kb SOD-1 mRNA translates 2 fold more efficiently than 0.93-kb SOD-1 mRNA by ultilizing two different promoters. Post-meiotic SOD mRNA undergoes adenylation changes, and one of the post-meiotic SOD-1 mRNAs is transcribed during mid-spermiogenesis and translated days later in a partially deadenylated form (Gu et al, 1995) **(Fig.30.4)**. Although the sizes and relative amounts of the multiple SOD-1 mRNAs vary as male germ cells enter meiosis and proceed into the postmeiotic stages of spermatogenesis, the amount of SOD-1 protein and enzyme activity does not fluctuate significantly, suggesting a precise control of SOD-1 activity in male germ cells (Gu and Hecht, 1997).

Developmental expression of Mn-SOD or SOD-2 mRNAs during spermatogenesis in the

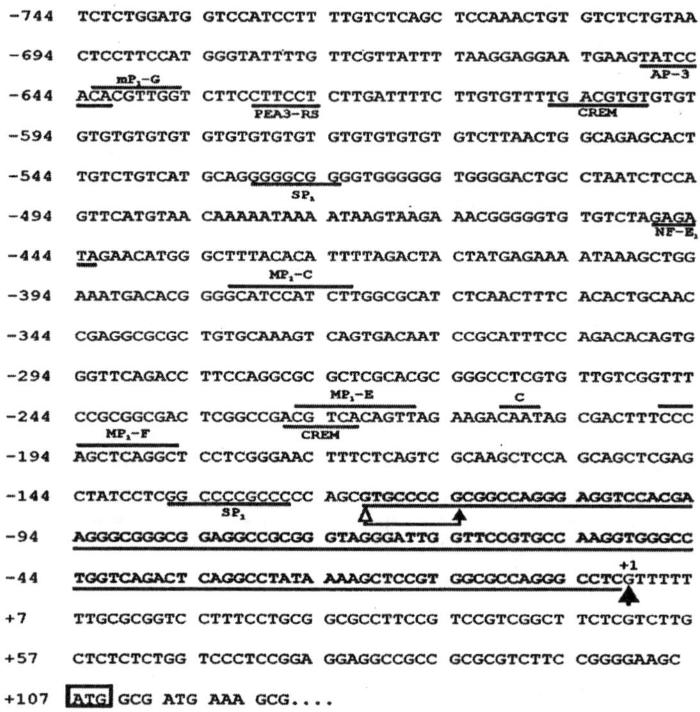

Fig.30.4 Sequence analysis of mouse testicular SOD-1 cDNAs gene. The underlining between the arrows indicates the additional RNA segment located at the 5'-terminus of the testis-specific SOD-1 mRNA. The somatic transcription start site (large solid arrow) and transcription start sites (open-solid arrow) of the post-meiotic 0.93-kb SOD-1 mRNA are shown. Translation start ATG of SOD-1 is boxed. Conserved sequences from a mouse genomic fragment of SOD-1 gene are included. CAAT (C), SP1, CRE, AP-3, and NF-E transcription factor elements are shown. The conserved sequences present in the promoters of mouse protamine 1, protamine 2, and transition protein 1 are indicated. Reprinted with permission from W. Gu et al. J Biol Chem 270; 236-43: 1995 © American Society for Biochemistry and Molecular Biology.

mouse was studied by Gu and Hecht. In RNA preparation derived from prepuberal and adult testes and from isolated populations of meiotic and post-meiotic germ cells three SOD-2 mRNA of about 2.2, 1.2 and 1.0-kb were present in testis. The SOD-2 mRNA levels are developmentally and translationally regulated with maximal expression in early post-meiotic germ cells. Comparative study suggested that translational regulation plays a more prominent role for SOD-2 expression than for GPx or catalase expression in the mammalian testis (Gu and Hecht, 1996) **(Fig. 30.4).**

Characterization and Localization of Germ Cell SOD: The entire cDNA sequences (Cu/Zn-SOD) (SOD 1) and (Mn-SOD SOD2) have been determined from adult testes. The equine Cu/Zn-SOD revealed a protein coding region with 465-bp accompanied by an estimated 154 amino acid residues. Coding region of equine Mn-SOD contained a total of 669-bp and an estimated 222 residues of amino acids (Ishida et al, 1999). Activity of CuZn-SOD is found in both the cytoplasm and the nucleus of the spermatogonia of the seminiferous tubules, spermatocytes and further differentiated germ cells and Sertoli cells. The principal cells in the ductus epididymis showed weak activity except for the stereocilary region, while epididymal basal cells had intense reactivity. It appears that CuZn-SOD in the male genital organs is localized where it could play an important role in cell differentiation, including spermatogenesis (Nonogaki et al, 1992).

Regulation of SOD-1 expression: As stated SOD-mRNA-1 is transcribed mainly in post-meiotic germ cells in mouse and is translationally regulated during spermiogenesis. Testis specific SOD-1 is translated only slightly less efficiently than somatic SOD-1. A testicular cytoplasmic protein (Cu/Zn-SOD RNA-binding protein [SOD-RBP]) of about about 65-kDa specifically binds to the extended 5'-UTR of testis specific SOD-1. The SOD-RBP can repress the in vitro translation of testis form of SOD-1 mRNA but not somatic SOD-1 mRNA. Thus SOD-RBP acts as a repressor in the translation of testis SOD-1 mRNA during spermatogenesis and thereby fine-tunes the level of Cu/Zn SOD in maturing germ cells (Gu and Hecht 1996).

Following TP administration to intact adult rats, SOD content of various subcellular fractions declined. This is accompanied by an elevation in lipid peroxidation of various fractions suggesting the role of SOD in induction of oxidative stress. This was associated with decline in activities of three enzymes related to the metabolism of superoxide radical (SOD) and hydrogen peroxide (catalase and GPx) in response to TP (Chainy et al, 1997). In both cryptorchid and contralateral testes and after efferent duct ligation, the germ cell-specific 0.9-kb SOD and PHGPX mRNA transcripts are significantly lowered suggesting that the overall decline in testicular mRNA transcripts after efferent duct ligation and cryptorchidism is primarily due to germ cell depletion (Zini and Schlegal, 1997).

Sertoli Cell SOD: A 68-kDa testicular variant polypeptide from primary Setoli cell-enriched culture medium consisted two monomers of 35 (α chain) and 33 (ss chain). Partial N-terminal amino acid sequence analysis of these two monomers revealed sequences of NH2-DXGESGVDLADRL (SOD-α) and NH2-XXDTGESGVDLADXL (SOD Ex-ss), which are identical to rat extracellular SODEx with the exceptions that SODEx-α and SODEx-ss do not contain, respectively, four (Trp-Thr- Met-Ser) and two / (Trp–Thr) amino acids from their N-termini, compared to rat SODEx. It suggested that the cleavage sites of the SODEx gene in the testis are different from that of other organs. Germ cell derived factors seem to regulate Sertoli cell SODEx in testis. Steroids including testosterone do not help in its expression whereas FSH is able to induce SODEx expression. Thus, Sertoil cells as well as germ cells synthesize and/or secrete a testicular variant of SODEx that may provide essential clues to understanding superoxide radical-mediated damage in testis (Mruk et al, 1998, Mruk and Cheng, 2000).

Epididymal SOD / Extracellular SOD: Extracellular SOD (SODex), a tetrameric glycoprotein found predominantly in the extra cellular matrix of tissues and to a lesser degree in extra-cellular fluids, has a high affinity for heparin and is bound to heparin sulfate proteoglycan in the glycocalyx on cell surfaces as well as in the connective tissue matrix (Ookawara etal., 1998; Williams et al, 1998). The epididymis is known to be a major site of expression in the rat. However, the properties of SOD_{EX} in the rat are different from those of other mammals. In the rat, SOD_{EX} is dimeric, has a low affinity for heparin, and does not bind to heparin sulfate in vivo. Furthermore, a single amino acid mutation from Asp to Val at position 28 from the N-terminus can convert the rat SOD_{EX} to a tetramer with a high affinity for heparin sulfate. The sequence of a cDNA encoding a secreted form of SOD indicated that the corresponding transcript is expressed principally in the cauda region of the epididymis, consistent with the high levels of SOD enzyme activity found in cauda-epididymidal plasma. Lower levels of an identical size transcript exist in other tissues, including placenta (Perry et al, 1993). SODEx could be detected in seminal plasma of vasectomized men and precludes its use as diagnostic indicator. Epididymal-SOD mRNA is primarily detected in the corpus. With the exception of PHGPX, the relative mRNA levels of the antioxidant enzymes remains unchanged after efferent duct ligation (Zini and Schlegel, 1997).

30.2.8. Search for Other Enzymes of Antiperoxidative Pathway

Glutathione reductase: Glutathione reductase (GR) has been observed in tesis and sperm of rat (Latchoumycandane et al., 2002; Tramer et al., 1998), mice (Mahboob et al., 2001) and human (Institoris et al., 1995). However, studies on tissue specificities of GR or its isozymes are few. During course of data search, it was not possible to find any information relating to GR characterization from testis and sperm.

Catalase: Catalase activates the decomposition of H_2O_2 into water and oxygen, thus removing an initiator of free radical chain reactions which are responsible for lipid peroxidation. After ejaculation, mammalian spermatozoa migrate to the oviduct which is recognized as the site for sperm storage in the mouse, hamster, rabbit, sheep, pig and cow. During storage period, spermatozoa closely interact with oviduct epithelial cells by means of their rostral plasma membranes and acquire oviductal proteins at their surfaces. The adhesion of some of these proteins to spermatozoa suggested a role in affecting sperm physiology. However, our knowledge on characterization of germ cell catalase is yet to begin. Catalase prevents the decreased motility of bovine sperm incubated in presence of specific amino acids. H_2O_2 can be a major player in the death of sperm in the female genital tract, since most mammalian sperm have little or no endogenous catalase. Catalase activity has also been detected in porcine oviductal fluid, human oriductal fluid, and cervical mucus. Western blots of oviduct fluid revealed two major bands at 60 and 40 kDa. The 60 kDa purified protein bound to bovine, boar, and human sperm, and to sperm membranes. However, bovine liver catalase did not bind to sperm. This enzyme may play an important role in sperm survival within the female tract against H_2O_2. The bovine oviductal catalase, which preferentially binds to the acrosomal surface of some mammalian spermatozoa has been sequenced (Lapointe et al, 1998).

γ-Glutamyl Transpeptidase: γ-glutamyltranspeptidase is an important enzyme involved in glutathione metabolism. GGT is present in epididymal epithelial cells and luminal fluids. Though testis is a poor source of GGT, epididymal GGT seems to function to protect spermatozoa from oxidative stress in the epididymal duct and/or to recover extracellular cysteine for the synthesis of epididymal proteins (Lieberman et al, 1993; Palladino et al, 1994). In the rat, multiple forms of mRNA are encoded by a single GGT gene using different promoters. Three forms of GGT mRNAs (II, III and IV) are expressed in the rat epididymis. GGT mRNA-II and GGT mRNA-III are mainly expressed in the initial segment. After unilateral efferent duct ligation, the level of GGT mRNA-IV decreased in the intial segment but not GGT mRNA-II and GGT mRNA-III.

Testicular factors including spermatozoa may be involved in the regulation of the expression of GGT. Ovine rete testis fluid (RTF) contains a number of growth factors produced by the testis. It seems that the development of the epididymis and the expression of GGT mRNA-IV in the rat initial segment is under the regulation of testicular factors (Palladino et al., 1994).

30.3. NITRIC OXIDE SYNTHASE

Nitric oxide (NO) is a unique signaling molecule which is endothelium-derived relaxing factor (EDRF). The initial step in the biosynthesis of NO is catalyzed by nitric oxide synthase (NOS).

```
  1 MEDHMFGVQQ IQPNVISVRL FKRKVGGLGF LVKERVSKPP VIISDLIRGG AAEQSGLIQA
 61 GDIILAVNGR PLVDLSYDSA LEVLRGIASE THVVLILRGP EGFTTHLETT FTGDGTPKTI
121 RVTQPLGPPT KAVDLSHQPP AGKEQPLAVD GASGPGNGPQ HAYDDGQEAG SLPHANGLAP
181 RPPGQDPAKK ATRVSLQGRG ENNELLKEIE PVLSLLTSGS RGVKGGAPAK AEMKDMGIQV
241 DRDLDGKSHK PLPLGVENDR VFNDLWGKGN VPVVLNNPYS EKEQMRKLRI TEGFGVQRGS
301 HNHPPPQENS PPQRMAAPPS VHASSRSRTG RLRWFSLTPS TLRAHWKRDA LSTSAWAPSC
361 ILLSMQGGLK TSAQKDSSSL SPKSLLINTI HQLKDLAPKP TWKGWKR
```

Fig.30.5. Amino acid sequence of Human truncated form of nNOS related protein with tex 2 insertion between exons 3 and 4. A GLGF motif is underlined. The bold underlined M letter indicates starting of novel sequence. Reproduced with permission from Y. Wang et al. J Biol Chem 272: 11392-401; 1997 © American Society for Biochemistry and Molecular Biology.

Two types of enzymes that catalyse the NO production are known: a Ca^{2+} dependent constitutive enzyme or cNOS, and a Ca^{2+} independent inducible enzyme (iNOS). The cNOS exists atleast in two isoforms: eNOS present in endothelial cells and nNOS present in neuronal cells. The iNOS from different cell types demonstrates variable Ca sensitivity. However, nNOS differs from the other two isoforms in that it exhibits a unique -NH2 terminal extension containing a PDZ/GLGF motif of`~100 amino acids. The NOS is regulated through Ca^{2+}/calmodulin, both in neurotransmission and vasodilation and is inhibited by antagonists like trifluoroperazine (Shinde et al, 1998). The NOS is regulated by phosphorylation, which decreases the NOS activity. Consensus sequences for phosphorylation by cAMP-dependent protein kinases have- been identified in nNOS and eNOS, but not in iNOS. Other kinases including protein kinase C, cGMP-dependent protein kinase, have been also implicated in phosphorylation of nNOS.

NOS in Testis: Within the testis, eNOS protein is localized in Leydig cells and Sertoli cells and all stages of spermatogenesis. Within the epididymis and vas deference, NOS is localized to the epithelium. The eNOS is also localized to endothelial cells in all tissues; it was not detectable in normal germ cells. In addition, prematurely shed spermatocytes and spermatids had intense eNOS expression. Rat Sertoli cells express an iNOS isoform in response to the combined action of IFNγ, TNFα, 1L-1α and LPS. Addition of cytokines and LPS to cultured peritubular cells resulted in high nitrite and iNOS mRNA levels, indicating the induction of an iNOS isoform. It appears that NOS activity in the somatic cells of the semniferous tubules is induced and regulated by multiple factors that act in combination, and that nitric oxide may participate in the endocrine and paracrine control of testicular function (Stephan et al, 1995; Bauche et al 1998; Middendorff et al., 1997). Recent evidences suggest that mouse and human spermatozoa contain constitutive cNOS and can synthesize nitric oxide. The inhibition of human sperm cNOS could affect sperm-oocyte fusion and sperm binding to the zona pellucida. A study provides an evidence that cNOS plays a role in the human sperm's capacity to fuse with oocyte but not in the ZP binding (Francavilla et al, 2000). Sperm NOS is active after oocyte activation and acrosome reaction. The NOS and nitric oxide-related bioactivity satisfy the primary criteria of an egg activator and fertilization (Kuo et al, 2000).

Germ Cell Specific NOS: The structural organization of the human nNOS gene shows that *NOS1* is a complex locus consisting of 29 exons and 28 intrones, localized to 12q24.2 spanning a region greater than 160 kb as a single copy in the haploid human genome. The full-length open reading frame of 4302-bp encodes a protein of 1434 amino acides. Analysis of cDNA isolated from human testis identified a testis-specific nNOS (TnNOS) mRNA transcript **(Fig30.5)**. A predicated 3294-bp open reading frame encodes an NH2-terminal truncated protein of 1098 amino acids. This protein is a calcium dependent nitric-oxide synthase with catalytic activity comparable to that of fulllength nNOS. The TnNOS transcripts exhibit novel 5'mRNA sequences encoded by two unique axons spliced to axon 4 of the fulllength nNOS. Characterization of the genomic structure indicates that exonic regions used by TnNOS are expressed from intron 3 of NOS gene. Although lacking canonical TATA and CAAT boxes, the 5' flanking region of the TnNOS exon 1 contains multiple putative *cis* elements including those implicated in testis specific gene. The downstream promoter of the human nNOS gene, which directs testis specific expression of a novel NH2- terminal truncated nitric oxide synthase, represents the first reported example in the NOS gene family of transcriptional diversity producing a variant NOS protein (Wang et al., 1997).

30.4. HEME OXYGENASES

The heme oxygenase isozymes, HO-1 and HO-2, oxidatively degrade the heme molecule to produce antioxidants (the bile pigments), CO (the gaseous cellular messenger), and iron, (a regulator of transferrin, ferritin, and nitric oxide synthase gene expression). The HO-1 and HO-2 differ in primary structure regulation and tissue distribution. The 32kDa HO-1 (hsp 32), induced by heat and variety of other factors is found in tissues with great heme catabolism e.g. liver and spleen. HO-2, on the other hand, a 36 kDa protein is constitutively expressed. In rat it is present in testis and brain (McCoubrey et al, 1992; Rublevskaya and Maines, 1994). Ewing and Maines (1995) described the cellular distribution of HO-1 and HO-2 in the testes of normal and heat shocked rats and defined the cell specific expression of the isozymes and a stage-specific expression of HO-2 in the tissue. In normal testis, HO-1 is present at a low level in the Sertoli cells and could not be detected in germ or Leydig cells. HO-2 on the other hand was most prominently expressed in residual bodies and was not detected in sprematogonia. Modest levels of HO-2 were observed in spermatocytes, sprematids, and select Leydig cells. In contrast, expression of HO-2 messenger RNA was detected in spermatogonia, as well as spermatocytes, spermatids, and residual bodies of the seminiferous epithelium. The expression pattern of HO-2 protein and transcription in testes of heat-stressed rats remained unaffected, whereas, the expression pattern of HO-1 resulted in an increase in the testicular HO-1 mRNA. Activation of soluble guanylyl cyclase (sGC) by CO has been found to show elevated levels of the second messenger cGMP (Maines, 1997). Whether CO is also involved in testicular function, was addressed by Middendorff et al (2000), who confirmed that the CO-generating enzyme, HO-1 is present in Sertoli cells, whereas sGC is distributed exclusively in basal compartment. Seminiferous tubules on treatment with sodium arsenite (inducer of HO-1) or hematin (an HO substrate), gave increased cGMP production. The CO scavenger hemoglobin significantly reduced c-GMP generation. These observations demonstrate a link between HO-1 activity in Sertoli cells and sGC-dependent cGMP production in seminiferous tubules, and a functional role of CO in the human testis.

30.5. SELENOPROTEINS (NON-ENZYMATIC)

Knowledge of selenoproteins has increased markedly within the last several years. The cDNAs encoding a number of mammalian selenoproteins have been sequenced. Selenocysteine is encoded by the codon, UGA in all these selenoproteins. Although UGA is usually a stop codon, it also directs the incorporation of selenocysteine into selenoprotein. A unique tRNA serves as the site of selenocysteine synthesis from serine and recognizes the UGA in the mRNA. In eukaryotic mRNAs, interpretation of UGA as selenocysteine codon requires a

SECIS element, which forms a conserved stem loop structure in the 3' untranslated region of the mRNA. Spacing between UGA codons and the SECIS element determines whether a particular UGA specifies selenocysteine or a stop. Evidence for a low molecular weight protein containing selenium, now called selenoprotein W was presented a number of years ago. This selenoprotein was purified and characterized from rat skeletal muscle and the corresponding cDNA was determined (Whanger et al., 1997). The selenoprotein W increases in tissues after intake of dietary selenium. Its gene contains five introns and six exons. The coding sequence and SECIS element, which are both important for funcion of the protein, are much more highly conserved than other regions of the gene.

Selenoprotein P: Selenoprotein P was identified as a selenium-containing protein distinct from glutathione peroxidase in 1977. Since that early report, the protein has been shown to contain selenium in the form of selenocysteine. Consistent with the cDNA sequence, the amino acid composition of the purified protein shows the presence of multiple selenocysteine residues. At least two isoforms of the selenoprotein P do exist. One isoform is a truncated protein with termination of protein synthesis occurring at the second selenocysteine codon. A high level of expression of selenoprotein P has been also observed in testes. In plasma, selenoprotein P is thought to be a key protein because more than 60% of the plasma Se is found as apart of this protein. The encoding gene is unique, since its cDNA contains 10 TGA codons within the open reading frame, each of which supposedly encodes a selenocyteine residue (Hill et al 1991). Injections of Se into Se-deficient rats induced protection against diquat-induced liver necrosis and lipid peroxidation. Moreover, its protecting ability was positively correlated with a rise in selenoprotein P plasma concentration before a rise in GPx activity. To determine its physiological function, Koga et al (1998) investigated the expression of selenoprotein P mRNA in the rat testis and showed that selenoprotein P is exclusively and predominantly expressed in the Leydig cells. Selective degeneration of Leydig cells by EDS treatment resulted in disappearance of selenoprotein-P mRNA from the testis. Upon recovery of regenerative differentiation of Leydig cells, selenoprotein-P mRNA reappeared.

34-kDa DNA Bound Selenoprotein: Several investigations have shown that Se plays an important role in spermatogenesis. Rats and mice fed low Se diets over long period of time produce spermatozoa with morphological anomalies, which are most frequently observed in the midpiece region of the tail but are also found in the sperm head. Severe depletion of Se led to disruption of spermatogenesis. Among several selenium containing proteins, a 34 kDa-protein was only found in rat testis (Behne et al, 1997). The 34-kDa-protein is localized in the nuclei of the spermatids in the developmental stages 12-15. These nuclei are characterized by their resistance to sonication. Other proteins with molecular masse of 32, 30, 26, 24, 22 and 20-kDa were also found, but the 34-kDa-protein was the most strongly labeled ^{75}Se compound and contained about 80% of the tracer present. After its extraction from the spermatid nuclei, 34kDa band disintegrated into 20-kDa protein. This transformation also seems to take place as physiological process during sperm maturation, since in the nuclei of epididymal spermatozoa, the 34-kDa-band was only weakly labeled and instead replaced by a strongly labeled 20kDa-band. After dissolution of the nuclei, the 34-kDa compound was tightly bound to the DNA and could not be removed by treatment with several detergents and other treatments. The fact that it was not removed by RNAse indicated that it is not associated with the RNA. As its location in the chromatin of the spermatid nuclei is a singular characteristic of a Se compound, it was named as "DNA-bound spermatic selenoprotein".

Mitochondrial Capsule Selenoprotein (MCS): A keratinous structure, known as the mitochondrial capsule that exhibits the characteristic cresent shape of the outer membranes of mitochondria in the helix, has been isolated from bull and rat sperm. The most abundant protein in purified bull sperm mitochondrial capsules is a 20-kDa protein with a very unusual composition including >20% cysteine and proline and virtually all of the selenium in sperm. The most abundant protein in rat sperm mitochondrial capsules is a 17-kDa selenoprotien. The 17k-Da rat capsular protein is identical to the 20-kDa bull capsular protein. Mouse testes cDNAs encodes a protein with a predicted cysteine- and proline-rich composition that is very similar to the 20kDa bull sperm capsule protein. Raveri et al. (1992, 1994) have proposed that the selenoprotein in the sperm mitochondrial capsule is the enzyme, phospholipid hydroperoxide glutathione peroxidase.

A protein, associated with outer membrane of sperm mitochondria is synthesized in elongated spermatids. This protein has been also named as mitochondrial capsule selenoprotein (MCS), and sperm mitochondria associated cysteine rich protein (SMCP). The seleinium content of MCS/MCP has been questioned in recent years. The SMCP has been recognized as an auto-antigen (Herr et al., 1999). An antibody to a synthetic oligopeptide based on the predicted sequence of mouse cyteine rich protein recognizes a 24kDa protein in epididymal sperm tails of mice. However, the 24kDa protein does not appear to be a selenoprotein because: (1) it is not labled with Se^{75} in seminiferous tubule culture; (2) In vitro transcripts indicate that the translation start site is located downstream of potential UGA selenocysteine codons in the cysteine-rich mRNA; (3) the reading frame encoding the cysteine-rich protein in rat lacks in-phase UGA selenocysteine codons. The cysteine-rich protein first appears during step 11 of spermiogenesis in the mouse demonstrating that the cysteine-rich protein mRNA is under temporal translational control. Cysteine-rich protein is evenly distributed in the cytoplasm of spermatids in steps 11 through early step 16 and that it is associated with the outer mitochondrial membranes of sprematids in late step 11 through early step 16 in mouse and epidermal spermatozoa (Cataldo et al, 1996). MCS has been known as a structural protein of the mitochondrial sheath in spermatozoa. A full-length cDNA encoding the MCS has been isolated from the testes of golden hamsters and its mRNA expression was investigated in the hamster tissues. Hamster MCS cDNA was 820 bp long, including 24 bp of the 5' untranslated region (UTR) and 243 bp of the 3'UTR, and showed identity of 75.6% and 73.9% with mouse and rat MCS. The amino acid sequence analysis shows hamster MCS as a polypeptide of 184 amino acids, including a cysteine and proline-rich domain which is characteristic sequence of MCS, and contained 2 in-frame UGA codons for selenocyteine. Hamster MCS also shared amino acids identity of 64.4% with mouse MCS and contained an Arg-Lys-Ser-Thr- rich region in the N-terminus similar to the mitochondrial targeting signal. On the other hand, hamster MCS-mRNA was found-expressed in various tissues as well as the testes. This indicates that MCS in hamster may have more function than just a function of mitochondrial sheath formation of sperm (Nam et al, 1998).

30.6. METALLOTHIONEINS

30.6.1. Metallothioneins

Metallothioneins (MTs) are a group of low-molecular weight (6000-7000 D), cysteine rich (30%) intracellular proteins with a high affinity for transitional metals such as essential zinc, copper and toxic cadmium. MT is expressed in variety of species and in most of the tissues.

MT-I and MT-II are the two of the four isoforms of MT that are found in testis. Altough a high basal level of MT has been demonstrated in the testis, however, hepatic like MT in testis cannot be induced by injection of metals such as Zn and Cd, which are major inducers of hepatic MT synthesis. MT plays important role in Cu and Zn homeostasis and Cd detoxification. Cd sensitivity to testicular injury varies among mouse strains. Liu et al, (2001) suggested that genetic background but not metallothioneine phenotype dictates the sensitivity to Cd induced testicular injury. Mammalian testes contain hight amount of MT. However, aminoacid content of monkey testis MT is different from that of rodent testis. In human testis, Kaur et al (1993) reported a tissue specific testis metal binding protein (TMBP), which is different from liver MT. The molecular weight of TMBP based on aminoacid composition was 6-kDa. The MT significantly increases in the testis of allograft recipients. Testicular Zn and Cu levels, but not Fe level, were significantly higher in testis with allograft kidney than that with isograft kidney. This suggested that allogenic stimuli may induce MT synthesis in the allograft recipients and offer a protective action from oxidative damage in the testis (Cai et al., 2000).

Due to a high content of cysteine, MT is not only involved in the detoxification of heavy metals but also acts as a free radical scavenger both and in vitro and in vivo. The protective effect of MT against radiation and transient metal induced oxidative damage has been demonstrated. It has also been considered as an acute-phase or stress- response protein since it is induced readily by a variety of stress-related factors such as glucocorticoids, bacterial LPS, interferon, alkylating agents, and irradiation. Testis contains high levels of both MT mRNA (10 fold higher than those in adult liver). MT mRNA levels are age dependent, being low for the first 2 weeks after birth and increases slowly thereafter to maximal levels in the adult. In the adult testis, seminiferous tubules contain high levels of MT mRNA that accumulates after the initial differentiation of primary spermatocytes and maintained in spermatids. Pachytene spermatocytes and round spermatids contain both MT-I and MT-II mRNAs, equivalent to those found in zinc treated hepatocytes, whereas low levels of MT mRNAs were present in Sertoli cells (ST) with basal levels in interstitial, spermatogonial, and mature sperm cells at developmental stages. Primary spermatocytes and round spermatids actively synthesize MT-I and MT-II. The MT mRNA in adult testes not increased after cadmium, zinc or LPS administration, in marked contrast to increased levels of hepatic and ovarian MT mRNA. The MT genes are actively expressed in mouse male germ cells in a developmental way (De et al, 1991).

30.6.2. Tesmin-60

Four mouse tesmin cDNAs so far reported encode a testis-specific, metallothionein-like, 30-kDa protein (tesmin-30). But a protein of 60-kDa (tesmin-60) is present in the mouse testis. The relationship between the two proteins indicated that tesmin-30 is not full-length but is part of the C-terminal half of tesmin-60. Tesmin-60 is a testis specific gene, which is responsive to induction by heavy metals. It shares a pair of cysteine-rich regions reported to be similar to metallothionein (Sugihara et al., 1999). Two tesmin related transcripts (2.2 and 1.8-kb) in human testis encode a cysteine-rich 32-kDa protein that contains a metallothionein-like motif. Tesmin is specifically expressed in pachytene spermatocytes and shows its appearance as early as day 8 when germ cells enter meiosis. The adult W/W^v sterile mice with c-kit mutation lacked tesmin expression. The tesmin is located to mouse chromosome 19B and undergoes spatial regulation through G2/M transition in meiosis (Matsuura et al., 2002; Sugihara et al., 1999). The full-length cDNA (2.2-kb) encoding tesmin-60 (475 aa residues) and its genomic DNA (23-kb) indicated that tesmin is a member of the CXC-hinge-CXC family. The tesmin shows dynamic subcellular localization that changes its pattern during spermatogenesis. The temporally and spatially dynamic localization of tesmin suggested that tesmin is involved in multiple stages of spermatogenesis and spermiogenesis (Sutou et al., 2003).

30.7. REFERENCES

Aitken RJ. The human sprermatozoon – a cell in crisis. J Reprod Fertil 1999; 115: 1-7.

Andonian S, Hermo L. Immunocytochemical localization of the Ya, Yb1, Yc, Yf, and Yo subunits of glutathione S-Transferaes in the cauda epididymidis and vas deferens of adult rats. J Androl 1999; 20: 145-57.

Anuradha D, Reddy KV, Kumar TC, Neeraja S, Reddy PR, Raddanna P. Purification and characterization of rat testicular glutathione S-transferases: role in the synthesis of eicosanoids. Asian J Androl 2000; 2: 277-82.

Bauche F, Fouchard MH, Jegou B. Antioxdant system in rat testicular cells. FEBS Lett 1994; 349: 392-6.

Bauche F, Stephan JP, Touzalin AM, Jegou B. In vitro regulation of an inducible-type NO synthase in the rat seminiferous tubule cells. Biol Reprod 1998; 58: 431-38.

Behne D, Kyriakopoulos A, Kalchlosch M, Weiss-Nowak C, Pfeifer H, Gessner H, Hammel C. Two new selenoproteins found in the prostatic glandular epithelium and in the spermatid nuclei. Biomed Environ Sci 1997; 10: 340-5.

Borchert A, Savaskan NE, Kuhn H. Regulation of expression of the phospholipid hydroperoxide/sperm nucleus glutathione peroxidase gene. Tissue-specific expression pattern and identification of functional cis- and trans-regulatory elements. J Biol Chem 2003; 278: 2571-80.

Boyer TD, Kenney WC. Acidic glutathione S-transferases of rat testis. Biochem J 1985; 230: 125-32.

Brigelius-Flohe R. Tissue-specific functions of individual glutathione peroxidases. Free Radical Biol Med 1999; 27: 951-65.

Cai L, Deng DX, Jiang J, Chen S, Zhong R, Cherian MG, Chakrabarti S. Induction of metallothinein synthesis with preservation of testicular function in rats following long term renal transplantation. Urol Res 2000; 28: 97-103.

Campbell E, Takahashi Y, Abramovitz M, Peretz M, Listowsky I. A distinct human testis and brain mu-class glutathione S-transferase. Molecular cloning and characterization of a form present even in individuals lacking hepatic type mu isoenzymes. J Biol Chem 1990; 265; 9188-93.

Cataldo L, Baig K, Oko R, Mastrangelo MA, Kleene KC. Developmental expression, intracellular localization, and selenium content of the cysteine-rich protein associated with the mitochondrial capsules of mouse sperm. Mol Reprod Dev 1996; 45: 320-31.

Chainy GB, Samantaray S, Samanta L. Testosterone-induced changes in testicular antioxidant system. Andrologia 1997; 29: 343-9.

Comstock KE, Johnson KJ, Rifenbery D, Henner WD. Isolation and analysis of the gene and cDNA for a human Mu class glutathione S-transferase GSTM4. J Biol Chem 1993; 268: 16958-65.

De Lamirande E, Gagnon C. A positive role for the superoxide anion in the triggering of human sperm hyperactivation and capacitation. Int J Andol 1993a; 16: 21-25.

De Lamirande E, Gagnon C. Human sperm hyperactivation and capacitation as parts of an oxidative process. Free Radical Biol Med 1993b; 14: 157-63

De Lamirande E, Jiang H, Zini A, Kodama H, Gagnon C. Reactive oxygen species and sperm physiology. Rev Reprod 1997; 2: 48-54.

De Lamirande E. Harakat A, Gagnon C. Human sperm capacitation induced by biological fluids and progesterone, but not by NADH or NADPH, is associated with the production of superoxide anion. J Androl 1998; 19: 215-25.

De SK, Enders GC, Andrews GK. High levels of metallothionein mssenger RNAs in male germ cells of the adult mouse. Mol Endocrinol 1991; 5: 628-36.

Drevet JR, Lareyre JJ, Schwaab V, Vernet P, Dufaure JP. PEA3 protein of the Ets oncogene family is a putative transcriptional modulator of the mouse epididymis-specific glutathione peroxidase gene *gpx5*. J Reprod Dev 1998; 49: 131-40.

Ewing JF, Maines MD. Distribution of constitutive (HO-2) and heat inducible heme oxygenase (HO-1) isoenzymes in rat testes: HO-2 displays stage-specific expression in germ cells. Endocrinology 1995; 136: 2294-302.

Ford WC, Whittington K, Williams AC. Reactive oxygen species in human sperm suspensions: production by leukocytes and the generation of NADPH to protect sperm against their effects. Int J Androl 1997; 20: 44-9.

Foresta C, Flohe L, Garolla A, Roveri A, Ursini F, Maiorino M. Male fertility is linked to the selenoprotein phospholipid hydroperoxide glutathione peroxidase. Biol Reprod 2002; 67: 967-71.

Francavilla F, Santucci R, Macerola B, Ruvolo G, Romano R. Nitric oxide synthase inhibition in human sperm affects sperm-oocytes fusion but not zona pellucida binding. Biol Reprod 2000; 63: 425-29.
Fulcher KD, Welch JE, Klapper DG, O'Brien DA, Eddy EM. Identification of a unique mu-class glutathione S-transferase in mouse spermatogenic cells. Mol Reprod Dev 1995; 42: 415-24.
Giannattasio A, Girotti M, Williams K, Hall L, Bellastella A. Puberty influences expression of phospholipid hydroperoxide glutathione peroxidase (GPX4) in rat testis: probable hypophysis regulation of the enzyme in male reproductive tract. J Endocrinol Invest 1997; 20: 439-44.
Godeas C, Sandri G, Panfili E. Distribution of phospholipid hydroperoxide glutathione peroxidase in rat testis mitochondria. Biochim Biophys Acta 1994; 1191: 147-50.
Godeas C, Tramer F, Micali F, Roveri A, Maiorino M, Nisii C, Sandri G, Panfili E. Phospholipid hydroperoxide glutathione peroxidase (PHGPx) in rat testis nuclei is bound to chromatin. Biochem Mol Med 1996; 59: 118-24.
Godeas C, Tramer F, Micali F, Soranzo M, Sandri G, Panfili E. Distribution and possible novel role of phospholipid hydroperoxide gluathione peroxidase in rat epididymal spermatozoa. Biol Reprod 1997; 57: 1502-8.
Gopalakrishnan B, Aravinda S, Pawshe CH, Totey SM, Nagpal S, Salunke DM Shaha C. Studies on glutathione S-transferases important for sperm function: evidence of catalytic activity independent functions. Biochem J 1998a; 329: (Pt 2) 231-41.
Gopalakrishnan B, Shaha C. Inhibition of sperm glutathione S-transferase leads to functional impairment due to membrane damage. FEBS Lett 1998b; 422: 296-300.
Gu W, Hecht NB. Developmental expression of glutathione peroxidase, catalase, and managanese superoxide dismutase mRNAs during spermatogenesis in the mouse. J Androl 1996; 17: 256-62.
Gu W, Hecht NB. Translation of a tesis-specific Cu/Zn superoxide dismutase (SOD-1) mRNA is regulated by a 65-kilodalton protein, which binds to its 5' untranslated region. Mol Cell Biol 1996; 16: 4535-43.
Gu W, Morales C, Hecht NB. In male mouse germ cells, copper-zinc superoxide dismutase utilizes alternative promoters that product multiple transcripts with different translation. J Biol Chem 1995: 270: 236-43.
Hall L, Williams K, Perry AC, Frayne J, Jury JA. Majority of human glutathione peroxidase type 5 (*GPX5*) transcripts are correctly spliced: implications for the role of GPX5 in the male reproductive tract. Biochem J 1998; 333: 5-9.
Hayes JD, Pulford DJ. The glutathione S-transferase supergene family: Regulation of GST and the contribution of the isoenzymes to cancer chemoprotection and drug resistance. Crit Rev Biochem Mol Biol 1995; 30: 445-600.
Hayes JD. Purification and characterization of glutathione S-transferases P, S and N. Isolation from rat liver of Yb1 Yn protein, the existence of which was predicted by subunit hybridization in vitro. Biochem J 1984; 224: 839-52.
Hermo, Papp S, Robaire B. Developmental expression of the Yf subunit of glutathione S-transferase P in epithelial cells of the testis, efferent ducts, and epididymis of rat. Anat Rec 1994; 239: 421-40.
Herr JC, Thomas D, Bush LA, Coonrod S, Khole V, Howards SS, Flickinger CJ. Sperm mitochondria-associated cysteine-rich protein (SMCP) is an autoantigen in lewis rats. Biol Reprod 1999; 61: 428-35.
Hill KE, Lloyd RS, Yang JG, Read R, Burk RF. The cDNA for rat selenoprotein P contains 10 TGA codons in the open reading frame. J Biol Chem 1991; 266: 10050-53.
Hsieh CH, Tsai SP, Yeh HI, Sheu TC, Tam MF. Mass spectrometric analysis of rat ovary and testis cytosolic glutathione S-transferases (GSTs): Identification of a novel class-αGST, rGSTA6*, in rat testis. Biochem J 1997; 323(Pt2): 503-10.
Hussey AJ, Hayes JD. Human Mu-class glutathione S-transferases present in liver, skeletal muscle and testicular tissue. Biochim Biophys Acta 1993; 1203: 131-41.
Ichikawa T, Oeda T, Ohmori H, Schill WB. Reactive oxygen species influence the acrosome reaction but not acrosin activity in human spermatozoa. Int J Androl 1999; 22: 37-42.
Imai H, Suzuki K, Ishizaka K, Ichinose S, Oshima H, Okayasu I, Emoto K, Umeda M, Nakagawa Y. Failure of the expression of phospholipid hydroperoxide glutathione peroxidase in the spermatozoa of human intertile males. Biol Reprod 2001; 64: 674-83.
Institoris E, Eid H, Bodrogi I, Bak M. Glutathione related enzymes in human testicular germ cell tumors and normal testes. Anti Res 1995; 15: 1371-4.
Ishida N, Katayama Y, Sato F, Hasegawa T, Mukoyama H. The cDNA sequences of equine antioxidative enzyme genes Cu/Zn-SOD and Mn-SOD, and their expressions in equine tissues. J Vet Med Sci 1999; 61: 291-4.
Jow WW, Schlegel PN, Cichon Z, Phillips D, Goldstein M, Bardin CW. Identification and localization of copper zinc superoxide dismutase gene expression in rat testicular development. J Androl 1993; 14: 439-47.

Kaur G, Nath R, Gupta GS. Purification and characterization of low molecular weight metal binding protein from human testis. Trace Elem Expt Med 1993; 6: 1-11.
Knopp EA, Arnodt TL, Eng KL, Caldwell M, LeBoeuf RC, Deeb SS, O'Brien KD. The phospholipid hydroperoxide glutathione peroxidase: cDNA sequence, tissue expression and mapping. Mamm Genome 1999; 10: 601-5.
Koga M, Tanaka H, Yomogida K, Tsuchida J, Uchida K, Kitamura M, Sakoda S, Matsumiya K, Okuyama A, Nishimune Y. Expression of Selenoprotein-P-messenger ribonucleic acid in the rat testis. Biol Reprod 1998; 58: 261-65.
Kuo RC, Baxter GT, Thompson SH, Stricker SA, Patton C, Bonaventura J, Epel D. NO is necessary and sufficient for egg activation at fertilization. Nature 2000; 406: 633-35.
Lahti PP, Shariatmadari R, Penttinen JK, Drevet JR, Haendler B, Vierula M, Parvinen M, Huhtaniemi IT, Poutanen M. Evaluation of the 5'-flanking regions of murine glutathione peroxidase five and cysteine-rich secretory protein-1 genes for directing transgene expression in mouse epididymis. Biol Reprod 2001; 64: 1115-21.
Lapointe S, Legare C, Gaudreault C, Sullivan R, Sirard MA. cDNA sequence and deduced amino acid sequence of bovine oviductal fluid catalase. Mol Reprod Dev 1998; 51: 265-73.
Latchoumycandane C, Chitra KC, Mathur PP. The effect of 2,3,7,8-tetrachlorodibenzo-p-dioxin on the antioxidant system in mitochondrial and microsomal fractions of rat testis. Toxicology 2002; 28: 127-35.
Lei XG, Ross DA, Parks JE, Combs GF Jr. Effects of dietary selenium and vitamin E concentrations on phospholipid hydroperoxide glutathione peroxidase expression in reproductive tissues of pubertal maturing male rats. Biol Trace Elem Res 1997; 59: 195-206.
Lieberman MW, Wiseman AL, Shi Z-Z, Carter BZ, Barrios R et al., Growth retardation and cysteine deficiency in γ-glutamyl transpeptidase-deficient mice. Proc Natl Acad Sci USA 1996; 93: 7923:26.
Listowsky I, Rowe JD, Patskovsky YV, Tchaikovskaya T, Shintani N, Novikova E, Nieves E. Human testicular glutathione-S-transferases : insights into tissue-specific expression of the diverse subunit classes. Chem Biol Interact 1998; 24: 111-12.
Liu J, Corton C, Dix Dj, Liu Y, Waalkes MP, Klaassen CD. Genetic background butnot metallothionenin phenotype dictates sensitivity to cadmium-induced testicular injury in mice. Toxicol Appl Pharmacol 2001; 1: 1-9.
Lopes S, Jurisicova A, Sun JG, Casper RF. Reactive oxygen species : potential cause for DNA fragmentation in human spermatozoa. Reproduction 1998; 13: 896-900.
Mahboob M, Shireen KF, Atkinson A, Khan AT. Lipid peroxidation and antioxidant enzyme activity in different organs of mice exposed to low level of mercury. J Environ Sci Health B 2001; 36: 687-97.
Mains MD. The heme oxygenase system: a regulator of second messanger gases. Annu Rev Pharmacol Toxicol 1997; 37: 517-54.
Maiorino M, Wissing JB, Brigelius-Flohe R, Calabrese F, Roveri A, Steinert P, Ursini F, Flohe L. Testosterone mediates expression of the selenoprotein PHGPx by induction of spermatogenesis and not by direct transcriptional gene activation. FASEB J 1998; 12: 1359-70.
Matsuura T, Kawasaki Y, Miwa K, Sutou S, Ohinata Y, Yoshida F, Mitsui Y. Germ cell-specific nucleocytoplasmic shuttling protein, tesmin, responsive to heavy metal stress in mouse testes. J Inorg Biochem 2002; 88: 183-91.
McCoubrey WK Jr, Ewing JF, Maines MD. Human heme oxygenase-2: chacterization and expression of a full-length cDNA and evidence suggesting that the two HO-2 transcripts may differ by choice of polyadenylation signal. Arch Biochem Biophys 1992; 295: 13-20.
Middendorff R, Kumm M, Davidoff MS, Holstein AF, Muller D. Generation of cyclic guanosine monophosphate by heme oxygenases in the human testis-a regulatory role for carbon monoxide in sertoli cells. Biol Reprod 2000; 63: 651-57.
Middendorff R, Muller D, Wichers S, Holstein AF, Davidoff MS. Evidence for production and functional activity of nitric oxide in seminiferous tubules and blood vessels of the human testis. J Clin Endocrinol Metab 1997; 82: 4154-61.
Mizuno K, Hirata S, Hoshi K, Shinohara A, Chiba M. Analysis of the phospholipid hydroperoxide glutathione peroxidase mRNA in the rat spermatozoon and effect of selenium deficiency on the mRNA. Biol Trace Elem Res 2000; 74: 71-9.
Mruk D, Cheng CH, Cheng YH, Mo MY, Grima J, Silvestrini B, Lee WM, Cheng CY. Rat testicular extracellular dismutase: its purification, cellular distribution, and regulation. Biol Reprod 1998; 59: 298-308.
Mruk DD, Cheng CY. In vitro regulation of extracellular superoxide dismutase in Sertoli cells. Life Sci 2000; 67: 133-45.
Mueller A, Hermo L, Robaire B. The effects of aging on the expression of glutathione S-transferes in the

testis and epididymis of the Brown Norway rat. J Androl 1998; 19: 450-65.
Mukherjee SB, Aravinda S, Gopalakrishnan B, Nagpal S, Salunke DM, Shaha C. Secretion of glutathione S-transferase isoforms in the seminiferous tubular fluid, tissue distribution and sex steroid binding by rat GSTM1. Biochem J 1999; 340(Pt 1): 309-20.
Nam S, Nakamuta N, Kurohmaru M, Hayashi Y. Cloning and sequencing of the mouse cDNA encoding a phospholipids hydroperoxide glutathione perioxidase. Gene 1997; 198: 245-9.
Nam SY, Fujisawa M, Kim JS, Kurohmaru M, Hayashi Y. Expression pattern of phospholipid hydroperoxide glutathione peroxidase messenger ribonucleic acid in mouse testis. Biol Reprod 1998a; 58: 1272-6.
Nam SY, Maeda S, Gujisawa M, Kurohmaru M, Hayashi Y. Cloning and expression of mitochondrial capsule selenoprotein gene in the golden hamster. J Vet Med Sci 1998b; 60: 1113-8.
Nonogaki T, Noda Y, Narimoto K, Shiotani M, Mori T, Matsuda T, Yoshida O. Localization of CuZn-superoxide dismutase in the human male gental organs. Hum Reprod 1992; 7: 81-5.
O'Flaherty CM, Beonegui NB, Beconi MT. Reactive oxygen species requirements for bovine sperm capacitation and acrosome reaction. Theriogenology 1999; 15: 289-301.
Okamura N, Iwaki Y, Hiramoto S, Tamba M, Bannai S, Sugita Y, Syntin P, Dacheux F, Dacheux JL. Molecular cloning and characterization of the epididymis-specific glutathione peroxidase-like protein secreted in the porcine epididymal fluid. Biochim Biophys Acta 1997; 1316: 99-109.
Ookawara T, Imazeki N, Matsubara O, Kizaki T, Oh-Ishi S, Nakao C, Sato Y, Ohno H. Tissue distribution of immunoreactive mouse extracellular superoxide dismutase. Am J Physiol 1998; 275 (3 Pt 1):C 840-7.
Palladino MA, Hinton BT. Expression of multiple-γ-glutamyl transpeptidase messenger ribonucleic acid transcripts in the adult rat epididymis is differentially regulated by androgens and testicular factors in a region-specific manner. Endocrinology 1994; 135: 1146-56.
Papp S, Robaire B, Hermo L. Developmental expression of the glutathione S-transferase Y_0 subunit in the rat testis and epididymis using light microscope immunocytochemistry. Anat Rec 1994; 240: 345-57.
Papp S, Robaire B, Hermo L. Immunocytochemical localization of the Ya, Yc, Yb1 and Yb2 subunits of glutathione S-transferases in the testis and epididymis of adult rats. Microsc Res Tech 1995; 30: 1-23.
Perry AC, Jones R, Hall L. Isolation and characterization of a rat cDNA clone encoding a secreted superoxide dismutase reveals the epididymis to be a major site of its expression. Biochem J 1993; 293 (Pt 1): 21-5.
Pharmacol Exp Therap 1994; 269: 230-37.
Puspha-Rekha TR, Burdsall AL, Leksa LM, Chisolm GM, Driscoll DM. Rat phospholpid hydroperoxide glutathione peroxidase: cDNA cloning and identification of multiple transcription and translation start sites, J Biol Chem 1995; 270: 26993-99.
Rao AV, Shaha C. Role of glutathione S-transferases in oxidative stress-induced male germ cell apoptosis. Free Radical Biol Med 2000; 29: 1015-27.
Roberts KP. Y chromosome deletions and male infertility: state of the art and clinical implications. J Androl 1998; 19: 255-59.
Roveri A, Casasco A, Maiorino M, Dalan P, Calligaro A and Ursini F. Phospholipid hydroperoxide glutathione peroxidase of rat tesis. Gonadotropin dependence and immunocytochemical indentification. J Biol Chem 1992; 267: 6142-46.
Roveri A, Maiorino M, Nisii C, Ursini F. Purification and characterization of phospholipid hydroperoxide glutathione peroxidase from rat testis mitochondrial membranes. Biochim Biophys Acta 1994; 1208: 211-21.
Rowe JD, Nieves E, Listowsky I. Subunit diversity and tissue distribution of human glutathione S-transferases: interpretations based on electrospray ionization-MS and peptide sequence-specific antisera. Biochem J 1997; 325 (Pt2): 481-6.
Rowe JD, Patskovsky YV, Patskovska LN, Novikova E, Listowsky I. Rationale for reclassification of a distinctive subdivision of mammalian class Mu glutathione S, transferases that are primarily expressed in testis. J Biol Chem 1998; 273: 9593-601.
Rublevskaya IN, Maines MD. Interaction of Fe-protoporphyrin IX and heme analogues with purified recombinant heme oxygenase-2, the constitutive isoenzyme of the brain and testes. J Biol Chem 1994; 269: 26390-95.
Schwaab V, Faure J, Dufaure JP, Drevet JR. The plasma-type glutathione peroxidase is expressed under androgenic control in the mouse epididymis and vas deferens. Mol Reprod Dev 1998; 51: 362-72.
Schwaab V, Lareyre JJ, Vernet P, Pons E, Faure J, Dufaure JP, Drevet JR. Characterization, regulation of the expression and putative roles of two glutathione peroxidase proteins found in the mouse epididymis. J Reprod Fertil Suppl 1998; 53: 157-62.
Sharma RK, Agarwal A. Role of reactive oxygen species in male infertility. Urology 1996; 48: 835-50.
Sheehan D, Ryle CM, Mantle TJ. Selective induction of glutathione S-transferase D in rat testis by phenobarbital. Biochem J 1984; 219: 687-8.

Shinde UA, Mehta AA, Goyal RK. Nitric oxide: a molecule of the millennium. Ind J Expt Biol 2000; 38: 201-10.

Stephan JP. Guillemois C, Jegou B, Bauche F. Nitric oxide production by Sertoli cells in response to cytokines and lipopolysaccharide. Biochem Biophys Res Commun 1995; 213: 218-24.

Sugihara T, Wadhwa R, Kaul SC, Mitsui Y. A novel testis-specific metallothionein-like protein, tesmin, is an early marker of male germ cell differentiation. Genomics 1999; 57: 130-6.

Sutou S, Miwa K, Matsuura T, Kawasaki Y, Ohinata Y, Mitsui Y. Native tesmin is a 60-kilodalton protein that undergoes dynamic changes in its localization during spermatogenesis in mice. Biol Reprod 2003; 68: 1861-9.

Tam MF, Hsieh CH, Tsai SP and S.Tam TC. Amino acid sequence of glutathione S-transferase rGSTM5* from rat testis. Biochem J 1998; 333: 735-39.

Thundathil J, de Lamirande E, Gagnon C. Nitric oxide regulates the phosphorylation of the threonine-glutamine-tyrosine motif in proteins of human spermatozoa during capacitation. Biol Reprod 2003; 68: 1291-8.

Ufer C, Borchert A, Kuhn H. Functional characterization of cis- and trans-regulatory elements involved in expression of phospholipid hydroperoxide glutathione peroxidase. Nucleic Acids Res 2003; 31: 4293-303.

Ursini F, Heim S, Kiess M, Maiorino M, Roveri A, Wissing J, Flohe L. Dual function of the selenoprotein PHGPx during sperm maturation. Science 1999; 285 (5432): 1393-6.

Veri JP, Hermo L, Robaire B. Immunocytochemical localization of glutathione S-transferase Yo subunits in the rat testis and epididymis. J Androl 1994; 15: 415-34.

Veri JP, Hermo L, Robaire B. Immunocytochemical localization of the Yf subunit of glutathione S-transferase P shows regional variation in the staining of epithelial cells of the testis. J Androl 1993; 14: 23-44.

Vernet P, Rock E, Mazur A, Rayssiguier Y, Dufaure JP, Drevet JR. Selenium-independent epididymis-restricted glutathione peroxidase 5 protein (GPX) can back up failing Se-dependent GPXs in mice subjected to selenium deficiency. Mol Reprod Dev 1999; 54: 362-70.

Wang Y, Goligorsky MS, Lin M, Wilcox JN, Marsden PA. A novel testis-specific mRNA transcript encoding an NH_2-terminal truncated nitric oxide synthase. J Biol Chem 1997; 272: 11392-401.

Whanger PD, Vendeland SC, Gu QP, Beilstein MA, Ream LW. Selenoprotein W cDNAs from five species of animals. Biomed Environ Sci 1997; 10: 190-97.

Williams K, Frayne J, Hall L. Expression of extracellular glutathione peroxidase type 5 (GPX5) in the rat male reproductive tract. Mol Hum Reprod 1998; 4: 841-8.

Yang Y, Cheng JZ, Singhal SS, Saini M, Pandya U, Awasthi S, Awasthi YC. Role of glutathione S-transferases in protection against lipid peroxidation: Overexpression of hGSTA2-2 in K562 cells protects against hydrogen peroxide-induced apoptosis and inhibits JNK and caspase 3 activation. J Biol Chem 2001; 276: 19220-30.

Yeh JY, Vendland SC, Gu Q, Butler JA, Ou BR, Whanger PD. Dietary selenium increases selenoprotein W levels in rat tissues. J Nutr 1997; 127: 2165-72.

Zhang H, Zheng RL. Promotion of human sperm capacitation by superoxide anion. Free Radical Res 1996; 24: 261-68.

Zini A, de Lamirande E, Gagnon C. Reactive oxygen species in semen of infertile patients : levels of superoxide dismutase- and catalase-like activities in seminal plasma and spermatozoa. Int J Androl 1993; 16: 183-88.

Zini A, Schlegel PN. Identification and characterizatio of antioxidant enzyme mRNAs in the rat epididymis. J Androl 1997; 20: 86-91.

Chapter 31

QUALITY CONTROL OF GERM CELL PROTEINS

Functional proteins in a single conformation state are obtained by many intracellular events. Nonetheless, best quality control of protein levels and 3-D structure is achieved by two complementary processes: i) molecular degradation by ubiquitination and ii) molecular chaperonining. Intracellular protein degradation has important roles in the modulation of the levels of specific proteins and the elimination of damaged proteins and cells. The selective degradation of many proteins in eukaryotic cells is carried out by the ubiquitin (Ub) mediated pathway. This system requires a heat stable polypeptide for the activity of an ATP-dependent proteolytic degradation. Molecular chaperones are defined as a family of unrelated classes of proteins that mediate the correct assembly of other polypeptides but they are not components of the functional assembled structures. In recent years, there is an increasing appreciation of the relationship between ATP-dependent proteases (e.g. the proteasome) and chaperonins (e.g. GroEL), both of which act to isolate the protein intermediate from the cellular environment and promote the unfolded state; in one case to allow for complete degradation and, in the other, to allow for refolding into the native state. Thus it is clear that molecular chaperones and the proteasome play complementary roles in the maturation and quality control of proteins, in the regulation of protein function and in the elimination of damaged proteins and cells (Kirschner, 1999).

I. UBIQUITINATION AND PROTEOLYSIS

31.1. UBIQUITIN SYSTEM

Intracellular proteolysis of proteins through ubiquitination has been implicated in a variety of processes such as the heat shock response, DNA repair, cell cycle progression, the modification of histones and of receptor genes, silencing and the possible pathogenesis of some neurodegenerative diseases. The ubiquitin-proteasome pathway exerts control over the half-life of cyclins, transcription factors, and components of signal transduction pathways. Early steps in mitosis, such as spindle formation, chromosome congregation and chromosome condensation are susceptible to reversible blockage in response to DNA and spindle damage and employ reversible modifications. The ubiquitin (Ub) mediated pathway requires a heat stable polypeptide for the activity of an ATP-dependent proteolytic degradation. This polypeptide was subsequently identified as ubiquitin, a 76-amino-acid residues containing highly conserved protein present in all eukaryotes. In this process, ubiquitin is covalently ligated to protein substrates in an ATP-dependent manner, and the ubiquitin ligation commits proteins for degradation. The system is well conserved and is present in all cells in high amount. In addition, there are many non-proteasome dependent functions of the ubiquitin system, such as a role in endocytosis of integral membrane proteins, a chaperone function in protein, (re)folding, and functions in regulation of chromatin structure and gene expression.

The processive nature of proteasome mediated degradation assures that, in most cases, all domains of a protein would be eliminated simultaneously, avoiding the possibility that partial degradation would leave behind other domain that could interfere with regulatory pathway. A ubiquitin-dependent, sperm quality control mechanism that resides in the mammalian epididymis has been described. Most of the defective sperm are phago-cytosed by the epididymal epithelial cells through ubiquitin pathway. The presence of cell surface ubiquitination in defective sperm provides a possible mechanism for sperm quality control in mammals and a new marker for semen abnormalities in men and animals (Sutovsky et al., 2001). It is important that the proteasome is not limited in its ability to degrade all proteins, because it could then afford an opportunity for viruses or parasites to evade immune surveillance. The important feature of proteolysis is that it is irreversible and obliterate. Therefore, lack of fidelity in the process may have grave consequences for chromosome loss that can promote tumorogenesis (Kirschner, 1999).

Information on the enzymatic reactions of the ubiquitin pathway is summarized in **Figure 31.1.** Mono- or poly-ubiquitination of proteins leads to protein degradation. The ligation of ubiquitin to target proteins is initiated by the activation of its C-terminal Gly residue, catalyzed by a specific Ub-activating enzyme, E1 (Step 1) and achieved in multi-steps, which involve Ub-conjugating enzymes (E2), and Ub-ligating enzymes (E3) **(Fig.31.1)**. The system follows a hierarchy, in which a single E1 interacts with several E2s. In turn, one E2 can function together with different E3s (or independent of an E3), whereby the E3 plays a role in determining substrate specificity. Only a few E1 enzymes are known in the human, and in other animal species. Among E2 enzymes, 13 different E2 enzymes have been identified in yeast, and many different E2 enzymes are known in human (reviews- Hershko and Ciechanover, 1998; Barrends et al., 1999; 2000). The function of these E2-like proteins is still unknown. E3 enzymes show an even greater variation of proteins and protein complexes than the E2 enzymes. Different families of E3 enzymes have been described. The largest E3 family is characterized by the so-called HECT-domain. A much smaller family is the UBR family (named after Ubr1, encoding N-recognin, the recognition component of the N-end rule pathway), of which only three members have been identified in mammals.

Repeated additions of ubiquitin to a single substrate through the E1-E2-E3 machinery results in polyubiquitination. This often marks proteins for selective destruction by the 26S proteasome, a large mutlisubunit ATP-dependent protease. Ubiquitin can be released from a polyubiquitin chain, or from a protein substrate, through the action of deubiquitinating enzymes. This function is carried out by a large number of enzymes from two families: ubiquitin carboxyl-terminal hydrolases (UCHs), and ubiquitin specific processing proteases (UBPs). The UCHs are capable of removing ubiquitin only from peptides or small proteins. The UBPs have a broader specificity, and some enzymes of this family are most likely also involved in deubiquitination of intact proteins, thereby conunteracting the E1-E2-E3 machinery (review - Wilkinson, 1997).

The ubiquitin systems are used in other evolutionary conserved processes also. There are several ubiquitin related analogous enzymatic pathways, which have been referred to as human/ rat/mouse SUMO-1 (also called UBL1, sentrin, and several other names), SUMO-2 and 3, and mouse Nedd-8 and UCRP. Similar to ubiquitin, these proteins can be coupled to cellular substrates

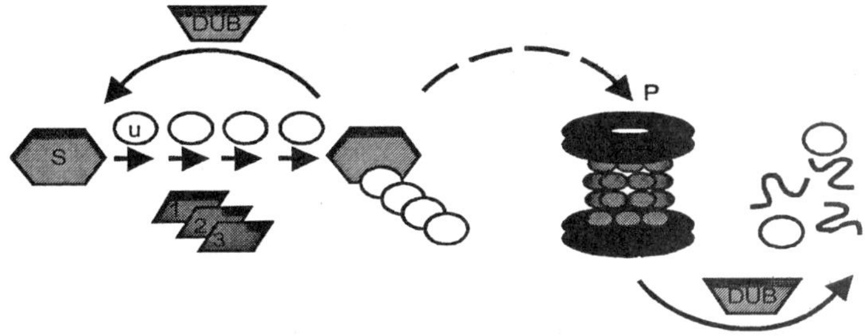

	Gonad/meiosis-enhanced expression	Gonad/meiosis-specific expression	(candidate) Fertility/ sporulation factor
= Substrate	*sp*Rad21	*sp*Rec8	*sp*Rec8
(u) = Ubi/Ubi-like	*sc*poli-Ubi *sc*RAD23 *ch/mam/*Ubi *h*UBL1		*sc*poli-Ubi
= E1		*m*Ube1y	*m*Ube1y
= E2	*sc*RAD6/ *h-m-r-*HR6A/B *sc*UBC9/ *h-m*UBC9 *dm*UbcD1 *r*E217k *h*UbcH-ben	*r*8A isoform	*sc*Rad6/ *h-m-r-*HR6A/B *dm*UbcD1
= E3	*sc*SCF- and *sc*APC- components *m*Rjs	*sc*SCF-components scSPO70(APC-component)	
= proteasome		*dm-x-h* testis-specific subunits	*m*Rjs
DUB = UCH	*dm*Uch-D *r*PGP9.5 *h*BAP1		
UBP	*m-h*DFFRX *h*DFFRY *h*ISOT3	*m*-DFFRY	*m*-hDFFRY

Fig. 31.1. Ubiquitin system in gametogenesis. Ubiquitination of substrate (S) occurs in presence of ubiquitin (Ubi/u) activating enzyme (E1), ubiquitin-conjugating enzyme (E2) and ubiquitin ligase (E3). Polyubiquitinated proteins are recognized by 26S proteasome (P) and degraded. Table summarizes which components of Ub system in different eukaryotes are enhanced or show specific expression during gametogenesis/meiosis. UBP, ubiquitin specific processing protease; UCH, ubiquitin C-terminal hydroxylase; sp, *S. pombe*; sc, *S. cereviseae*; h, human; m, mouse; r, rat; dm, *D. melanogaster*; ch, chicken; mam, mammalian; x, *xenopus*. Reprinted with permission from G. Baarends et al. Mol Cell Endocrinol. 151:5-16:1999© Elsevier.

by an analogous cascade of enzymatic coupling reactions. Ubiquitin itself is one of the most strongly conserved protein in evolution, differing only 3 amino acids between the budding yeast *Saccharomyces cerevisisae* and human. It has become clear that certain components of the ubiquitin system that are involved in the regulation of the mitotic cell cycle, are also induced during the middle phase of sporulation when the meiotic divisions take place (Chu et al., 1998; Patton et al., 1998; Townsley and Ruderman, 1998).

A functional connection with the ubiquitin system has been described for a number of proteins involved in DNA repair and meiosis/sporulation. For example, the E2 enzyme Rad6/UBC2 is involved in DNA repair, sporulation, and several other processes, such as gene silencing. There are indications that ubiquitin-mediated proteasomal degradation of Rad21 plays a role in sister chromatid separation at the onset of mitotic anaphase. The Rec8 might be involved in successful completion of meiosis. Search of these components in testis is the subject of recent investigations (Santti et al., 2003; Tanaka et al., 2004; Yan et al., 2003)

31.2. UBIQUITIN SYSTEM IN VERTEBRATE GONADS

31.2.1. Ubiquitin

Watson et al., 1978 identified ubiquitin as an abundant component of trout testis chromatin. Later high level of ubiquitin was found in chicken and mammalian testes. Ubiquitin-like protein (UBL1) is elevated in human testis, but also highly expressed in other organs including the ovary (Shen et al., 1996). The cDNAs encoding polyubiquitin were cloned from human granulosa cells. Experimental findings support a link between the ubiquitin dependent pathway of protein degradation and spermatogenesis. The main features are: (a) The existence of testis-specific isoforms of E1, E2 and E3. (b) A Y-linked gene is transcribed in mouse spermatogenic cells and encodes E1. (c) Members of the E2 family are dominantly present in mammalian testis. Two E2 homologues (HR6A and HR6B) are expressed during spermatogenesis and one of them (HR6B) is particularly abundant in elongating spermatids. (d) Transgenic male mice lacking a functional E2 are sterile (Roest et al., 1996). (e) An increase in the ubiquitinated histones is observed in elongating spermatids (rat: Chen et al., 1998; mouse: Baarends et al., 1999) at the time when somatic histones are replaced by more stable basic proteins. The cDNA clones corresponding to ubiquitin have been characterized in testis (Mezquita et al., 1997).

The ubiquitin specific proteolytic system in human tissues is extremely complex and diverse (Quesada et al., 2004). In human and mouse, two homologs of UBP gene, *DFFRX (Dffrx)* and *DFFRY (Dffry)*, are located on the X- and Y-chromosomes respectively (Jones et al., 1996 Brown et al., 1998). The X- and Y-chromosomal genes share 89% base pair identity. The X-chromosomal gene is ubiquitously expressed, but with enhanced expression in both human and mouse testis (Jones et al., 1996). The Y-chromosomal gene is ubiquitously expressed in the human, but in a testis specific manner in the mouse (Brown et al., 1998). In the human, DFFRY gene maps to azoospermia factor region AZFa. In addition, *Dffry is located* in the *Sxrb* region of the mouse Y chromosome (Brown et al., 1998; Mazeyrat et al., 1998; reviewed in Barrends et al., 1999; 2000). Quesada et al., (2004) identified and cloned 22 human cDNAs encoding novel members of the ubiquitin-specific protease (UBP) family. Eighteen of the identified proteins contain all structural features characteristic of cysteine proteinases, whereas four of them were classified as non-peptidase homologs. The identified UBPs are broadly and differentially distributed in human tissues, some of them being especially abundant in skeletal muscle or testis.

Two testis specific isoforms of ubiquitin specific processing proteases (UBP) (UBPt1, and UBPt2) are present in germ cells of testis. Both contain identical core regions but distinct N-termini, thereby permitting dissection of the functions of these two regions **(Fig 31.2)**. Both isoforms are germ cell specific and developmentally regulated. The UBP-tI was induced in step 16 to 19 spermatids while UBP-t2 is expressed in step 18 to 19 spermatids. TheUBP-t1 is found in the nucleus while UBP-t2 is extranuclear and found in residual bodies. The UBP-t2 colocalizes with anti-γ-tubulin, indicating that like several other components of the ubiquitin system, a

1	MSQLSSTLKR	YTESSRYTDA	PYAKSGYGTY	TPSSYGANLA	ASFLEKEKLG	FKPVSPTSFL
61	PRPRTYGPSS	ILDCDRGRPL	LRSDITGGSK	RSESQTRGNE	RPSGSGLNGG	SGFPYGVTSN
121	SLSYLPMNAR	DQGVTLGQKK	SNSQSDLARD	FSSLRTSDSY	RTSDGYRASD	GSRIDPGNLG
181	RSPMLARTRK	ELCALQGLYQ	AASRSEYLTD	YLENYGRKGS	APQVLTQAPP	SRVPEVLSPT
241	YRPSGRYTLW	EKNKGQASGA	SRSTSPGRDT	MNSKSAQGLA	GLRNLGNTCF	MNSILQCLSN
301	TRELRDYCLQ	RLYMRDLGHT	SSAHTALMEE	FAKLIQTIWT	SSPNDVVSPS	EFKTQIQRYA
361	PRFVGYNQQD	AQEFLRFLLD	GLHNEVNRVA	ARPKPSPESL	DHLPDEEKGR	QMWRKYLERE
421	DSRIGDLFVG	QLKSSLTCTD	CGYCSTVFDP	FWDLSLPIAK	RGYPEVTLMD	CMRLFTKEDV
481	LDGDEKPTCC	RCRARKRCIK	KFSVQRFPKI	LVLHLKRFSE	SRIRTSKLTT	FVNFPLRDLD
541	LREFASENTN	HAVYNLYAVS	NHSGTTMGGH	YTAYCRSPVT	GEWHTFNDSS	VTPMSSSQVR
601	TSDAYLLFYE	LASPPSRM				

Fig. 31.2. Amino acid sequence of Ubiquitin specific processing protease t2 from rat testis. Source: http// www.ncbi.nlm.nih.gov. (Accession 446226).

deubiquitinating enzyme is associated with the centrosome (Lin et al., 2000).

31.2.2. E1-E2-E3 Enzymes

Conjugation of ubiquitin to proteins, during spermatogenesis, is mediated by ubiquitin-activating enzyme (E1), ubiquitin-conjugating enzymes (UBCs or E2s), and ubiquitin protein ligases (E3s). The activated ubiquitination is UBC4 dependent.

E1: Several components of the ubiquitin system often show enhanced or specific expression of isoforms, in testis and/or ovary. Beginning with the E1-E2-E3 machinery, the X-chromosomal gene *Ube1x*, encoding an E1 enzyme, is expressed in many tissues in the mouse, but the homologous Y-chromosomal gene *Ube1y* is expressed only in the testis. *Ube1y* lost from the Y chromosome during evolution of the primate lineage, and is thus absent from the human Y chromosome. Loss of *Ube1y* may represent an example of an evolutionary process on the sex chromosomes that leads to X chromosome specific gene localization. The expression of *Ube1x* and *Ube1y* mRNAs starts in spermatogonia A. Post-meiotically both genes are highly expressed. However, not much is known about specific functions of the ubiquitin system in meiosis in higher eukaryotes. In isolated germ cell populations of the mouse, the ubiquitin conjugating activity and proteasome activity are higher in spermatids than in spermatocytes. Several proteasome sub-units are highly expressed in the testis, and testis specific subunits have been identified in *Drosophila* and humans testis.

E2: Several ubiquitin conjugating (E2/UBC) enzymes show exclusive expression in testis of rat (HR6a and HR6b by Koken et al., 1996; E217kb by Wing and Jain, 1995; and the isoform 8A by Wing et al., 1996) and human (hUBC9 by Kovalenko et al., 1996; Yasugi and Howley, 1996 and UbcH-ben by Yamaguchi et al., 1996). Using the *S. cerevisisae* and the UBC4 sequence as primer in a PCR reaction, a 390 bp DNA fragment predicted the amino acid sequence of E2 that was 83% identical to yeast UBC4. Rat testes cDNA library identified a family of cDNAs, which predicted two very similar proteins with basic pls and molecular masses of 16,700 Da (E217 kb) (**Fig 31.3**). Isoform E2 expressed in E.coli was expressed in a broad spectrum of tissues, but at a high level in the testis. Fractionation of a testis extract identified several putative ubiquitin protein ligase activities with which this E2 could interact in promoting conjugation of ubiquitin to proteins. One of these activities supported conjugation of ubiquitin to histone H2A, a substrate degraded in the ubiquitin system by a non-N-end rule mechanism (Wing and Jain 1995). Another isoform, 8A, which despite having 91% amino acid identity with the other

```
E2_17kB (2E)       MAL---KRIHKELNDLARDPPAQCSAGPVGDDMFHWQATIMGPNDSPYQG   47
E2_17kB (4A/10A)   ----------N---S-----------------------------------   47
UBCD1              ----------N---Q--G--------------L----------P------   47
ATHUBC8            --S-------L---K--QK---TS-----AE------------AE---S-   47
UBC4               -SSS------A---S--E----TS--------LY----S----A----A-   48
E2_14k             -STPARR-LMRDFKR-QE---VGV-GA-SENNIMQ-N-V-F--EGT-FED   50

E2_17kB (2E)       GVFFLTIHFPTDYPFKPPKVAFTTRIYHPNINSNGSICLDILRSQWSPAL   97
E2_17kB (4A/10A)   --------------------------------------------------   97
UBCD1              --------------------------------------------------   97
ATHUBC8            ---LV----P-----------R-KVF----------------KE------   97
UBC4               -----S-------------IS---K-------A--N------KD------   98
E2_14k             -T-K-V-E-SEE--N---T-R-LSKMF---VYAD--------QNR---TY  100
RABE2F1                IE-N--AE-------      ------DEK

E2_17kB (2E)       TISKVLLSICSLLCDPNPDDPLVPEIARIYKTDREKYN-RIAREWTQKYA  146
E2_17kB (4A/10A)   ----------------------------------D------S--------  146
UBCD1              ---------------------------------------EL-----R---  146
ATHUBC8            -------------T-------------HM-----A--E-AT--N------  146
UBC4               -L-----------T-A-----------H------P--E-AT-----K---  147
E2_14k             DV-SI-T--Q---DE---NS-ANSQA-QL-QENKRE-EKRVSAIVEQSWN  150
RABE2F1             TDQ-IQ-LIA-VN--Q-EH--R

E2_17kB (2E)       M-   147  (100%/100%)
E2_17kB (4A/10A)   --   147  (97%/99%)
UBCD1              --   147  (95%/97%)
ATHUBC8            -G   148  (80%/87%)
UBC4               V-   148  (80%/88%)
E2_14k             DS   152  (38%/54%)
RABE2F1                      (52%/67%)
```

Fig. 31.3. Deduced amino aid sequences for clones 2E, 4A and 10 A of ubiquitin-conjugating enzyme $E2_{17kb}$ and comparison with E2s from *Drosophila* (UBCD1), A thaliana (ATHUBC8), *S. cereviseae* (UBC4) and rabbit reticulocytes ($E2_{14K}$ and RABE2F1). Reprinted with permission from S.S.Wing and P. Jain. Biochem J. 305;125-32:1995 © Portland Press, Ltd.

isoforms, shows several other features. The 8A isoform is restricted to testis and is induced during puberty. Hypophysectomy reduced the expression of the 8A isoform. The 8A mRNA is expressed mainly in round spermatids and is found not only in subfractions of germ cells enriched in round spermatids but also in subfractions containing residual bodies indicating that the protein possesses a longer half-life than its mRNA. Unlike all identified homologues of UBC4, which possesses a basic pI, the 8A isoform is unique in having an acidic pI (Wing et al, 1996). Studies also evaluated contribution of various UBC4 isoforms during testis development. UBC4-1 mRNA is expressed in most cells whereas UBC4-2 mRNA is restricted to germ cells with high levels of expression in round and elongated spermatids. Induction of various UBC4 isoforms activates overall conjugation and plays an important role in the cellular remodeling and protein loss occurring during spermatogenesis (Rajapurohitam et al., 1999).

The hUBC9 is another E2 enzyme that localizes to the synapotonemal complex (SC) during the pachytene stage of the prophase of meiosis in mouse spermatocytes (Kovalenko et al., 1996). Many proteins involved in regulation of meiotic recombination localize to these complexes, and a role of hUBC9 in control of this process and/or in the breakdown of the SC has been suggested (Kovalenko et al., 1996; Tarsounas et al., 1997). The Hsubc9, a human gene encoding an ubiquitin-conjugating enzyme codes a 18 kDa protein. The HsUbc9 protein is homologs to the ubiquitin conjugating enzymes Hus5 of *S. pombe* and Ubc9 of *S. cerevisisae*. The *Hsubc9* gene complements a ubc9 mutation of *S. cerevisisae* and mapped to chromosome 16p13.3. It is expressed in many human tissues, with the highest levels in testis and thymus. The *HsUbc9* protein interacts with human recombination protein Rad51. A mouse homolog, Mmubc9, encodes an amino acid sequence that is identical to the human protein. In mouse spermatocytes, MmUbc9 protein, like Rad51 protein, localizes in synaptonemal complexes, which suggests that Ubc9 protein plays a regulatory role in meiosis (Kovalenko et al., 1996).

NEDD8 (a ubiquitin-like protein) controls vital biological events through its conjugation to target proteins. A negative regulator of the NEDD8 conjugation system, named NEDD8 ultimate buster-1 (NUB1) recruits NEDD8 and conjugates to the proteasome for degradation. NUB1 interacts with ubiquitin precursor UbC1 (UbC1) through its UBA domain. The UBA domain of NUB1 is a specific acceptor for the linear ubiquitin precursor. NUB1 seems to form a protein complex with the unidentified ubiquitin C-terminal hydrolase and recruit UbC1 to this complex. This might allow the ubiquitin C-terminal hydrolase to hydrolyze UbC1, in order to generate ubiquitin monomers. Both mRNAs were strongly expressed in seminiferous tubules of the testis. This implies that the UbC1 hydrolysis mediated by NUB1 is involved in cell functions during spermatogenesis (Tanaka et al., 2004).

E3: Most of the known E3 enzymes show widespread expression, including expression in testis. Often, testis-specific mRNA species have been observed (Kwon et al., 1998; Perry et al., 1998), but testis specific E3 enzymes have not yet been identified. A large protein named Rjs that contains a HECT-domain as well as several RCC1 repeats, is mainly expressed in mouse brain and testis. The *rjs* gene has been identified as part of a genomic region that, when disrupted is responsible for a complicated phenotype including male/female infertility (Barrends et al., 1999, 2000). A UBC4-dependent E3 is expressed in the testis. Rat100 is a 300-kDa protein expressed mainly in the brain and testis and is similar to the human E3. Rat100 is induced during postnatal development of the rat testis. It is localized only in germ cells and is highly expressed in spermatocytes, moderately in round and slightly in elongating spermatids. In contrast to UBC4, Rat100 has a limited subset of substrates. Rat100 is the homolog of human EDD and likely of *Drosophila* hyperplastic discs. This homology with other results (Oughtred et al., 2003) suggests that induction of this E3 results in ubiquitination of specific substrates, some of which are important in male germ cell development.

31.2.3. Deubiquitination Enzymes

Specific control of protein deubiquitination takes part in the ubiquitin system. One of the enzymes which catalyzes protein deubiquitination is PGP 9.5, an UCH (ubiquitin carboxyl-terminal hydrolase) found at high level in testis (Kajimoto et al., 1992). Protein gene product 9.5 (PGP9.5), originally isolated as a neuron specific protein, belongs to a family of ubiquitin carboxyl terminal hydrolases that play important roles in the nonlysosomal proteolytic pathway. In the developing testes, PGP9.5 was only localized on the spermatogonia, where as in adult mice it appeared not only on spermatogonia, but also on Sertoli cells. In the testis of the male sterile W/Wv mutant, very little, but strong reactivity was present in Sertoli cells. PGP9.5 is a useful marker for activated Sertoli cells, playing an important role in degradation of abnormal proteins (Kon et al., 1999). Among mammalian species, the human enzyme UCH (BAP1), which physically interacts with BRCA1 (a protein involved in breast cancer) shows enhanced expression in testis and ovary (Jensen et al., 1998). Human ISOT-3 encodes an enzyme, which belongs to the UBP group of deubiquitinating enzymes and is highly expressed in ovary and testis, with a low level expression in other tissues (Timms et al., 1998).

31.2.4. Multiubiquitin Chain Binding Protein (Mcb1)

The selective protein degradation mediated by the ubiquitin proteasomal pathway has recently attracted much attention as a mechanism for controlling the levels of regulatory peptides with a rapid turnover, e.g. cyclin, p53, c-Myc, c-Fos, and c-Jun. The binding of the multiubiquitin

1	MVLESTMVCV	DNSEYMRNGD	FLPTRLQAQQ	DAVNIVCHSK	TRSNPENNVG	LITLANDCEV
61	LTTLTPDTGR	ILSKLHTVQP	KGKITFCTGI	RVAHLALKHR	QGKNHKMRII	AFVGSPVEDN
121	EKDLVKLAKR	LKKEKVNVDI	INFGEEEVNT	EKLTAFVNTL	NGKDGTGSHL	VTVPPGPSLA
181	DALISSPILA	GEGGAMLGLG	ASDFEFGVDP	SADPELALAL	RVSMEEQRQR	QEEEARRAAA
241	ASAAEAGIAT	PGTEDSDDAL	LKMTINQQEF	GRPGLPDLSS	MTEEEQIAYA	MQMSLQGTEF
301	SQESADMDAS	SAMDTSDPVK	EEDDYDVMQD	PEFLQSVLEN	LPGVDPNNAA	IRSVMGALAS
361	QATKDGKNDK	KEEEKK				

Fig. 31.4. Amino acid sequence of mouse testis multiubiquitin chain binding (Mcb1) protein predicted from c-DNA. Source: http//www.ncbi.nlm.nih.gov. (Accession JC 6535).

substrate to the 26S proteasome is mediated by the 5S 'a' subunit, which is encoded by the gene for the multiubiquitin chain binding protein (Mcb1) **(Fig 31.4)**.

The mouse homologue of Mcb1 was identified from a testis cDNA library where its expression correlates with active spermatogenesis (Pusch et al., 1998). The encoded Mcb1 protein is highly conserved between mouse and human, and its transcripts are expressed in all mouse tissues. In the testis, however, there is additionally a second, longer Mcb1 transcript in wild type mice that is absent in the azoospermic W/Wv mutant mice, suggesting expression of this transcript in association with germ cell differentiation.

31.2.5. Proteasome

Many investigators had noticed that eukaryotic cells contain a novel protease with an unusually large size and multiple catalytic functions, which is distributed widely in all eukaryotes. A variety of terminology covering over 10 names was proposed for the enzyme, such as high molecular weight protease or multi-catalytic proteinase complex. Finally, in 1988, the term proteasome, based on its structural features, was proposed. Rechesteiner group showed that two large proteases were present in cells, one being the 20S proteasome and the other, an ATP activated protease with an apparent sedimentation coefficient of 26S **(Fig 31.5)**. The larger 26S proteasome was shown to contain the 20S proteasome together with multiple components containing potential ATPase. This protease was called the 26S proteasome to distinguish it from the original 20S proteasome. The 26S proteasome has been established as a eukaryotic ATP dependent protease that appears to be responsible for the process. Thus the eukaryotic ATP dependent system for proteolysis consists of two distinct processes: first, recognition of target proteins by Ub and then degradation of Ub ligated proteins by the 26S proteasome, with metabolic energy being required for both the steps (Review -Tanaka, 1998; Mochida et al., 2000).

Using anti-TBP Ab, Mochida et al., (2000) isolated the 26S proteasome from rat testis and sperm tail and visualized it by electron microscopy. During the purification of Sak57, a keratin filament present in outer dense fibers from epididymal sperm, was associated with a substantial amount of 26S proteasomes. Intact 26S proteasomes from rat testis displayed a rod shaped particles about 45nm in length and 11-17 nm in diameter. Each particle consists of a 20S barrel shaped component formed by four rings (αββα), capped by two polar 19S regulatory complexes, each identified by an element known as the Chinese dragon head motif. The TBP-1 appeared as an ATPase containing subunit of the 19S regulatory cap. Rat sperm preparations displayed both dissociated 26S proteasomes and Sak57 filaments. It was hypothesized that 26S proteasomes in the perinuclear arranged manchette are in a suitable location for recognition, sequestration, and degradation of accumulating ubiquitin conjugating somatic and transient testis specific histones during spermiogenesis (Mochida et al., 2000).

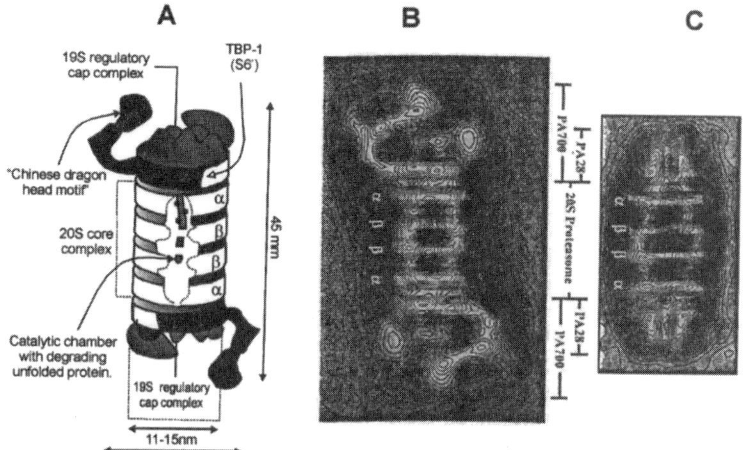

Fig. 31.5. (A). Diagramatic representation of 26 S Proteasome. (B) Image of dumbbell-like PA 700-20 S proteasome and (C) football like PA 28-20S proteasome. The alpha and beta rings of 20S proteasome are indicated. (A) reprinted with permission from K. Mochida et al. Mol Reprod Develop 57; 176-84:2000© John Wiley & Sons Inc http://www3.interscience.wiley.com/cgi-bin/jabout/37692/ProductInformation.html. (B, C) reprinted with permission from S. Tanaka. J Biochem (Japan) 123; 195-204:1998 © Japanese Biochemical Society.

31.2.6. Tat Binding Protein 1 and Proteasome

The rat TBP-1 (tat binding protein-1) (rt TBP-1) is a new member of the family of putative ATPase associated with 26S proteasome complex. The 1.63 kb rtTBP-1 cDNA encodes a 49 kDa protein with 99% amino acid identity to human TBP-1 protein. The rtTBP1 protein contains a heptads repeat of six leucine type zipper fingers at the amino terminal end and highly conserved ATPase and DNA/RNA helicase motifs towards the carboxyl terminal region. The encoded product consists of a protein triplet with a molecular mass range of 52-48 kDa and acidic pl (5.0-5.9). An identical immunoreactive triplet was detected in fractioned pachytene spermatocytes, round spermatids and epididymal sperm. The rtTBP-1 immunoreactive sites co-localizes with - tubulin decorated manchettes of elongating spermatids, in addition to its activity in fibriller and granular cytoplamic bodies observed in spermatocytes and spermatids as well as in association with paraaxonemal mitochondria and outer dense fibers of the developing spermatid tail. The rtTBP-1 is a highly evolutionary conserved TBP-1 like subfamily of putative ATPase, sharing region of identity including ATP binding sites (Rivkin et al., 1997).

31.2.7. Significance of Ubiquitination in Gametogenesis

DNA Repair : The regulation of the first and second meiotic division in testes depends on many components that are also present in mitotic cells. Now it is an established fact that ubiquitin-mediated protein degradation, for example of the different cyclins, is essential for the cascade of events that mediate cellular division. The chromatin rearrangement also could pose a high demand on ubiquitin dependent proteolysis, carried out by the 26S proteasome. In *Drosophila* and *Xenopus*, the proteasome contains testis-specific high expressed subunits (Yuan et al., 1996; Belote et al., 1998). The Rad6 is required for a variety of cellular functions in yeast, including DNA repair, sporulation, and gene silencing. It acts as a Ub-conjugating E2 enzyme in yeast. Two homologs of the *Rad6* gene have been identified in mouse and man: X chromosomal HR6A and autosomal HR6B (Roest et al., 1996) (Chapter 13). Both genes are expressed in many tissues, with the highest mRNA expression in brain, heart, testis and low expression mHR6B in ovary. The yeast RAD6 and its human homolgs hHR6A and hHR6B are implicated in post-replication repair and damage induced mutagenesis. A parallelism between yeast sporulation and mammalian spermatogenesis strongly implicates hHR6-dependent ubiquitination in chromatin remodeling. Since heterozygous male mice and even knockout female mice are completely normal and fertile and thus able to transmit the defect, similar hHR6B mutations may cause male infertility in man (Roest et al., 1996). The HR6B, possibly together with the ubiquitin ligating enzyme mRAD18Sc, is most likely involved in chromatin re-organization during the meiotic and post-meiotic phases of spermatogenesis (Baarends et al., 2000).

The E2 mutation leads also to infertility as found in *Drosophila*. The E2 enzyme UbcD1, suggested for proper telomere behaviour in all cells, has an unknown additional function that is essential for post-meiotic germ cell development. The UbcD1 enzyme is a homolog of the *S. cerevisisae* UBC4/UBC5 proteins, that are essential for degradation of short lived and abnormal proteins. The UbcD1 is most similar to the rat $E2_{17kb}$ (Wing and Jain, 1995) and its isoform 8A, both of which are highly expressed in rat testis. The *rjs* gene, encoding a E3-like proteins is involved in regulation of a number of substrates that function in different specific processes (Lehman et al., 1998).

Targetting Repressor Proteins : Human Cdc34 and Rad6B ubiquitin conjugating enzymes target repressors of cyclic AMP-induced transcription for proteolysis. The Cdc34, Rad6B, and the Cullin proteins are expressed in a developmentally regulated manner, with distinctly different patterns for Cdc34 and the Cullin proteins in germ cells. The Cdc34 and Rad6B protein are significantly elevated in meiotic and postmeiotic haploid germ cells when chromatin modifications occur. Thus, the stability of specific mammalian transcription factors is the result of complex targeting by multiple ubiquitin conjugating enzymes and may have an impact on cAMP inducible gene regulation during both meiotic and mitotic cell cycles (Pati et al., 1999).

Ubiquitin Mediated Protein Degradation : Now it is established that ubiquitin mediated protein degradation, e.g. of different cell cyclins and testicular histones is essetional for cascades of events that mediate spermatogenesis. Another example for the need of protein breakdown during gametogenesis, is the turnover of nuclear proteins during the post-meiotic histone-to-protamine transition in male germ cells (Baarends et al., 1999). Ubiquitination of histones is a general phenomenon found in mammalian cell types, and thought to function in the regulation of nucleosome structure. Also, during spermatids elongation in the mouse, just prior to replacement of the histones, an increase in the amount of ubiquitinated histones has been observed (Baarends et al., 1999). Since RAD6 and its mammalian homologs can ubiquitinate in vitro, Baarends et al (1999) investigated the pattern of histone ubiquitination in mouse testis. A high amount of ubiquitinated H2A (uH2A) was detected in pachytene spermatocytes in wild mouse testis. It was not detected in round spermatids, but then increased again in elongating spermatids. No other ubiquitinated histones were observed. In the HR6B knockout mice, there was no overt defect in the overall pattern of histone ubiquitination. The most intense uH2A signal in pachytene spermatocytes was detected in the sex body, an inactive nuclear structure that contains the heterochromatic X and Y-chromosomes. The postmeiotic uH2A in elongation spermatids indicates that nucleosome destabilization induced by histone ubiquitination might play a facilitating role during histone to protamine replacement (Barrends et al., 1999, Chen et al., 1998). Mutation of the ubiquitination site of H2A does result in a modest negative effect upon the sporulation efficiency (Swerdlow et al., 1990).

II. MOLECULAR CHAPERONS

Molecular chaperones are defined as a family of unrelated classes of proteins that mediate the correct assembly of other polypeptides but they are not components of the functional assembled structures. It is likely that most, if not all, cellular proteins interact with a chaperone at some stage of their lifetime. The function of chaperone proteins is to assist polypeptides to self-assemble by inhibiting alternative assembly pathways that produce nonfunctional structures. There exist several families of chaperones, many of which are heat shock proteins. In ATP dependent chaperones, binding of ATP triggers a critical conformational change thus releasing the substrate. Although the main function of the HSP70/HSP40 chaperone system is to minimize aggregation of newly synthesized polypeptide, some chaperones, like HSP60 facilitate the actual folding process by providing an environment for individual protein, and may also promote the unfolding and refolding of misfolded intermediates (review -Fink, 1999; Ellis, 1991).

The molecular chaperones involved in the folding of newly synthesized proteins recognize non-native substrate proteins predominantly via their exposed hydrophobic residues to form stable complexes. In most cases, these complexes are dissociated by the binding and hydrolysis of ATP. Thus, cells have solved the problem of misfolding and aggregation to a considerable extent at least, through the participation of molecular chaperones in the in vivo folding process (Fink, 1999). The major chaperone classes are small, 40-kDa heat shock protein (HSP40); the DnaJ family), 60-kDa heat shock protein [HSP60; including GroEL and the T-complex polypeptide 1 (TCP-1) ring complexes], 70kDa heat shock protein (HSP70), and 90-kDa heat shock protein (HSP90), calreticulin and calnexin. All these chaperones can prevent the aggregation of at least some unfolded proteins and the extent depends on the compatibility of molecular chaperon and the amino acid sequence of the protein to be folded. In addition to preventing aggregation, HSP90 may permit misfolded structures to unfold and refold.

31.3. HEAT SHOCK PROTEINS IN SPERMATOGENESIS

Heat shock proteins (HSP) are highly conserved cellular stress proteins present in every organism from bacteria to man. Many stimuli induce heat shock like response and generate heat shock proteins. In 1962 the salivary gland chromosomes of the Drosophila melanogaster exhibited a characteristic puffing after exposure to heat. The gene product of this chromosomal puffing was identified in 1974 when the term heat shock proteins was introduced (Tissiers et al. 1974). Since then the chromosomal location of the genes coding for many HSP proteins have been identified and the genes sequenced. The conformation and the mechanism of gene activation by nuclear heat shock transcription factor have been characterized (Westwood et al. 1991). Since the heat shock response is a vital cellular survival mechanism, attention on HSP has been attracted in every medical field including reproductive medicine (Mizzen, 1998;Review-Neuer et al., 2000).

31.3.1. Properties of Heat Shock Proteins

All organisms respond to an increase in temperature by switching off the synthesis of most proteins and commencing large-scale synthesis of a few HSP. The amino acid composition of

HSP has not changed very much during evolution. The HSPs of highly divergent organisms are very similar. Due to essential role in quality control, HSPs are among the most highly conserved gene products in nature. The term HSP refers to inducible protein products while HSC describes constitutively expressed HSP. The HSPs serve two major functions: firstly, under physiological conditions, they act as molecular chaperones, involved in mediating the folding and transport of other intracellular proteins and in some cases HSP play a role in their assembly into oligomeric structures from monomers. They fulfill crucial roles in intracellular transport, the maintenance of proteins in an inactive form and the prevention of protein degradation. In addition, two HSP70 proteins are products of genes expressed specifically in spermatogenic cells and are widely distributed in ciliary and flagellar components, suggesting that they play a potential role in axonemal protein dynamics.

Heat shock proteins are induced in response to various cellular stresses, which include change in temperature, the presence of free oxygen radicals, viral and bacterial infection, heavy metals, ethanol, and ischaemia. The stress-elicited activation of heat shock genes is called the heat shock response. This heat shock response is frequently found in clinical conditions, e.g. ischaemia, infection and circulatory and haemorrhagic shock. Cellular stress disturbs the tertiary structure of proteins and has adverse effects on cellular metabolism. However, pretreatment of cells with a mild stress, just sufficient to induce the expression of HSP, results in protection to subsequent insults. This phenomenon has been called 'stress tolerance'. It has been suggested that cellular structure like microfilaments and centrosomes and cellular functions like transcription and translation are more stabilized during a second stressful event in stress tolerant cells. (review Neuer et al., 2000).

31.3.2. 'Small' Heat Shock Proteins

Small HSP 12- to 43 kDa from crytallin family produced under stress are known to function in vitro as chaperons. But their role in vivo is not very clear. Among recently described, eighth small heat shock protein (sHspB8) is present in heart and skeletal muscle (Smith et al., 2000). The other, named HspB9, is specifically expressed in testis, notably in the spermatogenic cells from late pachytene spermatocyte stage till elongate spermatid stage. While mammalian sHsps are generally highly conserved, mouse HspB9 shows 38% sequence difference with human HspB9, which may confirm its sex-related role (Kappe et al., 2001). Expression of Hsp25 was not detected in mice testis in the early postnatal days, i.e., before the onset of spermatogenesis. The Hsp25 mRNA was detected at around 10 days postpartum and began to be expressed in coincidence with the onset of the first wave of spermatogenesis. Throughout the testis development, Hsp25 mRNA was localized exclusively to germ cells. The W/Wv mutant mice, exhibited no Hsp25 mRNA expression. The most intense signal of Hsp25 mRNA was localized to the spermatocytes at leptotene, zygotene and early pachytene phases and decreased in the late pachytene and diplotene spermatocytes and was not detected in spermatids. Hsp25 plays some specific role in the meiotic prophase of the testicular germ cell (Wakayama and Iseki, 1999). HSP-27 is present in the cytoplasm of mouse spermatogenic cells and reallocates to nucleus after heat shock (Biggiogera et al., 1996)

31.3.3. Heat Shock Protein - 40 or DnaJ

The HSP40 or DnaJ family consists of over 100 members, defined by the presence of a highly conserved J domain of ~78 residues. Proteins in this family typically consist of several domains, e.g., DnaJ contains at least four conserved regions representing potential functional domains : the J domain, which is linked by a Gly/Phe-rich region to a domain of unknown function, followed by a zinc-finger region, and ending with the COOH-terminal domain, also of unknown function. The best defined role thus far for the HSP40 is to act as a co-chaperone for HSP70; however, even this function is not well understood, and there is evidence to indicate that DnaJ and other members of the HSP40 family are chaperones in their own right, binding to at least some unfolded protein and nascent chains. Little is known about the structural features of DnaJ that are involved in its interaction with DnaK and unfolded proteins. A member of the DnaJ protein family, named MSJ-1, is 242 amino acid residues long that contains the finger-printing J domain in the NH2 terminus and Gly/Phe-rich domain in the middle of protein, which are characteristic structural domains of the DnaJ protein family. Analysis from different organs revealed that MSJ-1 is expressed only in the testis, while in-situ hybridization analysis demonstrates that the mRNA is first transcribed in spermatids. The antiserum developed against a MSJ-1 GST fusion protein recognizes a protein of 30-kDa in germ cell protein extracts (review Berruti et al., 1998a, b). The predicted sequence of DnaJ from spermatogenic cells is similar to that of bacterial DnaJ proteins in the amino terminal portion since it contains the highly conserved J domain, which is present in all DnaJ like proteins and is considered to have a critical role in DnaJ protein interactions. In contrast, the middle and carboxyl-terminal regions of the protein are not similar to any other DnaJ proteins, with the exception of the human neuronal HSJ-1, which displays a 48% identity in a 175 amino acid overlap. The mRNA is first transcribed in spermatids. A similar pattern of expression is exhibited also in rat testis. This mouse gene was named MSJ-1, for mouse sperm cell specific DnaJ first homologue (Berruti et al, 1998). MSJ-1 gene was called MFSJ1, for spermatogenic cell-specific DNAJ homolog in the Japanese monkey. The MFSJ1 gene is specifically expressed in testis and is active at adult and senile stages. The MFSJ1 gene is expressed mainly in spermatids and was found to have high similarity (71% identity) with MSJ-1, mouse spermatogenic cell-specific DnaJ homolog (Yu et al., 2003).

The MSJ-1 interacted with the testis specific Hsp70-2 protein and precipitated with Hsp70-2 from spermatogenic cells; binding of these two charperones is consistent with the presence of a third component, which is so far unknown. The MSJ-1 is weakly detected in early round spermatids, and its content increased in differentiating spermatids where it colocalizes with the developing acrosome and their post-nuclear region. The Hsp70-2, which is highly expressed in meiotic cells, shows a sub-cellular localization in late differentiating spermatids and sperm and overlap with MSJ-1. Findings are consistent with a role for MSJ-1 in acrosome formation and centrosome adjustment during spermatid development, whereas its presence in mature spermatozoa suggests a special function during fertilization (Berruti and Martegani, 2001).

31.3.4. Heat Shock Protein-60

Many of the HSP60 chaperones are also known as chaperonins (cpn60) and are ring-shaped oligomeric protein complexes with a large central cavity in which non-native proteins can bind. Members of HSP60 are also involved in assembly of large multi-protein complexes. During developmental stages of spermatogenesis, it is not surprising that this process is accompanied by the expression of different HSPs. The HSP60 is one of the best characterized among molecular chaperones. It is an essential chaperon protein involved in transport, folding and assembly of protein subunits. It is highly conserved and its production is elevated in response to environmental stress. Immune responses to conserved regions of HSP60 have been implicated in autoimmune phenomena. Pregnancy regulation and final out come of HSP60 has been reviewed (Neuer et al, 2000). The HSP60 group of proteins that is highly expressed in a constitutive manner is moderately stress inducible. The HSP60 and 70 are important in testis germ cells. During mouse and rat spermatogenesis, the constitutive form of HSP70 accumulates. Also mRNA coding for proteins related to HSP60 was found in rat and human testis in infertile men. The seminiferous epithelium showed a cell specific expression of HSP60 with localization in Sertoli and Leydig cells. In germ cells, mitochondria of spermatogonia and early primary spermatocytes were immunoreactive for HSP60; other germ cell types were completely negative. High concentrations of HSP60 were found in stages I-V and IX-XIV, and low levels were detected in the other stages, i.e., VI-VIII. In stages with high HSP60 expression, spermatogonia divide mitotically whereas in stages lacking mitosis, the HSP60 level was much weaker. The main site of HSP60 is located in the mitochondria; other loci include cell surface exposition of HSP60. In seminiferous epithelium, two different types of mitochondria are present. The crista type of mitochondria (e.g., in Sertoli cells and spermatogonia) reacted with the antibody against HSP60, whereas HSP60 was negative in so-called "condensed"-type mitochondria found in mid-pachytene spermatocytes and more advanced germ cells. The presence of HSP60 in stages with mitotic activity suggests a very active mitochondrial protein import and protein assembly machinery that generates further mitochondria for the dividing cells (Meinhardt et al., 1995, 1999). It has been demonstrated that the number of HSP60 expressing spermatogonia paralleles the loss of spermatogenic function, and that a low level of HSP60-expressing spermatogonia might lead to a decreased level of protection, which in turn could be involved in low spermatogenic efficiency (Meinhardt et al., 1999).

31.3.5. Heat Shock Protein-70

The HSP70 family comprises several proteins, at least a dozen, that are localized in distinct cellular compartments. The constitutively expressed HSP70 is found in the cytosol and the nuclei of cells. It is only moderately stress inducible and is most conserved among HSP super-family. The HSP70 chaperones are composed of two major functional domains. The NH2 terminus with highly conserved ATPase domain binds ADP and ATP very tightly and hydrolyzes ATP, whereas the COOH-terminal domain is required for polypeptide binding. Co-operation of both domains is needed for protein folding (Fink, 1999). In rat testis high levels of *Hsp70* RNA were found in stages XII-XIV and I to early VII of the spermatogenic cycle, and low levels were found in other stages, i.e., late VII (VII-d) through VIII-XI of the cycle. The in situ hybridization revealed that the *Hsp70* gene was activated in late pachytene primary spermatocytes during stage XII of the cycle, and that mRNA was then present in cells during differentiation through diakinesis, meiotic divisions, and early spermiogenesis (Krawczyk et al., 1988). An antibody originally raised against human sperm proacrosin (designated C3) was found to immunostain Hsp70 like protein in rat (Roab et al., 1995). In mice, the HSP70 family contains at least seven different proteins, including the constitutively expressed 70-kDa heat shock cognate (HSP70) and 75-kDa and 78-kDa glucose-regulated proteins GRP75 and GRP78. In addition, environmental or physiological stress induces expression of two additional HSP70s, HSP70-1 and HSP70-3, which protect cells and help them recover from stress induced damage. Studies demonstrated that Hsp70.1 and Hsp70.3 have an essential role in maintaining genomic stability under stressed conditions (Hunt et al., 2004). Hsp70-1 mRNA and protein are equally expressed in all segments of normal epididymis and ductus deferens and was specific to basal cells in the epididymal epithelium (Legare et al., 2004). In addition, spermatocyte-specific HSP70-2 is expressed at high levels in pachytene spermatocytes during the meiotic phase of spermatogenesis, whereas testis-specific HSC70 (HSC70-t) is expressed in postmeiotic spermatids (Dix et al., 1996). Mouse somatic tissues contain low levels of transcripts homologous to heat shock-inducible and cognate members of hsp70 gene family. A unique sized *hsp70* mRNA of 2.7-kb is present in testes in the absence of exogenous stress. Its expression is restricted to germ cells and is developmentally regulated. The 2.7-kb transcript first appears during the haploid phase of spermatogenesis and is stable throughout the morphogenic stages of spermiogenesis. A 2.7-kb *hsp70* mRNA is present in rat and human testes.

The developmental expression of HSP70-2 during spermatogenesis implies that it performs a specialized function during meiosis that cannot be accomplished by other HSP70 (Dix et al., 1996). To determine if this unique stress protein has a critical role in meiosis, Dix et al (1996) disrupted Hsp70-2 gene in mice. Male mice homozygous for the mutant allele (*Hsp70-2-/-*) did not synthesize Hsp70-2, lacked postmeiotic spermatids and mature sperm, and were infertile. However neither meiosis nor fertility was affected in female *Hsp70-2-/-* mice. It was found that Hsp70-2 was associated with synaptonemal complex in the nucleus of meiotic spermatocytes from mice and hamsters. While synaptonemal complexes assembled in *Hsp70-2-/-* spermatocytes, structural abnormalities became apparent in these cells by late prophase, and development rarely progressed to the meiotic divisions. Furthermore, analysis of nuclei and genomic DNA indicated that the failure of meiosis in *Hsp70-2-/-* mice was coincident with a dramatic increase in spermatocyte apoptosis. It suggests that Hsp70-2 participates in synaptonemal complex function during meiosis in male germ cells and is linked to mechanisms that inhibit apoptosis (Dix et al., 1996).

In juvenile homozygous mutant mice (*Hsp70-2-/-*), spermatogenesis proceeded normally until day 15 following which pachytene spermatocytes became apoptotic and differentiation of cells beyond the pachylene stage began to falter. Synaptonemal complexes assembled in *Hsp70-2-/-* mice and spermatocytes developed through the final pachytene substage, however, SCs failed to desynapse and normal diplotene spermatocytes were not observed. Metaphase spermatocytes were not seen in testis sections of *Hsp70-2-/-*mice, and expression of mRNAs and antigens, characteristic of late pachytene spermatocytes (e.g., cyclin A1) and development of spermatids did not occur. Thus, HSP70-2 is required for synaptonemal complex desynapsis, and its absence severely impairs the transition of spermatogenic cells through the late meiotic stages and results in apoptosis beginning with the first wave of germ cell development in juvenile mice (Dix et al., 1997). The TUNEL on testis sections indicated that chromosomal fragmentation was occurring in late pachytene spermatocytes in *Hsp70-2-/-* mice. The role of HSP70-2 in preventing apoptosis in pachytene spermatocytes was relatively specific.

HSP70-2 is required for CDC2 to form a heterodimer with cyclin B1, suggesting that it is a chaperone necessary for the progression of meiosis in the male germ cells. It appeared that one cause of failure to complete meiosis during spermatogenesis in *Hsp70-2 (-/-)* mice is disruption of CDC2/cyclin B1 assembly required as CDC2 kinase to trigger G2/M-phase transition. This provides in vivo evidence for a link between an HSP70 molecular chaperone and CDC2 kinase activity essential for the meiotic cell cycle during spermatogenesis (Zhu et al., 1997). In addition to CDC2, other proteins may require chaperoning by HSP70-2. The SC proteins that might be associated with HSP70-2 include those with structural roles, such as SCP1/COR1. In addition, HSP-70 proteins are associated with the SC that has roles in DNA recombination and repair, including RAD51, MLH1, DNA topoisomerase II, BRCA1, ATM, ATR and UBC9 (review Eddy, 1999).

A cDNA from spermatids that encoded HSP 70 showed only one continuous open reading frame capable of coding 630 amino-acid protein and matched with sequences of previous HSP70. A 2.7kb transcript derived from this gene was expressed in spermatids but not in other testicular germ cells and somatic tissues (Matsumoto and Fujimoto, 1990) **(Fig 31.6)**. A 0.8-kb upstream region from endogenous rat hsp70 contains sufficient information to function as an active spermatogenesis specific promoter (Wisniewski et al., 1993). The 252 bp of the upstream region of the hst70 gene is sufficient to direct the expression of the reporter gene with high specificity to testis of adult transgenic as well as sufficient to activate the transgene in a developmentally regulated way (Widlak et al., 1995) . A 165 bp-long fragment of the *Hst70* gene

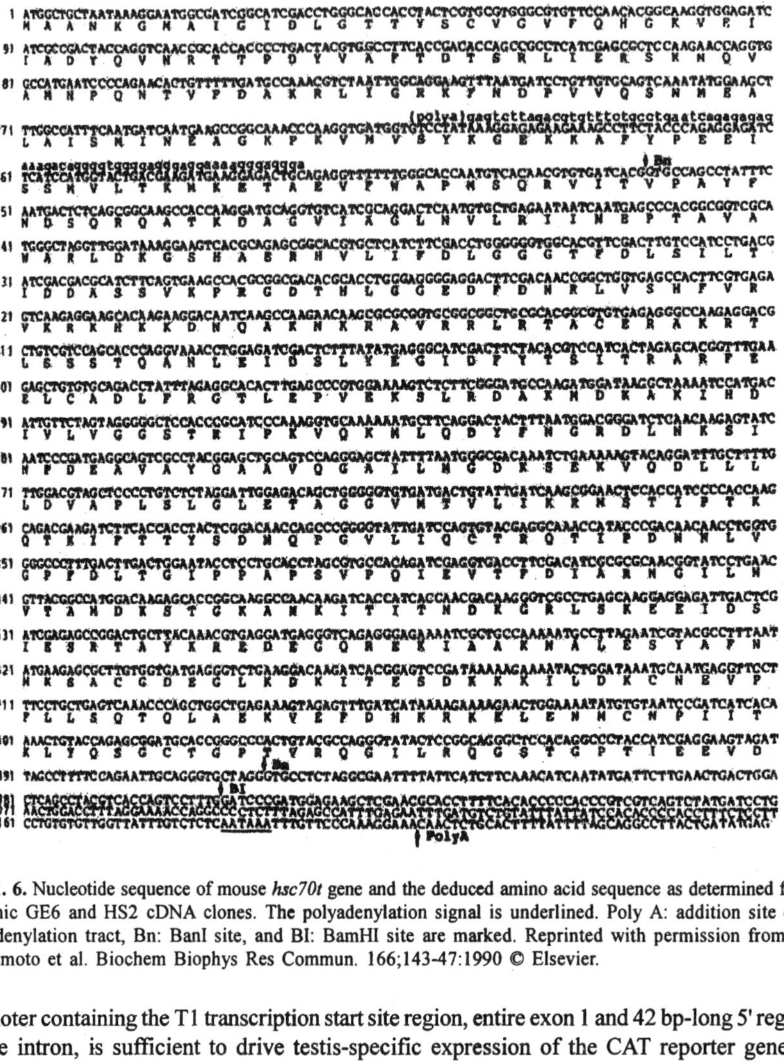

Fig.31. 6. Nucleotide sequence of mouse *hsc70t* gene and the deduced amino acid sequence as determined from genomic GE6 and HS2 cDNA clones. The polyadenylation signal is underlined. Poly A: addition site of a polyadenylation tract, Bn: BanI site, and BI: BamHI site are marked. Reprinted with permission from M. Matsumoto et al. Biochem Biophys Res Commun. 166;143-47:1990 © Elsevier.

promoter containing the T1 transcription start site region, entire exon 1 and 42 bp-long 5' region of the intron, is sufficient to drive testis-specific expression of the CAT reporter gene in transgenic mice with the same developmentally regulated pattern as the endogenous *Hsp70.2* gene (Scieglinska et al., 2004).

A mouse gene, containing extensive similarity to a heat shock-inducible *Hsp70* gene within the coding region but diverging in both 3-' and 5'-non-translated regions, was isolated by Zakari et al (1988). The gene appeared to be activated uniquely in the male germ line. An isoform of Hsp70 called 73T, with a Mr of 73-kDa was shown to be associated with germ cells, and found only in adult mouse testicular cells, but not in testes from animals that lacked meiotic germ cells. The 73T was synthesized in enriched cell populations of both meiotic prophase and postmeiotic cells, but was not inducible by in vitro heat shock. It was suggested that hsp72 is inducible in the somatic compartment and possibly in the pre-meiotic germ cells, but not in germ cells, which have entered meiosis and which are expressing members of the HSP70 in a developmentally regulated fashion (Zakeri et al., 1990). The 5' flanking region of *Hsp70* related gene, which is specifically expressed in rat and mouse spermatocytes contains GGTCAnnnCGACC sequence closely resembling the palindromic consensus estrogen responsive element (ERE) GGTCAnnnTGACC. However, in transient expression assay, GGTCAnnnCGACC sequence did not respond to hormone stimulation, showing the importance of T nucleotide in ERE for activation of transcription (Krawczyk et al., 1992).

The *HspA2* gene is the human homologue of the murine *Hsp70-2* gene with 91.7% identify in the nucleotide coding sequence. HSPA2 occurs in testes with normal spermatogenesis, whereas a low amount of HSPA2 was expressed in testis with Sertoli cell-only syndrome. The HSP-A2 was present in spermatocytes and spermatids. The mRNA expression of HSPA2 in testes of infertile men with azoospermia demonstrates that HspA2 gene expression is down regulated with abnormal spermatogenesis, which in turn suggests that the HspA2 might play a specific role during meiosis in human testes (Son et al., 2000).

Birds, however, are unique among homeothermic animals in developing spermatogenesis at the elevated avian internal body temperature of 40-410 C. To shed light on the mechanisms that maintain an efficient avian spermatogenesis at elevated temperatures, expression of HSP70 and ubiquitin was compared in mouse and chicken testicular cells. While the expression of Hsp70 and ubiquitin did not change upon heat shock in mouse testicular cells, both the amount and polyadenylation of Hsp70 and ubiquitin transcripts increased when male germ cells from adult chicken testis were exposed to elevated temperatures (Mezquita et al., 1998).

31.3. 6. Heat Shock Protein-90

In mouse testis, HSP90 heat shock protein is present in the cytoplasm and to a lesser extent in the nucleus throughout spermatogenesis. It is absent from mitochondria but disappears from the cell nucleus at the stage of elongating spermatids. After heat shock, both HSP90 and HSP27 increase and partly relocate to the nucleus. Following hypothermic treatment to testis, a sharp increase in the labeling of HSPs occurs in chromatid bodies of round spermatids (Biggiogera et al., 1996). The 105-kDa protein is testis-specific and related to HSP90 (90-kDa heat-shock protein). During development of rat testis, the 105-kDa protein does not appear until the age of five weeks, while HSP90 could be detected at three weeks. In the spermatozoa, the 105-kDa protein was more abundant than in Leydig tumor cells of rat testis (Itoh and Tashima, 1991).

31.3.7. Heat Shock Protein-110

The Apg-1 encodes a heat shock protein belonging to HSP110 family. The apg-1 transcripts are constitutively expressed in the testicular germ cells and to some extent in most tissues. The apg-1 is abundantly expressed in testis and responds to a low temperature heat shock rather than the traditional elevated temperatures (Kaneko et al, 1997) . In a Sertoli cell line (TAMA26) of mouse, apg-1 transcripts were induced by a temperature shift from 32 to 390C, but not by a shift from 37 to 420C, the traditional heat stress, or a shift from 32 to 42 0C. The heat response pattern of hsp110 expression was similar to that of apg-1. It was suggested that the expression of apg-1, hsp60 and hsp70 could be induced at lower temperature than the traditional elevated temperatures and that the mechanisms regulating the transcript levels of apg-1 and hsp110 are different from those of hsp70 (Kaneko et al., 1997a).

The 105-kDa HSP105 and HSP105 are mammalian heat shock proteins that belong to the HSP105/ HSP110 family. Both HSP105 and HSP110 consist of acidic and basic isoforms. The acidic isoforms are serine phosphorylated. Furthermore, HSP105 or HSP105 seem to act as substrates for casein kinase II, which phosphorylates HSP105 and HSP105. Since phosphorylated HSP105 is prominently present in the brain of mice and rats, the phosphorylation of HSP105 by casein kinase II may be biologically important (Ishihara et al., 2000). Matsumori et al., (2002), however, identified a testis and brain-specific 105-kDa protein, which cross-reacted with an anti-bovine HSP90 antibody. The protein was induced in germ cells by heat stress and inhibited the aggregation of citrate synthase as a molecular chaperone in vitro. It was able to bind to ATP-Sepharose like HSP70. A partial amino-acid sequence (24 amino-acid residues) showed identity with mouse testis- and brain-specific APG-1 and osmotic stress protein 94 (OSP94). The 105-kDa testis-brain specific protein was also detected in the medulla of the rat kidney similar to OSP94. It appeared as if APG-1 and OSP94 were identical to 105-kDa protein (Matsumori et al., 2002).

The expressions of the apg-1 transcripts were detected in germ cells at specific stages of development including spermatocytes and spermatids. Although heat induction of apg-1 transcripts was observed in W/Wv mutant testis lacking germ cells, it was not detected in wild type testis nor in the purified germ cells. Thus, the apg-1 expression is not heat regulated but developmentally regulated in germ cells, suggesting that APG-1 plays role in normal development of germ cells (Kaneko et al. 1997b).

31.4. HEME OXYGENASES AS CHAPERONES

The heme oxygenase (HO) isozymes HO-1 and HO-2 oxidatively cleave the heme molecule to produce the bile pigments, CO and iron (See Chapter 30). The HO-1 (Hsp32) is a stress inducible enzyme, whereas HO-2 is constitutively expressed at high levels in testis and brain. The cellular distribution of HO-1 and HO-2 in the testis of normal and heat shocked rats indicated the cell specific expression of these isozymes. In normal testis HO1 was not detected in germ or Leydig cells but could be detected at low levels in Sertoli cells. The HO-2 on the other hand, was most prominently expressed in residual bodies and was not detected in spermatogonia. Modest levels of HO-2 were observed in spermatocytes, spermatids, Leydig cells. In contrast, prominent expression of HO-2 mRNAs was detected in spermatocytes, spermatids, and residual bodies of the seminiferous epithelium (Ewing et.,1995).

31.5. CALRETICULIN AND CALNEXIN

Calnexin is a transmembrane molecular chaperone that resides in the endoplasmic reticulum (ER). Calreticulin, which has sequence homology with calnexin, is a soluble ER chaperone. Both proteins are involved in folding and assembly of nascent proteins in the ER in a calcium-dependent manner and play an important role in glycoprotein maturation and quality control.

31.5.1. Calreticulin

The endoplasmic reticulum is considered one of the most important relevant source of cellular Ca^{2+}. Ca^{2+} is released from the ER by the inosital triphosphate (IP3) receptor and /or ryanodine receptor (RyR) Ca^{2+}-release channels. Calreticulin is a unique ER luminal Ca^{2+} binding chaperone implicated in playing its role in many cellular functions, including lectin like chaperoning, Ca^{2+} storage and signaling, regulation of gene expression, cell adhesion and autoimmunity. The cDNA and genes encoding calreticulin have been isolated from several vertebrates and invertebrates. The calreticulin gene consists of nine exons and spans approx. 3.6- kb or 4.6- kb of human or mouse genomic DNA respectively. Human and mouse genes have been localized to chromosomes 19 and 8 respectively. The exon-intron organization of the human and mouse gene is almost identical. The nucleotide sequences of the mouse and the human gene show more than 70% identity, with the exception of introns 3 and 6. The promoter of the mouse and human calreticulin genes contain several putative regulatory sites, including AP-1 and AP-2 sites, GC rich areas including an Sp1 site, an H4TF-1 and four CCAAT sequences. AP2 and H4TF-1 recognition sequences are typically found in genes that are active during cellular proliferation. There are several poly (G) sequences, including GGGNNGGG motif in the promoter regions of calreticulin and other ER luminal proteins.

Calreticulin is a 46kDa protein with an N-terminal cleavable amino acid signal sequence and a C-terminal KDEL ER retrievel signal. Depending on species, calreticulin may have one or more potential N-linked glycosylation sites. Heat shock may trigger glycosylation of calreticulin. Three highly conserved cysteine residues are located in the N-domain of the protein. Two of three cysteine residues found in the protein form a disulphide bridge (Cys120 - Cys146), which may be important for proper folding of the N-terminal region of calreticulin. Structural predictions of calreticulin suggest that the protein has at least three domains (Michalak et al., 1999). Besides its localization in ER/SR membranes, the protein has also been localized to the cytoplasmic granules of the cytotoxic T-cell, sperm acrosomes and tick's cell surface. Calreticulin has been implicated to participate in many cellular functions. However two major functions of calreticulin in chaperoning and regulation of Ca^{2+} homeostasis are well documented. Calreticulin modulates cell adhesion integrin dependent Ca^{2+} signaling and steroid sensitive gene expression both in vitro and in vivo. Calreticulin from spermatogenic cells has been identified and found to be present in the acrosome of both round spermatids and mature sperm. However, under immunoelectron microcopy, gold-particles were seen over other sub-cellular structures of spermatocytes, spermatids, and Sertoli cells. The acrosome was found to be most heavily labeled. The amino acid sequence of calreticulin, deduced from the cDNA sequence shares a high degree of identity with mouse protein. The cDNA encodes a protein of 416 amino acids, including a 17-residue NH2- terminal signal sequence. The mature protein contains a KDEL sequence as an ER signal at the COOH terminus. Sperm calreticulin contained no glycosyl moiety. Calreticulum mRNA was present in both pre- and post-meiotic germ cells of rat. Rat calreticulin contained two Ca2+ binding sites, a low affinity/high capacity site and another high affinity/low capacity site. The calreticulin, which is not specific to testis, is closely associated during spermiogenesis, and may play an important role in the regulation of sperm motility and acrosome reaction (Nakamura et al., 1993).

31.5.2. Calmegin (Calnexin-t)

Calnexin was first discovered as one of the proteins consisting signal sequence receptor complexes on the ER membrane. Subsequently it was found that calnexin functions as a molecular chaperone. It associates transiently with many membrane glycoproteins during their maturation in the ER, including MHC I heavy chain, MHC class II and other proteins (Ohsako et al., 1998). Partially trimmed, monoglycosylated, N-linked oligosaccharides have been shown to be recognized by calnexin. Calnexin has a long internal domain, which resides in the luminal compartment of the ER. The domain is composed of proline rich repeat sequences bearing a high homology to a similar region in calreticulin. The proline rich repeat sequence of calreticulin has been identified as a high affinity and low capacity Ca^{2+} binding domain **(Fig. 31.7)**. In constrast to that of calreticulin, the cytoplasmic region of calnexin is relatively short and has several potential casein kinase II phosphorylation sites. In vivo casein is isolated as a phosphorylated protein and casein kinase II was identified as the probable kinase for calnexin.

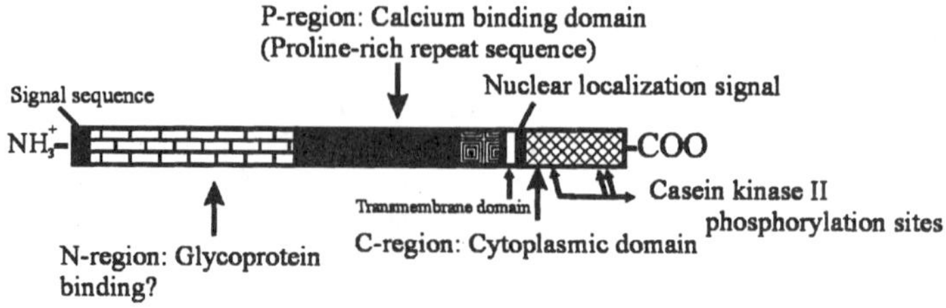

Fig. 31.7. Domain structure of testis specific, type 1 transmembrane protein calnexin-t. Calnexin-t has a Ca2+ binding domain (P-region), a glycoprotein binding domain (N-region), and a cytoplasmic domain (C-region), which has three phosphonylation sites for casein kinase II.

The molecular cloning of a male germ cell specific Ca^{2+} binding protein showed it to be initially expressed in middle pachytene spermatocytes and persisted until maturation phase spermatids. The male germ cell calnexin from testis was called calnexin-t (or calmegin called by watanabe et al, 1994) as a testis specific variant. Deletion of the gene resulted in complete infertility in male mice in spite of the production of sperm with normal morphology and motility, suggesting that calmegin /calnexin-t is indispensable for sperm function (Ikawa et al., 1997; c/r Ohsako et al., 1998). Sperm from homozygous null males failed to adhere to the egg extracellular matrix zona pellucida in vitro. This suggested that calmegin functions as a chaperone for one or more sperm surface proteins that mediate the interactions between sperm and egg (Ikawa et al., 1997) However, sperm from Calmegin (-/-) and (+/+) mice did not show any difference in antibody staining of acrosome or in strength of antigen in sperm as measured by flow cytometry (Ikawa et al,1997) (**Fig 31.8**). Besides selective localization on ER, calmegin was present on nuclear envelope of spermatogenic cells. During the maturation phases, a dramatic reduction in calmegin occurred in the ER of the spermatids, suggesting that the major function of calmegin has been completed by the time spermatids reached step 14. Although the immunoreactivity is absent in the calmegin deficient mouse testis, mature sperm from the knockout mice were ultrastructurally normal (Yoshinaga et al., 1999). Calmegin is required for the heterodimerization of fertilin and the appearance of fertilin- on the sperm surface. The sperm defect in calmegin-null animals is not at the level of sperm-egg binding but rather may involve either sperm-ZP binding and/or sperm transit to the oviduct. Because fertilin- is absent from calmegin-deficient mice, results indicated that the role of fertilin in sperm-egg interaction be re-evaluated (Yamagata et al., 2002).

A cDNA representing calmegin from mouse testis cDNA expression library, showed that this transcript of 2.3 kb in length, was expressed only in the testis and not in other somatic tissues or in the ovary. The expression of the mRNA was first detected at the pachytene spermatocyte stage, which correlated with the expression of the protein. Sequence analysis of the cDNA predicted a protein of 611 amino acids, including a hydrophobic -NH2 terminus characteristic of a signal peptide, two sets of internal repetitive sequences of IPDPSAVKPEDWDD and GEWXPPMIPNPXYQ, and a hydrophilic -COOH terminus.

Protein extracts from developing mice testes revealed calnexin as a 101-kDa protein. The101-kDa form of calnexin-t is prevalent at all stages pf spermatogenesis. The ER luminal domain of calnexin-t is structurally different in middle pachytene versus haploid germ cell phases. The 45Ca2+ overlay assay showed that the internal proline rich repeat region has Ca^{2+} binding ability and contains an epitope recognized by specific mAb 1C9 and as a 101 kDa protein band during mouse testicular development. The 101-kDa form of calnexint (named as Calmegin by Watanabe, 1994) is present at all stages of spermatogenesis, and is the phosphorylated type of

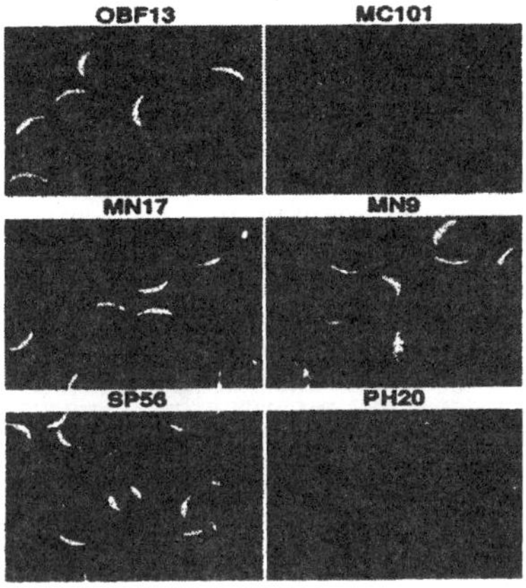

Fig. 31.8. Sperm from calmegin-deficient and wild type mice, stained with anti-acrosomal monoclonal and polyclonal antibodies against various sperm antigens (OBF13, MN17, SP56, MC101, MN9, and PH20). Reprinted with permission from M. Ikawa et al. Nature 387;607-11:1997 © Macmillan Magazines, Ltd. http//www.nature.com/nature.

calnexin. Mouse calnexin-t contains a high affinity calcium binding domain as present in both calnexin and calreticulin and one transmembrane domain similar to that of calnexin.

DNA of calmegin promoter region contained GC rich sequences and potential binding sites for AP2 and Sp1, but lacked the TATA sequence. The 330 bp5' flanking sequence of calmegin genomic DNA fused with the CAT gene indicated that the 30-bp calmegin 5' sequence was sufficient for the testis specific expression. The existence of testicular nuclear factors specifically bound to the putative promoter sequence was also demonstrated (Watanabe et al., 1995). In addition, the cytoplasmic domain of calnexin-t is highly phosphorylated immediately after protein synthesis and constitutively phosphorylated during spermatogenesis. Analogous to the function of somatic calnexin, calnexin-t has a role in quality control in the ER for the folding and assembly of glyco-proteins that end up in the acrosomal matrix or on the surface of spermatozoa (Watanabe et al., 1994; 1995).

The human homolog of the mouse germ cell specific calmegin showed 80% identity with the mouse calmegin. The predicted acid sequence showed strong conservation of the two sets of internal repetitive sequences (Ca^{2+} binding motif), and the hydrophilic COOH terminus, which corresponds to the putative ER retention motif. The transcript of 3-kb length was expressed exclusively in the testis **(Fig. 31.9)**. The human calmegin gene was mapped to chromosome 4q28.3-q31.1 (Tanaka et al. 1997). The deduced amino acid sequence has 58% homology with dog calnexin and significant partial homology with calreticulin.

31.6. IMMUNOPHILINS AND TETRATRICOPEPTIDE REPEATS

FKBP59: FKBP59 is a high molecular weight immunophilin that was originally cloned and sequenced in the rabbit. It corresponds to the human FKBP52 and the mouse p59. As with all other immunophilins, FKBP59 has peptidylprolyl isomerase ["rotamase"] activity potentially involved in conformational changes of proteins through the cis trans isomerization of the proline amidic peptide bonds. The protein, is associated with Hsp90 and hence acts as a chaperon. The FKBP59 has a ubiquitous distribution in mammals and colocalizes with microtubules during interphase and mitotic spindle. Among structural domains involved in the direct protein interactions is tetratricopeptide repeat (TPR) motif. The TPRs are found in diverse proteins with common biochemical functions. Thus, proteins with TPR motif include those

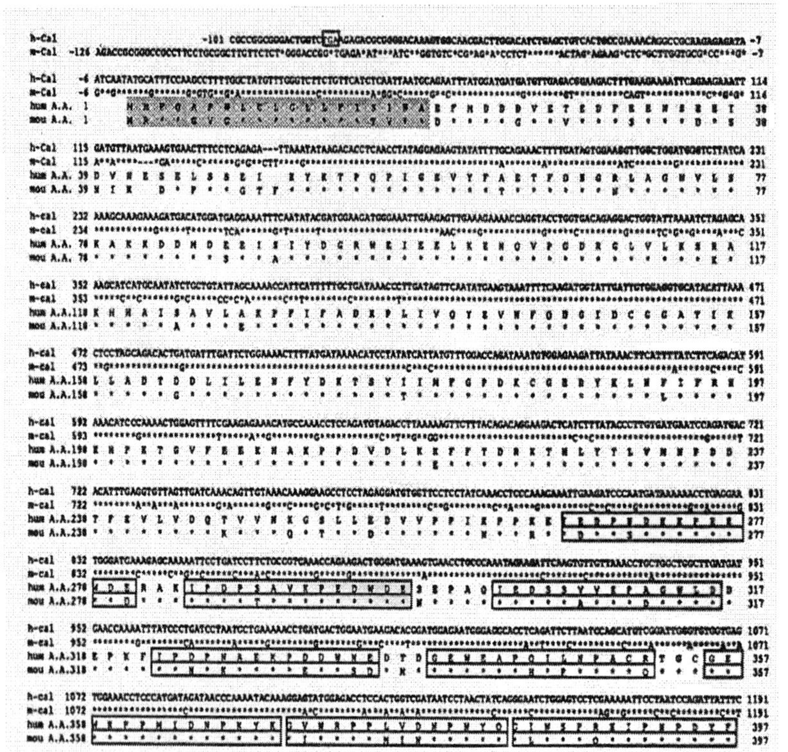

Fig. 31.9. Comparison of nucleotide and deduced amino acid sequence of h-calmegin with mouse calmegin. Amino acid residues 1-19 represent hydrophobic leader sequence. The ER-retention motif (aa 604-610) and transmembrane domain residues (aa 471-492) are shown. Polyadenylation signal is underlined. Reprinted with permission from H. Tanaka et al. Gene 204;159-63:1997 © Elsevier.

involved in cell cycle regulation, transcriptional control, mitochondria and peroximsomal protein transport, neurogenesis, and protein kinase inhibition. Proteins with a characteristic 34 amino acid repeat unit TPR have been discovered in more than 20 genes from various organisms.The TPRs have been identified in proteins involved in regulating transcription and protein transport and in several cell division cycle gene products. The FKBP59 also includes a 3-unit TPR domain through which it binds hsp90. Hence, FKBP59 might intervene in various cellular processes. In situ hybridization experiments revealed an over expression of FKBP59 mRNA in rabbit and rat testes in comparison with other organs. The high level of expression was restricted to germ cells of seminiferous epithelium. The strongest signal was observed in the late pachytene, diplotene, and diakinesis primary spermatocytes and declined in postmeiotic early spermatids. High expression of FKBP59 mRNA did not occur in earlier and later germ cell stages. The possibility that FKBP59 plays some role in the cell division process is strengthened by the fact that several proteins essential in the cell division cycle have a TPR domain (Sananes et al., 1998).

TPIS in Testis: Nucleotide sequence homology search identified a human 75-kDa infertility-related sperm protein, which was named TPIS after the acronym of TPR-containing protein involved in spermatogenesis. The transcript of tpis coding TPR in testis was longer than that observed in other tissues, and the deduced amino acid sequence contained 8 tetratricopeptide repeat (TPR) domains. The TPIS from mouse embryonic skin was deduced to possess a 529 amino acid sequence, containing 5 TPRs (Takaishi and Huh, 1999). TPIS from testis is longer, encoding 372 additional amino acid residues at the 5'-side with 3 more TPRs. In situ hybridization revealed specific expression of tpis at a distinct differentiation stage of spermatogenesis Chromosomal localization of tpis gene showed at 18. 10 cM of chromosome 15.

Three of the 8 TPRs of tpis , present in the extra region, indicate that the testis type TPIS may interact with broader spectrum of proteins. The amino acid sequence of TPIS showed limited homology with a number of proteins with diverse structures and functions, but containing TPR domains, which are conserved relatively well. Thus, the fifth and sixth TPRs of TPIS showed 73% identity with the second and third TPRs of human 34-kDa translocase (hTOM34) of outer mitochondrial membrane. Spermatogonia were negative for TPIS, while stronger signal was seen at intermediate layer of each tubule.In this respect, it may be noteworthy that some types of heat shock proteins, HSP70-2 and HSC70t, are present at specific stage of differentiating spermatogenic cells, and targeted disruption of Hsp70-2 gene leads to male infertility due to blockade of spermatogenesis. Clarification of biological functions of TPIS as well as the nature of its interacting targets may provide a cue to proceed in the direction of control of infertility.

31.7. SPECIALIZED CHAPERONES IN GERM CELLS

Some molecular chaperones are highly specific since they interact with only one, or a very limited number of target proteins. The 47-kDa HSP (HSP47) is an ER-resident chaperone found in collagen-producing cells, where it interacts with pro-collagen. Receptorassociated protein (RAP) is another example of a specialized molecular chaperone for the low-density lipoprotein receptor-related protein (LRP). The major role of RAP is to facilitate correct folding of LRP.

Template Activating Factor1 and Chromatin Decondensation: Remodeling of chromatin structure contributes to various aspects of gene activity, such as transcription, replication, repair and recombination. Factors involved in chromatin remodeling include histone modifying enzymes, multi-subunit ATP dependent chromatin remodeling factors and histone chaperones.

The TAF-1 has histone binding and nucleosome assembly and remodeling activities, and is a novel histone chaperone. The cDNA cloning revealed that TAF-1 is a very acidic factor consisting of a homo- or heterodimer of the 41 kDa TAF-1 and the 39kDa TAF-1ß proteins. Recombinant human and *Xenopus* TAF-1ß decondenses *Xenopus* sperm chromatin. TAF-1ß releases basic proteins from the sperm chromatin through direct interaction, and remains bound to the decondensed sperm chromatin after releasing the basic proteins. Since TAF-1ß preferentially binds to core histones H3/H4 in solution, it was suggested that TAF-1 associates with the chromatin through its interaction with histones H3/H4 (Matsumoto et al., 1999).

Cofactor A (p14): The correct folding of tubulins and the generation of functional tubulin heterodimers requires the participation of a series of molecular chaperones and CCT (or TRiC), the cytosolic chaperonin containing TCP-1. Cofactor A (p14) is a highly conserved protein that forms stable complexes with tubulin. The p14 is more abundantly expressed in testis than in other adult mammalian tissues. It is progressively up-regulated from the onset of meiosis through spermiogenesis, being more abundant in differentiating spermatids. The close correlation between the mRNA expression waves for p14 and testis specific tubulin isotypes 3 and 3/7, together with site of expression suggest that p14 would presumably be associated to tubulin processing rather than meiosis itself. It seems that p14 plays a double role in tubulin folding, enhancing the dimerization of newly synthesized tubulin isotypes as well as capturing excess tubulin monomers. Evidences suggest that p14 is a chaperone required for the actual tubulin folding process in vivo in excess of tubulin in testis, where excessive microtubule remodeling could lead to a disruption of the - balance (Fanarraga et al., 1999)

THEG : An insertional transgenic mouse mutant (termed Kisimo mouse) has been isolated that results in abnormal germ-cell development, showing abnormal elongated spermatids in the lumina of seminiferous tubules. Yanaka et al, cloned the disrupted locus of Kisimo and identified the testis-specific gene, THEG, which is specifically expressed in spermatids and was disrupted in the transgenic mouse. THEG protein strongly interacts with chaperonin containing t-complex polypeptide suggesting that the THEG protein functions as a regulatory factor in protein assembly. Findings indicated that the Kisimo locus is essential for the maintenance of spermiogenesis and that a gene expression disorder in the locus may be involved in male infertility (Yanaka et al., 2000).

31.8. REFERENCES

Baarends WM, Hoogerbrugge JW, Roest HP, Ooms M, Vreeburg J, Hoeijmakers JH, Grootegoed JA. Histone ubiquitination and chromatin remodeling in mouse spermatogenesis. Dev Biol 1999; 15: 322-33.

Baarends WM, Roest HP, Grootegoed JA. The ubiquitin system in gametogenesis. Mol Cell Endocrinol 1999; 151: 5-16.

Baarends WM, van der Laan R, Grootegoed JA. Specific aspects of the ubiquitin system in spermatogenesis. J Endocrinol Invest 2000; 23: 597-604.

Belote JM, Miller M, Smyth KA. Evolutionary conservation of a testis specific proteasome subunit gene in Drosophila. Gene 1998; 215: 93-100.

Berruti G, Martegani E. MSJ-1, a mouse testis-specific DnaJ protein, is highly expressed in haploid male germ cells and interacts with the testis-specific heat shock protein Hsp70-2. Biol Reprod 2001; 65: 488-95.

Berruti G, Perego L, Borgonovo B, Martegani E. MSJ-1, new member of the DNAJ family of proteins, is a male germ cell-specific gene product. Exp Cell Res 1998; 15: 430-41.

Berruti G, Perego L, Martegani E. Molecular cloning and developmental pattern of expression of MSJ-1, a new male germ cell-specific DNAJ homologue. Adv Exp Biol 1998; 444: 145-50.

Biggiogera M, Tanguay RM, Marin R, Wu Y, Martin TE, Fakan S. Localization of heat shock proteins in mouse male germ cells: an immunoelectron microscopical study. Exp. Cell Res 1996; 25: 77-85.

Brown GM, Furlong RA, Sargent CA, Erickson RP, Longepied G, Mitchell M, Jones MH, Hargreave TB,

Cooke HJ, Affara NA. Characterization of the coding sequence and fine mapping of the human DFFRY gene and comparative expression analysis and mapping to the Sxrb interval of the mouse Y chromosome of the Dffry gene. Hum Mol Genet 1998; 7: 97-107.
Brown P, H. Goulding E, D. Strong B, and Eddy EM. HSP70-2 is required for desynapsis of synaptonemal complexes during meiotic prophase in juvenile and adult mouse spermatocytes. cells and interacts with the testis-specific heat shock protein Hsp70-2. Biol Reprod 2001; 65: 488-95.
Chen HY, Sun JM, Zhang Y, Davie JR, Meistrich ML. Ubiquitination of histone H3 in elongating spermatids of rat testes. J Biol Chem 1998; 273: 13165-69.
Chu S, DeRisi J, Eisen M, Mulholland J, Botstein D, Brown PO, Herskowitz I. The transcriptional program of sporulation in budding yeast. Science 1998; 282: 699-05.
Dix DJ, Allen JW, Collins BW, Mori C, Nakamura N, Pooman-Allen P, Goulding EH, Eddy EM. Targeted gene disruption of Hsp70-2 results in failed meiosis, germ cell apoptosis, and male infertility. Proc Natl Acad Sci USA 1996: 99: 3264-8.
Dix DJ, Allen JW, Collins BW, Poorman-Allen P, Mori C, Blizard DR, Brown PR, Goulding EH, Strong BD, Eddy EM. HSP70-2 is required for desynapsis of synaptonemal complexes during meiotic prophase in juvenile and adult mouse spermatocytes. Development 1997; 124: 4595-603.
Eddy EM. Role of heat shock protein HSP70-2 in spermatogenesis. Rev Reprod 1999; 4: 23-30.
Ellis RJ, van der Vies SM. Molecular Chaperons. Ann Rev Biochem 1991; 60: 321-47.
Ewing JF, Maines MD. Distribution of constitutive (HO-2) and heat-inducible (HO-1) heme oxygenase isozymes in rat testis : HO-2 displays stage-specific expression in germ cells. Endocrinology 1995: 136: 2294-302.
Fanaaaga ML, Parraga M, Aloria K, del Mazo J, Avila J, Zabala JC. Regulated expression of p14 (cofactor A) during spermatogenesis. Cell Motil Cytoskeleton 1999; 43: 243-54.
Fink AL. Chaperone-mediated protein folding. Physiol Rev 1999;79:425-49.
Hershko A, Ciechanover A. The ubiquitin system. Annu Rev Biochem 1998; 67: 425-479.
Hunt CR, Dix DJ, Sharma GG, Pandita RK, Gupta A, Funk M, Pandita TK. Genomic instability and enhanced radiosensitivity in Hsp70.1- and Hsp70.3-deficient mice. Mol Cell Biol 2004; 24: 899-911.
Iida H, Doiguchi M, Yamashita H, Sugimachi S, Ichinose J, Mori T, Shibala Y. Spermatid-specific expression of 1bal, an calcium binding adapter molecule-1, in rat testis. Biol Reprod 2001: 64: 1138-46.
Ikawa M, Wada I, Kominami K, Watanabe D, Toshimori K, Nishimune Y, Okabe M. The putative chaperone calmegin is required for sperm fertility. Nature 1997; 387: 607-11.
Ishihara K, Yasuda K, Hatayama T. Phosphorylation of the 105-kDa heat shock proteins, HSP 105 and HSP105 by casein kinase II. Biochem Biophys Res Comm 2000; 270: 927-31.
Itoh H, Tashima Y. Different expression time of the 105-kDa protein and 90-kDa heat-shock protein in rat testis. FEBS Lett 1991; 2: 110-2.
Jensen DE, Proctor M, Marquis ST, Gardner HP, Ha SI, Chodosh LA, Ishov AM, Tommerup N, Vissing H, Sekido Y, et al. A novel ubiquitin hydrolase which binds to the BRCA1 RING finger and enchances BRCA1 mediated cell growth suppression. Oncogene 1998; 16: 1097-12.
Jones MH, Furlong RA, Burkin H, Chalmers IJ, Brown GM, Khwaja, et al, The Drosophila developmental gene fat facets has a human homologue in XpII.4 which escapes X-inactivation and has related sequences on Yq 11.2. Hum Mol Genet 1996; 5: 1695-01.
Kajimoto Y, Hashimoto T, Shirai Y, Nishino N, Kuno T, Tanaka. cDNA cloning and tissue distribution of a rat ubiquitin carboxyl-terminal hydrolase Pgp9.5. J Biochem 1992; 112: 28-32.
Kaneko Y, Kimura T, Nishiyama H, Noda Y, and Fujita J. Developmentally regulated expression of APG-1, a member of heat shock protein 110 family in murine male germ cells. Biochem Biophy Res Comm 1997; 233: 113-116.
Kaneko Y, Nishiyama H, Nonoguchi K, Higashitsuji H, Kishishita M and Fujita J. A hsp110- related gene, apg-1, that is abundantly expressed in the testis responds to a low temperature heat shock rather than the traditional elevated temperatures. J Biol Chem 1997;272:2640-45.
Kappe G, Verschuure P, Phillipsen RL, Staldunen AA, Van de Boogaart P, Boelens WC, De Jong WW. Characterization of two novel human small heat shock proteins : protein kinase related HspB8 and testis-specific HspB9. Biochim Biophys Acta 2001; 30: 1520: 1-6.
Kirschner M. Intracellular protelysis. TIBS 1999; 24: M42-M46.
Koken MHM, Hoogerbrugge JW, Jqaspers-Dekker I, de Wit J, Willemsen R, Roest HP, Grootegoed JA, Hoeijmakers JHJ. Expression of the ubiquitin-conjugating DNA repair enzymes HHR6A and B suggests a role in spermatogenesis and chromatin modification Dev Biol. 1996; 173: 119-132.
Kon Y, Endoh D, Iwanaga T. Expression of protein gene product 9.5, a neuronal ubiquitin C-terminal hydrolase, and its developing change in Sertoli Cells of Mouse Testis. Mol. Reprod Dev 1999;54:333-41.
Kovalenko OV, Plug AW, Haaf T, Gonda DK, Ashley T, Ward DC, Radding CM, Golub EI. Mammalian ubiquitin-conjugating enzyme Ubc9 interacts with RadS 1 recombination protein and localizes in

synaptonemal complexes. Proc Natl Acad Sci USA 1996; 93: 2958-63.
Krawczyk Z, Mali P, Parvinen M. Expression of a testis-specific hsp70 gene-related RNA in defined stages of rat seminiferous epithelium. J Cell Biol 1988: 107: 1317-23.
Krawczyk Z, Schmid W, Harkonen P, Wolnirzek P. The ERE-like sequence from the promoter region of the testis specific hsp70 related gene is not estrogen responsive. Cell Biol Int Rep 1992: 16: 937-48.
Kwon YT, Reiss Y, Fried VA, Hershko A, Yoon JK, Gonda DK, Sangan P, Copeland NG, Jenkins NA, Varshavsky A. The mouse and human genes encoding the recognition component of the N-end rule pathway. Proc Natl. Acad Sci USA 1998; 95: 7898-03.
Legare C, Thabet M, Sullivan R. Expression of heat shock protein 70 in normal and cryptorchid human excurrent duct. Mol Hum Reprod 2004; 10: 197-202.
Lehman AL, Nakatsu Y, Ching A, Bronson RT, Oakey RJ, Keiper-Hrynko N, Finger JN et al,. A very large protein with diverse functional motifs is deficient in rjs (runty, jerky, sterile) mice. Proc Natl Acad Sci USA 1998; 95: 9436-41.
Lin H, Keriel A, Morales CR, Bedard N, Zhao Q, Hingamp P, Lefrancois S, Combaret L, Wing SS. Divergent N-terminal sequences target an inducible testis deubiquitinating enzyme to distinct subcellular structures. Mol Cell Biol 2000; 20: 6568-78.
Matsumori M, Itoh H, Toyoshima I, Komatsuda A, Sawada K, Fukuda J, Tanaka T, et al. Characterization of the 105-kDa molecular chaperone. Identification, biochemical properties, and localization. Eur J Biochem 2002; 269: 5632-41.
Matsumoto M, Fujimoto H. Cloning of a hsp70-related gene expressed in mouse spermatids. Biochem Biophys Res Commun 1990; 166; 43-7.
Mazeyrat S, Saut N, Sargent CA, Grimmond S, Longepied G, Ehrmann IE et al. The mouse Y chromosome interval necessary for spermatogonial proliferation is gene dense with syntenic homology to the human AZFa region. Hum Mol Genet 1998; 7: 1713-24.
Meinhardt A, Parvinen M, Bacher M, Aumuller G, Hakovirta H, Yagi A, Seitz J. Expression of mitochondrial heat shock protein 60 in distinct cell types and defined stages of rat seminiferous epithelium. Biol Reprod 1995; 52: 798-07.
Meinhardt A, Wilhelm B, Seitz J. Expression of mitochondrial marker proteins during spermatogenesis. Hum Reprod Update 1999, 5: 108-119.
Mezquita B, Mezquita C, Mezquita J. Marked differences between avian and mammalian testicular cells in the heat shock induction and polyadenylation of Hsp70 and ubiquitin transcripts. FEBS Let 1998; 436: 382-86.
Mezquita J, Pau M, Mezquita C. Characterization and expression of two chicken cDNAs encoding ubiquitin fused to ribosomal proteins of 52 and 80 amino acids. Gene 1997; 195: 313-19.
Michalak M, Corebett EF, Mesaeli N, Nakamura K, Opas M. Calreticulin one protein, one gene, many functions. Biochem J 1999; 344: 281-92.
Mizzen L. Immune responses to stress proteins. Applications to infectious diseae and cancer. Bioessays 1998; 10: 171-89.
Mochinda K, Tres LL, Kierszenbaum AL. Structural features of the 26S proteasome complex isolated from rat testis and sperm tail. Mol Reprod Dev 2000; 57: 176-84.
Nakamura M, Moriya M, Baba T, Michikawa Y, Yamanobe T, Arai K, Okinaga S, Kbayashi T. An endoplasmic reticulum protein, calreticulin, is transported into the acrosome of rat sperm. Exp Cell Res. 1993: 205: 101-10.
Neuer A, Spandorfer SD, Giraldo P, Dieterle S, Rosenwaks Z, Witkin SS. The role of heat shock proteins in reproduction. Human Reprod Update 2000; 6: 149-159.
Ohsako S, Hayashi Y, Bunick D. Molecular cloning and sequencing of calnexin-t. An abundant male germ cell-specific calcium binding protein of the endoplasmic retriculum. J Biol Chem 1994: 13: 14140-8.
Ohsako S, Janulis L, Hayashi Y, Bunick D. Characterization of domains in mice of calnexin-t, putative molecular chaperone required in sperm fertility, with use of glutathione S-transferase-fusion proteins. Biol Reprod 1998: 59: 1214-23.
Oughtred R, Bedard N, Adegoke OA, Morales CR, Trasler J, Rajapurohitam V, Wing SS. Characterization of rat100, a 300-kilodalton ubiquitin-protein ligase induced in germ cells of the rat testis and similar to the Drosophila hyperplastic discs gene. Endocrinology 2002; 143: 3740-7.
Patton EE, Willems AR, Tyres M. Combinatorial control in ubiquitin dependent proteolysis: don't Skp the F-box hypothesis. Trends Genet 1998; 14: 236-43.
Perry WL, Hustad CM, Swing DA, TN OS, Jenkins NA, Copeland NG. The itchy locus encodes a novel ubiquitin protein ligase that is disrupted in A18H mice. Nat Genet 1998; 18: 143-46.
Pusch W, Jahner D, Ivell R. Molecular cloning and testicular expression of the gene transcripts encoding the murine mutliubiquitin-chain-binding protein (Mcb1). Gene 1998; 207: 19-24.
Quesada V, Diaz-Perales A, Gutierrez-Fernandez A, Garabaya C, Cal S, Lopez-Otin C. Cloning and enzymatic analysis of 22 novel human ubiquitin-specific proteases. Eur J Biochem 2004; 271: 339-48.

Raab LS, Polakoski KL, Hancock LW, Hamilton DW. Characterization of the heat shock protein P70 in rat spermatogenic cells. Mol Reprod Dev 1995: 40: 186-95.
Rajapurohitam V, Morales CR, El-Alfy M, Lefrancois S, Bedard N, Wing SS. Activation of a UBC-4 dependent pathway of ubiquitin conjugation during postnatal development of the rat testis. Dev Biol 1999;12:217-28.
Rivkin E, Cullinan EB, Tres LL, Kierszenbaum AL. A protein associated with the manchette during rat spermiogenesis is encoded by a gene of the TBP-1 like subfamily with highly conserved ATPase and protease domains. Mol Reprod Dev 1997; 48: 77-89.
Roest HP, Klaveren van J, Wit de J, Gurp van CG, Koken MHM, Vermey M et al,. Inactivation of the HR6B ubiquitin-conjugating DNA repair enzyme in mice causes a defect in spermatogenesis associated with chromatin modification. Cell 1996; 86: 799-810.
Sananes N, Baulieu EE, and Le Goascogne C. Stage-specific expression of the immunophilin FKBP59 messenger ribonucleic acid and protein during differentiation of male germ cells in rabbits and rats. Biol Reprod 1998: 58: 353-60.
Santti H, Mikkonen L, Hirvonen-Santti S, Toppari J, Janne OA, Palvimo JJ. Identification of a short PIASx gene promoter that directs male germ cell-specific transcription in vivo. Biochem Biophys Res Commun 2003; 308: 139-47.
Scieglinska D, Vydra N, Krawczyk Z, Widlak W. Location of promoter elements necessary and sufficient to direct testis-specific expression of the Hst70/Hsp70.2 gene. Biochem J 2004; Pt.
Shen Z, Pardington-Purtymun PE, Comeaux JC, Moyzis RK, Chen DJ. UBL1, a human ubiquitin like protein associating with human RAD51/RAD52 proteins. Genomics 1996;36, 271-79.
Son WY, Han CT, Hwang SH, Lee JH, Kim S, Kim YC. Repression of hspA2 messenger RNA in human testes with abnormal spermatogenesis. Fertil Steril 2000; 73: 1138-44.
Son WY, Hwang SH, Han CT, Lee JH, Kim S, Kim YC. Specific expression of heat shock protein HspA2 in human male germ cells. Mol Hum Reprod 1999; 5: 1122-6.
Swerdlow PS, Schuster T, Finley D. A conserved sequence in histone H2A which is a ubiquitination site in higher eukaryotes is not required for growth in Saccharomyces cerevisiae. Mol Cell Biol 1990;10:4905-11.
Takaishi M, Huh NH. A tetratricopeptide repeat containing protein gene, tpis, whose expression is induced with differentiation of spermatogenic cells. Biochem Biophys Res Commun 1999; 264: 81-5.
Tanaka H, Ikawa M, Tsuchida J, Nozaki M, Suzuki M, Fujiwara T, Okabe M, Nishimune Y. Cloning and characterization of the human calmegin gene encoding putative testis-specific chaperone. Gene 1997; 204: 159-63.
Tanaka K. Proteasomes : Structures and Biology. J Biochem 1998; 123: 195-04.
Tanaka T, Yeh ET, Kamitani T. NUB1-mediated targeting of the ubiquitin precursor UbC1 for its C-terminal hydrolysis. Eur J Biochem 2004; 271: 972-82.
Tarsounas M, Pearlman RE, Gasser PJ, Park MS, Moens PB. Protein-protein interactions in the synaptonemal complex. Mol Biol Cell 1997; 8: 1405-14.
Timms KM, Ansari-Lari MA, Morris W, Brown SN, Gibbs RA. The genomic organization of Isopeptidase T-3 (ISOT-3), a new member of the ubiquitin specific protease family (UBP). Gene 1998; 217: 101-6.
Tissiere A, Mitchell HK, Tracy U. Protein synthesis in salivary glands of Drosophila melanogaster: relation to chromosomal puffs. J Mol Biol 1974; 84: 389-98.
Townsley FM, Ruderman JV. Proteolytic ratchets that control progression through mitosis. Trends Cell Biol. 1998; 8: 238-44.
Wakayama T, Iseki S. Specific expression of the mRNA for 25 kDA heat shock protein in the spermatocytes of mouse seminiferous tubules. Anat Embyol 1999; 199: 419-25.
Watanabe D, Okabe M, Hamajima N, Morita T, Nishina Y, Nishimune Y. Characterization of the testis-specific gene ' calmegin' promoter sequence and its activity defined by transgenic mouse experiements. FEBS Lett 1995: 24: 509-12.
Watanabe D, Yamada K, Nishina Y, Tajima Y, Koshimizu U, Nagata A, Nishimune Y. Molecular cloning of a novel Ca2+ binding protein (calmegin) specifically expressed during male meiotic germ cell development. J Biol Chem 1994; 11: 7744-9.
Watson DC, Levy WB, Dixon GH. Free ubiquitin is a non-histone protein of trout testis chromatin. Nature 1978; 276: 196-98.
Westwood JT, Clos J, Wu C. Stressed induced oligomerization and chromosomal relocalization of heat shock factor. Nature 1991; 353: 822-27.
Widlak W, Markkula M, Krawczyk Z, Kananen K, Huhtaniemi I. A252 bp upstream region of the rat spermatocyte specific hst70 gene is sufficient to promote expression of the hst70-CAT hybrid gene in testis and brain of transgenic mice. Biochim Biophys Acta 1995: 1264: 191-200.
Wilkinson KD. Regulation of ubiquitin-dependent processes by deubiquitinating enzymes. FASEB J. 1997; 11: 1245-56.

Wing SS, Bedard N, Morales C, Hingamp P, Trasler J. A novel rat homolog of the Saccharomyces crevisiae ubiquitin conjugating enzymes UBC4 en UBC5 with distinct biochemical features is induced during spermatogenesis. Mol Cell Biol 1996; 16: 4064-72.
Wing SS, Jain P. Molecular cloning expression and characterization of a ubiquitin conjugation enzyme (E2(17)kB) highly expressed in rat testis. Biochem J 1995: 305: 125-32.
Wisniewski J, Malezewski M, Krawczyk Z, Gedamu L. An upstream region of the rat spermatogenesis-specific heat-shock-like Hst70 gene confers testis specific expression in transgenic mice. Eur J Biochem 1993; 15: 137-43.
Yamagata K, Nakanishi T, Ikawa M, Yamaguchi R, Moss SB, Okabe M. Sperm from the calmegin-deficient mouse have normal abilities for binding and fusion to the egg plasma membrane. Dev Biol. 2002; 250: 348-57.
Yamaguchi T, Kim NS, Sekine S, Seino H, Osaka F, Yamao F, Kato S. Cloning and expression of cDNA encoding a human ubiquitin-conjugating enzyme similar to the Drosophila bendless gene product. J Biochem 1996; 120; 494-97.
Yan W, Santti H, Janne OA, Palvimo JJ, Toppari J. Expression of the E3 SUMO-1 ligases PIASx and PIAS1 during spermatogenesis in the rat. Gene Expr Patterns 2003; 3: 301-8.
Yanaka N, Kobayashi K, Wakimoto K, yamada E, Imahie H, Imai Y, Mori C. Insertional mutation of the murine kisimo locus casued a defect is spermatogenesis. J Biol Chem 2000; 275: 14791-4.
Yasugi T, Howley PM. identification of the structural and functional human homolog of the yeast ubiquitin conjugating enzyme UBC9. Nucleic Acids Res 1996; 24: 2005-2010.
Yoshinaga K, Tanii I, Toshimori K. Molecular chaperone calmegin localization to the endoplasmic reticulum of meiotic and post-meiotic germ cells in the mouse testis. Arch Histol Cytol 1999; 62: 283-93.
Yu SS, Takenaka O. Molecular cloning, structure, and testis-specific expression of MFSJ1, a member of the DNAJ protein family, in the Japanese monkey (Macaca fuscata). Biochem Biophys Res Commun. 2003; 301: 443-9.
Yuan X, Miller M, Belote JM. Duplicated proteasome subunit genes in Drosophila melanogaster encoding testis-specific isoforms. Genetics 1996; 144: 147-57.
Zakeri ZF, Welch WJ, Wolgemuth DJ. Characterization and inducibility of hsp 70 proteins in the male germ line. J Cell Biol 1990; 111: 1785-92.
Zakeri ZF, Wolgemuth DJ, Hunt CR. Identification and sequence analysis of a new member of the mouse HSP70 gene family and characterization of its unique cellular and developmental pattern of expression in the male germ line. Mol Cell Biol 1988; 8: 2925-32.
Zakeri ZF, Wolgemuth DJ. Developmental-stage-specific expression of the hsp70 gene family during differentiation of the mammalian male germ line. Mol Cell Biol 1987: 7: 1791-6.
Zhu D, Dix DJ, Eddy EM. HSP70-2 is required for CDC2 kinase activity in meiosis I of mouse spermatocytes. Development 1997; 124: 3007-14.

Chapter 32

CANCER ASSOCIATED TESTIS ANTIGENS

32.1. TUMOR SUPPRESSOR ANTIGENS IN SPERMATOGENIC CELLS

32.1.1. P53: Regulation of Spermatogenesis and Tumorigenesis

p53 and Cell Cycle Control: The p53 is a tumor suppressor that protects the cell from a variety of genotoxic and environmental stresses. Induction of double strand breaks in DNA by ionizing radiation or genotoxic chemotherapy leads to the rapid stabilization of p53 through phosphorylation of its N-terminus. This stabilization results either in cell cycle arrest via transcriptional induction of the p21 CDK inhibitor or apoptosis (Chapter 5 and 11). The p53 is one of the most commonly mutated genes in human cancer, being inactivated in almost half of all tumors. The p53 has been dubbed as the guardian of genome and hence p53-mutant cells are much more prone to accumulate amplified DNA copies. Such increased mutability accelerates the genetic changes that seem to be the rate-limiting steps in tumor progression. The p53 has multifaceted activities. One of its important effectors is p21. Both being coordinately activated, there is a chance they have overlapping pathways that signal senescence and DNA damage. Most work on p53 has centered on its control of the G1 phase of the cell cycle that falls between the end of M phase and the start of S phase. But the coordination of the S and M phases of the cycle may also fall within the purview of p53. Since cell-cycle receives and responds a wide variety of growth–controlling signals through its growth cycle, by differentiating or by entering into a quiescent growth state, G_0, it seems that the cell cycle has to have a quality control. In most of the vertebrate organisms this is achieved through p53, which shuts off the cell cycle until damage in corrected. Under alarming situations, p53 with other proteins may initiate a more drastic response resulting into apoptosis. Otherwise p53, the watchman induces the synthesis of p21 and other proteins capable of shutting down the CDKs activities and the cell cycle. Unfortunately in absence or inactivation of p53 mechanisms as in human tumors, the cell cycle is in danger and loses its control. In p53 mutant animals, cells continue to proliferate due to failure of guarding activity of p53. Mutations in p53 gene have long been associated with the process of immortalization, in which cells escape apoptosis and go on to create tumors(**Fig.32.1**). Further studies indicated that the mortality of normal cell populations is triggered by collapse of the cell's telomeres (c/r Jacks and Weinberg, 1996).

p53 and Telomere Function: Telomeres are nucleoprotein structures that serve a critical function in protecting the ends of linear chromosomes. Maintenance of telomere length and function requires telomerase, a specialized reverse transcriptase, as well as a complex of telomere associated proteins (see Chapter 12). In order to explain the telomere loss, Chin et al (1999) reviewed a possible role of p53 deficiency in rescuing adverse effects of teleomere loss. Strikingly, as telomerase deficient (mTR-/-) mice grow with age and generation, an increase in cancer incidence is observed presumably due to increased genetic instability caused by telomere shortening, repeated chromosomal fusion bridge-breakage cycles, and chromosomal loss (Ruldolph et al., 1999). The contrasting effect of telomere loss on cellular transformation phenotypes in cells in DNA damaged pathway suggested that the p53 pathway may alter the cellular response to telomere loss in both normal and neoplastic cells. Chin et al (1999) examined the functional interactions between telomere dysfunction and p53 in cells and organs of late-generation mTR-/- mice. Telomere loss in late-generation mTR-/- testis culminates in sterility due to an apoptotic elimination of germ cells. In mTR-/- testis, p53 status did not affect testicular volume or significantly alter the cytoarchitecuture of seminiferous tubules. But a decrease in testis size correlating depletion of germ cells was evident in G5 mTR-/- mice. As telomere function worsens in later generations of mTR-/- deficiency, germ cell death by p53-independent mechanisms became evident. Whereas all G5 mTR-/- P53-/- were histologically normal, 40% of testes showed varying degree of pathology. P53 independent cell death was more consistent in G7 and G8.

Coincident with severe telomere shortening and associated genomic instability, p53 is activated, leading to growth arrest and/or apoptosis. Deletion of p53 significantly attenuated the adverse cellular and organ effects of telomere dysfunction. Both the apoptotic response and cell cycle arrest induced by telomere dysfunction are mediated by p53. Thus telomere dysfunction acts synergistically with p53 deficiency to initiate malignant transformation, presumably by enhancing genetic instability. Studies established a key role for p53 in the cellular response to telomere dysfunction in both normal and neoplastic cells (Chin et al., 1999).

p53 and Spermatogenesis: Spermatogenesis offers a convenient system to explore the possibility that p53 plays a role in cell differentiation in vivo. The number of cells entering meiosis is held constant, since excess of spermatogonia or premeiotic spermatocytes are eliminated by apoptosis. Mice with reduced levels of p53 protein exhibit the testicular giant-cell degenerative syndrome. The giant-cell degenerative syndrome was also observed in some genetic strains of homozygous p53 null mice (Rotter et al., 1993). The p53 is expressed during meiosis in normal rat spermatogenesis and its expression is localized to the preleptotene-early pachytene spermatocytes. The most prominent expression is in zygotene. *In situ* hybridization experiments indicated that the p53 gene is expressed in tetraploid primary spermatocytes at the meiotic pachytene stage of the first round of spermatogenesis in young mice. Thus p53 confined to the tetraploid primary spermatocytes of the meiotic pachytene phase was considered a potential member of the meiosis control protein family (Schwartz et al., 1993). Western blot analysis showed that p53 is enriched in the nuclear envelope of germ cells. Nuclear envelope is the appropriate site for rapid entry of p53 into nucleus and binding to DNA for apoptosis and cell cycle arrest in response to an appropriate stimulus. Exposure of the testis to heat stress induced translocation of p53 into the nucleus. The nuclear envelope sequestration of p53 also provides a framework to understand how mitosis and meiosis in the testis may proceed despite high intracellular concentration of p53 (Yin et al., 1997). After irradiation, p53 levels increased in a time and a dose-dependent manner. This increase occurs in the same cells that normally express elevated levels of p53. This supports the view that p53 is involved in meiosis of the male rat and p53 has a role in recombination processes and/or formation of the synaptonemal complex (Sjoblom and Lahdetie 1996). Normally, p53 is not detected in spermatogonia but after irradiation at 4Gγ, p53 can be found in these cells, with a possible function since irradiation induces apoptosis of differentiating spermatogonia **(Fig.32.1).** Beumer et al (1997) showed

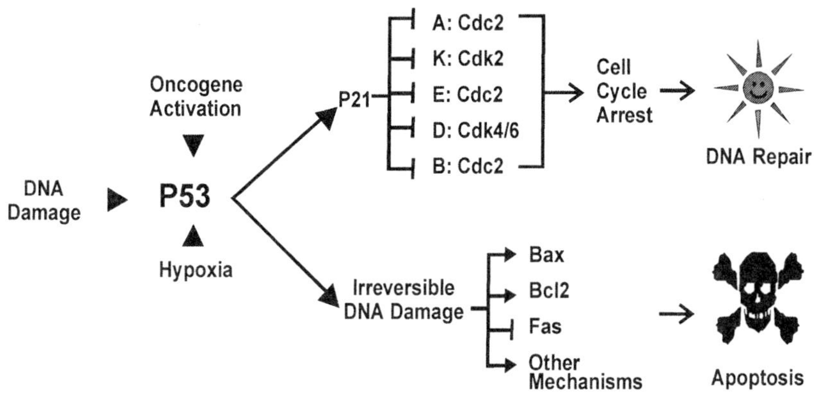

Fig. 32.1. Mechanism of action of P53 in DNA repair and Apoptosis. Arrows (→ or ▶) indicate activation of various steps, whereas ⊣ indicates inhibiting steps.

that p21 is not expressed in spermatogonia before or after irradiation but describes time course of spermatogonial apoptosis after a 4Gy dose of X-rays. Hence role of p53 in DNA damage induced spermatogonial apoptosis is not affected by inducing the expression of its target gene *p21*. Though loss of expression of p21 by p53 associated induction was confirmed by West and Lahdetie (1997), these authors demonstrated expression of p21 mRNA in pachytene spermatocytes and round spermatids in a dose dependent manner. On the contrary, Stephan et al., (1996) showed strong reactivity of p53 in the nuclei of a number of spermatogonia, some pre-meiotic spermatocytes and probably in all spermatids and appeared to run parallel to that of DNase I. In contrast, p53 was absent in mature spermatozoa in testicular tubules or ductus epididymidis suggesting that p53 is either degraded or retained in residual bodies during later stage of spermatid maturation. In p53 deficient mice, however testis contains more spermatogonia than in wild type. This suggests that p53 in testis is involved in cell cycle regulation of undifferentiated spermatogonia and that p53 induces apoptosis in these cells (Stephan et al., 1996).

Wip1 Gene and p53: The *Wip1* gene of a serine/threonine phosphatase is induced in a p53-dependent manner by DNA-damaging agents. *Wip1* message is expressed at very high levels in the testes, particularly in the post-meiotic round spermatids. Though, *Wip1* null mice are viable but show a variety of postnatal abnormalities, including variable male runting, male reproductive organ atrophy, reduced male fertility, and reduced male longevity besides diminished T and B-cell function. Fibroblasts derived from *Wip1* null embryos have decreased proliferation rates and appear to be compromised in entering mitosis (Choi et al., 2002).

Feedback between p53 and TR2: The testicular orphan receptor-2 (TR2) expression, which modulates many signal pathways, was completely repressed in the surgery induced cryptorchidism of the rhesus monkey. Further studies on linked TR2 repression to the induction of p53 suggested that induced p53 could repress TR2 expression via the p53—>p21—>CDK—>Rb—>E2F signal pathway. In return, TR2 could also control the expression of p53 and retinoblastoma supressor (Rb) through the regulation of human papilloma virus 16E6/E7 genes.

This suggests a feedback control mechanism between TR2 and p53/Rb tumor suppressors, which might play important role in male infertility associated with cyrptorchidism (Xm et al., 2000).

32.1.2. Retinoblastoma Family of Proteins

The retinoblastoma tumor suppressor (Rb) protein promotes G1/S transition in cell cycle. Retinoblastoma family proteins pRb, p107 and p130 are differentially expressed in the rat testis. They function in specific cell types during testicular development and spermatogenesis. Their expression levels and phosphorylation status are modulated during germ cell cycle progression and apoptosis. Hyperphosphorylated states and elevated levels of p107 are correlated with cell cycle progression, whereas hypophosphorylated states and reduced levels are associated with suppression of proliferation and apoptosis in germ cells. These proteins may also serve as markers of cell cycle status of germ cells during spermatogenesis (Toppari et al., 2003).

The p19, a member of the INK4 family of cyclin-dependent kinase inhibitors, negatively regulates the proto-oncogenic cyclin D/CDK4/6 complexes whose ability to phosphorylate the Rb promotes G1/S transition. As the Rb pathway is commonly targeted in cancer, the mouse model suggests a role for p19 in spermatogenesis. The p19 protein is abundant in spermatocytes of normal human adult testes, whereas virtually no p19 was detectable in testicular cancer. Together with the deficiency of p19 in human foetal germ cells, these results supported the concept of foetal origin of the testicular germ cell tumors, and help to better understand the emerging role of the Rb pathway in spermatogenesis and tumorigenesis in the human testis (Bartkova et al., 2000) (see Chapter 11).

32.1.3. C9orf11

The chromosome 9p21 region is frequently deleted in several neoplasias. The cyclin dependent kinase inhibitor 2A (CdkN2A or p16) gene is present in this region and identified as a tumour suppressor gene. Much evidence indicates the existence of another tumor suppressor gene, located proximal to the CdkN2A gene, and involved in cutaneous malignant melanoma (CMM) initiation. A gene within it *(C9orf11)* shares no similarities to any known gene or predicted protein representing a novel human gene. Nevertheless, a putative leucine zipper pattern is located at the C-terminal end of the predicted protein, suggesting that it could dimerise. The C9orf11 encodes for a protein of 294 amino acids with a predicted molecular mass of 32.8-kDa. The *C9orf11* is organized in eight exons that encompass a region of approx. 13-kb. Expression analysis demonstrated that C9orf11 is highly expressed in testis, although minor expression was also seen in other tissues. Mutations in the *C9orf11* gene were not detected in CMM families that were negative for CdkN2A mutations. Two SNPs for the *C9orf11* gene have been identified, which could be used in segregation or association studies for other disorders (Ruiz et al., 2000).

32.1.4. hH-Rev107-3 cDNA

The *hH-Rev107-3* cDNA identical to *hH-Rev107-1* cDNA is a class II tumor suppressor gene. Two transcripts (1 and 0.8-kb) were detected in all human tissues, except in thymus. The strongest expression was found in testis, skeletal muscle and heart. These two mRNAs are probably transcribed from only one gene that was mapped to the q12-q13 region of the chromosome 11. In human testis, *hH-Rev107* gene expression was localized within the round spermatids. A strong *hH-Rev107* gene expression was observed in normal testis as well as in samples with preinvasive carcinoma in situ but was completely absent in overt tumors, both seminomas and non-seminomas. Assessment of *hH-Rev107* mRNA level in testicular germ cell tumors exhibited a ninefold decrease in the gene expression. Thus down regulation of hH-Rev107 may be associated with invasive progression of testicular germ cell tumors (Siegrist et al., 2001).

32.1.5. Testisin –A Serine Protease

The recently characterized human serine protease, Testisin, is expressed on pre-meiotic testicular germ cells and is a candidate type II tumor suppressor for testicular cancer. Mouse Testisin, *Prss21* gene comprises six exons and five introns and spans approximately 5-kb of genomic DNA with an almost identical structure to the human Testisin gene, *PRSS21*. The gene was localized to murine chromosome 17 A3.3-B, a region syntenic with the location of *PRSS21* on human chromosome 16p13.3.3. A 1.3-kb mRNA transcript corresponding to testisin is present only in testis. The human and murine Testisin cDNA shares 65% identity and encodes a putative pre-pro-protein of 324 amino acids with 80% similarity. The predicted amino-acid sequence includes an N-terminal signal catalytic domain typical of a serine protease with trypsin like specificity, and a C-terminal hydrophobic extension, which is predicted to function as a membrane anchor. It is present in the cytoplasm and on the plasma membrane of round and elongating spermatids of mouse testis. The onset of murine Testisin mRNA expression occurs corresponding to the appearance of spermatids in contrast to the expression of human Testisin in spermatocytes (Scarman et al., 2001).

32.1.6. Tumor Suppressor Gene: *PTEN*

The tumor suppressor gene *PTEN*, which is frequently mutated in human cancers, encodes a lipid phosphatase for phosphatidylinositol 3, 4 , 5-triphosphate [PtdIns (3,4,5)P3] and antagonizes phosphatidylinositol 3 kinase. Primordial germ cells (PGCs), which are the embryonic precursors of gametes, are the source of testicular teratoma. Male mice that lacked PTEN exhibited bilateral testicular teratoma, which resulted from impaired mitotic arrest and outgrowth of cells with immature characters. The PTEN appears to be essential for germ cell differentiation and an important factor in testicular germ cell tumor formation (Kimura et al., 2003).

32.2. CANCER-TESTIS ANTIGENS

Cancer/testis antigens (CTA) are tumor-associated antigens expressed during oncogenesis, in a number of solid tumors but not in normal tissues except testis. Most of these CTA are highly immunogenic, eliciting a humoral and cellular immune response in the patients with advanced cancer, and are useful for tumor-specific immunotherapy. During the last decade, many of these immunotherapy candidate genes have been discovered using various approaches. Zendman et al., (2003) have reviewed outlines of the CT antigen family and focussed on the expression and putative function during male germ cell development and melanocytic tumor progression. Most of these genes are localized on the X-chromosome, often as multigene families. Methylation status seems to be the main, but not the only regulator of their specific expression pattern. In testis, CTAs are exclusively present in cells of the germ cell lineage, though there is a lot of variation in the moment of expression during different stages of sperm development.

Likewise, there is also a lot of heterogeneity in the expression of CTAs in melanoma samples. Clues regarding functions for many of these proteins point to a role in cell cycle regulation or transcriptional control. Better insights in the function of these genes may shed light on the link between spermatogenesis and tumor growth and could be of use in anti-tumor therapies. Evidences regarding functions of CTAs point to a role in cell cycle regulation or transcriptional control (Zendman et al., 2003).

Location of Cancer-Testis Antigen Genes

Brun et al., (2003) analysed the genomic structure and transcriptional activity of a 2.3-Mb genomic sequence in the juxtacentromeric region of human chromosome 21. This region comprises two different chromosome domains, 1.5-Mb and 0.8-Mb. The 1.5-Mb proximal domain, besides other characteristics harbors two genes (TPTE and BAGE2) that belong to gene families, which have a cancer and/or testis expression profile. The TPTE gene family was generated by intra- and interchromosome duplications of the ancestral TPTE gene mapping to phylogenetic chromosome XIII. By contrast, the 0.8-Mb distal domain has a chromosome 21-specific sequence and harbors six genes including housekeeping genes. These structural and transcriptional features suggested that the proximal domain has heterochromatic properties, whereas the distal domain has euchromatic properties (Brun et al., 2003).

Hypomethylation of C/T Antigens Genes in Tumor Progression

Studies on molecular mechanisms for the expression of cancer/testis antigens have been few (Cho et al., 2003). Most of the genes of C/T antigens are localized on the X-chromosome, often as multigene families. Methylation status seems to be the main, but not the only regulator of their specific expression pattern. In testis, CTAs are exclusively present in cells of the germ cell lineage, though there is a lot of variation in the moment of expression during different stages of sperm development. Likewise, there is also a lot of heterogeneity in the expression of CTAs in melanoma samples. Cancer/testis antigen gene, $CAGE_4$ expresses in various histological types of tumors, but not in normal tissues, with the exception of the testis. In some cancer cell lines *CAGE* did not express. But the expression of *CAGE* could be restored in these cell lines by treatment with 5'-aza-deoxycytidine, suggesting that the expression of CAGE is mainly suppressed by hypermethylation. The methylation of the CpG sites inhibited the binding of transcription factors. It has been suggested that the methylation status of the CpG sites of *CAGE* determines its expression. The hypomethylation of *CAGE* precedes the development of gastric cancer and hepatocellular carcinoma, and the high frequencies of hypomethylation of *CAGE*, in various cancers can be a cancer diagnostic marker (Cho et al., 2003) (see Chapter 14) **(Fig.32.2)**. MAGE-encoding genes are expressed in various tumour types via demethylation of their promoter CpG islands, which are silent in all non-neoplastic tissues except for the testis and placenta. Demethylation of MAGE-A1 and -A3 occurs during progressive stages of gastric cancer, and may be associated with aggressive biological behavior of gastric cancer also (Honda et al., 2004). The demethylating agent, 5' Aza-deoxycytocytidine (DAC) induces MAGE-1 and other antigens in normal and malignant lymphoid cells and also in myeloid monocyte cell lines. Demethylation alone is a sufficient stimulus to induce MAGE tumor rejection antigens in both normal and malignant lymphoid cells in most cases.

32.2.1. Breast Cancer Antigens (BRCA-1 and BRCA-2)

The *Brca1* and *Brca2* breast cancer susceptibility genes encode proteins, the normal cellular functions of which are complex and multiple, and germ line mutation in individuals predispose both to breast and to ovarian cancer. The BRCA1 is a nuclear phosphoprotein and expressed in a broad spectrum of tissues during cell division. The 17q-linked breast and ovarian cancer susceptibility gene *(Brca1)* is a tumor suppressor gene. The inheritance of a mutant BRCA1 allele dramatically increases a woman's lifetime risk for developing both breast and ovarian cancers. Measurements of BRCA1 mRNA levels in highly purified somatic cells of the testis and in staged germ cells showed that the high level BRCA1 mRNA expression is limited to the germ cells. Rat *Brca1* mRNA was expressed in most tissues with the highest level in testis, consistent with studies in human and mouse. The splicing sites between exon 6 and exon 7 are conserved between rat and human. Partial sequencing of the rat *Brca1* intron 6 revealed a polymorphism of a pentanucleotide TTTTG repeat between the WKY and WF strains. The rat *Brca1* gene mapped to chromosome 10 (Chen et al., 1996).

The mouse homologues of the breast cancer susceptibility genes, *Brca1* and *Brca2,* are expressed in a cell cycle-dependent fashion in vitro and appear to be regulated by similar or overlapping pathways. Both of these genes are expressed concordantly in proliferating cells of embryos, and the mammary gland undergoing morphogenesis and in most adult tissues. There is nevertheless substantial evidence linking BRCA1 and BRCA2 to homologous recombination and DNA repair, to transcriptional control and to tissue proliferation (see chapter 12). There is controversy regarding the localization of BRCA1 and BRCA2 proteins to either nucleus or cytoplasm and whether the expression is present in pre-meiotic germ cells or can still be expressed in mitotic spermatogonia (Bernard-Gallon et al., 2001). The expression pattern of *Brca1* and *Brca2* correlates with the localization of proliferating cell nuclear antigen, an indicator of proliferative activity. In the testes, *Brca1* and *Brca2* are expressed in mitotic spermatogonia and early meiotic prophase spermatocytes. The time course of *Brca2* expression was delayed in spermatogonia to *Brca1*. Thus, while *Brca1* and *Brca2* share concordant cell-specific patterns of expression in most proliferating tissues, observations suggested that they may have distinct roles during meiosis. Brca2 is required for efficient DNA repair, and a study suggested that the loss of the p53 checkpoint may be essential for tumour progression triggered by mutations in *Brca2* (Cressman et al., 1999). Within the germ cell lineage, the high level expression was detected in meiotic cells, specifically pachytene spermatocytes and in post-meiotic round spermatids. This is in contrast to pre-meiotic germ cells, which were found to express little or no BRCA1 mRNA. These observations argued against a function for BRCA1 early progenitor cells in both tissues and cells (Cressman et al., 1998; Zabludoff et al., 1996). Most of the *Brca1* mutant mice died within 10 days. But the survival of remaining mutant mice was likely due to additional unknown mutations or epigenetic effects. Analysis of the *Brca1(-/-) p53(-/-)* animals indicated that BRCA1 is not required for the development of most organ systems. However, these mice are growth retarded, males are infertile due to meiotic failure, and the mammary gland of the female mouse is underdeveloped.

32.2.2. PLU-1: A Nuclear Protein

The PLU-1, a large multi-domain nuclear protein with strong transcriptional repression activity is a member of the ARID family of DNA binding proteins. In humans and mice, the expression of nuclear protein (PLU-1) is restricted in adult tissues, with high expression being seen in the testis, ovary and transiently in the mammary gland of the pregnant female, while the protein

```
   1 MGLILSTRLN FFRAGTLIHS KWSIKVGSTE TLRINSIHSG RARPRPIRGR GGEPRAEGGS
  61 MGVGWDSFSS PWRRRRTAWA CGETSSSEAE KAQGAAVART TRTCCCSSRV CTGLGPSGAR
 121 SLGPGAHLRL ALAQPAVMEA ATTLHPGPRP ALPLGGPGPL GEFLPPPECP VFEPSWEEFA
 181 DPFAFIHKIR PIAEQTGICK VRPPPDWQPP FACDGDKLHF TPRIQRLNEL EAQTRVKLNF
 241 LDQIAKYWEL QGSTLKIPHV ERKILDLFQL NKLVAEEGGF AVVCKDRKWT KIATKMGFAP
 301 GKAVGSHIRG HYERILNPYN LFLSGDSLRC LQKPNLTTDT KDKEYKPHDI PQRQSVQPSE
 361 TCPPARRAKR MRAEAMNIKI EPEETTEART HNLRRRMGCP TPKCENEKEM KSSIKQEPIE
 421 RKDYIVENEK EKPKSRSKKA TNAVDLYVCL LCGSGNDEDR LLLCDGCDDS YHTFCLIPPL
 481 HDVPKGDWRC PKCLAQECSK PQEAFGFEQA ARDYTLRTFG EMADAFKSDY FNMPVHMVPT
 541 ELVEKEFWRL VSTIEEDVTV EYGADIASKE FGSGFPVRDG KIKLSPEEEE YLDSGWNLNN
 601 MPVMEQSVLA HITADICGMK LPWLYVGMCF SSFCWHIEDH WSYSINYLHW GEPKTWYGVP
 661 GYAAEQLENV MKKLAPELFV SQPDLLHQLV TIMNPNTLMT HEVPVYRTNQ CAGEFVITFP
 721 RAYHSGFNQG FNFAEAVNFC TVDWLPLGRQ CVEHYRLLHR YCVFSHDEMI CKMASKADVL
 781 DVVVASTVQK DMAIMIEDEK ALRETVRKLG VIDSERMDFE LLPDDERQCV KCKTTCFMSA
 841 ISCSCKPGLL VCLHHVKELC SCPPYKYKLR YRYTLDDLYP MMNALKLRAE SYNEWALNVN
 901 EALEAKITKK KSLVSFKALI EESEMKKFPD NDLLRHLRLV TQDAEKCASV AQQLLNGKRQ
 961 TRYRSGGGKS QNQLTVNELR QFVTQLYALP CVLSQTPLLK DLLHRVEDFQ QHSQKLLSEE
1021 TPSAAELQDL LDVSFEFDVE LPQLAEMRIR LEQARWLEEV QQACLDPSSL TLDDMRRLID
1081 LGVGLAPYSA VEKAMARLQE LLTVSEHWDD KAKSLLKARP RHSLNSLATA VKEIEEIPAY
1141 LPNGAALKDS VQRARDWLQD VEGLQAGGRV PVLDTLIELV TRGRSIPVHL NSLPRLETLV
1201 AEVQAWKECA VNTFLTENSP YSLLEVLCPR CDIGLLGLKR KQRKLKEPLP NGKKKSTKLE
1261 SLSDLERALT ESKETASAMA TLGEARLREM EALQSLRLAN EGKLLSPLQD VDIKICLCQK
1321 APAAPMIQCE LCRDAFHTSC VAVPSISQGL RIWLCPHCRR SEKPPLEKIL PLLASLQRIR
1381 VRLPEGDALR YMIERTVNWQ HRAQQLLSSG NLKFVQDRVG SGLLYSRWQA SAGQVSDTNK
1441 VSQPPGTTSF SLPDDWDNRT SYLHSPFSTG RSCIPLHGVS PEVNELLMEA QLLQVSLPEI
1501 QELYQTLLAK PSPAQQTDRS SPVRPSSEKN DCCRGKRDGI NSLERKLKRR LEREGLSSER
1561 WERVKKMRTP KKKKIKLSHP KDMNNFKLER ERSYELVRSA ETHSLPSDTS YSEQEDSEDE
1621 DAICPAVSCL QPEGDEVDWV QCDGSCNQWF HQVCVGVSPE MAEKEDYICV RCTVKDAPSR
1681 K
```

Fig.32.2. Amino acid sequence of PLU-1 (Accession NP_006609). Source: http//www. ncbi.nlm.nih.gov.

and its mRNA are specifically up-regulated in breast cancer. PLU-1 protein contains a small domain of amino acids (168-209), found in the jumonji family of transcription factors, in addition to a BRIGHT, ARID region (A/T-rich interaction Domain) at aa 235-324. It also has three PHD zinc finger regions at amino acids: 448-494, 1316-1359, and 1629-1673 besides a C5HC2 zinc finger at 829-882. The *Plu-1* gene containing highly conserved putative DNA/chromatin binding motifs is specifically up-regulated in breast cancer (Lu et al, 1999) **(Fig.32.2)**. The expression in mammary tumors from c-neu transgenic mice is high. *Plu-1* is also differentially expressed in the adult mammary gland. In the developing embryo *Plu-1* is expressed in a restricted fashion with tissue specific expression being limited to parts of the developing brain, whisker follicle, mammary bud, thymus, limbs, intervertebral disc, olfactory epithelium, teeth, eye, and stomach. The mouse homologue (Plu-1) shows 94% overall homology with human PLU-1 at the protein level, with almost 100% identity in the conserved domains. Due to its high expression in the testis and its relationship to cancer, PLU-1 has been proposed to belong to the family of testis-cancer antigens. In testis, *Plu-1* mRNA and PLU-1 protein are both highly expressed in the mitotic spermatogonia. Following reduced expression in the early prophase I stages (zygotene, leptotene), PLU-1 reappears at pachytene and is still detectable in diplotene cells. PLU-1 localizes over the nucleus and is present in the chromatin. It points to a role of PLU-1 in meiotic transcription, which may be restricted to certain meiotic stages and mediated by the ability of this protein to associate with the chromatin. In addition, the temporal and spatial expression patterns of the transcription factors Bf-1 and Pax9, which bind PLU-1 through the PLU domain, suggest that Plu-1 plays an important role in mouse embryonic development that may involve interaction with Pax9 and Bf-1 (Madsen et al., 2002; 2003). Like-wise human PLU-1 has transcriptional repression properties and interacts with the developmental transcription factors BF-1 and PAX9 (Tan et al, 2003). Gene targeting of Desrt, a novel ARID class DNA-binding protein, causes growth retardation and abnormal development of reproductive organs (Lahoud et al., 2001).

32.2.3. MAGE Gene Products

A body of data has shown that human cancer cells express tumor-rejection antigens recognized by CTL on the appropriate MHC class I antigens. Genes that code for tumor specific antigens on melanoma cell lines tested by CTL include *MAGE, BAGE* and *GAGE.* van der Bruggen et al. reported the identification of a human gene *MAGE-1* (melanoma associated gene-1), which directs the expression of a tumor-rejection antigen. The gene *MAGE-1*, encoding an antigen MZ2-E, was identified by a gene transfection. Several new genes encoding melanoma-related antigens recognized by CTL have been identified. These tumor-rejection antigens are not truly foreign, but rather they are differentiation antigens expressed by no other normal tissue except testis. However, little is known about the function of human cancer/testis antigens (CTAs). The *MAGE-1* gene comprises of three exons spread over 4.5-kb and shows no homology to any reported gene. The *MAGE* gene family consists of at least 12 closely related genes located on the long arm of chromosome X (Xq28). The *MAGE-1*, -2, -3, -4, (-4a/-4b), -6, and –12 genes are preferentially expressed at the mRNA level in many different cancers, including melanomas, lung cancers, head and neck cancer, esophageal cancers, ovarian cancers, and breast cancers. In contrast to preferential for cancer cells, no normal cells other than testicular cells express the *MAGE* gene (Itoh et al., 1996). Recent studies indicated that expression of MAGE genes is regulated by DNA methylation. The demethylating agent, 5' Aza-deoxycytidine induces MAGE-1 and other antigens in normal and malignant lymphoid cells and also in myeloid monocyte cell lines. Demethylation alone is a sufficient stimulus to induce MAGE tumor rejection antigens in both normal and malignant lymphoid cells in most cases. Because of the preferential expression in different types of cancer cells, the MAGE gene products are the most appropriate target molecules for development of a new cancer vaccine

MAGE Gene in Germ Cells: Normal adult tissues and some tissues from fetuses over 20 weeks old were negative for MAGE gene at the mRNA level, with the exception of testis and placenta. The testis expresses all the MAGE genes except the MAGE-7 gene, while the placenta expressed MAGE-3, -4, and MAGE-8 to –11 genes. Polyclonal anti-MAGE-1 Ab was reactive to the testes from two donors. This Ab recognized a 38kDa protein as a major band and the additional 42 and 45-kDa proteins in the testes. The molecular weight of cellular MAGE-1 protein in melanomas was reported as a 46-kDa. Both the anti-MAGE4 polyclonal and mAb were reactive to the testes and recognized a 45-kDa protein as a major band and the additional -28, -35, and 40-kDa proteins in the testes, and a 45-kDa protein in the RPMI-1788 B cell line. These various bands could be the MAGE-1 or MAGE-4 protein by itself, or their various processed products although a number of MAGE genes have been identified in recent years (Osterlund et al., 2000). The MAGE-1 and –4 proteins were observed in the nucleus and the cytoplasm of spermatogonia, but undetectable in spermatids and Sertoli's cells as well as in the primary spermatocytes adjacent to the basement membrane but not in the primary spermatocytes far from the basement membrane with relatively small nuclei. Therefore, MAGE proteins are normal tissue antigens found in testicular germ cells, playing an important role in the early phase of the spermatogenesis (Itoh et al., 1996; Takahashi et al., 1995).

The testis germ cells undergo a process of gene demethylation process as in cancer cells. Characterization studies of a murine MAGE homologue, MAGE-b4 revealed that MAGE–b4 is

```
  1 MSSEQKSQHC KPEEGVEAQE EALGLVGAQA PTTEEQEAAV SSSSPLVPGT LEEVPAAESA
 61 GPPQSPQGAS ALPTTISFTC WRQPNEGSSS QEEEGPSTSP DAESLFREAL SNKVDELAHF
121 LLRKYRAKEL VTKAEMLERV IKNYKRCFPV IFGKASESLK MIFGIDVKEV DPTSNTYTLV
181 TCLGLSYDGL LGNNQIFPKT GLLIIVLGTI AMEGDSASEE EIWEELGVMG VYDGREHTVY
241 GEPRKLLTQD WVQENYLEYR QVPGSNPARY EFLWGPRALA ETSYVKVLEH VVRVNARVRI
301 AYPSLREAAL LEEEEGV
```

Fig.32.3. Amino acid sequence of melanoma-associated antigen-4 (MAGE-4 antigen) (Accession P43358). Source: http//www. ncbi.nlm.nih.gov.

specifically expressed in fetal and adult gonads. Double labeling experiments using antibodies against the meiosis specific SCP3 protein and the MAGE-b4 protein showed that MAGE-b4 is down regulated as the germ cells enter meiosis in adult testis. In contrast, MAGE-b4 was expressed in female germ cell throughout meiosis, and the protein was also found in dormant primary oocytes (Osterlund et al., 2000) **(Fig.32.3).** Like *MAGE-1 to MAGE-12, MAGE-C1* gene also expresses in testis. *MAGE-C1* is expressed in significant proportion in tumors of various histological types (De Smet et al., 1997). Because of the preferential expression in different types of germ cells, the *MAGE* gene products are the most appropriate target molecules for development of a male contraceptive.

32.2.4. LAGE-1

A cDNA clone specific for tumors and the corresponding gene named *LAGE-1* was sequenced. Using *LAGE-1* as a probe, a closely related gene, identical to NY-ESO-1, was reported. The *LAGE-1* codes for an antigen recognized by autologous antibodies in an esophageal squamous cell carcinoma. Gene *LAGE*-1 maps to Xq28. It comprises 3 exons. Alternative splicing produces 2 major transcripts encoding polypeptides of 210 and 180 residues, respectively. LAGE-1 is expressed in 25-50% of tumor samples of melanomas, non-small-cell lung carcinomas, bladder, prostate and head and neck cancers. The only normal tissue that expressed the gene was testis. As for *MAGE-A1*, the expression of *LAGE-1* is induced by 5'-deoxyazacytidine in lymphoblastoid cells, and suggested that tumoral expression is due to demethylation. The expression of *LAGE-1* is strongly correlated with that of NY-ESO-1 and clearly correlated with the expression of *MAGE* genes (Lethe et al., 1998).

32.2.5. CAGE-1

CAGE is expressed in a variety of cancers but not in normal tissues except testis, and might play a role in cellular proliferation. Results suggest that the methylation status of the CpG sites of CAGE determines its expression, and the hypomethylation of *CAGE* precedes the development of gastric cancer and hepatocellular carcinoma (Cho et al., 2003). Park et al., (2003) characterized a novel gene with expression pattern similar to that of C/T antigens. Its open reading frame is 1920-bp in size and encodes for putative protein composed of 639 amino acids. This gene exists as single copy and encodes a protein of 73- kDa in size. It is composed of 11 exons and 10 introns and displays testis-specific expression among normal tissues, and wide expression among various cancer tissues and cancer cell lines. Because of its association with cancer, this gene was named cancer-associated gene-1 (CAGE-1). The CAGE-1 is main protein, whose expression is relatively higher in cancer tissues. It appeared that over expression of CAGE-1 might be associated with the progression of tumor. The exact functional role of CAGE-1 in tumorigenesis remains to be seen (Park et al., 2003) **(Fig.32.4A, B).**

A

```
  1 MSHWAPEWKR AEANPRDLGA SWDVRGSRGS GWSGPFGHQG PRAAGSREPP LCFKIKNNMV
 61 GVVIGYSGSK IKDLQHSTNT KIQIINGESE AKVRIFGNRE MKAKAKAAIE TLIRKQESYN
121 SESSVDNAAS QTPIGRNLGR NDIVGEAEPL SNWDRIRAAV VECEKRKWAD LPPVKKNFYI
181 ESKATSCMSE MQVINWRKEN FNITCDDLKS GEKRLIPKPT CRFKDAFQQY PDLLKSIIRV
241 GIVKPTPIQS QAWPIILQGI DLIVVAQTGT GKTLSYLMPG FIHLDSQPIS REQRNGPGML
301 VLTPTRELAL HVEAECSKYS YKGLKSICIY GGRNRNGQIE DISKGVDIII ATPGRLNDLQ
361 MNNSVNLRSI TYLVIDEADK MLDMEFEPQI RKILLDVRPD RQTVMTSATW PDTVRQLALS
421 YLKDPMIVYV GNLNLVAVNT VKQNIIVTTE KEKRALTQEF VENMSPNDKV IMFVSQKHIA
481 DDLSSDFNIQ GISAESLHGN SEQSDQERAV EDFKSGNIKI LITTDIVSRG LDLNDVTHVY
541 NYDFPRNIDV YVHRVGSLDG QERLHISSLI TQRDSKMAGE LIKILDRANQ SVPEDLVVMA
601 EQYKLNQQKR HRETRSRKPG QRRKEFYFLS
```

B

```
  1 MTEKPEFQSQ VYNYAKDNNI KQDSFKEENP METSVSANTD QLGNEYFRQP PPRSPPLIHC
 61 SGEMLKFTEK SLAKSIAKES ALNPSQPPSF LCKTAVPSKE IQNYGEIPEM SVSYEKEVTA
121 EGVERPEIVS TWSSAGISWR SEACRENCEM PDWEQSAESL QPVQEDMALN EVLQKLKHTN
181 RKQEVRIQEL QCSNLYLEKR VKELQMKITK QQVFIDVINK LKENVEELIE DKYKIILEKN
241 DTKKTLQNLE EVLANTQKHL QESRNDKEML QLQFKKIKAN YVCLQERYMT EMQQKNKSVS
301 QYLEMDKTLS KKEEEVERLQ QLKKELEKAT ASALDLLKRE KEAQEQEFLS LQEEFQKLEK
361 ENLEERQKLK SRLEKLLTQV RNLQFMSENE RTKNIKLQQQ INEVKNENAK LKQQVARSEE
421 QNYVPKFETA QLKDQLEEVL KSDITKDTKT THSNLLPDCS PCEERLNPAD IKRASQLASK
481 MHSLLALMVG LLTCQDIINS DAEHFKESEK VSDIMLQKLK SLHLKKKTLD KEVIDCDSDE
541 AKSIRDVPTL LGAKLDKYHS LNEELDFLVT SYEEIIECAD QRLAISHSQI AHLEERNKHL
601 EDLIRKPREK ARKPRSKSLE NHPKSMTMMP ALFKENRNDL D
```

Fig.32.4.A. Amino acid sequence of cancer-associated gene (CAGE) product, called DEAD (Asp-Glu-Ala-Asp) box polypeptide 53. Region 53-110, "K homology RNA-binding domain", binds single-stranded RNA, found in a wide variety of proteins including ribosomal proteins, transcription factors and post-transcriptional modifiers of mRNA". Region 231-428 = "DEAD-box helicases". Region 441-556 = "Helicase superfamily c-terminal domain" (Cho et al, 2002). Source: http//www. ncbi.nlm.nih.gov.(Accession NP_874358 XP_291343). **B.** Amino acid sequence of cancer/testis antigen 3 isoform 2; cancer/testis antigen gene 1 CAGE-1 Source: http//www. ncbi.nlm.nih.gov (Accession NP_995586) (Parks et al, 2003).

32.2.6. Testis Specific Cyclin A1 in Testicular and Ovarian Tumors

Cyclin A1 is a tissue-specific A-type cyclin that is essential for spermatogenesis (see Chapter 11). Over-expression of cyclin A1 was found in acute myeloid leukemia and cyclin A1 induced leukemia in a transgenic mouse model. Cyclin A1 expression was very low in breast cancer, non-small cell lung cancer and in cervical carcinoma, whereas substantial expression of cyclin A1 was found in testicular and ovarian cancer and in endometrial cancer. In testis specimens, cyclin A1 expression was much higher in testicular tumors compared to Sertoli cell only syndrome that lacks spermatogenesis. However, compared to normal spermatogenesis testicular cancers expressed on average lower levels of cyclin A1. Cyclin A1 is expressed in selected solid tumors. Its known oncogenic function and the high expression levels in aggressive testicular tumors suggest a role for cyclin A1 in germ cell tumorigenesis (Muller-Tidow et al., 2003).

32.2.7. Testis-Specific Protein Y-encoded (TSPY)

Genetic mapping studies have identified a gonadoblastoma locus on the human Y chromosome (GBY) that predisposes the dysgenetic gonads of XY sex-reversed patients to tumorigenesis. The testis-specific protein Y-encoded (*TSPY*) gene that locates on the GBY critical region, has been demonstrated to express preferentially in tumor cells in gonadoblastoma and testicular germ cell tumors. The TSPY shares high homology to a family of cyclin B binding proteins and is considered to play a role in cell cycle regulation or cell division. The TSPY is expressed at low levels in normal epithelial cells and benign prostatic hyperplasia (BPH), but at elevated levels in tumor cells of prostate cancers at various degrees of malignancy. Sequence analysis of TSPY mRNA revealed a complex pattern of RNA processing of the TSPY transcripts involving cryptic intron splicing and/or intron skipping. The variant transcripts encode a variety of polymorphic isoforms or shortened versions of the TSPY protein, some of which might possess different biochemical and/or functional properties. The shortened versions of transcripts were more abundant in prostatic cancer tissues than the testicular ones. The differential expression of TSPY gene might be involved in the multi-step prostatic oncogenesis besides its putative role in gonadoblastoma and testicular seminoma (Lau et al., 2004).

32.2.8. Markers of Seminomas

Spermatocytic seminoma is a rare germ cell derived old age testis tumor. Rajpert-De Meyts et al., (2003) studied the expression of several recently discovered markers for germ cell differentiation and the mitosis-meiosis transition in order to define the antigen profile for diagnostic purposes and to clarify the biology and histogenesis of spermatocytic seminoma. The panel of markers included Chk2, p19INK4d, p53, MAGE-A4, KIT, TRA-1-60, neuron-specific enolase and placental-like alkaline phosphatase. Proteins highly expressed in gonocytes and spermatogonia, such as Chk2, MAGE-A4 and neuron-specific enolase were consistently present in spermatocytic seminoma. Antigens of embryonic germ cells (TRA-1-60) and a proto-oncogene p19INK4d were not detectable. But the expression of p53 protein was demonstrated in 80% of cases. This pattern of expression was consistent with the origin of spermatocytic seminoma from a pre-meiotic germ cell, which has lost embryonic traits but had committed to spermatogenic lineage (Rajpert-De Meyts et al, 2003). Compared to normal spermatogenesis, testicular cancers also expressed lower levels of cyclin A1. Cyclin A1 is expressed in selected solid tumors where its high expression level suggests a role for cyclin A1 in germ cell tumorigenesis (Muller-Tidow et al., 2003).

The NKX3.1 is a homeobox gene, which exhibits prostate and testis specific expression. Loss of NKX3.1 expression has been implicated in prostate development and tumorigenesis. The NKX3.1 expression is dramatically down-regulated in testicular cancer of germ cell origin. NKX3.1 is expressed at high levels in normal germ cells and in carcinoma *in situ*, but is sharply decreased or absent in most seminomas and all embryonal carcinomas. However, NKX3.1 is expressed in a subset of the more differentiated non-seminomas. These changes in NKX3.1 protein levels are mainly due to transcriptional effects and suggested that NKX3.1 is essential for normal testis function (Skotheim et al., 2003).

A peanut agglutinin-binding tumor antigen, gp200, is a surface membrane glycoprotein expressed on human embryonal carcinoma, a malignant stem cell testicular tumor. Gp200 is remarkably similar to another embryonal carcinoma antigen, GCTM-2, a differentiation antigen that is also detected in blood of testis cancer patients. Based on sequence similarity, the gp200 is a podocalyxin. Therefore, gp200 is a testicular tumor form of podocalyxin, a surface membrane glycoprotein that was originally discovered as a scaffolding extracellular matrix protein of kidney podocyte cells. Podocalyxin is also present on hematopoietic cells where it has a putative function as a cell adhesion protein (Schopperle et al., 2003).

Given the importance of growth signals in normal testicular development and their acquired deregulation in most human cancers, role of growth factors and signaling molecules has been implicated in the genesis of testicular germ cell tumors This suggests the existence of specific paracrine functions during male germ cell differentiation and development of male germ cell tumors (Devouassoux-Shisheboran et al., 2003; Franke et al., 2003). A somatic mutation of SMAD4 gene results in the development of seminomas. This mutational inactivation may affect the activity of members of TGF beta superfamily (TGF beta, activin, inhibin, BMP). tHE VEGF expression has been associated with metastasis in seminomas, whereas HST-1 expression, a member of fibroblast growth factors, has been linked with the nonseminomatous phenotype and with tumour stage. Despite intense expression of c-kit in almost all seminomas, activating mutation of *c-kit* gene is seldom reported. The RET (GDNF receptor) expression is known in human seminomas, and not in nonseminomatous tumours. However, the molecular bases of these alterations in germ cell tumours are not known. Signaling molecules such as peptide hormones are also involved in testicular carcinogenesis.

Franke et al., (2003) reviewed the current knowledge on ACE and its potential substrates with special emphasis on the differentiation-restricted ACE expression during spermatogenesis and pre-spermatogenesis. In addition to testis specific tACE, the somatic isoform (sACE) is also present in human germ cells. Similar to other oncofoetal markers, sACE exhibits a transient expression during foetal germ cell development and appears to be a constant feature of intratubular germ cell neoplasm, in particular, of classic seminoma (Franke et al., 2003). A specific pattern of somatostatin (sst) receptors expression was shown in testicular germ cell tumours, with a loss of sst3 and sst4 in seminomas and loss of sst4 and expression of sst1 in nonseminomas only. This suggested an antiproliferative action of somatostatin that may be involved in the differentiation of germ cell tumours (Devouassoux-Shisheboran et al., 2003).

32.2.9. Other Testis Antigens Expressed in Cancer

The HCA587 Antigen: The *HCA587* gene from a hepatocellular carcinoma (HCC) patient, encodes a new member of cancer-testis antigens. The HCA587 mRNA expression in normal tissues and cancers is known. HCA587 protein was not shown in normal tissues except germ cells in testis and Purkinji cells in cerebellum. In HCC specimens the HCA587 protein was expressed in 37.1% (26 of 70) samples. Being located at the nucleus or cytoplasm, there appeared a correlation between the tumor differentiation of HCC and HCA587 protein expression. HCA587 antigen was also expressed in different proportions in melanoma, lymphoma, pancreatic cancer, and lung cancer (Li et al., 2003).

SP17 as Cancer Testis Antigen: Although SP17 was originally described as a testis-specific antigen, emerging evidence indicates that it may be more ubiquitously expressed than was previously thought. With the use of a specific antiserum, SP17 was found on the surface of malignant lymphoid cells, including B- and T-lymphoid cell lines, and on the surface of primary cells of B-lymphoid tumors. The role of SP17 in promoting lymphoid cell adhesion was addressed with the use of recombinant SP17 (rSP17). Findings suggested that SP17 promotes heparin sulfate-mediated cell aggregation and thereby plays a role in regulating adhesion and migration of normal and malignant lymphocytes. The SP17 is detectable in tumor cells from myeloma patients. Since a high proportion of normal individuals develop antibodies against SP17 following vasectomy, Sp17 is likely to be a highly immunogenic protein in vivo. The SP17 is therefore, a member of the cancer testis antigen family and could be an ideal target for immunotherapy of multiple myeloma (Lim et al., 2001; Lacy and Sanderson, 2001) (see Chapter 24).

LDH-C: As evidenced, ectopic activation of testis-specific genes in cancer is a frequent phenomenon. Lactate dehydrogenase-C (LDH-C), the germ cell-specific protein escapes from transcriptional repression, resulting in significant expression in virtually all tumor types tested (Koslowski et al., 2002). Surprisingly, aberrant splicing of LDH-C was restricted to cancer cells, resulting in four novel tumor-specific variants displaying structural alterations of the catalytic domain. LDH-C expression in tumors is not mediated by gene promotor demethylation, as described for other germ cell-specific genes activated in cancer (Koslowski et al., 2002). Since LDH-C has a preference for lactate as a substrate, LDH-C activation in cancer may depend on lactate for ATP production (see Chapter 28).

Besides several antigens expressed in spermatogenic cells, a spermatogenesis-associated factor (SPAF) was found to be aberrantly expressed at malignant conversion stage in a epidermal model of chemical carcinogenesis. Sequence analysis revealed SPAF to belong to AAA-protein family (ATPase associated with diverse activities) (Liu et al., 2000). Two additional tumor specific antigens, HOM-MEL-40 and NY-ESO-1, encoded by testis specific genes activated in tumors, have been identified (De Smet et al., 1997).

32.3. DIAGNOSTIC AND THERAPEUTIC POTENTIAL OF C/T ANTIGENS

32.3.1. MAGE in Squamous Cell Carcinoma and Childhood Astrocytoma

MAGE antigens are expressed with frequencies ranging between 22.7% (larynx) and 50% of cases (lung) in squamous cell carcinomas from different anatomic areas and in large cell carcinomas of the lung (37.9%). Bolli et al., (2002) provided the first description of MAGE antigen expression in basalioma (48.1%). MAGE expression in a frequently positive tumor type, a non-small-cell lung cancer (TMA) showed that 43.2% of tumors were 57B positive. In squamous cell carcinoma, MAGE TAA positivity correlated with a shorter tumor-specific survival. In lung squamous cell carcinoma, the significant association of MAGE TAA expression with poor prognosis suggests that patients with 57B-positive tumors may benefit from early immunotherapy procedures (Bolli et al., 2003). Among childhood astrocytomas, the MAGE-1 CT-antigen was observed only in anaplastic, high-grade ASTRs (100%), including glioblastomas (Bodey et al., 2002).

To determine the expression of 12 different CT antigens in Japanese primary lung cancers and cell lines, MAGE-3 (41%, 19/46), and SSX-4 (35%, 16/46) showed highest expression. The expression frequency of NY-ESO-1 in Japanese was drastically different from Caucasians. Vaccination with MAGE-3 and SSX-1 could cover 57% of all patients, while combination of three antigens, MAGE-3, SSX-1, and MAGE-4, would cover 65%, and with four antigens, MAGE-3, MAGE-4, SSX-1, and SSX-4, would cover 70%. Polyvalent CT antigen vaccines may be effective to reduce a chance of emergence of antigen loss variants, thus preventing tumors from escaping from the immune system. For this purpose, combination of vaccinogens of MAGE-3 with MAGE-6, SSX-4, MAGE-1 or BAGE may be effective for a quarter of Japanese lung cancer patients. A novel gene, TES101RP, expressed only in some small cell lung cancers (SCLC) and in testis was also identified (Tajima et al., 2003).

Sarcevic et al., (2003) investigated the expression of cancer/testis antigens in cervical squamous cell carcinomas. HeLa cervical cancer derived cell line expresses MAGE-A1, MAGE-A2, MAGE-A3, MAGE-A4, MAGE-A6, MAGE-A12, GAGE-3/6, LAGE-1, and PRAME genes, encoding defined C/T antigens). In contrast, the expression of MAGE-A10, BAGE, GAGE-1/2, or NY-ESO-1 genes was not observed. A group of 60 patients was studied, and 57B positivity was detectable to different extents in 33% of the cases (20/60). Considering the high tumor specificity of C/T antigens, data provided the rationale for the design of immunotherapy targeting these antigens in cervical cancers (Sarcevic et al., 2003). In an effort to examine the potential of NY-ESO-1 and LAGE-1 CT antigens for immunotherapy in epithelial ovarian cancer (EOC),

Odunsi et al., (2003) indicated that NY-ESO-1 and LAGE-1 are attractive targets for antigen-specific immunotherapy in EOC (Odunsi et al., 2003).

32.3.2. NY-ESO

Medullary thyroid carcinoma (MTC) is a neoplasm derived from the parafollicular cells of the thyroid and occurs in either a sporadic or a familial form. Expression of a number of genes encoding C/T antigens in serological samples of MTC indicated humoral antibody response to NY-ESO-1 that can provide a new attractive target for vaccine-based immunotherapy of MTC (Maio et al., 2003). Juretic et al., (2003) reviewed immunohistochemical evidence of the expression of C/T antigens, known to be recognized by specific cytotoxic T lymphocytes. The emerging picture was consistent with a mostly heterogeneous expression in human cancers. The wide range of tumors in which C/T TAA have been detected urges further efforts to develop effective specific immunotherapeutic procedures. Synovial sarcomas are striking with regard to CT antigen expression, with >80% of specimens expressing NY-ESO-1 and MAGE-A3. Although highly expressed in sarcoma, CT antigens do not induce frequent humoral immune responses in sarcoma patients. A novel CT antigen, NY-SAR-35, mapping to chromosome Xq28 was identified among the cancer-related antigens, that encodes a putative extra-cellular protein. In addition to testis-restricted expression, NY-SAR-35 mRNA was expressed in sarcoma, melanoma, esophageal cancer, lung cancer and breast cancer. NY-SAR-35 is therefore a potential target for cancer vaccines and monoclonal antibody-based immunotherapies (Lee et al., 2003). Cellular immune responses have also been studied for NY-ESO-1 peptide. The CD_8^+ T cell response to a NY-ESO-1 peptide vaccine composed of the two defined peptides 157-165 and 157-167, administered with GM-CSF as a systemic adjuvant. The NY-ESO-1 peptide vaccine elicited a CD_8^+ T cell response directed against multiple distinct epitopes in the 157-167 region. However, only a minor fraction of the elicited CD_8^+ T cells, namely those recognizing the peptide 157-165 with sufficiently high functional avidity, recognized the naturally processed target on NY-ESO-1 (+) tumor cells. Because of the complexity of the CD_8^+ T cell repertoire that can be elicited by vaccination with synthetic peptides, a precise definition of the targeted epitope, and hence, of the corresponding peptide to be used as immunogen, is required to ensure a precise tumor targeting (Dutoit et al., 2002).

32.3.3. Semenogelin 1 and HAGE in Leukemia

Adams et al (2002) investigated the expression of a range of CT genes (MAGE-A1, -A3, -A6, -A12, BAGE, GAGE, HAGE, LAGE-1, NY-ESO-1 and RAGE) for their expression in a cohort of acute and chronic myeloid leukemia patients. Cancer-testis antigens expression was not detected in normal bone marrow or peripheral blood stem cell samples. On the other hand, in acute myeloid leukemia (AML) nine of the 26 (35%) samples expressed one or more of the CT genes with six of the samples (23%) expressing HAGE. In chronic myeloid leukemia (CML), 24 of 42 (57%) patient samples expressed one or more CT antigen with 23 expressing HAGE. Thus HAGE is frequently expressed in CML, and to a lesser extent in AML patients (Adams et al., 2003).

Niemeyer et al., (2003) investigated the expression of 5 CT genes (SSX-1, HOM-MEL-40/SSX-2, HOM-TES-14/SCP-1, SCP-3 and NY-ESO-1) in leukemic blood samples obtained from patients with either acute lymphatic leukemia (ALL) or myelocytic leukemia (AML). A majority of the ALLs might be amenable for specific immunotherapeutic interventions. However, the identification of additional antigens with a frequent expression in leukemias is warranted to allow the development of widely applicable polyvalent leukemia vaccines (Niemeyer et al., 2003). Semenogelin 1, a major protein of human semen coagulum thought to be highly specific to seminal vesicles, is expressed in leukemic cells. However, Semenogelin 1 expression is normally confined to the testis, suggesting that it is a novel Cancer-Testis antigen (Zhang et al., 2003). Translation of the mRNA to Semenogelin 1 protein confirmed that Semenogelin 1 is a novel CT antigen capable of inducing B-cell responses in vivo in chronic leukemias (Zhang et al., 2003). The expression of a panel of CT antigenic proteins was examined in patients with malignant gammopathies using the following monoclonal antibodies (mAbs): mAb MA454 to MAGE-A1, mAb M3H67 to MAGE-A3, mAb 57B to MAGE-A4, mAb CT7-33 to CT7/MAGE-C1 and mAb ES121 to NY-ESO-1. The CT antigens have important biological implications in malignant gammopathies and that CT-7 may be a suitable target for T cell-based and possibly antibody-mediated immunotherapy of myeloma (Dhodapkar et al., 2003).

32.4. REFERENCES

Adams SP, Sahota SS, Mijovic A, Czepulkowski B, Padua RA, Mufti GJ, Guinn BA. Frequent expression of HAGE in presentation chronic myeloid leukaemias. Leukemia 2002; 16: 2238-42.

Bartkova J, Thullberg M, Rajpert-De Meyts E, Skakkeback NE, Bartek J. Lack of p19INK4d in human testicular germ-cell tumours contrasts with high expression during normal spermatogenesis. Oncogene 2000; 14: 4146-50.

Bernard-Gallon DJ, Dechelotte P, Vissac C, Aunoble B, Cravello L et al., BRCA1 and BRCA2 protein expression in an ovotestis of a 46, XX true hermaphrodite. Breast Cancer Res 2001; 3: 61-5.

Beumer TL, Roepers-Gajadien HL, Gademan IS, Rutgers DH, deRooij DG. $P21^{cip1/waf1}$ expression in mouse testis before and after X-irradiation. Mol Reprod Dev. 1997; 47: 240-47.

Bodey B, Siegel SE, Kaiser HE. MAGE-1, a cancer/testis-antigen, expression in childhood astrocytomas as an indicator of tumor progression. In Vivo 2002; 16: 583-8.

Bolli M, Kocher T, Adamina M, Guller U, Dalquen P, Haas P, Mirlacher M, Gambazzi F et al., Tissue microarray evaluation of Melanoma antigen E (MAGE) tumor-associated antigen expression: potential indications for specific immunotherapy and prognostic relevance in squamous cell lung carcinoma. Ann Surg 2002; 236: 785-93.

Brun ME, Ruault M, Ventura M, Roizes G, De Sario A. Juxtacentromeric region of human chromosome 21: a boundary between centromeric heterochromatin and euchromatic chromosome arms. Gene 2003; 312: 41-50.

Chen KS, Shepel LA, Haag JD, Heil GM, Gould MN. Cloning, genetic mapping and expression studies of the rat Brca1 gene. Carcinogenesis 1996; 17: 1561-6.

Chin L, Artandi SE, Shen Q, Tam A, Lee SL, Gottlieb GJ, Greider CW. p53 deficiency rescues the adverse effects of telomere loss and cooperates with telomere dysfunction to accelerate carcinogenesis. Cell 1999; 97: 527-38.

Cho B, Lee H, Jeong S, Bang YJ, Lee HJ, Hwang KS, Kim HY, Lee YS, Kang GH, Jeoung DI. Promoter hypomethylation of a novel cancer/testis antigen gene CAGE is correlated with its aberrant expression and is seen in premalignant stage of gastric carcinoma. Biochem Biophys Res Commun 2003; 307: 52-63.

Cho B, Lim Y, Lee DY, Park SY, Lee H, Kim WH, Yang H, Bang YJ, Jeoung DI.Identification and characterization of a novel cancer/testis antigen gene CAGE.Biochem Biophys Res Commun 2002; 292 : 715-26.

Choi J, Nannenga B, Demidov ON, Bulavin DV, Coone A et al.,. Mice deficient for the wild-type p53-induced phosphatase gene (Wip-1) exhibit defects in reproductive organs, immune function, and cell cycle control. Mol Cell Biol 2002; 22 : 1094-105.

Connor F, Bertwistle D, Mee PJ, Ross GM, Swift S et al,. Tumorigenesis and a DNA repair defect in mice with a truncating Brca2 mutation. Nat Genet 1997; 17: 423;30.

Cressman VI, Backlund DC, Avrutskaya AV, Leadon SA, Godfrey V, Koller BH. Growth retardation, DNA repair, and lack of spermatogenesis in BRCA1-deficient mice Mol Cell Biol 1999; 19: 7061-75.

De Smet C, Martelange V, Lucas S, Brasseur F, Lurquin C, Boon T. Identification of human testis-specific transcripts and analysis of their expression in tumor cells. Biochem Biophys Res Commun 1997:241: 653-57.

Devouassoux-Shisheboran M, Mauduit C, Tabone E, Droz JP, Benahmed M. Growth regulatory factors and signalling proteins in testicular germ cell tumours. APMIS 2003; 111: 212-24.

Dhodapkar MV, Osman K, Teruya-Feldstein J, Filippa D, Hedvat CV, Iversen K, Kolb D et al., Expression of cancer/testis (CT) antigens MAGE-A1, MAGE-A3, MAGE-A4, CT-7, and NY-ESO-1 in malignant gammopathies is heterogeneous and correlates with site, stage and risk status of disease. Cancer Immun. 2003: 23 ; 3:9.

Dutoit V, Taub RN, Papadopoulos KP, Talbot S, Keohan ML et al., Multiepitope CD_8^+ T cell response to a NY-ESO-1 peptide vaccine results in imprecise tumor targeting. J Clin Invest. 2002; 110: 1813-22.

Franke FE, Pauls K, Metzger R, Danilov SM. Angiotensin I-converting enzyme and potential substrates in human testis and testicular tumours. APMIS. 2003; 111: 234-43.

Honda T, Tamura G, Waki T, Kawata S, Terashima M, Nishizuka S, Motoyama T. Demethylation of MAGE promoters during gastric cancer progression. Br J Cancer 2004; 90: 838-43.

Itoh K, Hayashi A, Nakao M, Hashino, Seki N, Schichyo S. Human Tumor rejection antigens MAGE. J Biochem 1996; 119: 385-90.

Jacks T, Weinberg RA. Cell-cycle control and its watchman. Nature 1996; 381: 643-44.

Juretic A, Spagnoli GC, Schultz-Thater E, Sarcevic B. Cancer/testis tumour-associated antigens: immunohistochemical detection with monoclonal antibodies. Lancet Oncol. 2003; 4: 104-9.

Kimura T, Suzuki A, Fujita Y, Yomogida K, Lomeli H, Asada N, Ikeuchi M, Nagy A, Mak TW, Nakano T. Conditional loss of PTEN leads to testicular teratoma and enhances embryonic germ cell production. Development 2003; 130: 1691-700.

Koslowski M, Tureci O, Bell C, Krause P, Lehr HA, Brunner J, Seitz G, Nestle FO, Huber C, Sahin U. Multiple splice variants of lactate dehydrogenase C selectively expressed in human cancer. Cancer Res 2002; 62: 6750-5.

Lacy HM, Sanderson RD. Sperm protein 17 is expressed on normal and malignant lymphocytes and promotes heparan sulfate-mediated cell-cell adhesion. Blood 2001; 98: 2160-65

Lahoud MH, Ristevski S, Venter DJ, Jermiin LS,Bertoncello I, Zavarsek S, Hasthorpe S, Drago J, de Kretser D,Hertzog PJ, Kola I. Gene targeting of Desrt, a novel ARID class DNA-binding protein,causes growth retardation and abnormal development of reproductive organs. Genome Res. 2001;11:1327-34.

Lau YF, Lau HW, Komuves LG. Expression pattern of a gonadoblastoma candidate gene suggests a role of the Y chromosome in prostate cancer. Cytogenet Genome Res. 2003;101(3-4): 250-60.

Lee SY, Obata Y, Yoshida M, Stockert E, Williamson B, Jungbluth AA, Chen YT, Old LJ, Scanlan MJ. Immunomic analysis of human sarcoma. Proc Natl Acad Sci U S A. 2003; 100: 2651-6.

Lethe B, Lucas S, Michaux L, De Smet, Godelaine D et al., LAGE-1, a new gene with tumor specificity. Int J Cancer 1998; 76: 903-8.

Li B, Qian XP, Pang XW, Zou WZ, Wang YP, Wu HY, Chen WF. HCA587 antigen expression in normal tissues and cancers: correlation with tumor differentiation in hepatocellular carcinoma. Lab Invest. 2003; 83: 1185-92.

Lim SH, Wang Z, Chriva-internati M, Xue Y. Sperm protein 17 is a novel cancer-testis antigen in multiple myeloma. Blood 2001; 97: 1508-10

Liu Y, Black J, Kulessz-Martin MF. SPAF, a new AAA-protein specific to early spermatogenesis and malignant conversion. Oncogene 2000; 19: 1579-88.

Lu PJ, Sundquist K, Baeckstrom D, Poulsom R, Hanby A, Meier-Ewert S, Jones T, Mitchell M, Pitha-Rowe P, Freemont P, Taylor-Papadimitriou J. A novel gene (PLU-1) containing highly conserved putativDNA/ chromatin binding motifs is specifically up-regulated in breast cancer.J Biol Chem 1999;274:15633-645.

Madsen B, Spencer-Dene B, Poulsom R, Hall D, Lu PJ, Scott K, Shaw AT, Burchell JM, Freemont P, Taylor-Papadimitriou J. Characterisation and developmental expression of mouse Plu-1, a homologue of a human nuclear protein (PLU-1) which is specifically up-regulated in breast cancer. Gene Expr Patterns. 2002; 2: 275-82.

Madsen B, Tarsounas M, Burchell JM, Hall D, Poulsom R, Taylor-Papadimitriou J. PLU-1, a transcriptional repressor and putative testis-cancer antigen, has a specific expression and localisation pattern during meiosis. Chromosoma 2003; 112: 124-32.

Maio M, Coral S, Sigalotti L, Elisei R, Romei C, Rossi G, Cortini E, Colizzi F, Fenzi G, Altomonte M, Pinchera A, Vitale M.Analysis of cancer/testis antigens in sporadic medullary thyroid carcinoma: expression and humoral response to NY-ESO-1. J Clin Endocrinol Metab. 2003; 88: 748-54.

Muller-Tidow C, Diederichs S, Schrader MG, Vogt U, Miller K, Berdel WE, Serve H. Cyclin A1 is highly expressed in aggressive testicular germ cell tumors. Cancer Lett 2003; 190: 89-95.

Niemeyer P, Tureci O, Eberle T, Graf N, Pfreundschuh M, Sahin U. Expression of serologically identified tumor antigens in acute leukemias. Leuk Res 2003; 27: 655-60.

Odunsi K, Jungbluth AA, Stockert E, Qian F, Gnjatic S et al. NY-ESO-1 and LAGE-1 cancer-testis antigens are potential targets for immunotherapy in epithelial ovarian cancer. Cancer Res 2003; 63:6076-83.

Osterlund C, Tohonen V, Forslund KO, Nordqvist K. Mage-b4, a novel melanoma antigen (MAGE) gene specifically expressed during germ cell differentiation. Cancer Res 2000; 60: 1054-61.

Park S, Lim Y, Lee D, Cho B, Bang YJ, Sung S, Kim HY, Kim DK, Lee YS, Song Y, Jeoung DI. dentification and characterization of a novel cancer/testis antigen gene CAGE-1. Biochim Biophys Acta 2003; 625: 73-82.

Rajpert-De Meyts E, Jacobsen GK, Bartkova J, Aubry F, Samson M, Bartek J, Skakkebaek NE. The immunohistochemical expression pattern of Chk2, p53, p19INK4d, MAGE-A4 and other selected antigens provides new evidence for the premeiotic origin of spermatocytic seminoma. Histopathology 2003; 42: 217-26.

Rotter V, Schwartz D, Almon E, Goldfinger N, et al. Mice with reduced levels of p53 protein exhibit the testicular giant-cell degenerative syndrome. Proc Natl Acad Sci 1993; 90: 9075-9.

Rudolph L, Chang S, Lee HW, Gottlieb GJ et al. Longevity, stress response and cancer in aging telomerase deficient mice. Cell 1999; 96: 701-12.

Ruiz A, Pujana MA, Estivill X. Isolation and characterization of a novel human gene (C9orf11) on chromosome 9p21, a region frequently deleted in human cancer. Biochim Biophys Acta 2000; 1517: 128-34.

Sarcevic B, Spagnoli GC, Terracciano L, Schultz-Thater E, Heberer M, et al. Expression of cancer/testis tumor associated antigens in cervical squamous cell carcinoma. Oncology 2003; 64: 443-9.

Scarman AL, Hooper JD, Boucaut KJ, Sit ML, Webb GC, Normyle JF, Antalis TM. Organization and chromosomal localization of the murine Testisin gene encoding a serine protease temporally expressed during spermatogenesis. Eur J Biochem 2001; 268: 1250-8.

Schopperle WM, Kershaw DB, DeWolf WC. Human embryonal carcinoma tumor antigen, Gp200/GCTM-2, is podocalyxin. Biochem Biophys Res Commun. 2003; 300: 285-90.

Schwartz D, Goldfinger N, Rotter V. Expression of p53 protein in spermatogenesis is confined to the tetraploid pachytene primary spermatocytes. Oncogene 1993; 8: 1487-94.

Siegrist S, Feral C, Chami M, Solhonne B, Mattei MG et al., . hH-Rev107, a class II tumor suppressor gene, is expressed by post-meiotic testicular germ cells and CIS cells but not by human testicular germ cell tumors. Oncogene 2001; 20: 5155-63.

Sjoblom T, Lahdetie J. Expression of p53 in normal and gamma-irradiated rat testis suggests a role for p53 in meiotic recombination and repair. Oncogene 1996; 12: 2499-505.

Skotheim RI, Korkmaz KS, Klokk TI, Abeler VM, Korkmaz CG, Nesland JM, Fossa SD, Lothe RA, Saatcioglu F. NKX3.1 expression is lost in testicular germ cell tumors. Am J Pathol. 2003; 163: 2149-54.

Stephan H, Polzar B, Rauch F, Zanotti S, Ulke C, Mannherz HG. Distribution of deoxyribonuclease I (DNase I) and p53 in rat testis and their correlation with apoptosis. Histochem Cell Biol 1996; 106: 383-93.

Tajima K, Obata Y, Tamaki H, Yoshida M, Chen YT et al., Expression of cancer/testis (CT) antigens in lung cancer. Lung Cancer 2003; 42: 23-33.

Takahashi K, Shichijo S, Noguchi M, Hirohata M, Itoh K. Identification of MAGE-1 and –4 proteins spermatogonia and primary spermatocytes of testis. Cancer Res 1995; 55: 3478-82.

Tan K, Shaw AL, Madsen B, Jensen K, Taylor-Papadimitriou J, Freemont PS. Human PLU-1 Has transcriptional repression properties and interacts with the developmental transcription factors BF-1 and PAX9.J Biol Chem 2003;278:20507-13.

Toppari J, Suominenf JS, Yan W. The role of retinoblastoma protein family in the control of germ cell proliferation, differentiation and survival. APMIS 2003;111:245-51

West A, Lahdetie J. P21 ^{WAF1} expression during spermatogenesis of normal and X-irradiated rat. Int J Radiat Biol 1997; 73: 283-91.

Xm MU, Liu Yx, Collins LL, Kim E, Chang C. The p53/retinoblastoma-mediated repression of testicular orphan receptor-2 in the rhesus monkey with cyrptorchidism.. J Biol Chem 2000; 275: 2387-83.

Yin Y, DeWolf WC, Morgentaler A. p53 is associated with the nuclear envelope in mouse testis. Biochem Biophys Res Commun 1997; 235: 689-94.

Zabludoff SD, Wright WW, Harshman K, Wold BJ. BRCA1 mRNA is expressed highly during meiosis and spermiogenesis but not during mitosis of male germ cells. Oncogene 1996; 13: 649-53

Zendman AJ, Ruiter DJ, Van Muijen GN. Cancer/testis-associated genes: identification, expression profile, and putative function. J Cell Physiol 2003; 194 :272-88.

Zhang Y, Wang Z, Liu H, Giles FJ, Lim SH. Pattern of gene expression and immune responses to Semenogelin 1 in chronic hematologic malignancies. J Immunother 2003; 26: 461-7.

Chapter 33

SELECTIVE GROUP OF GERM CELL SPECIFIC PROTEINS

33.1 POLYAMINES AND ORINITHINE DECARBOXYLASE

33.1.1 Polyamines

Polyamines (puterscine, spermidine, and spermine)) are necessary components for normal cell growth. The polyamine content of cells is elaborately regulated by biosynthesis, degradation, uptake and excretion. The proliferative effects of polyamines on cell growth are thought to be due to stimulation of nucleic acid and protein synthesis. Extra-cellular spermine has multiple effects at the N-methyl-D-aspartate (NMDA) subtype of glutamate receptor, including stimulation that increases the size of NMDA receptor currents, and voltage-dependent block. The interactions between polyamines and DNA are weak (0.4 to 1.3mol/100 phosphate of DNA). So, the structural change of DNA caused by polyamines may be small compared with the structural change of RNA caused by polyamines. Since polyamines preferentially bind to GC rich regions of RNA or DNA, therefore, polyamine effects in GC rich regions of DNA in cell free system reflect the effects *in vivo*. Polyamine mediated condensation is inhibited by histone hyperacetylation and hence polyamines are repressors of transcription in *vivo,* and that one role of histone hyperacetylation is to antagonize the ability of polyamines to stabilize highly condensed states of chromosomal fibers. Polyamines, especially spermine make a ternary complex with ATP-Mg^{2+} that can affect phosphorylation by protein kinases. Casein kinase II is strongly enhanced by polyamines. It seems that puterscine has strictly selective local stimulatory and inhibitory actions during spermatogenic DNA synthesis, and that its excess amount ultimately may lead to decrease in fertility (Hakovirta et al., 1993). The polyamine levels are strictly regulated in male germ cells, by ornithine decarboxylase (ODC) perhaps in the following manner: ODC catalyses polyamine biosynthesis, and polyamine stimulates synthesis of antizyme OAZ-t through the programmed +1 frame shift mechanism. In turn, the synthesized OAZ-t facilitates the degradation of ODC. In this way, OAZ-t functions as a negative regulator of ODC activity to regulate polyamine concentrations in elongated spermatids, and plays an important role in controlling spermiogenesis (Tosaka et al., 2000). Orinithine decarboxylase - overproducing transgenic mice showed a reduction in sperm count, malformation of mature spermatozoa, and impaired motility.

33.1.2. Orinithine Decarboxylase

Orinithine decarboxylase catalyzes the conversion of orinithine to putresine and is the first rate-controlling enzyme in polyamine biosynthesis in mammalian cells. Since polyamines are essential for normal cell growth but cytotoxic in excess, therefore, cells have developed several ways to regulate the levels of polyamines rapidly and transiently. The cell content and the level of polyamine biosynthetic enzymes change dramatically in response to hormones, growth factors, and turnover of promoters. The ODC levels increase under conditions of rapid cell growth, proliferation, and differentiation. The ODC is regulated at the level of transcription, protein stability and possibly also at a post-translational level. The carboxy terminal end of ODC with antizyme (AZ) plays a critical role for being trapped by the 26S proteasome. Thus, the degradation pathway of ODC proceeds as a sequence of multiple distinct processes, and ultimate degradation mediated by the 26S proteasome (Murakami et al., 2000).

Several unique features for molecular mechanisms underlying the polyamine regulated ODC degradation in animal cells have been documented: (i). polyamine-enhanced degradation is mediated exclusively by antizyme (AZ), a 26.5-kDa protein induced by polyamines. (ii) Polyamines promote biosynthesis of AZ by raising programmed, ribosomal frame shifting. (iii). AZ-mediated ODC degradation is cancelled by AZ inhibitor, which is the product of an ODC related gene and binds to AZ more firmly than ODC does. (iv) ODC associated with AZ is degraded by the 26S proteasome with ubiquitination. AZ inhibitor apparently stabilizes newly synthesized ODC during the phase of ODC increase after growth stimuli, whereas AZ accelerates ODC degradation during the phase of ODC decline. In addition to the role in the polyamine-regulated ODC degradation, AZ also suppresses cellular uptake of polyamines. These dual functions of AZ effectively prevent excessive accumulation of cellular polyamines.

Transgene mouse lines (K2 and K15) over-express ODC and hence contain high level of puterscine. In K2 mice with 30 fold testicular ODC over-expression showed a specific stimulation of DNA synthesis in type A4, intermediate, and type B spermatogonia. The K15 mice that had about 70-fold ODC over-expression, showed an elevation of DNA synthesis only at stage V of the cycle, suggesting a specific dependence of type B spermatogonia on puterscine. In these animals [3H] thymidine incorporation at stage VIII of tubule segments was decreased, suggesting that excess amounts of puterscine selectively inhibited meiotic DNA synthesis. The level of two ODC transcripts (2.2 and 2.7-kb) was low but detectable in the testes of 6-16-day-old mice and increased substantially as the first spermatogenic wave proceeded into spermiogenesis. In agreement with the developmental studies, ODC mRNA levels increased substantially in enriched populations of pachytene spermatocytes, round spermatids, and residual bodies/ cytoplasts isolated from mature testes. The ODC mRNA is distributed in both non-polysomal and polysomal fractions, which suggested that ODC is translationally regulated in the mouse testis.

It seems that only little testicular ODC is mainly associated with Sertoli cells and activity is seen in fractions containing other cell types, including spermatids. FSH induced ODC activity in Sertoli cells obtained from 7- to 14-day-old calves. However FSH decreased the activity of ODC in Sertoli cells of 18 days old rat. The two ODC mRNAs of 2.2 and 2.7-kb are present in mouse and rat testes. Considering the ubiquitous distribution and multiple functions of polyamines, it is not surprising that ODC mRNA is present throughout the mammalian testis. One can only speculate that the increased levels of ODC during later stages of spermatogenesis may well reflect hormonal effects or essential roles the polyamines play in chromatin alteration or condensation during meiosis and spermiogenesis (Alcivar et al., 1989).

33.1.3. Ornithine Decarboxylase Antizyme

A testis specific antizyme, ornithine decarboxylase antizyme (OAZt) has been described by various authors (Tewari et al., 1994; Ivanov et al., 2000; Tosaka et al., 2000) **(Fig.33.1)**. The

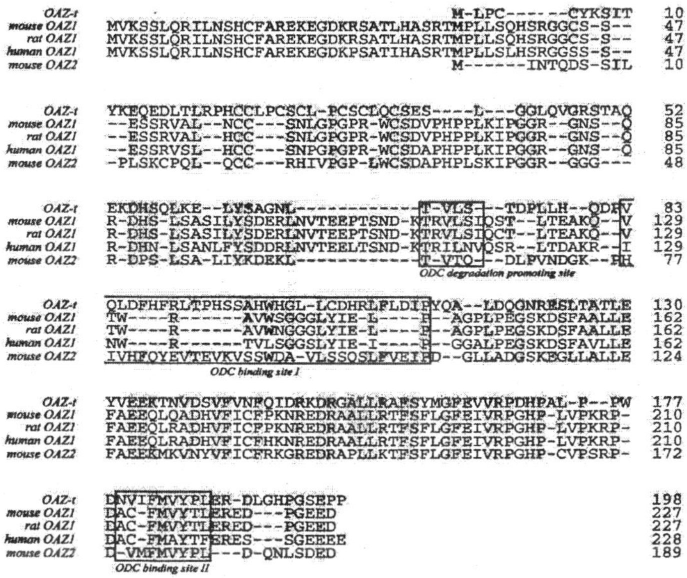

Fig.33.1. The amino acid sequence of OAZ-t is aligned with the somatic OAZ1 of mouse, rat and human and the mouse somatic OAZ2. The somatic OAZ1 regions to bind to ODC (ODC binding sites I and II) and to promote degradation of ODC (ODC degradation promoting site) are boxed, respectively. Printed with permission from Y. Tosaka et al. Genes Cells 5;265-76:2000© Blackwell Science.

originally discovered antizyme (termed antizyme1) binds ODC, inactivates it in a catalytic manner, targets it for degradation by 26S proteasome. The structure and mapping of antizyme 1 have been described (c/r Ivanov et al., 2000). Recently antizyme 2 was identified and its encoding gene identified. Both antizyme 1 and 2 are present in most of the mammalian tissues. The mRNA encoding the antizyme in testis (OAZ-t) is specifically expressed in haploid germ cells. In contrast, the mRNA level of somatic ODC antizyme 1 (OAZ1) decreased markedly at the late stages of haploid germ cell differentiation. The OAZ-t mRNA was first observed on day 23 whereas the OAZ-t protein was detected at 35 days after birth. Polyamines were capable of inducing a frame-shift at the frame-shift sequence of OAZ-t mRNA, resulting in the translation of OAZ-t, as was the case with the somatic OAZ1. Transfection of OAZ-t cDNA inactivated the ODC activity in the HEK293 cells. Thus expression of OAZ-t is controlled at both transcriptional and translational levels, and that OAZ-t likely plays a key role in regulating the intracellular concentration of polyamines in haploid germ cells. The 357-bp region, which includes a TATA-less promoter and an untranslated region, is sufficient for *OAZt* gene expression in the spermatids of transgenic mice. The 10-bp element that contains an initiator (Inr) plays a central role as the core promoter, in combination with a downstream element, while two CRE-like sites in the upstream region also contribute to promoter activity (Ike et al., 2004). Ivanov et al, (2000) described the presence of a 3rd antizyme present exclusively in testicular germ cells. Like antizyme 1 and 2, antizyme 3 cDNA has same arrangement of reading frames and a potential shift site with definite, although limited homology to antizyme 1 and 2. In testis, antizyme 3 expression starts in early spermiogenesis and finishes in the late spermatids phase (Ivanov et al., 2000; Tosaka et al., 2000).

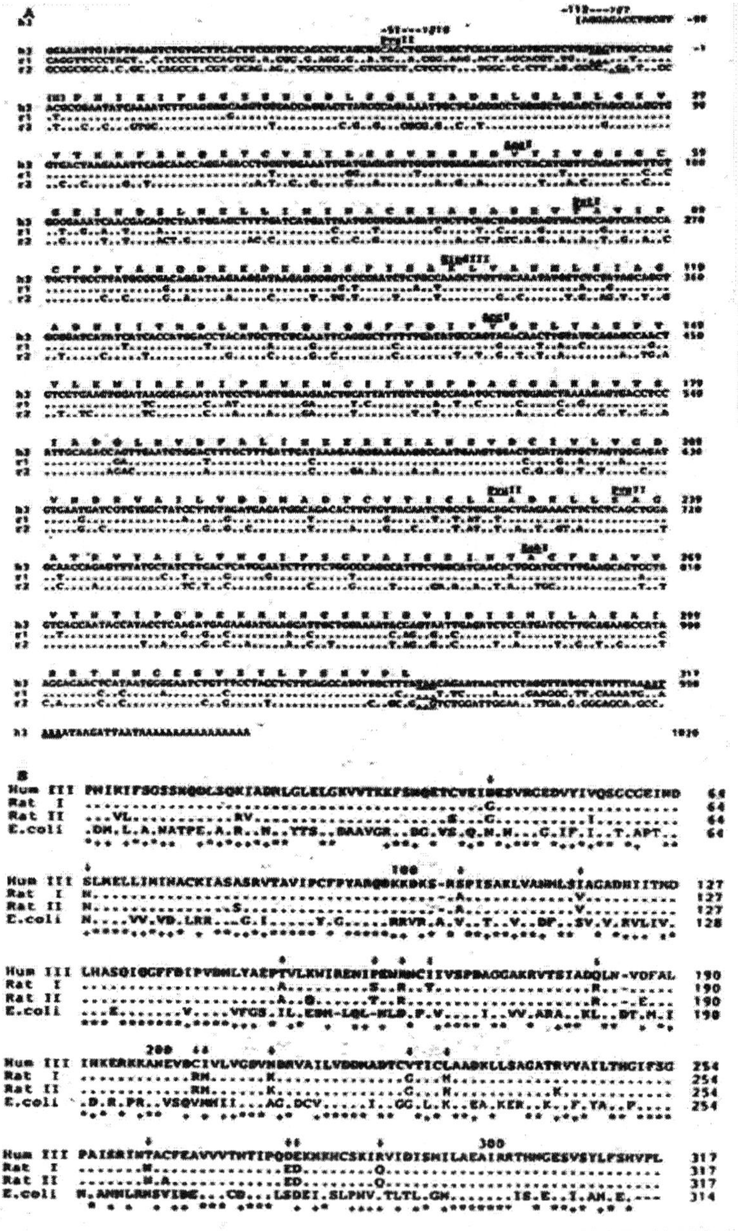

Fig.33.2. Nucleotide and deduced amino acid sequence of human PRPS3 cDNA. (A) Nucleotide sequence of PRPS3 cDNA (human) and its comparison with PRPS3, rat PRPS1, and rat PRPS2, cDNAs. Comparison of perdicted amino acid sequences of testis specific PP-Rib-P (PRS III) subunit (Hum III) and its comparisan with rat PRSI (Rat I) and PRSII (Rat II), and E.Coli enzyme (E.Coli). Dots and asterisk (*) denote identical residues among four. + indicates conservative substitutions. Printed with permission from M. Taira et al, J Biol Chem 265; 16491-97:1990 © American Society for Biochemistry and Molecular Biology.

33.2. SELECTIVE GROUP OF GERM CELL SPECIFIC PROTEINS

33.2.1. Isoaspartyl Methyltransferase

Spermatogenic cells possess high levels of L-isoaspartyl (D-asparty)-O-methyltransferase (PIMT, E.C. 2.1.1.77), which modifies altered aspartyl residues that arise from spontaneous deamidation or isomerization processes in cells. The PIMT has been shown to initiate the repair of age-damaged proteins by catalyzing the conversion of L-isoaspartyl residues to L-aspartyl residues. A unique transcript-encoding PIMT is expressed in mouse testes that is distinguished from other transcripts by the lack of splicing in the 3'-untranslated region (3-UTR). Northern blot analysis revealed that a unique 1.1-1.3-kb PIMT mRNA transcript is present in round spermatid and pachytene spermatocyte poly (A)+ RNA. Small PIMT transcript is four times higher in round spermatids than in pachytene spermatocytes, indicating that the PIMT gene is actively transcribed during the haploid phase of spermatogenesis. The germ cell-specific PIMT transcripts are distributed between the polysomal fraction and the non-polysomal fractions of testis RNA, suggesting that translational controls also contribute to the high concentrations of PIMT in mammalian sperm. PIMT function is not essential for spermatogenesis, since normal spermatozoa were recovered from transgenic testis lacking PIMT. The protein repair function of the PIMT may be more important in maintaining the fertilization competence of translationally-inactive mature sperm during the prolonged period of epididymal transit and storage in the male reproductive tract (Chavous et al., 2000 also see c/r).

33.2.2. Phosphoribosyl Pyrophosphate (PP-Rib-P) Synthetase

Phosphoribosyl pyrophosphate (PP-Rib-P) synthetase (ATP: D-ribose 5 phosphate pyrophosphotransferase, EC 2.7.6.1) catalyzes a step in the utilization of ribose 5-phosphate during pentose phosphate pathway to form purine, pyrimidine, and pyridine nucleotides. The PP-Rib-P is an oligomeric complex composed of 34-kDa subunits. Rat enzyme shows the presence of two distinct types of subunits, PRS I and PRS II. Either PRS I or PRS II or both mRNAs are detected in almost all tissues of the rat. However, in the testes of rats, mice and humans, an additional mRNA has been detected. Human gene mapping showed that these genes, referred to as *PRPSI* and *PRPS2*, are located on different regions of the X chromosome, while two additional PRPS I-related sequences were present on autosomes. The human cDNAs, cloned for the testis specific autosomal PRPS3-mRNA revealed that the human testis-specific PRPS3 mRNA was encoded by an autosomal gene, termed *PRPS3*. The expression of this gene began at 4 weeks of age in rats, coinciding with the reported appearance of primary spermatocytes. A cDNA clone of *PRPS3* encodes a predicted product of 317 amino acids, which was highly homologous to those of *PRPS1* and *PRPS2* (94.3% and 91.2% identities, respectively). However, the *PRPS3* cDNAs lacked an ATG initiator for translation at the expected position, and instead contained an ACG triplet. Analysis of amino-terminal sequence of the radio-labeled *PRPS3* product supported the predicted sequence starting with Pro-1. Study suggested that the synthesis of the nascent polypeptide could initiate with methionine at the position corresponding to the ACG codon (Taira et al, 1990) **(Fig.33.2)**. Three genes identified for subunits of PP-Rib-P synthetase (X-linked PRPS1 and PRPS2 and autosomal PRPS3) showed differential developmental expression in testis. It is speculated that the expression of autosomal PRPS3 may be associated with X-chromosome inactivation in the spermatogenic cells, as has been postulated for X-linked PGK-1 and autosomal/testis-specific PGK-2 (Taira et al, 1990).

33.2.3. Glucosamine-6-Phosphate Deaminase

During mammalian fertilization, the oocyte is dependent on the sperm and its ability to induce intracellular Ca^{2+} oscillations. A cytosolic sperm factor from hamster spermatozoa, called oscillin seemed to correlate with the ability of sperm to cause calcium oscillations and activation when injected into mouse oocytes. Oscillin showed a high sequence homology to bacterial glucosamine 6-phosphate deaminase (GNPDA). Initially found to be present in only hamster testis and spermatozoa, Wolosker et al. (1998) reported a ubiquitous distribution of mammalian (rat) GNPDA. However, the role of sperm GNPDA is still unresolved. The GNPDA catalyzes the formation of fructose 6-phosphate and ammonia from glucosamine 6-phosphate. Mouse glucosamine phosphate deaminase (GNPDA/Oscillin) has been cloned. A polyclonal rabbit anti-GNPDA antibody recognized a 33-kDa protein in soluble extracts of various tissues including testis and sperm. Localization of GNPDA in male reproductive tissue revealed its presence in spermatids and in spermatozoa. In spermatids, GNPDA localized close to the developing acrosome vesicle and in spermatozoa close to the acrosomal region. The GNPDA plays a role within the cascade, which leads to the initiation and completion of the acrosome reaction (Montag et al., 1999). However, Amireault and Dube (2000) suggested that GNPDA/oscillin is ubiquitously expressed in rat tissue and that a recombinant hamster GNPDA/oscillin protein does not exhibit oscillin activity when injected into oocytes. In the mouse, the nature and role of such a GNPDA/oscillin is not clear whereas another protein, tr-kit, has been proposed as a sperm factor causing oocyte activation. Mouse GNPDA/oscillin shows over 96% identity with the hamster and human homologs. The GNPDA is localized in the equatorial and neck regions of the human spermatozoa and the post-acrosomal region of the hamster spermatozoa (Amireault and Dube, 2000).

33.2.4. Carnitine Transferases

Carnitine: Spermatozoa produced in the testis, undergo post-gonadal modifications in the epididymis to acquire fertilizing ability. In epididymis high molecular-weight proteins and molecules as small as free-L carnitine convert the male gamete into "competent" and functional cells. In mammals, including man, free L-carnitine is taken from blood plasma and concentrated in the epididymal lumen. This epididymal secretion is beneficial for sperm and is not merely an excretory waste. Free L-carnitine goes through the sperm plasma membrane by passive diffusion, where it is acetylated in mature sperm. The excess acetyl-CoA from the mitochondria is probably stored as acetyl-L-Carnitine and modulates the reserves of free CoA essential to the function of the tricarboxylic acid cycle. These properties of L-carnitine of buffering CoA in the mitochondrial matrix are known in somatic cells but are also accentuated in the male germinal cells. The relation between the endogenous pools of free and acetylated L-carnitine and the percentage of progressive sperm motility indicates a more important metabolic function related to flagellar movement. The uptake of "cytoplasmic" free L-carnitine in mature spermatozoa may be a protective form of mitochondrial metabolism, useful to the survival of this isolated cell (Jeulin and Lewin, 1996).

Carnitine Transferases: Carnitine palmitoyltransferases I and II (CPT I and II), and carnitine acetyltransferase (CAT) are key enzymes in the mitochondrial transport of long-chain fatty acids. Acetyl-CoA derived from the oxidation of long-chain fatty acids can react with carnitine under the influence carnitine acetyltransferase to yield acetylcarnitine. The CAT activity also develops during spermatogenesis, reaches high levels by spermatids and is fully retained by the spermatozoa. During their transit through the epididymis, the maturing sperm encounter extremely high levels of carnitine in the surrounding fluid from which they generate a large reservoir of acetylcarnitine, which is thought to provide a ready source of energy during their post-ejaculation activity.

The CPT I and II are key enzymes in the mitochondrial transport of long-chain fatty acids. The outer membrane enzyme, CPT I is inhibited by malonyl-coenzyme A (CoA). Whereas CPT II (inner membrane and malonyl-CoA insensitive) appears to be the product of a single gene, located on chromosome1p32 in humans, CPT I exists in at least two isoforms L (Liver) and M (muscle) types, with different kinetic properties and sensitivity to malonyl CoA. Adult rat testis is rich source of mRNA for M form of mitochondrial CPT I, where as L-CPT I is the sole isoform of immature testis. The human genes for L and M-CPT I reside on chromosomes 11q13 and 22q13, respectively (McGrarry and Brown, 1997). Expression of the *M-CPT I* gene was associated only with meiotic and post-meiotic germ cells. Although the adult testis contains a mixture of the L-and M-CPT I enzymes, the L and M form dominate the extra-tubular cells and spermatids, respectively. Mature epididymal spermatozoa appeared to be deficient in CPT I activity, while CPT II and CAT were abundantly present. Five days of dietary etomoxir treatment at a dose that resulted in complete inhibition of CPT I in somatic tissues was totally without effect on either the L- or M-type enzyme in the testis of mature rats. This points to an important role for transient expression of M-CPT I, coupled with sustained activity of CAT, in the maturation and/or function of rat sperm. It suggests that at least one form of testis CPT is unusually resistant to inhibitor (etomoxir), when given orally (Adams et al., 1998).

33.2.5. Organic Cation/Carnitine Transporters

Carnitine transporter (CT2) represents a novel transporter family, specifically expressed in human testis (Enomoto et al., 2002). It is phylogenetically located between the organic cation transporter (OCT/OCTN) and anion transporter (OAT) families. Hence, when expressed in *Xenopus oocytes* CT2 mediates the high affinity transport of L-carnitine but does not accept mainstream OCT/OCTN cationic or OAT anionic substrates. The CT2 protein is located in the luminal membrane of epididymal epithelium and within the Sertoli cells of the testis (Enomoto et al., 2002). Spermatozoal maturation depends upon the progressive increase in epididymal and spermatozoal carnitine, which is critical for mitochondrial fatty acid oxidation, as sperm pass from the caput to the cauda region of the epididymis. The organic cation/carnitine transporters, OCTN1, OCTN2, and OCTN3, are expressed in sperm as three distinct proteins with an expected molecular mass of 63-kDa. Identification of individuals with defective sperm carnitine transport may provide potentially treatable etiologies of male infertility, responsive to L-carnitine supplementation (Xuan et al., 2003).

33.2.6. NM23-H5 (Nucleoside Diphosphate Kinase)

Nucleoside diphosphate (NDP) kinases, responsible for the synthesis of nucleoside triphosphates and produced by the *nm23* genes, are involved in numerous regulatory processes associated with proliferation, development, and differentiation. A new *Nm23-H5* is specifically expressed in testis germ cells and located on chromosome 5q23-31. The cDNA encoding a 212 amino acid protein is 27-31 % identical to the other human NDP kinase gene family and is transcribed as 1.1-kb transcript. Nm23-H5 was present in sperm extract, together with the ubiquitous A and B NDP kinases (but not the C and D isoforms). Nm23-H5 is located in the flagella of spermatids and spermatozoa, adjacent to the central pair and outer doublets of axonemal microtubules. High levels of NDP kinases A and B were observed at specific locations

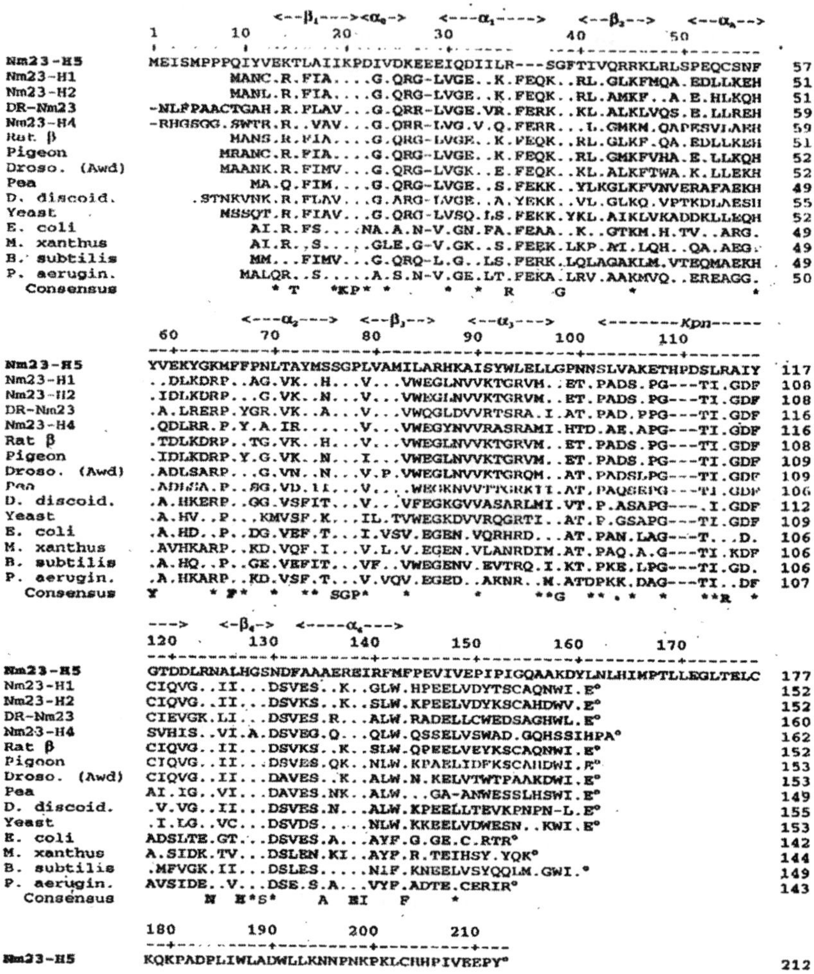

Fig.33.3. Alignment of the amino acid sequence of human Nm23-H5/NDP kinase of human (Nm23-H1, Nm23-H2, DR-Nm23 and Nm23-H4), eukaryotic (rat, pigeon, Drosophila, pea, slime mold and yeast) or prokaryotic (E. coli M. xanthus, B. subtilis and P. aeruginosa) origin. The numbering at the top refers to the Nm23-H5 sequence. Reproduced with permission from Munier A et al,FEBS Lett 434;289-94:1998© Elsevier.

in post-meiotic germinal cells. NDP kinase A was transiently located in round spermatid nuclei and became asymmetrically distributed in the cytoplasm at the nuclear basal pole of elongating spermatids. The distribution of NDP kinase B was reminiscent of the microtubular structure of the manchette (Munier et al., 1998; 2003).

33.2.7. Farensyl Transferase

Changes in protein prenyltransferase activity with levels of prenylated protein, and the type of isoprenoid modification are known in cells of rat seminiferous epithelium and correlate with

```
A ATT CGG TTT TTT TTG CGC GGC GTT TCA GTC TCT CTT GAG TTG GGC TGT  49

  TGT CGA GTC GCG GCG AGA CCG GCG CAC CGA GCC ATG ACA GAG GCT GAT  97
                                               M   T   E   A   D    5

  GTG AAT CCG AAG CCT ATC CCC TCG CAG ATG CCC ACC TCA CCA AGA AGC 145
   V   N   P   K   P   I   P   S   Q   M   P   T   S   P   R   S   21

  TGC TGG ACC TTG TTC AAC AGT CAT GTA ACT ACA AGC AGC TTC GGA AAG 193
   C   W   T   L   F   N   S   H   V   T   T   S   S   F   G   K   37

  GAG CCA ATG AAG CCA CCA AAA CCC TCA ACA GAG GCA TCT CTG AGT TAC 241
   E   P   M   K   P   P   K   P   S   T   E   A   S   L   S   S   53

  ATT GTG ATG GCA GCA GAC GCT GAG CCC TTG GAG ATC ATC CTG CAC CCT 289
   L   V   M   A   A   D   A   E   P   L   E   I   I   L   H   P   69

  CCC TCT GCT GTG CGA AGA CAA GAA TGT CCC CTA CGT ATT TGT GCG CTC 337
   P   S   A   V   R   R   Q   E   C   P   L   R   I   C   R   L   85

  CAA GCA GGC TTT AGG ACG GCG TGT GGC GTC TCC AGG CCA GTC ATC GCC 385
   Q   A   G   F   R   T   A   C   G   V   S   R   P   V   I   A  101

  TGT TCT GTT ACC ATC AAA GAA GGC TCT CAA CTA AAG CAG CAG ATC CAG 433
   C   S   V   T   I   K   E   G   S   Q   L   K   Q   Q   I   Q  117

  TCC ATT CAG CAG TCT ATT GAA AGG CTC TGG TGT AGG CTG TGG CCT CTG 481
   S   I   Q   Q   S   I   E   R   L   W   C   R   L   W   P   L  133

  CCC TTT CCT GTC TAC TGC CAC CCT GTT GTG TAC ATA TTG CTG TTA GCA 529
   P   F   P   V   Y   C   H   P   V   V   Y   I   L   L   L   A  149

  TGT ACT ATT TCA ACC AGT TAC TAT TAT AAT TGT ACT ACA TCT CGT TAA 577
   C   T   I   S   T   S   Y   Y   Y   N   C   T   T   S   R   *  164
                                      -----------
  AAA AAA AAA AAA AAA AAA AAA AAA AAA AAA AAA AAA AAA AAA AAA AAA 625

  AAA AAA AAA AAA AAA AAC CGA ATT                                 649
```

Fig 33.4. Nucleotide and deduced amino acid sequences of Fertilization antigen-1. Reproduced from X. Zhu and R.K. Naz. Proc Natl Acad Sci USA. 1994;4704-09:1997© National Acadamy of Sciences,USA.

differentiative events of spermatogenesis. The activity of protein farensyltransferase (FTase) is at least 10-fold higher than that for protein geranylgeranyltransferase-I (PGGT-I) in seminiferous epithelium and spermatogenic cells of pre-pubertal rats of different ages. These activities increase during the meiotic stages of differentiation and reach peak at 23 days of age. The highest activity was associated with intermediate sized spermatocytes appearing late during meiosis, suggesting that independent changes in prenyltransferase activity and protein prenylation accompany the differentiation events during the pre-meiotic and meiotic stages of spermatogenesis. The expression pattern and distribution of protein FTase in hamster showed only weak activity of FTase in round spermatids, whereas elongating spermatids exhibited a high level of FTase expression that was segregated to the cytoplasmic lobe surrounding the anterior flagellum. Although FTase was released with the residual body, mature spermatids retained FTase within the mid-piece and cytoplasmic droplets. In epididymal spermatozoa, FTase remained associated with the cytoplasmic droplet during its migration to the mid-piece-principal piece junction. Both the α- and β-subunits of FTase were present in sperm lysate (Olson et al., 1997).

33.3. OTHER DOMINANT PROTEINS IN GERM CELLS

Fertilizing Antigens: A group of sperm specific antigens has been studied by Naz and associates, with a search of their application in contraceptive technology. Fertilization antigen-1 (FA-1) is involved in human immunologic infertility in both men and women. FA-1 interacts with ZP3 receptor on ZP of ovum, which is involved in acrosome reaction (Kadam et al., 1995; Zhu and Naz, 1997) **(Fig 33.4)**. The FA-2 antigen from sperm is another candidate, which was identified on immunoaffinity column with a Mr of 95-kDa (Naz et al., 1993). Another antigen, designated NZ-3 is located on acrosome and tail regions of human sperm (Naz and Zhu, 2001).

A c-DNA encoding a testis specific antigen-1 (TSA-1) was reported by Santhanam and Naz (2001). The TSA-1 has an ORF of 471 bp encoding a protein of 156 aa and a molecular mass of 17.4. The recombinant TSA-1 is present on acrosomal, equatorial, mid-piece and tail regions of human sperm

Phosphatidyl Choline Transfer Protein: Yamanaka et al, (2000) isolated a cDNA clone specifically expressed in mouse testis during spermiogenesis. The cDNA consisted of 1392 nt and had an ORF of 873 nucleotides encoding a protein of 291 amino acid residues. The deducted amino acid sequence had similarity to mouse phosphatidylcholine transfer protein (PCTP). This newly PCTP-like protein revealed a 1.4-kb mRNA expressed in the testis, kidney, liver, and intestine with the highest level in the testis. The protein, having a molecular weight of approximately 40-kDa, was detected at the tail of the elongated spermatids and sperm (Yamanaka et al., 2000). One of the clones, TISP-81 showed specific expression in haploid cells and significant homology with PCTP.

Sulfoglycolipid Immobilizing Protein Slip: Sulfogalactosylglycerolipid (SGG) is the major sulfoglycolipid of mammalian male germ cells. Like other sulfoglycolipids, SGG is believed to be involved in cell-cell/extracellular matrix adhesion. Sperm SGG plays a role in sperm-egg interaction. Anti SGG-IgG localized SGG to the convex and concave ridges and the postacrosome of the mouse sperm head. Pretreatment of sperm with anti-SGG-IgG/Fab inhibited sperm-zona pellucida binding in vitro. Sperm treated with anti-SGG-IgG underwent the spontaneous and ZP-induced acrosome reaction suggesting an involvement of sperm SGG in direct binding to the ZP (White et al., 2000). A sulfoglycolipid immobilizing protein, termed SLIP1, is a conserved germ cell membrane protein that has been shown in vitro to bind specifically to the mammalian germ-cell-specific SGG. The SLIP1, present on the acrosomal plasma membrane is absent from ZP. Low concentrations of exogenous purified SLIP1decreased the number of sperm bound per egg. Sperm SLIP1 is involved in sperm-egg interactions (Tanphaichitr et al., 1993).

Testis Specific Phosphatase: Testis contains a germ cell specific phosphatase (GCAP), which resembles with placental alkaline phosphatase. The enzyme is a marker for seminomas and is utilized for clinical use (Otto et al 1998). An embryonic alkaline phosphatase (EAP) expresses at the two-cell to blastocyst stage of preimplantation development in the mouse. The isozyme is re-expressed in trace amounts in the thymus, intestine and testis during adult life. The EAP transcripts can be detected, in very low amounts in the testes of newborn mice, but at 24 days of age, EAP mRNA levels reach the highest concentrations. No positive cells were recognized in the testis of newborns (day 0) and 8-day old mice. Immunoreactive positive cells were first observed at day 15 and, at 24 days of age, corresponding to spermatocytes in mid- to late prophase of meiotic division I. Positive M-phase cells were also observed at 40 days and 151 days of age. Transgenic mice expressing the human GCAP isozyme in a tissue-specific manner in the testis showed equivalent stages of M-phase staining (Narisawa et al., 1992).

P125/6.5 Nuclear Antigen: A p125/6.5 nuclear matrix antigen displaying a marked increase in mitotic cells was identified by Hadyiolova et al (1989) at different stages of postnatal development: In Sertoli cells, the p125/6.5 antigen parallels [3H] thymidine incorporation: present in newborn and absent in sexually mature testes. The antigen is present also in some prespermatogonia of the newborn rat testies, which do not incorporate [3H] thymidine. At later stages of development, the p125/6.5 antigen is present also in first meiotic prophase spermatocytes displaying an extra-chromosomal nucleoplasmic distribution, while absent in spermatids and spermatozoa. Thus the p125/6.5 antigen increases not only during mitosis, but also during meiosis.

Gametogenetins: The germ cell-deficient (gcd) mutation is a recessive transgenic insertional mutation leading to a deficiency of primordial germ cells (PGCs). The gene underlying this mutation is *Pog*, which is necessary for normal proliferation of PGCs. Meiosis is impaired in Pog-deficient mice. POG interacted with GGN1 and GGN3, two proteins formed by alternate splicing of the same gene, gametogenetin (*Ggn*). This gene has more than 10 different splice variants giving rise to three proteins (GGN1, GGN2, and GGN3), which show different cell localization. The expression of *Ggn* was confined to late pachytene spermatocytes and round spermatids, a time window concomitant with the occurrence of meiosis. Mouse *Ggn* and *Pog* are both expressed in primary spermatocytes. The GGN1 and GGN3 are spatially and temporally regulated proteins, which interact with and regulate the localization of POG, and play a role in spermatogenesis (Lu et al., 2003).

MHC Class I Molecule Mamu-AG: The nonclassical MHC class I locus HLA-G is expressed primarily in the placenta, although other sites of expression have been noted in normal and pathological conditions. In addition, HLA-G isoforms are present in the serum of pregnant and non-pregnant women as well as men. The rhesus monkey placenta expresses a nonclassical MHC class I molecule Mamu-AG, which has features remarkably similar to those of HLA-G. The Mamu-AG5 mRNA was readily detected in rhesus testis but not in sperm. Immunochemically, Mamu-AG was consistently stained the cells within the seminiferous tubules. While primary spermatocytes were negative, Sertoli cells, spermatocytes, and spermatids were consistently positive for immunostaining. This demonstrated that a specific soluble nonclassical MHC class I molecule is expressed in the rhesus monkey placenta and testis (Ryan et al., 2003).

Phosphatidylethanolamine Binding Proteins: The phosphatidylethanolamine binding proteins (pebps) comprise an evolutionarily conserved family of proteins recently implicated in mitogen-activated protein (MAP) kinase pathway regulation, where they are called raf kinase inhibitory proteins. A new member, pebp-2, with potential roles in male fertility has been described. The pebp-2 is a testis-specific 21-kDa protein found within late meiotic and haploid germ cells in a stage-specific pattern that is temporally distinct from that of pebp-1. The pebp-2 forms a distinct subset of the pebp family within mammals. Database analyses revealed the existence of a third subset. The specificity of the distinct pebps subsets is likely to be determined by the amino terminal 40 amino acids or the 3'-untranslated region, where the majority of sequence differences occur. However, protein modeling suggests that pebp-2 protein is topologically similar to other pebps and composed of Greek key fold motifs, a dominant β-sheet formed from five anti-parallel β strands forming a shallow groove associated with a putative phosphatidylethanolamine binding site. The *pebp-2* gene is intronless and it is a retrogene derived from pebp-1. The pebp-2 co-localizes with members of the MAP kinase pathway in late spermatocytes and spermatids and on the midpiece of epididymal sperm (Hickox et al.,2000; 2002).

Phosducin-Like Protein: A new member of the phosducin-like (PhLP) protein family is predominantly, if not exclusively, expressed in male and female germ cells. Spermatogenic cell fractions evidenced a stage-specific expression with high levels of RNA and protein in pachytene spermatocytes and round spermatids. Three mRNA species were detected, which correspond to different polyadenylation sites and vary in abundance during germ cell maturation (Lopez et al., 2003).

Phtfl Gene Product: Phtfl is a gene evolutionarily conserved from *Drosophila* to human that is abundantly expressed in testis. In adult rat, transcripts were abundant in germinal meiotic and postmeiotic cells. From weak activity in a juxtanuclear region of early pachytene spermatocytes, activity progressively extended and persisted until the end of spermiogenesis, when it was eliminated in the residual bodies. Phtfl protein showed the properties of an integral membrane protein. In haploid cells of rat seminiferous epithelium, it co-localized with calnexin and calmegin. Phtfl is a potential new marker for Golgi modifications as well as for many of the obscure transformations undergone by the endoplasmic reticulum (Oyhenart et al., 2003).

Sperm-Associated Glutamate (E)-Rich Protein: A family of genes, indicating an ORF, which codes 200-220 amino acid proteins, has been identified. These proteins have high proportion of α-helical secondary structure and comprise approximately 15% glutamate residues. Because of this property, the family has been named sperm-associated glutamate (E)-Rich protein (SPEER). Three members of this family, SPEER-1 (pseudogene), SPEER-2, and SPEER-4D, are expressed specifically in the testis of mice, with only very weak expression evident in the rat testis but in no other species tested (Spiess et al., 2003). All three transcripts are expressed at high levels in haploid germ cells at the spermatocyte-spermatid transition. While SPEER-1 mRNA is present in the cytoplasm as a sense transcript, SPEER-2 appears to be mostly as an antisense transcript, whereas SPEER-4D appears to be localized within a subcellular compartment as a conventional sense transcript. While there are no true homologies to other proteins in the genome databases, some motifs are present that suggested a relationship of these proteins to nuclear matrix proteins ((Spiess et al., 2003).

Serine-Rich Spermatogenic Protein-1: By means of mRNA differential display and cDNA library screening, Geisinger et al., (2002) characterized a novel gene of the rat that is differentially expressed during spermatogenesis. Its mRNA reaches the highest level during the first meiotic prophase. The transcript appears to be testis-specific as it was not detectable in any of the nine other tissues tested. The full length ORF encodes a putative phosphoprotein containing a serine stretch and a bipartite nuclear localization signal, which was called as SRSP1 (serine-rich spermatogenic protein-1. Comparison of cDNA of SRSP1 with genomic sequences in databases identified the presence of some putative regulatory sequences (Geisinger et al., 2002).

TEPP: A gene that is specifically expressed in testis, prostate, and placenta was identified. The gene has one major transcript of 1.0kb in size and encodes for a protein of 30.7kDa molecular weight. This gene was named as TEPP (expressed in testis, prostate, and placenta). The amino acid sequence analysis of TEPP using Signal P program shows that it has a signal peptide with a predicted cleavage site between amino acids 19 and 20, indicating that it might be a secreted protein. These proteins are highly conserved in chordates. In addition, a splice variant of TEPP, which encodes a 37kDa protein was also present (Bera et al., 2003).

Other Clones and Related Proteins: Several mouse and human testis-specific genes have been identified using antibodies raised against testicular germ cells. The relaxin like factor gene *(Rinl)* was mapped to chromosome (Chr) 8, and collapsin response mediator protein 1 *(Crmp1)* was mapped to Chr5. Three other genes encoding testis-specific proteins *A2 (Tsga2), A8*

(Tsga8), and *A12 (Tsga12)* were mapped to chromosomes 3, X, and 10, respectively (Taketo et al., 1997). Following genes express during spermatogenesis and seem to be involved in testicular germ cell differentiation and sperm formation:

i). *Gsg 1-3* are expressed specifically in testicular germ cells, in both 24-day-old and adult testes, but they are not expressed in 4-10-, or 16 day old testes. They are expressed in the haploid germ cells after meiotic division. Because the product of *Gsg3* has a 37% amino acid sequence identity to an actin-capping protein, Cap Z and *Gsg3* protein might interact with actin filaments during the morphogenesis of spermatids. Both *Gsg1* and *Gsg3* are mapped on Chr 6. *Gsg1* has been mapped close to *Fgf6* that belongs to the fibroblast growth factor gene family, some of which are involved in testicular functions. Putative positions of human homologs of *Gsg1* and *Gsg3* are 12p11-12 and 12p13, respectively. In the human, more than 80% of adult testicular germ cell tumors (TGCTs) have an isochromosome of the short arm of Chr 12, i(12p). This isochromosome is highly specific. Since chromosomal abnormality is found in human germ cell tumors, it is conceivable that *Gsg1* or *Gsg3* over-expression is involved in the tumorigenesis. The *Gsg2* gene was mapped to mouse Chr 11(Matsui et al., 1997).

ii). A testis-specific protein, termed LRTP, was identified by screening both human and mouse testis and mouse pachytene spermatocyte cDNA libraries. LRTP contains an amino terminus leucine-rich repeat domain. There are several acidic regions rich in glutamic acid in the C-terminus. The sequence, shows similarities to LANP and SDS22+ leucine-rich repeat proteins, ,localized to the nucleus and involved in the regulation of protein phosphatases. In mouse, the mRNA is first detected presumably in mid-pachytene spermatocytes. The protein is most abundant in the cytoplasm of pachytene and diplotene cells, corresponding to late prophase of meiosis I suggesting a functional association of LRTP with meiosis (Xue and Goldberg, 2000).

iii). Chen et al (1997) isolated two types cDNA clones specifically expressed in mouse testes. Type A cDNA (2,071 nt) is predicted to encode 347 amino acids, whereas type B cDNA (1,536 nt) has a deletion of 535 bp from nucleotides 1,206 to 1,740 of type A cDNA, probably due to alternative splicing. Both types of mRNAs were specifically expressed in testis; type B mRNA was more abundant than type A mRNA. These two mRNAs also expressed in immature testes until 24 days after birth, specifically in spermatogonia, Sertoli cells and Leydig cells. In the W/W^v mouse testis, which lacks c-kit activity and spermatogonia, but contains Sertoli and Leydig cell, both mRNAs were found to be expressed in the latter two types of cells. These clones were termed *tsec-1*, testis-specifically expressed cDNAs (Chen et al., 1997).

iv). A monoclonal antibody, MC301, against rat testicular extracts corresponding to the onset of meiosis recognized a 90-kDa glycoprotein, GP90-MC301. The glycoprotein was ubiquitously expressed in various tissues and localized predominantly in the Golgi area of epithelial cells. In testes, GP90-MC301 was shown in the cytoplasm of meiotic spermatogenic cells from the preleptotene to mid-pachytene stages. Glycoprotein was localized in spermatocytes on organelles such as Golgi apparatus, endoplasmic reticulum, and nuclear envelope and in the extracellular fluid of the testes (Tani et al., 2000).

v) A 2.5-kb cDNA *(hss)* codes for a human sperm surface protein (HSS), which contains five direct repeats, three mirror repeats, three possible stem loop structures and three unique palindromic motifs located at functionally relevant positions. Open reading frame encodes a protein of 766 amino acid. NH_2 terminus extra-cellular domain had a leucine zipper (LZ) motif, a transmembrane helical domain and a cytoplasmic domain at the COOH terminus. *Hss* transcript is present only in testis. Cell type expression of *hss* mRNA showed it only on the round spermatids suggesting haploid germ cell expression if *hss* (Shankar et al., 1998)

vi) Iba1: During rat testis development a cDNA fragment isolated by differential display method turned out to be Iba1, an ionized calcium binding adapter molecule-1, that contains two EF hand like motifs. The Iba1 mRNA is present in the rat testis first at 4 week and then increased up to adulthood. The Iba 1 protein was specifically expressed in the cytoplasm of elongate spermatids as well as in residual bodies that are ultimately engulfed by Sertoli cells. On the other hand, Iba1 mRNA is present in round spermatids as well as early elongate spermatids (steps 1 – 12) but not in late spermatids, suggesting that Iba1 mRNA undergoes post-transcriptional regulation. It seems that Iba1 may be involved in the final stages of spermiogenesis (Iida et al., 2001).

vii) RSD-3 gene: A novel full-length cDNA composed of 2228 bp, designated as RSD-3 is expressed in germ cells of testis. The reading frame encodes a polypeptide consisting of 526 amino acid residues, containing a number of DNA binding motifs and phosphorylation sites for PKC, CK-II, and p34cdc2. The initial expression of the RSD-3 gene was detected in the testis on the 30th postnatal day and attained adult level on the 60th postnatal day; its expression is restricted to primary spermatocytes, undergoing meiosis division I. The human testis homolog of RSD-3 cDNA, designated as HSD-3.1 consisted of 12 exons that span approximately 52.8 kb of the genome sequence and was mapped to chromosome 14q31.3 (Zhang et al., 2003).

viii) PIASx and PIAS1, two members of the conserved protein inhibitor of activated STAT (PIAS) family, are able to interact with and modulate activities of several distinct nuclear proteins, including androgen receptor. PIASx gene encodes two SUMO E3 ligases that are highly expressed in the testis. The expression of both PIASx and PIAS1 was low or undetectable in newborn rats. PIASx mRNA started to accumulate after day 20 of postnatal life, whereas expression of PIAS1 mRNA increased around day 30 after birth. In the adult rat testis, both PIASx and PIAS1 mRNA were present in Sertoli cells and in germ cells in the seminiferous epithelium. However, PIASx mRNA was more abundant in spermatocytes than in other cell types, whereas higher levels of PIAS1 mRNA were detected in late spermatocytes and round spermatids than in early spermatocytes. Since PIASx and PIAS1 accumulate in developing male germ cells, their regulatory functions are not only restricted to androgen receptor in Sertoli cells, but they also participate in molecular processes during meiosis. The proximal promoter of the murine PIASx gene from testicular nuclear extracts forms a major DNA-protein complex containing Sp1, Sp2, and Sp3 transcription factors. A fragment comprising a segment from -168 to +76 nt relative to transcription start site is sufficient for basal promoter activity in cultured cells, but these in vitro assays failed to reveal clear differences in promoter activity between testis- and non-testis-derived cell lines (Santti et al., 2003; Yan et al., 2003).

ix) Bv8, a 8-kDa protein from skin secretions of *Bombina variegata*, reacts with mammalian brain and intestine. The murine and human homologues of Bv8 have identical amino-terminal sequences and contain 10 cysteines. In mouse testis, two forms of Bv8 mRNA have been characterized, of which one contains an additional exon, which codes for 21 mostly basic amino acids. The mouse Bv8 gene is most active in mid-late pachytene spermatocytes. The mouse and human precursor proteins are very similar but considerably different from Bv8 from *Bombina variegata*. However, the N-terminal of Bv8 from three different sources was very similar, suggesting its importance in biological functions (Wechselberger et al., 1999).

33.4. REFERENCES

Adams SH, Esser V, Brown NF, Ing NH, Johnson L, Foster DW, McGarry JD. Expression and possible role of muscle-type carnitine palmitoyltransferase I during sperm development in the rat. Biol Reprod 1998; 59: 1399-405.

Alcivar AA, Hake LE, Mali P, Kaipia A, Parvinen M, B.Hecht N. Developmental and differential expression of the ornithine decarboxlyase gene in rodent testis. Biol Reprod 1989; 41: 1133-42.

Amireault P, Dube F. Cloning, sequencing and expression analysis of mouse glycosamine 6-phosphate deaminase (GNPDA/oscillin). Mol Reprod Dev 2000; 56: 424-35.

Bera TK, Hahn Y, Lee B, Pastan IH. TEPP, a new gene specifically expressed in testis, prostate, and placenta and well conserved in chordates. Biochem Biophys Res Commun 2003; 312: 1209-15.

Chavous DA, A. Hake L, E Lynch RJ, O'Connor CM. Translation of a unique transcript for protein isoaspartyl methyltransferase in haploid spermatids : implications for protein storage and repair. Mol Reprod Dev 2000; 56: 139-44.

Chen L, Sato M, Inoko H, and Kimura M. Molecular cloning and analysis of novel cDNAs specifically expressed in adult mouse testes. Biochem Biophy Res Comm 1997; 240: 261-8.

Enomoto A, Wempe MF, Tsuchida H, Shin HJ, Cha SH, Anzai N, Goto A, Sakamoto A, Niwa T, Kanai Y, Anders MW, Endou H. Molecular identification of a novel carnitine transporter specific to human testis. Insights into the mechanism of carnitine recognition. J Biol Chem 2002; 277: 36262-71.

Geisinger A, Dos Santos A, Benavente R, Wettstein R. Identification and characterization of SRSP1, a rat gene differentially expressed during spermatogenesis and coding for a serine stretch-containing protein. Cytogenet Genome Res 2002; 98: 249-54.

Hadjiolova KV, Marlinova YS, Yankulov KY, Davidov V, Kancheva LS, Hadjiolov AA. An immunocytochemical study of the proliferating cell nuclear matrix antigen. J Cell Sci 1989; 93(pt1): 173-7

Hakovirta H, Keiski A, Toppari J, Halmekyto M, Alhonen L, Janne J, Parvinen M. Polyamines and regulation of spermatogenesis : selective stimulation of late spermatogonia in transgenic mice overexpressing the human ornithine decarboxylase gene. Mol Endocrinol 1993; 7: 1430-6.

Helmekyoto M, Hyttinen JM, Sinervirta R et al. Transgenic mice aberrantly expressing human ornithine decarboxylase gene. J Biol Chem 1991; 266: 19746-51.

Hickox DM, Gibbs G, Morrison JR, Sebire K, Edgar K, Keah HH, Alter K, Loveland KL, Hearn MT, de Kretser DM, O'Bryan MK. Identification of a novel testis-specific member of the phosphatidylethanolamine binding protein family, pebp-2. Biol Reprod. 2002; 67: 917-27.

Hickox DM, Gibbs G, Morrison JR, Sebire K, Edgar K, Keah HH, Alter K, Loveland KL, Igarashi K, Kashiwagi K. Polyamines : mysterious modulators of cellular functions Biochem Biophys Res Comm 2000; 271: 559-64.

Igarashi K, Kashiwagi K. Polyamines; mysterious modulaors of cellular functions. Biochem Biophys Res Commun 2000; 271: 559- 64.

Iida H, Doiguchi M, Yamashita H, Sugimachi S, Ichinose J, Mori T, Shibala Y. Spermatid-specific expression of 1bal, an calcium binding adapter molecule-1, in rat testis. Biol Reprod 2001: 64: 1138-46.

Ike A, Ohta H, Onishi M, Iguchi N, Nishimune Y, Nozaki M. Transient expression analysis of the mouse ornithine decarboxylase antizyme haploid-specific promoter using in vivo electroporation. FEBS Lett 2004; 559 : 159-64.

Ivanov IP, Rohrwasser A, Terreros DA, Gesteland RF, Atkins F. Discovery of a spermatogenesis stage-specific ornithine decarboxylase antizyme : Antizyme 3. Proc Natl Acad Sci USA 2000; 97: 4808-13.

Jeulin C, Lewin LM. Role of free L-carnitine and acetyl-L-carnitine in post-gonadal maturation of mammalian spermatozoa. Hum Reprod update 1996; 2: 87-102.

Kadam AL, Fatch M, Naz RK. Fertilization antigen-1 (FA-1) completely blocks human sperm binding to human zona-pellucida: FA-1 may be sperm a sperm receptor for zona pellucida in humans. J Reprod Immunol 1995; 29: 19-30.

Lopez P, Yaman R, Lopez-Fernandez LA, Vidal F, Puel D, Clertant P, Cuzin F, Rassoulzadegan M. A novel germ line-specific gene of the phosducin-like protein (PhLP) family. A meiotic function conserved from yeast to mice. J Biol Chem 2003; 278: 1751-7.

Lu B, Bishop CE. Mouse GGN1 and GGN3, two germ cell-specific proteins from the single gene Ggn,interact with mouse POG and play a role in spermatogenesis. J Biol Chem 2003;278:16289-96.

Matsui M, Ichihara H, Kobayashi S, Tanaka H, Tsuchida J et al mapping of six germ cell specific genes to mouse chromosomes Mamm Genome 1997; 8: 873-74.

McGarry JD, Brown NF. The mitochondrial carnitine palmitoyltransferase system. From concept to molecular analysis. Eur J Biochem 1997; 244: 1-14.

Montag M, van der Ven K, Dorbecker C, van der Ven H. Characterization of testicular mouse glucosamine 6-phosphate deaminase. FEBS Lett. 1999; 458: 141-44.

Munier A, Feral C, Milon L, Pinon VP-B, Gyapay G, Capeau J, Guellaen G, Lacombe ML. A new human nm23

homologue (Nm23-H5)/ specifically expressed in testis germinal cells. FEBS Lett 1998; 434: 289-94.
Munier A, Serres C, Kann ML, Boissan M, Lesaffre C, Capeau J, Fouquet JP, Lacombe ML. Nm23/NDP kinases in human male germ cells: role in spermiogenesis and sperm motility? Exp Cell Res 2003; 289: 295-306.
Murakami Y, Matsufuji S, Hayashi S-I, Tanahashi N, Tanaka K. Degradation of ornithine decarboxylase by the 26S proteasome. Biochem Biophys Res Commun 2000; 267: 1-6.
Narisawa S, Hofmann MC, Ziomek CA, Millan JL. Embryonic alkaline phosphatase is expressed at M-phase in the spermatogenic lineage of the mouse. Development 1992; 116: 159-65.
Naz RK, Morte C, Garcia-Framis V, Kaplan P, Martinez P. Characterization of a sperm specific monoclonal antibody and isolation of 95-kilodalton fertilization antigen-2 from human sperm. Biol Reprod 1993; 49: 1236-44.
Naz RK, Zhu X. Molecular cloning and sequencing of a novel cDNA encoding for a protein involved in human sperm function. Biochem Biophys Res Commun 2001; 284: 911-17.
Olson GE, Nagdas SK, Winfrey VP. Temporal expression and localization of protein famesltransferase during spermiogenesis and post-testicular sperm maturation in the hamster. Mol Reprod Dev 1997; 48: 71-6.
Otto VI, Schar BK, Sulser T, Hanseler E. Specific determination of germ cell alkaline phosphatase for early diagnosis and monitoring of seminoma: performance and limitations of different analytical techniques. Clin Chim Acta 1998; 273: 131-47.
Oyhenart J, Le Goffic R, Samson M, Jegou B, Raich N. Phtfl is an integral membrane protein localized in an endoplasmic reticulum domain in maturing male germ cells. Biol Reprod 2003; 68: 1044-53.
Ryan AF, Grendell RL, Geraghty DE, Golos TG. A soluble isoform of the rhesus monkey nonclassical MHC class I molecule Mamu-AG is expressed in the placenta and the testis. J Immunol. 2002;169:673-83.
Santhanam R, Naz RK. Novel human testis-specific cDNA: Molecular cloning, expression and immunological effects of the recombinant protein. Mol Reprod Dev 2001; 60:1-12.
Santti H, Mikkonen L, Hirvonen-Santti S, Toppari J, Janne OA, Palvimo JJ. Identification of a short PIASx gene promoter that directs male germ cell-specific transcription in vivo. Biochem Biophys Res Commun. 2003; 308: 139-47.
Shankar S, Mohapatra B, Suri A. Cloning of a novel human testis mRNA specifically expressed in testicular haploid germ cells, having unique palindromic sequences and encoding a leucine zipper dimerization motif. Biochem Biophy Res Comm 1998; 243: 561-5.
Spiess AN, Walther N, Muller N, Balvers M, Hansis C, Ivell R. SPEER—a new family of testis-specific genes from the mouse. Biol Reprod 2003; 68: 2044-54.
Taira M, Iizasa T, Shimada H, Kudoh J, Shimuzu N, Kudoh J, Tatibana M. A human testis-specific mRNA for phosphribosylpyrophosphate synthetase that initiates from a non–AUG codon. J Biol Chem 1990; 265: 16491-97
Tanii I, Yoshinaga K, Toshimori K. A 90-kDa glycoprotein exhibited stage-specific expression in meiotic germ cells in rat testis. Histochem Cell Biol 2000; 114: 181-9.
Tanphaichitr N, Smith J, Mongkolsirikieart S, Gradil S, Lingwood CA. Role of a gamete-specific sulfoglycolipid immobilizing protein on mouse sperm-egg binding. Dev Biol 1993; 156: 164-75.
Tewari DS, Qian Y, Thornton RD et al. Molecular cloning and sequencing of human cDNA encoding ornithine decarboxylase antizyme Biochim Biophys Acta 1994; 1209: 293-95.
Tosaka Y, Tanaka H, Yano Y, Masai K, Nazoki M, Yomogida K, Otani S, Nojima H, Nishimune Y. Identification and characterization of testis specific ornithine decarboxylase antizyme (OAZ-t) gene; expression in haploid germ cells and polyamine induced frameshifting. Genes Cells 2000; 5: 265-76.
Wechselberger C, Puglisi R, Engel E, Leeperdinger G, Boitani C, Kreil G. The mammalian homologues of frog Bv8 are mainly expressed in spermatocytes. FEBS Lett 1999; 462: 177-88.
White D, Weerachatyanukul W, Gadella B, Kamolvarin N, Attar M, Tanphaichitr N. Role of sperm sulfogalactosylglycerolipid in mouse sperm-zona pellucida binding. Biol Reprod 2000; 63: 147-55.
Wolosker H, Kline D, Bian Y, Blackshara S, Cameron AM et al. FASEB J 1998; 12: 91-99.
Xuan W, Lamhonwah AM, Librach C, Jarvi K, Tein I. Characterization of organic cation/carnitine transporter family in human sperm. Biochem Biophys Res Commun 2003;306:121-8.
Xue JC, Goldberg E. Identification of a novel testis-specific leucine-rich protein in humans and mice. Biol Reprod 2000; 62: 1278-84.
Yamanaka M, Koga M, Tanaka H, Nakamura Y, Ohta H, et al. Molecular cloning and characterization of phosphatidylcholine transfer protein-like protein gene expressed in murine haploid germ cells. Biol Reprod 2000; 62: 1694-01.
Yan W, Santti H, Janne OA, Palvimo JJ, Toppari J. Expression of the E3 SUMO-1 ligases PIASx and PIAS1 during spermatogenesis in the rat. Gene Expr Patterns. 2003; 3: 301-8.
Zhang X, Liu H, Zhang Y, Qiao Y, Miao S, Wang L, Zhang J, Zong S, Koide SS. A novel gene, RSD-3/HSD-3.1, encodes a meiotic-related protein expressed in rat and human testis. J Mol Med 2003; 81: 380-7.
Zhu X, Naz R. Fertilization antigen-1: cDNA cloning, testis specific expression, and immunocontraceptive effects. Proc Natl Acad Sci USA 1997; 94: 4704-7.

Chapter 34

SPERM MATURATION IN EPIDIDYMIS

34.1. ROLE OF EPIDIDYMIS

During their transit in the epididymis, spermatozoa acquire the capacity to move and to fertilize eggs; they are then stored in the cauda-end of the lumen in a quiescent state prior to ejaculation. The ability of sperm to undergo the events of fertilization is developed as a result of interaction with certain epididymal secretions, besides a complex cascade of changes involving the remodeling of the sperm surface, the induction of chromatin condensation, the acquisition of movement, and development of the potential for capacitation. Although secretions of the epididymal epithelium are clearly important for sperm maturation and survival, their role in this process has yet to be fully determined. Alterations in epididymal sperm membranes may result from the incorporation of protein, sugar and lipid determinants. Being a dynamic structure, the sperm surface is altered even after the spermatozoon leaves the male tract. For example, in the female reproductive tract, the spermatozoa undergo capacitation and the acrosome reaction that enable them to fertilize the egg. Both of these processes are accompanied by re-localization of many proteins. During sperm maturation in the epididymis, the acrosome undergoes a structural change in morphology. Biochemically, several intra-acrosomal proteins have been shown to exhibit modifications in their molecular form and antigenicity. Cell surface alterations in epididymis as well as the sperm maturation in vivo are also dependent on the action of androgen although the biochemical basis of its action is largely unknown (Cooper 1998; Jones, 1998; Sivashanmugam et al., 1999). Epididymal epithelium provides possible mechanism through ubiquitination for quality control of sperm in mammals. Ubiquitin binds to the surface of defective sperm, and most of the ubiquitinated sperm are subsequently phagocytosed by the epididymal epithelial cells (Sutovsky et al., 2001). Sperm maturation in epididymis also involves DNA stability. Mature sperm from the cauda epididymis are also able to incorporate foreign DNA in a buffer containing only salts and calcium, whereas immature spermatozoa are unable to bind DNA (Carballada and Esponda 2001). Cyclic AMP levels increase in the spermatozoa of ram, bovine, boar and goat during epididymal transit. The level of cAMP is determined by the relative rates of cyclic AMP synthesis and degradation i.e., by the relative activities of adenylate cyclase and its cyclic AMP phosphodiesterase (CPD) respectively. The CPD has an important role in epididymal sperm maturation. The intrasperm pH has been found to increase 0.4 unit from caput to cauda in goat indicating thereby that intrasperm pH may have a role in the regulation of sperm motility initiation in epididymis (Nath and Majumder; Uma Devi et al, 1997).

34.2. SPERM SURFACE ALTERATIONS

Membrane fluidity: Little is known about the lipid asymmetry and fluidity of sperm plasma membrane. It is characterized by an unusually high proportion of polyunsaturated phospholipids that give it special physical characteristics and compartmentalization of many of its component proteins and lipids into discrete domains on the sperm head and tail. During passage of spermatozoa through the epididymis remodeling of the plasma membrane is commensurate with acquisition of motility and fertilizing capacity. Remodeling processes include uptake of secreted epididymal glycoproteins, removal or utilization of specific phospholipids from the inner leaflet of the bilayer, processing of existing or acquired glycoproteins by endoproteolysis and re-positioning of both protein and lipid components to different membrane domains (Jones, 1998). The fluidity of the isolated goat sperm plasma membrane (PM) decreases significantly during the epididymal transit of sperm

Lipid Profile: Major sperm membrane lipids comprise phospholipid (approx. 75% w/w), neutral lipid (approx. 15% w/w) and glycolipid (approx. 10% w/w). During epididymal maturation there is a significant decline in the total lipids, phospholipid and glycolpid contents of sperm membrane. On the contrary, the mature cauda-sperm membrane showed greater neutral lipid content than that of the immature caput sperm. Phoshatidylcholine (PC), phosphatidylethanolamine (PE) and sphingomyelin are the phospholipids of the sperm membrane, the former two being the major lipids. Phosphatidylethanolamine decreased most strikingly in sperm maturation. The sperm maturity is associated with marked increase of sterol and steryl ester, and with decrease of the other membrane bound neutral lipids. The fatty acid profile of the various membrane lipids undergoes marked alteration during the epididymal transit of the male gametes. Cholesterol/phospholipid and saturated/unsaturated fatty acid ratio increases greatly in the maturing sperm membrane. The altered lipid profile of the mature sperm membrane leads to changes in its fluidity that plays an important role in determining the structure and functions of the sperm (Rane et al., 1991). Rat spermatozoa undergo significant modifications in lipid content from the caput to the cauda epididymis. A depletion of phosphatidylcholine and phosphatidylethanolamine, concomitant with a virtually unchanged amount of the corresponding plasmalogens, are the major alterations, plasmenylcholine thereby becoming the major phospholipid. Concerning the fatty acids, the proportions of oleate (C18:1, n-9) and linoleate (C18:2, n-6) in most lipids decreased on movement of sperm from caput to cauda, augmenting in turn the proportions of longer-chain (C20 to C24) and more unsaturated fatty acids. These fatty acids are found to undergo the most significant changes during sperm maturation.

Redox status: Aitken and Vernet (1998) reviewed that the changes in the redox status of mammalian spermatozoa from all regions of the epididymis exhibit a spontaneous capacity for superoxide anion production, which can be enhanced by the exposure to NADPH, in the caput region. This spontaneous free radical generating activity is mediated by a membrane-bound NADPH oxidase, the function of which is to generate the peroxides that are needed to serve as hydrogen acceptors for phospholipid hydroperoxide glutathione peroxidase during chromatin condensation in sperm. As spermatozoa enter the cauda epididymidis they also express a capacity for hydrogen peroxide (H2O2) when released into a defined culture media.

34.3. PROCESSING OF SPERM PROTEINS IN EPIDIDYMIS

Several types of glycoprotein changes including their appearance, loss, alteration of staining intensity, and alteration of electrophoretic mobility are noted. Some maturation dependent sperm surface glycoproteins co-migrated with glycoproteins present in epididymal fluid. Proteins on the sperm surface, localized to specific domains. such as fertilin β, PH-20, rat orthologue 2B1, rat posterior tail protein CE9, rat sperm mannosidase and various ADAM proteins with a potential role in fertilization, are processed during sperm maturation in the epididymis (Blobel, 2000; Frayne et al., 1998; Linder et al., 1995; Yuan et al., 1997).

PH20: The PH-20 is initially present on whole of the membrane and subsequently becomes restricted to a particular domain by some mechanism, not yet defined. The 2B1 on rat spermatozoa is a homologue of mouse/man PH-20. The 2B1 is expressed postmeiotically in the testis as a precursor glycoprotein of approximately 60kDa that first appears on the plasma membrane of stage 6 to 8 round spermatids. As spermatozoa pass through the caput epididymidis, 2B1 is endoproteolytically cleaved at a specific arginine residue (Arg 312) to produce a heterodimeric glycoprotein (40-kDa and 19-kDa) containing intra-molecular disulphide bridges (Jones et al., 1996). The equine PH-20 is modified during epididymal transit. Significant differences occur in electrophoretic mobility of PH-20 from caput and cauda epididymal equine sperm. N-deglycosylation resulted in the loss of hyaluronidase activity of sperm from both epididymal regions, whereas O-deglycosylation or trypsinization did not affect hyaluronidase activity. PH-20 distribution during epididymal maturation is dependent on proteolytic trypsin like mechanisms and, possibly, on complementary membrane associated factors (Rutilant and Meyers, 2001).

Glycoconjugates: Lectin binding studies provide evidence that glycan moieties of sperm plasma membrane glycoproteins undergo extensive modification during epididymal transit. The luminal fluid fucosyltransferase and β-D-galactosidase of epididymis are also responsible in the modification of sperm surface glycoproteins during sperm maturation (Tulsiani et al., 1995, 1998). Sperm surface proteins are modified by proteolytic processing in epididymis. For examples, rat sperm plasma membrane mannosidase is synthesized in the testis in an enzymatically inactive or less active prescursor form of 135-kDa, which is proteolytically processed to an enzymatically active mature form of 115-kDa during development in the testis and maturation in the epididymis. (Tulsiani et al, 1995). Similarly, guinea-pig sperm plasma membrane proteins PH20 and PH 30, mouse sperm antigen M42, rat sperm glycoprotein CE9, and antigen 2B1 undergo proteolysis during sperm maturation. In addition to the modification of sperm plasma membrane proteins/ glycoproteins, spermatozoa undergo intra-acrosomal modifications for acrosin and acrogranin (Tulsiani et al., 1998). Evidence presented demonstrate that at least one sperm surface glycoprotein (86-kDa) is fucosylated in vitro when caput spermatozoa are incubated with GDP fucose. Mouse testes have high activity of fucosyl transferase (FT), which is mainly on sperm surface. Spermatozoa from the caput epididymis have significantly more FT activity per cell than do spermatozoa from the cauda epididymis (Ram et al., 1989).

ADAM Proteins: In guinea pig, fertilin is distributed on testicular sperm head plasma membrane, but is found only on posterior head once sperm have passed through the epididymis. This can be mimicked if testicular sperm are briefly treated with trypsin. Fertilin from seven regions of epididymis showed migration of the protein from anterior head to posterior head after proteolytic processing of full length fertilin β processor (85 kDa pro -βform) to 75 kDa intermediate proβ* to 25-28 kDa mature form (Hunnicutt et al., 1997). One or more serine protease activities associated with testicular sperm can process fertilin β in vitro and in a similar fashion in epididymis (Blobel 2000; Lum and Blobal, 1997; Hunnicutt et al., 1997; Hellsten et al., 2001). Inositol polyphosphate 5-phosphatase deficient mice showed lack of fertilin-β processing and disrupted sperm function (Hellsten et al., 2001).

Acrins: Intra-acrosomal antigen, Acrin1 (MN7), shows not only a specific localization in the apical segment of the guinea pig sperm acrosome, but also a distal alteration during maturation. Acrin1 (MN7) was exclusively found both at the dorsal matrix and on the acrosome membrane (OAM) matrix associated materials in the apical segment. Acrin1 was initially distributed throughout dorsal matrix in immature sperm. But, during maturation, it became more restricted in the spherical bodies (Yoshinaga et al., 2000). The acrosomal location of acrin2 in caput epididymidal sperm was almost identical to that observed in the final step spermatids. But during maturation it was more restricted in area, until in distal cauda epididymidal sperm it remained only in the dorsal region. Acrin2 was recognized as a 165-kDa protein in the mature sperm and showed a reduction in molecular weight during sperm passage through the epididymis (Yashinaga e al., 2001).

Cytoskeleton Proteins: The actin regulatory proteins (thymosin β10, destrin and a testis specific actin capping protein) involved in controlling the balance between actin monomers (G-actin) and actin filaments (F-actin) are localized to acrosomal domain of bull sperm but during epididymal maturation they become confined to the equatorial segment. Following ejaculation, however, they extend back into the acrosomal region. The redistribution of actin and actin-regulatory proteins, coupled with changing levels of actin polymerization, suggests a continuing role for actin in both post-testicular sperm maturation and acrosomal exocytosis (Howes et al., 2001). In the tammer wallaby. The assembly and disassembly of F-actin within the acrosome as the spermatozoa transit the epididymis, suggested that actin filaments might play a significant role in the acrosomal transformation (Scarlet et al., 2001).

CE9: CE 9 is a domain specific integral plasma membrane protein of rat sperm. On testicular spermatozoa, it was concentrated within the posterior tail domain of the plasma membrane, whereas on vasa deferens spermatozoa, CE9 was concentrated within the anterior tail domain. N-terminal sequencing of CE9 suggested that the cleavage occurred on the carboxy terminal side of arginine 74 in the primary sequence of CE9, resulting in the loss of approximately 40% of the amino acids in the extracellular domain of this transmembrane glycoprotein (Petruszak et al., 1991).

34.4. PROTEINS ADSORBED BY SPERM DURING EPIDIDYMIS TRANSIT

In addition to processing of various proteins, proteins secreted from the epididymis are apparently added to the sperm surface in transit through the epididymis (McLaughlin et al., 1997; Waters and White, 1997; Jury et al., 1998). Most of these proteins participate in egg-sperm interaction and ovum fertilization. A comparison of the different protein patterns, in epididymis showed 61 spots, of which 11 were secreted only in the presence of hormones, 3 appeared to show hormone-related changes, and 25 were region-specific. Most of these secreted proteins were low-molecular-weight acidic proteins.

34.4.1 Mouse and Rat

Mouse epididymal proteins 7, -8 and –9: Two classes of epididymal protein-sperm interactions are distinguished on the basis of regional synthesis and secretion. The major class consists of proteins that are synthesized, secreted, and bound to spermatozoa in the caput epididymidis. In this class, however, the binding of protein to the spermatozoa is variable. Mouse MEP 25-kDa protein is a member of the phospholipid binding protein family and may have a role in lipid metabolism during sperm maturation (Araki et al., 1992). The 25–kDa mouse epididymal secretory protein (MEP-9) is similar to mouse testicular protein MTP-9 in several characteristics including molecular mass (25 kDa), isoelectric point (pl 6.0) and immunoreactivity. Although the 25–kDa antigens from the epididymal fluid and testicular extract are quite similar, they may have different immunological conformations. The amino acid composition and partial amino acid sequence of the testicular antigen showed substantial homology (> 80%) with a phosphatidylethanolamine binding protein from bovine brain. Another class of proteins (54, 44, 29 kDa) are synthesized and secreted from all epididymal regions but bound only to spermatozoa. Most of the epididymal proteins appeared to be tightly bound to the sperm. Whereas the activity of 25-kDa protein in mouse testis was found in the residual cytoplasm of the elongated spermatids, in the residual bodies, and in the cytoplasmic droplets of spermatozoa, the epithelial principal cells in epididymis were stained from the distal caput to the distal cauda. The activity seen in the principal cell is due to origin of synthesis rather than endocytosis of the testicular proteins from disrupted cytoplasmic droplets suggesting that the 25-kDa protein is synthesized independently by both testis and epididymis (Vierula et al., 1992). Among the eleven major MEPs identified in caudal fluid, three proteins were characterized: MEP 9, secreted from the proximal to the distal caput epididymidis; MEP 10, secreted from mid- and distal caput and MEP 7, secreted from the distal caput, corpus, and cauda. The secretion of these proteins is influenced by androgens. MEPs 7, 9 and 10 appeared to be associated with spermatozoa.

MEP-10 was localized within the cytoplasm of the principal cells, the stereocilia, and the lumen of the epididymis at the junction of the distal caput and corpus and dissociated completely from caudal spermatozoa under low salt conditions, indicating that it was not firmly bound to spermatozoa. The related rat protein migrated at approximately 20kDa. Amino acid sequence analysis of MEP-10 revealed an 86% sequence similarity with rat proteins B and C. MEP7 and MEP-10, are homologues of rat proteins D and E, and B and C, respectively and are likely to be functionally equivalent (Rankin et al., 1992). Mouse epididymal protein 7 was localized initially within the supranuclear regions of some principal epithelial cells in the proximal corpus. Sequential extraction of caudal spermatozoa indicated that MEP7 was predominantly loosely associated with sperm. Examination of other rodent caudal fluids revealed a MEP7 related protein of 32 kDa in rat caudal fluid and has a 68% sequence similarity with rat proteins AEG and D/E.

Mouse Sperm Maturation Antigen 4 (SMA-4): Mouse sperm maturation antigen 4 (SMA-4) is a surface component of sperm tail, and may be secreted by principal cells of the distal caput epididymidis and bound to spermatozoa as they pass through that region of the duct. A band corresponding to 54kDa was PAS positive and further processed on sperm surface. During transit through epididymis, several sperm antigens, produced in testis, are masked by secretary products of epididymis. Maturation antigen produced by epididymal epithelium is initially masked by sialic acid residues and finally is desialyted. The SMA4 shows WGA reactivity and binds to the flagellar surface of sperm. It contains disulfide bonds, which stabilize it on sperm surface and its mobility. The primary function of SMA4 is to prevent tail-to-tail agglutination of sperm during storage in epididymis (Feuchter et al., 1988; Toshimori et al., 1988). An intra-acrosomal 155kDa protein increases the antigenicity of mouse sperm antigens (Toshimori et al., 1995)

Rat Sperm Maturation Antigen: The major maturation antigen associated with sperm membrane is the prominent glycopeptide, which appears on the surface of rat spermatozoa during post – testicular sperm maturation. This phenomenon is not restricted to the rat. Rather in situ RNA hybidization and genomic DNA fragments made it evident that homolgous counterparts exist in other mammalian species, including human. The major maturation glycoproetein found on cauda epididymal sperm is linked to membrane via phosphatidylinositol (Moore et al., 1994). A 105kDa sperm membrane antigen (SMA2) is localized in the anterior head overlying the acrosome and absent from sperm flagellum (Sarkar and Chatterjee, 1997).

Prostaglandin D_2 Synthase: Glutathione independent prostaglandin (PG) D_2 synthetase (prostaglandin H_2 D-isomerase, EC 5.3.99.2) (PGD-S) is the enzyme that catalyzes the conversion of the cyclooxygenase derived intermediate PGH_2 to PGD_2 in the presence of sulfhydryl compounds. However, under different conditions, PGH_2 can also be converted to PGD_2 by glutathione dependent PGD synthetase, glutathione S-transferase, and even albumin. PGD-S has been detected in many mammalian organs. The PGD_2 is known to be involved in diversified physiological functions. The PGD-S is an important molecule in testicular and epididymal function and sperm maturation. The regionalization of PGH synthase in epididymis suggests that PGs play a role in fluid and ion transport (Marshburn et al., 1989). The PGD-S was shown to be a member of the lipocalin family, which is composed of secretory proteins that bind and transport small lipophilic molecules. PGD-S has been sown to be a bifunctional molecule. The concentration of PGD-S is highest in the epididymis-about 6 and 80 fold greater than that in the brain and testis, respectively (Lareyre et al, 2000b).

E-3: A Defensin Like Protein: The E-3, epididymis-specific secretory glycoprotein, is a sperm-associated isoantigen containing defensin- and lectin-like motifs. The E-3 was detected in rat epididymal fluid and in sperm with a spot of 28-kDa with a (pI) of 3.5. A 449-bp consisting of a complete ORF of 111 amino acids showed similarity to the defensin and lectin families. The E-3 mRNA was only predominantly expressed in the corpus and cauda of the epididymis, but not in caput. Indirect immunofluorescence localized E-3 on the entire tail, and with less intensity on the head of the sperm. The E-3 may play a role in protecting the sperm from microbial infections in the epididymis and in the female reproductive tract (Rao et al., 2003).

Other Proteins in Sperm Maturation in Rodents: Acidic epididymal glycoprotein (AEG) has a slight stimulatory effect on the motility of spermatozoa of rats. These studies emphasized the importance of the effects of AEG protein, but did not provide any clear evidence for a specific effect of AEG. A glycoprotein of 32kDa was isolated from the rat caudal epididymal fluid. The 32-kDa protein interacts with epididymal spermatozoa and shows charge heterogeneity having pl values of 5.4 and 4.9, with the major components (Zehab and Orr 1984). In mouse GP-83 and GP-49, sperm maturation related glycoproteins are secreted de novo in the epididymis, and their secretion is developmentally regulated and androgen dependent (Liu et al., 1992).

34.4. 2. Guinea pig/Rabbit/Hamster

HEP64: HEP64, a 64 kDa glycoprotein is secreted by principal cells of hamster corpus and proximal cauda epididymidis that specifically binds to dead spermatozoa. Its monomer contains approximately 12kDa carbohydrate. Following deglycosylation, HEP64 migrates as a 52-kDa polypeptide. Both soluble (luminal fluid) and sperm associated HEP64 are assembled into disulfide linked high molecular weight oligomers that migrate as a doublet band of 260/280kDa in non-reducing SDS-PAGE. In the epididymal lumen HEP64 is concentrated in aggregates of structurally abnormal or degenerating spermatozoa and forms a shroud like coating surrounding abnormal spermatozoa (Nag Das et al., 2000).

P26h: P26h abundant in the caput epididymal fluid of hamster and is taken up by sperm during epididymal transit. This protein is present on mature spermatozoa and thus, could be considered a marker of epididymal sperm maturation (c/r Legare et al., 2001). P26h is associated to epididymal prostasomes. Phospholipase C treatment of epididymal prostasomes indicated that P26h is anchored to these vesicles via a phosphatidylinositol. Epididymal sperm maturation involves a cell-to-cell transfer of a phosphaditylinositol-anchored protein and that prostasomes may be implicated in this process (Legare et al., 1999a). A 34-kDa human epididymal sperm protein, P34H is homologous to P26h and has been shown to be involved in sperm zona pellucida interaction. The predicted amino acid sequence of P34H has 65% identity with P26h, the hamster counterpart of the P34H (see Chapter 24).

Rabbit Epididymal Secretory Protein: A protein designated as BE-20, is present in cauda epididymal fluid of the rabbit with an estimated molecular weight of 20kDa. The initial eight amino acid residues were His-Gly-Ala-Asp-Lys-Pro-Gly-Val. The BE-20 cDNA consisted of 499bp with an open reading frame of 285 bp encoding a deduced polypeptide composed of 95 amino acids. The BE-20 cDNA had 78.5% identity with 479 bp human epididymis specific HE4 cDNA. The N-terminus of the purified protein showed the sequence: N-His-Gly-Ala-Asp-Lys-Pro-Gly-Val-Cys-Pro-Gln-Leu-Ser-Ala-Asp-Leu-Asn-Cys-Thr-Gln-Asp-Cys-Arg-Ala-Asp-Gln-Asp-Cys-Ala-Glu.

The deduced polypeptide contained 16 cysteine residues and had partial sequence homology with proteins belonging to the disulfide core family of extracellular proteinase inhibitors (Xu et al., 1996). The full length cDNA of BE-20 consisted of 585-bp with a poly(A) tail of 26 residues and an open reading frame of 369 bp encoding a polypeptide of 123 amino acids with a calculated molecular mass of 13 kDa. The N-terminus contained a leucine rich segment. BE-20 cDNA has about 76.8% homology with the HE4 gene of human epididymis. Northern blot analysis showed that BE-20 is expressed only in the epididymis (Fan et al., 1999).

Rabbit epididymal protein-38 of 38-kDa (REP-38) is localized in principal cells of cauda region and does not express any where in the body. The REP-38 is predominantly localized over the acrosomal and post acrosomal regions of the head and middle piece. It inhibited sperm-egg interaction in vitro. REP-38 is homologous to rat testicular Odf2 and KTT4, which putatively encode outer dense fiber proteins. The cDNA of REP38 encodes an ORF, which encodes a polypeptide of 666 amino acids. Cleavage of a 22-amino acid N-terminal signal peptide revealed a mature protein with a theoretical molecular mass of 74.5-kDa. Two cross-hybridizing transcripts of 1.3 and 2.5-kb appeared to result from alternative mRNA splicing. This finding could explain the discrepancies between the observed (38-kDa) and deduced molecular mass of REP38. The secretion of REP-38 is under androgen control (Nixon et al., 2002).

34.4.3. Ram/Goat/Bull

More than 150-200 proteins are secreted into epididymis lumen. Most of them are secreted in relatively small amounts. In ram for example fewer than 10 proteins contribute 90% of the total secretions and only two contribute 52% of the total protein secreted (Dacheux et al., 1998). The radiolabeling patterns of ram sperm surface showed an overall shift in the predominant labeled glycoproteins from the zone 78-115-kDa in testicular spermatozoa to relatively low molecular weights of between 15 and 95-kDa in cauda epididymal or ejaculated spermatozoa. Labeling procedures specific for glycoproteins and sialoglycoproteins revealed additional complexities in transformation patterns of ram sperm surface and suggested that cauda epididymal spermatozoa exposed to accessory sex gland secretions adsorb or produce a component of high molecular weight (approx. 350 kDa) (Voglmayr et al, 1985). In another study (Weaver et al., 1993), a monoclonal antibody identified an antigen of 18 kDa that is present on the surface of ram ejaculated sperm, but is absent from testicular sperm.

Bands corresponding to 17-111-kDa with different regionalization have been confirmed in the epididymis of ram (Gatti et al, 2000). The strongest epitopes at 17-kDa and 23-kDa were restricted to the cauda epididymidis. These 17 and 23-kDa proteins disappeared after orchidectomy, but they reappeared in the same regions after testosterone supplementation, indicating that they are secreted by the epithelium. The 17- and 23-kDa peptides had a low isoelectric point and were glycosylated. Their N-terminal sequences did not match any protein in current database. A polyclonal antibody against the fluid 17-kDa protein localized it on the sperm flagellum membrane and at the luminal border of all cells in the cauda epididymal epithelium. The secreted proteins with hydrophobic properties could be directly integrated in a specific domain of the sperm plasma membrane (Gatti et al., 2000). The ESA152 antigen is another highly hydrophobic integral membrane protein that resists aqueous extraction but partitions in detergents. ESA152 is an anchored protein, whose anchoring is unaffected by phosphatidylinositol specific phospholipase C. The antigen is absent from extracts of caput and corpus epididymidis but appears abruptly in the first segment of the cauda. Ecto-PPase of goat epididymal sperm outer surface showed a significant increase in its activity at the initial stage of epididymal sperm maturation followed by a sharp fall towards the terminal phase of the maturation event. The isolated PPase has a molecular mass of approximately 36-kDa and an isoelectric point of 5.95. The ecto-PPase is localized on the external surface of viable sperm (Barua et al., 2001).

34.4.4. Porcine Epididymal Proteins

Electrophoretic analysis of boar epididymal fluids has demonstrated the presence of proteins specific to the epididymis. Some of the proteins interact with boar spermatozoa and may alter the sperm membrane properties. Some of these may modify pre-existing compounds through the action of proteases, glucosidases, glucosasminidases, or glycosyltransferases. A total of 146 epididymal proteins covering 220 spots, were found to be secreted by the boar epididymis. The distal caput showed the highest number of spots; the lowest number of proteins secreted being found in the proximal caput and cauda. Most of the epididymal proteins analyzed are highly polymorphic in terms of both isoelectric and molecular mass. Of the various major proteins, clusterin, glutathione peroxidase, retinal binding protein, lactoferrin, EP4, β-N-acetyl-hexosamindidae, α-mannosidase, and procathepsin L were identified and localized along the organ (Syntin et al., 1996; Dacheux et al., 1998).

Boar epididymal anti-agglutinin, from cauda epididymal plasma shows the presence of sialic acid residues. Removal of sialic acid residues from antiagglutinin greatly reduced its immunoreactivity with the specific antiserum. The head to head agglutinability of boar spermatozoa gradually increases during their transit through the caput to the proximal corpus epididymidis, but declines by the time they reach cauda epididymidis. A boar "anti-agglutinin", which inhibits head to head agglutination of spermatozoa is a 25kDa sialoprotein contained in epididymal and seminal plasma (Harayama et al., 1996, 1999). A specific 135kDa protein has been purified from porcine cauda epididymal fluid and localized on acrosome (Okamura et al., 1992). A porcine homologue of the major secretory protein of human epididymis, HE1, is secreted in the epididymal fluid as a 19-kDa glycoprotein, whose sugar moiety was gradually processed to form a 16-kDa protein during transit through the epididymis. The HE1 homologue mRNA was detected only in the caput and corpus epididymis among the porcine tissues examined. The purified HE1 homologue specifically bound to cholesterol and is the major cholesterol binding protein in the porcine epididymal fluid (Okamure et al., 1999).

34.4.5. Stallion

FITC-lectins confirmed the presence of macromolecules containing carbohydrate residues in the epithelial cells of stallion epididymis with a distribution and relative density that was dependent on the epididymal region analyzed. The most prominent bands correspond to 66, 55, 45 and 14-kDa proteins, present in different concentrations, in the three major regions. A major band of 36-kDa was observed in the cauda epididymis. The stallion epididymis contains both O-and N-glycocunjugates, probably in the N-acetyl O-diacetyl form. Although sperm maturation is an androgen dependent process, striking differences were not detected in animals in breeding and non-breading seasons (Retamal et al., 2000).

34.4.6. Primate Epididymis Secretory Proteins

116-kDa Glycoprotein: Monkey spermatozoa acquire the capacity to move and to fertilize eggs while they are in transit rhrough epididymis (Yeung et al., 1996). The major maturation dependent sperm membrane glycoproteins identified from five segments of primate epididymis, include O-linked 170, 150, 86 and 60/58 kDa glycoproteins; N-linked 68, 56, 48 and 38-kDa glycoproteins and N and O-linked 116-kDa glycoprotein, all of which exhibited marked differences in the degree of glycosylation between immature and mature sperm surfaces. The 100% inhibition of fertility in female rats and rabbits immunized with major maturation dependent 116-kDa glycoprotein showed significance of glycosylation changes occurring in the maturation status of epididymal spermatozoa (Srivastav, 2000).

27-kDa EP-1: Perry et al (c/r Perry et al., 1999) identified several cDNA clones in a cynomologus macaque (*Macaca fascicularis*) epididymal cDNA library. Among the known encoded proteins are the homologues of CD52 and others. Perry et al. (1995) isolated a cDNA clone, ESP14.6, which is homologue to HE1. Epithelia I-V of the efferent ducts did not show characteristic anti-EP1 binding. EP1 appears to be synthesized and secreted primarily in the caput region of the ductus epididymis and may be reabsorbed non-selectively across epithelia with apical microvilli (Smithwick and Young, 1999). The 27-kDa (EP-1) protein identified in chimpanzee cauda epididymal fluid is maturation related protein that may be involved in the initiation of motility or in the attainment of fertilizing capacity of sperm.The 27kDa-glycoprotein comprises approximately 20% of the total protein in chimpanzee *(Pan troglodytes)* cauda epididymal fluid. 27-and 25-kDa immunoreactive components are present in cauda epididymal fluid and in seminal fluid of human and sub-human primates. The 27-and 25-kDa components (chimpanzee EP-1) are identical in amino acid sequencing to the putative human epididymal protein HE1. At the cDNA level, chimpanzee EP-1 and human HE1 are 99% identical. The derived amino acid sequence of rhesus monkey EP-1 is identical to chimpanzee WPI-1 and to human HE1. Northern

```
1.  Variant B    (Accession NP_478113)

            1 MKVFFLFAVL  FCLVQTNSVH  ISHQEARGPS  FRICVDFLGP  RWARGCSTGN

2.  Variant E    (Accession NP_478114)

            1 MKVFFLFAVL  FCLVQTNSGD  VPPGIRNTIC  RMQQGICRLF  FCHSGEKKRD
           51  ICSDPWNRCC  VSNTDEEGKE  KPEMDGRSGI

3.  Variant G    (Accession NP_478107)

            1 MRQRLLPSVT  SLLLVALLFP  GSSQARHVNH  SATEALGELR  ERAPGQGTNG
           51  FQLLRHAVKR  DLLPPRTPPY  QGDVPPGIRN  TICRMQQGIC  RLFFCHSGTG
          101 QQHRQRCG

4.  Variant C    (Accession NP_478110)

            1 MRQRLLPSVT  SLLLVALLFP  GSSQARHVNH  SATEALGELR  ERAPGQGTNG
           51  FQLLRHAVKR  DLLPPRTPPY  QEPASDLKVV  DCRRSEGFCQ  EYCNYMETQV
          101 GYCSKKKDAC  CLH

5.
    EP2              (Accession Q08648)
            1 MRQRLLPSVT  SLLLVALLFP  GSSQARHVNH  SATEALGELR  ERAPGQGTNG
           51  FQLLRHAVKR  DLLPPRTPPY  QVHISHQEAR  GPSFRICVDF  LGPRWARGCS
          101 TGN
```

Fig.34.1. Human SAGA11 gene products. Out of several (9 already known) variants, five variants with their accession numbers are shown. Source: http//www. ncbi.nlm.nih.gov

analysis localized rhesus monkey EP-1 mRNA to the distal caput and the proximal corpus epididymidis and expresses specifically in the epididymis (Frochlich and Young, 1996). The EP-1 protein was not detected in caput epididymis and hence caput epididymal sperm do not show motility, whereas caudal sperm showed forward progressive motility (Young et al., 1987 c/r).

EP2: The SPAG11/*EP2* gene codes for at least nine message variants that are all specifically expressed in the epididymis. Alternative splicing of this gene results into seven transcript variants encoding different isoforms. These variants putatively encode small secretory proteins, which differ in their N- and C-termini, resulting in proteins that can have little or no sequence similarity to each other. The *EP2* gene has two promoters, eight exons, and seven introns. Exons 3 and 6 encode protein sequences homologous to β-defensins, a family of antimicrobial peptides. This sequence homology and the arrangement of promoters and defensin encoding exons suggest that the *EP2* gene originated from two ancestral β defensin genes arranged in tandem, each contributing a promoter and two exons encoding a leader sequence and defensin peptide. The *EP2* gene is located on chromosome 8p23 near the defensin gene cluster and is separated by 100 kb or less from *DEFB2*, the gene for β defensin-2 (Frohlich et al., 2001). In chimpanzee epididymis, several mRNA HE2/EP2 variants are produced. Of these three cDNAs code for two different protein sequences that have a cysteine distribution characteristic for β-defensins. For example variant B has an alternative 5' UTR compared to variant A. The N-terminus of encoded isoform (B) is identical to that of isoform E, but the remainder of the protein is identical to isoform A. Amino acid sequences of some of the isoforms from database are given **(Fig.34.1).**

ESC42: Thirty-six cDNAs from a subtracted *Macaca mulatta* epididymis library were analyzed (Liu et al., 2001). These clones encode proteins with a range of motifs characteristic of protein-modifying enzymes, protease inhibitors, hydrophobic ligand-binding and transport proteins, extracellular matrix – interacting proteins, and transcription regulatory factors, but more than

```
  1 MKLLLLALPM LVLLPQVIPA YSGEKKCWNR SGHCRKQCKD GEAVKDTCKN LRACCIPSNE
 61 DHRRVPATSP TPLSDSTPGI IDDILTVRFT TDYFEVSSKK DMVEESEAGR GTETSLPNVH
121 HSS
```

Fig.34.2. Amino acid sequence of ESC42 (beta defensin) from human epididymis (Liu et al, 2001). Zinc finger motif is italics faced. Stars indicate presence of trefoil like motif. Cysteine residues are underlined. Source: http//www. ncbi.nlm.nih.gov (Accession Q96PH6).

half showed no relationship to any known proteins. The most abundant clone, ESC42, contains a cysteine rich region similar to the signature-binding domain of the trefoil family of motogenic wound repair proteins. The monkey and human proteins are nearly 90% identical **(Fig.34.2)**. The protein is most abundant in the epithelium of the caput and is also present in the lumen and bound to sperm. The *ESC42* gene, located on chromosome, 20q11, contains two exons encoding two nearly identical predicted signal peptides and a third exon encoding the rest of the protein. The predicted proteins have a calculated molecular mass of 11.2-kDa after signal peptide removal. Several sites for post-translational processing are present, including an asparagine glycosylation site, four PKC phosphorylation sites and two casein kinase C phosphorylation sites. ESC42 protein showed an apparent mol wt of about 20kDa, somewhat larger than the 11.2 kDa predicted for the mature peptide and consistent with post-translational processing such as phosphorylation or glycosylation. The cysteine rich region of ESC42 is 50% identical to N-terminus of another epididymis – specific protein, ESP13.2 (Perry et al., 1999). Though the function of this protein is not known, the cysteine-rich region similar to domains present in defensins suspects its function in innate immunity (Li et al., 2001).

34.5. HUMAN EPIDIDYMIS PROTEINS AND THEIR ANIMAL ORTHOLOGUES

Epididymal secretory proteins that interact with the human sperm surface are reported to be characterized by a low molecular mass (15-35kDa), a high degree of glycosylation, and high species-specific variation. The cDNA sequencing suggested that most of these proteins represented novel human gene products. In situ mRNA hybridization showed that the human epididymal (HE) proteins HE1, HE2, HE4, and HE5 expressed by the human epididymal epithelium exhibit region-specific variation of expression. At least seven of SAGA11 genes are known to be secreted by human epididymis **(Fig. 34.1)**.

GP-83: The WGA binding proteins synthesized by the corpus epididymal epithelial cells and GP-83 present in seminal fluid may play an important role in sperm maturation. The GP-83 and GP-39 found on mature spermatozoa may be secreted by the principal cells of corpus and cauda epididymis and bound to spermatozoa during their transit in human epididymis (Liu et al., 2001). The GP-83 from human seminal fluid showed the pI of 6.57 and was found in fluid tissue and sperm extracts of corpus and cauda epididymis, but not in the caput. The GP-83 was found on the anterior acrosome in ejaculated spermatozoa, and shifted to the equatorial region after capacitation and the acrosome reaction (Sun et al., 2000).

34.5.1. Human Epididymal (HE1-HE4) Proteins

Kirchhoff et al. (1998a, 1999) characterized a series of human epididymal (HE) sperm proteins, which are important part of sperm maturation. Human post-testicular proteins, named HE1-

```
             -19                 -1+1                                30
Porcine      MHFLAAAFL  LLTLSASALA EPVHFRDCGS GVGVIKEVNV NPCPTQPCQL
Dog          MRLLVAAFL  LLALGDLGPG GAVHFKDCGS AVGVIKELNV NPCP-QPCQL
m.fas.       MRFLAATFL  LLALSTAAQA EPVQFKDCGS VDGVIKEVNV SPCF-QPCQL
Human        MRFLAATFL  LLALSTAAQA EPVQFKDCGS VDGVIKEVNV SPCF-QPCQL
Chimp.       MRFLAATFL  LLALSTAAQA EPVQFKDCGS VDGVIKEVNV SPCF-QPCQL
                                                                 80
Porcine     HKGQSYSVNV TFTSNTQSKG SKAVVHGIVM GVPIPFPIPD PDGCKSGINC
Dog         HKGQSYSVNV TFTSNIPSQS SKAVVHGIVL GVAVPFPIPE ADGCKSGINC
M.fas.      SKGQSYSVNV TFTSNIPSKS SKAVVHGILM GVPVPFPIPE PDGCKSGINC
Human       SKGQSYSVNV TFTSNIPSKS SKAVVHGILM GVPVPFPIPE PDGCKSGINC
Chimp       SKGQSYSVNV TFTSNIPSKS SKAVVHGILM GVPVPFPIPE PDGCKSGINC
                                                                130
Porcine     PIQKDQTYSY LNKLPVKAEY PSIKLVVEWK LQDDNDQCLF CWQIPVQIES --
Dog         ---KDKTYSY LNKLPVKNEY PSIKLVVQWM LLGDNNQHLF CWEIPVQIEG --
M.fas.      ---KDKTYSY LNKLPVKSEY PSIKLVVEWQ LQDDKNQSLF CWEIPVQIVS HL
Human       ---KDKTYSY LNKLPVKSEY PSIKLVVEWQ LQDDKNQSLF CWEIPVQIVS HL
Chimp       ---KDKTYSY LNKLPVKSEY PSIKLVVEWQ LQDDKNQSLF CWEIPVQIVS HL
```

Fig.34.3 Amino acid sequences of HE1 homologues from porcine, dog, macaque, human, and chimpanzee. -1 and +1 show the suggested signal sequence cleavage site. Reproduced with permission from N. Okamura et al. Biochim Biophys Acta 1437; 377-87: 1999 © Elsevier.

HE6, are derived from abundant epididymal mRNAs. With the exception of HE5, which turned out to be identical to the lymphocyte surface antigen CD52, they represented completely novel human gene products. The predicted proteins have been localized in the human epididymal epithelium, within the lumen of the epididymal duct and vas deferens, and also on the surface of ejaculated spermatozoa. Sperm association has been reported for at least four HE proteins, ARP, HE2, HE4, and HE5 (CD52). However, as is largely the case in other species, a link to a specific function for any luminal component of the human epididymis, including the cloned secretory glycoproteins, to either sperm maturation or storage has not been demonstrated convincingly (Kirchoff 1998b).

The HE1 represented the most frequent epididymal cDNA. The HE1 mRNA is abundant in the epididymis but not detectable in other human tissues including testis. Unlike the other HE products HE1 seemed to be well conserved among mammals on the nucleic acid level suggesting a common functional role in the mammalian epididymis. The HE1, encoded by a single gene, predicted the peptide identical to that of EPI-1 from Chimpanzee epididymis and to ESP14.6 from the macaque homologue. HE1 related antigen was shown to be present within the epithelium and lumen of the human corpus epididymidis, and also in the lumen of the cauda epididymidis and efferent duct. Moreover, HE1 seemed to be associated with epididymal spermatozoa and its association with ejaculated spermatozoa remained doubtful. Antiserum against the synthetic HE1 epitope reacted, specifically, with 25-27-kDa proteins present in human seminal plasma (Kirchoff et al., 1998a). A porcine homologue of the major secretory protein of human epididymis, HE1, is secreted into the epididymal fluid as a 19-kDa glycoprotein, whose sugar moiety was gradually processed to form a 16-kDa protein during transit through the epididymis (Okamura et al., 1999). Amino acid sequences of HE1 from epididymis of different sources are shown in figure (**Fig.34.3**).

The amino acid sequence of a major human epididymis specific protein, HE4 deduced from the nucleotide sequence showed characteristics of a secretory protein, with a signal peptide followed by a small (approximately 10-kDa), acidic (pI 4.3), and cysteine-rich polypeptide. The positions of half-cysteines suggested that it is two-domain member of the family of "four-disulfide core" proteins to which a number of proteinase inhibitors belong. The transcripts originated from a single copy gene. Northern blot and in situ hybridization specifically localized the HE4 (human epididymis gene product) mRNA to the epithelial cells of the epididymal duct. The protein functions in sperm maturation as indicated by amino acid similarities to extracellular proteinase inhibitors of genital tract mucous secretions (Kirchoff et al., 1999).

The localization of HE2 to acrosomal and equatorial regions of ejaculated sperm head, suggests the involvement of this protein is sperm maturation. The HE2 contains potential sites for signal peptide cleavage and glycosylation. HE2 mRNA (700bp) and protein (107 amino acids) has no sequence homology with known sequences (Osterhoff et al., 1994). HE2 mRNA is androgen dependent. Multiple classes of HE2 related sequences have been discovered in human caput corpus cDNA library and three have been assigned structural sub-types α, β, and γ. The HE2 transcripts code for identical 71-amino acid residues N-terminus, different C-termini, and 5'- and 3'- untranslated regions. Compared with the original HE2, HE2β and HE2γ proteins have a 25 amino acid deletion. These frame shifting deletions result in C-termini differing in length, amino acid sequence, including number of cysteines, and isoelectic point resulting into HE2 isoforms, which are derived from alternative splicing of 8 or more exons of a single gene. Northern hybribization revealed that the 0.9kb mRNA is most abundant in human caput, less of it (20%) in corpus and least (<5%) in cauda. The β1 form is highly expressed in principal cells of the initial segment of caput, secreted into the lumen and binds to the sperm surface in the postacrosomal and neck regions. The β2 form is expressed in principal cells primarily in efferent ducts (Hamil et al., 2000).

Canine Orthologs of HEs: The dog expresses a relatively high number of abundant, epididymis-specific gene products, which are also expressed in the human, and offers an acceptable source of tissue suitable for in vitro studies (Ivell et al., 1998). The mature sperm of dog from cauda epididymis and freshly ejaculated sperm demonstrated a functional membrane-bound progesterone receptor while less matured spermatozoa from testis and corpus epididymis did not express this reactivity. It was suggested that acrosome reaction in dog sperm is dependent on epididymal maturation (Srivaidyapong et al., 2001). Using cDNA probes specific for human epididymal genes established that the canine equivalents of the HE1, HE4 and HE5 mRNAs are expressed in the canine epididymis. The cDNA cloning and sequencing confirmed that the canine gene products, CE1, CE4 and CE5 were indeed true structural homologues of their human counterparts (Ellerbrock et al., 1994). The cloning of a human gene product, HE12, by homology to its canine counterpart, (CE12) has been described. Sequence analyses showed that these gene products belong to a family of epididymis-specific cDNAs encoding novel fibronectin type II (Fn2)-module proteins, which are similar but not homologous to the major proteins of bovine seminal plasma (called BSP proteins) **(Fig.34.7).**

34.5.2. CD52 (HE5) and Its Orthologs

The CD52 is a 12 amino acid glycopolypeptide, which is linked to the plasma membrane via a GPI (Glycosylphosphatidyl inositol) linkage. The human antigen CD52, encoded on chromosome 1, is expressed in lymphocytes, monocytes and eosinophils. In male genital tract CD52 has been found in epididymal cells, seminal vesicles, seminal plasma and mature sperm. Sequencing of HE5 from human epididymis cDNA suggested it to be identical to that of human leucocyte differentiation antigen CD52. Both products are transcribed by same gene. HE5 gene originates from epithelial cells or epididymal and deferent duct (Kirchhoff et al., 1993). The secretory protein in men has a collinear cDNA sequence with lymphoctye CD52, a sialylated glycoprotein. Flow cytometric detection of cynomolgus monkey sperm, during epididymal maturation showed

A

```
 1 MKRFLFLLLT ISLLVMVQIQ TGLSGQNDTS QTSSPSASSN ISGGIFLFFV
51 ANAIIHLFCF S
```

B

```
 1 MKSFLLFLTI ILLVVIQIQT GSLGQATTAA SGTNKNSTST KKTPLKSGAS
51 SIIDAGACSF LFFANTLMCL FYLS
```

Fig.34.4. Amino acid sequences of secretory protein from human epididymis. A) CAMPATH-1 antigen precursor (CD52 antigen) (CDW52) (Cambridge pathology 1 antigen) (Epididymal secretory protein E5) (Accession P31358). B) CD52 antigen from mouse [Mus musculus] epididymis (Accession NP_038734).

increase in CD52 from 20% stained sperm in caput to 85% sperm in corpus with double staining intensities. Unlike the sperm coating antigens, CD52 binds firmly to the sperm membrane via its GPI anchor during epididymal passage (see Chapter 29). From both in vivo and in vitro studies, it was concluded that androgen and temperature are principal factors synergistically modulating epididymal CD52 expression (Kirchhoff et al., 1998a; 1999) **(Fig.34.4)**. The monkey epididymal secreted CD52 on sperm underwent changes in antigenic characteristics during sperm maturation, which were reversed under capacitation conditions (Yeung et al., 2000). During incubation in capacitating conditions, there is an increasing accessibility of some sialic acid residues and of the core peptide, particularly the GPI anchor (Yeung et al., 2001).

SAGA-1: In one study, Isojima et al., (1987) generated a human mouse heterohybridoma with the peripheral blood lymphocytes of an infertile woman whose serum exhibited high sperm inhibitory antibodies. The human IgM mAb, H6-3C4, so produced had sperm agglutinating and sperm immobilizing activities in vitro. The H6-3C4 mAb was shown to recognize N-lined oligosaccharide epitopes on a sperm surface glycoprotein showing as a series of 15 to 25 kDa immunoreactive bands on Western blots. Tsuji et al., (1988) demonstrated that the epitope reacting with H6-3C4 mAb contained repetitive *N*-acetyl lactosamine. However, the cognate H6-3C4 antigen remained unknown. In another study sperm agglutination antigen – 1 (SAGA-1), a polymorphic (~15-25 kDa), highly acidic hydrophobic glycoprotein was found localized over the entire surface of the human spermatozoa (Diekman et al, 2000). Competitive inhibition studies, as well as similarities in molecular weight suggested that SAGA-1 and the H6-3C4 antigen represent the same sperm glycoprotein (s) (Diekman et al., 1999a,b).

Treatment with phosphatidylinositol specific phospholipase C showed that SAGA-1 is anchored in the sperm plasmalemma via a GPI-lipid linkage. The core peptide of the SAGA-1 glycoprotein is identical to the sequence of CD52. Comparison of anti-SAGA-1 and anti-CD52 immunoreactivities revealed that the sperm form of CD52 exhibits N-linked glycan epitopes, including the epitope recognized by the infertility associated H6-3C4 mAb, which are not detected on lymphocyte CD52. Thus, the two forms of CD52 glycoprotein on lymphocytes and spermatozoa represent glycoforms of glycoprotein with the same core amino acid sequence but different carbohydrate structures. The mAbs to the unique carbohydrate epitopes on sperm CD52 give multiple inhibitory effects on sperm function, including a cytotoxic effect on spermatozoa in the presence of complement. These results implicate unique carbohydrate moieties of a sperm CD52 glycoform as target epitopes in the anti-sperm immune response of an infertile woman (Diekman et al., 1999a,b). The CD-52 homologue is present on entire chimpanzee spermatozoan and offers potential animal model for development of immunocontraceptive vaccine (McCauley et al., 2002) **(Fig.34.4)**.

gp20 isforms: Though CD52 structure is largely unknown but there is some evidence that it differs from that of the immune system in glycan chains and GPI-anchor structures of Diekman et al., (1999a,b) and Schroter et al., (1999). A human surface sialoglycoprotein (gp20) is homologous to CD52. Analysis of release of sialic acid residues in the capacitation medium showed that sialylglycoconjugates are abundant in freshly ejaculated sperm but are lost during capacitation. The anti-gp20 antiserum, highly specific for gp20, did not detect a similar antigen in murine sperm, rat sperm or any other tissue (c/r Focarelli et al, 1999a). On the uncapacitated sperm surface, sialylglycoconjugates were localized from the post-acrosomal region of the sperm head to the tail middle-piece, whereas after in vitro capacitation these molecules were only found in a small area of the post-acrosomal region. The N-terminus sequence of gp20 was GLY-GLN-ASN-ASP-THR-SER-GLN, which showed 100% identity with N-terminus of CD52. Like CD52, gp20 behaves as a glycosylphosphatidylinositol (GPI)-anchored protein and that anti-gp20 antiserum reacts with an antigen on leukocytes of the same molecular weight as CD52. However, the effect of CAMPATH-1 antibody was different from that of anti-gp20, since Western blot revealed a 20 kDa band with anti-gp20, whereas a 14-20-kDa band was detected with CAMPATH-1. Anti-gp20 stained the equatorial region of the sperm head, whereas CAMPATH-1 stained the tail region of capacitated spermatozoa. While CAMPATH-1 agglutinated motile spermatozoa, this was not seen with anti-gp20. This suggests that the epitopes recognized by the two antibodies are different (Hale et al., 1996; Focarelli et al., 1999a). The anti-gp20 antibody intensely stained the head and the mid-piece. However, on capacitated spermatozoa the antigen was restricted to a sharp zone in the equatorial region. The antibody did not bind to differentiating germ cells but antigen was present in epididymal epithelial cells. Anti-gp20 exerted a blocking effect in a test for sperm penetration of zona free hamster eggs, thus suggesting that gp20 is involved in the early stages of fertilization. Immunodepletion experiments revealed that CAMPATH and anti-gp20 interact with the same antigen, but that anti-gp20 had a much higher avidity for the antigen than CAMPATH. Anion exchange fractionation analysis of the antigen revealed three differently charged gp20-CD52 glycoforms; the least charged was largely without a GPI-anchor. All three glycoforms are associated with freshly ejaculated sperm, whereas capacitated sperm only contained the two GPI-anchored, more charged forms, which were also present in prostasome fraction of seminal plasma and in leukocytes (Giovampaola et al, 2000).

In a similar study, immunoblot analysis with anti-gp20 antibody showed that the sperm and seminal plasma antigens were similar and appeared to consist of two components, whereas the leukocyte antigen is unique in showing a singlet. Evidence for the presence of two components of the sperm antigen at about 19 and 21 kDa, was obtained by analyzing the purified antigen. Both components had an isoelectric point (pI) between 3 and 6. However, in contrary to immunoblots, MALDI confirmed the presence of two components and indicated masses (Mr) of 8243 and 10908 lower than apparent molecular weight. Since antigens of freshly ejaculated sperm and seminal plasma were identical and appeared as a doublet whereas that of leukocytes as a singlet, it indicated that of the CD52 antigen of the reproductive system is more complex than that the immune system (Focarelli et al., 1999b).

HE6: Human Epididymis-specific protein 6 [HE6 (GPR64)] is a highly conserved, tissue-specific seven-transmembrane receptor of the human epididymis. In rodent counterparts, downstream from the highly conserved signal peptide-coding sequence, the 5'-regions contained at least six mini-exons of less than 50 nucleotides. The tissue distribution of the mRNA was very similar in human and rodents. The human HE6 gene was assigned to the X chromosome in a region, which is syntenic to the mouse. The HE6 sequence predicted a two-subunit receptor of the LNB-TM7 subfamily. Two-subunit proteins were abundant in human and rodents, comprising an approximately 180 kDa hydrophilic ectosubunit and a <40 kDa hydrophobic endosubunit. The large ectosubunits were highly glycosylated, the carbohydrate side chains dramatically increasing the apparent molecular mass. Both subunits were associated with apical membranes of efferent ductule and proximal epididymal duct epithelia (Obermann et al., 2003).

34.5.3. P34H and its Orthologs

A 34-kDa (P34H) human epididymal sperm protein that shows homologies with the hamster P26h is localized on the acrosomal cap of human sperm and has been proposed to be involved in the interaction with the zona pellucida. P34H first appeared in the caput epididymidis and was restricted to the acrosomal cap. Signal intensity increased considerably from the proximal corpus to the cauda region of the epididymis. The P34H appeared to be strongly anchored to the sperm plasma membrane during epididymal transit.The P34H has been shown to be involved in sperm-zona pellucida interaction. The cloning and characterization of the full-length cDNA encoding human P34H, revealed 65% identify in amino acid sequence with P26h. The deduced P34H amino acid sequence revealed a 71% similarity with a pig lung tetrameric carbonyl reductase. The P34H mRNA was highly expressed in the human epididymis principally in the corpus region. A single 912-bp transcript, P34H mRNA was predominantly expressed in the proximal and distal sections of the corpus epididymis by the principal cells of the epididymal epithelium. Based on its pattern of expression and its function in one of the key steps leading to fertilization, P34H can be considered as a marker of epididymal sperm maturation in humans (Legare et al., 1999b). Using P34H as a marker, results showed that vasectomy alters the pattern of gene expression along the human epididymis, and that the vas deferens can be a major contributor to sperm maturation in certain situations (Legare et al., 2001).

34.5.4. Cysteine Rich Secetory Protein (CRISP) Family

Crisp-1/ -2/ -3: The cysteine rich secretory proteins (CRISPs) represent an evolutionarily highly conserved family, originally described in rodent male reproductive tract and recently in salivary glands. The CRISPs derive their name from a cluster of cysteine residues residing in their C-terminal portion, which apparently forms a discrete, compact domain. In the mouse CRISP-1 is mainly found in the cauda epididymis where it is synthesized by the principal cells and secreted into lumen, accounting for about 15% of the protein content of epididymal fluid. Data on mouse epididymal protein 7 (MEP 7A), which is probably identical to CRISP-1 report a weak binding to spermatozoa and indicate that the protein is taken up by the clear cells of the epididymis **(Fig 34.5).**

CRISP transcripts are also present in mouse salivary gland, where besides low *CRISP-1* level, abundant expression of the closely related protein, *CRISP-3* has been observed. The corresponding CRISP-3 shares 77% sequence identity with *CRISP-1,* with all 16 cysteines conserved. Expression of both the *CRISP-1* and *CRISP–3* genes in the salivary gland is controlled by androgens. The structure of the *CRISP-3* gene has been identified and putative androgen– respective elements were detected in the promoter region (Hinton and Palladino, 1995). The CRISP is abundantly present in equine seminal plasma and can also be detected in extracts of testicular, epididymal and ejaculated spermatozoa in increasing amounts. CRISP antigens are localized on the midpiece, and the postacrosomal and equatorial region of the sperm head. Tissue distribution and localization of CRISPs on equine spermatozoa point to a

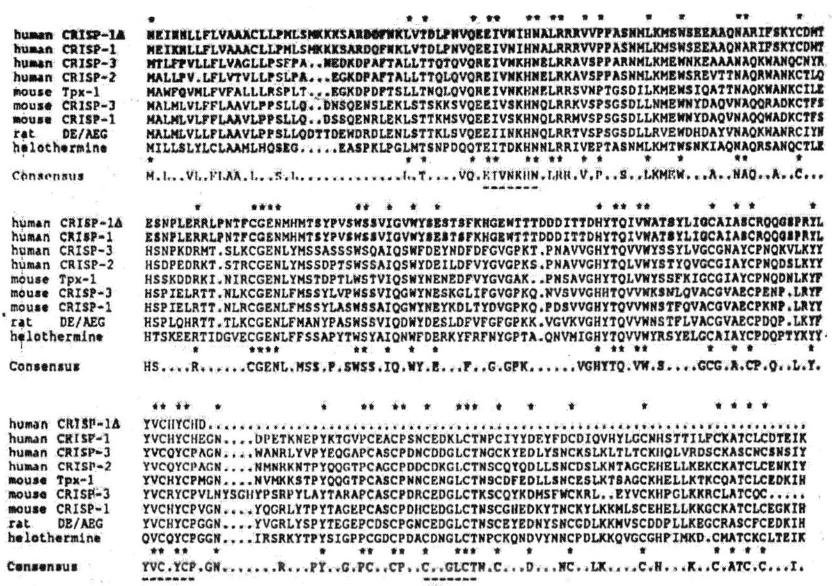

Fig.34.5 Amino acid sequences of human and murine epididymal CRISP1,-2, and -3. The sequences have been aligned to obtain similarities and identities among various CRISPs.Consensus amino acid residues are shown and underlined for at least six of the nine sequences. Complete conservation is shown by asterisks. Adapted with permission from J. Kratzschmar et al. Eur J Biochem 236; 827-36: 1996 © Blackwell Science.

role of these proteins in epididymal sperm maturation and equine reproduction (Schambony et al., 1998).

The CRISP-1 exists in five cDNA subtypes differing by the presence or absence of a stretch coding for a C-terminal cysteine rich domain so for found in all members of the family, and by the length of their 3'-untranslated region. A protein reacting with an anti-mouse CRISP-1 antibody from human epididymal extracts revealed N-terminal sequencing that corresponded to the CRISP-1 cDNA. The CRISP-2 cDNA corresponds to the previously described TPX1 form, with so far unreported 5'-untranslated sequence heterogenities while CRISP-3 codes for a unique protein. The *Tpx-1* renamed as *CRISP-2* in view of its high sequence identity to *CRISP-1* is expressed specifically in male haploid germ cells. The acrosomal auto-antigen AA1, the guinea pig counterpart of Tpx-1 makes 64% of the acrosomal proteins of guinea pig sperm. The CRISP-1 transcripts are epididymis specific whereas CRISP-2/TPX1 transcripts are detected mainly in the testis and also in the epididymis. The CRISP-3 transcripts are more widely distributed and found predominantly in the salivary gland, pancreas and prostate, and in less abundance in the epididymis, ovary, thymus and colon. Furthermore in cell-free system, helothermine another CRISP, blocks the ryanodine receptors, a ryanodine sensitive sarcoplasmic calcium release channel. This is the only biological effect known of any of the four CRISPs from humans (Kratzschmar et al., 1996; Giese et al., 2002a) **(Fig 34.5)**

Acidic Epididymal Glycoprotein: In contrast to findings on its rat counterpart epididymal protein D/E or AEG, association of CRISP-1 with human spermatozoa has not been observed (Kratzschmar et al., 1996). Acidic epididymal glycoprotein (AEG), the rat counterpart of CRISP-1, has been studied in greater detail. It shares 70% sequence identity with mouse CRISP-1 including all 16 cysteine residues and predominantly found in the cauda epididymis **(Fig.34.5)**.

It has been reported to bind to the heads of spermatozoa, possibly via a specific receptor. The involvement of DE/AEG in rosette formation and the presence of complementary sites on oocytes have also been documented. However, a clear biological function of CRISP/AEG has so far been eluded.

The deduced human AEG-like molecule from a cDNA is made up of 230 amino acids, excluding a signal peptide, and contained one potential N-linked glycosylation site. All cysteinyl residues of human are conserved between the human AEG-like molecule and the AEG molecules of rats and mice, and testis specific protein TPX1 of human and mice (approximately 40% amino acid sequence similarity). The human AEG-like gene is expressed specifically in the epididymis. The antibody against AEG detected a major band of 30 kDa and a minor band of 26 kDa in the caput, corpus, and cauda regions of the epididymis, the ductus deferens, the sperm, and the seminal plasma. The human AEG-like molecule is located in the lumen and epithelium of distal ductus efferentes and epididymal ducts, and on the postacrosomal region of the sperm head (Giese et al., 2002a) (see Chapter 24). The cloning and DNA sequence of 90kb of horse genomic DNA from equine chromosome 20q22 contains the complete equine AEG1 gene. The equine AEG1 gene consists of eight exons spanning 31-kb. Although equine AEG1 transcripts did not reveal any evidence for alternative splicing, but three different transcription start sites were identified. The intronless equine *Pgk2* gene for the testis-specific phosphoglycerate kinase is located approximately 39-kb downstream of *AEG1* (Giese et al., 2002b).

D/E proteins: Some of the sperm surface components are fucosylated during sperm maturation. In rat epididymis, two proteins (D and E of M_r 27 and 28-kDa respectively) are associated with spermatozoa during epididymal transit. Proteins D and E belong to the CRISP family and may be associated with surface receptors responsible for gamete recognition. Glycoprotein D is a major androgen sperm binding protein. Regional differences in the concentration of protein D suggests that its expression may be regulated at the level of mRNA within the corpus epididymidis (Hall and Reddy 1992). The antigen apparently a glycoprotein, retained on a Ricinus communis agglutinin I is present in epididymal fluid, which has an apparent Mr(s) of 38-26 kDa, whereas the membrane associated form of the molecule has a M_r of 26 kDa and released by PI-PLC. The D/E molecules interact with specific membrane proteins, and are subsequently covalently bound to the surface of spermatozoa via a glycosyl-phosphatidyl inositol linkage.The characterization of 26-37kDa glycoprotein associated with maturation of rat sperm revealed the presence of large percentage of carbohydrate, which is primarily attached through O-glycosidic bonds to 18 threonine residues. The molecular weight of the protein component was estimated to be 16600-Da (Moore et al., 1994; Eccleston et al., 1994). Cloning and sequencing of D/E revealed a lack of hydrophobic domains and the presence of 16 cysteine residues in the molecule. While most of the protein is removable from sperm by mild ionic strength, a low amount of D/E, resistant to even 2 M NaCI, can be completely extracted by agents that remove integral proteins. This supports the existence of two populations of D/E: a major, loosely bound population that is released during capacitation, and a minor strongly bound population that remains after capacitation, migrates to the equatorial segment (ES) with the acrosome reaction, and thus corresponds to the one with a role in gamete fusion (Cohen et al., 2000; Ellerman et al., 2002). It has been suggested that carbohydrates in D/E do not have any role in gamete fusion, whereas disulfide bridges hold the full biological activity of the protein (Ellerman et al., 2002). Amino acid sequencing of protein D and E confirmed that these proteins are nearly identical and differ only by presence of the 4E9 epitope on protein E. Protein D associates with the sperm head and that a portion of this protein may be proteolytically processed in the carboxy terminal region (Roberts et al., 2002).

These molecules remain covalently bound to spermatozoa even after deposition in the female reproductive tract, an observation, which focuses its physiological function in the fertilization process (Tubbs et al., 2002). Immunization of male and female rat with D/E results in 90% infertility. The exposure of sperm to immune sera prior to uterine insemination resulted in a significant reduction in the percentage of fertilized eggs. Immunization with epididymal protein DE does not produce orchitis, epididymitis or vasitis but specifically interferes with the sperm fertilizing ability, supporting the use of epididymal protein DE for contraceptive vaccine development (Ellerman et al., 1998).

Allurin: Allurin, a 21-kDa sperm chemoattractant from Xenopus egg jelly, is related to mammalian sperm-binding proteins. The protein consists of 184 amino acids having a molecular mass of 21,073 Da. The protein shares homology with the mammalian cysteine rich secretory protein (CRISP) family that includes testes-specific spermatocyte protein 1, a cell adhesion protein, which links spermatocytes to Sertoli cells, and acidic epididymal glycoprotein (AEG) that bind to sperm and have been implicated in sperm egg fusion (Olson et al., 2001).

34.5.5. Cystatin-Related Epididymal Spermatogenic Protein

The cystatin-related epididymal spermatogenic (CRES) and recently identified testatin and cystatin T proteins define a new subgroup within the family-2 cystatins of cysteine protease inhibitors (see Chapter 24). CRES, however, inhibits the serine protease prohormone convertase 2 (PC2), a protease involved in prohormone processing in the neuroendocrine system, whereas cystatin C showed no inhibition. CRES did not inhibit subtilisin, trypsin, or the convertase family members, PC1 and furin, indicating that it selectively inhibits PC2 (Cornwall et al., 2003). However, CRES lacks sequences important for cysteine protease inhibitory activity and is specifically expressed in reproductive and neuroendocrine tissues. Thus, CRES is distinct from cystatins and may perform unique tissue-specific functions. The *CRES* gene expression is highly restricted to the caput epididymidis with low levels of expression in the testis and without detectable expression in the other 24 tissues examined (Cornwall and Hann 1995). Its expression in the epididymis appears to be regulated by factors present in testicular fluid, other than androgens. CRES protein is present in elongating spermatids in the testis and is synthesized and secreted by the proximal caput epididymal epithelium. The presence of CRES protein in developing germ cells and in the luminal fluid surrounding maturing spermatozoa prompted Syntin and Cornwall (1999) to examine the presence of CRES in spermatozoa. The human ortholog of mouse CRES is highly expressed in the human testis, specifically within clusters of round spermatids. Human CRES is predominantly 19-kDa protein and a minor 14-kDa protein. However, in contrast to the acrosomal localization of CRES in mouse spermatozoa, human CRES was strictly localized to the equatorial segment (Wassler et al., 2002). While the 19 and 14-kDa CRES proteins were present in testicular and proximal caput epididymal spermatozoa, the 14-kDa CRES protein was the predominant form present in mid-caput to cauda epididymal spermatozoa. Within the testis, CRES expression is stage specific during spermatogenesis and is exclusively expressed by the round spermatids of Stages VII-VIII and the early elongating spermatids of Stages IX and X. Within the testis the protein was localized to the elongating spermatids, whereas within the epididymis CRES protein was exclusively synthesized by the proximal caput epithelium and then secreted into the lumen. Surprisingly the secreted CRES protein is absent from distal caput epididymidis. Testis and epididymal proteins showed that the CRES antibody specifically recognized a predominant 19 kDa CRES protein and a less abundant 14-kDa form suggesting that CRES protein performs a specialized role during sperm development and maturation (Cornwall and Hann, 1995).

Two novel genes, designated *Cres2* and *Cres3* encode proteins with four conserved cysteine residues and predicted molecular weights characteristic of family 2 cystatins but have divergent cystatin inhibitory sequences. Furthermore, the genes exhibited reproductive-specific expression with *Cres2* exclusively expressed in the epithelial cells of the proximal and midcaput epididymal regions and Cres3 expressed in the proximal caput epididymal epithelium, Sertoli cells of the testis, and early follicles and corpora lutea in the ovary. Like other *Cres*, both *Cres2* and *Cres3* genes are dependent on testicular factors for epididymal expression. Thus CRES2 and CRES3 represent new members of a subgroup of cystatin family 2 proteins (Hsia and Cornwall, 2003).

Two members of the mouse cystatin multigene family named cystatin SC (cystatin-related gene expressed in Sertoli cells) and cystatin TE-1 (cystatin-related gene highly expressed in testis and epididymis) have been isolated (Li et al., 2002). The full-length cDNA of cystatin SC contains an ORF that encodes a putative signal peptide of 20 amino acids and a mature protein of 110 amino acids, whereas that of cystatin TE-1 encodes a 128 amino acid protein with a predicted signal peptide of 21 amino acids. Both of these protein sequences contain four highly conserved cysteine residues in alignment with other cystatin family members. However, they lack some of the specific, highly conserved motifs assumed to be necessary for cysteine proteinase inhibition activity. Cystatin SC mRNA was detected only in the testis, whereas the cystatin TE-1 gene was highly expressed in testis and epididymis with very low expression in ovary and prostate. Cystatin SC mRNA was localized mainly to Sertoli cells with an obvious stage-dependent expression, while cystatin TE-1 mRNA was mainly expressed in Sertoli cells without apparent stage-dependent expression. It appears that cystatins SC and TE-1 play a specialized role in the testis and epididymis (Li et al., 2002).

34.5.6. Clusterin

Clusterin (SGP-2) is a heterodimeric glycoprotein produced by a wide array of tissues and found in most biological fluids. A number of physiological functions have been proposed for clusterin based on its distribution and in vitro properties. A prominent and a puzzling feature of clusterin is its induction in disease states (Rosenberg et al., 1995). The *TRPM-2/clusterin* gene is induced de novo during the regression of the prostate and other hormone-dependent tissues after hormone ablation, and is over-expressed in several human neurodegenerative diseases including Alzheimer's disease, epilepsy and retinitis pigmentosa. In the rat reproductive tract, clusterin/SGP-2 is present in the ventral prostate, seminal vesicle, testis and epididymis. In the epididymis, the expression of SGP-1 is not under the control of testicular or pituitary factors, as is also the case for cathepsin D expression by principal, narrow, and basal cells (Hermo and Andonian, 2003). In the ventral prostate, SGP-2 is associated with the programmed cell death (apoptosis), while in the testis and epididymis a role for SGP-2 in sperm maturation has been proposed. Information suggests that there are both inter and intra-organ variations in SGP-2 localization, molecular forms, and response to androgen ablation. The epididymis demonstrates a non-uniform staining pattern of SGP-2. The caput shows strong reaction over the apical membrane and sterocilia of all principal cells. Molecular forms of SGP-2 differed between testis and epididymis but were similar between ventral prostate and seminal vesicle. Prostate and seminal vesicle forms of SGP-2 differed from those of both testis and epididymis. The presence of multiple molecular forms of SGP-2 in various organs of the male reproductive tract of rats suggests a possible variation in functional activity and/or half life of SGP-2 in these organs (Sensibar et al., 1993).

```
                                   I
  1 MQKKNYARIR NKRPVEMNPW SSYLLGWTTF LLYFYETSGK IPNLSSLGKH EFTKPWISIK
 61 EDQKDSCVFP FVYKGSSYFS CIKTNSFSPW CATRAVYNGQ WKFCMADDYP RCIFPFIFRG
                II                                               III
121 KSHNSCITEG SFLRRLWCSV TSSFDENQQW KYCETNEYGG NSFSKPCIFP SIFRNSTIFE
                                                              IV
181 CMEDENNKLW CPTTENMDED GKWSLCADTR ISSLVPGFPC HFPFSYKNKN YYNCIGKGTK
241 ENLTWCATSY NYDRDHTWVY C
```

Fig.34.6. Secretory protein from dog epididymis with fibronectin activity (Saalmann et al, 2001). The four fibronectin type 2 domains are shown. Source: http//www. ncbi.nlm.nih.gov (Accession NP_001002931).

The genomic structure of the human *TRPM-2/ clusterin* gene is organized into nine exons, ranging in size from 47-bp (exon 1) to 412-bp (exon V), spanning a region of 16, 580 bp. It has extensive homology to human SP-40. Clusterin from bull rete testis fluid (RTF), cauda epididymal fluid (CEF) and octyl-β-D-glucopyranoside extract of cauda epididymal sperm (CES) showed that clusterin had dimeric and monomeric molecular weights of approximately 94-kDa (bovine RTF), 42-kDa (CEF) and 43-kDa (CES), respectively. Clusterin in CEF and CES had similar dimeric structure of M_r(s) (74 kDa). Reduced CEF clusterin appeared as three monomers (M_r = 40, 39, and 38-kDa), whereas reduced CES clusterin appeared only at M_r = 40-kDa. Enzymatic deglycosylation resulted in similar M_r (5-kDa) of clusterin from RTF, CEF, and CES. The three clusterins from RTF, CEF and CES showed carbohydrate content of 45, 31 and 32%, respectively, suggesting two isoforms of the herterodimeric protein. Thus it seems that several isoforms of clusterin occur in the bull reproductive tract and that the variation in carbohydrate content among these isoforms may affect the biological or functional activity of the protein (Ibrahim et al., 1999; Wong et al., 1994).

34.5.7. Other Epididymal Proteins in Sperm maturation

Fibronectin: Fibronectin (FN) is present along the entire length of the epididymis, and increases 12-fold from proximal caput to distal corpus. The greater number of FN positive sperm coincident with FN accumulation in distal regions of the epididymis supports the role of FN in human sperm maturation (Miranda and Tezon, 1992). Fibronectin type II (Fn2) is the first known example of proteins with four tandemly arranged Fn2-domains. The encoding mRNAs were abundant products of the epididymal duct epithelium, but not detectable in other tissues. Homologous mRNAs have been identified in the epididymidis of various mammals, representing members of this family of epididymal origin. Within the Fn2-module-encoding stretches, species homologues displayed >85% sequence identity, but showed high variability at their predicted N-termini. Western blots detected 30-35kDa immunoreactive protein bands in epididymal tissue, cauda epididymidal fluid, and sperm membrane protein preparations. The tandem arrangement of increasing numbers of Fn2-modules might functionally correspond to the tendency to form oligomers that have been described for lipid-binding proteins. Despite a considerable similarity to the BSP proteins, which comprise only two tandemly arranged Fn2-domains, the predicted epididymal proteins appeared much larger, the vast majority comprising four such motifs (Saalmann et al., 2001) **(Fig 34.6)**

Epididymal Retinoic Acid Binding Protein: Epididymal retinoic acid-binding protein (ERABP) is the major androgen-dependent protein present in the lumen of the epididymis and is thought to be involved in sperm maturation. Observations provide support that the charged residues near the open end of the binding pocket are responsible for restricting the specificity of ERABP for retinoic acid (Sundaram et al., 1998).

FLB: An antibody CA6 recognized the epididymal antigen, which was referred to as FLB1. The FLB1 was found to be secreted by the epididymis and to bind specifically to a human, macaque, and rodent subacrosomal sperm region. It is present in human epididymal extracts with a molecular mass of 94-kDa and in human, macaque, mouse, rat, and hamster sperm with a Mr of 100-kDa, suggesting modification of 94 kDa protein after its binding to sperm. The FLB1 specific to epididymis is formed of two subunits with identical mass of 46-kDa and slightly different pI (5.8-5.9). The protein is partially homologous with human cytokeratins 1 and 10 suggesting that FLB1 is an epididymis specific cytokeratin like protein that is involved in the sperm-oocyte interaction (Boue et al., 1995).

34.6. PROSTASOME

The seminal plasma of several mammals contains membraneous vesicles of different anatomic derivation. In humans, these particles derive from the prostate by exocytosis and diacytosis, whereas in bulls they originate from seminal vesicles. Therefore, these particulate elements are named prostasomes in human and vesiculosomes in bovine ejaculates. At ejaculation, these prostasomes are expelled with prostate secretions and are to be found in seminal plasma as seminal prostasomes, which facilitate sperm function in various ways. These vesicles are ascribed to have many physiologiocal functions, the primary function being enhancement of sperm motility. The several enzyme systems, small signaling molecules, and neuroendocrine markers associated with prostasomes reveal the complex nature of these vesicles in regulating sperm viability and motility. Other functions of prostasome include the liquefaction of semen and immunosuppression. The functional significance of molecules that regulate complex pathways in these small vesicles is still a matter of debate. Prostasomes are rich in Ca^{2+} and GDP. Many proteins at their surface possess catalytic activity or are involved in the immune response (Kravets et al., 2000). Average diameter of prostasome is about 150 nm and they appear as a lipoprotein membrane surrounding organized material. Their lipid composition is peculiar, having much cholesterol and sphingomyelin. On the other hand, many of their proteins possess catalytic activity and are involved in the immune response. The fatty acid pattern of prostasome is completely different from that of sperm membrane lipid. Polyunsaturated phosphatidylcholines common in sperm are rare in prostasome. The fusion between prostasome and sperm seems to stabilize sperm plasma membrane by enriching it in cholesterol, sphingomyelin, and saturated glycerophospholipid. This would prevent the untimely occurrence of the acrosome reaction (Arienti et al., 1998). Proteins present in human seminosomes include the seminal vesicle– derived fibronectin, prostate derived 5'-nucleotidase and an unidentified 100kDa membrane protein from epididymis, seminal vesicle and prostate respectively. Fibronectin also secreted from the seminal vesicles participates in the formation of the seminal clot. Fibronectin immunoreactivity on spermatozoa from different donors shows a relatively broad distribution and depends on Ca^{2+} flux and thus points a possible post-testicular regulatory function of seminal fibronectin. Addition of 5'nucleotidase inhibitor and adenosine channel antagonist inhibit sperm motility indicating a significant role for ecto–5' nucleotidase in the regulation of sperm motility. However immunodepletion of 100 kDa antigen from seminosomes had no effect on sperm motility and not in agreement with earlier reports on prostasomes.

34.7. REFERENCES:

Aitken RJ, Vernet P. Maturation of redox regulatory mechanisms in the epididymis. J Reprod Fertil Suppl 1998; 53: 109-18.

Araki Y, Vierula ME, Rankin TL, Tulsiani DR, Orgebin–Crist MC. Isolation and characterization of a 25-kilodalton protein from mouse testis: sequence homology with a phospholipids-binding protein. Biol Reprod 1992; 47: 832-43.

Arienti G, Carlini E, Polci A, Cosmi EV, Palmerini CA. Fatty acid pattern of human prostasome lipid. Arch Biochem Biophys 1998; 358: 391-5.

Barua M, Nath D, Majumder GC. Alteration of goat sperm ecto-phosphoprotein phosphatase activity and its distribution on the sperm surface during epididymal maturation. Reprod Fertil Dev 2001; 13: 443-50.

Blobel CP. Functional processing of fertilin: evidence for a critical role of proteolysis in sperm maturation and activation. Rev Reprod 2000; 5; 75-83.

Boue F, Duquenne C, Lassalle B, Lefevre A, Finaz C. FLB1, a human protein of epididymal origin that is involved in the sperm-oocyte recognition process. Biol Reprod 1995; 52: 267-78.

Carballada R. Esponda P. Regulation of foreign DNA uptake by mouse spermatozoa. Exp Cell Res 2001; 262: 104-13.

Cohen DJ, Rochwerger L, Ellerman DA, Morgenfeid MM. Busso D, Cuasnicu PS. Relationship between the association of rat epididymal protein "DE" with spermatozoa and the behaviour and function of the protein. Mol Reprod Dev 2000; 56: 180-8.

Cooper TG. Interactions between epididymal secretions and spermatozoa. J Reprod Fertil (Suppl) 1998; 53: 119-36.

Cornwall GA, Cameron A, Lindberg I, Hardy DM, Cormier N, Hsia N. The cystatin-related epididymal spermatogenic protein inhibits the serine protease prohormone convertase 2. Endocrinology 2003; 144: 901-8.

Cornwall GA, Hann SR. Transient appearance of CRES protein during spermatogenesis and caput epididymal sperm maturation. Mol Reprod Dev 1995; 41: 37-46.

Dacheux JL, Druart X, Fouchecourt S, Syntin P, Gatti JL, Okamura N, Dachexu F. Role of epididymal secretory proteins in sperm maturation with particular reference to the boar. J Reprod Fert (Suppl) 1998; 53: 99-107.

Diekman AB, Norton EJ, Klotz KL, Westbrook VA, Nabby-Hansen S, Herr JC. Evidence for a unique N-linked glycan associated with human infertility on sperm CD52: a candidate cotraceptive vaccinogen. Am J Reprod Immunol 1999b; 43; 134-43.

Diekman AB, Norton EJ, Klotz KL, Westbrook VA, Shibahara H, Naaby-Hansen N, J. Flickinger C and C. Herr J. N-linked glycan of a sperm CD52 glycoform associated with human infertility. FASEB J 1999a; 13: 1303-13.

Diekman AB, Norton EJ, Westbrook VA, Klotz KL, Nabby-Hansen S, Herr JC. Antisperm antibodies from infertile patients and their cognate sperm antigens: a review. Identity between SAGA-1 the H6-3C$_4$ antigen and CD52. Am J Reprod Immunol 2000; 43: 134-43.

Eccleston ED, White TW, Howard JB, Hamilton DW. Characterization of a cell surface glycoprotein associated with maturation of rat spermatozoa. Mol Reprod Dev 1994; 37: 110-19.

Ellerbrock K, Pera I, Hartung S, Ivell R. Gene expression in the dog epididymis: a model for human epididymal function. Int J Androl 1994; 17: 314-23.

Ellerman DA, Brantua VS, Penz-Martinez S, Cohen DJ, Conesa D, Cuasnicu PS. Potential contraceptive use of epididymal proteins. Immunization of male rats with epididymal protein DE inhibits sperm fusion ability. Biol Reprod 1998; 59 1029-36.

Ellerman DA, Da Ros VG, Cohen DJ, Busso D, Morgenfeld MM. Expression and structure-function analysis of DE, a sperm cysteine-rich seretory protein that mediates gamete fusion. Biol Reprod 2002; 67: 1225-31.

Fan HY, Miao SY, Wang LF, Koide SS. Expression and characterization of an epididymis specific gene. Arch Androl 1999; 42: 63-9.

Feuchter FA, Tabet AJ, Green MF. Maturation antigen of the mouse sperm flagellum. I. Analysis of its secretion, association with sperm, and function. Am J Anat 1988; 181: 67-76.

Focarelli R, Francavilla S, Francavilla F, Giovampaola CD, Santucci A, Rosati F. A sialoglycoprotein, gp20, of the human capacitated sperm surface is a homologue of the leukocyte CD52 antigen: analysis of the effect of anit-CD52 monoclonal antibody on capacitated spermatozoa. Mol Hum Reprod 1999a; 5: 46-51.

Focarelli R, Giuffrida A, Capparelli S, Schibona M et al. Specific localization in the equatorial region of gp20, a 20 kDa sialyglycoprotein of the capacitated human spermatozoon acquired during epididymal transit which is necessary to penetrate zona-free hamster eggs. Mol Human Reprod 1998; 4: 119-25.

Frayne J, Hall L. The gene for the human tMDC 1 sperm surface protein is non-functional: implications for its proposed role in mammalian sperm-egg recognition. Biochem J 1998; 334 (PEI): 171-6.
Frohlich O, Po C, Young LG. Characterization of human gene encoding the epididymis specific EP2 protein variants and its relationship to defensin genes. Biol Reprod 2001 64: 1072-9.
Frohlich O, Young LG. Molecular cloning and characterization of EPI-1, the major protein in chimpanzee (pan troglodytes) cauda epididymal fluid. Biol Reprod 1996; 54: 857-64.
Gatti JL, Druart X, Syntin P, Guerin Y, Dacheux JL, Dacheux F. Biochemical characterization of two ram caude epididymal maturation – dependent sperm glycoproteins. Biol Reprod 2000; 62: 950-8.
Giese A, Jude R, Kuiper H, Piumi F, Schambony A, Guerin G, Distl O, Topfer-Petersen E, Leeb T. Molecular characterization of the equine AEG1 locus. Gene. 2002a; 292: 65-72.
Giese A, Jude R, Kuiper H, Raudsepp T, Piumi F, Schambony A, Guerin G, Chowdhary BP, Distl O, Topfer-Petersen E, LeebT. Molecular characterization of the equine testis-specific protein 1 (TPX1) and acidic epididymal glycoprotein 2 (AEG2) genes encoding members of the cysteine-rich secretory protein (CRISP) family. Gene 2002b; 299: 101-09.
Giovampaola CD, Flori F, Sabatini L, Incerti L, LA Sala GB, Rosati F, Focarelli R. The surface of human sperm bears three differently charged CD52 glycoforms two of which remain stably bound to sperm after capacitation. Mol Reprod Dev 2000 (Submitted).
Hale G, Bartholomeco M, Taylor V, Stables J, Topley P, Tile J, Recognition of CD52 allelic gene products by CAMPATH-1 H antibodies. Immunology 1996; 88: 183-90.
Hall JC, Killian GJ. Changes in rat sperm membrane glycosidase activities and carbohydrate and protein contents associated with epididymal transit. Biol Reprod 1987; 36: 709-18.
Hall JC, Reddy NG. Protein D is differentially expressed and regulated in the rat epididymis. Biochem Biophys Res Commun 1992; 31: 1109-16.
Hamil KG, Sivashanmugam P, T. Richardson RT, Grossman G, Ruben SM et al. HE2β and HE2γ, new members of an epididymis-specific family of androgen-regulated proteins in the human. Endocrinology 2000; 141: 1245-53
Harayama H, Kato S, Hammerstedt RH. Electrophoretic characterization of boar epididymal antiagglutinin. Biol Reprod 1996; 55: 325-32.
Harayama H, Miyake M, Kato S, Immunolocalization of anti-agglutinin for spermatozoa in boars. Mol Reprod Dev 1999; 52: 269-76.
Hellsten E, Evans JP, Bernard DJ, Janne PA, Nussbaum RL. Disrupted sperm function and fertilin beta processing in mice deficient in the inositol polyphosphate 5-phosphatase Inpp5b. Dev Biol 2001; 240: 641-53.
Hermo L, Andonian S. Regulation of sulfated glycoprotein-1 and cathepsin D expression in adult rat epididymis. J Androl 2003; 24: 408-22.
Hinton BT, Lan ZJ, Rudolph DB, Labus JC, Lye RJ. Testicular regulation of epididymal gene expression. J Reprod Fertil 1998; 53: 47-57.
Hinton BT, Palladino MA. Epididymal epithelium: its contribution to the formation of adluminal fluid microenvironment. Microsc Res Tech 1995; 30: 67-81.
Howes EA, Hurst SM, Jones R. Actin and actin-binding proteins in bovine spermatozoa : potential role in membrane remodeling and intracellular signaling during epididymal maturation and the acrosome reaction. J Androl 2001; 22: 62-72.
Hsia N, Cornwall GA. Cres2 and Cres3: new members of the cystatin-related epididymal spermatogenic subgroup of family 2 cystatins. Endocrinology 2003; 144: 909-15.
Hunnicutt GR, Koppel DE, Myles DG. Analysis of the process of localization of fertilin to the sperm posterior head plasma membrane domain during sperm maturation in the epididymis. Dev Biol 1997; 191: 146-59.
Ibrahim NM, Troedsson MH, Foster DN, Loseth KJ, Farris JA, Blaschuk O, Crabo BG. Reproductive tract secretions and bull spermatozoa contain different clusterin isoforms that cluster cells and inhibit complement-induced cytolysis. J Androl 1999; 20: 230-40.
Isojima S, Kameda K, Tsuji Y, Shigeta M, Ikeda Y, Koyama K. Establishment and chacterization of a human hybridoma secreting monoclonal antibody with high titers of sperm immobilizing and agglutination activities against human seminal plasma. J. Reprod Immunol 1987; 10: 67-78.
Ivell R, Pera I, Ellerbrock K, Beiglbock A, Gebhardt K, Osterhoff C, Kirchhoff C. The dog as a model system to study epididymal gene expression. J Reprod Fertil (Suppl) 1998; 53: 33-45.
Jones R, Ma A, Hou ST, Shalgi R, Hall L. Testicular biosynthesis and epididymal endoproteolytic processing of rat sperm surface antigen 2B1. J Cell Sci 1996; 109: 2561-70.
Jones R. Maturation in the epididymis. J Reprod Fertil (Suppl) 1998; 53: 73-84.
Jury JA, Frayne J, Hall L. Sequence analysis of a variety of primate fertilin a genes: evidence for non-functional genes in the gorilla and man. Mol Reprod Dev 1998; 51: 92-7.
Kirchhoff C, Krull N, Pera I, Ivell R. A major mRNA of the human epididymal principal cells, HE5, encodes

the leucocyte differentiation CDw52 antigen peptide backbone. Mol Reprod Dev 1993; 34: 8-15.

Kirchhoff C, Osterhoff C, Pera I, Schroter S. Function of human epididymal proteins in sperm maturation. Andrologia 1998a; 30: 225-32.

Kirchhoff C, Osterhoff C, Young L. Molecular cloning and characterization of HE1, a major secretory protein of human epididymis Biol Reprod 1998b; 54: 847-56.

Kirchhoff C. Gene expression in the epididymis. Int Rev Cytol 1999; 188: 133-202.

Kratzschmar J, Haendler B, Eberspaecher U, Roosterman D, Donner P, Schleuning WD. The human cysteine-rich secretory protein (CRISP) family. Primary structure and tissue distribution of CRISP-1, CRISP-2 and CRISP-3. Eur J Biochem 1996; 15: 827-36.

Kravets FG, Lee J, Singh B, Trocchia A, Pentyala SN, Khan SA. Prostasomes: current concepts. Prostate 2000; 43: 169-74.

Lareyre JJ, Winfrey VP, Kasper S, et al. Gene duplication gives rise to a new 17-kilodalton lipocalin that shows epididymal region-specific expression and testicular factor(s) regulation. Endocrinology 2000; 20: 261-74.

Legare C, Berube B, Boue F, Lefievre L, Morales CR, B-Alfy M, Sullivan R. Hamster sperm antigen P26h is a phosphatidylinositol-anchored protein. Mol Reprod Dev 1999a; 52: 225-33.

Legare C, Gaudreault C, St-Jacques S, Sullivan R. P34H sperm protein is preferentially expressed by the human corpus epididymidis. Endocrinology 1999b; 140: 3318-27.

Legare C, Thabet M, Picard S, Sullivan R. Effect of vasectomy on P34H messenger ribonucleic acid expression along the human excurrent duct: a reflection on the function of the human epididymis. Biol Reprod 2001; 64: 720-7.

Li P, Chan HC, He B, et al. An antimicrobial peptide gene found in the male reproductive system of rats. Sci 2001; 291: 1783- 85.

Li Y, Friel PJ, Robinson MO, McLean DJ, Griswold MD. Identification and characterization of testis- and epididymis-specific genes: cystatin SC and cystatin TE-1. Biol Reprod 2002; 67: 1872-80.

Linder B, Bammer S, Heinlein UA. Delayed translation and posttranslational processing of cyritestin, an integral transmembrane protein of the mouse acrosome. Exp Cell Res 1995; 221: 66-72.

Liu HW, Sun GH, Shy SR, Shyu HY. Postnatal development and testosterone-dependence of GP-83 and GP-49, two sperm maturation related glycoprotions in BALB/c mouse epididymis. Cell Tissue Res 1992; 269: 189-94.

Liu Q, G. Hamil K, Sivashanmugam P, Grossman G, Soundararajan R, Jagannadha Rao A et al,. Primate epididymis specific proteins characterization of ESC42, a novel protein containing a trefoil like motif in monkey and human. Endocrinology 2001; 142: 4529-39.

Lum L, Blobel CP. Evidence for distinct serine protease activities with a potential role in processing the sperm protein fertilin. Dev Biol 1997; 191: 131-45.

Marshburn PB, Clark MR, Shabanowitz RB. Immunohistochemical localization of prostaglandin H synthase in the epididymis and vas deferens of the mouse. Biol Reprod 1989; 41: 491-7.

McCauley TC, Kurth BE, Norton EJ, et al., Analysis of a human sperm CD52 glycoform in primates: Identification of an animal model for immunocontraceptive vaccine development. Biol Reprod 2002; 66: 1681-88.

McLaughlin EA, Frayne J, Barker HL, Jury JA, Jones R, Ford WC, Hall L. Cloning and sequence analysis of rat fertilin α and β developmental expression, processing and immunolocalization. Mol Hum Reprod 1997; 3: 801-9.

Miranda PV, Tezon JG. Characterization of fibronectin as a marker for human epididymal sperm maturation. Mol Reprod Dev 1992; 33: 443-50.

Moore A, Ensrud KM, White TW, Frethem CD, Hamilton DW. Rat epididymis specific sperm maturation antigens. I. Evidence that the 26 kD 4E9 antigen found on rat caudal epididymal sperm tail is derived from a protein secreted by the epididymis. Mol Reprod Dev 1994; 37: 181-94.

NagDas SK, Winfrey VP, Olson GE. Identification of a hamster epididymal region-specific secretory glycoprotein that binds nonviable spermatozoa. Biol Reprod 2000; 63: 1428-36.

Nath D, Majumder GC. Maturation dependent modification of the protein phosphorylation profile of isolated goat sperm plasma membrane. J Reprod Fertil 1999; 115: 29-37.

Nixon B, Hardy CM, Jones RC, Andrews JB, Holland MK. Rabbit epididymal secretory proteins. III. Molecular cloning and characterization of the complementary DNA for REP38. Biol Reprod 2002b; 67: 147-52.

Nixon B, Jones RC, Clarke HG, Holland MK. Rabbit epididymal seretory proteins.II. Immunolocalization and sperm association of REP38. Biol Reprod 2002; 67: 140-46.

Obermann H, Samalecos A, Osterhoff C, Schroder B, Heller R, Kirchhoff C. HE6, a two-subunit heptahelical receptor associated with apical membranes of efferent and epididymal duct epithelia. Mol Reprod Dev 2003; 64:13-26.

Okamura N, Dacheux F, Venien A, Onoe S, Huet JC, Dacheux JL. Localization of a maturation- dependent epididymal sperm surface antigen recognized by a monoclonal antibody raised against a 135-kilodalton

protein in porcine epididymal fluid. Biol Reprod 1992; 47: 1040-52.
Okamura N, Kiuchi S,, Tamba M, Kashima T, Hiramoto S et al. A porcine homolog of the major secretory protein of human epididymis HE1, specifically binds cholesterol. Biochim Biophys Acta 1999; 1438: 377-87.
Olson JH, Xiang X, Ziegert T, Kittelson A, Rawls A, Bieber AL, Chandler DE. Allurin, a 210kDa sperm chemoattractant from Xenopus egg jelly, is related to mammalian sperm-binding proteins. Proc Natl Acad Sci USA 2001; 98: 11205-10.
Osterhoff C, Kirchhoff C, Krull N, Ivell R. Molecular cloning and characterization of a novel human sperm antigen (HE2), specifically expressed in the proximal epididymis. Biol Reprod 1994; 50: 576-25.
Perry ACF, Jones R, Hall L. The monkey ESP14.6 mRNA, a novel transcript at high levels in the epididymis. Gene 1995; 153: 291-92.
Perry ACF, Jones R, Moisyadi S, Coadwell J, Hall L. The novel epididymal secretory protein ESP13.2 in Macaca fasicularis. Biol Reprod 1999; 61: 965-72.
Petruszak JA, Nehme CL, Bartles JR. Endoproteolytic cleavage in the extracellular domain of the inegral plasma membrane protein CE9 precedes its redistribution from the posterior to the anterior tail of the rat spermatozoon during epididymal maturation. J Cell Biol 1991; 114: 917-27.
.Ram PA, Cardullo RA, Millette CF. Expression and topographical localization of cell surface fucosyltransferase activity during epididymal sperm maturation in the mouse. Gamete Res 1989; 22: 321-22.
Rane APS, Majumder GC, Misra S, Ghosh A. Lipid changes of goat sperm plasma membrane during epididymal maturation. Biochim Biophys Acta 1991; 1061: 185-96.
Rankin TL, Tsuruta KJ, Holiand MK, Griswald MD, Orgebin Crist MC. Isolation, immunolocalization and sperm-association of three proteins of 18, 25, and 29 kilodaltons secreted by the mouse epididymis. Biol Reprod 1992; 46: 747-66.
Rao J, Herr JC, Reddi PP, Wolkowicz MJ, Bush LA, Sherman NE, Black M, Flickinger CJ. Cloning and characterization of a novel sperm-associated isoantigen (E-3) with defensin- and lectin-like motifs expressed in rat epididymis. Biol Reprod 2003; 68: 290-301.
Retamal C, Urzua J, Alves EW, Lopez ML. Cytochemical and electrophoretic study of the stallion epididymal glycoproteins. J Submicrosc Cytol Pathol 2000; 32: 117-30.
Roberts KP, Ensrud KM, Hamilton DW. A comparative analysis of expression and processing of the rat epididymal fluid and sperm-bound forms of proteins D and E. Biol Reprod 2002; 67: 525-33.
Rosenberg ME, Silkensen J. Clusterin : physiologic and pathophysiologic considerations. Int J Biochem Cell Biol 1995; 27: 633-45.
Rutllant J, Meyers SA. Posttranslational processing of PH-20 during epididymal sperm maturation in the horse. Biol Reprod 2001; 65: 1324-31.
Saalmann A, Munz S, Ellerbrock K, Ivell R, Kirchhoff C. Novel sperm-binding proteins of epididymal origin contain four fibronectin type II-modules. Mol Reprod and Dev 2001; 58: 88-100.
Sarkar M, Chatterjee T. Distribution of 105-kDa sperm unique antigen on goat epididymal mature spermatozoa. Biochem Biophys Res Commun 1997; 231: 662-6.
Scarlett CJ, Lin M, Aitken RJ. Actin polymerization during morphogenesis of the acrosome as spermatozoa undergo epididymal maturation in the tammar wallaby (Macropus eugenii). J Anat 2001; 198; 93-101.
Schambony A, Hess O, Gentzel M, Topfer-Petersen E. Expression of CRISP proteins in the male equine gential tract. J Reprod Fertil (Suppl) 1998; 53: 67-72.
Schroter S, Derr P, Conradt HS, Nimtz M. Hale G, Kirchhoff C. Male specific modification of human CD52. J Biol Chem 1999; 274: 29862-73.
Sensibar JA, Qian Y, Griswold MD, Sylvester SR, Bardin CW, Cheng CY, Lee C. Localization and molecular heterogeneity of sulfated glycoprotein-2 (clusterin) among ventral prostate, seminal vesicle, testis, and epididymis of rats. Biol Reprod 1993; 49: 233-42.
Sirivaidyapong S, Bevers MM, Gadella BM, Colenbrander B. Induction of the acrosome reaction in dog sperm cells is dependent on epididymal maturation. Mol Reprod Dev 2001; 58: 451-9.
Sivashanmugam P, Richardson RT, Hall S, Hamil KG, French FS, O,Rand MG. Cloning and characterization of an androgen-dependent acidic epididymal glycoprotein/CRISP-1 like protein from the monkey. J Androl 1999; 20: 384-93.
Smithwick EB, Young LG. Immunohistochemical localization of epididymal secretory glycoprotein EP1 in the adult male chimpanzee. Tissue Cell 1999; 31: 54-65.
Srivastav A. Maturation dependent glycoproteins containing both N- and O-linked oligosacharides in epididymal sperm plasma membrane of rhesus monkeys (Macaca Mulatta). J. Reprod Fert 2000; 119: 241-52.
Sun GH, Lin YC, Guo YW, Chang SY, Lin HW. Purification of GP-83, a glycoprotein secreted by the human epididymis and conjugated to mature spermatozoa. Mol Hum Reprod 2000; 6: 429-34.
Sundaram M, Sivaprasadarao A, Aalten DM, Findlay JB. Expression, characterization and engineered specificity of rat epididymal retinoic acid-binding protein. Biochem J 1998; 334: 155-60.

Sutovsky P, Moreno R, Ramalho-Santos J, Dominko T, Thomoson WE, Schatten G. Putative ubiquitin dependent mechanism for the recognition and elimination of defective spermatozoa in mammalian epididymis. J Cell Sci 2001; 114 (Pt a): 1665-75.

Syntin P, Cornwall GA. Immunolocalization of CRES (Cystatin-Related Epididymal Speratogenic) protein in the acrosomes of mouse spermatozoa. Biol Reprod 1999; 60: 1542-52.

Syntin P, Dacheux F, Druart X, Gatti JL. Characterization and identification of proteins secreted in the various regions of the adult boar epididymis. Biol Reprod 1996; 55: 956-974.

Toshimori K, Araki S, Oura C. Masking of sperm maturation antigen by sialic acid in the epididymis of the mouse. An Immunohistochemical study. Histochem 1988; 90: 195-200.

Toshimori K, Tanii I, Araki S. Intra-acrosomal 155,000 dalton protein increases the antigenicity during mouse sperm maturation in the epididymis: a study using a monoclonal antibody MC101. Mol Reprod Dev 1995; 42: 72-9.

Tsuji Y, Clausen H, Nudelman E, Kaizu T, Hakomori SI, Isojima S. Human sperm carbohydrate antigens defined by an antrisperm human monoclonal antibody derived from an infertile woman bearing antisperm antibodies in her serum. J Exp Med 1988; 168: 343-56.

Tubbs CE, Hall JC, Scott RO, Clark VP, Hermon TL, Bazemore-Walker C.Binding of protein D/E to the surface of rat epididymal sperm before ejaculation and after deposition in the female reproductive tract. J Androl 2002; 23: 512-21.

Tulsiani Dr, NagDas SK, Skudlarek MD, Orgebin-Crist MC. Rat sperm plasma membrane mannosidase: localization and evidence for Proteolytic processing during epididymal maturation. Dev Biol 1995; 167: 584-95.

Tulsiani DRP, Orgebin-Crist MC, Skudlarek MD. Role of luminal fluid glycosyltransferases and glycosidases in the modification of rat sperm plasma membrane glycoproteins during epididymal maturation. J Reprod Fert (Suppl). 1998; 53,85-97.

Uma Devi, Basheer Ahmad M, Shivaji S. A maturation related differential phosphorylation of the plasma membrane proteins of the epididymal spermatozoa of hamster by endogenous protein kinases. Mol Reprod Dev 1997; 47: 341-50.

Vierula ME, Araki Y, Rankin TL, Tulsiani DR, Orgebin-Crist MC. Immunolocalization of a 25-kilodalton protein in mouse testis and epididymis. Biol Reprod 1992; 47: 844-56.

Voglmayr JK, Sawyer RF Jr, Dacheux JL. Glycoproteins : a variable factor in surface transformation of ram spermatozoa during epididymal transit. Biol Reprod 1985; 33: 165-76.

Wassler M, Syntin P, Sutton-Walsh HG, Hsia N, Hardy DM, Cornwall GA.Identification and characterization of cystatin-related epididymal spermatogenic protein in human spermatozoa: localization in the equatorial segment. Biol Reprod 2002; 67: 795-803.

Waters SI, White JM. Biochemical and molecular characterization of bovine fertilin a and b (ADAM1 and ADAM2): A candidate sperm-egg binding/fusion complex. Biol Reprod 1997; 56: 1245-54.

Weaver FE, Dino JE, Germain BJ, Wolf DE, Fairbanks G. Biochemical characterization and epididymal localization of the maturation dependent ram sperm surface antigen ESA152. Mol Reprod Dev 1993; 35: 293-301.

Wong P, Taillefer D, Lakins J, Pineault J, Chander G et al etc. Molecular characterization of human TRPM-2/clusterin, a gene associated with sperm maturation, apoptosis and neurodegeneration. Eur J Biochem 1994; 221: 917-25.

.Xu WD, Wang LF, Miao SY, Zhao M, Fan HY, Zong SD, Wu YW, Shi XQ, Koide SS. Identification of a rabbit epididymal protein gene. Arch Androl 1996; 37: 135-41.

Yeung CH, Cooper TG, WeinBauer GF. Maturation of monkey spermatozoa in the epididymis with respect to their ability to undergo the acrosome reaction. J Androl 1996; 17: 427-32.

Yeung CH, Perez-Sanchez F, Schroter S, Kirchhoff C, Cooper TG. Changes of the major sperm maturation associated epididymal protein HE5 (CD52) on human ejaculated spermatozoa during incubation. Mol Hum Reprod 2001; 7: 617-24.

Yeung CH, Schroter S, Kirchhoff C, Cooper TG. Maturational changes of the CD52 like epididymal glycoprotein on cynomologus monkey sperm and their apparent reversal in capacitation conditions. Mol Reprod Dev 2000; 57: 280-9.

Yoshinaga K, Tanii I, Oh-oka T, Toshimori K. Changes in distribution and molecular weight of the acrosomal protein acrin2 (MC41) during guinea pig spermiogenesis and epididymal maturation. Cell Tissue Res 2001; 303: 253-61.

Yoshinaga K, Tanii I, Oh-oka T, Toshimori K. Transport and rearrangement of the intra-acrosomal protein acrin1 (MN7) during spermiogenesis in the guinea pig testis. Anat Rec 2000; 259: 131-40.

Yuan R, Primakoff P, Myles PG. A role for the disintegrin domain of cyritestin, a sperm surface protein belonging to the ADAM family, in mouse sperm-egg plasma membrane adhesion and fusion. J Cell Biol 1997; 137: 105-12.

Zeheb R, Orr GA. Characterization of a maturation-associated glycoprotein on the plasma membrane of rat caudal epididymal sperm. J Biol Chem 1984; 259: 839-48.

INDEX

About the Author

Dr. Gupta is a former Professor and Chairman of the Department of Biophysics, Panjab University, Chandigarh, India. His primary areas of teaching are molecular and cellular biology/ biophysics, biophysical chemistry and radiation biology. His primary areas of research are molecular and cellular biology, reproductive biology/reproductive immunology, enzymology/enzyme technology, and radiation biology.

As a visiting researcher, he has worked at many institutions, including Northwestern University in Evanston, IL, the Center of Immunopathology and Experimental Immunology, INSERM, and the Center of Cytogenics and Immunogenetics, INSERM.

He has been honored by various Indian agencies such as Council of Scientific and Industrial Research (CSIR), University Grants Commission (UGC), Indian Council of Medical Research (ICMR), and Department of Science and Technology (DST).

Professor Gupta has made significant research contribution in the field of Reproductive Molecular Biology and Molecular Radiation Biology and published 120 original research articles and reviews in international journals.

He is the recipient of several awards and international fellowships including a WHO fellowship and the INSERM fellowship. Indian Council of Medical Research has honored him by conferring Swaran Kanta Dingley Oration Award of 1993 for his research contribution and extending knowledge in male reproduction. Associated with several national and international scientific societies, he has chaired scientific sessions and delivered invited lectures at national and International conferences.

Professor G.S. Gupta holds master's degrees in Physical Chemistry and Biochemistry from Lucknow University, Lucknow, and a Ph.D. in Biophysics from Panjab University, Chandigarh (India).